2013 9th IEEE International Symposium on Diagnostics for Electric Machines, Power Electronics and Drives

(SDEMPED 2013)

Valencia, Spain
27 – 30 August 2013

IEEE Catalog Number: CFP13SDE-POD
ISBN: 978-1-4799-0024-4

Copyright © 2013 by the Institute of Electrical and Electronic Engineers, Inc
All Rights Reserved

Copyright and Reprint Permissions: Abstracting is permitted with credit to the source. Libraries are permitted to photocopy beyond the limit of U.S. copyright law for private use of patrons those articles in this volume that carry a code at the bottom of the first page, provided the per-copy fee indicated in the code is paid through Copyright Clearance Center, 222 Rosewood Drive, Danvers, MA 01923.

For other copying, reprint or republication permission, write to IEEE Copyrights Manager, IEEE Service Center, 445 Hoes Lane, Piscataway, NJ 08854. All rights reserved.

***This publication is a representation of what appears in the IEEE Digital Libraries. Some format issues inherent in the e-media version may also appear in this print version.**

IEEE Catalog Number: CFP13SDE-POD
ISBN 13:   978-1-4799-0024-4

Additional Copies of This Publication Are Available From:

Curran Associates, Inc
57 Morehouse Lane
Red Hook, NY 12571 USA
Phone: (845) 758-0400
Fax: (845) 758-2633
E-mail: curran@proceedings.com
Web: www.proceedings.com

Introduction, Message from the Symposium Chairs

These proceedings include 90 papers which were presented during the 9th IEEE International Symposium on Diagnostic of Electrical Machines, Power Electronics and Drives (SDEMPED 2013) that was held in Valencia, Spain, between 27th and 30th September, 2013.

These papers were presented in 14 regular sessions and 4 special sessions. The regular sessions deal with classical topics in the diagnostic area as:

- Rotor Faults (4 sessions).
- Stator Faults (2 sessions).
- Power electronics faults (3 sessions devoted to Power electronics Devices, Adjustable Speed Drives and Power Converters).
- Tools for diagnostics (3 sessions devoted to Signal Analysis tools, Artificial Intelligence tools and Diagnostic tools applied to Mechanical faults).
- Permanent magnet and Synchronous machines.
- Test for predictive maintenance and partial discharge tests.

The Special Sessions deal with emerging subjects in the diagnostic area. They were proposed by specialists and approved by the Steering Committee. There were Special Sessions on:

- Fault tolerant Power Converters, Electrical machines and drives (2 sessions).
- Advanced artificial intelligence approaches applied to fault characterization and classification for electrical machines diagnostic purposes (1 session).
- Failure prognosis methods in electrical drives (1 session).

All these papers were selected among 130 submitted papers, after a review process in which every paper was revised by at least three independent reviewers.

The proceedings are organized maintaining the structure of the conference; the papers can be easily localized through the Table of Contents and author list.

The conference also included three tutorials and three Plenary Sessions that were carried out by invited speakers. Details on Tutorials and Plenary Sessions are given in the Technical Program summary.

We want to acknowledge the support of many volunteers who contributed to the success of the Symposium. We thank the Steering Committee and Local Committee members, the Track Chairs, the Special Sessions Organizers, the Tutorial Organizers, the Keynote Speakers, the Session Chairs and the reviewers, the Sponsors and other collaborating Institutions, each of them developing different but important tasks. And, last but not least, we specially thank the authors, for their effort in their research and preparation of their works that reached a high technical level; they are the main motivation of this Symposium.

Martin Riera-Guasp   Jose A. Antonino-Daviu

SDEMPED 2013 General Co-Chairs

Table of Contents

Regular Sessions

RS1 Power Electronics (I)

RS2 Rotor Faults (I)

RS3 Tools for Diagnostics (I). Signal Analysis Techniques

RS4 Rotor Faults (II)

RS5 Tools for Diagnostics (II). AI Techniques

RS6 Stator Faults (I)

RS7 Adjustable Speed Drives and Power Converters

RS8 Stator Faults (II)

RS9 Rotor Faults (III)

RS10 Permanent Magnet and Synchronous Machines

RS11 Tools for Diagnostics (III). Mechanical faults

RS12 Test for predictive maintenance. Partial Discharge tests

RS13 Rotor Faults (IV)

RS14 Power Electronics II, Power Converters

Special Sessions

SS1 Fault tolerant Power Converters, Electrical machines and drives (I)

SS2 Fault tolerant Power Converters, Electrical machines and drives (II)

SS3 Advanced Artificial Intelligence Approaches

SS4 Failure Prognosis Methods In Electrical Drives

RS1 Power Electronics (I)

TOP ^

On-Line Algorithm for Early Stage Fault Detection in IGBT Switches Jason M. Anderson, Robert W. Cox, Paul O'Connor	1
Onboard Condition Monitoring of Solder Fatigue in IGBT Power Modules B. Ji, V. Pickert, W. P. Cao, and L. Xing	9
Power MOSFET failure and degradation mechanisms in flyback topology under high temperature and high humidity conditions Ilkka Vaalasranta, Juha Pippola, and Laura Frisk	16
Remote Monitoring System of Electrical Machines via INTERNET O. Touhami, R. Sadoun, A. Belouchrani, S. Hamdani, A. Boukoucha and S. Ouaged	23
Low-Cost IC less Self Oscillating Boost PFC Converter S. Borekci, I. M. Luleci	28

RS2 Rotor Faults (I)

TOP ^

Evaluation of Different Broken Bar Fault Diagnostic Means in Double- Cage Induction Motors with FEM Konstantinos N. Gyftakis, Dimitrios K. Athanasopoulos and Joya Kappatou,	36
A Hybrid Kangaroo Algorithm to Assess the State Of Health of Electric Motors H. Razik, M.El.K. Oumaamar, G. Clerc	43
Optimal Wavelets for Broken Rotor Bars Fault Diagnosis Pu Shi, Zheng Chen, Yuriy Vagapov, Zoubir Zouaoui	49
New Quantitative Rotor Fault Evaluation in Wound Rotor Induction Machine Drives Under Time-Varying Conditions Y. Gritli, M. Mengoni, C. Rossi, F. Filippetti, D. Casadei	57
Induction Motor Broken Rotor Bar Detection Using Vibration Analysis – A Case Study Ž. Kanović, D. Matić, Z. Jeličić, M. Rapaić, B. Jakovljević, M. Kapetina	64

RS3 Tools for Diagnostics 1 . Signal Analysis Techniques

TOP ^

Diagnosis of Induction Motor Faults using a DSP and Advanced Demodulation Techniques M. Pineda-Sanchez, J. Perez-Cruz, J. Roger-Folch, M. Riera-Guasp, A. Sapena-Baño, R. Puche-Panadero	69
On the Use of Spectral Kurtosis for Diagnosis of Electrical Machines E. Fournier, A. Picot1, J. Régnier1, M. Tientcheu Yamdeu, J-M. Andréjak, P. Maussion	77
Air-gap Power and Rotor Loss Estimation for Induction Motor Efficiency Monitoring based on Kalman Filtering N. Jirasuwankul, C. Manop	85
Diagnosis of Induction Machines under Non-Stationary Conditions by means of the Spectral Filter F. Vedreño-Santos, M. Riera-Guasp, M. Pineda-Sánchez	91
Fault Detection and Classification in Permanent Magnet Synchronous Machines using Fast Fourier Transform and Linear Discriminant Analysis Reemon Z. Haddad, Elias G. Strangas	99

RS4 Rotor Faults (II)

TOP ^

Early Broken Rotor Bar Detection Techniques in VSD-fed Induction Motors at Steady-state R. J. Romero-Troncoso, D. Morinigo-Sotelo, O. Duque-Perez, P. E. Gardel- Sotomayor, R. A.,Osornio-Rios, A. Garcia-Perez	105
Use of Discrete and Optimized Continuous TFD Tools for Transient- Based Diagnosis in Controversial Fault Cases J. Pons-Llinares, J. Antonino-Daviu, M. Riera-Guasp, S.B. Lee, T-J. Kang , C. Yang	114
A New Method to Separate Broken Rotor Failures and Low Frequency Load Oscillations in Three-Phase Induction Motor T. Göktaş, M. Arkan, and Ö. F. Özgüven	122
Analysis of Mixed-Eccentric Induction Machine Rijaniaina Njakasoa Andriamalala, Hubert Razik, François-Michel Sargos, Bruno Francois	128
Motor Current Signature Analysis Apply for external Mechanical Fault and Cage Asymmetry in Induction Motors A. J. Fernández Gómez, T. J. Sobczyk	136
Dynamic Model of Induction Machine with Faulty Rotor in Field Reference Frame Vanja Ambrožic, Rastko Fišer, Mitja Nemec, Klemen Drobnic	142

RS5 Tools for Diagnostics (II). AI Techniques

TOP ^

Support Vector Machine for Diagnosis of Induction Motors: a Comparative Analysis in Terms of the Quantity and the Signal Processing Tool Used to Build the Feature Space A. Sapena-Bañó, M. Pineda-Sanchez, R. Puche-Panadero, J.Roger-Folch, J. Perez-Cruz, M. Riera-Guasp	150
Detection of Induction Machine Winding Faults Using Genetic Algorithm M. Alamyal, S. M. Gadoue and B. Zahawi	157
Broken bar condition monitoring of an induction motor under different supplies using a Linear Discriminant Analysis M. Fernandez-Temprano, P.E. Gardel-Sotomayor, O. Duque-Perez, D. Morinigo-Sotelo	162
Intelligent Sensor based on Acoustic Emission Analysis applied to Gear Fault Diagnosis Daniel Zurita, Miguel Delgado, Juan Antonio Ortega Redondo, Luis Romeral	169

17

RS6 Stator Faults (I)

TOP ^

Evaluation of the Applicability of FRA for Inter-Turn Fault Detection of in Stator Windings F. R. Blánquez, Carlos. A. Platero, E. Rebollo, F. Blázquez	177
Detection of Stator Slot Magnetic Wedge Failures for Induction Motors without Disassembly Kun Wang Lee, Jongman Hong, Doosoo Hyun, Sang Bin Lee, Ernesto Wiedenbrug, Mike Teska, Chaewoong Lim	183
Induction Motor Model Validation using Fast Fourier Transform and Wavelet tools F.J. Villalobos-Piña, R. Alvarez-Salas, Eduardo Cabal-Yepez, Arturo Garcia- Perez	192
A Novel Non-Invasive Method for Detecting Missing Wedges in an Induction Machine Maciej Orman, Agnieszka Nowak, J.R. Ottewill, C. T. Pinto	200
Stator circulating currents as media of fault detection in synchronous motors Pedro Rodriguez, Pawel Rzeszucinski, Maciej Sulowicz, Rolf Disselnkoetter, Ulf Ahrend, Cajetan T.Pinto, James R. Ottewill, Stephan Wildermuth.	207
An accurate and Fast Technique for Correcting Spectral Javier Martinez, François Philipp, Manfred Glesner, Antero Arkkio	215

RS7 Adjustable Speed Drives and Power Converters

TOP ^

Analysis of Electrical and Non-Electrical Causes of Variable Frequency Drive Failures Osama A. Al-Naseem, Mohamed A. El-Sayed	221
Detecting High-Resistance Connection Asymmetries in Inverter Fed AC Drive Systems G. Stojčić, T. M. Wolbank	227
FPGA-based Smart-sensor for Fault Detection in VSD-fed Induction Motors A.G. Garcia-Ramirez, R.A. Osornio-Rios, A. Garcia-Perez, R.J.,Romero- Troncoso	233
IGBT Fault Diagnosis using Adaptive Thresholds during the Turn-on Transient M. A. Rodríguez-Blanco, A. Vázquez-Pérez, L. Hernández-González, A. Pech-Carbonell, M. May-Alarcón.	241
Fault-Tolerant Converter for AC Drives using Vector-Based Hysteresis Current Control Nuno M. A. Freire, A. J. Marques Cardoso	249

RS8 Stator Faults (II)

TOP ^

Finite Element Investigation of the Short-Circuit Fault in the Stator Winding of Induction Motors and Harmonics of the Neighboring Magnetic Field V. Fireteanu, A-I. Constantin, R. Romary, R. Pusca, S. Ait-Amar	257
Modeling and Simulation of Stator Turn Faults. Detection Based on Stator Circular Current and Neutral Voltage Yassine Maouche, Abdelfettah Boussaid, Mohamed Boucherma, Abdelmalek Khezzar	263
Temperature Field Analysis of Winding Short-circuits in DFIGs Zheng Liu, Wenping Cao, Zheng Tan, Xueguan Song, Bing Ji, and Guiyun Tian	269
Induction Motor Stator Faults Diagnosis by Using Parameter Estimation Algorithms Fang Duan, Rastko Zivanovic	274
Diagnosis of Stator Winding Inter-turn Short Circuit in Three-Phase Induction Motors by Using Artificial Neural Networks P. J. Broniera, W. S. Gongora, A. Goedtel, W. F. Godoy	281
Naïve Bayes classifier for Temporary short circuit fault detection in Stator winding D. A. Asfani, M. H. Purnomo, D. R. Sawitri	288

RS9 Rotor Faults (III)

TOP ^

Analytical study of pulsating torque and harmonic components in rotor current of six-phase induction motor under healthy and faulty conditions Yassine Maouche, Abdelfettah Boussaid, Mohamed Boucherma, Abdelmalek Khezzar	295
The Zero-Sequence Current Spectrum as an On-Line Static Eccentricity Diagnostic Mean in Δ-Connected PSH-Induction Motors K. N. Gyftakis, J. C. Kappatou	302
Discriminating time-varying loads and rotor cage fault in Induction motors A. E. Mabrouk, S. E. Zouzou, M. Sahraoui and S. Khelif	309
Mathematical Modeling of Eccentricities in Induction Machines by the Mono-harmonic Model A. J. Fernández Gómez, A. Dziechciarz, T. J. Sobczyk	317
Winding Function Approach for Induction Machine Fault Detection Pu Shi, Zheng Chen, Yuriy Vagapov, Zoubir Zouaoui	323

RS10 Permanent Magnet and Synchronous machines

TOP ^

On-line Inter-Turn Short-Circuit detection in Permanent Magnet Synchronous Generators B. Aubert, J. Regnier, S.Caux, D. Alejo	329
Saturation Independent Detection of Dynamic Eccentricity Fault in Salient-Pole Synchronous Machines T. Ilamparithi, Subhasis Nandi	336
Coupled Magnetic Circuit Based Magnetic Vibrations Modeling of PMSM Guillaume Verez, Ouadie Bennouna, Yacine Amara, Ghaleb Hoblos, Georges Barakat	342
2-pole turbo-generator eccentricity diagnosis by split-phase current signature analysis Claudio Bruzzese	349

RS11 Tools for Diagnostics (III). Mechanical faults

TOP ^

Gear Tooth Surface Damage Fault Detection Using Induction Machine Electrical Signature Analysis Shahin Hedayati Kia, Humberto Henao, Gérard-André Capolino	358
Sensorless Speed Estimation and Diagnosis of Induction Motors Based on Purified Space Vectors Dongfeng Shi	365
Bearing Faults Detection in Induction Machines Based on Statistical Processing of the Stray Fluxes Measurements Ciprian Harlişca, Loránd Szabó, Lucia Frosini, Andrea Albini	371
Identification of variable mechanical parameters using extended Kalman filters M. Perdomo, M.Pacas, T. Eutebach , J. Immel	377

RS12 Test for predictive maintenance. Partial discharge tests

TOP ^

Partial Discharge measurements in Electrical Machines controlled by Variable Speed Drives: from Design Validation to permanent PD Monitoring Luca Fornasari, Andrea Caprara, Gian Carlo Montanari	384
An Applied Laboratory Characterisation Approach for Electric Machine Insulation D. F. Kavanagh, D. A. Howey and M. D. McCulloch	391
A Wideband Partial Discharge Meter using FPGA Radek Sedláček, Josef Vedral, Ján Tomlain	396

RS13 Rotor Faults (IV)

TOP ^

A Novel Methodology for the Broken Bar Fault Diagnosis in Single and Double Cage Induction Motors Fed by Asymmetrical Voltage Supply K. N. Gyftakis, D. K. Athanasopoulos , J. C. Kappatou	402
Broken Bar Detection Using Current Analysis - A Case Study Dragan Matić, Željko Kanović, Dejan Reljić, Filip Kulić, Đura Oros, Veran Vasić	407
A Novel Broken Rotor Bar Fault Detection Method Using Park's Transform and Wavelet Decomposition Ramin Salehi Arashloo, José Luis Romeral Martinez, Mehdi Salehifar	412
Analytical Evaluation of inductances for induction machine with dynamic eccentricity using MWFA and FE methods S. Hamdani, O. Touhami, R. Ibtiouen and M. Hasni	420

RS14 Power Electronics 2, Power Converters

TOP ^

Analysis of radiated EMI for power converters switching in MHz frequency range A. Majid, J. Saleem, F. Alam, K. Bertilsson	428
Design of Current Source DC/DC Converter and Inverter for 2kW Fuel Cell Application A. Andreiciks, I. Steiks, O. Krievs, F. Blaabjerg	433
An adaptive robust position control for induction machines using a sliding mode flux observer Oscar Barambones	439
Calculation of Stator Winding Parameters to Predict the Voltage Distributions in Inverter Fed AC Machines Oliver Magdun, Sébastien Blatt and Andreas Binder	447
Circulating Current Minimization and Current Sharing Control of Parallel Boost Converters Based on Droop Index Sijo Augustine, Mahesh K. Mishra, N. Lakshminarasamma	454

SS1 Fault tolerant Power Converters, Electrical machines and drives (I) TOP ^

A Simple and Robust Method for Open Switch Fault Detection in Power Converters Mehdi Salehifar, Ramin Salehi Arashloo, Manuel Moreno-Eguilaz, Vicent Sala, L. Romeral	461
Efficiency Optimization on Vector Controlled Six-Phase Induction Motor in Healthy and Faulted Mode M. Moghadasian, A. Sivert, A. Yazidi, F. Betin, G.A. Capolino	469
Magnetic Optimization of a Fault-Tolerant Linear Permanent Magnet Modular Actuator for Shipboard Applications M. Bortolozzi, C. Bruzzese, F. Ferro, T. Mazzuca, M. Mezzarobba, G. Scala, A. Tessarolo, D. Zito	477
Fault Tolerant High Voltage Resonant Power Converter Application A. Hultgren, S. Bui, J. Linnér, P. Ranstad, M. Lenells	485
Experimental Evaluation of Combined Reference Frames Transformation for Stator Fault Detection in Multi-Phase Machines C. Bianchini, F. Immovilli, E. Lorenzani, A. Bellini, E. Fornasiero	491
The Performance of a Three-Phase Induction Motor fed by a Three- Level NPC Converter with Fault Tolerant Control Strategies. B. R. O. Baptista, M.B. Abadi, A.M. S. Mendes, S. M. A. Cruz	497

SS2 Fault tolerant Power Converters, Electrical machines and drives (II) TOP ^

Full Detection of High Resistance Connection in Multiphase Induction Motor Drives L. Zarri, M. Mengoni, A. Tani, Y. Gritli, G. Serra, F. Filippetti, D. Casadei	505
Improved Fault Detection Based on Current Average in Multiphase Fault Tolerant Converters Mehdi Salehifar, Manuel Moreno-Eguilaz, Vicent Sala, Ramin Salehi Arashloo, L.Romeral	512
Inter-turn Fault Detection in Five-Phase PMSMs. Effects of the Fault Severity Harold Saavedra, Jordi-Roger Riba, Luís Romeral	520
Detection of Coupling Inductor Faults in Three-Phase Adjustable Speed Drives with Direct Power Control-Based Active Front-End Rectifiers Joaquín G. Norniella, José M. Cano, Gonzalo A. Orcajo, Carlos H. Rojas,Joaquín F. Pedrayes, Manés F. Cabanas Manuel G. Melero	527
Study of fault – tolerant inverter F.khelifi, B.Nadji	533

SS3 Advanced Artificial Intelligence Approaches Applied to Fault Characterization and Classification for Electrical Machines Diagnostic Purposes

TOP ^

Bearing Fault Detection Using Relative Entropy of Wavelet Components and Artificial Neural Networks Helder L. Schmitt, Lyvia B. Silva, Paulo R. Scalassara, Alessandro Goedtel	538
Dedicated Hierarchy of Neural Networks applied to Bearings Degradation Assessment Miguel Delgado, Giansalvo Cirrincione, Antonio Garcia Espinosa, Juan Antonio Ortega, Humberto Henao	544
A Dedicated Application of Artificial Ants for the Condition Monitoring of Induction Motors A. Soualhi, H. Razik, G. Clerc	552
Comparison of supervised classification algorithms combined with feature extraction and selection : Application to a turbo-generator rotor fault detection Alexandre Bacchus, Mélisande Biet, Ludovic Macaire, Yvonnick Le Menach and Abdelmounaïm Tounzi	558
Neural Approach for Bearing Fault Detection in Three Phase Induction Motors W. S. Gongora, H. V. D. Silva, A. Goedtel, W. F. Godoy, S. A. O. da Silva	566
Early Detection of Unbalance Voltage in Three Phase Induction Motor Based on SVM D. R. Sawitri, D. A. Asfani, M. H. Purnomo, I. K. E. Purnama, M. Ashari	573

SS4 Failure Prognosis Methods In Electrical Drives

TOP ^

Exploitation of Induction Machine's High-Frequency Behavior for Online Insulation Monitoring Peter Nussbaumer, Markus A. Vogelsberger, Thomas M. Wolbank	579
Long-Term Prediction of Bearing Condition by the Neo-Fuzzy Neuron A. Soualhi, G. Clerc, H. Razik, F. Rivas	586
Bar Breakage Mechanism and Prognosis in an Induction Motor Vicente Climente-Alarcon, Jose Alfonso Antonino-Daviu, Elias Strangas, Martin Riera-Guasp	592
Time-Frequency Complexity Based Remaining Useful Life (RUL) Estimation for Bearing Faults Rodney K. Singleton II, Elias G. Strangas, Selin Aviyente	600
Improvements on Lifespan Modeling of the Insulation of Low Voltage Machines with Response Surface and Analysis of Variance Antoine Picot, David Malec, Pascal Maussion	607
Diagnosis and Prognosis of In-Service Electric Machine in the Absence of Historic Data Related to Faults and Faults Progression Syed Sajjad H. Zaidi	615

Online Algorithm for Early Stage Fault Detection in IGBT Switches

Jason M. Anderson, Robert W. Cox, and Paul O'Connor
Department of Electrical and Computer Engineering
UNC Charlotte
Charlotte, NC 28223
Email: jmander1@uncc.edu and Robert.Cox@uncc.edu

Abstract— The early detection of incipient faults is desirable in mission-critical applications such as shipboard propulsion drives. This paper presents an online condition-monitoring approach for detecting early stage faults in IGBTs. The proposed algorithm extracts important device features (i.e. on-state resistance, gate charge, etc.) and compares them to healthy values recorded over a range of operating conditions. The algorithm is based on principal-components analysis (PCA). An experimental implementation in an IGBT-based drive is described, and results recorded with two different faults over a range of operating conditions are presented. The scheme integrates well with new FPGA-based gate drives and provides a powerful alternative to rules-based fault detection.

I. INTRODUCTION

Power electronic drives are becoming increasingly common in mission-critical applications. High-power, medium-voltage examples include propulsion motors aboard all-electric ships [1], large industrial motors such as those driving recirculation pumps in nuclear power plants [2], and large multi-megawatt, grid-tied inverters for wind turbines and photovoltaics [3]. This increasing dependence on power electronics in mission-critical applications has created a need for real-time techniques that can detect *early stage* faults.

The two components most prone to failure in switch-mode drives are electrolytic filtering capacitors and power semiconductors [4]. Approaches for online monitoring of capacitor health have been described in the literature, and are relatively easy to implement in digitally controlled drives [4], [5]. In the case of MOSFET and IGBT switches, however, the focus has been to develop fault-detection schemes that allow one to detect complete device failures such as shorted or open transistors [6], [7], [8], [9], [10]. In the event that catastrophic failure does not occur, various schemes allow operation in a degraded mode until service can be performed [7], [8], [11].

Although post-fault detection and fault tolerance will always be a necessary safety feature in any mission-critical drive, there is good reason to consider incipient fault detection and online condition monitoring. Consider, for instance, an IGBT that slowly wears over its lifetime because of thermal cycling. Ultimately, such a device will short circuit, which is desirable in a drive with $N + 1$ components. Eventually, however, the bond wires burn away, leading to an indeterminate failure state that may cause an arc flash and subsequent collateral damage to the rest of the circuit or to nearby humans [12].

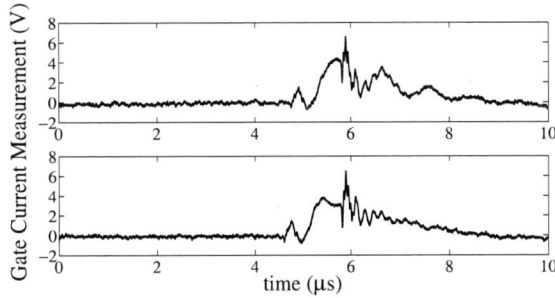

Fig. 1. Output of a differential amplifier in a prototype drive measuring the voltage across a 10Ω gate resistor during two different turn-on instances. Note that the gate current has a slightly different shape during the two instances as a result of switching noise.

The literature includes a number of articles describing the physics of failure in MOSFETs and IGBTs [13], [14], [15], [16]. Authors have described and verified the relationships between device health and device parameters (i.e. on-state resistance, threshold voltage, etc.). Until recently, however, very few works have described the use of such features for monitoring switch health in real-time in an operational circuit [17], [6].

A major limiting factor in the development of condition monitoring for switches has been the difficulty in obtaining the required signals, which are often very small and thus highly susceptible to corruption from switching noise [10]. Several useful health-related parameters are most easily accessed during switching when noise can be particularly problematic. Consider gate charge, for example, which can help to identify naturally occurring gate-oxide degradation [18], [19]. Figure 1 shows the noise corrupting the gate-current measurements in a 208V motor drive. Clearly, switching noise makes it difficult to consistently extract the injected gate charge. Another issue apparent in Fig. 1 is that the signals of interest change rapidly and thus must be sampled at very high rates and/or carefully conditioned using well-designed analog circuits. Given the noise issues, such sampling and conditioning must be performed near the switch.

Recent advancements in the use of digital signal processors (DSPs) and field-programmable gate arrays (FPGAs) close to the power semiconductor [7], [20], [21], [22] provide the

978-1-4799-0024-4/13 $31.00 © 2013 IEEE

opportunity to overcome the aforementioned issues to perform early stage fault detection in real time. Intelligent FPGA-based controllers close to the switches have been used to measure switch terminal variables at very high speeds [21], [22]. These measurements can be used to extract useful health-related features, including threshold voltage, gate charge, and on-state voltage and resistance.

This paper describes a robust fault-detection algorithm taken from the field of facial recognition that uses health-related features to detect multiple early stage faults in IGBTs. The algorithm is capable of overcoming variations in operating conditions and does not require excessive processing. Furthermore, this scheme avoids the need to develop tedious fault-detection rules. Such rules-based approaches often fail to include failure modes that are either overlooked or unknown.

The paper begins in Section II with a description of some of the common failure mechanisms in power transistors. Section III presents a prototype drive architecture and describes the proposed fault-detection algorithm. Section IV describes an experimental prototype and presents results for several different faults and operating conditions. Section V presents conclusions and directions for future work.

II. A REVIEW OF FAILURE MECHANISMS IN POWER TRANSISTORS

There are two different categories of failure mechanisms in power transistors. The first group includes intrinsic mechanisms related to the physics of the actual semiconductors. Some of the most prevalent examples are dielectric breakdown and electromigration [18], [23], [24]. The other group includes factors related to transistor packaging, such as contact migration, bond-wire lift, and die-solder degradation [25], [13], [26], [27]. Our focus here is placed on IGBT switches.

A. Example Intrinsic Failure Mechanisms

Dielectric breakdown occurs when a strong electric field creates a current channel in an insulating medium [28]. During conduction, breakdown can occur between the gate and collector terminals or between the gate and emitter terminals. Two different forms of breakdown are noted in the literature [18]. Catastrophic breakdown of the gate oxide typically results from severe thermal or electrical over-stress (i.e. electrostatic discharge, junction over-voltage, etc.). Time-dependent dielectric breakdown (TDDB), which occurs more gradually over time, refers to the natural breakdown of the gate oxide. TDDB is caused by chronic defect accumulation in the SiO_2 insulator during standard operation [19]. At least three defect-generation mechanisms have been identified [18]. These include impact ionization, hot-carrier injection, and so-called trap creation attributed to the redistribution of hydrogen within the device. Before causing a complete failure, these naturally occurring phenomena affect various device parameters [18]. For instance, they can change the gate leakage current. Similarly, any charges that become trapped in the gate oxide affect important device parameters, such as the threshold voltage V_T and the transconductance g_m [18]. Note that breakdown can also occur between the collector and emitter when the device is in a blocking state.

Electromigration is another intrinsic failure phenomenon [23]. This mechanism results when high current densities within the silicon cause adjacent metal connections to migrate. If any voids form in the interconnects as a result of this process, then the connection may open-circuit or the overall device resistance may increase [23], [29].

B. Example Extrinsic Failure Mechanisms

Various extrinsic failure phenomena have also been observed. Bond-wire lift is one of the most commonly occurring examples [13]. This phenomenon is a failure in the bond between the package wire and the silicon die. Thermal expansion mismatch between the bond solder and the attachment point is the primary cause. Bond-wire lift leads to higher junction temperatures (T_J), and thus it impacts parameters such as on-state resistance and on-state v_{CE} [30]. Changes in these parameters can increase power dissipation, thus causing further increases in T_J. The resulting positive feedback mechanism ultimately leads to a complete device failure [13].

Die-solder degradation is another extrinsic issue [26]. Solder attaching the silicon die to the package heat sink can develop cracks and voids due to dissimilar thermal expansion in the two materials [27]. The junction-to-case thermal impedance, θ_{jc}, thus increases, which leads to a higher T_J. A positive feedback mechanism is thus created once again. As in the case of bond-wire failures, this mechanism affects parameters such as the on-state resistance and on-state v_{CE} [13].

A third extrinsic failure mechanism is contact migration. This phenomenon, which is related to electromigration, occurs when voids between external metal contacts and silicon cause metal to diffuse into the semiconductor. Ultimately, this diffused metal can short-circuit internal pn junctions. Before causing a complete failure, this mechanism impacts parameters such as on-state resistance and on-state v_{CE} [13].

III. FAULT-DETECTION ALGORITHM

Real-time switch condition monitoring introduces new requirements into the design of the overall drive. First, one must obtain high-rate samples of switch terminal variables including, v_{CE}, i_C, v_{GE}, and i_G. Additionally, these signals must be processed and analyzed over time. This section considers a drive architecture that addresses these issues, and it presents an algorithm that can be included to track switch health. Note that the proposed algorithm can be used in any drive capable of measuring the required quantities.

A. Drive Architecture

Figure 2 shows a three-phase, full-bridge converter capable of sampling the required terminal variables. The key feature is the advanced gate-drive concept that has now been described in several works [7], [20], [21], [22]. FPGA-based gate drives have been used to acquire switch terminal variables

978-1-4799-0024-4/13 $31.00 © 2013 IEEE

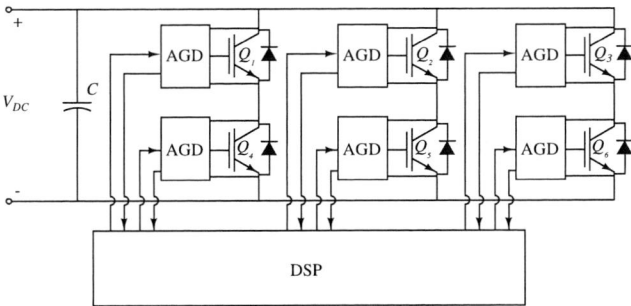

Fig. 2. A potential smart drive architecture including FPGA-based gate drives such as those discussed in [7], [20], [21], [22]. These devices are labeled here as advanced gate drivers (AGDs). The AGDs provide the controller with appropriate measurements. These connections would be fiber optic.

at rates as high as 100MHz. with emphasis on their use in optimizing turn-on and turn-off performance [20], [21]. The signals measured by such devices can also be used to extract meaningful health-related features, such as on-state voltage and resistance. Feature extraction can be performed locally at the gate-drive unit, and features can be transmitted back to a central controller over fiber-optic cables at a much lower data rate.

The advanced gate-drive concept is included here simply because it provides a feasible approach for monitoring the key signals of interest. It should be noted, however, that such signals could be obtained without the use of high-speed FP-GAs. The experimental implementation described in Section IV relies on measurements available to such devices, but it uses analog measurements and off-line processing for now. It would not be difficult to implement the proposed algorithm using an FPGA-based gate drive, so the off-line processing should not be viewed as a limitation of the approach. It was used here simply for ease of demonstration. It should be noted that the limited resolution of the high-speed data converters on the FPGA would likely require some level of analog preprocessing regardless of the format of the implementation.

B. Generalized Health-Monitoring Algorithm for Switches

Figure 3 shows a generalized health-monitoring algorithm for power switches. This algorithm is designed to distinguish between the effects of true faults and other naturally occurring phenomena such as changes in temperature and operating conditions (i.e. loading) without the need for cataloging all of the known failure modes. Inputs to the algorithm include the device terminal variables, i.e. v_{CE}, i_C, v_{GE}, and i_G. The first step in the process is to extract relevant health indicators, such as on-state voltage. Various algorithms can be used. In general, multiple samples of the input data are needed to calculate any feature. As a result, indicators are extracted at a rate well below the sampling frequency. Section IV presents algorithms for estimating key IGBT parameters. As shown in Fig. 3, various feature estimates are ultimately combined with measurements of the case and ambient temperature and passed to the remainder of the algorithm.

Feature vectors computed at the time t_k are grouped into a column vector $\boldsymbol{x}_k$. In our current implementation in the IGBT-based drive, this vector includes the following features, each measured once per line cycle:

- Voltage drops across the *pn* junction and drift regions, V_J
- On-state resistance of the MOSFET channel, R_{ON}
- Injected gate charge at turn-on, Q_G
- Root-mean-squared value of the collector current, I_C
- Average on-state gate-to-emitter voltage, $V_{GE.ON}$
- Ambient temperature, T_A
- Case temperature, T_C

Assuming that other parameters will be added in the near future, we make the general assumption that $\boldsymbol{x}_k$ has a length d. The basic approach of the principal-component-based algorithm is to compare each measurement $\boldsymbol{x}_k$ to an expectation. This expected vector is computed by projecting $\boldsymbol{x}_k$ onto a vector space created using "healthy" features. These healthy values are learned during a training phase in which the drive is new and presumably in good condition. During training, the healthy vectors are decomposed into a small set of characteristic vectors that best describes the distribution of the healthy parameters. During operation, each measured vector is projected onto this space. In the language of information theory, we are extracting the relevant information from the feature vector, encoding it efficiently, and then comparing the encoded result to a database of healthy features encoded in a similar manner. Any differences indicate that the device may be degrading. This approach is partially patterned after the facial recognition scheme presented in [31].

The training space includes features recorded over a range of expected operating conditions. In all, there are M such vectors and they are denoted as $\Gamma_1, \Gamma_2, \Gamma_3, ..., \Gamma_M$. These training vectors are subject to a principal-component analysis (PCA) in which one seeks a set of orthonormal vectors $\boldsymbol{e}_i$ that best describe the distribution of the data. The j-th training vector can thus be expressed as

$$\Gamma_j = \boldsymbol{m} + \sum_{i=1}^{d'} a_{j,i}\, \boldsymbol{e}_i, \qquad (1)$$

where $\boldsymbol{m}$ is the sample mean, i.e.

$$\boldsymbol{m} = \frac{1}{M} \sum_{j=1}^{M} \Gamma_j. \qquad (2)$$

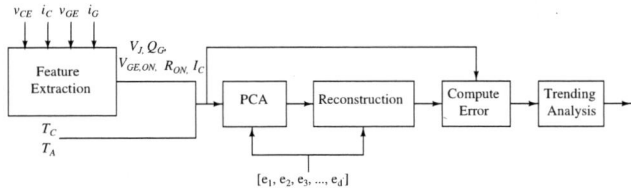

Fig. 3. Generalized condition-monitoring algorithm for an operational IGBT based on principal-components analysis (PCA). The $\boldsymbol{e}_i$ are based on healthy conditions. New input signals and parameters can be added as needed.

Note that the distribution of the data is best described using $d' \leq d$ orthonormal vectors [32], [31]. During the training phase, one calculates these by minimizing the squared-error criterion function

$$J = \sum_{j=1}^{M} \left\| \left(\boldsymbol{m} + \sum_{i=1}^{d'} a_{j,i} \, \boldsymbol{e_i} \right) - \boldsymbol{\Gamma}_j \right\|^2 . \tag{3}$$

References [32] and [31] show that the $\boldsymbol{e}_i$ correspond to the eigenvectors of the sample covariance matrix which is

$$S = \sum_{j=1}^{M} \left(\boldsymbol{\Gamma}_j - \boldsymbol{m} \right) \left(\boldsymbol{\Gamma}_j - \boldsymbol{m} \right)^T . \tag{4}$$

The actual $\boldsymbol{e}_i$ are the eigenvectors corresponding to the d' largest eigenvalues of S [32], [31]. Ultimately, this process yields a compact basis that efficiently encodes the relevant features of a healthy switch over a range of expected operating conditions.

During normal operation of the drive, the fault-detection algorithm in Fig. 3 projects the features measured at time t_k onto the space spanned by the $\boldsymbol{e}_i$. This projection is performed by the block labeled PCA, which computes the set of coefficients

$$a_{k,i} = \boldsymbol{e}_i^T \left(\boldsymbol{x}_k - \boldsymbol{m} \right) . \tag{5}$$

The next block uses these coefficients to reconstruct an approximation of $\boldsymbol{x}_k$. The resulting estimate is thus denoted as

$$\hat{\boldsymbol{x}}_k = \boldsymbol{m} + \sum_{i=1}^{d'} a_{k,i} \, \boldsymbol{e}_i. \tag{6}$$

Following reconstruction, the algorithm calculates the two-norm of the residual vector $\boldsymbol{r} = \hat{\boldsymbol{x}}_k - \boldsymbol{x}_k$. This quantity represents the error between the measured features and their projection onto the healthy features. If the error is small, the switch is healthy; if the error grows, a problem may be developing. The final block monitors for such variations. Changes in this so-called reconstruction error are used to detect developing faults.

IV. INCIPIENT FAULT DETECTION IN IGBTS

The condition-monitoring algorithm described in Section III has been tested using an IGBT-based drive of the form shown in Fig. 2. This custom-built system drives a 1hp induction motor and is controlled using a dsPIC30F6010A. The switching frequency is set to approximately 1kHz in order to mimic operation in a high-power, medium-voltage drive.

A. Signal Measurement and Feature Extraction

In our prototype implementation we did not use high-speed, FPGA-based gate drives, but we did rely on signals that could be acquired by such devices. Signal measurement and feature extraction are thus performed using a combination of analog hardware and off-line signal processing that could easily be adapted for online use. The use of off-line analysis and analog

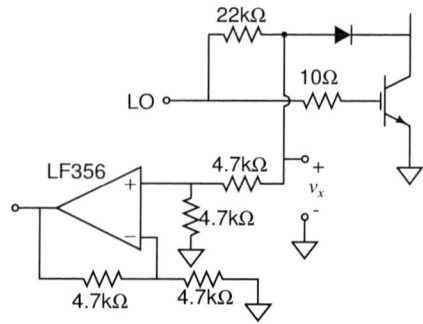

Fig. 4. Circuit used to measure the on-state collector-to-emitter voltage.

processing should not be viewed as a limitation, but rather as an approach used to demonstrate proof-of-concept. The data-acquisition system samples its inputs simultaneously at 50kHz. The system is connected to a PC with MATLAB and is referenced to the low-voltage ground of the control board. For demonstration purposes, only a single low-side switch is monitored; it would not be difficult to monitor the others.

Several key features are obtained by processing measurements of i_C and v_{CE}. The collector current is easily measured using a Hall-effect transducer. The collector-to-emitter voltage, on the other hand, is difficult to measure directly because the required amplifier would be exposed to common-mode swings on the order of several hundred volts and the subsequent analog-to-digital converter would have difficulty measuring the very low on-state voltage with adequate resolution. To overcome these issues, v_{CE} is measured in the on-state using the desaturation detection circuit included in the gate drive. Figure 4 shows the approach. When the IGBT is conducting, the signal v_X has the form

$$v_X = V_D + v_{CE}, \tag{7}$$

where V_D is the forward voltage drop across the diode in the desaturation circuit. Since the voltage at the output of the differential amplifier in Fig. 4 is referred to the high voltage bus, the circuit shown in Fig. 5 is used to transmit its output to the data-acquisition system. Similar isolation would be required in an FPGA-based design, but its exact location in the signal path would depend upon the specifics of the design.

Although the literature suggests that on-state v_{CE} can be directly applied in fault detection, this is only true under carefully controlled conditions that do not necessarily apply in the field. To understand this, consider the circuit model for a conducting IGBT. Since such devices are modeled as a diode in series with a power MOSFET, the on-state voltage is

$$v_{CE,ON} = V_J + i_C R_{ON}. \tag{8}$$

Given that most incipient faults have only a small impact on $v_{CE,ON}$, Eq. 8 suggests difficulty in distinguishing between changes in i_C and true fault conditions. This problem is further compounded by the fact that both V_J and R_{ON} are affected by temperature [33]. Measurement is particularly problematic

978-1-4799-0024-4/13 $31.00 © 2013 IEEE

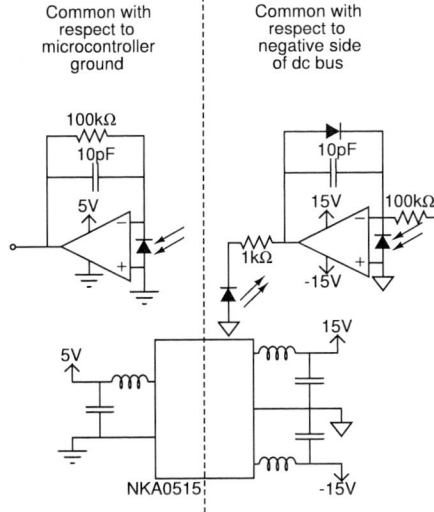

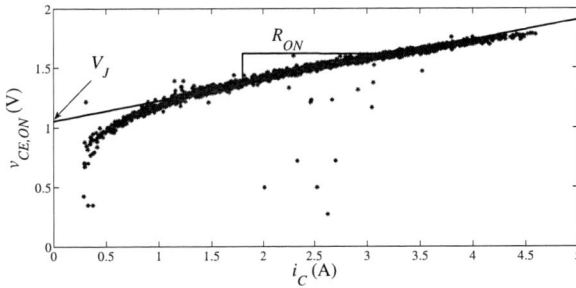

Fig. 7. Plot showing v_{CE} versus i_C over a line cycle. Note that a fit is performed to extract V_J and R_{ON}.

Fig. 5. Optical transmission circuit used to provide isolation between the low voltage data-acquisition system and measurements recorded with respect to the high voltage bus. The NKA0515 provides isolated DC supplies for all of the measurement circuits included here.

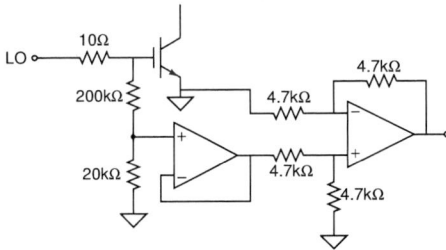

Fig. 8. Differential amplifier circuit used to measure the gate-to-emitter voltage.

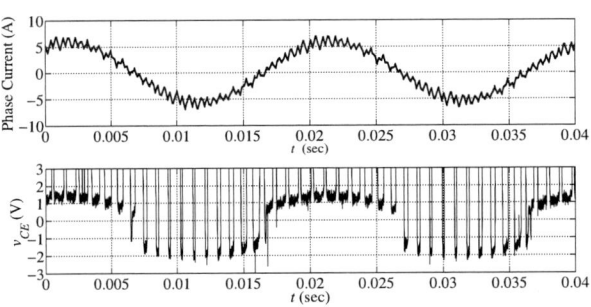

Fig. 6. Measured phase current and v_{CE} for one of the switches in the experimental drive. When v_{CE} is negative, the anti-parallel diode is conducting.

in variable-speed AC drives in which the current is subject to continuous variation.

Figure 6 illustrates the effect of Eq. 8 in an operational drive. Note that v_{CE} essentially follows the current when the IGBT is conducting. To isolate the effect of the current, we separately estimate the two parameters in Eq. 8 using a least-squares approach over each line cycle. Figure 7 shows an example. These parameters and the corresponding RMS current over the cycle are extracted and included in the feature vector $\boldsymbol{x}_k$ computed at time t_k

The next useful feature included in the present implementation is $V_{GE,ON}$. Figure 8 shows the circuit measuring the gate-to-emitter voltage, which is simply a differential amplifier. A circuit similar to the one in Fig. 5 transmits the output back to the data-acquisition system. Samples are averaged over each conduction interval to compute $V_{GE,ON}$ for that interval.

Figure 9 shows the circuit used to directly extract the parameter Q_G. Note that a high-bandwidth differential amplifier

(AD8055) measures the voltage across the gate resistor to obtain i_G. The subsequent signal is amplified and passed to an operational transconductance amplifier (OPA660) configured as an integrator. A sample-and-hold (AD781) triggered by the data-acquisition system acquires the output of this circuit. Figure 10 shows the output of the integrator as well as the trigger signal and the corresponding sample-and-hold output. Note that the integrator output rises during the device turn-on, i.e. while v_{GE} is rising. Once i_G has fallen to zero and v_{GE} has become steady, the integrator capacitor begins to slowly discharge through the output resistance of the OPA660. Note that a sample is acquired during this discharge in order to avoid the switching noise at device turn-on. Figure 11, which shows the integrator output signal during three different device turn-ons, provides the rationale. Note that each signal is clearly impacted differently by the common-mode switching noise. Once this has dissipated, however, each output is quite consistent. The output at the point labeled 'Sample' is taken as the feature Q_G. Further discussion follows in Section IV-B, which shows that the output at the sampling point varies significantly as the gate oxide exhibits signs of degradation.

A solid-state sensor measures the ambient temperature, and a thermocouple measures the case temperature. These measurements are also averaged over a conduction interval. Ultimately, estimates of T_C, T_A, Q_G. $V_{GE,ON}$, I_C, V_J, and R_{ON} over each line cycle are used to assemble a feature vector at time t_k. Threshold voltage and possibly other features will be included in the future.

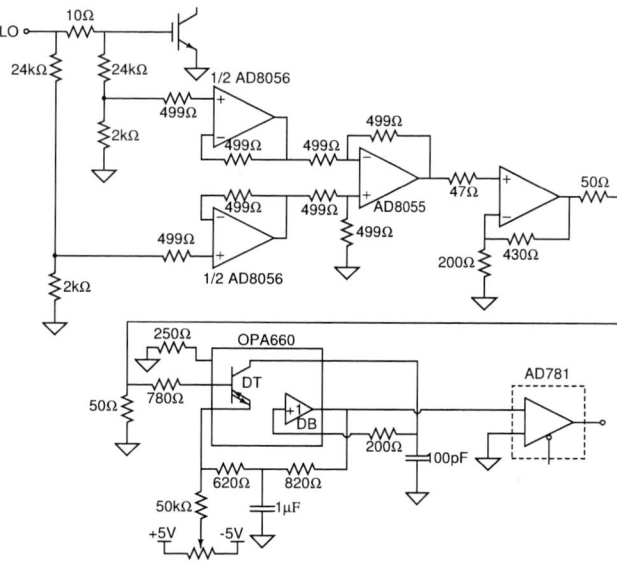

Fig. 9. Circuit used to extract the gate charge injected during turn-on. The OPA660 is an operational transconductance amplifier that integrates using the 100pF capacitor. The trigger signal for the AD781 sample-and-hold is provided from the data-acquisition system using the inverse of the circuit shown in Fig. 5

B. Experimental Demonstration

To illustrate the effectiveness of the proposed algorithm, two different early stage faults were introduced into a single IGBT

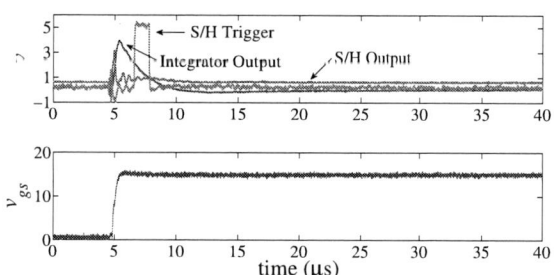

Fig. 10. Top trace: Integrator output, sample-and-hold output, and trigger signal during a device turn-on. Note the effect of switching noise. Bottom trace: The corresponding gate-to-emitter voltage.

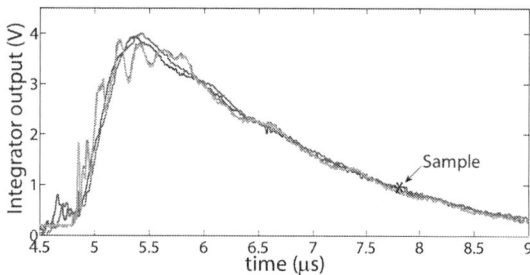

Fig. 11. Integrator output during three different device turn-ons. Note that when the signals are rising, i_G is being integrated. The sampling point is shown.

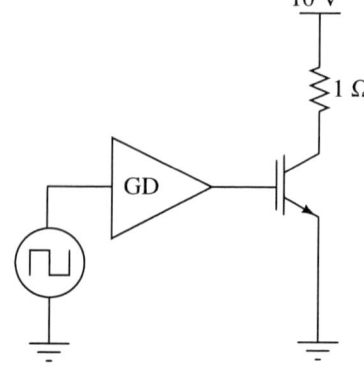

Fig. 12. Circuit used to age the IGBT. A 1kHz waveform with $D = 0.4$ drives the gate.

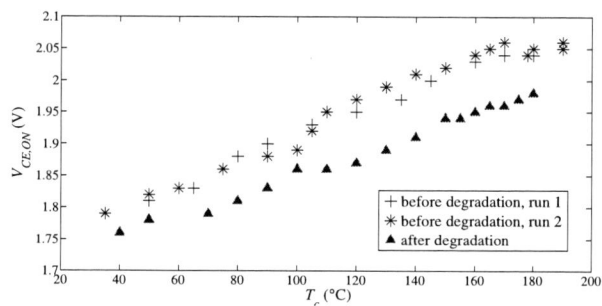

Fig. 13. Average on-state v_{CE} (labeled here as $V_{CE,ON}$) versus temperature for the IGBT, both before and after aging.

switch. First, the monitored switch was removed and subjected to an accelerated aging test to produce an early stage die-attach fault. To do so, the IGBT was placed in the circuit shown in Fig. 12. The gate signal in this circuit is a 1kHz square wave with a 40% duty ratio. The average current recorded during testing was approximately 3A, which is rather high given that the part is only rated for 10A and no heat sink was used. Case temperature was measured during the test using an infrared camera. The test was stopped once T_C reached approximately 190°C. This test was repeated for a number of hours. A similar approach was used in [34] to develop die-attach faults.

Figure 13 shows results recorded during the ageing process. Note the positive temperature coefficient; this is expected for IGBTs [34]. Additionally, note that the component has a lower average on-state v_{CE} at all temperatures once it has degraded. This result was also obtained in [34]. It is believed that the drop occurs because the degraded die attach increases θ_{jc} and thus creates a higher internal temperature at any given T_C. Since the *pn* junction has a negative temperature coefficient of resistance [35], an increase in the temperature of this junction lowers average on-state v_{CE}.

Figure 14 shows results obtained using the proposed condition-monitoring algorithm once the degraded part was returned to the drive. Note that the part was tested in both the healthy and degraded state, and that the load on the

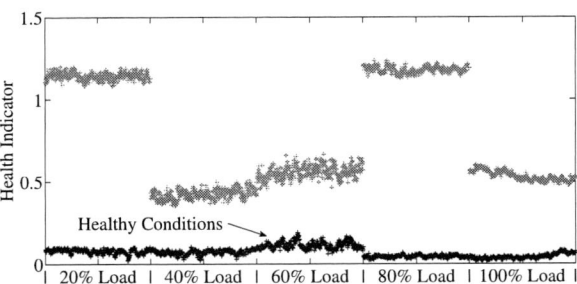

Fig. 14. The reconstruction error recorded by the health-monitoring algorithm before and after the die-attach fault was induced. Note that there are a number of samples recorded at each loading condition and that the higher error at each load step corresponds to the degraded transistor.

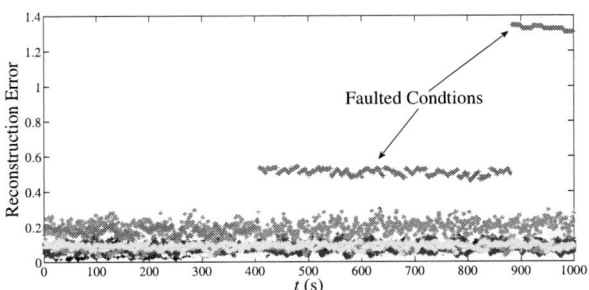

Fig. 15. The reconstruction error recorded by the health-monitoring algorithm in 10% load steps for a healthy part. Note that they are all reasonably similar. The high errors beginning at ∼ 400s and ∼ 900s correspond to increased gate leakage current, i.e. ∼ 3.2mA and ∼ 32mA, respectively, at 100% load.

machine was varied in order to simulate changes in operating conditions. The figure shows the reconstruction errors in both the healthy and degraded conditions, using data extracted at 20% load steps. Note that the degraded part caused a significantly higher reconstruction error at each load step and that changes in the loading condition had minimal impact. Note that drive operation appeared normal despite the fact that the device was nearing failure.

As another example, we simulated the effect of a breakdown in the gate oxide, which can result in a leakage current [18]. As in [6], we simulated this fault by placing resistance between the gate and the emitter of the monitored switch. To demonstrate the ability to detect early stage faults, we used different resistor values. Figure 15 shows the reconstruction error versus time for both healthy and faulted operation. Note that time series are shown for the healthy part in 10% increments from no load to full load. To simulate a faulted part, we added approximately $4.7k\Omega$ at $t \approx 400s$ and then 470Ω at $t \approx 900s$. Note that the corresponding leakage currents are small ($\sim 3.2mA$ and $\sim 32mA$, respectively), and that both conditions correspond to scenarios in which the drive maintains complete functionality despite the presence of the soon-to-fail switch.

V. CONCLUSION

Online methods for early stage fault detection will be essential as systems become more heavily dependent on power electronics. This paper has presented and demonstrated a health-monitoring algorithm for power switches that can distinguish between true faults and changes in operating conditions and does not require the development of rules. Ongoing work focuses on the inclusion of additional features and the completion of an exhaustive battery of tests. Note that this algorithm is for fault detection, and thus diagnostic and prognostic algorithms using the measured features are under development. These would operate on feature measurements after faults have been detected. Human operators could also observe the data.

VI. ACKNOWLEDGEMENTS

The authors would like to acknowledge comments and input from Adam Kabulski at Converteam Naval Systems.

REFERENCES

[1] T. Dalton, A. Boughner, C. Mako, and C. N. Doerry, "LHD 8: A step toward the all electric warship," in *ASNE Day*, 2002.

[2] A. N. Inc., "Variable frequency drives: More reliability, less houseload, more profit," 2008.

[3] J. Rodriguez, L. Franquelo, S. Kouro, J. I. Leon, R. C. Portillo, M. A. M. Prats, and M. A. Perez, "Multilevel converters: An enabling technology for high-power applications," *Proc. IEEE*, vol. 97, no. 11, pp. 1786–1817, Nov. 2009.

[4] A. M. Imam, T. G. Habetler, R. G. Harley, and D. Divan, "Failure prediction of electrolytic capacitors using dsp methods," in *Proc. of the 20th IEEE Applied Power Electronics Conference (APEC'05)*, vol. 2, 2005, pp. 965–970.

[5] J. M. Anderson, R. W. Cox, and J. Noppakunkajorn, "An on-line fault diagnosis method for power electronic drives," in *Proc of IEEE Electric Ship Technologies Symposium*, Alexandria, VA, Apr. 2011, pp. 492 – 497.

[6] L. Chen, F. Z. Peng, and D. Cao, "A smart gate drive with self-diagnosis for power MOSFETs and IGBTs," in *Proc. of IEEE Applied Power Electronics Conference and Exposition*, Austin, TX, Feb. 2008, pp. 1602 – 1607.

[7] P. Xiao, G. K. Venayagamoorthy, K. A. Corzine, and R. Woodley, "Self-healing control with multifunctional gate drive circuits for power converters," in *Conference Record of the 2007 IEEE Industry Applications Conference*, New Orleans, LA, Sept. 2007, pp. 1852 – 1858.

[8] T.-H. Liu, J.-R. Fen, and T. A. Lipo, "A strategy for improving reliability of field-oriented controlled induction motor drives," *IEEE Trans. Ind. Appl.*, vol. 29, no. 5, pp. 910–918, Sep/Oct 1993.

[9] R. Ribeiro, C. B. Jacobina, E. da Silva, and A. Lima, "Fault detection of open-switch damage in voltage-fed PWM motor drive systems," *IEEE Trans. Power Electron.*, vol. 18, no. 2, pp. 587–593, Mar. 2003.

[10] A. K. Kharegakar and P. Pavana Kumar, "A novel scheme for protection of power semiconductor devices against short circuit faults," *IEEE Trans. Ind. Electron.*, vol. 41, no. 3, pp. 344–351, Jun 1994.

[11] H. A. Toliyat, "Analysis and simulation of five-phase variable-speed induction motor drives under asymmetrical connections," *IEEE Trans. Power Electron.*, vol. 13, no. 4, pp. 748–756, Jul. 1998.

[12] N. D. Benavides, T. J. McCoy, and M. A. Chrin, "Reliability improvements in integrated power systems with pressure-contact semiconductors," in *Proc. ASNE Day*, Apr 2009.

[13] M. Ciappa, "Selected failure mechanisms of modern power modules," *Microelectronics Reliability*, no. 42, pp. 653–667, Jan 2002.

[14] J. Celaya, N. Patil, S. Saha, P. Wysocki3, and K. Goebel, "Towards accelerated aging methodologies and health management of power MOSFETs (technical brief)," in *Proc. of Annual Conference of the Prognostics and Health Management Society*, San Diego, CA, Sept./Oct. 2009.

[15] G. Sonnenfeld, K. Goebel, and J. R. Celaya, "An agile accelerated aging, characterization and scenario simulation system for gate controlled power transistors," in *Proc of IEEE AUTOTESTCON*, Salt Lake City, UT, Sept. 2008, pp. 208 – 215.

[16] M. Trivedi and K. Shenai, "Failure mechanisms of IGBTs under short-circuit and clamped inductive switching stress," *IEEE Trans. Power Electron.*, vol. 14, no. 1, pp. 108–116, Jan. 1999.

[17] J. Morroni, A. Dolgov, R. Zane, and D. Maksimovi, "Online health monitoring in digitally controlled power converters," in *Proc. IEEE Power Electron. Specialists Conf.*, Orlando, FL, Jun 2007.

[18] S. Lombardo, J. Stathis, B. Linder, K. Pey, F. Palumbo, and C. Tung, "Dielectric breakdown mechanisms in gate oxides," *Journal of Applied Physics 98*, 2005.

[19] G. Buh, H. Chung, and Y. Kuk, "Real-time evolution of trapped charge in a SiO_2 layer: An electrostatic force," *Applied Physics Letters*, vol. 79, no. 13, pp. 2010–2012, 2001.

[20] Y. Lobsiger and J. W. Kolar, "Voltage, current and temperature measurement concepts enabling intelligent gate drives," in *European Center for Power Electronics (ECPE) Workshop - Electronics Around the Power Switch: Gate Drives, Sensors, and Control*, 2011.

[21] H. Kuhn, T. Koneke, and A. Mertens, "Potential of digital gate units in high power applications," in *Proc. 13th Power Electronics and Motion Conference (EPE-PEMC)*, 2008, pp. 1458–1464.

[22] ——, "Considerations for a digital gate unit in high power applications," in *Proc. 2008 Power Electronics Specialists Conference (PESC)*, 2008, pp. 2784–2790.

[23] J. R. Black, "Electromigration - a brief survey and some recent results," *IEEE Trans. Electron Devices*, vol. 16, no. 4, pp. 338 – 347, April 1969.

[24] E. Ameraseka and F. Najm, *Failure Mechanisms in Semiconductor Devices*, 2nd ed. John Wiley & Sons Ltd, 1998.

[25] J. Celaya, A. Saxena, P. Wysocki, S. Saha, and K. Goebel, "Towards prognostics of power MOSFETs: Accelerated aging and precursors of failure," in *Proc of Annual Conference of the Prognostics and Health Management Society*, Portland, Oregon, Oct. 2010.

[26] W. Wu, M. Held, P. Jacob, P. Scacco, and A. Birolini, "Investigation on the long term reliability of power IGBT modules," in *Proceedings of International Symposium on Power Semiconductor Devices & ICs*, Yokohama, Japan, 1995, pp. 443 – 448.

[27] D. Katsis and J. van Wyk, "Void-induced thermal impedance in power semiconductor modules: Some transient temperature effects," *IEEE Trans. Ind. Appl.*, vol. 39, no. 5, pp. 1239 – 1246, Sept. / Oct. 2003.

[28] J. D. Kraus, *Electromagnetics*. McGraw-Hill Book Company, Inc., 1953.

[29] D. L. Goodman, "Prognostic methodology for deep submicron semi-conductor failure modes," *IEEE Trans. on Components and Packaging Technologies*, vol. 24, no. 1, pp. 109 – 111, March 2001.

[30] A. Hamidi, N. Beck, K. Thomas, and E. Herr, "Reliability and lifetime evaluation of different wire bonding technologies for high power IGBT modules," *Microelectronics Reliability*, no. 39, pp. 1153 – 1158, 1999.

[31] M. Turk and A. Pentland, "Eigenfaces for recognition," *Journal of Cognitive Neuroscience*, vol. 3, no. 1, pp. 71–86, 1991.

[32] R. O. Duda, P. E. Hart, and D. G. Stork, *Pattern Classification*. Wiley, 2006.

[33] N. Mohan, T. Undeland, and W. Robbins, *Power Electronics Converters, Applications, and Design*, T. Edition, Ed. John Wiley & Sons, 2003.

[34] N. Patil, D. Das, K. Goebel, and M. Pecht, "Identification of failure precursor parameters for insulated gate bipolar transistors (IGBTs)," in *Proc of International Conference of Prognostics and Health Management*, Denver, CO, Oct. 2008, pp. 1–5.

[35] J. Baliga, *Power Semiconductor Devi*. Boston, MA: PWS Publishing, 1996.

Onboard Condition Monitoring of Solder Fatigue in IGBT Power Modules

B. Ji, V. Pickert, W. P. Cao, and L. Xing

Abstract—This paper proposes a novel on-board condition monitoring of the aging of solder layers in IGBTs for electric vehicle applications. The diagnostic technique makes use of the chip itself as a temperature sensor while current sensors are already in place for control purposes. An auxiliary power supply unit which can be created from the 12V battery and an in situ data-logger circuit is developed for condition monitoring. The novel aspect of the proposed technique relates to monitoring IGBTs in situ when the electric vehicle is operating during stop-and-go traffic conditions or at routine services. The accelerated aging tests are performed on the test vehicles and the condition monitoring system is validated using simulation and thermo-electrical experimentation. The thermal performance of the thermal resistance/impedance and junction temperature of the IGBTs demonstrates the effectiveness of the proposed technique for IGBT health monitoring.

Index Terms—Monitoring, insulated gate bipolar transistors, power electronics, prognostics and health management, semiconductor device reliability, thermal management

I. INTRODUCTION

Due to the superior performance and rapid development towards high power density, high efficiency, and low cost, insulated gate bipolar transistor (IGBT) power modules have been widely used in the consumer and industrial power electronic systems such electric vehicles (EVs) [1]-[3], ships

---

B. Ji, V. Pickert and L. Xing are with the School of Electrical and Electronic Engineering, Newcastle University, Newcastle upon Tyne, NE1 7RU, United Kingdom, England. (bing.ji@ncl.ac.uk)

L. Xing is with the School of Chemical Engineering &Advanced Materials, Newcastle University, Newcastle upon Tyne, NE1 7RU, United Kingdom, England.

[4], aircraft [5], wind turbines [6], smart grid [7] and industrial drives [8]. IGBT modules are switching elements designed for high-power levels and constitute the central component of modern power electronic converters. Their reliability and operational lifetime are of great concern, especially for safe critical applications where a sudden failure may result in a catastrophic accident and high penalty costs.

The failure rate has traditionally been employed as a reliability index to electrical apparatus and devices, which generally follow the so-called bathtub curve. As shown in Fig. 1, it can be divided into three stages including early failure period, random failure period and wear-out failure period. A product manufacture defect and improper design can always lead to early failures while random failures are mainly due to the intermittent excessive stress (electrical, thermal, mechanical, etc) over the maximum rating of the devices. With continuous improvements within quality control, semiconductor chip and packaging technologies, system protection and optimum control algorithms, such failures can be largely reduced. On the other hand, the wear-out failure occurring towards the end of lifetime is generally caused by the chronic environmental and operational loadings. Solder fatigue is a commonly observed wear-out failure mode, which is susceptible to IGBT failures if left untreated, and is therefore the focus of this paper.

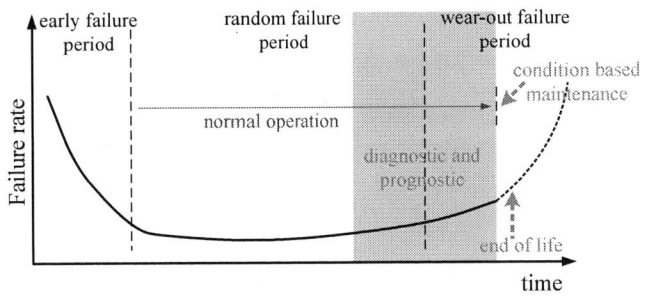

Fig. 1. Change in failure rate over time (bathtub curve)

In this paper, an in situ condition monitoring (CM) circuit was proposed for a traction motor drive in EV applications. The on-board diagnostic and prognostic test was conducted to demonstrate the solder layer condition monitoring capability. The CM circuitry can be embedded in an advanced gate driver unit (GDU) and each IGBT in the inverter can be monitored sequentially within a short period of time when the EV is operating during stop-and-go traffic conditions or at routine services. In this case, many of the catastrophic failures as a result of solder fatigue could be avoided.

This paper is organized with an introduction of the contemporary IGBT packaging and review of solder fatigue detection methods in Section II. Section III then presents the on-board condition monitoring system with considerations for real applications. Section IV shows the simulation and experimental results. Section V gives the conclusion.

II. IGBT MODULES AND SOLDER FATIGUE

Power electronic converters EV applications are required to adapt to harsh environments and diverse mission profiles. This will burden the IGBT power modules which undergo substantial power cycles and thermal cycles [9-12].

A. Solder Fatigue of IGBT Modules

The schematic diagram of the cross-section view of a standard IGBT module is shown in Fig. 2. It is assembled with different materials in a multilayered structure. Typically, the IGBT chip is soldered onto a direct copper-bonded (DCB) substrate, which is composed of ceramic and metallized copper films. The DCB is then soldered onto a copper baseplate. In common, chip surface is connected to copper tracks by wire bonding methods. The assembly is housed by a plastic case and encapsulated with silicon gel and epoxy. Thermal interface material (TIM) is always inserted between the baseplate and heat sinks for cooling purpose.

Due to the coefficient of thermal expansion (CTE) mismatch between base plate and DCB substrate, as well as temperature gradients, repetitive expansion and contraction stresses are built up in the solder layer, resulting in failure mechanisms such as creep [13], voids [14], cracks [15,16] and delamination [15,16]. Consequently, the heat conduction

capability is diminished resulting in increased thermal impedance and junction temperature. This may not generate a sudden failure but can deteriorate the device performance and eventually lead to ultimate failures (i.e. hot spot, latch up, burn-out, etc.). In addition, device power losses are always increased as a result of the increased junction temperature, which becomes a positive feedback mechanism and will accelerate deteriorations.

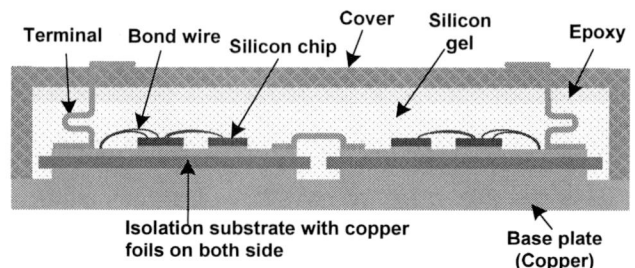

Fig. 2.Schematic diagram of a standard IGBT module

B. Condition Monitoring of Solder Fatigue

Recently, extensive research activities in achieving high IGBT reliability with CM methods are going on world-wide, which can be generally divided into model-driven and data-driven methods. Model-driven estimation deduces the temperature cycle in real-time from case temperature, device losses, and thermal network [17-20]. Apart from the difficulty of loss estimation, assumptions for simplicity are always made to neglect the coupling effects from other forms of fatigue mechanisms. Indeed, an increase of thermal resistance always comes with the solder aging. An alternative is to use data-driven method where diagnosis and condition monitoring are performed entirely depending on the measured signals. Various non-destructive testing (NDT) techniques were studied to correlate featuring parameters with solder layer fatigue in the literature. Scanning acoustic tomography or microscopy (SAT/SAM) techniques have been widely used and a real-time based failure detector was claimed [21]. Active thermography based strategies are also investigated and the detection of the degraded solder joints after a 2mm thick Cu layer was demonstrated [22]. A time-domain reflectometry technique by monitoring RF impedance stimulated by high-frequency signal (GHz) was also introduced which shows superior effectiveness over traditional dc resistance based monitoring method [23]. All above methods rely on sophisticated measurement systems which are normally costly and space consuming. In addition, necessary physical charges of the conventional power converter assembly are required to accommodate these systems, which are not feasible to be implemented in practice. These methods are yet immature and have only been used for laboratory test purpose. A simplified CM method was proposed to estimate the thermal resistance increase as a result of case-above-ambient temperature rise, but the effect of degradation upon thermal interface material and heat sink cannot be separated [24].

III. IMPLEMENTATION OF IN SITU CONDITION MONITORING

A. Condition Monitoring with Thermal Characterization

The junction-to-case thermal impedance is characterized and monitored here to enable IGBT solder failure detection. The transient thermal impedance (TTI) is generally defined as the time dependent temperature difference between silicon chip and reference point divided by the constant power loss (P).

$$Z_{thjr}(t) = \left[T_j(t) - T_r(t) \right]/P \qquad (1)$$

Heating and cooling curves are complementary for linear thermal system and can both be used for TTI measurement. Generally, cooling curve is recorded when heating power is removed after reaching a thermal steady state. It normally requires a long period of time (i.e. tens or hundreds of seconds) which is depending on the packaging and cooling properties. Heating curve, on the other hand, can be implemented with defined heating power prior to thermal steady state and allows fast in situ monitoring. In order to meet EVs' requirement, the heating method is used and the length of the heating pulse is controlled to focus on the DCB solder layers. Since several factors have been reported in the literature that can affect the TTI measurement result including power dissipation magnitude, type of temperature response, environmental conditions (mounting techniques, heat sink temperature, etc) and selection of temperature sensors (for both junction and reference temperature measurement), careful considerations are required in order to assure accurate and meaningful results under actual applications.

B. Hardware Setup

Generally, fault diagnostic functions implemented in conventional gate drive units (GDUs) are used for overstress detections and post-fault protections (over-voltage, over-current, over temperature, etc). Up to now it was almost impossible to diagnose the solder layer health and predict its lifetime. One of the clear technical challenges in current solder CM method is measurement accuracy. It requires the extraction of small temperature (a few °C) and electrical signals (a few millivolts) from larger ones (hundreds of volts) meaning that conventional methods can not apply to in situ measurements. That is, the variations in prognostic parameters resulting from ageing are relatively small and it can be overwhelmed by noise or disturbances in the EV power network which are associated with changes in operating conditions (temperature, loading or control). As a result, a dedicated data-acquisition method is developed to improve the measurement accuracy.

Fig. 3 shows photographs and schematic diagrams of the complete experimental setup. In Fig. 3(a), the DUTs (IGBTs) are placed in the thermal chamber to maintain a required testing environment and tested using the proposed on-board CM circuit. In Fig. 3(b), T1-T6 are six identical IGBTs, D1-D6 are six FWDs and M is an electric motor. The measurement circuitry consists of: 1) an auxiliary power supply unit (PSU); 2) a gate drive and protection circuit; 3) a measurement circuit with digital isolation; and 4) selector relays.

(a) Photographs of the experimental setup

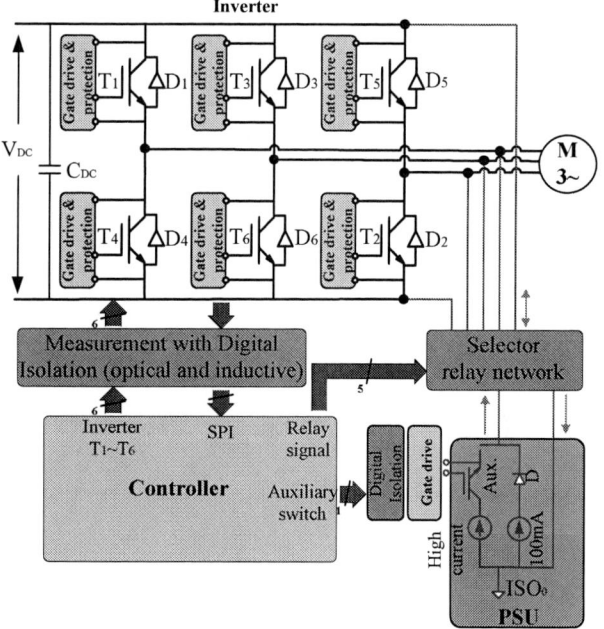

(b) The condition monitoring circuit within the EV power inverter

Fig. 3. The proposed in situ measurement circuitry

IV. SIMULATION AND EXPERIMENTAL RESULTS

A. FEM Simulation Result

Finite element method (FEM) is performed using COMSOL Multiphysics software to evaluate the thermal characteristics of the device under different solder conditions and their results are later compared with practical measurement results. It is assumed that conduction is the only heat transfer mechanism and all heat generated from IGBT chip flows through the DCB and base plate and out of the heat sink. Side walls are assumed to be adiabatic, i.e. the flow of heat is totally constricted across the edges. The contact thermal resistance between two adjacent layers is also excluded. The temperature dependence of other material properties is neglected except for the silicon thermal conductivity, which is sufficiently large. This is approximated based on Table I [25,26]. A dissipative power step is applied to the active chip volume, i.e. 100um from the chip top surface. The ambient temperature is set to be 21°C and a defined heat transfer coefficient is applied to the bottom of the heat sink to simulate the forced air cooling. The heat flow in each layer is calculated with the partial differential equation

$$\rho C_p \frac{\partial T}{\partial t} + \nabla . (-k\nabla T) = Q + q_s T \qquad (2)$$

where k, C_p, ρ, q_s and Q are the thermal conductivity, heat capacity, density, absorption/production coefficient and heat, respectively.

The dimension of each layer and material properties of the corresponding layer within the IGBT module are given in Table II and III. The applied power corresponds to the power loss based on practical measurements when it conducts current of 60A with on-state voltage measured in real time. The switching losses are neglected due to the low switching frequency and high duty ratio. The same loading is applied when solder fatigue are introduced. The effect of progressive solder fatigue was modeled by introducing a perfect 3 um thick 'crack' layer (infinite thermal resistance) located within the DCB to baseplate solder joint and their results are compared in Fig. 4. It is clear that the heat in the 'crack' pocket is difficult to remove and thus give rise to its temperature.

TABLE I

TEMPERATURE DEPENDENT THERMAL CONDUCTIVITY OF SILICON

Temperature, T (K)	Thermal Conductivity, k (W/mK)
250	191
300	148
350	119
400	98.9
500	76.2

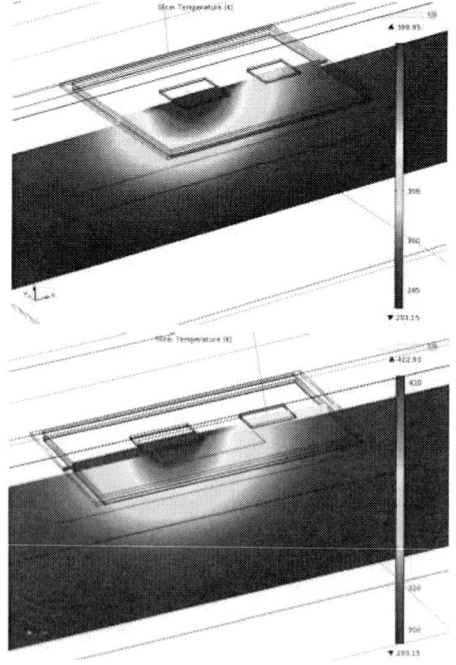

Fig.4 FEM analysis using COMSOL for (a) healthy (100%) and (b) faulty (36%) solder layers between DCB and base plate

978-1-4799-0024-4/13 $31.00 © 2013 IEEE

TABLE II

MATERIAL PROPERTIES FOR IGBT POWER MODULE CONSTRUCTION

Material	Thermal conductivity, λ (W/(m*K))	Specific heat,c (J/(kg*K))	Mass density ,ρ (kg/m³)
silicon	163	703	2330
Solder	70	250	8000
copper	400	385	8700
alumina	27	900	3900
copper	400	385	8700
Dow corning 340	0.54	250	2140
aluminium	160	900	2700

TABLE III

DIMENSIONS OF THE IGBT POWER MODULE

	IGBT	Die attach	DCB copper	ceramic	Base plate solder	Base plate
L(mm)	6.5	6.5	28.44	30.60	28.44	91.48
W(mm]	6.5	6.5	26.12	28.44	26.12	31.40
H(mm)	0.22	0.08	0.3	0.38	0.08	3

B. Power Loss Measurement

Power dissipation from the device junction is developed by injecting constant heating current during IGBT forward conduction state. Ambient temperature is controlled with the thermal chamber and heat sink with forced air cooling is used. A controlled heating current pulse train is shown in Fig. 5(a). The heating current and on-state voltage are measured with their multiplication representing the instantaneous power loss. Only conduction losses are considered and the switching losses are neglected due to low switching frequency of the heating pulses. Since the on-state voltage $V_{CE(on)}$ gradually increases with the elevated junction temperature, an increased power dissipation is observed as illustrated in Fig.5(b). In addition, the on-state voltage, as well as the power dissipation, is also influenced by ambient temperature. Although fixed heating power can be achieved by adjusting the gate voltage or collector current via feedback control, this may add system cost and complexity [27].

As a consequence, fluctuated power dissipation is segmentally averaged for quasi-TTI measurement. The power dissipation waveform is represented approximately by m sequential pulses with averaged amplitudes of P_1, P_2, $\cdots$, P_m. The amplitude of each pulse can be calculated based on the equation below with N being the total samples within the pulse.

$$P_{av} = \frac{1}{N}\sum_{u=1}^{N} V_{CE(on)}(u)\cdot I_C(u) \qquad (3)$$

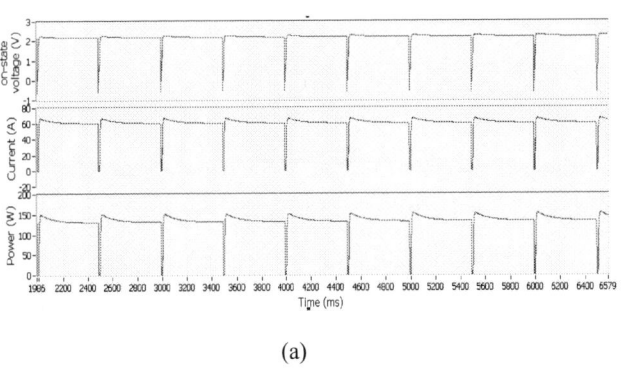

(a)

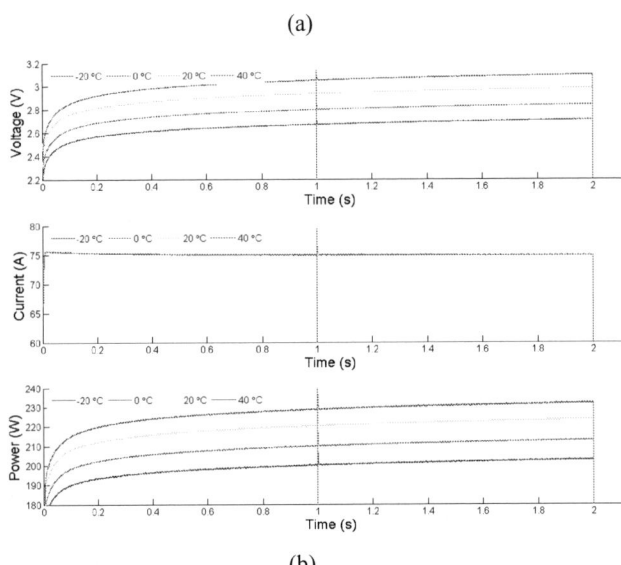

(b)

Fig.5 heating pulse for TTI measurement (a) and (b) under different ambient temperature

Since the thermal resistance of DCB solder contributes only a small portion to the total junction-to-case thermal resistance, high dissipation power is required which enables larger temperature gradient at given thermal resistance. Consequently, more accurate junction-to-case thermal impedance can be determined by increasing the signal-to-noise ratio (SNR).

The heating power can be assumed to follow a conductive heat flow path from the heat dissipating junction of the IGBT to the environment. A finite amount of time is required for the heat generated from device to propagate outward. This offers a spatial inspection of the thermal performance of the multilayered power module. By controlling the length of the applied heating pulse, the depth of the heat diffusion can be regulated. This allows the solder layer condition to be focused within the power assembly. However, a compromise has to be

978-1-4799-0024-4/13 $31.00 © 2013 IEEE

made between maximising the variation due to the thermal property change in the layer of interest and minimising the interference due to changes from subsequent layers in the measurement of the thermal impedance.The length of the heating pulse is defined based on the transient dual interface (TDI) measurement [28] and 2s is selected as the end of TTI measurement.

C. Junction temperature measurement

The IGBT junction temperature is required for thermal performance test and it is normally determined by temperature sensitive electrical parameters (TSEPs) [29,30]. The on-state voltage drop is used here for its simple implementation and a linear temperature dependency. Prior to the thermal characterization, the TSEP of each DUT shall be calibrated individually. Since TSEP measurement can not be achieved at the same time as the heat generation, an approach referred to as the switched method is used during heating response. The IGBT is switched from a heating condition when heating current is applied to the TSEP measurement condition when only calibration current is applied (as shown in Fig. 5a) [30]. The thermal impedance calculation error resulting from the cooling effect due to measurement delay is corrected by extrapolation.

D. Results

Accelerated passive thermal aging test is performed on devices with the air-to-air thermal shock chamber. The temperature varies from -50°C to 160°C. The dwell time is 10 min with a 2 min transition time. The thermal cycling is interrupted at 800 and 1300 cycles. TTI measurement are performed and the thermal resistance at t =1s, 2s are compared as shown in Fig. 6. This is also compared with simulation result.

The IGBT chip surface temperature under healthy and faulty solder conditions is demonstrated by the thermal images shown in Fig. 7(a). An increase in the maximum chip temperature goes from 117°C in initial state to 129°C in aged state according to the DCB solder changes from healthy state to degraded state after 1300 thermal cycles. The temperature gradient over the IGBT chip surface did not exceed 50°C in the initial state but reached more than 55°C in the aged state and this is illustrated in Fig. 7(b).

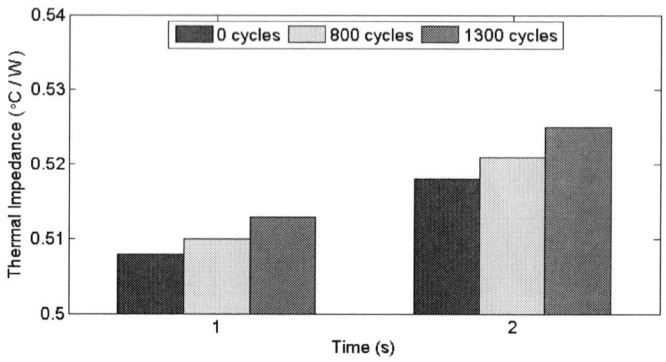

Fig.6 Heating curve TTI measurement at 1s and 2s for different solder conditions

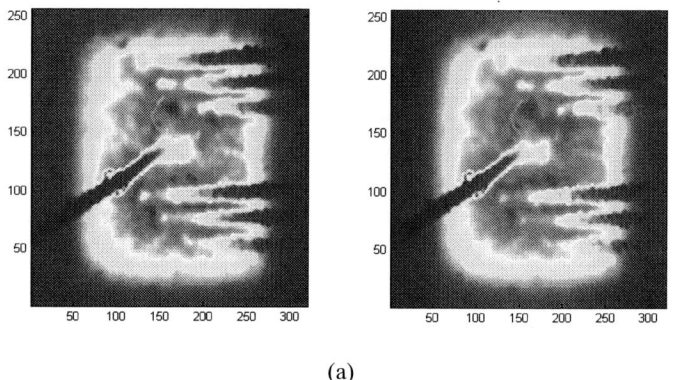

(a)

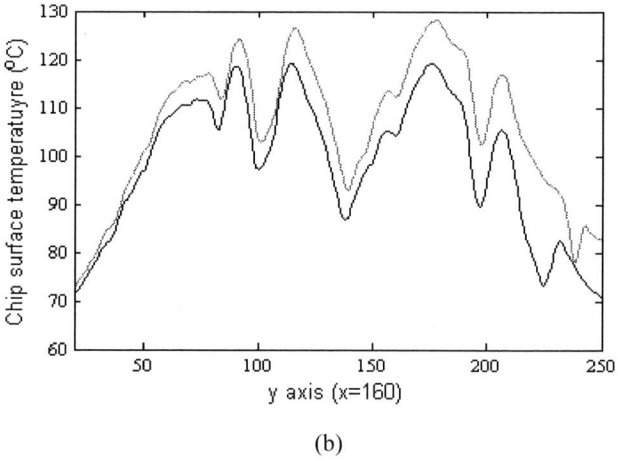

(b)

Fig. 7 Thermal images of an IGBT surface temperature with healthy and faulty solders (a) and its 1-dimentional temperature plot (b)

V. CONCLUSION

This paper has demonstrated importance of the diagnostic and prognostic capability for IGBT solder layers. Research has shown that with the solder layer degradation, junction-to-case thermal impedance and chip surface temperature variation also increase. The principle of the proposed method for in situ condition monitoring was explained. A novel on-board D/P technique with protection and sensor circuitry for IGBTs has been proposed in this paper, which can be embedded in an advanced gate drive circuit to improve system reliability. The condition monitoring system was verified using computer simulation and thermo-electrical experimentation on an in situ diagnostic and prognostics prototype.

ACKNOWLEDGMENT

The authors gratefully acknowledge the financial support of EPSRC project EP/K008552/1 on novel calorimeters for developing high-efficiency power converters and electrical machines.

REFERENCES

[1] Z. Yilu, G. W. Gantt, M. J. Rychlinski, R. M. Edwards, J. J. Correia, and C. E. Wolf, "Connected Vehicle Diagnostics and Prognostics, Concept, and Initial Practice," *Reliability, IEEE Transactions on,* vol. 58, pp. 286-294, 2009.

[2] S. Ofsthun, "Integrated vehicle health management for aerospace platforms," *Instrumentation & Measurement Magazine, IEEE,* vol. 5, pp. 21-24, 2002.

[3] K. P. Logan, "Intelligent Diagnostic Requirements of Future All-Electric Ship Integrated Power System," *Industry Applications, IEEE Transactions on,* vol. 43, pp. 139-149, 2007.

[4] Zahedi, B.; Norum, L.E., "Modeling and Simulation of All-Electric Ships With Low-Voltage DC Hybrid Power Systems," *Power Electronics, IEEE Transactions on ,* vol.28, no.10, pp.4525,4537, Oct. 2013

[5] Wenping Cao; Mecrow, B.C.; Atkinson, G.J.; Bennett, J.W.; Atkinson, D.J., "Overview of Electric Motor Technologies Used for More Electric Aircraft (MEA)," *Industrial Electronics, IEEE Transactions on ,* vol.59, no.9, pp.3523,3531, Sept. 2012

[6] Blaabjerg, F.; Liserre, M.; Ke Ma, "Power Electronics Converters for Wind Turbine Systems," *Industry Applications, IEEE Transactions on ,* vol.48, no.2, pp.708,719, March-April 2012

[7] Popović-Gerber, J.; Oliver, J.A.; Cordero, N.; Harder, T.; Cobos, J.A.; Hayes, M.; O"Mathuna, S.C.; Prem, E., "Power Electronics Enabling Efficient Energy Usage: Energy Savings Potential and Technological Challenges," *Power Electronics, IEEE Transactions on ,* vol.27, no.5, pp.2338,2353, May 2012

[8] Chaal, H.; Jovanovic, M., "Toward a Generic Torque and Reactive Power Controller for Doubly Fed Machines," *Power Electronics, IEEE Transactions on ,* vol.27, no.1, pp.113,121, Jan. 2012

[9] E. Wolfgang, "Examples for failures in power electronics systems," EPE Tutorial 'Reliability of Power Electronic Systems'April 2007.

[10] Y. Shaoyong, A. Bryant, P. Mawby, X. Dawei, R. Li, and P. Tavner, "An Industry-Based Survey of Reliability in Power Electronic Converters," Industry Applications, IEEE Transactions , vol. 47, pp. 1441-1451, 2011.

[11] Smet, V.; Forest, F.; Huselstein, J.; Rashed, A.; Richardeau, F., "Evaluation of V_{ce} Monitoring as a Real-Time Method to Estimate Aging of Bond Wire-IGBT Modules Stressed by Power Cycling," *Industrial Electronics, IEEE Transactions on,* vol.60, no.7, pp.2760,2770, July 2013

[12] Scheuermann, Reliability challenges of automotive power electronics, Microelectronics Reliability, Volume 49, Issues 9–11, September–November 2009, Pages 1319-1325

[13] W.D. Zhuang, P.C. Chang, F.Y. Chou, R.K. Shiue, Effect of solder creep on the reliability of large area die attachment, Microelectronics Reliability, Volume 41, Issue 12, December 2001, Pages 2011-2021

[14] Katsis, D.C.; Van Wyk, J.D., "Void-induced thermal impedance in power semiconductor modules: some transient temperature effects," *Industry Applications, IEEE Transactions on ,* vol.39, no.5, pp.1239,1246, Sept.-Oct. 2003

[15] Bouarroudj, M.; Khatir, Z.; Ousten, J. -P; Lefebvre, S., "Temperature-Level Effect on Solder Lifetime During Thermal Cycling of Power Modules," *Device and Materials Reliability, IEEE Transactions on ,* vol.8, no.3, pp.471,477, Sept. 2008

[16] E. Herr, T. Frey, R. Schlegel, A. Stuck, R. Zehringer, Substrate-to-base solder joint reliability in high power IGBT modules, Microelectronics Reliability, Volume 37, Issues 10–11, October–November 1997, Pages 1719-1722

[17] Hui Huang; Mawby, P.A.; , "A Lifetime Estimation Technique for Voltage Source Inverters," *Power Electronics, IEEE Transactions on ,* vol.28, no.8, pp.4113-4119, Aug. 2013

[18] Musallam, M.; Johnson, C.M.; , "Real-Time Compact Thermal Models for Health Management of Power Electronics," *Power Electronics, IEEE Transactions on ,* vol.25, no.6, pp.1416-1425, June 2010

[19] Musallam, M.; Johnson, C.M.; , "An Efficient Implementation of the Rainflow Counting Algorithm for Life Consumption Estimation," *Reliability, IEEE Transactions on ,* vol.61, no.4, pp.978-986, Dec. 2012

[20] Bin Du; Hudgins, J.L.; Santi, E.; Bryant, A.T.; Palmer, P.R.; Mantooth, H.A.; , "Transient Electrothermal Simulation of Power Semiconductor Devices," *Power Electronics, IEEE Transactions on ,* vol.25, no.1, pp.237-248, Jan. 2010

[21] A. Watanabe, I. Omura, Real-time failure imaging system under power stress for power semiconductors using Scanning Acoustic Tomography (SAT), Microelectronics Reliability, Volume 52, Issues 9–10, September–October 2012, Pages 2081-2086

[22] Christiane Maierhofer, Mathias Röllig, Henrik Steinfurth, Mathias Ziegler, Marc Kreutzbruck, Christian Scheuerlein, Simon Heck, Non-destructive testing of Cu solder connections using active thermography, NDT & E International, Volume 52, November 2012, Pages 103-111

[23] Daeil Kwon; Azarian, M.H.; Pecht, M.; , "Nondestructive Sensing of Interconnect Failure Mechanisms Using Time-Domain Reflectometry," *Sensors Journal, IEEE ,* vol.11, no.5, pp.1236-1241, May 2011

[24] Dawei Xiang; Li Ran; Tavner, P.; Bryant, A.; Shaoyong Yang; Mawby, P.; , "Monitoring Solder Fatigue in a Power Module Using Case-Above-Ambient Temperature Rise," *Industry Applications, IEEE Transactions on ,* vol.47, no.6, pp.2578-2591, Nov.-Dec. 2011

[25] M.P. Rodriguez, N.Y.A. Shammas, A.T. Plumpton, D. Newcombe, D.E. Crees, Static and dynamic finite element modelling of thermal fatigue effects in insulated gate bipolar transistor modules, Microelectronics Reliability, Volume 40, Issue 3, 17 March 2000, Pages 455-463

[26] Shammas, N. Y A; Rodriguez, M. P.; Plumpton, A. T.; Newcombe, D., "Finite element modelling of thermal fatigue effects in IGBT modules," *Circuits, Devices and Systems, IEE Proceedings - ,* vol.148, no.2, pp.95,100, Apr 2001

[27] Xiao Cao; Tao Wang; Ngo, K.D.T.; Guo-Quan Lu; , "Characterization of Lead-Free Solder and Sintered Nano-Silver Die-Attach Layers Using Thermal Impedance," *Components, Packaging and Manufacturing Technology, IEEE Transactions on ,* vol.1, no.4, pp.495-501, April 2011

[28] "JESD51-14, EIA / JEDEC Standard, Transient Dual Interface Test Method for the Measurement of the Thermal Resistance Junction to Case of Semiconductor Devices with Heat Flow Trough a Single Path," 2010.

[29] Dupont, L.; Avenas, Y.; Jeannin, P., "Comparison of junction temperature evaluations in a power IGBT module using an IR camera and three thermo-sensitive electrical parameters," *Applied Power Electronics Conference and Exposition (APEC), 2012 Twenty-Seventh Annual IEEE ,* vol., no., pp.182,189, 5-9 Feb. 2012

[30] Blackburn, D.L., "Temperature measurements of semiconductor devices - a review," *Semiconductor Thermal Measurement and Management Symposium, 2004. Twentieth Annual IEEE ,* vol., no., pp.70,80, 9-11 Mar 2004

978-1-4799-0024-4/13 $31.00 © 2013 IEEE

Power MOSFET failure and degradation mechanisms in flyback topology under high temperature and high humidity conditions

Ilkka Vaalasranta, Juha Pippola, and Laura Frisk

Abstract -- **This article describes observations about power MOSFET failures and degradation experienced during accelerated testing involving high temperature and high humidity stresses. The examined power MOSFETs were operated in commercial variable speed drives in flyback transformer topology as power switches. Known power MOSFET failure mechanisms are summarized and electrical stresses typical for the flyback topology are reviewed. In addition, effects of electrical stresses due to power interruptions were studied. The power MOSFET failure analysis results are presented. The visual appearance of the samples with catastrophic damage was examined with such analysis methods as X-ray, acoustic microscopy (SAM) and optical microscopy. The samples with no obvious failures were also analyzed in research for electrically measurable failure precursor parameters to characterize the physical degradation of the devices. Under these test circumstances, the power MOSFET channel off-resistance R_{DS-off} was discovered to have degraded. This resistive leakage phenomenon was also visualized with backside OBIRCH technique.**

Index Terms--**Accelerated aging, Failure analysis, Flyback transformers, Industrial electronics, Industry applications, Power MOSFET, Semiconductor device breakdown, Variable speed drives**

I. INTRODUCTION

Variable speed drives are commonly used devices in industrial applications requiring electric motor control. Modern variable speed drives are power electronics devices offering sophisticated motor control features such as accurate control of speed, frequency, torque and also safety features such as safe torque off (STO). Modern variable speed drives also operate with high power efficiency.

A simplified structure of a general motor drive is shown in Fig. 1. The power is input from a 3-phase AC voltage network and then rectified to DC voltage for the intermediate circuit with large DC capacitors to store energy. The inverter circuit converts the DC voltage to AC voltage to operate and control the electric motor. Power electronics components, mainly IGBT's (Insulated Gate Bipolar Transistor) or power MOSFET's (Metal-Oxide-Semiconductor-Field-Effect-Transistor), are used as switches for DC-AC conversion. Common feature for these power electronics devices are that

they can be switched on and off with a low voltage control signal [1].

Auxiliary power circuit with smaller DC voltages is required to provide supply voltages for control and measurement electronics. There are several possible methods to implement the DC-DC power conversion from higher voltage in the range of hundreds of volts down to few tens of volts. One possible and widely used method is flyback transformer topology with a power MOSFET component as a power switch.

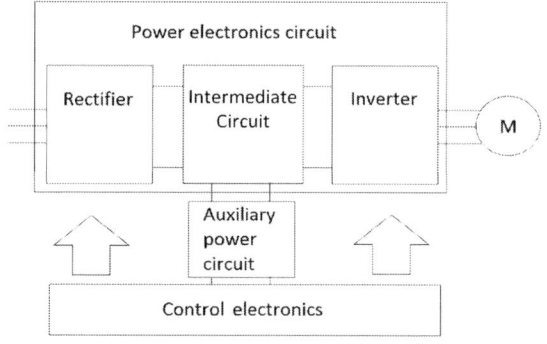

Fig. 1. Block diagram of a general motor drive

The electrical drive is often exposed to various environmental stresses during its service life. With increasing reliability demands, efficient accelerated testing methods are needed to ensure reliable system operation.

In this study, reliability of commercial variable speed drives with 0.75kW rated power and 4 kHz switching frequency was investigated using multiple stresses. The drives were operating motors in the test conditions with 85°C temperature and 85%RH (relative humidity). In addition, further stress was applied to a set of the tested units through input power interruptions.

Several failures were observed during testing. The failures could be roughly categorized to catastrophic MOSFET failures, IGBT failures, and other soft failures to cause electrical drive to stop functioning. The overview of the main failure modes and mechanisms, experienced in this study, is presented in earlier research article by J. Pippola. [2]

In this paper, more detailed analysis results about the power MOSFET degradation and failures are presented.

This work was supported in part by TEKES, the Finnish Funding Agency for Technology and Innovation, and ABB Drives.

I. Vaalasranta, J. Pippola, and Laura Frisk are with the Department of Electrical Engineering, Tampere University of Technology, Tampere, Finland. (e-mail: ilkka.vaalasranta@fi.abb.com, juha.pippola@tut.fi, laura.frisk@tut.fi).

Additionally, characterization of aged devices and the effects of degradation on the eventual failures are discussed.

II. FLYBACK TRANSFORMER AND POWER MOSFET

A. Flyback Topology

The flyback converter is widely used topology in industrial applications due to its relative simplicity and cost-effectiveness. It is generally used in multiple output power supplies for implementing DC-DC power conversion up to 100W power levels. One of the advantages in the flyback topology is that secondary output inductors are not required, which then results in savings in PCB surface area, volume, and cost. The topology also provides electrical isolation between the primary side and the secondary side, which can be considered as a safety feature. A flyback transformer and a power switch are the essential components in the flyback topology. Additionally, snubber circuit components are typically utilized for voltage transient filtering. Typical design topology with snubber components R_S, C_S, and D_S is presented in Fig. 2. Also the parasitic leakage inductance L_{lk}, the effect of the transformer component non-ideal characteristics, is shown in the figure.

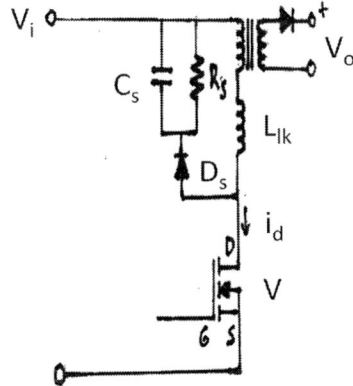

Fig. 2. Flyback topology with RCD snubber [5]

B. Flyback Transformer

The flyback transformer is a coupled inductor on a single magnetic core with multiple windings on the secondary side. When the power switch is turned on, electrical current starts to flow through the primary side windings of the transformer. The increasing current induces increasing magnetic flux through the magnetic core, storing energy in the magnetic field. Turning off the power switch causes the magnetic field to decrease resulting in negative magnetic flux, which in turn induces voltages to the secondary windings enabling energy transfer.

The transformer is not ideal, which means that the input energy does not transfer entirely to the secondary side. Some energy is stored in the non-magnetic gap in series with the magnetic core. Also, there is always some physical separation between multiple windings causing unequal coupling to the magnetic core. In addition, some energy is also stored within and between the windings. These non-ideal characteristics of the transformer are represented in the circuit as leakage inductance.

In the flyback topology, the characteristic leakage inductance of the transformer results in large transient voltage at the power switch drain terminal and at the secondary side rectifier diodes. The presence of these transient voltages, occurring instantly after the power switch is turned off, is the major disadvantage of the flyback topology. Therefore, appropriate snubber circuitries are generally implemented in the design topology to control and reduce the transient voltage stresses at the component terminals [3]-[5]. Voltage and current waveforms, typical for the flyback topology, are illustrated in Fig. 3.

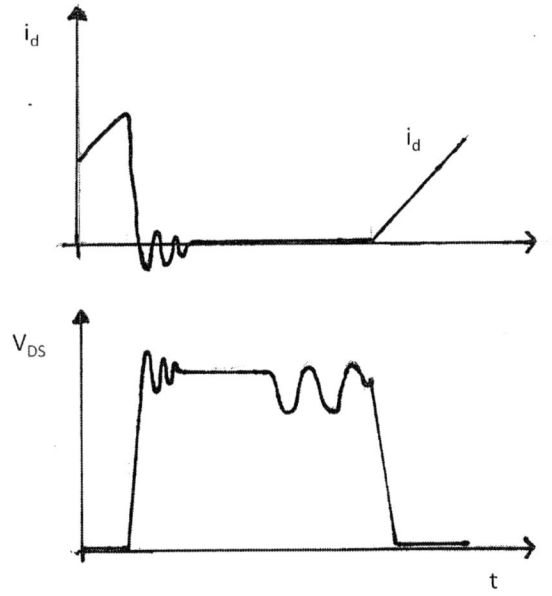

Fig. 3. Typical flyback waveforms in discontinuous conduction mode [5]

C. Power MOSFET

Power MOSFETs are used extensively as switches in power converter designs in modern industrial applications. A power MOSFET is a field-effect-transistor (FET) where the electrical current flows through the die vertically. Hence, the structure can also be called as vertical MOSFET. Power MOSFET's are unipolar devices, which means that the current transportation is carried out by majority carriers (electrons in n-channel MOSFET's or holes in p-channel MOSFET's) [6]. A cross-section of a power MOSFET structure is illustrated in Fig. 4.

To provide high current handling capability in a single power MOSFET component, this structure is repeated thousands of times in parallel connection. In an n-channel MOSFET discussed here, the MOSFET channel is in non-conductive state without external voltage on gate terminal.

The structure becomes conductive, when a gate voltage V_{GS} larger than the threshold voltage V_{TH} is applied. In this state, the electric field in the gate oxide is strong enough to form inversion layer in the p-base region, allowing the electron current to flow from the source terminal metallization on the top of the chip to the drain terminal metallization on the bottom of the chip. [7]

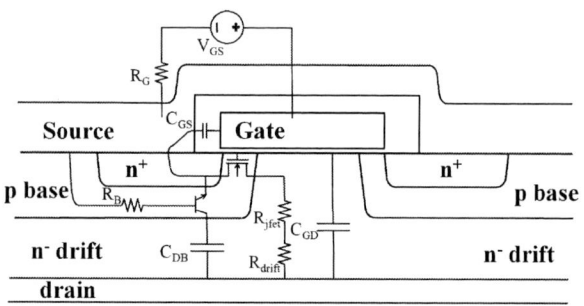

Fig. 4. Power MOSFET structure

In non-conductive state with voltage V_{DS} applied between drain and source, minor leakage current is generated. This small leakage current is consequential to electric field induced electron-hole pair generation in the deserted space charge region in the n⁻ drift zone. The leakage current increases with applied voltage and temperature [6]. However, the leakage current can be considered to be negligible in terms of system operation.

III. POWER MOSFET FAILURE AND DEGRADATION MECHANISMS

A. Parasitic bipolar transistor (BJT) turn-on

Various MOSFET failure modes have been identified. In switch mode power supply applications, the MOSFET failures typically occur due to integral parasitic bipolar transistor (BJT) turn-on, even though the power MOSFET designs have been continually optimized to improve the parasitic BJT turn-on immunity. Normally this BJT turn-on does not happen because p-base and n⁺-source are connected to each other. However, these failures may occur as consequence of high rate of change in the drain-source voltage dV_{DS}/dt. In the dynamic situations of this kind, displacement current may flow laterally into the p-base region due to presence of parasitic capacitances C_{DB} and C_{GD}, causing voltage drop across its resistance R_B. If the voltage drop V_{BE} exceeds ~0.65V, the parasitic BJT is activated.

In the parasitic BJT structure, emitter is the n⁺ source contact, the base is the p-base region, and the collector is the n⁻ drift region. Once the parasitic BJT turns on, the nMOS conductivity can no longer be controlled with the gate voltage. With high voltage applied across the drain and the source terminals, the increasing current through the nMOS channel destroys the device eventually. [7]-[9]

B. The MOSFET Body Diode Reverse Recovery Failure Mechanism

The MOSFET body diode reverse recovery failure mechanism may occur particularly in designs where the MOSFET is exposed to negative drain-source voltages V_{DS}.

In this setting, the MOSFET becomes reverse biased causing body diode to become forward biased and thus to conduct current. When the body diode forward current becomes switched off with high rate of dI/dt, the charge carrier recombination occurs in the depletion layer of the p-n junction. This may induce high local electric fields internally, leading to voltage drop in the p-base region and eventual catastrophic MOSFET failure. [7]-[9]

C. Single Event Breakdown Phenomenon

Single event breakdown phenomenon may occur if the specified forward breakdown voltage of the MOSFET component is insufficiently derated in relation to application demands. In long term operation, a proportion of the MOSFETs may fail even if the drain-source voltage V_{DS} does not exceed the specified forward breakdown voltage. Moreover, the forward breakdown voltage has temperature dependence, decreasing in value as temperature increases. [7]

D. TDDB (Time Dependent Dielectric Breakdown)

The gate oxide can degrade and eventually break down, if the power MOSFET is exposed to a high current stress for a long period of time. The degradation takes place due to gradual charge accumulation in the gate oxide caused by hot electron injection. The high energy electrons in the MOSFET channel may collide with the lattice atoms, causing injection to the oxide dielectric. There will be threshold voltage V_{TH} shift as the degradation propagates. The hot electron injection (HEI) phenomenon has negative correlation to temperature increase. In cold temperatures, the electrons have greater probability of travelling sufficiently longer distances without colliding to crystal atoms, thus gaining more energy and increasing the likelihood of injection. [7]-[8]

E. Die Delamination Induced $R_{DS\text{-}on}$ Increase

In high temperature operation with high currents, the interface between the die and the component lead frame may deteriorate due to temperature cycling. Delaminated interface can no longer conduct any current, increasing the current density across still undamaged interface. This accelerates the delamination propagation further, causing increase in the channel on-resistance $R_{DS\text{-}on}$ as well as in the channel temperature during the operation of the device. [9]

IV. FAILURE PRECURSORS

Failure precursor is a data event or trend that indicates impending failure. Usually, the precursor is a measurable variable or parameter that can be associated with the eventual failure. The precursor parameter can be selected based on experimental information about the application or analyzing potential failure modes and mechanisms. The change in the precursor parameter can be monitored to

978-1-4799-0024-4/13 $31.00 © 2013 IEEE

evaluate the remaining useful lifetime of the product. In accelerated aging test settings, the eventual failures can be analyzed to confirm the correlation between the parameter change and the physical degradation of the sample. [10]

In vertical MOSFET devices, gate leakage current and resistance along with drain-source leakage current or resistance have been identified as critical parameters to be considered as indicators of incipient failure. [11]-[12]

V. OBIRCH

Precise component die level fault localization is necessary in failure and yield analysis. The failure must often be described accurately in detailed level before corrective actions can be determined to remove the failure causes. To pinpoint a failure location in a failed device, it is often necessary to perform the analysis under electrical stress. Moreover, additional sample excitation may be needed as well. For example, such techniques as liquid crystal thermography and emission microscopy along with scanning electron microscopy (SEM) or laser irradiation can be used. [13]

The fault localization is a major challenge also when analyzing leakage current failures. The traditional front-side analysis methods are generally unable to localize the failure if the fault location is covered up with one or several metal layers. However, optical beam induced resistance change (OBIRCH) technique has been used successfully to analyze leakage current failures. A significant advantage is that OBIRCH analysis can be performed both from the front side and from the backside of the sample. In the back side OBIRCH analysis, the sample preparation must be carried out before analysis. The component lead frame has to be ground out from the bottom side to reveal the semiconductor substrate. With vertical semiconductor devices, where the current flows vertically through the chip, the preparation process temporarily disconnects the bottom terminal (drain), and the functionality must be restored with applying electrically conductive substance on the substrate corner.

In OBIRCH analysis, the sample under test is powered with a constant voltage or current source. The sample is then scanned with a laser beam, which generates localized heat on the sample surface. Once the sample temperature increases on a location with a resistive failure, it affects the conductivity of the sample. This resistance change can then be measured. On areas with no defects, heat is allowed to transmit evenly in the sample, and no changes in resistance will be observed. [14]-[15]

VI. EXPERIMENTAL

Commercial variable speed drives with 0.75kW rated power and 4 kHz switching frequency were operated in 85°C temperature and 85%RH (relative humidity) test conditions. Further stress was applied to a set of the tested units with input power interruptions. Several failures were observed during testing. The failures could be roughly categorized to catastrophic MOSFET failures, IGBT failures, and other soft failures to cause electrical drive to stop functioning. [2]

VII. RESULTS

The failure analysis was performed to selected power MOSFETs which were exposed to the accelerated test setting. The used power MOSFET components are commercially available with 1500V/2A voltage/current rating. Initially, the samples were characterized with electrical measurements. The summary of the results is presented in Table I.

TABLE I
THE ELECTRICAL CHARACTERIZATION SUMMARY OF THE ANALYZED POWER MOSFET SAMPLES

Sample #	I_D (U_{DS}=20V)	$R_{DS\text{-off}}$	test time in 85°C /85%RH test conditions	power interruptions
#2	170μA	120kΩ	1750h	no
#14	1700μA	12kΩ	1053h	yes
#16	650μA	31kΩ	1223h	yes
#17	short	0Ω	750h	yes
#20	1700μA	12kΩ	1695h	yes
#24	32μA	630kΩ	958h	yes
#25	280μA	71kΩ	962h	yes
#26	880μA	23kΩ	1099h	no
#27	690μA	29kΩ	1210h	yes
ref#1	<1μA	-	0h	n/a

The visual appearance of the samples with catastrophic damage (such as sample #17) was examined with X-ray, scanning acoustic microscopy (SAM), and optical microscopy. The aged samples with no obvious failures were characterized in more detail in research for electrically measurable failure precursor parameters to correlate with the physical degradation of the devices.

After exposing the devices to these test circumstances, the power MOSFET channel off-state resistivity $R_{DS\text{-off}}$ was discovered to have degraded. Liquid crystal thermography analysis showed no anomalies on the samples. However, the resistive leakage locations could be visualized successfully with backside OBIRCH technique.

A. Electrical Failure Analysis

In the electrical measurements, a subgroup of samples was observed to be short circuited between all three terminals indicating catastrophic damage in the MOSFET die. In addition, the aged samples were measured to conduct low amount of current when a moderate voltage V_{DS} was applied across the MOSFET channel while maintaining the structure in the off-state with gate voltages V_{GS} below MOSFET threshold voltage V_{TH}. The measured leakage currents were low compared to on-state currents but significantly larger than what was observed in unused reference samples used for comparison. The corresponding leakage resistance values for these samples were in the range of ~10 kΩ to ~700 kΩ,

978-1-4799-0024-4/13 $31.00 © 2013 IEEE

measured after approximately 1000 hours of testing. This leakage current could not be influenced by the gate voltage V_{GS}, as illustrated in Fig. 5. Moreover, there is no shift in the threshold voltage V_{TH}, compared with unused reference device, indicating that the behavior is not originating from charge accumulation in the gate oxide.

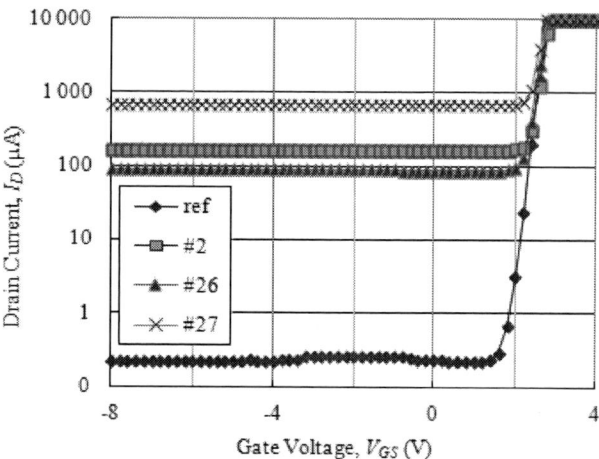

Fig. 5. Aged power MOSFET drain current I_D in relation to gate voltage V_{GS}. ($V_{DS} = 20V$, Current is limited to 10mA.)

B. Visual imaging

The samples with catastrophic damages as well as the aged samples with measurable leakage current were investigated with X-ray and scanning acoustic microscope (SAM) and the results were compared to those with an unused reference sample. Using these non-destructive methods, an aged sample (sample #2) did not reveal any significant differences, but both methods were capable of distinguishing the sample with the catastrophic damage. In the X-ray analysis, a minor deviation could be observed, but with SAM imaging, clear evidence of internal delamination becomes visible. The samples were then de-encapsulated to expose the die for visual inspection. The sample with the catastrophic damage (sample #17) revealed severely destroyed die while the aged sample showed no deviation, as shown in Fig. 6.

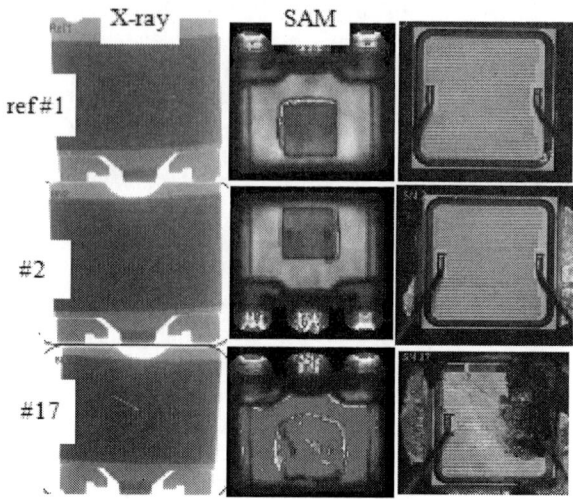

Fig. 6. A severely damaged sample (#17) can be exposed with non-destructive failure analysis methods while an aged sample (#2) shows no deviation in comparison to an unused sample (ref#1) used for reference.

C. OBIRCH Analysis

The OBIRCH (Optical Beam Induced Resistance CHange) analysis, when performed from the front side of the die, reveals some deviation on the edge of the die, but the observation is almost completely obscured by the source and gate metallization layers. The back side OBIRCH analysis reveals evenly distributed hot spot locations indicating leakage current, as illustrated in Fig. 7.

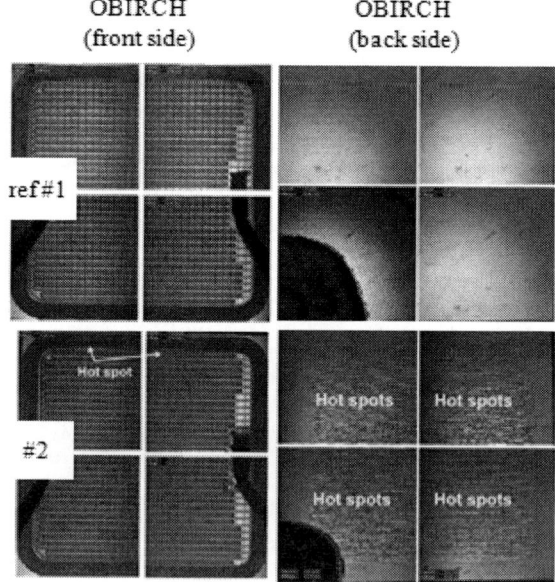

Fig. 7. Back side OBIRCH technique reveals multiple evenly distributed hot spot locations in the n⁻ drift region of the aged power MOSFET sample.

D. Further electrical characterization

The aged samples with measurable leakage resistance R_{DS-off} were characterized in more detail. The resistive character of the leakage was confirmed, as shown in Fig. 8. Also the temperature dependence of the phenomenon was studied, as illustrated in Fig. 9. The leakage resistance is observed to increase as the temperature increases, which means the temperature dependence of the observed leakage current is opposite to the temperature dependence of the initial leakage current [6].

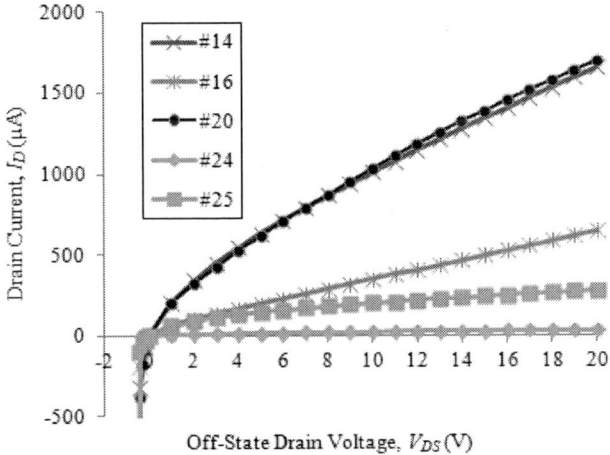

Fig. 8. Power MOSFET off-state channel leakage current I_D in relation to drain-source voltage V_{DS}. (@V_{GS} = 0V.)

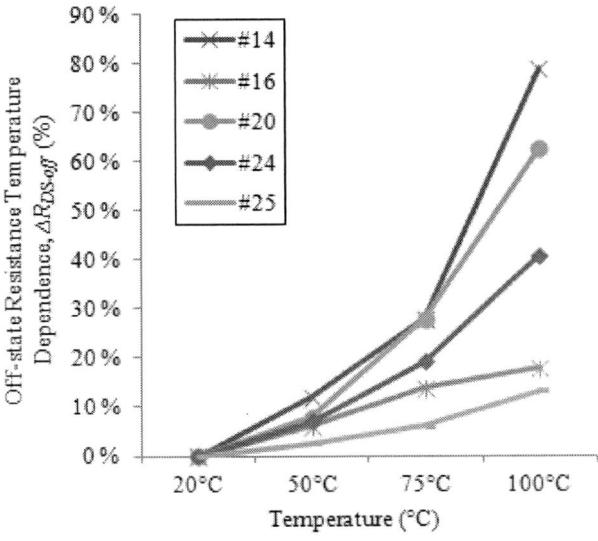

Fig. 9. Temperature dependence of drain-source off-state resistance R_{DS-off}.

VIII. CONCLUSIONS AND DISCUSSION

It can be concluded from the experiment results that the input power cycling has a clear accelerating effect on the eventual catastrophic failure mechanism. This MOSFET degradation may be caused by the increased current stress due to increased MOSFET duty cycle, which is consequential to temporarily lower supply voltage in the intermediate circuit. Power interruptions may also cause higher dV_{DS}/dt conditions at the power MOSFET drain terminal. The leakage current is also increased as the component temperature decreases, which will increase the likelihood of the MOSFET inherent BJT turn-on (and therefore the risk of catastrophic damage) after the device has cooled down temporarily.

The leakage current, observed in the samples, causes temperature increase in the MOSFET structure, which accelerates charge carrier generation in the p-base region, which in turn will increase likelihood of BJT turn-on. However, the actual mechanism for the leakage current phenomenon could not be concluded and therefore there may be also other mechanisms (than those induced by the temperature increase) to accelerate the failure.

The temperature dependence of the observed leakage current phenomenon is opposite to the initial small leakage current, consequential to electric field induced electron-hole pair generation in the deserted space charge region in the n⁻ drift zone. Therefore, the leakage mechanism should also be different. For more comprehensive evaluation, the R_{DS-on} characteristics and the switching losses could be investigated as well.

The conducted experiment does not provide enough data to statistically evaluate the leakage propagation and what would be an appropriate threshold level for significantly increased risk for failure. However, continuous monitoring of the leakage current could be implemented in the flyback topology with little effort. During the device operation, the MOSFET gate driver circuit is constantly monitoring MOSFET current during the duty cycle to control the pulse width modulation. The current is measured utilizing the current sensing resistors located between the ground and the MOSFET source terminal. Same resistors could be used for leakage current sensing during the PWM sub-interval when the MOSFET is switched off. Practically, the temperature dependence will cause challenges in accurate data interpretation. Still, the leakage phenomenon can be used as a failure precursor parameter to indicate increased failure risk in challenging environments.

IX. ACKNOWLEDGMENT

The authors would like to thank TEKES, the Finnish Funding Agency for Technology and Innovation, and ABB for collaboration. The authors would also like thank IC Failure Analysis Lab for support in the component level failure analysis.

X. REFERENCES

[1] A. Hughes, Electric Motors and Drives, Fundamentals, Types and Applications, 3rd ed., Burlington, MA: Elsevier Ltd., 2006, pp. 72.

[2] J. Pippola, L. Frisk, K. Kokko, J. Kiilunen, T. Marttila, "Effect of Input Power Interruptions on Power Electronics Reliability," in *Proc. 2012 IEEE 7th International Conference on Integrated Power Electronics Systems (CIPS)*, paper 06.5 pp. 1-6.

[3] K. Patel, "Voltage Transient Spikes Suppression in Flyback Converter Using Dissipative Voltage Snubbers," in *Proc. 2008 3rd IEEE Conference on Industrial Electronics and Applications*, pp. 897-901.

[4] A. Rahnamaee, J. Milimonfared, K. Malekian, "Reliability Considerations for a High Power Single Switch Flyback Power Supply," in *Proc. 2008 14th IEEE Mediterranean Electrotechnical Conference (MELECON 2008)*, pp. 527-533.

[5] G-B Koo, "Design Guidelines for RCD Snubber of Flyback Converters," Fairchild Semiconductor Corporation. Application Note AN-4147. [Online]. Available: http://www.fairchildsemi.com/an/AN/AN-4147.pdf

[6] R. Perret, Power Electronics Semiconductor Devices, London/ New Jersey: ISTE Ltd./Wiley, 2009, pp. 4, 28-29.

[7] P.Singh, "Power MOSFET Failure Mechanisms," in *Proc. 2004 26th Annual International IEEE Telecommunications Energy Conference (INTELEC,)* pp. 499-502.

[8] J.R. Celaya, N. Patil, S. Saha, P. Wysocki, K. Goebel, "Towards Accelerated Aging Methodologies and Health Management of Power MOSFETs (Technical Brief)," in *Proc. 2009 Annual Conference of the Prognostics and Health Management Society*, pp. 1-8.

[9] B. Khong, M. Legros, P. Tounsi, Ph. Dupuy, X. Chauffleur, C. Levade, G. Vanderschaeve, E. Scheid, "Characterization and Modelling of Ageing Failures on Power MOSFET Devices," in *Proc. 2007 18th Microelectronics Reliability European Symposium on Reliability of Electron Devices, Failure Physics and Analysis*, vol. 47, issues 9-11, pp. 1735-1740

[10] N. Vichare, M. Pecht, "Prognostics and Health Management of Electronics," IEEE Trans. Components and Packaging Technologies, vol. 29, no. 1, pp. 222-229. Mar, 2006.

[11] F.H. Born, R.A. Boenning, "Marginal Checking – Technique to Detect Incipient Failures," in *Proc. 1989 IEEE National Aerospace and Electronics Conference (NAECON,)* pp. 1880-1886.

[12] H. Zhang, R. Kang, M. Luo, M. Pecht, "Precursor Parameter Identification for Power Supply Prognostics and Health Management," in *Proc. 2009 IEEE 8th International Conference on Reliability, Maintainability and Safety (ICRMS)*, pp. 883-887.

[13] M. Ohring, Reliability and Failure of Electronic Materials and Devices, Academic Press, 1998, pp.621-635.

[14] F. Beaudoin, G. Imbert, P.Perdu, C. Trocque, "Current Leakage Fault Localization Using Backside OBIRCH," in *Proc. 2001 IEEE International Symposium on the Physical and Failure Analysis of Integrated Circuits (IPFA)*, pp. 121-125.

[15] K. Nikawa, S. Inoue, K Morimoto, S. Sone, "Failure Analysis Case Studies Using the IR-OBIRCH (Infrared Optical beam Induced Resistance Change) Method," in *Proc. 1999 Eighth Asian Test Symposium. (ATS '99)*, pp. 405-409.

Remote Monitoring System of Electrical Machines via INTERNET

O. Touhami, R. Sadoun, A. Belouchrani, S. Hamdani, A. Boukoucha and S. Ouaged

Abstract -- **Electrical machine condition monitoring is of particular important to factory efficiency and worker safety. A variety of signal analysis techniques exists for machine status diagnosis. This paper deals with remote diagnosis of electrical systems. Herein, we developed a remote monitoring system based on the Internet technology. The use of such system gives many advantages including minimum intervention on the client side, and simplicity to expand the system. The users can check the machine status data through the internet and mobile terminals. The automatic alarm part can actively send alerting messages to the engineers' mobile phones and call these phones to make sure they get alert when machine's status is abnormal. Another advantage is the reduction of the maintenance cost.**

Index Terms--**Remote monitoring, Remote diagnosis, spectral analysis, induction motor, and rotor bar faults**

I. INTRODUCTION

PROGRESS in control and technological advances in the field power electronics and microelectronics have made possible the implementation of efficient controls for the induction machine, making it a competitor to some other electrical machines in areas of variable speed and torque control fast. Ensure continuity of operation requires the implementation of programs preventive and corrective maintenance. Indeed, the reliability and safety of their operation can partially ensure the safety, the quality of service and profitability of facilities. Unfortunately, the new constraints and integration of these machines in complex systems of energy conversion makes diagnosis more difficult. Traditionally the maintenance procedure of electrical machines is made naturally by the repair or replacement of equipment following the judgment. However, this procedure has considerable economic losses, what is commonly called corrective maintenance. Users are increasingly focused on the maintenance because it can contribute significantly to the overall performance of the company and the security company personal.

With the development of technology, a new approach is gaining ground rapidly in the maintenance management is predictive maintenance which involves the detection and localization of defects to intervene early against different types of electrical or mechanical fault [1]. The activity of plant maintenance calls for more than twenty years in diagnostic techniques more sophisticated. Several techniques for detecting defects are now available to engineers, [2-4]. Among these, one can note: the analysis of mechanical vibration, flux as well as the analysis of stator current [5, 6]. Various techniques for analyzing signals, for example, the

Fourier transform and the wavelet transform, have long been used in the fault diagnosis of such machines in the context of spectral analysis [7,8].

The current tendency of manufacturing moves towards regionalization. Therefore, the production unit may be far from experts, whereas machines must be monitored by such experts. Using techniques based on Internet and mobile communications, monitoring of machines and remote fault diagnosis becomes feasible. Therefore, quick decisions can be taken. Contingency plans can be provided to compensate for production losses due to untimely breakdowns of production machines. Finally, control of remote machines is critical to companies involved in the competition.

Remote diagnostics and remote monitoring of electrical systems are increasingly used in many industrial fields, hence the growing interest in this subject.

Once the Web became a more stable, researchers have begun to exploit the systems fault diagnosis of electrical machines, [9]. Caldwell et Al [10] have used the internet to diagnose a scanning electron microscope.

Today, wireless communication systems are widely deployed together with Machine to Machine communications. The WAP (Wireless Application Protocol) was also developed for mobile devices such as mobile phones and PDAs (Personal Digital Assistants) to access the internet. The development of these technologies allows for continued monitoring of electrical machines anywhere any time [11].

In this paper, we use the Fast Fourier Transform (FFT) of stator current analysis for fault diagnosis of squirrel cage induction machine. Defects are considered the breaking of the bars and the end-ring portion. The current trend of diagnosis is now facing the monitoring is to say that a plant may be located far from experts while the machine should be monitored.

Using techniques based on INTERNET and mobile communications, monitoring of machines and remote faults diagnosis becomes feasible. As a result, decisions "just in time" can be taken away. Contingency plans can be provided to compensate for the loss of production due to the unexpected failure of electrical machines. So the control of electrical machines remotely, becomes essential for companies involved in the competition.

II. REMOTE MONITORING SYSTEM OF ELECTRICAL MACHINES BY INTERNET

In our work we adopt the system of Fig.1. The system can be divided into three parts: acquisition, processing and sharing and the client side. The sensors gather the signals of

978-1-4799-0024-4/13 $31.00 © 2013 IEEE

different sizes (current, voltage, speed….) of the monitored machine and transmit amplifiers. Amplifiers condition the signals in formats acceptable so that the DAQ can convert them to digital signals. In our laboratory we have a card for signal conditioning connected to a data acquisition card type IOTEQ/DAQ series 1005, with a sampling frequency of 200 kHz, 16 analog inputs, 16 digital.

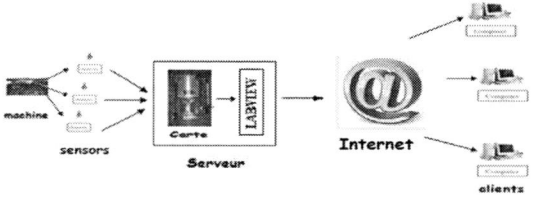

Fig.1: System using an acquisition card connected to LABVIEW DAQ

Adapting our LABVIEW card, the data will be processed and analyzed more rigorously by simple visual programming in LABVIEW. The combination of DAQ 1005, LABVIEW and network outputs, which are guaranteed by the latter as a "server LABVIEW" equivalent to a DATA-LOGGER and will share the results of treatment (messages displays signals…) via the internet.

It is sufficient for the client (or expert) install an application Matlab or C++, which have LABVIEW software installed and properly configured in the terminal for access to diagnostic data, and consequently a decision on the condition of the machine.

To adequately explain the method to follow such a system diagnostics, we include the following steps:

- *Model and test the implementation of the DAQ.*

For this we opted for Fig.2. For the adaptation of the card LABVIEW DaqIO with some instructions, then we have done some testing and acquisition process to ensure that the card is integrated in LABVIEW.

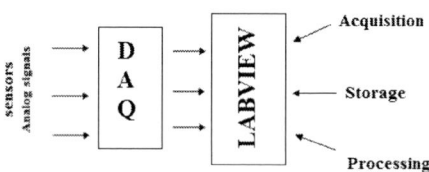

Fig.2: Implementation of the DAQ

- *Model of processing and storage.*

Once the card is adapted to LABVIEW, appropriate treatment is performed for the diagnosis, by saving in databases and by sharing via internet. Figure 3 summarizes this process.

- Expertise part

After treatment outcomes are communicated and shared on

the internet, the client (or expert) must have access to our system. It must also have:

- An application MATLAB, C++, or other languages,
- Or LABVIEW installed and configured in the terminal.

It is recommended to have the second case. Figure 4 shows the tasks offered to the customer.

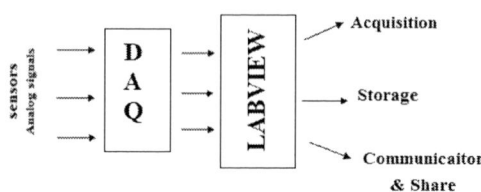

Fig.3: A method of processing and storage of data

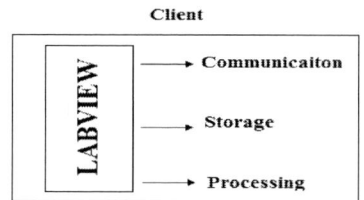

Fig.4: Client-side

By combining the three models described above, we reconstructed the general model structure of our station. LABVIEW also offers the ability to store data in databases for possible later use.

III. EXPERIMENTAL RESULTS AND ANALYSIS

We apply the method of spectral analysis of the stator current for the diagnosis of broken rotor bars and end-ring portion broken. The study focuses on faults and rotor bar breakages and end-ring portion broken. Acquisition of a duration of 10s at a frequency of 10 kHz (N=100.000) of the three currents, phase voltages and speed has been performed on this test bed laboratory research in electrical engineering, Fig.5. Four squirrel cage induction machines whose main characteristics are given in Appendix.

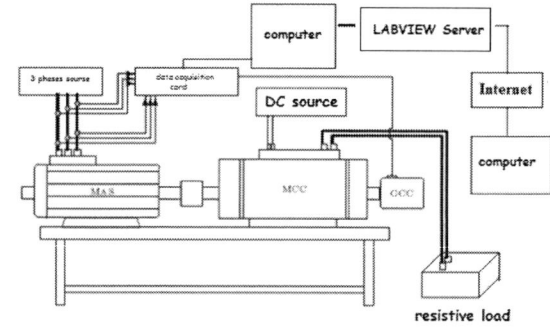

Fig.5: Bench testing of the induction machine

978-1-4799-0024-4/13 $31.00 © 2013 IEEE

For signal acquisition, the laboratory has a card for signal conditioning connected to a data acquisition card IOTQ/DAQ series 1005, with a sampling frequency of 200kHz, 16 analog inputs, 16 digital inputs; all inputs can be used as outputs.

To detect defects in the rotor, the surveillance system is designed for the acquisition of the stator current signature analysis (MCSA). The stator current is initially taken in the time domain, the power spectral density is computed and analyzed in order to detect frequencies defects with those of the same machine in healthy state. Based on the amplitudes in dB, it is possible to determine the extent of fault. In the system described, the map data acquisition is used to take the current engine operating vacuum and then at 70% of rated load. The current signals is then transformed into the frequency domain using Fast Fourier Transform (FFT) based on the power spectral density. The block-diagram used to obtain the power spectral density using DaqView is shown in Fig.6.

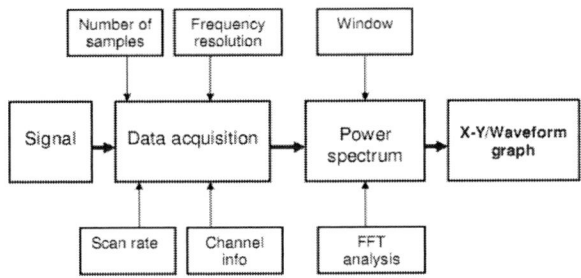

Fig.6: Fault detection and diagnosis system

We will analyze the various quantities, namely voltage, speed and current. Figure 7 shows the voltage across the stator windings.

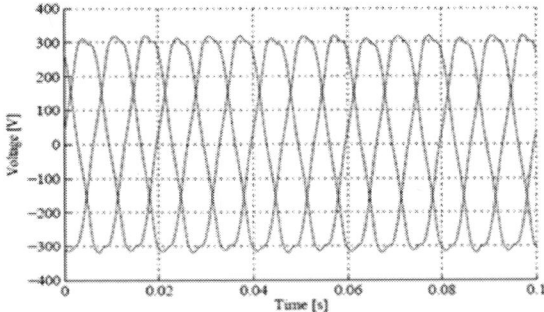

Fig.7: stator voltage

Breakage of the rotor bars and the end-ring portion has the effect of increasing the transient speed (see Fig.8). This increase grows with the severity of the defect, in the case of a broken bar the mechanical time constant is 0.053s and for two broken bars, it is 0.11s, while for the sound machine it is 0.047s. The time constant in the case of end-ring portion broken, is 0.054s which is virtually identically to that of a broken bar.

Faults cause the undulations of the stator current, as can even in Fig.9 which shows that the current ripple of the machine with two broken bars is greater that of one rotor bar or end-ring portion broken which is virtually identical.

- *Stator current analysis of the induction machine*

Currents in the rotor cage produce a three phase magnetic field with the same number of poles and the same direction as the electromagnetic field with the stator but with the slip frequency $f_s.s$.

Asymmetric due to a defect of the cage (bars or end-ring portion broken) generates a rotating field direction to the slip frequency. As a result, an additional current induced in the stator winding frequency $f_{sb} = f_s (1 - 2s)$.

This will produce a cyclic variation of the current will produce a torque oscillation frequency $2f_s.s$ and oscillation of speed with the same frequency will depend on the inertia. This oscillation speed creates two currents of the same amplitude, frequency $f_s (1 - 2s)$ and $f_s (1 + 2s)$.

Other harmonics of stator current can be created by the same phenomenon due to the current of frequency $f_s (1 \pm 2s)$

So the influence of broken bars on the stator current spectrum is the appearance of harmonics frequencies:

$$f_{sb} = f_s (1 \pm 2.k.s) \qquad (1)$$

Where,

f_s is the frequency of stator field; f_{sb} the frequency of broken rotor bars and s the slip.

Fig.10 gives an overview of the harmonics generated by the rotor defects.

For operation at 70% load, the harmonics characterizing defects are visible to only k=1 and the others pics are low in amplitude (case of one broken rotor bar). It is the same for the case of a machine with an end-ring portion broken. The spectral density values are given in table1. For two broken rotor bars, the harmonics are visible for k=1, 2 and 3.

Table1: Spectral density values

Characteristics frequency of the defects 70% of load and Slip is 0.028					
k=1		k=2		k=3	
LSB (Hz)	USB (Hz)	LSB (Hz)	MSB (Hz)	LSB (Hz)	MSB (Hz)
47.2	52.8	41.5	58.5	38.6	61.4
Observable for two broken bars		Observable for two broken bars		Observable for two broken bars	
LSB (Hz)	USB (Hz)	LSB (Hz)	MSB (Hz)	LSB (Hz)	MSB (Hz)
47.2	52.8	41.5	58.5	38.6	61.4
Observable for one broken bar		Not visible for one broken bar		Not visible for one broken bar	
LSB (Hz)	USB (Hz)	LSB (Hz)	MSB (Hz)	LSB (Hz)	MSB (Hz)
47.2	52.8	41.5	58.5	38.6	61.4
Observable for an end-ring broken		Not visible for an end-ring broken		Not visible for an end-ring broken	

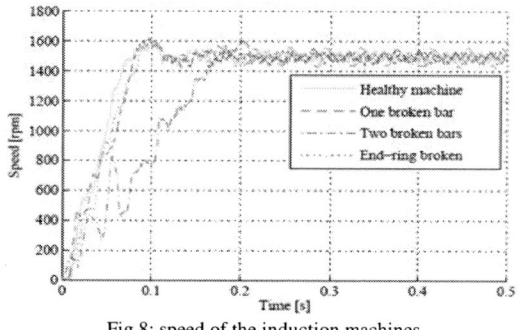

Fig.8: speed of the induction machines

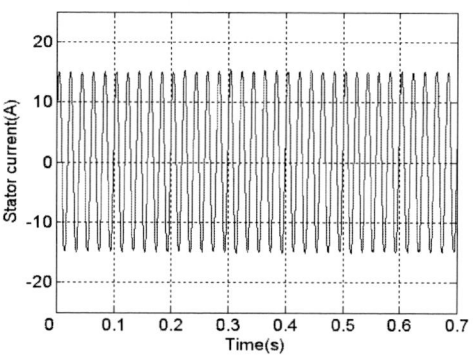

a/ Healthy machine

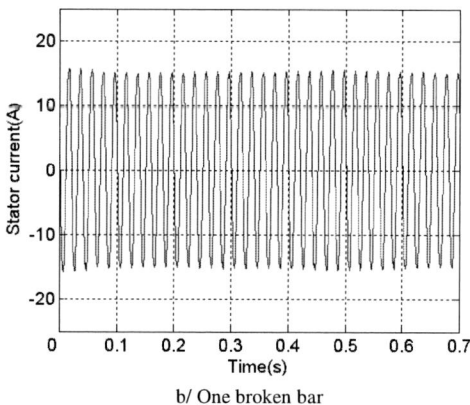

b/ One broken bar

c/ Two broken bars

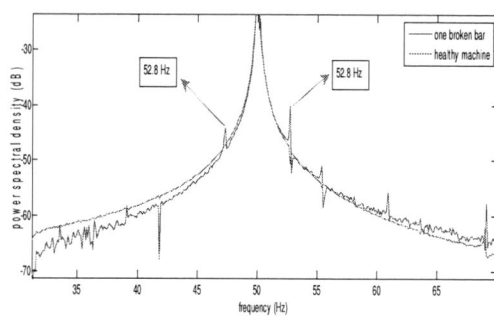

d/ end-ring portion broken
Fig. 9: Stator current for the induction machines

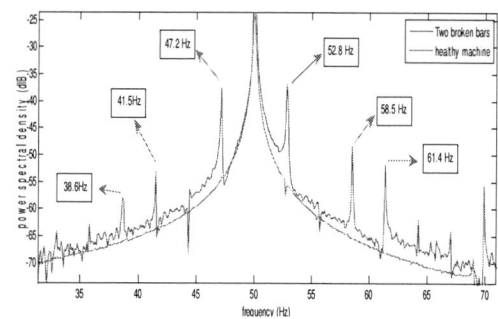

a/ one broken rotor bar

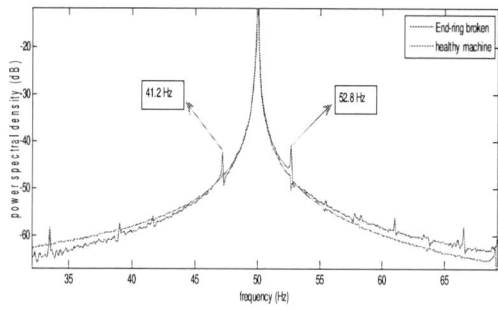

b/ Two broken rotor bars

c/ A end-ring portion broken

Fig.10: Spectral analysis of the stator current at 75% of load

978-1-4799-0024-4/13 $31.00 © 2013 IEEE   26

IV. CONCLUSION

- We used the internet as a new technology for data exchange in the remote diagnosis of electrical machines. At first, we proposed an architecture well advanced, using intelligent software. We took as an example the SMAS (Smart Asset Maintenance System). A highly recommended and widely used in the industry, thanks to the simplicity of its architecture and the benefits it highlights. The use of such a system diagnostics aims to reduce the maintenance costs by replacing the traditional round by monitoring in real time and distance. The decision support tool with the machine learning becomes possible.

- The system we developed is the use of an acquisition card DAQ 1005 we adapted to LABVIEW. Together they form one server to ensure data sharing internet. This system provides a large space for the treatment and diagnosis, uses software developed and evolved the LABVIEW and provides the ability to remotely communicate the results of its integrated network outputs.

- For the evaluation of diagnostic systems remotely, we can mention some advantages:
- Setting safety
- The staff is available for problem solving
- The possibility to reduce the time to understand more rapidly changing behavior of electrical machines
- The analysis is performed by experienced personnel
- Increased productivity with the location and the rapid removal of defects

V. APPENDIX

The nameplate data of induction machines are:

P_n = 4kW; V_n/U_n=220/380 V; I_n (Δ) =15.3A; $\cos\varphi_n$=0.83; p (pair-pole number)=2; number of stator slots=36; number of rotor bars=28.

VI. REFERENCES

[1] J. Salsona, "Effects of rotor bar and end ring faults over the signals of a position estimation strategy for induction motors, " *IEEE Trans. On Energy Conversion, Vol. 03, no. 04,* pp. 873-879, December 1988. Applications, 0-7803-7817, 2003.

[2] MEH. Benbouzid, "Induction motors' faults detection and localization using stator current advanced signal processing techniques," *IEEE Trans. On Industry Applications, Vol.14, no. 1,* January 1999.

[3] B. Yazici, "An adaptive statistical time-frequency method for detection of broken rotor bars and bearing faults in motors using stator current," *IEEE Trans. On Industry Applications, Vol.35* pp.442_452, Mar./Apr. 1999.

[4] R. R. Shoen, Brian K. Lin, Th. G. Halbetler, J H. Schlag, and S. Farag, "An unsupervised, on-line system for induction motor fault detection using stator current monitoring," *IEEE Trans. On Industry Applications, Vol. 31, No. 6,* pp. 1280-1286, 1995.

[5] D.G., Dorrell and W.T., Thomson "Analysis of air-gap flux, current, and vibration signals as a function of the combination of static and dynamic air-gap eccentricity in 3-phase induction motors". *IEEE Trans. On Industry Applications,* vol.33, No1, pp.24-34, Jan/Feb. 1997.

[6] Li Heming, Wan Shuting, Li Yonggong, Wang Aimeng, "Condition monitoring of generator stator winding inter-turn short circuit fault based on electrically excited vibration". *Electric Machines and Drives, 2005 IEEE International Conference On,* IEMDC,2005, pp.1-4, San Antonio, TX

[7] A. Bouzida, O. Touhami, R. Ibtiouen, M. Fadel, A. Rezzoug, A. Belouchrani, "Fault Diagnosis in Industrial Induction Machines through Discrete Wavelet Transform*," Trans. On Industrial Electronics,* vol.58, No9, pp. 4385-4395, 2011.

[8] O. Touhami, N. Lahcène, R. Ibtiouen, M. Fadel, "Modeling of the Induction Machine for the Diagnosis of Rotor Defects. Part. I: An Approach of Magnetically Coupled Multiple Circuits, " *IEEE-IECON, Industrial Electronics Society,* pp.1580-1587, 2005.

[9] Ong K, An N, Nee "A Web-based fault diagnostic and learning system", *Int. J. Adv. Manuf. Technoly 18(7):502–511,* (2001)

[10] Caldwell NH, Breton BC, Holburn DM, "Remote instrument diagnosis on the Internet" *IEEE Intel. Syst. App. 13 (3):70–76,* (1998)

[11] Peter Tavner, Li Ran, Jim Penman and Howard Sedding. *"Condition monitoring of Rotating Electrical Machines".* IET Power and Energy Series 56, London, United Kingdom © 2008 The Institution of Engineering and Technology First published 2008, pp.1-306

Low-Cost IC less Self Oscillating Boost PFC Converter

S. Borekci, I. M. Luleci

Abstract -- **Discontinuous conduction mode (DCM) boost power factor correction (PFC) converters have the advantages of zero-current turn-on for the boost-switch and no reverse recovery stress on boost-diode. When the converter switching frequency is constant, it has better EMI performance as well. In the proposed study, a novel approach is introduced. By using the new technique, IC'less PFC solution with less number of components is achieved. The proposed design is in DCM with constant duty and fixed frequency. The boost mosfet is driven by a transistor RC phase shift oscillator (T-RC PSO). In addition to that, IEC 61000-3-2 regulation limits have been met and the presented converter have lower bill of material (BOM) cost which is from 0.5 to 1 USD cost-down for per product compared with the IC solution.**

Keywords -- *SMPS, switching mode power supply, PFC, power factor correction, transistor RC phase shift oscillator, boost converter, DC/DC converter, DCM, discontinuous conduction mode.*

I. INTRODUCTION

In parallel with the constantly developing technology and increasing consumer demands, a better quality of electronic products has become a necessity. Presently, many products require fixed and stable DC power supply voltage such as PC/Laptop, LCD/LED TV, power LED lamp and communication devices. They draw sine wave current from the main electrical network, thanks to PFC techniques. Circuits without PFC stage, current drawing from the electrical network is too far from the sine waveform with low power factor (PF) and high total harmonic distortion (THD) [1], [2], [7]. The methods of achieving PFC can be classified into two. First one is the passive PFC approach, and the second is the active [1]-[11], [13]-[36]. The passive one is a type of L-C filter which is cheap and simple. However, they are heavy, bulky, and provide only poor regulation [6] and they will not be examined in this paper.

There are many different topologies for achieving active PFC techniques [1]-[36]. Boost converters are the widely used topology because they offer many advantages such as low ripple in the input current, high PF, small size in the output capacitor and being simple. In addition, the boost converter is one of the cheapest methods for the line-harmonics reduction circuit [6], [7].

Boost PFC converters can be designed and operates in three modes: continuous conduction mode (CCM), critical conduction mode (CRM) and discontinuous conduction mode (DCM) [3], [4]. For low and middle power requirements, DCM active PFC topology is one of the most used and the least expensive one [8], [9].

Moreover, DCM is less complicated than CCM is reported in [28].

All known active PFC circuits except Valley-fill and [12] can be achieved by a controller IC. Valley-fill is a simple solution but its PF is not greater than 0.85 [11]. [11] has only constant current performance which is 0.970 PF and 26% THD for different line voltage. Constant duty fixed frequency DCM boost PFC converter has higher performance especially for low line input voltage. In [12], the converter has two mosfets and two inductors are utilized. That increases the BOM.

IC manufactures offer many different types of PFC ICs to industry. Some of ICs drive the converter with constant duty and fixed switching frequency [2], [8], [20]-[22]. Some of them drive converters with variable duty and constant switching frequency [3], [4], [9], [13]-[17]. With some modifications, the second type ICs can be operated with constant duty but variable switching frequency [18].

The objective of this research is to propose a new method of achieving a high PF for the DCM boost PFC converter with no IC and low BOM. Instead of using an IC, the transistor RC phase shift oscillator (T-RC PSO) drives the boost converter at constant duty and fixed frequency to meet the load requirement. The proposed design mathematically analyzed, simulated in PSpice and implemented. As a prototype, a 75W boost PFC converter has been built and tested.

In section II, DCM boost PFC converters working with constant duty cycle and fixed switching frequency are mathematically analyzed. In section III, the proposed circuit is introduced. In section IV, simulation and experimental results are presented. In addition to that, in the section, cost and performance comparison among the converters are illustrated.

II. DCM BOOST PFC CONVERTER WITH CONSTANT DUTY CYCLE AND FIXED FREQUENCY

A. Two-Stage SMPS Approach

Typical two-stage SMPS is shown in Figure 1, and as reported in [10], [23], [24], compared with single-stage approach, two-stage approach has many advantages. Better efficiency, lower weight and volume, good performance with protection can be mentioned [25]. In [24]-[27] universal-line two-stage approach with the same size as the conventional single-stage one is presented and two-stage approach has higher efficiency than single-stage approach across the entire line.

978-1-4799-0024-4/13 $31.00 © 2013 IEEE

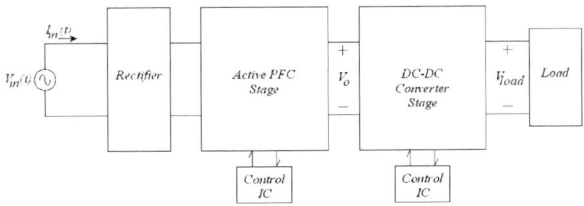

Figure 1. Typical two-stage SMPS

At the V_o line, generally about 400V DC bus voltage is expected. 400V DC bus voltage converted to the required lower DC voltage by suitable DC/DC converter with its peripheral circuits. The suitable DC/DC converter, for example; flyback, half-bridge, full-bridge, LLC resonant, or etc. selected by the designer with considering load. On the load side, single or multiple outputs may be needed, so that the DC/DC converter may have single or multi-output [25]. In this paper, second stage of the SMPS will not be examined.

B. Boost PFC Converter

Figure 2 shows the circuit of a boost PFC converter. For simple calculation, fallowing approaches are made. All circuit elements are ideal and lossless, output voltage and current ripple are neglect able and the switching frequency of the mosfet is much higher than the electrical network line frequency.

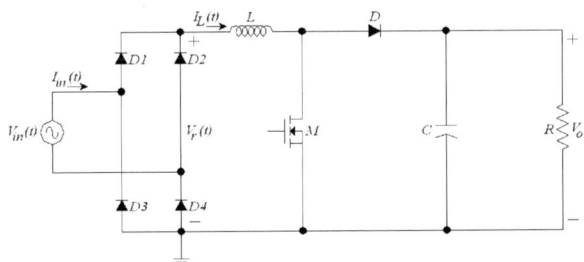

Figure 2. The circuit of a boost PFC converter

The electrical network line voltage means input voltage is defined as:

$$V_{in}(t) = V_m \sin(2\pi f t) \tag{1}$$

where, f is the line frequency; V_m is the line voltage amplitude.

The rectified voltage is:

$$V_r(t) = |V_m \sin(2\pi f t)| \tag{2}$$

When the boost PFC converter operates at DCM, the inductor current waveform in a line cycle looks like in Figure 3.

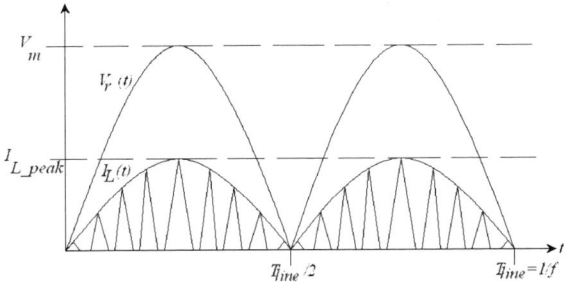

Figure 3. The inductor current waveform in a line cycle (DCM)

If it is looked at this signal in a boost switching frequency, it is seen a portion of this signal as shown in Figure 4. This is the inductor current waveform in a few switching cycle. In a boost switching cycle, the inductor peak current $I_{L\_peak}(t)$ is:

$$I_{L\_peak}(t) = \frac{V_r(t)}{L} D_{rise} T_s \tag{3}$$

where D_{rise} is the duty cycle and T_s is the switching cycle of the boost converter.

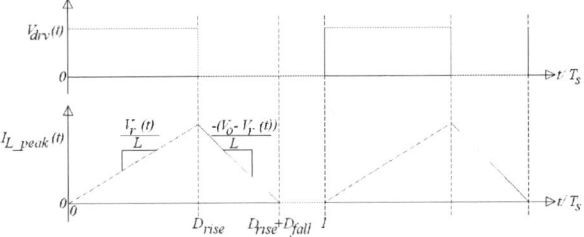

Figure 4. DCM Boost inductor current waveform in a switching cycle

In each switching cycle, voltage-second balance has been established and:

$$V_r(t) D_{rise} T_s = (V_o - V_r(t)) D_{fall} T_s \tag{4}$$

where, D_{fall} is the discharging duty of the inductor and V_o is the output voltage on the load side.

In DCM, the time between $(D_{rise}+D_{fall})/T_s$ and the $1/T_s$ is called dead duty and it is not interested this part because of the voltage-second balance.

In a switching cycle T_s, average inductor current can be calculated from (3) as;

$$I_{L\_avg}(t) = \frac{1}{2} I_{L\_peak}(t)(D_{rise} + D_{fall}) \tag{5}$$

where;

$$D_{fall} = \frac{V_r(t)}{V_o - V_r(t)} D_{rise} = \frac{|V_m \sin(2\pi f t)|}{V_o - |V_m \sin(2\pi f t)|} D_{rise} \tag{6}$$

978-1-4799-0024-4/13 $31.00 © 2013 IEEE

comes from (4). The inductor current in a switching cycle (5) can be rewritten as:

$$I_{L\_avg}(t) = \frac{V_m D_{rise}^2}{2Lf_{sw}} \frac{|\sin(2\pi ft)|}{1 - \frac{V_m}{V_o}|\sin(2\pi ft)|} \tag{7}$$

where, f_{sw} is the boost switching frequency that is equal to the $1/T_s$.

So the electrical network line current means input current is equal to the not-absolute value of the (7) is that:

$$I_{in}(t) = \frac{V_m D_{rise}^2}{2Lf_{sw}} \frac{\sin(2\pi ft)}{1 - \frac{V_m}{V_o}|\sin(2\pi ft)|} \tag{8}$$

According to (3) and (7), the inductor current for a half line cycle plotted in Figure 5.

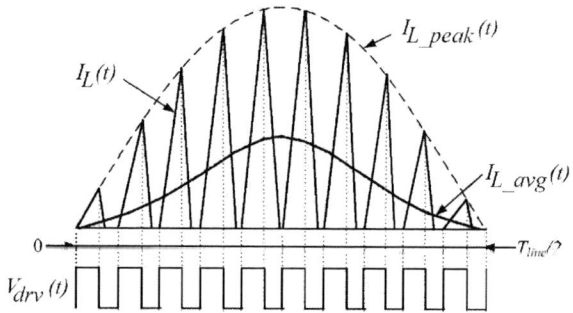

Figure 5. The inductor current waveform in a half-line cycle (DCM)

Using assumption in Section II that is all circuit elements are ideal and lossless, $P_{in}=P_{out}$ and efficiency of the converter is 100%, and then the input power in a half-line cycle is:

$$P_{in} = \frac{1}{T_{line}/2} \int_0^{Tl_{ine}/2} V_{in}(t) I_{in}(t) dt \tag{9}$$

$$P_{in} = \frac{V_m^2 D_{rise}^2}{2Lf_{sw}} \frac{1}{\pi} \int_0^\pi \frac{\sin^2(2\pi ft)}{1 - \frac{V_m}{V_o}|\sin(2\pi ft)|} d(2\pi ft) \tag{10}$$

From (10), the inductor current discharge duty D_{rise} can be written as:

$$D_{rise} = \frac{1}{V_m} \sqrt{\frac{2\pi Lf_{sw} P_{in}}{\int_0^\pi \frac{\sin^2(2\pi ft)}{1 - \frac{V_m}{V_o}|\sin(2\pi ft)|} d(2\pi ft)}} \tag{11}$$

The RMS value of the input current is;

$$I_{in\_rms} = \sqrt{\frac{1}{\pi} \int_0^\pi (I_{in}(t))^2 d(2\pi ft)} \tag{12}$$

Then the Power Factor (PF) is calculated as:

$$PF = \frac{P_{in}}{S} = \frac{\frac{V_m^2 D_{rise}^2}{2Lf_{sw}} \frac{1}{\pi} \int_0^\pi \frac{\sin^2(2\pi ft)}{1 - \frac{V_m}{V_o}|\sin(2\pi ft)|} d(2\pi ft)}{\frac{V_m}{\sqrt{2}} \sqrt{\frac{1}{\pi} \int_0^\pi (\frac{V_m D_{rise}^2}{2Lf_{sw}} \frac{\sin(2\pi ft)}{1 - \frac{V_m}{V_o}|\sin(2\pi ft)|})^2 d(2\pi ft)}}$$

$$= \frac{\frac{2}{\sqrt{\pi}} \int_0^\pi \frac{\sin^2(2\pi ft)}{1 - \frac{V_m}{V_o}|\sin(2\pi ft)|} d(2\pi ft)}{\sqrt{\int_0^\pi (\frac{\sin(2\pi ft)}{1 - \frac{V_m}{V_o}|\sin(2\pi ft)|})^2 d(2\pi ft)}} \tag{13}$$

According to (13), input PF can be plotted as in Figure 6. It can be seen in Figure 6 that, if the V_m/V_o is larger, than the PF is lower. For 220V AC input and 400V DC output, PF is about **0.96** and it is high enough for the IEC 61000-3-2 limit.

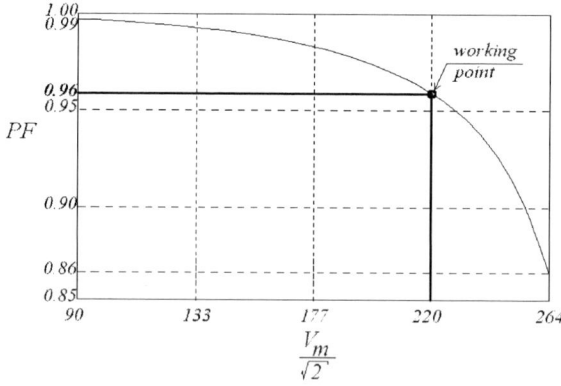

Figure 6. PF for constant duty cycle control for DCM boost converter and proposed working point illustration

In this paper, proposed circuit for PFC is a cost-effective solution for two-stage SMPS. Although this solution does not have a PFC IC and its peripheral components, it offers active PFC solution.

To improve the PFC stage of the SMPS, many different solutions are introduced. For example, to eliminate the high frequency component of the current and to meet with the EMI standards, an input filter can be used [10], [28]. The other example is that, to minimize

978-1-4799-0024-4/13 $31.00 © 2013 IEEE   30

the stress and power losses of switching devices, snubber circuits are used [29]-[33]. For more improvements, instead of the PFC IC, a DSP or PLU solution has also introduced [34]-[36]. But these types of solutions need more complicated circuit and more expensive components.

Some of the improvements given above can be adapted to the solution given in this paper.

C. Transistor RC Phase Shift Oscillator (T-RC PSO)

In Figure 7, transistor RC phase shift oscillator with amplifier and high-pass filter is illustrated. V_{OUT} is the oscillator signal and its frequency can be adjusted by the oscillator stage component R_{OSC}, C_{OSC} as described in [37] or another methods.

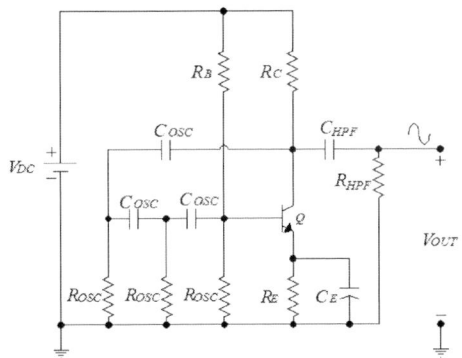

Figure 7. Transistor RC phase shift oscillator (T-RC PSO) with amplifier and high-pass filter

The shifted phase and oscillator frequency also depends on the number of the RC stage. Without mathematical details, from [38], the oscillation frequency of the transistor RC phase shift oscillator is:

$$f_{osc} = \frac{1}{2 R_{OSC} C_{OSC} \sqrt{6}} \quad (14)$$

where, the circuit in Figure 7 has *three* RC stages.

III. IMPLEMENTING PROPOSED CIRCUIT

The proposed solution is an integration of a boost PFC converter and a transistor RC phase shift oscillator used as a driver illustrated in Figure 8.

Transistor RC phase shift oscillators produce high frequency signal which modulate the input current of the boost PFC converter. To drive the mosfet in switching mode, the output signal of the oscillator must be higher than the threshold voltage (V_{GS}) which is commonly about 4V.

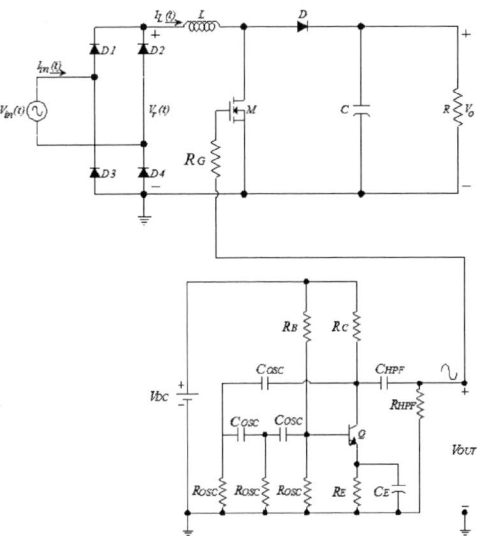

Figure 8. The proposed circuit (The boost PFC converter + Transistor RC phase shift oscillator)

The oscillator output signal is in sinusoidal waveform. The mosfet of the boost PFC converter starts to conduct when the signal reaches the threshold level of V_{GS}. Then it keeps being on till the driving signal decreasing to threshold level of V_{GS}. This is illustrated in Figure 9.

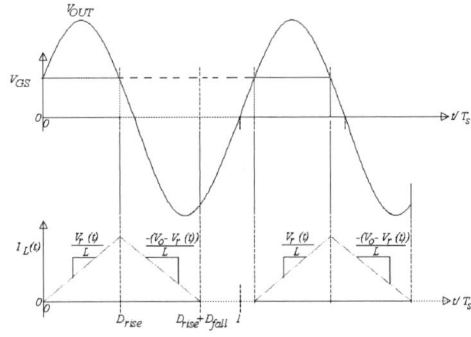

Figure 9. DCM Boost PFC converter's inductor current waveform in a switching cycle which is driving by transistor RC phase shift oscillator

where, V_{OUT} is the output signal of the transistor RC phase shift oscillator. V_{GS} is the threshold voltage of the mosfet.

$V_{OUT} > V_{GS}$ Mosfet starts conducting.
$V_{OUT} < V_{GS}$ Mosfet finishes conducting.

In this application, D_{rise} duty is quite different than the traditional DCM boost PFC converter and it can be calculated below equation. V_{out_m} is the peak value of V_{OUT}.

$$D_{rise} = \frac{\pi - 2 \sin^{-1}(V_{GS} / V_{out\_m})}{2\pi} \quad (15)$$

978-1-4799-0024-4/13 $31.00 © 2013 IEEE   31

IV. SIMULATION AND EXPERIMENTAL RESULTS

As a prototype, 75W boost PFC converter with 220Vrms @ 50Hz input, 400V DC output is designed. The circuit is simulated and analyzed in PSpice program and then it is realized. The specifications of the proposed circuit are given in Table I.

TABLE I
THE SPECIFICATIONS OF THE PROPOSED CIRCUIT

f:50Hz	$V_{in}(t)$:$V_m\sin(2\pi ft)$	L:1.2mH
$T_{line}=1/f$:20ms	$V_{in\ rms}$:220V	M:IRFP460
V_m:311V	$V_r(t)$:$V_m|\sin(2\pi ft)|$	D:STTA3006P
$D1,D2,D3,D4$: KBPC1010	Vo:400V DC	R_B:3.3kΩ
C:220µF 450V	V_{DC}:60V DC	R_C:1kΩ
R:2120Ω 100W	Q:BD139 /16S	R_E:470Ω
C_E:4.7µF	R_{OSC}:1kΩ	$T_s=1/f_{osc}$:41.96µs
R_{HPF}:10kΩ	C_{OSC}:2.2nF	V_{out_m}:4.62V
C_{HPF}:4.7nF	$f_{osc}=f_{sw}$:23830Hz	V_{OUT}: $V_{out\_m}\sin(2\pi f_{osc}t)$
V_{GS}:4V(max.) from IRFP460 datasheet	R_G:10Ω	P_{out}:75W

From (15), the rising duty D_{rise} is equal to 16.68% and duty time $D_{rise}T_s$ is equal to 7µs. Mathematical, simulation and experimental results of the proposed 75W boost PFC converter are given in Table II.

TABLE II
MATHEMATICAL, SIMULATION AND EXPERIMENTAL RESULTS OF THE PROPOSED 75W BOOST PFC CONVERTER

	Mathematical	Simulation	Experimental
Vin(rms) (V)	220	220	220
IL_peak(π/4) (A)	1.814	1.822	1.815
Iin_rms (A)	0.362	0.379	0.389
fosc=fsw (Hz.)	23830	23830	23830
Ts (us.)	41.96	41.96	41.96
Duty (%)	16.68	16.68	16.68
Duty time (us.)	7	7	7
S (VA)	79.56	83.42	85.50
Pin (W)	76.45	80.12	81.95
Vo (V) DC	402	401	400
Pout (W)	76.45	75.85	75.47
PF (Pin/S)	0.961	0.960	0.959
THD (%)	28.38	29.73	30.28
Effeciency (Pout/Pin) (%)	100	94.68	92.09

In Table II, the results are in harmony. However, there are minor differences in results come from losses on circuit components and their tolerances.

In Figure 10, the current harmonics of the proposed converter and limits which is defined by IEC 61000-3-2 standard for 75W are illustrated. The line current harmonics are in allowable limits. IEC THD limit is 32.28% and also achieved with 30.28%.

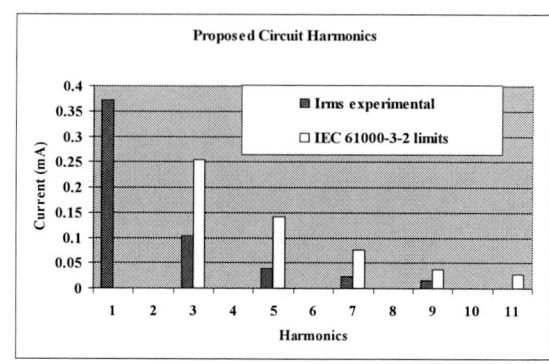

Figure 10. Proposed circuit's experimental results-line current harmonics and IEC 61000-3-2 limits

Mathematical, simulation and experimental results for line current waveform are given in Figure 11.

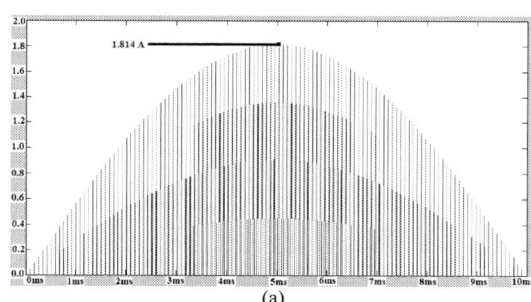

(a)

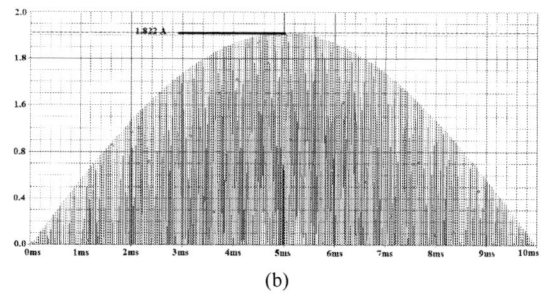

(b)

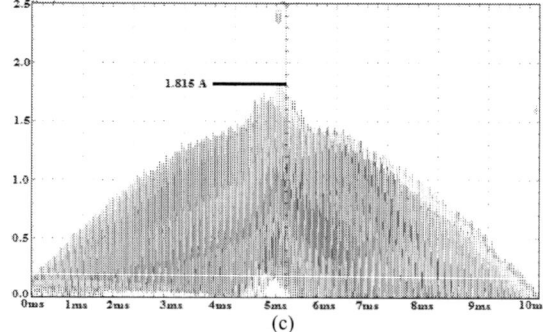

(c)

Figure 11. Line current waveform for a-half line cycle. Y-axis: 0.5A/div, X-axis: 1ms./div. a) Mathematical-MATLAB, b) Simulation-PSpice c) Experimental

In Figure 12, one cycle line current and its zoomed portion are illustrated. It can be seen that PFC converter working in DCM.

978-1-4799-0024-4/13 $31.00 © 2013 IEEE

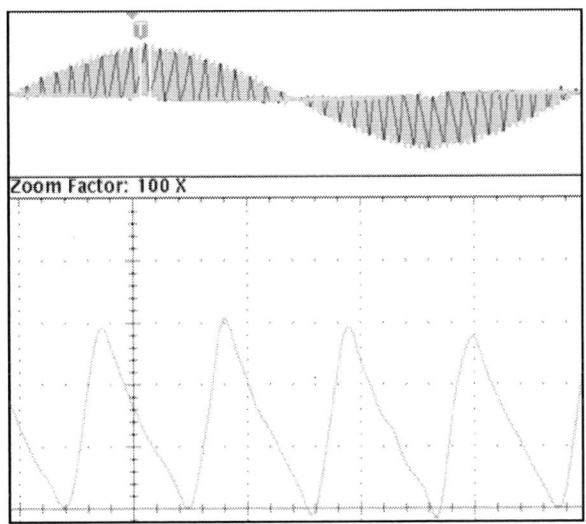

Zoom Factor: 100 X

Figure 12. Line current waveform for a line cycle. In zoomed part Y-axis: 0.5A/div, X-axis: 40μs./div., 100x zoomed (Experimental result)

In Table III, BOM and performance comparison among commonly used converters in the literature are demonstrated. The converters with PFC ICs have better PF and THD. However, their BOM cost is high. It is obvious that the solution given by this paper is not high performance as PFC IC solution but is in the IEC 61000-3-2 limits. On the other hands, constant duty and fixed frequency solution has higher performance, when the input voltage is 110 Vac. For example, from [2], 0.990 PF and 9.8% THD were achieved. That is already mentioned in Figure 6.

TABLE III
COMPONENT AND PERFORMANCE COMPARISON OF THE BOOST PFC CONVERTERS AND WITHOUT PFC RESULTS

Component / Performance	No PFC	With PFC IC			With PFC IC			
	Without PFC	Varible Duty Fixed Freq Boost PFC Converter	Const. Duty Varible Freq. Boost PFC Converter	Const. Duty Fixed Freq Boost PFC Converter	Valley Fill PFC	Valley Fill + Forward Converter Ref.[11]	Self-Oscillating PFC Ref.[12]	Proposed Boost PFC Converter
Mosfet	No	Yes	Yes	Yes	No	Yes	Yes (x 2)	Yes
Inductor	No	Yes	Yes	Yes	No	Yes	Yes (x 2)	Yes
Capacitor	Yes	Yes	Yes	Yes	Yes (x2)	Yes (x 2)	Yes	Yes
Others (R,C,D,etc.)	Yes	Yes	Yes	Yes	Yes	Yes	Yes	Yes
PFC IC	No	Yes	Yes	Yes	No	No	No	No
Transistor	No	No	No	No	No	No	No	Yes
PF	0.600	>0.995	>0.995	0.960 Ref.[4]	0.850	0.970	0.992	0.959
THD	134.00%	<5.00%	<5.00%	20.59% Ref.[9]	32.00%	26.00%	13.00%	30.28%
Vo Ripple	Large	Too Small	Too Small	Too Small	Large	Too Small	Too Small	Too Small

When the BOM cost of the converter with IC and proposed circuit compared, the proposed solution has less component and less cost. The comparison is given in Table IV. The prices are just for to give an idea. They are for one unit order and may change by order quantity. (www.digikey.com)

TABLE IV
PRICE DIFFERENCE BETWEEN THE PFC SOLUTIONS WITH IC AND PROPOSED CIRCUIT

WITH PFC IC			
COMPONENT	QUANTITY	UNIT COST (USD)	TOTAL COST (USD)
PFC IC (L6561)	1	0.92	0.92
PFC IC's PERIPHERAL COMPONENT			
Resistor	4	0.09	0.36
Zener Diode	1	0.42	0.42
Diode	2	0.42	0.84
Capacitor	4	0.33	1.32
Total			*3.86*

PROPOSED CIRCUIT			
COMPONENT	QUANTITY	UNIT COST (USD)	TOTAL COST (USD)
Transistor (BD139 /16S)	1	0.48	0.48
TRANSISTOR's PERIPHERAL COMPONENT			
Resistor	7	0.09	0.63
Capacitor	6	0.33	1.98
Total			*3.09*
Difference			*0.77*

http://www.digikey.com

V. CONCLUSIONS

DCM PFC circuits have some advantages over critical and continuous mode converters such as less power loss and better EMI performance. In addition to that IC-less solution offers low BOM cost and simplicity. Proposed technique which works in DCM gives an active PFC solution without using a PFC IC. 75W prototype circuit is assembled and tested. It gives 0.959 PF, 30.28% THD, and 92.09% efficiency with less BOM which is about 0,77 USD cost-down. Also, the IEC 61000-3-2 regulation limits have been achieved. Minor differences in results come from losses on circuit components and their tolerances.

VI. REFERENCES

[1] F. Beltrame, L. Roggia, L. Schuch, and J. R. Pinheiro, "A Comparison of High Power Single-Phase Power Factor Correction Pre-Regulators," *Industrial Technology (ICIT), 2010 IEEE International Conference*, pp. 625-630, March 2010.

[2] Y.-C. Chuang, and H.-L. Cheng, "Single-Stage Single-Switch High-Power-Factor Electronic Ballast for Fluorescent Lamps," *Industrial and Commercial Power Systems Technical Conference, IEEE*, pp. 1-7, 2006.

[3] K. Yoa, X. Ruan, X. Mao, and Z. Ye, "DCM Boost PFC Converter with High Input PF," *Applied Power Electronics Conference and Exposition (APEC), 2010 Twenty-Fifth Annual IEEE*, pp. 1405-1412, Feb 2010.

[4] K. Yoa, X. Ruan, X. Mao, and Z. Ye, "Variable-Duty-Cycle Control to Achieve High Input Power Factor for DCM Boost PFC Converter," *Industrial Electronics, IEEE Transactions*, vol. 58, no. 5, pp. 1856-1865, May 2011.

[5] D. C. Hopkins, and D. W. Root, "Synthesis of a New Class of Converters That Utilize Energy Recirculation," *Power Electronics Specialists Conference, PESC '94 Record., 25th Annual IEEE*, vol. 2, pp. 1167-1172, Jun 1994.

[6] R. Redl, "Reducing Distortion in Boost Rectifiers with Automatic Control," Applied Power Electronics Conference and Exposition, 1997. APEC '97 Conference Proceedings 1997., Twelfth Annual, vol. 1, pp. 74-80, Feb 1997.

[7] E. Maset, E. Dede, G. Hau, and F. C. Lee, " 100 kHz, 2 kW Boost ZVT-PWM Converter For Power Factor Correction," Power Electronics Congress, 1993. Technical Proceedings. CIEP 93., 2nd International, pp. 102-106, Aug 1993.

[8] W. Chen, and F. C. Lee, "Single Magnetic, Unity Power Factor, Isolated Power Converter With Ripple Free Input Current," Power Electronics Specialists Conference, 1998. PESC 98 Record. 29th Annual IEEE, vol. 2, pp. 1450-1455, May 1998.

[9] H. S. Athab, and P. K. S. Khan, "A Cost Effective Method of Reducing Total Harmonic Distortion (THD) in Single-Phase Boost Rectifier," Power Electronics and Drive Systems, PEDS '07. 7th International Conference, pp. 669-674, 2007.

[10] K.-H. Liu, and Y.-L. Lin, "Current Waveform Distortion In Power Factor Correction Circuits Employing Discontinuous-Mode Boost Converters," Power Electronics Specialists Conference, 1989. PESC '89 Record, 20th Annual IEEE, vol. 2, pp. 825-829, Jun 1989.

[11] N. Y. Choi, and C. H. Lee, "A New Single-Stage Converter for Improving THD," Power Electronics and Drive Systems, 2003. PEDS. The Fifth International Conference, vol. 2, pp. 1476-1479, Nov. 2003.

[12] L. C. Gomes de Freitas, E. E. A. Coelho, J. B. Vierira Jr., and L. C. de Freitas, "A Single-Stage PFC Converter Applied as an Electronic Ballast for Fluorescent Lamps," Applied Power Electronics Conference and Exposition, APEC '04. Nineteenth Annual IEEE, vol. 1, pp. 164-169, 2004.

[13] Y.-K. Lo, S.-Y. Ou, and T.-H. Song, "Varying Duty Cycle Control for Discontinuous Conduction Mode Boost Rectifiers," Power Electronics and Drive Systems, Proceedings, 4th IEEE International Conference, vol. 1, pp. 149-151, Oct. 2001.

[14] Y. Hongxiang, L. Min, and J. Yanchao, "An Advanced Harmonic Elimination PWM Technique for AC Choppers," Power Electronics Spectalists Conference, PESC 04, IEEE 35th Annual, vol. 1, pp. 161-165, June 2004.

[15] H. S. Athab, "A Duty Cycle Control Technique for Elimination of Line Current Harmonics in Single-Stage DCM Boost PFC Circuit," TENCON 2008 - 2008 IEEE Region 10 Conference, pp. 1-6, Nov. 2008.

[16] L. S. C. Silva, F. J. M. de Seixas, and M. A. G. de Brito, "Bridgeless Interleaved Boost PFC Converter With Variable Duty Cycle Control," Power Electronics Conference (COBEP), Brazilian, pp. 397-402, Sept. 2011.

[17] K. Taniguchi, and Y. Nakaya, "Analysis and Improvement of Input Current waveforms for Discontinuous-Mode Boost Converter with Unity Power Factor," Power Conversion Conference - Nagaoka, Proceedings, pp. 397-402, Aug 1997.

[18] Y.-K. Lo, J.-Y. Lin, and S.-Y. Ou, "Switching-Frequency Control for Regulated Discontinuous-Conduction-Mode Boost Rectifiers," Industrial Electronics, IEEE Transactions, vol. 54, no 2, pp. 760-768, April 2007.

[19] Y.-L. Chen, Y.-M. Chen, and C.-N. Wu, "The Time-Domain Analysis for Constant On-Time Critical Mode Boost-Type PFC Converters," Energy Conversion Congress and Exposition (ECCE), IEEE, pp. 4643-4648, Sept. 2012.

[20] M. Ponce, J. Arau, A. Lopez, and J. M. Alonso, "A Novel High-Power-Factor Single-Switch Electronic Ballast for Compact Fluorescent Lamps," Power Electronics Congress, CIEP. VII IEEE International, pp. 194-198, 2000 .

[21] J. Cardesín, J. Ribas, M. Rico , J. M. Alonso, A. Calleja, E. Corominas, and J. García, "A Low Cost Electronic Ballast for 250W High Pressure Sodium Vapour Lamps Using the CC/CC Buck Converter as Power Factor Preregulator," Industry Applications Conference, 37th IAS Annual Meeting. Conference Record, vol. 3, pp. 1847-1851, 2002.

[22] Y.-C. Chuang, and H.-L. Cheng, "Single-Stage Single-Switch High-Power-Factor Electronic Ballast for Fluorescent Lamps," Industry Applications, IEEE Transactions, vol. 3, no 6, pp. 1434-1440, Nov-Dec 2007.

[23] S. V. Mollov, A. J. Forsyth, and D. R. Nuttall, "Performance/Cost Comparison between Single-Stage and Conventional High Power Factor Correction Rectifiers," Power Electronics and Drives Systems, PEDS. International Conference, vol. 2, pp. 876-881, Nov 2005.

[24] C. Shin-Young, L. Il-Oun, P. Jeong-Eon and M. Gun-Woo, "Two-stage Configuration for 60W Universal-line AC-DC Adapter," IECON 2012 - 38th Annual Conference on IEEE Industrial Electronics Society, pp. 1445-1450, Oct. 2012.

[25] M.S. Göksu, and I. Alan, "250W Flyback SMPS Design for a Big Size CTV," Consumer Electronics, IEEE Transactions, vol. 49, no 4, pp. 911-916, 2003.

[26] J.-E. Park, J.-W. Kim, B.-H. Lee, and G.-W. Moon, "Design on Topologies for High Efficiency Two-Stage AC-DC Converter," Power Electronics and Motion Control Conference (IPEMC), 7th International, vol. 1, pp.257-262, June 2012.

[27] T. K. Jappe, and S. A. Mussa, "Discrete-Time Current Control Techniques Applied In PFC Boost Converter At Instantaneous Power Interruption," Power Electronics Conference, COBEP '09. Brazilian, pp.1118-1123, Sept-Oct 2009.

[28] W. Grigore, J. Kyyra, and J. Rajamaki "Input Filter Design for Power Factor Correction Converters Operating in Discontinuous Conduction Mode," Electromagnetic Compatibility, IEEE International Symposium, vol. 1, pp.145-150, 1999.

[29] I. Matsuura, K. M. Smith, and K. M. Smedley, "A Comparison of Active and Passive Soft Switching Methods for PWM Converters," Power Electronics Specialists Conference,PESC 98 Record, 29th Annual IEEE, vol. 1, pp.94-100, May 1998.

[30] B. T. Irving, and M. M. Jovanović, "Analysis, Design, and Performance Evaluation of Flying-Capacitor Passive Lossless Snubber Applied to PFC Boost Converter," Applied Power Electronics Conference and Exposition, APEC. Seventeenth Annual IEEE, vol. 1, pp.503-508, 2002.

[31] Y. Jang, D. L. Dillman, and M. M. Jovanovic´, "A New Soft-Switched PFC Boost Rectifier With Integrated Flyback Converter for Stand-by Power," Power Electronics, IEEE Transactions, vol. 21, no 1, pp.66-72, Jan 2006.

[32] R. Zhao, J. Pan, and J. Hui, "A Novel Soft-Switching Boost PFC with a Passive Snubber," Industrial Electronics and Applications, ICIEA. 2nd IEEE Conference, pp.1473-1476, May 2007.

[33] M. Mahesh, and A. K. Panda, "A High Performance Single-Phase AC-DC PFC Boost Converter with Passive Snubber Circuit," Energy Conversion Congress and Exposition (ECCE), IEEE, pp. 2888-2894, Sept. 2012.

[34] K. De Gusseme, D. M. Van De Sype, and J. A. A. Melkebeek, "Design Issues for Digital Control of Boost Power Factor Correction Converters," Industrial Electronics, ISIE, Proceedings, IEEE International Symposium, vol. 3, pp. 731-736, 2002.

[35] M. Ferdowsi, Z. Nie, and A. Emadi, "A New Estimative Current Mode Control Technique for DC-DC Converters Operating in Discontinuous Conduction Mode," Power Electronics and Motion Control Conference, IPEMC. The 4th International, vol. 2, pp. 497-501, Aug 2004.

[36] Z. Z. Ye, M. M. Jovanovic´, and B. T. Irving, "Digital Implementation of a Unity-Power-Factor Constant-Frequency DCM Boost Converter," Applied Power Electronics Conference and Exposition, APEC. Twentieth Annual IEEE, vol. 2, pp. 818-824, March 2005.

[37] R. W. Johnson, "Extending the Frequency Range of the Phase-Shift Oscillator," Proceedings of the IRE, vol. 33, no 9, pp. 597-603, Sept. 1945.

[38] R. W. Johnson, "Generalized Equations for RC Phase-Shift Oscillators," Proceedings of the IRE, vol. 42, no 7, pp. 1169-1172, July 1954.

VII. BIOGRAPHIES

Selim Borekci received the M.Sc. and Ph.D. degrees in Electrical Engineering from New Mexico State University,Las Cruces, in 1997 and 2003, respectively.

978-1-4799-0024-4/13 $31.00 © 2013 IEEE

From 2001 to 2004, he was a Design Engineer with Howard Industries, Laurel, MS. Since 2004, he has been with the Department of Electrical and Electronic Engineering, Akdeniz University, Antalya, Turkey, where he is currently an Assistant Professor. His research interest includes applications of power electronics such as power filters, reactive power control techniques, power supplies, multilevel inverters, electronic ballasts, resonant circuits, and induction heating systems.

Ihsan Murat Luleci was born in Nevsehir, Turkey. He received the B.E.E. degree from Ege University (E.U), in 2003. He worked as a technical product manager between 2005 and 2010 at the Vestel Durable Goods Marketing Company. He has been working for government for three years and simultaneously continuing M.Sc thesis at Akdeniz University. His research interests are power electronics, SMPS systems and their improvements'.

Evaluation of Different Broken Bar Fault Diagnostic Means in Double-Cage Induction Motors with FEM

Konstantinos N. Gyftakis, *Member, IEEE*, Dimitrios K. Athanasopoulos and Joya Kappatou, *Member, IEEE*

Abstract -- **The aim of the present paper is to compare and evaluate different diagnostic means for the identification of the broken bar fault in delta connected double-cage induction motors. The work is carried out with FEM. The motors are studied at nominal load as well as under low load operation. The traditionally used diagnostic means such as: the current, torque and apparent power waveforms' spectrums are thoroughly studied with respect to prior work. Moreover, in this work the zero-sequence current spectrum is also investigated. The advantages and disadvantages of each diagnostic mean will be discussed and evaluated, offering an important insight of the broken bar fault diagnosis in the double-cage induction motor.**

Index Terms-- **Broken bar, Double-cage induction motor, FEM, Fault diagnosis.**

I. INTRODUCTION

THE double-cage induction motor belongs to NEMA design class C [1]. In this motor, one can see that the rotor contains two separate conducting cages, from different materials. This means that, in each rotor slot there exist two conducting bars: the upper and the inner. Usually, in order to improve the starting behavior of the motor, the upper bar is characterized by greater resistivity than the inner bar. As a consequence, the upper bar is small and made from aluminum or bronze, whereas the inner bar from copper [2].

Double-cage induction motors are characterized by higher starting torque, lower starting current and normal nominal efficiency compared to the standard NEMA's class A induction motor [3]. Due to those characteristics and despite the greater manufacturing cost, the double-cage induction motor is used in applications such as: conveyors, crushers, compressors, loaded pumps, stirrers etc [3]-[4].

The health condition of the operating motors in an industrial environment is of crucial importance, since it affects the effectiveness of the production line and the motors' life cycle. The main faults which can appear during

motor operation are the following: stator faults, eccentricity and bearing faults and the broken or cracked bar/end-ring faults [5]. Around 5-10% of total induction motor failures are rotor failures and, more specifically, broken rotor bars and end-ring faults [5].

Many methods can be found in the literature on the induction motor's broken bar fault detection in single cage induction motors. Most common is the MCSA method which is based on the study of the current's frequency spectrum around the basic harmonic component [6]-[7]. Alternatively, it is proposed in [8]-[9] the detection of the broken bar fault, by observing the lower sideband frequency of the fifth current harmonic. The instantaneous power [10]-[12], and the air gap torque signature [12]-[16] have also been used, when frequencies at $2ksf_s$, near the dc component, are indicating broken bar/end-ring faults. Also, in [17] the authors have shown that it is possible to detect the fault at the frequency area close to 300Hz in the torque spectrum. Moreover, in [12] it is reported that it is possible to detect the fault around the 100Hz harmonic of the power spectrum. Finally, experimental testing has shown in [18], that the zero-sequence current spectrum can also reveal the broken bar fault at the frequency area close to the saturation related, third harmonic, in delta connected induction motors. All the above mentioned works deal with the broken bar fault detection in conventional induction motors. Except from the FFT, other methods have also been proposed for the broken-bar fault detection such as Wavelet [19]-[20] and MUSIC [21]-[22].

Concerning the broken bar fault in double-cage induction motors, the literature is quite poor. Generally, it has been reported that this specific fault type is very difficult to be detected [2]. This is due to the fact that, usually the fault occurs at an upper bar and as a consequence, the total initial fault impact on the motor operation is low. Lately, it is proposed to identify this type of fault through a dynamic model [23] of double cage induction motors and the combination of current signature and vibration analysis [24]. The offline methods of DWT analysis under startup transient [4] and Wavelet analysis under time-varying condition [25] were also recommended.

In this work, the aim is to study, with the application of the FFT, the frequency spectrum of various diagnostic means of the double-cage induction motor and evaluate their effectiveness at nominal operation and low load operation as well.

The work is carried out with FEM analysis with the use of

This work was supported by the research program: "K. Karatheodori 2010", of the Research Committee of the University of Patras, Greece.

K. N. Gyftakis is with the Laboratory of Electromechanical Energy Conversion, Dept of Electrical and Computer Engineering, University of Patras, Greece (phone: 0030-2610-996413; e-mail: kosgyftak@upatras.gr).

D. K. Athanasopoulos is with the Laboratory of Electromechanical Energy Conversion, Dept of Electrical and Computer Engineering, University of Patras, Greece (E-mail: athanasd@upatras.gr).

J. C. Kappatou is with the Laboratory of Electromechanical Energy Conversion, Dept of Electrical and Computer Engineering, University of Patras, Greece (E-mail: joya@ece.upatras.gr).

978-1-4799-0024-4/13 $31.00 © 2013 IEEE

the software OPERA. The simulated motor's stator geometrical and electromagnetic characteristics are taken from a real stator of a 4kW, 400V, 3-phase induction motor. The non-linear B-H magnetic characteristics of the rotor and stator iron core are taken into consideration. The simulated rotor has the upper bars from aluminum, the inner from copper and the rotor slot middle area from iron [26]. All the motors are considered un-skewed due to the high computational time. The analysis carried out is transient under constant speed. The application of the FFT on the motors' signals waveforms is implemented with the use of MATLAB.

II. SIMULATIONS AT 1460RPM

In this paragraph, two cases are studied. The motor operates under healthy condition and with one broken upper bar. The motor operates under constant speed 1460rpm.

A. Spatial Electromagnetic Characteristics

In Fig. 1, it can be seen that, the location of the selected upper broken bar. Moreover, in Fig. 2-3 the current density versus the geometrical angle for the upper and inner rotor bars is presented for healthy and faulty condition respectively. It has been shown [17] that, in conventional cage induction motors the breaking of the bar affects strongly its neighboring bars which draw increased current and are expected to break next. A different situation is met in the case of the broken bar fault in double cage induction motors. It is shown in Fig. 3-b, that in the double-cage induction motor, the breaking of the upper bar affects the inner bar of the same rotor slot, significantly more than its neighboring bars.

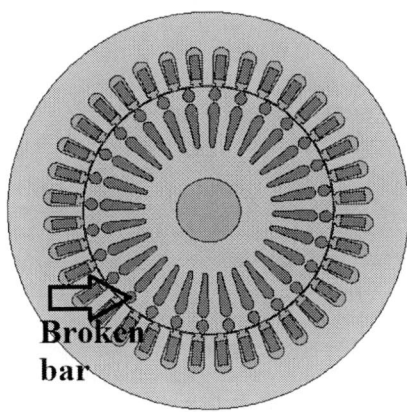

Fig. 1. The simulated double-cage induction motor. The arrow indicates the location of the broken upper bar.

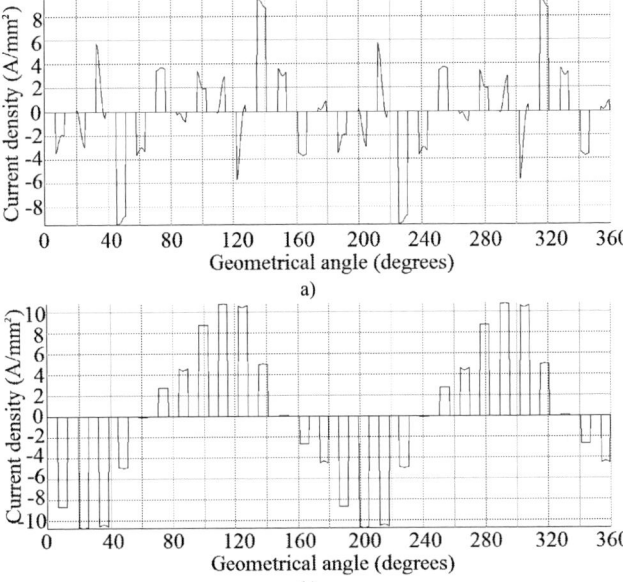

Fig. 2. The current density distribution versus the geometrical angle in the: a) upper bars and b) inner bars for the healthy motor.

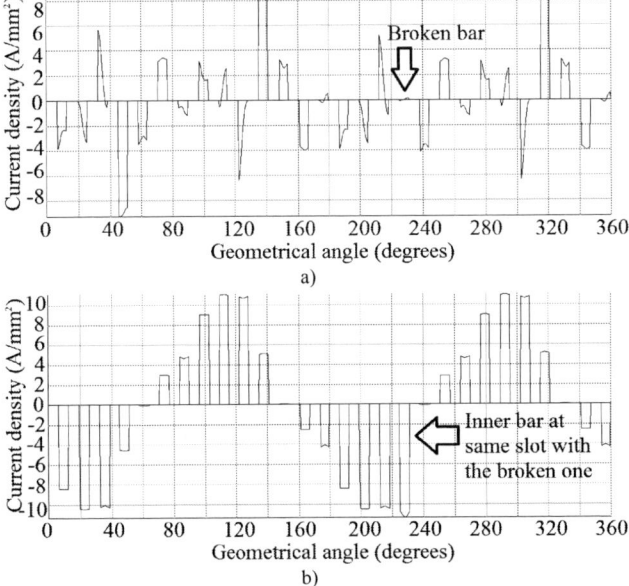

Fig. 3. The current density distribution versus the geometrical angle in the: a) upper bars and b) inner bars for the motor with the broken bar.

B. Simulation Results for Speed 1460rpm

The current, torque, apparent power and zero-sequence current waveforms have been extracted for healthy and faulty cases and the FFT is applied to each one of them. The signals' spectrums are presented in Fig. 4. For each case, the spectrums are normalized to the basic harmonic of the healthy signal. Blue color stands for the healthy motor whereas red for the motor with the broken bar. The frequency components, which are fault related, are marked for all cases with arrows.

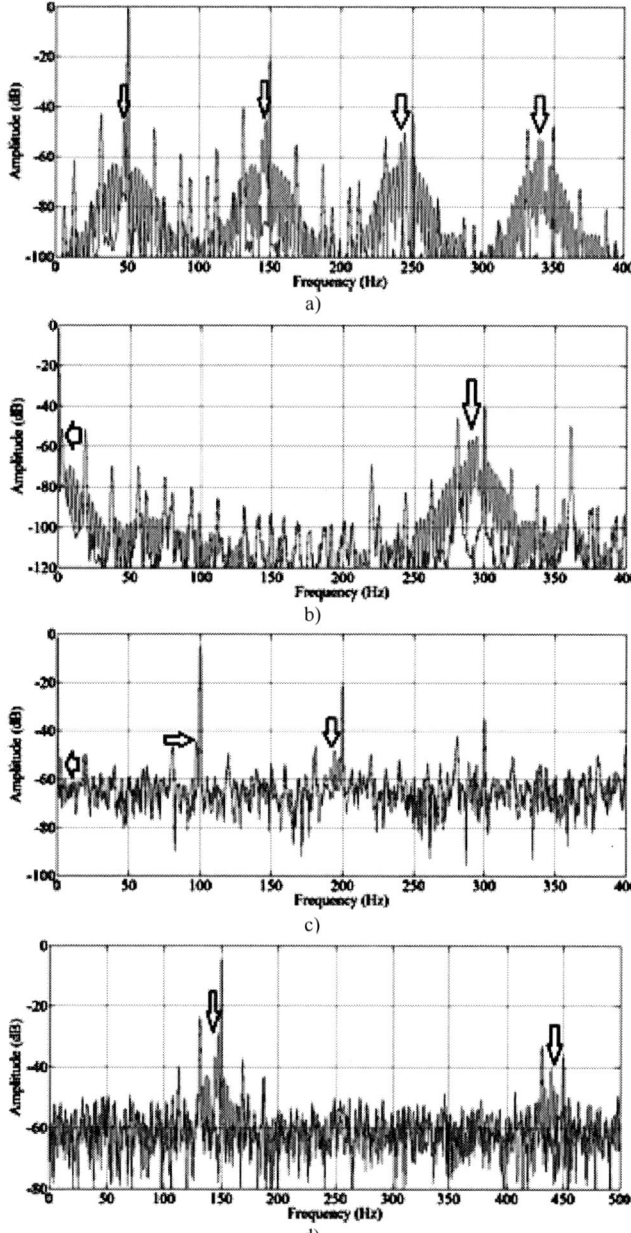

Fig. 4. Frequency spectrum of: a) current, b) torque, c) apparent power and d) zero-sequence current for motor operation at 1460rpm. Blue is for the healthy and red for the faulty motor.

C. Observations and Remarks on the Simulation Results for Speed 1460rpm

In this section, the results of Fig. 4 will be thoroughly discussed.

Firstly, in Fig. 4-a, it can be observed that, the current spectrums of the healthy and faulty motors. Next to the basic 50Hz harmonic, one can see the traditional broken bar fault signature at 47.5Hz. The amplitude of this harmonic is -45.6dB, and has increased about 20dB compared to the healthy motor. Moreover, two fault signatures exist close to

the saturation related 150Hz harmonic. More specifically, the signatures at 147.5Hz and 144.4Hz have amplitudes -45.4dB and 52.8Hz respectively, having both increased about 40dB compared to the healthy motor. Also, as it can be seen, close to 250Hz there exist significant fault related harmonics, complying with previous knowledge [8],[9]. The 244.4Hz and 241.9Hz harmonics present -50.2dB and -53.7dB amplitudes respectively, in the faulty motor. This means an important amplitude increase of more than 50dB compared to the healthy motor. Finally, close to the 7th harmonic at 350Hz there exist important fault related signatures. More specifically, both harmonics at 341.9Hz and 339.4Hz present amplitudes about -52dB in the faulty motor, having increased by 59dB and 57dB respectively compared to the healthy motor.

In Fig. 4-b, the torque spectrums of the healthy and faulty motors are presented. The traditional low frequency fault-related component at 2.5Hz has increased by 29dB, having an amplitude -50.7dB in the faulty motor. Moreover, in the frequency area close to 300Hz, one can detect three fault signatures with important amplitudes. The component at 295Hz, 291.9Hz and 289.4Hz are characterized by -55.2dB, -56.3dB and -56.3dB amplitudes in the case of the broken bar fault. This means an increase of more than 50dB for all signatures, compared to the healthy motor. The existence of those harmonics close to the 300Hz component in the torque spectrum complies with previous knowledge [17]. The difference pointed in the double-cage induction motor, has to do with the much greater amplitude increase in the three components close to 300Hz, compared to conventional NEMA's class A induction motors.

Furthermore, in Fig. 4-c the apparent power spectrums for healthy and faulty motors are presented. The traditional low frequency fault signature at 2.5Hz is characterized by an amplitude -50.9dB in the motor with the broken bar, having increased by 11.5dB compared to the healthy one. By examining the component close to 100Hz, at 97.5Hz, it is shown that, it presents greater amplitude than the traditional one since it is -44.2dB in the case of the fault. This implies a difference amplitude of 22dB compared to the healthy motor. Finally, two more fault related signatures can be observed close to the 200Hz harmonic. The 197.5Hz and 194.4Hz harmonics have amplitudes -51.33dB and -48dB respectively, having increased 13.3dB and 11dB respectively, compared to the healthy motor.

In Fig. 4-d, the zero-sequence current spectrums of the healthy and faulty motors are presented. More specifically, close to the 150Hz harmonic there exist two broken bar fault signatures at 147.5Hz and 144.4Hz with amplitudes -28.3dB and -36.5dB respectively. Those harmonics have increased by 50dB and 25dB respectively, compared to the healthy motor. Furthermore, a wide frequency area exists close to the 9th saturation related harmonic at 450Hz, in which many fault signatures appear. To be specific, the harmonics at 444.4Hz, 441.9Hz, 439.4Hz and 433.8Hz present important amplitudes -48.5dB, -45.9dB, -40dB and -46.5dB respectively in the motor with the broken bar fault. Those harmonics' amplitude increase is 7-27dB compared to the healthy motor.

The previously described results are shown in Table I-IV, for a better overview.

TABLE I
FAULT RELATED HARMONICS AMPLITUDES FOR THE HEALTHY AND FAULTY MOTORS IN THE CURRENT SPECTRUM AT 1460RPM

Signatures	healthy (dB)	faulty (dB)
47.5Hz	-74.89	-45.6
147.5Hz	-85.65	-45.36
241.9Hz	-103.7	-53.74
244.4Hz	-107.6	-50.16
339.4Hz	-109.3	-52.03
341.9Hz	-116	-52.91

TABLE II
FAULT RELATED HARMONICS AMPLITUDES FOR THE HEALTHY AND FAULTY MOTORS IN THE TORQUE SPECTRUM AT 1460RPM

Signatures	healthy (dB)	faulty (dB)
2.5Hz	-79.51	-50.7
289.4Hz	-113.6	-56.29
291.9Hz	-108.4	-56.3
294.4Hz	-103.6	-54.79

TABLE III
FAULT RELATED HARMONICS AMPLITUDES FOR THE HEALTHY AND FAULTY MOTORS IN THE APPARENT POWER SPECTRUM AT 1460RPM

Signatures	healthy (dB)	faulty (dB)
2.5Hz	-62.68	-50.9
97.5Hz	-67.59	-44.25
194.4Hz	-59.04	-48

TABLE IV
FAULT RELATED HARMONICS AMPLITUDES FOR THE HEALTHY AND FAULTY MOTORS IN THE ZERO-SEQUENCE CURRENT SPECTRUM AT 1460RPM

Signatures	healthy (dB)	faulty (dB)
144.4Hz	-61.3	-36.49
147.5Hz	-78.11	-28.28
439.4Hz	-63.94	-39.97

D. Discussion

The results which occurred for motor operation under constant speed 1460rpm lead to several important derivations. First of all, the broken bar fault signatures, discussed in prior works, exist in all diagnostic means. Moreover, the zero-sequence current proves to be diagnostically a significant mean, if one examines the frequency areas close to the 3rd and 9th harmonics. It is also shown that, the strongest diagnostic signatures are offered in the zero-sequence current spectrum compared to all diagnostic means studied in this work. Finally, it is shown that in the apparent power spectrum the fault can also be identified at signatures close to 200Hz. The diagnostic signatures close to 100Hz and 200Hz present greater amplitudes than the traditional low frequency signature close to the DC component in the apparent power spectrum.

III. SIMULATIONS AT 1490RPM

In this paragraph, two cases are studied. The motor operates under healthy condition and with one broken upper bar. The motor operates under constant speed 1490rpm. It is important to evaluate the different diagnostic means for low load operation.

A. Simulation Results for Speed 1490rpm

The current, torque, apparent power and zero-sequence current waveforms have been extracted for healthy and faulty cases and the FFT is applied to each one of them. The signals' spectrums are presented in Fig. 5. For each case, the spectrums are normalized to the basic harmonic of the healthy signal. Blue color is for the healthy motor whereas red is for the motor with the broken bar.

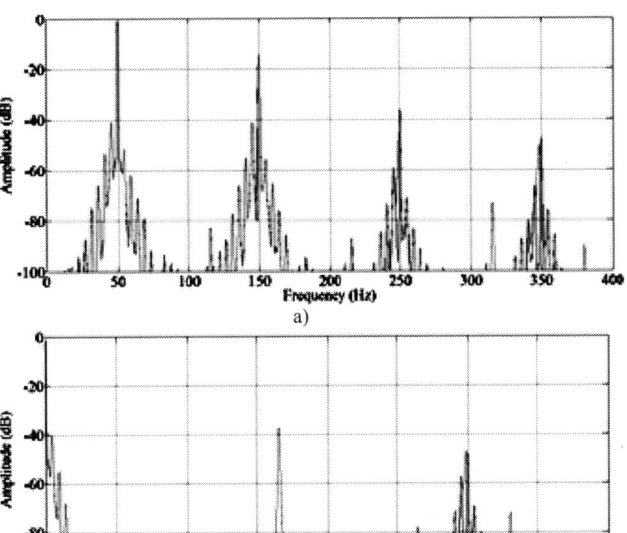

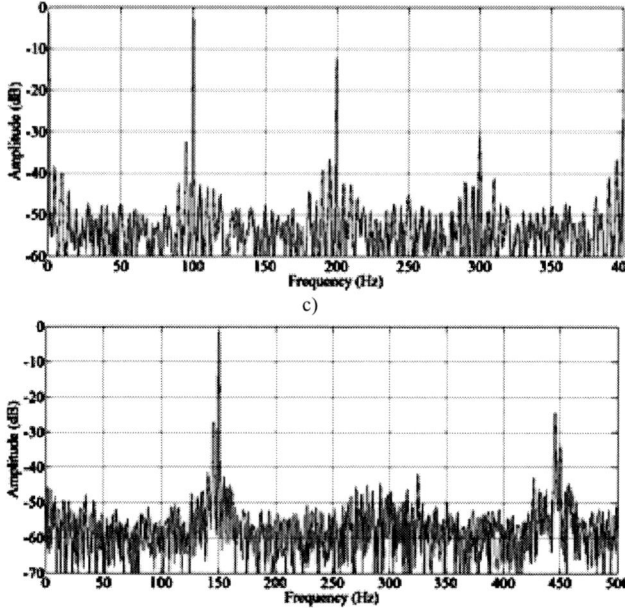

c)

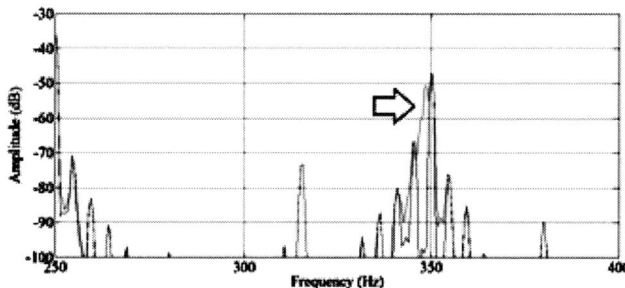

d)

Fig. 5. Frequency spectrum of: a) current, b) torque, c) apparent power and d) zero-sequence current for motor operation at 1490rpm. Blue is for the healthy and red for the faulty motor.

A first look in Fig. 5 reveals that the traditional diagnostic means and methods are incapable of detecting the broken bar fault in double cage induction motors at low load operation. Instead, a closer look in higher frequencies spectrum reveals accurate diagnostic signatures, which are presented in Fig. 6-Fig. 8.

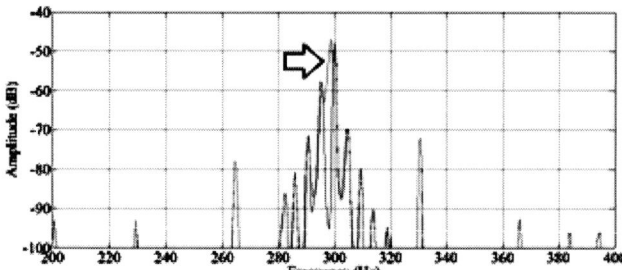

Fig. 6. The current spectrum close to 350Hz for motor operation under speed 1490rpm. Blue is for the healthy and red for the faulty motor.

Fig.7. The torque spectrum close to 300Hz for motor operation under speed 1490rpm. Blue is for the healthy and red for the faulty motor.

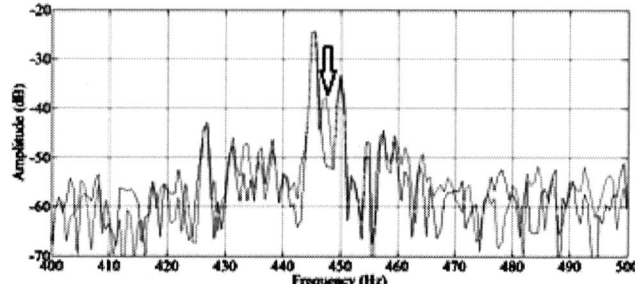

Fig. 8. The zero-sequence current spectrum close to 450Hz for motor operation under speed 1490rpm. Blue is for the healthy and red for the faulty motor.

B. Observations and Remarks on the Simulation Results for Speed 1490rpm

In this section, the simulation results presented in the previous section A of Paragraph III will be thoroughly discussed.

Firstly, it is clear from Fig. 5-c that the apparent power spectrum is completely insufficient and unreliable towards the broken bar fault detection at low load operation of the double cage induction motor, since it does not contain any fault related signatures.

Secondly, it is clear that the traditional signature close to 50Hz, as well as the signatures close to 150Hz and 250Hz do not exist in the current spectrum, as one can see from Fig. 5-a. On the other hand, a fault signature appears close to the 7th current harmonic at 350Hz (Fig. 6). This component exists at 348.1Hz and has an amplitude -50.7dB in the faulty motor, which is 48dB greater than in the healthy motor.

Furthermore, the traditional low frequency signature close to the DC component of the torque spectrum does not exist for low load operation. Instead, in Fig. 7 it can be observed that, there exists a broken bar fault signature close to the 300Hz harmonic. This component can be found at 298.8Hz and is characterized by an amplitude -46.8dB, having increased about 42dB, compared to the healthy motor.

Finally, while observing the zero-sequence current spectrum of Fig. 5-d, it seems that the fault cannot be detected close to the 3rd saturation related 150Hz component. On the other hand, Fig. 8 shows that close to the 9th saturation related 450Hz harmonic, there exists a fault signature at 447.5Hz. Its amplitude is -37.8dB and it greater than the healthy motor by 14dB.

C. Discussion

The results obtained for low load operation of the double cage induction motor reveal significant derivations. Traditional broken bar fault detection methods are completely unreliable to detect the fault in the case of double cage induction motors.

The fault can be identified reliably with the use of the fault signature close to the 7th current harmonic.

Moreover, the fault signature in the frequency area close to the 300Hz in the torque spectrum reveals the fault with reliability. Notably, this indication has already been commented in [17], in which the authors suggested that the monitoring of the torque spectrum at the frequency area close to 300Hz could be applied for the broken bar fault identification in double cage induction motors.

Furthermore, the broken bar fault signature close to the 9th saturation related 450Hz harmonic in the zero-sequence current spectrum, clearly reveals the fault. It is worthy to note that, this specific signature's amplitude is the greatest compared to the other above two. Undoubtedly, it has to be mentioned that, the monitoring of the zero-sequence current spectrum demands for three current sensors, instead of one, needed in the MCSA method.

IV. CONCLUSION

This paper studies and evaluates four diagnostic means for the broken bar fault detection in Δ-connected double-cage induction motors with FEM. While the motor operates at 1460rpm, the current, torque and apparent power spectrums' analysis agrees with previous work, concerning conventional induction motors.

Moreover, double-cage induction motor operation under low load at 1490rpm reveals different approaches concerning the broken bar fault detection. Firstly, it is shown in the paper that the apparent power waveform's spectrum is completely unreliable to detect the broken bar fault. On the other hand, the fault can be detected while observing the frequency area of the 7th current harmonic, while it remains completely undetectable at lower frequencies. Furthermore, the broken bar fault can be reliably detected close to the 300Hz torque harmonic, while it is unnoticeable at the area close to the traditional DC torque component.

Finally, the analysis results reveal that, the strongest fault signatures can be found in the zero-sequence current spectrum close to the 150Hz and 450Hz saturation related harmonics, for nominal and low load operation respectively.

REFERENCES

[1] T. Gönen, "Electrical Machines with Matlab," 2nd Ed., CRC Press, pp. 251, 2012.
[2] Jongbin Park, Byunghwan Kim, Jinkyu Yang, Sang Bin Lee, Ernesto J. Wiedenbrug, Mike Teska, Seungoh Han, "Evaluation of the Detectability of Broken Rotor Bars for Double Squirrel Cage Rotor Induction Motors", Energy Conversion Congress and Exposition(ECCE), pp 2493 – 2500, IEEE 2010.
[3] Motors and generators, NEMA standards pub. MG 1-2006, 2006.
[4] Antonino-Daviu, J.; Riera-Guasp, M.; Pons-Llinares, J.; Jongbin Park; Sang Bin Lee; Jiyoon Yoo; Kral, C., "Detection of Broken Outer-Cage Bars for Double Cage Induction Motors Under the Startup Transient", IEEE Trans. on Industry Applications, Vol. 48, No. 5, pp 1539-1548, September/October 2012.
[5] S. Nandi, H. Toliyat and X. Li, "Condition Monitoring and Fault Diagnosis of Electrical Motors-A Review," IEEE Trans.Ener.Conv., Vol. 20, No. 4, pp. 719-729, Dec 2005.
[6] J. Milimonfared, H. M. Kelk, S. Nandi, A. D. Minassians and H. A. Toliyat, "A Novel Approach for Broken-Rotor-Bar Detection in Cage Induction Motors," IEEE Trans.Ind.Applicat., Vol. 35, pp. 1000-1006, Sep/Oct 1999.
[7] Pinjia Zhang, ,YiDu , Thomas G. Habetler, Bin Lu, "A Survey of Condition Monitoring and Protection Methods for Medium-Voltage Induction Motors", IEEE Trans. Ind. Applicat. Vol.47, No.1, pp. 34-46, Jan/Feb. 2011.

[8] Humberto Henao, Hubert Razik, Gerard-Andre Capolino, "Analytical Approach of the Stator Current Frequency Harmonics Computation for Detection of Induction Machine Rotor Faults", IEEE Trans. Ind. Applicat. Vol.41,No 3, pp. 801-807, May/June 2005.
[9] J. Cusido, J. Rosero and E. Aldabas, "New Fault Detection Techniques For Induction Motors," Electrical Power Quality and Utilisation, Magazine Vol. II, No. 1, 2006.
[10] Andrzej M. Trzynadlowski, Ewen Ritchie," Comparative Investigation of Diagnostic Media for Induction Motors: A Case of Rotor Cage Faults" IEEE Trans. Indust. Electronics, Vol.47, No. 5, pp. 1092-1099, Oct. 2000.
[11] Zhenxing Liu, Xianggen Yin, Zhe Zhang, Deshu Chen, Wei Chen, "Online Rotor Mixed Fault Diagnosis Way Based on Spectrum Analysis of Instantaneous Power in Squirrel Cage Induction Motors", IEEE Trans. Energy Convers., Vol.19, No. 3, pp. 485-490, Sep.2004.
[12] Mario Eltabach, Ali Charara,, Ismail Zein, "A Comparison of External and Internal Methods of Signal Spectral Analysis for Broken Rotor Bars Detection in Induction Motors", IEEE Trans. Indust. Electronics, vol. 51, No. 1, pp. 107-121, Feb.2004.
[13] J. S. Hsu, "Monitoring of defects in induction motors through air-gap torque observation," IEEE Trans. Ind. Appl., Vol. 31, No. 5, pp. 1016–1021, Sep./Oct. 1995.
[14] V. V. Thomas, K. Vasudevan, and V. J. Kumar, "Online cage rotor fault detection using air-gap torque spectra," IEEE Trans. Energy Convers., vol. 18, no. 2, pp. 265–270, Jun. 2003.
[15] ST. J. Manolas and J. A. Tegopoulos, "Analysis of Squirrel CageInduction Motors with Broken Bars and Rings", IEEE Trans. Energy Conv., vol. 14, no. 4, pp. 1300–1305, Dec. 1999.
[16] M. Eltabach, A. Charara, I. Zein and M. Sidahmed, "Detection of broken rotor bar of induction motors by spectral analysis of the electromagnetic torque using Luenberger observer", IECON'01, 27th Annual Conference of the IEEE IES, Denver, CO, pp.658-663, 29 Nov.-02 Dec. 2001.
[17] K. N. Gyftakis, D. V. Spyropoulos, J. Kappatou and E. D. Mitronikas, " A Novel Approach for Broken Bar Fault Diagnosis in Induction Motors through Torque Monitoring," IEEE Trans. Ener. Conv., Vol. 28, No. 2, pp. 267-277, Jun. 2013.
[18] F. Briz, M. W. Degner, P. Garcia and A. B. Diez, "Induction machine diagnostics using zero sequence components," Conference Record of the 2005 Industry Applications Conference, 2005. Fourtieth IAS Annual Meeting, Vol. 1, pp.34-41, Oct. 2005.
[19] J. Pons-Linares, J. A. Antonino-Daviu, M. Riera-Guasp, M. Pineda-Sanchez and V. Climente-Alarcon, "Induction Motor Diagnosis Based on a Transient Current Analytic Wavelet Transform vio Frequency B-Splines," IEEE Trans. Ind. Electr., Vol. 58, No. 5, pp. 1530-1544, May 2011.
[20] A. Bouzida, O. Touhami, R. Ibtiouen, A. Belouchrani, M. Fadel and A. Rezzoug, "Fault Diagnosis in Industrial Induction Machines Through Discrete Wavelet Transform," IEEE Trans. Ind. Electr., Vol. 58, No. 9, pp. 4385-4395, Sep. 2011.
[21] F. Cupertino, E. de Vanna, L. Salvatore and S. Stasi, "Analysis techniques for detection of IM broken rotor bars after supply disconnection," IEEE Trans. Ind. Appl., Vol. 40, No. 2, pp. 526-533, Mar.-Apr. 2004.
[22] Y.-H. Kim, Y.-W. Youn, D.-H. Hwang and J.-H. Sun, "High-Resolution Parameter Estimation Method to Identify Broken Rotor Bar Faults in Induction Motors," IEEE Trans. Ind. Electr., Vol. 60, No. 9, pp. 4103-4117, Sep. 2013.
[23] Lorenzani, E.; Salati, A.; Bianchini, C.; Immovilli, F.; Bellini, A.; Lee, S.B.; Yoo, J.; Kwon, C., "Dynamic Modeling of Double Cage Induction Machines for Diagnosis of Rotor Faults", Energy Conversion Congress and Exposition (ECCE), pp 1299 – 1305, IEEE 2012.
[24] Gritli, Y.; Di Tommaso, A.O.; Filippetti, F.; Miceli, R.; Rossi, C.; Chatti, A., "Investigation of Motor Current Signature and Vibration Analysis for Diagnosing Rotor Broken Bars in Double cage Induction Motors", International Symposium on Power Electronics, Electrical Drives, Automation and Motion (SPEEDAM), pp 1360-1365, IEEE 2012.
[25] Gritli, Y.; Sang Bin Lee; Filippetti, F.; Zarri, L., "Advanced Diagnosis of Outer Cage Damage in Double Squirrel Cage Induction Motors Under Time-Varying Condition Based on Wavelet Analysis", Energy Conversion Congress and Exposition (ECCE), pp 1284 – 1290, IEEE 2012.
[26] K. N. Gyftakis, D. Athanasopoulos and J. Kappatou, "Study of Double Cage Induction Motors with Different Rotor Bar Materials", IEEE

International Conference on Electrical Machines XXth ICEM 2012, Marseille, France, 2-5 Sep. 2012.

Konstantinos N. Gyftakis was born in Patras, Greece, in May 1984. He received the diploma in Electrical and Computer Engineering from the University of Patras, Patras, Greece in 2010. He is a PhD Candidate in the Department of Electrical and Computer Engineering, University of Patras. His research activities are in FEM design, fault diagnosis and optimization of electrical machines. He is an IEEE member, member of IEEE PES, IAS, IES and Magnetics Society, member of the HELIEV (Hellenic Institute of Electric Vehicles) and finally member of the Technical Chamber of Greece. (E-mail: **kosgyftak@upatras.gr**)

Dimitrios K. Athanasopoulos was born in Patras, Greece, in November 1989. He received the diploma in Electrical and Computer Engineering from the University of Patras, Patras, Greece in 2012. He is a PhD Candidate in the Department of Electrical and Computer Engineering, University of Patras. His research activities are in optimization of electrical machines' design with FEM, power electronics and fault diagnosis. (E-mail: **athanasd@upatras.gr**)

Joya C. Kappatou was born in Argostoli, Greece. She received the diploma in Electrical Engineering from the University of Patras, Patras, Greece and the PhD from the same University in 1991 in the field of Electrical machines and Power Electronics. She is Assistant Professor in the Electrical and Computer Engineering Department of the University of Patras. Her teaching and research activities are in electrical machines, power electronics, modeling and design using FEM, faults diagnosis in electrical machines. Dr. Kappatou is a member of IEEE and the Technical Chamber of Greece. (University of Patras, Electrical and Computer Engineering Department, 26500 Rion-Patras, Greece, Tel: +30 2610/996413, Fax: +30 2610/997362, E-mail: **joya@ece.upatras.gr**)

A Hybrid Kangaroo Algorithm to Assess the State Of Health of Electric Motors

H. Razik, *Senior Member, IEEE*, M.El.K. Oumaamar, *Member, IEEE*, G. Clerc, *Senior Member, IEEE*

Abstract—Even if the asynchronous machine is widely used in industrial process under speed variable, we have to monitor the process which is composed of an inverter and a motor in an enclosed space. Thus, the stator current is not accessible and the AC supply line current is only available to monitor the system in comparison to a motor connected directly to the power supply. In this paper, we focus our attention on the tracking of faulty lines rising up in the supply line current due to a rotor bar defect. We suggest to use a meta-heuristic method to tackle this problem. In this paper a hybrid kangaroo and a non-linear great deluge are taken into consideration to track these defective lines thanks to the analysis of the supply line. Experiments results show the effectiveness of this approach.

Index Terms—Diagnostic, induction motor, great deluge, kangaroo.

NOMENCLATURE

f_b	the fault frequencies
f_s	the electrical supply frequency
s	the slip in per unit
f_c	the fundamental frequency of the converter
s_o	the current solution
s_o^*	the new solution
$f(s_o)$	the cost function
$N(s_o)$	the neighborhood solution
$\triangle L$	the decay rate ($\triangle L > 0$)
L	tevel limit (water-level)
$\psi()$	a chaotic function

I. INTRODUCTION

Due to safety and economical reasons, monitoring and diagnosis are of growing interest in complex industrial processes. The reputation of roughness and robustness of the induction motor are the life-motive and conduct to its widely use in electromechanical processes. In sensitive industrial processes, knowing that failures sometimes appear, early fault detection is necessary to avoid or to limit the cost as well as the dangerousness of the process in the whole. Consequently, diagnostic and monitoring systems are naturally of growing interest in sensitive industrial production lines, spreading all over the world.

H. Razik and G. Clerc are with the Université Lyon 1; Laboratoire Ampère, Ampère UMR CNRS 5005, Bât Omega, 43 Bd du 11 Novembre 1918, F-69622 Villeurbanne, Cedex, France, France (e-mail: [hubert.razik; guy.clerc]@univ-lyon1.fr)

M.El.K. Oumaamar is with the Université Mentouri, Laboratoire d'électrotechnique de Constantine (LEC), 25000-Constantine, Algeria (e-mail: oumaamarkamel@yahoo.fr)

Since 82, ever-fast researches on the study of defects on induction motor have brought significant impacts on the scientific community. Deleroi [1] and Kliman [2]–[4] have studied physical phenomena and have suggested to analyze the spectra component of the stator current which involves the research of specific lines around the supply frequency. Usually the stator current measure provides with suitable information about the state of health of the asynchronous motor. That is why a huge part of research have been made with diagnosis technique based on the motor current signature analysis (MCSA) [5], [6]. To be strictly accurate, the success of this approach depends both on the simplicity of the measure and the ability to discriminate various faults between themselves, in a large range of frequencies. Moreover the stator current is an accessible low-cost measure source.

Among all approaches described in the literature, one can find at least two types of methods: the off-line process and the on-line process. Although basic monitoring and analysis are done thanks to the Fast Fourier transform (FFT) of the stator current. A huge amount of information could be obtained through it [4]. Nonetheless, most of the methods are based on steady state operating condition. Moreover, the induction motor can be fed either by the electrical supply network, or by a voltage inverter. Anyway, electrical drives, operating under a field oriented control or scalar vector control, are now largely present in industrial plants. In case of incipient failure, the sliding discrete Fourier transform technique is wonderfully adapted to the on-line analysis. An upgraded spectrum is given at each sample period and this is advocated for the monitoring of electrical machines at distance [7], [8]. So, a general purpose software for the monitoring of electrical machines at distance was suggested. This approach consisted in a software allowing the acquisition of data far from the process and the transfer for analysis purpose via internet. Consequently, new horizon are coming up for the monitoring of process.

As the induction motor is more and more fed by an inverter due to the necessity to has speed variable systems, this induction motor have to support voltage having great harmonics. Thus, even if this kind of motor is robust, faults can occur due to rotor defects coming from vibrations, torque ripple and so on, which accelerate the ageing of the process. The goal of this paper is to investigate a non intrusive method consisting in monitoring the AC supply and line current of the induction motor fed by an inverter. It puts in front a novel standalone approach to track the faulty lines in a process [9]. Among several optimization approaches,

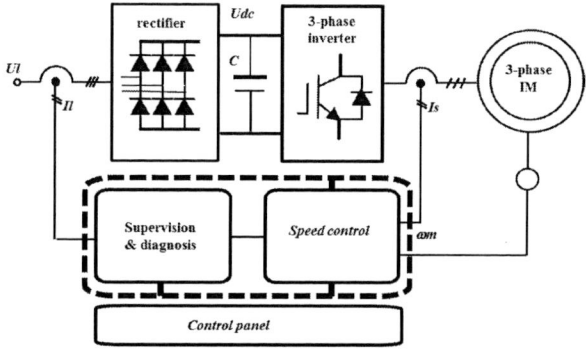

Fig. 1. Inverter drive system

we focus our attention on a stochastic method based on the kangaroo technique. As all techniques, it starts with an initial solution which is improved by a succession of moves in the neighborhood of the current solution in order to optimize the solution of the problem. In order to prove the efficiency of this approach, the well known reference dealing with the rotor bar defect of the induction motor is made. Oumaamar [10] has suggested a generalized formula for covering signatures for this type of application using an inverter and experimental results confirm the effectiveness of this study.

II. FAULTS SIGNATURE DUE TO A DEFECT

The motor drive system studied is composed by a 3-phase induction machine fed by a 3-phase inverter. Fig. 1 shows the system in the whole. As we can see, the induction motor operated under field oriented control or scalar control. The access of currents are possible thanks to to the power line supply in order to monitor the process.

A. The diagnosis principle

Faults can occur in this type of process. The former is due to the induction machine and the latter due to the power electronic devices.

1) Repartition of faults in induction motor: Induction motors have at least 4 causes of faults. The rotor faults and stator faults are due to a stress or to a combination of stress [11]. For the stator like the rotor, the effects emerge from a problem which can be either/or: thermal (overload, ...); electrical (insulation, ...); mechanical (winding, ...); environmental (aggressive, ...). Bearing faults account for almost 51%, stator winding faults account for almost 16%, environmental faults account for almost 16 %, rotor bar faults for almost 5%, shaft faults for almost 2% and the others for about 10% [12]. Consequently, we propose to put forward the main faults of induction motors which create lines in the stator line current spectrum which components are explained by some mathematical expressions.

2) Repartition of faults in power device: Faults can occur at large power systems and can be dangerous, causing a breakdown of a process. Fuchs claims that 38% of faults are due to power devices, 53% due to the control circuits and the rest due to external auxiliaries circuits [13]. These faults can originate from the rectifier, the inverter or the gate interface for the power devices. Nevertheless, studies are made on fault tolerant control strategies in order to improve the reliability. Many researches have been conducted to analyze the component failure impact in a three-phase voltage-source inverter but they mainly consider the case of an open circuit but in order to get rid of a general and catastrophic failure, the process has to be monitored. Different types of faults can occur in a converter and can be classified as:

- a problem of short-circuit from the dc link to the ground;
- a problem of short-circuit in the dc link capacitor;
- a problem of damage (open or short circuit) of switch;
- a problem connection of line of the induction motor to the ground.

So, protections are taken into consideration in order to prevent the damage of the inverter. Naturally, the consequence of a fault is the possibility of a total breakdown of the process as an emergency intervention with the altered process.

3) Spectral components due to Broken Bars: Under abnormal conditions, it is well known that the rotor current produces forward and backward rotating fields. This non-zero backward rotating field induces a first harmonic in the stator at the frequency $(1 - 2s)f_s$ [2]. This is the widely starting point of the motor fault theory. Sidebands due to the broken bars are given by the relation:

$$f_b = (1 \pm 2ks)f_s \qquad (1)$$

where: $k =$1, 2, 3 ..., knowing that k is an integer. Lines induced by this type of defect are close to the supply frequency. The stator current presented in Fig. 2 is made with an induction motor fed by an inverter at $f_c = 15Hz$. This current is a little bit different from an induction motor directly connected to the supply lines. The switching frequency at $10kHz$ is manifest.

4) Spectral components in the lines: For the diagnosis of induction motor fed by an inverter converter as depicted in Fig. 1, the line current is represented in Fig. 3. The line current frequency is modulated by the twice of the product of slip by the converter frequency. Thus, in case of the appearance of a broken bar, faulty lines rise up with the expression given as follows [10]:

$$f_b = (h.f_s \pm 2ks.f_c) \qquad (2)$$

where h an odd integer (1, 5, 7, 11, 13 ...), k any integer.

The diagnosis of an induction is based on the evolution of sides band. In order to quantify the fault severity, we have to monitor the amplitude of lines close to the supply frequency ($k = 1$). Lines described by (2) with: $h =$1, 5, 7 ... and $k = 1$ will be monitored. Thus, the standalone diagnosis consists in

978-1-4799-0024-4/13 $31.00 © 2013 IEEE

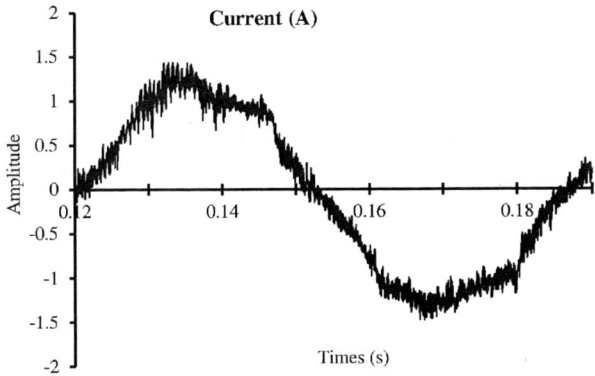

Fig. 2. Stator current of the induction machine fed by an inverter

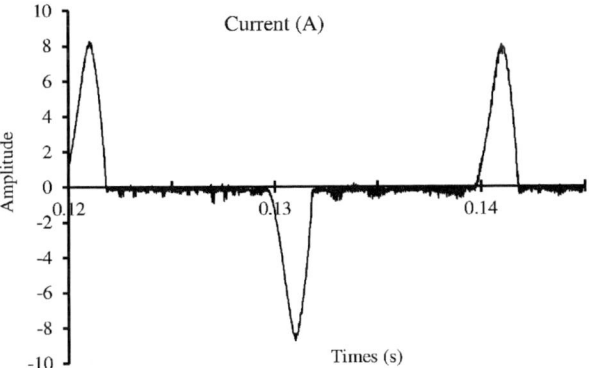

Fig. 3. Line current of the induction machine fed by an inverter

monitoring the amplitude of these faulty lines and we suggest to use the hybrid kangaroo algorithm to track and evaluate the amplitude of these faulty lines and to alert the operator of the fault severity.

III. A HYBRID KANGAROO ALGORITHM

The hybrid kangaroo algorithm is a meta-heuristic method for the optimization process solving large scale problems. At the beginning of the meta-heuristic procedure, the initial solution is set which evolves to an optimal solution. In this paper, a Hybrid Kangaroo Non-Linear Great Deluge Algorithm (also known as Degraded Ceiling) is suggested as an optimization approach to solve complex systems. Duerke in 1993 [14] has introduced and developed the Great Deluge which is a meta-heuristic algorithm. This one accepts worse candidate solutions than the current one. Burke in 2004 [15] has extended the Great Deluge Algorithm by integrating the acceptance of all better moves.

A. The Extended Great Deluge Algorithm

As written by Baykasoglu [16], this approach can be illustrated by analogy like depicted in Fig. 4. A person hopes to find a way up as the water-level (L) rises during a constant rain. Among all solutions, S_a is not an acceptable solution, S_b and S_c are acceptable solutions. As Dueck wrote: "Our

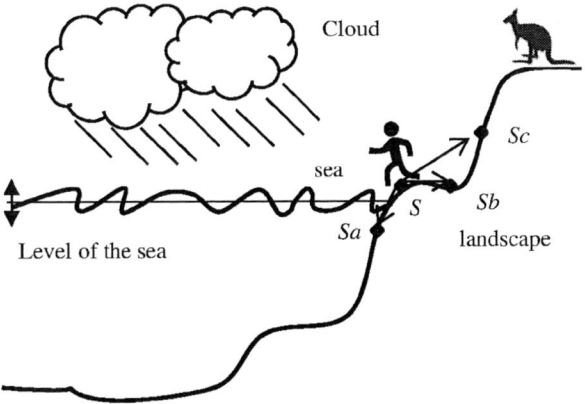

Fig. 4. Representation of Extended Great Deluge

idea is that in the end the GDA (Great Deluge Algorithm) 'gets wet feet' when it has reached one of the very highest points in the country so that is has found a point close to the optimum". As a consequence, knowing that it rains continuously, the algorithm continues its exploration hoping to improve the solution.

The algorithm of the Extended Great Deluge is presented in Alg. 1.

As we can see, at every iterations a new solution s_o^* replaces the last one while a stopping criterion is satisfied. The new solution is taken in the neighborhood $N(s_o)$. The accepted solution could be worse than the other which characterizes this extended great deluge algorithm to others. It explains the cost function could be higher than the previous which is in contrast with the "hill climbing rule".

The initial value of the level limit is a slightly higher value of the initial cost function, the decay rate is unique and monotonically decreases this level limit. This linear function in the conventional Great Deluge Algorithm is given by:

$$L = L - \triangle L. \qquad (3)$$

The quality of the solution depends on the decay rate $\triangle L$ (also called "rain speed") [14], so if this one is large, the algorithm is faster than if the decay rate would be small.

Algorithm 1 The Extended Great Deluge algorithm

Set the initial solution s_o
Calculate initial cost function $f(s_o)$
Initial level $L = f(s_o)$
Set initial decay rate $\triangle L$
While stopping criterion is not satisfied
    Define neighborhood $N(s_o)$
    Randomly select the candidate solution $s_o^* \in N(s_o)$
    If $f(s_o^*) \leq f(s_o)$ or $f(s_o^*) \leq L$ then
        Accept $s_o = s_o^*$
        Decrease L ($L = L - \triangle L$)
Return the best solution s_o^*

B. The neighborhood

As the neighbor generation is not unique, we use a chaotic process as suggested by R. May in 1976 [17] and later for example by B. Liu in 2005 [18] as it is a "well known Logistic map". This approach is frequently used in chaotic search algorithm as for example [19]. The chaotic process enlarges the searching region whatever the solution is local or global.

The chaotic function is based on this equation with the constraint $C_n \in [0 - 1]$:

$$C_n = \mu C_{n-1} \left(1 - C_{n-1}\right). \tag{4}$$

Nevertheless, the dynamic is chaotic only if the following conditions are respected: $\mu = 4$, and $C_{n=0} \neq \{0, 0.2, 0.75, 1\}$. In this particular case, the periodicity p of this chaotic function when the periodicity is huge is given as follows:

$$C_{n=0} \approx \frac{\pi^2}{4p^2}. \tag{5}$$

The neighborhood equation is based on:

$$x_i = x_i + \alpha \left(2\psi\left(\right) - 1\right) \tag{6}$$

with $\psi\left(\right) = C_n$ and α the amplitude of the disturbance.

C. The fitness

The objective function used in order to find the major components in the current spectrum is based on (2). To do that, Gaussian functions are used as a window function.

$$w(f) = \prod_h \prod_k \exp\left(-\frac{1}{2}\frac{(f - [h.f_s \pm 2ks.f_c])^2}{\sigma^2}\right) \tag{7}$$

these window functions depend on the supply frequency f_s and the frequency $2ks.f_c$. Consequently, only two parameters have to be find in the spectrum to analyze: f_s and $2s.f_c$. As the diagnosis is based on the evolution of lines depending of $2ks.f_c$ and thanks to our experience, we limit ourselves to the tracking of lines to $h = 1, 5, 7, 11$ and $k = 0, 1$. Thus, 12 Gaussian functions are used in the fitness function we have to minimize, which are based on the expression (8) of the product of the spectral components to analyze by the spectral window (7).

$$fitness = \sqrt{\frac{1}{N}\sum_{i=0}^{N}[y(i) - y(i).w(i)]^2}. \tag{8}$$

D. The Non-Linear Great Deluge Algorithm

The non-linear great deluge has a decay rate which is not constant. In order to decrease the level of water the expression, we suggest this expression as follows:

$$L = (L - \beta)\exp^{-\delta.iter} + \beta + rand(.). \tag{9}$$

Algorithm 2 The Kangaroo subroutine

If $f(s_o^*) \leq f(s_o)$ or $f(s_o^*) \leq L$ then
    Accept $s_o = s_o^*$
    Decrease L $(L = L - \triangle L)$
else $counter+ = 1$
end
If $counter > Stagnation\,max$ then
    generate new neighbors from the best solution
    $counter = 0$
end

This equation has several parameters in order to modify the slope of the water-level decay and its decay velocity. The value of β is equal to zero if we don't know the final value of the objective function expected. The parameters δ and the random function limited in the range $[min, max]$ modify the decay velocity and the search process as well.

E. The Kangaroo effect

As the fitness evolves with the number of iterations, this one can stagnate and there is no improvement. In order to continue to explore the neighborhoods of the best solution, one suggest to use a neighborhood search method based on the characteristic of the Kangaroo algorithm. So, using a counter of the fitness stagnation, if this one is greater than a predetermined value, we proceed to another neighbor generation. It allows to continue to explore the search space during a lot of iterations so as to avoid to feel trapped in a local optimum. After that we return to the search process.

The effect of the Kangaroo algorithm on the Non-Linear Extended Great Deluge is presented in Alg. 2.

IV. THE EXPERIMENTS

The power plant is composed of a $3kW$ asynchronous motor (Tab. I) fed by a three-phase inverter converter operating at $10kHz$ and coupled with a DC motor as represented in Fig. 5. The system has a data acquisition board and a DSpace DS1104 which generate and control the signal of the system.

The signal to establish a diagnosis was done thanks to a data board acquisition. The current was recorded at the sampling frequency of $20kHz$ during a little bit more than 10s. The data length is 2^{18}, and consequently the resolution of the spectral analysis based on the FFT is better than $0.01Hz$. Tests were made thanks to two types of experiments. In the

TABLE I
NAMEPLATE RATING OF THE INDUCTION MOTOR

Manufacturer	SEW USOCOME
Power	3 kW
Voltage	230/400 V
Current	10.4/5.9 A
Number of Poles	2
Number of rotor slots	28
Number of stator slots	36
Rated speed	2800 rpm

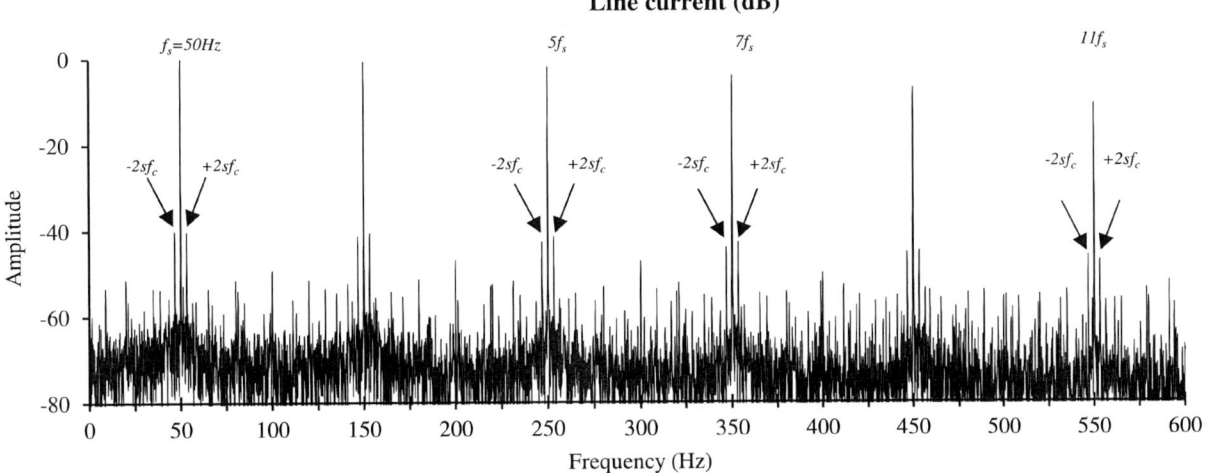

Line current (dB)

Fig. 6. Normalized FFT spectrum of experimental supply line current (before the rectifier) of a induction motor operating under broken bars defect

TABLE II
MAGNITUDE OF THE DEFECTIVE LINES

Formula	f_b (Hz)	Healthy	Defective	$\triangle$ (dB)
$f_s - 2sf_c$	46.74	-48.39 dB	-40.11 dB	8.28
$f_s + 2sf_c$	53.23	-47.50 dB	-40.28 dB	7.22
$5f_s - 2sf_c$	246.68	-50.11 dB	-42.68 dB	7.43
$5f_s + 2sf_c$	253.17	-49.42 dB	-41.44 dB	7.98
$7f_s - 2sf_c$	346.65	-57.39 dB	-44.08 dB	13.31
$7f_s + 2sf_c$	353.14	-50.81 dB	-42.92 dB	7.89
$11f_s - 2sf_c$	546.59	-54.53 dB	-46.03 dB	8.50
$11f_s + 2sf_c$	553.08	-53.17 dB	-47.14 dB	6.03

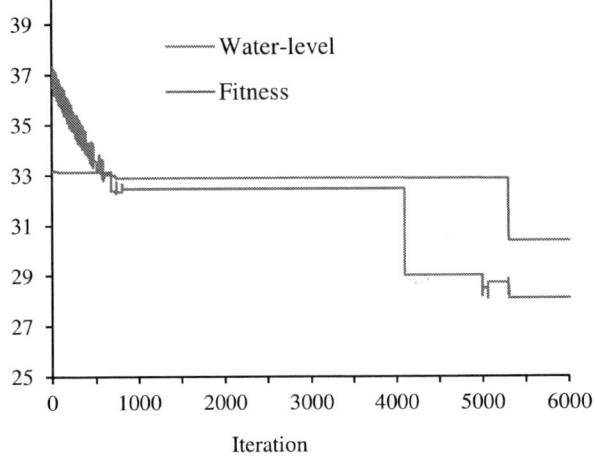

Fig. 7. Best fitness and water-level evolution: Non-Linear decay

first one, the induction motor is healthy and operates at full load. In the second one, the induction motor has a broken bar and has operated strictly in the same condition. The induction motor was fed by an inverter. The supply frequency is $f_s = 50Hz$ for the line and the motor is fed at $f_c = 16Hz$. The analysis is made thanks to the access of the supply line current (before the rectifier).

In order to test this algorithm, we have used the spectrum above presented in Fig. 6 where faulty lines have to be detected. Tab. II summurizes the amplitudes of lines in case or not of a defect. Thus, this first tested algorithm uses the Non-Linear decay and the second algorithm uses the Kangaroo technique.

Fig. 7 shows the evolution both of the best fitness and the water-level. The decay rate of the water-level is Non-Linear.

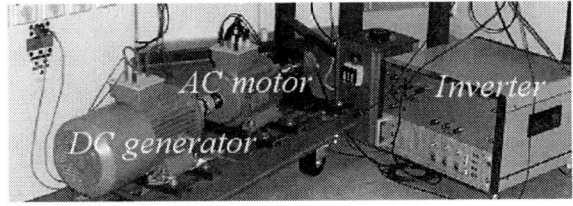

Fig. 5. The test-bed

We have restricted the number of iteration to 6000 because the best solution is found before this number.

As we can see, at the beginning, the fitness is the highest and consequently the water-level also. The evolution of both variables are not linear. This is due to the test in the algorithm "If $f(s_o^*) \leq f(s_o)$ or $f(s_o^*) \leq L$" then accepts or not the solution and decreases or not the water-level. We have to notice that this approach minimizes the fitness.

Fig. 8 shows the evolution once again of both the best fitness and the water-level. The decay rate of the water-level is not Non-Linear once again and is based on the (9). One can see on this figure that the initials conditions are strickly the same than the previous test. The initial decreasing value of the water are close to each other techniques. But sometimes the water-level is disturbed by the amplitude of waves. As one can see near to the iteration 2200 and 2500, the effect of the Kangaroo technique allows to decrease significantly the value of the best fitness. Consequently, such a Kangaroo

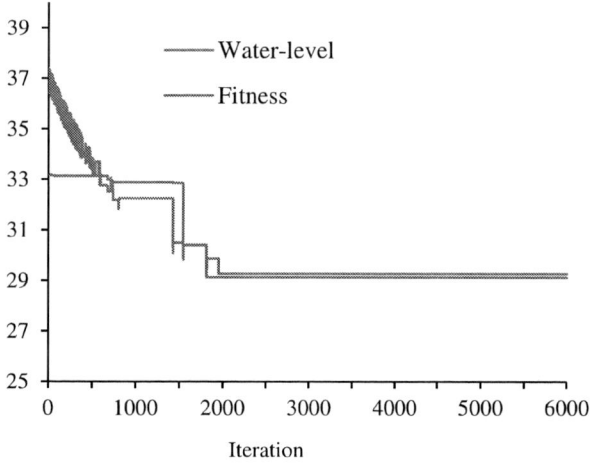

Fig. 8.   Best fitness and water-level evolution: Kangaroo technique

disturbance is welcome to minimise quickly the fitness.

The initial position of the person in the landscape is a little bit higher than the water-level in both examples. Nevertheless, each final value converges to the same place. Both approaches need a high number of iterations. In our own personal computer (Intel I7-2600 CPU 3.40 GHz), this algorithm (written in C language) consumes around 0.35s using the Kangaroo technique and 1.32s without it.

V. Conclusion

This paper has introduced a novel approach for the tracking of faulty lines in induction motor in order to monitor it using the signal from the power line supply. The difficulty is that the monitoring as well as the diagnosis are based on the spectral analysis of the current, knowing that the induction has to operate in steady-state condition during the acquisition time. The other difficulty is the complexity to distinghish faulty lines from each other in high power systems. The effective signature to diagnose a broken defect is based on this expression "$f_b = h.f_s \pm 2ks.f_c$".

The approach illustrated in this paper does not require any special technique. The Non-Linear Great Deluge is simple to implement and does not require a huge computational process to track the faulty lines. The first approach is based on the Non-Linear Great Deluge, the second one is an extention using the Kangaroo technique. In both cases, the solution is found but the Kangaroo technique allows to explore a new solution in order to quickly find the final solution. Nevertheless the number of iterations is determined by suuccessive test.This technique is adequate for off-line diagnostic and the monitoring of faulty lines is made easier.

References

[1] W. Deleroi, "Squirrel-cage motor with broken bar in rotor - Physical phenomena and their experimental assesment," *in Proc. ICEM'82*, pp. 767–770.

[2] G.B. Kliman, R. A. Koegl, J. Stein, R. D. Endicot and M. W. Madden, "Noninvasive detection of broken rotor bars in operating induction motors,"*IEEE Trans. Energy Convers.*, vol. 3, no. 4, pp. 873–879, Dec. 1988.

[3] G.B. Kliman and J. Stein, "Induction motor fault detection via passive current monitoring," *in Proc. ICEM'90*, vol. 1, pp. 13–17.

[4] G.B. Kliman and J. Stein, "Methods of motor current signature analysis," *IEEE Electrical Machines and Power Systems*, vol. 20, no. 5, pp. 1463–474, Sept. 1992.

[5] H.A. Toliyat and S. Nandi, "Condition monitoring and fault diagnosis of electrical machines—A review," *in Proc. 1999 IEEE Industry Applications Conference*, pp. 197–204

[6] M. Benbouzid, "Bibliography on induction motors faults detection and diagnosis," *IEEE Trans. on Energy Conv.*, vol. 14, pp. 1065–1074, Dec. 1999.

[7] M. Artioli, G.A. Capolino, F. Filippetti, A. Yazidi, "A general Purpose Software for distance monitoring and diagnosis of electrical machines," *in Proc. SDEMPED'03*, pp. 272–276.

[8] H. Razik, G. Didier, M.B.R. Correa, E.R.C. da Silva, "The remote surveillance device in monitoring and diagnosis of induction motor by using a PDA.," *in Proc. POWERENG 2007*, pp. 84–89.

[9] H. Razik, M.B.R. Correa and E.R.C. Da Silva, "A Novel Monitoring of Load Level and Broken Bar Fault Severity Applied to Squirrel-Cage Induction Motors Using a Genetic Algorithm," *IEEE Trans. on Industry Electronics*, vol. 56, no. 11, pp. 4615-4626, Nov. 2009.

[10] M.El.K. Oumaamar, H. Razik, A. Rezzoug, H. Chemali and A. Khezzar, "Experimental Investigation of New Indices of Broken Rotor Bar in Induction Motor," *in Proc. ACEMP-Electromotion 2011*, pp. 305–308.

[11] A.H. Bonnet and G.C. Soukup, "Analysis of Rotor Failures in Squirrel-Cage Induction Motors," *IEEE Transactions on Industry Applications*, vol. 24, no. 6, pp. 1124-1130, Nov./Dec. 1988.

[12] A.H. Bonnet and C. Yung, "Increased efficiency versus increased reliability," *Industry Applications Magazine*, vol. 14, no. 1, pp. 29-36, Jan/Feb 2008.

[13] F.W. Fuchs, "Some diagnosis methods for voltage source inverters in variable speed drives with induction machinesa survey," *in Proc. 2003 IEEE-IES Conf.*, pp. 1378–1385.

[14] G. Dueck, "New optimization heuristics. The great deluge algorithm and the record-to-record travel," *Journal of Computational Physics*, vol. 104, no. 1, pp. 86–92, 1993.

[15] E.K. Burke, Y. Bykov, J.-P. Newall and S. Petrovic, "A time predefined local search approach to exam timetabling problems," *Journal of IIE*, vol. 36, no. 6, pp. 509–528, June 2004.

[16] A. Baykasoglu, "Design optimization with chaos embedded great deluge algorithm," *Journal Applied Soft Computing*, vol. 12, issue: 3, pp. 1055–1067, March 2012.

[17] R. May, "Simple mathematical models with very complicated dynamics," *Nature 261*, pp. 459–467, June 1976.

[18] B. Liu, L. Wang, Y-H Jin, F. Tang and D.X. Huang, "Improved particle swarm optimization combined with chaos," *Elsevier Chaos, solitons and Fractals*, vol. 25, Issue 5, pp. 1261-1271, Sept. 2005.

[19] H. Razik, M.B.R. Correa and E.R.C. Da Silva, "The tracking of induction motor's faulty lines through particle swarm optimization using chaos," *in Proc. IEEE International Conference on Industrial Technology (ICIT 2010)*, pp. 1245–1250.

VI. Biographies

Hubert Razik was graduated from the Ecole Normale Supérieure, Cachan, France, in 1987. He received the Ph.D. degree in Electrical Engineering from the Institut Polytechnique de Lorraine, Nancy, France, in 1991. Since 2009, he is professor of electrical engineering at the "Université Claude Bernard Lyon I" and he carried out researches on diagnosis on multiphases induction motors.

M.E.K. Oumaamar received the M. Sc. degree in 2004 from Mentouri University- Constantine, Algeria. Currently, he is with the Electrical Engineering Institute of the University of Constantine. He received the Ph.D. degree in 2012 on the diagnosis of faults of the induction machines.

Guy Clerc was born in Libourne, France, on November 30, 1960. He received the Engineer's degree and the PhD in electrical engineering from the Ecole Centrale de Lyon, France. He is Professor of Universities. He teaches electrical engineering at the "Université Claude Bernard Lyon I" in France. He carried out researches on control and diagnosis of induction machines.

Optimal Wavelets for Broken Rotor Bars Fault Diagnosis

Pu. Shi, Zheng. Chen, Yuriy. Vagapov, Zoubir. Zouaoui

Abstract – **This paper discusses the use of an optimal wavelets approach for broken rotor bars diagnosis. In this paper, discrete wavelet coefficients (DWT) of stator current in a specific frequency band are derived and analyzed. Wavelets db8, db9, db10, sym7 and sym8 are compared to select a most optimal one. This approach facilitates the detection and diagnosis of broken rotor bar occurrence or even number of broken bars under load variation. Simulation and experimental tests on inductions motors with 1, 2, and 3 broken bars are conducted, which demonstrate the effective fault detection of the proposed approach.**

Index Terms-- **Wavelets, Broken Rotor Bars, Induction Machine**

I. INTRODUCTION

NOWADAYS, induction machines are widely used in a variety of diverse industries product line, ranging from mining industry, process manufacturing, automation applications, heating and air conditioning, transportation, aerospace and marine propulsion applications to the health care industry.

Even though, the reliability of induction machines is good. However, internal motor faults (e.g., leads and interturn short circuits, ground faults, bearing and gearbox failures, broken rotor bars and cracked rotor end-rings), as well as external motor faults (e.g., phase failure, asymmetry of main supply and mechanical overload), are expected to happen sooner or later [1]. Therefore, an effective approach for monitoring electrical machines components condition has received considerable attention for many years.

In the past decades, Fast Fourier Transforms (FFT) and Wavelet Transforms (WT) are the two most widely used approaches for fault diagnosis of induction machines [2][3][4]. Although the FFT is effective for steady state analysis of the stator current at rated load and speed, it is not suitable for analyze the transient stator current signal under a varying load condition. Also it is impossible to estimate the time of the fault occurrence using the FFT [5]. For FFT analysis, the elementary functions are complex exponentials producing the same results for a particular waveform being analyzed. However, in wavelet analysis, the fundamental functions could be any permissible mother wavelets and the results produced are unique to the selected wavelet. In Douglas and Pillay, it has been shown that use of high-order wavelets can improve the accuracy of diagnosis of the

P. Shi, Z. Chen, Y. Vagapov, and Z. Zouaoui are with Engineering Department, Glyndwr University, Wrexham, UK (e-mail: s07001957@mail.glyndwr.ac.uk).

broken rotor bars [6]. In Ye, et al., wavelet packet decomposition of the stator current is used for online noninvasive detection of broken rotor bars [7]. In Daviu, et al., the evolution of certain frequency components associated with broken rotor bars, during the start-up transient mode has been extracted [8].

In this paper, the proposed approach is focused on the study of different wavelet coefficients derived from the DWT analysis [9][10]. By comprehensive analysis these coefficients, a best representative wavelet is closed for rotor fault detection. This allows a good interpretation of the phenomenon due to the variation of these signals reflects clearly the evolution of the harmonics associated with broken rotor bars during the transient. Moreover, the use of the wavelet signals (approximation and high-order details) resulting from the DWT constitutes an interesting advantage because these signals act as filters, according to Mallat algorithm, allowing the automatic extraction of the time evolution of the low frequency components that are presented in the signal during the transient [11]. The computation time required for the analysis is usually negligible due to the proposed technique does not use any intricate algorithm for the extraction of the evolution of the signal components.

The approach for the diagnosis of broken rotor bars is described and applied to industrial induction machines. This new approach is also compared with the well-known method, based on the FFT analysis of the stator current in steady state. Several experiments are developed for different fault cases and operating conditions such as one-bar breakage, two-bar breakages and also variation rated load. For the purposes of testing, the bar breakages were forced in the workshop in industrial motors to compare machine behavior in healthy and faulty conditions.

II. EXPERIMENTAL SETUP

In this section, an on-line experimental rig is developed based on the proposed approach in order to test and verify the performance of the diagnosis system. The on-line current monitoring system is shown in Figure 1. The experiment is carried out under the self-designed test rig which is mainly composed a set of three phase induction machines, DC generate, current transducer, A/D converter, and computer. Firstly, transient stator current signals are collected from tested motors and signal preprocessing is conducted which contains smoothing and subtraction. Moreover, Matlab & Wavelet toolbox is used to decompose the acquired time domain signal into time-frequency domain. Then, fault features frequency band is extracted from all specified wavelet transform level. Finally, the individual diagnosis results are used to validate the developed model.

The tested motors are three identical three-phase, 2-pole, 36 stator slots and 28 rotor slots induction machines. The specifications of the proposed induction motors used in our experiment are summarized in Table I. The tests are carried out on a healthy motor and a motor with bored bars. The rotor bar breakages were broken deliberately by drilling holes in workshop.

Stator currents of the motor are sampled by a current Hall Effect sensor which is placed in one of the phase line current cables. The stator current is sampled with a 1.92KHz rate and interfaced to a PC by an ADC-11 data acquisition system. The quantities have been measured for healthy and three broken rotor bars at varied load.

The motor load is controlled through the generator shaft speed. A DC generator is coupled to the motor as the load. The excitation current of the generator has been adjusted in order to regulate the output voltage. A high power resistance box has been connected to the terminals of the generator. The resistance of this box can be selected step by step by a selector on the box. By regulating the output voltage of the generator inserted in the excitation current pass, the load level can be regulated precisely.

TABLE I
INDUCTION MOTOR CHARACTERISTICS USED IN THE EXPERIMENT

Description	Value
Power	5.5KW
Input Voltage	240/380 V
Full load current	20.6/11.9 A
Supply Frequency	50Hz
Number of poles	2
Number of stator slots	36
Number of rotor slots	28
Speed	3000 rpm

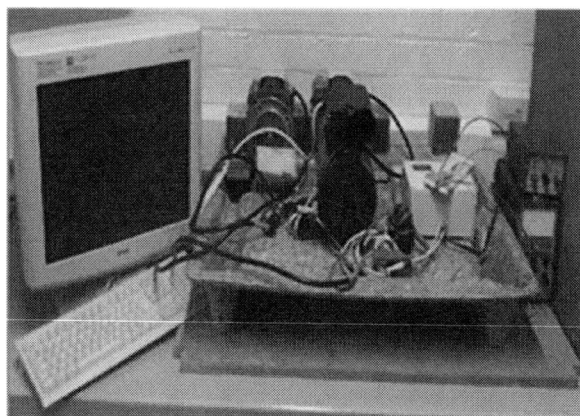

Fig. 1.   View of the experimental setup

TABLE II
FREQUENCY LEVELS OF WAVELET COEFFICIENTS

Wavelet	Frequency(Hz)
A5	0-30
D5	30-60
D4	60-120
D3	120-240
D2	240-480
A1	480-960

III.   DISCRETE WAVELET TRANSFORM

A.   Bases of DWT

The main idea that underlies the application of DWT is the dyadic bandpass filtering process carried out by this transformation. Provided a certain sampled signal S = (i_1, i_2, . . . , i_N), the DWT decomposes it onto several wavelet signals (an approximation signal an and n detail signals d_j) [8][12]. A certain frequency band is associated with each wavelet signal; the wavelet signal reflects the time evolution of the frequency components of the original signal S, which are contained within its associated frequency band as shown in Figure 2 [8].

The decomposition coefficients can be determined through convolution and implemented by using a filter [11]. The filter, LPF, is a low-pass filter and HPF is a high-pass filter. The decomposition process can be iterated, with successive approximations being decomposed in turn, so that one signal is broken down into many lower resolution components.

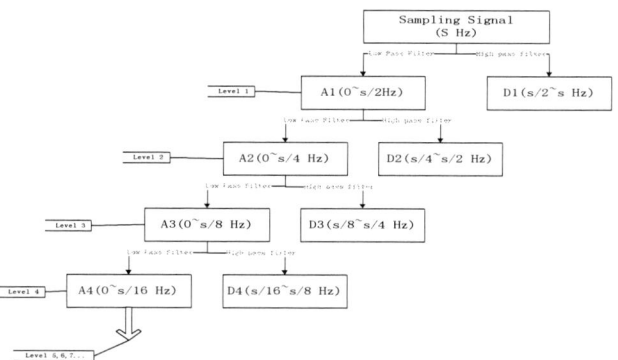

Fig. 2.   Demonstration of DWT

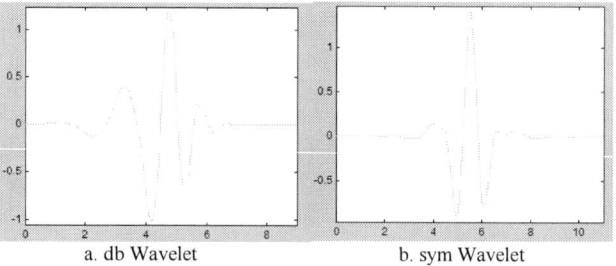

a. db Wavelet   b. sym Wavelet
Fig. 3.   Demonstration of wavelet family

More concretely, if f_s (in samples per second) is the sampling rate used for capturing S, then the detail d_j contains the information concerning the signal components with frequencies included in the interval

$$f\left(d_j\right) \in \left[2^{-(j+1)} \cdot f_s, 2^{-j} \cdot f_s\right] Hz \quad (1)$$

The approximation signal an includes the low-frequency components of the signal belonging to the interval

$$f\left(a_n\right) \in \left[0, 2^{-(n+1)} \cdot f_s\right] Hz \quad (2)$$

Due to the automatic filtering performed by the wavelet transform, this tool provides a very attractive flexibility for the simultaneous analysis of the transient evolution of rather different frequency components present in the same signal. At the same time, in comparison with other tools, the computational requirements are low. In addition, the DWT is available in standard commercial software packages, so no special or complex algorithm is required for its application.

B. Some Considerations about DWT Parameters

In this paper, Matlab Wavelet Toolbox is used, although other software packages could perfectly be suitable for applying the methodology. Prior to the application of the DWT, some considerations have to be done regarding the different parameters of the DWT decomposition, such as the type of mother wavelet, the order of the mother wavelet, or the number of decomposition levels.

1) Selection of Mother Wavelet

With regarding to the DWT, an important step is the selection of the mother wavelet to carry out the analysis. The selected mother wavelet is related to the coefficients of the filters used in the filtering process inherent to the DWT [12]. During these last decades, several wavelet families with rather different mathematical properties have been developed. Infinite-supported wavelets (Gaussian, Mexican Hat, Morlet, Meyer, etc.) and wavelets with compact support (orthogonal wavelets such as Daubechies or Coiflet, and biorthogonal wavelets) have been proposed [13]. In some fields of science, some families have shown better results for particular applications.

Nevertheless, regarding the transient extraction of fault components, the experience achieved after the development of multiple tests shows that a wide variety of wavelet families can lead to satisfactory results.

However, it has to be remarked that, in the case of compactly supported wavelets, once the wavelet family is selected, it is advisable to carry out the DWT using a high-order mother wavelet, this is, a wavelet with an associated filter with a large number of coefficients. If a low-order wavelet is used, the frequency response gets worse, and the overlap between adjacent frequency bands increases. Daubechies or Symlet with high orders has shown satisfactory results. In addition, dmeyer, within the infinite support wavelets, has behaved very well.

In this paper, Daubechies 8, 9, 10 and Symlet 7, 8 have been selected as the mother wavelets used for the DWT

analyses. Figure 3 shows wavelet family of Daubechies or Symlet.

2) Specification of the Number of Decomposition Levels

The number of decomposition levels is determined by the low frequency components to be traced. The lower the frequency components to be extracted, the higher the number of decomposition levels of the DWT [11]. So, the evolution of these components will be reflected through the high-level signals resulting from the analysis.

Typically, for the extraction of the frequency components caused by rotor asymmetries or even eccentricities, the number of decomposition levels should be equal or higher than that of the detail signal containing the fundamental frequency.

Finally, the number of decomposition levels n_d is related to the sampling frequency of the signal being analyzed (i.e., f_s). This parameter has to be chosen in such a way that the DWT supplies at least three high-level signals (i.e., two details and an approximation) with frequency bands below the supply frequency f; the following condition applies [11]:

$$n_d \geq n_f + 2 \quad (3)$$

n_f being the level of the detail that contains the supply

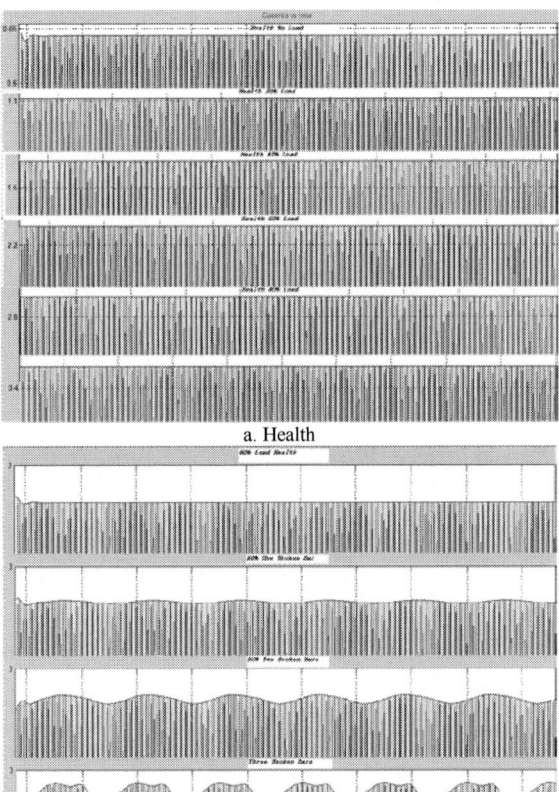

a. Health

b. Two broken bars

Fig. 4. Demonstration of stator current under varying load for health and two broken rotor bar (60% Load).

frequency, which can be calculated using (4), i.e.,

$$2^{-(n_d+1)} \cdot f_s < f \quad (4)$$

This condition means that the lower limit of the frequency band of the n_f level detail is lower than the supply frequency. Thus

$$n_d > \frac{\log(f_s / f)}{\log(2)} + 1 \quad \text{(Integer)} \quad (5)$$

Figure 2 illustrates the implementation procedure of a Discrete Wavelet Transform (DWT), in which S is the original signal, LPF and HPF are the low-pass and high-pass filters respectively. It must be considered, when capturing the transient signal, that the sampling frequency f_s plays an important role. Taking into account the Nyquist criterion, a very high sampling frequency is not mandatory for the application of the method, since most of the important fault components are usually in the low-frequency region [8]

Another practical remark is that, due to the nonideal filtering carried out by the wavelet signals, it is advisable not to set the limits of the band of the wavelet signal containing the fundamental frequency f_s very close to this frequency. Otherwise, this component could partially be filtered within the adjacent bands, masking the evolution of other components within these bands due to its much higher amplitude. Sampling frequencies of 1920 samples/s which enable good resolution analyses and, according to (1.16) and

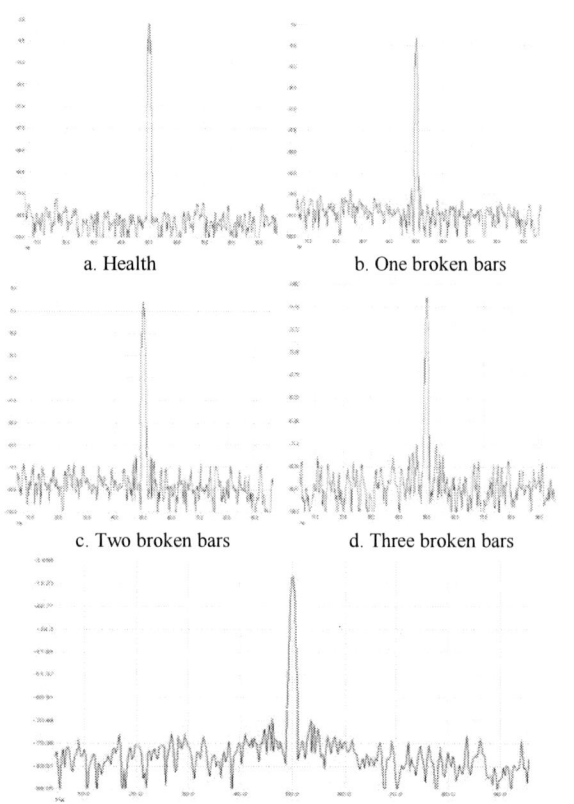

a. Health   b. One broken bars

c. Two broken bars   d. Three broken bars

e. Three broken bars Experimental

Fig. 5.   Line current spectra analysis of a induction motor.

(1.17), Table II shows the frequency levels of the wavelet function coefficients.

Once the mother wavelet and the number of decomposition levels have been selected, it is possible to carry out the DWT of the analyzed signal, obtaining wavelet decomposition graphics as those shown in Figures 5–9.

IV. SPECTRUM ANALYSIS OF STATOR CURRENT UNDER VARYING LOAD

When an induction motor under normal operation, frequency of the rotor induced current is equal to sf_s. This current generates a forward rotating magnetic field with respect to the rotor. If the rotor operates under one broken bar, the induced rotor current with frequency sf_s consists of two forward and backward magnetic components that generate the magnetic fields with speeds $\pm sf_s$ regarding to the rotor. If the speeds of these fields are added up with the rotor speed equal to $(1-s)f_s$ the speed of these fields with respect to the stator winding is as follows [11]:

$$sf_s + (1-s)f_s = f_s \quad (6)$$

$$-sf_s + (1-s)f_s = (1-2s)f_s \quad (7)$$

Considering (1.7), a current with frequency $(1-2s)f_s$ is induced in the stator winding. The interaction between the backward field of the rotor and the stator field leads to some ripples on the stator current as shown in Figure 4. The relative speed between the rotor and stator fields is equal to $2sf_s$ as follows:

$$f_s - (1-s)f_s = 2sf_s \quad (8)$$

This oscillation in the rotor speed produces a current with frequency $3sf_s$ in the rotor and similar with the previous case. According to (1.9) and (1.10), a current with frequencies $(1+2s)f_s$ and $(1-4s)f_s$ are induced in the stator.

$$3sf_s + (1-s)f_s = (1+2s)f_s \quad (9)$$

$$-3sf_s + (1-s)f_s = (1-4s)f_s \quad (10)$$

Therefore, when rotor bars broken happen, harmonic components of stator current having patterns as: $f_s \pm 2f_s$, $f_s \pm 4f_s$, $f_s \pm 6f_s$ …or in general harmonic with frequencies $(1 \pm 2K)f_s$ or $f_s \pm 2Kf_s$ are induced in the stator current. Therefore, considering the increase of the amplitudes of the

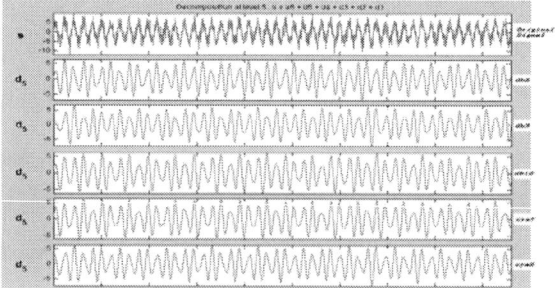

Fig. 6.   Wavelets coefficients for healthy motor.

978-1-4799-0024-4/13 $31.00 © 2013 IEEE

harmonic components in $(1\pm2Ks)f_s$ due to fault and its extension. The quantities have been measured for 1, 2 and 3 broken rotor bars and *3* broken bars in practical. Moreover, the frequency bands around the fundamental component of current must be concentrated on for diagnosis of the broken rotor bars.

By taking into account the above facts and Nyquist sampling theorem, the chosen sampling frequency of the signal in this research was chosen to be 1920 Hz. Frequency spectrums of the stator current for a healthy motor and a motor with 1, 2 and 3 broken bars were presented in Figure 5 (a-e). Taking comparison from Figure 5(a) to Figure 5 (e) indicates that 3 broken bars increase the harmonic components $(1\pm2s)f_s$ from -89 dB in the healthy to -71 dB in the faulty motor under load variation. Table III shows the measured amplitudes of harmonic components of $(1\pm2s)f_s$ for healthy and faulty motor with 1, 2 and 3 broken rotor bars and 3 broken bars in experimental.

Referring to Figures 5(a-e) the higher broken bar numbers increase the amplitude of harmonic components $(1\pm2s)f_s$. An interesting point in Table III that broken bar numbers not only affect the amplitude of the harmonic components but also influence the frequency of the harmonic components, as such that difference between the frequencies of harmonic components due to broken bars increase comparing with the fundamental frequency. The reason is that higher broken bars numbers decreases the speed of the motor and slip and difference of frequencies $(1\pm2s)f_s$ in respect to the fundamental frequency rises. Table III summarizes the frequency of harmonic components $(1\pm2ks)f_s$ for k=1, 2 of healthy and faulty motor with 1,2 and 3 broken bars under rate load.

V. INTRODUCING INDICES FOR DIAGNOSIS OF FAULT AND NUMBER OF BROKEN ROTOR BARS

As previously discussed, Rotor broken bars generate side band components around the fundamental frequency. Moreover, referring to Table II indicates that wavelet coefficients in D5 consist of side band components around the fundamental frequency. Therefore, in this paper, a different wavelet (db8, 9, 10, sym 7, 8) coefficient in D5 has been used to diagnose the fault and number of the broken bars. By comparing these coefficients in D5, the optimal wavelet was selected for diagnosis of rotor fault and number of broken bars.

In Figure 6, the wavelets coefficients in D5 for a healthy motor are presented. Figure 7 presents the wavelet coefficients in D5 for a motor with one broken rotor bar. Figure 8 demonstrates the wavelets coefficients in D5 for a motor with two broken rotor bars. Figure 9 shows the wavelets coefficients in D5 for a motor with three broken rotor bars. To validate the simulation results, experimental results was presented in Figure 14.

Figure 6-10 and Table 5-9 shows the different wavelet coefficients in D5 for a healthy motor by using db8, 9, 10, sym 7, 8 wavelets. Normal data analysis approaches, such mean stator current, standard deviation (STD) and average of D5 wavelet coefficients fluctuations were calculate to represent the features in a numerical way. In the end, two novel indexes were introduced as criterion function to

TABLE III
AMPLITUDES OF HARMONIC COMPONENTS FOR HEALTHY AND FAULTY MOTOR IN EXPERIMENTAL

No of broken bars	Amplitude of $(1-2s)f_s$	Amplitude of $(1+2s)f_s$	Frequency of $(1-2s)f_s$	Frequency of $(1+2s)f_s$
0	-89	-89	49Hz	51Hz
1	-77	-77	48Hz	52Hz
2	-75	-75	47Hz	53Hz
3	-71	-71	46Hz	54Hz
3(Test)	-70	-70	46Hz	54Hz

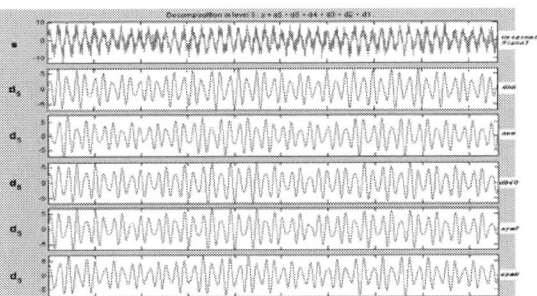

Fig. 8. Wavelets coefficients for two broken rotor bars.

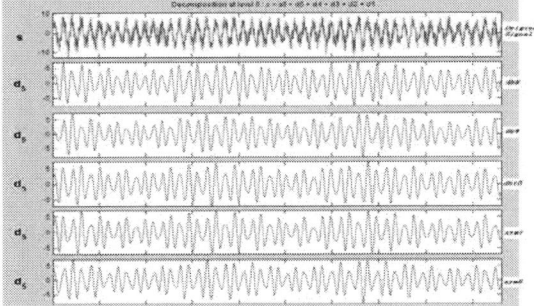

Fig. 9. Wavelets coefficients for three broken rotor bars.

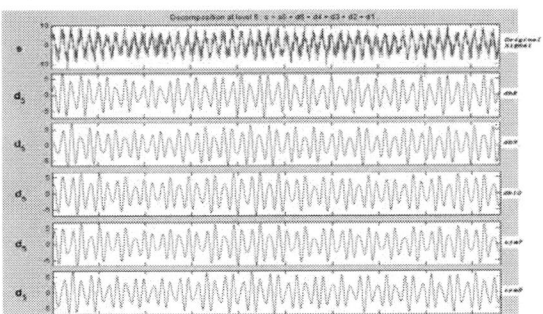

Fig. 7. Wavelets coefficients for one broken rotor bar.

diagnose the breakage in rotor bars.

For mean stator current analysis, comparison of wavelet coefficients of its amplitude in D5 for the healthy motor and motor with 1, 2 and 3 broken rotor bar confirms the increase of the oscillation due to the harmonic components $(1\pm2Ks)fs$ in the stator current. Figure 6-10 and Table V-IX indicate that the average value of the current amplitude is reduced due to the broken rotor bars, such that the current decreases from 4.3939A in db8, (4.3998A in db9, 4.4215A in db10, 4.4326A in sym7, 4.4628A in sym8), in the healthy motor to 4.1395A, (4.1563A, 4.1632A, 4.2356A, 4.2521A) in the motor with one broken bars under same wavelet analysis. By increasing the number of broken bars to two, this value is decreased to 3.9355A, (3.9856A, 3.9875A, 3.9921A, 3.9987A). Finally for a motor with 3 broken bars this reduces to 3.7793A, (3.7919A, 3.7965A, 3.8016A, 3.8109A). The trend of this reduction continues by increasing the number of the broken bars.

In addition, the standard deviation (STD) value of wavelet coefficients in D5 increases from 2.1131, 2.1157, 2.1161, 2.1177, 2.1186 in the healthy motor to 2.1323, 2.1325, 2.1357, 2.1362, 2.1368 in a motor with one broken rotor bar under db8, db9 db10, sym7, sym8. By increasing the number of broken rotor bars to two, this STD increases to 2.1679, 2.1686, 2.1689, 2.1697, 2.1712. This change relates to the

harmonic components in the stator current, which causes oscillation in D5. The STD value of the wavelet coefficients in D5 increases to 2.1869, 2.1889, 2.1896, 2.1908, 2.1911 for 3 broken bars.

Furthermore, average of fluctuations in absolute value of the wavelet coefficients in D5 increases from 0.0769, 0.0782, 0.0782, 0.0783, 0.0789 in the healthy motor to 0.0897, 0.0895, 0.0899, 0.0901, 0.0913 in a faulty motor with 1 broken bar of rated load in db8, db9 db10, sym7, sym8. Referring to Table VII, an increase is seen in fluctuations average of absolute value 0.1235, 0.1296, 0.1311, 0.1357, 0.1388 of the wavelet coefficients in D5, by increasing the broken number to 3 this fluctuations average in absolute value of the coefficients increases to 0.1635, 0.1696, 0.1711, 0.1757, 0.1788.

Through the numerical analysis of mean, STD and fluctuations, the change trends can be developed when broken rotor bars occurring in induction motor. However, these trends are not easy for operator to judge due to the values are too small to tell the difference. To solve this problem, according to the above mentioned facts the following indexes is proposed for diagnosis of the broken bar:

$$\text{Criterion function1} = \frac{\text{STD}}{\text{mean}} \quad (11)$$

$$\text{Criterion function2} = \frac{\text{Fluctuation coefficients}}{\text{Mean}} \quad (12)$$

Where the average of absolute value of the wavelet coefficients in D5 is expressed in per-unit (p.u.) with respect to the average amplitude of the corresponding currents.

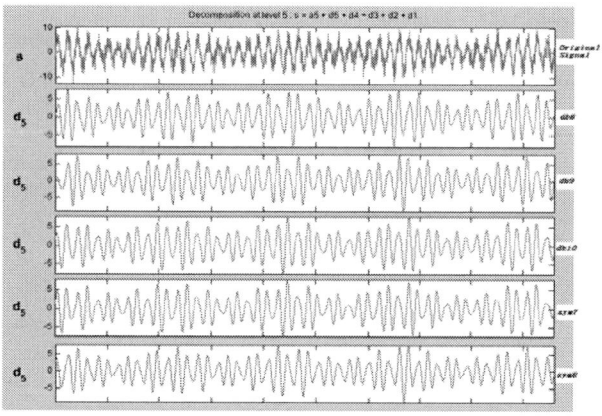

Fig. 10. Wavelets coefficients for three broken rotor bars under experimental results.

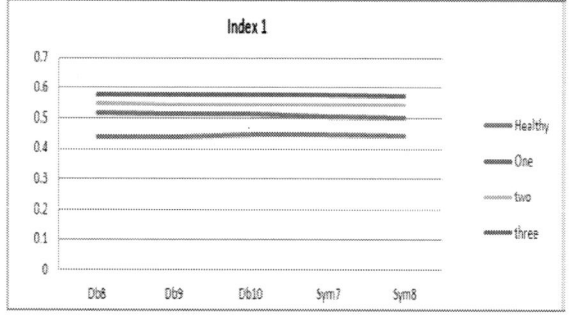

Fig. 11. Criterion function of index 1.

TABLE IV
CRITERION FUNCTION FOR DIFFERENT WAVELETS UNDER HEALTHY CONDITION

Wavelets	Mean	STD	Mean fluctuations	Index1	Index2
Db8	4.3939	2.1131	0.0769	0.4809	0.0175
Db9	4.3998	2.1157	0.0782	0.4808	0.0177
Db10	4.4215	2.1161	0.0782	0.4785	0.0176
Sym7	4.4326	2.1177	0.0783	0.4777	0.0176
Sym8	4.4628	2.1186	0.0789	0.4747	0.0176

TABLE V
CRITERION FUNCTION FOR DIFFERENT WAVELETS UNDER ONE BROKEN BAR

Wavelets	Mean	STD	Mean fluctuations	Index1	Index2
Db8	4.1995	2.1212	0.1097	0.5051	0.0261
Db9	4.1963	2.1225	0.1095	0.5058	0.0260
Db10	4.1732	2.2057	0.1221	0.5285	0.0292
Sym7	4.1956	2.1362	0.1001	0.5091	0.0238
Sym8	4.1821	2.2035	0.1173	0.5268	0.0280

TABLE VI
CRITERION FUNCTION FOR DIFFERENT WAVELETS UNDER TWO BROKEN BARS

Wavelets	Mean	STD	Mean fluctuations	Index1	Index2
Db8	3.9955	2.1679	0.1235	0.5425	0.0309
Db9	3.9856	2.1686	0.1296	0.5441	0.0325
Db10	3.8975	2.1689	0.1311	0.5564	0.0336
Sym7	3.9921	2.1697	0.1357	0.5434	0.0339
Sym8	3.9887	2.1712	0.1388	0.5443	0.0347

TABLE VII
CRITERION FUNCTION FOR DIFFERENT WAVELETS UNDER THREE BROKEN BARS

Wavelets	Mean	STD	Mean fluctuations	Index1	Index2
Db8	3.7793	2.1869	0.1635	0.5786	0.0432
Db9	3.7919	2.1889	0.1696	0.5772	0.0447
Db10	3.7965	2.1896	0.1711	0.5767	0.0450
Sym7	3.8016	2.1908	0.1757	0.5762	0.0462
Sym8	3.8109	2.1911	0.1788	0.5749	0.0469

TABLE VIII
CRITERION FUNCTION FOR DIFFERENT WAVELETS FOR THREE BROKEN BARS UNDER EXPERIMENTAL RESULTS

Wavelets	Mean	STD	Mean fluctuations	Index1	Index2
Db8	4.1793	2.4369	0.1935	0.5830	0.0462
Db9	4.1919	2.4389	0.1996	0.5818	0.0476
Db10	4.1965	2.4396	0.2011	0.5813	0.0479
Sym7	4.2016	2.4408	0.2057	0.5809	0.0489
Sym8	4.2109	2.4411	0.2088	0.5797	0.0495

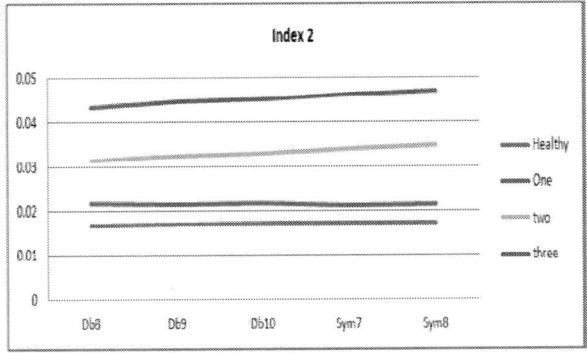

Fig. 12.   Criterion function of index 2.

The value of index 1 increases by occurring the breaking in the bars of the rotor and has ascending trend as shown in Figure 11. According to Table IV-VIII, the value of index 1 for a healthy motor is 47-48% and for the motor with one broken rotor bar is 50-51%. A noticeable difference between these values makes the proposed index enable to diagnose faulty motor from healthy one. On the other hand, comparison of this index for one and two broken bars indicates that this index is also a very good criterion for the diagnosis of the broken rotor bars. It is a convenient index for diagnosis of the number of broken bars.

In Table IV-VIII, values of Index 2 for different number of broken bars have been mentioned in Figure 12. It is observed that index 2 increases from 2.1% in one broken bar to 3.1-3.4% in two broken bars. Also this index increases from 3.1-3.4% in 2 broken bars to 4.3-4.6% in 3 broken bars.

Figure 10 shows the experimental for 3 rotor broken bars which are in good agreement with the simulation results.

Although by introducing these two indexes, boundaries between healthy and faulty even numbers of broken bars clearly showed in Figure 11, 12. There does still exist a problem, gap values between these boundaries are too small to diagnosis. To increase the accuracy of diagnosis, the most suitable wavelet must be select. However, there is no definite rule to guide how to choose the right wavelet until now due to sampled waves are unexpectable. Most of the researches are based on try and error approach. In this report, db8, db9, db10, sym7, sym8 were used to diagnosis broken rotor bars. According to Figure 11, 12, wavelet sym8 has the most obvious gap under 1, 2 and 3 broken bars.

VI. CONCLUSIONS

In this paper, the proposed algorithm is applied to the stator current of a healthy and faulty induction motor. Two novel criterion functions are introduced to select the optimal wavelet to diagnose the broken rotor bars fault in induction machines. FFT based spectrum analysis is used to compare with wavelet analysis result to demonstrate these criterion functions have the ability to detect fault in induction motors under rate load. Moreover, even the number of broken bars can be diagnosed accurately. Both simulations and experiments can prove that increase of the load and broken bars is followed by a growth in harmonic components amplitude of a faulty induction motor.

VII. REFERENCES

[1] M. Y. Chow, "Motor fault detection and diagnosis," *IEEE Industrial Electronics Newsletter*, Vol.44, No. 4, pp. 4-7, 1997.

[2] S. Radhika, G. R. Sabareesh, G. Jagadanand, and V. Sugumaran, "Precise wavelet for current signature in 3φ 1M," *Expert Systems with applications*, Vol. 37, No.1, pp. 450-455, 2010.

[3] M. Riera-Guasp, J. Antonino-Daviu, M. Pineda-Sanchez, R. Puche-Panadero, and J. Perez-Cruz, "A General Approach for the Transient Detection of Slip -Dependent Fault Components Based on the Discrete Wavelet Transform," *IEEE Transactions on Industrial Electronics*, Vol. 55, No. 12, pp. 4167-4180, 2008.

[4] J. Siau, A. Graff, W. L. Soong, and N. Ertugrul, "Broken Bar Detection in Induction Motors using Current and Flux Spectral Analysis", *Australasian Universities Power Engineering Conference* (AUPEC), Christchurch, New Zealand, 2003.

[5] M. E. H. Benbouzid, "A Review of Induction Motor Signature Analysis as a Medium for Faults Detection," *IEEE Transactions on Industrial Electronics*, Vol. 47, No. 5, pp. 984-993, 2000.

[6] H. Douglas, and P. Pillay, "The impact of wavelet selection on transient motor current signature analysis," *IEEE International Conference on Electric Machines and Drives*, San Antonio, Texas, USA, pp.80–85, 2005.

[7] Z. Ye, B. Wu, and A. Sadeghian, "Current signature analysis of induction motor mechanical faults by wavelet packet decomposition", *IEEE Trans on Industrial Electronics*, Vol.50, No. 6, pp. 1217–1227, 2003.

[8] J. A. Daviu, M. R. Guasp, J. R. Folch, and M. P. Molina, "Validation of a new method for the diagnosis of rotor bar failures via wavelet transformation in industrial induction machines," *,IEEE Trans. Ind. Appl.*, Vol. 42, No. 4, pp. 990–996, 2006.

[9] R. S. Arashloo, and A. Jalilian, "Design, Implementation and Comparison of Two Wavelet Based Methods for the Detection of Broken Rotor Bars in Three Phase Induction Motors," in 2010 1st *Power Electronic & Drive Systems & Technologies Conference*, PEDSTC, pp. 345-350, 2010.

[10] H. Sadri, "Induction Motor Broken Rotor Bar Fault Detection Using Wavelet Analysis," Master's Thesis, Iran University of Science & Technology, 2004.

[11] F. Ponci, A. Monti, L. Cristaldi, and M. Lazzaroni, "Diagnostic of a Faulty Induction Motor Drive via Wavelet Decomposition," *IEEE*

Transactions on Instrumentation and Measurement, Vol. 56, No. 6, pp. 2606-2615, 2007.

[12] C. S. Burrus, R. A. Gopinath, and H. Guo, "Introduction to Wavelets and Wavelet Transforms. A Primer," Englewood Cliffs, NJ: Prentice-Hall, 1998.

[13] J. W. Zhang, N. H. Zhu, L. Yang, Q. Yao, and Q. Lu, "A Fault Diagnosis Approach for Broken Rotor Bars Based on EMD and Envelope Analysis," *Journal of China University Mining & Technology*, Vol.17, No. 2, pp.205-209, 2007.

VIII. BIOGRAPHIES

Pu Shi was born in Henan, China, in 1982. He received the M.Eng. degree from Glyndwr University, Wrexham, UK, in 2009 and the B.Eng. degree from Southwest University for Nationalities, Chengdu, China in 2007.

In September 2009, he started his PhD degree in Glyndwr University, UK. He currently is focusing his research on induction machine fault detection, applying artificial intelligence technology such as neural network, expert system and fuzzy logic.

New Quantitative Rotor Fault Evaluation in Wound Rotor Induction Machine Drives Under Time-Varying Conditions

Y. Gritli, M. Mengoni, C. Rossi, *Member, IEEE,*
F. Filippetti, *Member, IEEE,* D. Casadei, *Senior Member, IEEE.*

Abstract -- **This paper presents a new diagnosis approach for quantitative rotor faults evaluation in three-phase wound rotor induction machines. In the considered application, the rotor windings are supplied by a static converter for the control of active and reactive power flows exchanged between the machine and the electrical grid. Under speed or fault-varying conditions, the typical rotor fault frequency components, which appear in rotor voltages, are spread in a bandwidth proportional to the speed which complicates its detectability. The proposed diagnosis approach is based on the use of wavelet analysis improved by a pre-processing of the rotor voltage commands. Thus, the time evolution of fault components can be effectively analyzed. Experimental results show the validity of the proposed method under speed and intermittent fault varying conditions, leading to an effective diagnosis procedure for rotor electrical faults in wound rotor induction machines.**

I. INTRODUCTION

THE use of wound rotor induction machines (WRIMs) is nowadays the preferred technology for modern wind turbines. Monitoring the induction machine is crucial to ensure safe operation, timely maintenance and increased operation reliability for maximizing wind energy extraction. Investigations on different failure modes in induction motors have revealed that 19% of the overall motor faults are related to the rotor part [1]. Some faults are not destructive and can evolve to more serious levels. Specifically, electrical faults that occur in the rotor windings of a WRIM, such as resistance variations or short-circuits, produce a phase dissymmetry because the phase impedances are not equal anymore [2]. The increase in a phase resistance, commonly referred to as "high resistance connection" in the literature, is a common problem that can occur in any power connections of industrial electrical machines [2]-[4]. Winding connections of WRIMs are subjected to corrosion, abrasion and fretting, leading generally to local heating, which can propagate to insulation damages [2]. Consequently, the detection of the first anomalies, such as resistance changes and winding unbalances, can prevent serious damages from happening. Different diagnostic methods have been proposed for WRIM [5]-[8]. In particular, some authors have developed a comprehensive diagnostic technique based on the Fourier analysis (FA) of the rotor voltage set-points at constant speed [5]. Unluckily,

induction machines operate mostly in time-varying conditions. Hence, slip and speed vary unpredictably and the classical application of FA for processing the voltage set-points or the measured currents fails. This shortcoming in the FA-based techniques can be reduced by analyzing a small interval of the signal by means of the Short-Time Fourier Transform (STFT). This method has been widely used to detect rotor failures in induction motors. However, the fixed width of the window and the high computational cost required to obtain a good resolution still remain major drawbacks of this technique [9]-[10]. The Hilbert-Huang Transform (HHT) was proposed for motor diagnosis, and has shown quite interesting performances in term of fault severity evaluation [11]-[13]. Being a linear decomposition, the Wavelet Transform (WT) provides a good resolution in time for high frequency components of a signal and a good resolution in frequency for low frequency components. In this sense, wavelets have a window that is automatically adjusted to give the appropriate resolution. WT was used with different approaches for monitoring motors. The related techniques are the un-decimated discrete WT [14], the wavelet ridge method [15], the wavelet coefficients analysis [16] or the direct use of wavelet signals [7]-[8], [17] for rotor fault detection. Recently, fractional Fourier transform (FrFT) was proposed in [18], providing an innovative graphical representation of rotor-fault components issued from discrete wavelet transform in time–frequency domain. More intensive research efforts have been focused on the usage of both approximation and detail signals for tracking different failure modes in motors such as broken bars [9], [17], [19]-[20], mixed eccentricity [17], [21], and increasing rotor phase resistance [5], [7]-[8], [10]. Most of the reported contributions are based on wavelet analysis of currents during the start-up phase or during any load variation. In this context, the frequency components are spread in a wide bandwidth as the slip and the speed change considerably. The situation is more complex under rotor fault conditions due to the proximity of the fault components to the fundamental frequency. These facts justify the usage of multi-detail or/and approximation signals resulting from the wavelet decomposition [9], [16]-[17], [19]. Moreover, the different decomposition levels are imposed by the sampling frequency. In fact, the dependency on the choice of the sampling frequency and on the capability of tracking multiple-fault frequency components makes difficult to interpret the fault pattern coming from wavelet signals and increases the diagnosis complexity. Typically, and for rotor faults detection, the amplitudes of the sideband around the fundamental at frequencies *(1±2s)f* are monitored. As

---

Y. Gritli, L. Zarri, M. Mengoni, C. Rossi, F. Filippetti, and D. Casadei are with the Dipartimento di Ingegneria dell'Energia Elettrica e dell'Informazione « Guglielmo Marconi », University of Bologna, 40126 Bologna, Italy (e-mail: yasser.gritli@unibo.it; michele.mengoni3@unibo.it; claudio.rossi@unibo.it; fiorenzo.filippetti@unibo.it; domenico.casadei@unibo.it).

explained in [2] and [21]-[22], the best diagnostic index for diagnosing rotor faults is the sum of the amplitudes of the two side bands. Such index is independent of inertia, avoids the underestimation of the fault and has an improved sensitivity. Recently, this technique was extended to time-frequency domain in [7], under speed transients in open-loop conditions for stationary rotor unbalance. Since in closed-loop drives, the control itself has an impact on the magnitude of the fault frequency components issued from currents [8], the same approach based on the chain of rotor fault components $(f_{krr} = \pm(1+2k)sf)_{k=0,1,2,...}$ issued from rotor voltages is here investigated. The paper is organized as follows. Section II presents the WRIM control system when the machine is mounted as generator in a wind turbine. The analysis of rotor fault frequencies is described in Section III. Experimental results are presented and commented in Sections IV.

II. WRIM CONTROL SYSTEM

In this paper the WRIM is supposed to be fed by a back-to-back converter on the rotor side, whereas the stator side is directly connected to the power grid. This configuration is common in wind energy plants. An independent control of active and reactive powers on the stator side is achieved by using a stator flux oriented control [5]. The relationships between the active and reactive stator power and the rotor current components are:

$$P_s \cong -\frac{3}{2} \left| \overline{v_s} \right| \frac{L_m}{L_s} i_{qr} \tag{1}$$

$$Q_s \cong \frac{3}{2} \left| \overline{v_s} \right| \frac{L_m}{L_s} \left(\frac{\left| \overline{v_s} \right|}{2\pi f L_m} - i_{dr} \right). \tag{2}$$

Equations (1) and (2) show that the stator active power is proportional to $-i_{qr}$, whereas the stator reactive power is proportional to $-i_{dr}$ since the stator voltage magnitude v_s and the frequency f are imposed by the power grid. The complete control system consists of two cascaded control loops based on two proportional-integral (PI) controllers (Fig. 1).

The outer loop controls the stator active and reactive powers, whereas the inner loop controls the d and q components of the rotor current. The matrices D and D^{-1} represent the Clarke and the inverse Clarke transformations, which convert a three phase system into an equivalent two phase system and vice versa, whereas T and T^{-1} represent direct and inverse rotations of the reference frame. The angle θ denotes the phase angle of the stator flux vector, whereas θ_r is the rotor position in electrical radians.

Consequently, $T(\theta)$ is the transformation matrix from the stationary reference frame to the stator-flux oriented reference frame, and $T(\theta-\theta_r)$ from the rotor reference frame to the stator-flux oriented reference frame [5]. The phase θ of the stator flux vector is computed by using the components of the magnetizing current vector,

$$\theta = \arg \overline{i}_{ms} \tag{3}$$

whereas the rotor angle θ_r is measured by a mechanical

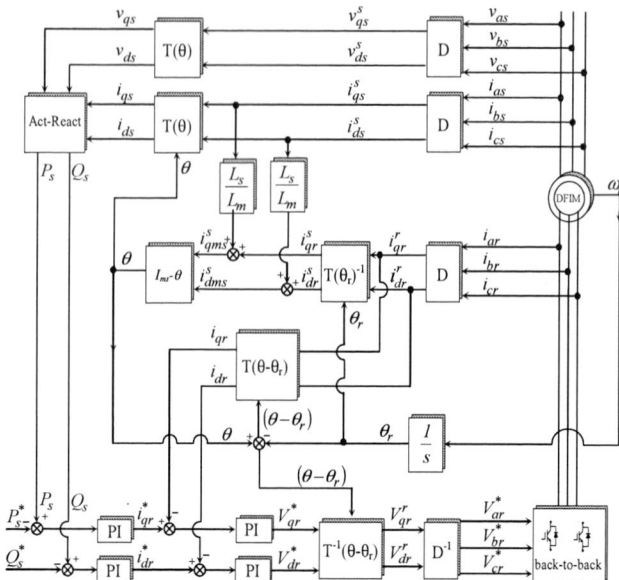

Fig. 1. Block-scheme representation of the complete WRIM control system

sensor such as an incremental or an absolute encoder.

III. ROTOR FAULT FREQUENCY PROPAGATION– THE PROPOSED CONCEPT

A. Rotor fault components propagation

The WRIM is subject to both electromagnetic and mechanical stresses, which are symmetrically distributed. In healthy conditions, the three stator and rotor impedances are identical and the currents are balanced. Under these normal conditions, only the normal frequency components at f and sf exist respectively in the stator and rotor currents/voltages (f denotes the power grid frequency, s the slip). If the rotor part is damaged in such a way that the phase windings are unbalanced, the rotor symmetry of the machine is lost and the rotor currents become unbalanced. The rotor body is subjected to oscillations and asymmetric heating, which may produce a thermal bow in the rotor.

These undesired effects can lead to a further propagation of the damage. Figure 2 illustrates the propagation of the harmonic components of a rotor fault. As can be seen, the first and the most relevant fault components that occur on rotor and stator currents are at frequencies $-sf$ and $(1-2s)f$ respectively. The negative-sequence frequency $-sf$ generates the component at frequency $(1-2s)f$, which in turn produces a pulsating torque and speed oscillations at frequency $2sf$. This speed frequency causes both a reaction stator current component at frequency $(1-2s)f$ and a new component at frequency $(1+2s)f$.

The new stator component at frequency $(1+2s)f$ interacts with the arising torque and speed oscillation at frequency $2sf$ and produces the new rotor current harmonics at the frequencies $\pm 3sf$, which in turn induce a reaction on the stator part and generate a new frequency component at $(1-4s)f$. Globally, these interactions leads to the appearance of

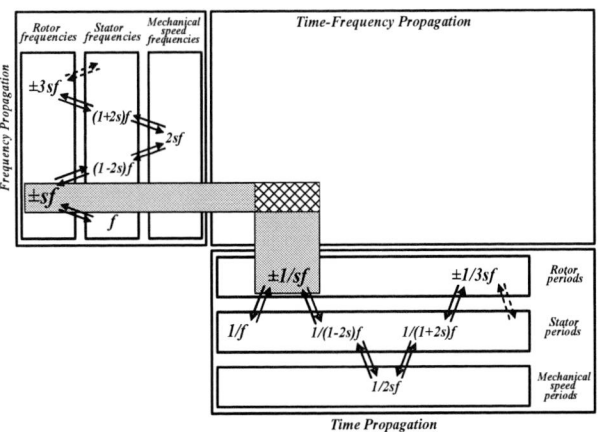

Fig. 2. Time-frequency propagation of a rotor fault. The minus sign (–) identifies current inverse sequence components.

the following chains of fault components on the rotor and on the stator currents, respectively at frequencies $(f_{kra}=\pm ksf)_{k=1,3,5,...}$ and $f_{ksa}=(1\pm 2ks)f)_{k=1,2,3,...}$.

As aforementioned in this system operating in closed loop, the rotor fault frequencies are mitigated by the control action consequently the same fault frequencies are reflected in the rotor voltages [5]. For this reason in the present work, the focus is exclusively on tracking the most relevant fault frequencies in the rotor voltages. Due to the damping effect of the machine-load inertia on higher order fault harmonics, only the signature of the side bands $f_l=-sf$ and $f_r=3sf$ are monitored.

B.   The proposed rotor fault detection strategy

Wavelet analysis consists in signal decomposition, using successive combinations of approximation signals and detail signals. With the well known dyadic down sampling procedure, frequency bands of the jth level decomposition are strictly related to the sampling frequency f_s. These bands, given by $[0, 2^{-(J+1)}f_s]$ or $[2^{-(J+1)}f_s, 2^{-J}f_s]$, cannot be changed unless a new acquisition with a different sampling frequency is made [23]. This fact complicates any fault detection based on DWT, particularly in speed-varying condition.

In this paper, an efficient solution to overcome this limitation is proposed. Hereafter the sampling frequency f_s is assumed equal to 3.2 kHz, and an eight level decomposition ($J=7$) is chosen to cover the frequency bands in which the fault component is tracked. With regard to the type of mother wavelet, a 10th order Daubechies family is chosen, although other families (Dmeyer and Coiflet) also allow a clear detection of the phenomenon.

Under healthy conditions, only the fundamental component sf exists in the rotor voltages. If the rotor part is damaged, the most notable fault components are at frequencies $-sf$ and $3sf$ (Fig. 2). In time-varying conditions

the magnitudes of these fault components cannot be detected through a classical frequency analysis since they are spread in a wide frequency range. A simple processing of the stator phase current allows shifting the fault components $-sf$ and $3sf$ to a prefixed frequency band.

TABLE I
FREQUENCY BANDS AT EACH LEVEL

Approximations «a_j»	Frequency bands (Hz)	Details «d_j»	Frequency bands (Hz)
a_7	: [0– 12.5]	d_7	: [12.5– 25]
a_6	: [0– 25]	d_6	: [25– 50]
a_5	: [0– 50]	d_5	: [50– 100]
a_4	: [0– 100]	d_4	: [100– 200]

In such a way, all the information related to the fault is isolated and confined to a fixed frequency band. More in details, a frequency sliding f_{sl} is applied to the rotor voltage space vectors at each time slice, so that the harmonic component of interest ($-sf$ or $3sf$) is moved to a frequency band in which the corresponding fault signature can be easily tracked. Then the real part of the shifted signal is analyzed by means of DWT. In conclusion, the processed signal is as follows:

$$v_{sl}(t) = \mathrm{Re}\left[v_r(t) \ e^{-j2\pi f_{sl}t} \right] \qquad (4)$$

The evaluation of f_{sl} is made in such a way that the contribution of the fault frequency component, for the whole considered slip range, is shifted into one of the intervals $[0, 2^{-(J+1)}f_s]$ or $[2^{-(J+1)}f_s, 2^{-J}f_s]$ and consequently the DWT can be applied to analyze only the frequency band of interest. In this way, the fault detection and quantification become easier. For this reason, approximation a_7 has been chosen for extracting the contribution of the fault components $-sf$ and $3sf$. Therefore the choice of f_{sl} is made to shift these fault components inside the frequency interval $[0, 12.5]$ Hz corresponding to approximation a_7 (see Table I).

The proposed approach was applied to the rotor voltage space vectors with $f_{sl}=11.25$Hz and then with $f_{sl}=-8.75$ Hz to isolate respectively the contribution of the fault components $-sf$ and $3sf$. Once the state of the machine has been qualitatively diagnosed by the approximation signals a^l_7 and a^r_7, issued from the wavelet analysis related respectively to the fault components at frequencies $-sf$ and $3sf$, a quantitative evaluation of the fault degree is necessary.

For extracting the best fault signature, it is necessary to define a fault indicator that incorporate both the signatures issued from the $-sf$ and $3sf$ components. For this purpose the following multiresolution mean power indicator mPa_j at the resolution level j is used to quantify the signature of each fault component (f_l and f_r):

$$mPa_j(v_{sl}) = \frac{1}{\Delta n}\sum_{n=1}^{\Delta n}\left|a_j(n)\right|^2 \qquad (5)$$

where $a_j(n)$ is the approximation signal of interest (d^l_7 or d^r_7 in our case), Δn denotes the number of samples and j the decomposition level. The fault indicator is periodically calculated over time, every 400 samples inside a window of 6400 samples [7]-[8]. These values were regulated experimentally to reduce variations that can lead to false alarms in healthy operating conditions of the motor. The fault indicator is defined in (6) and normalized by mPa^{fun}_7, which denotes the energy related to the fundamental component.

$$\% F_{ind} = \frac{mPa^l_7 + mPa^r_7}{mPa^{fun}_7} \times 100 \qquad (6)$$

When the fault occurs, the energy distribution of the signal is changed in the resolution levels related to the characteristic frequency bands of the default. Hence, the energy excess localized in the approximation signal of interest (a_7 in our case) is considered as an anomaly indication in case of rotor unbalance.

IV. Experimental Results

The experimental set-up (Fig. 3) consists of a 5.5kW, 220V/380V, 50Hz, 4 poles WRIM (see appendix) connected to a PWM back-to-back power converter on the rotor side. The generator is driven by a 9kW DC motor fed by a commercial speed-controlled DC/DC converter. The field oriented control for the WRIM has been implemented on a dSPACE DS1103 board. Since the speed of the WRIM is not fixed by the control system (because the WRIM set points are the active and reactive powers), the rotor speed is determined by the DC motor, and can be varied by acting on the corresponding set-point.

For the assessment of the fault diagnosis algorithm, the grid voltages, the rotor-side line currents, the reference rotor voltages, the rotor and stator currents, and the rotor speed are sampled at 3.2 kHz with a data acquisition window of 10 seconds. Finally, the frequency bandwidth of the current loops has been set to 100 Hz. Since the common effect of rotor faults is the unbalance of the considered winding impedances, the fault configurations have been emulated in a non-destructive way by increasing the rotor phase resistances with additive components (R_{add}). During the tests, active and reactive power set-points at the stator side are kept equal to 5.5 kW and 4.1 kVA respectively. Initially, the performance of the WRIM has been verified in healthy condition during a speed transient from 1395 rpm to 1305 rpm. With a sampling frequency f_{sam} of 3.2 kHz, a 7th level decomposition ($J=7$) has been chosen to track the fault components $-sf$ and $3sf$ in the time domain.

A. Frequency domain analysis

In this section, FFT is used to carry out a preliminary study to detect experimentally the contribution of the fault components $-sf$ and $3sf$ under healthy and faulty ($Radd=Rr$) cases. Fig. 4 shows the experimental rotor voltage space-

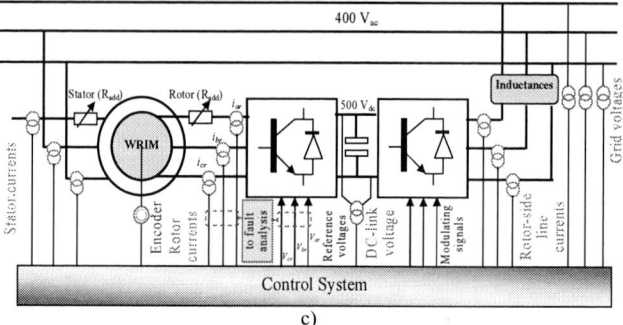

Fig. 3. Experimental set-up. a) Power converter cabinet. b) WRIM and prime mover. c) Schematic diagram of the test bench and position of the current and voltage sensors.

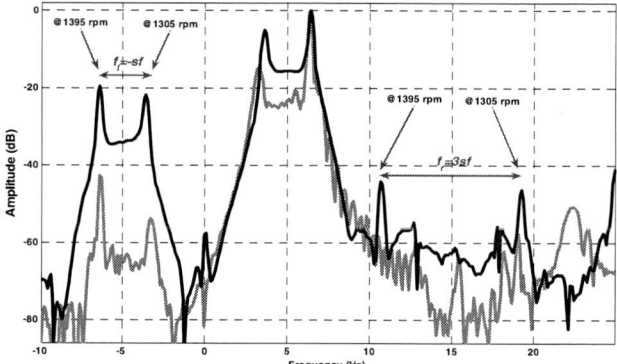

Fig. 4. FFT rotor voltage space-vector spectra under speed-varying conditions, for healthy *(Red)* and rotor fault *(Black)* conditions.

vector spectra under healthy (Red) and rotor fault (Black) conditions. For both analysis, it is evident that a rotor asymmetry produces a negative sequence harmonic component in the spectrum of the voltage space-vector at frequency $-sf$, and a fault component at $3sf$.

Comparing the contribution of each component, it is possible to note that the rotor fault signature coming from $-sf$ is more relevant than the one related to the component at frequency $3sf$. This is mainly due to the damping effect of the machine-load inertia. However, being dependent on slip values, under speed varying conditions, the magnitude of these fault components are spread in a bandwidth proportional to the speed variation. Consequently, the use of Fourier transform in these conditions can lead to an erroneous diagnosis. Then, a more appropriate technique is needed for providing a reliable diagnosis procedure.

978-1-4799-0024-4/13 $31.00 © 2013 IEEE

B. Time-Frequency domain analysis

As explained in the previous section, it is possible to track over time the signature of these two fault components by means of wavelet multiresolution analysis. More specifically, according to the proposed approach, the frequency band of interest for tracking the rotor fault signature of the side bands is the approximation signal a_7. The first rotor fault signature tracked is related to the fault component at frequency $-sf$. The corresponding experimental results, under healthy and rotor fault condition, are presented respectively in Fig. 5 and Fig. 6. It is possible to notice that the 7th approximation signal, issued from wavelet decomposition for the healthy machine (Fig. 5-c), and does not show any variations. This indicates the absence of the fault component $-sf$ in the rotor voltage space vector, allowing diagnosing the healthy condition of the motor under time-varying operating condition. On the contrary, in faulty condition ($R_{add}=R_r$) approximation a_7 shows significant variation in amplitude, as can be seen from Fig.6-c. Effectively, the large amplitude variations reflects the presence of the fault frequency component $-sf$. Despite the speed transient, the oscillations reproducing the fault frequency evolution $-sf$ are still significant. The second rotor fault signature tracked is related to the fault component at frequency $3sf$. The corresponding simulation results, under healthy and rotor fault condition, are presented respectively in Figs. 7 and Fig. 8. The same observations as for the signature of the side band $-sf$ under the considered rotor unbalance, can be made.

By comparing the approximation signal a_7 calculated in healthy and unbalance conditions (Figs. 7-c and 8-c), one understands clearly the amplitude evolution of the fault component $3sf$ dynamically over time. If the attention is focused on the approximation signal a_7 depicted in Fig. 6-c and Fig. 8-c, it is possible to note that the rotor fault signature coming from the component at frequency $-sf$ is more relevant than the one related to the component at frequency $3sf$. This is mainly due to the damping effect of the machine-load inertia. Anyway, the sum of their amplitudes is still a reliable diagnostic index. As introduced in the previous section, a fault indicator based on the sum of the cyclic energy calculation related to the fault components $-sf$ and $3sf$ is proposed to evaluate quantitatively the rotor fault extent. The cyclic evolution of the fault indicator F_{ind} obtained from, under healthy and rotor fault condition, are depicted in Fig. 9. In healthy condition, and under large speed variations, mPa_7 does not show any significant change. Consequently, the indicator values for the healthy motor are considered as a baseline to set the threshold to discriminate between healthy and faulty conditions. Under rotor unbalance condition, the calculated fault indicator F_{ind} shows a noteworthy increase. The large energy deviation observed in faulty condition proves the effectiveness of the proposed approach, since the transient speed motor operation does not disturb the fault assessment in comparison with the healthy case. Finally, the results that have been obtained prove the robustness and the effectiveness of the developed approach for the diagnosis of rotor faults under time-varying operating conditions.

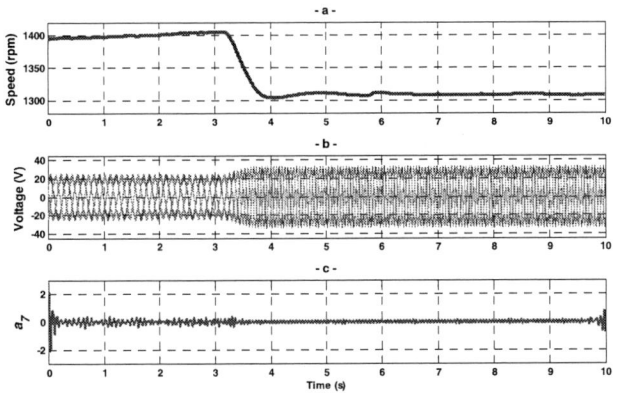

Fig. 5. DWT of the computed signal v_{sl} under speed-varying condition, for tracking the fault component $-sf$ under healthy conditions.

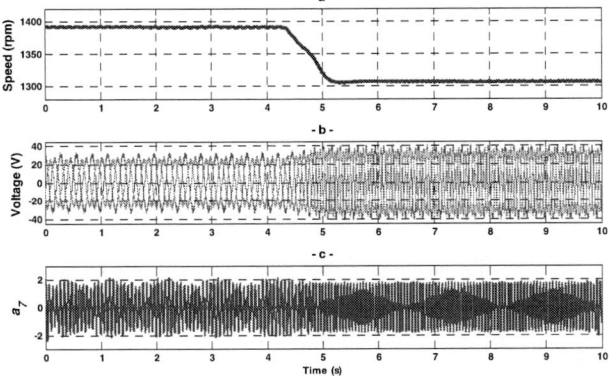

Fig. 6. DWT of the computed signal v_{sl} under speed-varying condition, for tracking the fault compoent $-sf$ under rotor unbalance *(Radd=Rr)*.

V. CONCLUSION

A diagnosis procedure for detection and quantification of rotor faults, in speed-varying condition has been proposed. The developed approach is based on the DWT after a bi-sliding frequency pre-processing of the stator phase current. Once the state of the machine has been qualitatively diagnosed, a time-frequency fault indicator is used to quantify the fault extent.

The proposed approach allows the calculation of the fault signatures related to the typical side bands at $-sf$ and $3sf$, which are typical in rotor faults of induction machines. The proposed technique has been also tested for low level rotor unbalance, and its sensitivity has been proved.

The effectiveness of this new approach can be extended for the diagnosis of other types of faults under time–varying conditions.

VI. APPENDIX

WRIM Parameters

Parameter		Value
Rated power	kW	5.5
Rated stator/ rotor voltage	V	380/186
Rated frequency	Hz	50
Rated speed	rpm	1400
Stator phase resistance	Ω	0.531
Rotor phase resistance	Ω	0.310
Stator inductance	H	0.083
Rotor inductance	H	0.019
Mutual inductance	H	0.038
Pole pairs		2

VII. REFERENCES

[1] A.H. Bonnett, C. Yung,, "Increased Efficiency Versus Increased Reliability", *IEEE Ind. App. Mag.*, Vol. 14, Issue 1, Jan./Feb. 2008.

[2] A. Bellini, F. Filippetti, C. Tassoni, G.A. Capolino, "Advances in diagnostic techniques for induction machines," *IEEE Trans. Ind. Elec.*, vol. 55, no. 12, pp. 4109–4126, Dec. 2008.

[3] J. Yun, J. Cho, S.B. Lee, J. Yoo, "On-line detection of High-Resistance Connections in the incoming electrical circuit for induction motors, " *IEEE Trans. Ind. Appl.*, vol. 45, no. 2, pp. 694–702, Mar./Apr. 2009.

[4] J. Yun, K. Lee, K.W Lee, S.B. Lee, J.Y. Yoo, "Detection and classification of stator turn faults and High-Resistance electrical connections for induction machines," *IEEE Trans. Ind. Appl.*, vol. 45, no. 2, pp. 666–675, Mar./Apr. 2009.

[5] A. Stefani, A. Yazidi, C. Rossi, F. Filippetti, D. Casadei, G.A. Capolino, "Doubly fed induction machines diagnosis based on signature analysis of rotor modulating signals, " *IEEE Trans. Ind. Appl.*, vol. 44, no. 6, pp. 1711–1721 , Nov./Dec. 2008.

[6] V. Dinkhauser, F.W. Fuchs, "Rotor turn-to-turn faults of doubly-fed induction generators in wind energy plants-modelling, simulation and detection, " in *Proc. EPE-PEMC*, Posnan, Poland, Sept. 1-3, 2008, pp. 1819–1826.

[7] Y. Gritli, C. Rossi, L. Zarri, F. Filippetti, A. Chatti, D. Casadei, A. Stefani, "Double frequency sliding and wavelet analysis for rotor fault diagnosis in induction motors under time-varying operating condition, " in *Proc. IEEE Int. Symp. Diagnostics Electr. Mach., Power Electron. Drives*, Bologna, Italy, Sep. 5–8, 2011, pp. 676–683.

[8] Y. Gritli, L. Zarri, C. Rossi, F. Filippetti, G. Capolino, and D. Casadei, "Advanced diagnosis of electrical faults in wound rotor induction machines," *IEEE Trans. on Ind. Elec.*, vol. 60, no. 9, pp. 4012–4024, Sept. 2013.

[9] J. Cusido, L. Romeral, J.A. Ortega, J.A. Rosero, A.G. Espinosa, "Fault detection in induction machines using power spectral density in wavelet decomposition, " *IEEE Trans. Ind. Elec.*, vol. 55, no. 2, pp. 633–643, Feb. 2008.

[10] I.P. Tsoumas, G. Georgoulas, E.D. Mitronikas, and A.N. Safacas, "Asynchronous machine rotor fault diagnosis technique using complex wavelets, " *IEEE Trans. Energy Conv.*, vol. 23, no. 2, pp. 444–459, Jun. 2008.

[11] J. Antonino-Daviu, P. J. Rodriguez, M. Riera-Guasp, A. Arkkio, J. Roger-Folch, and R. B. Perez, "Transient detection of eccentricity related components in induction motors through the Hilbert–Huang transform," *Energy Convers. and Manage.*, vol. 50, no. 7, pp. 1810–1820, Jul. 2009.

[12] A. Espinosa, J. Rosero, J. Cusido, L. Romeral, and J. Ortega, "Fault detection by means of Hilbert–Huang transform of the stator current in a PMSM with demagnetization," *IEEE Trans. Energy Conv.*, vol. 25, no. 2, pp. 312–318, Jun. 2010.

[13] R. Puche-Panadero, M. Pineda-Sanchez, M. Riera-Guasp, J. Roger-Folch, E. Hurtado-Perez, J. Peres-Cruz, "Improved resolution of the MCSA method via Hilbert transform, enabling the diagnosis of rotor asymmetries at very low slip, " *IEEE Trans. Energy Conv.*, vol. 24, no. 1, pp. 52–59, Mar. 2009.

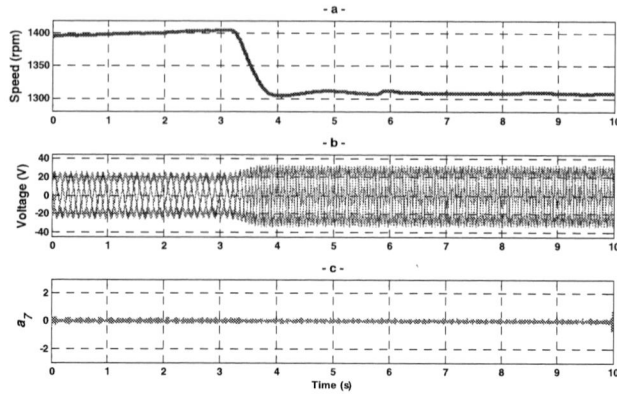

Fig. 7. DWT of the computed signal v_{sl} under speed-varying condition, for tracking the fault component *3sf* under healthy conditions.

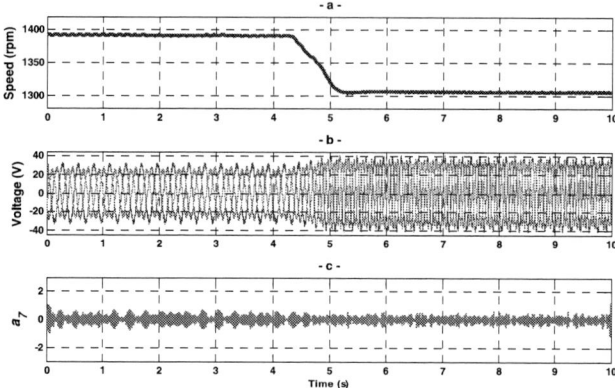

Fig. 8. DWT of the computed signal v_{sl} under speed-varying condition, for tracking the fault component *3sf* under rotor unbalance *(Radd=Rr)*.

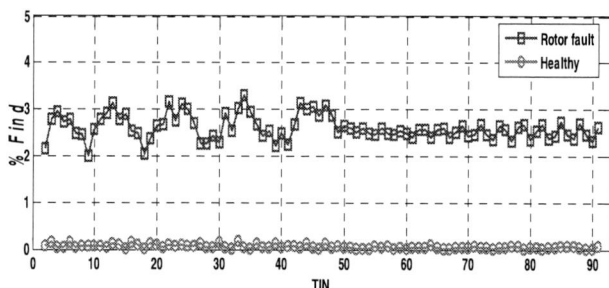

Fig. 9. Cyclic values of the fault indicator F_{ind}, resulting from the 7th wavelet decomposition level of the signals v_{sl} under healthy and rotor unbalance ($R_{add}=R_r$) for time-varying operating condition.

[14] W.G. Zanardelli, E.G. Strangas, S. Aviyente, "Identification of intermittent electrical and mechanical faults in permanent-magnet AC drives based on time–frequency analysis, " *IEEE Trans. Ind. Appl.*, vol. 43, no. 4, pp. 971–980. Jul./Aug. 2007.

[15] S. Rajagopalan, J.M. Aller, J.A. Restrepo, T.G. Habetler, R.G. Harley, "Analytic wavelet ridge-based detection of dynamic eccentricity in brushless direct current (BLDC) motors functioning under dynamic operating conditions, " *IEEE Trans. Ind. Elec*, vol. 54, no. 3, pp. 1410–1419, Jun. 2007.

[16] A. Ordaz-Moreno, R.J. Romero-Troncoso, J.A. Vite-Frias, J.R. Rivera-Gillen, A. Garcia-Perez, Automatic online diagnosis algorithm for broken-bar detection on induction motors based on discrete wavelet transform for FPGA implementation, ," *IEEE Trans. Ind.*

Elec., vol. 55, no. 5, pp. 2193–2202, May 2008.

[17] M.R. Guasp, J.A. Daviu, M.P. Sanchez, R.P. Panadero, J.P. Cruz, "A general approach for the transient detection of slip-dependent fault components based on the discrete wavelet transform," *IEEE Trans. Ind. Elec.*, vol. 55, no. 12, pp. 4167–4180 Dec. 2008.

[18] M. Pineda-Sanchez, M. Riera-Guasp, J. A. Antonino-Daviu, J. Roger-Folch, J. Perez-Cruz, and R. Puche-Panadero, "Diagnosis of induction motor faults in the fractional Fourier domain," *IEEE Trans. Inst. and Meas.*, vol. 59, no. 8, pp. 2065–2075, Aug. 2010.

[19] A. Bouzida, O. Touhami, R. Ibtiouen, A. Belouchrani, M. Fadel, A. Rezzoug, "Fault diagnosis in industrial induction machines through discrete wavelet transform, " *IEEE Trans. Ind. Elec.*, vol. 58, no. 9, pp. 4385–4395. Sept. 2011.

[20] J. Pons-Llinares, J. A. Antonino-Daviu, M. Riera-Guasp, M. Pineda-Sanchez, V. Climente-Alarcon, "Induction motor diagnosis based on a transient current analytic wavelet transform via frequency B-splines, " *IEEE Trans. Ind. Elec.*, vol. 58, no. 5, pp.1530–1544, May 2011.

[21] W. Yang, P. J. Tavner, C. J. Crabtree, M. Wlikinson, "Cost-effective condition monitoring for wind turbines, " *IEEE Trans. on Ind. Elec.*, vol. 57, no. 1, pp. 263–271, Jan. 2010.

[22] Bellini, A., "Quad Demodulation: A Time-Domain Diagnostic Method for Induction Machines", *IEEE Tran. on Ind. Appl.*, Vol. 45 , Issue 2, pp. 712 – 719, 2009.

[23] T.K. Sarkar, C. Su, R. Adve, M.S. Palma, L.G. Castillo, R. Boix, "A tutorial on wavelets from an electrical engineering perspective, Part 1: discrete wavelet techniques, " *IEEE Ant. and Prop. Mag.*, vol. 40, no. 6, pp. 36–49, Oct. 1998.

Induction Motor Broken Rotor Bar Detection Using Vibration Analysis – A Case Study

Ž. Kanović, D. Matić, Z. Jeličić, M. Rapaić, B. Jakovljević, M. Kapetina

Abstract -- **Early fault diagnosis and condition monitoring can reduce the consequential damage and breakdown maintenance, prolong the machine life and increase the performance of industrial systems. This paper describes a real fault detection problem of a high-power (3.2 MW) induction motor driving pumps in a heating plant. Steady-state vibration signals were collected and some characteristic low- and high-frequency features were observed, resulting in broken rotor bar diagnosis. The obtained results were verified by disassembling the motor, which proved that this particular technique can be successfully applied in induction motor fault detection.**

Index Terms—**Fault detection and diagnosis, frequency domain analysis, induction motor, vibration analysis.**

I. INTRODUCTION

Induction motors are the primary movers of the industries because of their rugged configuration, low cost, versatility, reasonably small size and capability to operate with an easily available power supply. In the practical applications, however, they are subjected to the unavoidable stresses, such as electrical, environmental, mechanical and thermal stresses, which create failures in different motor parts [1]. These failures disturb the safe operation of induction motors, threaten the normal manufacturing and, accordingly, result in the substantial cost penalties.

The issue of robustness and reliability is very important to guarantee the good operational condition of the entire industrial system. Therefore, condition monitoring of induction motors has received considerable attention in recent years. Early fault diagnosis and condition monitoring can reduce the consequential damage, breakdown maintenance and reduce the spare parts of inventories. Moreover it can prolong the machine life and increase the performance and the availability of machine.

The condition of an induction motor is mostly examined from the signals acquired through the sensors and supporting instrumentation methods. Signals can manifest fault signature using an appropriate signal processing technique. A number of induction motor condition monitoring techniques have been developed, which monitor a certain parameter of the motor and determine its health. These techniques are based on monitoring various variables, such as acoustic emission, air-gap torque, electromagnetic field, instantaneous power, vibration, voltage, stator current and other [2 - 8].

Rotor is the most inner part of the induction motor, which is rotated by an electromagnetic field induced in its coils from the stator field. The rotational force is then applied to the external equipment. Induction motor rotors are very rugged, but still, rotor defects such as broken bar, cracked end-ring, bent shaft and eccentricity do occur. These failures do not initially cause motor to fail, but they bring about secondary effects that lead to a serious malfunction, so detection of these faults plays an important role.

This paper deals with detection of broken rotor bar using vibration signal analysis, which is one the oldest condition monitoring techniques, characterized by easy measurability, high accuracy and reliability. A case study is presented, where a faulty 3.2 MW induction motor was examined and the presence of broken rotor bar was detected. The paper is organized as follows: in section 2 some basics of broken rotor bar detection using vibration analysis are presented, in section 3 the observed system, data acquisition equipment and the results obtained using signal processing techniques are described, and section 4 concludes the paper.

II. BROKEN ROTOR BAR DETECTION USING VIBRATION ANALYSIS

Rotor bars can be partially or completely cracked during the operation of induction motor, due to stresses, improper rotor geometry design or some imperfection in material or rotor production process. The bar breakage is the major fault in the rotor of induction motor. Once a bar breaks, the condition of the neighboring bars also deteriorates progressively due to the increased stresses. To prevent such a cumulative destructive process, the problem should be detected early, that is, when the bars are beginning to crack [9].

Vibration signal analysis is a fault detection technique which is generally used for mechanical faults diagnosis, such as bearing problems, gear mesh defects, rotor misalignment and mass unbalance [10]. However, it can also be successfully applied to detect broken rotor bar, since this fault excites the electromagnetic field disturbance and thus intensifies the torque modulations and vibrations of the motor, which can be measured by placing vibration sensors on motor housing [11-14].

The This paper is a result of the research within the project TR32018, supported by the Ministry of Education, Science and Technology, Republic of Serbia.

The authors are with the Computing and Control Department, Faculty of Technical Science, University of Novi Sad, Trg Dositeja Obradovica 6, 21000 Novi Sad, Serbia (phone: +381 21 485 2449, e-mails: kanovic@uns.ac.rs, dmatic@uns.ac.rs, jelicic@uns.ac.rs, rapaja@uns.ac.rs, bjakov@uns.ac.rs, mirna.kapetina@uns.ac.rs)

Analyzing the obtained vibration signals in frequency domain, using for example Fast Fourier Transformation (FFT), one can note some fault-specific features in signal spectrum, which imply the presence of the particular fault.

If a broken rotor bar exists, no current will flow in that bar. As a result, the field in the rotor around that particular bar will not exist. Therefore the force applied to that side of the rotor would be different from that on the other side of the rotor, creating an unbalanced magnetic force that rotates at one times rotational speed and modulates at a frequency equal to slip frequency times the number of poles, which is known as pole pass frequency. It means that in vibration spectrum increased amplitudes will occur at the rotation frequency f_r and its sidebands

$$f_{brb1} = f_r \pm f_p \qquad (1)$$

where f_p is pole pass frequency defined as

$$f_p = \left(f_{sync} - f_r \right) \cdot P \qquad (2)$$

with P being number of poles and f_{sync} being synchronous speed [4]. These sidebands occur also in higher harmonics of rotation speed ($2f_r, 3f_r,...$). Detection of broken bar based on vibration analysis and observation of features (1) and (2) is common in literature and widely used in practical applications [6].

One must note that pole pass frequency defined in (2) is highly slip-dependent. The low value of slip causes the rotation frequency f_r to be close to synchronous speed f_{sync}, and sidebands to be closer to the central frequency, which makes them less distinctive in vibration signal spectrum. This fact can cause difficulties if the motor operates under low load, and consequentially with low slip, since in that case fault detection reliability is questionable. Therefore, vibration signals must be acquired when motor operates close to full load in order to have sidebands that can be clearly distinguished from central frequencies. This is the main flaw and limitation when features (1) and (2) are used for fault detection, since the applicability and reliability of the results depends on motor load, meaning that low slip is undesirable.

If we still need to examine an induction motor operating under low load, some additional features can be considered. Namely, broken rotor bar will also result in increased vibration amplitude at rotor bar pass frequency $RBPF$

$$RBPF = f_r \cdot Nr \qquad (3)$$

and its sidebands modulated at two times the frequency of the power source f

$$f_{brb2} = RBPF \pm 2f . \qquad (4)$$

In (3), Nr is number of rotor bars [10]. This characteristic feature is not slip-dependent and, combined with previously mentioned features, it can provide more reliable fault detection results in case of low load condition. However, since number of rotor bars varies from 16 to 60 or more, the value of $RBPF$ can be very high, implying the examination

of high frequency domain of vibration spectrum. This demands high sampling rates for signal acquisition which is the reason why $RBPF$ is a parameter not frequently used in fault detection due to hardware-imposed limitations. Still, when motor operates with very low slip and the reliability of low frequency features (1) and (2) is questionable, the observation of vibration amplitudes at $RBPF$ and its sidebands can confirm broken bar diagnosis and augment the overall reliability of the results [10 - 12].

III. SYSTEM DESCRIPTION AND OBTAINED RESULTS

The subject of the present research was to detect fault(s) on a 3.2 MW high-voltage induction motor with fabricated rotor consisting of 56 bar, driving a low- and high-pressure pump in a heating plant, depicted in Fig.1.

Fig. 1. A 3.2 MW high-voltage induction motor driving low- and high-pressure pumps in heating plant, which was the subject of fault detection.

The faulty operation of this particular motor was manifested through high level of vibration and acoustic noise and decreased momentum when operating under load, which implied the presence of broken bar. However, the costs of motor disassembling and rotor removal were considerably high, especially having in mind that periodic maintenance was not in schedule, so it was necessary to determine the nature of the fault and to recommend the procedure for its elimination.

Vibration signals were collected through two shear accelerometers, attached by magnetic mount, as shown in Fig. 2.

Radial and axial vibration signals were recorded, containing 100.000 samples collected with 10 kHz sampling rate. The signals were acquired from faulty motor and from the other motor which was assumed to be healthy, being the same type as the faulty one and driving the same type of pumps.

Fig. 2. Two shear accelerometers with magnetic mounting used for radial and axial vibration signal collection.

The research was conducted in the summer period, when the plant was not in operation. The signals were collected at two different operation point, one with lower load, and the other with higher load level, approximately 35% and 60% of nominal load, respectively. The nominal load could not be achieved, since the heating plant was not in operation. The slip value, measured using stroboscope, was 0.0010 (0.10%) for the first operation point and 0.0016 (0.16%) for the second operation point. The rotation frequencies were calculated to be 49.95 Hz and 49.9 Hz for the first and second operation point, respectively. The number of poles P equals two and the power source frequency f is 50 Hz, which is also equal to synchronous motor speed f_{sync}. Using these data, characteristic features defined by (1) - (4) can be calculated.

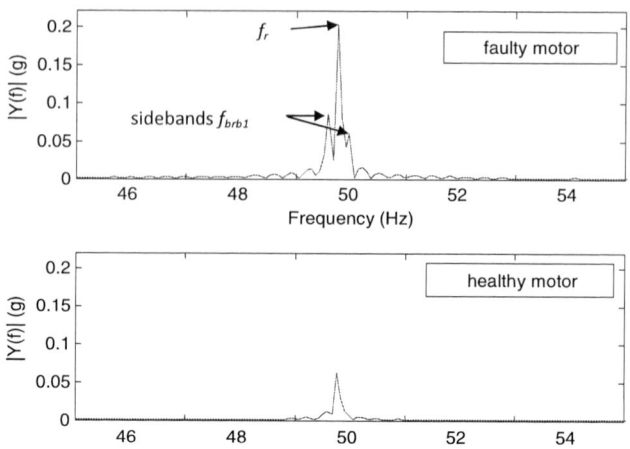

Fig. 4. Vibration spectra of faulty and healthy motor – low frequency domain, higher load level (slip 0.16%)

The time-domain signals were processed using FFT and signal spectra were obtained for both healthy and faulty motor, depicted in Fig. 3, Fig. 4 and Fig. 5.

In Fig. 3, showing low frequency domain vibration spectra of both healthy and faulty motor at the first operation point, one can only note the increased vibration level of faulty motor at rotation frequency. Since the load level is low, sidebands f_{brb1} are not distinguished.

The increased vibration amplitude at the rotation frequency of the faulty motor is notable also in Fig. 4, depicting low frequency domain vibration spectra at the second operation point. Although the slip value is still very low, rotation frequency sidebands f_{brb1} for faulty motor are distinctive, implying the presence of broken bar.

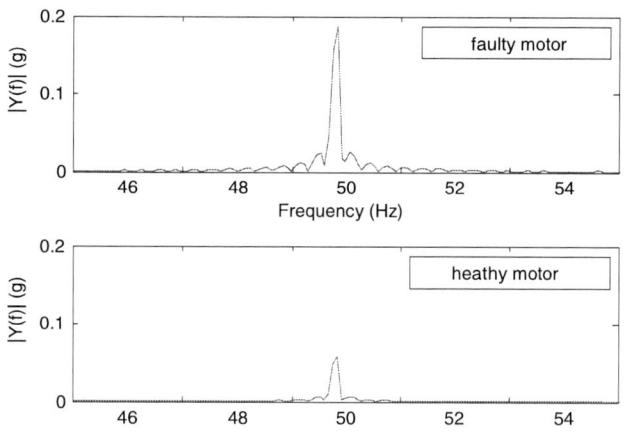

Fig. 3. Vibration spectra of faulty and healthy motor – low frequency domain, lower load level (slip 0.10%)

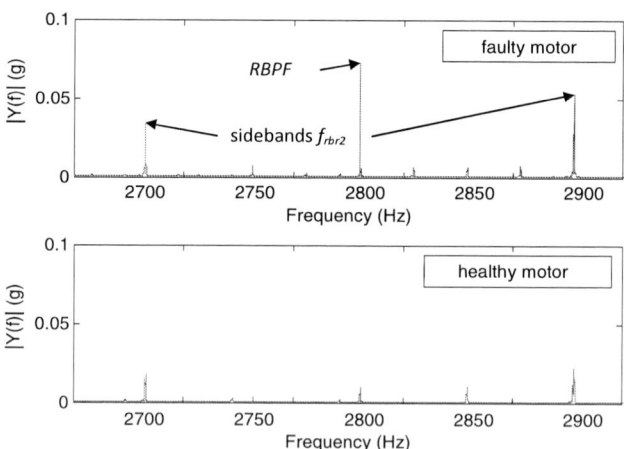

Fig. 5. Vibration spectra of faulty and healthy motor – high frequency domain, higher load level.

Since in both operating points the slip was very low, in order to increase diagnosis reliability, high-frequency domain features were also considered. Observing Fig. 5, one can note the increased vibration amplitude of faulty motor at rotor bar pass frequency *RBPF* and its sidebands f_{brb2}, which confirmed the fault diagnosis and led to the conclusion that at least one rotor bar is broken.

Based on the presented results of fault detection, the suggestion was to disassemble the motor and to remove the rotor in order to be examined in detail, despite high costs of

Fig.6. The rotor after disassembling; two broken rotor bars at the end rings can be observed.

this operation. As shown in Fig. 6, after rotor removal it was determined that two rotor bars were completely broken at their endpoints, which confirmed the results of fault detection procedure.

IV. CONCLUSION

This paper presents a case-study where vibration signal analysis was applied to detect presence of broken rotor bar in high power induction motor. The vibration signals were collected and processed and some characteristic spectral features were observed. Since the motor operates with very low slip, beside some well-known and widely used low-frequency features, some additional features in high-frequency domain were also considered. Based on the results of fault detection procedure, it was suggested to remove the rotor and inspect its condition, which confirmed the fault diagnosis, proving that vibration analysis can be successfully applied in detection of induction motor rotor faults.

V. REFERENCES

[1] A.H. Bonnett, G.C. Soukup, "Analysis of rotor failures in squirrel cage induction motors", *IEEE Transactions on Industry Applications* 24 (6) 1124–1130, 1988.

[2] N. Arthur and J. Penman, "Induction Machine Condition Monitoring with Higher Order Spectra," *IEEE Transactions on Industrial Electronics*, vol. 47, no. 5, 1031-104, 2000.

[3] M.E.H. Benbouzid, "A Review of Induction Motors Signature Analysis as a Medium for Faults Detection," *IEEE Transactions on Industrial Electronics*, vol. 47, no. 5, 984-993, 2000.

[4] W.R. Finley, M.M. Hodowanec and W.G. Holter, "An Analytical Approach to Solving Motor Vibration Problems," *IEEE Transactions on Industry Appications*, vol. 36, no. 5, 1467-1480, 2000.

[5] F. Filippetti, G. Franceschini, G. Gentile, S. Meo, A. Ometto, N. Rotondale and C. Tassoni, "Current Pattern Analysis to Detect Induction Machine Non-Rotational Anomalies," In *Proceedings of International Conference on Electrical Machines*, vol. 1, 448-453.

[6] M. R. Mehrjou, N. Mariun, M. H. Marhaban, N. Misron, "Rotor fault condition monitoring techniques for squirrel-cage induction machine—A review", *Mechanical Systems and Signal Processing* 25, 2827–2848, 2011.

[7] M. Pineda-Sanchez,M. Riera-Guasp, J. A. Antonino-Daviu, J. Roger-Folch, J. Perez-Cruz, R. Puche-Panadero, "Instantaneous Frequency of the Left Sideband Harmonic During the Start-Up Transient: A New Method for Diagnosis of Broken Bars", IEEE Transactions on Industrial Electronics, vol. 56 , iss. 11, 4557-44570, 2009.

[8] D. Matić, F. Kulić, M. Pineda-Sanchez, I. Kamenko, "Support vector machine classifier for diagnosis in electrical machines: Application to broken bar", *Expert Systems with Applications*, vol.39, no.10, 8681-8689, 2012.

[9] S. Nandi, R. Bharadwaj, H.A. Toliyat, A.G. Parlos, "Study of three phase induction motors with incipient rotor cage faults under different supply conditions", In *Proceedings of the IEEE Industry Applications Conference, 34th IAS Annual Meeting 1999*, 1922–1928.

[10] G.K. Singh, S. A. S. Al Kazzaz, "Induction machine drive condition monitoring and diagnostic research - a survey", *Electric Power Systems Research* 64, 145 – 158, 2003.

[11] M. Neale, et al., *A Guide to the Condition Monitoring of Machinery*, London: HMSO, 1979.

[12] P.J. Tavner, J. Penman, *Condition Monitoring of Electrical Machines*, UK: Wiley and Sons, 1987.

[13] V. Climente-Alarcon, J. Antonino-Daviu, F.Vedreño-Santos, R. Puche-Panadero, "Vibration Transient Detection of Broken Bars by PSH Sidebands", In *Proceedings of XXth International Conference on Electrical Machines (ICEM) 2012*, 2517-2521.

[14] Ž. Kanović, M. R. Rapaić, Z. D. Jeličić, "Generalized Particle Swarm Optimization Algorithm - Theoretical and Empirical Analysis with Application in Fault Detection" *Applied Mathematics and Computation* 217, 10175-10186, 2011.

VI. BIOGRAPHIES

Željko Kanović was born on July 18th, 1976, in Sombor, Serbia. He received the B.Sc. degree in Mechanical Engineering in 2000, and the M.Sc. and Ph.D. degree in Electrical Engineering in 2007 and 2012, all from the University of Novi Sad, Serbia. Currently, he is a Teaching Assistant at the Computing and Control Department at the same University. His research interest is in the field of global optimization and its application in control, modeling, fault detection, artificial intelligence and engineering design. He is a member of IEEE.
Dragan Matić was born in Novi Sad, Serbia in 1978. He received PhD from Faculty of technical science, University of Novi Sad in 2012. He is involved in lecturing as an assistant professor and currently works on several projects. Fields of interests are: artificial intelligence, system control and fault detection. He is a member of IEEE.
Zoran Jeličić was born in Serbia in 1971. He received the Dipl. Ing., M.Sc., and Ph.D. degrees in electrical and computer engineering from University of Novi Sad, Serbia, in 1995, 1999, and 2002, respectively.
He is currently a Associate Professor in the Department of Computing and Control, and he is Head of Automatic Control Group.
His research interests include optimal control, nonlinear control, optimization, and fractional calculus. He has published more than 100 scientific papers in fields of interest.
Milan Rapai was born in Ruma, Serbia, on November 17, 1982. In 2006 he received Master's degree in electrical and computer engineering from the University of Novi Sad in Novi Sad, Serbia. He received PhD degree from the same university in 2011. Currently he holds the position of Assistant

professor at Computing and Control Department, with the Group for Automatic Control. His primary research interests are global optimization, optimal control theory and fractional systems. However, recently he has also done research in the field of self-adaptive systems for fault detection and isolation.

Boris Jakovljević was born in Novi Sad, Serbia, on July22, 1982. He received his MSc degree in automatic control and system engineering from the Faculty of Technical Science, University of Novi Sad, Serbia, in 2007. The same year he enrolled in PhD studies at the same university.

In 2007, he joined the Department of Automatic and Control System, University of Novi Sad, as a Teaching Assistant, where he is currently employed.

His current research interests include optimization, control systems and fractional calculus, with over 10 technical publications in the fields of interest. He is currently IEEE student member.

Mirna Kapetina was born in Sarajevo, Bosnia and Herzegovina, on November 27, 1988. She received her Bachelor's degree in electrical and computer engineering from the University of Novi Sad, Novi Sad, Serbia, in 2011. and her MSc degree in automatic control and system engineering from the Faculty of Technical Science, University of Novi Sad, Serbia, in 2012. The same year she enrolled in PhD studies at the same university.

In 2012, she joined the Department of Automatic and Control System, University of Novi Sad, as a Teaching Assistant.

Diagnosis of Induction Motor Faults using a DSP and Advanced Demodulation Techniques

M. Pineda-Sanchez, J. Perez-Cruz, J. Roger-Folch, M. Riera-Guasp, A. Sapena-Baño, R. Puche-Panadero

Abstract – On-line diagnosis of induction motors faults requires special, high speed hardware, such as DSP or FPGAs. Practical implementation of diagnosis algorithms in such a device must take into account the limited amount of memory available for storing sampled data, and for performing spectral analysis using the FFT. Another practical problem is the need to filter the mains component, whose leakage can hide fault harmonics, prior to compute the FFT of the current's signal. This requires the use of digital filters, that must be tuned in case of using variable speed drives that can operate the motor at different speeds. In this paper, an advanced demodulation technique that is able to eliminate the mains component with an extremely low memory requirement, based on the Teager-Kaiser energy operator, is presented. The demodulated current is footprint is down sampled, so that only 2kb of memory are needed to perform the diagnosis process. The proposed method is implemented in a DSP commercial board online diagnosis system and tested on commercial induction motors with broken bars. Finally, the results are compared with the results obtained offline using conventional Motor Current Signature Analysis method.

Index Terms—Broken bar rotor faults, demodulation, DSP, signal analysis, fault diagnosis, induction machine, motor current signature analysis.

I. INTRODUCTION

INDUCTION machines are the horse-work of the industry; they are spread in all of the processes and the faults in these electrical machines could cause extremely costly losses. So, diagnosis system that are able to assess the motor condition working online are continuously growing in number and importance.

The reference method used to detect faults in electrical machines is the Motor Current Signature Analysis method (MCSA), which has been extensively treated in the technical [1-2]. It based on the detection of spectral components that appear in the stator current caused by the specific faults. The characteristic frequencies of these harmonic have been obtained analytically for different types of motor faults. This method has been applied successfully to the diagnosis of rotor bar breakages [3-5], eccentricity [6-8], bearing damages [9-12], inter-turn short circuits [13-16] gearbox and load faults and power supply asymmetries [17-18].

One of the main reasons of the success of MCSA is its simplicity: only one phase current of the motor, measured with an external clamp, is needed to implement the diagnostic of the machine. Besides, a standard FFT suffices to detect the spectral components associated to each type of fault. The diagnosis can be implemented on-line, continuously monitoring the appearance of fault harmonics, or off-lien, storing the sampled current and performing its spectral evaluation on a computer running signal analysis software.

However, MCSA has practical drawbacks. One of the main problems is the spectral leakage, which is the result of the use of a finite-time window for sampling the motor's current. The energy of the mains frequency spreads over the other frequencies and can hide the sideband component. In order to avoid this problem different approaches have been proposed. In [19-21] a smoothing window is applied to the signal before performing the spectral analysis. Although window functions reduce side lobe leakage, their application results in a decrease of the spectral resolution, due to the broadening of the spectral characteristics. Notch filters [22-23], band-pass filters [24] or Wiener adaptive filters [25] have been also used to filter out the main supply harmonic, which is considered as "noise" from the point of view of fault detection.

A different approach to solve the leakage problem has been to analyze the current's envelope instead of the current signal. The envelope retains all the necessary diagnostic information, and transforms the supply harmonic into a dc component which can be easily removed [4]. But this approach requires either a sampling or computing intensive process, departing from the simplicity that has made MCSA an industrial choice.

In this paper, a new and simple way to eliminate the influence of the main supply harmonic before performing the spectral analysis of the current is presented. The proposed method is based on an extremely short non-linear filter, the Teager-Kaiser Energy Operator (TKEO) [26-27]. It operates on just three consecutive samples of the current, which makes it virtually instantaneous. This operator, transforms the main supply harmonic into a constant component, which can be easily eliminated and, in addition, compared to [4], [28-29], the computational cost and hardware requirements are substantially reduced. This feature makes this technique especially well suited for its implementation in low-memory

---

This work was supported by the Spanish "Ministerio de Ciencia e Innovación" in the framework of the "Programa Nacional de proyectos de Investigación Fundamental" (project reference DPI2011-23740).

M. Pineda-Sánchez (mpineda@die.upv.es), J. Pérez-Cruz (juperez@die.upv.es), J. Roger-Folch (jroger@die.upv.es), M. Riera-Guasp (mriera@die.upv.es), A. Sapena-Baño (ansaba2@upvnet.upv.es), R. Puche-Panadero(rupucpa@die.upv.es) are with the Electrical Engineering Department of Universitat Politècnica de València (UPV), Spain.

978-1-4799-0024-4/13 $31.00 © 2013 IEEE

devices, such as standard DSP components. Another advantage of the proposed approach is its ability to display the fault harmonics right at their characteristic frequencies, instead of bands around the supply frequency, as in classical MCSA, which improves the fault detection capabilities of the diagnosis system.

This paper is structured as follows. Section III introduces the mathematical definition of the Teager-Kaiser energy operator. In Section IV, the practical test bed is shown. In Section V, the DSP algorithm is outlined. In section VI, the experimental verification of the proposed system is made, and the experimental results are shown. Section VII presents the conclusions

II. THE TEAGER-KAISER ENERGY OPERATOR

The first used of the TKEO operator is linked to the studies on speech [30-32] , in which modulation plays an important role [31]. It has been applied mainly in communications [33], for estimation of the instantaneous frequency of transient signals [34-35], for detection of power supply oscillations [36], and also, in the field of electrical machines diagnostic, for the diagnosis of bearing faults [37-38].

A. Mathematical definition

The continuous form of the TKEO, applied to a continuous time signal $x(t)$, is

$$\psi(x(t)) = \dot{x}(t)^2 - x(t) \cdot \ddot{x}(t) \qquad (1)$$

where $\dot{x}(t) = dx / dt$.

To apply (1) to a discrete signal (such as the sampled motor current), which is obtained by sampling a continuous one with a frequency f_s,

$$x[n] = x(n\Delta t) \quad with \ \Delta t = 1/f_s, \ n = 0,1,2,..... \qquad (2)$$

backward approximation of the time derivatives in (1) is used, so that

$$\psi(x[n]) = \left(\frac{x[n]-x[n-1]}{\Delta t} \right)^2 - x[n] \cdot \frac{x[n]-2 \cdot x[n-1]+x[n-2]}{\left(\Delta t\right)^2} =$$

$$= \frac{1}{\left(\Delta t\right)^2} \left(x[n-1]^2 - x[n-2] \cdot x[n] \right). \qquad (3)$$

The expression (3) is normally scaled and centered, yielding

$$\psi(x[n]) = x[n]^2 - x[n-1] \cdot x[n+1]. \qquad (4)$$

Since only three samples are required in (4), the response of the TKEO is almost instantaneous. Besides, it can be easily implemented in DSP processors, because of its extremely low requirements of memory storage.

B. The effect of the TKEO on the current of a healthy and faulty machine

In (5) the phase current of an ideal, healthy machine is purely sinusoidal

$$i_{Healthy}(t) = I_m \cos(\omega t) = I_m \cos(2\pi f t) \qquad (5)$$

where f is the supply frequency (50/60 Hz). Direct application of the TKEO (1) to (5) gives:

$$\psi(i_{Healthy}(t)) = I_m^2 \omega^2 \qquad (6)$$

which is a constant value, corresponding to the current's signal envelope.

In the case of periodic disturbances, like the one produced by a broken bar, the amplitude of the phase current is modulated by the principal frequency f_0 characteristic of the fault; thus the phase current of an ideal machine suffering one of these faults can be characterized by:

$$i_{Faulty}(t) = i_{Healthy}(t) \cdot [1 + \beta \cos(\omega_0 t)] \qquad (7)$$

where β denotes the modulation depth (modulation index) and $\omega_0 = 2\pi f_0$. The value of f_0 has been established theoretically $f_0 = 2 \cdot s \cdot f$, where f is the current main component frequency, s is the slip.

The current of the faulty machine (7) can also be expressed, using (5) as

$$i_{Faulty}(t) = I_m cos(\omega t)$$
$$+ \frac{I_m\beta}{2} [cos(\omega + \omega_0)t + cos(\omega - \omega_0)t] \qquad (10)$$

which shows the fault harmonics appearing as characteristic sideband spectral lines around the main supply harmonic.

The application of the TKEO (1) to the current in the faulty machine (10) gives

$$\psi\left(i_{Faulty}(t)\right) = I_m^2 \omega^2 \cdot \left[1 + \beta \left[\frac{1}{2} + \frac{1}{4}a[cos(2\omega - \omega_0 t + cos 2\omega + \omega_0 t + 2 + a2 cos \omega_0 t + \beta 2[a \cdot cos 2\omega t + cos 2\omega_0 t]\right]\right] \qquad (11)$$

where a is $(\omega_0/\omega)^2$.

The amplitude of the fault harmonics is usually very low compared to the amplitude of the main current component ($\beta \ll 1$). For example, the harmonic corresponding to a bar breakage is around 35-45 dB lower than the fundamental. Therefore, neglecting in (11) the terms which are multiplied by β^2, it gives:

$$\psi\left(i_{Faulty}(t)\right) = I_m^2\omega^2 \cdot \left[1 + \frac{\beta}{4}[2 + a[cos(2\omega - \omega_0)t + cos(2\omega + \omega_0)t] + (8 + 2a)cos(\omega_0 t)]\right] \qquad (12)$$

In (12) three components appear regarding to:

- The harmonic produces by the main supply like a constant term.
- The fault frequency which is a oscillating term.
- Two sideband terms around twice the supply frequency.

C. Application of the TKEO to the diagnosis of motor faults

The dominant term in the TKEO demodulated current (12) is the constant term that corresponds to the supply harmonic. However, it can be removed very easily, simply subtracting to (12) its mean value. Therefore, the proposed diagnostic methodology relays on the analysis of a new signal, $i_{TK}(t)$, derived from the original signal $i(t)$ through the TKEO and defined as:

$$i_{TK}(t) = \frac{\psi(i(t)) - \overline{\psi(\iota(t))}}{\overline{\psi(\iota(t))}} \tag{13}$$

where $i_{TK}(t)$ is the AC component of the function $\psi(i(t))$, normalized dividing it by its DC component.

For a healthy machine, taking into account (6), it results that $i_{TK,Healthy}(t) = 0$, whereas in the case of faulty machine, from (12) results:

$$i_{TK,Faulty}(t) = \frac{\beta}{4} [a[\cos(2\omega - \omega_0)t + \cos(2\omega + \omega_0)t] + (8 + 2a)\cos(\omega_0 t)] . \tag{14}$$

As it is shown in equation (14), the diagnostic function $i_{TK}(t)$ corresponding to a faulty machine contains:

- a component at the characteristic frequency of the fault ω_0, which is not perturbed by the spectral leakage of the main harmonic, and whose amplitude is proportional to the amplitude of the original fault component in the stator current, so it can be used to characterize the presence and the severity of the fault.
- two symmetric sideband components around a frequency that is twice that of the supply, which can be used to further asses the fault diagnostic.

As the fundamental component has been suppressed in the diagnostic function $i_{TK}(t)$, the fault related harmonics can be detected with the spectrum of $i_{TK}(t)$ more easily than using the spectrum of the original signal $i(t)$.

Finally, based on the previous reasoning, an improved MCSA approach using the TKEO is proposed. The approach consists of 3 steps (see Fig. 1):

1) The TKEO algorithm (4) is applied to the tested signal $i(t)$ and then the diagnostic signal $i_{TK}(t)$ is calculated through (13)
2) The spectrum of $i_{TK}(t)$ is calculated through the conventional FFT
3) In the last step, the spectrum of $i_{TK}(t)$ is evaluated, looking for the characteristic fault components given in Table I, which are present in (12) with frequencies f_0 and $(2f \pm f_0)$.

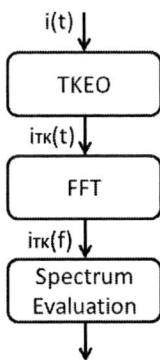

Fig. 1. Proposed approach: demodulation of the phase current by TKEO, followed by the FFT computation and spectrum evaluation.

Diverse methods based on the analysis of the current's envelope suppressed easily the DC component which represents the transformation of the main supply harmonic, prior to the spectral analysis.

III. TEST BED

The proposed method in this paper has been implemented in a DSP-system and it is experimentally validated on five commercial motors with the same characteristics (given in the Appendix). Two of the motors are healthy ones, and the other three have different number of broken bars (one, four and seven). Fig. 2 shows rotors with the one and four broken bars, which have been artificially broken by drilling the bars.

Fig. 2. Different rotors with one broken bar and four broken bars.

The test bed consists in two motors of the same reference coupled by the shaft. One of them, the motor to be diagnosed, operated in motor regime, and the other one in generator regime, thus acting as a regulated load. Both of them are fed by a variable speed drive (VSD). However, the generator needs a brake resistor in its VSD for dissipating the regenerative energy. The block diagram of the test-bed is shown in Fig. .

In all the tests, one of the healthy machine acts like a generator and the other four motors (healthy, one, four and seven broken bars) act like a motor in each case. In Fig. 1 the experimental test-bed is shown.

Besides the test bed is prepared for measuring and recording in parallel the currents and the speed of the machine in a digital oscilloscope, Yokogawa DL750. Thus, the traditional methods to analyze offline could be implemented in a personal computer and later it could be compared with the previous online results obtained with the DSP. The effectiveness of the algorithm implemented in a DSP is possible to verify with this system, which it is only necessary in the lab for checking the system based on a DSP. So, this parallel system is using to calibrate the DSP system.

At the end, more time is necessary to acquire with the traditional methods because when the test motor is unloaded or with low load, the traditional methods in steady state needs 100 seconds to detect the faulty, in other case is possible not detecting the fault.

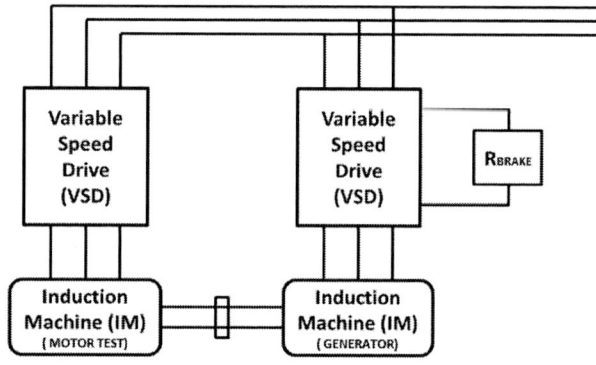

Fig. 3. Block diagram of the test bed.

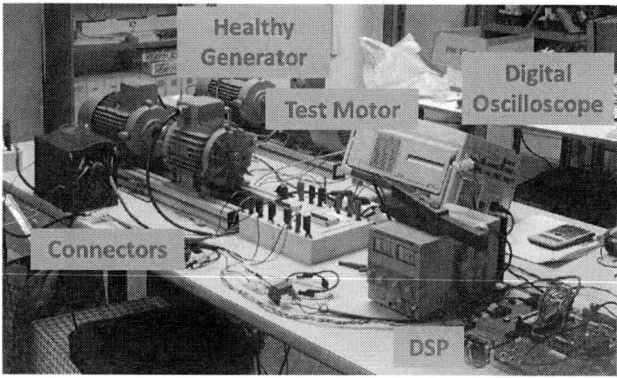

Fig. 1- The test-bed used for the experimental work.

Finally, the DSP board is shown in Fig. and it incorporate a digital output as a led which indicates the status of the induction machine tested. The red light is brightening when the motor has a broken bar.

The test cases evaluated are from 10 to 50 Hz in the part of the motor and from the part of the load from 0 to 6-7% of slip. And in every case at least 100 seconds of a phase current have been sampled at 10 kHz in the case of the digital oscilloscope and in the part of the online DSP system three cycles are evaluating.

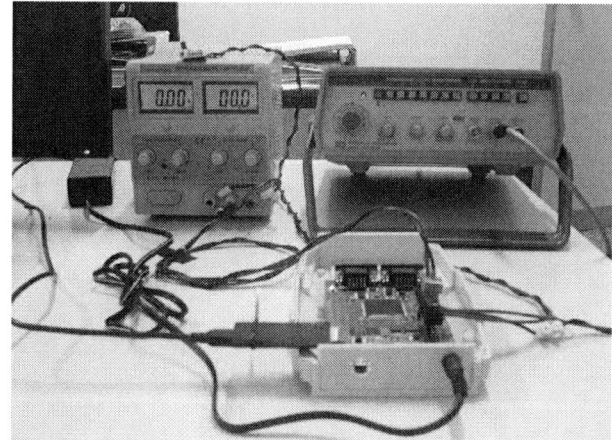

Fig. 5. eZdsp™ platform for TMS320F28335 development, used for the TKEO-based diagnostic method implementation.

IV. DSP ALGORITHM

The DSP algorithm implemented is shown in Fig. 6. This algorithm is very fast and it executes at least three times before signalling the induction motor as a faulty one, to increment the reliability of the system.

The whole process is running on-line during the normal work of the induction machine in steady state regime and independent of the frequency of the supply (the grid or the VSD).

A. Pre-processing stage

The pre-processing stage consists of configuring the analogical channel of the DSP, for adapting the levels of voltage from the clamp to the values supported by the DSP. Also the digital input that is dedicated to the measurement of the motor's speed, using a 200 pulse/rev incremental encoder, is configured at this stage.

B. Processing stage

The processing stage is based on sampling and scaling the input signals:

- The speed must be computed by translating the pulses of the incremental encoder to rpm values.
- The current signal sampled from the current's clamp is scaled and filtered, using the TKEO algorithm presented in section II to eliminate the

fundamental component. This signal is further down sampled to reduce the need of memory storage, so that 10 seconds of sampled current can be stored in 2 Kb of the internal DSP memory.

C. Analysis stage

At this stage, the slip of the motor is calculated from the real speed and from the fundamental component of the current. At the same time, the current samples are processed with the FFT and converted into the frequency domain for its analysis.

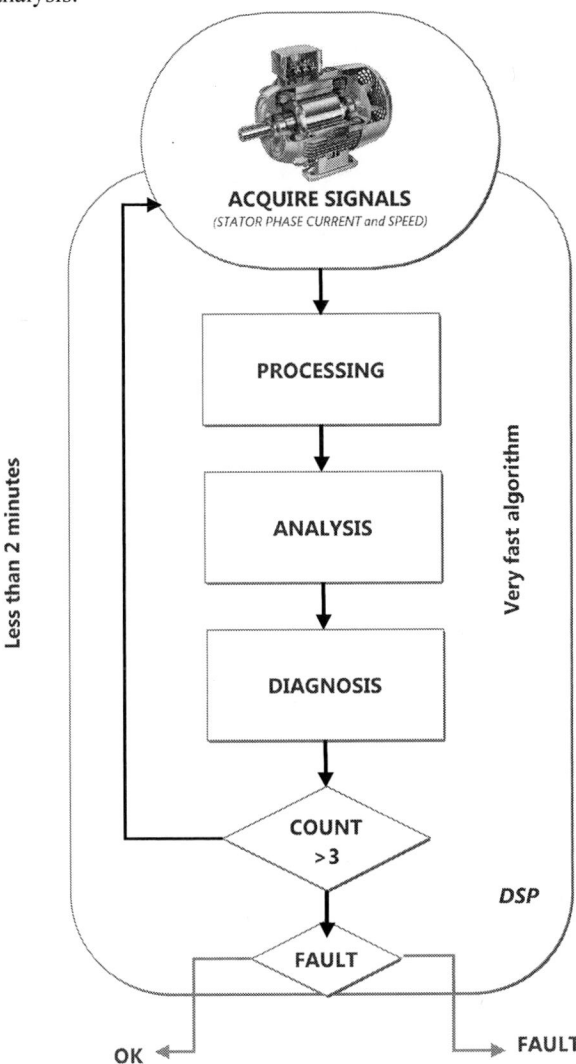

Fig. 6. Algorithm implemented in the DSP.

D. Diagnosis stage

When the Fourier transform of the TKEO current is implemented and the slip is obtained, the diagnosis stage starts. First, the theoretical position of the peak associated to the fault is computed. Second, the spectrum is scanned to identify the presence of such fault harmonics, focusing the

search in a narrow frequency band around the theoretical position of the harmonic.

V. EXPERIMENTAL RESULTS

This system has been validated experimentally using the test bed described in Section III, under different working conditions. Different supply frequencies have been applied to the IM, in the range of 10-50 Hz, working under different load levels, from no-load test up to rated load. The tests performed are shown in Table 1.

In all the tests, the DSP system diagnosed accurately the motor condition, either faulty or healthy, giving the indication through two leds connected to the digital outputs of the development board.

TABLE 1
TEST PERFORMED TO VALIDATE THE DSP BASED DIAGNOSIS SYSTEM

Test	Reference Speed Motor (Hz)	Reference Speed Generator (Hz)	Speed (rpm)
A1	50	OFF	1496
A2	50	49	1480
A3	50	48	1463
A4	50	47	1447
A5	50	46	1424
B1	40	OFF	1196
B2	40	39	1182
B3	40	38	1166
B4	40	37	1150
B5	40	36	1130
C1	30	OFF	897
C2	30	29	882
C3	30	28	865
C4	30	27	846
C5	30	26	825
D1	20	OFF	595
D2	20	19	581
D3	20	18	563
D4	20	17	542
D5	20	16	520
E1	10	OFF	293
E2	10	09	278
E3	10	08	258
E4	10	07	237
E5	10	06	214

To further assess the performance of the developed system, a parallel analysis has been performed off-line. Fig. 7 shows the off-line analysis of the sampled current in the test A4, obtained with the proposed method, but using all the current samples (10 seconds of the motor's current, sampled at 10 kHz). Finally, Fig. 8 shows the spectrum obtained with the DSP running the algorithm of Fig. 6, using the down sampled i_{TK} signal. The fault harmonic is clearly visible in this spectrum, as confirmed by the output of the diagnosis system.

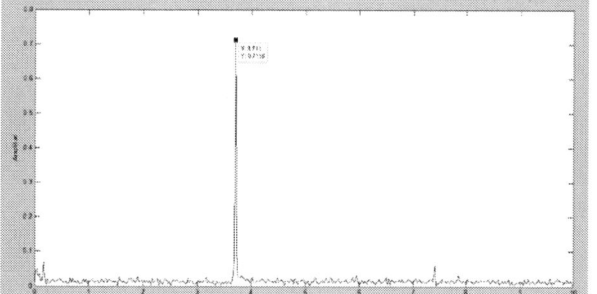

Fig. 7. Spectrum of the TKEO current of the faulty machine, processed by the off-line system.

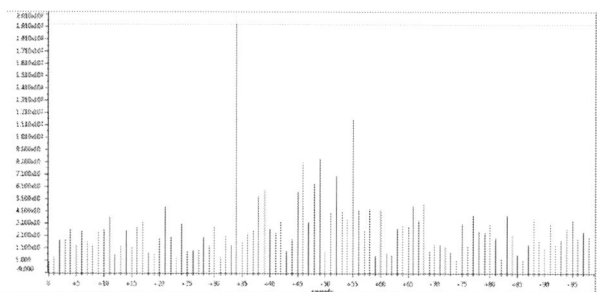

Fig. 8. Spectrum of the TKEO current of the faulty machine, processed by the DSP system.

VI. CONCLUSIONS

The algorithm implemented in the DSP, presented in this paper, is very simple and fast. The use of a Teager-Kaiser for demodulation of the current offers two major advantages: on the one hand, it filters the mains component, thus avoiding the leakage effect, independently of the supply's frequency, which makes it especially suitable for motors fed from a VSD; on the other hand, the memory space that the DSP needs to operate and storage data acquire from the speed and stator current can be reduced only 2kb, due to the extremely low requirements of the proposed algorithm, because he demodulation technique only needs three consecutive samples to operate.

This system has been experimentally validated with different tests using healthy and faulty machines, operating at different frequencies and load levels.

Future works will address the validation of the proposed method to detect other types of faults such as eccentricities, damage bearings, and inter-turn short circuits.

VII. FUTURE WORKS

Due to the good results obtained in this work the following step would be extended to other working regime, for example in a transient regime (start-up).

Finally the integration with the frequency converter seems

inmediately because the algorithm is not heavy from the programming point of view.

VIII. APPENDIX

The characteristics of the motors tested are: Star connected, rated voltage (U_n) 400V, rated power (P_n) 1.1 kW, 2 pole pairs, stator rated current (I_{1n}) 2.55A, rated speed (n_n) 1415 rpm, cos φ =0.81. The motor model and series are: Siemens 1LA7090-4AA10 (Fig. 9).

Fig. 9. Plate of the motor tested.

IX. REFERENCES

[1] O. Poncelas, J. Rosero, J. Cusido, J. Ortega, and L. Romeral, "Motor fault detection using a Rogowski sensor without an integrator," Industrial Electronics, IEEE Transactions on, vol. 56, no. 10, pp. 4062–4070, Oct. 2009.

[2] A. Bellini, F. Filippetti, C. Tassoni, and G. A. Capolino, "Advances in diagnostic techniques for induction machines," Industrial Electronics, IEEE Transactions on, vol. 55, no. 12, pp. 4109–4126, Dec. 2008.

[3] M. Riera-Guasp, M. Cabanas, J. Antonino-Daviu, M. Pineda-Sanchez, and C. Garcia, "Influence of nonconsecutive bar breakages in motor current signature analysis for the diagnosis of rotor faults in induction motors," Energy Conversion, IEEE Transactions on, vol. 25, no. 1, pp. 80–89, March 2010.

[4] R. Puche-Panadero, M. Pineda-Sanchez, M. Riera-Guasp, J. Roger-Folch, E. Hurtado-Perez, and J. Perez-Cruz, "Improved resolution of the MCSA method via Hilbert transform, enabling the diagnosis of rotor asymmetries at very low slip," Energy Conversion, IEEE Transactions on, vol. 24, no. 1, pp. 52–59, March 2009.

[5] A. Bellini, A. Yazidi, F. Filippetti, C. Rossi, and G. A. Capolino, "High frequency resolution techniques for rotor fault detection of induction machines," Industrial Electronics, IEEE Transactions on, vol. 55, no. 12, pp. 4200–4209, Dec. 2008.

[6] T. Ilamparithi and S. Nandi, "Detection of eccentricity faults in three-phase reluctance synchronous motor," Industry Applications, IEEE Transactions on, vol. 48, no. 4, pp. 1307–1317, 2012.

[7] C. Concari, G. Franceschini, and C. Tassoni, "Toward practical quantification of induction drive mixed eccentricity," Industry Applications, IEEE Transactions on, vol. 47, no. 3, pp. 1232–1239, 2011.

[8] D. Morinigo-Sotelo, L. Garcia-Escudero, O. Duque-Perez, and M. Perez-Alonso, "Practical aspects of mixed-eccentricity detection in PWM voltage-source-inverter-fed induction motors," Industrial Electronics, IEEE Transactions on, vol. 57, no. 1, pp. 252–262, Jan. 2010.

[9] E. Bouchikhi, V. Choqueuse, and M. Benbouzid, "Current frequency spectral subtraction and its contribution to induction machines' bearings condition monitoring," Energy Conversion, IEEE Transactions on, vol. 28, no. 1, pp. 135–144, 2013.

[10] L. Frosini and E. Bassi, "Stator current and motor efficiency as indicators for different types of bearing faults in induction motors,"

Industrial Electronics, IEEE Transactions on, vol. 57, no. 1, pp. 244–251, Jan. 2010.

[11] A. Knight and S. Bertani, "Mechanical fault detection in a medium-sized induction motor using stator current monitoring," *Energy Conversion, IEEE Transactions on*, vol. 20, no. 4, pp. 753–760, Dec. 2005.

[12] M. Blodt, M. Chabert, J. Regnier, and J. Faucher, "Mechanical load fault detection in induction motors by stator current time-frequency analysis," Industry Applications, IEEE Transactions on, vol. 42, no. 6, pp. 1454–1463, Nov.-Dec. 2006.

[13] A. Gandhi, T. Corrigan, and L. Parsa, "Recent advances in modeling and online detection of stator interturn faults in electrical motors," Industrial Electronics, IEEE Transactions on, vol. 58, no. 5, pp. 1564–1575, 2011.

[14] S. Grubic, J. Aller, B. Lu, and T. Habetler, "A survey on testing and monitoring methods for stator insulation systems of low-voltage induction machines focusing on turn insulation problems," Industrial Electronics, IEEE Transactions on, vol. 55, no. 12, pp. 4127–4136, Dec. 2008.

[15] J. Seshadrinath, B. Singh, and B. Panigrahi, "Single-turn fault detection in induction machine using complex-wavelet-based method," Industry Applications, IEEE Transactions on, vol. 48, no. 6, pp. 1846–1854, 2012.

[16] S. Das, P. Purkait, D. Dey, and S. Chakravorti, "Monitoring of inter-turn insulation failure in induction motor using advanced signal and data processing tools," Dielectrics and Electrical Insulation, IEEE Transactions on, vol. 18, no. 5, pp. 1599–1608, 2011.

[17] S. M. A. Cruz, "An active-reactive power method for the diagnosis of rotor faults in three-phase induction motors operating under time-varying load conditions," Energy Conversion, IEEE Transactions on, vol. 27, no. 1, pp. 71–84, 2012.

[18] W. Sleszynski, J. Nieznanski, and A. Cichowski, "Open-transistor fault diagnostics in voltage-source inverters by analyzing the load currents," Industrial Electronics, IEEE Transactions on, vol. 56, no. 11, pp. 4681–4688, Nov. 2009.

[19] M. Aiello, A. Cataliotti, and S. Nuccio, "An induction motor speed measurement method based on current harmonic analysis with the chirp-z transform," Instrumentation and Measurement, IEEE Transactions on, vol. 54, no. 5, pp. 1811–1819, Oct. 2005.

[20] S. Kia, H. Henao, and G. A. Capolino, "Torsional vibration effects on induction machine current and torque signatures in gearbox-based electromechanical system," Industrial Electronics, IEEE Transactions on, vol. 56, no. 11, pp. 4689–4699, Nov. 2009.

[21] B. Ayhan, H. Trussell, M.-Y. Chow, and M. H. Song, "On the use of a lower sampling rate for broken rotor bar detection with DTFT and AR-based spectrum methods," Industrial Electronics, IEEE Transactions on, vol. 55, no. 3, pp. 1421–1434, March 2008.

[22] P. Zahradnik and M. Vlcek, "Fast analytical design algorithms for FIR notch filters," Circuits and Systems I: Regular Papers, IEEE Transactions on, vol. 51, no. 3, pp. 608–623, 2004.

[23] F. Costa, L. A. L. De Almeida, S. Naidu, and E. R. Braga-Filho, "Improving the signal data acquisition in condition monitoring of electrical machines," Instrumentation and Measurement, IEEE Transactions on, vol. 53, no. 4, pp. 1015–1019, Aug. 2004.

[24] S. A. S. A. Kazzaz and G. Singh, "Experimental investigations on induction machine condition monitoring and fault diagnosis using digital signal processing techniques," Electric Power Systems Research, vol. 65, no. 3, pp. 197–221, 2003.

[25] W. Zhou, T. Habetler, and R. Harley, "Bearing fault detection via stator current noise cancellation and statistical control," Industrial Electronics, IEEE Transactions on, vol. 55, no. 12, pp. 4260–4269, Dec. 2008.

[26] D. Dimitriadis, A. Potamianos, and P. Maragos, "A comparison of the squared energy and Teager-Kaiser operators for short-term energy estimation in additive noise," Signal Processing, IEEE Transactions on, vol. 57, no. 7, pp. 2569–2581, July 2009.

[27] D. Vakman, "On the analytic signal, the Teager-Kaiser energy algorithm, and other methods for defining amplitude and frequency," Signal Processing, IEEE Transactions on, vol. 44, no. 4, pp. 791–797, Apr. 1996.

[28] A. da Silva, R. Povinelli, and N. A. O. Demerdash, "Induction machine broken bar and stator short-circuit fault diagnostics based on three-phase stator current envelopes," Industrial Electronics, IEEE Transactions on, vol. 55, no. 3, pp. 1310–1318, March 2008.

[29] S. M. A. Cruz and A. J. M. Cardoso, "Stator winding fault diagnosis in three-phase synchronous and asynchronous motors, by the extended Park's vector approach," Industry Applications, IEEE Transactions on, vol. 37, no. 5, pp. 1227–1233, Sep/Oct. 2001.

[30] H. Teager, "Some observations on oral air flow during phonation," Acoustics, Speech and Signal Processing, IEEE Transactions on, vol. 28, no. 5, pp. 599–601, Oct. 1980.

[31] P. Maragos, J. Kaiser, and T. Quatieri, "Energy separation in signal modulations with application to speech analysis," Signal Processing, IEEE Transactions on, vol. 41, no. 10, pp. 3024–3051, Oct. 1993.

[32] A. O. Boudraa, J. C. Cexus, and K. Abed-Meraim, "Cross ΨB-energy operator-based signal detection," The Journal of the Acoustical Society of America, vol. 123, no. 6, pp. 4283–4289, 2008.

[33] A. O. Boudraa, S. Benramdane, J. C. Cexus, and T. Chonavel, "Some useful properties of cross ΨB-energy operator," AEU - International Journal of Electronics and Communications, vol. 63, no. 9, pp. 728 – 735, 2009.

[34] A. O. Boudraa, "Instantaneous frequency estimation of FM signals by ψB-energy operator," Electronics Letters, vol. 47, no. 10, pp. 623–624, 2011.

[35] M. Pineda-Sanchez, M. Riera-Guasp, J. Antonino-Daviu, J. Roger-Folch, J. Perez-Cruz, and R. Puche-Panadero, "Instantaneous frequency of the left sideband harmonic during the start-up transient: A new method for diagnosis of broken bars," Industrial Electronics, IEEE Transactions on, vol. 56, no. 11, pp. 4557 –4570, nov. 2009.

[36] I. Kamwa, A. Pradhan, and G. Joos, "Robust detection and analysis of power system oscillations using the Teager-Kaiser energy operator," Power Systems, IEEE Transactions on, vol. 26, no. 1, pp. 323–333, 2011.

[37] H. Li and H. Zheng, "Bearing fault detection using envelope spectrum based on EMD and TKEO," in Fuzzy Systems and Knowledge Discovery, 2008. FSKD '08. Fifth International Conference on, vol. 3, Oct. 2008, pp. 142–146.

[38] P. H. Rodríguez, J. B. Alonso, M. A. Ferrer, and C. M. Travieso, "Application of the Teager-Kaiser energy operator in bearing fault diagnosis," ISA Transactions, vol. 52, no. 2, pp. 278 – 284, 2013.

X. Biographies

Manuel Pineda-Sanchez (M'02) was born in 1962 in Albacete (Spain). He received his Dipl. Ing. and Dr. Ing. degrees in electrical engineering from the Universitat Politecnica de Valencia (Spain) in 1985 and 2004, respectively. He joined the Universitat Politecnica de Valencia in 1987 as Associate Professor in the Department of Electrical Engineering, in the area of theory and control of electrical machines. His research interests include electrical machines and drives, induction motor diagnostics, numerical simulation of electromagnetic fields, and software development.

Juan Perez-Cruz (M'09) obtained his M.Sc. in electrical engineering in 1997 and his Ph.D. in 2006 from the Universitat Politecnica de Valencia. From 1970 to 1992 he worked in the electrical industry in the field of industrial systems maintenance and automation. In 1992 he joined the Universitat Politecnica de Valencia and is currently Associate Professor of electrical installations and machines. His research interests focus on induction motor diagnostics and maintenance, numerical modelling, and automation.

José Roger-Folch (M'03) received the M.Sc. degree from the Universidad Politecnica de Cataluña, Barcelona, Spain, in 1970, and the Ph.D. degree from the Universitat Politecnica de Valencia, Valencia, Spain, in 1980, both in electrical engineering. From 1971 to 1978, he worked in the electrical industry as a Project Engineer. Since 1978, he has been with the Department of Electrical Engineering, Universitat Politecnica de Valencia, where he is currently a Professor of Electrical Installations and Machines. His main research areas are numerical methods (finite-element method and others) applied to the design and maintenance of electrical machines and equipment.

Martín Riera-Guasp (M'94-SM'12) received the M.Sc. degree in Industrial Engineering and the Ph.D. degree in Electrical Engineering from the Universitat Politecnica de Valencia (Spain) in 1981 and 1987, respectively. Currently he is an Associate Professor in the Department of Electrical Engineering of the Universitat Politecnica de Valencia. His research interests include condition monitoring of electrical machines and electrical systems efficiency.

Angel Sapena-Baño obtained his M.Sc in Electrical Engineering and his M.Sc on Renewable and Efficient Energy from the Universitat Politecnica de Valencia (Spain) in 2009 and in 2012, respectively. From 2008 he works as a Researcher in the Electrical Department of Universitat Politecnica de Valencia where he is developing its Ph.D.His research interests focus on induction motor diagnostics and the maintenance based on the condition monitoring, numerical modeling of electrical machines, and advanced automation processes and electrical installation.

Rubén Puche-Panadero (M'09) received his M.Sc. in Automatic and Electronic Engineering in 2003 from the Universitat Politecnica de Valencia. He received his Ph.D. in 2008 in the field of Condition Monitoring of Electrical Machines. From 2003 to 2006 he worked as a PLC programmer and as a developer of the SCADA programs. He joined to the Universitat Politecnica de Valencia in 2006 and he is currently Associate Professor of Control Electrical Machines. His research interests focus on induction motor diagnostics and the maintenance based on the condition monitoring, numerical modeling of electrical machines, and advanced automation processes and electrical installations.

On the Use of Spectral Kurtosis for Diagnosis of Electrical Machines

E. Fournier[1,2], A. Picot[1], J. Régnier[1], M. Tientcheu Yamdeu[2], J-M. Andréjak[2] and P. Maussion[1]

Abstract—This paper explores the efficiency of Spectral Kurtosis (SK) in the area of electrical machines diagnosis. In the literature, Spectral Kurtosis is mainly presented as a tool used to detect non-stationary components in a signal. However, classical use of SK is unsuitable for detection of new stationary components or slow evolutions in a spectrum. In order to detect different types of faults, three indicators are designed from the original definition of the Spectral Kurtosis. These indicators are first tested and compared on synthetic signals. Then, their performance are demonstrated for unbalance detection in a Induction Machine (IM) using current signal.

Index Terms—Spectral Kurtosis, Electrical Machine Diagnosis, Statistical Analysis

I. INTRODUCTION

UNEXPECTED failures in electromechanical systems may lead to important losses of production, safety issues and additional costs. Thus, diagnosis and condition monitoring of electrical machines is an important topic among industrial and academic research. Developing efficient diagnosis systems is more and more essential for manufacturers of electrical machines.

Numerous methods are based on the analysis of vibratory signals [1], [2] and [3]. Their aim is to diagnose electrical, mechanical or load faults occurring in electromechanical systems. However, accelerometers are expensive devices. Therefore, these methods are only profitable for high power systems where accelerometers represent a negligible part of the total cost. So, research has focused on diagnosing machine failures through the analysis of current signals, because these signals are often already available for control purposes. These methods are mainly based on the Motor Current Signature Analysis (MCSA). A review can be found in [4]. Failures happening in electrical machines are generally listed in two groups :

- Localized faults which appears in current spectrum (and in some other quantities) at specific frequencies. This category includes unbalances, eccentricity faults, localized bearing faults (ball, cage and races faults), etc.
- Generalized faults which are not associated to any particular frequencies. This category includes generalized roughness bearing faults.

MCSA focuses on localized faults occurring in an electrical machine. Indeed, these faults produce specific signatures in electro-mechanic quantities (vibration, speed, torque, flux, etc.) which can be tracked in the current spectrum, often computed with the Fast Fourier Transform (FFT), at particular frequencies defined in MCSA. These characteristic frequencies depend on many variables such as rotor speed or bars numbers for example, as shown in [5] and [6].

Recent works have shown that Spectral Kurtosis (SK) is an interesting indicator to monitor and diagnose electrical machines. This statistical tool has been first presented in [7] as a complement of Power Spectrum Density (PSD). The aim of SK is to detect non-stationary components of a signal. Thus, SK is representative of the behavior of the signal spectral components whereas PSD is representative of their energies. This property makes SK an efficient indicator for several types of faults. Indeed, many failures such as localized bearing faults and gear faults are characterized by impulses at fault frequencies predicted in MCSA. These impulses may be considered as non-stationarities and can be detected using SK as presented in [8], [9] and [10] where SK is used to detect localized faults such as bearing faults and gearbox failures. Moreover, SK can also be used to detect generalized-roughness bearing fault, as shown in [11].

Several definitions of SK can be found in the literature. R. Dwyer introduces, in [7], the SK of a signal $x(t)$ as the kurtosis of its spectral components in order to detect non-stationary components. Vrabie et al. in [12], have completed this definition by proposing an unbiased estimator of the SK. Antoni and Randall have proposed a formal definition with a Short Time Fourier Transform (STFT)-based estimator in [13]. More recently, a modified STFT has been presented in [14] in order to use the SK as an efficient Gaussianity test.

However, SK is mostly used on current or vibration signals to detect bearing or gearbox faults which are characterized by non-stationary components. SK, in its classical form, is unsuited to study long terms evolutions in a signal. A different approach is presented in [15] with a new protocol of computation of the SK. Indeed, in [15] the calculation of SK uses several recordings of a signal in order to detect bearing faults in a Permanent Magnet Synchronous Machine (PMSM). Nevertheless there is no study comparing the different definitions of SK and their characteristics.

In this context, this article presents several protocols of computation of SK in order to detect slow dynamic faults in current spectra. In Section II, definition and properties of the classical SK are presented to understand its working. In section III, two protocols of computation of SK are presented to over-

---

1: Université de Toulouse ; INPT, UPS ; LAPLACE (Laboratoire Plasma et Conversion d'Energie) ; ENSEEIHT, 2 rue Charles Camichel, BP 7122, F-31071 Toulouse cedex 7, France. CNRS ; LAPLACE ; F-31071 Toulouse, France. Email: name.surname@laplace.univ-tlse.fr

2: Moteurs LEROY SOMER, 16000 Angoulème, France. Email: name.surname@emerson.fr

---

978-1-4799-0024-4/13 $31.00 © 2013 IEEE

come the limitations of the classical form. The performance of the indicators is tested in Section IV on the detection of a developing unbalance.

II. SPECTRAL KURTOSIS

A. Kurtosis

Let x be a random variable with mean μ and standard deviation σ. The centered reduced variable x_{CR} associated to x is defined as:

$$x_{CR} = \frac{x - \mu}{\sigma} \qquad (1)$$

In high order statistics, the kurtosis of the random variable x is defined in (2) as the fourth *moment* of the centered reduced variable x_{CR}.

$$\tilde{K}(x) = \mathbb{E}[x_{CR}^4] = \mathbb{E}[(\frac{x-\mu}{\sigma})^4] = \frac{\mu_4}{\mu_2^2} \qquad (2)$$

with μ_i the ith central moments of x.

In order to have a coefficient equal to 0 for a Gaussian distribution, a standardized kurtosis may be defined in terms of *cumulants* rather than in terms of *moments*. Cumulants are statistical parameters linked to moments by the following recurrence formula

$$\kappa_n = \mu_n - \sum_{k=0}^{n-1} \binom{k-1}{n-1} \kappa_k \mu_{n-k} \qquad (3)$$

Standardized kurtosis calculated from cumulants is given by

$$K(x) = \frac{\kappa_4}{\kappa_2^2} = \frac{\mu_4 - 3\mu_2^2}{\mu_2^2} = \tilde{K}(x) - 3 \qquad (4)$$

with κ_i the ith cumulants of x.

The standardized kurtosis is used in the rest of this paper because it is much more used in statistics literature [16].

In statistics, the kurtosis is the second shape factor (after the skewness) and reflects the sharpness of a distribution. The sharper is the distribution of x, the higher is its kurtosis value and reciprocally, the flatter is the distribution, the lower is the kurtosis. Fig. 1 shows three examples of distributions and their associated kurtosis.

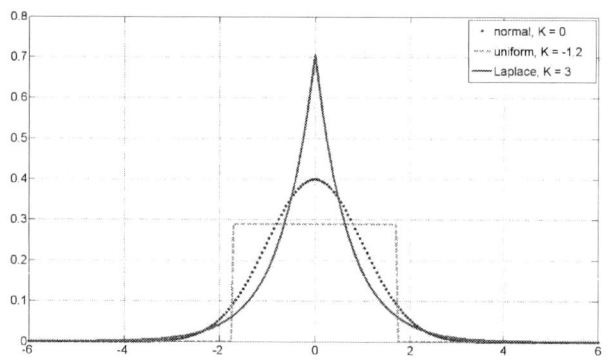

Figure 1. Probability density function of a normal distribution (dotted line), uniform distribution (dashed line), Laplace distribution (solid line)

The uniform distribution, plotted with a light blue dashed line, is very flat and has a kurtosis of -1.2. The normal distribution, plotted with blue dotted line, has a kurtosis of 0. Finally, the Laplace distribution, plotted with a purple solid line, is very sharp and has a kurtosis of 3.

As shown in (1), the kurtosis of a random variable x is defined as the fourth moment of x_{CR}, the centered reduced variable associated to x. Therefore its value is independent from the mean μ and the standard deviation σ of x. It is only dependent of the kind of distribution of x. Thus, two random variables with the same type of distribution (normal, uniform, etc.) but with different mean and standard deviation have the same kurtosis.

B. Spectral Kurtosis

1) Definition: The Spectral Kurtosis, noted SK, of a signal is defined as the kurtosis of its spectral components. It has been shown in [12] that SK is a complement of the classical Power Spectrum Density (PSD) to detect non-stationary components of a signal and that it can be applied as well on the real part, the imaginary part or on the modulus of the signal's spectrum. However, a minimum number of spectra is required to correctly estimate the SK of a signal. In practice, this number is reached using the STFT. The principle of STFT is to split a signal into N segments and compute the FFT on each segment. Thus, a set of local spectrums is obtained on which SK is computed.

2) Example: Let $x(t)$ be a temporal signal and $x[n]$ its associated discrete signal sampled at frequency f_e. Firstly, the discrete signal $x[n]$ is divided in N segments of length L (with or without overlap). Then, FFT is computed on each segment i of the signal in order to create a Time-Frequency representation $X_i(f_k)$ for every frequency slot f_k. The principle of this method is presented in Fig. 2.

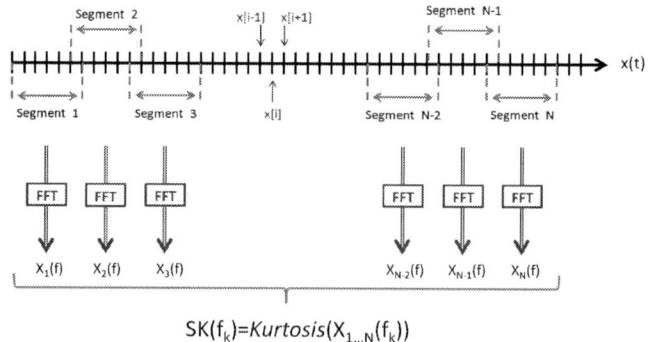

Figure 2. Principle of Spectral Kurtosis computation using STFT

Once all spectra have been computed, the SK of x at frequency f_k is given by

$$SK_x(f_k) = K(|X_1(f_k)|, |X_2(f_k)|, ..., |X_N(f_k)|) \qquad (5)$$

To understand how SK works, we consider a signal x defined by

$$x(t) = sin(2*\pi*0.25*t) + m(t)*sin(2*\pi*0.1*t) + b(t) \qquad (6)$$

with

- $sin(2 * \pi * 0.25 * t)$ is a stationary component of x,
- $m(t) * sin(2 * \pi * 0.1 * t)$ is a non-stationary component where $m(t)$ worth 1 on 2% of the N segments of x and 0 on the rest of the segments,
- $b(t)$ is a Gaussian noise with mean 0 and standard deviation 0.1.

The spectrum of this signal is presented in Fig. 3.a.

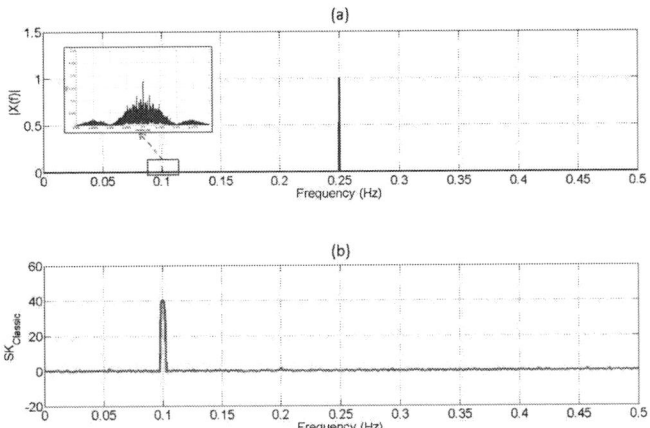

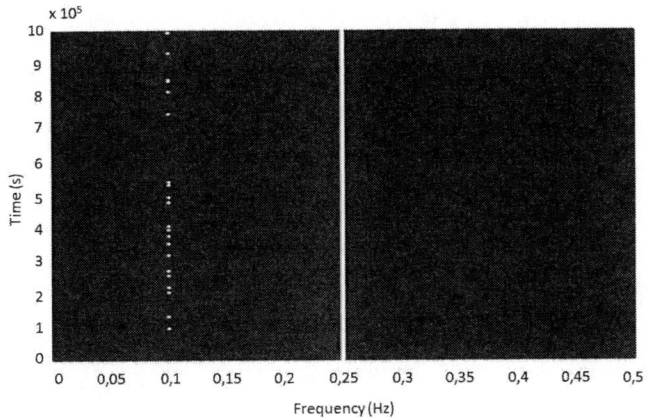

Figure 3. FFT of $x(t)$ (a), SK of $x(t)$ (b)

The stationary sinusoidal component of $x(t)$ is clearly visible at frequency 0.25 Hz with an amplitude of 1 whereas the non-stationary sinusoidal component is barely noticeable at 0.1 Hz with an amplitude about equal to 0.02. FFT computes the mean intensity of spectral components of a signal over a certain amount of time. So, it smoothes components with a low percentage of non-stationary components. A first approach to deal with this problem is to display the Spectrogram of the signal $x(t)$ using STFT. The Spectrogram of $x(t)$ using 1000 windows with no overlap is presented on Fig. 4.

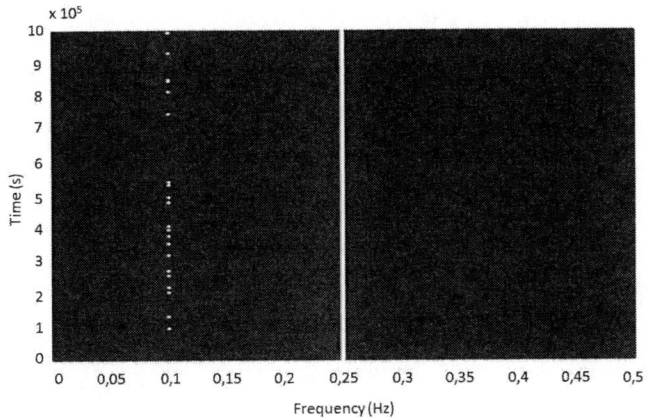

Figure 4. Spectrogram of $x(t)$

It can be seen in this figure that the non-stationary character of the component at 0.1 Hz is clearly visible with this Time-Frequency representation.

Then, it is possible to calculate the SK of x from the STFT computed. The result is represented on Fig. 3.b. As expected,

the SK presents a high value at frequency 0.1 Hz, which is a non-stationary component, and a low value for the other frequencies, which are stationary components. Fig. 5 shows the histograms of the non-stationary component $X(0.1Hz)$ and a stationary component, $X(0.15Hz)$ for example. We notice

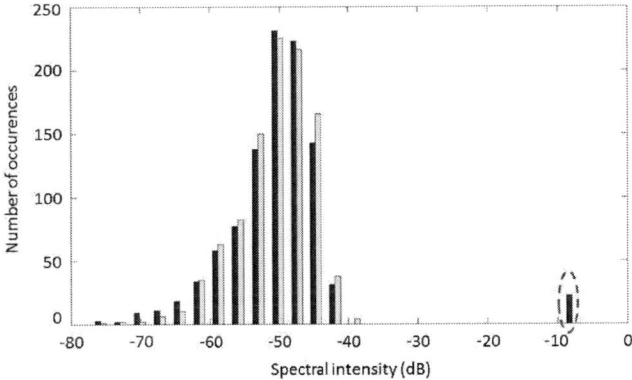

Figure 5. Histogram of $\{20log_{10}(|X_i(0.15Hz)|)\}_{i=1..N}$ (light blue) and $\{20log_{10}(|X_i(0.1Hz)|)\}_{i=1..N}$ (dark blue)

from Fig. 5 that the two histograms have the same overall shape centered on -50 dB. The only difference is the presence of extreme and rare values around -9 dB in the distribution of the non-stationary component (circled in red in Fig. 5). These rare values are responsible for the increase of the Spectral Kurtosis at this particular frequency.

III. SPECTRAL KURTOSIS APPLIED TO ELECTRICAL MACHINE DIAGNOSIS

Let consider the monitoring of an electrical machine in order to assess its health state. Phase currents (or other useful measures) are recorded for a certain amount of time T at regular intervals, once a day for example. Each current recording contains L samples. SK can be computed on each recording to detect non-stationary components over time T. This protocol is represented on Fig. 6.

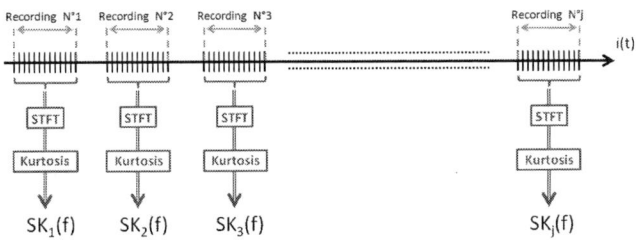

Figure 6. Protocol of computation of classical SK

As shown in section II, this method enlightens non-stationary components which can appear in current signals because of a bearing fault for example. Nevertheless, SK could also be used to detect other kinds of faults such as a developing unbalance or a misalignment, happening after a maintenance operation. In order to simulate such degradations, let consider $x(t)$ a synthetic signal representing the current of an electrical machine. The effect of a degradation occurring in the machine on $x(t)$ can be modelized by

$$
\begin{aligned}
x(t) &= A * cos(2\pi f_s t) \\
&+ A * m(t) * cos(2\pi(f_s - f_r)t + \phi_1) \\
&+ A * m(t) * cos(2\pi(f_s + f_r)t + \phi_2) \\
&+ b(t)
\end{aligned} \tag{7}
$$

This signal is composed of a fundamental component at frequency f_s, two side components at frequencies $f_s - f_r$ and $f_s + f_r$ and a Gaussian noise. The two side components can be representative of an AM or a PM modulation of the current fundamental with a modulation index $m(t)$.

The signal $x(t)$ is composed of 400 $5s$-recordings. There are three different modulation profiles whose are depicted in Fig. 7. These three profiles are representative of a sudden, a slow or a very slow evolution of the spectral components of x at frequencies $f_s \pm f_r$.

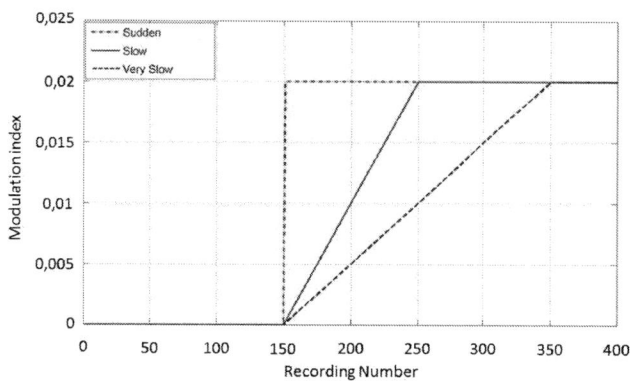

Figure 7. Profiles of modulation index m(t)

SK is computed on each profiles using STFT windows of $1000pts$ with 50 % overlap. The results obtained are presented in Fig. 8.

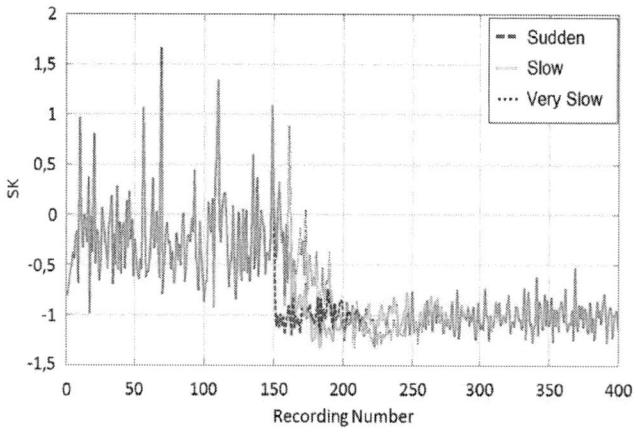

Figure 8. SK for three profiles of $m(t)$

We observe in Fig. 8 that SK decreases from about 0 to -1 after the increase of the modulation index $m(t)$. This result is explained in [12] and shows the changing of a Gaussian noise

to a sine component. However, the variation of SK between a *healthy* and a *faulty* signal is weak. Indeed, SK focuses on a time horizon T and can only detects non-stationarities occurring during this interval of time. A change in the spectrum occurring over a long time or between two successive recordings is a global non-stationarity but the resulting signal will be considered as locally stationary on each recording. Thus, the implementation of SK has to be improved in order to detect long-term evolutions in a spectrum.

A. Spectral Kurtosis with Past

SK can not reflect the non-stationarity of a component over a large scale of time because it focuses on one recording at a time. To overcome this problem, the SK calculation needs to include the previous recordings to be able to react on slow variations of spectral components. This form of SK is called "Spectral Kurtosis with past" and noted SK_{Past}. Past spectrum coefficients are stored in memory. For each new spectrum, the SK_{Past} is calculated on all the spectrums computed from the beginning. The computation principle of SK_{Past} is presented in Fig. 9.

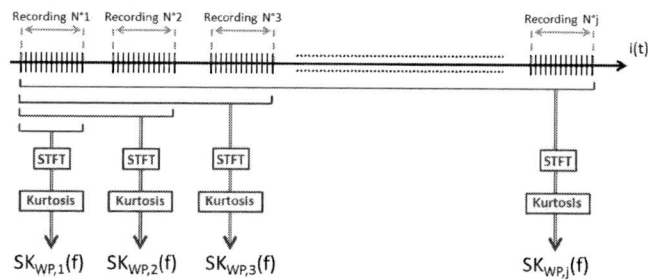

Figure 9. Protocol of computation of SK with Past

The SK_{Past} has been calculated for the three different profiles according to the principle presented in Fig. 9. The length of each STFT window is $L = 5000pts$ with no-overlap. The results of SK_{Past} are presented in Fig. 10.

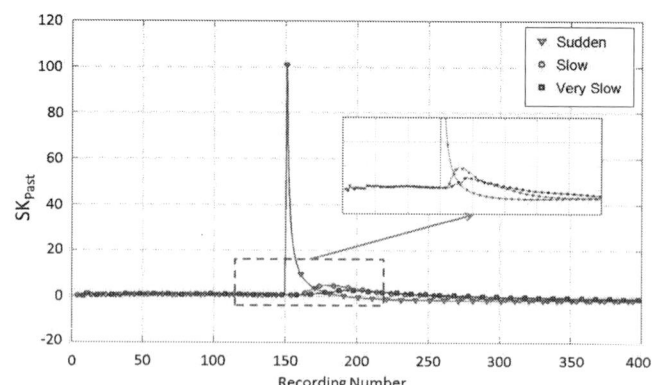

Figure 10. SK_{Past} for three profiles of $m(t)$

First, it is clearly visible in Fig. 10 that SK_{Past} reacts on the sudden, the slow and the very slow evolution of the spectral component $|X(f_s \pm f_r)|$. For the three kinds of faults,

SK_{Past} first increases because *"abnormal"* values appear in the distribution of $|X(f_s \pm f_r)|$. Indeed, the new sample values are *extremes* because they are out of the distribution of $|X(f_s \pm f_r)|$ during the *"normal"* recordings. They are also *rare* because they are much less numerous than all the past *"normal"* samples of $|X(f_s \pm f_r)|$. Then the value of SK_{Past} decreases because these *"abnormal"* components gradually become more and more frequent.

It is obvious in Fig.10 that the more slowly $m(t)$ increases, the lower the maximum value of SK_{Past} is. Another point is that SK_{Past} also depends on the time when the fault occurs. Indeed, let consider two signals with a sudden degradations. In the first signal the evolution occurs after the 150th recording whereas it occurs after the 200th recording in the second signal. SK_{Past} is applied on these two signals with the same protocol as before and results are presented in Fig. 11.

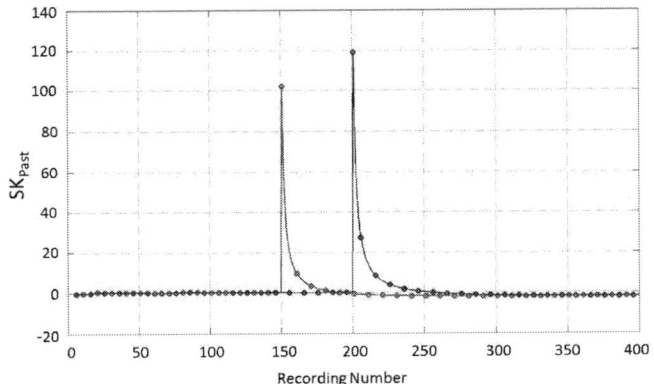

Figure 11. SK_{Past} for sudden evolutions occurring at different times

As shown in Fig. 11 the later the degradation is, the higher the maximum value of SK_{Past} is. Indeed, the new component appearing at the 200th recording is more infrequent than the one appearing at the 151th recording. Thus, its SK_{Past} has a higher value. So, SK_{Past} depends not only on the final value of the modulation index $m(t)$ but also on its dynamic and on the time when the change starts.

All these points make SK_{Past} difficult to interpret. Moreover, the computation time of SK_{Past} strongly increases with the number of recordings and it calculation requires a high storage capacity.

B. Spectral Kurtosis with Reference

In order to create an indicator easier to interpret than the SK with past and to reduce computational cost, a second way of computing spectral kurtosis is proposed in this section. The principle of this spectral kurtosis is to calculate the kurtosis on N recordings composed of :

- N_{ref} recordings of the signal considered as a reference set,
- N_{mov} recordings of the signal,

The first set composed with N_{ref} recordings is fixed whereas the second is a sliding set. These two sets have fixed sizes. Thus, the computational cost of the Spectral Kurtosis

with Reference, noted SK_{Ref}, is set by the length of N_{ref} and N_{mov} and does not increase in time. The computation principle of SK_{Ref} at the jth recording is presented in Fig. 12.

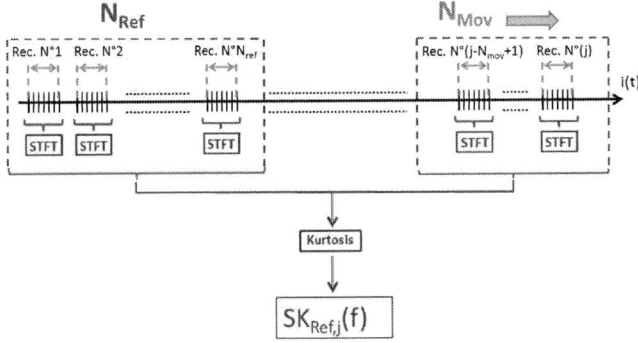

Figure 12. Principle of Spectral Kurtosis with Reference

SK_{Ref} is computed on $x(t)$ for the three modulation profiles. N_{ref} is set to 20 and N_{mov} to 1. The results are presented in Fig. 13. It can be seen in Fig. 13 that SK_{Ref}

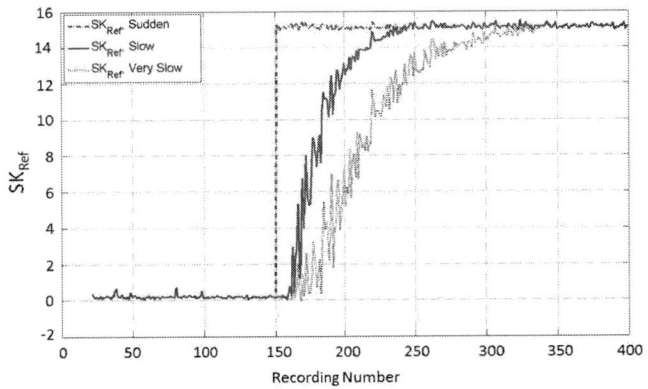

Figure 13. SK_{Ref} at f_r for a sudden degradation (red dash-dotted line), slow degradation (green solid line) and very slow degradation (blue dashed line)

reacts for the three types of faults and its final value is the same in the three cases. This last point shows that SK_{Ref} only depends on the value of the modulation index $m(t)$ and not on its dynamic or on the time when the degradation happens. Furthermore, value of SK_{Ref} does not decrease once the degradation appears but follows the evolution of modulation index $m(t)$.

However, SK_{Ref} has two parameters, $N_{reference}$ and N_{moving}, whose significantly modify its behavior. Several curves of SK_{ref} are plotted in Fig. 14 for different couples $\{N_{ref}, N_{mov}\}$.

It is visible that the behavior of SK_{ref} only depends on the ratio $R = \frac{N_{ref}}{N_{mov}}$. Two couples with the same ratio have the same influence on SK_{Ref}. Furthermore, it can be noticed in Fig.14 that the higher R is, the better SK_{Ref} reacts. Indeed, a change in the distribution appears *marginal* for a small value of N_{mov} whereas if N_{mov} is too high the change goes unnoticed because the new samples are not considered as infrequent.

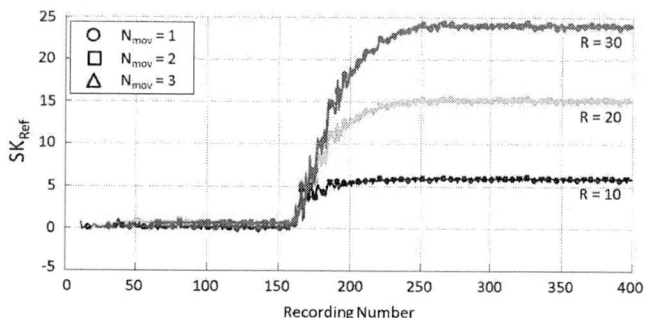

Figure 14. $SK_{Ref}(f_r)$ for a slow degradation with $R = 10$ (black lines), $R = 20$ (green lines) and $R = 30$ (red lines)

Thus, the ratio $R = \frac{N_{ref}}{N_{mov}}$ determines the efficiency of the indicator. A high value of R is required to obtain a good behavior of SK_{Ref}. In order to reduce computational costs and storage capacity required, N_{mov} has to be chosen equal to 1. For example, $R = 20$ can be obtained with different couples $\{N_{ref}, N_{mov}\}$ but the couple minimizing the number of data to be stored is $\{N_{ref} = 20, N_{mov} = 1\}$.

IV. EXPERIMENTAL RESULTS

A. Test Bench Description

An experimental test bench has been set up in order to illustrate the efficiency of the different SK indicators for diagnosis of electrical machines. The experimental system is composed of :

- a 5.5 kW *Leroy Somer* induction machine (IM) with two poles pairs,
- a direct current generator used to regulate the load level of the IM,
- an iron disk placed on the shaft and associated to several weights in order to create different unbalance levels.

An overall view of the test bench is presented in Fig. 15.

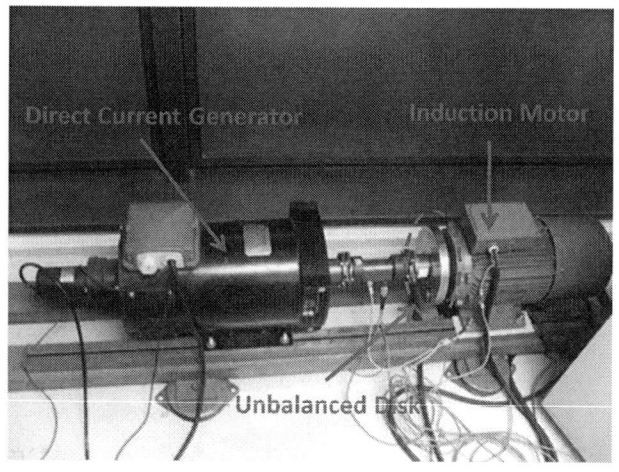

Figure 15. Experimental test bench

The IM is fed by an open-loop inverter at 40 Hz and operates at a low load level. Moreover, an acquisition system is used to measure phase currents of the IM with a sample frequency of $f_e = 10kHz$.

B. Spectral Kurtosis indicators for unbalance monitoring

250 5s-recordings of the IM currents (filtered and resampled at 200 Hz) have been measured on the machine. These measurements include :

- 100 recordings with no unbalance corresponding to the healthy state of the machine,
- 50 recordings with a low level of unbalance, noted B_1, created by a small weight (77.5 g) fixed to the iron disk. This unbalance produces load torque oscillations at f_r with an amplitude of 0.15% of the IM rated torque.
- 50 recordings with a medium level of unbalance, noted B_2, created by a medium weight (133 g). This unbalance produces load torque oscillations of 0.27% of the IM rated torque.
- 50 recordings with a high level of unbalance, noted B_3, created by a heavy weight (274 g). This unbalance produces load torque oscillations of 0.55% of the IM rated torque.

Let $X(f)$ be the spectral component of the IM current at frequency f. It is shown in [5] that an unbalance produces a Phase Modulation (PM) of the current fundamental (of frequency f_s) at the rotation frequency f_r. So, unbalance is visible in current spectrum on components $X(f_s \pm kf_r)$. In order to tolerate small changes of the system operating point while monitoring spectral components, we define the variable $\hat{X}(f)$ defined as

$$\hat{X}(f) = max_{k=0..2}\{X(f \pm k\Delta f)\} \quad (8)$$

with Δf the frequency resolution of the current spectrum.

$\hat{X}(f_s - f_r)$ and $\hat{X}(f_s + f_r)$ are computed on the IM current and plotted in Fig. 16.

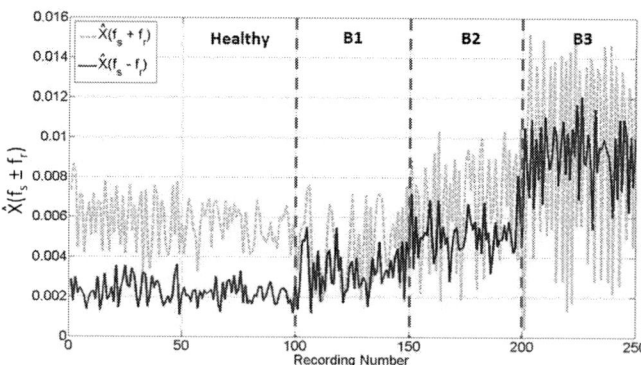

Figure 16. $\hat{X}(f_s - f_r)$ (blue solid line) and $\hat{X}(f_s + f_r)$ (green dashed line)

It is obvious from Fig. 16 that the effect of the different unbalance levels is visible on the two components $\hat{X}(f_s - f_r)$ and $\hat{X}(f_s + f_r)$. It can be noticed that $\hat{X}(f_s - f_r)$ better reflects the evolution of the unbalance level than $\hat{X}(f_s + f_r)$. However the values of $\hat{X}(f_s - f_r)$ and $\hat{X}(f_s + f_r)$ vary with the type of machine and with the operating point. Thus, it may be difficult to predict a fault threshold for these energy indicators.

To overcome this problem, SK_{Past} is first computed on $\hat{X}(f_s - f_r)$ and $\hat{X}(f_s + f_r)$ as presented in section III-A. SK_{Past} calculation uses one spectrum per current recording in order to have a sufficient frequency resolution ($\Delta f = 0.2 Hz$). The results are presented in Fig. 17.

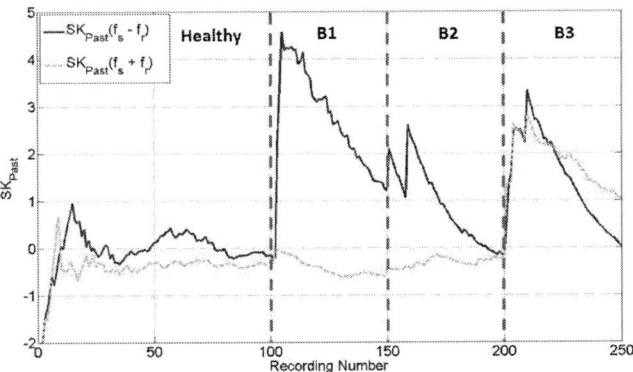

Figure 17. $SK_{Past}((f_s - f_r))$ (blue solid line) and $SK_{Past}((f_s + f_r))$ (green dashed line)

It is clearly visible in Fig. 17 that $SK_{Past}(f_s - f_r)$ increases for each new unbalance level whereas its level remains very low during the healthy state of the IM. However, it is higher during the unbalance level B1 than during the unbalance levels B2 and B3. Thus, SK_{Past} does not reflect the severity level of the unbalance although it reacts to every change of unbalance.

In order to obtain a better indicator of the unbalance level, SK_{Ref} is computed on $\hat{X}(f_s - f_r)$ and $\hat{X}(f_s + f_r)$ according to the protocol presented in section III-B. SK_{Ref} is computed using one spectrum per current recording and with $\{N_{ref} = 50; N_{mov} = 1\}$. The results are presented in Fig. 18.

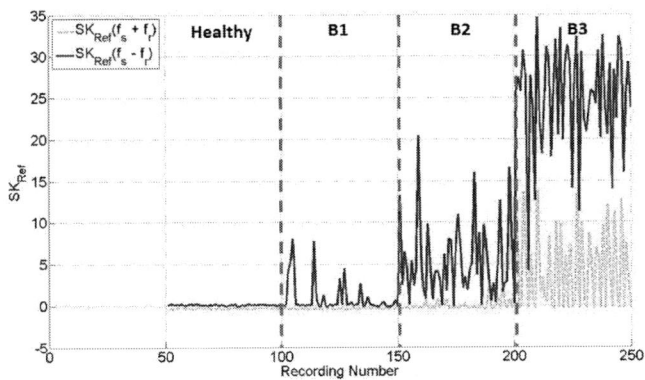

Figure 18. $SK_{Ref}((f_s - f_r))$ (blue solid line) and $SK_{Ref}((f_s + f_r))$ (green dashed line)

It is visible in Fig. 18 that SK_{Ref} is an efficient indicator of the unbalance level, especially at frequency $f_s - f_r$. Its value remains very low during healthy recordings and increases with the unbalance level. At frequency $f_s + f_r$, SK_{Ref} only reacts for the higher unbalance level B3 but it always has a very low value during healthy recordings. Unlike $\hat{X}(f)$, SK_{Ref} evolves in the same range of values for both $f_s - f_r$ and $f_s + f_r$ and its healthy level remains really close to 0.

It has been shown in Section III that the sensitivity of SK_{Ref} increases with the ratio $R = \frac{N_{ref}}{N_{mov}}$. In order to check this property on experimental current signals, SK_{Ref} has been

computed on $\hat{X}(f_s - f_r)$ for different ratios R. The results are shown in Fig. 19.

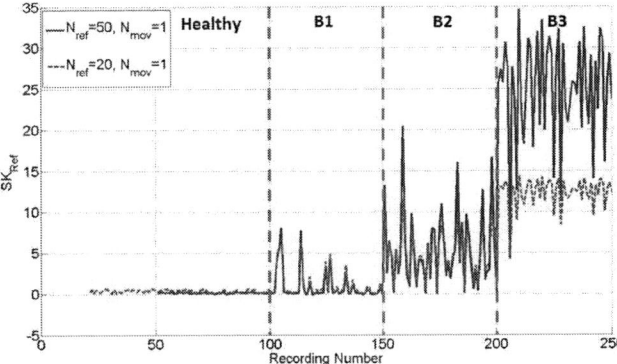

Figure 19. $SK_{Ref}((f_s - f_r))$ with $R = 50$ (blue solid line) and $R = 20$ (purple dashed line)

The sensitivity of $SK_{Ref}(f_s - f_r)$ is clearly improved using a high value of R, according to the results obtained on synthetic signals in Section III.

V. CONCLUSION

Different protocols of computation of Spectral Kurtosis have been presented in this paper. Original approaches, noted SK_{Past} and SK_{Ref}, have been developed in order to detect stationary faults such as unbalance or misaligned shaft. The relevancy of these indicators has first been shown on synthetic signal. Then their performance have been illustrated on the current recording of an IM for the diagnosis of unbalance faults. It has been shown that both SK_{Past} and SK_{Ref} react with unbalance changes. SK_{Ref} is a particularly interesting indicator as it increases proportionally with the unbalance level. The advantage of such a method compared to classical MCSA techniques is that the level of the indicator has a statistical meaning as it based on the statistical Kurtosis. So, its level should not depend on the application and it would be easier to define a fault threshold. Indeed, the indicator $SK_{Ref}(i)$ applied on a spectral component $X(f)$ only depends on the relative value of $X_i(f)$ compared to the reference set $\{X_1(f), ..., X_N(f)\}$. Thus, this indicator is independant of the type of machine monitored and of the operating point chosen to realise the diagnosis. It only requires a learning period in order to create a representative reference set of the monitored spectral components.

Future work should concentrate on the tolerance of these indicators toward small changes in the operating point. Moreover, this statistical-based approach will be tested for others types of faults and for different kinds of machine.

REFERENCES

[1] G. Betta, C. Liguori, A. Paolillo, and A. Pietrosanto, "A DSP-based FFT-analyzer for the fault diagnosis of rotating machine based on vibration analysis," *IEEE Transactions on Instrumentation and Measurement*, vol. 51, no. 6, pp. 1316–1322, Dec. 2002.

[2] G. Maruthi and K. Panduranga Vittal, "Electrical fault detection in three phase squirrel cage induction motor by vibration analysis using MEMS accelerometer," in *International Conference on Power Electronics and Drives Systems*, vol. 2, Nov. 2005, pp. 838–843.

[3] M. Ahmed, M. Baqqar, F. Gu, and A. Ball, "Fault detection and diagnosis using principal component analysis of vibration data from a reciprocating compressor," in *UKACC International Conference on Control*, Sep. 2012, pp. 461–466.

[4] M. E. H. Benbouzid, "A review of induction motors signature analysis as a medium for faults detection," in *Proceedings of the 24th Annual Conference of the IEEE Industrial Electronics Society*, vol. 4, Sep. 1998, pp. 1950–1955.

[5] M. Blodt, J. Regnier, and J. Faucher, "Distinguishing load torque oscillations and eccentricity faults in induction motors using stator current wigner distributions," *IEEE Transactions on Industry Applications*, vol. 45, no. 6, pp. 1991–2000, 2009.

[6] M. Sahraoui, S. Zouzou, A. Ghoggal, S. Guedidi, and H. Derghal, "An improved algorithm for detection of rotor faults in squirrel cage induction motors based on a new fault indicator," in *XXth International Conference on Electrical Machines (ICEM)*, 2012, pp. 1572–1578.

[7] R. Dwyer, "Detection of non-gaussian signals by frequency domain kurtosis estimation," in *Acoustics, Speech, and Signal Processing, IEEE International Conference on ICASSP*, vol. 8, Apr. 1983, pp. 607–610.

[8] V. Vrabie, P. Granjon, C.-S. Maroni, and B. Leprettre, "Application of spectral kurtosis to bearing fault detection in induction motors," in *Proceedings of the 5th International Conference on acoustical and vibratory surveillance methods and diagnostic techniques*, Senlis, France, 2004.

[9] J. Antoni and R. Randall, "The spectral kurtosis: application to the vibratory surveillance and diagnostics of rotating machines," *Mechanical Systems and Signal Processing*, vol. 20, no. 2, pp. 308–331, 2006.

[10] W. Taiyong and L. Jinzhou, "Fault diagnosis of rolling bearings based on wavelet packet and spectral kurtosis," in *2011 International Conference on Intelligent Computation Technology and Automation (ICICTA)*, vol. 1, Mar. 2011, pp. 665–669.

[11] F. Immovilli, M. Cocconcelli, A. Bellini, and R. Rubini, "Detection of generalized-roughness bearing fault by spectral-kurtosis energy of vibration or current signals," *IEEE Transactions on Industrial Electronics*, vol. 56, no. 11, pp. 4710–4717, Nov. 2009.

[12] V. Vrabie, P. Granjon, and C. Servière, "Spectral kurtosis: from definition to application," in *Proceedings of the 6th IEEE International Workshop on Nonlinear Signal and Image Processing*, Grado-Trieste, Italie, 2003.

[13] J. Antoni, "The spectral kurtosis: a useful tool for characterising non-stationary signals," *Mechanical Systems and Signal Processing*, vol. 20, no. 2, pp. 282–307, 2006.

[14] F. Millioz and N. Martin, "Circularity of the STFT and spectral kurtosis for time-frequency segmentation in gaussian environment," *IEEE Transactions on Signal Processing*, vol. 59, no. 2, pp. 515–524, Feb. 2011.

[15] Z. Obeid, A. Picot, S. Poignant, J. Regnier, O. Darnis, and P. Maussion, "Experimental comparison between diagnostic indicators for bearing fault detection in synchronous machine by spectral kurtosis and energy analysis," in *38th Annual Conference on IEEE Industrial Electronics Society, IECON*, Oct. 2012, pp. 3901–3906.

[16] A. Bellini, M. Cocconcelli, F. Immovilli, and R. Rubini, "Diagnosis of mechanical faults by spectral kurtosis energy," in *34th Annual Conference of IEEE Industrial Electronics, IECON*, Nov. 2008, pp. 3079–3083.

Etienne Fournier Etienne Fournier got his MSc in Electrical Engineering from SUPELEC Gif-sur-Yvette, France in 2012. He is now a PhD Student at the Laboratory of Plasma and Energy Conversion (LAPLACE) in Toulouse, France. His Scholarship is funded under an industrial agreement with Moteurs LEROY SOMER in Angoulême, France. His work focuses on preventive diagnostic of electrical machine under variable-speed control.

Antoine Picot (IEEE Member) Antoine Picot graduated from the Telecom Department of the Institut National Polytechnique (INP) Grenoble, France in 2006. He received the MSc degree in signal, image, speech processing and telecommunications in 2006 and his PhD in automatic control and signal processing in 2009 from the INP Grenoble. He is actually an associate professor at the INP Toulouse. He is also a Researcher with the Laboratory of Plasma and Energy Conversion (LAPLACE). His research interests are in monitoring and diagnosis of complex systems with signal processing and artificial intelligence techniques.

Jérémi Régnier (IEEE Member) Jérémi Régnier received the Ph.D. degree in electrical engineering from the Institut National Polytechnique de Toulouse (INP Toulouse), Toulouse, France, in 2003. Since 2004, he has been an Assistant Professor with the Electrical Engineering and Control Systems Department, INP Toulouse. He is also a Researcher with the Laboratoire Plasma et Conversion d'Energie, Université Paul Sabatier-Institut National Polytechnique de Toulouse, Toulouse. His research interests include modeling and simulation of faulty electrical machines and drives as well as the development of monitoring techniques using signal-processing methods.

Pascal Maussion (IEEE Member) Pascal Maussion got his MSc and PhD in Electrical Engineering in 1985 and 1990 from Toulouse Institut National Polytechnique (France). He is currently full Professor with the University of Toulouse and with LAPLACE, Laboratory for PLAsma and Conversion of Energy. His research activities deal with control and diagnosis of electrical systems (power converters, drives, lighting) and with the design of experiments for optimisation in control and diagnosis. He is currently Head of Control and Diagnosis group in LAPLACE. He teaches control and diagnosis in a school of engineers.

978-1-4799-0024-4/13 $31.00 © 2013 IEEE

Air-gap Power and Rotor Loss Estimation for Induction Motor Efficiency Monitoring based on Kalman Filtering

N. Jirasuwankul, C. Manop

⊕Abstract **– This paper presents a technique of induction motor's efficiency monitoring based on air-gap power and rotor loss estimation by Kalman filtering. A simplified model of three phase induction motor, with equivalent circuit of five elements, has been tested by the proposed technique with varying load torque and power to represent practical operations. Good agreement between the simulation and experimental results are found in a normal operating range of load torque and power.**

Index Terms--: **Air-gap power, Efficiency, Estimation, Kalman filtering, and Rotor loss**

INTRODUCTION

INDUCTION motors are universal equipment which have been widely used in modern industrial plants and contribute moreover 70% of total electrical energy consumption [1]. In a present realm of industrial machine operation and control, i.e., squirrel cage induction motor with or without variable speed drive, efficient energy utilization is found to be able to accountability and measurable [2]. In order to know efficiency of a running motor accurately, many approaches have been defined and applied as standardization tools [3]-[5].

Determination of efficiency of a running induction motor, it is clearly seen from theoretical point of view that, we need to know the ratio between mechanical power output and electrical power input [6]. However, since efficiency of the induction motor is not an explicit parameter and impractical or uneconomical to measure in-field directly [7], therefore the possible approach is alternatively based on parameters estimation techniques [8]-[9].

In this paper, the technique of states and parameters estimation by Kalman filtering from [10]-[12] has been applied to estimate air-gap power simultaneously with rotor loss, and thus efficiency of the machine can be determined.

ESTIMATION OF INDUCTION MACHINE'S EFFICIENCY

Machine Model and State Equation

A machine model of three phase induction motor which has been used in this paper is the steady-state equivalent

---

N. Jirasuwankul is with the Department of Electrical Engineering, Faculty of Engineering, KMITL Bangkok 10520 Thailand
(e-mail: nitudh@yahoo.com).
C. Manop is with the Department of Electrical Engineering, Faculty of Engineering, KMITL Bangkok 10520 Thailand
(e-mail: kmchaler@hotmail.com)

circuit of 5 elements, and shown in Fig.1. From the model, there are four states and two parameters to be determined, i.e., stator currents and flux linkage in stationary reference frame, rotor resistance and inductance which denoted by $i_{\alpha s}$, $i_{\beta s}$, λ_α, λ_β, R_r and $1/L_m$ respectively.

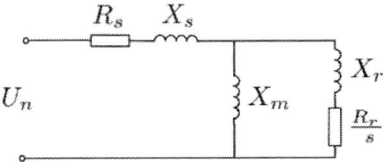

Fig.1. Equivalent circuit per phase of the induction motor

State equations of the model are described by the following,

$$\dot{x} = Ax + Bu + w \qquad (1)$$

$$y = Cx + v \qquad (2)$$

$$x = [i_{\alpha s}\ i_{\beta s}\ \lambda_\alpha\ \lambda_\beta\ R_r\ 1/L_m]^T \qquad (3)$$

$$u = [v_{\alpha s}\ v_{\beta s}]^T \qquad (4)$$

$$A = \begin{bmatrix} A_{11} & A_{12} \\ A_{21} & A_{22} \end{bmatrix} \qquad (5)$$

Where elements of the matrix A are sub-matrices, i.e.,

$$A_{11} = \begin{bmatrix} -\dfrac{(R_s + R_r)}{l} & 0 & \dfrac{R_r}{L_m l} & \dfrac{\omega_r}{l} \\ 0 & -\dfrac{(R_s + R_r)}{l} & -\dfrac{\omega_r}{l} & \dfrac{R_r}{L_m l} \\ R_r & 0 & -\dfrac{R_r}{L_m} & -\omega_r \\ 0 & R_r & \omega_r & \dfrac{R_r}{L_m} \end{bmatrix}$$

$$A_{12} = \begin{bmatrix} 0 & 0 & 0 & 0 \\ 0 & 0 & 0 & 0 \end{bmatrix}^T, \quad A_{21} = A_{12}^T, \quad A_{22} = \begin{bmatrix} 0 & 0 \\ 0 & 0 \end{bmatrix},$$

and elements of sub-matrix A_{11} are of the following,

$$l = L_s - L_m, \quad L_m = M'^2 / L_r, \quad L_s = l_1 + 3L_1 / 2,$$
$$L_r = l_2 + 3L_2 / 2, \quad R_r = r_2 L_m / L_r, \quad M' = 3M / 2,$$

Where l_1, L_1, R_s and l_2, L_2, R_r are leakage inductance, self-inductance and resistance of the stator and that of the rotor respectively. Finally, the matrix B is defined as,

$$B = \begin{bmatrix} \dfrac{1}{l} & 0 & 0 & 0 & 0 & 0 \\ 0 & \dfrac{1}{l} & 0 & 0 & 0 & 0 \end{bmatrix}^T \qquad (6)$$

Where w in (1) and v in (2), are the zero-mean Gaussian noise of the measurements and of the input with covariance R and Q respectively.

Kalman Filtering

The prescribed state-space equations of a steady-state model have then been used by the Kalman filter where structure of the estimator is shown in Fig.2.

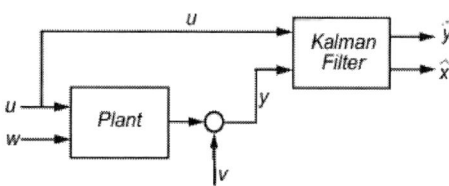

Fig.2. Plant and the Kalman estimator

From the Kalman estimator model, the new state variables vector z is defined as a vector that estimates the plant states,

$$z = \hat{x} \qquad (7)$$

And the plant output by,

$$\hat{y} = Cz + Bu + v \qquad (8)$$

Thus a state equation of the Kalman filter is written by

$$\dot{z} = Az + G(y - \hat{y}) \qquad (9)$$

Where the filter gain matrix, G, is optimally solved by an

algebraic matrix Riccati equation with the known inputs, u and the difference between the measurements and estimated output $(y - \hat{y})$.

Air-gap Power and Rotor Loss Estimation

The estimated state parameters which primarily resulted from the filter can then be further manipulated to determine other parameters, i.e., air-gap power, developed torque, rotor loss and rotational speed.

In case of efficiency, it is theoretically to calculate a ratio between mechanical power output and electrical power input. Therefore, the developed torque and rotational speed of rotor are two parameters that is needed to be estimated by,

$$\hat{T} = \frac{P}{2}(\hat{\lambda}_\alpha \hat{i}_{\beta s} - \hat{\lambda}_\beta \hat{i}_{\alpha s}) \qquad (10)$$

$$\hat{T} = \frac{P}{2}(z_2 z_3 - z_1 z_4) \qquad (11)$$

$$\omega_r = \frac{1}{J} \int \hat{T} dt \qquad (12)$$

Where ω_r is approximated over a short time interval of sampling period, i.e., $\omega_r(k\Delta t) = s(k\Delta t)/J$, $s(k\Delta t)$ is an integral of the developed torque from Eq.(10) over time interval $k\Delta t$, and J is the machine's moment of inertia. And power input of the induction motor is calculated by,

$$P_{input} = \frac{3}{2}(v_{\alpha s} \hat{i}_{\alpha s} + v_{\beta s} \hat{i}_{\beta s}) \qquad (13)$$

$$P_{input} = \frac{3}{2}(u_1 z_1 + u_2 z_2) \qquad (14)$$

Determination of Efficiency

From a given set of known parameters of the induction motor under test and the estimate parameters which resulted by (10) and (11), it is obviously to evaluate efficiency of the running motor by calculate the ratio of mechanical power output to electrical power input directly. An alternative way is to obtain by applying the estimated air-gap power together with rotor loss as the following,

$$P_{airgap} = P_{mech} + P_{rotor loss} \qquad (15)$$

$$\hat{T}\omega_s = \hat{T}(\omega_r + s\omega_s) \qquad (16)$$

$$P_{rotor\ loss} = s\,\hat{T}\,\omega_s \qquad (17)$$

$$\hat{\eta} = \frac{P_{output}}{P_{input}} \times 100\% \qquad (18)$$

Simulation Results

The proposed technique was tested by simulation on a system in Fig.2 which incorporated the plant model of the induction motor from (1) to (5). By applying a given set of known parameters of a machine as shown in Table I, the simulation results of 5-second running period are illustrated by set of states and parameters as shown in Fig.3 to Fig.4, and the machine's power and efficiency in Fig.5.

TABLE I
INDUCTION MOTOR PARAMETERS FOR SIMULATION

Parameters	Rated Value / Unit
Power	11 kW
Speed	1450 rpm
Frequency	50 Hz
Voltage	220 V/380 V
Rated Current	21.5 A
Poles	4
R_s	0.435 Ω
R_r	0.816 Ω
L_s	0.002 H
L_r	0.002 H
M	0.069 H
J	0.089 kg.m^2

In Fig.3(a), the induction motor starts to run from stand-still and speeds up to rated speed to a vicinity of 1,450-1500 rpm with the applied periodic load torque of 20 N-m., and then it speeds up close to 1,500 rpm when the load torque absents.

Fig.3(b) illustrates an occurrence of the developed air-gap torque that fluctuates periodically according to the presence or absence of the load torque with the same duty cycle. The periodic load torque was set to form 50:50 duty cycle of 2 seconds duration as shown in Fig.3(c).

Some states and parameters pertaining to rotor are shown in Fig.3 (d) to Fig.3 (f). Those represent the flux linkage between stator and rotor, rotor current, and rotor inductance respectively.

Fig.4(a) to Fig.4(c) represents states and parameters of the running plant model with Kalman estimator. There are the estimated total power loss that determined by (13) and (15), rotor loss by (17), and rotor resistance respectively.

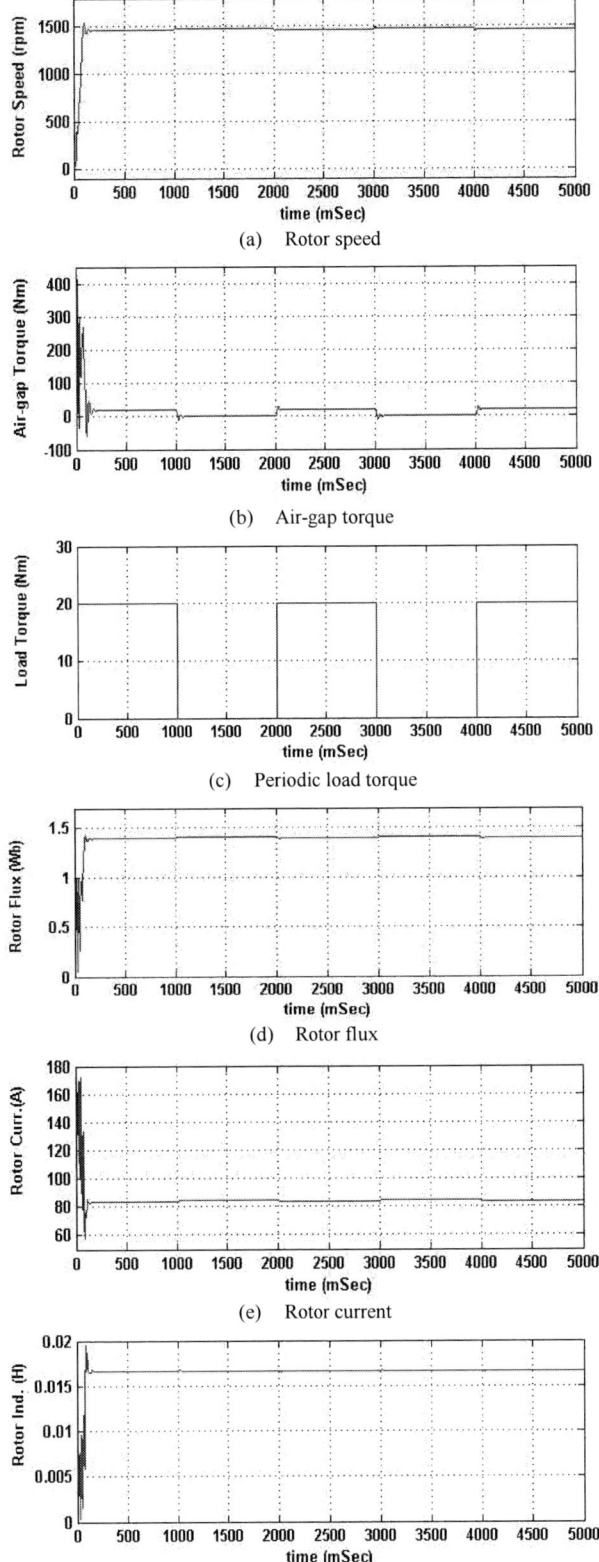

(a) Rotor speed

(b) Air-gap torque

(c) Periodic load torque

(d) Rotor flux

(e) Rotor current

(f) Rotor inductance

Fig.3 Estimated states and parameters of the induction motor model

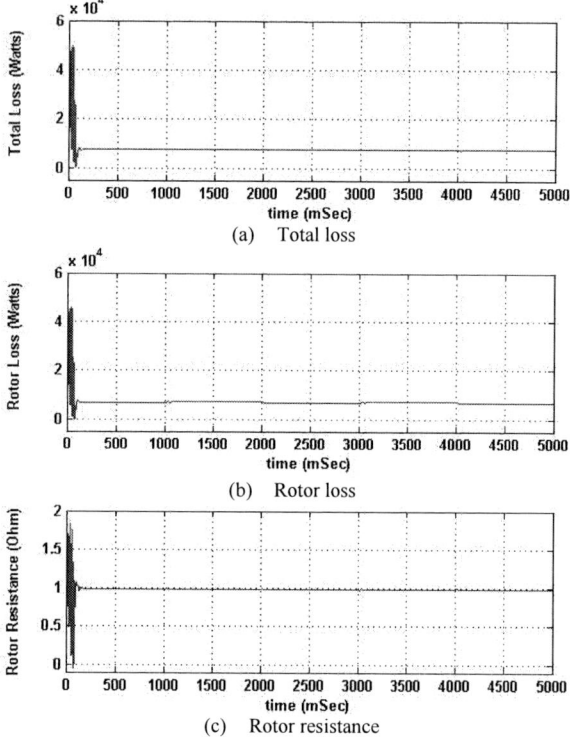

Fig.4 Estimated loses of the induction motor model

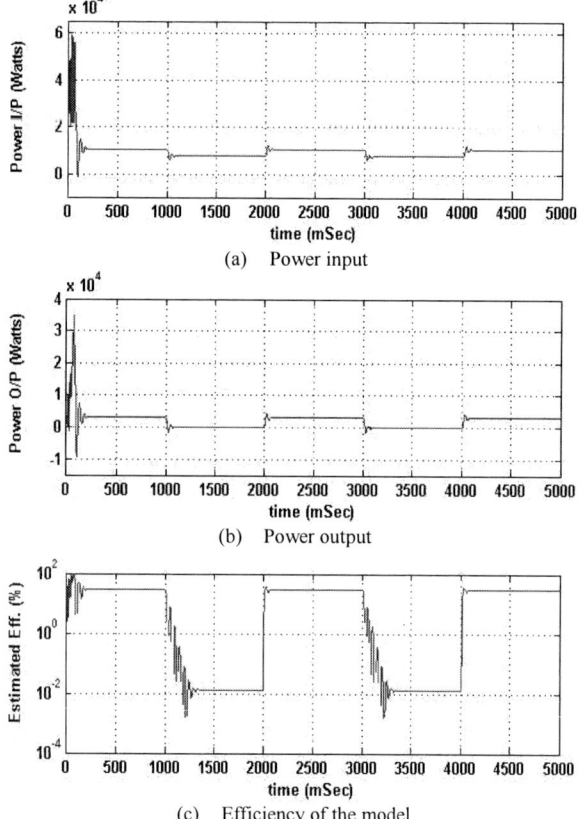

Fig.5 Estimated power and efficiency of the induction motor model

In Fig.5(a), input power of the running machine is determined by the estimated states of stator currents from an arbitrary reference α-β axis in (13), while Fig.5(b) shows mechanical output power of the machine which determined by a product between the estimated air-gap torque and speed.

Since the states and parameters of the machine have been predetermined by the Kalman filter, therefore the machine's efficiency can then be calculated by a ratio between output and input power in (18) and the result is shown in Fig.5(c).

From those graphical illustration of all signal profiles. Most of them consists of two parts, firstly the beginning time, which refered to as a transient period of the machine when it starts, and secondly a steady state period that proceeds after succeeding of the starting process. In this period, the machine responses, i.e., input power drawn by the motor, output power counteracts to load torque, and the estimated efficiency, vary precisely to the applied periodic load torqe.

Experimental Results

To validate the proposed technique, the experiment was set up and tested by a test rig which is depicted by Fig.6.

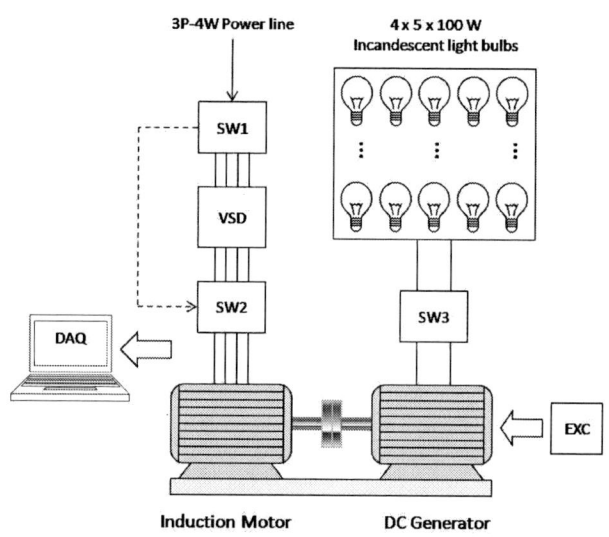

Fig.6. Experimental setup

In Fig.6, the system comprises of an induction motor, i.e., 5.5kW 4 poles 220/380V 50Hz, 2.4kW DC generator with separated or shunt exciter (EXC), a set of electrical load of 20x100W-220V incandescent light bulbs with control switches, two bypass control switches (SW1,SW2), a variable speed drive (VSD) module, and data acquisition system (DAQ). The induction motor can either be configured to run in variable speed mode through variable speed drive (VSD) or in a mode of single speed from power line frequency with direct on line start. Those have been managed by two bypass switches, SW1 and SW2, and all relevant signals, i.e., line voltages, currents, and shaft speed of the rotor, are measured by corresponding transducers and recorded in the DAQ module.

The experiment throughout this work had been carried out only for the case of single speed from power line frequency. By putting on appropriate adjustment of the field current in the exciter, the DC generator is thus set to generate output voltage of 220V at no-load and synchronous speed conditions. To apply the step-wise mechanical load torque periodically, the electrical load is scheduled to connect on or disconnect off the output terminals of the DC generator by main single circuit interrupter, SW3, whereas all 20 light bulbs can be grouped individually into 5 sub-circuits of 4 bulbs, i.e., 5 steps of 400 watts each.

The test was conducted by, firstly stating up the motor and left it running up to rated speed for 5 seconds, secondly applying the virtual mechanical load torque by turn on the switch SW3 and run the motor for next 5 seconds, and finally stop it at a time of 10 seconds. All necessary signals which are required by the model had been measured and recorded and then used to test the proposed technique in off-line manners. The test results are shown in Fig.7.

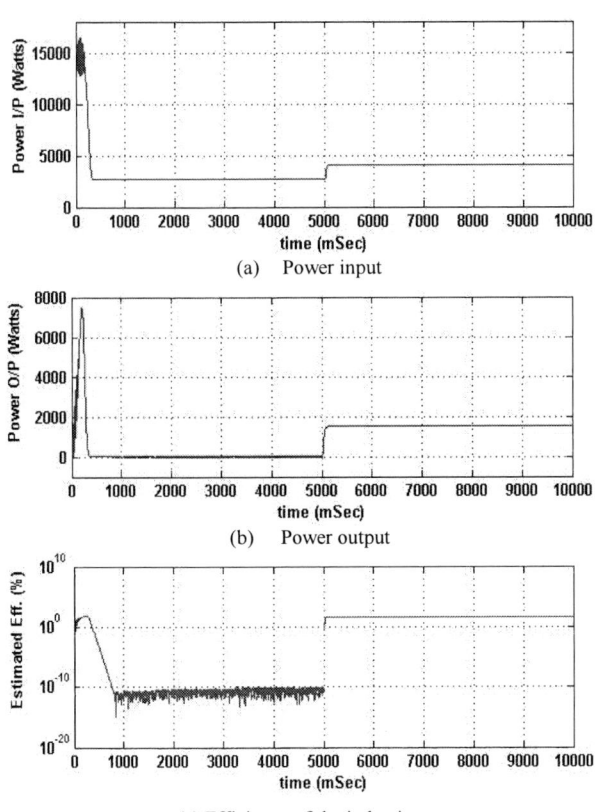

(a)   Power input

(b)   Power output

(c) Efficiency of the induction motor

Fig.7 Estimated power and efficiency of the motor under test

Fig.7 (a) shows the estimated input power to the motor, which is determined by states and transformed signals in α-β reference axis as defined in (14). It is obviously seen that the peak power occurs during a transient period of starting up

the motor. In the post start-up period, the motor run no-load for 5 seconds and then loaded for the next 5 seconds with mechanical load torque of 12 Nm, which approximates to 1,800 watts of incandescent light bulbs. During a period of stepwise loading, the induction motor draws power directly from utility line the same as to a load profile.

Fig.7 (b) shows mechanical power output of the motor that is estimated by (15). It is also found that the overall signal profiles of mechanical power output appears likely to that of electric power input. Therefore, from those two estimated states, efficiency of the machine is thus calculated by (18) and the result is shown in Fig.7(c).

At a first glance in Fig.7(c), significant peak of efficiency rise high during a startup time caused by peak starting torque of the motor, even thought there was no load. In the post starting period, efficiency decreases and converges to a vicinity of zero as long as there was no load. When the load torque was applied, efficiency increases scalable to a level of rated torque.

In particular to this experimental results and test conditions, efficiencies of the motor are approximated to 42.8 % by measurement and 44.7% by the Kalman estimator under the same load torque of equivalent to 1,800 watts. In addition, a slightly difference between two numerical values could resulted by the machine model, calculation process and measurements. To obtain the best result, further improvement in refining the model together with enhancing validation technique is required.

Conclusion

This paper presents a technique to estimate efficiency of a three phase induction motor based on air-gap power and rotor loss estimation by Kalman filtering technique. By including the Kalman filter to a simplified machine model, states and parameters can be estimated. In order to test the technique, the system model with varying load torque has been simulated and the experiment has been conducted. Good agreement between the simulation and experimental results are found in a normal operating range of load torque and power.

References

[1] H. Zhang, P.Zanchetta, C.Gerada, K. Bradley, and J. Liu, "Performance Evaluation of Induction Motor Efficiency and In-Service Losses Measurement Using Standard Test Methods", *Proc. 2011 IEEE International Electric Machines and Drives Conference (IEMDC)*, pp.913-917

[2] S. Manoharan, N. Devarajan, S.M. Deivasahayam, and G. Ranganathan, "Review on Efficiency Improvement in Squirrel Cage Induction Motor by Using DCR Technology", *Journal of Electrical Engineering*, vol.60, no.4, pp.227-236, 2009

[3] *IEEE Standard Test Procedure for Polyphase Induction Motors and Generators*, IEEE Standard and 112-2004, Nov. 2004

[4] H. M. Mzungu, A. B.Sebitosi, and M. A. Khan, "Comparison Standards of Determining Losses and Efficiency of Three-Phase Induction Motors", *IEEE-PES Power Africa 2007 Conference and Exposition*, pp.1-6, Johannesburg, South Africa, 16-20 July 2007

[5] A. Boglietti, A. Cavagnino, M. Lazzari, and M. Pastorelli,"International Standards for the Induction Motor Efficiency Evaluation: A Critical Analysis of the Stray-Load Loss Determination", *IEEE Trans. on Industry Applications*, vol.40, no.5, September/October 2004, pp.1294-1301

[6] R. Razzali, A. N. Abdalla, R. Ghoni, and C. Venkataseshaiah, "Improving Squirrel Induction Motor Efficiency: Technical Review", *International Journal of Physical Sciences*, vol.7, no.8, February 2012, pp.1129-1140

[7] E. B. Agamloh, "A Comparison of Direct and Indirect Measurement of Induction Motor Efficiency", *Proc. 2009 IEEE International Electric Machines and Drives Conference (IEMDC)*, pp.36-42

[8] J. Bacher, and F. Waldhart, " Effiviency Determination of Standard Asynchronous Machines from Start-Up Data", *IEEE-ICEM International Conference on Electrical Machines*, pp.1-6, Rome, 2010

[9] K. Banan, M.B.B. Sharifian, and J. Mohammadi, "Induction Motor Efficiency Estimation using Genatic Algorithm", *World Academy of Science, Engineering and Technology 3*, 2007, pp.688-692

[10] T. Kataoka, S. Toda, and Y. Sato, "On-line Estimation of Induction Motor Parameters by Extended Kalman Filter", *Proc. 1993 The European Power Electronic Association (EPE)*, pp.325-329

[11] S. Aksoy, A. Muhurcu, and H. Kizmaz, "State and Parameter Estimation in Induction Motor Using the Extended Kalman Filtering Algorithm", *Proc. 2010 Modern Electric Power Systems (MEPS'10)*, paper P13

[12] K. Radhakrishnan, A. Unnikrishnan, and K.G. Balakrishnan, "Joint State and Parameter Estimation of Squirrel Cage Induction Motor: A System Identification Approach using Expectation Maximisation based Extended Kalman Filter", *International Journal of Computer Application (0975-8887)*, vol. 27, no.1, August 2011, pp.24-29

BIOGRAPHIES

Nirudh Jirasuwankul received the B.Eng. and M.Eng. degrees in electrical engineering from King Mongkut's Institute of Technology Ladkrabang (KMITL), Thailand, in 1991 and 1998 respectively, and the Ph.D. degree in energy technology from the Joint Graduate School of Energy and Environment, King Mongkut's University of Technology Thonburi (JGSEE-KMUTT), Bangkok, Thailand, in 2008. He is currently a professor of the Department of Electrical Engineering, Faculty of Engineering, KMITL, Bangkok, Thailand. His interest area is energy efficiency both in electric power and combustion systems.

Chalermchat Manop received the B.Ind.Ed degree in electrical engineering from King Mongkul's University of Technology Thonburi (KMUTT), Bangkok, Thailand, the M.Eng. and the D.Eng. degrees in electrical engineering from King Mongkut's Institute of Technology Ladkrabang (KMITL), Bangkok, Thailand, in 2001. He is currently an Assistant Professor with the Department of Electrical Engineering, KMITL. His research interests are energy conversion, motor condition monitoring and diagnosis.

978-1-4799-0024-4/13 $31.00 © 2013 IEEE

Diagnosis of Induction Machines under Non-Stationary Conditions by means of the Spectral Filter

F. Vedreño-Santos, M. Riera-Guasp, M. Pineda-Sánchez

Abstract -- **This paper introduces a new way to extract the fault components of a signal by means of a new proposed filter based on the Sampling Theorem and the properties of Discrete Fourier Transform. The proposed method is able to extract fault components either in steady-state or in transient conditions. This paper also proposes a new way to compute the energy of a signal in the frequency domain by means of the Plancherel's Theorem. Unlike the Discrete Wavelet Transform, which makes the filtering process in the time domain, the Discrete Fourier Transform allows the filtering process in the frequency domain. The use of the Discrete Fourier Transform makes the filtering process possible without applying any pre-treatment to the current as it is done in the former works. As a consequence of the filtering process in the frequency domain and in order to reduce useless computations, in the current paper is also demonstrated the possibility of the computation of the energy of the extracted fault component in the frequency domain. The results are compared with the former and already published works in the area showing the good reliability of the new filtering process and energy computation in the frequency domain.**

Index Terms -- **discrete Fourier transform, discrete wavelet transform, fault diagnosis, filtering processes, fluctuating load conditions, frequency domain, energy computation, Hilbert transform, induction machine, instantaneous frequency, mixed eccentricity, rotor asymmetry, sliding frequency, squirrel cage induction machine, stator asymmetry**

I. Nomenclature

a_n	Wavelet Approximation of order n
f_{Low}	Lower Bandwidth Filter Limit
f_{max}	Maximum fault frequency
f_{min}	Minimum fault frequency
f_{RAL}	Lower Side Band Harmonic Frequency in a Rotor Asymmetry
f_{raS}	Fault Frequency for a Rotor Asymmetry
f_{RAU}	Upper Side Band Harmonic Frequency in a Rotor Asymmetry
f_s	Supply Frequency
f_{samp}	Sampling Frequency
f_{Upp}	Upper Bandwidth Filter Limit
j	Wavelet order for Details
n	Wavelet order for Approximations
p	Number of pole pairs

This work was supported by the Spanish "Ministerio de Ciencia e Innovación" in the framework of the "Programa Nacional de proyectos de Investigación Fundamental" (project reference DPI2011-23740)

F. Vedreño-Santos, M. Riera-Guasp and M. Pineda-Sánchez are with the Instituto de Ingeniería Energética, Universitat Politècnica de València, Camino. de Vera s/n, 46022, Valencia, SPAIN.
(e-mails: fravedsa@etsii.upv.es, mriera@die.upv.es, mpineda@die.upv.es).

s	Slip
E	Energy of a given signal
F	Frequency of a given signal
N	Number of samples of a digitalized signal
DFT	Discrete Fourier Transform
DWT	Discrete Wavelet Transform
MCSA	Motor Current Signature Analysis
WRIM	Wound Rotor Induction Machine

II. Introduction

The scope of application of electrical machines is continuously growing since the end of XIX century when their industrial use began. Nowadays, new applications as wind generation or electrical vehicles, in which machines work under non-stationary conditions and undergoing continuous changes in load and supply conditions, are quickly increasing their importance. In [1], [2] different techniques for carrying out diagnosis in non-stationary conditions are reviewed. For such applications, conventional diagnostic methodologies based on steady-state analysis –as MCSA– usually lead to unsatisfactory results [1]. As a consequence, new diagnostic approaches focused on non-steady state operation have recently been proposed, based on the application of different kind of signal analysis tools as Continuous Time-Frequency representations (such as Wigner Ville Distribution [3]-[5], Short-Time Fourier Transform [6], Continuous Wavelet Transform [7]) or Discrete Time Frequency transforms (as the Discrete Wavelet Transform [8]-[12]).

Essentially, the basis of these diagnostic methods can be understood as an extension of conventional steady-state methodology.

In steady state, diagnostic methodologies based on MCSA rely on detecting fault related components through the appearance of beans at specific frequencies in the current spectrum. For instance, 0 correspond to a diagnostic process of a wound rotor induction generator working in steady state. 0a depicts the slip evolution (constant in steady operation), 0b the tested current and 0c-d the current spectrum for the cases of healthy and faulty machine respectively. In this case the fault consisted of a rotor asymmetry. In 0d, peaks at frequencies:

$$f_{RAU} = (1 + 2 \cdot s) \cdot f_s \qquad (1)$$

$$f_{RAL} = (1 - 2 \cdot s) \cdot f_s \qquad (2)$$

reveal the existence of the fault and enable to quantify it through the amplitude of the peaks [10].

Fig 2 shows the same diagnostic process applied to the same machine but now working under non-stationary condition,

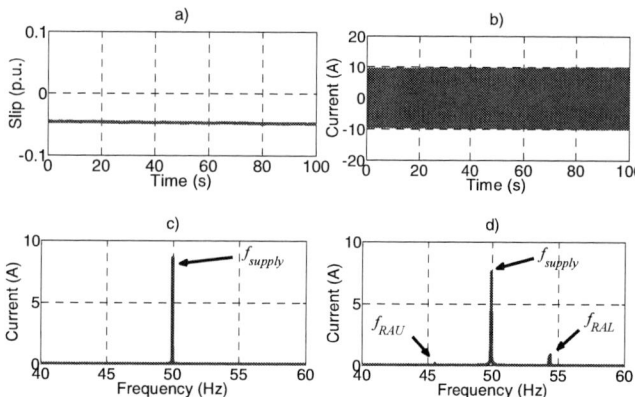

Fig 1. Diagnostic process of a wound rotor induction machine in steady-state by means of MCSA methodology a) Slip evolution b) Current c) Healthy Machine Current Spectrum d) Faulty Machine Current Spectrum

characterized by a slip (Fig 2a) fluctuating between two limits s_{min}, s_{max}; Fig 2b shows the corresponding tested current. In this case the frequency of the main fault component varies as the slip does, as indicated in (2), taking values into the intervals [f_{min}, f_{max}], which are obtained by substituting s_{min}, s_{max} into (2). Fig 2d shows the spectrum of the non-stationary current of the faulty machine. In this case the fault does not produce identifiable peaks as in 0d,

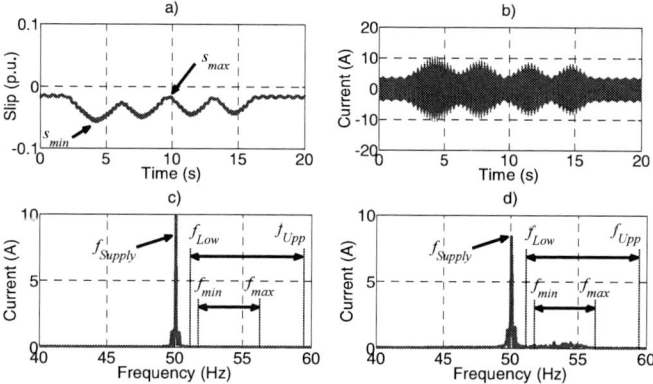

Fig 2. Diagnostic process of a wound rotor induction machine under non-steady state by means of MCSA methodology a) Slip evolution b) Current c) Healthy Machine Current Spectrum d) Faulty Machine Current Spectrum

since the energy of the fault component spread into the frequency interval [f_{min}, f_{max}]. Nevertheless, the comparison of the spectra shown in Fig 2c (healthy machine) and Fig 2d (faulty machine) reveals that the increase in the energy of the signal in the frequency interval in which the fault components evolves is detectable and can be used for diagnosing the fault.

Detecting the increase in signal's energy produced by the fault components is the basis of recent approaches for diagnosing under non stationary conditions [7]-[12]. These methods basically consist of applying a filter to the diagnosed signal to extract the components which belong to a bandwidth [f_{Low}, f_{Upp}], which includes a faulty frequency interval [f_{min}, f_{max}], as it is shown in Fig 2d; when fault happens, an increase in the signal's energy arises. It is remarkable that, if the filter is set up in such a way that the fault component is the most relevant component in the

bandwidth [f_{Low}, f_{Upp}], then the resultant signal reproduces approximately the evolution in time of the fault component. For this reason, this filtering process is designated as *Fault component extraction*.

Until now, most of the non-stationary diagnostic methods described in the literature as [8]-[12] use the DWT as a filtering tool for extracting the fault components. The worldwide use of the DWT comes from its easiness to be set and its efficiency when performing a filtering process with a reasonable computational cost. In [8] the authors introduce an adaptive algorithm which extracts the fundamental current component from transient waves and then it is analyzed the residual by using the DWT, then rotor faults are diagnosed through the increase in some wavelet coefficients. Papers [9]-[12] deal with the diagnosis of stator and rotor faults in Wound Rotor Induction Machines (WRIG) under non stationary operation. In [9] an approach for detecting stator and rotor asymmetries in WRIG, valid under load transient conditions, is presented. The approach is based on detecting increases in the energy in the monitored signals (stator current space vector magnitude, instantaneous magnitude of stator current, rotor current) into specific frequency bands, determined by the detail or approximation signals of the DWT. In [10] the proposed method is based on the current frequency sliding pre-processing, and the DWT. After applying the pre-processing, the mean power calculation of wavelet signals, at different resolution levels, is introduced as a dynamic fault indicator for quantifying the fault extents. This methodology is applied in [11] to a double fed induction generator in which the rotor windings are supplied by an electronic converter; the use of rotor voltages as diagnostic signal is proposed to diagnose the machine in closed-loop operation. In [12] the use of approximation signals of different levels and their combination is proposed to extract the fault component; a pre-processing treatment through the Hilbert transform is applied to the stator currents in the case of the diagnosis of rotor asymmetries. In [12] the extraction and quantification of the fault component under non-stationary conditions through DWT is applied to diagnose mixed eccentricity faults in squirrel cage machines working under motor or generator mode.

In the previous commented papers, a critical issue is the filtering process of extracting the related fault component, previous to the computation of its energy. The more accurate the filtering process is (i.e., the more similar the extracted frequency band [f_{Low}, f_{Upp}] is to the frequency range [f_{min}, f_{max}] in which the fault component moves, see Fig 2d) the more sensitive the method for detecting an increase in the signal's energy caused by the fault is.

As it has already been pointed out, the DWT has been the standard tool used for the fault component extraction up to now. Nevertheless this technique involves two important drawbacks when it is used as a filtering tool: (i) its stiffness to choose an arbitrary frequency band according to the needs of the diagnosed fault and (ii) the large width of the extracted frequency band [f_{Low}, f_{Upp}] supplied by the DWT filter compared with the objective frequency interval [f_{min}, f_{max}].

Both inconveniences provoke the need of applying

978-1-4799-0024-4/13 $31.00 © 2013 IEEE

complex pre-treatments on the acquired currents –as those used in the already developed techniques [8]-[12]– to carrying out the right performance of diagnosing electrical machines in transient conditions.

The current paper deals with the filtering process problem associated to the issue of the diagnosis under non-stationary conditions; a new way to extract the fault components from the current signals of an electrical machine when it is undergoing non-stationary conditions is proposed. This method enables a more accurate extraction of the fault components than the methods based on DWT and also avoids the need of applying complex pre-treatments to the current.

The new proposed filtering tool is based on the Sampling Theorem and in the Discrete Fourier Transform (DFT) properties, filtering the signal in the frequency domain unlike the previous works which perform the filtering process in the time domain.

The propounded method can be applied to improving the diagnosis of different kind of faults such as rotor asymmetry, stator asymmetry and mixed eccentricity as explained in the following sections; nevertheless, in this paper the validation of the method is focused on the case of the diagnosis of rotor asymmetries which are the most complex case.

The performance of the filter in the frequency domain also arises the problem of computing the energy of the extracted component whether in the time domain or in the frequency domain. To solve such inconvenience, in the current paper, it is also presented a new way to compute the energy of a signal in the frequency domain by means of the Parseval's Theorem. The proposed method has the same accuracy in the computation of the energy that those which does it in the time domain, but requiring less computational resources.

The current paper is structured as follows: section III characterizes the frequencies of the fault components that will be used for diagnostic purposes; section IV introduces both filtering tools used in the current paper to filter and extract the fault components from the current signals; section V shows how the energy is computed in the time and in the frequency domain; section VI presents how the filtering process has to been carried out depending on the chosen filter to extract the fault component; section VII and VIII show the experimental validation of the proposed filtering method under non-stationary conditions and under severe transients as startup, respectively. In both cases the spectra filtering is compared with the classical DWT filtering process. Finally, section IX summarizes the conclusions of the present work.

III. PHYSICAL BASIS

The fault components used in the current paper rely on the well established MCSA methodology, as they do in the former works [9]-[15]. It is well known that every kind of fault causes a characteristic perturbation in the air gap field which produces specific families of harmonics which appear (or greatly increase in their amplitude) in the currents circulating through the windings of the machine.

A stator asymmetry may be diagnosed by means of the rotor current in WRIM due to the amplitude of a series of components in the rotor current spectrum which rises when a stator asymmetry takes place. The frequency of the main harmonics is given by [14]

$$f_{saR}(s) = (2 - s) \cdot f_s \qquad (3)$$

where f_{saR} is the frequency of the main fault component produced by a stator asymmetry in the rotor current, s is the slip and f_s is the supply frequency.

A Rotor Asymmetry may be diagnosed by means of the stator current in Induction Machines, squirrel cage and wound rotor, due to the increase in the amplitude of the lower and upper sideband components of it, which frequencies are given by [16]

$$f_{raS}(s) = (1 \pm 2 \cdot s) \cdot f_s \qquad (4)$$

where f_{raS} is the frequency of the main fault component produced by a rotor asymmetry in the stator current.

A Mixed Eccentricity can be diagnosed by means of the stator current in IM, at the frequency given by [16]

$$f_{eccS}(s) = f_s - \frac{f_s}{p}(1 - s) \qquad (5)$$

where f_{eccS} is the frequency of the main fault component produced by a mixed eccentricity in the stator current and p the number of pair of poles.

Equations (3), (4) and (5) give the frequencies, in the current spectrum, where the fault components appear when a fault happens, in a non-stationary regime in which the slip fluctuates between two values $[s_{min}, s_{max}]$. The aforementioned expressions enable to compute the limits of the bandwidth $[f_{min}, f_{max}]$ in which the frequency of the fault component is included.

IV. FILTERING TOOLS

A. DWT Filter

The DWT is a filtering tool which performs an efficient low pass filtering process.

Provided a certain sampled signal $I = (i_1, i_2,..., i_N)$, the DWT decomposes it as the sum of $n+1$ wavelet signals, an approximation signal a_n and n detail signals d_j, where n is the decomposition level number [17], which can be set freely:

$$I = d_1 + d_2 + ... + d_n + a_n \qquad (6)$$

The practical procedure for the application of DWT is known as Mallat's algorithm [18] or Subband Coding algorithm. Mallat's algorithm shows that each wavelet signal is associated with a certain frequency band. Given a f_{samp}, sampling frequency in Hz, used for acquiring I, then the a_n approximation signal includes the low frequency components of the signal, belonging to the interval $[0, 2^{-(n+1)} \cdot f_{samp}]$ Hz and d_j includes the frequency components contained in the frequency band $[2^{-(j+2)} \cdot f_{samp}, 2^{-(j+1)} \cdot f_{samp}]$.

Therefore, to extract a component from I which contains a maximum frequency f_{max}, the highest level approximation n_{max} is characterized by [12]:

$$n_{max} = \operatorname{int}\left[\frac{\log\left(f_{samp}/f_{max}\right)}{\log 2} - 1\right] \qquad (7)$$

Approximation a_{nmax} brings the strongest low pass filter of the signal but keeping in all frequencies lower than f_{max}.

On the other hand, the highest level approximation n_{min} which does not contain higher frequencies than f_{min} is characterized by

$$n_{min} = \text{int}\left[\frac{\log\left(f_{samp}/f_{min}\right)}{\log 2}\right] \quad (8)$$

Approximation a_{nmin} brings the strongest low pass filter of the signal but keeping out all frequencies greater than f_{min}.

Subtracting (8) from (7) it is possible to extract any Frequency Band (FB) according to

$$F.B. = \left[2^{-(n_{min}+1)} \cdot f_{samp}, 2^{-(n_{max}+1)} \cdot f_{samp}\right] = \left[f_{Low}, f_{Upp}\right] \quad (9)$$

Note that (9) is the generalization of the expression for the DWT details based on the use of simple approximation signals.

Fig 3 graphically shows the DWT filtering process based on approximation signals.

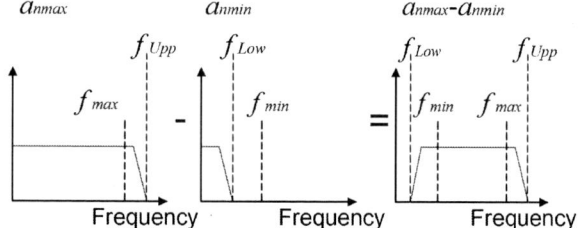

Fig 3. DWT filtering process based on approximations signals.

B. Spectral Filter

Any real signal when it is sampled by an acquisition system has a finite bandwidth. As a consequence, the acquired signal's spectrum cannot have any frequencies over a maximum value F_{max} [19].

The Sampling Theorem shows that a continuous-time band limited signal may be represented perfectly by its samples at uniform intervals of T seconds if T is small enough or, in other words, the continuous-time signal may be reconstructed perfectly from its samples if the signal is sampled at a high enough rate.

The Sampling Theorem states that "if $x(t)$ is a waveform (i.e. a finite energy function) that is band-limited to F_{max} Hz, then $x(t)$ at all times t can be recovered from its samples at integer multiples of $T=1/(2 \cdot F_{max})$ sec" [19], [20], as

$$x(t) = \sum_{n=-\infty}^{\infty} x(nT) \cdot \text{sinc}\left(\frac{t-nT}{T}\right) \quad (10)$$

where the 'sinc' function is

$$\text{sinc}(t) = \frac{\sin(\pi t)}{(\pi t)} \quad (11)$$

The Spectral filtering process of extracting the bandwidth$[f_{min}, f_{max}]$ is performed in three steps:

1.- The first step is to acquire N sample values, $x[n] = (x_0, x_1, x_2, ..., x_{N-1})$ of a signal $x(t)$ and compute its spectrum. Given a series of N samples the spectrum is computed by means of

$$X\left(\frac{2 \cdot \pi \cdot k}{N}\right) = \sum_{n=0}^{N-1} x(n) \cdot e^{-j \cdot \frac{2 \cdot \pi \cdot k \cdot n}{N}} \quad (12)$$

2.- Once the spectrum of the signal has been computed it is needed to set the maximum and minimum frequency which has to be extracted by means of

$$f_{Upp} = \text{int}\left(N \cdot \frac{f_{max}}{F_{max}}\right)$$

$$f_{Low} = \text{int}\left(N \cdot \frac{f_{min}}{F_{max}}\right) \quad (13)$$

3.- The last step is to reconstruct the filtered signal by

$$x(n) = \frac{1}{N} \sum_{k=f_{Low}}^{k=f_{Upp}} X(k) \cdot e^{j \cdot \frac{2 \cdot \pi \cdot k \cdot n}{N}} \quad (14)$$

V. ENERGY COMPUTATION

In all of the previously referenced works [9]-[15] the computation of the energy of signal is done in the time domain, but it can also be computed in the frequency domain as a direct consequence of the Parseval's Theorem [21].

A. Time Domain

The computation of the energy of a signal on the time domain as in [9]-[15] can be carried out by means of the expression

$$E = \int_{-\infty}^{\infty} |x(t)|^2 \, dt \quad (15)$$

In the case of discrete signals, (15) is computed as

$$E = \sum_{0}^{N-1} |x[n]|^2 \quad (16)$$

B. Frequency Domain

Parseval's Theorem is often written as [21]:

$$E = \int_{-\infty}^{\infty} |x(t)|^2 \, dt = \int_{-\infty}^{\infty} |X(f)|^2 \, df \quad (17)$$

where $X(f) = F\{x(t)\}$ represents the continuous Fourier transform and f the frequency component.

The interpretation of this form of the theorem is that the total energy contained in a waveform $x(t)$ summed across all of time t is equal to the total energy of the waveform's Fourier Transform $X(f)$ summed across all of its frequency components f.

In the case of discrete signals, the relation (17) becomes

$$E = \sum_{n=0}^{N-1} |x[n]|^2 = \frac{1}{N} \sum_{k=0}^{N-1} |X[k]|^2 \quad (18)$$

where $X[k]$ is the DFT of the $x[n]$, both of length N.

VI. FILTERING PROCESS

The current section describes how the filtering process is carried out by means both filtering techniques: the DWT [9]-[15], and the new proposed Spectral filter.

A. DWT Filter

The filtering process in [9]-[15] is carried out by means of the DWT filter in order to extract the fault component of the current signal.

Due to the aforementioned inconveniences in section II, some pre-treatments have to be done to the current in order to extract the fault component when a rotor asymmetry is diagnosed. The different used pre-treatments are explained in detail in [10]-[12], [14], and [15].

In [10], [11], [14], and [15] the Frequency Sliding technique is applied to the current signal to move the fault components to the low frequency region of the spectrum where the DWT filter performs a better extraction.

In [12] the Hilbert Transform is the chosen technique to set the fault components in the low frequency region of the spectrum.

When a stator asymmetry is diagnosed [9], [12] there is no need of a pre-treatment but the width of the extracted band is too wide. Thus, the extracted band may contain unwanted frequency components.

In the case of an eccentricity fault [12], the width of the extracted band may be also too wide as it happens when a stator asymmetry is diagnosed. Therefore, the extracted band may also contain unwanted frequency components.

In [9], [12], and [12] can be seen the stiffness of the DWT filter for extracting a specific bandwidth and its dependence on the supply and sampling frequency.

B. Spectral Filter

To extract a bandwidth of current by means of the new proposed filter, the Spectral filter, based on the DFT, it has to be followed the three aforementioned steps described in section IV.B.

Note that the only restriction on establishing the bandwidth limits $[f_{Low}, f_{Upp}]$ of the extracted signal is the length N of the sampled signal. The length of the sampled signal depends on the sampling frequency f_{samp} and the amount of time in which the signal is acquired.

Therefore, given a signal with enough length (N), achieved either by a high sampling frequency rate, f_{samp}, or a long sampling time, the bandwidth limits $[f_{Low}, f_{Upp}]$ can be set equals to the frequency fault bandwidth limits $[f_{min}, f_{max}]$ in which the fault component evolves as it is stated in (19). As a consequence of the previous statements, the maximum sensitivity of noticing increases in the energy is reached.

$$F.B = \left[N \cdot \frac{f_{min}}{\frac{f_{samp}}{2}}, N \cdot \frac{f_{max}}{\frac{f_{samp}}{2}} \right] = \left[f_{Low}, f_{Upp} \right] \approx \left[f_{min}, f_{max} \right] \quad (19)$$

VII. EXPERIMENTAL VALIDATION UNDER NON-STATIONARY FUNCTIONING

In order to validate the new filtering process the extraction of the fault component, when a rotor asymmetry is present in a machine, is carried out by testing a WRIG directly coupled to the network but undergone to random speed fluctuations around the rated speed. Under these conditions, the asymmetry related component of the stator current is extracted by means of the DWT filter, applying the same pre-treatment as it is done in [12] and also by means of the new Spectral filter.

To make easier the comparison between both filters, firstly the pre-treatment will be applied to the extraction of the fault component for each filter, although the Spectral filter does not need any pre-treatment.

Secondly, the extraction of the fault component will be done without applying any pre-treatment to the current. The aim of this second section is to show the main advantage of the Spectral filter which is the skill at extracting the fault component without applying any pre-treatment.

The electrical scheme that is set up for the test carried out to get the current from a faulty machine is shown in Fig 4.

Due to practical restrictions, the rotor windings had to be used as primary winding, connected to a three-phase supply source of 160 V.

The stator windings were short-circuited working as secondary winding. An additional resistance Ras = 4.15 Ω (which produces an increase of 94% in the winding phase resistance) is connected in series to a stator phase for simulating a secondary winding asymmetry fault.

The induction machine is directly coupled to a DC machine. The field winding of the DC machine is fed through a motorized rheostat, which is controlled by a PLC. This assembly enables the control of the load set on the tested machine and, as a consequence, it also allows the programming of fixed load fluctuations.

The primary current is measured through a current clamp (10 A - 100 mV), and the speed through a 360 pulse/turn encoder, connected to a Yokogawa DL 750 digital oscilloscope (16 bit AD converter). The oscilloscope is connected, via an intranet network, to a PC where the acquired signal is processed. A sampling frequency f_{samp} = 5 kHz is set.

A transient load is imposed to the machine and the current (Fig 5) and the speed (Fig 6) are captured. The capture of the speed is only necessary to calculate accurately the slips s_{min}, s_{max} which allows the knowledge of the minimum and maximum frequency of the fault component by means of (4) [10]-[15]. Nevertheless, s_{min}, s_{max} can be estimated without measuring the speed, as the slips corresponding to the no- load and full load regimes, without a significant lack of sensitivity.

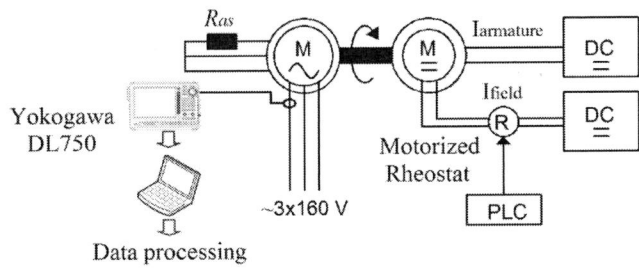

Fig 4. Rotor asymmetry setup scheme

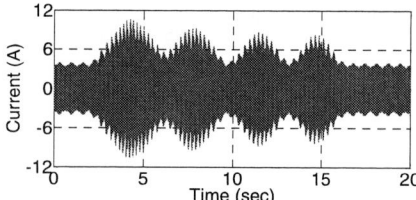

Fig 5. Acquired current from an electrical machine with rotor asymmetry undergoing a transient load

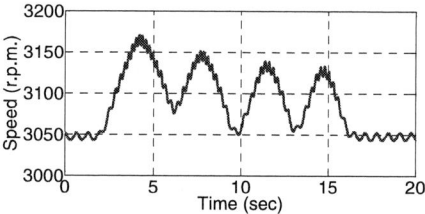

Fig 6. Acquired speed from an electrical machine with rotor asymmetry undergoing a transient load

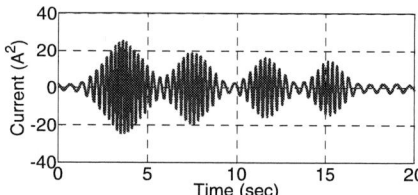

Fig 7. Fault component extracted by a DWT filter after applying the pre-treatment presented in [12].

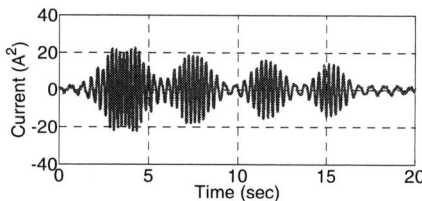

Fig 8. Fault component extracted by the Spectral filter after applying the pre-treatment presented in [12].

TABLE I.

ENERGY OF THE FAULT COMPONENT EXTRACTED VIA DWT AND SPECTRAL FILTER (NON STATIONARI FUNCTIONING, PRETREATMEN)

Filter	Energy	
	Faulty	Healthy
DWT	4.13 %	0.0053 %
Spectral Filter	4.02 %	0.0053 %

A. Extraction of the fault component applying a pre-treatment to the current

The extraction of the fault component by means of the DWT is performed in a similar way as it is done in [12]: First a new diagnostic signal is computed as the magnitude of the analytical signal obtained through the Hilbert Transform of the current. In this new signal the network supply frequency component (50 Hz) and the fault component are shifted to the low frequency region of the frequency spectrum. Then the extraction of the fault component is carried out through a filtering process based on the DWT. The extracted fault component is shown in Fig 7. If the extraction procedure described in [12] is followed but Spectral filter is applied to extracting the fault component

instead of the classical DWT filter, the achieved result is shown in Fig 8.

The energy of the fault component is computed in the time domain by (16). Table I shows the energy for the fault component extracted by the DWT and the Spectral filter versus the total energy of the transient current signal. According to the results (4.13% of the energy when the fault component is extracted by DWT filter versus 4.02% when the fault component is extracted by Spectral filter for the faulty machine and 0.0053% for both filters when the machine is healthy), it can be concluded that both filters perform rightly the extraction of the fault component.

Obviously the values of the energy for the Spectral filter will be equal or smaller than the values of the energy for the DWT filter since the extracted bandwidth for the Spectral filter is narrower than the DWT bandwidth.

B. Extraction of the fault component without applying a pre-treatment to the current

In the current section no pre-treatment is applied to the transient current for extracting the fault component from it.

The waveform of the extracted component by means of a DWT filter is shown in Fig 9, whereas in Fig 10 it is shown the fault component extracted by means of the new proposed Spectral filter. After the extraction of the fault component by means of the DWT filter (Fig 9) and by means of Spectral filter (Fig 10), the energy of the extracted components is computed.

The filtering process carried out by means of the DWT filter is done in the time domain, therefore, the computation of the extracted fault component by the DWT filter is computed in the time domain.

On the other hand, the filtering process carried out by means of the Spectral filter is done in the frequency domain. As a consequence, there is no need of changing the signal domain, from the frequency domain to the time domain, in order to compute the energy of the extracted fault component as it was stated in section V.B due to direct consequence of the Plancharel's Theorem.

The values of the computed energies for the extracted fault component by means of the DWT filter and the Spectral filter are shown in Table II.

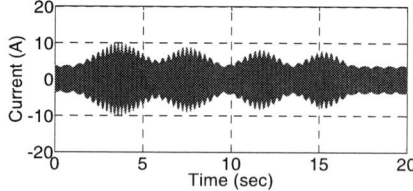

Fig 9. Fault component extracted by a DWT filter without applying any pre-treatment to the transient current.

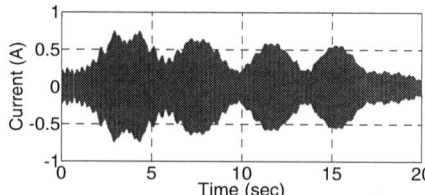

Fig 10. Fault component extracted by the Spectral filter without applying any pre-treatment to the transient current.

TABLE II.
ENERGY OF THE FAULT COMPONENT EXTRACTED VIA DWT AND SPECTRAL FILTER (NON STATIONARI FUNCTIONING, NO PRETREATMEN)

Filter	Energy	
	Faulty	Healthy
DWT	99.75 %	99.89 %
Spectral Filter	4.72 %	0.217 %

The results agree to the former results presented in [9]-[12], [14], and [15]where it is said that a pre-treatment is mandatory in order to extract a rotor fault component from a current signal due to the closeness of the main component of the current, otherwise the performance of the DWT filter is unable to extract the fault component from the main component of the current and the last consequence, the energy of the extracted signal is almost equal to the total energy of the original signal.

Table II shows that if no pre-treatment is applied to the current for extracting the rotor fault component with a DWT filter, the energy of the fault component is 99.75% of the total energy of the current signal for the faulty machine and 99.89% for the healthy machine. Therefore, it implies that the filtering process has been unsuccessful.

On the other hand, if the Spectral filter is chosen for the extraction of the fault component without applying any pre-treatment and although the energy is computed in the frequency domain, the value of the extracted fault component for the faulty machine is 4.72%, similar to the values of the energies (4.13% with DWT and 4.02% with Spectral filter) of the extracted fault components in the previous section where a pre-treatment is set to the current before extracting the fault component, and 0.217% for the healthy machine (0.053% when a pre-treatment is applied to the diagnostic process).

Therefore, it has been verified that the use of the Spectral filter makes possible the extraction of the fault component without applying any complex pre-treatment to the current.

VIII. EXTRACTION OF THE LSH DURING A STARTUP TRANSIENT USING SPECTRAL FILTERING. EXPERIMENTAL VALIDATION

The extraction of the LSH from the stator current by means of the DWT was proposed in [22] as a method for diagnosing broken bars. There, it is demonstrated that under bar breakage conditions, the evolution in time of the LSH during a startup follows a characteristic pattern, as shown in Fig.11 [22]. It is also demonstrated that a specific approximation signal resulting of the DWT practically fits the time evolution of the LSH, and consequently can be used for diagnostic purposes.

In this section the spectral filtering is applied to the transient startup current of a machine with a
broken bar, with the aim to extract the LSH. It will be seen that this method enables for a more precise extraction of the LSH than the classic approach based on the DWT. Fig. 12 a shows the startup stator phase current measured during the startup of a 1.1kW, two pole pairs cage machine, with one broken bar [17]. The current is tested using a sampling frequency f_{samp}= 5000 samp/seg. For this sampling rate, the

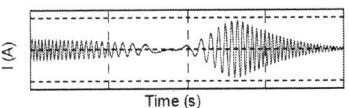

Fig 11 Theoretical evolution of the LSH during a startup [22]

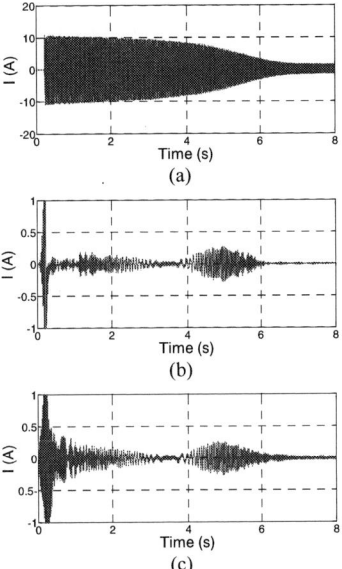

Fig 12 (a) Startup current of a cage motor with a broken bar[17]. (b). LSH extracted using the DWT (extraction band [0,39.6]Hz.)[17]. (c). LSH extracted using the spectral filtering (extraction band [0,47]Hz.)

approximation of level 6, depicted in Fig.12 b reproduces the evolution of the LSH in the frequency interval [0, 40]Hz. The shape of this subsignal fits quite well the theoretical pattern of the LSH shown in Fig.11.

On the other hand, Fig.12c shows the result of extracting the LSH from the startup current of Fig.12 a, but using the spectral filtering technique. With this technique there is no stiffness in the selection of the limits of the extraction band. In this case the LSH has been extracted in the band [0,47] Hz, wider than the band allowed by DWT, and thus covering almost all the evolution of the LSH during the startup; in this way a better fit to the theoretical shape is reached than when the DWT was used.

IX. CONCLUSIONS

This paper introduces a new filter based on the Sampling Theorem and the Discrete Fourier Transform properties and also a new way to compute the energy of a signal in the frequency domain by means of the Plancharel's Theorem.

Unlike the previous works in the field, the new proposed filter allows the accurate extraction of the fault components of a current signal without applying, previously the extraction, any complex pre-treatment.

The use of the new Spectral filter improves the reliability of the filtering process by means of the DFT due to its own features, reducing the stiffness of the bandwidth selection imposed by the use of the DWT filters and thus, enabling for an optimal selection of the extraction band.

On the other hand, the computation of the energy in the frequency domain, instead of the classical computation of the energy in the time domain as a consequence of the filtering process which is carried out in the frequency domain, reduces the number of computations which are needed to perform the diagnostic process in the former works improving the already developed techniques.

The improvement achieved has been validated by means of the comparison of the achieved results by the new proposed Spectral filter with the classical methodology based on the DWT filters in the case of rotor asymmetries.

X. References

[1] Shahin Hedayati Kia, Humberto Henao, Gérard-André Capolino, "Efficient Digital Signal Processing Techniques for Induction Machines Fault Diagnosis", *Proceedings 2013 IEEE Workshop on Electrical Machines Design, Control and Diagnosis (WEMDCD)*, pp. 230-244, Paris, France ,11 - 12 March, 2013

[2] M. Riera-Guasp, J. Pons-Llinares, V. Climente-Alarcón, F.Vedreño-Santos, M. Pineda-Sánchez, J. Antonino-Daviu, R.Puche-Panadero, J. Perez-Cruz, IEEE, J. Roger-Folch "Diagnosis of Induction Machines under Nonstationary Conditions: Concepts and Tools" *Proceedings 2013 IEEE Workshop on Electrical Machines Design, Control and Diagnosis (WEMDCD)*, pp. 218-229, Paris, France ,11 - 12 March, 2013

[3] M. Blödt, D. Bonacci, J. Regnier, M. Chabert, J. Faucher, "On-line monitoring of mechanical faults in variable-speed induction motor drives using the Wigner distribution," *IEEE Trans. Industrial Electronics*, vol. 55, no. 2, pp. 522-533, Feb. 2008.

[4] S. Rajagopalan, J.A. Restrepo, J.M. Aller, T.G. Habetler, R.G. Harley, "Nonstationary motor fault detection using recent quadratic time-frequency representations," *IEEE Trans. Industry Applications*, vol. 44, no. 3, pp. 735-744, May/June 2008.

[5] Climente-Alarcon, V.; Riera-Guasp, M.; Antonino-Daviu, J.; Roger-Folch, J.; Vedreno-Santos, F.; "Diagnosis of rotor asymmetries in wound rotor induction generators operating under varying load conditions via the Wigner-Ville Distribution," *Power Electronics, Electrical Drives, Automation and Motion (SPEEDAM)*, 2012 International Symposium on, pp.1378-1383, 20-22 June 2012

[6] J. Cusidó, L. Romeral, J.A. Ortega, J.A. Rosero, A.G. Espinosa, "Fault detection in induction machines using power spectral density in wavelet decomposition," *IEEE Transactions Industrial Electronics*, vol. 55, no. 2, pp. 633-643, Feb. 2008.

[7] Simon Jonathan Watson, Beth J. Xiang, Wenxian Yang, Peter J. Tavner, and Christopher J. Crabtree, "Condition Monitoring of the Power Output of Wind Turbine Generators Using Wavelets", *IEEE Transactions on Energy Conversion*, vol. 25, no. 3, pp. 715-721,september 2010.

[8] H. Douglas, P. Pillay, A.K. Ziarani, "A new algorithm for transient motor current signature analysis using wavelets," *IEEE Trans. Industry Applications*, vol. 40, no. 5, pp. 1361-1368, Sept./Oct. 2004.

[9] Kia, Shahin Hedayati; Henao, Humberto; Capolino, Gerard-Andre; "Windings monitoring of wound rotor induction machines under fluctuating load conditions ", *IECON 2011 - 37th Annual Conference on IEEE Industrial Electronics Society* , Melbourne , Australia, November 2011 , Page(s): 3459 - 3465

[10] Y. Gritli, A. Stefani, C. Rossi, F. Filippetti, A. Chatti, "Experimental validation of doubly fed induction machine electrical faults diagnosis under time-varying conditions", *Journal of Electric Power Systems Research*, Vol. 81, Issue 3, pp. 751-766, March 2011.

[11] Gritli, Y.; Zarri, L.; Rossi, C.; Filippetti, F.; Capolino, G.; Casadei, D.; , "Advanced Diagnosis of Electrical Faults in Wound Rotor Induction Machines," *Industrial Electronics, IEEE Transactions on* , doi: 10.1109/ TIE.2012.2236992

[12] F. Vedreño-Santos, M. Riera-Guasp, H. Henao, M. Pineda-Sanchez, "Diagnosis of faults in induction generators under fluctuating load conditions through the instantaneous frequency of the fault components", *XX International Conference on Electrical Machines (ICEM)*, 2012. Marseille

[13] F.Vedreño-Santos, M. Riera-Guasp, H. Henao, M. Pineda-Sanchez, J.A.Antonino-Daviu, "Diagnosis of eccentricity in induction machines working under fluctuating load conditions, through the instantaneous frequency", *38th International Conference on Industrial Electronics, Control, and Instrumentation (IECON)* 2012, Montréal

[14] Gritli, Y.; Rossi, C.; Zarri, L.; Filippetti, F.; Chatti, A.; Casadei, D., "Double frequency sliding and wavelet analysis for rotor fault diagnosis in induction motors under time-varying operating condition," *Diagnostics for Electric Machines, Power Electronics & Drives (SDEMPED), 2011 IEEE International Symposium on*, vol., no., pp.676,683, 5-8 Sept. 2011

[15] Gritli, Y.; Sang Bin Lee; Filippetti, F.; Zarri, L., "Advanced diagnosis of outer cage damage in double squirrel cage induction motors under time-varying condition based on wavelet analysis," *Energy Conversion Congress and Exposition (ECCE), 2012 IEEE*, vol., no., pp.1284,1290, 15-20 Sept. 2012

[16] Thomson, W.T.; Fenger, M.; , "Current signature analysis to detect induction motor faults," *Industry Applications Magazine, IEEE*, vol.7, no.4, pp.26-34, Jul/Aug 2001

[17] M. Riera-Guasp, J. A. Antonino-Daviu, M. Pineda-Sanchez, R. Puche-Panadero, and J. Perez-Cruz, "A General Approach for the Transient Detection of Slip-Dependent Fault Components Based on the Discrete Wavelet Transform," *IEEE Trans. Ind. Electron.*, vol. 55, pp. 4167-4180, 2008

[18] C.S. Burrus, R.A. Gopinath, H. Guo, "Introduction to Wavelets and Wavelet Transforms: A Primer", Prentice-Hall, Englewood Cliffs,NJ,

[19] Nyquist, H., "Certain topics in telegraph transmission theory," *Proceedings of the IEEE*, vol.90, no.2, pp.280,305, Feb 2002

[20] Harold S. Black, "Modulation Theory", 1953

[21] Plancherel, Michel "Contribution a l'etude de la representation d'une fonction arbitraire par les integrales définies," *Rendiconti del Circolo Matematico di Palermo*, vol. 30, pages 298–335, 1910.

[22] M. Riera-Guasp, J. Antonino-Daviu, J. Roger-Folch, and M. P. Molina,"The use of the wavelet approximation signal as a tool for the diagnosis and quantification of rotor bar failures," *IEEE Trans. Ind. Appl.*, vol. 44, no. 3, pp. 716–726, May/Jun. 2008.

XI. Biographies

Francisco Vedreño-Santos received the M.Sc degree in electrical engineering from Universidad Politecnica de Valencia, Spain, in 2008, where he is currently working toward the Ph.D. degree in electrical engineering in the Departamento de Ingeneria Eléctrica. His research interests include electric-machine diagnostics, condition monitoring of electric machines and windmills.

Martin Riera-Guasp received the M.Sc.degree in industrial engineering and the Ph.D. degree in electrical engineering from the Universitat Politècnica de València (UPV), València, Spain, in 1981 and 1987, respectively. He is currently an Associate Professor in the Eléctrical engineering Department, UPV. His research interests include condition monitoring and diagnostics of electrical machines, applications of signal analysis to electrical engineering, and efficiency in electric power applications.

M. Pineda-Sanchez received the Dipl. Ing. And Dr. Ing. degrees in electrical engineering from the Universidad Politécnica de Valencia, Valencia, Spain, in 1985 and 2004, respectively. He joined the faculty of the Universidad Politéc nica de Valencia in 1987 as an Associate Professor in the area of theory and control of electrical machineswith the Department of Electrical Engineering. His research interests include electrical machines and drives, induction motor diagnostics, numerical simulation of electromagnetic fields, and software development

978-1-4799-0024-4/13 $31.00 © 2013 IEEE

Fault Detection and Classification in Permanent Magnet Synchronous Machines using Fast Fourier Transform and Linear Discriminant Analysis

Reemon Z. Haddad, Elias G. Strangas

Abstract-- **The main objective of this paper is to propose a method to detect the presence of a fault in Permanent Magnet Synchronous Machines (PMSMs), determine the type of that fault and estimate the severity in the case of eccentricity fault. In this paper, three types of faults are discussed: static eccentricity, inter-turn short circuit, and demagnetization faults. The machine is controlled using a three phase current source and the harmonics of the stator voltage are used as detailed features for the classifier for fault detection. Two dimensional (2-D) Finite Element Analysis (FEA) is used to model and simulate the machine under healthy and faulty conditions. Fast Fourier Transform (FFT) analysis is performed to the phase voltage signal to detect the frequency spectrum and Linear Discriminant Analysis (LDA) is chosen as a classification method. To validate the method, two different types of PMSMs are tested: the first with a concentrated winding and the second with a series distributed winding. The tests are applied for different operation loads.**

Index Terms-- **Demagnetization, Eccentricity, Inter-turn short circuit, Linear discriminant analysis classification, Permanent magnet synchronous machine.**

I. INTRODUCTION

Permanent magnet synchronous machines play a major role in many industrial applications because of their high efficiency, reliability, wide operation range, and high torque density. These applications include power steering in electric/hybrid vehicles, robotics and wind generation. Detecting a fault in PMSMs is important because each fault requires its own mitigation action (either interruption in the operation or change in the controller). The main three faults that we are concerned with in this paper are eccentricity, inter-turn short circuit, and partial demagnetization faults.

Many methods have been used to detect and estimate the type and the degree of these faults in PMSM. These methods can be categorized as time domain methods, frequency analysis methods, using Fast Fourier Transform (FFT), [1] and time scale analysis methods, using discrete and continuous wavelet transform [2], [3]. The motor signature current analysis (MSCA) is one of the most common online methods for fault detection since it doesn't require any additional connections or hardware. This method uses spectral analysis techniques (FFT) or time frequency analysis techniques, like wavelet transform applied to the stator current or voltage signal, and by comparing the healthy to the faulty case, faults can be detected.

In this paper, a method is proposed to detect whether a machine is healthy or faulty, detect the type of the fault in the case of faulty machine (static eccentricity, inter-turn short circuit or demagnetization), and estimate the severity of the fault in the case of static eccentricity. Two machines are tested using this method. The model and simulation of these machines, under healthy and faulty conditions, is performed using two dimensional finite element analysis software (FLUX-2D). The machines are controlled using a three phase current source. FFT analysis is applied to the phases voltage signal, since the machines are operated at steady state conditions, and the generated harmonics are the features for the classifier. Linear Discriminant Analysis (LDA) classification is used to detect the type of fault and estimate the severity of static eccentricity fault. The classification is applied using MATLAB.

Section II discusses the characteristics of the three faults, and the methods used for fault detection. Section III shows the main parameters of the two tested machines and how finite element analysis can be used to apply different faults. Section IV talks about the main idea of LDA and how it is performed for fault classification. Section V shows the LDA classification results for fault detection and estimation. And section VI is the conclusion.

II. CHARACTERISTICS OF THE FAULTS

A. Eccentricity fault

Eccentricity faults are one of the most common mechanical faults in machines. In the case of a healthy machine the air gap between the stator and the rotor is uniformly distributed. In the cases of an eccentricity fault the air gap is no longer uniform, which cause the flux distribution to be asymmetric and creates a radial force between the stator and the rotor. This force increases with the degree of eccentricity and will cause several effects on

---

This work was partially funded by National Science Foundation grant ECCS1102316, and Jordan University of Science and Technology.

R. Z. Haddad is with the Department of Electrical and Computer Engineering, Michigan State University, East Lansing, MI 48824 USA. (e-mail: haddadre@msu.edu).

E. G. Strangas is with the Department of Electrical and Computer Engineering, Michigan State University, East Lansing, MI 48824 USA. (e-mail: strangas@egr.msu.edu).

978-1-4799-0024-4/13 $31.00 © 2013 IEEE

the machine, like vibration, noise, and possibly wear of the bearing, which, in time, might increase the eccentricity further and cause the rotor and the stator to rub. Therefore, detecting eccentricity fault while it is still in the early stage is very important to protect the machine from severe damage.

There are three different types of eccentricity. The first one is static eccentricity (SE), in which the stator geometric axis center is different than the rotor and the rotation axis center. This may be caused by incorrect positioning during assembling of the machine or stresses applied to the machine stator. The second type of eccentricity is dynamic eccentricity (DE), in which the rotor geometric axis center is different than the stator and the rotation axis center. The main reasons for dynamic eccentricity are a bent machine shaft, bearing wear, stresses applied to the shaft (e.g. thermal stresses) and mechanical resonance at critical speed. The third type of eccentricity is mixed eccentricity (ME). Here, the rotation axis center is different than the stator and the rotor geometric axis center. Fig. 1a shows a cross sectional area of a machine with dynamic eccentricity fault and Fig. 1b shows the machine with static eccentricity fault.

In [1], MSCA is used by applying FFT to the machine stator current. According to this paper, in the case of eccentricity fault a side band frequency pattern will appear at frequencies

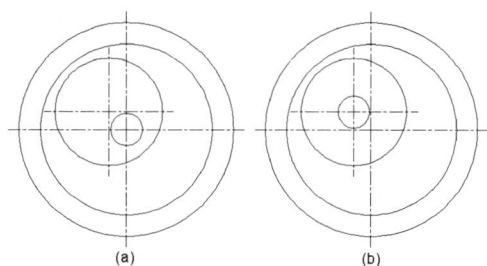

(a)   (b)

Fig. 1. (a) dynamic eccentricity, (b) static eccentricity

$$(1 \pm (2k-1)/P) * fs \qquad (1)$$

where fs is the supply frequency, P is the number of pole pairs, and k is an integer. The amplitude of these side band components can be used to detect the type of eccentricity fault (static or dynamic) and also to estimate the severity of that fault. In [3], discrete wavelet transform (DWT) is applied to the stator current, and is used to estimate eccentricity fault in the PMSM by calculating the energy in their detailed signal. It is shown that the details chosen for eccentricity detection depend on the speed and the sampling frequency. In [4], air gap flux density sensors are used to detect static eccentricity fault in a synchronous generator for different loads, FEA simulation is used to compare and validate the experimental results. According to [5], the minimum eccentricity fault that needs to be detected is about 10% and the maximum eccentricity fault allowed is 60%. Any eccentricity that is less than 10% can be neglected and

any eccentricity fault that is higher than 60% requires an immediate fix to prevent any rub between the stator and the rotor. For this paper, only static eccentricity is discussed because it is more common than dynamic eccentricity.

B. Inter-turn Short circuit

Of the many types of stator winding faults, the inter-turn short circuit fault is one of the most common faults. The causes an inter-turn fault can be classified as mechanical, electrical and thermal stresses applied to the stator winding. These stresses may lead to an insulation break down of the coil conductor which will lead to shorting some of the turns in that coil. In the case of an inter-turn fault, the shorted turn will cause an extra high current path that is coupled with the winding current and flux circuit path. This current will heat the shorted turns and the turns that are near, which will cause insulation damage and might expand the fault to other windings. That's why detecting the inter-turn fault while it is still in the early stage is important in order to protect the machine winding from severe damage. Fig. 2 shows a series connection of a three phase winding with inter-turn short circuit fault at phase A. The fault is modeled by a small resistance r_f connected across the shorted turns.

In [6], a dynamic circuit model was developed, based on the winding function theory, to represent DC brushless motors under the case of inter-turn short circuit fault, and to analyze the behavior and the performance of the faulty machine and compare it with a healthy model. In [7], the MSCA method is used to detect the inter-turn fault by analyzing the machine stator current spectrum using FFT

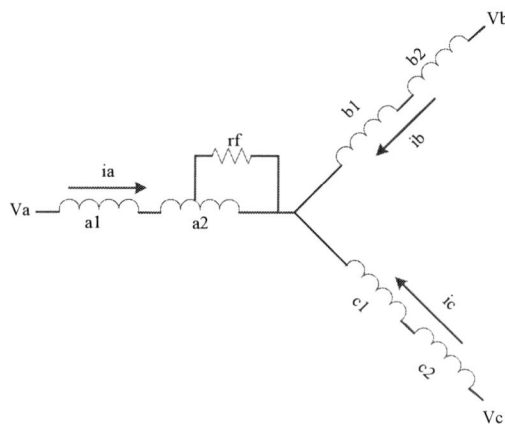

Fig. 2. Series winding with shorted turn.

analysis. According to [8], in the case of inter-turn fault, a sub harmonic frequency pattern will appear in the frequencies similar to that in eqn. (1). The appearance of this pattern can be used as an indicator for a short circuit fault and the amplitude of this pattern is used to estimate the number of shorted turns. K-nearest neighbor classifier (K-

NN) is used to detect the short circuit fault and the support vector machines (SVMs) classifier is used to estimate the number of shorted turns. In [8], the stator short circuit fault is detected by injecting a high frequency signal to the stator current and using the negative sequence impedance as an indicator for detecting the winding short circuit fault.

C. Demagnetization

Demagnetization is also a common fault in PM machines. In the case of a demagnetization fault a high current will flow in the stator winding. This current will weaken the insulation of the winding and affect the machine performance and parameters. Many factors may cause a magnet to demagnetize such as aging of the magnet, the arrangement of the magnets inside the rotor, the number of magnets in each pole and the shape of the magnet. All of these parameters will affect the magnet and may cause partial demagnetization with time.

In [9], MSCA method is used to detect demagnetization fault by applying FFT analysis to the stator current. The generated zero sequence current is used to determine the demagnetization fault at high and low speed. In [10], the continuous wavelet transform (CWT) was applied to the stator current to detect and analyze demagnetization by comparing the CWT of a healthy case with a faulty machine, and the discrete wavelet transform (DWT) is used to evaluate demagnetization by calculating the energy of the detailed DWT. In [11], the Hilbert Huang transform (HHT) is applied to the stator current as a processing tool to detect demagnetization faults in PMSMs. The tests show that HHT can be used for both static and dynamic operating conditions.

A few methods have been used to separate between different faults. In [12] the d-axis incremental inductance is used to detect eccentricity faults and distinguish between eccentricity and demagnetization faults, by the change of the inductance pattern. However, in [13], the harmonic current was used to distinguish between inter-turn short circuit and rotor dynamic eccentricity faults for a permanent magnet synchronous generator. It is shown that side band frequency currents at 25 and 75 Hz can be given by the following equation:

$$(1 \pm k / P) * fs \qquad (2)$$

The 17th and the 19th order stator components can be used to detect a fault and distinguish between dynamic eccentricity and inter-turn fault. From previous reviews it is shown that a fault can be detected in machines, but the main problem is that there is no specific way to distinguish between these faults since they have similar effects on the frequency spectrum.

III. NUMERICAL EXPERIMENTS

Finite element analysis [14] is a powerful and flexible technique for solving differential equation using numerical methods. In this paper, two types of PMSMs are modeled and tested using finite element analysis. Both machines are 3 phase, 300V. The first machine has 16 poles and a concentrated winding with 8 parallel branches per phase, while the second has 12 poles and a distributed winding with 2 parallel branches per phase. Fig. 3a shows the geometric cross sectional area of the concentrated winding machine and Fig. 3b shows the geometric cross section of the distributed winding machine. The specifications and the parameters of the two PMSMs are summarized in Table I.

FEA can be a useful tool for machine analysis and design; it can be used to analyze any machine design topology with any materials, parameter or winding distribution. It is also useful to apply faults and detect the effects of faults on the machine performance. To apply static eccentricity fault in FEA, the axis center for the stator geometry should be different than that for both the rotor geometry and the

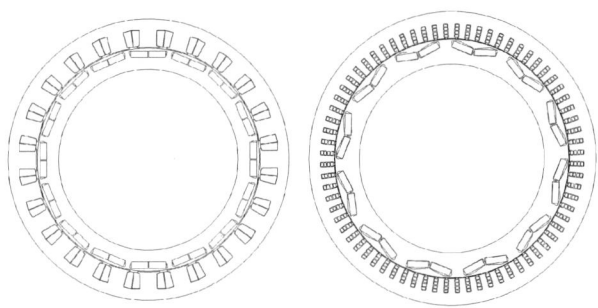

Fig. 3. Geometry cross section (a) concentrated winding, (b) distributed winding

TABLE I
PARAMETERS FOR THE TWO TESTED MACHINES

	Concentrated winding	Distribution winding
Number of phases	3 phase	
Rated voltage	300 V	
Effective length	55 mm	
Max speed	6000 rpm	9500 rpm
Number of slots	24	48
Number of poles	16	12
Magnet remanent Flux	1.15 T	1.2 T
Turns per conductor	46	1
Air gap length	0.8 mm	0.75 mm

rotational axis center. A separate coordinate system is assigned to the stator geometry that is different than the rotor and the rotational coordinate system; this will allow controlling the direction and the degree of eccentricity without affecting the rotor geometric or the rotation center.

In the case of partial demagnetization fault, a new material with lower remanent flux density value is assigned to the faulty magnet to represent the partial demagnetization. For the inter-turn short circuit fault, since each slot consists of many turns, a new region needs to be created in the winding slot. This region will represent the fault and the number of shorted turns will be assigned to that region. A change also needs to be made to the circuit. A new coil conductor corresponding to the new faulty region needs to be added to the circuit to represent this fault in the circuit, a resistance is connected across the coil conductor. Fig. 4 shows the modified circuit used in the case of inter-turn short circuit fault for the concentrated winding machine, where R represents the winding resistance, L represents the end turn inductance, C represents the coil conductor and Rf is used to represent the inter-turn short circuit fault. Fig. 5 shows part of the geometric cross section for the concentrated winding machine with the inter-turn short circuit area applied to phase A of the machine.

IV.   LINEAR DISCRIMINANT ANALYSIS

Three phase currents are used to control and operate the machine in steady state condition (i.e. no variation on the speed or the torque of the machine). To detect the frequency

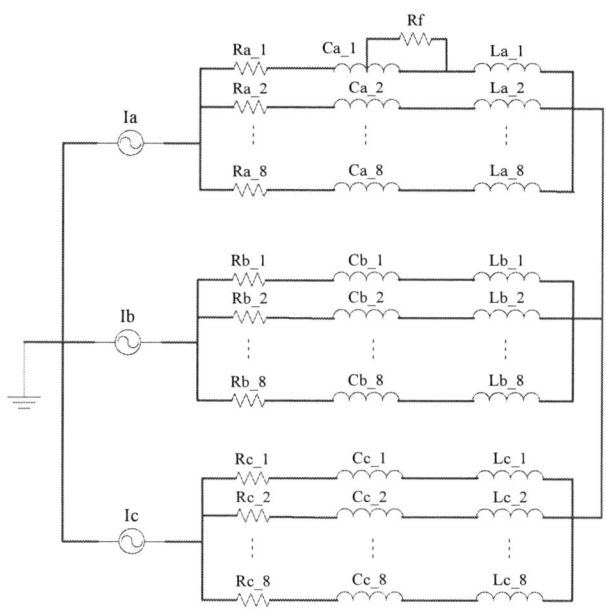

Fig. 4. Modified circuit with inter-turn short circuits fault applied to phaseA for the concentrated winding machine

components, FFT analysis is applied to the machine phase voltage signal. Comparing the frequency spectrum for healthy and faulty cases is not an adequate way to determine which harmonic has the most detailed information for detecting a fault, distinguishing between them or the side band pattern. Therefore, the first 14 harmonics were selected

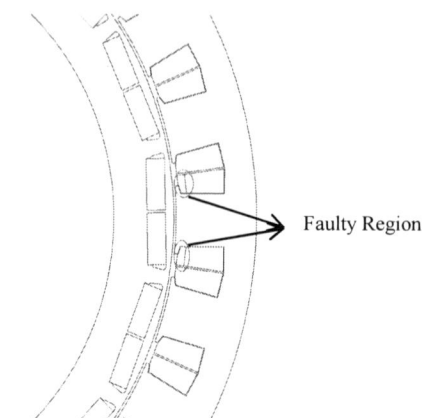

Fig. 5. Modified geometry for concentrated winding with inter-turn fault applied for phase A

as features for the classifier. Linear Discriminant Analysis (LDA) is chosen as a method to classify the three different faults and it is also used to estimate the severity of static eccentricity faults. To validate the classification, two different PMSMs were used and the tests were applied for different operating loads.

The main concept of LDA [15] is based on maximizing the ratio between the class-variance and the within-class variance, to achieve a maximum separation between the feature sets in each class. The main procedure of LDA is to divide the sample space into K number of classes, where each class consists of a specific number of samples corresponding to the same state. These classes are associated through weighting coefficients and each class has its own coefficients that are used to calculate the corresponding linear discriminant function for that class. The linear discriminant function for class k is given by the following formula:

$$C_k(x) = a_{1k}x_1 + a_{2k}x_2 + \ldots + a_{Nk}x_N + a_{N+1,k} \qquad (3)$$

where x is the N dimensional sample vector and $[a_{1k} \ldots a_{N+1,k}]$ is the coefficient matrix for the kth class.

The weighting coefficient matrix is determined using an iterative process called the training phase. During this phase, since we know the proper classification for each sample, the weighting matrix will keep changing until each sample is classified into its correct class. To classify an unknown sample, the measured coefficients from the training phase are used in eqn. (3) to calculate the discriminant functions for this sample. A sample vector belongs to a particular class if the linear discriminant function for that sample is greater than any other linear discernment function. For example, a sample vector i belongs to a class j if:

$$C_j(x) \geq C_k(x) \quad \forall \ j \neq k \qquad (4)$$

V.   SIMULATION RESULTS

FFT analysis is applied to the phase voltage signal, to

detect the frequency components in the signal. Table II shows the mean for the first, fifth and seventh harmonics of the stator voltage for the healthy case and the three different faults (12% static eccentricity, 2 shorted turns and 80% demagnetization).

From the harmonic table it can be noted that all three faults show similar effects in the frequency harmonic spectrum. It is hard to distinguish between the three faults or determine the severity of the fault by simply observing the frequency spectrum. Therefore, a classification method is needed to distinguish between these different faults and to estimate the severity of the fault.

TABLE II
FIRST, FIFTH AND SEVENTH HARMONICS MEAN AMPLITUDE FOR THE THREE DIFFERENT FAULTS

	Harmonic number		
	1st	5th	7th
Healthy	0.2933	0.01186	0.00779
12% Static	0.2937	0.01173	0.00794
2 shorted turns	0.2949	0.01096	0.00851
demagnetization	0.2912	0.008575	0.00661

A. *Identifying the fault type*

LDA classification is used to detect the type of the fault (static eccentricity, inter-turn short circuit or demagnetization). Since only the first 14 harmonics are used as features for the classification, a total number of samples higher than 14 is needed for the LDA classification to converge. In this experiment, the sample space contains 44 samples that correspond to 4 different classes. Each class represents a specific state as follows: class 0 corresponds to healthy case, class 1 corresponds to 12% static eccentricity, class 2 corresponds to 2 shorted turns and class 3 corresponds to 80% demagnetization for one magnet. Each class contains 11 samples generated by varying the speed from 1000 rpm to 2000 rpm in steps of 100 rpm at two different operating loads. The first 14 harmonics from the voltage spectrum are chosen as features for classification. To validate the classification results, the leave one out method is used, in which one sample was not used during the training phase to determine the coefficient matrix. This sample was categorized to its corresponding class using the coefficient determined from the other samples; this method is repeated for every sample in the sample space. Table III shows the classification results for both machines at two different operating loads of 30% and at 60% of full load. It can be seen that LDA is able to classify the type of the fault correctly and distinguish between different faults.

B. *Identifying the degree of static eccentricity*

After detecting the type of the fault, it is necessary to detect its severity. Here, it's assumed that the type of the fault is correctly detected to be eccentricity fault from the above mentioned method. LDA is used again to estimate the severity of the eccentricity fault. For the second classification, the sample space consists of 33 samples and corresponds to static eccentricity for three different severities; 12%, 25% and 45%. Each sample corresponds to a specific speed from 1000rpm to 2000 rpm in steps of 100rpm for the same load. So a total of 3 classes are assigned as follows: class 0 corresponds to 12% static eccentricity, class 1 corresponds to 25% static eccentricity and class 2 corresponds to 45% static eccentricity. The "leave one out" method was used again to validate the results. Table IV shows the classification results for eccentricity severities for both machines for two different loads at 30% and 60% of full load. From this, it is seen that LDA can be used to estimate the type and the degree of eccentricity in PMSM.

From the classification results, it is noted that some of the samples related to 12% static eccentricity fault were not classified correctly, even though only simulation experiments were used. With low eccentricity faults, most of the harmonics amplitude are close, so the LDA classification cannot distinguish between healthy and the 12% static eccentricity fault for a few number of samples.

TABLE III
LDA CLASSIFICATION RESULT FOR FAULT DETECTION

	Correct classification			
	Concentrated winding		Distributed winding	
	30% full load	60% full load	30% full load	60% full load
Class 0 Healthy	100%	100%	100%	91%
Class 1 12% ECC	100%	91%	100%	91%
Class 2 2 short	100%	100%	100%	100%
Class 3 Demag.	100%	100%	100%	100%

TABLE IV
LDA CLASSIFICATION RESULT TO DETECT THE SEVERITY OF STATIC ECCENTRICITY

	Correct classification			
	Concentrated winding		Distributed winding	
	30% full load	60% full load	30% full load	60% full load
Class 0 12% ECC	91%	100%	100%	100%
Class 1 25% ECC	91%	100%	91%	100%
Class 2 45% ECC	100%	100%	100%	100%

VI. Conclusion

In this paper the harmonics of the phase voltage are used as features for fault detection in PMSMs. LDA is used as a classification method to detect the type of the fault. Three types of faults are discussed: eccentricity, inter-turn short circuit and demagnetization. LDA is also used to detect the degree of static eccentricity. The classification results show that the proposed method can be used to detect the types and estimate the severity of the fault correctly for operation at different loads. All the simulations were performed using FEA and the classification was done using MATLAB.

VII. References

[1] Torkaman, H.; Afjei, E.; Yadegari, P., "Static, Dynamic, and Mixed Eccentricity Faults Diagnosis in Switched Reluctance Motors Using Transient Finite Element Method and Experiments," Magnetics, IEEE Transactions on , vol.48, no.8, pp.2254-2264, Aug. 2012

[2] Rosero, J.; Romeral, J.L.; Cusido, J.; Ortega, J.A.; Garcia, A., "Fault detection of eccentricity and bearing damage in a PMSM by means of wavelet transforms decomposition of the stator current," Applied Power Electronics Conference and Exposition, 2008. APEC 2008. Twenty-Third Annual IEEE , vol., no., pp.111-116, 24-28 Feb. 2008

[3] Georgakopoulos, I.P.; Mitronikas, E.D.; Safacas, A.N.; Tsoumas, I.P., "Detection of eccentricity in inverter-fed induction machines using wavelet analysis of the stator current," Power Electronics Specialists Conference, 2008. PESC 2008. IEEE , vol., no., pp.487-492, 15-19 June 2008

[4] Iamamura, B. A T; Le Menach, Y.; Tounzi, A.; Sadowski, N.; Guillot, E.; Jacq, T.; Langlet, J., "Study of synchronous generator static eccentricities — FEM results and measurements," Electrical Machines (ICEM), 2012 XXth International Conference on , vol., no., pp.1829-1835, 2-5 Sept. 2012

[5] Thomson, W.T.; Barbour, A., "On-line current monitoring and application of a finite element method to predict the level of static airgap eccentricity in three-phase induction motors," Energy Conversion, IEEE Transactions on , vol.13, no.4, pp.347-357, Dec 1998

[6] Taehyung Kim; Hyung-Woo Lee; Sangshin Kwak, "The Internal Fault Analysis of Brushless DC Motors Based on the Winding Function Theory," Magnetics, IEEE Transactions on , vol.45, no.5, pp.2090,2096, May 2009

[7] Ebrahimi, B.-M.; Faiz, J., "Feature Extraction for Short-Circuit Fault Detection in Permanent-Magnet Synchronous Motors Using Stator-Current Monitoring," Power Electronics, IEEE Transactions on , vol.25, no.10, pp.2673-2682, Oct. 2010

[8] Briz, F.; Degner, M.W.; Zamarron, A.; Guerrero, J.M., "Online stator winding fault diagnosis in inverter-fed AC machines using high-frequency signal injection," Industry Applications, IEEE Transactions on , vol.39, no.4, pp.1109-1117, July-Aug. 2003

[9] Rosero, J.A.; Cusido, J.; Garcia, A.; Ortega, J.A.; Romeral, L., "Study on the Permanent Magnet Demagnetization Fault in Permanent Magnet Synchronous Machines," IEEE Industrial Electronics, IECON 2006 - 32nd Annual Conference on , vol., no., pp.879-884, 6-10 Nov. 2006

[10] Rosero, J.; Romeral, L.; Cusido, J.; Ortega, J.A., "Fault detection by means of wavelet transform in a PMSMW under demagnetization," Industrial Electronics Society, 2007. IECON 2007. 33rd Annual Conference of the IEEE , vol., no., pp.1149-1154, 5-8 Nov. 2007

[11] Espinosa, A.G.; Rosero, J.A.; Cusido, J.; Romeral, L.; Ortega, J.A., "Fault Detection by Means of Hilbert–Huang Transform of the Stator Current in a PMSM With Demagnetization," Energy Conversion, IEEE Transactions on , vol.25, no.2, pp.312-318, June 2010

[12] Jongman Hong; Sanguk Park; Doosoo Hyun; Tae-june Kang; Sang Bin Lee; Kral, C.; Haumer, A., "Detection and Classification of Rotor Demagnetization and Eccentricity Faults for PM Synchronous Motors," Industry Applications, IEEE Transactions on , vol.48, no.3, pp.923-932, May-June 2012

[13] Xiao Zhaoxia; Fang Hongwei, "Stator Winding Inter-Turn Short Circuit and Rotor Eccentricity Diagnosis of Permanent Magnet Synchronous Generator," Control, Automation and Systems Engineering (CASE), 2011 International Conference on , vol., no., pp.1-4, 30-31 July 2011

[14] Jacek F. Gieras, Mitchell Wing. Permanent Magnet Motor Technology: Design and Applications, second edition, New York.Basel

[15] T. Y. Young and T. W. Calvert, Classification, Estimation and Pattern Recognition. American Elsevier Publishing Co., Inc., 1974.

Early Broken Rotor Bar Detection Techniques in VSD-fed Induction Motors at Steady-state

R. J. Romero-Troncoso, D. Morinigo-Sotelo, O. Duque-Perez, P. E. Gardel-Sotomayor, R. A. Osornio-Rios, A. Garcia-Perez

Abstract -- Condition monitoring has become necessary to detect failures in induction motors (IM), where the detection of incipient faults is of great concern. However, the detection of partially-broken rotor bars at an early stage is not so easily achieved. Therefore, it is necessary to use suitable condition monitoring accompanied with signal processing techniques to detect partially-broken rotor bar. This paper presents a comparative study of various condition monitoring methods accomplished for IM, with the aim of early detection of one partially-broken rotor bar by steady-state current spectrum analysis and different supply conditions, such as two different variable speed drives providing three fundamental supply frequencies, and the line supply case. The study includes three different load conditions for each case. Results show that the most accurate and robust analysis methodology for early detection of broken rotor bars under different supply conditions, fundamental supply frequencies and load conditions during steady-state analysis, are the subspace methods.

Index Terms-- Condition monitoring, Fault diagnosis, Induction motors, Multiple signal classification, Spectral analysis, Wavelet transforms.

I. INTRODUCTION

INDUCTION motors (IM) are the most popular rotating electrical machines consuming more than 50% of the electrical energy used in industry. However, in industrial applications, IM is subject to inevitable stress that produces failures in its different parts. An IM failure may yield unexpected interruptions at the industry plant, with consequences in costs, product quality, and safety. Hence, condition monitoring has become necessary to detect failures in IM and it has been a subject of research for the last three decades, as it can be verified in [1-8]. During the last decade, there has been much interest in the development of early fault detection and diagnosis techniques for use in condition-based maintenance (CBM), being the key for its success the effective early fault condition detection [9]. Thus, the

---

Partially supported by project SEP PIFI-2012 Universidad de Guanajuato.

A. Garcia-Perez is with HSPdigital CA-Procesamiento Digital de Señales at DICIS, University of Guanajuato, Salamanca, Gto. 36885 Mexico (corresponding author; e-mail: arturo@ugto.mx).

R. J. Romero-Troncoso is with HSPdigital CA-Telematica at DICIS, University of Guanajuato, Salamanca, Gto. 36885 Mexico (e-mail: troncoso@hspdigital.org).

D. Morinigo-Sotelo and O. Duque-Perez are with the Department of Electrical Engineering, University of Valladolid, UVa. 47011 Valladolid, Spain (e-mail: daniel.morinigo@eii.uva.es, oscar.duque@eii.uva.es).

P. Gardel is with the University of Valladolid, UVa, 47011 Valladolid, Spain, and with the National University of Asuncion, Paraguay (e-mail: pedroesteban.gardel@alumnos.uva.es).

R. A. Osornio-Rios is with HSPdigital CA-Mecatronica at the Faculty of Engineering, Autonomous University of Queretaro, San Juan del Rio, Qro. 76806 Mexico (e-mail: raosornio@hspdigital.org).

detection of incipient faults is of great concern. Full broken-rotor bars in squirrel-cage IM are easy to detect using steady-state current monitoring [10-11]. This is based on monitoring the amplitudes of the two slip frequency sidebands of the fundamental supply frequency in the current spectrum [12]. It has been shown that the greater the rotor-bar fault severity, the higher is the amplitude of these sidebands. However, the detection of partially-broken rotor bars at an early stage is not so easily achieved. A partially-broken rotor bar can lead to larger failures, or even be catastrophic, and yet may not be detectable under full load conditions. Therefore, it is necessary to use suitable condition monitoring accompanied with powerful signal processing techniques to detect the broken rotor bar at an early stage.

Several methods and techniques have been used to detect one full or a partially-broken rotor bar in an IM under different load conditions, or even when the IM is fed by a variable speed drive (VSD) at different fundamental frequencies. For instance, Garcia-Escudero *et al.* [13] proposed an expert system to detect incipient broken rotor bars in VSD-fed IM. Razik *et al.* [14] presented a diagnosis of a growing broken bar using a genetic algorithm. Yazidi *et al.* [15] examined the detection of one half-broken rotor bar using flux signature analysis. Garcia-Perez *et al.* [16] proposed an experimental study of one partially broken-rotor-bar time-frequency evolution effects using the short-time multiple signal classification (MUSIC) method, but the computational load is very high compared to the FFT. Wolbank *et al.* [17] proposed monitoring a rotor bar defect in a VSD-fed induction machine by using the VSD to establish a voltage pulse excitation and the built-in current sensors to extract the fault indicator. Rangel-Magdaleno *et al.* [18] examined the methodology for one half-broken rotor bar detection, which combines current and vibration analysis. Kowalski and Wolkiewicz [19] proposed the early stator and rotor fault detection of induction motors supplied from a VSD by the FFT algorithm, used for the spectral analysis of the instantaneous active power, the instantaneous reactive power and the estimated electromagnetic torque. Soualhi *et al.* [20] examined an approach to intelligent fault detection and diagnosis of three-phase induction motor fed by a VSD and using signal-based method. Duque-Perez *et al.* [21] presented a statistical study for condition monitoring of induction motors fed by different VSD at different load conditions. The Wigner-Ville and Hilbert-Huang methods are also applied to condition monitoring accomplished for induction motor; however, these techniques are best suited for transient signals. Some of these methods are non-deterministic; others require some kind of training, but they have in common the loss of analytical information on the fault-related frequencies. Few of these references treat the early fault detection problem; yet, some of the proposed analysis techniques are sensitive enough to be used in the detection of partially broken rotor bars. Despite the potentiality of some analysis techniques, there is not a comparative study that assesses the effectiveness of the proposed analysis techniques in the early detection of partially-broken rotor bars under different operating conditions such as: severity of the fault, line-fed or VSD-fed IM, VSD brand, VSD operating frequency, and different load conditions. The introduction of VSD-fed motors has introduced significant challenges in the field of diagnostics needing further research in order to overcome some inconveniences such as noise that reduces the possibility of true fault signature recognition using line current spectrum [22], [23], dynamically changing excitation frequency and the fact that signatures produced by a fault can significantly change from open-loop to closed-loop VSD operation. It is also important to notice that manufactures have adopted different techniques to generate supply voltage that produces different current line harmonic content. This has been considered in this paper, analyzing two different VSD brands.

This paper focuses on a comparative study and evaluation of various condition monitoring methods accomplished for induction motor, with the aim of early detection of one partially-broken rotor bar by steady-state current spectrum analysis and different supply conditions, such as two different VSD providing three fundamental supply frequencies at 35 Hz, 50 Hz and 65 Hz and the line supply case at 50 Hz. The study includes three different load conditions for each case. The goal is to determine the most suitable and accurate steady-state analysis methodology for early detection of broken rotor bars under different conditions on the power supply, fundamental frequency, and motor load. The methodologies considered in the study are FFT, Wavelet plus FFT, MUSIC, Empirical Mode Decomposition (EMD) plus FFT, and EMD plus MUSIC.

II. BACKGROUND

A. Broken Rotor Bar Fault

The detection of broken rotor bar faults can be done by the observation of the space harmonics (f_{BB}) components as a fault indicator:

$$f_{BB} = \left[k\left(\frac{1-s}{p}\right) \pm s \right] f_s \qquad (1)$$

where s is the per-unit motor slip, p is the number of pole pairs of the motor, $k/p = 1,3,5,\ldots$ are the characteristic values of the motor, and f_s is the electrical supply frequency [16]. The characteristic frequencies of broken rotor bars are very close to the supply frequency, because the slip of an induction motor at rated load is small and even smaller under light or no load.

B. Broken Rotor Bar Fault Detection Techniques

A brief description of the Fast Fourier Transform (FFT), Wavelet, MUSIC, Empirical Mode Decomposition (EMD) with FFT and EMD with MUSIC spectral analysis techniques applied to steady-state current is presented next.

1) Fourier Based Analysis:

The FFT algorithm is a spectrum analysis technique that has been successfully used for the detection of broken rotor bars in induction motors requiring low computational effort, based on the steady state motor current signature analysis (MCSA). However, this method suffers from some serious drawbacks such as: it is applicable only in the steady-state regime and not for transient regime; it is a difficult task the detection at the no-load or light load condition of the motor; and at light load condition, it is quite complicated to distinguish between healthy and faulty rotors because the faulty frequencies of a broken rotor bar are very close to the fundamental component and their amplitudes are smaller in comparison. Besides, FFT has high sensitivity to noise, making difficult the analysis when the acquired signal is noisy. As a result, detection of the fault and classification of the fault severity under light load is arduous work with this technique. However, it is possible to apply the FFT in the torque waveform in the area close to 300Hz; where it is possible to detect the broken bar fault reliably even at low-load operation [24].

2) Wavelet Analysis

Wavelet Transform (WT) provides a time-scale representation of a signal with a multi-resolution characteristic, providing flexibility in describing signals that include regions of different frequency contents [25]. The WT at frequency scale l is given by,

$$W(a^l, \tau) = \sum_{n=0}^{N-1} x[n] \psi_l^*[n-\tau] \qquad (2)$$

where

$$\psi_l[n] = \frac{1}{\sqrt{a^l}} \psi\left(\frac{n}{a^l}\right) \qquad (3)$$

and where $\psi(\cdot)$ is the mother wavelet which satisfies a number of conditions. The wavelet transform method is the DWT (Discrete wavelet transform) and the number of decomposition levels is three. For the extraction of the frequency components caused by rotor broken bars, the number of decomposition levels should be equal or higher than that of the detail signal containing the fundamental frequency. The FFT algorithm is applied then to the level where faulty frequencies are located.

3) MUSIC Analysis

978-1-4799-0024-4/13 $31.00 © 2013 IEEE

The subspace methods are known as high-resolution methods, which detect frequencies with low signal-to-noise ratio. The MUSIC algorithm is a class of spectral techniques based on eigen analysis of the autocorrelation matrix. The subspace methods assume that the discrete-time signal $x[n]$ can be represented by m complex sinusoids in noise $e[n]$ [12], i.e.

$$x[n] = \sum_{i=1}^{m} \overline{B_i} e^{j2\pi f_i n} + e[n], \qquad n = 0,1,2,....,N-1 \quad (4)$$

where B_i is the complex amplitude of the i-th complex sinusoid, f_i is its frequency, and $e[n]$ is a sequence of white noise with zero mean and a variance σ^2. The MUSIC spectrum Q is given by:

$$Q^{MUSIC}(f) = \frac{1}{\left| \mathbf{e}(f)^H \mathbf{v}_{m+1} \right|^2} \quad (5)$$

where $\mathbf{e}^H(f_i)$ is the signal vector, and $\mathbf{v}_{m+1}$ is the noise eigen-vector. This expression exhibits the peaks that are exactly at frequencies of principal sinusoidal components where $\mathbf{e}(f)^H \mathbf{v}_{m+1} = 0$.

The advantage of the MUSIC algorithm is that with short-time samples, the spectrum estimation has a higher resolution compared with the FFT algorithm, thus the effects of stator current fluctuation and the noise to the detection of broken rotor bar fault can be decreased and the estimated spectrum is very clean.

4) Empirical Mode Decomposition with FFT Analysis

The EMD method decomposes a signal into oscillating components obeying some basic properties, called Intrinsic Mode Functions (IMF). The principle of EMD method is to decompose any signal $s(t)$ into a set of band-limited functions $Cn(t)$, which are zero mean oscillating components, called the IMF. The idea of finding the IMF relies on subtracting the highest oscillating components from the signal with a step-by-step process, which is called the shifting process. The signal $s(t)$ is represented as a sum of n IMF signals plus a residue signal $r_n(t)$ [26],

$$s(t) = \sum_{i=1}^{n} C_i(t) + r_n(t) \quad (6)$$

After the decomposition of the signal into IMF by the EMD method, each IMF is analyzed by the FFT algorithm to identify the related fault frequencies.

5) EMD with MUSIC Analysis

In this case, after the signal is decomposed in IMF by the EMD method, the second IMF is analyzed by the MUSIC algorithm to identify the related fault frequencies. This technique reduces the computational load compared to the MUSIC algorithm alone, because the order required to obtain the spectrum is much lower.

III. EXPERIMENTAL SETUP

The steady-state current signal is used to realize the comparative study that assesses the effectiveness of the proposed analysis techniques in the early detection of partially-broken rotor bars under different operating conditions. Fig.1 shows the general experiment setup with a 1.1-kW three-phase induction motors (model Siemens 1LA7090-51). The tested motor has two pairs of poles and receives a power supply of three different sources, line supply (LS) at 230 V_{AC} and 50 Hz, a VSD (PowerFlex 40) by Allen-Bradley (AB), and a VSD (MicroMaster 420) by Siemens (SM). Both VSD are tested supplying at 35, 50, and 65 Hz of fundamental frequency. The mechanical load consists in a Lucas-Nülle SE 2662-5R electromagnetic brake, with its control unit, which sets a constant-torque load to provide three different slips: low, medium and high. The current signal is acquired by using a proprietary board with Honeywell CSNE151 Hall-effect current sensors. A National Instruments NI cDAQ-9174 base platform with a 16-bit NI 9215 acquisition module is used for data acquisition. The instrumentation system uses a sampling frequency f_o=4 KHz, obtaining 40,000 samples in 10 s during the induction motor steady state. Tests were performed first with the motor in healthy condition (HLT). Then, a hole was progressively drilled at near the end ring into one of the rotor bars drilling 1 mm of depth to produce an early fault (EAR); after that, a drilling of 12 mm is done to produce a severe (SEV) partially-broken rotor bar; and finally drilling until the rotor bar is fully broken at 18 mm (BRB); where a total of twenty tests were realized for each case.

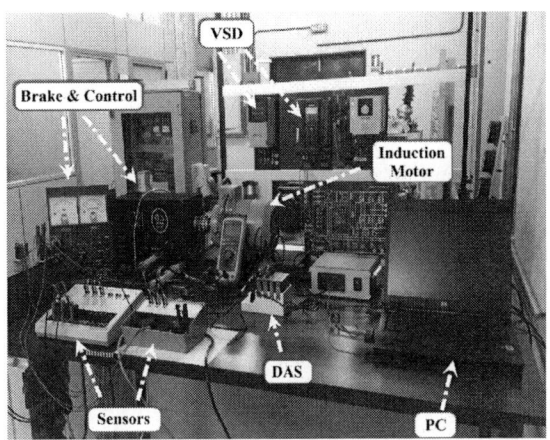

Fig. 1. General aspect of the experimental setup.

IV. EXPERIMENT RESULTS

The comparative study of techniques is implemented in the Matlab Digital Signal Processing Toolbox. The Power Spectral Density (PSD) is estimated from the signal itself applying the PSD estimators studied in this work. In order to reduce the computation time and to optimize the spectrum estimation, after the data-acquisition stage, a low-pass filter with a cutoff frequency of 100 Hz is used to limit the frequency region. This limited band corresponds to the detection region where the sideband fault-frequency components of a rotor with broken bars are observed, so the model order of the subspace methods and their computation time are considerably reduced. The next figure considers a special case called "false positive" and it is defined for the analysis of a healthy motor; where the obtained spectrum has some spectral components induced by the VSD, so this spectrum is indistinguishable from the spectrum of a faulty motor; then the used algorithm is not able to discern between the two motor conditions.

Fig. 2 shows three FFT spectra for the following conditions: a) A healthy motor with the line supply at 50 Hz, b) A healthy motor fed with the Allen-Bradley (AB) VSD at 65 Hz, where the two side components are not equidistant from the main 65 Hz frequency due to the VSD induced effects, c) An early fault condition in the rotor bar with a VSD from AB at 65 Hz, and d) An early fault condition in the rotor bar with a VSD from Siemens (SM) at 65 Hz; all cases with a mechanical load for medium slip level. Fig. 2a depicts the spectrum of a healthy motor where no other frequency than the line supply at 50 Hz is observed. Fig. 2b shows the spectrum of a healthy motor where a false positive (FP) error in the detection of a broken rotor bar fault is present; this condition occurs because the spectrum is highly influenced by the specific VSD used [16].Fig. 2c shows the spectrum of an early fault condition, where this fault is not deteted by the FFT method. Fig 2d depicts the spectrum of an early fault condition in a different VSD-fed motor where the early fault is not detected by the FFT method.

Fig. 3 depicts the spectra of the EMD+MUSIC methodology, showing that it is able to correctly detect all conditions: healthy, early fault, severe fault and fully-broken rotor bar, when the motor is directly fed with the power line at a medium slip.

Fig. 4 shows the spectra for an early fault condition in the rotor bar, under the following methods and VSD models: a) FFT and AB VSD; b) EMD+FFT and AB VSD; c) Wavelet+FFT and AB VSD; d) MUSIC and AB VSD; and e) EMD+MUSIC and AB VSD. All these cases have a low slip level and the supply frequency is 35 Hz. The early fault condition of partially-broken rotor bar is not detected with FFT, Wavelet+FFT or EMD+FFT methods; however, the MUSIC and EMD+MUSIC methods are able to recognize this early-fault condition, even when the motor is VSD fed and with a low slip level; but only the right sideband is detected because the VSD-induced effects.

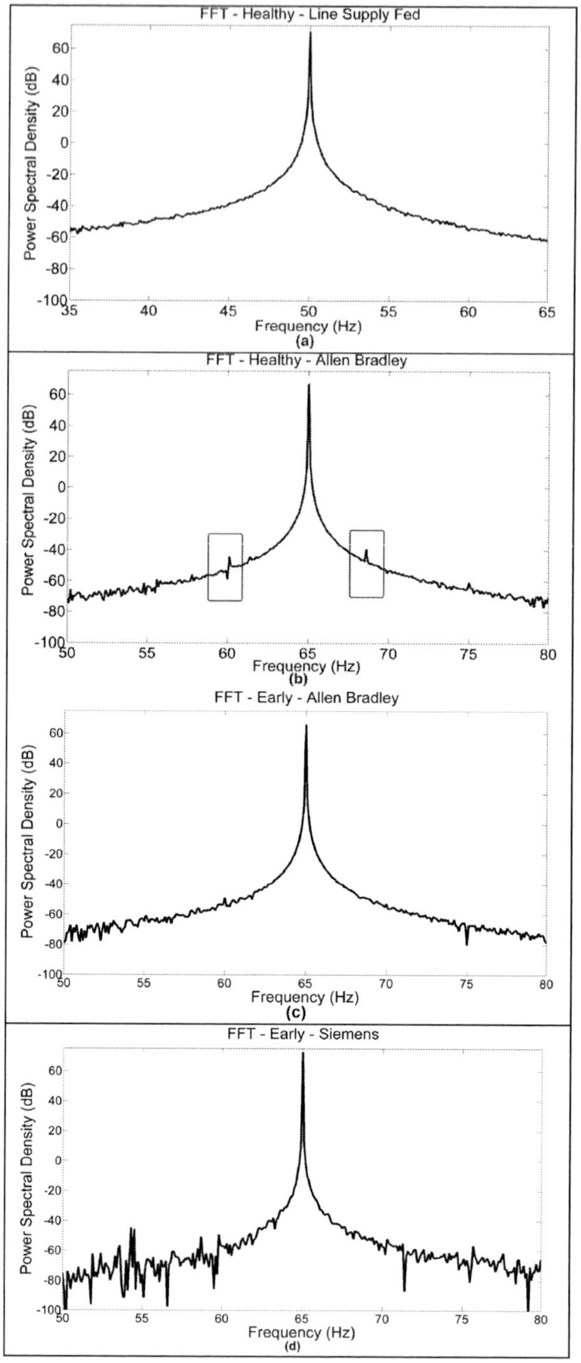

Fig. 2. FFT spectra of: a) Healthy motor with the line supply at 50 Hz, b) Healthy motor fed with the Allen-Bradley (AB) VSD; c) Early fault condition in the rotor bar for the motor fed with a VSD from AB d) Early fault condition in the rotor bar for the motor fed with a VSD from Siemens (SM).

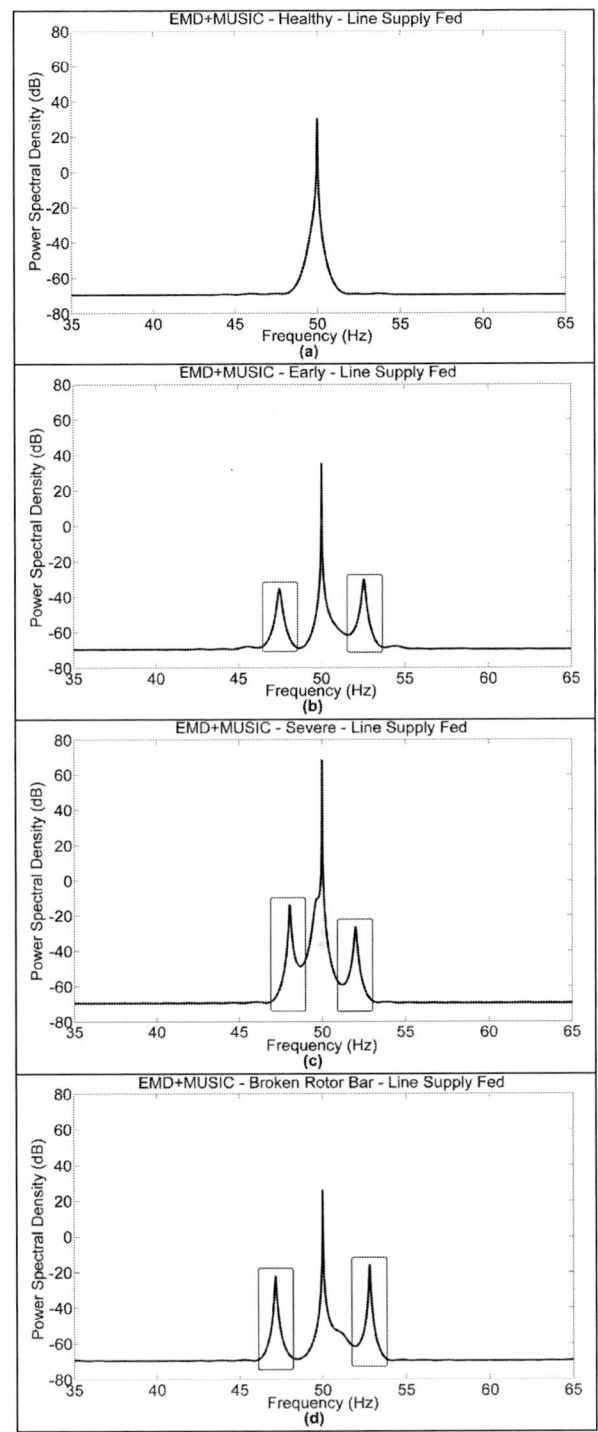

Fig. 3. EMD+MUSIC spectra of: a) A healthy motor; b) An early fault condition in the rotor bar, c) Severe fault condition in the partially-broken rotor bar, and d) A fully-broken rotor bar.

Table I and Table II show the detectability comparison in decibels for all methodologies with Table I showing results for the VSD-fed motor, and Table II depicting results for the power-line-fed motor. The detectability is calculated as the amplitude ratio average between the peak amplitude of both side frequencies present in the faulty condition. The comparative study to assess the effectiveness of the proposed analysis techniques in the early detection of partially-broken rotor bars is done under the following combination of operating conditions:

a) Severity of the fault:
    1. Healthy motor (HLT),
    2. Early partially-broken rotor bar (EAR),
    3. Severe partially-broken rotor bar (SEV),
    4. Fully-broken rotor bar (BRB),
b) VSD brand and operating frequencies:
    1. Allen-Bradley (AB), f_s=35, 50 and 65 Hz
    2. Siemens (SM), f_s=35, 50 and 65 Hz
c) Load condition for a slip level:
    1. Low
    2. Medium (Med)
    3. High
d) Noise level:
    1. Low-noise level (E). For this condition, the analyzed signal is directly taken from the data acquired in the experimentation.
    2. Noisy signal (N). This condition is produced by adding Gaussian noise to the original acquired signal.

The notation presented in Tables is as follows:
- (H) means that the methodology is able to rightly detect a Healthy condition of the IM, because it is perfectly distinguishable from faulty conditions. This assessment is done by establishing a threshold, that clearly differentiates the healthy from the faulty condition.
- (FP) indicates that the methodology detects a False Positive condition when a healthy IM is tested, which is indistinguishable from some faulty conditions.
- (---) indicates that the methodology is unable to detect the faulty condition because it is indistinguishable from other conditions.
- The numbers represent the detectability values for the different cases studied in this work, indicating that the faulty condition can be detected correctly.

For better understanding, the detectability values obtained in both tables, two cases are reviewed. The first case (highlighted as FP in yellow in Table I) considers a false positive (FP) condition when a healthy Allen-Bradley (AB) motor is VSD-fed at a supply frequency of f_s=65 Hz. The acquired signal has low-noise contents (E), the slip condition is Medium (Med), and the PSD estimation is based on the FFT method corresponding to the spectrum depicted in Fig. 2b. For this case, the FP indicates that the methodology is unable to distinguish from a healthy condition and an early fault.

The second case (highlighted as 15 in cyan in Table I) considers the case of an early broken-rotor bar condition with a VSD from Siemens (SM), at a supply frequency of f_s=35 Hz.

978-1-4799-0024-4/13 $31.00 © 2013 IEEE

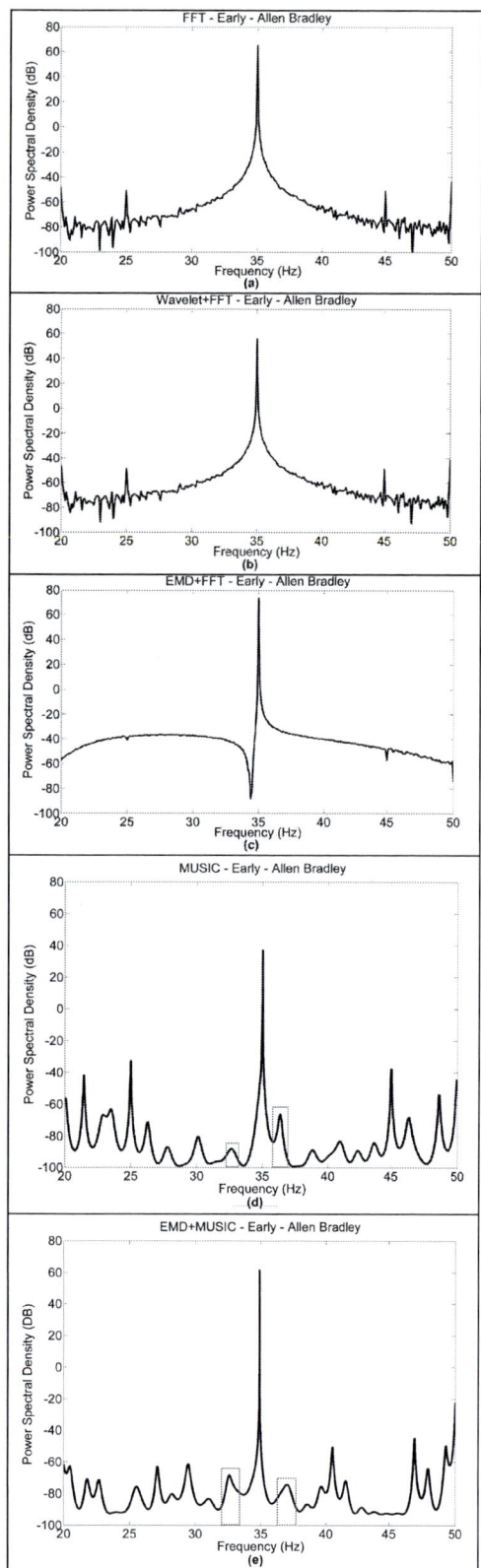

Fig. 4. Spectra for an early fault condition in the rotor bar, under the following methods and VSD models: a) FFT and AB VSD fed; b) EMD+FFT and AB VSD fed; c) Wavelet+FFT and AB VSD fed; d) MUSIC and AB VSD fed, and e) EMD+MUSIC and AB VSD fed.

The acquired data has low-noise contents (E), with a condition of Low slip, and the PSD estimation is obtained with the MUSIC method as depicted in Fig. 4c. The detectability value of 15 dB is calculated as the amplitude average between the peak amplitude of both faulty side-frequencies around the supply frequency (10 dB and 20 dB, respectively in this case). The 15 dB value indicates that the methodology is able to distinguish from the healthy and early fault conditions. The same procedure is made for all possible cases studied in this work.

V. DISCUSSION

From results in both tables, some remarks can be made:

a) The use of VSD to feed a motor can result in false-positive errors in the detection of the broken rotor bar condition when the motor is healthy. This condition occurs because the spectrum is highly influenced by the specific VSD-induced harmonics and sub-harmonics.

b) The VSD-induced harmonics make some algorithms such as the FFT, unable to identify broken rotor bars fault under conditions of light load and incipient fault; however, the fault can be detected for a severe fault and full load conditions.

c) The detectability, when the IM is fed from the line supply, depends on the PSD estimation method used; where the best results are obtained with MUSIC and EMD+MUSIC methods. The MUSIC order was determined heuristically by adjusting the order manually as a function of fault condition; where the order was between 15 and 35. The order in the EMD+MUSIC method is fixed at 20.

d) The FFT and Wavelet+FFT methods have a similar behavior for most cases. These methods are suited for detecting severe or fully-broken rotor bar conditions with a high slip level.

e) The detectability of FFT method is worse when the signal has additive Gaussian noise (N) for most cases. The other methodologies are not affected by noise as the FFT is.

f) The worst analysis condition occurs when the supply frequency is at 35 Hz and the slip level is low; nevertheless, MUSIC and EMD+MUSIC methods are sensitive enough to detect most of the faulty conditions.

MUSIC method presents better results than EMD+MUSIC. However, MUSIC method alone requires a fine hand-tuning of the optimal order and it has a higher computational load compared to the combination EMD+MUSIC, where the order of MUSIC in this combination is fixed and lower than the MUSIC alone method.

TABLE I
Detectability In Decibels For The Comparative Study of Techniques for Two VSD-fed IM.

	Slip	f_s	FFT				Wavelet + FFT				MUSIC				EMD + FFT				EMD + MUSIC			
			AB		SM		AB		SM		AB		SM		AB		SM		AB		SM	
			E	N	E	N	E	N	E	N	E	N	E	N	E	N	E	N	E	N	E	N
HEALTHY	Low	35	H	H	H	H	H	H	H	H	H	H	H	H	H	H	H	H	H	H	H	H
		50	H	H	H	H	H	H	H	H	H	H	H	H	H	H	H	H	H	H	H	H
		65	H	H	H	H	H	H	H	H	H	H	H	H	H	H	H	H	H	H	H	H
	Med	35	H	H	H	H	H	H	H	H	H	H	H	H	H	H	H	H	H	H	H	H
		50	H	H	H	H	H	H	H	H	H	H	H	H	H	H	H	H	H	H	H	H
		65	FP	FP	H	H	FP	H	H	H	H	H	H	H	FP	H	FP	H	H	H	H	H
	High	35	FP	FP	H	H	FP	H	H	H	H	H	H	H	FP	FP	H	H	H	H	H	H
		50	H	H	H	H	H	H	H	H	H	H	H	H	H	H	H	H	H	H	H	H
		65	H	H	H	H	FP	H	H	H	H	H	H	H	FP	H	H	H	H	H	H	H
EARLY	Low	35	---	---	---	---	---	---	---	---	13	10	15	5	---	---	---	---	10	15	15	10
		50	---	---	---	---	---	---	---	---	6	15	10	20	---	---	---	---	10	30	7	20
		65	---	---	---	---	---	---	---	---	30	12	15	15	---	---	---	---	30	10	15	20
	Med	35	---	---	---	---	---	---	---	---	10	15	25	21	---	---	---	---	20	10	25	15
		50	---	---	5	---	---	---	3	---	20	13	10	20	---	---	---	---	22	20	10	30
		65	---	---	---	---	---	---	---	---	27	24	12	20	----	---	----	---	30	20	12	12
	High	35	---	---	---	---	---	---	---	---	30	28	15	20	---	---	---	---	29	25	15	10
		50	3	---	25	---	4	---	50	---	19	17	13	25	---	---	---	---	25	25	30	25
		65	20	---	---	---	---	---	40	---	26	15	25	15	---	---	---	---	25	30	23	15
SEVEVE	Low	35	---	---	---	---	---	---	---	---	15	13	14	14	---	---	---	---	13	17	10	20
		50	---	---	---	---	---	---	---	---	40	18	25	15	---	---	---	---	13	12	20	15
		65	---	---	---	---	---	---	---	---	50	15	16	20	4	---	---	---	22	13	12	17
	Med	35	10	---	30	---	10	---	40	---	50	20	50	15	8	---	2	---	60	33	50	20
		50	18	---	24	---	20	---	30	---	50	20	50	25	3	---	4	---	60	28	50	25
		65	10	---	18	---	12	---	20	---	60	16	35	25	---	---	4	---	72	24	40	25
	High	35	15	---	35	---	26	---	50	---	40	30	48	19	8	4	2	---	55	35	55	45
		50	24	---	40	---	25	---	60	---	30	28	50	20	4	4	3	3	30	26	54	30
		65	20	---	40	---	20	---	40	---	65	26	34	25	12	15	6	---	70	40	50	25
BROKEN ROTOR BAR	Low	35	---	---	---	---	---	---	---	---	6	5	10	15	---	---	---	---	7	7	15	24
		50	---	---	---	---	---	---	---	---	25	9	10	12	---	---	---	---	12	12	12	25
		65	---	---	---	---	---	---	---	---	12	17	10	10	---	---	---	---	10	10	10	10
	Med	35	25	---	30	---	24	---	36	---	44	25	40	15	4	---	3	---	55	29	46	22
		50	15	---	22	---	35	---	24	---	50	18	48	24	5	---	3	---	50	12	50	18
		65	20	---	40	---	20	---	40	---	60	20	48	15	4	---	4	---	70	16	54	25
	High	35	23	---	50	---	24	---	60	---	47	20	35	27	4	---	3	---	55	16	45	30
		50	38	---	40	---	38	---	45	---	50	20	50	28	4	---	5	---	60	40	62	30
		65	24	---	50	---	40	---	50	---	65	30	50	30	5	---	5	6	75	30	58	30

Spectral Analysis Techniques, VSD-fed by Allen-Bradley (AB) and by Siemens (SM) and noise condition

TABLE II
DETECTABILITY IN DECIBELS FOR THE COMPARATIVE STUDY OF TECHNIQUES FOR LINE SUPPLY-FED IM.

			Spectral Analysis Techniques, line-fed (LS) and noise condition								
	Slip	f_s	FFT		Wavelet + FFT		MUSIC		EMD + FFT		EMD + MUSIC	
			E	N	E	N	E	N	E	N	E	N
HLT	Low	50	H	H	H	H	H	H	H	H	H	H
	Med	50	H	H	H	H	H	H	H	H	H	H
	High	50	H	H	H	H	H	H	H	H	H	H
EAR	Low	50	---	---	---	---	22	22	---	---	35	5
	Med	50	---	---	---	---	10	24	---	---	15	30
	High	50	---	---	1	1	20	26	---	---	25	28
SEV	Low	50	---	---	---	---	10	10	---	---	12	26
	Med	50	---	---	4	---	32	20	2	---	50	25
	High	50	---	---	3	---	48	28	2	---	50	30
BRB	Low	50	---	---	---	---	4	12	---	---	8	15
	Med	50	15	---	20	---	40	15	8	---	46	12
	High	50	10	---	16	14	46	28	7	---	44	33

VI. CONCLUSIONS

This work presents a comparative study and evaluation of various condition monitoring methods accomplished for induction motors, with the aim of early detection of one broken rotor bar by steady-state current spectrum analysis and different supply conditions, such as two different VSD providing three fundamental supply frequencies at 35 Hz, 50 Hz and 65 Hz and the line supply case at 50 Hz. The study includes three different load conditions for all cases. Gaussian noise was added to the signal in order to evaluate the methodologies under low- and high-noise conditions.

When a motor is fed through a VSD, several methodologies for detecting early broken rotor bars such as FFT, Wavelet+FFT, and EMD+FFT, fail in the task because the VSD-induced harmonics and sub-harmonics make the spectrum indistinguishable from a healthy and a faulty condition. Yet, MUSIC and the combination EMD+MUSIC methodologies are able to detect most faulty conditions, even in the early stage, when the motor is fed through VSD or the line supply.

Therefore, it could be determined that the most accurate and robust analysis methodology for early detection of broken rotor bars under different supply conditions, fundamental supply frequencies and load conditions during steady-state analysis are the MUSIC and EMD+MUSIC methods where the twenty tests showed repeatability in each one of the results. From the computational point of view, if one of those methodologies is to be automated, EMD+MUSIC is better than MUSIC alone. This is because MUSIC requires fine hand-tuning when selecting the optimal order and, in general, it requires a high order for analysis. Meanwhile EMD+MUSIC requires lower computational load because the MUSIC part can be fixed to a lower order than MUSIC alone. If the order of the methodology is fixed, it is possible to develop an automated hardware/software unit for early detecting partially-broken rotor bars in induction motors.

For further development, the study can be extended to test these and other methodologies in transient-state analysis.

VII. REFERENCES

[1] M. Riera-Guasp, J. A. Antonino-Daviu, J. Roger-Folch, M. P. M. Palomares, "The Use of the Wavelet Approximation Signal as a Tool for the Diagnosis of Rotor Bar Failures," *IEEE Trans. Ind. Electron.*, vol. 44, pp. 716–726, Jun. 2008.

[2] A. Yazidi, H. Henao, G.-A. Capolino, F. Betin, F. Filippetti, "A Web-Based Remote Laboratory for Monitoring and Diagnosis of AC Electrical Machines," *IEEE Trans. Ind. Electron.*, vol. 58, pp. 4950–4959, Jun. 2008.

[3] M. Delgado, A. Garcia, J. A. Ortega, J. Urresty, J. R. Riba, "Bearing diagnosis methodologies by means of Common Mode Current," in *Proc. 2009 IEEE European Power Electronics and Applications Conf.*, pp. 1-10.

[4] M. Bouzid, G. Champenois, N. Bellaaj, K. Jelassi, "Automatic and Robust Diagnosis of Broken rotor bars fault in Induction Motor," in *Proc. 2010 IEEE Electrical Machines International Conf.*, pp. 1-7.

[5] W.T. Thomson and M. Fenger, "Current Signature Analysis to Detect Induction Motor Faults," *IEEE Trans. on IAS Magazine*, Vol. 7, pp. 26–34, July 2001.

[6] G.B. Kliman, J. Stein, R. D. Endicott and R. A. Koegl, "Noninvasive Detection of Broken Rotor Bars in Operating Induction Motor," in *IEEE Trans. on Energy Conversion*, Vol. 3, pp. 873–879, Dec. 1998.

[7] R. Schoen and T. Habetler, "A new method of current-based condition monitoring in induction machines operating under arbitrary load conditions", Taylor & Francis, *Electric Machines and Power Systems*, vol. 25, pp. 141-152, Sep. 1997.

[8] H. Henao, H. Razik, G.-A. Capolino, Analytical approach of the stator current frequency harmonics computation for detection of induction machine rotor faults, IEEE Transactions Industry Applications, Vol. 41, pp. 801–807, May 2005.

[9] K. Kim and A.G. Parlos, "Induction Motor Fault Diagnosis Based on Neuropredictors and Wavelet Signal Processing," *IEEE/ASME Trans. Mechatronics*, vol. 7, pp. 201–219, Jun. 2002.

[10] A. Soualhi, G. Clerc, H. Razik, A. Lebaroud, "Fault detection and diagnosis of induction motors based on hidden Markov model," in *IEEE 2012 International Conference on Electrical Machines*, pp. 1693-1699.

[11] J. Cusido, J. Rosero, E. Aldabas, J. A. Ortega, L. Romeral; "New fault detection techniques for induction motors", *Electrical Power Quality and Utilization Magazine*, Vol. 2, pp.39-46, 2006.

[12] A. Garcia-Perez, R. J. Romero-Troncoso, E. Cabal-Yepez, and R. A. Osornio-Rios, "The Application of High-Resolution Spectral Analysis for Identifying Multiple Combined Faults in Induction Motors," *IEEE Trans. Ind. Electron.*, vol. 58, pp. 2002–2010, May 2011.

[13] L.A. Garcia-Escudero, O. Duque-Perez, D. Morinigo-Sotelo, M. Perez-Alonso, "Robust multivariate control charts for early detection of broken rotor bars in an induction motors fed by a voltage source inverter," in *Proc. 2011 IEEE Power Engineering, Energy and Electrical Drives Int. Conf.*, pp. 1-6.

[14] H. Razik, M.B.R. Correa and E.R.C. Silva, "A Novel Monitoring of Load Level and Broken Bar Fault Severity Applied to Squirrel-Cage Induction Motors Using a Genetic Algorithm," *IEEE Trans. Ind. Electron.*, vol. 56, pp. 4615–4626, Nov. 2009.

[15] A. Yazidi, H. Henao and G.-A. Capolino, "Broken Rotor Bars Fault Detection in Squirrel Cage Induction Machines," in *Proc. 2005 IEEE International Symposium on Diagnostics for Electric Machines, Power Electronics and Drives*, pp. 741-747.

[16] A. Garcia-Perez, R. J. Romero-Troncoso, E. Cabal-Yepez, R. A. Osornio-Rios, J. J. Rangel-Magdaleno, H. Miranda, "Startup Current Analysis of Incipient Broken Rotor Bar in Induction Motors using High-Resolution Spectral Analysis," in *Proc. 2011 IEEE International Symposium on Diagnostics for Electric Machines, Power Electronics and Drives*, pp. 657-663.

[17] T. M. Wolbank, P. Nussbaumer, H. Chen, P. E. Macheiner,

"Monitoring of rotor-bar defects in inverter-fed induction machines at zero load and speed," *IEEE Transactions on Industrial Electronics*, Vol. 58, No. 5, pp. 1468-1478, May 2011.

[18] J.J Rangel-Magdaleno, R.J. Romero-Troncoso, R.A. Osornio-Rios, E. Cabal-Yepez and L.M. Contreras-Medina, "Novel Methodology for Online Half-Broken-Bar Detection on Induction Motors," *IEEE Trans. Instrum. Meas.*, vol. 58, pp. 1690-1698, May 2009.

[19] C. T. Kowalski, M. Wolkiewicz, "Converter-fed induction motor diagnosis using instantaneous electromagnetic torque and power signals," in *Proc. 2011 IEEE EUROCON*, pp. 811-816.

[20] A. Soualhi, G. Clerc, H. Razik, A. Lebaroud, O. Ondel, "Detection of induction motor faults by an improved artificial ant clustering," in *Proc. 2011 IEEE Industrial Electronics Society Conference*, pp. 3446-3451.

[21] O. Duque-Perez, L. A. Garcia-Escudero, D. Morinigo-Sotelo, P. E. Gardel, M. Perez-Alonso, "Condition monitoring of induction motors fed by Voltage Source Inverters. Statistical analysis of spectral data," in *IEEE 2012 International Conference on Electrical Machines*, pp. 2479-2484.

[22] B. Akin, U. Orguner, H.A. Toliyat and M. Rayner, "Low order PWM inverter harmonics contributions to the inverter-fed induction machine fault diagnosis," IEEE Trans. Ind. Electron., vol. 55, no. 2, pp. 610–619, 2008.

[23] F. Briz, M. W. Degner, P. Garcia and A. B. Diez, "High-frequency carrier-signal voltage selection for stator winding fault diagnosis in inverter-fed ac machines," IEEE Trans. Ind. Electron., vol. 55(12), pp. 4181–4190, 2008.

[24] K.N. Gyftakis, D.V. Spyropoulos, J.C. Kappatou, E.D. Mitronikas, "A Novel Approach for Broken Bar Fault Diagnosis in Induction Motors Through Torque Monitoring", *IEEE Transactions on Energy Conversion*, Vol. 28, pp. 267-277, June 2013.

[25] Antonino-Daviu, J.A.; Riera-Guasp, M.; Folch, J.R.; Palomares, M.P.M., "Validation of a new method for the diagnosis of rotor bar failures via wavelet transform in industrial induction machines," *IEEE Transactions on Industry Applications*, Vol. 42, pp. 990-996.

[26] Antonino-Daviu, J.; Aviyente, S.; Strangas, E.G.; Riera-Guasp, M.; Roger-Folch, J.; Perez, R.B., "An EMD-based invariant feature extraction algorithm for rotor bar condition monitoring," in *Proc. 2011 IEEE International Symposium on Diagnostics for Electric Machines, Power Electronics and Drives*, pp. 669-675.

VIII. BIOGRAPHIES

Rene de J. Romero-Troncoso (M'07–SM'12) received the Ph.D. degree in mechatronics from the Autonomous University of Queretaro, Queretaro, Mexico, in 2004. He is a National Researcher level 2 with the Mexican Council of Science and Technology, CONACYT. He is currently a Head Professor with the University of Guanajuato and an Invited Researcher with the Autonomous University of Queretaro, Mexico. He has been an advisor for more than 180 theses, an author of two books on digital systems (in Spanish), and a coauthor of more than 90 technical papers published in international journals and conferences. His fields of interest include hardware signal processing and mechatronics. Dr. Romero–Troncoso was a recipient of the 2004 Asociación Mexicana de Directivos de la Investigación Aplicada y el Desarrollo Tecnológico Nacional Award on Innovation for his work in applied mechatronics, and the 2005 IEEE ReConFig Award for his work in digital systems.

Daniel Morinigo-Sotelo (M'04) received the B.S. and Ph.D. degrees in Electrical Engineering from the University of Valladolid, Valladolid, Spain, in 1999 and 2006, respectively. He is currently with the Research Group in Predictive Maintenance and Testing of Electrical Machines, Department of Electrical Engineering, University of Valladolid and is a Research Collaborator in Electromagnetic Processing of Materials with the Light Alloys Division, CIDAUT Foundation, Valladolid. His current research interests also include condition monitoring of induction machines, optimal electromagnetic design, and heuristic optimization techniques.

Oscar Duque-Perez received the B.S. and Ph.D. degrees in Electrical Engineering from the University of Valladolid (UVA), Spain, in 1992 and 2000, respectively. In 1994, he joined the E.T.S. de Ingenieros Industriales, UVA, where he is currently Full Professor with the Research Group in Predictive Maintenance and Testing of Electrical Machines, Department of Electrical Engineering. His main research fields are power systems reliability, condition monitoring, and heuristic optimization techniques.

Pedro Esteban Gardel-Sotomayor received the B.S. degree in Electromechanical Engineering from the National University of Asuncion (UNA), Paraguay, in 2006. He is currently a Ph.D. student at the University of Valladolid, Spain. His research interests are monitoring of induction machines and electric power system optimization.

Roque A. Osornio-Rios (M'10) received the B.E. degree from the Instituto Tecnologico de Queretaro, Queretaro, Mexico, and the M.E. and Ph.D. degrees from the University of Queretaro, Queretaro, Mexico, in 2007. He is a National Researcher with CONACYT. He is currently a Head Professor with the University of Queretaro. He was an Advisor of over 30 theses, and a coauthor of over 40 technical papers in international journals and conferences. His fields of interest include hardware signal processing and mechatronics. Dr. Osornio-Rios received the "2004 ADIAT National Award on Innovation" for his works in applied mechatronics.

Arturo Garcia-Perez (M'10) received the B.E. and M.E. degrees in electronics from the University of Guanajuato, Salamanca, Mexico, in 1992 and 1994, respectively, and the Ph.D. degree in electrical engineering from the University of Texas at Dallas, Richardson, in 2005. He is currently a Titular Professor with the Department of Electronic Engineering, University of Guanajuato. He is a National Researcher with the Consejo Nacional de Ciencia y Tecnología level 1. He was an Advisor of over 50 theses. His fields of interest include digital signal processing for applications in mechatronics.

978-1-4799-0024-4/13 $31.00 © 2013 IEEE

Use of Discrete and Optimized Continuous TFD Tools for Transient-Based Diagnosis in Controversial Fault Cases

J. Pons-Llinares, J. Antonino-Daviu, M. Riera-Guasp, S.B. Lee, T-J. Kang and C. Yang

Abstract – **Transient-based diagnosis of electromechanical failures in induction motors has gained an increasing attention over recent years. The diagnostic in some specific situations (presence of load toque oscillations, light loading conditions) or of specific failures may be difficult when using the classical MCSA approach. In this context, the transient-based methodologies have been proven to become valuable informational sources for the diagnosis, either confirming the MCSA results or avoiding its possible false positives. The application of these transient methodologies requires the use of modern signal processing tools that are in continuous evolution. This work proposes the application of an advanced tool; the recently developed Adaptive Slope Transform. The paper compares the performance of this continuous transform and that of a discrete counterpart, the Discrete Wavelet Transform, when applied to different controversial fault cases in which the classical MCSA may not lead to correct results: outer bar breakages in double cage motors and motors with rotor axial duct influence. The results show the potential of the continuous transforms for the transient tracking of high-order fault-related components as well as for the improved discrimination between fault components.**

Index Terms—**fault diagnosis, induction motors, signal processing, Fourier transforms, transient analysis, wavelet transforms**

I. INTRODUCTION

TRANSIENT Motor Current Signature Analysis (TMCSA) [1] consists in the diagnosis of electromechanical failures in electrical machines (induction motors [1-4], induction generators [5], DC motors [6]) through the analysis of transient signals (currents [1-4], voltages [7], vibrations [8]).

TMCSA has proven to be a valuable source of information in some real situations in which the classical MCSA may have difficulties to reach a correct diagnostic. For instance, under the presence of load torque oscillations, supply voltage oscillations, and use of transmission elements (such as pulleys, straps, or multi stage speed reduction gearboxes), the classical MCSA may lead to false positives that can be discarded when applying the TMCSA [1]. In this regard, it can be said that MCSA is only strictly suitable under pure

stationary conditions, that are, on the other hand, quite unusual in the industrial environment, where applications in which the machine load continuously changes (sewage treatment plants, compressors, or coal mills in thermal plants...) and variable speed applications (pumps, machine tools, conveyor belts...) are rather frequent.

Especially interesting is the recently reported difficulty of MCSA to diagnose rotor faults in some specific controversial cases in which the TMCSA has proven to provide satisfactory results:

- *Breakages in the outer bars of double cage rotors*: the breakage effect on the steady-state stator current is reduced, since most of the current is confined in the inner cage during stationary operation [9-10]. The application of TMCSA over the startup current has proven to enhance the sensitivity of the diagnostic since, during the startup, most of the current circulates through the outer cage and the effects of the breakage are much more noticeable [10].

- *Motors with rotor axial ducts (in which the number of axial ducts, N_d, and poles, N_p, are identical)*: this constructive characteristic may introduce frequency components in the FFT spectrum identical to the breakage-related ones, leading to false positives [11]. TMCSA has proven to provide satisfactory results through a startup analysis, enabling to avoid false positives, due to the fact that the presence of axial ducts has no influence at high slip [12].

TMCSA is based on the use of suitable signal processing tools enabling the time-frequency decomposition (TFD) of non-stationary quantities. There are a variety of these time-frequency (t-f) transforms that have been recently applied in the fault diagnosis area [13]. These can be classified in two main groups: continuous and discrete, each of them having its particular advantages and drawbacks.

Among the continuous transforms, there are also two main groups: the Wigner Ville Distributions (WVD) [8], based on the signal correlation with a time and frequency translation of itself, and the atom based transforms, based on the signal correlation with a family of time-frequency atoms. In this latter group we find the Short Time Fourier Transform (STFT) [17], the Continuous Wavelet Transform (CWT) [2],[6], or the recently developed Adaptive Slope Transform (AST) [14].

This paper proposes the application of the AST, a continuous transform enabling an optimized capture of the faulty components complete evolutions, to some of the aforementioned controversial cases barely treated in the literature (outer cage bar breakages in double cage rotors and

---

This work was supported by the Conselleria d'Educació, Formació i Ocupació of the Generalitat Valenciana, in the framework of the "Ayudas para la Realización de Proyectos de I+D para Grupos de Investigación Emergentes", project reference GV/2012/020.

J. Pons-Llinares, J. Antonino-Daviu and M. Riera-Guasp are with Institute for Energy Engineering, Universitat Politecnica de València, Camino de Vera s/n, 46022, Valencia, SPAIN (e-mails: jpons@die.upv.es, joanda@die.upv.es, mriera@die.upv.es).

S.B. Lee, T-J. Kang and C. Yang are with the Department of Electrical Engineering, Korea University, Seoul, Korea (e-mails: sangbinlee@korea.ac.kr).

978-1-4799-0024-4/13 $31.00 © 2013 IEEE   114

motors with rotor axial ducts with $N_d = N_p$). In each case, the AST is compared with a discrete transform extensively applied in the field, the Discrete Wavelet Transform (DWT). Results confirm the potential of the continuous transforms for the transient-based diagnostics and their suitability in these controversial cases in which the classical MCSA may not behave satisfactorily.

II. CONTINUOUS VS DISCRETE TRANSFORMS

The most common approach to solve the problem of electric machines diagnosis in transient regime is the analysis of a transient current, using a t-f transform, in order to obtain the time evolution of its components frequencies (Fig. 1).

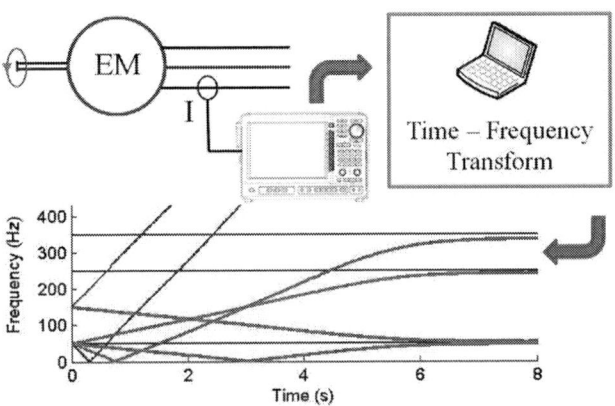

Fig. 1 TMCSA common approach.

If the electric machine has a certain fault, it causes the appearance of specific components in the currents, which will be reflected in the transform result, enabling the fault detection, such as the bar-breakage related components in an induction motor startup current (depicted in red color in Fig. 1). The differences between the different fault diagnosis techniques hitherto proposed rely on the time-frequency transform used to analyze the current.

A. Discrete Transforms

The DWT is used to analyse a transient current (as the stator startup current of an induction motor plotted in red in Fig. 2 (a)), decomposing it in a set of sub-signals (rest of Fig. 2 (a)), each one covering a different frequency band (depicted beside). If the fault appears (as 3 bar breakages in the induction motor: Fig. 2 (b)), the oscillations caused by the faulty components (Lower Sideband Harmonic, LSH) can be quantified in the sub-signals covering the frequency band where the faulty components evolve (a_9, d_9, and d_8, showing the characteristic V (or Λ) pattern [10]). In short, Fig. 2 (a) and (b) show the high-level wavelet signals resulting from the application of the DWT to the first example analyzed in the paper: stator startup current of an induction motor with a single-cage deep-bar Al-die-cast rotor (from now on called Sample 1) under healthy condition (a) and with 3 broken bars (b).

As observed in Fig. 2, the DWT signals clearly show the evolution of the LSH when the machine is faulty [10]. However, it has notable difficulties in tracking the evolution of the high-order fault harmonics, as the Upper Sideband Harmonic (USH) in the case of bar breakages, which in this example are contained in signal d_6. As it can be observed in the figure, the frequency band covered by the wavelet signals increases as the wavelet signal covers higher frequencies. This provokes that the wavelet signals containing the high-order fault-related harmonics (signal d_6) cover very wide frequency bands where many other components can be present or can evolve: Winding Harmonics (WHs), Principal Slot Harmonics (PSHs), other high-frequency harmonics, noises, etc... This fact, combined with the usual low amplitude of the high-order fault-related harmonics, explains the difficulty of the DWT to track with enough precision the high-frequency harmonics evolutions.

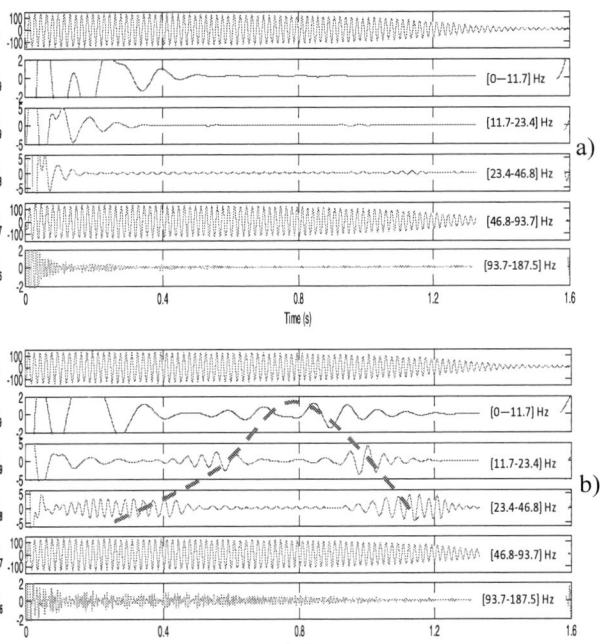

Fig. 2 Startup current, s, and high-level wavelet signals, a_9, d_9, d_8, d_7 and d_6 resulting from the DWT of the startup current of Sample 1 for: (a) healthy machine, (b) machine with 3 broken bars.

B. Continuous Transforms

While the FFT is intended to obtain the distribution of the signal energy in the frequency domain (energy spectral density), the continuous transforms obtain the distribution of the signal energy in the t-f plane (energy t-f density). The classical MCSA is based on the detection of failure-related frequency 'peaks' in the current FFT spectrum; while the underlying idea of the TMCSA, is the identification of the time evolutions of these frequency peaks in the t-f map, during a transient.

In order to obtain the time evolutions of the components frequencies, the t-f energy density of the current analyzed must be obtained. This can be done correlating the signal with a family of t-f atoms (Short Time Fourier Transform or Continuous Wavelet Transform, which have an inherent tradeoff between time and frequency resolutions, with little more freedom in the case of the CWT) or correlating the signal with a time and frequency translation of itself (Wigner-Ville Distributions, which introduces cross terms). In this paper, in order to avoid the mentioned drawbacks of the classic continuous transforms, the use of the AST is proposed: a t-f transform based on using the appropriate atom in each point of the t-f plane analyzed, in order to obtain the best t-f resolution [14]. Fig. 3 (a) and (c) depict the result of analyzing with the AST the same currents whose DWT results are shown in Fig. 2 (a) and (b). In order to avoid the influence of the Fundamental Component (FC), instead of using a logarithmic scale, a different color reference has been used at each frequency analyzed.

Fig. 3 (b) shows the typical t-f evolution of the main healthy (blue) and bar breakage related (red) components in the stator startup current of an induction motor. First, the healthy components (blue) with horizontal evolutions are the WHs, with different orders: WH1 (FC), WH3, WH5, and WH7. Second, the healthy components (blue) with variable frequencies are the PSHs, whose frequencies are given by the following formula:

$$f_{PSH} = \left[kR(1-s)/p \pm v \right] f_m \qquad (1)$$

where in this case $k=1$ (positive integer), $R=44$ (rotor slots), $p=2$ (pole pairs), $v=1,3,5$ with a previous $+$, $-$, and $-$ sign respectively (order of the supply voltage harmonic), $f_m = 60$ Hz (supply frequency) and the slip s evolves from 1 to nearly zero as in a usual startup. The PSHs are named indicating the value and sign (positive or negative) assigned to v: P$v$sign.

Third, the frequencies of the bar breakage-related components (red) are given by the following two formulas ($k \in \mathrm{N}$, $k/p = 1,3,5,...$):

$$f_{bb} = \left[1 \pm 2ks \right] f_m \qquad (2)$$

$$f_{bb} = \left[k/p(1-s) \pm s \right] f_m \qquad (3)$$

In the case of (2), the broken bar related components are named with a "B" indicating the value assigned to k and the sign used: B$k$sign; while in the case of (3) are named with a "b" indicating the value assigned to k/p and the sign used: bk/psign. In Fig. 3 (b) there is a bar breakage component that is not predicted by the previously mentioned theoretical formulas, but it appears on the experimental results: the one evolving from 180 Hz to 300 Hz.

The differences between Fig. 4 (a) and (c) are clearly noticeable: they do not rely only on a single fault harmonic (LSH, that, on the other hand, is captured with much higher precision than with the DWT) but on a wide set of fault-components evolutions (USH as well as other high-order fault related components). Their detection can substantially increase the reliability of the diagnosis in some cases.

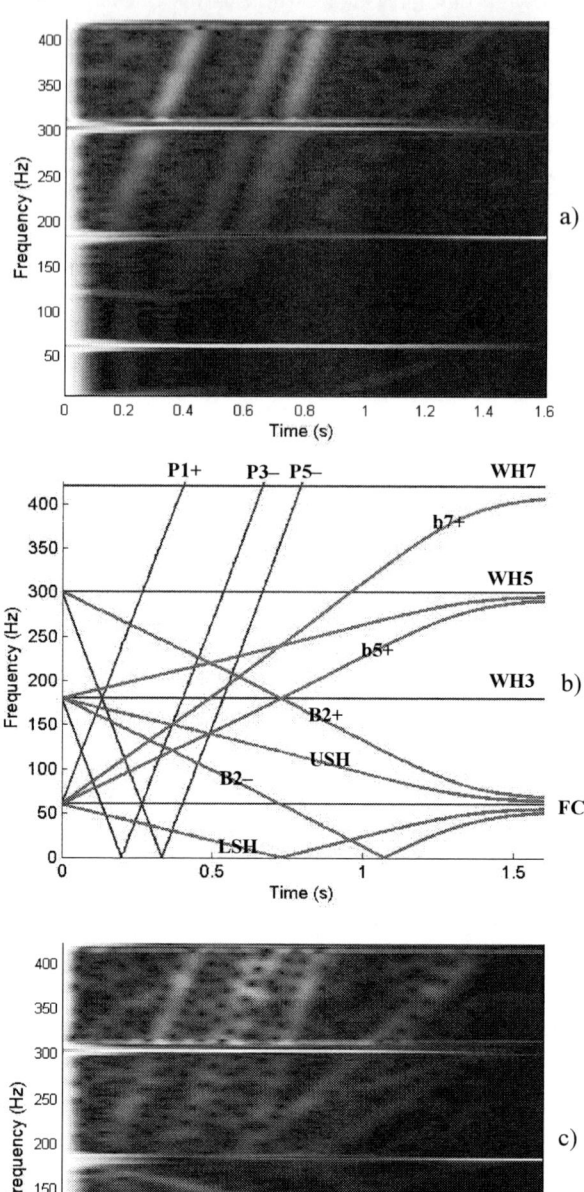

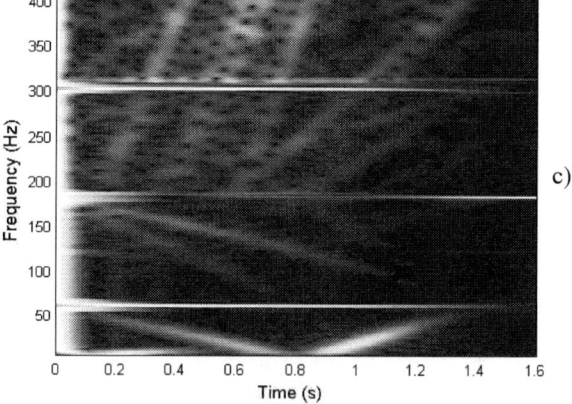

Fig. 3 AST of the startup current of Sample 1 for: (a) healthy machine, (c) machine with 3 broken bars. Theoretical t-f evolutions of the components (blue: healthy, red: bar breakage-related) in the stator startup current of the induction motors analyzed (b).

As the results reveal, continuous transforms such as the AST avoid some of the problems of the DWT. The nature of this tool leads to a continuous picture of the time-frequency map, where the exact frequencies present at each time can be observed (of course, considering some limitations, as the ones fixed by the Uncertainty Principle). Furthermore, proper selection of the dimensions of the time-frequency atoms enables a high-precision tracking of the high-order harmonics [14]. Indeed, TMCSA takes advantage of the characteristic variation during the transient of the fault components frequencies over time, following very specific trajectories, the identification of which is a reliable indicator of the presence of the failure.

C. Comparison

Both the continuous and the discrete transforms have particular advantages and drawbacks, though the former group shows an especial suitability, mainly for off-line diagnosis, where the computational burden is not a crucial constraint. On the one hand, the DWT can be fast and easily applied, and the resulting wavelet signals are also easily managed. On the other hand, among other, continuous TFD tools provide the following advantages over their discrete counterparts in the context of transient-based fault diagnosis:

- *Clearer extraction of the low-frequency fault components evolutions.* The continuous tools provide like a 'photo' of the machine that enables to visualize the whole continuous evolution of the fault-related components during their trips across the transient. On the other hand, the DWT shows this evolution as a finite number of oscillations in the wavelet subsignals; this makes the identification and tracking of the fault related components more difficult.

- *Tracking of the high-order harmonics evolutions.* The continuous tools, if properly optimized, enable to extract the evolutions of the fault components that evolve in the high frequency region. These components have often low amplitudes, a fact complicating sometimes their visualization in the FFT spectrum. On the other hand, some discrete TFD tools (such as the DWT) are better suited for the extraction of the low-frequency components, since their frequency resolution substantially drops at high frequencies. High-order fault components may provide very interesting information for the diagnostic, mainly in those cases in which the tracking of the low-frequency ones is not conclusive.

- *Easier fault discrimination.* The fact that these tools enable the visualization of which specific frequencies are present in the signal at each time (fulfilling the restrictions of the Uncertainty Principle) enables to better segregate the different fault components present in the signal or even to distinguish fault components evolutions and other phenomena.

III. TRACKING HIGH-ORDER HARMONICS IN DOUBLE CAGE MOTORS AND MOTORS WITH AXIAL DUCTS

This section compares the application of the DWT and the AST to the startup currents corresponding to the second and third samples, under different faulty conditions:

- *Sample* 1: Single-cage, deep-bar, Al-die-cast rotor.
- *Sample* 2: Double cage, Cu rotor, with common end ring.
- *Sample* 3: Motor with 4 rotor axial ducts and 4 poles.

The analyses of Samples 2 and 3 are especially interesting, since these are two situations in which the classical FFT analysis may not work well, as proven in [9-12]: Fig. 4 (top) represents the evolution of the classical MCSA indicator for all three previous samples under different faulty conditions ranging from healthy machine up to machine with 3 broken bars. In the case of Sample 1, as expected, the MCSA measurements reveal an increase of the LSH peak when the number of broken bars increases (increment of ~38dB from 0 to 3 broken bars). For the case of double cage motor (Sample 2), the sensitivity of the MCSA analysis is substantially reduced (increment of ~18dB from 0 to 3 broken bars); as proven in [9], this is due to the reduced current circulating through the outer cage in steady-state. Analysis of the startup current was proven to increase the sensitivity [10], based on the fact that during the startup most of the current is indeed confined in that cage. Finally, Fig. 4 (top) also shows the MCSA measurements corresponding to Sample 3 for 0, 1, and 2 broken bars located at ϕ_e=45° and ϕ_e=135° (electric angles, see Fig. 4 (bottom)). It proves that, if N_d=N_p, the presence of axial ducts may lead to false positives when the machine is healthy (-51 dB for healthy condition) and that the LSH peak can even decrease with broken bars (ϕ_e=135°) [11,12]. Again, TMCSA may help to avoid these constraints [12].

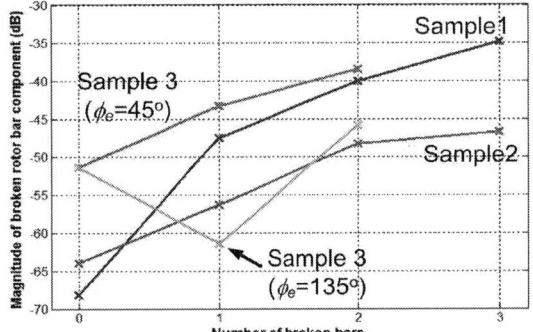

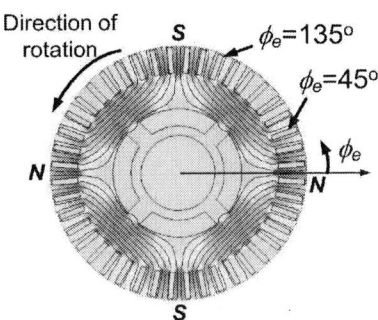

Fig. 4 MCSA measurements of LSH peak for Samples 1, 2, and 3 and different number of broken bars (top); considered reference frame (bottom).

A. Double cage rotor motor

Fig. 5 shows the DWT analyses of a phase startup current corresponding to Sample 2 for the cases of healthy machine (Fig. 5 (a)) and machine with 3 broken bars in the outer cage (Fig. 5 (b)). Fig. 6 (a) and (c) shows the equivalent analysis using the AST.

DWT analysis (Fig. 5) reveals the evolution of the LSH when the machine is faulty. However, the amplitude is much lower than for the single-cage motor (compare Fig. 5 and Fig. 2), a fact making its detection difficult even during startup conditions. This problem is aggravated in the case of double cage motors with separate end-ring, as shown in [10]: despite the LSH startup evolution is detected using the DWT, its low amplitude forces to adopt reduced scales in order to enhance its visibility. Moreover, the very difficult tracking of higher order harmonics using the DWT, makes it difficult to confirm the diagnostic by using those harmonics.

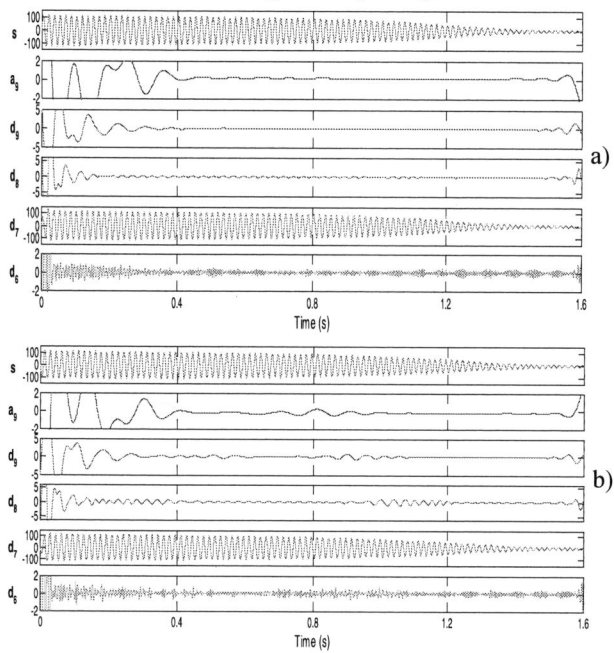

Fig. 5 Startup current, s, and high-level wavelet signals, a_9, d_9, d_8, d_7 and d_6 resulting from the DWT of the startup current of Sample 2 for: (a) healthy machine, (b) machine with 3 broken bars.

AST analysis provides much more information for the diagnosis, as observed in Fig. 6. Again, clear differences are observed between healthy and faulty conditions. These differences do not only rely on the presence of the LSH (much more clearly observed), but also on the identification of other fault harmonics whose presence enables to ratify the presence of the failure, hence increasing the reliability of the diagnostic. The usefulness of the continuous transforms is therefore also confirmed for these cases where the reduced amplitude of the low frequency harmonics may make the diagnostic uncertain.

B. Motor with rotor axial ducts ($N_d=N_p$).

Fig. 7 shows the DWT analysis of a phase startup current for Sample 3 in healthy condition (Fig. 7 (a)) and with 2 broken bars (Fig. 7 (b)). Fig. 8 (a) and (c) are equivalent but for the AST analysis.

DWT analysis (Fig. 7) proves that the possible drawbacks of the MCSA can be avoided with transient analysis. That figure shows how, for the healthy machine, even if the motor has axial ducts and $N_d \neq N_p$, no pattern related to the LSH

evolution appears (Fig. 7 (a)). Of course, if the machine has broken bars (Fig. 7 (b)) the classical pattern associated with the LSH is present, indicating the existence of the failure.

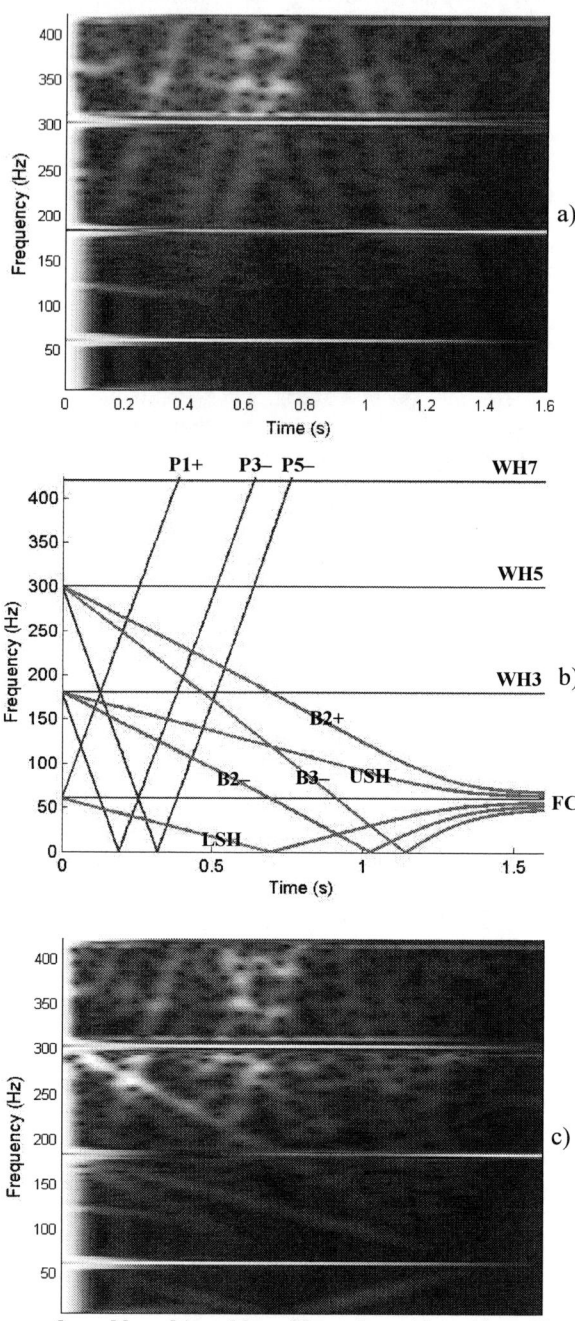

Fig. 6 AST of the startup current of Sample 2 for: (a) healthy machine, (c) machine with 3 broken bars. Theoretical t-f evolutions of the components (blue: healthy, red: bar breakage-related) in the stator startup current of the induction motors analyzed (b)

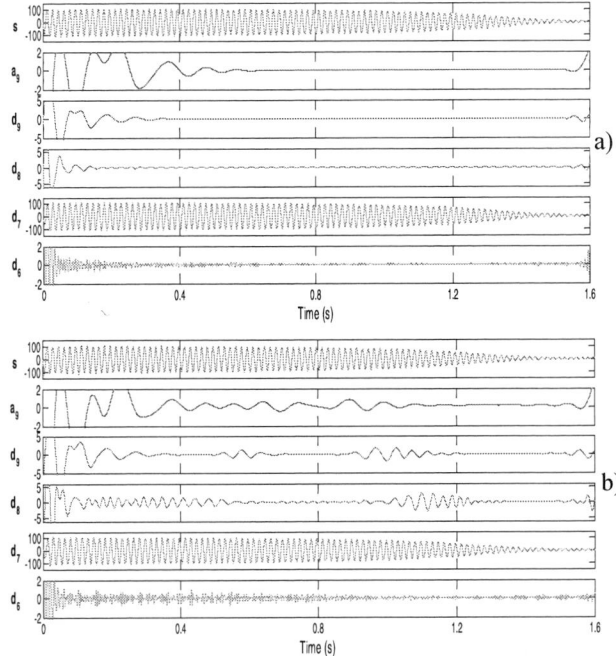

Fig. 7 Startup current, s, and high-level wavelet signals, a_9, d_9, d_8, d_7 and d_6 resulting from the DWT of the startup current of Sample 3 for: (a) healthy machine, (b) machine with 2 broken bars.

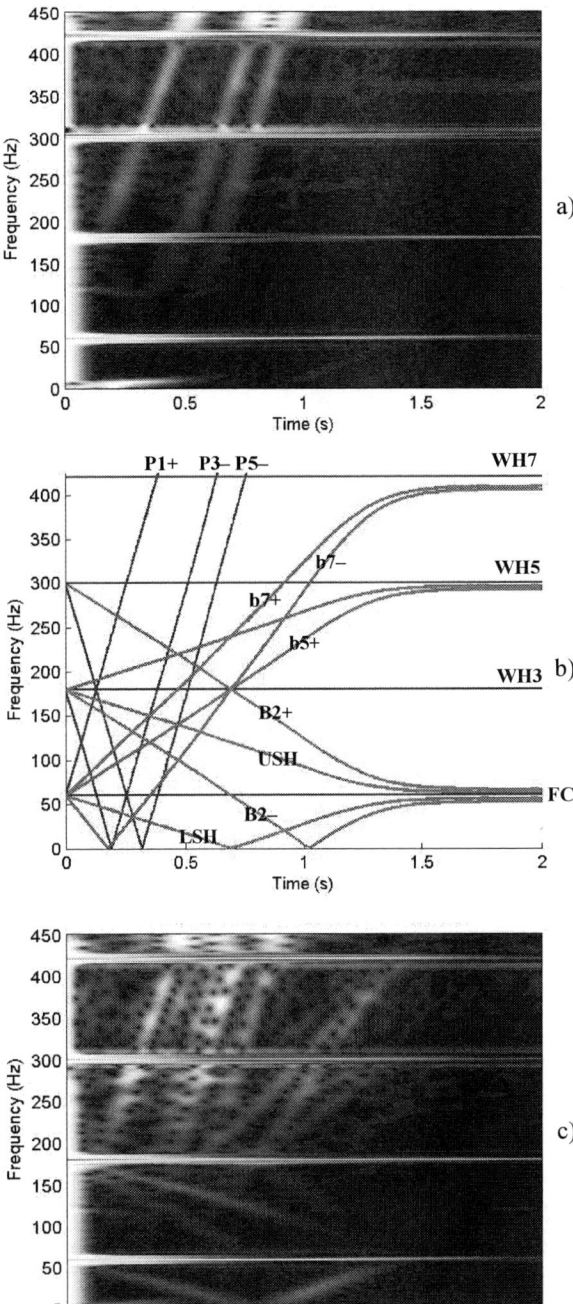

Fig. 8 AST of the startup current of Sample 3 for: (a) healthy machine, (b) machine with 2 broken bars. Theoretical t-f evolutions of the components (blue: healthy, red: bar breakage-related) in the stator startup current of the induction motors analyzed (b).

The AST analysis leads to similar conclusions. The LSH is clearly detectable if the machine is faulty, but it is not present when the machine is healthy. Hence, we can discard possible false positives in this latter case, which would be obtained in the case of applying the conventional MCSA.

The additional advantage is that the AST analysis, as happened in the previous cases, enables to ratify the diagnostic by using the evolutions of the high-order fault harmonics. As observed in Fig. 8 (a), the high-order components are absent, hence confirming the healthy condition of the machine. In the case of the faulty one (Fig. 8 (b)), it is possible to track the evolutions of the USH as well as other fault harmonics which ratify the diagnostic conclusion reached with the LSH.

C. Importance of high-frequency harmonics

The identification of the evolution of these high-frequency harmonics may make the diagnostic conclusion clearer in some situations in which the tracking of the low order ones is not clear; for instance, when load torque oscillations are present, these may partially distort the evolution of the LSH [15] being much less likely to affect the evolutions of the high-order fault harmonics [16]. Also, when other failures are simultaneously present, the evolutions of some of their low-frequency related components may mutually interfere [15]. Tracking of the high frequency harmonics may help to determine which faults are present. Finally, presence of inter-bar currents have been reported to mainly affect the amplitude of the low-frequency fault components, being less likely to reduce the amplitudes of the high-frequency ones [16].

IV. DISCRIMINATION BETWEEN FAULT COMPONENTS

This section shows the ability of the continuous TFD transforms to discern the components introduced by different failures. This fact is not always possible in the case of the classical MCSA, in which frequencies due to different

978-1-4799-0024-4/13 $31.00 © 2013 IEEE

failures can appear very near in the FFT spectrum of the steady-state current.

A. Eccentricity and rotor asymmetry

Fig. 9 (a) shows the DWT of the startup current for a machine with an uncertain condition (AST revealed later a certain slight rotor asymmetry). Fig. 9 (b) corresponds to the same machine with an additional 40% static + 40% dynamic eccentricity. Figs. 10 (a) and (c) are equivalent but for the AST application.

The DWT analyses reveal how, in the former case, no clear pattern is obtained. Some slight oscillations appear in the DWT signals, some of them resembling the LSH pattern during the startup. For the second case, the detection is not clear at all. This is due to the fact that the evolutions of the components due to the inherent asymmetry and the eccentricity overlap within the different frequency bands associated with the wavelet signal, so it becomes impossible its separation.

The application of the AST enables to overcome these problems. Fig. 10 (a) shows clear traces corresponding to the LSH evolution that follow the characteristic pattern. The evolutions of the eccentricity components are not observed since the machine has no eccentricity. On the other hand, Fig. 10 (b) clearly shows the evolution of the LSH associated with the rotor asymmetry but, additionally, it also shows the characteristic evolutions of the components related to the eccentricity (E– evolving from 60 Hz to 30 Hz and E+ evolving from 60 Hz towards 90 Hz). Hence, the continuous transform enables to distinguish the evolutions of components created by different simultaneous failures, reducing the uncertainty in the diagnostic.

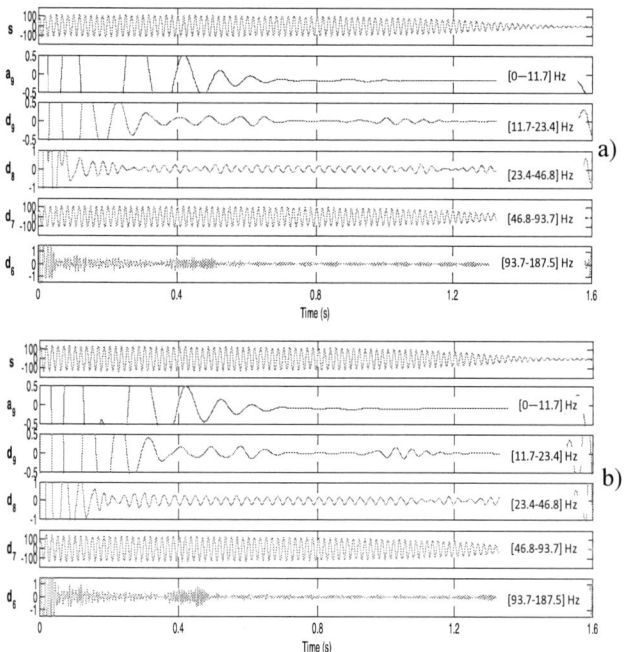

Fig. 9 Startup current, i, and high-level wavelet signals, a_{10}, d_{10}, d_9, resulting from the DWT for a machine with: (a) certain rotor asymmetry, (b) rotor asymmetry + 40% static and 40% dynamic eccentricity.

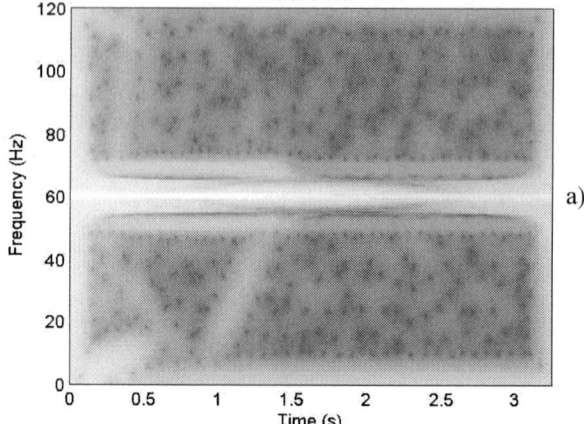

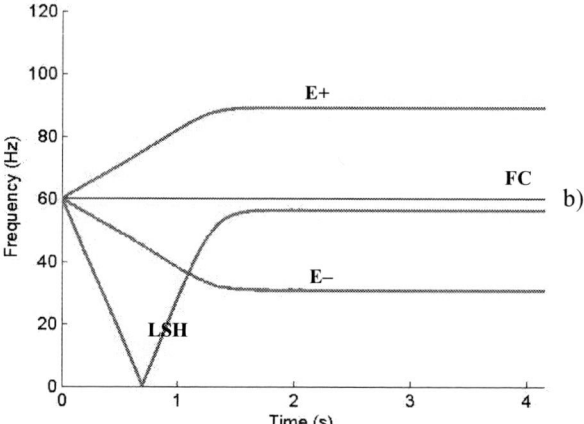

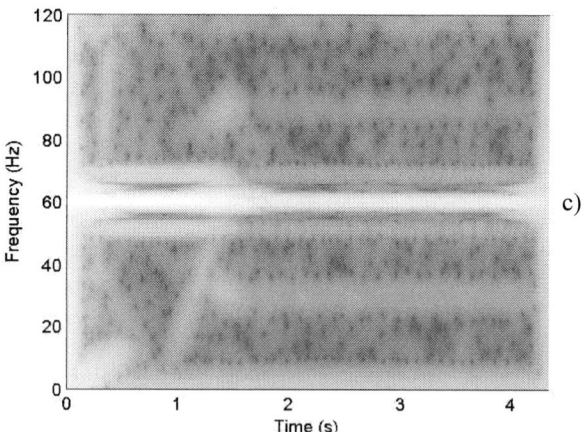

Fig. 10 AST of the startup current and subsequent steady state of an induction motor with an inherent rotor asymmetry (a) and adding 40% static and 40% dynamic eccentricity (c). Theoretical t-f evolutions of the components (blue: healthy, red: bar breakage and eccentricity related) in the stator startup current and subsequent steady state of the induction motors analyzed (b).

V. CONCLUSION

This paper proposes the application of continuous TFD tools to fault diagnosis in induction motors. In certain controversial situations these tools may become an important informational source that can ratify, when not correct, the conclusions reached with MCSA.

The work compares the performance of a specific continuous transform, the AST, with that of a classical discrete transform such as the DWT. The comparison is carried out in different controversial cases, namely, the detection of outer bar breakages in double cage motors and the diagnosis of bar breakages in motors with rotor axial ducts. Moreover, a case of combined failures (eccentricities + rotor asymmetries) is analyzed.

Results prove the ability of the continuous transforms for an accurate tracking of the high-order fault harmonics as well as for a better distinction between coexisting components related to different failures.

VI. REFERENCES

[1] M. Riera-Guasp, J. A. Antonino-Daviu, M. Pineda-Sanchez, R. Puche-Panadero, and J. Perez-Cruz, "A General Approach for the Transient Detection of Slip-Dependent Fault Components Based on the Discrete Wavelet Transform," *IEEE Trans. Ind. Electron.*, vol. 55, pp. 4167-4180, 2008.

[2] F. Briz, M.W. Degnert, P. Garcia, D. Bragado, "Broken rotor bar detection in line-fed induction machines using complex wavelet analysis of startup transients," *IEEE Trans. Ind. Appl.*, Vol. 44, No. 3, May-June 2008, pp. 760-768.

[3] A.Ordaz-Moreno, R.Romero-Troncoso, J.A.Vite-frías, J.Riviera-Gillen, A.García-Pérez, "Automatic online diagnostic algorithm for broken-bar detection on induction motors based on Discrete Wavelet Transform for FPGA implementation", *IEEE Trans. Ind. Electron.*, vol.55, no.5, pp.2193-2200, May. 2008.

[4] S.H. Kia, H. Henao and G.A. Capolino, "Diagnosis of Broken-Bar Fault in Induction Machines Using Discrete Wavelet Transform Without Slip Estimation," *IEEE Trans. Ind. Appl.*, Vol. 45, No. 4, July/August 2009, pp. 1395-1404.

[5] S. Rajagopalan, J.M. Aller, J.A. Restrepo, T.G. Habetler and R.G. Harley, "Analytic-Wavelet-Ridge-Based Detection of Dynamic Eccentricity in Brushless Direct Current (BLDC) Motors Functioning Under Dynamic Operating Conditions", *IEEE Trans. Ind. Electron.*, vol. 54, no. 3, pp. 1410-1419, June 2007.

[6] Y. Gritli, L. Zarri, C. Rossi, F. Filippetti, G. Capolino, D. Casadei, "Advanced Diagnosis of Electrical Faults in Wound Rotor Induction Machines" , *IEEE Trans. Ind. Electron.*, To be published.

[7] N. M. Elkasabgy, A. R. Eastham y G. E. Dawson, "Detection of broken bars in the cage rotor on an induction machine," *IEEE Trans. Ind. Appl.* vol. 28, nº 1, pp. 165-171, 1992.

[8] V. Climente-Alarcón, J. Antonino-Daviu, F. Vedreño-Santos and R. Puche-Panadero, "Vibration Transient Detection of Broken Rotor Bars by PSH Sidebands," *IEEE Trans. Ind. Appl.*, To be published.

[9] J. Park, B. Kim, J. Yang, K. Lee, S.B. Lee, E.J. Wiedenbrug, M. Teska, and S. Han, " Evaluation of the Detectability of Broken Rotor Bars for Double Squirrel Cage Rotor Induction Motors," *Proc. of IEEE ECCE*, pp. 2493-2500, Sept. 2010.

[10] J. Antonino-Daviu, et al., "Detection of broken outer cage bars for double cage induction motors under the startup transient," *IEEE Trans. on Ind. Appl.*, vol. 48, no. 5, pp. 1539-1548, Sept./Oct. 2012.

[11] J. Hong, et al., "Evaluation of the Influence of Rotor Axial Air Duct Design on Condition Monitoring of Induction Motors," Proc. of IEEE ECCE 2012, pp. 3016-3023, 2012.

[12] Sang Bin Lee, et al. "Reliable Detection of Induction Motor Rotor Faults under the Rotor Axial Air Duct Influence," IEEE ECCE 2013.

[13] J. Pons-Llinares et al., "Electric machines diagnosis techniques via transient current analysis", Proc. IEEE IECON 2012, pp. 3893-3900.

[14] J. Pons-Llinares, "A methodology for fault diagnosis in induction motors through the analysis of transient stator currents using time-frequency atoms", PhD dissertation, Universitat Politècnica de València, February 2013.

[15] J. Antonino-Daviu, P. Jover Rodriguez, M. Riera-Guasp, M. Pineda-Sánchez, A. Arkkio, "Detection of Combined Faults in Induction Machines with Stator Parallel Branches through the DWT of the startup current" *Mechanical Systems and Signal Processing*, vol. 23, no. 7, October 2009, Pages 2336-235.

[16] M. Fernandez-Cabanas et al. "Maintenance and diagnosis techniques for rotating electric machinery", Marcombo, Barcelona 1999.

VII. BIOGRAPHIES

Joan Pons-Llinares received the M.Sc. degree in Industrial Engineering and the Ph.D. degree in Electrical Engineering from the Universitat Politècnica de València (UPV, Spain), in 2007 and 2013, respectively. He is currently an Assistant Professor in the Electric Engineering Department of the UPV. His research interests include time-frequency transforms, condition monitoring and diagnostics of electrical machines.

Jose Antonino-Daviu (S'04–M'08-SM'12) received his M.S. and Ph. D. degrees in Electrical Engineering, both from the Universitat Politècnica de València, in 2000 and 2006, respectively. He was working for IBM during 2 years, being involved in several international projects. Currently, he is Associate Professor in the Department of Electrical Engineering of the mentioned University. He has been invited professor in Helsinki University of Technology (Finland) in 2005 and 2007, Michigan State University (USA) in 2010 and Korea University (South Korea) in 2013. He has over 90 publications between international journals, conferences and books. His primary research interests are condition monitoring of electric machines, wavelet theory and its application to fault diagnosis.

Martín Riera-Guasp (M'04-SM'12) received the M. Sc. degree in industrial engineering and the Ph.D. degree in electrical engineering from the Universitat Politècnica de València, Valencia (Spain), in 1981 and 1987, respectively. Currently, he is an Associate Professor with the Department of Electrical Engineering, Polytechnic University of Valencia. His research interests include condition monitoring of electrical machines, applications of the Wavelet Theory to electrical engineering and efficiency in electric power applications.

Sang Bin Lee (S'95-M'01-SM' 07) received the B.S. and M.S. degrees from Korea University, Seoul, Korea in 1995 and 1997, respectively, and the Ph.D. degree from Georgia Institute of Technology, Atlanta, GA in 2001, all in electrical engineering. From 2001 to 2004, he was with the Electric Machines and Drives Laboratory, General Electric Global Research Center (GRC), Schenectady, NY. At GE GRC, he developed an inter-laminar core fault detector for generator stator cores, and worked on insulation quality assessment for electric machines. From 2010 to 2011, he was with the Electric Drive Technologies, Austrian Institute of Technology, Vienna, Austria, as a Research Scientist where he worked on condition monitoring of PM synchronous machines. Since 2004, he has been a professor of Electrical Engineering at Korea University, Seoul, Korea. His current research interests are in protection, monitoring and diagnostics, and analysis of electric machines and drives. Dr. Lee was the recipient of 6 Prize Paper Awards from the IEEE Power Engineering Society, the Electric Machines Committee of the IEEE Industry Applications Society, and the Technical Committee on Diagnostics of the IEEE Power Electronics Society. He serves as an Associate Editor for the IEEE Transactions on Industry Applications for the IEEE IAS Electric Machines Committee.

Tae-june Kang (S'11) received the B.S. degree in electrical engineering from Korea University, Seoul, Korea, in 2011. He is currently pursuing the M.S. degree in electrical engineering at Korea University, Seoul, Korea. His research interests are in stator insulation testing for rotating electric machines.

Chanseung Yang (S'12) received the B.S. degree in electrical engineering in 2012 from Korea University, Seoul, Korea, where he is currently working toward the M.S degree in electrical engineering. In 2012, he was an Intern with the Polytechnic University of Valencia(UPV) ,Valencia, Spain, where he worked on analysis of the startup transient on induction machines. His research interests are in condition monitoring of induction machines, and analysis of electric machinery.

A New Method to Separate Broken Rotor Failures and Low Frequency Load Oscillations in Three-Phase Induction Motor

T. Göktaş, M. Arkan, and Ö. F. Özgüven

Abstract— For a time-varying loads, the presence of load fluctuation may sometimes have the same effect as broken rotor failure in a stator current of induction motors. In recent years, many methods, which are used to distinguish these two effects, have been published. In this study, a new method, which is based on Analytical Signal Angular Fluctuation (ASAF) signal, is developed to separate effects of load oscillation from broken rotor bar, especially when load oscillation is close to twice slip frequency. The simulation results are presented for the proposed method to show that, discerning broken rotor bar faults from low frequency load oscillation is possible. The developed method is independent of motor parameters.

Index Terms— Induction motor, broken rotor bar, Hilbert transform, load torque oscillation, fault diagnosis, fault detection.

I. INTRODUCTION

INDUCTION motors are the most dominant ones when compared to the other types of motors used in industry, due to their advantages. There are huge demands for early fault detection techniques to prevent system from unexpected interrupt by industry. Otherwise, halt of system can cause serious financial losses, and increase production costs. For industry, in addition to fault determination, it is also important to detect the failures as early as possible [1]. Many methods are developed to detect faults. Motor Current Signature Analysis (MCSA) technique is commonly used for stator, rotor, eccentricity, and bearing faults [2].

Despite induction motors are used in more harsh environments; cage rotor design has undergone a little change [2]. Broken rotor bar failure accounts for around 5%-10% of total induction motors failures [2]. Even though broken rotor failure does not directly affect the motor, overheating occurs in broken bar due to this failure. Besides, broken rotor failure cause high voltage in stator core at high speeds. This condition causes serious isolation issues, e.g. burn of coils, expensive maintenance and production issues.

Many signal processing techniques are used for rotor fault detection based on MCSA. Phase current's analytic signal modulus obtained by Hilbert transform has been used in [3] for MSCA to detect broken rotor bar fault at very low slip. Phase modulation techniques based on analytical and space

transformation signals is used for rotor fault diagnosis [4]. In [5], to improve fault diagnosis accuracy of broken rotor bar a method based on adaptive notch filter and Hilbert-Huang Transform is presented. A fault detection based on Hilbert-Park pattern analysis by using low sampling data is given in [6]. Hilbert transform (HT) also has been used for diagnosis of induction motors fed by frequency converters [7]. It is appears that the readability and resolution of HT is better than other techniques at very low slip.

Broken rotors bar failure and load torque oscillation at low frequency are especially confused with each other [8]. For time-varying loads, the presence of load fluctuation may sometimes have the same effect as broken rotor failure in stator current spectrum. In recent years, many methods, which are used to separate these two effects, have been proposed. In [9, 10], load effect is decoupled from rotor fault by using model based approach. In [11], the instantaneous active and reactive powers are used for discriminating broken rotor bar from oscillating load effects. While sideband components appear only in instantaneous active power spectrum for load torque oscillations, broken rotor failure is appears in instantaneous reactive power spectrum. Load torque oscillation and broken rotor failures can be separated by looking instantaneous active and reactive currents too [12]. In [13], angular displacement of space vector is used for the same purpose.

A method based on combined analysis of the amplitude and phase spectra of the instantaneous active and reactive powers is suggested in [14]. Quantification numbers of broken bars are calculated by using phase modulation, while fault determination is provided with amplitude modulation. In [15] active and reactive power media signals and their derived signals- the instantaneous power factor and phase angle, are used to identify rotor fault, and separate them from abnormal load effects.

In this study, a new scheme is proposed for discriminating low frequency load torque oscillation and broken rotor bar failure. The analytical signal is obtained from sampling one phase stator current by using Hilbert transform. By using the analytical signal angular fluctuation spectrum, discerning broken rotor failure from load torque oscillation is shown with presented simulation results.

T. Göktaş is with Osmancık Ömer Derindere Vocational School of Higher Education, Hitit University, Çorum, Turkey (e-mail: tanergoktas@hitit.edu.tr).

M. Arkan and Ö. F. Özgüven are with the Department of Electrical and Electronics Engineering, Inonu University, Malatya, Turkey (e-mail: muslum.arkan@inonu.edu.tr; ofozguven@inonu.ed.tr).

978-1-4799-0024-4/13 $31.00 © 2013 IEEE

II. EFFECTS OF BROKEN ROTOR BARS AND OSCILLATING LOADS

Broken rotor fault reflects on sidebands harmonics of the fundamental stator current. These harmonics are defined as in (1) [16].

$$f_b = (1 \mp 2ks) f_s \qquad k = 1, 2, 3 \ldots \ldots \tag{1}$$

where f_s is the supply frequency, s is the slip of motor, and k is an integer constant. The amplitude of the sideband harmonics is proportional to the number of broken rotor bars and motor load. The sideband harmonics distance to the main harmonic is proportional to the motor slip. In case of broken bar, an oscillation at twice slip frequency appears in speed and torque signals.

There are two types of mechanical load abnormalities. First type mechanical faults, which cause a rotor eccentricity inside the motor, are unbalanced load, angular and radial shaft misalignments. Second type mechanical faults are pulsating load such as reciprocating compressors that have the same effect as rotor fault on the stator current [8,15]. The oscillating load torque fluctuating at single frequency can be define as

$$T_{load} = T_a \left(1 + T_{osc} \sin \left(2\pi f_{osc} t \right) \right) \tag{2}$$

where f_{osc} is the oscillating torque frequency, T_{osc} is the oscillating torque percentage and T_a is the constant average load torque.

Torque oscillation at twice slip frequency appears at the same sideband harmonics of fundamental frequency given by (1), and generally they can be confused with rotor fault. A typical sample of the spectrum of stator current is given in Fig 1, for one broken rotor bar fault and 6% load torque oscillation. As it is seen from the figure, if load fluctuation is at twice slip frequency, determination of the source of the fault is difficult by using only these sideband harmonics on the current spectrum. So, additional processing need to be done for distinguishing these two types of faults from each other.

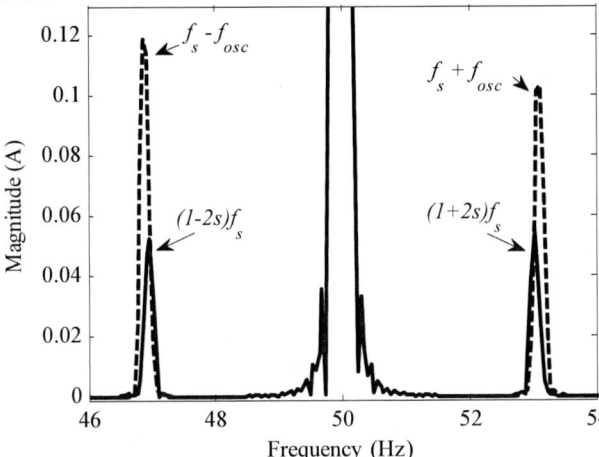

Fig. 1. Spectrum of stator current in the case of one broken rotor bar (solid line) and 6% load oscillation (dashed line) ($2sf_s$=3.1Hz, f_{osc}=3.1Hz).

III. DEVELOPED FAULT DETECTION METHOD

Hilbert transform is a mathematical conversion process, which shift the phase of original signal by 90° without changing its amplitude. Hilbert transform of the actual signal is defined by its convolution with function $1/\pi t$ [17].

$$HT(x(t)) = \frac{1}{\pi t} * x(t) = \frac{1}{\pi} \int_{-\infty}^{+\infty} \frac{x(\tau)}{t - \tau} d\tau \tag{3}$$

Analytic signal $\overline{x}(t)$ can be defined from actual signal $x(t)$ and the Hilbert transform of this signal $HT(x(t))$, as in (4).

$$\overline{x}(t) = x(t) + j\, HT(x(t)) \tag{4}$$

From these mathematical expressions, it can be seen that spectrum of $HT(x(t))$ is the same as the actual signal $x(t)$ except phase of each frequency component is shifted 90°. Negative frequency values are zero in FFT of analytical signal $\overline{x}(t)$. So, there are only positive frequency components in FFT of analytical signal.

Angular fluctuation of analytical signal can be obtained by similar procedure of space vector angular fluctuation techniques (SVAF) [18]. The method of space vector angular fluctuation reveals oscillation with respect to the reference vector in case of faulty condition. In three-phase system, if six current space vectors samples are taken for a one period, angle between each vector is equal to 60° for 50Hz supply as shown in Fig. 2.

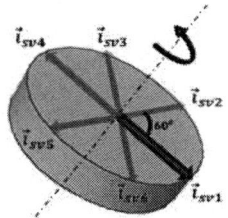

Fig. 2. Representation of space vectors in balanced system.

By using two consecutive space vector samples and reference degree, the space vectors angular fluctuation can be find as in (5):

$$\Delta\theta_i = \theta_i - \theta_{i-1} - 60^0 \tag{5}$$

In the SVAF spectrum, the dominant fundamental frequency is eliminated. Fluctuating harmonics, which are results of any abnormalities in motor or load, will be present at their exact frequencies.

The analytical signal angular fluctuation method proposed in this study, can be used to obtain similar information as the SVAF signal by using only one phase stator current. To obtain the analytical signal angular fluctuation (ASAF), firstly instantaneous vector positions of analytic signal are calculated. Then, a sampling frequency is selected similar to the SVAF (e.g. six samples in a period, i.e. 300Hz). Depending on the selected sampling frequency, in ideal condition, reference angle is calculated between two

978-1-4799-0024-4/13 $31.00 © 2013 IEEE

consecutive samples (60° for 50Hz). Angular fluctuation of the analytical signal is calculated by using (5). The ASAF algorithm is given in Fig. 3.

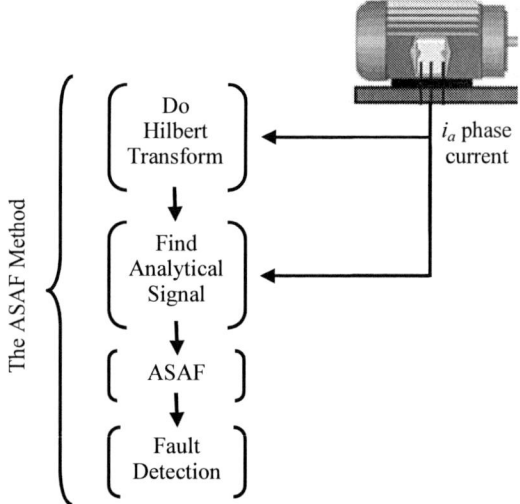

Fig. 3. The ASAF algorithm

Like the SVAF method, the ASAF spectrum has some advantages over traditional method's spectrum analysis. Since fundamental frequency is absent from the ASAF spectrum, its leakage is prevented and linear axis can be used instead of logarithmic axis. Spectral data give components at frequencies corresponding to their origin, that is, rotor fault and load oscillation will appear at $2sf_s$ and f_{osc} frequencies respectively. By using stator current spectrum, the fault detection is very difficult when fault related component is close to fundamental frequency at very low slips. However, with this technique, fault can be detected even at very small slip only by using one phase stator current.

When motor is supplied with unbalanced source, the three phase motor current contain both positive and negative sequences. Negative sequence current rotates at $-\omega_s$ and so does their analytic vectors. Thus, spectrum of the ASAF will contain frequency at $2f_s$ under unbalance source. When both rotor fault and supply unbalance exist at the same time, the spectrum of the ASAF will have additional spectral components at $(1-s)2f_s$ like the SVAF spectrum [18]. In case of load torque oscillation, simulation carried out by authors' showed that this component is almost negligible.

IV. SIMULATION RESULTS

Simulations are carried out to demonstrate that rotor bar fault and low frequency load torque fluctuation can be separated from each other by using the ASAF method, when motor is line connected and there is some degree of supply unbalance.

Oscillating load torque can be modeled by using (2) with

oscillating frequency close to rotor fault frequency.

Different models have been used in the literature for broken rotor bar fault. Modeling of squirrel cage rotor with broken rotor bar is complex. Broken rotor bar failure can be modeled by adding an extra resistor, which its value is proportional to the number of broken rotor bar, to one phase of the rotor resistance [19].

When the contribution of the end-ring is neglected, in case of consecutive broken bars, the extra resistor value is given by

$$\Delta R = \frac{n}{N/3 - n} R_r \qquad (6)$$

where R_r is the rotor phase resistance, n is the number of broken bar, and N is the total number of rotor bars.

Simulations are carried out by using $4kW$ standard wound type rotor induction motor model in Matlab/Simulink. The motor parameters are given in Table I. Simulations are done for healthy motor, broken rotor bar, load oscillation, and when both rotor fault and load oscillation are present. For all conditions, it is assumed that negative to positive voltage ratio is 0.4%, which is very common for laboratory condition.

TABLE I
INDUCTION MOTOR DATA

Nominal Power	4 kW
Line-Line Voltage	400 V
Frequency	50 Hz
Stator Resistance	1.405 Ω
Rotor Resistance	1.395 Ω
Mutual Inductance	0.1722 H
Total Number of Rotor Bars	45

Fig. 4 shows the spectrum of stator current (Fig. 4a) and the ASAF signal (Fig. 4b and Fig. 4c) for 2 broken bars when motor is loaded with 75% of constant load. The motor slip is 0.031. The ASAF spectrum is dived into two sections for clear presentation. As it can be seen from the current spectrum rotor fault related sidebands harmonics appear at 46.9Hz and 53.1Hz (Fig. 4a). The rotor fault frequency ($2sf_s$) and its related harmonics ($1-s$)$2f_s$ appear directly in the ASAF spectrum.

Fig. 5 shows the spectrum of stator current (Fig. 5a) and the ASAF signal (Fig. 5b and Fig. 5c) for 6% oscillation load at 3.1Hz frequency. The average motor load is 75% of full-load. Since oscillating load frequency is at rotor fault frequency, sidebands harmonics appear at 46.9Hz and 53.1Hz in the current spectrum (Fig. 5a). While the oscillating load frequency f_{osc} clearly seen in the ASAF spectrum, its related harmonic ($2f_s$-f_{osc}) amplitude, however is very small when compared to the rotor fault case (Fig. 5c).

978-1-4799-0024-4/13 $31.00 © 2013 IEEE   124

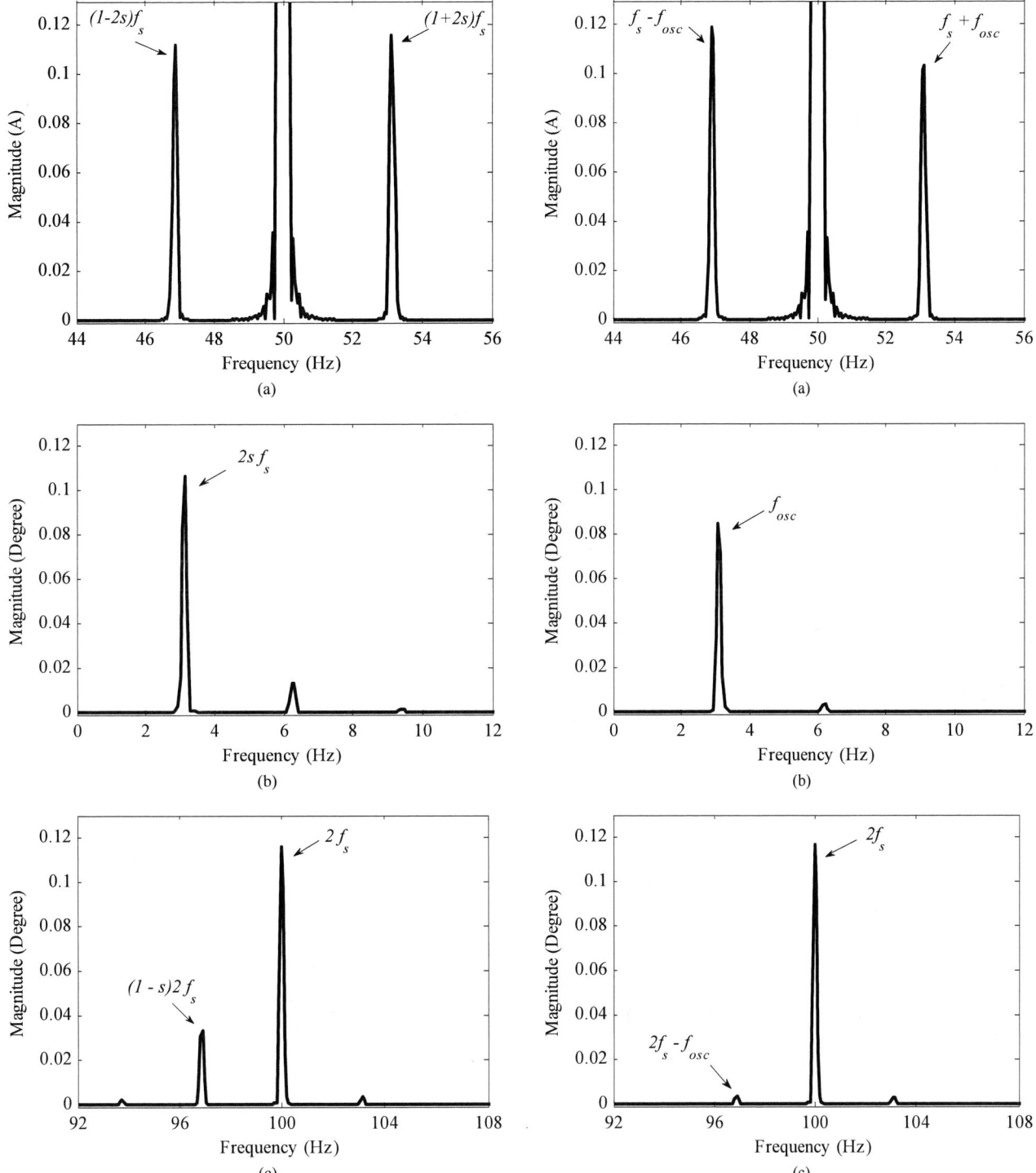

Fig. 4. Simulation with 2 broken rotor bars at 75% full-load ($2sf_s = 3.1$Hz). (a) Stator current spectrum. (b) ASAF spectrum (0Hz-to-12Hz section). (c) ASAF spectrum (92Hz-to-108Hz section).

Fig. 5. Simulation with 6% load oscillation ($f_{osc} = 3.1$ Hz, which is equal to $2sf_s$) at 75% full-load. (a) Stator current spectrum. (b) ASAF spectrum (0Hz-to-12Hz section). (c) ASAF spectrum (92Hz-to-108Hz section).

To show the performance of the proposed method, the motor was simulated with 2 broken rotor bars and 6% fluctuation load torque, as well. Average load torque was 75% of full-load torque. Fig. 6 shows the ASAF spectrum. As it can be seen from Fig. 6b, both effects can be clearly discerned from each other by using $2f_s$ side harmonics. Even though, load oscillation f_{osc} component's amplitude is bigger than the rotor fault $2sf_s$ component (Fig. 6a), the rotor fault sideband component $(1-s)2f_s$ amplitude is much higher than load oscillating sideband $(2f_s.-f_{osc})$ component (Fig. 6b).

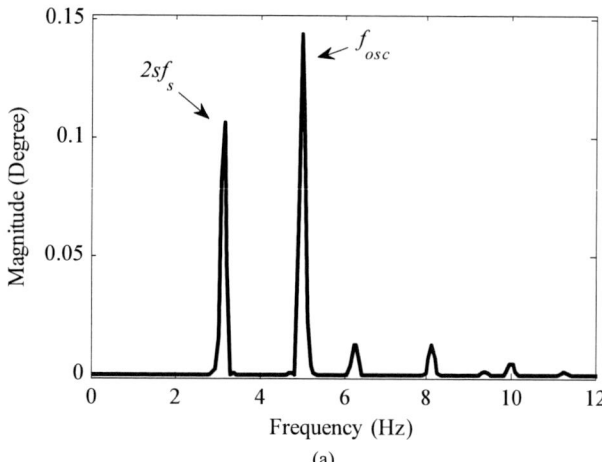

(a)

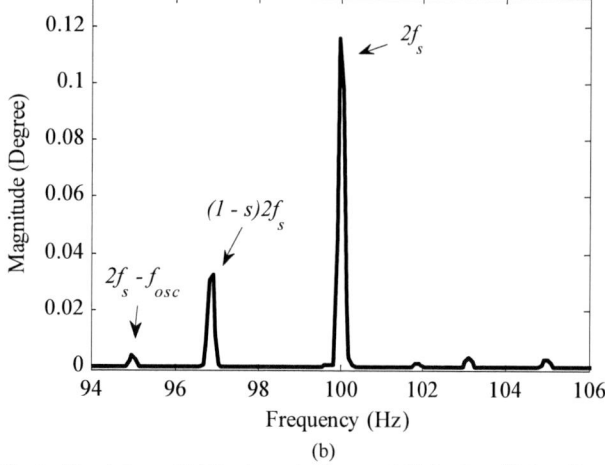

(b)

Fig. 6. Simulation with 2 broken rotor bars and 6% load oscillation (f_{osc} = 5Hz) at 75% full-load ($2sf_s$ = 3.15Hz). (a) The ASAF spectrum (0Hz-to-12Hz section). (b) The ASAF spectrum (92Hz-to-108Hz section).

These results show that broken rotor bar fault and fluctuating load torque effects can be separated by using the presented method.

V. CONCLUSIONS

A new method is presented for fault detection in line fed induction motor when load oscillation and broken rotor bar faults occur simultaneously at the same frequency. The analytical signal angular fluctuation (ASAF) method is a combination of Hilbert transform and the SVAF. From presented simulation results, it has been demonstrated that, while load oscillation appears only at $2sfs$ in the ASAF spectrum, rotor fault components appear both at $2sfs$ and $2fs(1-s)$. Thus, by using these two components, source of fault can be specified.

The advantages of presented method are fault detection even at very low slips (low load), requirement of only one phase current and low sampling frequency. To use this method, there must be an unbalance in the supply. Authors' experiences suggest that there is always an unbalance in real systems.

VI. REFERENCES

[1] J. J. Magdaleno, J. R. Troncoso, R. A. O. Rios, E. C. Yepez, and L. M. C Medina, "Novel methodology for online half-broken-bar detection on induction motors", *IEEE Trans. Instr. and Measur.*, vol. 58, no. 5, pp. 1690-1698, 2009.

[2] S. Nandi, and H. A. Toliyat, "Condition monitoring and fault diagnosis of electrical machines—A review", *IEEE Trans. Energy Convers.*, vol. 20, no. 4, pp. 719–729, Dec. 2005.

[3] R. Punche-Panadero, M. Pineda-Sanchez, M. Riera-Guasp, J. Roger-Folch, E. Hurtado-Perez, and J. Perez-Cruz, "Improved Resolution of the MCSA Method Via Hilbert Transform, Enabling the Diagnosis of Rotor Asymmetries at Very Low Slip", *IEEE Transactions On Energy Conversion*, Vol. 24, No. 1, pp. 52-59, March. 2009.

[4] I. Jaksch, and J. Zalud, "Rotor fault detection of induction motors by sensorless irregularity revolution analysis", *XIX International Conference on Electrical Machines (ICEM)*, pp. 1-6, Sept.2010.

[5] W. Xin, and Z. Zhike, "Novel adaptive filter method and application in broken rotor bar fault diagnosis of induction motor", *Second International Conference on Digital Manufacturing & Automation(ICDMA)*, pp. 780-783, Aug. 2011.

[6] S. B. Salem, K. Bacha, and M. Gossa, "Induction motor fault diagnosis based on a Hilbert current space vector pattern analysis", *16th IEEE Mediterranean Electrotechinical Conference (MELECON)*, pp. 818-823, March 2012.

[7] R. Puche-Panadero, V. Sarkimaki, and P. Rodriguez, "Detection of broken rotor bar fault in induction machine fed by frequency converter", *International Symposium on Power Electronics, Electrical Drives, Automation and Motion (SPEEDAM)*, pp. 1027-1032 June 2012.

[8] W. Long, T.G. Habetler, and R.G. Harley, "A review of separating mechanical load effects from rotor faults detection in induction motors", in Proc. *IEEE SDEMPED*, Gracow, Poland, 2007, pp. 221-225.

[9] R.R. Schoen and T.G. Habetler, "Effects of time-varying loads on rotor fault detection in induction machines", *IEEE Transactions on Industry Applications*, vol. 31, no. 4, pp. 900-906, July-Aug. 1995.

[10] R.R. Schoen and T.G. Habetler, "Evaluation and implementation of a system to eliminate arbitrary load effects in current-based monitoring of induction machines," *IEEE Transactions on Industry Applications*, vol. 33, no. 6, pp. 1571-1577, Nov.-Dec. 1997.

[11] C. H. Angelo, G. R. Bossio, and G. O. Garcia, "Discriminating broken rotor bar from load oscillation using active and reactive powers components", *IEEE IET Electr. Power Appl.*, vol. 4, Iss. 4, pp 281-290, April 2010.

[12] G. R. Bossio, C. H. De Angelo, J. M. Bossio, C. M. Pezzani, and G. O. García, "Separating broken rotor bars and load oscillations on IM fault diagnosis through the instantaneous active and reactive currents", *IEEE Transactions On Industrial Electronics*, vol. 56, no. 11, Nov. 2009

[13] C. Concari, G. Franceschini, and C. Tassoni, " Induction machine current space vector features to effectively discern and quantify rotor faults and external torque ripple", *IET Electric Power Applications*, vol. 6, No. 6, July 1012.

[14] S. M. A. Cruz, "An active-reactive power method for the diagnosis of rotor faults in three-phase induction motors operating under time-

varying load conditions", *IEEE Transactions On Energy Con.*, vol. 27, no. 1, March 2012.

[15] M. Drif, and A. J. Marques Cardoso, "Discriminating the simultaneous occurrence of three-phase induction motor rotor faults and mechanical load oscillations by the instantaneous active and reactive power media analyses", *IEEE Transactions on Industrial Electr.*, vol. 59, no. 3, March 2012.

[16] W. T. Thomson and I. D. Stewart, "On-line current monitoring for fault diagnosis in inverter fed induction motors", *in Proc. Inst. Elect. Eng., 3rd Int. Conf. Power Electronics Drives*, London, U.K., 1988, pp. 432–435.

[17] M. Feldman. *Encyclopedia Of Vibration, Chapter Hilbert Transforms*, pages 642–648. Academic press, 2002.

[18] D. Kostic-Perovic, M. Arkan, and P. J. Unsworth, "Induction motor fault detection by space vector angular fluctuation", *In: Proceedings of the IEEE Industry Applications Conference. 34th Annual Meeting – World Conference on Industrial Applications of Electrical Energy*, vol. 1, Rome, Italy, pp 388–394, October2000.

[19] A. Bellini, F. Filippetti, G. Francheschini, C. Tassoni, and G.B. Kliman, "Quantitative evaluation of induction motor broken bars by means of electrical signature analysis", IEEE Trans. Ind. Appl. 37 (September/October (5)) (2001) 1248–1255.

VII.   BIOGRAPHIES

Taner Göktaş was born in Elazığ, Turkey, in 1983. He received M.Sc. degree in Electrical and Electronics Engineering from Fırat University, Elazığ, Turkey, in 2010. Now, He is a lecturer in Hitit University and also currently working toward the Ph.D degree in Electrical and Electronics Engineering at Inonu University, Turkey. His research interests are in the field of electric machines drives and condition monitoring and diagnostic techniques.

Müslüm Arkan was born in Bozova, Turkey on January 1, 1970. He received B.Sc. degree in Electrical and Electronics Engineering from University of Gaziantep, Gaziantep, Turkey, and his DPhil degree in Electrical Engineering from University of Sussex, Brighton, UK, in 1994 and 2000 respectively.

Since 1994, he has been with the Electrical and Electronics Engineering Department, Inonu University, Malatya, Turkey, where he is currently an Associate Professor. His research activities include the use of digital signal analysis for diagnostic, condition monitoring and motor failure prediction by sensorless methods and modeling of electrical machines for diagnosis purpose.

Ömer Faruk Özgüven was born in Malatya, Turkey, in 1963. He received B.Sc., M.Sc. and Ph.D. degrees in Electrical Engineering from Yıldız Technical University, Istanbul, Turkey, in 1985, 1988 and 1993 respectively.

Since 1994, he has been with the Electrical and Electronics Engineering Department, Inonu University, Malatya, Turkey, where he is currently an Assistant Professor. His research activities include intelligent control systems, fractional order control, and applications of microprocessors.

Analysis of Mixed-Eccentric Induction Machine

Rijaniaina Njakasoa Andriamalala, Hubert Razik, *Senior Member, IEEE*, François-Michel Sargos, *Member, IEEE*, and Bruno Francois, *Member, IEEE*

Abstract—**This work presents a model of a mixed-eccentric induction machine (IM) by considering space harmonics. Two forms of permeance function are presented and compared. The mixed-eccentricity fault is studied because it generalizes better the case encountered in any machine. IM inductances are calculated by using the modified winding function and including space harmonics of winding functions. It is also analytically detailed the genesis of the eccentricity signatures allowing the diagnosis. Coherence of simulated and experimental results in term of frequency content proves a validity of the study.**

Index Terms—**Mixed-eccentricity, permeance function, Legendre polynomial, space harmonics, fault signatures, induction machine (IM).**

I. INTRODUCTION

Various kinds of failures may occur within an IM during its working life time. Most of this faults includes stator inter-turn short circuit, a rotor misalignment, an eccentricity [1] or worn out bearings. Besides, some failures often induce other kinds of breakdowns, so the users should have to detect these problems from their initial and incipient levels. The rotor eccentricity really encountered in any motor is the mixed-one . Therefore, study will be especially focused on this kind of eccentricity which is the combination of the static and dynamic eccentricity [2], [3], [4]. Such defect is mostly caused by the bearing ageing and mechanical problem in the shaft. It may also occurs during the manufacturing and assembly processes but very rarely. An eccentric-rotor is to be monitored permanently in an electric drive because it results in acoustic noise, unbalanced magnetic pull, mechanical vibration and ultimately a contact between the rotor and the stator armatures [5], [6], [7]. A standard method of rotor eccentricity assessment consists of building an adequate model permitting to identify fault signatures. There are several indicators but harmonic content of stator current is among the most reliable one. One of the modeling key is the determination of a reliable permeance function under eccentricity condition.

R.N. Andriamalala is with the Ecole Centrale de Lille, L2EP Laboratoire, CS 20048, F-59651 Villeneuve d'Ascq, Cedex, France (email: Rijaniaina-Njakasoa.Andriamalala@ec-lille.fr);

H. Razik, *Senior member IEEE*, is with the Université Lyon 1; Laboratoire Ampère, Ampère UMR CNRS 5005, Bât Omega, 43 Bd du 11 Novembre 1918, F-69622 Villeurbanne, Cedex, France (e-mail: hubert.razik@univ-lyon1.fr);

F-M. Sargos is with the Ecole Nationale Supérieure d'Electricité et de Mécanique de Nancy, Laboratoire GREEN UMR-7037, 2 avenue de la Foret de Haye, F-54516, Vandœuvre-lès-Nancy, Cedex, France (email: Francois.Sargos@univ-lorraine.fr);

B. Francois, *Senior member IEEE*, is with the Laboratoire L2EP, Ecole Centrale de Lille, CS 20048, F-59651 Villeneuve d'Ascq, Cedex, France (email: bruno.francois@ec-lille.fr).

The present work presents two forms of permeance function: the first one is a non-classical permeance function generating Legendre polynomials [8] while the second one is the classical form.

A former work has dealt with the Legendre polynomials but this work appears limited because it stays only on the static eccentricity case [9]. Moreover, the range of the validity of such non-classical permeance function is not analyzed.

This paper is organized as follows: section II provides mathematical formulas that will be helpful for building the model of eccentric machine; section III presents two forms of permeance function and their comparison: those using the Legendre polynomials and the classical one. Comparison of these permeance functions is undertaken briefly. In section IV will be highlighted all analytical calculations of motor inductances. Section V shows also analytically how eccentricity-related harmonics rise in the machine meaning the model. Confrontation between simulated and experimental testings in term of eccentricity related harmonics will be presented in section VI.

II. USEFUL MATHEMATICAL RELATIONS

Analytical modeling of an eccentric IM uses stator and rotor Ampere-turn functions which depend on the stator and rotor mechanical angles ϕ_s and ϕ_r measured around the stator internal periphery and the rotor external periphery. It is therefore necessary to define an accurate relation between these angles to achieve the inductance calculations.

But before defining such angular relation between ϕ_s and ϕ_r, this section determines first instantaneous position of mixed eccentricity and its severity.

A. Mixed-eccentricity position

The axis of the mixed-eccentric rotor (O_r) turns around an another axis (O_c) that is displaced from the stator axis (O_s) (fig.1). The static eccentricity severity is defined as:

$$\varepsilon_s = \frac{O_s O_c}{e_0} \qquad (1)$$

The dynamic eccentricity degree is given by relation:

$$\varepsilon_d = \frac{O_c O_r}{e_0} \qquad (2)$$

e_o stands for the air-gap of the balanced machine. For a general situation, the direction ($O_s O_c$) is not parallel with the stator first phase axis but rather displaced by a constant angle δ_0.

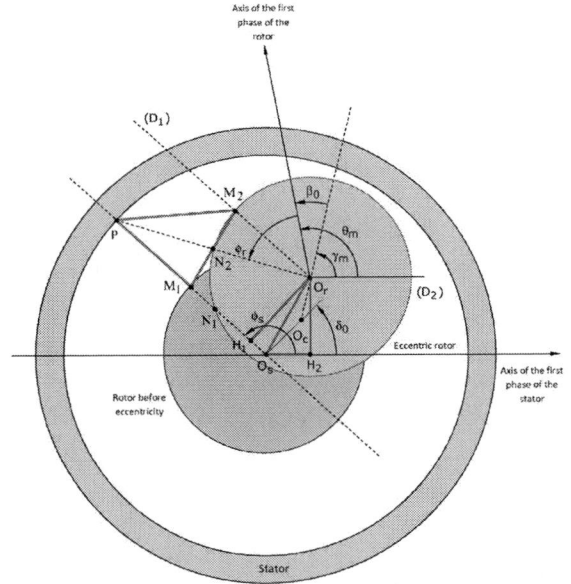

Fig. 1. Highlighting eccentricity phenomenon

Consider the line (D_1) parallel with the stator internal periphery radius O_sP (fig.1) and passing by the rotor center O_r, the line O_sP intersects with the rotor periphery on the point N_1. This notice yields to the following angular relation:

$$\widehat{(\overrightarrow{N_1O_s}, \overrightarrow{N_1O_r})} = \widehat{(\overrightarrow{O_rN_1}, \overrightarrow{O_rM_2})} \quad (3)$$

Considering the parallelism between (D_2) and the axis of the stator first phase:

$$\phi_s = \phi_r + \theta_m - \widehat{(\overrightarrow{O_rN_1}, \overrightarrow{O_rM_2})} \quad (4)$$

where θ_m designates the rotor angular position.

Let H_1 and H_2 be the orthogonal projections of O_r on the stator radius O_sP and on the axis of the stator first phase, respectively. Let Θ be also the angle:

$$\Theta = \widehat{(\overrightarrow{O_sH_2}, \overrightarrow{O_sO_r})} \quad (5)$$

Considering the triangle (O_s, H_2, O_r), the tangent of angle Θ can be calculated as follows:

$$\tan\Theta = \frac{O_rH_2}{O_sH_2} = \frac{O_sO_c \sin\delta_o + O_cO_r \sin\gamma_m}{O_sO_c \cos\delta_o + O_cO_r \cos\gamma_m} \quad (6)$$

γ_m designates the instantaneous angular position of the line (O_cO_r). Note also that the shift angle β_0 between the rotor first phase axis and this line is constant i.e:

$$\theta_m - \gamma_m = C^{te} = \beta_0 \quad (7)$$

θ_m is the rotor angular position. Previous works often assume that $\beta_0 = 0$ but it is a particular case. Hence, equations (6) and (7) yield to the following relation:

$$\Theta = \arctan\left[\frac{\varepsilon_s \sin\delta_0 + \varepsilon_d \sin(\theta_m - \beta_0)}{\varepsilon_s \cos\delta_0 + \varepsilon_d \cos(\theta_m - \beta_0)}\right] \quad (8)$$

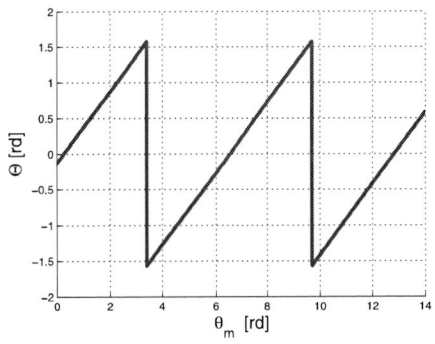

(a) $\varepsilon_s = 0.10$ and $\varepsilon_d = 0.10$

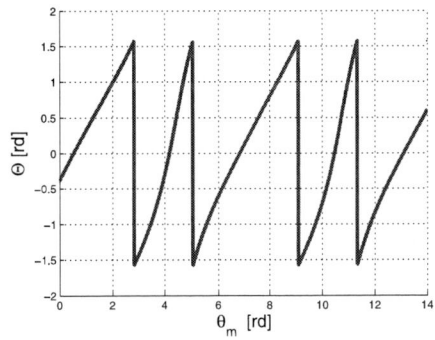

(b) $\varepsilon_s = 0.10$ and $\varepsilon_d = 0.20$

Fig. 2. Evolution of the angle Θ

Physically, Θ defines the instantaneous position of the rotor axis (O_r) under mixed-eccentricity condition or merely the mixed eccentricity position. More the eccentricity is severe more the evolution of this angle is complex rendering also electromagnetic phenomenon in the eccentric IM more complex (fig.2).

B. Mixed-eccentricity severity

We can write:

$$\overrightarrow{O_sO_r} = \overrightarrow{O_sO_c} + \overrightarrow{O_cO_r} \quad (9)$$

Therefore:

$$O_sO_r^2 = O_sO_c^2 + O_cO_r^2 + 2.O_sO_c.O_cO_r.\cos\widehat{(\overrightarrow{O_sO_c}, \overrightarrow{O_cO_r})} \quad (10)$$

Considering relations (1) and (2), equation (10) becomes:

$$O_sO_r^2 = \varepsilon_s^2 + \varepsilon_d^2 + 2.\varepsilon_s.\varepsilon_d.\cos\widehat{(\overrightarrow{O_sO_c}, \overrightarrow{O_cO_r})} \quad (11)$$

Further inspection of (fig.1) permits to state that:

$$\widehat{(\overrightarrow{O_sO_c}, \overrightarrow{O_cO_r})} = \theta_m - \delta_0 - \beta_0 \quad (12)$$

By substitution, relation (11) becomes :

$$O_sO_r = e_o\sqrt{\varepsilon_s^2 + \varepsilon_d^2 + 2\varepsilon_s\varepsilon_d \cos(\theta_m - \delta_0 - \beta_0)} \quad (13)$$

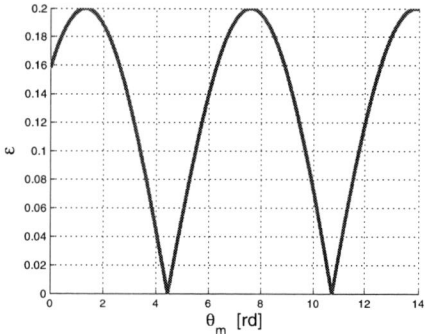

(a) $\varepsilon_s = 0.10$ and $\varepsilon_d = 0.10$

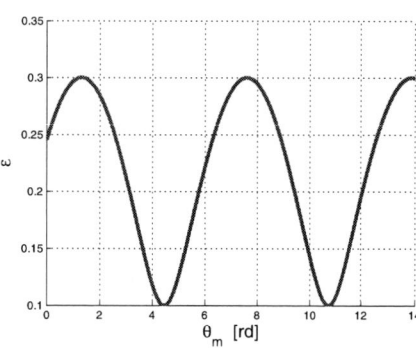

(b) $\varepsilon_s = 0.10$ and $\varepsilon_d = 0.20$

Fig. 3. Evolution of eccentricity severity ε

or:

$$O_s O_r = \varepsilon e_o \qquad (14)$$

where ε is given by:

$$\varepsilon = \sqrt{\varepsilon_s^2 + \varepsilon_d^2 + 2\varepsilon_s\varepsilon_d \cos(\theta_m - \delta_o - \beta_o)} \qquad (15)$$

ε is the mixed-eccentricity severity that depends on static and dynamic eccentricity severities and rotor position. As illustrated in fig.(3), ε is a periodic function. Minimal value of ε is $|\varepsilon_d - \varepsilon_s|$ (when $\cos(\theta_m - \delta_o - \beta_o) = -1$) while maximal value is $|\varepsilon_d + \varepsilon_s|$ (when $\cos(\theta_m - \delta_o - \beta_o) = 1$) (verified by a particular case in fig.(3)).

C. Angular relation between ϕ_s and ϕ_r

$$\sin(\phi_s - \Theta) = \frac{O_r H_1}{O_s O_r} = \frac{O_r H_1}{\varepsilon e_o} \qquad (16)$$

Further geometrical considerations in (fig.1) lead to:

$$(\overrightarrow{N_1 O_s}, \overrightarrow{N_1 O_r}) = \arcsin\left(\frac{O_r H_1}{r}\right) = -(\overrightarrow{O_r N_1}, \overrightarrow{O_r M_2}) \qquad (17)$$

r indicates the rotor radius. By taking into account By deducing $O_r H_1$ from relation (16) and substituting in (17):

$$-(\overrightarrow{O_r N_1}, \overrightarrow{O_r M_2}) = \arcsin\left[\frac{\varepsilon e_o}{r}\sin(\phi_s - \Theta)\right] \qquad (18)$$

By replacing this expression in relation (4), a relation between ϕ_s and ϕ_r can be deduced:

$$\phi_s = \phi_r + \theta_m + \arcsin\left[\frac{\varepsilon e_o}{r}\sin(\phi_s - \Theta)\right] \qquad (19)$$

For usual motors, the rotor radius r is much greater than the rotor displacement εe_o:

$$\frac{\varepsilon e_o}{r} << 1 \Longrightarrow \frac{\varepsilon e_o}{r} \longmapsto 0 \qquad (20)$$

and finally, equation (19) reduces to:

$$\phi_s \simeq \phi_r + \theta_m \qquad (21)$$

III. PERMEANCE FUNCTION

This section is dedicated two forms of the permeance function: the first one uses the Legendre polynomials while second one is the classical form

A. Use of the Legendre polynomials

Always in the fig.(1), N_2 stands for the intersection of the line $(O_r P)$ and the rotor periphery. Line segments as $N_1 P$ or $N_2 P$ may be chosen as acceptable approximations of the air-gap function under eccentric condition but their expressions seem very complicated. Inspection parallelism between the rotor radius $(O_r M_2)$ and the stator radius $(O_s P)$ permits to state:

$$N_2 P < M_2 P < N_1 P \qquad (22)$$

Hence, $M_2 P$ may be an another approximation of the air-gap function. Let M_1 be the intersection of the stator radius $O_s P$ and the non-eccentric rotor:

$$\overrightarrow{M_1 M_2} = \overrightarrow{O_s O_r} \qquad (23)$$

therefore:

$$\begin{cases} M_1 M_2 = O_s O_r = \varepsilon e_o \\ \overrightarrow{M_1 M_2} // \overrightarrow{O_s O_r} \end{cases} \qquad (24)$$

As M_1 belongs to the line segment $[O_s P]$:

$$(\overrightarrow{M_1 M_2}, \overrightarrow{M_1 P}) = (\overrightarrow{O_s O_r}, \overrightarrow{O_s P}) = \phi_s - \Theta \qquad (25)$$

Since:

$$\overrightarrow{M_2 P}^2 = \overrightarrow{M_1 P}^2 + \overrightarrow{M_1 M_2}^2 - 2.\overrightarrow{M_1 P}.\overrightarrow{M_1 M_2} \qquad (26)$$

Then:

$$M_2 P^2 = e_o^2 + \varepsilon^2 e_o^2 - 2.\varepsilon e_o.e_o.\cos(\phi_s - \Theta) \qquad (27)$$

Another possible expression of the air-gap function is therefore:

$$e(\phi_s) = e_o\sqrt{1 - 2\varepsilon\cos(\phi_s - \Theta) + \varepsilon^2} \qquad (28)$$

which corresponds to the permeance function :

$$\lambda(\phi_s) = \frac{1}{e_o}\frac{1}{\sqrt{1 - 2\varepsilon\cos(\phi_s - \Theta) + \varepsilon^2}} \qquad (29)$$

978-1-4799-0024-4/13 $31.00 © 2013 IEEE

Let $P_n(t)$ ($n \in \mathbf{N}$) be the Legendre polynomials:

$$P_n(t) = \sum_{k=0}^{n'} \frac{(-1)^k (2n - 2k)! t^{n-2k}}{2^n k! (n-k)! (n-2k)!},$$

$$\text{with} \quad n' = Int\left[\frac{n}{2}\right]$$

(30)

Then:

$$\frac{1}{\sqrt{1 - 2tx + x^2}} = \sum_{n=0}^{+\infty} P_n(t) x^n$$

$$\text{with} \quad |t| \le 1, |x| < 1 \quad (31)$$

Furthermore:

$$|P_n(t)| < 1 \quad (32)$$

So the permeance function development is:

$$\lambda(\phi_s) = \frac{1}{e_o} \sum_{n=0}^{+\infty} P_n[\cos(\phi_s - \Theta)] \varepsilon^n \quad (33)$$

Hence, the development of permeance function of the static eccentricity in [9] can be generalized for the mixed eccentricity by using the angle Θ instead of δ_0. For easier inductance calculations, Legendre polynomials in $\cos(\phi_s - \Theta)$ have to be linearized. Hence, the permeance function takes an form of Fourier series development:

$$\lambda(\phi_s) = \frac{1}{e_o} \sum_{k=0}^{+\infty} \lambda_k \cos[k(\phi_s - \Theta)] \quad (34)$$

Approximations of the first eight of the coefficients λ_k in [9] abide true by replacing δ_0 by Θ.

B. Classical form the permeance function

The classical form of the permeance function is given by [11]:

$$\Lambda(\phi_s) = \frac{1}{e_o[1 - \varepsilon \cos(\phi_s - \Theta)]} \quad (35)$$

$$= \frac{1}{e_o} \sum_{k=0}^{+\infty} \Lambda_k \cos[k(\phi_s - \Theta)] \quad (36)$$

Expressions of coefficients of Fourier series development Λ_k ($k = 0, 1, 2, ...$) are available in literature [10], [11]. Fig.4(a) illustrating both permeance functions for $\varepsilon_s = 0.10$ and $\varepsilon_d = 0.10$ (20% of eccentricity) shows clearly that they are equivalent. Nonetheless, inspection of fig.4(b) confirms a consequent difference between them for $\varepsilon_s = 0.10$ and $\varepsilon_d = 0.20$ (30% of eccentricity). Therefore, the permeance function using Legendre polynomials is valid for incipient or non-exceeding eccentricity. Such permeance function is rather useful for analysis of the eccentricity prognosis.

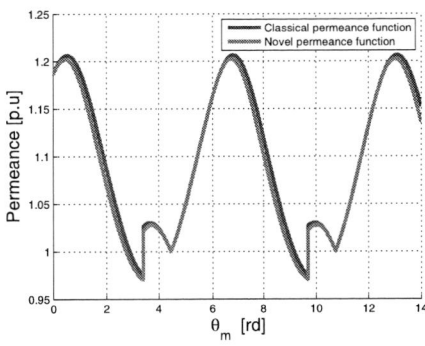

(a) $\varepsilon_s = 0.10$ and $\varepsilon_d = 0.10$

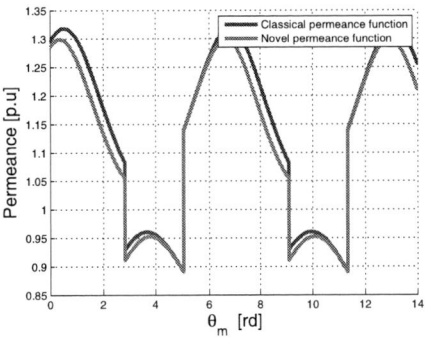

(b) $\varepsilon_s = 0.10$ and $\varepsilon_d = 0.20$

Fig. 4. Comparison of two permeance functions

IV. CALCULATIONS OF THE MOTOR INDUCTANCES

This section determines inductances of a mixed-eccentric IM by using the classical expression of permeance function. The model is can be applied for IM whose phase number is superior or equal to 3. The other specificity lies on the fact of taking into account space harmonics of winding functions. Other time harmonics indicating the presence of eccentricity are therefore expected. It is assumed that eccentricity does not affect leakage inductances. The self-magnetic field created by one phase may be determined as follows [12], [13]:

$$H_i = \left[n_i(\phi) - \frac{< \Lambda(\phi) n_i(\phi) >}{< \Lambda(\phi) >} \right] \Lambda(\phi_i) i_i(t) \quad (37)$$

$< f >$ designates the average value of the function f while n_i designates the Ampere turns of the ith stator or rotor phase winding. Λ indicates the classical permeance function previously defined [11]. $i_i(t)$ is the current in the ith phase.

A. Stator inductances

Relation (37) allows to calculate the flux density produced of the ith stator phase. Such flux density permits to express the elementary flux linkages $d\psi_{s_{ij}}$ induced in the jth phase:

$$d\psi_{s_{ij}} = \mu_o H_{s_i}(\phi_s) n_{sw_j}(\phi_s) ds_s \quad (38)$$

978-1-4799-0024-4/13 $31.00 © 2013 IEEE

μ_o, n_{sw_j} and ds stand for the free space permeability, the winding function of the jth phase and a very small area on the internal stator periphery, respectively. ds_s can be expressed as:

$$ds_s = R l \, d\phi_s \tag{39}$$

R and l designate the average radius of air-gap and the axial stack length of the machine. After integration and using orthogonality property of the cosine function [11], [10] expression of the mutual-magnetic flux linkages $\psi_{s_{ij}}$ is obtained. Therefore expression of the mutual inductance $L_{s_{ij}}$ may be calculated as:

$$L_{s_{ij}} = \frac{\psi_{s_{ij}}}{i_{s_i}(t)} \tag{40}$$

or more precisely:

$$L_{s_{ij}} = L_{s_0} \left\{ \sum_{k=0}^{+\infty} A_{s_k} \cos\left[(2k+1)(\varphi_{s_i} - \varphi_{s_j}) \right] \right.$$
$$\left. + \sum_{k=0}^{+\infty} B_{s_k} \cos\left[(2k+1)p(2\Theta + \varphi_{s_i} + \varphi_{s_j}) \right] \right\} \tag{41}$$

Therein L_{s_0} is expressed as:

$$L_{s_0} = \frac{4}{\pi} \frac{\mu_o R l}{e_o} \left(\frac{N_s}{k_{sw_1} p} \right)^2 \tag{42}$$

N_s indicates the number of the stator turns. The coefficients A_{s_k} and B_{s_k}, $(k \in N)$ are given:

$$\begin{cases} A_{s_k} = \left(\frac{k_{sw_{2k+1}}}{2k+1} \right)^2 \left(\Lambda_0 - \frac{\Lambda_{(2k+1)p}^2}{2\Lambda_0} \right) \\ B_{s_k} = \frac{1}{2} \left(\frac{k_{sw_{2k+1}}}{2k+1} \right)^2 \left(\Lambda_{2(2k+1)p} - \frac{\Lambda_{(2k+1)p}^2}{2\Lambda_0} \right) \end{cases} \tag{43}$$

A_{s_k} and B_{s_k} mainly depend on harmonics of winding coefficients $k_{sw_{2k+1}}$ and Fourier series coefficients $\Lambda_{(2k+1)p}$ of the permeance function [11]. As afore-mentioned, coefficient $\lambda_{(2k+1)p}$ associated to the permeance function using the Legendre polynomial can be resorted to in case of small eccentricity Referring to the relation (15), the mixed-eccentricity ratio depends on angle Θ (that is function of the rotor position). Therefore, A_{s_k}, B_{s_k} are also time-depending and make all stator inductances time-depending. As it is observed in fig.5, oscillation occurs in each stator inductances after the presence of eccentricity. This will cause additional ripples in electromagnetic torque. This explains partially why an eccentric-motor is subject of vibration.

B. Stator to rotor mutual inductances matrix

The rotor is assumed to be a three-phase winding system. The mutual flux linkage between the ith stator phase and the jth rotor phase is given by:

$$d\psi_{s_i r_j} = \mu_o H_{s_i}(\phi_s) n_{rw_j}(\phi_r) ds_r \tag{44}$$

n_{rw_j} and ds_r depict the winding function of jth rotor phase and a very small surface on the rotor periphery. By using angular relation (21) and after integration, one obtains the

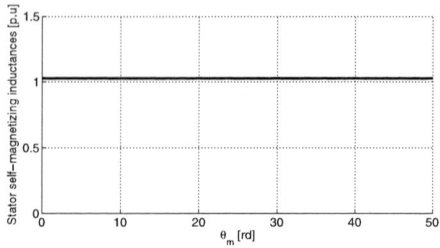

(a) balanced machine

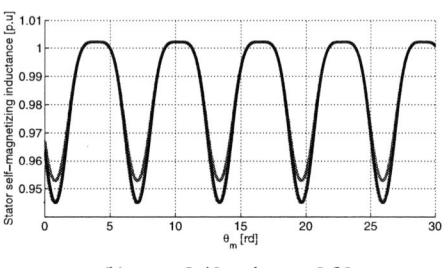

(b) $\varepsilon_s = 0.40$ and $\varepsilon_d = 0.30$

Fig. 5. Evolution of stator self-magnetizing inductances $L_{s_{11}}$ (red), $L_{s_{22}}$ (blue) and $L_{s_{33}}$ (black)

mutual flux $\psi_{s_i r_j}$. By dividing it by the instantaneous current $i_{s_i}(t)$, any mutual inductance between stator and rotor can be deduced by relation:

$$L_{s_i r_j} = \frac{\psi_{s_i r_j}}{i_{r_i}(t)} \tag{45}$$

or:

$$L_{s_i r_j} =$$
$$M_{sr_0} \left\{ \sum_{k=0}^{+\infty} A_{sr_k} \cos\left[(2k+1)(p\theta_m + \varphi_{s_i} - \varphi_{r_j}) \right] \right.$$
$$\left. + \sum_{k=0}^{+\infty} B_{sr_k} \cos\left[(2k+1)(2p\Theta - p\theta_m + \varphi_{s_i} + \varphi_{r_j}) \right] \right\} \tag{46}$$

where the constant inductance M_{sr_0} is given by the relation:

$$M_{sr_0} = \frac{4}{\pi} \frac{\mu_o R l}{e_o} \frac{N_s N_r}{k_{sw_1} k_{rw_1} p^2} \tag{47}$$

N_r indicates the turn number of any rotor phase and:

$$\begin{cases} A_{sr_k} = \frac{k_{sw_{2k+1}} k_{rw_{2k+1}}}{(2k+1)^2} \left(\Lambda_0 - \frac{\Lambda_{(2k+1)p}^2}{2\Lambda_0} \right) \\ B_{sr_k} = \frac{k_{sw_{2k+1}} k_{rw_{2k+1}}}{(2k+1)^2} \left(\Lambda_{2(2k+1)p} - \frac{\Lambda_{(2k+1)p}^2}{2\Lambda_0} \right) \end{cases} \tag{48}$$

Coefficients A_{sr_k}, B_{sr_k} $(k \in N)$ are mainly function of harmonics of stator and rotor winding coefficients and the Fourier series development of the permeance function. Hence, these coefficients cause additional time-dependence of such mutual inductances. As it is observed in fig.6, oscillations also occur in each stator to rotor mutual inductances after the presence of eccentricity.

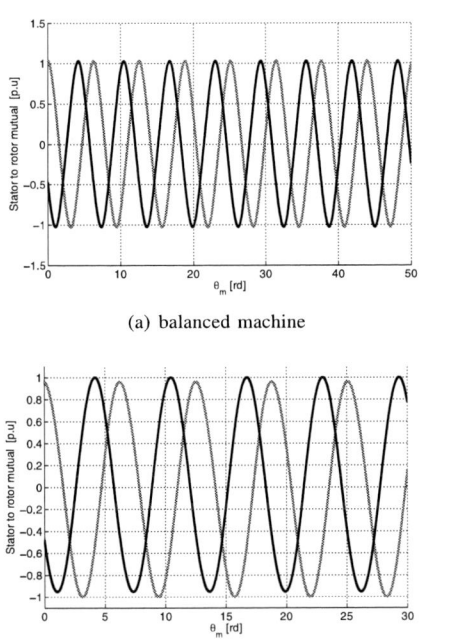

(a) balanced machine

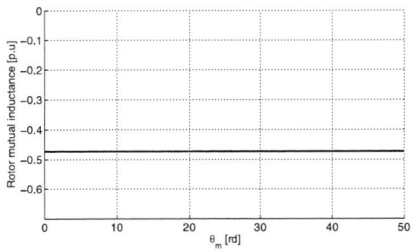

(a) balanced machine

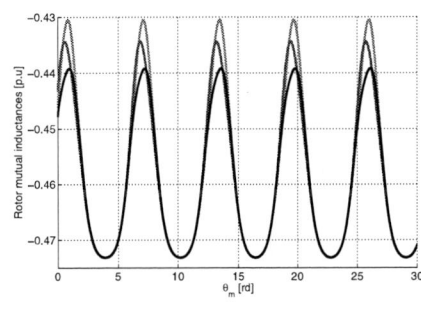

(b) $\varepsilon_s = 0.40$ and $\varepsilon_d = 0.30$

Fig. 6. Evolution of stator to rotor mutual inductances $L_{s_1 r_1}$ (red) and $L_{s_1 r_2}$ (black)

(b) $\varepsilon_s = 0.40$ and $\varepsilon_d = 0.30$

Fig. 7. Evolution of rotor mutual inductances $L_{r_{12}}$ (red), $L_{r_{13}}$ (blue) and $L_{r_{23}}$ (black)

C. Rotor inductances matrix

By adopting analogous way, all rotor inductance can be calculated by the following relation:

$$
\begin{aligned}
L_{r_{ij}} = L_{r_0} \Bigg\{ & \sum_{k=0}^{+\infty} A_{r_k} \cos\left[2(k+1)(\varphi_{r_i} - \varphi_{r_j})\right] \\
& + \sum_{k=0}^{+\infty} B_{r_k} \cos\left\{(2k+1)[2p(\Theta - \theta_m) + \varphi_{r_i} + \varphi_{r_j}]\right\} \Bigg\}
\end{aligned} \tag{49}
$$

where the constant inductance L_{r_0} is:

$$
L_{r_0} = \frac{4}{\pi} \frac{\mu_o R l}{e_o} \left(\frac{N_r}{k_{rw_1} p}\right)^2 \tag{50}
$$

Like previous kinds of inductances, each rotor inductance is also subject of ripples caused by eccentricity (fig.7).

V. GENESIS OF DEFECT HARMONICS

This section defines mixed-eccentricity signatures in stator currents from the built model. Referring to relation (41), stator mutual inductance includes the harmonic term $L_{s_0} B_{s_k} \cos\left[(2k+1)p(2\Theta + \varphi_{s_i} + \varphi_{s_j})\right] (k \in N)$ which also contains the term:

$$
\begin{aligned}
L'_{s_{ij}} = {} & L_{s_0} B_{s_k} \cdot \\
& \cos\Theta \cos\left[[2(2k+1)p - 1]\Theta + (2k+1)(\varphi_{s_i} + \varphi_{s_j})\right] \\
& = L_{s_0} B_{s_k} \cos\Theta f_{s_{ij}}(\Theta) \tag{51}
\end{aligned}
$$

In steady state $\theta_m = \frac{(1-s)}{p}\omega_s t, \omega_s = 2\pi f_s$, f_s is the fundamental frequency, s is the slip and p the pole pair number. Referring to fig.1, $\cos\Theta$ can be calculated by the formula:

$$
\cos\Theta = \frac{\varepsilon_s}{\varepsilon}\cos\delta_o + \frac{\varepsilon_d}{\varepsilon}\cos\left[\frac{(1-s)}{p}\omega_s t - \beta_o\right] \tag{52}
$$

By considering time harmonics caused by inverter, the line current of the jth phase current can be expressed as:

$$
i_{s_j}(t) = \sum_{h=1}^{+\infty} \sqrt{2} I_{s_h} \cos[h(\omega_s t + \varphi_{s_j} + \phi_0)]
$$
$$
h = 6m \pm 1 \qquad m \in N \tag{53}
$$

therein ϕ_0 is a constant shifted angle. Hence, the corresponding mutual flux linkage is expressed as follow:

$$
\begin{aligned}
\psi'_{s_{ij}} = {} & L_{s_0} B_{s_k} f_{s_{ij}}(\Theta) \\
& \Bigg\{ \frac{\varepsilon_s}{\varepsilon}\cos\delta_0 \sum_{h=1}^{+\infty} \sqrt{2} I_h \cos(h\omega_s t + \varphi_{s_j} + h\phi_0) \\
& + \frac{1}{2}\frac{\varepsilon_d}{\varepsilon} \sum_{h=1}^{+\infty} \sqrt{2} I_h \cos\left[\left(h + \frac{1-s}{p}\right)\omega_s t + \varphi_{s_j} + h\phi_0 - \beta_0\right] \\
& + \frac{1}{2}\frac{\varepsilon_d}{\varepsilon} \sum_{h=1}^{+\infty} \sqrt{2} I_h \cos\left[\left(h - \frac{1-s}{p}\right)\omega_s t + \varphi_{s_j} - h\phi_0 - \beta_0\right] \Bigg\} \tag{54}
\end{aligned}
$$

Based on the Faraday law, such flux induces the following back-emf:

$$E'_{s_{ij}} = -\frac{d\psi'_{s_{ij}}}{dt} = L_{s_0} B_{s_k} f_{s_{ij}}(\Theta)$$

$$\left\{ \frac{\varepsilon_s}{\varepsilon} \cos\delta_0 \sum_{h=1}^{+\infty} \sqrt{2} h\omega_s I_h \sin(h\omega_s t + \varphi_{s_j} + h\phi_0) \right.$$

$$+ \frac{1}{2}\frac{\varepsilon_d}{\varepsilon} \sum_{h=1}^{+\infty} \sqrt{2}\left(h + \frac{1-s}{p}\right)\omega_s I_h \sin\left[\left(h + \frac{1-s}{p}\right)\omega_s t + \Phi_{s_j}\right]$$

$$\left. + \frac{1}{2}\frac{\varepsilon_d}{\varepsilon} \sum_{h=1}^{+\infty} \sqrt{2}\left(h - \frac{1-s}{p}\right)\omega_s I_h \sin\left[\left(h - \frac{1-s}{p}\right)\omega_s t + \Phi'_{s_j}\right] \right\}$$

(55)

where $\Phi_{s_j} = \varphi_{s_j} + h\phi_0 - \beta_0$ and $\Phi'_{s_j} = \varphi_{s_j} - h\phi_0 - \beta_0$. Because of the Ohm law, this last expression shows that harmonics with frequencies of $\left[6m \pm 1 + \left(\frac{1-s}{p}\right)\right]f_s$ and $\left[6m \pm 1 - \left(\frac{1-s}{p}\right)\right]f_s$ will be generated in the stator current. These new harmonics interact again with the same inductance components to create the following frequencies $\left[6m \pm 1 + 2\left(\frac{1-s}{p}\right)\right]f_s$ and $\left[6m \pm 1 - 2\left(\frac{1-s}{p}\right)\right]f_s$. A recurrence reasoning allows to express mixed-eccentricity related frequency:

$$f_{s_{ecch}} = \left[6m \pm 1 \pm k\left(\frac{1-s}{p}\right)\right]f_s \qquad m, k \in N \quad (56)$$

These frequencies generalize the classical eccentricity signatures where only the time fundamental is considered. These new fault indicators occurs from interactions between all the time harmonic currents and the inductance harmonics. Further inspection of equation (55) states that the inherent dynamic eccentricity remains the main responsible of creating fault-related harmonics.

VI. SIMULATION AND EXPERIMENT

A confrontation between simulated and experimental testings is performed in this section. As it is mentioned in §.IV the model can be applied to a three-phase or multi-phase IM. Material possibility provides a 3kW, 127V, 50Hz, two-pole Dual-Stator Winding Induction Machine [3]. This motor is mechanically coupled to a DC-machine feeding a resistive load allowing to adjust the load level. All analysis are therefore performed basing upon this motor. In simulation, the inherent static severity is set at $\varepsilon_s = 0.4$ while the dynamic eccentricity is of $\varepsilon_d = 0.3$. The IM is loaded at 13% of the rated torque. The simulated slip is therefore of $s = 0.0485$ while the experimental one is of $s = 0.048$. FFT of the stator current is performed by subdividing frequency range at [0 200Hz] and [200Hz 500Hz] for more clarity.

Figs.8(a) and 8(b) depict FFT of simulated and experimental stator currents in [0Hz 200Hz]. Such comparison confirms a satisfactory coherence between theoretical prediction and practical result because all of sidebands met in simulation are observed in experiment in this range. Difference between sideband magnitude is observed because eccentricity in the experimental motor may not reach the level of 70%. Real physical phenomenon in motor is more complex than those

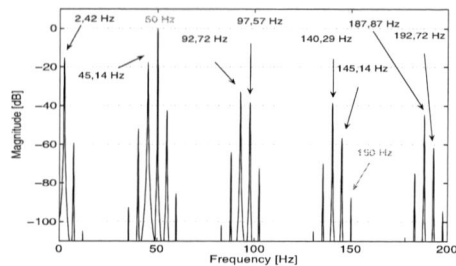

(a) Normalized FFT of simulated stator current $\varepsilon_s = 0.40$ and $\varepsilon_d = 0.30$

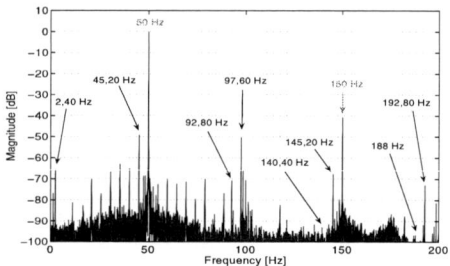

(b) Normalized FFT of experimental stator current

Fig. 8. Spectrum of stator current in [0Hz 200Hz]

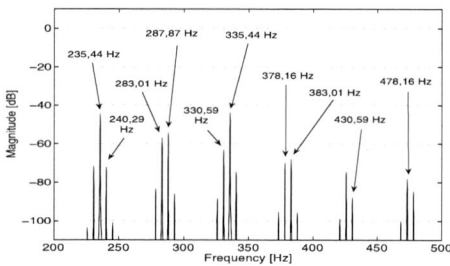

(a) Normalized FFT of simulated stator current $\varepsilon_s = 0.40$ and $\varepsilon_d = 0.30$

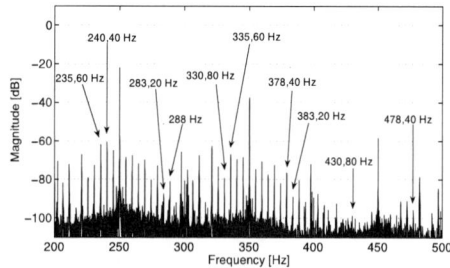

(b) Normalized FFT of experimental stator current

Fig. 9. Spectrum of stator current in [200Hz 500Hz]

in simulation that experimental spectrum appears more rich in harmonics. Similar notices can be carried out in figs.9(a) and 9(b). A good coherence is obtained again because both spectrum contain the same frequency in term of eccentricity-related harmonics. Such accordance confirms the validity of the investigation and the the realistic feasibility of eccentricity diagnosis in IM.

VII. CONCLUSION

This paper aims on the modeling, analysis and diagnosis of mixed-eccentricity fault in an IM. Study starts by building analytical expressions of the mixed-eccentricity instantaneous position, overall severity and an angular identity that are useful for the modeling. Comparison between permeance function based on Legendre polynomials and the classical one is undertaken. The non-classical permeance function constitutes an acceptable approximation of the classical form in the case of incipient or non-severe mixed-eccentricity. The modeling of eccentric IM takes into account to the space harmonics of winding and Ampre-turn functions. Such approach allows to characterize signatures due to interaction of eccentricity and space harmonics. All signatures got in simulation are effectively in observed in experimental testing. Difference is observed in theoretical and practical harmonics in term of magnitude because it is difficult to point out the exact severity of the real eccentricity in the motor. Negligible difference in term of frequency is due to small gap between theoretical and experimental slips. Satisfactory coherence between simulation and experiment constitutes a promising and realistic feasibility of eccentricity in industrial IM in spite of complexity of such defect. For industrial application, simultaneous presence of such predicted sidebands permits to state the presence of an eccentricity problem.

REFERENCES

[1] M. Berman "On the Reduction of Magnetic Pull in Induction Motors with Off-centre Rotor", in *Conf. Rec. IEEE IAS Annual Meeting*, pp. 343–350, Toronto, Canada, February 1993.

[2] G.M. Joksimović, "Dynamic simulation of cage induction machine with air gap eccentricity," in *IEE Proc. Electr. Power. Appl.*, vol. 152, no. 4, pp. 803–811, July 2005.

[3] R.N Andriamalala, H. Razik, L. Baghli and F.M. Sargos, "Eccentricity Fault Diagnosis of a Dual Stator Winding Induction Machine Drive Considering the Slotting Effects," in *IEEE Trans. on Indust. Electr.*, vol. 55, no. 12, pp. 42384251, December 2008.

[4] X. Huang, T.G. Habetler and R.G. Harley, "Analysis, Simulation and Experiments of Rotor Eccentricity in Closed-Loop Drive- Connected Induction Motors," *IEEE Inter. Symp. on Diag. for Elec. Mach. Power. Electr.& Drives, Conf. SDEMPED*, Austria, CDROM, 2005

[5] D.G. Dorrell, "Sources and Characteristics of Unbalanced Magnetic Pull in Three-Phase Cage Induction Motors With Axial-Varying Rotor Eccentricity," *IEEE Ind. Applicat. Soc.*, vol. 47, no. 1, pp. 12–24, 2011.

[6] D.G. Dorrell, J. Shek, M-F. Hsieh and M.A Mueller, "Unbalanced Magnetic Pull in Cage Induction Machines for Fixed-Speed Renewable Energy Generators," *IEEE Trans. on Magn.*, vol. 47, no. 10, pp. 4096–4099, October 2011.

[7] C. Bruzzese and G. Joksimovic, "Harmonic Signatures of Static Eccentricities in the Stator Voltages and in the Rotor Current of No-Load Salient-Pole Synchronous Generators," *IEEE Trans. on Indust. Electr.*, vol. 58, no. 5, pp. 1606–1624, May 2011.

[8] A.D. Poularikas, "Signals and Systems," *The Transforms and Applications Handbook: Second Edition*, Boca Raton: CRC Press LLC, 2000.

[9] R.N. Andriamalala, H. Razik, G. Didier, M.B.R. Corrêa and F.M. Sargos, "An Accurate Model by Using the Legendre Polynomial Functions of a Dual Stator Induction Machine Dedicated to The Static Eccentricity Diagnosis," in *IEEE Ind. Applicat. Soc.*, New-Orleans,-USA, Sept. 2007.

[10] A. Stavrou and J. Penman, "The on-Line Quantification of Air-gap Eccentricity in Induction Machines," *International Conference on Electrical Machines ICEM*, Paris, France, pp. 261–266, 1994.

[11] A. Stavrou and J. Penman, "Modelling dynamic eccentricity in smooth air-gap induction machines", *IEEE International Electric Machines & Drives Conf. IEMDC*, pp. 864–871, 2001.

[12] J. Faiz and I. Tabatabaei "Extension of winding function theory for nonuniform air gap of electric machinery," *IEEE Trans. on Magnetics*, vol. 38, pp. 3654-3657, 2002.

[13] J. Faiz, I.T. Ardekanei and H.A. Toliyat, "An Evaluation of Inductances of a Squirrel-Cage Induction Motor Under Mixed Eccentric Conditions," *IEEE Trans. on Energ. Conversions*, vol. 18, no. 2, pp. 252–258, June 2003.

Rijaniaina Njakasoa Andriamalala was born in Ambohitrolomahitsy, Madagascar. He was graduated of a PhD in electrical engineering from the Université Henri Poincaré, Nancy I, France in 2009. His PhD project focused mainly on modeling, diagnosis and control of multi-phase induction machine drive. In 2010, He was Research Assistant with the Laboratoire GREEN of the Université Henri Poincaré, Nancy I. Since September 2011, He is Research Fellow with the Laboratoire L2EP of the Ecole Centrale de Lille, France where He is involved to industrial projects (mainly with EDF, RTE and SNCF) covering power system analysis, wind power integration and PHIL.

Hubert Razik was graduated from the Ecole Normale Superieure, Cachan, France, in 1987. He received the Ph.D. degree in Electrical Engineering from the Institut Polytechnique de Lorraine, Nancy, France, in 1991, and received the Habilitation to Supervise Researches from the Université Henri Poincaré, France, in 2000. Since 2009, he is professor of electrical engineering at the Université Claude Bernard Lyon I and he is with the AMPERE laboratory - UMR 5005, Villeurbanne, France.

François-Michel Sargos was born in Talence, France, in 1947. He received the Engineer degree from the Ecole Nationale dElectricité et de Mécanique of Nancy, Nancy, France, in 1970. He is currently Professor of Electrical Engineering at the Ecole Nationale dElectricité et de Mécanique of Nancy. He is also with the Groupe de Recherche en Electronique et en Electrotechnique de Nancy (GREEN), CNRS UPRESA 7037, Institut National Polytechnique de Lorraine, Vandoeuvre-Lès-Nancy, France. His main research topics are analytical field calculation, reluctance and step motors, and machine modeling for simulation and control.

Bruno Francois (M'1996, SM'2006) received the PhD degree in electrical engineering in 1996 from the University of Science and Technology of Lille (USTL), France. His field of interest includes power electronics, renewable energy sources and power systems. He is currently working on renewable energy based active generators and on the design of advanced energy management systems. Bruno Francois is with the Laboratory of Electrical Engineering and Power Electronics of Lille (L2EP) and is a Professor at the department of Electrical Engineering of Ecole Centrale de Lille

978-1-4799-0024-4/13 $31.00 © 2013 IEEE

Motor Current Signature Analysis Apply for external Mechanical Fault and Cage Asymmetry in Induction Motors

A. J. Fernández Gómez, T. J. Sobczyk

Abstract -- The aim of this paper is to recognize if the differences between effects due to external mechanical faults and internal electrical faults in induction motors can be diagnosed through Motor Current Signature Analysis (MCSA) techniques applying Fourier analysis. For that purpose an alternative algorithm to traditional Fast Fourier Transform was used. Positive and negative frequency spectra have been studied. The paper shows a comparison between frequency spectra of the symmetrical and natural components of stator currents for internal electrical fault (rotor cage asymmetry) and external mechanical fault represented by a periodical oscillating torque. Cage asymmetry is defined by symmetrical and asymmetrical factors. The study is based on the classical model of induction machine considering only effects of Main Magneto-motive Forces. Result shown corresponds to the steady-state performance of the motor supplied directly from the net as a sinusoidal voltage source. The advantages of the algorithm use for Fourier analysis have also been discussed.

Index Terms-- Cage asymmetry, Condition Monitoring, External Fault, Modeling, Motor Current Signature Analysis, Squirrel Cage Motor.

I. INTRODUCTION

It is well-known that induction machine is the most common element existing in a large number of industries. This type of machine plays nowadays an important role in the control system of a wide number of processes.

Induction machines are characterised by an easy design and good properties such as reliability, low cost, robust construction or easy control through power electronic devices but on the other hand their functionality can be affected by plenty of factors. Research in fault field of electrical machines is still needed.

Many studies have dealt with diganosis of rotor faults and load torque oscillations [6]-[7]-[12] under different assumptions. Most of these studies agreed the use of

---

This work has the financial support from the Marie Curie FP7-ITN project "Energy savings from smart operation of electrical, process and technical equipment – ENERGY-SMARTOPS", Contract No: PITN-GA-2010-264940 is gratefully acknowledged.

T. J. Sobczyk is with Inst. of Electric Energy Conversion, Cracow University of Technology, Kraków, ul. Warszawska 24, 31-155 Poland (e-mail: pesobczy@cyf-kr.edu.pl).

A. J. Fernández Gómez is with Inst. of Electric Energy Conversion, Cracow University of Technology, Kraków, ul. Warszawska 24, 31-155 Poland (e-mail: afernandezpk@gmail.com).

additional components $(1 \pm 2s) \cdot f_0$ as indicators of the presence of the faults, e.g., some authors propose the creation of faults index [7] based on the relation of the amplitude of the additional components. In [14] the visualtization into the real and imagine axes of currents at frequencies $(1 \pm 2s) \cdot f_0$ has been proposed. Furthermore, others propose the study of the active and reactive instantaneous power [16] as an indicator of fault.

In the present paper a tehoretical study of internal electrical faults and external mechanical faults has been presented. Asymmetry and symmetry coefficients have been considered to represent internal rotor faults and an additional component of the load torque time dependant in form of periodic oscillations for mechanical fault. Equal efeccts on stator phase currents has been simulated defining the amplitude of the oscillating periodical torque as the ripple caused by the rotor cage asymmetry.

An alternative Fourier analysis of the stator currents is proposed in this paper. Results shown come from the analysis of the negative frequency espectra of symmetrical components of the stator phase currents.

The aim of this paper is to give an answer to the following question: is it possible to differentiate between external mechanical fault and internal electrical fault through the frequency spectra of stator phase currents?

Assumptions of main magneto-motive harmonics effect, steady-state performance of the induction machine with constant speed and small slips have been considered.

II. MODELING INTERNAL ELECTRICAL FAULTS

Induction machine can be represented by a set of four electrical equations and a set of two mechanical equations even under fault condition following the classical model of induction machines. In terms of internal electrical faults we meant rotor cage asymmetry. Despite the fact that statics indicating incipient broken bar and one broken bar as the common faults in rotor of induction machines, two neighbor broken bars have been studied as magnitude of their additional components is higher than in case of one broken bar providing a clear representation of the fault.

It is commonly known that additional components in the spectra of stator currents appear at a specific frequency band $(1 \pm 2s) \cdot f_0$ [2]-[7]-[10]. A fault in rotor cage creates oscillations in the rotor speed producing an alternating

---

978-1-4799-0024-4/13 $31.00 © 2013 IEEE   136

component in torque with frequency $2sf_0$ which is responsible for the appearance of the additional component $(1 + 2s) \cdot f_0$ in stator currents. For small slips the value of the component is also small, in consequence and due to the inertia of the rotor the speed becomes constant again so the oscillations disappear, as well as the component $(1 + 2s) \cdot f_0$. The component $(1 - 2s) \cdot f_0$ is the main effect produced by cage asymmetry fault at constant speed. The evaluation of this component has been accepted as an adequate measure of cage asymmetry.

The severity of the fault depends on how many elements of the rotor (ring segments and bars) are affected. The level of the component will be different for each case. Nevertheless, this component is not enough to diagnose which bar of the cage is damaged or how many of them are broken. For example: the level of the component decreases with the increasing number of broken bars or even it could be zero for a particular pair of broken bars [8]-[17]. Despite these disadvantages it is still considered a very useful method to detect an incipient broken bar in squirrel cage.

Cage asymmetry is modeled increasing the resistance of the element affected by the fault. The change of the resistance is typically estimated by an increase of 20 times the initial value, therefore, regarding the classical model of IM, an additional term of the rotor resistance will appear. On the other hand, the stator currents depend on the equivalent value of the rotor resistance, hence the level of the component $(1 - 2s) \cdot f_0$.

The rotor resistance $\mathbf{R_r}$ can be represented in a matrix form with size $(N + 1) \, x \, (N + 1)$ where "N" is the total number of bars (cage mesh currents) plus one extra term representing the current flowing in one of the rings, Fig. 1.

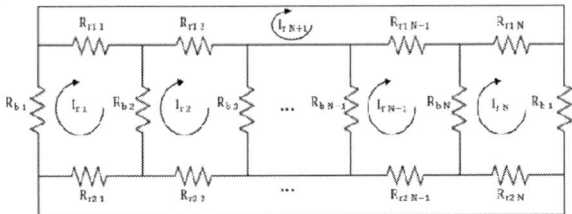

Fig. 1. Representation of cage mesh currents and resistances

The size can be reduced to an equivalent rotor matrix with dimensions (2 x 2). Changes of resistance values on any element of the rotor will be represented in that matrix. Differences between cage asymmetry and healthy machine have been represented by k_{as} k_s asymmetry and symmetry factors respectively [4]. Those factors become zero in case of healthy machine. It should be reminded that these factors will not be zero for the study case of two neighbor broken bars.

The quotient between backward I_s^2 and forward I_s^1 component of stator currents in symmetrical components k_{1-2s} gives us the relation of the component $(1 - 2s) \cdot f_0$ with the main frequency 50 Hz. As it has been proved in [15] the ratio k_{1-2s} can be approximated by the asymmetry factor k_{as}.

Summarizing, an internal electrical fault can be diagnosed by the level of the frequency component $(1 - 2s) \cdot f_0$ for analysis of symmetrical components of stator currents which depends on the value of asymmetry coefficient k_{as} measure of the cage asymmetry severity.

With the purpose of clarity the equations used has been transformed into components d-q.

$$\frac{d}{dt}\begin{bmatrix} \Psi_{sd} \\ \Psi_{sq} \\ \Psi'_{rd} \\ \Psi'_{rq} \end{bmatrix} + \begin{bmatrix} -\omega_s \cdot \Psi_{sq} \\ \omega_s \cdot \Psi_{sd} \\ -\omega_r \cdot \Psi'_{rq} \\ \omega_r \cdot \Psi'_{rd} \end{bmatrix} + R \cdot \begin{bmatrix} i_{sd} \\ i_{sq} \\ i'_{rd} \\ i'_{rq} \end{bmatrix} = \begin{bmatrix} u_{sd} \\ u_{sq} \\ u'_{rd} \\ u'_{rq} \end{bmatrix} \quad (1)$$

$$R = \begin{bmatrix} R_s & 0 & 0 & 0 \\ 0 & R_s & 0 & 0 \\ 0 & 0 & R_r'^p \cdot (1 + k_s + k_{as} \cdot \cos(2\gamma_r)) & R_r'^p \cdot k_{as} \cdot \sin(2\gamma_r) \\ 0 & 0 & R_r'^p \cdot k_{as} \cdot \sin(2\gamma_r) & R_r'^p \cdot (1 + k_s + k_{as} \cdot \cos(2\gamma_r)) \end{bmatrix} \quad (2)$$

$$\begin{bmatrix} \Psi_{sd} \\ \Psi_{sq} \\ \Psi'_{rd} \\ \Psi'_{rq} \end{bmatrix} = \begin{bmatrix} L_{\sigma s} + L_m & 0 & L_m & 0 \\ 0 & L_{\sigma s} + L_m & 0 & L_m \\ L_m & 0 & L'_{\sigma r} + L_m & 0 \\ 0 & L_m & 0 & L'_{\sigma r} + L_m \end{bmatrix} \cdot \begin{bmatrix} i_{sd} \\ i_{sq} \\ i'_{rd} \\ i'_{rq} \end{bmatrix} \quad (3)$$

$$\gamma_s = \gamma_r + p \cdot \varphi \quad (4)$$

where:

Ψ_{sd}, Ψ_{sq} stator-flux.

Ψ'_{rq}, Ψ'_{rd} rotor-flux referred to the stator.

ω_s, ω_r speed of stator and rotor, respectively.

i_{sd}, i_{sq} stator phase current.

i'_{rd}, i'_{rq} rotor currents referred to the stator.

u_{sd}, u_{sq} stator phase voltage.

u'_{rd}, u'_{rq} rotor voltage.

$R_s, R_r'^p$ stator and equivalent rotor resistances.

$L_{\sigma s}, L'_{\sigma r}$ leakage inductance of stator and rotor referred to stator.

L_m mutual inductance.

γ_s, γ_r geometrical angle of stator and rotor.

Finally, it needs to be remembered that the transformation between stator and rotor quantities (4) have to be fulfilled.

III. Modeling External Mechanical Faults

Electrical machines are usually the fundamental part of complex systems connected to a wide number of mechanical devices and they are also built using mechanical components. As it has been mentioned, their functionality can be affected by a fault in those mechanical elements. For example, an unbalanced rotation shaft due to a fault in a gearbox or a broken ball in a bearing creates an additional torque and in some cases an asymmetry in the air-gap.

The reaction of the induction motor due to this type of faults can be studied through the set of two mechanical equations adapting the additional torque component to the mechanical element damaged

$$J \frac{d\varphi_r{}^2}{dt} - D \frac{d\varphi_r}{dt} = elec - load \quad (5)$$

978-1-4799-0024-4/13 $31.00 © 2013 IEEE

$$\frac{d\varphi_r}{dt} = \omega_r \tag{6}$$

where J is the moment of inertia of the rotor and the load, D the dumping term, $_{elec}$ the electromechanical torque in the rotor created by the stator currents, $_{load}$ the resistive torque in shaft, φ_r is the geometrical angle of the rotor and ω_r the mechanical rotor speed. Usually the dumping term can be neglected.

An oscillating periodic torque at frequency $2sf_0$ [6] like in case of rotor cage asymmetry has been assumed. The amplitude of the torque has been defined as a percentage of the load torque and the main harmonic has been taken into account.

$$'_{load} = _{load} + _{AC} \cdot \sin(\omega_{AC} \cdot t) \tag{7}$$

where $_{AC}$ is the amplitude of the torque oscillations and $\omega_{AC} = 2s\pi f_0$.

Stator currents depend on the magnetic flux distribution in air-gap of the induction machine and the permeance function $\Box$ which is equal to the inverse of the air-gap length.

$$U = R \cdot I(t) + \frac{d}{dt}[L(t) \cdot I(t)] \tag{8}$$

$$\Psi = L(t) \cdot I(t) \tag{9}$$

In our study the permeance function will be constant because smooth air-gap has been considered and the Carter's factor neglected.

The flux density Ψ is calculated applying the integral to the magnetic field function $B(\theta, t)$ which depends on MMF function of stator and rotor.

On the other hand, when the motor is running a constant speed the electro-magnetic torque is equal to the load torque hence the speed of the rotor is equal to the integral of the oscillating torque component.

$$\omega_r = \frac{\partial \varphi_r}{\partial t} = -\int_{t_0}^{t} \frac{1}{J} \cdot _{AC} \cdot \sin(\omega_{AC} \cdot t) dt \tag{10}$$

$$\omega_r = \frac{1}{J} \cdot _{AC} \cdot \frac{1}{\omega_{AC}} \cdot \cos(\omega_{AC} \cdot t) - \omega_C \tag{11}$$

$$\omega_C = 2\pi f_r \tag{12}$$

and the mechanical angle of the rotor is the integral of the rotor speed

$$\varphi_r = \int_{t_0}^{t} \frac{1}{J} \cdot _{AC} \cdot \frac{1}{\omega_{AC}} \cdot \cos(\omega_{AC} \cdot t) - \omega_C \, dt \tag{13}$$

$$\varphi_r = \frac{1}{J} \cdot _{AC} \cdot \frac{1}{\omega_{AC}^2} \cdot \sin(\omega_{AC} \cdot t) - \omega_C \cdot t \tag{14}$$

MMF function depends on the angle reference frame which has to fulfill the constraint (4). Reference angle for stator quantities γ_s reminds constant but the reference angle for rotor quantities γ_r will be affected by the mechanical angle calculated in (14), accordingly the MMF function of the rotor will be affected by the alternating component of the load torque.

$$\gamma_r = \gamma_s - p \cdot [\alpha \cdot \sin(\omega_{AC} \cdot t) - \omega_C \cdot t] \tag{15}$$

where $\alpha = \frac{1}{J} \cdot _{AC} \cdot \frac{1}{\omega_{AC}^2}$

Therefore it has been proved that alternating component of the torque will affect the stator currents frequency spectra.

IV. ANALYSIS OF FREQUENCY SPECTRA OF STATOR CURRENTS

An iterative algorithm for determining Fourier spectra of stator currents of AC machine for steady-state analysis has been applied instead of usual techniques such as Fast Fourier Transform of Discrete Fourier Transform.

The algorithm can calculate the Fourier Spectra of a dynamic system described by a set of non-linear differential equations. The solving method allows more precise prediction of the Fourier spectra of stator phase currents.

Generally, currents at steady-states of AC machines are described by periodic or quasi-periodic time functions, whose Fourier spectra are determined by an infinite set of algebraic equations.

Let the vector $i_A(t)$ representation of the phase stator current A

$$i_A(t) = \sum_{s=-\infty}^{\infty} I_s^A \cdot e^{jsw_0 t} \tag{16}$$

which is periodic,

$$i_A(t) = i_A(t + T_p) \tag{17}$$

Limiting the function to a finite number $|s| \leq S$ for the set of points

$$w_0 \cdot t = j \cdot \alpha \quad -S \leq j \leq S \quad \alpha = \frac{2 \cdot \pi}{2 \cdot S + 1} \tag{18}$$

the Fourier coefficient I^A of the time domain current signal can be written in the form

$$I^A = T^1 \cdot i^A \tag{19}$$

where

$$i^A = [i_S^A, \dots, i_1^A, i_0^A, i_{S-1}^A, \dots i_{-S}^A] \tag{20}$$

$$I^A = [I_S^A, \dots, I_1^A, I_0^A, I_{S-1}^A, \dots I_{-S}^A] \tag{21}$$

and

$$T = \begin{bmatrix} a^{S^2} & \cdots & a^S & a^0 & a^{-S} & \cdots & a^{-S^2} \\ \vdots & \ddots & \vdots & \vdots & \vdots & \ddots & \vdots \\ a^S & \cdots & a^1 & a^0 & a^{-1} & \cdots & a^{-S} \\ a^0 & \cdots & a^0 & a^0 & a^0 & \cdots & a^0 \\ a^{-S} & \cdots & a^{-1} & a^0 & a^1 & \cdots & a^S \\ \vdots & \ddots & \vdots & \vdots & \vdots & \ddots & \vdots \\ a^{-S^2} & \cdots & a^{-S} & a^0 & a^S & \cdots & a^{S^2} \end{bmatrix} \quad a = e^{j\alpha} \tag{22}$$

$$T^{-1} = \frac{1}{(2S+1)} \cdot (conj(T))^T \tag{23}$$

The matrix T fulfills the condition

$$T \cdot (conj(T))^T = (2S + 1) \cdot I \tag{24}$$

where I is the identity matrix.

978-1-4799-0024-4/13 $31.00 © 2013 IEEE

Let's use a practical usage of the method to explain the advantages with respect to typical approach.

In the first place we have to define the characteristic of the stator current signal generated for our software regarding the resolution of the frequency band to be analyzed.

$$\Delta f = \frac{1}{T} = \frac{1}{k \cdot \Delta t} = \frac{f_s}{k} \qquad (25)$$

where Δf is the resolution of the frequency band, T is the period of the signal, k is the number of points of the signal, Δt step time and f_s is the sampling frequency.

Once the resolution is chosen the number of points has to be decided (set of points S in the above equations). It will depend on the desired quality of the signal analyzed. Period, step time and sampling frequency can be easily obtained by (25).

Vector I^A contains all Fourier coefficients from the sampling frequency $-f_s$ to f_s. The last step will consist in reduction of the system of equations to only those which allow calculation of the particular spectra of frequency wanted.

The advantage of this methodology in comparison to usual approaches is the possibility of study a specific frequency band with a high resolution and accuracy without the need of high calculation capacity requirements.

V. RESULTS OF SEPARATED FAULT SIMULATIONS

Comparison between frequency spectra of symmetrical and natural components of the stator phase currents is discussed in detail in this section. Currents in symmetrical components are defined as:

$$i_s^1 = \left(\frac{1}{\sqrt{3}}\right) \cdot (i_A + a \cdot i_B + a^2 \cdot i_C) \qquad (26)$$

$$i_s^2 = conj(i_s^1) \qquad (27)$$

$$a = e^{j\frac{2\pi}{3}} \qquad (28)$$

where $i^1(t)$ and $i^2(t)$ are the first and second symmetrical components.

Parameters of the machine used correspond to the model Sg112-M4 from Polish manufacture with rated data: $P_N = 1330\ W$, $N_N = 1492\ rpm$, $U_N = 400\ V$, $I_N = 2.86\ A$, $f_N = 50\ Hz$ with 28 bars and 2 pair of poles.

Both faults have been simulated under the same load condition; load torque - 9.35 Nm with rated slip s=0.07.

Two neighbor broken bars have been simulated. Asymmetry and symmetry factors have been calculated with resistance of the bar equal to $R_b = 0.0000641\ \Omega$.

With intention of simulating both faults under the same conditions, the ripple in case of internal electrical fault was firstly measured. Its percentage respect to the load torque applied was used to setup the amplitude of the oscillate torque to simulate external mechanical fault.

Resolution of the Fourier spectra has been set up for 0.1 Hz.

Comparison between stator phase currents in natural and symmetrical components is discussed in detail below.

Fig. 5.1, 5.2, 5.5, 5.6 show the frequency spectra of symmetrical components and Fig. 5.3, 5.4, 5.7, 5.8 show the frequency spectra of natural components for the case of healthy machine, cage asymmetry and mechanical fault respectability.

Values of asymmetry and symmetry coefficients are indicated in Table I. The amplitude of the oscillating torque rated is indicated in Table II.

A. Cage asymmetry

TABLE I
CAGE ASYMMETRY PARAMETERS

Broken Bars	k_{as}	k_s	Rated Slip	Frequency $(1 \pm 2s)f_o$
Two	0.649	0.698	0.075	42,5/57.5 Hz

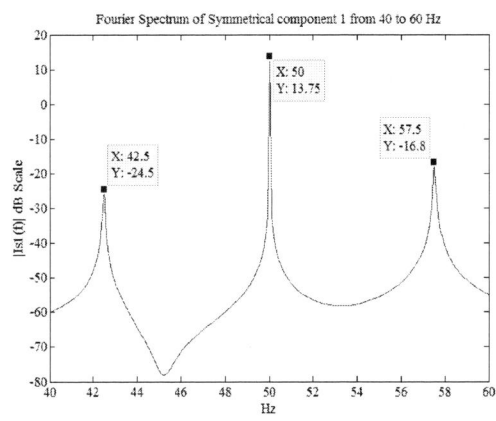

Fig. 5.1 Frequency spectra of Symmetrical Component 1 from 40 to 60 Hz

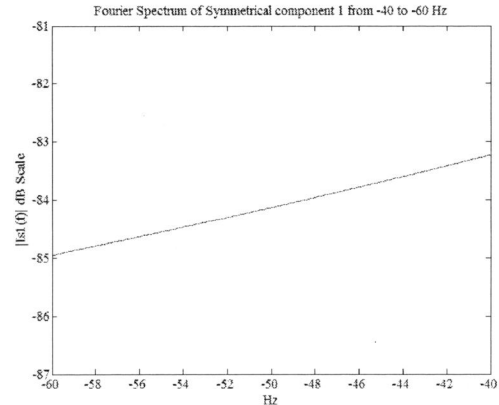

Fig. 5.2 Frequency spectra of Symmetrical Component 1 from -40 to -60 Hz

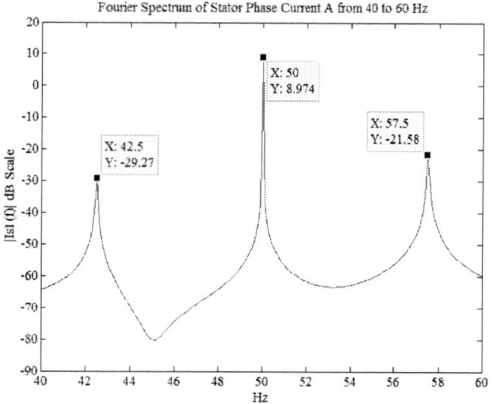

Fig. 5.3 Frequency spectra of Stator Phase Current A from 40 to 60 Hz

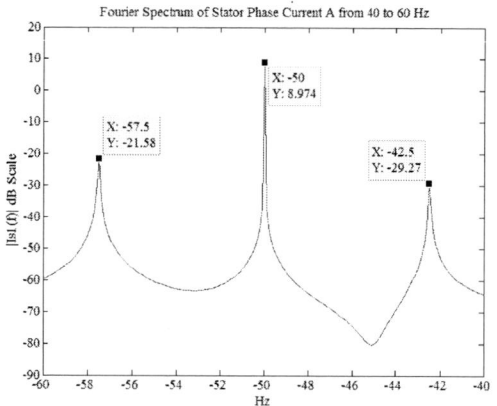

Fig. 5.4 Frequency spectra of Stator Phase Current A from -40 to -60 Hz

B. External mechanical fault

TABLE II
EXTERNAL MECHANICAL FAULT PARAMETERS

T_{AC} Percentage	Slip	Frequency $(1\pm2s)f_0$
6 %	0.073	42,7/57.3 Hz

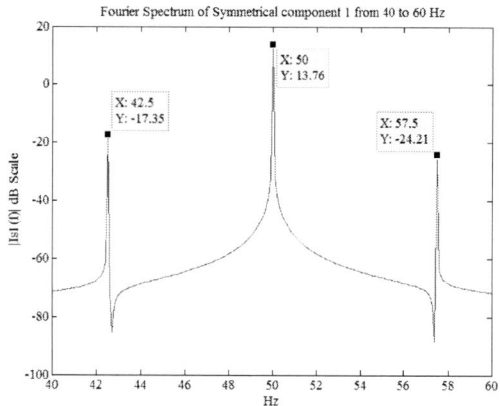

Fig. 5.5 Frequency spectra of Symmetrical Component 1 from 40 to 60 Hz

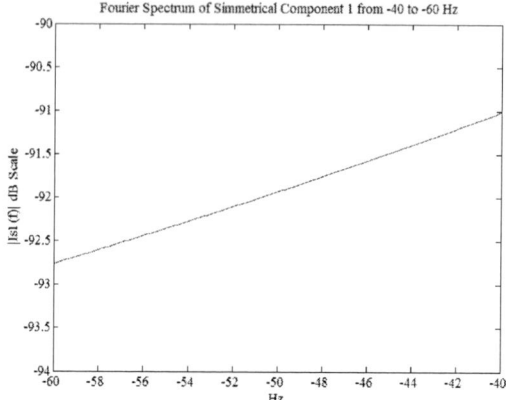

Fig. 5.6 Frequency spectra of Symmetrical Component 1 from -40 to -60 Hz

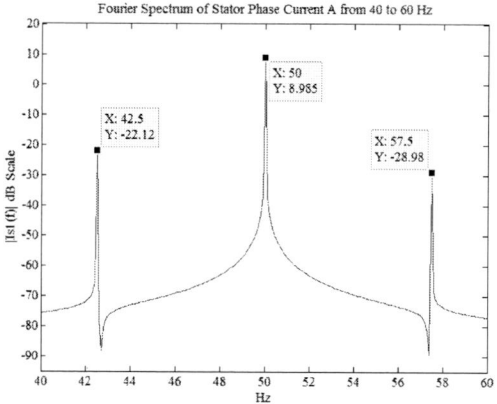

Fig. 5.7 Frequency spectra of Stator Phase Current A from 40 to 60 Hz

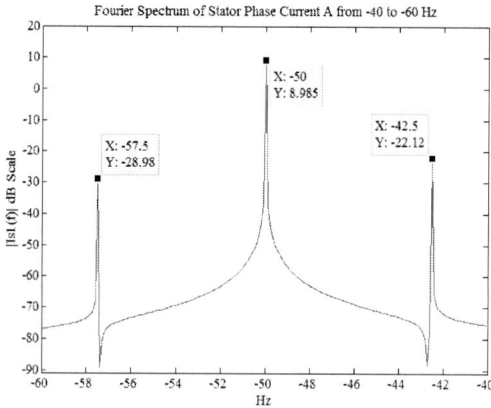

Fig. 5.8 Frequency spectra of Stator Phase Current A from -40 to -60 Hz

C. Discussion of Results

As expected, the additional components $(1 \pm 2s) \cdot f_0$ appear the positive spectrum of currents in natural and symmetrical components for both faults. In the case of cage asymmetry, the component $(1 - 2s) \cdot f_0$ has higher magnitude than $(1 + 2s) \cdot f_0$ and in the case of external mechanical faults the component $(1 + 2s) \cdot f_0$ has the highest magnitude.

In the negative spectra of symmetrical components do not emerge any frequency components in spite of the fact that a motor with cage asymmetry is asymmetrical and a motor with load perturbation is symmetrical.

VI. CONCLUSIONS

Classical model of induction machines has been used as a reference for faulty models development. Models for cage asymmetry and external mechanical faults have been presented. These models allow numerical computations of time-varying software package. Limitations of the models and consequences in the final results have been discussed.

The steady-state performance of the motor has been simulated under effect of main magneto-motive forces and small slips for constant rotor speed.

MCSA techniques have been applied to analyze the positive and negative frequency spectra of the instantaneous stator currents values in symmetrical and natural components frame. A special algorithm has been used instead of the typical Fast Fourier Transform approach.

Results of the negative frequency spectra for each fault have not proved useful to find differences between faults analyzed. Therefore, the question made at the beginning cannot be answered using suggested methods.

VII. REFERENCES

[1] T.J. Sobczyk, "Direct determination of two periodic solutions for non-linear dynamic systems", COMPEL International Journal for Computation and Mathematics in Electrical and Electronic Engineering, vol. 13, No. 3, pp. 509–529, 1994.

[2] F. Filippetti, G. Franceschini, C. Tassoni, P. Vas, "AI techniques in induction machines diagnosis including the speed ripple effect", IEEE Transactions on Industry Applications, vol. 34, no 1, pp. 98-108, Jan.-Feb. 1998.

[3] R. R. Obaid, T. G. Habetler, D. J. Gritter, "A simplified technique for detecting mechanical faults using stator current in small induction motors". In Industry Applications Conference, IEEE, vol. 1, pp. 479-483, Oct. 2000.

[4] T.J. Sobczyk, W. Maciołek, "Asymmetry factors of induction motor rotor cages", in Proc. 3rd IEEE Symp. On Diagnostics of Electric Machines, Power Electronics and Drives (SDEMPED), pp. 487-491, Sep. 2001.

[5] J. Faucher, B. Dagues, M. Chabert, "Mechanical load fault detection in induction motors by stator current time-frequency analysis", IEEE International Conference on Electric Machines and Drives, pp. 1881 - 1888, May. 2005.

[6] L. Wu, T.G. Habetler, R. G. Harley, "Separating load torque oscillation and rotor fault effects in stator current-based motor condition monitoring", IEEE International Conference on Electric Machines and Drives, pp. 1889-1894, May. 2005.

[7] A. Bellini, F. Filippetti, F. Franceschini, T.J. Sobczyk, C. Tassoni, "Diagnosis of induction machines by d-q and i.s.c rotor models", in Proc. 5th IEEE Symp. on Diagnostics of Electric Machines, Power Electronics and Drives (SDEMPED), pp.41-46, Sep. 2005.

[8] T.J. Sobczyk, W. Maciołek, "Does the component (1-2s)fo in stator currents is sufficient for detection of rotor cage faults?", in Proc. 5th IEEE Symp. on Diagnostics of Electric Machines, Power Electronics and Drives, (SDEMPED), pp. 175-179, Sep. 2005.

[9] A. Bellini, G. Franceschini, C. Tassoni, A. Toscani, "Assessment of induction machines rotor fault severity by different approaches", 31st Annual Conference of IEEE In Industrial Electronics Society, IECON, pp. 6-pp, Nov. 2005.

[10] S. Nandi, H. A. Toliyat, X. Li, "Condition monitoring and fault diagnosis of electrical motors-a review". Energy Conversion, IEEE Transactions on, vol. 20, no 4, pp. 719-72, Dec. 2005.

[11] G. Didier, E. Ternisien, O. Casparey, H. Razik, "Fault detection of broken rotor bars in induction motor using a global fault index", IEEE Trans. Industry Applications, vol. 42, pp. 79-88, Jan.-Feb. 2006.

[12] D. Basak, A. Tiwari, S.P. Das, "Fault diagnosis and condition monitoring of electrical machines - A Review", IEEE International Conference on Industrial Technology, pp. 3061-3066, Dec. 2006.

[13] M.R. Rao, "Estimation of parameters for induction motor's analytical model by direct search method", in Int. Conf. on Electrical Machines and Systems, ICEMS, pp. 8-10, Oct. 2008.

[14] C. Concari, G. Franceschini, C. Tassoni, "A MCSA procedure to diagnose low frequency mechanical unbalances in induction machines", XIX International Conference In Electrical Machines (ICEM), pp. 1-6, Sep. 2010.

[15] T.J. Sobczyk, W. Maciołek, "Influence of pole-pair number and rotor slot number on effects caused by cage faults", IEEE International Symposium on Diagnostics for Electric Machines, Power Electronics & Drives (SDEMPED), pp. 199-204, Sep 2011.

[16] S. M. Cruz, F. Gaspar, "A new PQ method to diagnose rotor faults in three-phase induction motors coupled to time-varying loads", IEEE International Symposium on Diagnostics for Electric Machines, Power Electronics & Drives (SDEMPED), pp. 16-23, Sep. 2011.

[17] T.J. Sobczyk, A.J. Fernández Gómez, "Influence of design data of induction motor on effects of cage asymmetry", XLVIII International Symposium on Electrical Machines SME, vol. I, pp. 357-364, Jun. 2012.

[18] J. Fraile Mora, Electrical Machines, edn.VI. Madrid: McGraw-Hill, 2003, pp. 259-378.

[19] G. M. Fitchtenholz, Differential and integral calculus, tom. 3, Warsaw: PWN (in Polish), 1980, pp. 390-408.

[20] T.J. Sobczyk, Metodyczne aspekty modelowania matematycznego maszyn indukcyjnych, Warsaw: WNT (in Polish), 2004Biographies

VIII. BIOGRAPHIES

Tadeusz J. Sobczyk is a professor at the Faculty of Electrical & Computer Engineering of the Cracow University of Technology. In the years 1993-1999, he was the Dean of this Faculty. Presently he is the Director of the Institute on Electromechanical Energy Conversion and the Head of Department of Electrical Machines in this Institute. Since 1991 is a member of the Committee of Electrical Engineering of the Polish Academy of Science. In 2000 he was awarded the honorary title Doctor Honorees Causa of the Russian Academy of Sciences. His main research fields are: electrical machines and drives, electromechanical systems, electrical energy conversion and transformation by power electronic systems.

Alejandro J. Fernández Gómez was born (1984) and educated in Spain. In 2010 he received M.Sc degrees in electrical engineering from the Faculty of Industrial Engineering at the University of Vigo. Since 2011 he has been working as early stage researcher in the Institute on Electromechanical Energy Conversion at the Faculty of Electrical & Computer Engineering at the Cracow University of Technology enrolled in ITN Energy Smartops project of Marie Curie Actions. His research field is electromechanical energy conversion and fault diagnosis of electrical machines.

Dynamic Model of Induction Machine with Faulty Rotor in Field Reference Frame

Vanja Ambrožič, Rastko Fišer, Mitja Nemec, and Klemen Drobnič

Abstract -- **This paper presents a control model of squirrel cage induction machine with broken rotor bars. The basis for model's equations is the fact that this type of faults inevitably alters the rotor parameters. Both the resistance and inductance take a sinusoidal shape, being a function of rotor angle, almost regardless of the configuration of fault. Differing from previous work, in this paper a model has been developed in rotor field coordinates, being this the preferable reference frame for machine control and thus easier for control engineers to grasp, especially in closed-looped systems. The approach has been tested on laboratory models of two different induction machines connected to the grid with completely different severities of faults. The results show a very good agreement between measurements and simulations.**

Index Terms-- **Control engineering, fault diagnosis, induction motors.**

I. NOMENCLATURE

Symbols

L, R, τ	inductance, resistance, time constant
v, i, ψ	voltage, current, flux
β, ε, ρ	slip, rotor and rotor flux angle
σ	leakage factor

Indexes

ab	stator reference frame (axes)
dq	rotor field reference frame (axes)
DQ	rotor reference frame (axes)
R, S	rotor, stator
sl, mR	slip, rotor magnetization

Abbreviations

FRF	(rotor) field reference frame
FOC	field oriented control
IM	induction motor (squirrel cage)
RRF	rotor reference frame
BRB	broken rotor bar

II. INTRODUCTION

ACCURATE and timely fault detection in electrical machines is a prerequisite for their reliable operation.

The ability to model various faults is an important tool when devising a new diagnostic method. The immediate benefits of using model-based studies over experimental tests are reduced economical cost, greater flexibility and readily available results. These models are mainly built upon healthy motor representations.

Finite element analysis (FEA) offers a detailed insight into field distribution of the induction machine (IM). FEA takes into account various secondary effects which accompany the primary fault expression. In case of broken rotor bars (BRBs), the FEA can predict an increase of magnetic field [1] and increased iron losses [2] in rotor laminations in immediate vicinity of the fault. In addition, a damping effect of skewing and local saturation onto typical frequency components in stator current spectrum can be ascertained. Computational burden of FEA is its major drawback, especially in case of transients and where asymmetric geometry has to be considered.

In multiple coupled circuits model (MCCM) N bars of rotor squirrel cage are replaced by $N+1$ magnetically coupled meshes. A separate electrical equation is introduced into each rotor mesh and end-ring. As stator windings are not point of interest their phenomena are generalized into one equation each. Then, mesh currents are calculated in order to determine the behavior of the machine [3, 4]. Even though basic MCCM does not take into account saturation and inter-bar currents it still offers a very detailed insight into motor operation [5].

Because of the conceptual simplicity a *dq* model of induction machine is well suited for simple and non-demanding drive analysis. The model itself enables simple implementation of other drive components (power converter, switching and modulation control, and mechanical system) into coherent simulation unit. The immediate advantage of the approach is fast execution time especially when comparing to FEA or MCCM [6-8].

In this paper, a *dq* model of induction machine with BRBs is presented. The reference frame is locked onto rotor magnetizing current. Field oriented control (FOC), most frequently used control scheme, depends on aforementioned reference frame representation as well. In order to understand behavior of closed-loop drive with faulty motor it is propitious to grasp the changes in internal signals. In this way the model becomes suitable for analysis of BRB fault in closed-loop schemes.

III. BASIS FOR TWO-PHASE FAULTY ROTOR MODELING

Two-phase models of the IM with BRB can be modeled on the fact that deviations in faulty rotor are expressed in both rotor resistance and inductance [9]. In that sense faulty rotor differs from the healthy one which preserves uniform

---

All authors are with University of Ljubljana, Faculty of Electrical Engineering, Trzaska 25, SI-1000 Ljubljana, Slovenia (e-mail: vanjaa@fe.uni-lj.si, rastko.fiser@fe.uni-lj.si, mitja.nemec@fe.uni-lj.si, klemen.drobnic@fe.uni-lj.si).

978-1-4799-0024-4/13 $31.00 © 2013 IEEE

parameters in both axes. In general, regardless of the configuration of the fault, both rotor parameters take a quasi-sinusoidal shape depending on rotor angle.

Fig. 1 shows several configurations of rotors with BRBs for Motor 1 (see Appendix) and corresponding dependence of resistance on rotor angle. This relation is obvious for faulty rotor, being its amplitude dependent on fault configuration. The measurement was made using pulsating (non-rotational) magnetic field in which successive rotor angular positions were set manually. At every angular position the input impedance was calculated using *U-I* method [9]. In this way, an angular dependence of rotor resistance is calculated. The shape of the graph is well defined for faulty rotor, being its amplitude dependent on

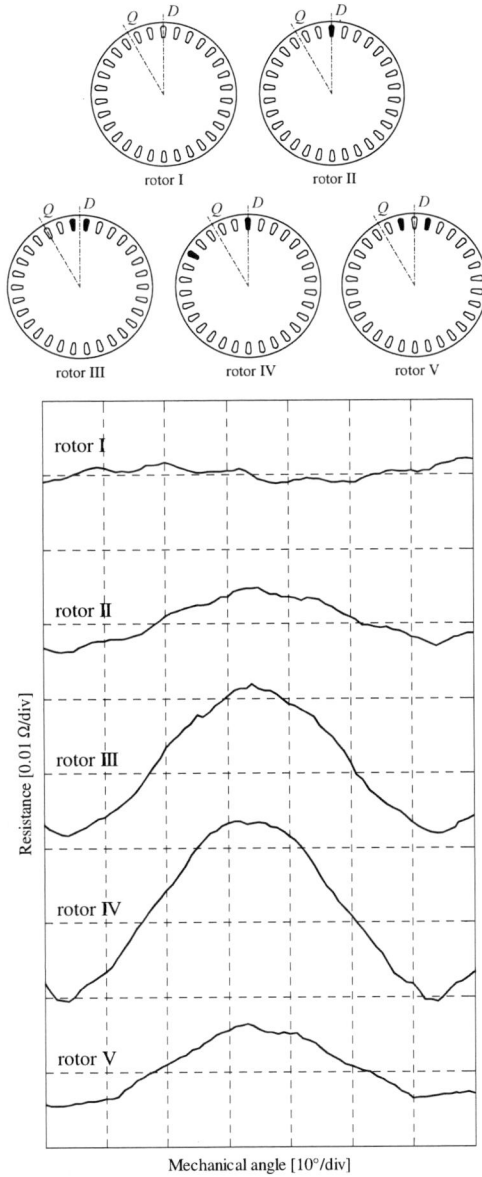

Fig. 1. Resistance vs. rotor angle (below) for different types of faults (above) – BRBs are shaded.

fault configuration. An interesting additional conclusion can be drawn from the measurements: a healthy rotor (denoted as "rotor I") also exhibits a non-constant resistance. This fact can be explained as consequence of manufacturing imperfection and is usually neglected in traditional modeling of a healthy machine. However, this manifestation is still small compared to faulty rotors with one or two BRBs (here in three different configurations).

This saliency effect allows for determining two distinctively different values in rotor's orthogonal *D* and *Q* axes. Consequently, models have been developed in stator coordinates [10]. However, these models suffer from usually complicated terms for transformation into rotor reference frame (RRF), in which the fault is best described.

The easiest way of modeling such a motor for control purposes is to develop the equations in RRF [11], as shown in Fig. 2. One of the benefits of this model is that instead of separate rotor parameters (resistance and inductance) these two merged into one parameter – rotor time constant τ_R. Of course, as said before, unlike the healthy motor, in faulty rotor actually two different rotor time constants exist – one for each rotor axis: τ_{RD} and τ_{RQ}. Additional influence of rotor inductance deviation and its different values in rotor coordinates is expressed through different rotor leakage inductances. Analysis of measurement procedure and sensitivity to measurement error of these parameters is not addressed in this paper.

Although simple, the mere fact that the model is developed in RRF cannot be much of a use for understanding the behavior of faulty machine for control purposes. Namely, the control engineers better understand the machine's behavior in rotor field reference frame (FRF), as field orientation control (FOC) is still the prevailing technique for machine control. Therefore, this paper proposes a model of IM developed in FRF showing how the equations of healthy motor change and what impact these changes have on the model itself. Consequently, in the next phase, this altered model can be used in order to understand the machine's behavior in a closed-loop system and offer an insight to alteration of control variables thus giving the opportunity for machine diagnosis.

Since in closed-loop systems the diagnostic footprint is somewhat compensated by the control loop [12], which intrinsically cause some complicated interactions, this paper, as the first step, will focus on the analysis of IM fed by the grid.

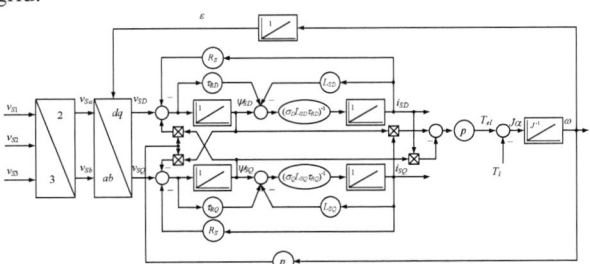

Fig. 2. Control model of IM in rotor reference frame.

978-1-4799-0024-4/13 $31.00 © 2013 IEEE   143

IV. Model of Faulty Motor in Field Reference Frame

Due to different nomenclature presented in literature, in this paper the notation in original works presenting field oriented control [13, 14] has been used. Voltage equations for stator (1) and rotor (2) winding of a 2-pole IM are defined in their own reference frames; DQ for rotor reference frame, as marked by the superscript and stator ones in stator reference frame ab (no superscript).

$$\mathbf{v}_s = R_s \mathbf{i}_s + L_s \frac{d\mathbf{i}_s}{dt} + L_m \frac{d}{dt}\left(\mathbf{i}_R^{DQ} e^{j\varepsilon}\right) \tag{1}$$

$$\mathbf{v}_R^{DQ} = 0 = R_R \mathbf{i}_R^{DQ} + L_R \frac{d\mathbf{i}_R^{DQ}}{dt} + L_m \frac{d}{dt}\left(\mathbf{i}_s e^{-j\varepsilon}\right) \tag{2}$$

A. Current model in FRF

Rotor current in its own reference frame is defined as (3) [13].

$$\mathbf{i}_R^{DQ} = i_{RD} + ji_{RQ} = \frac{\mathbf{i}_{mR} - \mathbf{i}_s}{(1+\sigma_R)} e^{-j\varepsilon} =$$
$$= \frac{i_{mR} e^{j\rho} - \left(i_{Sd} + ji_{Sq}\right)e^{j\rho}}{(1+\sigma_R)} e^{-j\varepsilon} = \tag{3}$$
$$= \frac{\left(i_{mR} - i_{Sd}\right) - ji_{Sq}}{(1+\sigma_R)} e^{j\beta}$$

where ρ, ε, and β are rotor field, rotor and slip angle with respect to stator reference frame ab, as shown in Fig. 3.

Here, the rotor magnetizing current and stator current are already presented in (rotor) field reference frame dq, as this is the system in which FOC is performed. Of course, magnetizing current has only a direct component in these coordinates.

As already mentioned in Chap. III, the influence of BRBs is manifested through different parameters in both rotor co-ordinates. For rotor current, this means that (3) has to be arranged properly by separating real and imaginary part and then consider different parameters (here, rotor leakage factor). Thus the equation for rotor current in RRF for a faulty motor is obtained

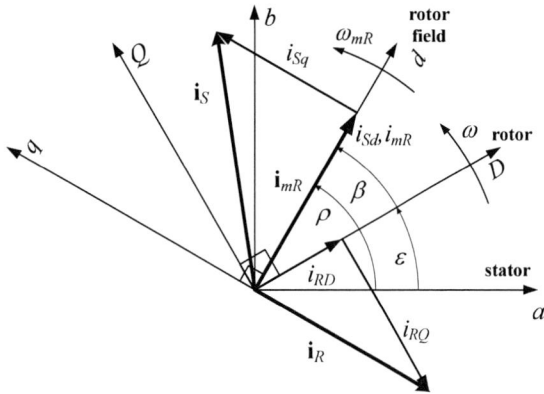

Fig. 3. Definition of space vectors and reference frames.

$$\mathbf{i}_R^{DQ} = \underbrace{\frac{\left(i_{mR} - i_{Sd}\right)\cos\beta + i_{Sq}\sin\beta}{(1+\sigma_{RD})}}_{i_{RD}} + j\underbrace{\frac{\left(i_{mR} - i_{Sd}\right)\sin\beta - i_{Sq}\cos\beta}{(1+\sigma_{RQ})}}_{i_{RQ}} \tag{4}$$

When inserting (4) into rotor voltage equation (2), influence of different rotor resistances also has to be considered (5)

$$0 = \left(\frac{R_{RD}}{1+\sigma_{RD}}\left(\left(i_{mR} - i_{Sd}\right)\cos\beta + i_{Sq}\sin\beta\right) + \right.$$
$$+ j\frac{R_{RQ}}{1+\sigma_{RQ}}\left(\left(i_{mR} - i_{Sd}\right)\sin\beta - i_{Sq}\cos\beta\right)\right) +$$
$$+ \frac{d}{dt}\left(L_{RD}\frac{\left(i_{mR} - i_{Sd}\right)\cos\beta + i_{Sq}\sin\beta}{(1+\sigma_{RD})} + \right. \tag{5}$$
$$\left. + jL_{RQ}\frac{\left(i_{mR} - i_{Sd}\right)\sin\beta - i_{Sq}\cos\beta}{(1+\sigma_{RQ})}\right)$$
$$+ L_m \frac{d}{dt}\left(\left(i_{Sd} + ji_{Sq}\right)e^{j\beta}\right)$$

For future manipulation we will consider different terms for rotor inductances in both axes

$$L_{RD} = L_m\left(1+\sigma_{RD}\right); \quad L_{RQ} = L_m\left(1+\sigma_{RQ}\right) \tag{6}$$

After dividing the equation (5) by L_m, the only parameters remaining are rotor time constants in the first term

$$\tau_{RD} = \frac{L_{RD}}{R_{RD}}; \quad \tau_{RQ} = \frac{L_{RQ}}{R_{RQ}} \tag{7}$$

The resulting equation

$$0 = \frac{1}{\tau_{RD}}\left(\left(i_{mR} - i_{Sd}\right)\cos\beta + i_{Sq}\sin\beta\right) +$$
$$+ j\frac{1}{\tau_{RQ}}\left(\left(i_{mR} - i_{Sd}\right)\sin\beta - i_{Sq}\cos\beta\right) +$$
$$+ \left(\frac{di_{mR}}{dt}\cos\beta - \omega_{sl}i_{mR}\sin\beta\right) + \tag{8}$$
$$+ j\left(\frac{di_{mR}}{dt}\sin\beta + \omega_{sl}i_{mR}\cos\beta\right)$$

is still defined in RRF. Note the definition of slip angular frequency

$$\omega_{sl} = \frac{d\beta}{dt} \tag{9}$$

In order to transform (8) to FRF (in which FOC is performed), the equation has to be multiplied by $e^{-j\beta}$. Consequently, after separating real and imaginary part, we get somewhat familiar terms that define rotor magnetizing current (10) and slip frequency (11)

$$\frac{di_{mR}}{dt} = f(\beta)\left(i_{Sd} - i_{mR}\right) - h(\beta)i_{Sq} \tag{10}$$

$$\omega_{sl} = g(\beta)\frac{i_{Sq}}{i_{mR}} + h(\beta)\left(1 - \frac{i_{Sd}}{i_{mR}}\right) \tag{11}$$

where

$$f(\beta) = \frac{1}{\tau_{RD}}\cos^2\beta + \frac{1}{\tau_{RQ}}\sin^2\beta \qquad (12)$$

$$g(\beta) = \frac{1}{\tau_{RD}}\sin^2\beta + \frac{1}{\tau_{RQ}}\cos^2\beta \qquad (13)$$

$$h(\beta) = \frac{1}{2}\left(\frac{1}{\tau_{RD}} - \frac{1}{\tau_{RQ}}\right)\sin(2\beta) \qquad (14)$$

Functions $f(\beta)$, $g(\beta)$ and $h(\beta)$ depend on parameters (rotor time constants) and slip angle. Note that in the healthy rotor, where $\tau_{RD} = \tau_{RQ} = \tau_R$, both equations change into a known form

$$\frac{di_{mR}}{dt} = \frac{1}{\tau_R}(i_{Sd} - i_{mR}) \qquad (15)$$

$$\omega_{sl} = \frac{i_{Sq}}{\tau_R i_{mR}} \qquad (16)$$

since $f(\beta) = g(\beta) = 1/\tau_R$ and $h(\beta) = 0$.

B. Torque equation in FRF

In order to grasp the influence of the rotor fault, electromagnetic torque will be defined through rotor currents, using generally known equation (17). Note that in that case, torque is negative. Vectors can be defined in arbitrary reference frame but in our case, for the reasons described before, RRF has been chosen

$$T_{el} = -\frac{2}{3}p\left|\boldsymbol{\psi}_R \times \mathbf{i}_R\right| = -\frac{2}{3}p\,\mathrm{Im}\left\{\boldsymbol{\psi}_R^{DQ*}\mathbf{i}_R^{DQ}\right\} \qquad (17)$$

In healthy motor, rotor flux in RRF is defined as
$$\boldsymbol{\psi}_R^{DQ} = L_R\left(i_{RD} + ji_{RQ}\right) + L_0\left(i_{Sd} + ji_{Sq}\right)e^{j\beta} \qquad (18)$$

In a faulty rotor, the complex conjugates of rotor flux, from (4) and (18) becomes

$$\boldsymbol{\psi}_R^{DQ*} = \left(L_{RD}i_{RD} - jL_{RQ}i_{RQ}\right) + L_0\left(\left(i_{Sd} + ji_{Sq}\right)e^{j\beta}\right)^* \qquad (19)$$

After some mathematical manipulations, a final term for electrical torque is obtained (20)

$$T_{el} = \frac{2}{3}pL_0\left(l(\beta)i_{mR}i_{Sq} + k(\beta)(i_{mR} - i_{Sd})i_{mR}\right) \qquad (20)$$

with

$$l(\beta) = \frac{\sin^2\beta}{1+\sigma_{RD}} + \frac{\cos^2\beta}{1+\sigma_{RQ}} \qquad (21)$$

$$k(\beta) = \frac{1}{2}\left(\frac{1}{1+\sigma_{RD}} - \frac{1}{1+\sigma_{RQ}}\right)\sin(2\beta) \qquad (22)$$

Again, for the healthy rotor, the rotor leakage factors in both axes become equal $\sigma_{RD} = \sigma_{RQ} = \sigma_R$, $l(\beta) = (1+\sigma_R)^{-1}$ and $k(\beta) = 0$, thus obtaining the known torque equation

$$T_{el} = \frac{2}{3}p\frac{L_0}{1+\sigma_R}i_{mR}i_{Sq} \qquad (23)$$

Equations (10), (11) and (20) form the basis for the current model of IM with BRB in rotor FRF, which is usually used for explaining the behavior of a controlled machine (Fig. 4).

C. Voltage model in FRF

Next, the voltage model is introduced, so as to encompass the limitations of voltage supply. In order to write the original stator voltage equation (1) in rotor field dq co-ordinates, the equation has to be multiplied by $e^{j\rho}$. Consequently, it takes the form of (24).

$$\left(v_{Sd} + jv_{Sq}\right) = R_S\left(i_{Sd} + ji_{Sq}\right) + j\omega_{mR}L_S\left(i_{Sd} + ji_{Sq}\right) +$$
$$+ L_S\frac{d}{dt}\left(i_{Sd} + ji_{Sq}\right) + L_m\left(\frac{d\mathbf{i}_R^{DQ}}{dt} + j\omega\mathbf{i}_R^{DQ}\right)e^{-j\beta} \qquad (24)$$

After taking into account (4) and a definition of synchronous or rotor flux angular frequency (25)

$$\omega_{mR} = \omega + \omega_{sl} \qquad (25)$$

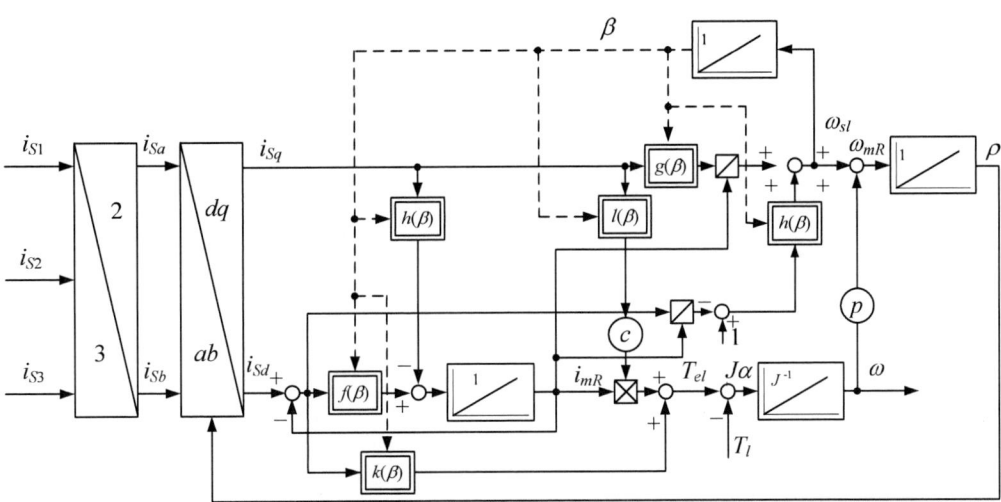

Fig. 4. Block scheme of current model of IM with faulty rotor in field reference frame.

978-1-4799-0024-4/13 $31.00 © 2013 IEEE

the final stator voltage equation, here split into real (26) and imaginary parts (27), becomes

$$v_{Sd} = R_S i_{Sd} - \omega_{mR} L_S i_{Sq} + L_S \frac{di_{Sd}}{dt} + p(\beta)\left(\frac{di_{mR}}{dt} - \frac{di_{Sd}}{dt}\right) +$$

$$+ r(\beta)\frac{di_{Sq}}{dt} + r(\beta)\left(\omega_{mR} - 2\omega_{sl}\right)\left(i_{mR} - i_{Sd}\right) + \qquad (26)$$

$$+ \left(s(\beta)m + t(\beta)n\right)i_{Sq}$$

$$v_{Sq} = R_S i_{Sq} + \omega_{mR} L_S i_{Sd} + L_S \frac{di_{Sq}}{dt} - r(\beta)\left(\frac{di_{mR}}{dt} - \frac{di_{Sd}}{dt}\right) -$$

$$- q(\beta)\frac{di_{Sq}}{dt} + r(\beta)\left(\omega_{mR} - 2\omega_{sl}\right)i_{Sq} + \qquad (27)$$

$$+ \left(t(\beta)m + s(\beta)n\right)\left(i_{mR} - i_{Sd}\right)$$

Terms in (26) and (27) being depended on the slip angle are

$$p(\beta) = m\cos^2\beta + n\sin^2\beta$$

$$q(\beta) = n\cos^2\beta + m\sin^2\beta$$

$$r(\beta) = \frac{1}{2}\sin 2\beta\,(m-n) \qquad (28)$$

$$s(\beta) = \left(\omega\sin^2\beta + \omega_{sl}\cos^2\beta\right)$$

$$t(\beta) = \left(\omega_{sl}\sin^2\beta + \omega\cos^2\beta\right)$$

where

$$m = \frac{L_m}{\left(1+\sigma_{RD}\right)}; \quad n = \frac{L_m}{\left(1+\sigma_{RQ}\right)} \qquad (29)$$

Interesting terms emerge in both (26) and (27) equal to twice the slip frequency $2\omega_{sl}$. Namely, this characteristic frequency component is known to be found in stator current spectrum of a faulty rotor (frequently employed). Both voltage components in FRF influence the stator current components that eventually form the output stator vector.

Equations (26) and (27) denoting components of stator voltage components in FRF are indeed complicated, but they comprise all the influences of rotor asymmetry due to BRB on stator voltage. Fig. 5 shows the block scheme of voltage model in FRF.

Again, in healthy rotor, with equal rotor leakage factors in both axes, the equations gets a much known form (30) [13].

$$v_{Sd} = R_S i_{Sd} + \sigma L_S \frac{di_{Sd}}{dt} - \sigma L_S \omega_{mR} i_{Sq} + (1-\sigma) L_S \frac{di_{mR}}{dt}$$

$$v_{Sq} = R_S i_{Sq} + \sigma L_S \frac{di_{Sq}}{dt} + \sigma L_S \omega_{mR} i_{Sd} + (1-\sigma) L_S \omega_{mR} i_{mR} \qquad (30)$$

after considering the definition of total leakage factor σ

$$\sigma = 1 - \frac{1}{\left(1+\sigma_s\right)\left(1+\sigma_R\right)} = 1 - \frac{L_m^2}{L_S L_R} \qquad (31)$$

since

$$p(\beta) = q(\beta) = \frac{L_m}{\left(1+\sigma_R\right)}$$

$$r(\beta) = 0$$

$$s(\beta)m + t(\beta)n = s(\beta)n + t(\beta)m = \frac{L_m}{\left(1+\sigma_R\right)}\omega_{mR}$$

V. Model Verification

In order to validate the model, it has been compared to the measurement made on laboratory set-ups with two different motors and different severity of rotor faults (number of BRBs). The first one (Motor 1) had only two BRBs out of 30 (rotor III from Fig. 1 – plausible scenario) while the second (Motor 2) has 7 consecutive BRBs out of 44. Although this severe type of fault is rather exaggerated, it has been used solely for purpose of demonstration the validity of the model. Data for both motor are given in the Appendix.

The proposed (voltage) model requires rotor parameters which describe the asymmetry of the cage. Rotor time constants $\tau_{RD,Q}$ and rotor leakage factors $\sigma_{RD,Q}$ determination is based on equivalent circuit of IM [10].

The results obtained from the simulation (voltage model of IM) have been compared with measurement results. In this paper, only behavior of both motors connected directly to the grid is presented. The torque has been measured with dynamometer whose signals have been filtered to eliminate the inherit noise.

First, the dynamic behaviors of torque and speed have been tested in "quasi" steady state, after speed transient.

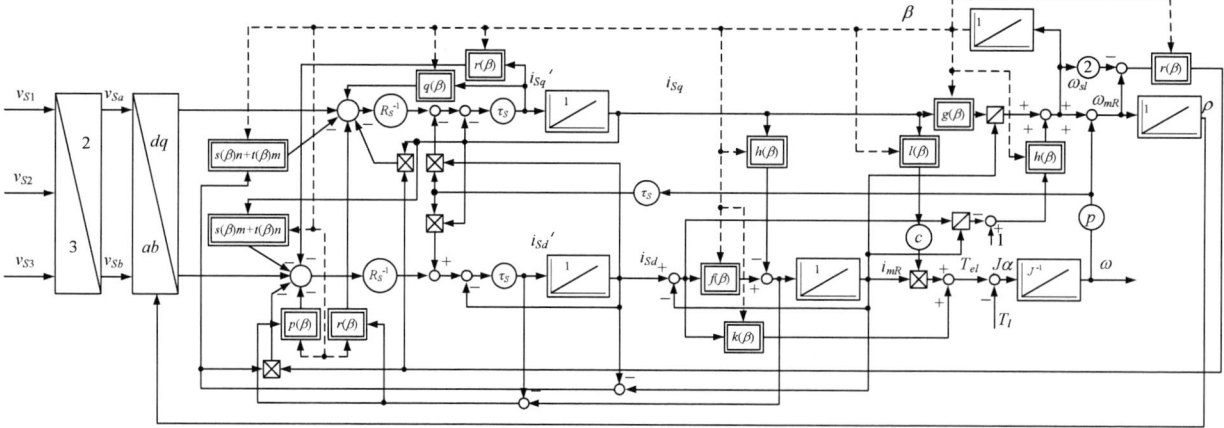

Fig. 5. Block scheme of voltage model of IM with faulty rotor in field reference frame.

However, as known, with faulty rotor, actual steady state (constant speed) never occurs, as even with constant (frequency and amplitude) supply voltage and constant load torque and speed oscillate.

Fig. 6 and Fig. 7 show the torque and speed of Motor 1 (with rotor III – two BRBs) for different load torques (33%, 66% and 100% of rated torque). In order to compare the model and experimental results more accurately, both the simulation and measurement speed and torque are depicted in detail. Please note, that in these two figures curves are separated on purpose (and not partially superimposed, as in reality) in order to better compare the frequency and amplitudes of the ripple.

Both comparisons show that the model has reconstructed the behavior of both quantities very well. As generally known [15], when fed with pure sinusoidal voltages and under constant load torque, machines with faulty rotor exhibit oscillations in both the speed and torque. Frequency of these oscillations equals twice the slip frequency. This effect is clearly visible in all of the presented figures.

Especially interesting is the behavior of speed and torque of a heavily damaged motor (Motor 2; seven BRBs) in Fig. 8 and Fig. 9, respectively. Again, measured and simulation results are intentionally separated for better view. Shapes and frequency of both variables match very well as before, thus proving the general validity of a model.

As in the previous case the rotor speed ripple has almost ideal sinusoidal form. However, the torque (Fig. 9) tends to have a distorted sinusoidal shape, which is reconstructed by the model in all details.

Of course, it has to be emphasized that the model of a faulty machine in FRF (as well as in its generalized form for the healthy rotor) is somewhat simplified and that rotor resistance and inductance dependence on rotor angle is just an approximation of sine function. Nevertheless the agreement between simulation and measurement is very good.

Now, the behavior of the simulation model and its internal values in FRF can be observed. For this purpose, machine has been first connected to the grid at no load. In order to analyze effect shown in Fig. 9, first a heavily damaged Motor 2 has been simulated.

After reaching steady state all model current components (stator current components in FRF and magnetizing current) reach constant values, additionally being $i_{Sq} = 0$ (Fig. 10). Thus, as known from the literature [16], no fault indicators are present at no-load, and diagnostics becomes impossible. The same condition can be observed on angular velocities (Fig. 11), where at no-load the rotor is rotating smoothly.

After impressing constant load at $t = 1$ s all currents start oscillating, as well as angular speeds. However, rotor speed is oscillating almost sinusoidally as in Fig. 8, but field velocity takes a form of distorted sine wave, analog to torque ripple in Fig. 9. This effect is obviously a consequence of slip frequency, which is related to torque. From the point of

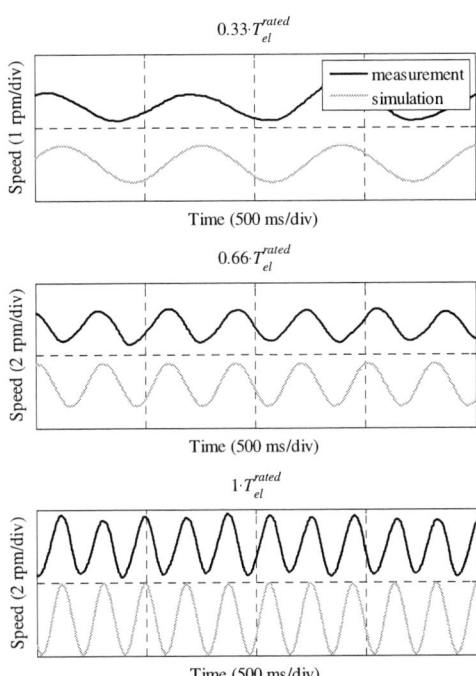

Fig. 6. Comparison of speed ripples for Motor 1 at different load torques.

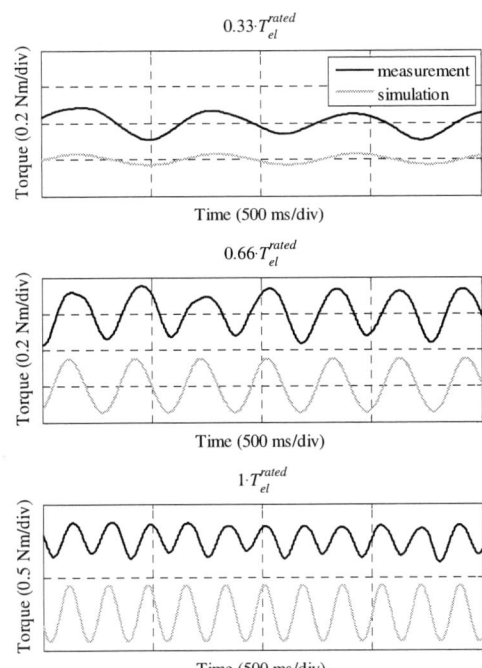

Fig. 7. Comparison of torque ripples for Motor 1 at different load torques.

view of stator reference frame, the rotor flux (or magnetizing current) vector in steady-state will rotate with constant speed and additional oscillating component.

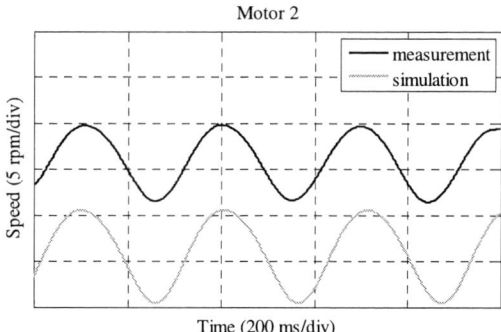

Fig. 8. Comparison of measured and simulated speed ripple at rated load torque (Motor 2).

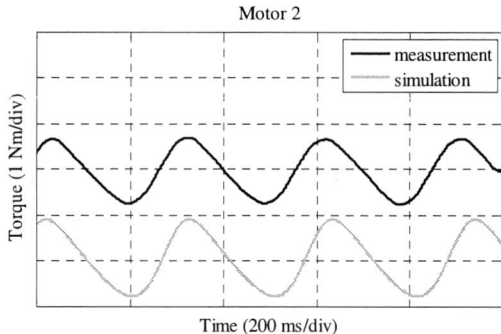

Fig. 9. Comparison of measured and simulated torque ripple at rated load torque (Motor 2).

On the other side described effects are not to be seen on current components (Fig. 12) and velocities (Fig. 13) of Motor 1, where fault extend does not affect sinusoidal shape.

VI. CONCLUSION

In this paper a dynamic model of induction motor with BRBs has been presented. Differing from the previous work, this model has been developed in rotor field co-ordinates. Thus, the motor dynamics becomes much clearer to control engineers working with electrical drives. Model performance has been compared with experimental results on two motors

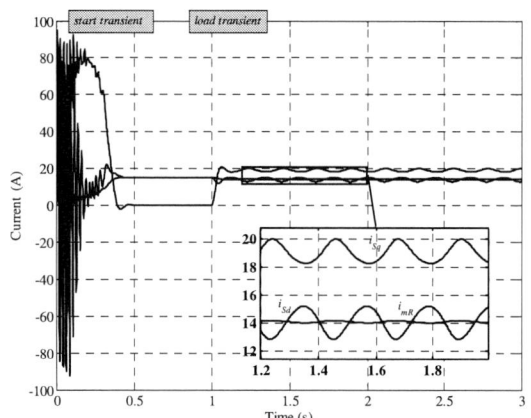

Fig. 10. Stator current components and magnetizing current in FRF (Motor 2).

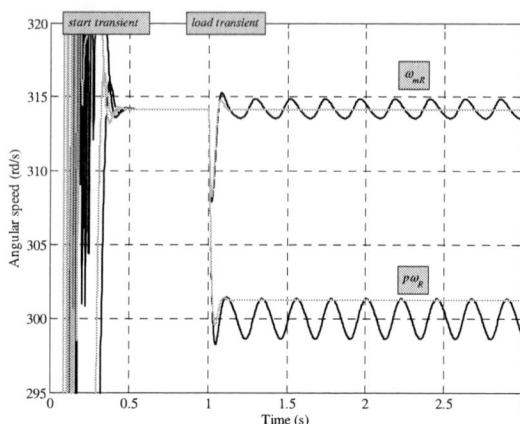

Fig. 11. Rotor and synchronous speed for faulty (black) and healthy (grey) case (Motor 2).

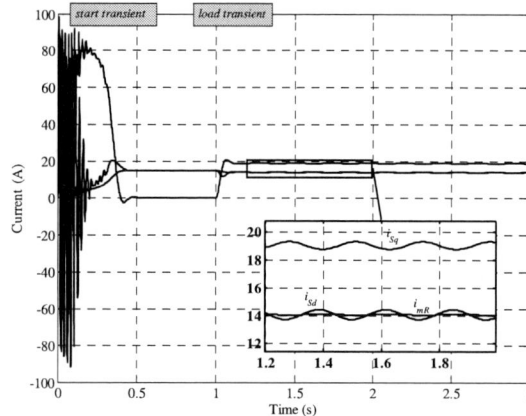

Fig. 12. Stator current components and magnetizing current in FRF (Motor 1).

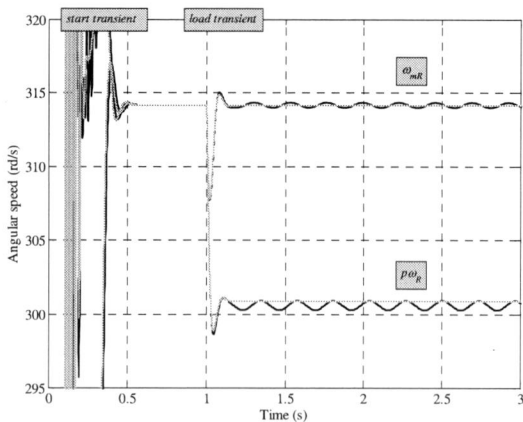

Fig. 13. Rotor and synchronous speed for faulty (black) and healthy (grey) case (Motor 1).

with different data and severity of the fault. In both cases the results are in a very good agreement thus showing a general validity of the approach.

Now it also become clearer how the machine model in FRF changes and how this change influences the behavior of internal motor quantities (stator current components in FRF, magnetizing current, slip frequency etc.).

Results in this paper have been obtained on machines connected to the grid. Following work will focus on analysis of the machine connected to converter and controlled in a closed-loop. In this way the mismatch between the machine model used by controller ("healthy motor") and distorted machine model presented here ("faulty motor") is expected to show the transition of fault signature otherwise invisible during the control.

VII. APPENDIX

MACHINE DATA

	Motor 1	Motor 2
Rated power	4.7 kW	3 kW
Rated torque	30 Nm	20 Nm
Rated speed	1500 rpm	1458 rpm
Rated current	13.0 A	14.6 A
Rated voltage	215 V	177 V
Number of pole pairs	3	2
Number of stator slots	36	36
Number of rotor slots	30	44
Moment of inertia	0.038 kgm^2	0.075 kgm^2
Stator resistance	0.687 Ω	0.214 Ω
Rotor resistance	0.550 Ω	0.231 Ω
Stator inductance	55.6 mH	35.4 mH
Mutual inductance	50.9 mH	31.3 mH
No. of broken rotor bars	2	7

VIII. REFERENCES

[1] N. M. Elkasabgy, A. R. Eastham, and G. E. Dawson, "Detection of broken bars in the cage rotor on an induction machine," *IEEE Trans. Ind. Appl.*, vol. 28, no. 1, p. 165–171, Jan/Feb 1992.

[2] J. F. Bangura and N. A. Demerdash, "Effects of broken bars/end-ring connectors and airgap eccentricities on ohmic and core losses of induction motors in asds using a coupled finite element-state space method," *IEEE Trans. Ener. Conv.*, vol. 15, no. 1, pp. 40–47, Mar 2000.

[3] X. Luo, Y. Liao, H. A. Toliyat, A. El-Antably, and T. A. Lipo, "Multiple coupled circuit modeling of induction machines," *IEEE Trans. Ind. Appl.*, vol. 31, no. 2, pp. 311–318, Mar/Apr 1995.

[4] S. D. Sudhoff, B. T. Kuhn, K. A. Corzine, and B. T. Branecky, "Magnetic equivalent circuit modeling of induction motors," *IEEE Trans. Ener. Conv.*, vol. 22, no. 2, p. 259–270, Jun 2007.

[5] H. A. Toliyat and T. A. Lipo, "Transient analysis of cage induction machines under stator, rotor bar and end ring faults," *IEEE Trans. Ener. Conv.*, vol. 10, no. 2, pp. 241–247, Jun 1995.

[6] H. Rodriguez-Cortes, C. N. Hadjicostis, and A. M. Stankovic, "Model-based broken rotor bar detection on an ifoc driven squirrel cage induction motor," in *Proceedings of the American Control Conference*, Jul 2004.

[7] M. Stocks, F. Rodyukov, and A. Medvedev, "Idealized two-axis model of induction machines under rotor fault," in *IEEE Conference on Industrial Electronics and Applications*, May 2006.

[8] T. J. Sobczyk and W. Maciolek, "On reduced models of induction motors with faulty cage," in *IEEE International Symposium on Diagnostics for Electric Machines, Power Electronics and Drives (SDEMPED)*, Aug. 31.-Sept. 3. 2009.

[9] D. Makuc, K. Drobnič, V. Ambrožič, D. Miljavec, R. Fišer, and M. Nemec, "Parameters estimation of induction motor with faulty rotor," *Przeglad Elektrotehniczny*, vol. 88, no. 1a, 2012.

[10] K. Drobnič, M. Nemec, D. Makuc, R. Fišer and V. Ambrožič, "Pseudo-salient model of induction machine with broken rotor bars," in *IEEE International Symposium on Diagnostics for Electric Machines, Power Electronics and Drives (SDEMPED)*, 2011.

[11] V. Ambrožič, K. Drobnič, R. Fišer, and M. Nemec, "Dynamic model of induction machine with faulty cage in rotor reference frame," in *2011 IEEE Ninth International Conference on Power Electronics and Drive Systems (PEDS)*, Singapore, 2011.

[12] K. Drobnič, M. Nemec, R. Fišer, and V. Ambrožič, "Simplified detection of broken rotor bars in induction motors controlled in field reference frame," *Control Engineering Practice*, vol. 20, no. 8, pp. 761-769, Aug. 2012.

[13] W. Leonhard, Control of electrical drives. Springer Verlag, 2001.

[14] F. Blaschke, "Das Prinzip der Feldorientierung, die Grundlage fuer die Transvektor-Regelung von Drehfeldmaschinen," *Siemens Zeitschrift*, vol. 45, no. 10, pp. 757-760, 1971.

[15] M. Nemec, K. Drobnič, D. Nedeljković, R. Fišer, and V. Ambrožič, "Detection of broken bars in induction motor through the analysis of supply voltage modulation," *IEEE Trans. Ind. Electr.*, vol. 57, no. 8, pp. 2879-2888, Aug. 2010.

[16] B. Mirafzal and N. A. O. Demerdash, "Effects of load magnitude on diagnosing broken bar faults in induction motors using the pendulous oscillation of the rotor magnetic field orientation," *IEEE Trans. Ind. Appl.*, vol. 41, no. 3, pp. 771-783, May-June 2005.

IX. BIOGRAPHIES

Vanja Ambrožič (M'92) received the B.S., M.S. and Ph.D. degrees in 1986, 1990, and 1993, respectively, from Faculty of Electrical Engineering, University of Ljubljana, Slovenia. In 1986 he joined the Laboratory of Control Engineering at the Faculty of Electrical Engineering, first as a Junior Researcher, then as Assistant and Assistant Professor at the Department of Mechatronics. He is currently Full Professor and head of the same department. His main research interests include control of electrical drives and power electronics.

Rastko Fišer (M'96) received the B.Sc., M.Sc., and Ph.D. degrees in electrical engineering from the Faculty of Electrical Engineering, University of Ljubljana, Slovenia, in 1984, 1989 and 1998, respectively. In 1986, he joined the Laboratory of Electrical Drives at the same institution, first as a Junior Researcher, then Assistant, Assistant Professor, and since 2009 he has been working as an Associate Professor at the Department of Mechatronics. He lectures on electrical machines, electrical drives and power electronics in undergraduate and postgraduate studies. Currently he is the Head of the Laboratory of Electrical Drives. His main research interests include condition monitoring and diagnostics of electrical drives, modeling, simulation, testing and control of electrical machines, power electronic converters and electrical traction systems.

Mitja Nemec (M'04) received the B.S. and Ph.D. from the Faculty of Electrical Engineering, University of Ljubljana, Slovenia, in 2003 and 2008, respectively. He is currently Senior Researcher at the same faculty in the area of power electronics and motion control. His main research interests include control of electrical drives, active power filters and application of power electronics in automotive industry.

Klemen Drobnič (M'08) received the B.S. and Ph.D. in electrical engineering from the Faculty of Electrical Engineering, University of Ljubljana in 2007 and 2012, respectively. In 2007, he joined the Department of Mechatronics as Junior Researcher and subsequently become an Assistant in 2013. His research interests include diagnostics of electrical machines and control of electrical drives.

Support Vector Machine for Diagnosis of Induction Motors: a Comparative Analysis in Terms of the Quantity and the Signal Processing Tool Used to Build the Feature Space

A. Sapena-Baño, M. Pineda-Sanchez, *Member, IEEE*, R. Puche-Panadero, *Member, IEEE*, J. Roger-Folch, *Member, IEEE*, J. Perez-Cruz, *Member, IEEE*, and M. Riera-Guasp, *Member, IEEE*,

Abstract—**The use of advanced diagnosis techniques for induction motor (IM) faults relies on the use of automated classifiers, such as those based on support vector machines (SVMs), which are able to assess the condition of the machine using a set of relevant features extracted either from the time domain or from the frequency domain machines signals. But the performance of such systems depends on two main factors: the quantity that is used to obtain the machine's condition, and the signal processing tool used for extract the features set. In this paper, a combination of the most used quantities and signal processing tools is used for diagnosis a set of machines with broken bars, fed from the mains and from variable speed drives, using the same SVM. In this way, the most efficient combination can be chosen, from the point of view of the performance of the automatic classifier system.**

I. Introduction

Motor current signature analysis (MCSA) is the reference method for diagnosis induction motor faults. The well-known MCSA approaches are based on the analysis of the Fourier spectrum of the steady state currents [1]–[3] ; theoretical studies show that different types of faults in electrical machines produce or amplify components in the stator current with characteristic frequencies associated to the fault. For example, in the case of induction motors, it has been demonstrated [3] that the rotor asymmetries produce, among others, harmonics in the stator current with frequencies:

$$f_b(s) = (k/p(1-s) \pm s)f_1 \quad \text{with } k/p = 1, 3, 5\dots \quad (1)$$

where f_1 is the power supply frequency, s is the slip and p is the number of pole pairs of the machine. In the case of mixed eccentricity, there are two different series of harmonics which can appear in the line current spectrum when this fault arises [4]: a high frequency series of harmonics, which appear as sidebands around the principal slot harmonics, and a low frequency series of harmonics, which appear as sidebands around the fundamental component, at frequencies given by

---

The authors are with the Department of Electrical Engineering, Universitat Politècnica de València, Camino de Vera s/n, 46022, Valencia, Spain. E-mails: sapena.angel@gmail.com,mpineda@die.upv.es, rupucpa@die.upv.es, jroger@die.upv.es, juperez@die.upv.es, mriera@die.upv.es, sapena.angel@gmail.com.

$f_{ME,k} = f_1 \pm (k \, (1-s)f_1/p), \quad k = 1, 2, 3 \dots$. Focusing on the most dominant component of this series, obtained with k = 1 [4], a mixed eccentricity (ME) could be characterized by the presence in the stator current spectrum of components with frequencies given by:

$$f_{ME}(s) = f_1 \pm (1-s)f_1/p = f_1 \pm f_r \quad (2)$$

where f_r is the rotational frequency of the motor. However, as stated in [4], harmonics with frequencies given by (2) can be found also in healthy machines; thus, it is advisable to complement the diagnostic based in (2) with analysis based on the high frequency components. In the case of cyclic faults in the outer or inner races of bearings, a new family of harmonics appears in the spectrum of the stator currents; the frequency of these harmonics depends on the fault localization (inner or outer race), on the supply frequency and on the frequencies produced by the defect in the spectrum of vibrations, which in turn depends on characteristics of the bearing (number of balls, ball diameter, bearing pitch diameter and the contact angle) [5]; in this reference approximate expressions for the vibration frequencies (suitable for most of the bearings with between six and twelve balls) are used, which enable estimating the frequencies that the cyclic bearing faults introduce in the spectrum of the stator currents as

$$f_{B,o} = |f_1 \pm (m \, 0.4 \, N_b(1-s) \, f_1/p)| = |f_1 \pm m \, 0.4 \, N_b \, f_r| \quad (3)$$

for the outer race, and

$$f_{B,i} = |f_1 \pm (m \, 0.6 \, N_b(1-s) \, f_1/p)| = |f_1 \pm m \, 0.6 \, N_b \, f_r| \quad (4)$$

for the inner race , where N_b is the number of bearing balls, and $m = 1, 2, 3 \dots$.

In MCSA the fast Fourier transform (FFT) is the mathematical tool commonly used for extracting the current spectrum. Once the stator current is sampled and the rotor speed is accurately measured, the diagnosis is carried out by comparing the amplitude of the harmonics whose frequencies are given by (1), (2), (3), (4), with threshold values that are associated to each frequency and type of fault. Nevertheless, the industrial application of MCSA presents some practical challenges, that may result in wrong diagnostics. Specifically, the peaks that

978-1-4799-0024-4/13 $31.00 © 2013 IEEE

reveal the presence of a given fault in the current's spectrum can be absent due to one of the following reasons:

- First, the amplitude of the mains harmonic is several order of magnitude greater than the amplitude of the fault harmonics. If these harmonics are close to the mains component, they can be buried under the spectral leakage generated by this component. This is the case, for example, of the fault harmonics generated by a broken bar fault when the machine is working under very low slip conditions. Low slip regime is the normal one for large motor, even at rated load, or can appear also when testing a given motor unloaded, irrespective of its rated power.

- Second, the signal analysis tool used to generate the spectrum can also hide the fault harmonics used for diagnostic. A high frequency resolution is needed for resolving harmonics which are closely spaced in the spectrum. It is also important that the peaks in the spectrum are properly isolated, which requires also a high definition of the spectral lines. Finally, due to the low amplitude of the fault harmonics, the noise level of the spectrum must be kept low enough to increase the reliability of the diagnosis process.

Several solutions have been proposed in the technical literature to overcome these difficulties. On the one hand, the leakage problem has been addressed using a quantity different than the raw current (windowed current, Hilbert transform of the current, extended Park vector, etc.). On the other hand, advanced signal processing tools have been used to improve the spectrum legibility (FFT, chirp-Z transform, Zoom-FFT, Welch spectrum, Cepstrum, etc.). But, which of these possible choices of both the quantity to be analysed and the signal processing tool used to analyse it is the best one? The aim of this work is to present these possibilities and to compare them from the point of view of their impact in the performance of an automatic classifier, based on a SVM. To achieve this goal, this paper has been structured in the following way. Section II presents the test bed and the tested motors used for developing the present work. In section III, some of the technical solutions that have been proposed in the literature to solve the leakage problem and to improve the ability of the spectrum to resolve fault harmonics are presented, both theoretically and using experimental currents obtained in the test bench; in section IV the basis of the SVM classifier used for detection of induction motor faults are presented.. The different solutions presented in section III, considering all the possible combinations of the quantity used to detect the fault and of the signal processing tool used for generating its spectrum, are used to feed the SVM. The results of this analysis are presented in section V. Finally, the most relevant conclusions of this analysis are presented in section VI

II. EXPERIMENTAL TESTS

The experimental part of this work has been carried out using the test bench depicted in Fig. 1. It allows both the testing of the IMs (Table of characteristics I) connected directly to the mains (50 Hz) or through a VSD, shown in Fig. 2.left

(ABB's ACS800-01-0005-3+E200+L503), to obtain tests at different supply frequencies, either in motor or in generator mode. The mechanical load is obtained through the use of a permanent magnet synchronous machine (PMSM), whose characteristics are given in Table II, which is controlled using a servodriver (ABB's ACSM1-04AS-024A-4+L516) and can be seen in Fig. 2.right. The use of a PMSM as motor load has several advantages. The most important is it is possible to manage the PMSM control either in speed or in torque control. If the control is selected to the torque this machine is able to offer a constant torque, independent of the speed. Besides, it can work in speed control mode, allowing the test of the IM at predefined speeds, or even driving the IM in the generator regime. Besides, to improve the efficiency of the test bed, both converters are linked through the DC bus, so reducing the power consumption of the system to the sum of the loses of both machines.

Fig. 1: Test Bench. Cage motor coupled to Permanent Magnet Synchronous Machine

TABLE I: Cage induction motor characteristics

Rated power	1.5 kW
Rated frequency	50 Hz
Rated primary current	3.25 A
Rated voltage	400 V
$\cos\phi$	0.85
Number of pole pairs	1
Rated speed	2860 r.p.m.
Rated slip	0.0467

In all the tested cases, two different type of motors has been used, all of the same reference: a healthy motor, and a motor with a broken bar, made artificially by drilling a hole in four bars. From the point of view of the work proposed in this paper, it is the most difficult fault among those cited in section I, because the fault harmonics can be very close to the

TABLE II: Permanent Magnet Synchronous Machine Characteristics

Rated current	14.4 A
Rated torque	15.5 Nm
Rated power	4.9 kW
Rated speed	3000 r.p.m.

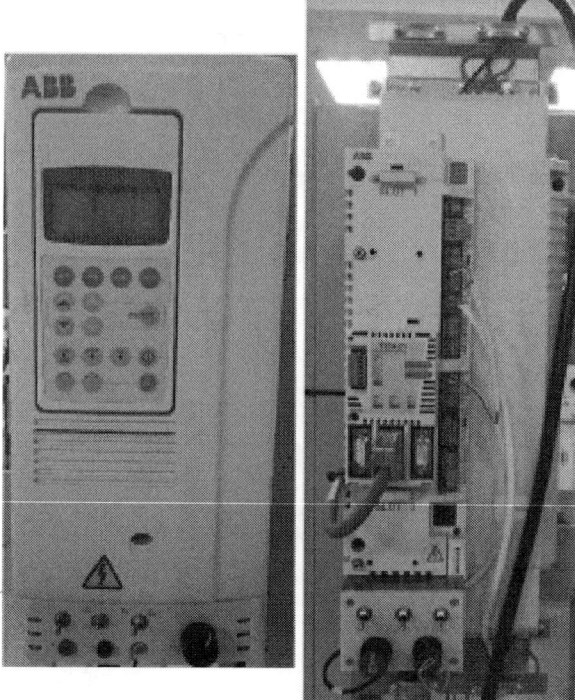

Fig. 2: ABB VSD (left). ABB Servodriver (right).

mains component, specially at very low slip, so that spectral leakage is specially problematic in these working conditions. The current obtained from this faulty machine during a test at rated load is shown in Fig. 3.

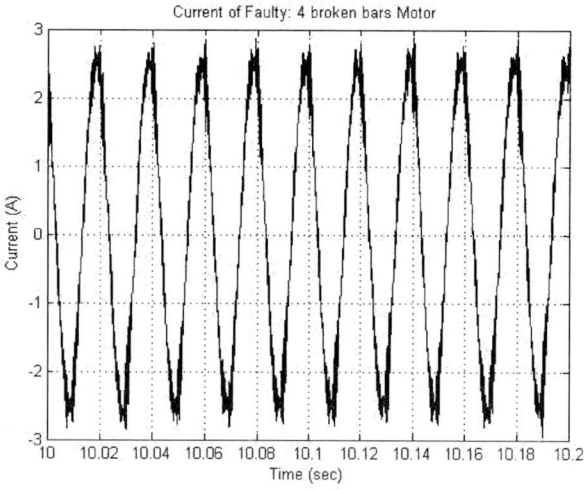

Fig. 3: Faulty current

The number of tests that have been made using de healthy and faulty motor at different load levels and using different

TABLE III: Number of tests used in this work

Number of tests	Direct connection	VSD in scalar mode	VSD in DTC mode
Healthy	15	24	24
Broken bar	15	45	45

types of supply, direct connection to the supply line, and VSD working in scalar mode or in direct torque control mode (DTC), is summarized in table III

III. TECHNICAL SOLUTIONS PROPOSED FOR SOLVING THE LEAKAGE PROBLEM AND TO IMPROVE THE SPECTRUM RESOLUTION

In this section, it is presented a brief review of some of the techniques used in the literature to solve the practical problems found in industrial application of MCSA highlighted in section I. This solutions can be grouped in two main areas: the use of a quantity that is more suitable than the raw current of the motor, and the use of advanced signal processing techniques to improve the legibility of the spectrum of the selected quantity.

A. Selection of the quantity to be analysed to minimize the leakage problem

To overcome the problem of spectral leakage, one of the first and more straightforward solutions is the use of a windowed current signal, obtained by multiplying the machine' current by a suitable window function (Hanning, Hamming, etc.). Fig. 4 shows the FFT of the current, and Fig. 5 shows the improvement of the spectrum obtained after applying a Hanning window to the original current' signal.

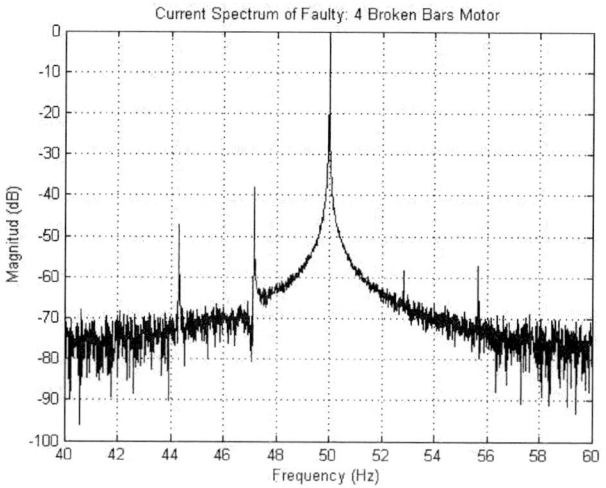

Fig. 4: Spectrum of the machine's current.

An improvement can be achieved by appending to the current's signal a mirror in the time domain, so that the end points of the signal coincide, which reduces the Gibb' phenomenon. Another approach is to obtain a signal derived from the original current, but free of the mains component,

978-1-4799-0024-4/13 $31.00 © 2013 IEEE

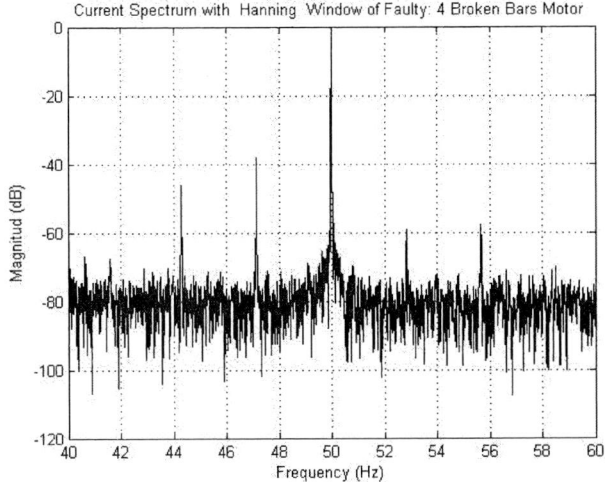

Fig. 5: Spectrum of the windowed current using a Hanning window.

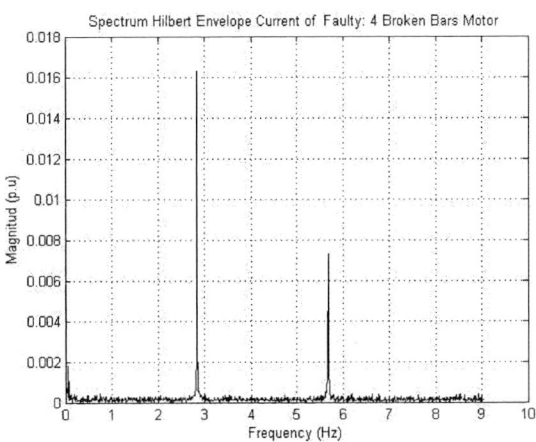

Fig. 6: Spectrum of the modulus of the AS obtained through the Hilbert transform.

which eliminates the leakage that this component introduces. The mains component can be eliminated using an appropriate notch filter [6]. The design of such a filter, with very sharp transitions between the stop and the pass bands, is a difficult task. Besides, in the case of use of a VSD, it must be recalculated as a function of the VSD output frequency. More advanced techniques to avoid the effect of the mains current rely on the analysis of the current's envelope. The Hilbert transform provides a mathematical basis for obtaining a complex signal, the so called analytic signal (AS), whose amplitude oscillates at the same frequency as the fault components, translating the mains component into a dc signal that can be easily eliminated just by substracting the mean value to the modulus of the AS [7]. Fig. 6 shows the spectrum of the modulus of the AS of the current shown in Fig. 3, after substracting its mean value. The effect of the mains component is absent of this spectrum, and the fault harmonics can be clearly distinguished even using a linear scale, instead a logarithmic one, as in Fig. 4. Besides, compared with this figure, the fault harmonics appear at the characteristic modulation frequency that the fault impresses on the machine's current, instead as sidebands around the supply frequency. And this property is independent of this frequency, which makes this technique specially useful in the case of varying supply frequencies, when the motor is fed from a VSD.

Similar to the method based on the Hilbert transform, using in this case the three currents of the machine, is to analyze the modulus of the extended Park's vector (EPV) [8]. This magnitude, as in the case of the AS, converts the mains component into a dc one, and keeps the information about the fault harmonics. Besides, and contrary to the use of the AS, the EVP can discriminate between the direction of rotation of these fault components, which is obtained through the sign of the instantaneous frequency.

The fault harmonics appear in the spectra obtained using the Hilbert transform or the EVP method right at the characteristics

frequencies of the fault (see Table IV, as can be seen for the broken bar fault in Fig. 6.

TABLE IV: VALUES OF THE FAULT FREQUENCY f_0 FOR DIFFERENT TYPES OF FAULTS

Fault	f_0
bar breakages	$2sf$
mixed eccentricity	$\dfrac{(1-s)}{p}f$
bearing inner race	$0.6N_b\dfrac{(1-s)}{p}f$
bearing outer race	$0.4N_b\dfrac{(1-s)}{p}f$

B. Signal processing tools used to improve the spectrum

The use of advanced signal processing tools can improve greatly the legibility of the current's spectrum, and so the ability to detect and isolate the fault-related harmonics given by (1), (2), (3) and (4). Spectral resolution is directly related to the total sampling time. Anyway, given a sampling time, the number of points used to represent the spectrogram can be increased just by appending zeros to the current's signal. This added number of points facilitates the identification and isolation of faul harmonics, but it can lead to extremely long sequences of data values to be processed by the FFT algorithms. An alternative is to use the Chirp-Z transform [9]. This transform can focus the output of the spectrum in a narrow range of values, but using all the available sampled data to compute the frequency bins comprised in the selected range, as shown in Fig. 7.

More advanced data processing techniques have been proposed in the literature to increase the spectrum resolution, such as the zoom-FFT [10] method (Fig. 8, the Esprit method [11], or the MUSIC method, which can achieve and extremely high frequency resolution at the cost of an added operational complexity.

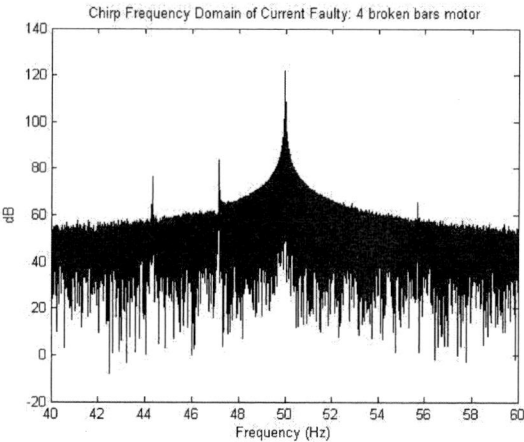

Fig. 7: Spectrum of the machine's current using the Chirp-Z transform.

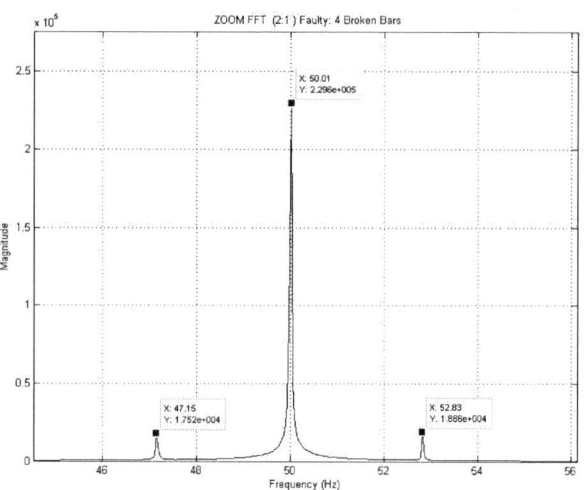

Fig. 8: Zoom FFT spectrum

A special feature of the harmonics generated by the fault is that, as stated in equations (1), (2), (3) and (4), these harmonics are not isolated spectral lines, but they appear as families of related spectral lines, that are repeated at integer multiples of the fault characteristic frequency. So, instead of detecting a single spectral line for assessing the machine' condition, it would be preferable to relay on the detection of a whole family of fault related harmonics. Unfortunately, taking into account several spectral lines complicates the task of automatic classifiers, so this technique is seldom used. Nevertheless, there is an extremely useful process that can detect the presence of repetitive harmonics in the spectrum, the cepstrum [12]. This process, which can be defined as taking the inverse Fourier transform of the log-magnitude Fourier spectrum, converts a whole family of equally spaced spectral

lines into a single line in the cepstrum, in the time domain. The position of this line is, in seconds, the inverse of the characteristic fault frequency, which separates two consecutive harmonics of a series generated by a given fault. The cepstrum of the machine's current shown in Fig. 3, whose spectrum is shown in Fig. 4, is shown in Fig. 9. The four spectral lines produced by the broken bar fault in Fig. 4, which are separated by a frequency given by (1) of 2.86 Hz, is transformed into a single cepstral line in Fig. 9, at a time of 1/2.86 = 0.35 seconds.

$$cepstrum = |IFFT\{log(|FFT\{i(t)\}|^2)\}|^2 \qquad (5)$$

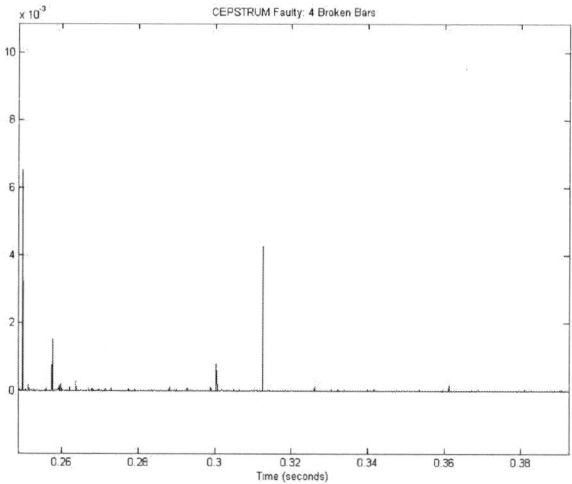

Fig. 9: Cepstrum of the machine's current in the case of a broken bar.

Other aspect of the spectral representation of the current that can difficult the identification of fault harmonics is the presence of a high noise level, which can bury these fault harmonics. A technique to lower this noise level is simply to average several consecutive spectra of the current, possibly overlapping, which reduces the variance of the resultant spectrum proportionally to the number of spectra averaged. Fig 10 shows the high reduction of the noise level, compare to the noise level in Fig. 4, that has been achieved just by averaging 8 consecutive spectra with an overlapping of 50% between two consecutive windows.

IV. SUPPORT VECTOR MACHINE CLASSIFIER

Support vector machines (SVM) is an automatic classifier based on the structural risk minimization (SRM), with foundations in statistical learning theory. SVM based classifiers are claimed to have better generalization properties in comparison with conventional classifiers. They rely on the identification of a subset of the samples data space, the support vectors, which contain sufficient information to define the classifier, so that the rest of the feature set is not needed to perform the classification

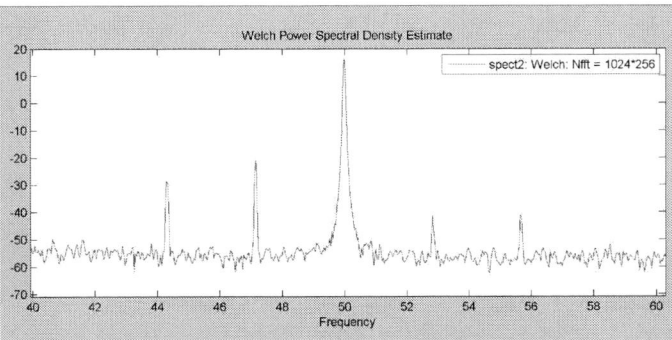

Fig. 10: Welch power spectral density estimate of the motor's current.

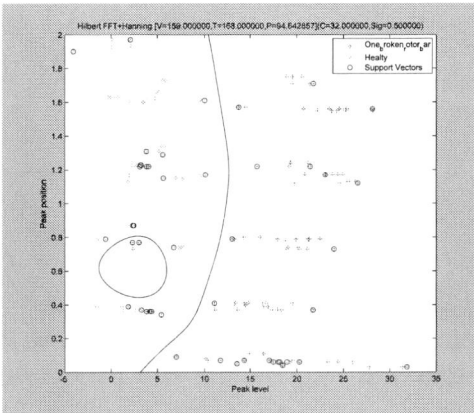

Fig. 12: Classifier using Hilbert transform.

tasks. This improves greatly the speed and efficiency of the classification process. Besides, with the use of appropriate kernels, non-linear problems can be easily addressed. In the present work, a radial basis function (RBF) has been chosen. A problem of the SVM classifier is the selection of the features used to establish the support vectors. Time-domain features (mean, variance, kurtosis, skewness) or frequency-domain features (frequency, amplitude of the spectral lines, etc.) can be used for performing the classification task. Several studies [13], based on the use of Principal Component Analysis (PCA) and Independent Component Analysis (ICA), has demonstrated that a SVM for detecting broken bars can be successfully implemented using just a two dimensional feature space: the frequency of the left sideband harmonic (LSH) ($k/p = -1$ in (1), relative to the supply frequency, and the amplitude of this harmonic.

In this work, the training set has a dimension of 2x105, the two features and 105 observations. Some results of the classifier, using the EPV and the AS, are shown in Fig. 11 and in Fig. 12

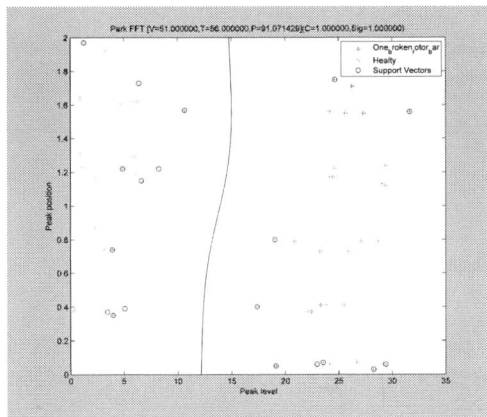

Fig. 11: Classifier using EPV modulus windowed with a Hanning window.

V. COMPARISON OF THE PERFORMANCE OF THE SVM

The combinations that have been analysed are related to the quantity that is analysed (current, windowed current, Hilbert transform, EPV) and the signal tool used to obtain the LSH (FFT, Chirp-Z transform, cepstrum, Welch). The comparison performed in table V gives information about the best combination of quantity-signal processing tool from the point of view of the SVM classifier.

TABLE V: Percentage of success of the SVM classifier as a function of the quantity selected (columns) and the signal processing tool used (rows)

	FFT	Welch	Chirp-Z	Cepstrum
Current	72.03%	71.02%	73.12%	62.05%
Current+Window	88.69%	85.12%	90.03%	56.55%
Modulus of the AS	94.64%	94.04%	93.55	47.53%
Modulus of the EPV	91.07%	91.07%	89.90%	57.32%

From the results presented in table V, the best combination between the quantity to be analyzed and the signal processing tool used for analysing it, from the point of view of the SVM classifier, is the combination Modulus of the AS - FFT. There are other combinations that perform as well as this one, but the optimal combination has, besides, some relevant features: it is more economic that the option that uses the EPV (only one current is needed in the Hilbert transform case, and the three currents are required to compute the EPV); It is also faster to process, because just the standard FFT algorithm is needed, which is faster than the Chirp-Z or the Welch signal processing tools.

VI. CONCLUSIONS

MCSA is the most extended method for diagnosing induction motors. Some practical problems that arise during the industrial application of MCSA, such as the problem of spectral leakage, or the need of high resolution spectra to detect closely spaced harmonics, have driven the proposal of several improvements to MCSA to address both problems, which have resulted in the proposal of new quantities and

978-1-4799-0024-4/13 $31.00 © 2013 IEEE

new signal processing tools. In this paper, a comparative study has been made in order to evaluate the best combination of chosen quantity - signal processing tool, from the point of view of an automatic classifier, based on a SVM. After extensive testing of a healthy machine and a machine with broken bars, tested at different load conditions and supply frequencies, the results show that the best combination is the analysis of the modulus of the AS via the FFT. Besides, this combination also outperforms close competitors in terms of easy of application, because just one current is needed, and in terms of speed of computation, because just a standard FFT is used to obtain the fault harmonics. Future work will extend this analysis to more combinations of quantities/signal processing tools, and to other types of faults different than the broken bar fault.

ACKNOWLEDGMENT

This work was supported by the Spanish "Ministerio de Ciencia e Innovación" in the framework of the "Programa Nacional de proyectos de Investigación Fundamental" (project reference DPI2011-23740).

REFERENCES

[1] V. F. Pires, M. Kadivonga, J. Martins, and A. Pires, "Motor square current signature analysis for induction motor rotor diagnosis," *Measurement*, vol. 46, no. 2, pp. 942 – 948, 2013. [Online]. Available: http://www.sciencedirect.com/science/article/pii/S0263224112003806

[2] A. Bellini, F. Filippetti, C. Tassoni, and G. A. Capolino, "Advances in diagnostic techniques for induction machines," *Industrial Electronics, IEEE Transactions on*, vol. 55, no. 12, pp. 4109–4126, Dec. 2008.

[3] A. Bellini, F. Filippetti, G. Franceschini, C. Tassoni, and G. Kliman, "Quantitative evaluation of induction motor broken bars by means of electrical signature analysis," *Industry Applications, IEEE Transactions on*, vol. 37, no. 5, pp. 1248 –1255, sep/oct 2001.

[4] S. Nandi, R. Bharadwaj, and H. Toliyat, "Performance analysis of a three-phase induction motor under mixed eccentricity condition," *Energy Conversion, IEEE Transactions on*, vol. 17, no. 3, pp. 392 – 399, sep 2002.

[5] R. Schoen, T. Habetler, F. Kamran, and R. Bartfield, "Motor bearing damage detection using stator current monitoring," *Industry Applications, IEEE Transactions on*, vol. 31, no. 6, pp. 1274 –1279, nov/dec 1995.

[6] M. El Hachemi Benbouzid, "A review of induction motors signature analysis as a medium for faults detection," *Industrial Electronics, IEEE Transactions on*, vol. 47, no. 5, pp. 984–993, 2000.

[7] R. Puche-Panadero, M. Pineda-Sanchez, M. Riera-Guasp, J. Roger-Folch, E. Hurtado-Perez, and J. Perez-Cruz, "Improved resolution of the MCSA method via Hilbert transform, enabling the diagnosis of rotor asymmetries at very low slip," *Energy Conversion, IEEE Transactions on*, vol. 24, no. 1, pp. 52–59, March 2009.

[8] S. M. A. Cruz and A. J. M. Cardoso, "Rotor cage fault diagnosis in three-phase induction motors by extended park's vector approach," *Electric Machines &Power Systems*, vol. 28, no. 4, pp. 289–299, 2000.

[9] M. Riera-Guasp, M. Pineda-Sanchez, J. Perez-Cruz, R. Puche-Panadero, J. Roger-Folch, and J. A. Antonino-Daviu, "Diagnosis of induction motor faults via gabor analysis of the current in transient regime," *Instrumentation and Measurement, IEEE Transactions on*, vol. 61, no. 6, pp. 1583–1596, 2012.

[10] S. H. Kia, H. Henao, and G. A. Capolino, "A high-resolution frequency estimation method for three-phase induction machine fault detection," *Industrial Electronics, IEEE Transactions on*, vol. 54, no. 4, pp. 2305–2314, 2007.

[11] B. Xu, L. Sun, L. Xu, and G. Xu, "An esprit-saa-based detection method for broken rotor bar fault in induction motors," *Energy Conversion, IEEE Transactions on*, vol. 27, no. 3, pp. 654–660, 2012.

[12] P. Gardel, D. Morinigo-Sotelo, O. Duque-Perez, M. Perez-Alonso, and L. A. Garcia-Escudero, "Neural network broken bar detection using time domain and current spectrum data," in *Electrical Machines (ICEM), 2012 XXth International Conference on*. IEEE, 2012, pp. 2492–2497.

[13] D. Matić, F. Kulić, M. Pineda-Sánchez, and I. Kamenko, "Support vector machine classifier for diagnosis in electrical machines: Application to broken bar," *Expert Systems with Applications*, vol. 39, no. 10, pp. 8681–8689, 2012.

Detection of Induction Machine Winding Faults Using Genetic Algorithm

M. Alamyal, S. M. Gadoue and B. Zahawi

Abstract -- **In this paper, an identification technique for fault detection of induction machines using Genetic Algorithm (GA) is investigated. The condition monitoring technique proposed in this paper indicates the presence of a winding fault and provides information about its nature and location. The data required for the proposed method are motor terminal voltages, stator currents and rotor speed obtained during steady state operation. The data is then processed off-line using an induction motor model in conjunction with GA to determine the effective motor parameters. The proposed technique is demonstrated using experimental data obtained from a 1.5 kW wound rotor three-phase induction machine with both stator and rotor winding faults considered. Results confirm the effectiveness of GA to properly identify the type and location of the fault without the need for knowledge of various fault signatures.**

Index Terms-- **Induction machine, genetic algorithm, condition monitoring.**

I. Introduction

ALTHOUGH induction motors are highly reliable, require low maintenance and have relatively high efficiency, they are subject to many electrical and mechanical types of faults. It is crucial to detect faults while they are still developing. A fault that is not identified in the initial stage may become catastrophic and the machine may suffer severe damage. Undetected faults can lead to serious machine failures. Fault identification is, therefore, essential in order to detect and diagnose potential failures in electrical motors.

Many methods have been developed for the purpose of detecting mechanical and electrical faults in induction motors, either directly or indirectly. Motor current signature analysis [1] is one of the most powerful methods used for online fault diagnosis due to its low cost and simplicity. This technique uses current spectrum of the machine for locating characteristic fault frequencies. Fast Fourier Transform and Wavelet transform are used to analyze motor current signature by identifying fault spectrum and extracting unique features for fault diagnosis. Vibration monitoring and analysis of the negative sequence components of the stator current can be also used.

Mohamoud Alamyal, Shady Gadoue and Bashar Zahawi are with the School of Electrical and Electronic Engineering, Newcastle University, Newcastle upon Tyne, NE1 7RU, England, United Kingdom. (e-mail: m.o.alamyal@ncl.ac.uk).

The monitoring and fault detection of electrical machines have moved in recent years from traditional techniques towards Artificial Intelligence (AI) techniques such as Artificial Neural Networks (ANNs) and Fuzzy Logic Systems (FLS) [2].

Stochastic optimization techniques have also been used in the fault detection of induction motors [3-5]. These techniques depend on the idea of neighborhood search. Every method uses a different search rule to find the optimal or near optimal solution. The space of all feasible solutions is called the search space. Each and every point in the search space represents one possible solution. Therefore each possible solution is represented by one point in the search space. In effect, this technique searches through the solution space and moves towards the direction of a known previous location until a solution is found. In [3] a Simulated Annealing (SA) algorithm has been used for induction machine fault detection. Particle Swarm Optimization (PSO) algorithm has also been used [4, 5] and compared with SA for induction machine stator and rotor winding fault identification. Similarly, a Tabu Search (TS) algorithm has been used in [6] for the vehicle routing problem.

Genetic Algorithm (GA) is one of the stochastic search techniques that has attracted much attention in the past few years as a powerful tool to solve many control problems. GA is an evolutionary optimization technique inspired by the mechanisms of evolution and natural selection. It has shown an efficient and effective way for optimization applications by searching global minimal without needing the derivative of the cost function.

The Fault identification technique proposed in this paper is based on adjusting the induction machine model parameters off-line using GA until the smallest error between the measured and calculated stator currents is achieved. The new set of model parameters defines the nature and location of the fault.

Experimental tests have been carried out to validate the proposed fault identification algorithm for a three-phase wound rotor induction machine with stator and rotor faults. Results confirm the capability of GA to identify and locate the fault without the need for a previous knowledge of different fault current signatures.

II. Induction Machine Mathematical Model

The mathematical ABCabc model of an induction motor is developed using Simulink software and used with GA to identify different machine winding faults. This ABCabc model is obtained from the standard machine voltage equations and represented by (1):

$$
\begin{bmatrix}
V_{sA} \\
V_{sB} \\
V_{sC} \\
V_{ra} \\
V_{rb} \\
V_{rc}
\end{bmatrix}
=
\begin{bmatrix}
R_{sA}+pL_{ss} & pM_{ss} & pM_{ss} & pM_{sr}\cos\theta_r & pM_{sr}\cos\theta_{r1} & pM_{sr}\cos\theta_{r2} \\
pM_{ss} & R_{sB}+pL_{ss} & pM_{ss} & pM_{sr}\cos\theta_{r2} & pM_{sr}\cos\theta_r & pM_{sr}\cos\theta_{r1} \\
pM_{ss} & pM_{ss} & R_{sC}+pL_{ss} & pM_{sr}\cos\theta_{r1} & pM_{sr}\cos\theta_{r2} & pM_{sr}\cos\theta_r \\
pM_{sr}\cos\theta_r & pM_{sr}\cos\theta_{r2} & pM_{sr}\cos\theta_{r1} & R_{ra}+pL_{rr} & pM_{rr} & pM_{rr} \\
pM_{sr}\cos\theta_{r1} & pM_{sr}\cos\theta_r & pM_{sr}\cos\theta_{r2} & pM_{rr} & R_{rb}+pL_{rr} & pM_{rr} \\
pM_{sr}\cos\theta_{r2} & pM_{sr}\cos\theta_{r1} & pM_{sr}\cos\theta_r & pM_{rr} & pM_{rr} & R_{rc}+pL_{rr}
\end{bmatrix}
\begin{bmatrix}
I_{sA} \\
I_{sB} \\
I_{sC} \\
I_{ra} \\
I_{rb} \\
I_{rc}
\end{bmatrix}
\quad (1)
$$

where (V_{sA}, V_{sB}, V_{sC}) ,(I_{sA}, I_{sB}, I_{sC}) are the stator winding voltages and currents, (V_{ra}, V_{rb}, V_{rc}), (I_{ra}, I_{rb}, I_{rc}) are the rotor winding voltages and currents, (R_{sA}, R_{sB}, R_{sC}), (R_{ra}, R_{rb}, R_{rc}) are the stator and rotor winding resistances respectively, L_{ss} and L_{rr} are the stator and rotor winding self-inductances respectively, M_{ss} and M_{rr} are the mutual inductance between pairs of stator and rotor windings respectively, M_{sr} is the peak value of the rotor position dependent mutual inductance between stator and rotor winding pairs, θ_r is the rotor position angle, $\theta_{r1}=\theta_r+2\pi/3$, $\theta_{r2}=\theta_r+4\pi/3$ and p is the differential operator.

III. PROPOSED CONDITION MONITORING TECHNIQUE USING GENETIC ALGORITHM

GA starts with a random population of potential individuals, each representing one possible solution to the problem. A population is made up of a set of individuals, and evolution from one generation to the next takes place by the deletion of existing individuals and the creation of new ones. During each generation, the individuals are evaluated according to the objective and fitness functions [7]. After obtaining the fitness of all individuals, a selection process is used to choose individuals for reproduction. Individuals with higher fitness should have a higher probability of being selected as parents so that the more successful individuals will have more chances to mate and generate offspring. The least fit individuals in each population are then replaced by the offspring so that the population size remains constant and another generation starts. Through an iterative process, the population evolves towards better regions of the search space. After many generations, the algorithm reaches convergence towards the best chromosome, or the individual which signifies the optimal solution or the nearest optimal solution to the problem [8]. The basic flow chart of GA is shown in Fig. 1.

The proposed condition monitoring technique is shown schematically in Fig. 2. Experimental measurements of the three-phase stator voltages (V_A, V_B, V_C), three-phase stator currents (I_A, I_B, I_C) and motor speed (ω_r) are recorded by using a digital oscilloscope and saved in the computer memory. The recorded three-phase voltages and the rotor speed are fed to the mathematical model in order to calculate the stator currents (I_{sA}, I_{sB}, I_{sC}). These stator currents are compared with the actual measured stator currents (I_A, I_B, I_C) to produce a set of current errors that are integrated and

summed to give an overall calculation error; the Integral Absolute Error (IAE) as defined in (2). This error function is the cost function to be minimized by GA.

$$
IAE = \sum \left(|i_A - i_{sA}| + |i_B - i_{sB}| + |i_C - i_{sC}| \right) \Delta T \quad (2)
$$

where ΔT is the sampling time.

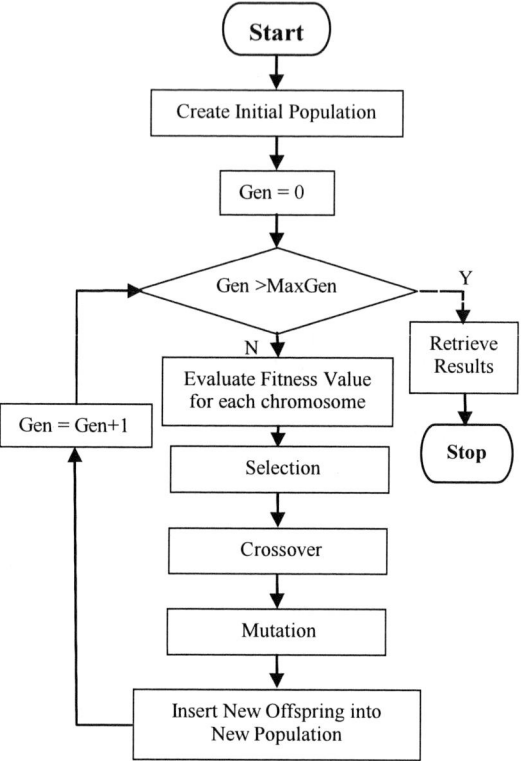

Fig. 1 GA flowchart

The GA algorithm is then implemented to minimize this cost function. This is achieved by adjusting the parameters of the machine model using GA until a minimum IAE between measured and calculated data is achieved. The new set of the model parameters indicates if the machine winding is healthy or if there is a fault and also the location and the nature of this fault. For a healthy machine the stochastic algorithm should identify the nominal parameters for the induction motor.

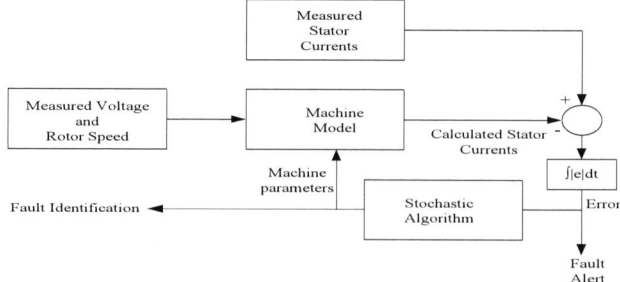

Fig. 2 Schematic representation of the fault identification technique

978-1-4799-0024-4/13 $31.00 © 2013 IEEE

In the case of a faulty machine, the final algorithm identified machine parameters would indicate the presence of a fault and its' location. For example, an increased value in the resistance of a rotor winding indicates a developing open-circuit fault in this circuit, and so on. This stochastic approach does not require any expert prior knowledge of the type of fault or its location.

IV. EXPERIMENTAL RESULTS

The experiment work is conducted on a 1.5kW, 50 Hz, 240 V, 2-pole wound rotor induction machine coupled to a 3 kW DC machine used as a generator to provide the necessary load torque. The induction motor has a star connected stator windings and a short circuited delta connected rotor winding. The nominal values of the induction machine equivalent circuit parameters were calculated using standard tests [9]. These are given in Table I.

TABLE I
INDUCTION MOTOR PARAMETERS

Stator resistances	$R_s = 5.88\ (\Omega)$
Rotor resistances	$R_r = 6.83\ (\Omega)$
Stator self-inductances	$L_{ss} = 0.729\ (H)$
Rotor self-inductances	$L_{rr} = 0.578\ (H)$
Mutual inductances between the stator windings	$M_{ss} = 0.25\ (H)$
Mutual inductances between the rotor windings	$M_{rr} = 0.7\ (H)$
Mutual inductance between stator and rotor winding pairs	$M_{sr} = 0.769\ (H)$ $M_{rs} = M_{sr}$

Tests are carried out emulating stator and rotor open-circuit winding fault conditions. In all tests the measured waveforms are the three terminal voltages, three stator currents and rotor speed. Voltage differential probes, current probe amplifier and a digital tachometer are used to measure these signals with a sampling interval of 1 ms.

Matlab/Simulink software environment is used to implement the GA algorithm. Before the GA could be used with the experimental test data, two important aspects should be considered: the coding of chromosomes and defining the evolution criteria. The chromosomes can be encoded as either binary or real values. In this study, the parameters are encoded with real values to alleviate errors in decimal-to-binary and binary-to-decimal conversions. The evolution criteria deal with evaluating each chromosome's fitness using an appropriate measure. In this study, performance is assessed by using the *IAE* (2) as the cost function to be minimized by GA where the best parameters are associated with the smallest *IAE*. The parameters of the GA algorithm used in this paper are shown in Table II.

TABLE II
GA PARAMETERS

Description	Value
Population size	12
Crossover rate	0.7
Mutation rate	0.05
Generation gap	0.9
Precision of variables	20
Number of generations	100

A. Stator fault test

This experiment is conducted by replicating a stator open-circuit fault. A developing stator open-circuit winding fault is emulated by connecting a 7-Ω resistor in series with a stator phase winding (winding B) as shown in Fig. 3.

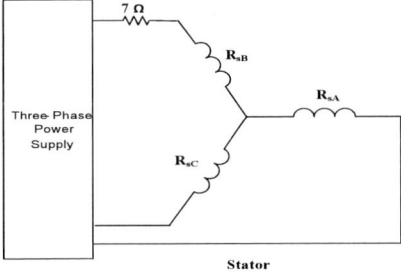

Fig. 3 Developing stator winding open-circuit fault; test circuit

The objective of the identification is to determine a vector of six parameters, which represent the six winding resistances (R_{sA}, R_{sB}, R_{sC}, R_{ra}, R_{rb}, R_{rc}) of the motor: Chromosome = (R_{sA}, R_{sB}, R_{sC}, R_{ra}, R_{rb}, R_{rc}).

The type and location of the fault are identified by adjusting the ABCabc model parameters. As shown in Fig. 4, there is a much higher estimated value of winding resistance (13.937 Ω) in stator phase B while the other estimated stator resistances are at approximately their nominal values, indicating the presence of a stator open circuit winding fault. At the same time, the estimated three rotor resistances are all at their nominal values as shown in Fig. 5, indicating a healthy state for the rotor winding. Fig. 6 shows the corresponding values of the *IAE* function during the optimization process.

The number of investigations of potential solutions required to obtain convergence with this data set was 16. The error falls from a under this rotor maximum value of 0.18 A.s to 0.007 A.s. The final values of the stator and rotor resistances obtained at the end of the GA optimization process are listed in Table III.

978-1-4799-0024-4/13 $31.00 © 2013 IEEE   159

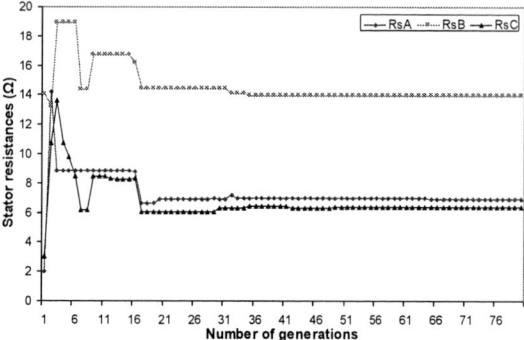

Fig. 4 Estimated stator resistances obtained using GA for operation of induction motor with stator open-circuit fault

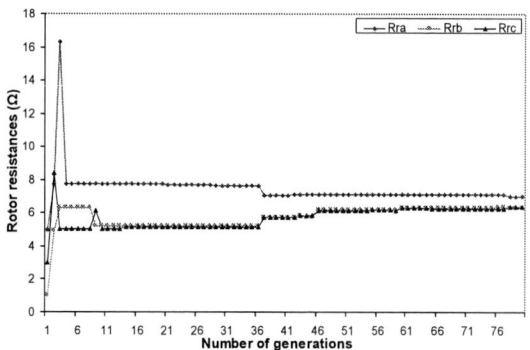

Fig. 5 Estimated rotor resistances obtained using GA for operation of induction motor with stator open-circuit fault

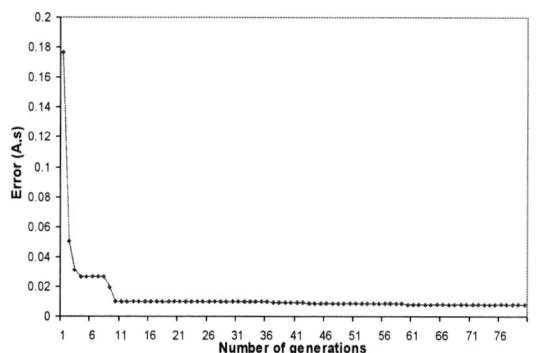

Fig. 6 IAE obtained using GA for operation of induction motor with stator open-circuit fault

TABLE III
FINAL VALUES OF WINDING RESISTANCES OBTAINED USING GA WITH
STATOR OPEN-CIRCUIT FAULT

R_{sA} (Ω)	R_{sB} (Ω)	R_{sC} (Ω)	R_{ra} (Ω)	R_{rb} (Ω)	R_{rc} (Ω)
6.937	13.964	6.369	6.991	6.311	6.387

Fig. 7 shows the measured (I_A, I_B, I_C) and calculated (I_{sA}, I_{sB}, I_{sC}) stator currents using the final parameter values obtained by the GA algorithm revealing good agreements between the two current waveforms. This gives confidence in the ability of the GA algorithm not only to identify the presence of the open-circuit winding fault but also to accurately estimate the resulting parameter values.

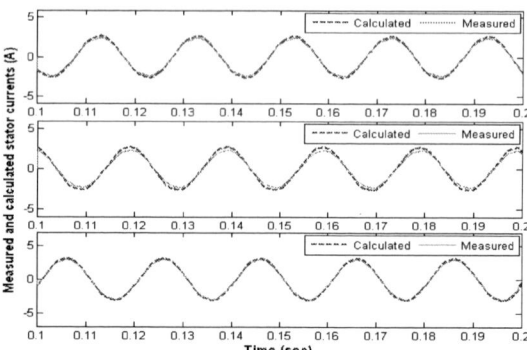

Fig. 7 Measured (I_A, I_B, I_C) and GA calculated (I_{sA}, I_{sB}, I_{sC}) stator current waveforms; stator open-circuit fault

B. Rotor fault test

A developing open-circuit rotor winding fault is emulated by connecting an external 7-Ω resistor in series with the line connected to the two ends of the R_{rb}–R_{ra} rotor delta windings as shown in Fig. 8.

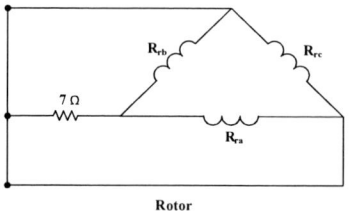

Fig. 8 Developing rotor winding open-circuit fault; test circuit

The GA algorithm is implemented to identify the presence of a developing rotor winding open-circuit fault based on the experimental measurements. In this test, the six winding resistances (R_{sA}, R_{sB}, R_{sC}, R_{ra}, R_{rb}, R_{rc}) are again the parameters to be optimized in order to minimize the IAE (2).

Fig. 9 shows the estimated stator resistances during the optimization process. It is clear from this figure that there is a rotor winding open-circuit fault as indicated by the high values of R_{ra} and R_{rb} compared with R_{rc}. On the other hand, the three stator resistances give nearly the same values; showing the healthy state of the stator (Fig. 10). The number of investigations of potential solutions required to obtain convergence with this data set was 20. Fig. 11 shows the error corresponding to the best solution open-circuit fault condition. The error falls from a under this rotor maximum value of 0.843 A.s to 0.449 A.s. The final estimated values of the stator and rotor resistances are given in Table IV. The measured (I_A, I_B, I_C) and calculated (I_{sA}, I_{sB}, I_{sC}) stator currents waveforms when using the final parameter values obtained by the GA algorithm are shown in Fig. 12.

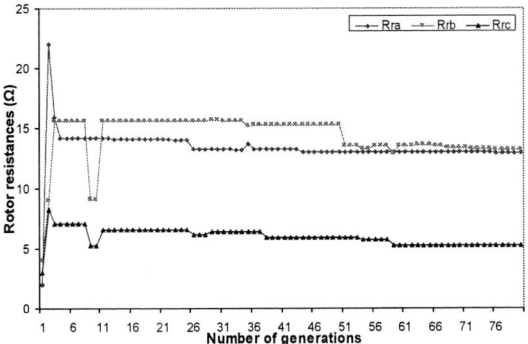

Fig. 9 Estimated rotor resistances obtained using GA for operation of induction motor with rotor open-circuit fault

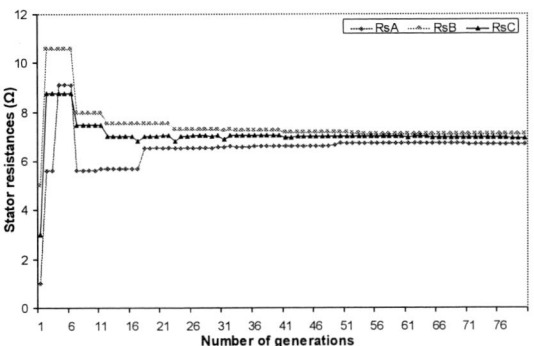

Fig. 10 Estimated stator resistances obtained using GA for operation of induction motor with rotor open-circuit fault

TABLE IV
FINAL VALUES OF WINDING RESISTANCES OBTAINED USING GA WITH
ROTOR OPEN-CIRCUIT FAULT

R_{sA} (Ω)	R_{sB} (Ω)	R_{sC} (Ω)	R_{ra} (Ω)	R_{rb} (Ω)	R_{rc} (Ω)
6.695	7.074	6.953	12.982	13.232	5.216

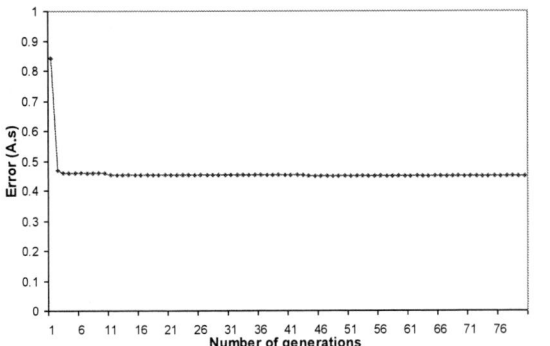

Fig. 11 IAE obtained using GA for operation of induction motor with rotor open-circuit fault

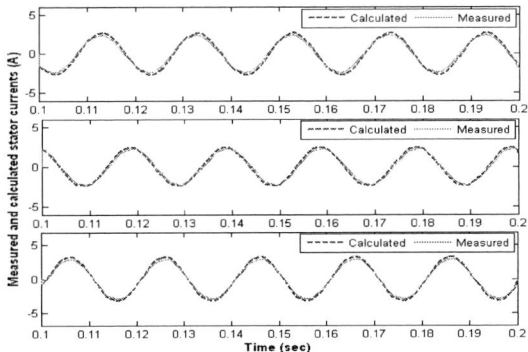

Fig. 12 Measured (I_A, I_B, I_C) and GA calculated (I_{sA}, I_{sB}, I_{sC}) stator current waveforms; rotor open-circuit fault

V. CONCLUSION

This paper has illustrated the ability of GA to identify and locate the presence of the stator and rotor winding faults for induction machines. Tests were carried out for a supply-fed wound rotor three-phase induction machine. The acquired data is then processed off-line using GA in conjunction with an induction motor mathematical model. The main objective of the stochastic optimization algorithm is to minimize an error function between measured and calculated stator currents by adjusting the parameters of the machine model. The new set of the model parameters indicates if the machine winding is healthy or if there is a fault and also the location and the nature of this fault.

REFERENCES

[1] M. E. H. Benbouzid, "Review of induction motors signature analysis as a medium for faults detection," in *IECON Proceedings (Industrial Electronics Conference)*, 1998, pp. 1950-1955.

[2] F. Filippetti, G. Franceschini, C. Tassoni, and P. Vas, "Recent developments of induction motor drives fault diagnosis using AI techniques," *IEEE Transactions on Industrial Electronics*, vol. 47, pp. 994-1004, 2000.

[3] K. P. Zakaria, P. P. Acarnley, and B. Zahawi, "Condition Monitoring of an Induction Machine using a Stochastic Search Technique," *3rd IET International Conference on, Power Electronics, Machines and Drives (PEMD2006)*, 2006, pp. 42-46.

[4] S. A. Etny, P. P. Acarlney, B. Zahawi, and D. Giaouris, "Induction Machine Fault Identification using Particle Swarm Algorithms," *International Conference on Power Electronics, Drives and Energy Systems (PEDES '06)*, 2006, pp. 1-4.

[5] S. A. Ethni, B. Zahawi, D. Giaouris, and P. P. Acarnley, "Comparison of particle swarm and simulated annealing algorithms for induction motor fault identification," *7th IEEE International Conference on Industrial Informatics (INDIN 2009)*, 2009, pp. 470-474.

[6] F. A. T. Montane and R. D. Galvao, "A tabu search algorithm for the vehicle routing problem with simultaneous pick-up and delivery service," *Computers and Operations Research*, vol. 33, pp. 595-619, 2006.

[7] O. Cordon, F. Gomide, F. Herrera, F. Hoffmann, and L. Magdalena, "Ten years of genetic fuzzy systems: Current framework and new trends," *Fuzzy Sets and Systems*, vol. 141, pp. 5-31, 2004.

[8] D. A. Coley, *An introduction to genetic algorithms for scientists and engineers*. Singapore: World Scientific Publishing Co. Pte. Ltd, 1999.

[9] *IEEE Standard Test Procedure for polyphase Induction Motors and Generators*. New York, USA, 2004.

Broken bar condition monitoring of an induction motor under different supplies using a Linear Discriminant Analysis

M. Fernandez-Temprano, P.E. Gardel-Sotomayor, O. Duque-Perez,
D. Morinigo-Sotelo, *Member, IEEE*

Abstract -- **This paper presents a procedure for broken rotor bar diagnosis in induction motors based in data extracted from stator current, which is calculated in the time and frequency domains. Data comes from a tested motor fed by different types of supply: direct line and two different Voltage Source Inverters. Diagnosis is always difficult in Voltage Source Inverter fed motors due to inherent noise level and the presence of additional non-related fault harmonics in the stator current spectrum. Moreover, the motor was tested under different load conditions, from no-load to full-load. Diagnosis is also more difficult at lower load levels. Previous to fault classification, a variable reduction was carried out using Principal Component Analysis. Fault classification was performed using Linear Discriminant Analysis. The motor was tested with different fault severities, what allowed us to perform an analysis oriented to different maintenance approaches, considering the criticality of the motor.**

Index Terms--**Fault diagnosis, Induction motors, Linear Discriminant Analysis, Principal Component Analysis.**

I. NOMENCLATURE

s: slip.
f_1 fundamental harmonic.
f_b frequency of the broken bar related sideband.

II. INTRODUCTION

F. Induction motors are essential in many industries, where they are the horsepower in many factories [1]. Although these machines are very rugged and their reliability is very high, they are not free from suffering faults when they are subjected to stresses greater to thresholds predetermined in the design stage [2]. These stresses can be classified in the following categories: electric, thermal, dielectric, mechanical, dynamic, residual, electromagnetic and environmental stresses. When a certain threshold is exceeded, or a combination of some of these stresses happens, one or more motor components fail, namely, rotor cage, stator winding, bearing, motor shaft, etc. [3], [4].

Of all possible induction motor faults, we considered in this work the failure through broken rotor bars, which is common in many industrial applications. This failure can be caused by large starting currents when cooling is at minimum, and this result in thermal and mechanical stresses being at a maximum. The impact of this failure mode is more significant when the start-up time is relatively long, and when frequent starts are required as part of a heavy-duty cycle [5]. Although a broken rotor bar is not a cause for a motor to stop, there can be serious secondary effects as torque reduction, inconsistent motor operation and safety concerns [6], [7]. In fact, broken rotor bars can be a major problem for some users [8].

A fault detection and diagnosis system is very important for any efficient industry in order to minimize costs of electrical machine failures and production downtime. To diagnose a faulty motor, many techniques are being used based in the analysis of vibrations, temperature, magnetic flux, but the most widespread one is based in the analysis of the stator current, and it is well known as Motor Current Signature Analysis [9]-[11]. The majority of these faults cause an asymmetry in the motor that is reflected as additional harmonics in the stator current spectrum. These fault-related harmonics appear at specific frequencies depending on the specific fault and motor rotation speed. There is an extensive bibliography giving formulas to calculate where these harmonics will appear in the stator current spectrum [6], [10], [12]. These harmonics amplitude, which also depends on the motor load, is indicative of the fault severity and is used as fault features to diagnose the motor condition [7]. These features are calculated in the frequency domain [13]. Although it is not usual, additional fault features can also be calculated in the time domain from the stator current signal [5], [14], [15]. This is more common in vibrational signal analysis, where some features are computed, such as root-mean-square, crest factor, skewness, kurtosis, and some cumulants and momentums. But, now they are also being used for induction motor diagnosis and calculated from the stator current in the time domain [16]-[18].

Nowadays, many induction motors are being fed by Voltage Source Inverters (VSI) due to the advantages provided. To name some of them, a VSI drive is more energy-efficient that one with a direct line supply and permits the motor to rotate at the exact speed demanded by the load.

But, despite these advantages, there are some drawbacks

P.E. Gardel-Sotomayor was granted an Erasmus-Mundus Doctoral Scholarship from E.U. This research is partially supported by the University of Valladolid.

M. Fernandez-Temprano is with the Department of Statistics and Operational Research, University of Valladolid, UVa, 47011 Valladolid, Spain (e-mail: miguelaf@eio.uva.es).

P.E. Gardel-Sotomayor is with the University of Valladolid, UVa, and with the National University of Asuncion, UNA (e-mail: pgardel@pol.una.py, pedroesteban.gardel@alumnos.uva.es).

O. Duque-Perez and D. Morinigo-Sotelo are with the Department of Electrical Engineering, University of Valladolid, UVa. 47011 Valladolid, Spain (e-mail: oscar.duque@eii.uva.es, daniel.morinigo@eii.uva.es).

978-1-4799-0024-4/13 $31.00 © 2013 IEEE

related to the diagnosis of the motor. A VSI fed motor shows a higher floor noise in the stator current than a direct line fed one. This makes more difficult to identify fault related harmonics, above all at low loads [19], [20]. Depending on the commutation technique and open-loop or closed-loop operation, VSIs also introduce additional harmonic content in the stator current that it is not fault related, and this also hinders true fault-related harmonics recognition [21]. VSI-fed motor faults have been analyzed and initial results are given in literature [22]–[26] but further investigation is still required [9], [19]. In these conditions, VSI fed motor and a wide range of motor loads, any fault diagnosis technique will face serious challenges.

The general procedure for a fault diagnosis technique has the following steps:

1. Data acquisition.
2. Feature extraction.
3. Feature set dimensional reduction and Classifier development.
4. Fault diagnosis.

In this case, we have adopted the following procedure. After the data acquisition step, where the stator current is recorded, a number of features are extracted from this signal. Some methods are available for feature extraction as Fast Fourier Transform, Wavelet Transform [27], [28], Wigner-Ville Transform [29], Bi-Spectrum [12], Cepstrum [30] or Hilbert Transform [31]. In our case, extracted features belong to the time and frequency domains, and will be explained later in Section IV. A large number of tests were carried out, with the motor fed by different power supplies (direct line and two VSIs) and at different load levels, from no load to full load. A broken bar failure was simulated by drilling a hole in one of the rotor cage bars. Four fault severities were considered (from healthy state to full broken bar) drilling different hole depths. As the number of extracted features is quite large, an initial analysis to avoid the existence of redundant information is required, since these redundancies may be inconvenient for the classification rule. In this work we have performed a Principal Component Analysis (PCA) to eliminate, from the initial data set, all features showing a strong correlation between them [32].

The next step is to develop a classification rule, based in the constructed data set. In this case, we choose the Linear Discriminant Analysis (LDA) technique in order to discriminate between normal and faulty operation conditions. This data-driven technique has not been used very frequently for induction motor diagnosis [33]-[35]. Other classifier methods that are being used in the literature are: Artificial Neural Networks [14], [36]-[39], Neuro-Fuzzy [40], [41], Robust Clustering Techniques [42], Support vector machine classifier [43]-[45], Nearest Neighbor classifiers [46], [47], Multiple Discriminant classifiers [48], [49].

As we consider that the diagnosis tool must be connected to the maintenance policy adopted, we evaluated the rules obtained using LDA in two different situations. First, we consider the scenario where we are interested in detecting whether a motor is faulty or not, no matter the fault severity. This would be useful when one is interested in diagnosing a critical motor, that is, if it is not acceptable that this motor can suffer any fault and must be replaced immediately, once a possible fault is detected.

In the second scenario, we try to distinguish between a healthy motor, or with an incipient fault, and a motor with a full broken bar, or almost full broken. This would be interesting when the motor is not so critical as in the first case, and it is only important to decide if the fault severity is high or not.

III. LABORATORY SETUP

The laboratory setup is shown in Fig. 1. In order to get data for this study, an induction motor with the following specifications was tested in a laboratory:

- Rated power: 1.1 kW
- Rated voltage: 400 V
- Rated current: 2.6 A
- Number of pole pairs: 2
- Rated speed: 1415 RPM.
- Rated voltage: 400 V.

The motor was loaded with a magnetic powder brake, and tested in all load conditions, from no load to full load. Three different supplies were considered, direct line and two different VSIs: Micromaster 420 by Siemens, and a PowerFlux 40 by Allen Bradley. The operating frequency was 50 Hz for the three supplies.

Stator current was acquired by a Hall Effect current transducer by LEM. A National Instruments NI cDAQ-9174 base platform with a NI 9215 acquisition module was used for data acquisition, with a sampling frequency of 80 kHz and sampling time of 10 s.

Fig. 1. General view of the laboratory setup.

The motor was tested first in a healthy condition. Faulty conditions were simulated by drilling a hole in one of the rotor bars. An incipient fault was obtained by drilling 6 mm

978-1-4799-0024-4/13 $31.00 © 2013 IEEE   163

depth hole in one of the bars. Next, a partially broken bar was obtained with a depth hole of 12 mm, and finally, a full-broken bar was obtained drilling an 18 mmm hole.

For each fault severity, the motor was tested with all the power supplies previously mentioned, and at a wide range of load conditions. The number of tests performed was:

- Healthy motor:
  - Direct line supply: 109
  - Micromaster 420 by Siemens at 50 Hz: 102
  - PowerFlex 40 by Allen Bradley at 50 Hz: 108
- Incipient fault state:
  - Direct line supply: 48
  - Micromaster 420 by Siemens at 50 Hz: 41
  - PowerFlex 40 by Allen Bradley at 50 Hz: 46
- Partially broken bar:
  - Direct line supply: 41
  - Micromaster 420 by Siemens at 50 Hz: 41
  - PowerFlex 40 by Allen Bradley at 50 Hz: 40
- Full broken bar:
  - Direct line supply: 46
  - Micromaster 420 by Siemens at 50 Hz: 41
  - PowerFlex 40 by Allen Bradley at 50 Hz: 46

IV. FEATURE EXTRACTION AND PRINCIPAL COMPONENT ANALYSIS

Once the stator current was acquired and registered, the following features were calculated.

A. Time domain statistical data

In previous works [5], [14], [15], time domain statistical features have been proposed as input data to different diagnosis techniques. In this work, 16 statistical features (including root mean square, RMS) were first considered and are shown in Table I.

B. Frequency domain data

Assuming a purely and balanced sinusoidal voltage supply and a symmetrical cage winding, there is only a forward rotating field at slip frequency, with respect to the rotor. When a rotor asymmetry happens, such as a broken bar, there will be a resultant backward rotating field at slip frequency with respect to the forward rotating rotor. This backward rotating field induces a voltage and a current in the stator winding at $(1-2s)f_1$ frequency, where s is the motor slip and f_1 is the main frequency. This induced current is the cause of torque and speed pulsations, which at the same time induce new electromotive forces in the stator. As a result of this force, new counter currents are produced at frequency $(1+2s)f_1$. This process goes on indefinitely, until it is damped and a pair of new sidebands appears around the main frequency f_1 [50]. Considering that sidebands also appear around some higher order harmonics as 5th or 7th, a more general equation of these sidebands frequencies is:

$$f_b = (1 \pm 2ks)k_1 f_1 \qquad (1)$$

where:

- $k = 1, 2, ...$
- $k_1 = 1, 5, 7, ...$

Amplitudes of the sidebands around the main, fifth and seventh harmonics are indicative of a broken bar fault, and are chosen as input data to the LDA. Table II shows the frequencies of the 8 sidebands selected for this paper.

C. Data reduction using Principal Component Analysis

PCA [51], [52] is a dimension reduction technique that is widely used in practice. Its main objective is to find a low-dimension linear subspace such that the projections of the original data onto that subspace retain as much information from the original data as possible. The vectors that generate that low-dimension linear subspace can be expressed as linear combinations of the original variables and are the new variables that contain as much information as possible from the original data. If the dimension of the subspace is low, there are a small number of these new variables and hence we have reduced the dimension of the set of data losing an amount of information as small as possible. This will be possible when there are high correlations among some of the original variables.

TABLE I
TIME DOMAIN STATISTICAL FEATURES

	Statistical feature	Equation
1	1st Order Moment (Mean)	$m_1 = \frac{1}{n}\sum(x)$
2	2nd Order Moment (Variance)	$m_2 = \frac{1}{n}\sum(x-\bar{x})^2$
3	3er Order Moment	$m_3 = \frac{1}{n}\sum(x-\bar{x})^3$
4	4th Order Moment	$m_4 = \frac{1}{n}\sum(x-\bar{x})^4$
5	Normalized 6th Order Moment	$m_6 = \frac{1}{(n*m_2)}\sum(x-\bar{x})^6$
	1st Order Cumulant	$c_1 = m_1$
6	2nd Order Cumulant	$c_2 = m_2 - m_1^2$
7	3er Order Cumulant	$c_3 = m_3 - 3m_1 m_2 + 2m_1^3$
8	4th Order Cumulant	$c_4 = m_4 + m_3 m_1 - 3m_2^2 + 12m_2 m_1^2 - 6m_1^4$
9	Skewness	$skew = \dfrac{m_3}{\left(\sqrt{m_2}\right)^3}$
10	Kurtosis	$kurt = \dfrac{m_4}{\left(\sqrt{m_2}\right)^4}$
11	Absolute mean	$|\bar{x}| = \frac{1}{n}\sum|x|$
12	Maximum peak value	$x_p = \max|x|$
13	Square root value	$x_r = \left(\frac{1}{n}\sum\sqrt{|x|}\right)^2$
14	Crest factor	$c_f = \frac{x_p}{x_{rms}}$
15	Shape factor	$s_f = \frac{x_{rms}}{|\bar{x}|}$
16	Root mean square	$x_{rms} = \sqrt{\frac{1}{N}\sum(x-\bar{x})^2}$

978-1-4799-0024-4/13 $31.00 © 2013 IEEE

Moreover, if the coefficients of the original values in these linear combinations are studied, it can be determined which variables are more relevant in each linear combination. Consequently, those variables having coefficients close to 0 in all linear combinations may be excluded from further analyses reducing the number of variables in the original set. These excluded variables will be the ones that have high correlations with some of the conserved ones so that they do not give much additional information and therefore may be dropped without substantial loss of information.

The algebra underlying PCA, which involves matrix diagonalization and eigenvector computation, is well known and may be found in many texts [51], [52]. For the problem at hand, we applied PCA to the two set of variables we are considering, namely the time domain statistical features appearing in Table I and the time spectrum data sidebands in Table II. There were no big associations among the spectrum variables and therefore all 8 variables of this type are considered in the next step. On the other hand, when the time domain statistical features were treated using PCA, several very strong associations were detected and only the first four moments, kurtosis, crest factor and shape factor (i.e. variables 1 to 4, 10, 14 and 15 in Table I) were selected to be considered in the second step of the procedure.

TABLE II
TIME SPECTRUM DATA SIDEBANDS

Sideband	Nomenclature	Frequency
1st harmonic, 1st left	BI	$f_b = (1 - 2s)f_1$
1st harmonic, 2nd left	BI2	$f_b = (1 - 4s)f_1$
1st harmonic, 1st right	BS	$f_b = (1 + 2s)f_1$
1st harmonic, 2nd right	BS2	$f_b = (1 + 4s)f_1$
5th harmonic, 1st left	B5I	$f_b = (1 - 2s)5f_1$
5th harmonic, 1st right	B5S	$f_b = (1 + 2s)5f_1$
7th harmonic, 1st left	B7I	$f_b = (1 - 2s)7f_1$
7th harmonic, 1st right	B7S	$f_b = (1 + 2s)7f_1$

V. LINEAR DISCRIMINANT ANALYSIS

Although many and more complex supervised classification techniques are being developed, LDA [52], originally developed by R. A. Fisher in 1936, is still being successfully employed in many applications to classify objects in $c \geq 2$ classes using a set of characteristics. LDA determines linear combinations of the explanatory variables such that the values of each linear combination for the objects in each class are as similar as possible, and the values for the objects in different classes are as different as possible. In statistical terms, it is said that the linear combinations minimize the variance inside the classes and maximize the variance between classes. An alternative, and perhaps more clear, formulation of LDA, is to consider the

squared estimated Mahalanobis distance. Assume that we have a set of objects $\{x_1,...,x_N\}$ where $x_i \in R_d$, that we call training sample, and that we know to which class $j=1,2,..,c$ each of the object in the training set belongs to. The squared estimated Mahalanobis distance between an object $y=(y_1,...,y_d)'$ and the estimated center of the jth class $\bar{y}_j$, $r^2(y, \bar{y}_j)$, that can be defined as:

$$r^2(y, \bar{y}_j) = (y - \bar{y}_j)'S^{-1}(y - \bar{y}_j) \qquad (2)$$

where $\bar{y}_j \in R^d$ is the mean of the objects in class j and $S \in R^{d \times d}$ is the pooled covariance matrix.

Now, the classification rule is to classify object y in the class j for which the squared Mahalanobis distance $r^2(y, \bar{y}_j)$ is smallest.

Since a single covariance matrix for all classes is considered, LDA assumes that the covariance between characteristics does not change between classes. (Notice that otherwise (2) would not be a linear function of y). The calculation of the Mahalanobis distance in (2) assumes that S is non-singular. However, if there are approximate linear relationships among the characteristics, the matrix will be ill-conditioned and the results would not be stable. This issue, called multicollinearity, is usual when a moderate to large number of characteristics is considered. The filter of characteristics made by our use of PCA allows guaranteeing that this sort of problem will not appear. In our case, we have made a second variable selection using stepwise methods inside the PCA procedure for selecting which characteristics are more useful to classify the objects.

Once the LDA has been performed, the classification power of the rule has to be evaluated computing the so-called generalization error (the capacity of the procedure to correctly classify future new observations). The most intuitive method of evaluation is called re-substitution. The re-substitution error, or apparent error (APPE), of the rule is just the percentage of objects in the training set that are incorrectly classified. Unfortunately, this error overestimates the generalization error. The reason for this is obvious. As the objects are being used to determine the rule (for computing the means in each class and the pooled covariance matrix) it is clear that they will be better classified than new independent objects that were not used in determining the rule.

In order to solve the problem of overestimation suffered by the apparent error, new procedures for estimating the generalization error have been defined. Two of the procedures more widely used in practice are k-fold crossvalidation and 632+ boostrap estimation.

In k-fold crossvalidation [52], the training sample is randomly partitioned in k subsamples. One of these subsamples is retained and the other *(k-1)* subsamples are used to determine the rule (compute the class means and the pooled covariance matrix). Then the observations in the retained subsample are classified with the rule based on the other *(k-1)* subsamples and the percentage of error is

978-1-4799-0024-4/13 $31.00 © 2013 IEEE

computed. This process is repeated retaining one of the k subsamples each time and using the rest for computing the rule. The k-fold error (KFE) is the mean of the k errors obtained with each of the k subsamples. The most commonly used value for k is $k=10$.

The procedure for obtaining the 632+ error is more elaborate. First, a bootstrap sample of size N (the training sample size) is obtained randomly and with replacement from the original training sample. Now the rule based on the bootstrap sample is computed, and the observations in the training sample not belonging to the boostrap sample are classified and the percentage of wrongly classified observations computed. This procedure is repeated B times (B is the number of boostrap samples extracted, in this work we have used $B=100$). The average of the B errors obtained is the bootstrap error (632+E) [53]. This bootstrap error (BE) tends to overestimate the generalization error so the 632+ bootstrap error is defined as follows [54]:

$$632+E = (1-\alpha)\,APPE + \alpha\,BE \qquad (3)$$

where $\alpha > 0.632$ [54].

TABLE III
MODELS AND THEIR PERFORMANCE
CLASSIFICATION ERRORS

Scenario	Variables Selected	APPE	KFE	632+E
1	1st and 4th order Moments, Shape Factor, Crest Factor, Kurtosis	0.342	0.352	0.349
1	BS, B5S	0.212	0.198	0.199
1	Kurtosis, Shape Factor, Crest Factor, BS, B5S	0.161	0.164	0.158
2	1st and 3rd order Moments, Shape Factor, Crest Factor	0.386	0.333	0.349
2	BS, B5S	0.100	0.099	0.096
2	Kurtosis, Shape Factor, Crest Factor, BS, B5S	0.095	0.093	0.095

Using this procedure on our motor failure data set, we obtained the results detailed in Table III. For each of the two diagnosis failure scenarios described in Section II, we selected three models, one initially considering only the time domain statistical features selected in the PCA step, a second one only considering the time spectrum sidebands and a third one considering both sets of features. In this way, we can check if one of the sets of features performs better than the other one and if advantages can be found combining both sets of features. For each of the models, we present the variables selected by the stepwise procedures that are finally present in the final model and the three errors we have just described, APPE, KFE and 632+E.

As expected, fault classification is more difficult in scenario 1 than in scenario 2, where an incipient fault must be detected even in extreme operation conditions, low load and VSI supply. Classification errors are lower in scenario 2.

It is also interesting to point out that classification errors are much lower when information from time and frequency domains are used together, being around 9.5% for scenario 2 and around 16% for scenario 1. In the analyzed cases, time domain data alone provided the worst classification results.

Table III also shows the variables selected to train the LDA procedure after applying PCA. In the time domain, the amplitudes of the upper-sidebands around the first and fifth harmonics were selected in the six cases analyzed. In the time domain, the number of selected variables was larger, including the kurtosis, shape and crest factors and the 1st, 3rd and 4th moments. The rest of the initial calculated variables or extracted features showed a strong correlation with these variables or contained little information for the diagnosis or classification purposes.

VI. CONCLUSIONS

This paper presents a fault classification method based in LDA applied for broken rotor bars diagnosis in induction motors. A data set is constructed to train this classification method using features extracted from the stator current of a motor. These features belong to the time and spectral domains. Initially, the number of features is very large, but it is reduced using PCA, eliminating features that show a strong correlation between them, but retaining the maximum information needed to perform a classification procedure. This data set was constructed testing an induction motor in a wide range of load conditions and in addition to the usual direct line motor supply, in this occasion, the motor was also tested fed by two different VSIs, at 50 Hz. The different load conditions and power supplies, plus the challenge of detecting incipient faults, make more difficult the fault diagnosis.

The motor was tested with four different fault states, which range from healthy to a full-broken bar condition. There were also two intermediate states, an incipient fault and almost full broken bar. This four fault conditions allowed us to apply the diagnosis technique in two different scenarios, each one oriented to a different maintenance policy. In scenario 1, we tried to distinguish when a motor is faulty, whatever the fault severity is. This approach can be used to monitor a critical motor, when it is important to detect that the motor is faulty, because it is not permitted its operation in this condition. In scenario 2, fault states were divided into two groups. Group 1 included healthy tests and incipient fault tests. Group 2 included the tests were the bar was completely broken or almost broken. In this second approach, the goal was to detect that the fault was completely developed since the motor may be allowed to operate with a non-fully developed fault.

VII. REFERENCES

[1]   D. J. T. Siyambalapitiya and P. G. Mclaren, "Reliability improvement and economic benefits of on-line monitoring systems for large

induction machines," *IEEE Trans. Ind. Appl.* vol. 26, pp. 1018-1025, 1990.

[2] B. Yanga, S. K. Jeonga, Y. M. Ohb, and A. C. C. Tanc, "Case-based reasoning system with Petri nets for induction motor fault diagnosis," Expert Systems with Applications, vol. 27, pp. 301–311, 2004.

[3] A.H. Bonnett and G.C. Soukup, "Cause and analysis of stator and rotor failures in three-phase squirrel-cage induction motors," *IEEE Trans. Industry Applications*, vol. 28, pp. 921-937, Jul./Aug. 1982.

[4] M. Perez-Alonso, O. Duque-Perez and D. Morinigo-Sotelo, "Characteristic faults in induction motors: stator faults (I)," *Ingeniería y Gestión de Mantenimiento*, vol. VII, n° 34, pp.79-84, March/Apr. 2004 (In Spanish).

[5] B.S. Payne, A. Ball and F. Gu, "Detection and Diagnosis on Induction Motor Faults using Statistical Measures," International Journal of Condition Monitoring and Diagnostics Engineering Management, vol. 5, no. 2, pp. 5-19, Apr. 2002.

[6] W. T. Thomson and M. Fenger, "Industrial application of current signature analysis to diagnose faults in 3-phase squirrel cage induction motors," in *Proc. 2000. Pulp and Paper Industry Technical Conference*, pp. 205-211.

[7] E. Germen, D.G. Ece and Ö.N. Gerek, "Self organizing map (SOM) approach for classification of mechanical faults in induction motors," in *Proc. 2007 9th international work conference on Artificial neural networks (IWANN'07)*, pp. 855-861.

[8] W. T. Thomson and D. Rankin, "Case Histories of Rotor winding Fault Diagnosis in Induction Motors", *in Proc. of the 2nd Int. Conf. on Condition Monitoring*, 1987.

[9] A. Bellini, F. Filippetti, C. Tassoni, and G.A. Capolino, "Advances in Diagnostic Techniques for Induction Machines," *IEEE Trans. Industrial Electronics*, vol.55, no.12, pp.4109-4126, Dec. 2008.

[10] M.E.H. Benbouzid, "A review of induction motors signature analysis as a medium for faults detection*," IEEE Trans. Ind. Electron.*, vol. 47, pp. 984–993, Oct. 2000.

[11] V. Climente-Alarcon; J. Antonino-Daviu; F. Vedreno-Santos, and R. Puche-Panadero " Vibration transient detection of broken bars by PSH sidebands" in *Proc. XXth International Conference on Electrical Machines (ICEM)*, Marseille, France, Sept. 2012.

[12] L. Saidi, H. Henao, F. Fnaiech, G.A. Capolino and G. Cirrincione, "Application of higher order spectral analysis for rotor broken bar detection in induction machines Diagnostics for Electric Machines, " in *2011 IEEE International Symposium on Diagnostics for Electric Machines, Power Electronics & Drives (SDEMPED)*, pp. 31-38.

[13] S. Hamdani, O. Touhami, R. Ibtiouen and M. Fadel, "Neural Network technique for induction motor rotor faults classification – Dynamic eccentricity and broken bar faults," in *2011 IEEE International Symposium on Diagnostics for Electric Machines, Power Electronics & Drives (SDEMPED)*, pp. 626-631.

[14] P. Gardel, D. Morinigo-Sotelo, O. Duque-Perez, M. Perez-Alonso, and L.A. Garcia-Escudero, "Neural network broken bar detection using time domain and current spectrum data" in *Proc. 2012 XXth International Conference on Electrical Machines (ICEM)*, Marseille, France.

[15] J. Zaeri, "Induction motors bearing fault detection using pattern recognition techniques" in *Expert Systems with Applications* vol. 39, Issue 1, pp. 68-73, Jan. 2012.

[16] J. Zhong, Z. Yang and S.F. Wong, "Machine condition monitoring and fault diagnosis based on support vector machine," in *Proc. 2010 International Conference on Industrial Engineering and Engineering Management*, pp. 2228–2233.

[17] V. Ghate and S. Dudul, "Optimal MLP neural network classifier for fault detection of three phase induction motor," *Expert Systems with Applications*, vol. 37, no. 4, pp. 3468-3481, Apr. 2010.

[18] V.T. Tran, B.-S. Yang, M.-S. Oh, and A.C.C. Tan, "Fault diagnosis of induction motor based on decision trees and adaptive neuro-fuzzy inference," *Expert Systems with Applications*, vol. 36, no. 2, pp. 1840-1849, March 2009.

[19] B. Akin, U. Orguner, H.A. Toliyat and M. Rayner, "Low order PWM inverter harmonics contributions to the inverter-fed induction machine fault diagnosis," *IEEE Trans. Ind. Electron.*, vol. 55, no. 2, pp. 610–619, 2008.

[20] F. Briz, M. W. Degner, P. Garcia, and A. B. Diez, "High-frequency carrier-signal voltage selection for stator winding fault diagnosis in inverter-fed ac machines," *IEEE Trans. Ind. Electron.*, vol. 55(12), pp. 4181–4190, 2008.

[21] R. Wieser, C. Kral, F. Pirker, and M. Schagginger, "On-line rotor cage monitoring of inverter-fed induction machines by means of an improved method," *IEEE Trans. Power Electronics*, vol. 14(5), pp. 858-865, 1999.

[22] O. Duque, M. Perez, and D. Moringo, "Detection of bearing faults in cage induction motors fed by frequency converter using spectral analysis of line current. in *Proc. 2005 IEEE International Electric Machines and Drives Conference*, pp. 17– 22.

[23] G. R. Bossio, C. H. D. Angelo, G. O. Garcia, J. A. Solsona, and M. I. Valla, "Effects of rotor bar and end-ring faults over the signals of position estimation strategy for induction motors," *IEEE Trans. Ind. Appl.*, vol. 41(4), pp. 1005–1012, 2005.

[24] J. Yang, S.B. Lee, J. Yoo, S. Lee, Y. Oh and C. Choi, "A stator winding insulation condition monitoring technique for inverter-fed machines," *IEEE Trans. Power Electron.*, vol. 22(5), pp. 2026–2033, 2007.

[25] O. Duque-Perez, M. Perez-Alonso, and D. Moringo-Sotelo, "Practical application of the spectral analysis of line current for the detection of mixed eccentricity in cage induction motors fed by frequency converter," in *Proc. 2004 16th International Conference on Electrical Machines*.

[26] D. Moringo-Sotelo, L. A. Garcia-Escudero, O. Duque-Perez, and M. Perez-Alonso, "Practical Aspects of Mixed Eccentricity Detection in PWM Voltage Source Inverter Fed Induction Motors," *IEEE Trans. Ind. Electron.*, vol. 57(1), pp. 252–262, 2010.

[27] T.W.S. Chow and S. Hai, "Induction machine fault diagnostic analysis with wavelet technique," *IEEE Trans. Ind. Electron.*, vol. 51, no. 3, pp. 558–565, Jun. 2004.

[28] M. Riera-Guasp, J.A. Antonino-Daviu, J. Roger-Folch, and M.P. Molina Palomares, "The use of the wavelet approximation signal as a tool for the diagnosis of rotor bar failures," IEEE Trans. Ind. Appl., vol. 44, no. 3, pp. 716–726, May/Jun. 2008.

[29] V. Climente-Alarcon, J.A. Antonino-Daviu, M. Riera, R. Puche-Panadero, and L.A. Escobar, "Wigner-Ville distribution for the detection of high-order harmonics due to rotor asymmetries," *2009 IEEE International Symposium on Diagnostics for Electric Machines, Power Electronics & Drives (SDEMPED).*, pp.1-6.

[30] Y.-R. Hwang, K.-K. Jen, and Y.-T. Shen, "Application of cepstrum and neural network to bearing fault detection," *Journal of Mechanical Science and Technology*, vol. 23, no.10, pp. 2730-2737.

[31] R. Puche-Panadero, M. Pineda-Sanchez, M. Riera-Guasp, J. Roger-Folch, E. Hurtado-Perez, and J. Perez-Cruz, "Improved Resolution of the MCSA Method Via Hilbert Transform, Enabling the Diagnosis of Rotor Asymmetries at Very Low Slip," *IEEE Transactions on Energy Conversion*, vol.24, no.1, pp.52-59, March 2009.

[32] Q. He, R. Yan, F. Kong, and R. Du, "Machine condition monitoring using principal component representations", *Mechanical Systems and Signal Processing*, vol. 23, pp. 446-466, 2009.

[33] B.B. Jakovljevic, Z.S. Kanovic, and Z.D. Jelicic, "Induction motor broken bar detection using vibration signal analysis, principal component analysis and linear discriminant analysis", *2012 IEEE International Conference on Control Applications*, pp.1686,1690.

[34] E.G. Strangas, S. Aviyente, and S.S.H. Zaidi, "Time–Frequency Analysis for Efficient Fault Diagnosis and Failure Prognosis for Interior Permanent-Magnet AC Motors", *IEEE Transactions on Industrial Electronics*, vol.55, no.12, pp.4191-4199, Dec. 2008.

[35] R. Casimir, E. Boutleux, and G. Clerc, "Fault diagnosis in an induction motor by pattern recognition methods," *2003 EEE International Symposium on Diagnostics for Electric Machines, Power Electronics and Drives, SDEMPED 2003*, pp.294-299, 24-26.

[36] C.T. Kowalski and T. Orlowska-Kowalska, "Neural networks application for induction motor faults diagnosis," *Mathematics and Computers in Simulation*, vol. 63, pp. 435-448, Nov. 2003.

[37] M.Y. Chow, P.M. Mangum, and S.O. Yee, "A neural network approach to real-time condition monitoring of induction motors," *IEEE Trans. Ind. Electron.*, vol. 38, no. 6, pp. 448–453, Dec. 1991.

[38] F. Filippetti, G. Franceschini, and C. Tassoni, "Neural networks aided online diagnostics of induction motor rotor faults, " *IEEE Trans. Ind. Appl.*, vol. 31, no. 4, pp. 892–899, Jul./Aug. 1995.

978-1-4799-0024-4/13 $31.00 © 2013 IEEE

[39] H. Su and K. T. Chong, "Induction machine condition monitoring using neural network modeling," *IEEE Trans. Ind. Electron.*, vol. 54, no. 1, pp. 241–249, Feb. 2007.

[40] M.S. Ballal, Z.J. Khan, H.M. Suryawanshi, and R.L. Sonolikar, "Adaptive Neural Fuzzy Inference System for the Detection of Inter-Turn Insulation and Bearing Wear Faults in Induction Motors," *IEEE Trans. Ind. Electron.*, vol. 54, no. 1, pp. 250-258, Feb. 2007.

[41] S. Altug, M.-Y. Chen, and H. J. Trussell, "Fuzzy inference systems implemented on neural architectures for motor fault detection and diagnosis," *IEEE Trans. Ind. Electron.*, vol. 46, no. 6, pp. 1069–1079, Dec. 1999.

[42] L. A. García-Escudero, O. Duque-Perez, D. Morinigo-Sotelo, and M. Perez-Alonso, "Robust condition monitoring for early detection of broken rotor bars in induction motors." *Expert Systems with Applications*, vol. 38(1), pp. 2653-2660, 2011.

[43] A. Widodo, B. S. Yang, and T. Han, "Combination of independent component analysis and support vector machines for intelligent faults diagnosis of induction motors," Expert Systems with Applications, vol. 32, pp. 299–312, 2007.

[44] Lu Shuang and Li Meng, "Bearing Fault Diagnosis Based on PCA and SVM," *2007 International Conference on Mechatronics and Automation, ICMA 2007*, pp.3503,3507, 5-8.

[45] Lu Shuang and Yu Fujin, "Fault Pattern Recognition of Bearing Based on Principal Components Analysis and Support Vector Machine," *2009 Second International Conference on Intelligent Computation Technology and Automation, ICICTA '09*, vol.2, pp.533-536.

[46] Han Sang-Bo, Don-Ha Hwang, Sang-Hwa Yi, and Dong-Sik Kang, "Development of diagnosis algorithm for induction motor using flux sensor," *2008 International Conference on Condition Monitoring and Diagnosis, CMD 2008*, pp.140-142.

[47] S.P. Santos and J.A.F. Costa, "A Comparison between Hybrid and Non-hybrid Classifiers in Diagnosis of Induction Motor Faults," *2008 11th IEEE International Conference on Computational Science and Engineering, CSE '08*, pp.301-306.

[48] B. Ayhan, Chow Mo-Yuen, and Song Myung-Hyun, "Multiple Discriminant Analysis and Neural-Network-Based Monolith and Partition Fault-Detection Schemes for Broken Rotor Bar in Induction Motors", *IEEE Transactions on Industrial Electronics*, vol.53, no.4, pp.1298-1308, June 2006.

[49] B. Ayhan, Chow Mo-Yuen, and Song Myung-Hyun, "Multiple signature processing-based fault detection schemes for broken rotor bar in induction motors," *IEEE Transactions on Energy Conversion*, vol.20, no.2, pp.336-343, June 2005

[50] F. Filippetti, G. Franceschini, C. Tassoni, and P Vas, "Impact of speed ripple on rotor fault diagnosis of induction machines," in *Proc. 1994 International Conference of Electric Machines and Drives (ICEM94)*, pp 452-456, 1994.

[51] G. H. Dunteman and H. George, *Brief description: principal component analysis*. London: Sage, 1989.

[52] R. O. Duda, P. E. Hart, and D. G. Stork, *Pattern classification*. John Willy and sons, Inc., 2001.

[53] B. Efron, "Estimating the error rate of a prediction rule: Improvement on cross-validation," *Journal of the American Statistical Association*, vol. 78, pp. 316-331, 1983.

[54] B. Efron and R. Tibshirani, "Improvements on cross-validation: The 632+ bootstrap method," *J. Amer. Statist. Assoc.*, vol. 92, pp. 548–560, 1997.

VIII. BIOGRAPHIES

M. Fernandez-Temprano received the B.S and Ph.D. degrees in Mathematics from the Universidad de Valladolid (UVA) in Spain in 1991 and 1995 respectively. He is currently with the Department of Statistics and Operations Research where he is Full Professor. His main research fields are reliability engineering, order restricted inference and circular data.

P.E. Gardel-Sotomayor received the B.S. degree in Electromechanical Engineering from the National University of Asuncion (UNA), Paraguay, in 2006. He is currently a Ph.D. student at the University of Valladolid, Spain. His research interests are monitoring of induction machines and electric power system optimization.

O. Duque-Perez received the B.S. and Ph.D. degrees in Electrical Engineering from the University of Valladolid (UVA), Spain, in 1992 and

2000, respectively. In 1994, he joined the E.T.S. de Ingenieros Industriales, UVA, where he is currently Full Professor with the Research Group in Predictive Maintenance and Testing of Electrical Machines, Department of Electrical Engineering. His main research fields are power systems reliability, condition monitoring, and heuristic optimization techniques.

D. Morinigo-Sotelo (M'04) received the B.S. and Ph.D. degrees in Electrical Engineering from the University of Valladolid, Valladolid, Spain, in 1999 and 2006, respectively. He is currently with the Research Group in Predictive Maintenance and Testing of Electrical Machines, Department of Electrical Engineering, University of Valladolid and is a Research Collaborator in Electromagnetic Processing of Materials with the Light Alloys Division, CIDAUT Foundation, Valladolid. His current research interests also include condition monitoring of induction machines, optimal electromagnetic design, and heuristic optimization techniques.

Intelligent Sensor based on Acoustic Emission Analysis applied to Gear Fault Diagnosis

Daniel Zurita, *Student Member, IEEE*, Miguel Delgado, *Member, IEEE*, Juan Antonio Ortega
Redondo, *Member, IEEE*, Luis Romeral, *Member, IEEE*

Abstract – The development of intelligent and autonomous monitoring systems applied to rotating machinery, represents the evolution towards the automatic industrial plants supervision. In this regard, an acoustic emission based intelligent sensor is presented in this work. The proposed sensor records regularly the acoustic emission signal generated by gearboxes. A time domain statistical analysis is applied in order to characterize the acquired data. Afterwards, a neural network based algorithm is applied to detect gear fault patterns. Finally, the diagnosis result is sent through a wireless transceiver to the central control unit. Moreover, in order to reach a real autonomous operation, the sensor power is approached by different energy harvesting solutions.

Index Terms--Acoustic Emission, Energy Harvesting, Gear Fault Diagnosis, Feature Analysis, Neural Networks, Preventive Maintenance, Rotating Machinery, Wireless Sensor System.

I. INTRODUCTION

Rotating machinery, such as electric motors, turbines, air compressors and others, are widely used in many industrial fields. Mechanical elements such as gears, bearings and shafts are critically important for a proper machinery operation. A high reliability operation for extended period of time in harsh environmental conditions is required. The most common failures in industrial machines are those related to the power transmission system as was reported by M. R. Wilkinson *et al.* [1]. As a result, the gearbox is found to be the most critical component, since its downtime per failure is higher in comparison to other components. For this reason, the gearbox condition monitoring is of significant importance. Unexpected faults of these components may lead to damages of the entire machine. Consequently, these unexpected faults will cause considerable machine repair/replacement costs, and associated labour and downtime losses [2]-[3]. Therefore, periodic non-destructive testing and condition monitoring are often used to conduct preventive maintenance in order to effectively diagnose and prevent further development of faults [4].

Classically, in order to perform preventive maintenance, highly experienced and skilled technicians are required. These technicians go around the plant periodically analyzing the machines in order to detect any malfunction sign. Hence, the most extended strategy is based on a pre-programmed downtime and substitution of the most critical machine parts. Consequently, a new concept of integrated smart sensor modules with the machines and processes, able to report automatically the status of the system under monitoring, represents the evolution towards a reliable distributed monitoring system in order to gain manufacturing performance and reduce maintenance costs.

The evolution of sensor networks technologies in the last decade provides a unique opportunity for the further development of a new concept of intelligent sensors. These sensors not only include the transducer, but also implement signal processing algorithms for asserting the machine status [5]-[6]. From a condition monitoring techniques point of view, classically, hand held devices are used by the industrial maintenance staff. These devices, based generally on vibration modes analysis, require and advanced degradation stage of the component under monitoring to generate a failure's warning. Therefore, an Intelligent Sensing Module (ISM), able to detect the fault in its early stage and report the machine status is a valuable tool in order to save time and money in maintenance operations.

Regarding to the physical magnitude, in early 2000, acoustic emission based methods have been proposed for preventive maintenance applications due to its early fault detection capabilities [7]-[8]. Acoustic emission in gear condition monitoring, are transient elastic waves produced by the interactions of two media of gears in relative motions. The main advantage of acoustic emission detection in front of traditional vibration detection is the better signal to noise ratio at very low frequencies [9]-[11], and in this way it is possible to detect the beginning of a failure at a very early phase [12]. However, unlike vibration based analysis, AE measurements and related methodologies need to be more deeply investigated in order to define standards and protocols for a proper signal interpretation. Different AE signal processing approaches have been carried out in this area as in [13], a simple AE analysis is performed to compute gear fault features, these features is introduced to a *k*-nearest neighbor algorithm for fault detection, in [14], a threshold-based denoising technique combined with a statistical feature extraction and reduction method is applied to detect faults in a notional split-torque gearbox.

However, in order to develop a reliable fault diagnosis procedure based on AE signals, a common approach based on statistical time domain features extraction is applied [12]. In these techniques, an AE signal is acquired, and then processed in order to extract a set of numerical features related to the current condition of the component [15], such as root mean square (RMS), crest factor or kurtosis among others. Once this set is obtained, it is time to analyze and select those features which are more related to the failure; then a diagnosis algorithm able to determinate the actual condition of the machine is applied. Unlike classical

978-1-4799-0024-4/13 $31.00 © 2013 IEEE

threshold based approaches, where sometimes is difficult to set a boundary between the healthy and faulty conditions, artificial intelligence methods are being applied with promising results [16]. For this porpoise, the Neural Networks (NN) represents one of the most used technique to learn different patterns and fusion all the information in order to obtain a diagnosis about the current state of the machine [16]-[17].

The contributions of this work are based on the presentation of an intelligent acoustic emission sensing module for gear condition monitoring. The proposed sensor is based on AE signal acquisition and processing, neural networks algorithms for fault detection and wireless technologies for communication. The effectiveness of this condition monitoring scheme has been verified by experimental results under laboratory conditions. The paper is organized as follows, Section II introduces the system architecture and the different subsystems involved in the ISM. Then, the AE signal analysis previous to the development of the diagnosis algorithm is introduced. In Section III the methodology to design the diagnosis algorithm is shown. Finally, in Section IV the experimental results and the diagnosis capabilities are analysed.

II. SENSING MODULE ARCHITECTURE

The new concept of smart sensor integrates different key modules for the diagnosis, which are: the signal conditioning module, in which the acquired signal from the transducer is pre-processed to fit the algorithm requirements. A signal processing module, in which different processing techniques are applied in order to highlight the failure presence of the system and give an estimation of the current state of the machine. A communication module which functions is to send the information to the user. And finally a user interface where all these indicators can be visualized.

Next, the main architecture and modules of the developed sensor module are detailed. The prototype system architecture is shown in Fig. 1. The system consists of four main blocks:

1) Sensor Layer (Amplifier/Filter/ADC).

2) MCU Layer.

3) Energy Harvesting Layer.

4) Communication Layer.

The sensor layer for AE signal sensing includes AE transducers and the signal pre-processing module. It is known that AE signals have a high frequency and low amplitude so an amplifying stage is required. When working with high frequency signals such as AE ones, the characteristics of the ADC are considered a critical design factor. In this sense, the ADC of the selected processor should be able to acquire data at a sampling frequency of 2MHz, because the characteristic frequency range of an AE signal goes from 20 to 1 MHz in order to avoid aliasing problems. As a result, an anti-aliasing filter is needed for a correct signal processing. Additionally, in order to avoid the

problem with the unknown sensor response takes place in frequencies under the sensor range, a high pass filter is also implemented. The cut frequency for the anti-aliasing filter is set to 400 kHz, and the cut frequency for the high pass filter is set to 100 kHz. As a result, the frequency range goes from 100 to 400 kHz.

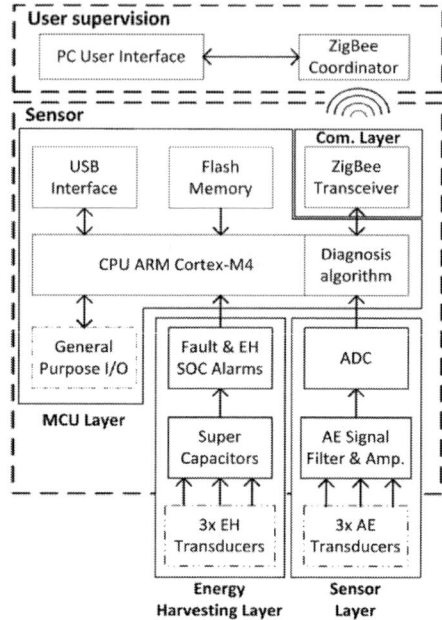

Fig. 1. Block diagram of the ISM's architecture. The ISM includes the Sensor Layer where the AE signal is acquired and conditioned, the Energy Harvesting Layer where the energy to power the system is collected, and the MCU layer which controls the whole system. The User supervision includes the wireless communication coordinator for communicating with different ISM's and the user interface where all the information is shown.

The ISM also has other general porpoise I/O ports. The processor consist of ARM Cortex-M4 MCU, the communication layer consists on a NXP JN5148 Zigbee wireless module. The flash memory is used to save the default/user configuration. Finally, the USB interface is used for further system configuration, possible upgrades and, additionally, on request, for raw data downloading directly to the PC.

The energy harvesting module is based on three different energy sources: the vibration of the rotating machinery converted in electric energy by means of the piezoelectric effect, the solar energy obtained from photovoltaic panels and thermoelectricity generators which take advantage of the heat dissipated by the machine. The harvesting module includes: rectifier and DC/DC converter for electromagnetic (vibration) energy harvester, ultra-low voltage DC/DC converter for thermoelectric generator, maximum power point tracking for indoor photovoltaic cells, super capacitor and solid state battery (thin film battery) as energy storage unit.

The data processing proposal is based on calculating several statistical time domain features and classifying the actual condition of the machine by means of a NN. The processed data is evaluated to obtain a diagnosis result which is sent to the user interface through Zigbee module. The user interface is considered as a common PC with a Zigbee coordinator, so it is able to send and receive messages with different sensors.

The proposed prototype consists of four subsystem layers as shown in Fig. 2: Energy harvester power management layer, Sensor layer (amplifier/filter/ADC), MCU layer and Zigbee communication layer. Three AE sensors can be connected to different channels of this system. The prototype's dimensions (with IP45 protection case) are 150mm×120mm×40mm. The ingress protection rating of the case is IP-68 making it suitable for harsh industrial environments.

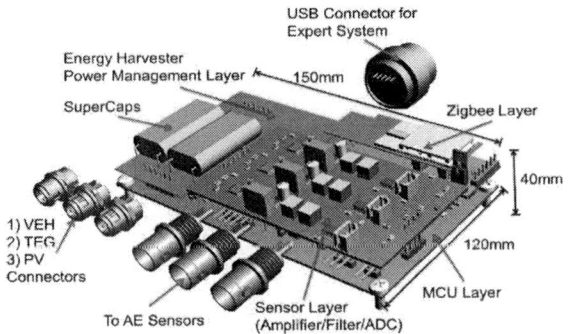

Fig. 2. Intelligent sensor module prototype 3D Illustration (the protection case is not shown in this illustration). The prototype has 4 different boards, the energy harvesting and power management board, a communication board with the Zigbee transceiver, a sensor board where the acquired signal is conditioned. All of them are connected to the main MCU board.

III. DIAGNOSIS METHODOLOGY

In this section the diagnosis algorithm, which is proposed to be implemented in the microprocessor, is explained. This section is divided in two different parts, the first one concern with the laboratory analysis of AE signals; this analysis is made in order to characterize the behavior of the gear's acoustic activity during a fracture process. The second part deals with the algorithm design and the main steps for performing the diagnosis.

A. Laboratory Analysis of Acoustic Emission Signals

In order to analyse the relation between the failure of the material and the acoustic emission activity, long term gear testing was conducted to identify the relation between the degradation procedure and AE signal characteristics.

Generally, during the fracture process, the acoustic activity of a material can be classified into three different phases depending on the development of the crack, which are: crack initiation, crack incubation, and crack propagation [18]. This principle has been validated in this work by means of fatigue test in gear surfaces. Different gears have been subjected to fatigue test while the AE activity (number of times the amplitude exceeds a threshold value), generated in

the gear's surface has been acquired. The results of this fatigue test are shown in Fig. 3. It can be clearly seen the evolution of the AE activity depending on the teeth degradation level. The resulting crack generated at the tensile face of the gear as a result of the test is shown in Fig. 4.

The obtained AE signals in each stage, present characteristic properties as can be seen in Fig. 5. It is known from the general AE theory that two different kinds of AE signals can be acquired: Continuous (Fig. 5 A) and discrete or Burst (Fig. 5 B) AE signals [12]. Discrete (or burst), AE signals are usual in electromechanical systems based on gears and contain more relevant information about the presence of failure in the system.

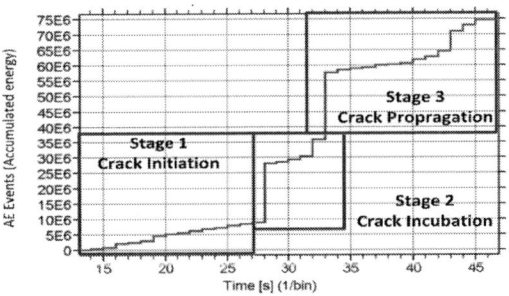

Fig. 3. Evolution of the AE signals activity during the fatigue test. In the graph three different stages could be easily identified. The time axis bin=0.1.

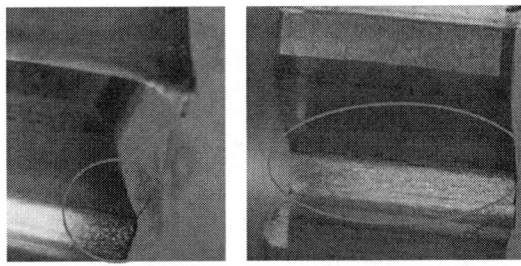

Fig. 4. Detail of the resulting crack generated within the gear under test during the experimental procedure.

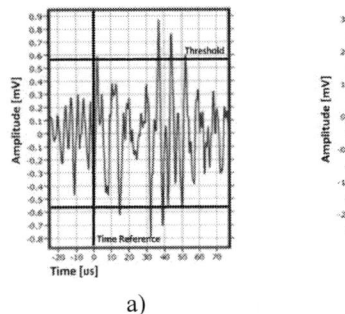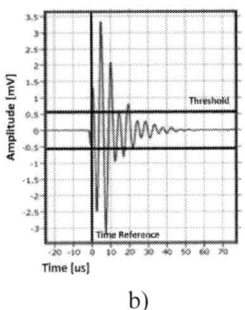

Fig. 5. Examples of AE signals obtained from the experimental test in different degradation stages. The vertical reference marks the same instant of time. The horizontal reference shows the same threshold value. a) Characteristic continuous AE signal obtained from stage 1. B) Characteristic discrete or burst AE signal obtained from stage 3.

The main conclusions extracted from this analysis are:

1) Fault presence implies a clear increase of the AE amplitude.
2) Density of "high energy" bursts increase during the fracture progress.
3) Spectral content of the AE signal increase with the presence of failure.

The developed diagnosis algorithm takes advantage of this fact to detect the health condition by calculating several statistical features; hence the apparition of a failure in the surface of the gear modifies the AE time domain signal shape and magnitudes. It is important to consider that all these features are not only related to the failure, they can be attached to the test conditions as it was reported by Y. He *et al.* [19]. The result of his works shows that the most negative influent test conditions are the speed, the defect size and the sensor location. In order to perform a proper diagnosis, the algorithm should consider this premise.

B. Definition of the diagnosis algorithm

This section covers the proposed diagnosis algorithm and the explanation of the different parts of it. The flow chart shown in Fig. 6 shows the different steps to perform a diagnosis. This algorithm proceeds as follows:

1. First of all, an AE signal (50 ms, 100k samples) is acquired by the ADC of the ISM.
2. Then, the envelope of the signal is calculated. This operation is necessary in order to improve the statistical information of the signal.
3. Feature calculation. Different statistical features from the envelope of the temporal AE signal are calculated.
4. A neural network, in base of the previously calculated features, gives a faulty condition if a certain pattern is found in these features. The probability of belonging to a class is also given.
5. A historic of 150 outputs is stored. The Machine Health Indicator (MHI) is calculated as the summation of the probability of having samples from the faulty class.
6. Finally, the diagnosis result is given to the user interface through the Zigbee module.

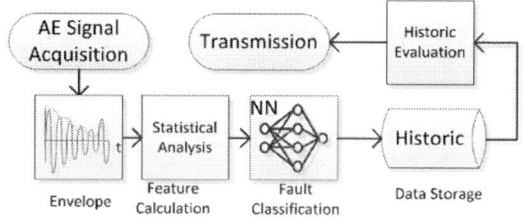

Fig. 6. Flow chart of the diagnosis algorithm. It starts in the signal acquisition by the MCU's ADC. Then several signal processing techniques are applied in order to extract suitable features for detecting the failure. These features are the input to a NN classifier which fills the historic with his outputs (class and probability). Then the historic is evaluated in order to send the MHI trough the communication module to the user interface.

Next, the different parts of the algorithm are detailed. The most important step at designing the algorithm is selecting the most suitable features for detecting the failure, this is a critical step because the performance of the NN highly depends on how much these features represents the current condition of the system. These features are calculated from the temporal form of the signal like have been the trend in the last decades [20]-[21]. Once the features are calculated for different experiments (fault severity, load and speed conditions, etc.) is time to observe the evolution or trends of them in function of the defects, and select those who are more related to the defect and less to the test conditions. Specific details of the fault detection algorithm are subjected to confidentiality terms of the project in which this sensor is being developed. The algorithm described in this paper is a proposal for the laboratory validation of the diagnosis capabilities of the ISM. The features selected for analyzing the signal in this paper are:

1) Variance

Information about how far the values of the signal are spread out from the mean. It highly increases with the presence of burst in the AE signal. The variance of a finite population (x_i) of size m is defined in (1).

$$\sigma^2 = \frac{1}{2} \cdot \sum_{i=1}^{m}(x_i - \bar{x})^2 \tag{1}$$

2) Shape Factor

It is the quotient between RMS and the mean. It gives information about the offset and symmetry of the signal. It present slightly differences between continuous and discrete AE signals. It is defined in (2).

$$SP = \frac{RMS}{\frac{1}{m}\sum_{i=1}^{m}|x_i|} \tag{2}$$

3) AE energy

The energy of an AE signal is defined as the root mean square (RMS) applied to the signal. This value has to be proportional to the area under the enveloping of the AE signal. This feature is related directly with the mechanical energy, the strain ratio and the deformation mechanism [12]. The RMS value of a finite population (x_i) of size m, is defined in (3).

$$RMS = \sqrt{\frac{\sum_{i=1}^{m}(x_i)^2}{m}} \tag{3}$$

Afterwards, the features selected in the previous section in addition with the speed of the machine, are the inputs to a neural network classifier. With these inputs, the NN estimates the condition of the gear by means of detecting a certain fault pattern within those features. The outputs of the

NN are the status of the machine R_{nn} (class: healthy or faulty), and the probability of belonging to that class P_{nn}.

According to the specifications of memory and energy consumption of the ISM, the AE signal acquisition cannot be made in real time. Therefore, a fixed time of 50 ms of AE signal is periodically acquired. As a result, with only one acquisition, a complete diagnosis of the actual machine status cannot be performed, since the acquired information is insufficient. In order to have significant data about the machine, the outputs of the NN are stored into a historic with 150 positions. Once the historic has been completely filled, the MHI is calculated as the summation of the probability of having samples from the faulty class, as shown in (4).

$$ MHI = \frac{\sum_{i=1}^{N} R_{nn}(i) \cdot P_{nn}(i)}{N} \cdot 100 \qquad (4) $$

Where R_{nn} is the estimated class of the acquired signal ($R_{nn}=1$ if faulty, $R_{nn}=0$ if healthy), P_{nn} is the probability of being in faulty condition, and N is the total length of the historic. The result is sent using the Zigbee module to the user interface. The configuration and the performance of this network will be explained in Section IV.

IV. EXPERIMENTAL RESULTS

In this section, the experimental results of applying the proposed diagnosis algorithm are shown. For this experiment a basic test bench has been designed. The aim of this test bench is to use the same gears who were degraded in the laboratory experiments. The setup of the test bench is shown in Fig. 7.

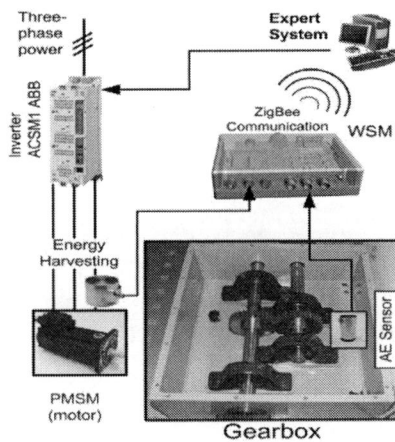

Fig. 7. Experimental setup for gear test.

It consists of a permanent magnet synchronous machine acting as a drive and a pair of gears in a simple 1:1 stage. The bearing supports for the shaft and gears are located inside a metallic box with a proper lubrication condition. A PMG FSH energy harvester from Perpetuum is located on the engine in order to transform the motor vibrations into electrical energy. A custom AE transducer designed by OPTEL is used for the test. This sensor is designed to cover a frequency band from 100 to 400 kHz. This sensor is located inside the gearbox, near to the gear under study, and connected to a BNC port of the ISM.

Three different speed conditions have been considered: the motor rotating at 150, 250 and 450 rpm. Three gears have been used to carry out the experiments, a healthy gear, a gear with a low severity failure (1 fracture), and a gear with high severity failure (8 fractures). The signals are acquired with the ADC of the ISM. Consequently, the sampling frequency is limited to 2MHz and a total amount of 100.000 samples are acquired, for avoiding memory limitations, the samples are stored in a 2MB external RAM. As a result, each acquisition corresponds to 50 ms of AE signal.

An example of the AE signals, obtained in experiments at different speeds and fault conditions, are shown in Fig. 8 and Fig. 9. The shape of both signals is analyzed in order to validate the previously mentioned statistical features used for the diagnosis algorithm.

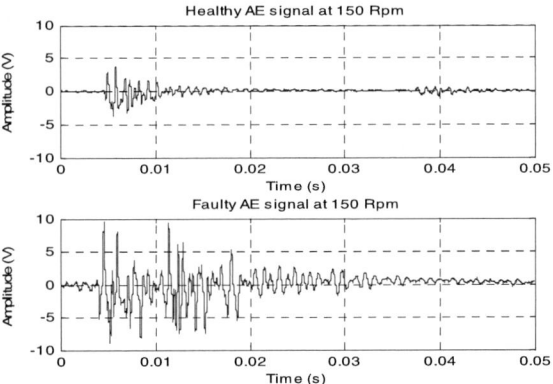

Fig. 8. Healthy (up) versus Faulty (down) AE Signal acquired from the gear based test bench at 150 Rpm.

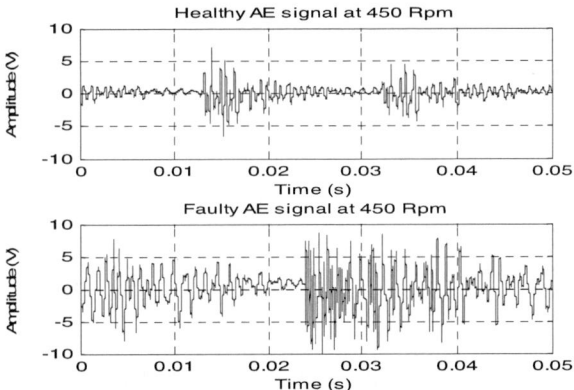

Fig. 9. Healthy (up) versus Faulty (down) AE Signal acquired from the gear based test bench at 450 Rpm.

Several differences in the signals in healthy versus faulty conditions are appreciable. In this sense, the time-domain statistical features considered for the algorithm show some recognizable trends as shown in Table 1. As can be seen in both figures, the RMS value can be used to evaluate the energy of the signal without taking in account any possible offset or peakedness. This happens because in every collision involving a faulty tooth, higher energy AE signals are generated, as was seen in Fig. 3 for the second and third degradation stages.

The Variance is also a suitable feature because it naturally increases due to the generation of more burst signals with the presence of the fracture. As much burst appear and higher are their amplitude higher the variance will be. The shape factor increases its value due to AE signal do not change its mean but it does in RMS, so apart from giving information about the symmetry of the signal it indicates the increasing of magnitude not consequence of an offset. As a conclusion from these acquired signals, the utilization of all the proposed features has been validated with the experimental results.

Table 1. Statistical features calculated from signals in healthy (H) and faulty (F) conditions under 150 and 450 Rpm (shown in Fig. 8 and Fig. 9).

Feature	150 H	150 F	450 H	450 F
RMS	0.559	1.742	1.162	2.675
Variance	0.024	3.036	1.347	7.153
Shape F	1.432	1.559	1.527	1.283

In Fig. 10 a three-dimensional feature space containing all the specified time-domain statistical features is shown. In this space, the x-axis is referred to the RMS value of the signal, the y-axis to the Shape Factor and the z axis to the Variance value. It should be noticed, that this space contains samples of all of the considered experiments, therefore it has samples from the faulty and healthy gears under three different speeds 150, 250 and 450 Rpm.

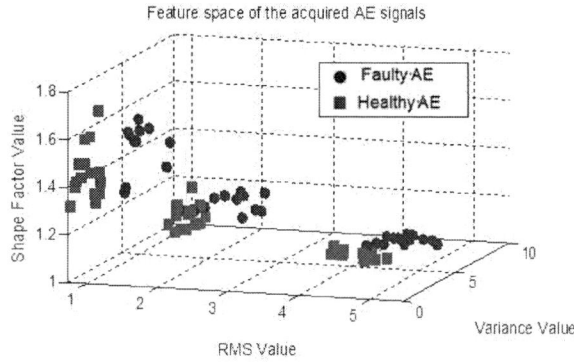

Fig. 10. Three-dimensional feature space containing samples of three different speeds 150, 250 and 450 Rpm, under faulty (●) and healthy (■) conditions.

Once the features have been calculated, the performance of the classification neural network can be tested. For this application, a multi-layer NN has been configured with one hidden layer which is composed of 5 hidden neurons. The neurons have been configured with a sigmoid activation function, which is recommended for classification problems involving multiple independent attributes. The network is trained by means of classical back propagation algorithm. Once the network has been defined and configured, it is trained with 300 inputs of the 3 features extracted from the experiments and tested with 150 other values (75 healthy and 75 faulty samples). The confusion matrix resulting from the test is shown in Table 2.

With the proposed NN configuration, a classification ratio of 76 % is achieved. Once the classification has been made, a historic of 150 positions is filled with the outputs of the NN. The historic saves the class and the probability of being to that class, and then calculates the MHI as shown in (4). Table 3 shows the MHI for different experiments performed under different gear fault severity and speed conditions.

Table 2. Confusion matrix extracted from the NN tested with 150 samples at different speed and failure conditions.

	Class Healthy	Class Faulty
Class Healthy	56	19
Class Faulty	17	58

Table 3. Machine health indicator calculated using the diagnosis algorithm under different speed and failure conditions.

Speed/ Condition	150 Rpm	250 Rpm	450 Rpm
Healthy	13.998	13.534	15.598
1 Tooth F	44.330	45.206	46.779
8 Tooth F	70.820	71.683	73.322

The results show that when the gear is in healthy condition, the MHI value is over 14 %, when a single fracture appears in the gear the value increases to 45%, and when the gear is under high failure conditions, the indicator increases up to 73 % clearly indicating the failure presence in the gear. As a result, the diagnosis capability of the proposed methodology has been correctly validated. This algorithm is capable of determining the actual condition of the machine, without being affected by the rotating speed of the experiment.

It should be considered, that the gears used for the experiments represent three discrete health conditions within the whole gear useful life. When working with real industrial applications, the gearboxes will suffer a progressive degradation. Under these degradation conditions, the MHI is expected to increase gradually (as shown in Fig. 11) , and it is precisely this increase which will early detect that the component is not working properly, and in order to prevent the failure, preventive maintenance actions should be applied regarding to the indicator value. Accordingly, several stages have been considered relating the MHI and the recommended maintenance actions that the user should performed. The different stages for the MHI, which are shown in Fig. 11, are:

1) Stage 1: Good condition

MHI value from 0 to 20 %. The machine is in Good Condition. No maintenance actions should be applied. A periodic inspection of the sensor's alarm is recommended.

2) Stage 2: Initial Degradation

MHI value from 21 to 50 %. The machine shows initial degradation signs. Common maintenance actions such as greasing should be applied. The MHI value should decrease.

3) Stage 3: Continuing Degradation

MHI value from 51 to 85 %. The machine is suffering a continuing degradation; a further analysis of the machine should be made. Consider replacing mechanical components.

4) Stage 4: Severe failure

MHI value from 86 to 100%. The machine has a severe failure. The mechanical components should be replaced immediately in order to avoid a critical failure.

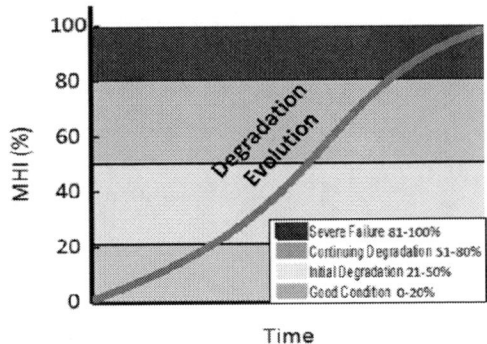

Fig. 11. Expected evolution of the machine health indicator (curve) among time. Different degradation stages regarding the MHI value.

It should be noticed that a classical approach, using threshold values for detecting the failure, could be also implemented, but the benefits of using a NN are higher. The NN introduces nonlinear relations between the different inputs, which improve the classification results. Additionally, it adds generalization ability, making the NN able to deal with situations not considered during the training step. Another important property is the fact that the NN could be easily enhanced with new variables and features, in order to improve the diagnosis results in some specific applications.

V. CONCLUSIONS

In this paper, an intelligent wireless sensor module, powered by an energy harvesting unit and based on acoustic emission analysis, has been proposed for gear fault diagnosis. The main architecture and components of this sensor has been explained in this work. Also the diagnosis capabilities of the module has been tested and validated under laboratory conditions.

The proposed sensor module integrates different key-technologies such as Zigbee transmission, energy harvesting, intelligent algorithms and acoustic emission analysis which work together to create an intelligent and autonomous sensor module. This device differs from the general offer from the market and it gives a new perspective to the diagnosis maintenance operations in large facilities.

The proposed diagnosis algorithm has been tested and validated in a real application under laboratory conditions. The classification ratio and with this the diagnosis performance achieved is over 76%. The algorithm itself requires a low computational cost, due to the simplicity of time-domain statistical features calculation, and the fact that the neural network would be trained off-line and only the weights would be introduced in the microprocessor.

As a further work, this sensor should be tested and validated in different industrial environments, in order to prove its diagnosis capabilities and improve its performance under real working conditions.

VI. ACKNOWLEDGMENT

This work was supported in part by the Research Executive Agency under the FP7-SME-2011 285848 Project.

VII. REFERENCES

[1] Wilkinson, M.R.; Spinato, F.; Tavner, P.J., "Condition Monitoring of Generators & Other Subassemblies in Wind Turbine Drive Trains," Diagnostics for Electric Machines, Power Electronics and Drives, 2007. SDEMPED 2007. IEEE International Symposium on , vol., no., pp.388,392, 6-8 Sept. 2007.

[2] Yanping Guo; Wenjun Yan; Zhejing Bao; , "Gear fault diagnosis of wind turbine based on discrete wavelet transform," Intelligent Control and Automation (WCICA), 2010 8th World Congress on , vol., no., pp.5804-5808, 7-9 July 2010.

[3] Bastianini, F., Sedigh, S., Pascale, G., and Perri, G. Cost-effective dynamic structural health monitoring with a compact and autonomous wireless sensor system. In Nondestructive Testing of Materials and Structures, pages 1065–1070. Springer. 2013.

[4] Berdinyazov, A.; Camci, F.; Sevkli, M.; Baskan, S., "Economic analysis of maintenance policies for a system," Diagnostics for Electric Machines, Power Electronics and Drives, 2009. SDEMPED 2009. IEEE International Symposium on , vol., no., pp.1,5, Aug. 31 20096-Sept. 3 2009

[5] Gungor, V.C.; Hancke, G.P.; Industrial Wireless Sensor Networks: Challenges, Design Principles, and Technical Approaches, IEEE Transactions on Industrial Electronics, Vol. 56, no. 10, pp. 4258 – 4265, Oct. 2009.

[6] Grosse, C. U. and Kruger, M. (2006). Wireless acoustic emission sensor networks for structural health monitoring in civil engineering. In Proc. European Conf. on Non-Destructive Testing (ECNDT), DGZfP BB-103-CD. Citeseer.

[7] Bohse, J. (2013). Acoustic emission. In Handbook of Technical Diagnostics, pages 137–160. Springer.

[8] De Silva, C. W. (2010). Vibration monitoring, testing, and instrumentation. CRC Press.

[9] Kral, C.; Habetler, T.G.; Harley, R.G.; Pirker, F.; Pascoli, G.; Oberguggenberger, H.; Fenz, C. -J M, "A comparison of rotor fault detection techniques with respect to the assessment of fault severity," Diagnostics for Electric Machines, Power Electronics and Drives, 2003. SDEMPED 2003. 4th IEEE International Symposium on , vol., no., pp.265,270, 24-26 Aug. 2003

[10] Wei Zhou; Habetler, T.G.; Harley, R.G., "Bearing Condition Monitoring Methods for Electric Machines: A General Review," Diagnostics for Electric Machines, Power Electronics and Drives, 2007. SDEMPED 2007. IEEE International Symposium on , vol., no., pp.3,6, 6-8 Sept. 2007

[11] Sidney Mindess, "Acoustic Emission Method", CRC Handbook on Non-destructive Testing of Concrete, CRC Press, 1991, p. 317.

[12] Christian U. Grosse, Masayasu Ohtsu. Acoustic Emission Testing. 2ª ed. Germany. Springer 2010. ISBN 978-3-540-69895-1.

[13] He, D.; Ruoyu Li; Bechhoefer, E., "Split torque type gearbox fault detection using acoustic emission and vibration sensors," Networking, Sensing and Control (ICNSC), 2010 International Conference on , vol., no., pp.62,66, 10-12 April 2010.

[14] Ruoyu Li; He, D., "Rotational Machine Health Monitoring and Fault Detection Using EMD-Based Acoustic Emission Feature Quantification," Instrumentation and Measurement, IEEE Transactions on , vol.61, no.4, pp.990,1001, April 2012

[15] Shen, G. T., Geng, R. S., and Liu, S. F., 2002, "Parameter Analysis of Acoustic Emission Signal," Chinese Journal of Non-destructive Testing, 24, pp. 72–77.

[16] Jianguo Cui; Xinqi Zheng; Ming Li; Zhonghai Li; Dong Liu, "Health Diagnosis for Aircraft Based on EMD and Neural Network," Measuring Technology and Mechatronics Automation, 2009. ICMTMA '09. International Conference on , vol.1, no., pp.677,680, 11-12 April 2009.

[17] Ghate, V.N.; Dudul, S.V., "Cascade Neural-Network-Based Fault Classifier for Three-Phase Induction Motor," *Industrial Electronics, IEEE Transactions on* , vol.58, no.5, pp.1555,1563, May 2011.

[18] Shi Z., Jarzynski J., Hurlebaus S., Jacobs LJ. Characterization of acoustic emission signals from fatigue fracture. Proceedings of the Institution of Mechanical Engineers 214-9 (2000) 1141-1149.

[19] Yongyong He, Xinming Zhang, and Michael I. Friswell, "Defect Diagnosis for Rolling Element Bearings Using Acoustic Emission" J. Vib. Acoust. 131, 061012 (2009).

[20] TANDON N. and MATA S., "Detection of defects in gears by acoustic emission measurements, Journal of acoustic emission" vol. 17, 1999.

[21] A. M. Al-Ghamd and D. Mba, "A comparative experimental study on the use of acoustic emission and vibration analysis for bearing defect identification and estimation of defect size," *Mechanical Systems and Signal Processing,* vol. 20, pp. 1537-1571, 10, 2006.

VIII. BIOGRAPHIES

Daniel Zurita Millan (S'13) received the B.S. and M.S. degrees in electronics engineering from the Universitat Politècnica de Catalunya (UPC), Barcelona, Spain in 2011 and 2013 respectively. Currently, he is a PhD student in the UPC, in the Motion and Industrial Control Group (MCIA) in Terrassa, Barcelona, Spain. His research interests include fault diagnosis and prognosis in electric machines and mechanical components, fault detection algorithms, machine learning and signal processing methods.

Miguel Delgado Prieto (S'08, M'12) received the M.S. degree in Electronics Engineering and the Ph.D. degree in Electronics Engineering from the Universitat Politècnica de Catalunya (UPC), Barcelona, Spain in 2007 and 2012 respectively. From 2004 to 2008 he was a Teaching Assistant in the Electronic Engineering Department of the UPC. In 2008 he joined the Motion and Industrial Control Group (MCIA), where he is currently a Post-Doc Researcher. His research interests include fault detection algorithms, machine learning, signal processing methods and embedded systems.

Juan Antonio Ortega Redondo received the M.S. Telecommunication Engineer and Ph.D. degrees in Electronics from the Technical University of Catalonia (UPC) in 1994 and 1997, respectively. In 1994, he joined the UPC Department of Electronic Engineering as a full time Associate Lecturer. In 1998, he obtained a tenured position as an Associate Professor. Since 1994 he has taught courses of microprocessors and signal processing. From 1994 to 2001 he was with Sensor Systems Group working in the areas of smart sensors, embedded systems, and signal conditioning, acquisition and processing. Since 2001 he belongs to the Motion Control and Industrial Applications research group working in the area of motor current signature analysis. His current research activities include: motor diagnosis, signal acquisition, smart sensors, embedded systems and remote labs. In the last years, he has participated in several Spanish and European funded research projects about these items.

Luis Romeral Martinez (M'98) received the M.S. degree in electrical engineering and the Ph.D. degree from the Universitat Politècnica de Catalunya (UPC), Barcelona, Spain, in 1985 and 1995, respectively. In 1988, he joined the Department of Electronic Engineering, UPC, where he is currently an Associate Professor and the Director of the Motion and Industrial Control Group, whose major research activities concern induction and permanent magnet motor drives, enhanced efficiency drives, fault detection and diagnosis of electrical motor drives, and improvement of educational tools. He has developed and taught postgraduate courses on programmable logic controllers, electrical drives and motion control, and sensors and actuators. Dr. Romeral is a member of the European Power Electronics and Drives Association and the International Federation of Automatic Control.

978-1-4799-0024-4/13 $31.00 © 2013 IEEE

Evaluation of the Applicability of FRA for Inter-Turn Fault Detection in Stator Windings

F. R. Blánquez, Carlos. A. Platero, *Member*, IEEE, E. Rebollo, F. Blázquez, *Member*, IEEE,.

Abstract -- In this paper, the applicability of the FRA technique is discussed as a method for detecting inter-turn faults in stator windings. Firstly, this method is tested in an individual medium-voltage stator coil with satisfactory results. Secondly, the tests are extended to a medium-voltage induction motor stator winding, in which inter-turn faults are performed in every coil end of one phase. Results of the frequency response in case of inter-turn faults are evaluated in both cases for different fault resistance values. The experimental setup is also described for each experiment. The results of the application of this technique to the detection of inter-turn faults justify further research in optimizing this technique for preventive maintenance.

Index Terms—Fault detection, inter-turn fault, stator winding.

I. INTRODUCTION

RELIABILITY of the power system is a primary concern nowadays. This fact involves not only the distribution system, but also power generation plants. AC generators and power transformers require precise protective and diagnostic techniques in order to ensure its correct operating condition and minimize the damage in case of internal defect.

Frequency Response Analysis (FRA) is a well-known technique for the diagnosis of power transformer windings. Since the first proposals of FRA [1], many improvements and new applications have been developed, resulting in new standards for the application of the FRA method, in both International Council on Large Electric Systems (CIGRE) and IEEE [2].

Regarding power transformers, FRA technique allows detecting small strains which may appear in the windings due to the effect of short-circuits or shocks during transportation [3], [4].

FRA technique is based on the evaluation of the equivalent impedance of the windings in the frequency domain. Since a winding can be modeled as an equivalent circuit with a complex network of capacitances, resistances and inductances, the frequency response is unique, and it can be used as a fingerprint of the winding under test. Thus, any physical alteration or electrical modification of the winding results in a variation of the frequency response, which is detected by comparing this test to the reference frequency response of the winding in healthy condition.

Although, FRA technique can be studied by the equivalent circuit, the interpretation of the frequency responses is sometimes complex due to the influence of exogenous situations such as the magnetization level of the core [5], or the connections of the measuring equipment [6].

Some authors suggest the applicability of FRA technique to the detection of insulation failures in its early stage [7], which extend the use of this method to predictive maintenance.

However, despite the great development of the FRA technique in power transformer, it is just being under research for its application in rotary machine [8], and it is practically unused currently in industry. This is because the high-frequency equivalent circuit of the windings in rotating machines is more complex than that of transformers: the stator winding is placed into slots, and there is a rotor winding.

Regarding the protection of generating units, several methods have been proposed in order to detect ground faults in the rotor [9], and in the stator winding. Advanced diagnostic methods based on the analysis of high frequency signals have been also proposed [10].

FRA technique has been presented as promising technique for detecting different faults in stator windings of both induction motors [11] and synchronous generator. In the latter case, some authors solved the need of removing the rotor winding for testing stator windings [12], which makes FRA technique very interesting for diagnoses of generating units.

This paper aims to continue laying the background for the future implementation of FRA technique as precise method for detecting inter-turn faults in rotary machines.

This paper is structured as follows: Section II presents the operating principle of FRA technique. Results of the applicability of FRA technique to the inter-turn detection in medium-voltage stator coils, and in an induction motor stator winding are discussed in Section III and Section IV. This paper finishes with the main conclusions.

II. FRA OPERATING PRINCIPLE

Any phase of the stator winding of an electrical machine can be represented as an equivalent circuit, composed by resistances, inductances and capacitors, in series or parallel connection. Concretely, a stator winding coil can be studied as a distributed winding circuit composed by n "pi" equivalents in series connection, as shown in Fig. 1, where $L \cdot n$ is the leakage inductance, Cs/n the series capacitance, $Cg \cdot n$ the shunt capacitance and Z is the impedance of the voltage measuring equipment (typically 50 Ω for FRA

F. R. Blánquez, C. A. Platero, E. Rebollo and F. Blázquez are with the Department of Electrical Engineering, ETSII, Polytechnic University, Madrid, 28006 Spain (e-mail: fr.blanquez@gmail.com).

978-1-4799-0024-4/13 $31.00 © 2013 IEEE

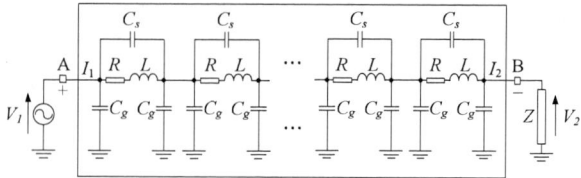

Fig. 1. Equivalent circuit of a stator winding coil.

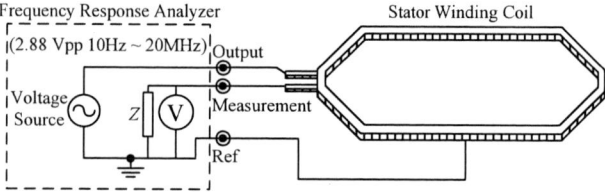

Fig. 2. Experimental setup for inter-turn fault detection test using FRA equipment.

equipment).

The technique is based on the application of an input sinusoidal voltage signal (V_1) of variable frequency to any terminal of the winding, and the measurement of the voltage in the free terminal (V_2). The gain of the frequency response is obtained using (1). This gain is commonly analyzed in dB units (2). The phase diagram is obtained applying (3).

$$\vec{H}(j\omega) = \frac{\vec{V_2}}{\vec{V_1}} \qquad (1)$$

$$H_{dB} = 20 \cdot \log_{10}\left|\vec{H}(j\omega)\right| \qquad (2)$$

$$\Phi = \frac{180}{\pi} \cdot \arg\left(\vec{H}(j\omega)\right) \qquad (3)$$

The FRA equipment used in this work is an Omicron FRAnalizer, which technical data is shown in Table I. The Omicron equipment generates a sinusoidal 2.88-V_{pp} signal, whose frequency grows from 10 Hz to 20 MHz. This input signal (V_1) is applied to any of the terminals of the winding coil, and the output voltage (V_2) signal is measured by the same equipment at the other terminal (Fig. 2). The voltage reference, which is common for V_1 and V_2, is connected to the grounding included in the external cover of the coil. A computer, connected to the FRA equipment, registers the frequency response.

The frequency response of a healthy winding is unique, so it can be used as a fingerprint of this winding. In this way, by comparing to this reference test, any fault or slight displacement in the winding coil can be detected.

The following research results provide conclusions in the using of FRA technique in the detection of inter-turn faults in medium-voltage stator coils and stator windings.

TABLE I
CHARACTERISTICS OF FRA EQUIPMENT USED IN THE EXPERIMENT

Frequency range	10Hz-20MHz
FRA Method	Sweep frequency
Output impedance	50 Ω
Input Impedance	50 Ω
Accuracy (down to -80 dB)	< 0.1 dB
Accuracy (down to -100 dB)	< 0.3 dB

TABLE II
CHARACTERISTICS OF THE MEDIUM-VOLTAGE STATOR COIL USED IN THE TEST

Rated power	560 kW
Rated voltage (± 5%)	6000 V
Frequency	50 Hz
Rated current	62 A
Pole pairs	2
Rated speed	2983 rpm
Rated Power Factor	0.91(FL)

III. RESULTS OF INTER-TURN FAULT DETECTION IN MEDIUM-VOLTAGE STATOR COILS

A. Experimental Setup for Inter-Turn Fault Detection Test

As described, the frequency response of the healthy coil is used as fingerprint of this concrete coil. By comparing to this reference test, any change in the frequency response is detected. The causes of these changes in the frequency response may be very diverse.

First, the reference test has to be correctly obtained, since any conclusions about the health of the coil are performed through the rigorous comparison of the frequency response at any time to this reference test. Since the connection and position of the cables, especially the grounding cable may influence the FRA results, this reference frequency response has to be confirmed by obtaining the identical response on several tests at healthy condition.

The characteristics of the stator coil under test are shown in Table II. For this medium-voltage stator coil, the reference frequency response is shown in Fig. 3. In electrical machine diagnosis, showing the phase diagram with the magnitude (dB) response is a typical practice, as some information can be obtained from the comparison of both diagrams. In the aforementioned figure, the absolute resonance is clearly identified in both diagrams. The db magnitude and the frequency of this resonance depend on the electrical parameter of the equivalent circuit [4]. This is a very characteristic point, as any internal change which affects to the electrical parameters of the coil, will modify this resonance magnitude or frequency.

For testing inter-turn faults, on the coil under test (Fig. 4 (a)), the insulation cover was removed (Fig. 4 (c)) in a zone near to the medium point (Fig. 4 (b)). In this way, rigid inter-turn faults, or inter-turn defects with different value of fault resistance was performed (Fig. 4 (d)). This coil has eleven turns, so the faults were tested at the ten different positions.

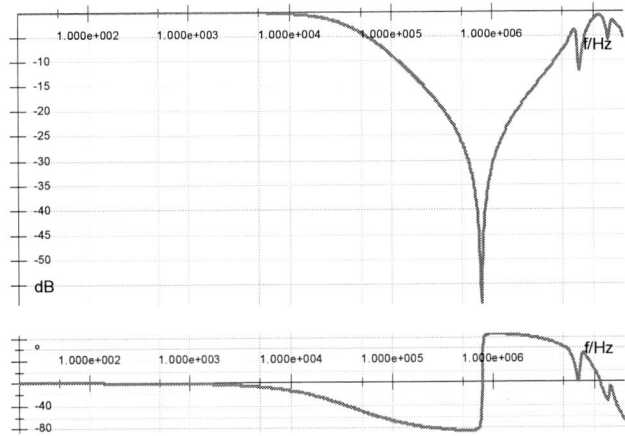

Fig. 3. Reference frequency response of the medium-voltage stator coil under test.

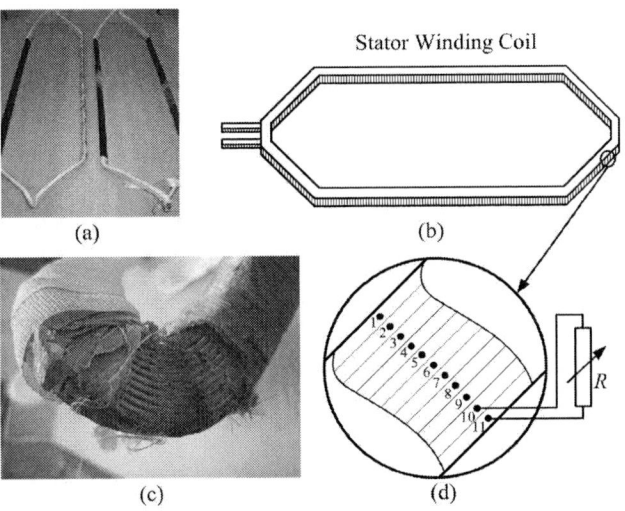

Fig. 4. (a) Medium-voltage stator winding coil under test. (b) Simplified scheme of the stator coil, whose insulation has been removed in the indicated point (d). (c) Picture of the point of insulation removal.

B. Results of zero resistance inter-turn fault tests

As described inter-turn faults were performed firstly with no fault resistance. In Fig. 5, the frequency responses of the coil in case of inter-turn fault are shown, where "INT_1_2", for instance, indicates that the inter-turn defect is test between the fault terminals 1 and 2 (Fig. 4 (d)). As observed, every fault implies a reduction of the magnitude of the main resonance. As the location of the defect moves from terminals 1 and 2, to terminals 5 and 6, the magnitude of this main resonance is also reduced. This effect is corresponded by the displacement of the location of the resonance in the phase diagram, as indicated in the same figure. In this diagram, as the fault is displaced to higher fault terminals, the resonance is displaced to the right, as indicated by the arrow.

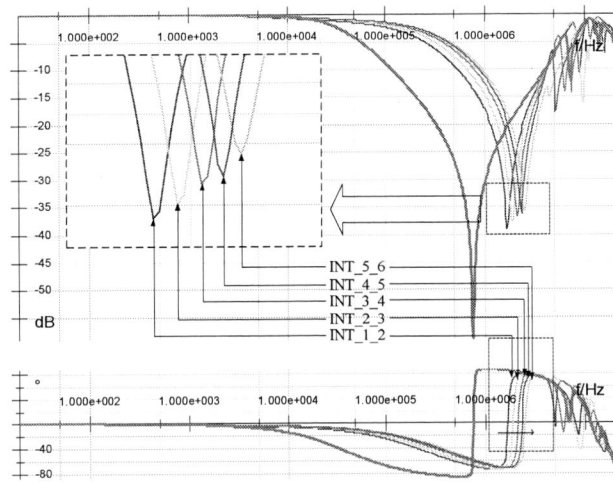

Fig. 5 FRA test for rigid inter-turn faults at different fault terminals (1 to 6).

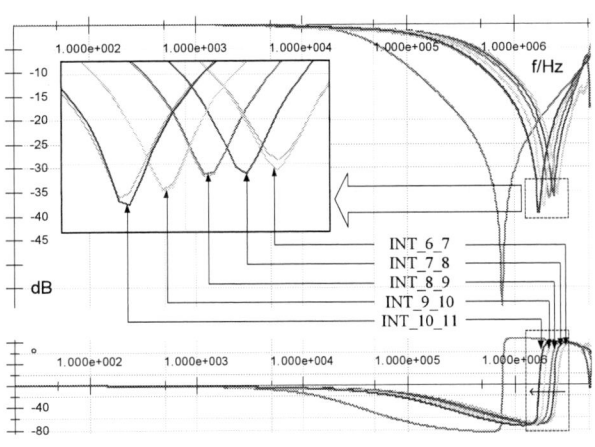

Fig. 6. FRA test for rigid inter-turn faults at different fault terminals (6 to 11).

These first results show that the inter-turn fault detection is achieved by de detection of a reduction of the main resonance magnitude. However, the conclusions about the fault location are discussed with the results shown in Fig. 6. In this figure, the results of rigid inter-turn faults at fault terminals 6 to 11 are shown. As observed, the defect also reduces the magnitude of the main resonance, but this point is displaced to the left as the fault location increases from terminal 6 to 11. This phenomenon is observed in the phase diagram, where the arrow indicates the movement of the main resonance. This effect on the frequency response could be justified by the symmetry of the n-stage ladder network equivalent impedance, as the inter-turn fault is moving from fault terminal 1 to 11. Although the fault can be clearly detected, the aforementioned phenomenon makes the fault location very complex, since the frequency response "INT_2_3" (fault at terminal 2 and 3) and "INT_9_10" are very similar.

978-1-4799-0024-4/13 $31.00 © 2013 IEEE

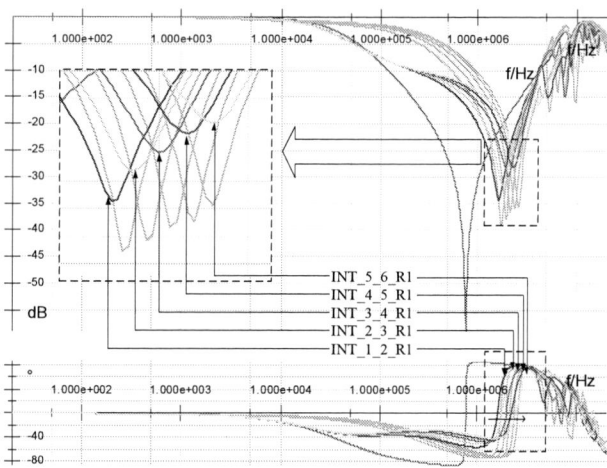

Fig. 7 FRA test for inter-turn faults with a fault resistance of 1 Ω at different fault terminals (1 to 6)

C. Effect of the fault resistance on the inter-turn faults

The influence of the value of the fault resistance is evaluated in Fig. 7. As observed, the inter-turn fault resistance affects to the magnitude of the main resonance and the frequency in which it happen.

Some general conclusions are pointed out: As expected, this type of faults is clearly detected in the frequency response. The influence of the fault resistance is also observable in the response. Moreover, inter-turn faults are detected in the high-frequency range of the response, while stator ground-faults are detected in the low-frequency range of the response [12].

The results in testing inter-turn fault detection in individual stator coils make this technique to be attractive for quality process in the manufacturing process of stator coils.

IV. RESULTS OF INTER-TURN FAULT DETECTION IN A MEDIUM-VOLTAGE INDUCTION MOTOR STATOR

Due to the clear results in the inter-turn fault detection at medium-voltage stator coil, an stator of a 520 kW induction motor (Fig. 8 (a)) was tested in order to evaluate the applicability of the FRA technique to detect internal inter-turn faults in stator windings. The characteristics of stator of the induction motor tested are summarized in Table III.

A. Experimental Setup

In this stator, each phase was composed by 16 coils (Fig. 9). In phase W of the stator winding, the external cover and insulation was removed at each coil end (Fig. 8 (b)), in order to test inter-turn faults. At each coil, up to three turns were available to perform inter-turn faults. The same FRA equipment was used for testing the described stator winding, where the reference was connected to the induction motor case in every case.

Tests were performed connecting the output signal to the

terminal W1, and the measurement cable was connected to

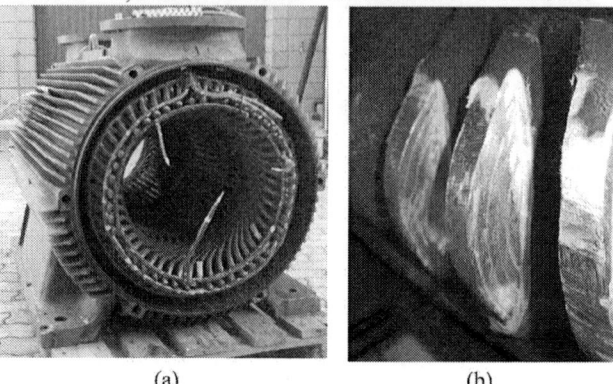

(a)   (b)

Fig. 8 Induction motor stator under FRA test (a). Cover and external insulation removal for fault testing (b).

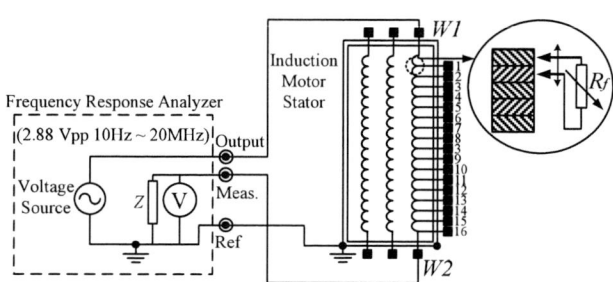

Fig. 9 Experimental setup for inter-turn fault test at the induction motor stator.

TABLE III
CHARACTERISTICS OF INDUCTION MOTOR STATOR USED IN THE TEST

Rated power	520 kW
Rated voltage (± 5%)	6300 V
Frequency	50 Hz
Pole pairs	1
Rated current	57 A
Rated speed	2985 rpm
Rated Power Factor	0.87
Number of slots	48
Coils per phase	16

the other terminal W2. The star point has been disconnected in this case in order to use the only healthy phase of this stator. If the star point is not accessible the test could be performed between two phases.

First the reference test was obtained (Fig. 10), and confirmed by the repetition of the FRA test with identical results, and also by inverting the connection of the phase terminals. In this latter case, the output signal was connected to the terminal W2, and the measurement cable was connected to the terminal W1. This is a typical practice for ensuring the veracity of the reference test.

As observed, the reference test of a stator phase winging "Ref_SW" is different to the individual coil reference frequency response "Ref". At least, three main resonances are now detected in this reference test, although the main conclusions are obtained by the absolute minimum of the response, as observed in the following figures.

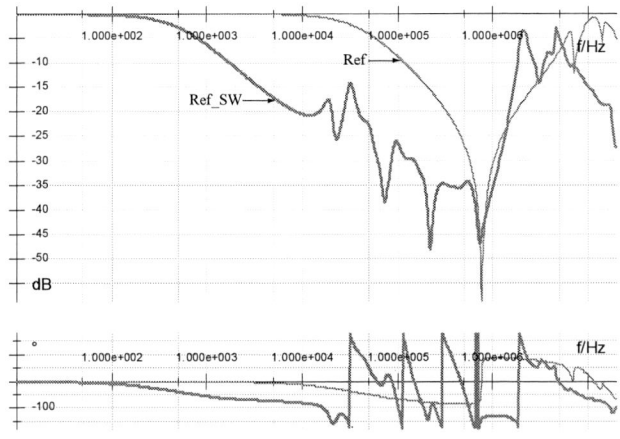

Fig. 10 Reference frequency response of the stator phase winding, "Ref_SW", compared to the reference test of the individual coil, "Ref".

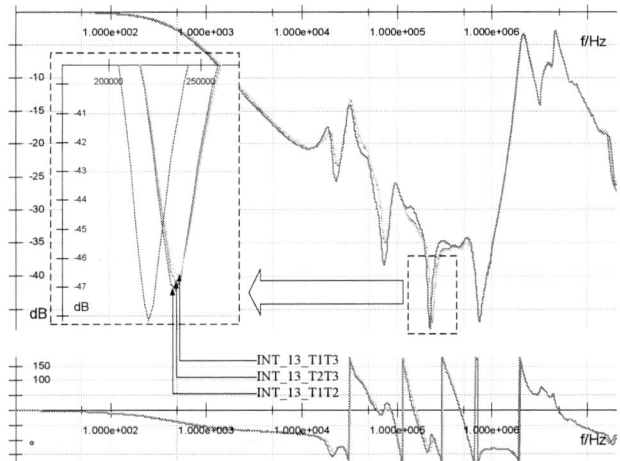

Fig. 11 FRA test of induction motor stator for inter-turn fault in different turns of a fixed fault terminal.

B. Result of FRA test for different inter-turn fault in the same coil.

First, the frequency response was obtained for inter-turn fault at the first three turns at each coil. Although, the changes in the frequency response are very slight, the differences between an inter-turn fault which involves the first turn and the second turn, and an inter-turn fault which involves the second turn and the third turn of the coil, are clearly detected. In Fig. 11, the FRA test of the phase W of the induction motor stator for inter-turn faults in coil 13 involving different turns, are shown, where "INT_13_T1T2", for instance, is the frequency response when the inter-turn fault is performed between turn 1 and 2 at fault terminal 13. The frequency responses for cases represented are very similar; however, slight changes can be detected in the main resonance peak (high frequency zone), as observed in the zoom of the same figure.

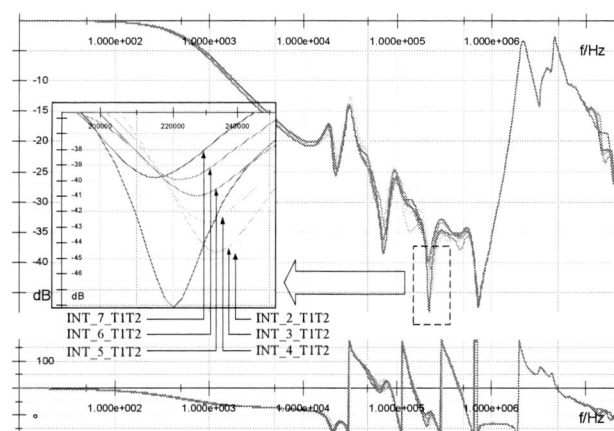

Fig. 12 FRA test of the induction motor stator for inter-turn fault in the first and second turns of every fault terminal with no fault resistance.

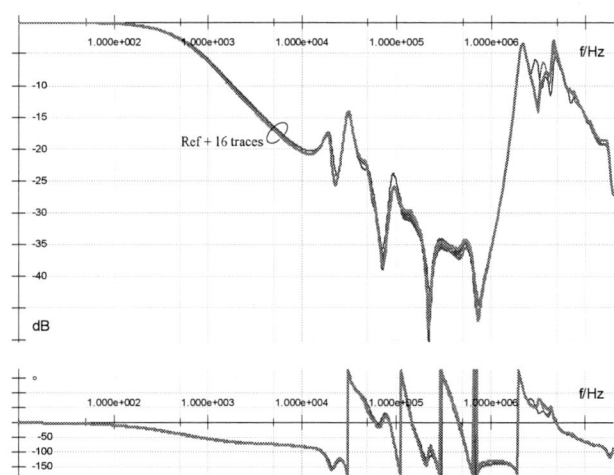

Fig. 13 test of the induction motor stator for inter-turn fault in the first and second turns of every fault terminal with a fault resistance of 10 Ω.

C. Result of FRA test for inter-turn fault at different coils

In Fig. 12, the frequency response of the test of inter-turn faults performed between the first and second turn at some coils of the phase W winding are shown. Again, the frequency responses are very similar, and no large differences are observed; However, in the main resonance peaks (high frequency zone) slight changes can be clearly detected, due to the different inter-turn fault location, as observed in the zoom of the same figure. In this way, the main resonance is affected not only by the existence of an inter-turn fault, but also by the fault resistance.

In Fig. 13, the frequency response of the test of inter-turn faults performed between the first and second turn at every coil with a fault resistance of 10 Ω is shown. Another important conclusion is obtained. As observed, as the fault resistance increases the changes in the response are slighter, and the inter-turn fault detection is not so clear. According to this, inter-turn fault detection is achievable for fault resistance value close to zero, which represents a clear inter-turn insulation failure.

V. CONCLUSIONS

In this paper, the applicability of the frequency response analysis for the inter-turn fault detection has been discussed, firstly for individual medium-voltage stator coils, and secondly for rotary machine stator windings.

Inter-turn fault are clearly detected in the frequency response of individual stator coils since the magnitude and phase of main resonance is affected. The fault resistance value also affects to the main resonance peak, since the influence of the fault resistance is detected in the frequency response. This main resonance is also affected by the fault location.

Tests in stator winding show, firstly that inter-turn faults present slight changes in the frequency response. However, this kind of fault may be detect due to they imply clear changes in the main resonance of the frequency response. The fault resistance value is a very dominant parameter, since as its value increases the changes in the frequency response are slighter.

The results obtained and shown in this work present another important conclusion; Inter-turn faults imply changes in the high frequency zone of frequency response. This fact defined the frequency range of the response where this kind of defects can be observed, considering that ground faults are generally detected in the low frequency zone of the frequency response.

Due to the presented results, further research is justified, focused not only on a defined criterion for inter-turn detection, but also on the future applicability of FRA in inter-turn fault location.

VI. ACKNOWLEDGMENT

The authors wish to acknowledge the technical support Mr. Gonzalez, Mr Hedgecock, Mr. Batlle of Omicron and Mr. Fernandez of ABB.

VII. REFERENCES

[1] P.T.M. Vaessen, E. Hanique, "A new frequency response analysis method for power transformers," *IEEE Trans. Power Delivery*, vol. 7, no. 1, pp. 384 – 391, 1992.

[2] M. Kraetge1, P. Krüger1, Fong. "Frequency Response Analysis – Status of the worldwide standardization activities", *International Conference on Condition Monitoring and Diagnosis*, April 21-24, 2008 Beijing, China

[3] M. Wang, A.J. Vandermaar, K.D. Srivastava, "Improved detection of power transformer winding movement by extending the FRA high frequency range," *IEEE Trans. Power Delivery*, vol. 20, no. 3, pp. 1930–1938, 2005.

[4] W. Zhongdong, Jie Li, D.M. Sofian, "Interpretation of Transformer FRA Responses - Part I: Influence of Winding Structure," *IEEE Trans. Power Delivery*, vol. 24, no. 2, pp. 703–710, 2009.

[5] N. Abeywickrama, Y.V. Serdyuk, S.M. Gubanski, "Effect of Core Magnetization on Frequency Response Analysis (FRA) of Power Transformers," *IEEE Trans. Power Delivery*, vol. 23, no. 3, pp. 1432 – 1438, 2008.

[6] J.A.S.B Jayasinghe, Z.D. Wang, P.N. Jarman, A.W. Darwin, "Winding movement in power transformers: a comparison of FRA measurement connection methods", *IEEE Trans. Dielectrics and Electrical Insulation*, vol. 13, no. 6, pp. 1342 – 1349, 2006.

[7] K.G.N.B. Abeywickrama, Y.V. Serdyuk, S.M. Gubanski, "Exploring possibilities for characterization of power transformer insulation by frequency response analysis (FRA)," *IEEE Trans. Power Delivery*, vol. 21, no. 3, pp. 1375 -1382, 2006.

[8] Lamarre, L.; Picher, P.; , "Impedance Characterization of Hydro Generator Stator Windings and Preliminary Results of FRA Analysis," Electrical Insulation, 2008. ISEI 2008. Conference Record of the 2008 IEEE International Symposium on , vol., no., pp.227-230, 9-12 June 2008

[9] C.A. Platero, F. Blázquez, P. Frías and M. Redondo, "A Novel Rotor Ground Fault Detection Technique for Synchronous Machines with Static Excitation", *IEEE Trans. Energy Convers.*, to be published.

[10] F. Perisse, P. Werynski, D. Roger, "A New Method for AC Machine Turn Insulation Diagnostic Based on High Frequency Resonances", *IEEE Trans. Dielectrics and Electrical Insulation*, vol. 14, no. 5, pp. 1308 – 1315, 2007.

[11] M. Florkowski, J. Furgal, "Detection of winding faults in electrical machines using the frequency response analysis method", *Measurement Science and Technology, IOP Publishing*, vol.15, 2004.

[12] Platero, C.A.; Blázquez, F.; Frías, P.; Ramírez, D.; , "Influence of Rotor Position in FRA Response for Detection of Insulation Failures in Salient-Pole Synchronous Machines," Energy Conversion, *IEEE Transactions on* , vol.26, no.2, pp.671-676, June 2011

VIII. BIOGRAPHIES

Francisco R. Blánquez was born in Spain in 1986. He received the Dipl. degree in electrical engineering from the Polytechnic University of Madrid, Spain, in 2010. He's currently pursuing the Ph.D degree in the Division of Electrical Engineering, Polytechnic University of Madrid. His research interests include protective relaying for power systems, electrical machine diagnosis and energy efficiency.

Carlos A. Platero (M'10) was born in Madrid, Spain, in 1972. He obtained the Dipl. degree and Ph.D. degree in electrical engineering from the Technical University of Madrid, Spain, in 1996 and 2007 respectively. Since 1996 to 2008 he has worked in ABB Generación S.A., Alstom Power S.A. and ENDESA Generación SA, always involved in design and commissioning of power plants. In 2002 he began teaching at the Electrical Engineering Department of the Technical University of Madrid, and joined an energy research group. Since 2008 he became full-time Associate Professor.

Emilio Rebollo was born in Spain in 1986. He received the Dipl. degree in electrical engineering from the Polytechnic University of Madrid Spain in 2010. He's a Ph.D student of the Electrical Engineering Department, Universidad Politécnica de Madrid, Spain. His research interests are Electrical machines design and Regulation of Power Plants.

F. Blázquez (M'07) was born in Toledo, Spain in 1972. He received the Dipl. degree in industrial engineering and the Ph.D. degree in electrical engineering from Polytechnic University of Madrid, in 1997 and 2004, respectively. Since 1999 he has been professor at the Electrical Engineering Department of the Polytechnic University of Madrid. His current research interests include electrical machine design and wind power generation.

Detection of Stator Slot Magnetic Wedge Failures for Induction Motors without Disassembly

Kun Wang Lee, Jongman Hong,
Doosoo Hyun, and Sang Bin Lee[*]

Department of Electrical Engineering,
Korea University, Seoul, Korea

Ernesto Wiedenbrug, and Mike Teska

Condition Monitoring Center,
SKF Corporation,
Fort Collins, CO U.S.A.

Chaewoong Lim

Hansung Electric Industrial Co.,
Dangjin, Chungnam, Korea

Abstract—The recent trend in large ac machines is to employ magnetic stator slot wedges for improving the motor efficiency, power factor, and power density. The mechanical strength of magnetic wedges is weak compared to the epoxy glass wedges, and many cases of loose and missing wedges have recently been increasingly reported. Magnetic wedge failure deteriorates the performance and reliability of the motor, but there is no method available for testing the wedge quality other than visual inspection after rotor removal. Monitoring of overall wedge condition without motor disassembly can help reduce the cost of maintenance and risk of degradation in performance. In this paper, a new off-line standstill test method for detecting magnetic wedge problems for ac machines without motor disassembly is proposed. An experimental study on 380 V, 5.5 kW and 6.6 kV, 3.4 MW motors with magnetic wedges is performed to verify the effectiveness of the new test method. It is shown that the new method can provide reliable monitoring of magnetic wedge problems over time, independent of other faults or motor design.

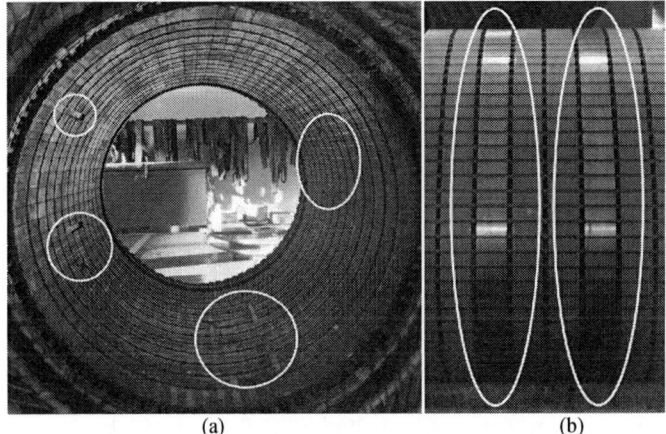

(a)   (b)

Fig. 1. Example of stator slot magnetic wedge damage on a 6.6 kV, 7.7 MW, 4 P induction motor; (a) magnetic wedges randomly missing or loose; (b) wear band mark on rotor surface due to magnetic wedge protrusion into airgap

I. INTRODUCTION

Magnetic stator slot wedges have become prevalent in large ac machines for improving the efficiency and power factor despite the higher cost compared to the epoxy glass wedges [1]-[5]. The overall trend is to use magnetic wedges for high output motors above 500 kW and for slow speed motors with 4 or more poles. They are typically used for large AC machines with form wound stators and open stator slot design, but have also been considered for performance improvement of small induction and permanent magnet synchronous motors [5]-[7].

Most magnetic slot wedges are made from iron powder (70%~75%), glass fabric, and epoxy resin binders, and therefore, have larger relative permeability ($\mu_r \leq 10$) compared to epoxy wedges. The increased μ_r in the slot section makes the flux concentrated on the tooth surface more uniformly spread over the tooth and wedge surface, and the behavior similar to that of semi-closed slot design [1]-[3]. This reduces the fluctuation of flux on the rotor core surface and decreases the effective airgap. The smoothened airgap flux distribution results in significant reduction in the surface core losses, and also helps reduce the acoustic noise and improve the torque characteristics. The decrease in effective airgap and closing of the slot results in increased magnetizing and leakage inductances, and decrease in the magnetizing current. The positive effect of reduced airgap is improvement in power factor and reduction in copper losses. Reduction of core and copper losses contributes to improved machine efficiency, reduced operating temperature, and increased power density of

the motor. With magnetic wedges, there can be cost and energy savings due to improved efficiency and reduction in material content and frame size, although the cost is increased. The price to pay for the advantages gained with magnetic wedges is the reduction in starting and pullout torque, increased maintenance requirements, and reliability risks [1]-[5].

Since the iron content in the magnetic wedge makes the mechanical strength weak, they are susceptible to failure. An example of a stator with loose or missing magnetic wedge segments that was recently found on a 6.6 kV, 7.7 MW boiler feed pump induction motor at a power plant is shown in Fig. 1. Many cases of wedge failures in medium-high voltage motors above 3.3 kV where up to 50% of the wedges were lost within 3 years of service are reported in [8]-[12]. For the case presented in [9], the decision was made to replace the magnetic wedges with non-magnetic epoxy wedges for 6.6 kV, 2.85 MW induction motor units with reoccurring wedge problems, since the cost of lost production and repair outweighed the efficiency and performance benefits. The examples presented in this section and in [8]-[12] show that missing magnetic wedge problems are very common and can have serious consequences.

II. DETECTION OF MAGNETIC WEDGE FAILURE

The main degradation mechanism behind magnetic wedge failure is wear, breakage, and/or disintegration due to mechanical stress [8]-[12]. In addition to the stress caused by vibration of the rotor and stator winding, electromagnetic force is induced in the wedge itself as it is an active component in the magnetic circuit. Therefore, a "wet" process such as epoxy gluing or vacuum pressure impregnation (VPI) is used in the wedging process to prevent wedge vibration. However, repeated electro-mechanical and thermal stress on the magnetic wedge and bonding epoxy makes the wedges loose [12]. Once the wedges are free to move and begin to vibrate, the wedge and core material are eroded and the wedge eventually falls out, breaks and disintegrates into small pieces. Evidence of magnetic wedge protrusion into the airgap and disintegration can be clearly observed in the wear band of the rotor surface in Fig. 1(b). It is shown in [9], [12] that wedge segments fall out randomly along the circumference of the bore, and are more likely to fail in the center of the slot in the axial direction. This is because wedge wear occurs during the wedge insertion process due to non-uniform lamination surface, and the wedge segments in the center travel the most. It has been observed that magnetic wedge failures are more likely to occur for the high-stress applications with frequent starts, long startup time, thermal overload, and large load oscillations [8]-[12].

Loose or missing magnetic wedges usually do not cause serious motor failure leading to a forced outage, since the wedges disintegrate into small pieces. However, inspection of the magnetic wedge tightness is a standard procedure of the stator inspection in most motor repair facilities, and partial or full re-wedging is performed if loose wedges are present since they can fall out and degrade motor performance. Missing stator slot wedges can degrade the efficiency and cause localized temperature rise in the slot teeth portion where the wedges are lost. If a considerable portion of magnetic wedges is missing in the slot, this can result in stator winding movement, especially during the startup transient [12]. Localized heating and winding movement are known to cause degradation of the stator winding insulation. The iron debris produced as a result of wedge wear contaminates the stator endwinding and can cause partial discharge [10] and increase surface leakage current for medium-high voltage motors.

Therefore, it is important to monitor the magnetic wedge condition regularly to minimize the risk of degradation in motor performance or reliability. However, there is currently no known field measurement technique for assessing the quality of magnetic wedges without disassembling the motor. Visual inspection is the only reliable method used in the field for magnetic wedge inspection [11]-[12]. Inspection of the exiting exhaust duct for iron/glass fabric debris [12], and borescope inspection of the rotor surface for wear bands [11] have been attempted for detecting missing wedges without disassembling the motor, but are not applicable to all motors.

The first attempt to detect missing wedges electrically without motor disassembly was made in [13]-[14] for inverter-fed machines. The current response to a high frequency voltage pulse (tens of μsec duration) is used to measure the transient leakage inductance, and the asymmetry in the inductance between phases is used as an indicator of missing wedges. It was shown that missing wedges can be detected with high sensitivity; however, it is possible that progressing wedge failures cannot be reliably monitored since the random distribution of missing wedges (Fig. 1) tends to cancel out the asymmetry. The ability to monitor the progress of missing

wedges with time is an important requirement considering that it is not cost effective to replace a small portion of missing wedges as the degradation in performance and reliability risks are not significant. Rotor cage problems produce an asymmetry in the inductance pattern is identical to that of missing wedges; therefore, testing must be performed at two different rotor positions to distinguish the two problems, if the test is performed off-line. There are also problems introduced by testing with high frequency pulses. It has been observed in [15]-[16] that testing of closed rotor slot induction motors with high frequency signals can be influenced by the residual flux in the rotor slot bridge distorting the test results. It is also possible that stator core inter-laminar insulation problems not observed with low frequency excitation can interfere in case of high frequency excitation [17].

It can be seen that loose or missing magnetic wedges is a common problem that can degrade motor performance and reliability. Although cases of magnetic wedge problems have recently increased, there is currently no practical test method available for monitoring the overall magnetic wedge condition other than visual inspection. There has not been much research effort on the detection of this issue, and research is currently in its initial stage [11]-[14]. Given the importance of testing magnetic wedge condition and the need for a reliable test method, the objective of this paper is to develop a new test method for magnetic wedge quality assessment. The requirement of the new method is the ability to monitor deterioration of wedge condition over time without motor disassembly, and independent of other faults or motor design.

III. PROPOSED METHOD FOR DETECTION OF MISSING WEDGES

A. Main Concept

The main concept of the new method is to use a portable power converter-type equipment to test the motor for missing magnetic wedges when the motor is at standstill. Testing can be performed at the motor control center (MCC) or at the motor terminals whenever the motor is shut down, as shown in Fig. 2. The test equipment is a low power, portable 3 phase inverter capable of injecting test signals for extracting parameters that are sensitive to missing wedges. Under the expectation that the stator leakage inductance will decrease with missing wedges, the motor is excited with pulsating magnetic fields at multiple circumferential locations, as shown in Fig. 2. It is shown in section III.B that the change in the stator leakage inductance

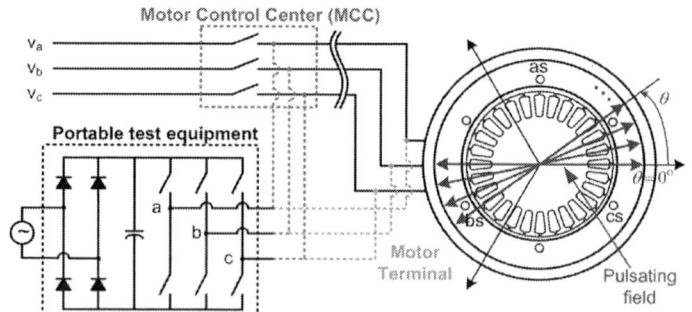

Fig. 2. Main concept: portable, low power 3 phase inverter used for testing magnetic wedges with pulsating field excitation at multiple circumferential positions, θ, at motor standstill

can be monitored with high sensitivity under pulsating field excitation. The rating of the test equipment is much lower than that of the motor since small test signals are sufficient for extracting information on wedge condition, as will be demonstrated in section V. The rotor does not rotate since the induced torque is zero with pulsating field excitation, and therefore, the proposed test method can provide non-invasive and remote testing whenever the motor is at standstill. The variation in the pattern and value of the equivalent resistance and inductance parameters of the motor as a function of pulsating field electrical angle, θ, is used as an indicator for detecting missing magnetic wedges.

If the electrical angle of the pulsating field vector is θ, the field pulsates between θ and $\theta+180°$ electrical degrees, as shown in Fig. 2. The pulsating vector at θ can be produced with sine-triangle PWM excitation with the voltage references,

$$v_{as}^*(\theta,\omega t) = V\cos(\theta)\cdot\sin(\omega t), \tag{1}$$

$$v_{bs}^*(\theta,\omega t) = V\cos(\theta - 2\pi/3)\cdot\sin(\omega t), \tag{2}$$

$$v_{cs}^*(\theta,\omega t) = V\cos(\theta + 2\pi/3)\cdot\sin(\omega t), \tag{3}$$

where V is the excitation voltage amplitude and ω is the excitation frequency. This results in current flow in the stator and rotor conductors, and the voltage and current vectors in the direction of θ, v_θ and i_θ, can be calculated from

$$v_\theta^*(\omega t) = k\cdot[\cos(\theta)\ \cos(\theta - 2\pi/3)\ \cos(\theta + 2\pi/3)]\cdot v_{abcs}^*, \tag{4}$$

$$i_\theta(\omega t) = k\cdot[\cos(\theta)\ \cos(\theta - 2\pi/3)\ \cos(\theta + 2\pi/3)]\cdot i_{abcs}, \tag{5}$$

where v_{abcs}, i_{abcs} represent the stator voltage and current matrices, and k is 2/3. The equivalent input impedance of the motor for excitation in the θ direction, $Z_{eq}(\theta)$, can be calculated from the phasors of the voltage and current vectors in the θ direction, V_θ and I_θ, as,

$$Z_{eq}(\theta) = \vec{V}_\theta / \vec{I}_\theta = R_{eq}(\theta) + jX_{eq}(\theta), \tag{6}$$

The electrical equivalent circuit of an induction motor excited with a pulsating field can be derived from the *double revolving field theory* where the field is resolved into two fields of equal magnitude rotating in opposite directions [18]. This circuit can be simplified since the slip, s, is equal to 1 when the motor is at standstill, as shown in Fig. 3. If the excitation frequency is not too low (e.g. 50 Hz), Z_{eq} can be approximated as (7) since $\omega L_m \gg R_r + j\omega L_{lr}$ at motor standstill ($s=1$).

$$Z_{eq}(\theta) \approx \left(R_s(\theta) + R_r(\theta)\right) + j\left(X_{ls}(\theta) + X_{lr}(\theta)\right), \tag{7}$$

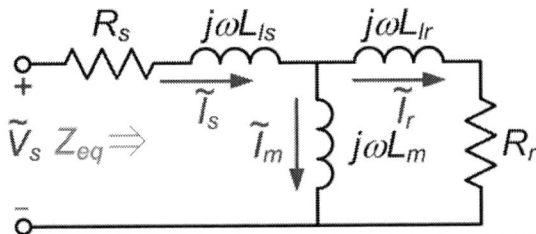

Fig. 3. Electrical equivalent circuit model of induction machine with pulsating magnetic field excitation at motor standstill ($s=1$)

If magnetic wedges are missing, the pattern and values of $R_{eq}(\theta)$ and $X_{eq}(\theta)$ as a function of θ are expected to change. The expected change in pattern and values are analyzed in III.B for deriving a sensitive indicator for detecting missing wedges. The behavior of $R_{eq}(\theta)$ and $X_{eq}(\theta)$ with other type of motor faults are also studied for distinguishing the faults.

B. Analysis of Equivalent Circuit Parameter Variation with Missing Magnetic Wedges

If the stator winding is excited with a 50 Hz pulsating field at motor standstill, the flux cannot penetrate into the rotor yoke due to eddy current flux rejection by the rotor cage. As a result, the flux is mainly distributed near the rotor surface, and the X_{eq} component mainly consists of leakage inductance, as shown in (7). This can be seen in the result of a 2 dimensional finite element (FE) simulation in Fig. 4, where a 4 pole motor was excited with a 50 Hz pulsating field in the direction of $\theta=0°$ at 10% rated current. The magnetic permeability, μ_r, of the magnetic wedge was assumed to be 10, which is typical of the wedges used. It can be clearly observed in Fig. 4 (upper figure) that the flux is concentrated on the stator and rotor surface and the magnetic wedge; therefore, X_{eq} mainly consists of zigzag and slot leakage components under pulsating field excitation.

The simulation was repeated for the case where the magnetic wedge of the slot in the $\theta=90°$ direction (mechanical angle $\theta_m=45°$) is missing, and the results are shown in the lower figure of Fig. 4. It can be seen that the difference in the flux strength and distribution is noticeable for the two cases under identical excitation conditions (10% I_{rated}), particularly in the vicinity of the slot with the missing wedge. For the case of

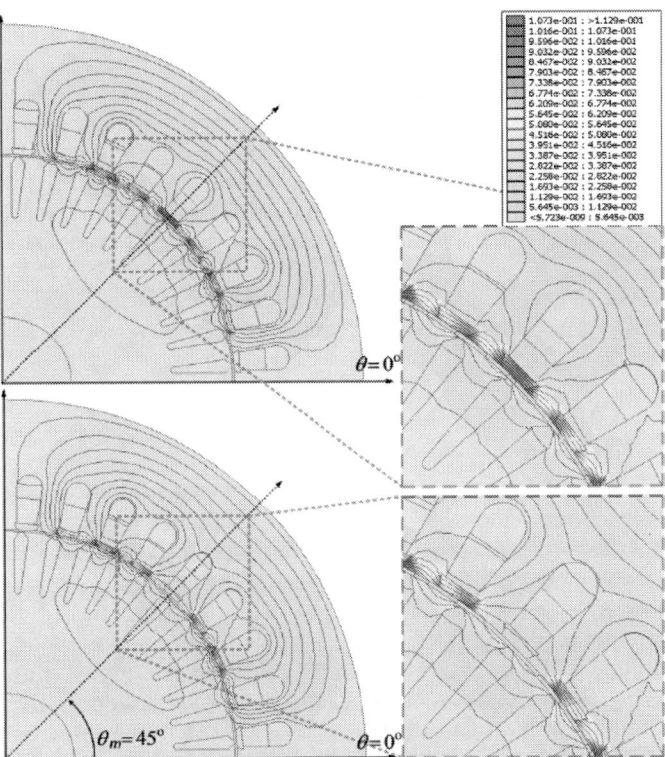

Fig. 4. Flux distribution in 4 P motor at standstill with 50Hz pulsating field excitation at $\theta=0°$ (10% of I_{rated}) for the case of no missing wedge (upper), and wedges missing in the slot at $\theta_m=45°$ location (lower)

θ=0° excitation shown in Fig. 4, the missing wedge increases the equivalent reluctance in the magnetic circuit, which results in decrease in the equivalent inductance. It can be clearly observed in Fig 4 that the magnetic wedge is the main path for the magnetic flux under the given excitation condition. 50 Hz pulsating field excitation is used in the proposed method because the variation in the value of X_{eq} obtained under this condition is sensitive to missing wedges. The resistance components, R_s and R_r, are independent of missing wedges, and only vary with temperature.

If the pulsating field excitation is in the direction of the missing wedge (θ=90°), the value of X_{eq} is not influenced since the magnetic wedge is not a part of the magnetic circuit, as shown in Fig. 5. Therefore, if X_{eq} is measured under pulsating field excitation for multiple circumferential positions between θ=0° and 180°, the expected pattern of $X_{eq}(\theta)$ for a number of cases are shown in Fig 6. For a healthy stator where no wedges are missing, $X_{eq}(\theta)$ is constant and independent of θ since the motor is symmetric. If the wedge is missing at the θ_m=45° location, X_{eq} decreases at θ=0° and 180°, and does not change at θ=90° resulting in the sinusoidal pattern of Fig. 6. If wedges are missing at the slots located at θ_m=45° and 135°, the value of X_{eq} at θ=0° and 180° decrease by a larger amount since the degree of increase in magnetic reluctance is larger. For the case where the wedge is missing at the θ_m=0° (or 90°) location, X_{eq} decreases at θ=90°, and does not change at θ=0° and 180° (Fig. 6). If the magnetic wedges of two slots at the θ_m=0° and 45° locations are missing, it can be predicted that $X_{eq}(\theta)$ would be constant at a lower value of X_{eq} since the asymmetry is canceled (Fig. 6).

It can be seen that the influence of missing wedges at slots 180° (electrical) apart (90° mechanical in this case) is additive, whereas missing wedges at slots 90° (electrical) apart (45° mechanical) tend to cancel out the asymmetry. The values of X_{eq} decrease with increase in the severity of wedge failure regardless of the location of the missing wedges. Magnetic wedge failure can therefore be detected by observing the change in the pattern and/or magnitude of $X_{eq}(\theta)$ over time. X_{eq} can serve as a reliable fault indicator provided that consistent excitation frequency and voltage level is used for X_{eq} measurement since it is independent of temperature, saturation, or skin effect under identical excitation conditions.

IV. IMPLEMENTATION AND PRACTICAL ISSUES

A. Fault Indicator

When the measurements of X_{eq} are obtained as a function of θ for a healthy motor, $X_{eq0}(\theta)$, it is not constant or independent of θ due to the inherent asymmetry in the motor. Experimental measurements of $X_{eq0}(\theta)$ were obtained on the test motor of section V under 50 Hz excitation with the rotor turned to different angles (0°,22.5°,45°,67.5°) with respect to a reference position. The measurements of $X_{eq0}(\theta)$ plotted in Fig. 7 show that it does not change significantly with rotor position. This implies that the fluctuations in $X_{eq0}(\theta)$ are due to the non-ideal inherent asymmetry in the stator winding, magnetic wedge, or core, and that it is independent of the rotor. The fluctuating

(ac) part of $X_{eq0}(\theta)$, $X_{eq0,ac}(\theta)$, can be separated from the average (dc part), $X_{eq0,dc}(\theta)$, as shown in (8), and stored to compensate for inherent stator asymmetries.

$$X_{eq0,ac}(\theta) = X_{eq0}(\theta) - X_{eq0,dc}(\theta), \qquad (8)$$

When determining the severity of magnetic wedge failure, it is desirable to have a fault indicator that can be used for all machines. Since the value of X_{eq} differs significantly from motor to motor, the normalized (and compensated) X_{eq}, $X_{eq,norm}(\theta)$, shown in (9) is suggested as a fault indicator.

$$X_{eq,norm}(\theta) = \left(X_{eq}(\theta) - X_{eq0,ac}(\theta)\right)/X_{eq0,dc}, \qquad (9)$$

If $X_{eq,norm}(\theta)$ is monitored over time, the degree of asymmetry and decrease in $X_{eq}(\theta)$ due to missing wedges can be observed.

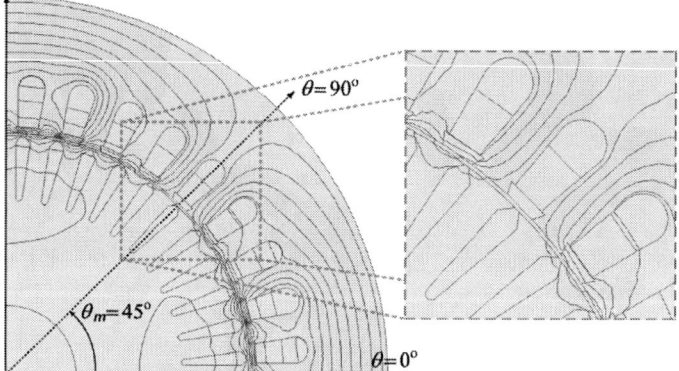

Fig. 5. Flux distribution in 4 P motor at standstill with 50Hz pulsating field excitation at θ=90° (10% I_{rated}) with missing wedge at θ_m=45°

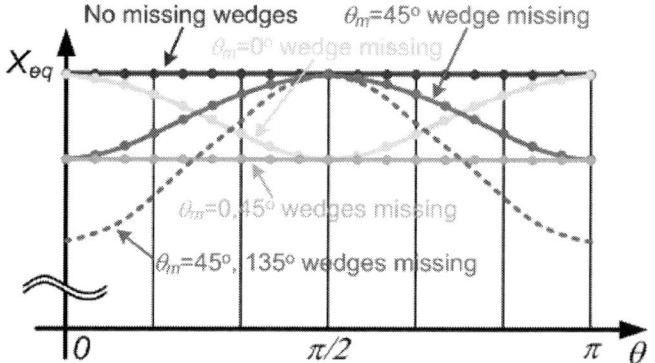

Fig. 6. Pattern of $X_{eq}(\theta)$ with magnetic wedges missing in 0, 1, and 2 stator slots (θ_m=0°, θ_m=45°, θ_m=45° & 135°, θ_m=0° & 45°)

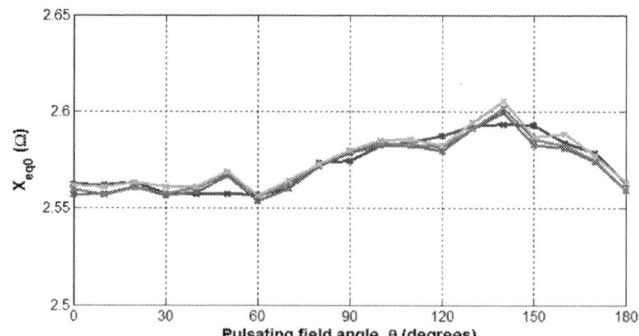

Fig. 7. Inherent asymmetry in $X_{eq}(\theta)$ measurements for healthy motor with no missing magnetic wedges with rotor positioned at 0°, 22.5°, 45°, 67.5° with respect to a reference position

The circumferential distribution of missing wedge segments is typically random, as shown in the example of Fig. 1, and the cases shown in [8]-[12]. If a small portion of the wedge is missing, corrective action will not be taken even if this can be detected, considering that it is not cost effective as the risk of motor performance degradation or failure is minimal. However, if the percentage of missing wedges increase significantly, as in the cases reported in [8]-[12], the wedges must be repaired to prevent performance degradation or failure. As the number of missing wedges increase, it is unlikely that they will all fall out from the same slot or from slots located 180 electrical degrees apart, and the "asymmetry" in the $X_{eq,norm}(\theta)$ pattern would not be observable. Therefore, the percent deviation in the average of (9) from unity, $\%\Delta X_{eq,norm,avg}$, is proposed as the quantitative fault indicator since $X_{eq,norm}$ decreases regardless of the distribution of missing wedges. The increase in $\%\Delta X_{eq,norm,avg}$ can be observed over time for monitoring missing magnetic wedges, and the pattern of $X_{eq,norm}(\theta)$ can be used as a reference for observing the asymmetry.

$$\%\Delta X_{eq,norm,avg} = (1 - avg(X_{eq,norm}(\theta))) * 100\% , \qquad (10)$$

The correlation between the change in $\%\Delta X_{eq,norm,avg}$ and the severity of missing wedges is expected to depend on the motor stator slot and wedge design/material, as in all condition monitoring methods. The fault threshold can be determined from a 2D FE analysis if the motor design is available.

B. Influence of Other Types of Faults and Motor Design

For every test method, it is very important to identify other conditions or factors that can cause false indications. The proposed test method is not influenced by source anomalies or problems in the bearing, shaft, coupling, or load since it is a standstill off-line test. If X_{eq} measurements are made under identical excitation conditions, it will not be influenced by temperature, or saturation; however, it can be influenced by other types of faults [17], [19]-[20]. The pattern of normalized $R_{eq}(\theta)$ and $X_{eq}(\theta)$ for stator core inter-laminar insulation failure, rotor cage failure, and airgap eccentricity measured under 50 Hz excitation are summarized in Fig. 8.

Stator core inter-laminar insulation failure causes increase in the local lamination eddy current and core loss since it is equivalent to having thicker insulation. Although this can result in change in $R_{eq}(\theta)$ and $X_{eq}(\theta)$ parameters under high frequency excitation, the influence is negligible under low frequency excitation as verified in [17]. The proposed method is not influenced by inter-laminar core insulation faults since 50 Hz pulsating field has very little impact on eddy current core loss. Airgap eccentricity causes variation in the airgap distribution and decrease in the minimum airgap. This results in a small amount of increase in the zigzag leakage inductance, where the increase in $X_{eq}(\theta)$ is uniform and independent of θ, as shown in Fig. 8 ($R_{eq}(\theta)$ is not influenced by airgap eccentricity) [19]. Since X_{eq} increases with airgap eccentricity, it can be easily distinguished from magnetic wedge failures. Rotor cage damage results in increase in the rotor resistance and leakage inductance when the position of the broken bar and pulsating field, θ, are 90 electrical degrees apart. The change in the pattern of the $R_{eq}(\theta)$ and $X_{eq}(\theta)$ measurements with a broken rotor bar are shown in Fig. 8 [20]. Since missing wedges result in a decrease only in the average of X_{eq}, the two faults can be easily distinguished. Rotor faults can also be distinguished from other faults since the $R_{eq}(\theta)$ and $X_{eq}(\theta)$ patterns shift in the x-axis direction depending on the rotor position, whereas all other faults are independent of rotor position.

It is reported in [15]-[16] that testing of closed rotor slot induction motors under low flux excitation can produce erroneous results because the residual flux in the bridge of the closed slot influences the test results. This occurs with high frequency excitation in the kHz ~10s of kHz range because the excitation flux is too low to overcome the residual flux. However, this is not a concern for the proposed method since 50 Hz excitation is 2~3 orders of magnitude smaller than the frequencies of concern (excitation flux is 2~3 orders of magnitude larger for the same excitation voltage). Therefore, the proposed method can be used for all motors whether they are of closed or open rotor slot design.

V. EXPERIMENTAL STUDY

A. Experimental Test Setup

To verify the validity of the proposed test method for magnetic wedge failure detection, an experimental study was performed on 380 V, 5.5 kW (4 pole, 36 stator slot) and 6.6 kV, 3.4 MW (6 pole, 72 stator slot) induction motors with magnetic wedges. An IGBT inverter was built and controlled with a commercial DSP for exciting the motor with variable voltage and frequency pulsating magnetic fields. The excitation voltage and frequency in (1)-(3) were set after comparing the X_{eq} measurements obtained under a number of excitation conditions. The consistency of the test results improved with higher voltage, and the voltage was set at the minimum voltage that provides consistent results. The excitation voltage was 18 V, 104 V for the 380 V, 6.6 kV units, respectively, which is lower than 5% and 2% of the rated voltage. The method was tested at the minimum voltage to verify that a portable inverter rated at much lower power than that of the motor can be used. The sensitivity of detection was improved as the excitation frequency was increased since flux penetration into the rotor decreases. 50 Hz was chosen because the sensitivity did not improve noticeably above 50 Hz, and because there is the risk of interference with core inter-laminar fault problems, and also with slot bridge saturation for closed slot rotor machines when high frequency excitation is used.

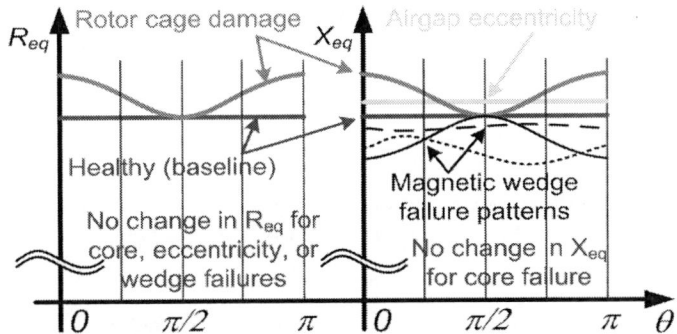

Fig. 8. Pattern of $R_{eq}(\theta)$ and $X_{eq}(\theta)$ for motor with stator inter-laminar core, airgap eccentricity, rotor cage, and magnetic wedge failures.

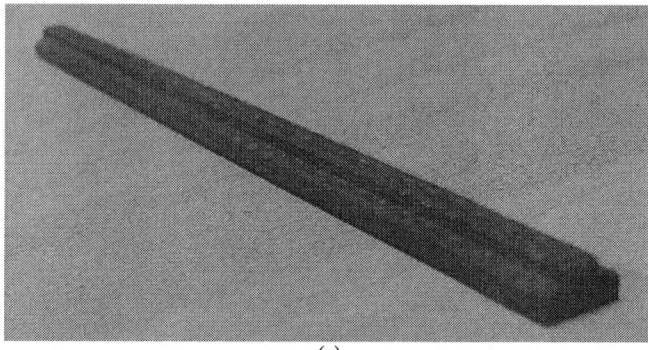

(a)

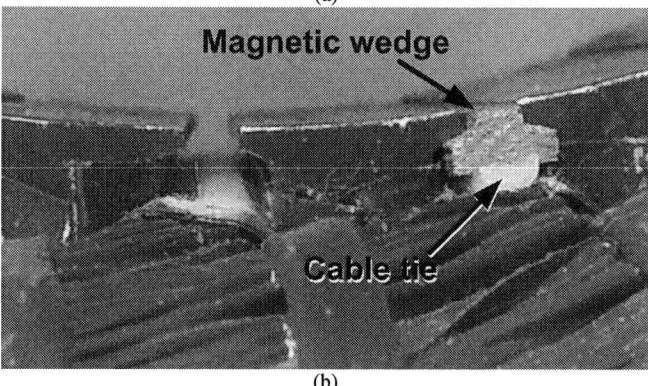

Magnetic wedge

Cable tie

(b)

Fig. 9. (a) Precision machined magnetic wedge sample for 380 V semi-closed slot test machine; (b) stator slot opening with and without magnetic wedges (cable tie used to fix magnetic wedge)

B. Experimental Results: 380 V, 5.5 kW lab motor

The 380 V test motor is a low voltage random wound motor that employs the semi-closed stator slot design with non-magnetic wedges. To test the proposed method under controlled conditions where the magnetic wedges can be inserted and removed, 3 mm thick magnetic wedges were precision machined to fit the slot opening, as shown in Fig. 9(a). The same magnetic material sheets used for high voltage motors were used for fabrication of the wedges. The stator of the test motor was rewound with windings of smaller diameter to make space for the wedges, and cable ties were placed between the magnetic wedge and stator winding, as shown in Fig. 9(b), to fix the magnetic wedges in place to prevent them from moving. Examples of slots with and without the magnetic wedges are shown in Fig. 9(b).

The values of R_{eq} and X_{eq} were measured under 18 V, 50 Hz excitation with the pulsating field electrical angle, θ, varied between $0°$ and $180°$ in $10°$ intervals. The $R_{eq}(\theta)$ and $X_{eq}(\theta)$ measurements are shown in Fig. 10 for the cases listed below.

- 0 missing wedges: healthy
- 1 missing wedge: $\theta_m = 45°$
- 2 missing wedges: $\theta_m = 45°, 135°$
- 3 missing wedges: $\theta_m = 45°, 135°, 225°$
- 4 missing wedges: $\theta_m = 45°, 135°, 225°, 315°$
- 36 missing wedges: all wedges removed

The results in Fig. 10 are the values of $R_{eq}(\theta)$ and $X_{eq}(\theta)$ obtained without any processing for compensation of inherent

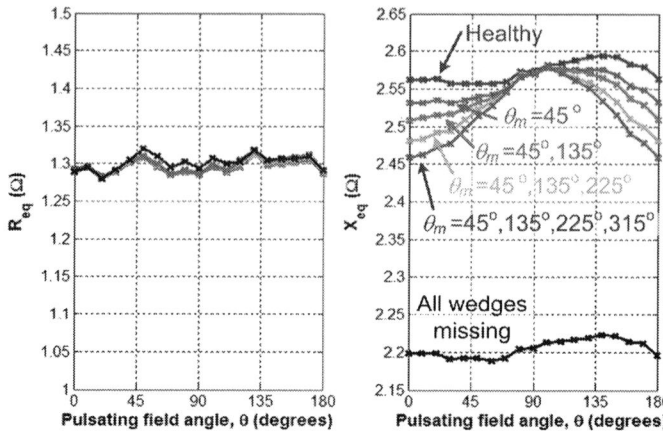

Fig. 10. Experimental measurements of $R_{eq}(\theta)$ and $X_{eq}(\theta)$ for 380 V motor stator with 0, 1 ($\theta_m=45°$), 2 ($\theta_m=45°,135°$), 3 ($\theta_m=45°,135°,225°$), 4 ($\theta_m=45°,135°,225°, 315°$), and 36 missing wedges.

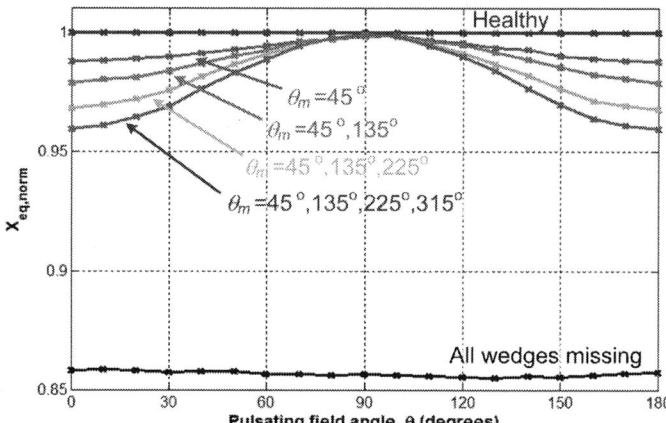

Fig. 11. Experimental measurements of $X_{eq,norm}(\theta)$ for 380 V motor stator with 0, 1 ($\theta_m=45°$), 2 ($\theta_m=45°,135°$), 3 ($\theta_m=45°,135°,225°$), 4 ($\theta_m=45°,135°, 225°,315°$), and 36 missing wedges.

asymmetry described in (8)-(10). It can be seen that the change in $R_{eq}(\theta)$ is negligible, and that $X_{eq}(\theta)$ decreases with increase in missing wedges. The change in the pattern of $X_{eq}(\theta)$ with missing wedges can be observed more clearly with the inherent asymmetry removed using (9). In addition, the decrease in $X_{eq}(\theta)$ with missing wedges is more evident, if it is normalized with respect to $X_{eq0,dc}(\theta)$ using (9). The normalized $X_{eq}(\theta)$, $X_{eq,norm}(\theta)$, is plotted in Fig. 11 for the same cases shown in Fig. 10. The increase in asymmetry, and decrease in magnitude of $X_{eq}(\theta)$ due to missing wedges can be observed with more clarity, as expected. Since the missing wedges are 180 electrical (90 mechanical) degrees apart, the influence is additive. The pattern of $X_{eq,norm}(\theta)$ for the cases with 1, 2, 3, and 4 missing wedges are in phase and the degree of asymmetry increases as the number of missing wedges increase. Therefore, the decrease in magnitude and asymmetry in the pattern of $X_{eq}(\theta)$ can serve as a good indicator for detecting missing wedges. The results with all 36 magnetic wedges removed shows how much the expected decrease in X_{eq} is for the worst case. If this information is available through FE analysis or testing, it could be used to determine the fault threshold, since the correlation between the decrease in $X_{eq,norm,avg}$ and fault severity can be found.

978-1-4799-0024-4/13 $31.00 © 2013 IEEE

The values of $X_{eq,norm}(\theta)$ calculated from (9) for the additional cases listed below are shown in Fig. 12.

- 0 missing wedges: healthy
- 1 missing wedge: $\theta_m = -5°$ and $\theta_m = 45°$
- 2 missing wedges: $\theta_m = -5°$, $45°$
- 3 missing wedges: $\theta_m = 45°$, $105°$, $165°$

The wedge fault conditions considered in Fig. 12 are cases where the missing wedge distribution in the slots is such that the asymmetry in X_{eq} is canceled, as shown in the example of Fig. 6 (wedges missing in the $\theta_m = 0°$ and $45°$ slots). To test the case where wedges are missing in two slots 90 electrical degrees apart, wedges were removed from slots located at $\theta_m = -5°$ and $45°$. Wedges at $\theta_m = 0°$ could not be removed as the test motor is a 36 slot machine with slots in $10°$ intervals. It can be seen in Fig. 12 that the pattern of $X_{eq,norm}(\theta)$ are roughly out of phase for the cases where the wedges are missing at the $\theta_m = -5°$ and $45°$ slots. The $X_{eq,norm}(\theta)$ pattern is almost constant and independent of θ for the case where both wedges are missing since the asymmetry is canceled. Similarly, if wedges are missing in three slots 120 electrical degrees apart ($\theta_m = 45°$, $105°$, and $165°$), the asymmetry is canceled, as shown in Fig. 12. The results of Fig. 12 clearly show that one cannot rely solely on monitoring the asymmetry of X_{eq} for progressing wedge failures since it can be misleading in case the multiple missing wedges cancel the asymmetry.

To observe the amount of decrease in $X_{eq,norm}$ with the percentage of missing wedges, the % decrease in the average of $X_{eq,norm}(\theta)$, $\%\Delta X_{eq,norm,avg}$, for all the cases considered in Figs 11-12 were calculated using (10), and shown in Fig 13. It can be seen that the fault indicator $\%\Delta X_{eq,norm,avg}$ increases with the % of missing wedges. It was observed that the % increase in $X_{eq,norm,avg}$ is different for the same number of missing wedges depending on whether the slot with the missing wedge contains windings from the same or different phases (test motor stator is a 7/9 fractional pitch double layer lap winding).

C. Experimental Results: 6.6 kV, 3.4 MW field motor

The 6.6 kV, 3.4 MW test motor (6 pole, 345 A) is a forced draft fan induction motor used at a power plant, and was brought into a motor repair shop for stator rewind. This high voltage motor has form wound stator bars with open stator slot design and employs magnetic wedges, as shown in Fig. 14(a). The authors were given a 4 hour window for testing the motor at the repair shop before the stator rewind. One segment of the magnetic wedge in one of the slots was missing when the rotor was removed, as shown in Fig. 14(a). The wear band on the rotor surface at the same axial location as the missing wedge (Fig. 14(b)) indicates that the magnetic wedge fell out into the airgap during motor operation. The missing wedge segment could not be found in the motor as it has disintegrated.

Testing was performed with the all the wedges removed in 0~4 slots, as shown in Fig. 15. Wedges were removed in slots

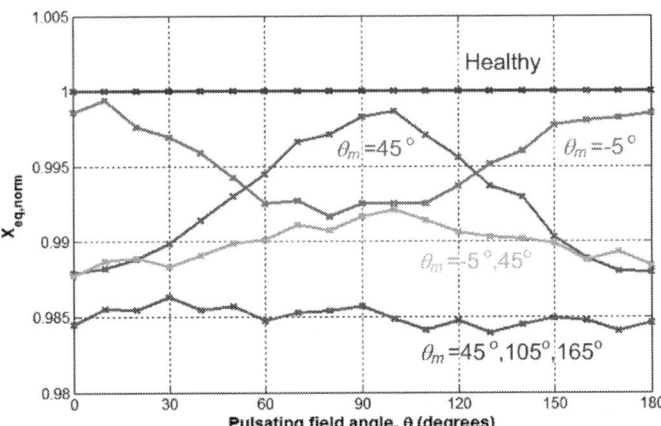

Fig. 12. Experimental measurements of $X_{eq,norm}(\theta)$ for 380 V motor stator with 0, 1 ($\theta_m = -5°$ & $\theta_m = 45°$), 2 ($\theta_m = -5°,45°$), 3 ($\theta_m = 45°,105°,165°$) missing wedges.

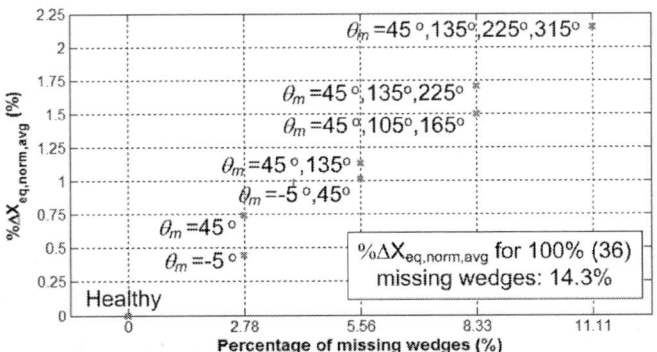

Fig. 13. Experimental measurements of $\%\Delta X_{eq,norm,avg}$ (380 V motor) as a function of the percentage of missing wedges for all cases considered in Figs 11-12

(a)

(b)

Fig. 14. 6.6 kV, 3.4 MW test motor with magnetic wedges; (a) one missing wedge segment; (b) wear band on rotor surface due to missing wedge

978-1-4799-0024-4/13 $31.00 © 2013 IEEE

60 degrees apart (180 electrical degrees) for the 6 P motor so that the influence is additive. The test conditions are listed below, where the mechanical angle θ_w is measured with respect to the missing wedge segment location, as shown in Fig. 14(a).

- No missing wedge: healthy
- 1 slot with missing wedge: $\theta_w = 0°$
- 2 slots with missing wedges: $\theta_w = 0°, 60°$
- 3 slots with missing wedges: $\theta_w = 0°, 60°, 120°$
- 4 slots with missing wedges: $\theta_w = 0°, 60°, 120°, 180°$

The reference (or healthy) case is the condition shown in Fig. 14(a), where there was 1 wedge segment missing at the $\theta_w = 0°$ slot. The test conditions listed above do not represent how the wedges randomly fall out in actual cases observed in the field, as shown in Fig. 1 and [8]-[12]. Testing was performed in this manner considering the short time window given for testing, since the purpose was to verify the validity of the proposed method. Testing was performed with the rotor removed, since insertion and removal of the rotor is time consuming and labor intensive, given the short time window. The result obtained with the rotor removed is expected to be more sensitive compared to when the rotor is inserted. This is because the alternate flux path through the rotor, shown in Fig. 4, is not present when the stator slot wedge is missing, and this is expected to make the "change" in the leakage inductance larger.

The values of X_{eq} were measured under 104 V, 50 Hz excitation under identical pulsating field excitation as the 380 V motor. The measurements of $X_{eq}(\theta)$ and $X_{eq,norm}(\theta)$ obtained for the five cases listed above for the 6.6 kV motor are shown in Fig 16. The results are similar to that of Figs. 10-11 where the missing wedges are 180 electrical degrees apart and the influence is additive. The pattern of $X_{eq}(\theta)$ and $X_{eq,norm}(\theta)$ for the cases with 1, 2, 3, and 4 missing wedges are in phase and the degree of asymmetry increases as the number of missing wedges increase. The values of $\%\Delta X_{eq,norm,avg}$ calculated for all the cases, and plotted in Fig 17, shows that the fault indicator $\%\Delta X_{eq,norm,avg}$ increases with the percentage of missing wedges. The results presented in Figs. 10-13, 16-17 verify that magnetic wedge failure can be monitored with the proposed method and fault indicator without disassembling the machine.

VI. CONCLUSION

An off-line test method for detecting magnetic wedge failures without disassembling the motor was proposed in this paper. The main concept is to use a low power portable inverter for injecting test signals into the motor from the MCC for extracting parameters indicative of missing wedges when the motor is at standstill. The pattern and decrease in the equivalent inductance measured as a function of pulsating field angle has been identified as a sensitive indicator of magnetic wedge failures since the stator leakage inductance decreases with missing wedges. An experimental study on 380 V, 5.5 kW, and 6.6 kV, 3.4 MW motors with magnetic wedges verified that the proposed fault indicator can provide reliable detection of missing wedges over time without motor disassembly. It was also shown that other types of faults such as stator core insulation, eccentricity, or rotor cage faults do not interfere with the proposed method.

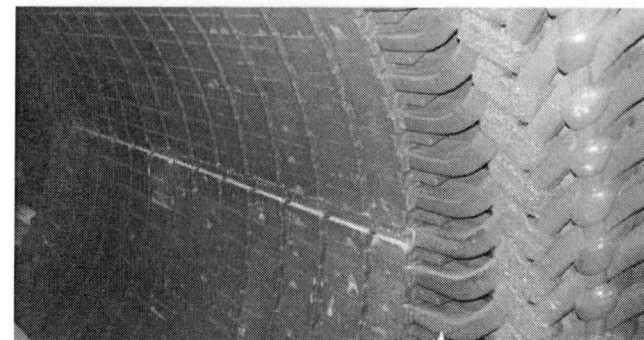

Fig. 15. 6.6 kV, 3.4 MW test motor with all the magnetic wedge segments intentionally removed in one slot for testing purposes

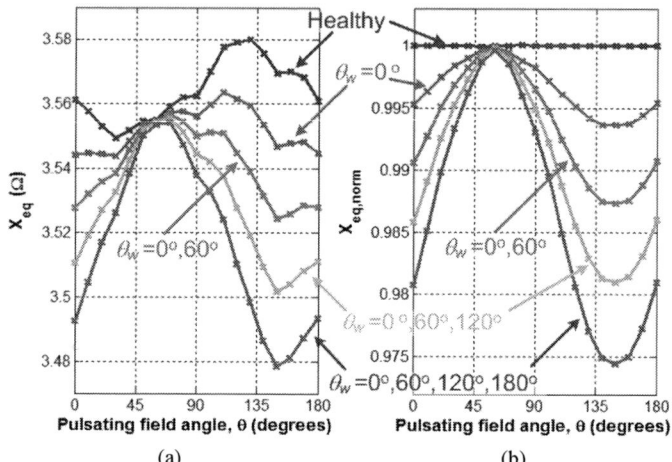

(a)   (b)

Fig. 16. Experimental measurements of (a) $X_{eq}(\theta)$; (b) $X_{eq,norm}(\theta)$ for 6.6 kV motor stator with 0, 1 ($\theta_w = 0°$), 2 ($\theta_w = 0°, 60°$), 3 ($\theta_w = 0°, 60°, 120°$), and 4 ($\theta_w = 0°, 60°, 120°, 180°$) missing wedges.

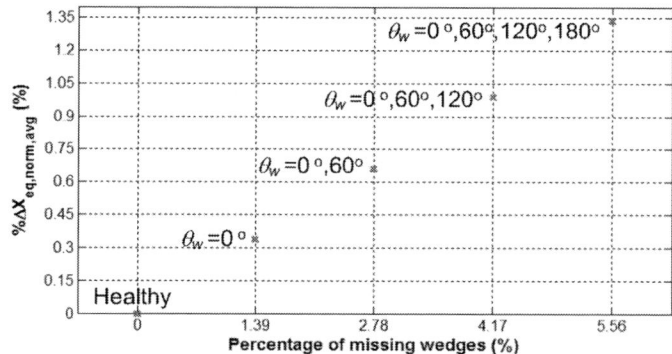

Fig. 17. Experimental measurements of $\%\Delta X_{eq,norm,avg}$ (6.6 kV motor) as a function of the percentage of missing wedges for the cases considered in Fig. 16

The proposed technique is meaningful considering that the only practical means of detecting wedge problems is with visual inspection after motor disassembly. It was shown that the proposed method can be developed as a small stand-alone off-line wedge condition test equipment as the required excitation power level is low. For inverter fed motors with magnetic wedges, the test algorithm can be implemented in the inverter as a built-in diagnostics feature without additional H/W for testing wedge automatically whenever the motor is stopped. Since missing magnetic wedges do not cause significant performance degradation or reliability risks unless it is severe, the proposed method can be used for regular monitoring of missing wedges for scheduling maintenance in an efficient manner.

ACKNOWLEDGMENT

The authors gratefully acknowledge Hansung Electric Industrial Company for sharing their experience on the inspection and repair of magnetic wedges, photographs of magnetic wedge failures, providing the custom made magnetic wedges and rewind for the test motor, and for the support in testing the 6.6 kV motor with magnetic wedge removal.

This work was supported by the Human Resources Development program (*No.* 201140102 03010) of the Korea Institute of Energy Technology Evaluation and Planning (KETEP) grant funded by the Korea government Ministry of Trade, Industry, and Energy.

REFERENCES

[1] B.J. Chalmers, and J. Richardson, "Performance of some magnetic slot wedges in an open slot induction motors," *IEE Proc.*, vol. 114, no. 2, pp. 258-260, Feb. 1967.

[2] H. de Swardt, Changing the wedge material in an electric motor (rev. 03), Marthinsen & Coutts, June 2007. [Online]. Available: http://www.rmwg. co.za/Presentations/Magnetic_Wedges/Magnetic_Wedges_Rev_03.pdf.

[3] S. Wang, Z. Zhao, L. Yuan, and B. Wang, "Investigation and analysis of the influence of magnetic wedges on high voltage motors performance," *Proc. of IEEE Vehicle Power & Propulsion Conf. (VPPC)*, Sept. 2008.

[4] R. Curiac, and H. Li, "Improvements in energy efficiency of induction motors by the use of magnetic wedges," *Proc. of IEEE IAS Petrol. & Chem. Ind. Conf. (PCIC)*, pp.1-6, Sept. 2011.

[5] Z. Milojkovic, D. Ban, M. Petrinic, J. Studir, Z. Maljkovic, and J. Polak, "Application of magnetic wedges for stator slots of hydrogenerators," *CIGRE Session*, Aug. 2010.

[6] A. Kaga, Y. Anazawa, H. Akagami, S. Watabe, and M. Makino, "A research of efficiency improvement by means of wedging with soft ferrite in small induction motors," *IEEE Trans. on Magn.*, vol. MAG-18, no. 6, pp. 1547-1549, Nov. 1982.

[7] F. Caricchi, F.G. Capponi, F. Crescimbini, and L. Solero, "Experimental study on reducing cogging torque and no load power loss in axial flux permanent magnet machines with slotted winding," *IEEE Trans. on Ind. Appl.*, vol. 40, no. 4, pp. 1066-1075, July/Aug. 2004.

[8] H. de Swardt, Electric motor failure prevention: wedge failures (rev. 08), Marthinsen & Coutts, Oct. 2007. [Online]. Available: http://www. mandc.co.za/pdfs/Electric_Motor_Failure_Wedges.pdf.

[9] R. Scollay, and W. Stewart, "The real cost of magnetic wedges in improved performance of induction motors," *Proc. of Coil Winding, Insulation & Electrical Manufacturing Exhibition (CWIEME)*, 2007.

[10] M. Davis, "Problems and solutions with magnetic stator wedges," *Proc. of Iris Rotating Machine Conf. (IRMC)*, 2007.

[11] R. Hanna, D.W. Schmitt, "Failure Analysis of Induction Motors: Magnetic Wedges in Compression Stations," *IEEE Ind. Appl. Mag.*, vol. 18, no. 4, pp. 40-46, July/August 2012.

[12] R.A. Hanna, W. Hiscock, and P. Klinowski, "Failure analysis of three slow-speed induction motors for reciprocating load application," *IEEE Trans. on Ind. Appl.*, vol. 43, no. 2, pp. 429-435, Mar./Apr. 2007.

[13] G. Stojicic, M. Samonig, P. Nussbaumer, G. Joksimovic, M. Vasak, N. Peric, and T. Wolbank, "A method to detect missing magnetic slot wedges in AC machines without disassembling," *Proc. of IEEE IECON*, pp. 1698-1703, Nov. 2011.

[14] G. Stojicic, M. Vasak, N. Peric, G. Joksimovic, and T.M. Wolbank, "Detection of partially fallen-out magnetic slot wedges in inverter-fed AC machines under various load conditions," *Proc. of IEEE Energy Conversion Congress and Exposition (ECCE)*, pp. 4015-4020, 2012.

[15] T. Kang, J. Hong, S.B. Lee, Y. Yoon, D. Hwang, and D. Kang, "The influence of the rotor on surge PD testing of low voltage AC motor stator windings," *IEEE Trans. on Dielec. & Elec. Ins.*, vol. 20, no. 3, pp. 762-769, June 2013.

[16] *Testing the rotor of a induction motor by measuring the inductance as a function of shaft position*, GET-8065 rev. 1, GE Industrial Control Systems, September 1997.

[17] K. Lee, J. Hong, K. Lee, S.B. Lee, and E.J. Wiedenbrug, "A stator core quality assessment technique for inverter-fed induction machines," *IEEE Trans. on Ind. Appl.*, vol. 46, no. 1, pp. 213-221, Jan./Feb. 2010.

[18] S.J. Chapman, *Electric Machinery Fundamentals (5th edition)*, Mcgraw Hill, 2011.

[19] D. Hyun, J. Hong, S.B. Lee, K. Kim, E.J. Wiedenbrug, M. Teska, S. Nandi, I.T. Chelvan, "Automated Monitoring of Airgap Eccentricity for Inverter-Fed Induction Motors Under Standstill Conditions," *IEEE Trans. on Ind. Appl.*, vol. 47, no. 3, pp. 1257-1266, May. 2011.

[20] S.B. Lee, J. Yang, J. Hong, B. Kim, J. Yoo, K. Lee, J. Yun, M. Kim, K. Lee, E.J. Wiedenbrug, and S. Nandi, "A New Strategy for Condition Monitoring of Adjustable Speed Induction Machine Drive Systems," *IEEE Trans. on Pwr. Elec.*, vol. 26, no. 2, pp. 389-398, Feb. 2011.

Induction Motor Model Validation using Fast Fourier Transform and Wavelet tools

F.J. Villalobos-Piña, Member, IEEE, R. Alvarez-Salas, Memeber, IEEE, Eduardo Cabal-Yepez,
Member, IEEE, Arturo Garcia-Perez, Member, IEEE.

Φ*Abstract* – This paper presents a comparative validation for electric stator and rotor faults through an induction-motor modeling utilizing Park Instantaneous Space Phasor (ISP) during stator current analysis and fast Fourier Transform (FFT2) to identify the fault spectrum and the band spectral density of wavelet coefficients using multi-resolution analysis (MRA). The spectral analysis identifies the fault signature modifying the sample frequency in the data acquisition system. The wavelet analysis maintains a constant sample frequency using MRA, which provides redundant information to identify the faults. Furthermore, the MRA analysis of ISP stator currents helps to identify small incipient faults choosing a threshold between the healthy and faulty machine. The cases considered are stator and rotor electric faults, which are modeled by means of parametric variations. In Spite of its simplicity, the model provides useful information for fault identification.

***Index Terms*— Fault detection, Frequency analysis, Induction motor, ISP, MRA, Wavelet analysis.**

I. NOMENCLATURE

V_s	Three-phase voltage supply
v_{si}	Stator phase i voltage, $i=a,b,c$
I_s	Stator current
I_r	Rotor current
i_{si}	Stator phase i current, $i=a,b,c$
i_{ri}	Rotor phase i current, $i=a,b,c$
Ψ_s	Stator flux
Ψ_r	Rotor flux
ψ_{si}	Stator phase i flux, $i=a,b,c$
ψ_{ri}	Rotor phase i flux, $i=a,b,c$
R_s	Stator resistance
R_r	Rotor resistance
R_{si}	Stator phase i resistance, $i=a,b,c$
ΔR_{si}	Stator phase i variable resistance, $i=a,b,c$
R_{ri}	Rotor phase i resistance, $i=a,b,c$
ΔR_{ri}	Rotor phase i variable resistance, $i=a,b,c$
L_s	Stator inductance
L_r	Rotor inductance
L_m	Mutual inductance
β_{m_max}	Maximum value of coefficient of viscous friction
L_{ms}	Stator mutual inductance
L_{mr}	Rotor mutual inductance
L_{lsi}	Stator phase i inductance, $i=a,b,c$
Δ_{lsi}	Stator phase i variable inductance, $i=a,b,c$
L_{lri}	Rotor phase i inductance, $i=a,b,c$
Δ_{lri}	Rotor phase i variable inductance, $i=a,b,c$
P	Number of poles
θ_m	Angular position
J	Moment of inertia
T_e	Torque
T_L	Load torque
β	Viscous friction coefficient
Δ_β	Variation of viscous friction coefficient
ω_m	Mechanical speed
$\theta_{m1} = 0^0$	Lower limit of angular position (bearing fault)
$\theta_{m2} = 10^0$	Upper limit of angular position (bearing fault)
F_L	Frequency of power supply
s	Slip
a	$1\angle 120^\circ$
ISP	Park's Instantaneous Space Phasor.
FFT	Fast Fourier Transform
IM	Induction Motor
MRA	Multi-resolution Analysis
DWT	Discrete Wavelet Transform
F_s	Sample Frequency
f_C	Frequency cut off

This work was supported in part by DEPI-ITA (Departamento de Posgrado e Investigación - Instituto Tecnológico de Aguascalientes), UASLP (Universidad Autónoma de San Luis Potosí México), Mexican Department of Education (SEP, PROMEP) and by project SEP PIFI-2012 Universidad de Guanajuato.

F.J. Villalobos-Piña, CA Electronica de Potencia y Control, Departamento de Ingeniería Electrónica, Instituto Tecnológico de Aguascalientes, Av. Adolfo López Mateos 1801 Ote., Col. Bona Genes, 20256, Aguascalientes, Aguascalientes, México (fvillalobos@mail.ita.mx).

R. Alvarez-Salas, CA Electronica de Potencia y Control, Facultad de Ingenieria, Universidad Autonoma de San Luis Potosi, Av. Manuel Nava 8, Zona Universitaria, 78290 San Luis Potosi, S. L. P., Mexico (ralvarez@uaslp.mx).

E. Cabal-Yepez, A. Garcia-Perez, HSPdigital - CA Telematica/Procesamiento Digital de Senales, Division de Ingenierias, Campus Irapuato-Salamanca, Universidad de Guanajuato, Carr. Salamanca-Valle km 3.5+1.8, Comunidad de Palo Blanco, 36700 Salamanca, Guanajuato, Mexico ({ecabal, agarcia}@hspdigital.org).

II. INTRODUCTION

Electric machines represent the principal source of movement in the industrial sector. Motors are critical components of electric utilities and industrial processes. The three-phase induction motor is widely recognized as the workhorse in industry. This kind of motor occupies a top position, almost exclusive, in converting electrical energy into mechanical energy, being responsible for nearly 90% of the electrical energy consumed by all electrical motors. It represents almost 60% of all used electric motors; the main reason is its practically null necessity of maintenance [1]. In recent years, direct current machines have gradually been replaced by induction motors in many industrial applications.

The success of the induction motor is due to its low cost, robustness, and high performance, which may be achieved thanks to developing new control laws and semiconductors devices. However, most control algorithms become ineffective and even dangerous when faults occur. Faults in the stator circuit of three-phase induction machines represent a significant percentage of motor failures. Faults in the stator windings, such as interturn and magnetic circuit short-circuits, are included in this category. In the first case, the internal asymmetry will cause circulating an extremely high current in the portion of the winding affected by the fault, thus contributing to degrading other portions of the winding. The lead time between the initial start of the fault and the complete failure of the machine depends on several factors, namely the initial number of shorted turns, winding configuration, rated power, rated voltage, and environmental conditions, among others [2]. Unfortunately, an induction motor can fail due to other defective mechanisms. For example, rotor cage faults, such as broken bars or cracked end-rings, can occur due to a combination of various stresses that act on the rotor [3], [8]. Another fault is the air-gap eccentricity, which can occur due to several causes, and leads to adverse effects in the induction machine if it overpasses a certain limit. One of the most obvious examples is the friction between the rotor and stator, which can result in serious damage to the stator core and windings. Another example is faulty bearings, which is the most common fault [4].

In this work time frequency analysis and wavelet tools are used to validate the mathematical modeling of a three-phase induction motor (IM) suitable for simulating the rotating-machine behavior under faulty conditions. The proposed model allows simulating stator, rotor, static eccentricity and bearing faults.

This paper is divided as follows: Section III presents the IM model. Section IV describes the phenomena produced for each kind of fault in IM. Simulation results are given in Section V. The validation model based on Park instantaneous space phasor (ISP), and frequency and wavelet analyses, for the different treated faulty conditions is assessed in Section VI. Some conclusions and future work are given in Section VII.

III. INDUCTION MACHINE DYNAMICAL MODEL UNDER FAULT CONDITIONS

In this section, a squirrel-cage IM dynamical model for faulty conditions analysis is proposed. The wound rotor machine analysis is not considered in this work. Since many assumptions for the IM d-q equivalent model in two phases are not longer valid when the motor undergoes some kind of fault, in this work the a-b-c three-phase model is presented [9], [10], [11]. Stator, rotor, static eccentricity and bearing faults are considered.

The IM dynamic model [20], [21] is given by

$$
\begin{pmatrix} \dot{\Psi}_s \\ \dot{\Psi}_r \end{pmatrix} = - \begin{pmatrix} R_s & 0 \\ 0 & R_r \end{pmatrix} \begin{pmatrix} L_s & T(\frac{P}{2}\theta_m) \\ T(\frac{P}{2}\theta_m) & L_r \end{pmatrix}^{-1} \begin{pmatrix} \Psi_s \\ \Psi_r \end{pmatrix} + \begin{pmatrix} V_s \\ 0 \end{pmatrix} \quad (1)
$$

where

$$
V_s = \begin{pmatrix} v_a \\ v_b \\ v_c \end{pmatrix}, \Psi_s = \begin{pmatrix} \psi_{sa} \\ \psi_{sb} \\ \psi_{sc} \end{pmatrix}, \Psi_r = \begin{pmatrix} \psi_{ra} \\ \psi_{rb} \\ \psi_{rc} \end{pmatrix} \quad (2)
$$

$$
R_s = \begin{pmatrix} R_{sa} + \Delta R_{sa} & 0 & 0 \\ 0 & R_{sb} + \Delta R_{sb} & 0 \\ 0 & 0 & R_{sc} + \Delta R_{sc} \end{pmatrix} \quad (3)
$$

$$
R_r = \begin{pmatrix} R_{ra} + \Delta R_{ra} & 0 & 0 \\ 0 & R_{rb} + \Delta R_{rb} & 0 \\ 0 & 0 & R_{rc} + \Delta R_{rc} \end{pmatrix} \quad (4)
$$

When an electric fault begins in the stator or rotor of the IM, a parametric variation of resistances and inductances occurs. From a previous work [?], models for stator faults like open or short-circuited windings are simulated, considering the parametric variations in stator resistances ΔR_{si} (3), and inductances ΔL_{si} (5) in a three-phase system ($i=a, b, c$). Similarly, rotor electrical faults like broken bars and cracked end-rings are considered, considering the parametric variations of rotor resistances ΔR_{ri} (4) and inductances ΔL_{ri} (6) in a three-phase system ($i=a, b, c$). The magnitude of the stator and rotor electric faults is modeled as a percentage of the parametric variation in stator and rotor resistances and inductances, which makes possible to introduce incipient or severe faults.

$$
L_s = \begin{pmatrix} L_{lsa} + \Delta_{lsa} + L_{ms} & -\frac{1}{2}L_{ms} & -\frac{1}{2}L_{ms} \\ -\frac{1}{2}L_{ms} & L_{lsb} + \Delta_{lsb} + L_{ms} & -\frac{1}{2}L_{ms} \\ -\frac{1}{2}L_{ms} & -\frac{1}{2}L_{ms} & L_{lsc} + \Delta_{lsc} + L_{ms} \end{pmatrix} \quad (5)
$$

$$
L_r = \begin{pmatrix} L_{lra} + \Delta_{lra} + L_{mr} & -\frac{1}{2}L_{mr} & -\frac{1}{2}L_{mr} \\ -\frac{1}{2}L_{mr} & L_{lrb} + \Delta_{lrb} + L_{mr} & -\frac{1}{2}L_{mr} \\ -\frac{1}{2}L_{mr} & -\frac{1}{2}L_{mr} & L_{lrc} + \Delta_{lrc} + L_{mr} \end{pmatrix} \quad (6)
$$

The currents generated by the model are

$$
\begin{pmatrix} I_s \\ I_r \end{pmatrix} = \begin{pmatrix} L_s & T(\frac{P}{2}\theta_m) \\ T(\frac{P}{2}\theta_m) & L_r \end{pmatrix} \begin{pmatrix} \Psi_s \\ \Psi_r \end{pmatrix} \quad (7)
$$

and

$$
I_s = \begin{pmatrix} i_{sa} \\ i_{sb} \\ i_{sc} \end{pmatrix}, I_r = \begin{pmatrix} i_{ra} \\ i_{rb} \\ i_{rc} \end{pmatrix} \quad (8)
$$

The proposed model considers the static eccentricity and the inner race bearing faults. Eccentricity faults affect the air gap permanence, and the bearing inner race fault is modeled by a variation in the friction viscous coefficient. In dynamical mechanics, the parameter $f(\theta_m)_{ec}$ allows to simulate the static eccentricity phenomenon, and it depends on the rotor angular position θ_m.

$$T\left(\frac{P}{2}\theta_m\right) = f(\theta_m)_{ec}\begin{pmatrix} \cos(\alpha) & \cos\left(\alpha+\frac{2\pi}{3}\right) & \cos\left(\alpha-\frac{2\pi}{3}\right) \\ \cos\left(\alpha-\frac{2\pi}{3}\right) & \cos(\alpha) & \cos\left(\alpha+\frac{2\pi}{3}\right) \\ \cos\left(\alpha+\frac{2\pi}{3}\right) & \cos\left(\alpha-\frac{2\pi}{3}\right) & \cos(\alpha) \end{pmatrix} \quad (9)$$

and

$$f(\theta_m)_{ec} = L_m(1-\text{Pr}) + \frac{L_m \text{Pr}(2-\cos(\theta_m))}{2} \quad (10)$$

In (10), *Pr* determines the static-eccentricity variation level as a percentage. The angular position α is given by

$$\alpha = \frac{P\theta_m}{2} \quad (11)$$

The generated torque is

$$T_e = \frac{P}{2}I_s^T T(\frac{P}{2}\theta_m + \frac{\pi}{2})I_r \quad (12)$$

Therefore, the angular position θ_m and speed ω_m are described by

$$\dot{\theta}_m = \omega_m \quad (13)$$

$$\dot{\omega}_m = \frac{1}{J}(T_e - \beta(\theta_m)\omega_m - T_L) \quad (14)$$

where T_L is the load torque and

$$\beta(\theta_m) = \beta\Delta_\beta \quad (15)$$

In (14), the bearing fault is modeled as the friction function $\beta(\theta_m)$, which describes the behavior of the inner-race section and depends on the angular position without considering the kind of fault. a air gap permanence variation. The parameters θ_{m1} and θ_{m2} in (16) represent the lower and upper boundaries in the angular position, respectively, and β_{m_max} in (17) is the maximum value of the friction function for the inner race bearing fault condition. The block diagram for the IM model is shown in Fig. 1.

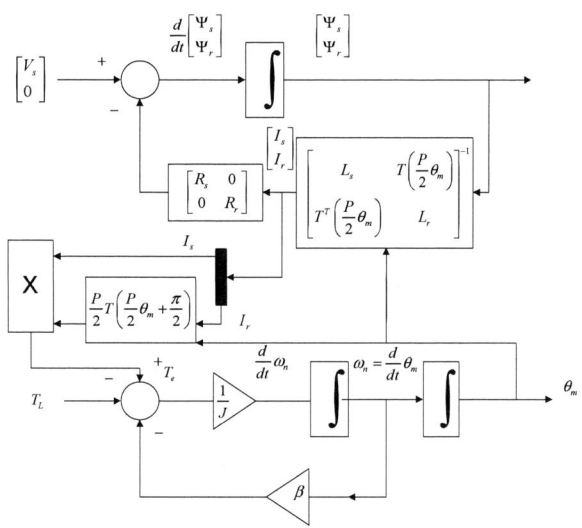

Fig. 1. Block diagram of the IM dynamical model.

The proposed model considers the electric stator and rotor faults simultaneously, by considering the corresponding resistance and inductance variations. For future work, it

could be considered modeling faults such as saturation, eddy current, resistance variation with temperature and so on, for the mechanical dynamics taking into account that eccentricity faults modify the air gap permanence, but this is not considered for the inner race bearing fault; however, it could be considered for future work.

$$\Delta_\beta = \begin{cases} \Delta_p, \theta_{m1} \le \theta_m \le \left[\theta_{m1}+\frac{\theta_{m2}-\theta_{m1}}{2}\right] \\ -\Delta_p, \left[\theta_{m1}+\frac{\theta_{m2}-\theta_{m1}}{2}\right] < \theta_m < \theta_{m2} \\ 1, \theta_m > \theta_{m2} \\ 1, \theta_m < \theta_{m1} \end{cases} \quad (16)$$

$$\Delta_p = \frac{\beta_{m\_max}}{\theta_{m2}-\theta_{m1}} \quad (17)$$

IV. FAULTS IN INDUCTION MACHINE

This section presents the Instantaneous Space Phasor (ISP), or Park-vector, and describes its properties and application for fault analysis in IM [7]. The spectral analysis of the ISP for each simulated fault in the IM model is also presented.

A. ISP

For a four wires non balanced system, the line currents in the *a, b* and *c* phases, and the neutral wire are

$$i_a = i_a^+ + i_a^- + i_a^0 \quad (18)$$

$$i_b = i_b^+ + i_b^- + i_b^0 \quad (19)$$

$$i_c = i_c^+ + i_c^- + i_c^0 \quad (20)$$

$$i_n = i_n^+ + i_n^- + i_n^0 \quad (21)$$

The measured ISP current in the *a, b,* and *c* phases are

$$\tilde{I} = \tilde{I}^+ + \tilde{I}^- + \tilde{I}^0 \quad (22)$$

$$\tilde{I}^+ = \frac{2}{3}(i_a^+ + ai_b^+ + a^2i_c^+) = \tilde{I}^+ e^{j(wt+\varphi^+)} \quad (23)$$

$$\tilde{I}^- = \frac{2}{3}(i_a^- + ai_b^- + a^2i_c^-) = \tilde{I}^- e^{j(wt+\varphi^-)} \quad (24)$$

$$\tilde{I}^0 = \frac{2}{3}(i_a^0 + ai_b^0 + a^2i_c^0) = 0 \quad (25)$$

Another important property of the ISP that helps to separate positive and negative components is its squared magnitude

$$|\tilde{I}| = |\tilde{I}^+ + \tilde{I}^-|^2 \quad (26)$$

$$|\tilde{I}^+ + \tilde{I}^-|^2 = (\tilde{I}^+)^2 + (\tilde{I}^-)^2 + 2(\tilde{I}^+\tilde{I}^-)\cos(2\omega t + \varphi^+ + \varphi^-) \quad (27)$$

And the mean value of the ISP squared magnitude is

$$\langle|\tilde{I}|^2\rangle = (\tilde{I}^+)^2 + (\tilde{I}^-)^2 \quad (28)$$

From (28), it is found that the value of $|\tilde{\mathbf{I}}|^2$ varies between a maximum

$$|\tilde{\mathbf{I}}|^2_M = (\tilde{I}^+ + \tilde{I}^-)^2 \quad (29)$$

and a minimum

$$|\tilde{\mathbf{I}}|^2_M = (\tilde{I}^+ - \tilde{I}^-)^2 \quad (30)$$

Solving above equations for the positive and negative sequence components, it is obtained

$$\tilde{I}^+ = (|\tilde{I}^+|_M + |\tilde{I}^+|_m)/2 \quad (31)$$

$$\tilde{I}^- = (|\tilde{I}^-|_M - |\tilde{I}^-|_m)/2 \quad (32)$$

This means that extreme value measurements in the ISP help to extract the positive $\tilde{I}^+$ and negative $\tilde{I}^-$ sequence current. The same holds for the voltages. The appropriate monitoring of these quantities enables measuring voltage and current unbalance

$$\%IU = 100\frac{\hat{I}^-}{\hat{I}^+} = 100\frac{|\tilde{I}|_M - |\tilde{I}|_m}{|\tilde{I}|_M + |\tilde{I}|_m} \quad (33)$$

B. Frequency Analysis

The frequency analysis in combination with ISP allows studying different kinds of faults in IM. Frequency analysis is normally based on FFT.

The stator electric fault appears when one or more windings are in short circuit. This produces an increase in the demanded current, and overheating in the winding and rotor cage. Stator faults are usually related to an insulation failure, which start as undetected turn-to-turn faults that evolve and culminate in major ones [5], [12], [13].

In this case, the parameters ΔR_{si} and ΔL_{si} in the IM model for $i = a, b, c$ have a negative value to indicate a decrease in the stator resistance and inductance under the fault condition (short-circuit) [14].

The spectral analysis of the ISP magnitude for electric stator fault gives a frequency component at

$$F_{stator\_fault} = 2F_L \quad (34)$$

where F_L is the power supply frequency.

The rotor fault appears when one or more bars in the rotor squirrel cage are broken [15], [16], [17].

In this case the parameters ΔR_{ri} and ΔL_{ri} in the IM model for $i = a, b, c$ have a positive value to indicate an increase in the rotor resistance and inductance under the fault condition (broken bars) for this case.

The broken bar fault generates a low frequency components related to rotor slip s and the power supply frequency F_L at

$$F_{rotot\_fault} = 2sF_L \quad (35)$$

The rotor slip s is a function of the IM number of poles P, the rotor speed ω_r, and the power supply frequency F_L

$$s = 1 - \frac{P\omega_r}{2(2\pi F_L)} \quad (36)$$

Bearing faults appear due to normal wear out, pollution, bad lubrication or bad installation of bearings. They can produce vibrations or mechanical sounds, and they are considered the easiest faults to identity. There are four basic motions that can be used to describe the dynamics of bearing fault movements. Each of them generates a unique frequency response [18].

These four different frequencies are defined as: inner race frequency, outer race frequency, cage frequency and rolling elements frequency. They are based on the bearing geometry, the number of rolling elements and the bearing rotational speed.

The bearing fault generates a characteristic frequency component in its spectrum at half of the power supply frequency F_L

$$F_{bearing\_fault\_n} = n\frac{F_L}{2} \quad (37)$$

Air gap eccentricity takes two basic forms: static and dynamic. Static eccentricity is characterized by a displacement of the rotating axis where the position of the minimal air gap length is fixed in space [19]. It can be caused by the stator oval shape or by the incorrect positioning of the rotor or stator. at the commissioning state. The non uniform air gap raises a radial force of electromagnetic origin, called unbalanced magnetic pull, which acts in direction of the minimum air gap. However, static eccentricity may also cause dynamic eccentricity [6].

The static eccentricity generates a similar spectrum to that of the bearing fault.

$$F_{eccentricity\_fault\_n} = n\frac{F_L}{2} - s \quad (38)$$

C. Wavelet analysis

The wavelet transform is a mathematical tool with a powerful structure and enormous freedom to decompose an ISP characterizing a stator current fault signature into several scales at a different level of resolution. Fig. 2 shows the MRA used for implementation of discrete wavelet transform.

In Fig. 2, x[n] is the acquired ISP fault signature of x(t), sampled at the rate of F_s Hz. The digitalized ISP signal x[n] is then decomposed into $a_1(n)$ and $d_1(n)$ using the low pass filter l(n) and high pass filter h(n), respectively, where $d_1(n)$ is called fluctuation, difference or detail function containing the high frequency terms, and $a_1(n)$ is called trend, average or approximation signal containing the low frequency terms. This is called first-scale decomposition. The second scale decomposition is now based on the coefficient $a_1(n)$ which gives $a_2(n)$ and $d_2(n)$. The next higher scale decomposition is now based on $a_2(n)$ and so on. At any level j, the approximation coefficient $a_j(n)$ will contain frequencies in the band 0-f_c Hz. Similarly, the detail coefficient $d_j(n)$ at any level j will include frequencies in the range f_c-2f_c Hz. The cut off frequency f_c for the approximation coefficients $a_j(n)$ in a given level j is given by

978-1-4799-0024-4/13 $31.00 © 2013 IEEE

$$f_c = \frac{F_s}{2^{j+1}} \quad (39)$$

Furthermore, the number of points in the decomposed detail and approximation coefficients decreases gradually through successive decimation.

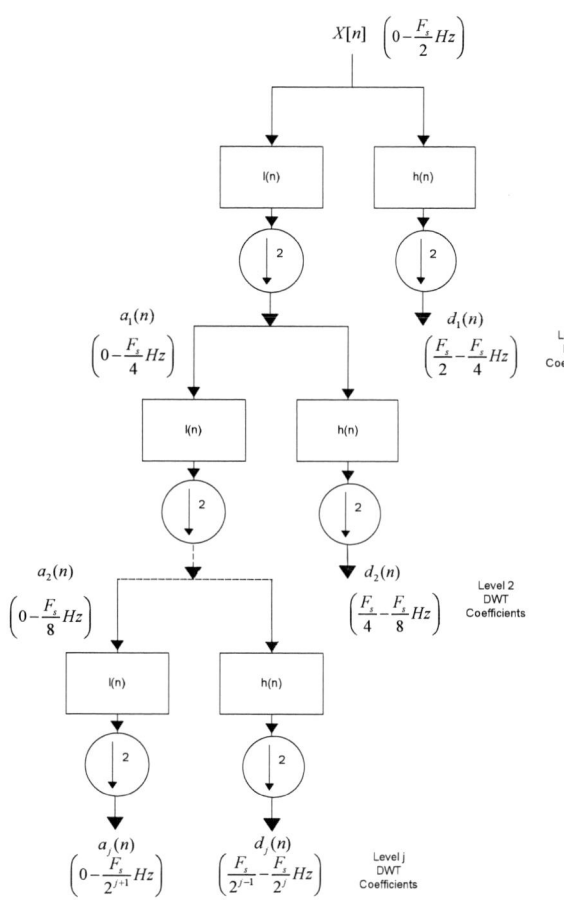

Fig. 2. ISP stator currents decomposition through MRA.

Therefore, the discrete wavelet transform can be computed through a bank of filters. The ISP signature signal is convolved with these filters. In contrast to the FFT, the time resolution becomes arbitrary good to high frequency, while the frequency resolution becomes arbitrary good to low frequencies.

The fault analysis using MRA is developed analyzing the coefficient-decomposition root medium square (RMS) spectral density (40) at each level. When the ISP corresponding to different stator current signatures at distinct fault scenarios are decomposed via wavelet, the contrast among wavelet coefficients can be seen. Therefore, the wavelet node value C_j at a decomposition level j is defined by

$$C_j = \sqrt{\frac{1}{N_j} \sum_{k=1}^{N_j} w^2_{j,k}} \quad (40)$$

In (40) N_j is the number of coefficients at each level j, $w_{j,k}$ is the kth coefficient calculated for jth level, C_j is the RMS value of the decomposed ISP for the stator current signature at level j. It measures the signal power contents in the specified frequency band indexed by the parameter j.

V. SIMULATION RESULTS

A. Spectral analysis results

Using the model proposed in a previous work [?]; a simulation study is performed for a 3 *KW* three-phase IM (Table I). The simulation results are given in terms of the ISP spectral analysis for each treated fault. Fig. 3 presents the results from the simultaneous-fault case when $F_L = 60$ *Hz* with rated load torque. It can be observed the presence of frequency components at $F_{stator\_fault} = 2F_L$ (stator fault), $F_{rotor\_fault} = 2s\ F_L$ (rotor fault), $F_{bearing\_fault\_n} = n\ F_L/2$ (bearing fault), and $F_{eccentricity\_fault} = n\ F_L/2\text{-}s$ (static eccentricity fault).

TABLE I
SIEMENS IM MODEL PARAMETERS

Parameter	Notation	Value	Units
Stator phase-a resistance	R_{sa}	1.16	Ω
Stator phase-b resistance	R_{sb}	1.16	Ω
Stator phase-c resistance	R_{sc}	1.16	Ω
Rotor phase-a resistance	R_{ra}	1	Ω
Rotor phase-b resistance	R_{rb}	1	Ω
Rotor phase-c resistance	R_{rc}	1	Ω
Stator phase-a inductance	L_{sa}	0.006	H
Stator phase-b inductance	L_{sb}	0.006	H
Stator phase-c inductance	L_{sc}	0.006	H
Rotor phase-a inductance	L_{ra}	0.003	H
Rotor phase-b inductance	L_{rb}	0.003	H
Rotor phase-c inductance	L_{rc}	0.003	H
Stator mutual inductance	L_{ms}	0.1	H
Rotor mutual inductance	L_{mr}	0.1	H
Mutual inductance	L_m	0.1	H
Moment of inertia	J	0.025	Kgm
Viscous friction coefficient	β	0.0045	Nms/rad

Number of poles	P	4	poles

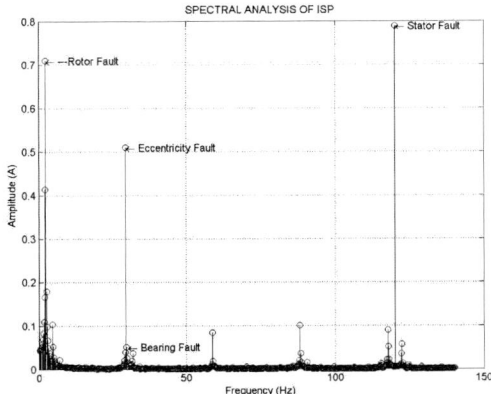

Fig. 3. Spectrum of the ISP magnitude for simultaneous faults (stator, rotor, bearing and static eccentricity) when $F_L = 60\ Hz$ with rated load torque - simulation.

B. Wavelet analysis results

The simulation-obtained wavelet node power for each stator and rotor electric fault case is shown in Table 2. For the analysis, a sampling frequency of 65536 Hz was used. For the stator electric fault, the RMS power spectrum magnitude of wavelet coefficients increased from 3.59 to 5.66 units, as shown in the frequency band 64-128Hz, whereas the magnitude in other bands stayed similar to the healthy case. For the rotor electric fault, the RMS power spectrum magnitude of wavelet coefficients increased from 0.56 to 31.64 units in the frequency band 0.5-1Hz. Both cases shown the wavelet tool capabilities for identifying the spectral band for each fault signature without modifying the sample frequency (F_s), different from windowed fast Fourier transform; furthermore, the wavelet analysis is a redundant tool to increase the reliability of the fault diagnosis.

TABLE 2
SIMULATED RMS WAVELET COEFFICIENTS

Level	Frequency band (Hz)	Healthy IM RMS	Stator fault RMS	Rotor fault RMS
1	16384.0-32768.0	4.186086	2.917641	4.264412
2	8192.0-16384.0	0.224049	0.220686	0.254327
3	4096.0-8192.0	0.402973	0.391107	0.456731
4	2048.0-4096.0	0.562348	0.497111	0.635194
5	1024.0-2048.0	0.389305	0.386403	0.439866
6	512.0-1024.0	0.740535	0.853333	0.835963
7	256.0-512.0	1.583877	2.018430	1.786359
8	128.0-256.0	3.134957	4.317916	3.519413
9	64.0-128.0	3.594713	5.679731	3.912542
10	32.0-64.0	3.640037	3.150577	3.877675
11	16.0-32.0	2.979974	3.008919	3.459305
12	8.0-16.0	1.702024	3.721670	2.936566
13	4.0-8.0	0.914715	1.351054	2.637007
14	2.0-4.0	0.800518	0.666765	6.863664
15	1.0-2.0	0.630051	0.770191	14.859729
16	0.5-1.0	0.566076	1.139732	31.648210
17	0.3-0.5	0.409553	0.180779	17.999383

VI. MODEL VALIDATION

This section presents additional results to validate the behavior of the IM model by comparing spectra and wavelet coefficients for ISP obtained through simulation against experimental results.

A. Spectral Validation without faults

Fig. 4 presents experimental spectra results for ISP from the real IM; these ones were taken as a benchmark to assess the spectra from the treated faulty conditions. It must be noted there exist a frequency component at $2F_L$ (120 Hz) and its amplitude increases proportionally with the load torque level. However, the maximum amplitude is small (0.4 A) even when the rated load torque is applied to the IM.

Consider the case when half of a winding is short circuited. Fig. 5 and 6 show the simulated and experimental spectra for the ISP, respectively. Both spectra are very similar; the fault frequency components have almost the same amplitudes for every scenario. It can be observed the components at $2F_L$ (120 Hz) are about ten times greater than those obtained from the healthy IM. This allows testing some diagnostic techniques with this model.

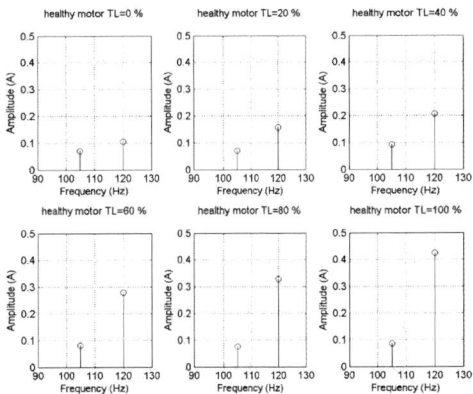

Fig. 4. Spectrum of the ISP magnitude for the healthy IM when $F_L = 60$ Hz and variable load torque - experimental.

B. Spectral Validation for rotor faults

For rotor failure, an incipient fault is generated by making a hole with 2 mm of depth and 1/8 inches of diameter between a bar and an end-ring. The bar thickness is 25mm, so the bar is cracked partially. The simulation and experimental spectra for the ISP, when $F_L = 60\ Hz$ and there is rated load torque, are depicted in figures 7 and 8, respectively. In both cases, the rotor fault component appears at 0.4 Hz with almost the same magnitude.

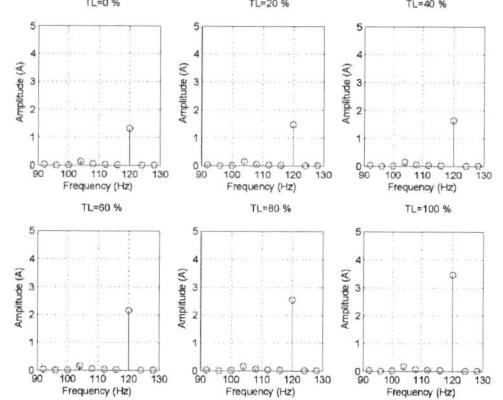

Fig. 5. Spectrum of the ISP magnitude for the stator fault (50% short-circuited winding) when $F_L = 60\ Hz$ and variable load torque – simulation.

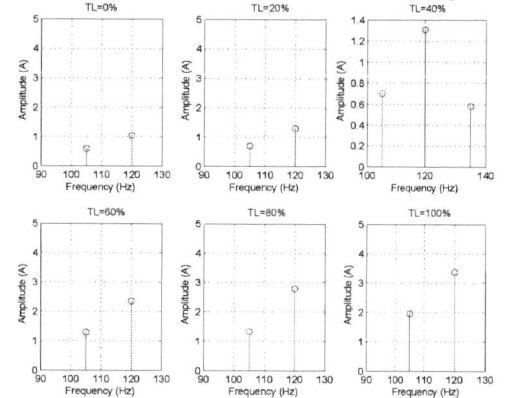

Fig. 6. Spectrum of the ISP magnitude for the stator fault (50% short-circuited winding) when $F_L = 60\ Hz$ and variable load torque – experimental.

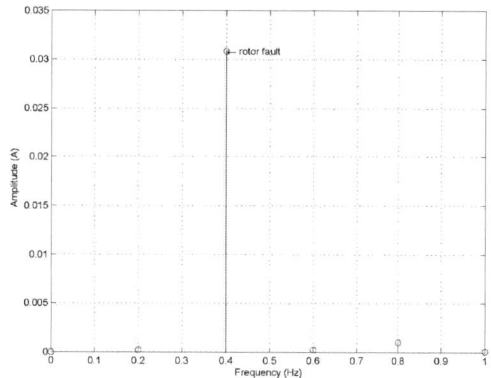

Fig. 7. Spectrum of the ISP magnitude for the incipient rotor fault when $F_L = 60\ Hz$ and the rated load torque – simulation.

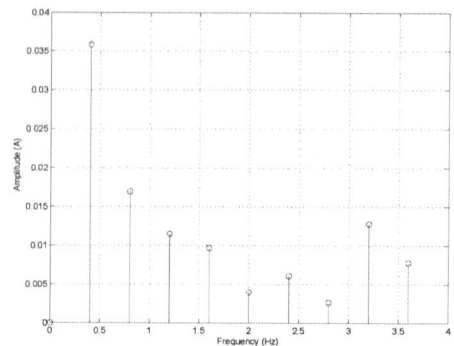

Fig. 8. Spectrum of the ISP magnitude for the incipient rotor fault when $F_L = 60\ Hz$ and the rated load torque - experimental.

C. Wavelet validation for stator and rotor faults

The simultaneous stator and rotor electric faults were validated using MRA wavelet analysis. For this case, a sampling frequency of 1920Hz was used. The experimental results are shown in Table 3. For the stator electric fault case, the RMS magnitude increases from 0.97 for the healthy case, to 2.15 units for the stator fault case in the frequency band of 60-120 Hz. For the rotor electric fault, the RMS magnitude increases from 0.62 units for the healthy case to 2.23 units for the faulty case in the frequency band of 0.23-0.46 Hz.

TABLE 3
EXPERIMENTAL RMS WAVELET COEFFICIENTS

Level	Frequency band (Hz)	Healthy IM RMS	Stator fault RMS	Rotor fault RMS
1	480.0-960.0	6.490875	6.226962	4.850153
2	240.0-480.0	0.780427	0.795897	1.173130
3	120.0-240.0	1.573381	0.842899	1.748283
4	60.0-120.0	0.976965	2.157393	1.596133
5	30.0-60.0	3.271422	2.129200	0.503343
6	15.0-30.0	0.258780	0.314066	0.745330
7	7.5-15.0	0.237494	0.251047	0.604578
8	3.75-7.5	0.308639	0.220973	0.510695
9	1.875-3.75	0.310048	0.388570	0.221874
10	0.9375-1.875	0.115363	0.506553	0.390122
11	0.4687-0.9375	0.806710	0.715985	3.219668
12	0.2343-0.4687	0.623901	0.658321	2.231234

VII. CONCLUSIONS

This paper carries out and validates a comparative between time-frequency spectral analysis and wavelet tools for a proposed three-phase dynamic model of an IM, which allows simulating different fault conditions, even simultaneously. The proposed model is capable to predict accurately the phenomena when stator, rotor, static eccentricity and bearing faults occur.

The comparative analysis shows that using both techniques, time-frequency analysis and wavelet tools, a reliability fault diagnosis of stator and rotor electric faults can be done in induction machines. Different from the windowed fast Fourier transform, which needs to change the sample frequency to identify different electrical and mechanical faults; the proposed techniques does not require of this change in the acquisition rate. Furthermore, for incipient faults, the spectra magnitudes present small changes, and wavelet analysis is a powerful tool to identify small changes in the magnitude of power spectra in different frequency bands, which allows utilizing a fixed sample frequency for different fault scenarios.

VIII. ACKNOWLEDGMENT

The authors wish to acknowledge the financial support of Instituto Tecnológico de Aguascalientes (ITA) México and Universidad Autónoma de San Luis Potosí (UASLP) México.

IX. REFERENCES

[1] S. Nandi, H.A. Toliyat, X. Li, Condition monitoring and fault diagnosis of electrical motors - A review, *IEEE Transactions on Energy Conversion, vol. 20, n. 4*, December 2005, pp. 719-729.

[2] S.M.A. Cruz, A.J. Marques Cardoso, Multiple reference frames theory: A new method for the diagnosis of stator faults in three-phase induction motors, *IEEE Transactions on Energy Conversion, vol. 20, n. 3*, September 2005, pp. 611—619.

[3] S.M.A. Cruz, A.J. Marques Cardoso, Rotor cage fault diagnosis in three-phase induction motors by extended Park's vector approach, *Electric Machines and Power Systems, vol. 28*, 2000, pp. 289-299.

[4] A.J. Marques Cardoso, E.S. Saravia, Computer-aided detection of airgap eccentricity in operating three-phase induction motors by Park's vector approach, *IEEE Transactions on Industry Applications, vol. 29, n. 5*, September 1993, pp. 897-901.

[5] S. M. A. Cruz, A. J. Marques Cardoso. Stator winding fault diagnosis in three-phase synchronous and asynchronous motors, by the extended Park's vector approach. *IEEE Transcations on Industruy Applications, vol 37, n. 5*, September 2001. pp. 1227-1233.

[6] M. Drif, A. J. Marques Cardoso, Airgap eccentricity fault diagnosis, in three-phase induction motors, by the complex apparent power signature analysis. *IEEE International Symposium on Power Electronics Electrical Devices, Automation and Motion SPEEDAM*, 2006.

[7] D. L. Milanez, A.E. Emanuel. The instanteneus-space-phasor a powerful diagnosis tool, *IEEE Transactions on Instrumentation and Measurement, vol. 52 n. 1*, February 2003. pp. 143-148.

[8] A. Bellini, F. Fillipeti, C. Tassoni, G.A. Capolino, Advances in diagnostic techniques for induction machines. *IEEE Transactions on Industrial Electronics, vol. 55, n. 12*. December 2008, pp. 4109-4126.

[9] R. Alvarez Salas, *Commande de machine asynchrone* (Éditions Universitaires Européennes, 2010).

[10] A. E. Fitzgerald, C. Kingsley, S. D. Umans, *Electric machinery* (McGraw Hill, 2003).

[11] P.C. Krause, O. Wasynczuk, S.D. Sudhoff, *Analsys of electric machinery and drive systems* (IEEE Press and Wiley Interscience, 2002).

[12] S. Grubic, J. M. Aller, B. Lu and T. H. Habelter. A Survey on testing and monitoring methods for stator insulation systems of low-voltage induction machines. *IEEE Transactions on Industrial Applications, vol. 55, n. 12*. December 2008. pp. 4181-4190.

[13] C. H. De Angelo, G. R. Bossio, S. J. Giaccone, M. I. Valla, J. A. Solsona, G. O. García. Online model-based stator-fault detection and identification in induction motors. *IEEE Transactions on Industrial Electronics, vol. 56, n. 11*. November 2006. pp. 4671-4680.

[14] G. M. Joksimovic, J. Penman. The detection of inter-turn short circuits in the stator windings of operating motors. *IEEE Transactions on Industrial Electronics, vol. 47, n. 5*. October 2000.

[15] G. Y. Sizov, A. S. Ahmed, C. C. Yeh, N. A. O. Demerdash. Analysis and diagnostics of adjacent and nonadjacent broken-rotor-bar faults in squirrel-cage induction machines. *IEEE Transactions on Industrial Electronics, vol. 56, no. 11*. November 2009. pp. 4627-4640.

[16] A. Bellini, A. Yazidi, F. Fillipeti, C. Rossi, G. A. Capolino. High frequency resolution techniques for rotor fault detection of induction machines. *IEEE Transactions on Industrial Electronics, vol. 55, n. 12*. December 2008, pp. 4200-4209.

[17] G. R. Bossio, C. H. De Angelo, J. M. Bossio, C. M. Pezzani, G. O. Garcia. Separating broken rotor bars and load oscillations on IM fault diagnosis through the instantaneous active and reactive currents. *IEEE Transactions on Industrial Electronics, vol. 56, n. 11*, November 2009. pp. 4571-4580.

[18] J. R. Stack, T. G. Habelter, R. G. Harley. Fault signature modeling and detection of inner-race bearing faults. *IEEE Transactions on Industrial Applications, vol. 42, n. 1*. January-February 200. pp. 61-68.

[19] G. M. Joksimovic. Dynamical simulation of cage induction machine with airgap eccentricity. *Proceeding on Instrumentation Electronics Engineer - Electric Power Applications, vol. 152, n. 4*, July 2005. pp. 803-811.

[20] F.J. Villalobos-Piña, R. Álvarez-Salas. A new Induction Motor Model for Fault Analysis. *International Review on Modelling and Simulation, vol. 4, n. 5*, October 2011. pp. 2145-2152.

[21] F.J. Villalobos-Piña, R. Álvarez-Salas. Diagnóstico de Fallas en Motores de Inducción, diagnóstico mediante el uso del ISP (Instanteneous Space Phasor). Vol. 1. Editorial Académica Española, 2011, pp. 43-48.

Francisco Javier Villalobos Piña received the electronics engineer and M. Sc. degree from Aguascalientes Institute of Technology, Mexico, in 1998 and 2001, respectively and Ph.D. degree in Electrical Engineering at the University of San Luis Potosí in 2011. He is also a professor in the Electronics Engineering Department, Aguascalientes Institute of Technology. His research areas are power electronics, fault diagnosis and control applied to electric machines.

Ricardo Álvarez Salas received the B.S. in 1993 from University of San Luis Potosi, Mexico. He received the M.S. from Scientific Research and Graduated Studies Center of Ensenada (CICESE), Mexico, in 1996 and the D.E.A. and Ph.D. degrees in control engineering from the Institute National Polytechnique of Grenoble, France, in 1998 and 2002, respectively. He is currently a professor in the Electrical Engineering Department, University of San Luis Potosí. His current research interests include control and fault diagnosis of electric machines and DSP applications. Dr. Alvarez is member of the IEEE, SIAM and the IET.

Eduardo Cabal-Yepez (M'09) received the B.Eng. and M.Eng. degrees from the University of Guanajuato, Salamanca, Mexico, and the D.Phil. degree from the University of Sussex, Brighton, U.K in 2007. Dr. Cabal-Yepez is an Associate Editor for the Journal of Computers and Electrical Engineering since 2011 and the Journal of Digital Signal Processing since 2012. He is Editor-in-Chief for the World Research Journal of Computer Architecture since 2012. He is currently a Titular Professor with the Division of Engineering, University of Guanajuato doing research work focused on hardware signal processing in field programmable gate arrays for applications in mechatronics. He has been an Adviser of over 15 theses.

Arturo Garcia-Perez (M'10) received the B.E. and M.E. degrees in electronics from the University of Guanajuato, Salamanca, Mexico, and the Ph.D. degree in electrical engineering from the University of Texas, Dallas, in 1994 and 2005, respectively. He is currently an Associate Professor with the Department of Electronic Engineering, University of Guanajuato. He has been an Adviser of over theses. His field of interest includes digital signal processing for applications in mechatronics.

A Novel Non-Invasive Method for Detecting Missing Wedges in an Induction Machine

Maciej Orman, Agnieszka Nowak, J.R. Ottewill, C. T. Pinto

Abstract -- **This paper presents a newly developed algorithm for evaluating the health of an induction machine. The proposed algorithm is based on spectrum analysis of an impedance calculated using measured stator current and voltage signals. The main idea is to calculate the frequency spectrum of the impedance for each power phase and compare specific differences between phases. Experimental investigations show that the method yields very accurate results and can form an important part of a machine monitoring system. In particular the presented method is shown to be successful in detecting missing wedges in electric motors.**

Index Terms-- **machine monitoring, impedance, induction machine, magnetic wedges**

I. INTRODUCTION

A missing or broken stator magnetic slot wedge is a well-known possible fault condition of an electrical machine. It is important to detect this failure at an early stage as it can lead to serious rotor or stator winding faults if left uncorrected. Magnetic wedges are brittle and are known for becoming loose and detaching from the slots because of the magnetic force acting in the air gap [1]-[3]. The most commonly used procedure for detecting failed magnetic wedges is visual inspection. Unfortunately such a procedure requires the machine to be stopped and dismantled; a particularly invasive and disruptive approach. In this paper a non-invasive method is proposed.

Current spectrum analysis is a method which today represents a standard for detecting faults in electric machines. The increasing availability of powerful data acquisition systems and suitable sensors such as current and voltage probes has led to the appearance of many condition monitoring systems based on these measurements. A variety of faults which can occur in induction machines, have been extensively studied [4], [5] and many monitoring methods have been proposed to detect problems [4]-[13]. However most of those methods are based on current signature analysis and do not utilize voltage signals.

Whilst it is true that in the case of a machine supplied direct-on-line the dynamic signatures in current signals owing to rotor faults are much more easily discerned than any equivalent signatures in the voltage, it is also true that by neglecting voltage measurements, some information relevant for condition monitoring is also neglected. This situation is particularly relevant in the case of electrical machines which are supplied by a drive, where controller actions can act to transfer information from the current signals to the voltage supplied to the machine. Thus, methods that combine both currents and voltages measured from power cables connecting the electrical rotating machines to the power source can ensure that no potentially useful diagnostic information is ignored. Examples of known methods for evaluating the health of electric machines using a combination of both measured current and voltage include the analysis of instantaneous power [9] and of estimated torques [14]. However, these methods are not without their own limitations for example methods of estimating the torque of a machine require accurate estimates of machine parameter values such as the stator resistance [14]. Such values are not always easily available and their accurate estimation is non-trivial.

Recently there have been some investigations into the diagnostic information contained with the admittance or impedance of a supply phase. For example, Ref. [15] describes a method of deriving and analyzing the admittance or impedance obtained for a single supply phase of an electrical machine. It is often the case that a greater amount of diagnostic information may be obtained by comparing the differences between the separate phases of a polyphase rotating electrical machine. An example of such a case is in the diagnosis of missing magnetic wedges in a three-phase electrical machine. In this situation, whilst certain characteristics of the impedance estimated from currents and voltages measured from two of the three phases will differ from the equivalent impedance estimates from a healthy machine, in the third phase the difference between the impedance estimated in the healthy case and in the case with missing magnetic wedges may be negligible. As a result, there is a risk of monitoring approaches based upon impedance estimated from currents and voltages measured from only one phase of an electrical machine being insensitive to certain developing faults, with the potential for many missed alarms.

There are various advantages of impedance based condition monitoring systems. In addition to being comparatively cheap to implement, there are inherent advantages associated with directly measuring signals from the power cables connecting the electrical rotating machines to the power source. Firstly, current and voltage sensors may be considered as non-invasive as the electric rotating machine forms part of the electromechanical system. Secondly, the influence of transmission path effects associated with the location of the transducer relative to a fault is less severe.

In the presented paper, as opposed to discarding the information available at the motor terminals, the method combines the information from both the voltage and current signals measurable at the motor terminals. Specifically, the measurements are combined to calculate the frequency spectrum of the impedance of the electric machine, impedance being the resistance to the flow of current that a circuit exhibits when a voltage is applied to it. The work presents an original method of calculating the impedance spectrum and a novel approach of analyzing the derived signals to obtain information pertaining to the missing magnetic wedges of the machine.

II. MAIN CONCEPT OF SPECTRAL IMPEDANCE CALCULATION

The proposed method for calculating the impedance spectrum requires that motor three phase currents and three phase voltages are measured by any known measurement tool(s). Initially, the current and voltage spectra of each phase are calculated using the discrete Fast Fourier Transform (FFT). The impedance spectrum, F_Z can be calculated as the ratio of the voltage spectrum of a single supply phase to the current spectrum of the same supply phase: $F_Z = F_U / F_I$, where F_I and F_U are the current and voltage spectrum respectively. The act of inverting the current spectrum serves to amplify noise inherent in typical measurements. As a result it is necessary to apply some further processing steps to the current spectra, using a modified version of the current spectrum in the impedance calculation.

Initially, frequencies connected with the supply frequency are removed from the current spectrum in order to minimize its, otherwise overbearing, influence on the impedance spectrum. Furthermore, in order to prevent the amplification of noise during the process of inverting the current spectrum, it is necessary to filter the spectrum so that frequencies, whose amplitudes are smaller than a given threshold ζ, are set to 1. This may be expressed as:

$$F_{If} = \begin{cases} 1 - \dfrac{F_I}{\max F_I}, & F_I \geq \zeta \, \overline{F_I} \\ 1 & F_I < \zeta \, \overline{F_I} \end{cases} \qquad (1)$$

where:
F_I – current spectrum with no harmonics,
$\overline{F_I}$ – average value of current spectrum with no harmonics.

After this first filtering operation, a further operation is applied. In order to prevent the overemphasis of low values close to zero, the values of the current spectrum that are below a certain threshold ξ are set to be equal to that threshold ξ. The resulting modified current spectrum, $\breve{F_I}$ can be expressed as:

$$\breve{F_I} = \begin{cases} F_{If}, & F_{If} > \xi \\ \xi, & F_{If} \leq \xi \end{cases} \qquad (2)$$

After the above computations, the impedance spectrum is estimated according to the formula:

$$F_Z = F_{Za} + F_{Zb} + F_{Zc} = \frac{F_{Ua}}{\widetilde{F_{Ia}}} + \frac{F_{Ub}}{\widetilde{F_{Ib}}} + \frac{F_{Uc}}{\widetilde{F_{Ic}}} \qquad (3)$$

where:
F_{Ua}, F_{Ub}, F_{Uc} – voltage spectrum of the phases (a, b and c respectively) normalized into [pu],
$\widetilde{F_{Ia}}, \widetilde{F_{Ib}}, \widetilde{F_{Ic}}$ – modified current spectrum of the phases (a, b and c respectively),
F_{Za}, F_{Zb}, F_{Zc} – impedance spectrum of the phases (a, b and c respectively).

III. DETECTION OF MISSING MAGNETIC WEDGES

The presented algorithm scheme assumes that the motor slip is known. There are various known methods of estimating motor slip on the basis of current measurements [16]-[18].

The purpose of inserting slot wedges is to protect windings in the core from becoming loose and being displaced into the path of the spinning rotor. The forces acting on the wedge may lead to the loss of the wedge and subsequently, after a short while may lead to the stator failing [19], [20]. Simulations performed at ABB Corporate Center showed that loose or missing magnetic wedges result in the appearance of components at specific fault frequencies in the spectrum of measured stator current. Specifically, it was identified that missing wedges produced amplitude components at:

$$f_{MW} = R \left(\frac{1-s}{p} \right) f_s \pm m f_s \qquad (4)$$

where:
f_{MW} – missing wedges fault frequencies
R – number of rotor bars
m – an integer (0, 1, 2 …)
p – number of pole pairs
f_s – supply frequency

Employing the dependency between the slip and the rotating frequency of the rotor, the above formula can be rewritten as

$$f_{MW} = R f_r \pm m f_s \qquad (5)$$

where:
f_r – rotor speed frequency.

IV. MEASUREMENTS AND NUMERICAL CALCULATIONS OF IMPEDANCE

For the purpose of this work current and voltage measurements were collected using the ABB MACHsense-P condition monitoring tool. ABB MACHsense-P is a walk around condition monitoring service tool provided by ABB which specifically focuses on electric motors. Three phase currents and voltages were measured by the MACHsense-P tool at a sampling frequency of 6.4 kHz. The methodology presented in this paper is an embedded functionality in the ABB MACHsense-P tool. The numerical calculations were carried out for two motors of the same type but with differing health states; one of the motors was considered as being healthy whilst the second motor was known to be missing magnetic wedges. Specific parameters of the motor are given in the appendix. In this section an example of how the previously described methodology is used to calculate the impedance spectrum is presented based on measurements of current and voltage of motor 1.

Fig. 1 presents the current spectrum of phase-a after the modifications described by Equations (1) and (2).

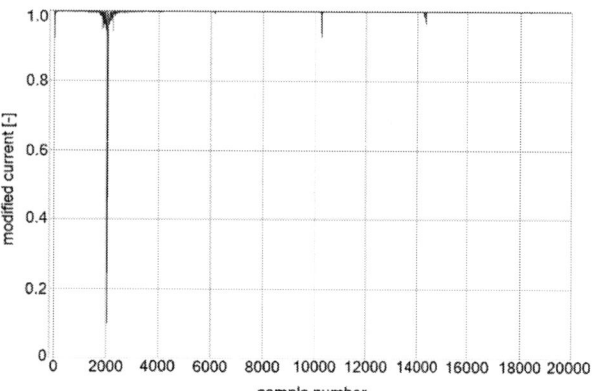

Fig. 1. Modified current spectrum calculated according to formula in (1) and (2)

It may be observed that the modified current spectrum does not contain any harmonics of power supply. After filtering out values of current spectrum close to zero according to (2) it is possible to calculate the impedance spectrum using (3). The impedance spectrum is shown in Fig. 2. Whilst the supply frequency and its harmonics dominate the spectrum, there are also peaks which are unrelated with the supply frequency visible. These additional peaks are related to the state of the machine.

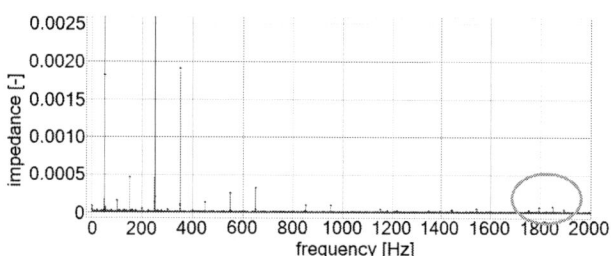

Fig. 2. Spectral impedance calculated according to formula (3), sample measurements of healthy motor under 75% loading.

Fig. 3 presents the impedance spectrum for the same loading case as in Fig. 2 but for a motor with missing magnetic wedges. Also highlighted in the figure is a region of the frequency spectrum where there are peaks visible which are not apparent in the equivalent impedance spectrum for the

healthy machine. These peaks are at frequencies predicted by (4).

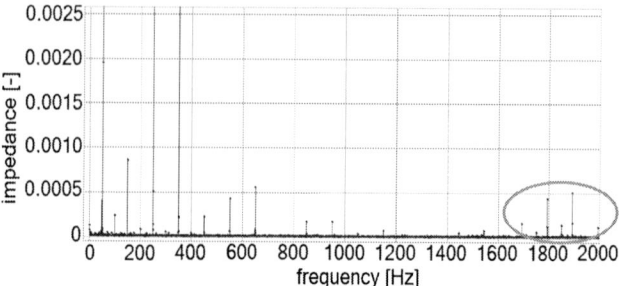

Fig.3. Spectral impedance calculated according to formula (3), sample measurements of the motor with missing wedges under 75% loading.

Fig. 4 and Fig. 5 show an enlarged view of the highlighted region of the impedance spectrum for the healthy and missing magnetic wedges case respectively.

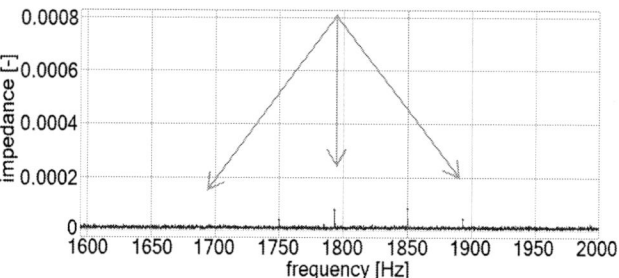

Fig. 4. Spectral impedance calculated according to formula (3), sample measurements of healthy motor under 75% loading, ZOOM.

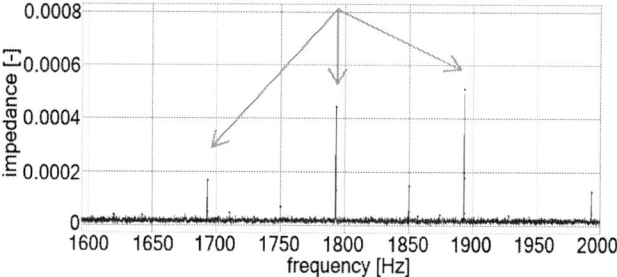

Fig. 5. Spectral impedance calculated according to formula (3), sample measurements of the motor with missing wedges under 75% loading, ZOOM

It may be observed in Fig. 4 and Fig. 5 that the frequency component with the largest amplitude in the highlighted range has an amplitude of almost 0.0006 [pu] for the case of missing magnetic wedges, whilst in the healthy motor case maximum amplitude is less than 0.0001 [pu].

Fig. 6 and Fig. 7 respectively show the impedance spectra for healthy (Fig. 6) and missing wedge (Fig. 7) cases for no load condition. In the case of missing magnetic wedges, the components previously identified in the case of 75% loading as being related to missing magnetic wedges are significantly larger than in the equivalent healthy case.

978-1-4799-0024-4/13 $31.00 © 2013 IEEE   202

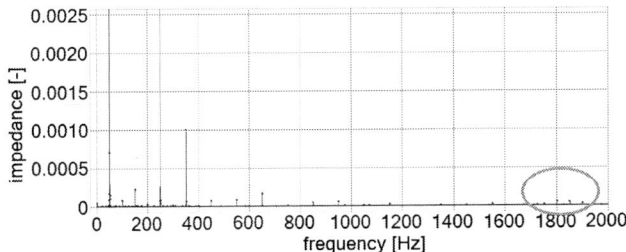

Fig. 6. Spectral impedance calculated according to formula (3), sample measurements of healthy motor under no load.

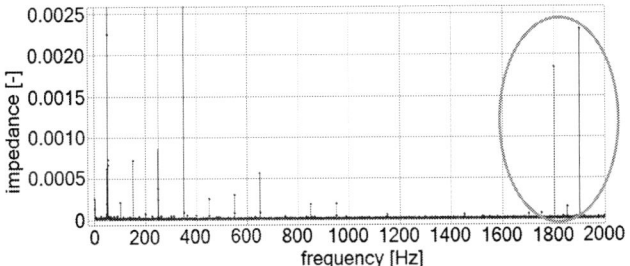

Fig.7. Spectral impedance calculated according to formula (3), sample measurements of the motor with missing wedges under no load

Fig. 8 and Fig. 9 present an enlarged view of the highlighted region of the impedance spectrum for the healthy and missing magnetic wedges case respectively for the no-load case. It may be observed that for the no-load case, the difference between amplitudes of certain frequencies is significant. The largest amplitude in this region is approximately 0.0025 [pu] in the case of missing wedges, whilst for the healthy case they are almost not visible at all. It is also noticeable is that this difference is much bigger than in the case of 75% load.

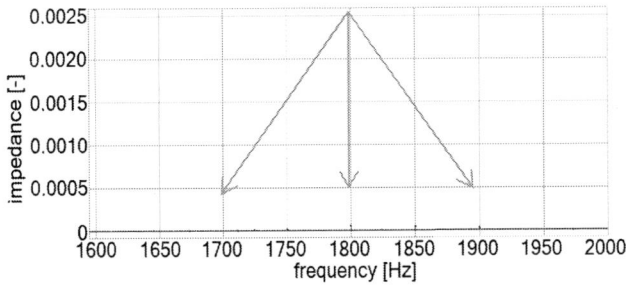

Fig. 8. Spectral impedance calculated according to formula (3), sample measurements of healthy motor under no load, ZOOM.

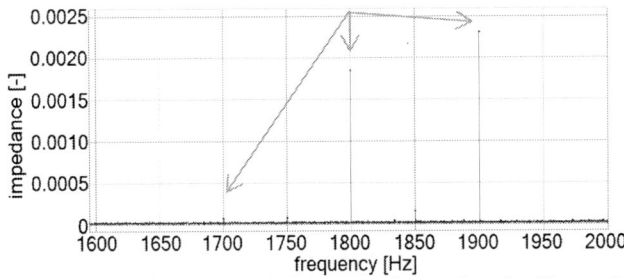

Fig.9. Spectral impedance calculated according to formula (3), sample measurements of the motor with missing wedges under no load, ZOOM.

The impedance spectrum for all phases and both motor cases is calculated in same way. For all the measurement cases the sampling frequency was set to 6.4 kHz while the signal length was equal to 10 s leading to a spectral resolution of 10 points per Hz. There are many known methods of interpolating between discrete points in a frequency spectra in order to improve the accuracy of amplitudes and frequencies required for spectral analyses [21]-[26]. In the presented approach the accuracy of the amplitudes of particular components relating to specific fault components is improved using such an interpolation method based on ratios involving three samples around any particular peak in the discrete impedance spectrum.

V. DIFFERENCES BETWEEN IMPEDANCE SPECTRA CALCULATED FOR DIFFERENT SUPPLY PHASES

The missing wedges fault frequencies are clearly visible in the impedance spectrum, calculated according to (3). The spectral impedance for each of the phases in the logarithmic scale of the machine AMA 500L4A with missing magnetic wedges under 75% loading is depicted in Fig. 10. Fault frequencies are indicated by the arrows.

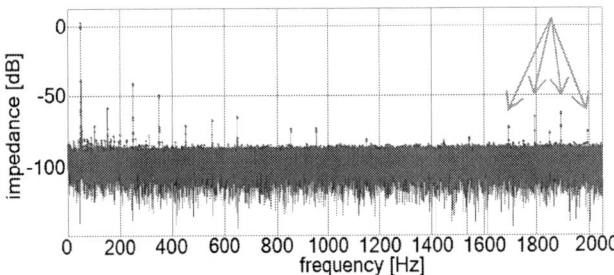

Fig. 10. Impedance spectra of each phase in logarithmic scale of the motor with missing magnetic wedges under 75% loading.

The height of the missing wedges fault frequencies amplitudes differ depending on the load condition. However, for every loading condition, there is always a high peak associated with the frequency equal to $Rf_r - 3f_s$ (for m = 3). Moreover, the amplitude of this particular component differs for each of the three phases of the machines.

Fig. 11 – Fig. 13 show comparisons between the amplitude of components in the impedance spectrum in the vicinity of the frequency $Rf_r - 3f_s$ for healthy machines and for machines with missing magnetic wedges. Three different load cases are considered, Fig. 11 shows the impedance spectra when the motors were loaded at 75% of nominal load, Fig. 12 shows the spectra when the motors were loaded at 50% of nominal load and Fig. 13 shows the spectra under no load conditions. It may be observed that for all loads the peak which has been identified as being related to a missing magnetic wedge fault is visible in the faulty motor spectra for at least one of the supply phases. Significant amplitude components at the identified frequency never appear in the equivalent healthy motor spectra for any of the supply phases. It may also be noted from the spectra of motors with missing wedges that the load has an influence on the amplitude of this component. Generally, the higher the load, the greater the amplitude of this peak relative to the other components in the spectrum which may be considered as background noise.

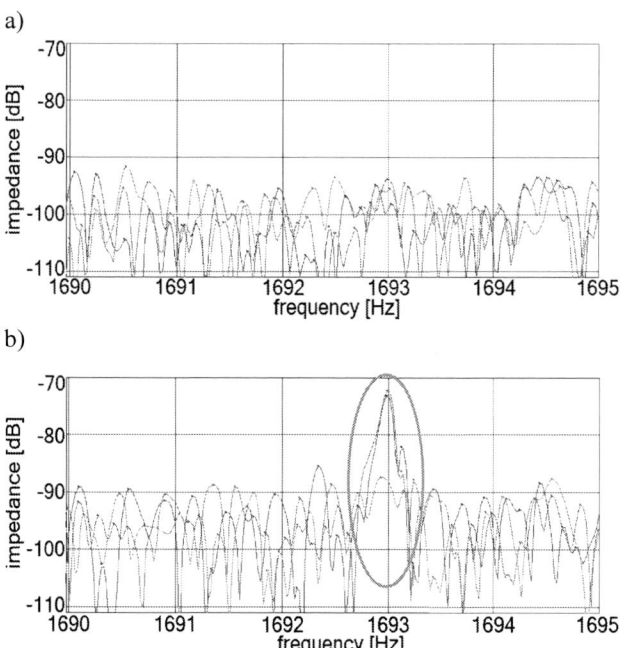

Fig.11. Comparison of the healthy motor (a) and motor with magnetic wedges loss (b) at impedance spectra of each phase in logarithmic scale, both under 75% loading

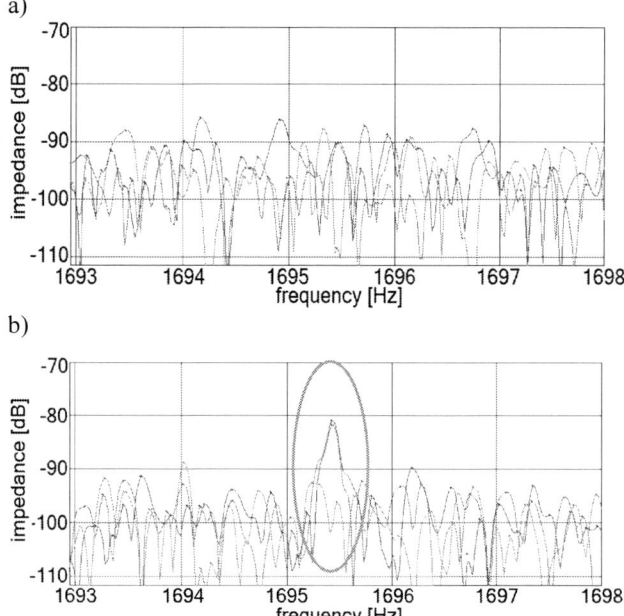

Fig.12. Comparison of the healthy motor (a) and motor with magnetic wedges loss (b) at impedance spectra of each phase in logarithmic scale, both under 50% loading

Finally and perhaps most interestingly, in the case of a defective motor, the amplitude of components associated with magnetic wedge problems differs between the three supply phases. In some cases the impedance spectrum of a particular supply phase shows no evidence of a magnetic wedge problem whilst the other two phases show clear evidence. This is likely to be due to the missing magnetic wedges allowing the stator windings to move within the space vacated by the wedge. Clearly geometry dictates that the effects of a missing magnetic wedge will be more significant in certain supply phases than in others. Thus the initial assertion of this paper that it is necessary to monitor multiple phases in order to reliably detect wedge problems, is shown to be valid

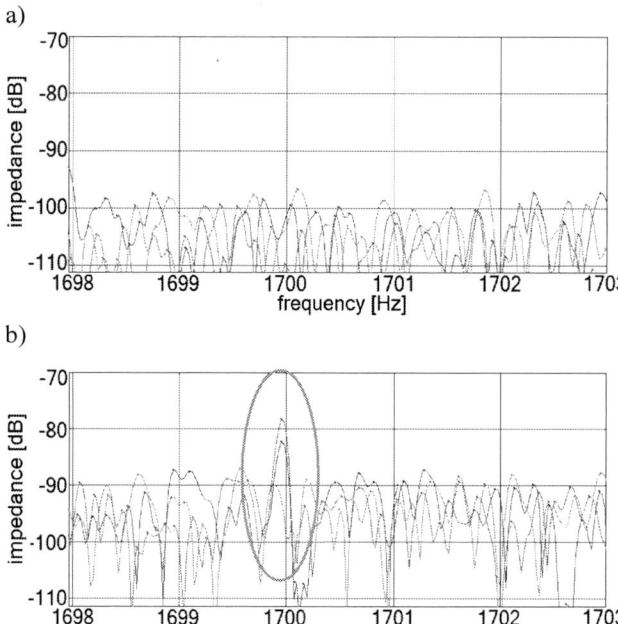

Fig.13. Comparison of the healthy motor (a) and motor with magnetic wedges loss (b) at impedance spectra of each phase in logarithmic scale, both under no load.

Armed with a knowledge of which frequency components may be analysed in order to identify whether or not a motor has problems associated with its magnetic wedges, it is possible to develop fault metrics in order to automate the detection of wedge problems. What follows is an example of a potential fault metric for diagnosing magnetic wedge problems. Two steps are performed in order to automatically detect problems associated with missing wedges. After the fault frequency is calculated, the amplitude of the peak at this component is estimated by comparing its amplitude with surrounding noise:

$$H_{MW} = \max\left(H(f_{MW,a}), H(f_{MW,b}), H(f_{MW,c})\right) \quad (6)$$

where:

$H(f_{MW,a}), H(f_{MW,b}), H(f_{MW,c})$ – height of the peak corresponding to the fault frequency of phase a, b and c, respectively

$|f_{MW,a}| = \text{Amp}(f_{MW,a}) - \text{RMS}(F_{Za})$

$\text{Amp}(f_{MW,a})$ – amplitude of the fault frequency

$\text{RMS}(F_{Za})$ – root mean square of the spectral impedance in the logarithmic scale of the phase A.

The first requirement to satisfy is that the peak height estimated using formula (6) has to be bigger than a given threshold value. Only if this condition is met, should further calculations be performed. If this condition is not met it is considered that there is no evidence of magnetic wedge loss.

In the second step, the difference in the amplitude of components associated with magnetic wedge problems between the three supply phases is estimated:

$$\text{Ind}_{MW} = \frac{(H(f_{MW,a}) - \overline{H})^2 + (H(f_{MW,b}) - \overline{H})^2 + (H(f_{MW,c})\,\overline{H})^2}{3} \quad (7)$$

Where:

$\overline{H}$ – average value of the height of the peak corresponding to the fault frequency of each phase.

As previously discussed, it is necessary to monitor more than a single supply phase as the amplitude of the component relating to a wedge fault differs between the various phases of a machine. The step taken in (7) removes the risks of missed alarms associated with considering the amplitude components of only a single phase. Investigations conducted by the authors, though not reported here, have indicated that values of $\text{Ind}_{\text{wedges}}$ greater than 15 (warning level) are considered as an indicator of possible lost magnetic wedges and values greater than 25 (alarm level) indicate evidence of the missing magnetic wedges fault. This methodology, which is based upon the results presented earlier in this paper, has been shown to be successful for a number of different test cases. Further verification studies are currently under way.

VI. CONCLUSION

In this paper, a method of missing magnetic wedges detection is presented. The proposed method is based on spectrum analysis of a 3-phase impedance calculated using measured stator current and voltage signals.

On the basis of measurements which have been conducted, it is possible to say that the proposed algorithm scheme gives satisfactory results. As presented in the results section, the method allows successful separation of healthy machine from one which has missing magnetic wedges. The results show that method is successful for different loading conditions of an electric motor, including the no-load case

The authors consider the presented method as a powerful tool which can significantly improve known diagnostic techniques and can be an important part of machine monitoring systems.

VII. APPENDIX

TABLE I
MOTOR NAMEPLATE DETAILS

Parameter	Value
Active power [KW]	2400
Nominal voltage [V]	750
Nominal current [A]	2070
Nominal power factor [-]	0.89
Rotor speed [rpm]	1490
Winding connection	Δ
Number of poles per phase winding [-]	2
Nominal frequency [Hz]	50
Number of rotor bars [-]	74

VIII. REFERENCES

[1] M. Davis, "Problem and solutions with Magnetic Stator Wedges", *Iris Rotating Machine Conference*, San Antonio, Texas, 5 June 2007.

[2] Scollay & Stewart, "The Real Cost of Magnetic Wedges in Improved Performance of Induction Motors", *CWIEME*, Berlin, Germany, 2007.

[3] R. A Hanna, Peter Klinowski, "Failure Analysis of Three-Slow Speed Induction Motors for Reciprocating Load Applications", *IEEE Transaction on Industry Applications*, Vol. 43, No. 2, April/March 2007.

[4] A.H. Bonnett, G.C. Soukup, "Analysis of rotor failures in squirrel-cage induction motors" *IEEE Transactions on Industry Applications*, Volume 24, Issue 6, Nov/Dec 1988 Page(s):1124 – 1130.

[5] A.H. Bonnett, G.C. Soukup, "Cause and analysis of stator and rotor failures in three-phase squirrel-cage induction motors" *IEEE Transactions on Industry Applications*, Volume 28, Issue 4, Jul/Aug 1992 Page(s):921 – 937.

[6] W. T. Thomson, M. Fenger, "Current signature analysis to detect induction motor faults" *IEEE Industry Applications Magazine*, Volume 7, Issue 4, Jul/Aug 2001 Page(s):26 – 34.

[7] Long Wu, "Separating Load Torque Oscillation and Rotor Faults in Stator Current Based-Induction Motor Condition Monitoring" Georgia Institute of Technology, May 2007.

[8] S. Nandi, S. Ahmed, H. Toliyat, "Detection of Rotor Slot and Other Eccentricity-Related Harmonics in a Three-Phase Induction Motor with Different Rotor Cages" *Power Engineering Review, IEEE*, Sept. 2001, Volume: 21, Issue: 9, Page(s): 62-62.

[9] P. Vas, "Parameter estimation, condition monitoring, and diagnosis of electrical machines" *Oxford: Calendar Press*, 1993.

[10] G. Didier, E. Ternisien, O. Caspary, H. Razik, "Fault Detection of Broken Rotor Bars in Induction Motor using a Global Fault Index", *IEEE Transactions on Industry Applications*, Volume 42, Issue 1, Jan.-Feb. 2006, Page(s): 79 – 88.

[11] S.M.A. Cruz, A.J.M. Cardoso, H.A. Toliyat, "Diagnosis of stator, rotor and airgap eccentricity faults in three-phase induction motors based on the multiple reference frames theory" : *Industry Applications Conference*, Publication Date: 12-16 Oct. 2003 Volume: 2, On page(s): 1340- 1346 vol.2.

[12] D.R. Rankin, "The industrial application of phase current analysis to detect rotor winding faults in squirrel cage induction motors" *Power Engineering Journal*, Volume 9, Issue 2, Apr 1995 Page(s): 77 – 84

[13] A. Bellini, et al, "On-field experience with online diagnosis of large induction motors cage failures using MCSA" *Transactions on Industry Applications*, Volume 38, Issue: 4, Aug 2002, Page(s): 1045-1053.

[14] D. Norman "Diagnosing Rotor Bar Issues with Torque and Current Signature Analysis" *SKF @ptitude Exchange*, October, 2009.

[15] Shih-Fu Ling, Lianyu Fu, "Method and Apparatus for Assessing Condition of Motor-Driven Mechanical System" US patent application US0282548A, 2007.

[16] Yoon-Ho Kim,Yoon-Sang Kook "Neural network based speed sensorless induction motor drives with Kalman filter approach" Industrial Electronics Society, 1998. Proceedings of the 24th Annual *Conference of the IEEE*, Volume 2, 4 Sep 1998, vol.2, Page(s): 997 - 1001

[17] Orman M., Orkisz M., Pinto C. T., "Parameter identification and slip estimation of induction machine". *Elsevier, Mechanical Systems and Signal Processing*, 18 November 2010

[18] Orman M., Orkisz M., Pinto C. T., Slip Estimation of a Large Induction Machine Based on MCSA, *Diagnostics for Electric Machines, Power Electronics & Drives (SDEMPED)*, IEEE International Symposium on, Bologna, Page(s): 568 – 572, August 2011

[19] Henk de Swardt "Electric Motor Failure Prevention: Wedge Failures". *Energizer*, February 2004

[20] Mike Davis "Problems and solutions with Magnetic Stator Wedges". *Iris Rotating Machine Conference,* San Antonio, TXBiographies 5 June 2007,

[21] B. G. Quinn, "Estimating Frequency by Interpolation Using Fourier Coefficients" *IEEE Trans. Signal Processing*, Volume 42, May 1994, Page(s): 1264-1268.

[22] B. G. Quinn, P. J. Kootsookos, "Threshold Behaviour of the Maximum Likelihood. Network: Comput. Neural Syst. Issue 13, 2002 , Page(s): 447–456

[23] B. G. Quinn, "Threshold Behavior of the Maximum Likelihood Estimator of Frequency" *IEEE Transactions on Signal Processing*, Volume 42, 1994 November, Page(s): 3291-3294.

[24] T. Grandke, "Interpolation Algorithms for Discrete Fourier Transforms of Weighted Signals" *IEEE Transactions on Instrumentation and Measurement*, Volume IM-32, June 1983, Page(s): 350-355.

[25] V.K. Jain et al, "High-Accuracy Analog Measurements via Interpolated FFT" *IEEE Transactions on Instrumentation and Measurement*, Volume IM-28, June 1979, Page(s): 113-122.

[26] D. C. Rife and R. R. Boorstyn, "Single-Tone Parameter Estimation from Discrete-Time Observations" *IEEE Transactions on Information Theory*, Volume IT-20, September 1974, Page(s): 591-598

Maciej Orman was born in Kraków in Poland, on October 27, 1983. He received MSc. Eng. degree in Automation and Robotics at AGH, University of Science and Technology in Krakow in 2008. He received his PhD degree at the same university in 2011. Since April 2009 he has worked at ABB's Corporate Research Center in Krakow, where he has dealt with condition monitoring of electric machines. Since December 2011 he has worked at ABB's Corporate Research Center in Shanghai were he continues working on condition monitoring methods of electric machines.

Agnieszka Nowak received her MSc. Eng. Degree in Automation and Robotics at AGH, University of Science and Technology in Krakow in 2010. Currently she is a Member of Research Staff at ABB's Corporate Research Center in Poland working of condition monitoring of electrical machines. Her main research interest include applying machine learning algorithms and techniques into diagnostic of electrical machine as well as modeling and analysis of non-linear systems.

James Ottewill received BEng and Ph.D. degrees in Mechanical Engineering from the University of Bristol, UK in 2005 and 2009, respectively. During his Ph.D. research, he investigated the non-linear problem of gear rattle. Currently, he is a Scientist at ABB's Corporate Research Center in Poland, working in the field of condition monitoring. His main research interests include dynamic testing, modeling and analysis of non-linear systems, signal processing and advanced diagnostic techniques.

Cajetan Pinto was born in Mumbai India on December 21, 1961. He graduated from VJTI, Mumbai with a Bachelors Degree in Electrical Engineering and from the University of California with a Masters degree Electrical Engineering. His employment experience included Lenzohm Electrical Engineering and ABB Ltd, and is currently R&D Manager for Motor and Generator Service at ABB. His special fields of interest included electrical insulation and electromagnetic behavior, analysis and diagnosis, electrical motor and generator service technology.

Stator Circulating Currents as Media of Fault Detection in Synchronous Motors

Pedro Rodriguez, Pawel Rzeszucinski, Maciej Sulowicz, Rolf Disselnkoetter, Ulf Ahrend, Cajetan T. Pinto, James R. Ottewill and Stephan Wildermuth.

Abstract -- **Often found in critical, high power applications, synchronous machines require reliable condition monitoring systems. Large synchronous machines are typically designed with parallel connected windings in order to split the currents in parallel paths, delivering the total power at the terminals. Under ideal symmetrical conditions, no current will circulate between parallel branches of the same phase. However, when a motor fault breaks this symmetry, currents circulate between the branches. Thus, due to the fact that they are only non-zero under faulty conditions, circulating currents potentially represent a sensitive indicator of faulty condition. In this paper, the advantages of using the circulating current between parallel branches of the stator of a synchronous motor as an early indicator of motor faults are shown. Analysis is conducted both through simulation, via the use of finite element methods (FEM), and through experimentation using a specially-designed synchronous machine which allows various fault conditions to be investigated. Through comparison between experiment and simulation, the simulation tool is validated. Furthermore, it is shown that the circulating current is better suited for fault detection than either the branch or the stator current. It is concluded that an improved condition monitoring and protection system for a synchronous machine may be achieved if these currents are monitored.**

Index Terms—**Circulating currents, synchronous machines, fault detection, eccentricity.**

I. INTRODUCTION

THIS work concerns the investigation of abnormal conditions in synchronous motors (SM). In addition to being used in applications where a fixed speed is desired, these machines are often found in high power applications where they are preferred over induction motors due to their higher efficiency. Synchronous motors represent large investments and typically drive processes where downtime results in significant capital losses. Thus, detecting faults at an early stage can help avoid catastrophic failures and be beneficial in the scheduling of maintenance actions. In contrast to induction motors (IM), the number of papers devoted to monitoring faults in SM is limited and reports on failure statistics are scarce. Most of the available literature relates to the monitoring of failures in synchronous generators or in permanent magnet SM [4]. There are various different physical quantities that can be monitored

and analyzed to ascertain motor condition, such as temperature, vibration, fluxes and currents [1]. The value of temperature monitoring is very limited due to its localized nature and slow response. Vibration monitoring is a well-established technique for monitoring both mechanical and electrical faults in IM. However, condition monitoring based on vibration analysis is often challenging in an industrial environment. This is due to the fact that pronounced background noise and complex electro-mechanical interactions modulate and mask the signals of interest. Additionally, not all faults may be related to a clear vibration signature. As an alternative, flux sensors are good indicators of SM failures but are limited by safety and accessibility issues. Due to their high reliability and because access is possible in most cases, current sensors may be considered as an appropriate part of a continuous monitoring system for a high power machine. Reference [2] showed through the application of support vector machines (SVM) that from among the different signals which may be measured and analyzed for diagnostic information, the circulating current clearly reacts to failures in induction motors. There have been some attempts to prove the usefulness of this indicator for the case of large generators [5]-[6]. However, no investigation has been found for the case of synchronous motors. These large machines are usually designed with parallel connected windings in order to split the currents in parallel paths, delivering the total power at the terminals. It is also known that parallel branches tend to reduce the unbalance magnetic pull produced by eccentricity problems in induction motors [3]. These currents act to smooth the distortion in the air-gap flux density. Thus, a motor with parallel branches is quieter than its counterpart. Ideally there is not any current circulating between parallel branches of the same phase. However, when ideal symmetry is lost due to a problem in the machine, such as eccentricity or short-circuits circulating currents can be induced between the two parallel connected windings. Thus, circulating currents between parallel branches represent a natural fault indicator, as shown in [8]-[9] for the case of IM.

In this paper, the usefulness of the circulating current as an indicator for detecting faults is proven through numerical simulation models and real measurements on a specially-designed synchronous motor. It is also shown that circulating current analysis has a higher sensitivity than the analysis of stator currents, giving the possibility of a more reliable monitoring system based on circulating current monitoring. This technique enables improved condition monitoring and protection of synchronous motors than an equivalent based on the monitoring of stator currents or stray fluxes.

P. Rodriguez is with ABB CRC Sweden (e-mail: pedro.rodriguez@se.abb.com).

P. Rzeszucinski and J. R. Ottewill are with ABB CRC Poland (e-mails: pawel.rzeszucinski@pl.abb.com, james.ottewill@pl.abb.com).

R. Disselnkoetter, U. Ahrend and S. Wildermuth are with ABB CRC Germany (e-mails:roll.disselnkoetter@de.abb.com, ulf.ahrend@de.abb.com, stephan.wildermuth@de.abb.com).

C. T. Pinto is with ABB Machine Service, India (e-mail: cajetan.t.pinto@in.abb.com).

M. Sulowicz is with Cracow Technical University (e-mail: pesoluwi@cyf-kr.edu.pl).

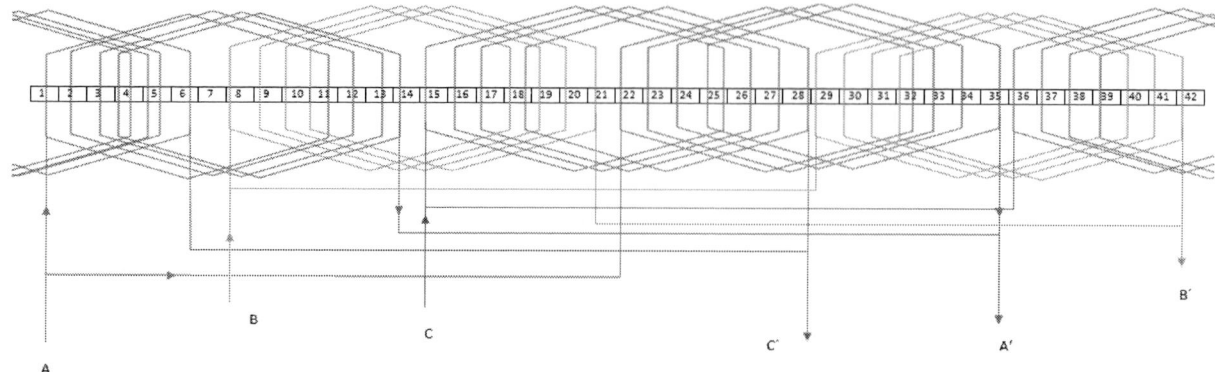

Fig. 1. Winding layout of the investigated SM.

II. CIRCULATING CURRENT ANALYSIS

A. Analytical fundamentals

The parallel connected stator winding layout of the motor under consideration is shown in Fig. 1. It may be observed that the stator layout corresponds to a 3-phase motor, with four poles and 42 slots in the stator. The parallel windings are represented by the same color. Geometrically, the parallel windings are 180 degrees apart from one another in a four pole motor and the induced electrical potential in these winding has the opposite direction. This is a general rule of winding design for any motor having two parallel windings. In other configurations the angle between parallel windings is a function of the number of poles and parallel windings.

If we derive the electrical circuit of phase A, the circuit depicted in Fig. 2 is obtained.

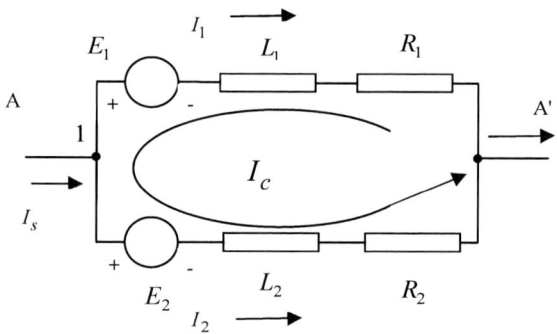

Fig. 2. Equivalent circuit of phase A.

By applying Kirchoff's first law and the superposition principle at node 1 we may obtain:

$$I_c = -I_s - I_1 \qquad (1)$$

$$I_c = I_s + I_2 \qquad (2)$$

Where I_c is the circulating current and I_s is the stator current and I_1, I_2 are the branch currents.

By combining (1) and (2), the following equation is obtained:

$$I_c = \frac{1}{2}(I_2 - I_1) \qquad (3)$$

If $I_1 = I_2$ there is no circulating current, which is the case of a symmetric machine. Kirchoff's second law allows the following equations, from which the branch currents may be calculated, to be obtained:

$$U_{AA'} = E_1 + I_1 \cdot (j\omega L_1 + R_1) \qquad (4)$$

$$U_{AA'} = -E_2 + I_2 \cdot (j\omega L_2 + R_2) \qquad (5)$$

Since the applied voltage is the same for the two parallel branches it is clear that variations in the the induced voltage, resistance and inductance of the branches would produce significant variations in the circulating currents. The induced voltage in each parallel branch is calculated for the case of two parallel branches [6]:

$$E_1 = 2qw_c k_{\omega\lambda} \tau \cdot f_r \cdot F_\lambda \cos(\lambda\omega t - \alpha_m - \beta) \cdot \frac{\mu_0}{g} \qquad (6)$$

$$E_2 = -2qw_c k_{\omega\lambda} \tau f_r F_\lambda \cos(\lambda(\omega t - \pi) - (\alpha_m + \pi) - \beta) \cdot \frac{\mu_0}{g} \qquad (7)$$

Where:

q is the slot number in each pole

w_c is turn number

$k_{\omega\lambda}$ the winding factor for the harmonic number λ

τ is the pitch of the coil

f_r is the rotation frequency of the rotor

F_λ is the mmf of the harmonic λ

α_m is the angle between two slots

β is the phase angle

μ_0 is the magnetic reluctivity of the air

g is the airgap between the rotor and stator

λ is the harmonic number

From equation (6) and (7), it is clear that airgap variation would produce substantial variations in the induced voltages.

The complexity involved in predicting the voltages increases in line with the number of parallel branches as well as with the complexity of the geometry of the machine. In the case of the machine under study, where the parallel branches are 180 degrees apart, eccentricity variation could be predicted from (6) and (7). However, for a machine with a higher number of poles it is rather difficult to calculate the induced voltages. Predicting the signatures in the circulating currents analytically is a complex task since it is necessary to take both the aigap permeance variation and the effect of every harmonic of the magnetic flux into account. This problem has been tackled in this investigation through the use of Finite Element Methods (FEM). The advantage of FEM over traditional analytical techniques is that it allows complex structures such as rotating machines to be modelled accurately, including material properties and non-linearity.

Regardless of the modelling approach employed it is worth noting that variation of the airgap permeance will produce substantial variation in the circulating current even if no variation is seen in the total stator current as shown in [9].

B. FEM model

In FEM, a complex problem represented by differential equations is transformed into a series of algebraic problems, which are easier to solve. The model utilized in this investigation solved the Maxwell equations in two dimensions. It is coupled with a circuit model to account for the three-dimensional end region fields, which are modelled approximately by constant end winding impedances in the circuit equations of the windings [7]. The calculation is made under the following assumptions: the magnetic field in the core is two-dimensional, the current density in the stator conductor is constant, the skin effect in the stator is negligible and the laminated iron core is treated as a non-conducting, magnetically non-linear medium, with the non-linearity being approximated to a single-value magnetisation curve. The details of the calculation method are presented in [2].

An internal ABB simulation tool is used in this investigation. As faulty conditions result in a loss of symmetry in the machine, in order to model the faulty conditions and calculate the circulating current in all branches, it is necessary to model the whole cross section. The machine data is shown in table I.

TABLE I. MACHINE DATA

PARAMETER	VALUE
Power [kW]	7.5
Voltage [V]	400
Phase winding voltage [V]	220
Current [A]	15.8
Rotational speed [rpm]	1500
Power factor	0.8
Frequency [Hz]	50
Field current [A]	8.56
Field voltage [V]	50

The simulation tool allows different types and severities of eccentricity to be modelled. As an example of the

simulation results which may be obtained from the FEM calculation, Fig. 3 shows the calculated flux lines at one instant of time when the motor has 40 % eccentricity.

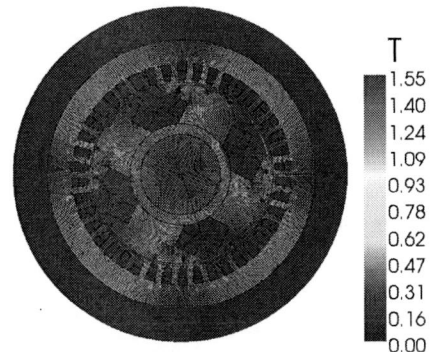

Fig. 3. Flux line when the motor is working in steady state with 40 % eccentricity.

C. Experimental setup and data acquisition

Fig. 4 shows the experimental arrangement with the synchronous motor investigated visible on the right hand side. The motor was connected via a Cardan coupling to a driven motor which acted as a generator and transferred the generated power to a large resistive load.

In order to properly start the synchronous machine, it was necessary to prepare a special set of electrical circuits, as shown in Fig. 5. In the initial stages of the starting process, the machine was powered by an autotransformer. As the angular speed of the rotor approached the synchronous speed of the machine, the excitation was switched to the field voltage. This resulted in the angular speed of the rotor reaching the synchronization stage. At this point in time the machine was powered by the power grid and was working without any load. The next step involved supplying voltage to the field winding of the DC generator. The value of the field voltage was regulated to obtain different values of the load. 12 kW resistance heaters were connected to the output of the DC generator and acted as the last stage of the system. This system of load regulation allowed different loading values ranging from 10 to approximately 120 percent of rated values to be achieved.

The field voltage was supplied to the rotor winding via a rectifier. Rectifier bridges SKBPC2504 with maximum DC current 25 A and maximum AC voltage 400 V were used. The input voltage of the rectifier was regulated by an autotransformer.

The values of current and voltage were measured using dedicated transducers. The transducers were mounted in the stator current circuit separately for each phase. Analog output signals from the current transducers LEM IT 60-S and voltage transducers LEM AV100-500 were transferred to a NI PXI 4495 data acquisition (DAQ) system. The DAQ system was connected to a PC with LabVIEW DAQ software. All analog signals were subject to anti-aliasing filtering and subsequently sampled at the rate of 20 kHz. The data was collected for a ten seconds time period.

Fig. 4. Driven motor-coupling-driving (evaluated) motor arrangement.

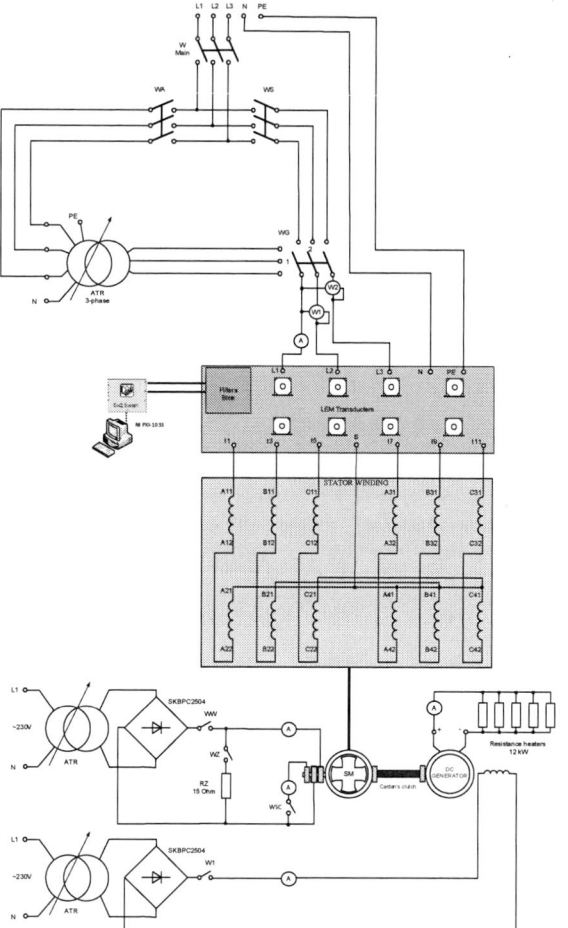

Fig. 5. Schematic diagram of electrical circuit of examined machine.

The static eccentricity fault was realized with the use of special end-rings which were placed in the lids on both sides of the machine. The axis of the inner diameter of the ring was offset from the axis of the outer diameter of the ring. The value of the offset determined the fault severity (marked as *r* in Fig. 6). The amount of offset determined the severity of static eccentricity which could be determined as a percentage with respect to the 3mm airgap of the machine.

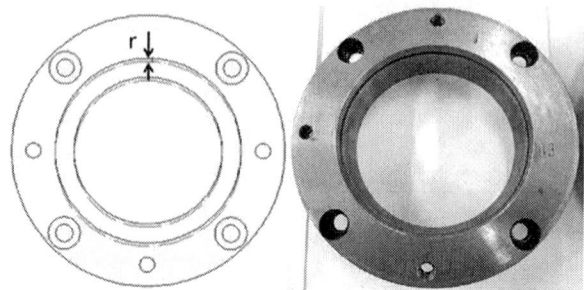

Fig. 6. Technical drawing and manufactured steel ring used for implementation of static eccentricity.

The dynamic eccentricity was introduced with the use of a special sleeve that was pushed onto both sides of the shaft of the rotor. Once in place, a bearing was installed on the sleeve. Similarly to the static eccentricity case, the axis of the sleeve was offset from the axis of the shaft (marked in as *r* in Fig. 7). The ring that supported the shaft did not have any offset between the axis of the inner and the outer diameter of the ring. The movement of the shaft operating in such a sleeve was artificially made to periodically oscillate around the central axis of the shaft.

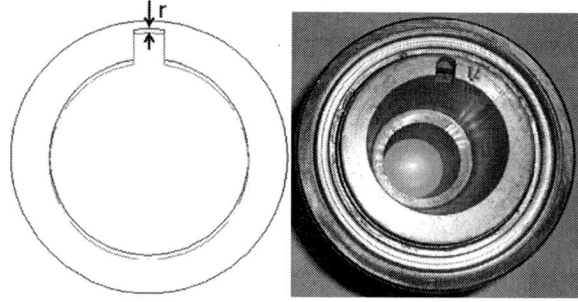

Fig. 7. Technical drawing and manufactured sleeve used for implementation of dynamic eccentricity.

III. FREQUENCY COMPOSITION OF CIRCULATING CURRENT

Circulating currents are calculated as a direct subtraction of two branch currents (3). Under ideal conditions it is expected that the energy in the circulating current would be close to zero. However the motor which was used in the experiments was assessed to have 8% static and 8% dynamic eccentricity for the nominally healthy condition. This fact was incorporated into the simulations and the resultant spectrum is shown in Fig. 8. One can see that the supply frequency (50Hz) and its harmonics are virtually invisible, however there is a clear presence of rotor speed (25Hz) related sidebands arising from the modulation caused by the movement of the rotor inside of the stator. Fig. 9 shows the reaction of frequency components to the onset of a much more severe static eccentricity case (20%). A very clear increase in the amplitude of the supply frequency and its third harmonic can be observed. This may be attributed to the fact that the rotor moves closer to one branch of a single phase and, at the same time, away from the second branch.

As a result the difference between the currents in both branches increases. Fig. 10 shows the changes in the circulating current caused by introduction of a severe dynamic eccentricity condition (30%). In this case a much stronger reaction can be observed in the amplitudes of the supply frequency sidebands. Some reaction in the supply frequency and its harmonics can also be observed, however these changes are not as distinct. The change in the amplitudes of these frequency components is caused by the modulating movement of the rotor within the stator frame which causes periodic oscillations at the rotor speed around the supply frequency.

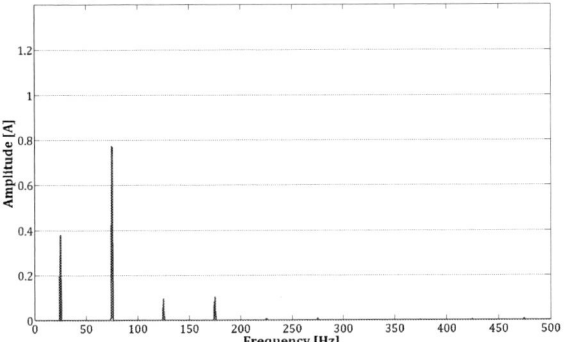

Fig. 8. Spectrum of circulating current under healthy condition – simulation.

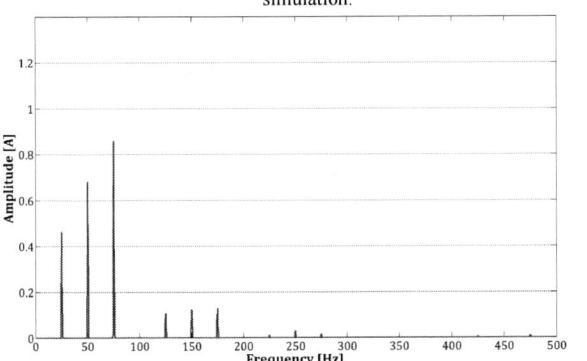

Fig. 9. Spectrum of circulating current under static eccentricity condition-simulation.

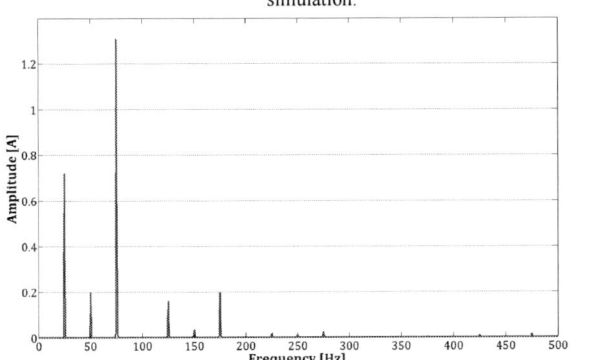

Fig. 10. Spectrum of circulating current under dynamic eccentricity condition – simulation.

The results obtained from simulations were verified by comparison with frequency content of circulating currents generated during experimental measurements under similar

conditions. It should be noted that, for a number of reasons, it is a very challenging task to faithfully simulate the exact physical behaviour of the motor, some of the more important reasons being the exact condition of the motor being unknown even for the supposedly healthy state, the assumption of ideal DC excitation and the fact that the system is modeled in 2D. What is expected from a model however is its ability to represent the general behavior of the motor under a number of different operating conditions and health states. Fig. 11 shows the spectrum of a circulating current recorded from the experimental machine under nominally healthy condition. When compared with the equivalent simulated spectrum in Fig. 8, in addition to the expected differences in the amplitudes of corresponding components, one observation stands out –the supply frequency of 50Hz and its third harmonic can be seen in the spectrum. Even though the amount of energy is low, their presence is clear when compared with the simulated case. This discrepancy appears mainly because of the presence of a voltage unbalance of approximately 1 % in the supply voltage of the experiment. On the other hand, rotor speed related sidebands around supply frequency components evident as was the case in the simulation case. Fig. 12 shows the spectrum of the circulating current recorded from machine with 20% static eccentricity fault. This spectrum may be compared with the equivalent simulation case given in Fig. 9. The reaction of the supply frequency components and their sidebands was verified to be of the same nature as in the simulations. Components at the supply frequency and its third harmonic increase in amplitude substantially, whereas the amplitudes of sidebands around these components remain relatively constant. Equally strong agreement is observed between simulated and experimental data for the case of 30% dynamic eccentricity, spectra for each being given in Fig. 10 and Fig. 13, respectively. The amount of energy contained in the rotor speed related sidebands around supply frequency harmonics clearly increases, whereas the energy in the supply frequency components retain the amount of energy observed in the healthy condition.

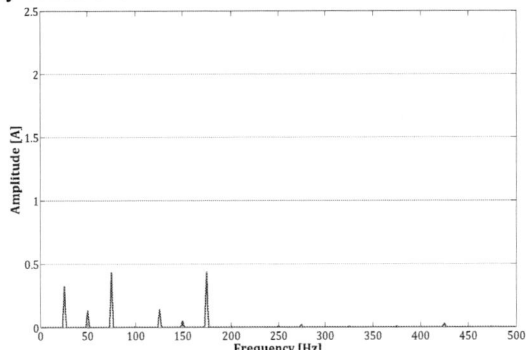

Fig. 11. Spectrum of circulating current under healthy condition – measurement.

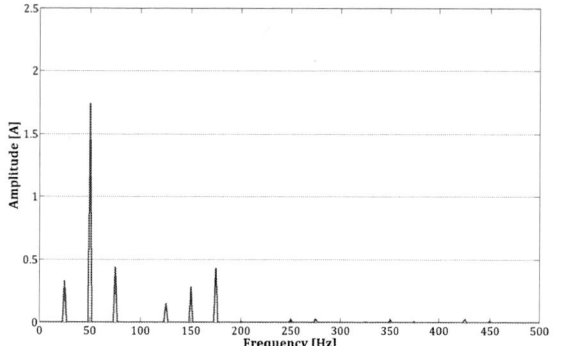

Fig. 12. Spectrum of circulating current under static eccentricity condition – measurement.

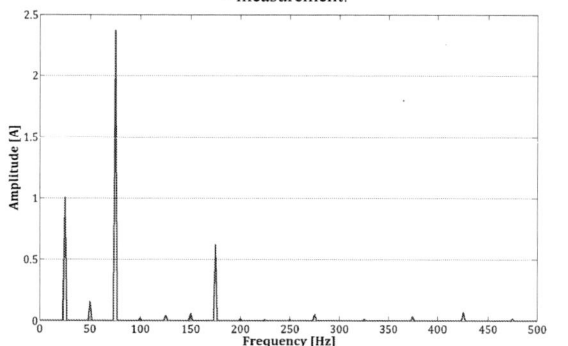

Fig. 13. Spectrum of circulating current under dynamic eccentricity condition – measurement.

IV. COMPARISON OF DIAGNOSTIC CAPABILITIES OF CIRCULATING CURRENT AND MOTOR CURRENT SIGNATURE ANALYSIS

The standard approach to condition monitoring of induction motors is to use Motor Current Signature Analysis (MCSA). This method is based on detecting signatures of specific motor faults by observing characteristic features that appear in the frequency spectrum of phase currents of the stator in the presence of a given fault [10]. The analysis of the circulating currents seems to be a natural advancement in the sense that they are only non-zero if the symmetry of the machine is broken, which typically indicates a fault in the machine. This section contains a comparison of the two methods showing the similarities and differences between them based on a series of measured currents spectra. Fig. 14 shows a spectrum of a phase current generated under the same healthy motor condition as was outlined in the previous section. There is a noticeable difference between the phase current spectrum and the equivalent circulating current spectrum (Fig. 8) in that the 50Hz component dominates the spectrum, decreasing the clarity of other frequency components. In addition to the supply frequency, three of its harmonics can be seen in the spectrum. Fig. 15 contains spectra of circulating currents generated both under nominally healthy condition and a condition of 20% static eccentricity. It can be seen that the reaction to the onset of the fault is visible solely in the supply frequency and its third harmonic, whereas the rotor speed sidebands retain constant amplitude values. Fig. 16 contains the same type of spectra

but generated from a phase current. The spectrum is limited on the y-axis in order to show the behavior of the corresponding frequencies that were described for the circulating current case. It can be seen that apart from the increase in the amplitudes of the frequencies related with the supply frequency, an increase in the amplitudes can also be seen for the rotor speed sidebands. Fig. 17 and Fig. 18 show spectra generated for the nominally healthy condition and a condition of 30% dynamic eccentricity fault, for circulating current and phase current respectively. By analyzing spectra in Fig. 17 one can see that the only reaction can be observed in the amplitudes of rotor speed sidebands around supply frequencies. The amplitudes of the supply frequency and its third harmonic remain at a very similar level. Conversely, Fig. 18 shows a similar reaction to that which was described in the static eccentricity case – apart from the supply frequency sidebands, the amplitude increases in the supply frequency and its harmonics (apart from the third harmonic). Fig. 19 presents magnified views on the previously discussed spectra from the machine operated under nominally healthy conditions (Fig. 11. and Fig. 14.). One can see that due to the subtraction of two branch currents the amount of noise in the baseline level of the circulating current is virtually zero, whereas the phase current spectra contain much greater amount of noise.

The comparison of spectra in Fig. 14 – 19 shows clear advantage of the diagnostic information contained in circulating current as opposed to the classical MCSA. The reaction of the signal to the onset of the static and dynamic fault is represented by an increase in frequencies of different type – supply frequency and its third harmonic in the case of the static eccentricity and rotor speed sidebands around the supply frequency and its third harmonic in the case of the dynamic eccentricity. In contrast, the reaction of the phase current could be seen in both groups of frequencies regardless of which type of fault was present. These properties favor the circulating current as a tool for automated monitoring of motor condition. In addition, noise in the baseline level of the spectrum (due to subtraction of the two branch currents) is strongly reduced. Therefore, circulating currents are a much more reliable and intuitive source of diagnostic information.

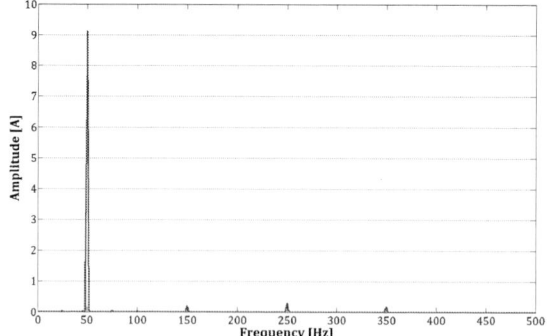

Fig. 14. Spectrum of a phase current under healthy condition.

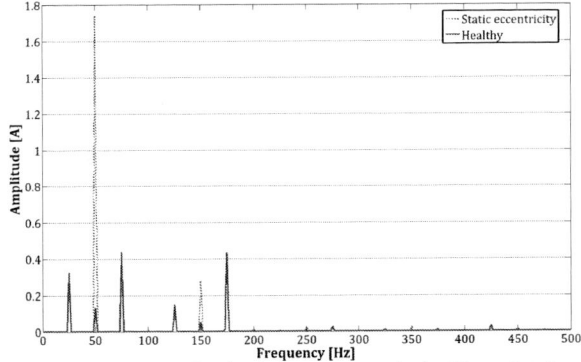

Fig. 15. Spectrum of a circulating current under healthy and static eccentricity condition (most of the frequency components are identical and obscured by the continuous line)

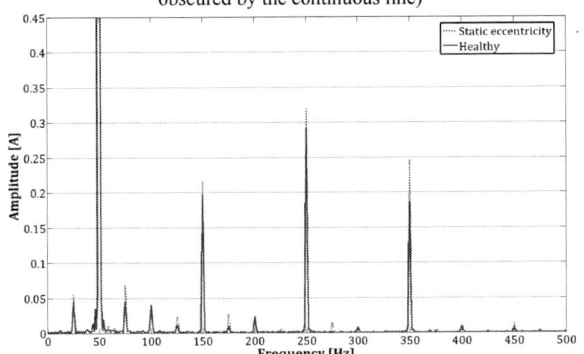

Fig. 16. Spectrum of a phase current under healthy and static eccentricity condition.

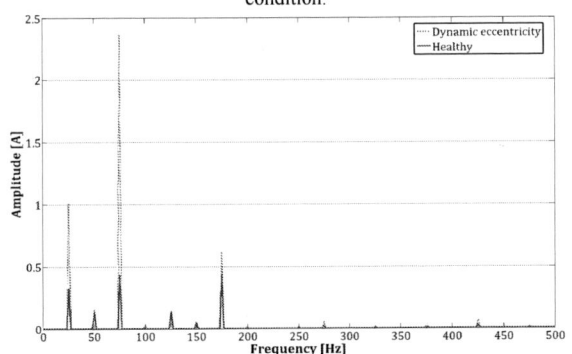

Fig. 17. Spectrum of a circulating current under healthy and dynamic eccentricity condition.

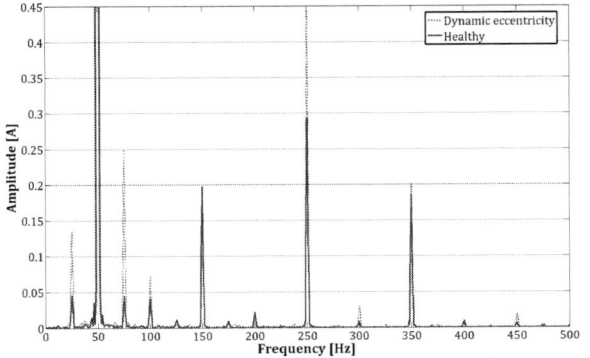

Fig. 18. Spectrum of a phase current under healthy and dynamic eccentricity condition.

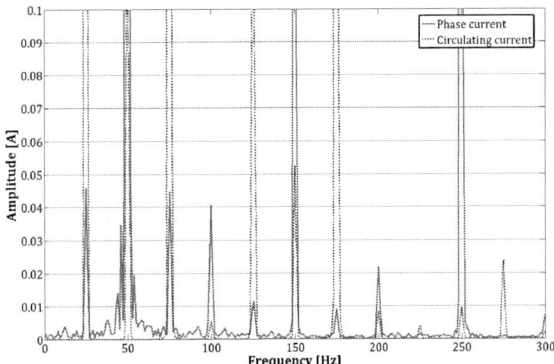

Fig. 19. Magnified view on signals from Fig. 8. And Fig. 9.

V. CONCLUSIONS

The comparison between frequency content of circulating currents obtained from numerical simulations and from experimental measurements validates the accuracy of the simulation in representing the behavior of the motor as well as in predicting the reaction of specific frequency components to the onset of different motor fault types. Thus, it is proven to be a useful tool to simulate electrical machines working under abnormal conditions. This is very important for the analysis of high-power machines where experimental investigations would be quite complex and expensive.

The concept of circulating current analysis as a reliable tool for assessing the condition of the monitored machine was shown through comparison with the standard phase current analysis technique – Motor Current Signature Analysis. The nature of diagnostic information contained in circulating current seems to indicate that this approach is much more intuitive and easier to automate in that specific frequency components react to the specific fault types. This in turn allows for unambiguous interpretation of results and may lead to reliable assessment of the exact type of fault present in the motor. In addition, due to subtraction of two branch currents the amount of background noise in circulating currents seems to be greatly reduced which may result in improved readability and precision of diagnosis. This type of system would likely find applications in high power synchronous motors and generators, where protection, condition monitoring and prognosis are very important to ensure the maximum availability of the power or drive system.

VI. REFERENCES

[1] Tavner P. J., "Review of Condition Monitoring of Rotating Electrical Machines", IET Electrical Power Applications, Vol. 2, Issues 4, pp. 215-257, 2008.

[2] Pöyhönen, S., Negrea, M., Jover, P., Arkkio, A., Hyötyniemi, H., 2003, "Numerical Magnetic Field Analysis and Signal Processing for Fault Diagnostics of Electrical Machines", COMPEL, The International Journal for Computation and Mathematics in Electrical and Electronic Engineering, Vol. 22, No. 4, pp. 969-981.

[3] Debertoli M. J., Salon S. J., Burow D. W., Slavik C. J., "Effects of Rotor Eccentricity and Parallel windings on Induction Machine Behaviuor: A Study Using Finete Element Analysis", IEEE Transactions on Magnetics, Vol. 29, March 1993.

[4] Biet M., Bijeire A., "Rotor Fault Diagnosis in Synchronous generators using feature selection and nearest neighbor rule", SDEMPED Conference Proceeding International Symposium on Diagnostic of Electrical Machines and Drives, Bologna, Italy 2011.

[5] Foggia A., Torlay E., Corenwinder C., Audoli A., Herigault J., "Circulating Current Analysis in the Parallel Connected Winding of Synchronous Generators under Abnormal Operating Conditions", Electric Machine and Drive, International Conference IEMD.

[6] Wang Shuting, Li Heming, Li Yonggang, Meng Fanchao, "Analysis of Stator Winding Parallel-Connected Branches Circulating Current and Its Application in Generator Fault Diagnosis", Industry Applications Conference, 2005. Fortieth IAS Annual Meeting. Conference Record of the 2005.

Dissertations:

[7] Arkkio A., "Analysis of induction motor based on numerical solution of the magnetic field and circuits equations", Acta Polytechn. Scand. Electri. Eng. Serie 1987, On. 59, pp. 97, available at http://lib.hut.fi/Diss/list.html#1980

[8] Jover Rodriguez, P. V., 2007, "Current-, Force-, and Vibration-Based Techniques for Induction Motor Condition Monitoring", Doctoral Dissertation, Helsinki University of Technology, Finland. http://lib.tkk.fi/Diss/2007/isbn9789512289387/isbn9789512289387.pdf

[9] Negrea, M., 2006, Electromagnetic Flux Monitoring for Detecting Faults in Electrical Machines, Thesis for the degree of Doctor of Science and Technology, Helsinki University of Technology, Laboratory of Electromechanics, October, Espoo, Finland, ISBN 951-22-8476-6, pp. 140.
http://www.tkk.fi/Units/Electromechanics/publications/index.html.

Books:

[10] Acton, A., 2012, Issues in Electronic Circuits, Devices, and Materials: 2011 Edition, Scholarly Editions, 2012

VII. BIOGRAPHIES

Pedro Vicente Jover Rodriguez was born in Santiago de Cuba, Cuba, on April 13, 1967. He graduated from the Orient University (ISPJAM), Santiago de Cuba in 1990 as Electronic Engineer. He graduated as Master of Science from Helsinki University of Technology (AALTO) in 2002 (electrical machines), where continued his studies and obtained the Ph. D in 2007. Since 2008, he is employed by ABB as a researcher in the field of rotating electrical machines (CRC Västerås, Sweden).

Pawel Rzeszucinski received MSc in Computer Science from Cranfield University (UK) and MSc in Electronics from Wroclaw University of Technology (PL). Subsequently he moved to The University of Manchester (UK) where he obtained Engineering Doctorate (EngD) for project sponsored by QinetiQ related to gearbox diagnostics. Currently, he is an Associated Scientist at ABB's Corporate Research Center in Poland working in the field of diagnostics and condition monitoring.

Maciej Sulowicz was born (1972) and educated in Poland. He received M.Sc degrees and Ph.D. degree in electrical engineering from the Faculty of Electrical and Computer Engineering at the Cracow University of Technology (Poland), in 1997 and 2005, respectively. Currently, he is an Associate Professor in the Department of Electrical Engineering and Computer Science, Institute Electromechanical Energy Conversion at the Tadeusz Kosciuszko Cracow University of Technology (Poland). His main research fields are data acquisition and fault diagnosis of electrical machines, drives and industrial process, neural networks, fuzzy logic, patterns recognition.

Rolf Disselnkoetter studied Physics at the University of Koeln with a focus on solid state physics and magnetics. During his Ph.D. work at the University of Goettingen he worked on dynamic laser light scattering on electrohydrodynamic flows in dielectric liquids. Currently, he is Senior Principal Scientist at ABB's Corporate Research Center in Germany. His main focus is on magnetic systems and on current sensors.

Cajetan Pinto studied Electrical Engineering at the University of Mumbai and at the University of California Irvine (USA). During his Masters of Science studies he worked on electromagnetic energy conversion systems at the University of California. Currently, he is Global Head of R&D for Motors and Generators Service at ABB Global & Industrial Services Ltd, Mumbai India. His main areas of focus include diagnostics, condition monitoring & reliability, motor and generator re-engineering, maintenance technologies, new service processes and materials for service.

James Ottewill received BEng and Ph.D. degrees in Mechanical Engineering from the University of Bristol, UK in 2005 and 2009, respectively. During his Ph.D. research, he investigated the non-linear problem of gear rattle. Currently, he is a Scientist at ABB's Corporate Research Center in Poland, working in the field of condition monitoring. His main research interests include dynamic testing, modeling and analysis of non-linear systems, signal processing and advanced diagnostic techniques.

Ulf Ahrend studied Physics at the Technical University Braunschweig and at Virginia Tech (USA). During his Ph.D. studies he worked on turbulent heat transfer at MIT and Technical University Braunschweig. Currently, he is Principal Scientist at ABB's Corporate Research Center Germany. His main focus is on sensor-based condition monitoring and autonomous sensing devices using energy harvesting.

Stephan Wildermuth received his Ph.D. degree in physics from the University of Heidelberg, Heidelberg, Germany, in 2005. From 2006-07, he was a Postdoctoral Research Fellow at the University of Otago, Dunedin, New Zealand, where his area of research was ultra-cold atom optics and two-species Bose Einstein Condensation. In 2007 he joined ABB Corporate Research where he is currently working as Principal Scientist. His main focus is on optical sensors for applications in medium and high voltage systems as well as sensor-based condition monitoring.

978-1-4799-0024-4/13 $31.00 © 2013 IEEE

An Accurate and Fast Technique for Correcting Spectral Leakage in Motor Diagnosis

Javier Martinez, François Philipp, *Member, IEEE*, Manfred Glesner, *Fellow, IEEE*, Antero Arkkio

Abstract—**This paper presents a technique for correcting the spectral leakage which occurs in windowed time series. The presented technique is aimed to be implemented on an autonomous embedded device based on reconfigurable hardware. This device is intended for being used in remote condition monitoring of motors. As a matter of example, the technique is used to analyse the common frequency components appearing in faulty motors suffering from inter-turn short circuits.**

I. INTRODUCTION

Induction motors are widely used in industry, for example, for traction purposes. Even though the reliability of induction motors is high compared to other type of machines such as DC motors, these machines are prone to failures. Among the possible sources of breakdowns, inter-turn short circuit is one of the most common. Thomson & al. reports that the origin of 38% of the faults is located in the stator [1]. Such a fault can be due to over-voltages and over-currents that can destroy the insulation of windings [1]. Many authors claim that inter-turn short circuit is the origin of a phase-to-ground fault in the motor [2].

Most novel diagnosis techniques based on MCSA (Monitoring Current Signature Analysis) are reported in [3]–[6]. Joksimovic & al. determine that a rise of the amplitude level in the Principal Slot Harmonics (PSH) in the induction motor takes place under an inter-turn short circuit [3]. Cruz & al. use the third harmonic current to detect a fault in a DTC-driven electric motor [4]. Kim & al. use the second harmonic appearing in the q component of the current to detect the combination of an opened switch and a turn short circuit in the stator of an inverter-driven motor [5] . Choqueuse & al. use PCA (Principal Component Analysis) together with an AM-FM demodulator to detect unbalances in the stator produced by short circuits in the stator [6]. The amplitude and phase demodulation parameters are estimated using statistical techniques.

Our implementation targets a small, portable, wireless device which can monitor the health of the motor online [7]. The high accuracy and robustness required by the utilisation of such devices in an industrial context, along with the complexity of the involved signal processing tasks, justify the choice of an FPGA-based target platform. The implementation of custom digital processing blocks improves both the performance and the accuracy of the monitoring system, as shown in [8] for the detection of broken bars. The wireless connectivity and the reconfigurability of the device make it highly flexible and usable for a large range of condition monitoring tasks. Related work includes the FPGA-based system developed by Medina &

al. for the online, vibration-based failure detection of induction motors [9]. They adopted Bartlett's method instead of a direct FFT to reduce the noise and the leakage of the vibration power spectrum. Ordaz-Moreno & al. implemented a wavelet-based diagnosis algorithm for broken bars on a low-cost FPGA platform [10].

This paper presents a technique to correct the effects of spectral leakage in windowed time series. This method is applicable to estimate the magnitude and the phase of any suitable frequencies considered in MCSA diagnosis techniques. The novelty of our approach arises in the application of a leakage correction function on the results of the FFT computed by the signal acquisition system. Thus, an FFT with a fixed sampling frequency and size is sufficient to accurately estimate all the different spectral components at the origin of the different faults. As a matter of example, we use the leakage correction function to analyse the different spectral signatures of inter-turn short circuits in induction motors.

The rest of this paper is divided as follows. Section II presents the the computation of the FFT for inter-turn short circuit and the corresponding leakage problematic. Section III presents the correction of phase and magnitude in FFT components suffering from leakage. Section IV analyses several electrical quantities of faulty motors having an inter-turn short circuit with the presented estimator. Section V concludes the paper and suggests future work to be done in the field of inter-turn detection.

II. FOURIER ANALYSIS FOR MCSA

A. Fourier Analysis of real-time series

The frequency spectrum of a real time series can be computed by the Fourier transform. Let $x(t)$ be a real time series representing a phase current

$$x(t) = \sum_{l=0}^{\infty} a_l \cos(\omega_l t) = \sum_{l=0}^{\infty} \frac{a_l}{2} \left(e^{j\omega_l t} + e^{-j\omega_l t} \right) \quad (1)$$

A coefficient $\{a_l\}_{l=0}^{\infty}$ can be computed by the inner product between $x(t)$ and a complex exponential $e^{j\omega t}$.

$$a_l = \left(x(t), e^{j\omega t} \right) = \begin{cases} \frac{a_k}{2} & \text{if} \quad \omega = \pm\omega_l \\ 0 & \text{otherwise} \end{cases} \quad (2)$$

The inner product is defined as

$$(a,b) = \int_0^{\infty} a \cdot b \, dt \quad (3)$$

The solution of Eq. (2) relies on the orthogonal properties of complex exponentials. Thus, the spectrum given by Eq. (2) shows that the spectrum is not vanishing when $\omega = \pm\omega_l$. This effect creates a problem in the analysis of electrical quantities in electric motors. For a single frequency supply, we obtain a set of two frequency components in the spectrum. The current signal is preprocessed with a Fortescue's transformation to overcome this problem.

B. Fourier Analysis of complex-time series

Fortescue's transformation can be seen as a projection of the space consisting in a set of three stator currents onto the space spanned by three orthogonal components. These components are a phasor rotating clockwise, a phasor rotating counter-clockwise and a pulsating sequence. The new components define a complex time series. Then, the location of the frequency component can be tracked by computing the Fourier transform of the Fortescue's transformation. We have

$$\mathbf{i_F} = \mathbf{A} \cdot \mathbf{i_s} \qquad (4)$$

where $\mathbf{i_F} = (i_o, i_d, i_i)^T$ is the vector of the Fortescue's components, $\mathbf{A}$ is the projection operator and $\mathbf{i_s}$ is the vector of symmetrical currents. $\mathbf{i_s}$ is given by

$$\mathbf{i_s} = a_l \begin{pmatrix} \cos(\omega_l t + \phi_l) \\ \cos\left(\omega_l t + \phi_l + \frac{2\pi}{3}\right) \\ \cos\left(\omega_l t + \phi_l + \frac{4\pi}{3}\right) \end{pmatrix} \qquad (5)$$

The projection matrix, A_F is defined as

$$A_F = \frac{1}{3}\begin{pmatrix} 1 & 1 & 1 \\ 1 & \alpha^2 & \alpha \\ 1 & \alpha & \alpha^2 \end{pmatrix} \qquad (6)$$

where $\alpha = e^{j\frac{2\pi}{3}}$. Substituting, Eq. (5) and (6) into (4), we obtain

$$\mathbf{i_F} = \left(0, e^{j\omega_l t}, e^{-j\omega_l t}\right)^T \qquad (7)$$

By taking the Fourier transform of the direct sequence, we obtain only a component located at $\omega = \omega_k$.

$$\mathcal{F}\{i_d\} = \sum_{l=0}^{\infty} i_l e^{-j\omega_l t} \qquad (8)$$

where the coefficients i_k are calculated by

$$i_l = \left(i_d(t), e^{j\omega t}\right) = \begin{cases} i_l & \text{if} \quad \omega = \omega_l \\ 0 & \text{otherwise} \end{cases} \qquad (9)$$

Eq. (8) and (9) show that the Fourier transform of the Fortescue component i_d is able to represent the sense of rotation of the electric drive. Thus, the analysis of this vector component is suitable to detect inter-turn short-circuit. Components located in the negative frequencies are analysed by Wu & al. [11]. When the asymmetry of the current supply vector $\mathbf{i_s}$ increases, the current components which are opposing the creation of the main flux wave will increase.

C. Digital computation considerations

On embedded systems, (8) and (9) are computed by a DFT (Discrete Fourier Transform). Our platform implements the DFT with a radix-2 Cooley-Tukey FFT algorithm mapped on a low-power reconfigurable hardware architecture [12]. Figure 1 illustrates the proposed hardware architecture used for the monitoring of the considered faults. The block FT implements the Fortescue's transformation as suggested in the section II-B. A 2^n-FFT has been prefered to the Goertzel algorithm where frequency bins can be computed individually. For a complete diagnosis application, the whole frequency spectrum is necessary to monitor other faults such as as broken bars and dynamic eccentricity.

The DFT of the i_d component can be expressed [13] as

$$\mathcal{DFT}\{i_d\} = I_d[k] = \sum_{n=0}^{N-1} i_d[n] \cdot w[n] e^{-j\frac{2\pi}{N}kn} \qquad (10)$$

where $w[n]$ represents the window function that bounds in time the current computation and N is the length of this window. Substituting Eq. (8) into Eq. (10), we obtain the following spectrum representation

$$I_d[k] = \sum_{l=0}^{\infty} \sum_{n=0}^{N-1} i_l e^{j\omega_l n} w[n] e^{-j\frac{2\pi}{N}kn} = \qquad (11)$$

$$= \sum_{l=0}^{\infty} \sum_{n=0}^{N-1} i_l e^{-j\left(2\pi\left(\frac{k}{N}-f_l\right)n\right)} w[n] =$$

$$= \sum_{l=0}^{\infty} \sum_{n=0}^{N-1} i_l w[n]\delta\left(2\pi\left(\frac{k}{N}-f_l\right)n\right) = \sum_{l=0}^{\infty} i_l W[k-f_l]$$

Eq. (11) presents the spectrum of the direct sequence component coming from the FT block. The spectrum of i_d is equivalent to a set of versions of the window $w[n]$ shifted in the frequency domain. Each shift is a frequency translation of value f_l. The effect of leakage appears when f_l and F_s are not having common factors. For this reason, we include the estimator block in Fig. 1 where the user selects the frequency of interest for the analysis of the faults. The advantage is twofold. Firstly, we do not need to change the sampling frequency or the number of points of the FFT for different fault analysis. Secondly, when a frequency bin is not suffering from leakage the estimator must be able to correct interferences from the leakage of neighbouring frequencies.

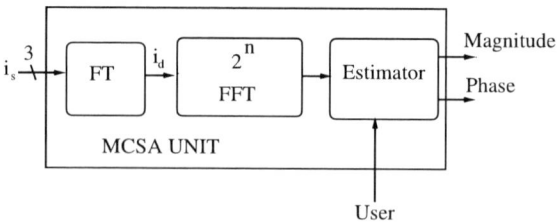

Figure 1.   Proposed architecture for our MCSA embedded core

III. Estimation of FFT spectral components

A. Introduction

This section derives expressions for the estimation of both amplitude and phase at the frequency of interest. An error estimate is also given. As a first approach of analysis, we assume that the frequency of supply is fixed. For this reason, the estimation of the line frequency is not treated in this work. We refer to [14] and [15] for a detailed treatment of the frequency computation. Prior to every leakage correction, we need to quantify the closest bin to the frequency of interest. The location can be found by the following relationships

$$df = \frac{F_s}{N} \qquad (12)$$

$$\frac{f_l}{df} = k + \delta_l \qquad (13)$$

where F_s corresponds to the sampling frequency, N is the length of the window and df is the frequency resolution of the FFT. k represents the index of the bin the closest to the frequency of interest in the DFT ratio $\frac{f_l}{df}$ and δ_l is the rounding factor such that δ_l is contained in the closed interval $[-0.5, 0.5]$.

B. Amplitude Estimation

Taking the closest bin to the frequency of interest, we can approximate the DFT as

$$I_d[k] \approx i_l W(k - f_l) + i_{l-1} W(k - f_{l-1}) + \qquad (14)$$
$$i_{l+1} W(k - f_{l+1}) = i_l W(k - f_l) + err(k)$$

The term $err[k]$ includes the interferences of the two closest frequencies around the frequency of interest.

We can easily estimate the amplitude by neglecting the error function if we assume that the closest frequency bins are far enough from the desired spectral component. The estimation can be simply expressed as

$$\widehat{i_l} = \frac{I_d[k]}{W(k - f_l)} = \frac{I_d[k]}{W(df\delta_l)} \qquad (15)$$

where $\widehat{i_l}$ is the estimated amplitude. When the error term cannot be neglected anymore, we can override the error with two simple properties. The first property relies on the dynamics of the spectral window in the frequency domain [15]

$$W_r(k + \delta) = \frac{\sin(\pi(k + \delta))}{N \sin\left(\frac{\pi k}{N}\right)} e^{-j\frac{N-1}{N}\pi k} \approx \qquad (16)$$
$$\frac{(-1)^k \sin(\delta\pi)}{\pi k} e^{-j\frac{N-1}{N}\pi k} = O\left(k^{-1}\right)$$

$$W_h(k + \delta) \approx \frac{1}{2} \frac{\sin(\pi(k + \delta))}{\pi(k + \delta)(1 - k^2)} e^{-j\frac{N-1}{N}\pi k} \approx \qquad (17)$$
$$\frac{1}{2} \frac{(-1)^k \sin(\pi\delta)}{\pi(k + \delta)(1 - k^2)} e^{-j\frac{N-1}{N}\pi k} = O\left(k^{-3}\right)$$

where H_r and H_h represent the frequency domain analytic expression for the rectangular and hanning windows respectively. The notation O represents the decay factor of the window. We can see in Eq. (16)-(17) that the sign of the expressions changes from sample to sample. This property allows us to form difference equations. The $j - th$ order finite difference equation can be written [14] as

$$\Delta^j x[n] = \sum_{i=0}^{j} \binom{n}{i} (-1)^i x[n + i] \qquad (18)$$

Applying the j-th sum of Eq. (18), without the changes of sign, to the left side of Eq. (14) and taking the modulus, we obtain the following relationship:

$$\left| \sum_{i=0}^{j} \binom{n}{i} I_d[k + i] \right| = \qquad (19)$$
$$\left| \sum_{i=0}^{j} \binom{n}{i} (i_l W(k - f_l + i) + err(k + i)) \right|$$
$$\leq i_l \left| \sum_{i=0}^{j} \binom{n}{i} W(k - f_l + i) \right| + \left| \sum_{i=0}^{j} \binom{n}{i} err(k + i) \right|$$

The inequality in Eq. (19) follows the Cauchy-Schwarz inequality. As a simple example, for k=1 and rectangular window, we obtain

$$|I_d[k + 1] + I_d[k]| \leq i_l |W_r(k + 1 - f_l) + W_r(k - f_l)| \\ + |err(k + 1) + err(k)| \qquad (20)$$

Eq. (20) shows that the sum of two frequency bins of the main lobe in the rectangular window is less than the sum of the window located at the two closest bins around the frequency of interest plus the deviation produced by the surrounding frequency components. By the change of sign property outside of the main lobe in the rectangular window, the addend $err[k + 1] + err[k]$ becomes the first order difference of the rectangular window. The first order difference of the rectangular window can be written as

$$\Delta W_r(k) = \frac{1}{\pi k} - \frac{1}{\pi(k + 1)} = \frac{1}{k(k + 1)} = O\left(k^{-2}\right) \qquad (21)$$

Eq. (21) shows that the decay function is one order of magnitude more with respect to Eq. (16). Thus, the error decreases. Finally, the estimation of the amplitude for a first order approximation gives

$$\widehat{i_l} = \frac{|I_d[k + 1] + I_d[k]|}{|W(df(\delta_l + 1)) + W(df\delta_l)|} \qquad (22)$$

In a similar manner, we can estimate the amplitude, $\widehat{i_l}$, computing the j-th order difference. The estimation can be written in a general form as

978-1-4799-0024-4/13 $31.00 © 2013 IEEE

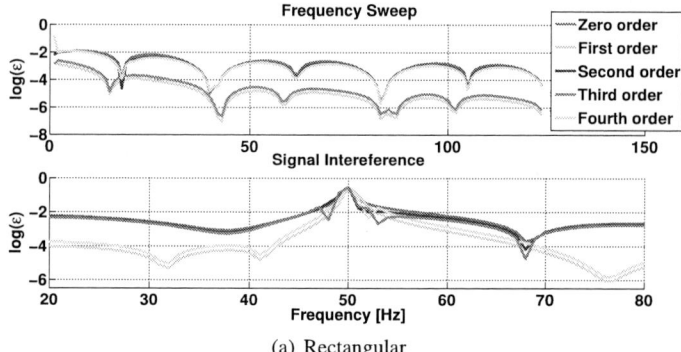

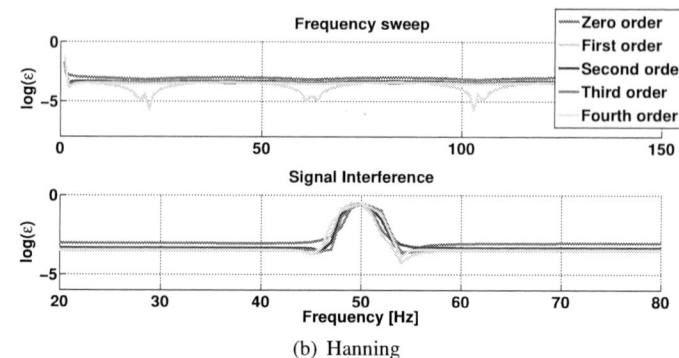

(a) Rectangular (b) Hanning

Figure 2. Estimation tests for rectangular and Hanning windows. N=1024.

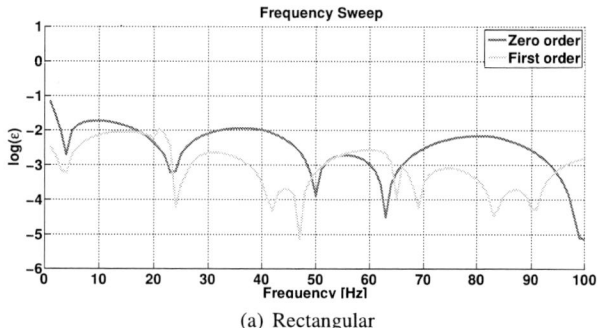

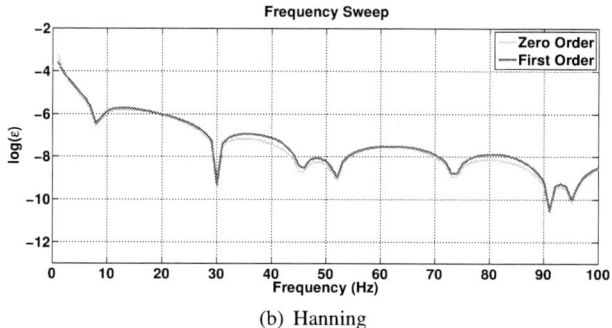

(a) Rectangular (b) Hanning

Figure 3. Estimation tests for Rectangular and Hanning windows. N=1024.

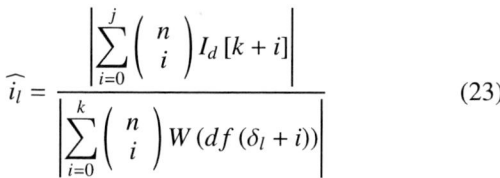

$$\widehat{i_l} = \frac{\left| \sum_{i=0}^{j} \binom{n}{i} I_d\,[k+i] \right|}{\left| \sum_{i=0}^{k} \binom{n}{i} W\,(df\,(\delta_l + i)) \right|} \qquad (23)$$

Two numerical tests have been performed in order to test accuracy of the amplitude estimator. Results are presented in Fig. 2. Firstly, the *Frequency Sweep* plot shows the error as a function of the desired frequency. Secondly, the *Signal Interference* plot shows the error committed while measuring the magnitude of the line frequency and we have another parasitic signal whose frequency ranges from 20 to 80 Hz. The error is quantified by $\epsilon = |\frac{\widehat{i_l} - i_l}{i_l}|$. According to Figure 2(a) and 2(b), we can see that the relative error in the Hanning window is more stable than in the rectangular case. In the case of the signal interference test, we observe a bigger error around the frequency of interest. For the case of Hanning windows, the area where the error becomes remarkable is wider than in the rectangular case. This effect can be explained by the width of the main lobe of the window. As the frequency of the interfering signal approaches to the main signal (50 Hz), the overlapping of the main lobes increases and the amplitude can not be evaluated properly.

C. Phase Estimation

The phase spectrum of a set of sinusoidal components can be easily written using Eq. (14). The phase spectrum located at the bin i_k can be written as

$$
\begin{aligned}
G\,(k) &= \sum_{l=m-1}^{m+1} i_l W\,(k - f_l)\,e^{j\phi_l} = \qquad (24)\\
&= i_l W\,(k - f_l)\,e^{j\phi_l} + err\,(k)
\end{aligned}
$$

Applying the phase operator in the previous equation, we obtain the expression for the k-th phase bin [16]

$$arg\,[G\,(k)] = \phi_l + \pi \frac{N-1}{N}\delta_l + \Delta err\,(k_l) \qquad (25)$$

When the error can be depreciated, the phase can be estimated with

$$\phi_l = arg\,[G\,(k)] - \pi \frac{N-1}{N}\delta_l \qquad (26)$$

As soon as the error increases, due to the leakage originated by neighbouring spectral components, we can override the situation by correcting the phase transition between the value k and the second closest bin with respect to the frequency of interest, f_l. Ideally, this transition should be a line whose slope is π. By calculating the current slope and applying the difference with the ideal slope, we can estimate the deviation produced by the surrounding frequencies. The geometric construction can be seen equivalent to a fixed-point iteration. In mathematical terms, this can be expressed as

$$\phi_l = arg\left[G\left(k\right)\right] - \pi \frac{N-1}{N}\delta_l + \Delta y \cdot \delta_l \qquad (27)$$

where $\Delta y = \pi - (arg\left[G\left(k\right)\right] - arg\left[G\left(k-1\right)\right])$. Fig. 3 shows the estimation of the relative phase error as a function of frequency. We can observe that the Hanning window performs better than the rectangular. For the rectangular case, we can observe a better estimation of phase using the first order algorithm around zero frequency. The origin of such error is related to the interference caused by a copy of the window located in the negative side of the spectrum.

IV. EXPERIMENTAL RESULTS

The presented method to estimate magnitude and phase of a specific frequency from FFT bins is applied to real measurements. The analysed motor has been short circuited by taking out several conductors and connecting them to set of external resistors. Every phase of the motor consists of two parallel branches. The short circuit, applied only in one turn, has been reproduced in both parallel branches of the phase B. Thus, the degree of freedom in our fault has been a change in the short circuit resistance. Rated values of the studied motor are presented in Table I. The motor has been loaded using the same type of motor by using the back-to-back configuration, Fig. 4. The motor used as a load is connected using a scalar control whereas the studied faulty motor is run by a grid supply. Both inverters are connected by a DC-link and the inverter which is connected to the device under test is in turn feeding a brake resistor by a DC-chopper. Consequently, the line-to-line voltages have been decreased in the same manner with respect to the nominal values in order to keep the flux voltage ratio constant. The electrical connection for both motors was star supply. The sampling frequency of the system was 3.3 kHz and the FFT length is N=2048. The MCSA unit is implemented on a Actel IGLOO AGL1000 FPGA tightly coupled to a TI CC2531 system-on-chip, combining a IEEE 802.15.4 radio transceiver with a microcontroller based on a 8051 CPU. Networking tasks are handled by the processor while the computationally intensive tasks from the estimator are assigned to the FPGA. The circuit is described in VHDL and the software running on the CPU is implemented in C on top of the Contiki operating system.

Table II compares the values of a healthy motor when the estimator block is working with the values taken directly from the FFT results together with a time-series where line frequency is not affected by spectral leakage. This time series has been sampled at 2.5 kHz and a register of 50 kSamples. First, the estimated peak value of currents presents a value that matches the expected RMS value of the motor at rated operating conditions (41 A). Secondly, the estimation of the phase is able to place the correct phase shift existing in the set of phase voltages and correct the sense of rotation of the drive. Finally, the estimation of phase current presents the motor as RL circuit and the phase difference between voltage and current matches the power factor of the motor.

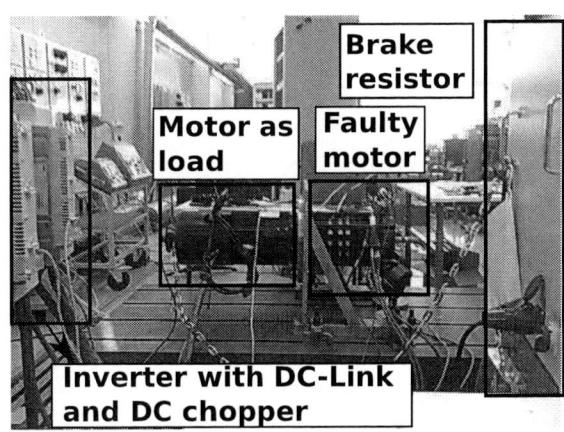

Figure 4.   Laboratory set-up for testing inter-turn short circuit fault.

The analysed frequencies for the faulty set-up are -50 Hz, 150 Hz and the Principal Slot Harmonic. The reasons for this choice of frequencies is studied in [3], [4], [11]. First, the increase of the frequency component located at -50 Hz in the space vector is related to the current flowing in the short circuited turn. This current opposes the main flux in the air-gap. Secondly, the authors in [4] and [11] study the increase of the harmonic located at 150 Hz. The rise is caused by the local saturation produced by the fault. Table III shows that this component increases with the degree of unbalance. It must be stated that [4] studies this fault under delta connection. This could explain the increase of this component in the model proposed by the authors. Finally, [3] studies the fault in the component located in the PSH, 996 Hz (Table III). This increase can be explained in the motor by considering the fault current flowing in the short-circuit loop. This loop can increase the coupling of waves having the same pair frequency and pole pair as the PSH. The aforementioned rises can be compared in Table III. Table III also shows that the current value at 50 Hz decreases since inter-turn loops creates a flux rotating oppositely to the main flux (component at -50Hz increases).

Table IV refers to the phase difference between the phase of the direct voltage, ϕ_v, and the direct current, ϕ_i. This is a way of estimating the impedance of the motor. The row for 50 Hz shows that the motor is seen as an RL circuit. The value of the difference decreases with the imbalance. The fault increases the losses in the drive and the inductance of the stator decreases due to the short circuit. The phase difference at -50 Hz has a clear effect in the fault. When the motor is healthy, the origin of this component is the voltage unbalance, and the phase is close to a pure inductance. This value can be explained by the inverse sequence model presented in [17]. The load resistance has a low value compared to the direct one. When the fault occurs, the phase reduces for the same reasons as the direct sequence. However, the effect seems to be clearer than in the direct frequency. Concerning the phase in the PSH, the effect has a similar trend than in the line frequency direct sequence. This harmonic component is a inter modulation product of the line frequency component and the air-gap function distribution of the rotor cage [3].

978-1-4799-0024-4/13 $31.00 © 2013 IEEE   219

Table I. NAMEPLATE OF THE STUDIED INDUCTION MOTOR

Pole pairs	2	
Parallel branches	2	
Stator Slots	48	
Rotor Slots	40	
Slip	5.5%	
Connection	Star	Delta
Rated Voltage [V]	380-415	380-415
Supply Frequency [Hz]	60	100
Rated Power [kW]	22	41
Power Factor	0.86	0.82

Table II. COMPARISON BETWEEN ESTIMATED, NON-ESTIMATED AND VALUES OF LINE FREQUENCY FOR THE PHASE CURRENTS AND VOLTAGES

	Est		Non-Est		No Leakage	
	Mag	Phase [rad]	Mag	Phase [rad]	Mag	Phase [rad]
V_a [V]	280.38	−0.53	219.22	1.41	280.47	−0.53
V_b [V]	278.23	1.59	214.19	−0.7156	278.34	1.58
V_c [V]	276.41	−2.62	214.93	−2.78	276.35	−2.61
I_a [A]	60.11	−1.1	53.38	1.98	60.11	−1.1
I_b [A]	59.62	1.09	52.44	−0.21	59.62	1.08
I_c [A]	55.01	3.12	48.4	−2.24	55.32	3.14

Table III. ESTIMATED VALUES OF THE MAGNITUDE OF DIRECT SEQUENCE CURRENT. UNITS : [A]

	Short circuit resistance		
Frequency	Healthy	1.5Ω	0.75Ω
50	58.47	59.54	61.15
150	0.0756	0.1864	0.23
-50	3.453	3.897	4.36
PSH	0.0015	0.0007	0.016

Table IV. ESTIMATED VALUES OF THE DIRECT VOLTAGE SEQUENCE AND DIRECT CURRENT SEQUENCE. UNITS: RADIANS

	Short circuit resistance		
Frequency	Healthy	1.5Ω	0.75Ω
50	1.15	1.14	1.11
150	1.58	1.5	0.74
-50	3.46	1.81	2.58
PSH	0.57	0.45	0.48

V. CONCLUSIONS

This paper introduces a phase and magnitude estimator with a degree of complexity which is suitable for on-line implementation in small and autonomous embedded system. Nevertheless, further studies should be carried before a final implementation. First of all, the study of the accuracy when fixed-point arithmetic is used must be evaluated. In addition to the errors committed by the arithmetic, the error in the estimation error induced by noise should also be considered. In practice, the accuracy of the estimator is restricted by the quality of the measured signal (signal to noise ratio), which depends on multiple factors like the quality of the sensor, the resolution of the analog-to-digital converter and the efficiency of the signal conditioning filters. Secondly, the authors suggest deeper studies for the diagnosis in inter-turn short-circuit. Phase relationships in the current/voltage spectrum can be further analysed, using FEA (Finite Element Analysis), in order to physically link the degree of short-circuit. This model will be used in turn to stablish the maximum errors tolerated by our digital platform.

REFERENCES

[1] W. Thomson and M. Fenger, "Current signature analysis to detect induction motor faults," *Industry Applications Magazine, IEEE*, vol. 7, no. 4, pp. 26 –34, jul/aug 2001.

[2] M. Sahraoui, S. Zouzou, A. Ghoggal, and S. Guedidi, "A new method to detect inter-turn short-circuit in induction motors," in *Electrical Machines (ICEM), 2010 XIX International Conference on*, sept. 2010, pp. 1 –6.

[3] G. Joksimovic and J. Penman, "The detection of inter-turn short circuits in the stator windings of operating motors," *Industrial Electronics, IEEE Transactions on*, vol. 47, no. 5, pp. 1078 –1084, oct 2000.

[4] S. Cruz and A. Cardoso, "Diagnosis of stator inter-turn short circuits in dtc induction motor drives," *Industry Applications, IEEE Transactions on*, vol. 40, no. 5, pp. 1349 – 1360, sept.-oct. 2004.

[5] K.-H. Kim, B.-G. Gu, and I.-S. Jung, "Online fault-detecting scheme of an inverter-fed permanent magnet synchronous motor under stator winding shorted turn and inverter switch open," *Electric Power Applications, IET*, vol. 5, no. 6, pp. 529 –539, july 2011.

[6] V. Choqueuse, M. Benbouzid, Y. Amirat, and S. Turri, "Diagnosis of three-phase electrical machines using multidimensional demodulation techniques," *Industrial Electronics, IEEE Transactions on*, vol. 59, no. 4, pp. 2014 –2023, april 2012.

[7] F. Philipp, F. Samman, and M. Glesner, "Design of an autonomous platform for distributed sensing-actuating systems," in *Rapid System Prototyping (RSP), 2011 22nd IEEE International Symposium on*, may 2011, pp. 85 –90.

[8] F. Philipp, J. Martinez, M. Glesner, and A. Arkkio, "A smart wireless sensor for the diagnosis of broken bars in induction motors," in *Electronics Conference (BEC), 2012 13th Biennial Baltic*, oct. 2012, pp. 119 –122.

[9] L. Medina, R. de Jesus Romero-Troncoso, E. Cabal-Yepez, J. de Jesus Rangel-Magdaleno, and J. Millan-Almaraz, "Fpga-based multiple-channel vibration analyzer for industrial applications in induction motor failure detection," *Instrumentation and Measurement, IEEE Transactions on*, vol. 59, no. 1, pp. 63 –72, jan. 2010.

[10] A. Ordaz-Moreno, R. de Jesus Romero-Troncoso, J. Vite-Frias, J. Rivera-Gillen, and A. Garcia-Perez, "Automatic online diagnosis algorithm for broken-bar detection on induction motors based on discrete wavelet transform for fpga implementation," *Industrial Electronics, IEEE Transactions on*, vol. 55, no. 5, pp. 2193 –2202, may 2008.

[11] Q. Wu and S. Nandi, "Fast single-turn sensitive stator interturn fault detection of induction machines based on positive- and negative-sequence third harmonic components of line currents," *Industry Applications, IEEE Transactions on*, vol. 46, no. 3, pp. 974 –983, may-june 2010.

[12] F. Philipp and M. Glesner, "Mechanisms and architecture for the dynamic reconfiguration of an advanced wireless sensor node," in *Field Programmable Logic and Applications (FPL), 2011 International Conference on*, sept. 2011, pp. 396 –398.

[13] G. Andria, M. Savino, and A. Trotta, "Windows and interpolation algorithms to improve electrical measurement accuracy," *Instrumentation and Measurement, IEEE Transactions on*, vol. 38, no. 4, pp. 856 –863, aug 1989.

[14] D. Belega and D. Dallet, "Frequency estimation via weighted multipoint interpolated dft," *Science, Measurement Technology, IET*, vol. 2, no. 1, pp. 1 –8, jan. 2008.

[15] D. Agrez, "Weighted multipoint interpolated dft to improve amplitude estimation of multifrequency signal," *Instrumentation and Measurement, IEEE Transactions on*, vol. 51, no. 2, pp. 287 –292, apr 2002.

[16] ——, "Improving phase estimation with leakage minimization," *Instrumentation and Measurement, IEEE Transactions on*, vol. 54, no. 4, pp. 1347 – 1353, aug. 2005.

[17] M. Arkan, D. Perovic, and P. Unsworth, "Online stator fault diagnosis in induction motors," *Electric Power Applications, IEE Proceedings -*, vol. 148, no. 6, pp. 537 –547, nov 2001.

978-1-4799-0024-4/13 $31.00 © 2013 IEEE

Analysis of Electrical and Non-Electrical Causes of Variable Frequency Drive Failures

Osama A. Al-Naseem and Mohamed A. El-Sayed

Electrical Engineering Department, Kuwait University

Abstract – **Variable frequency drives (VFD) are an integral part of many industrial plants and stations. Reliable operation and maintenance of these drives is vital to ensure sustained plant operation and availability. Understanding of the principles of operation of VFD systems as well as knowledge about their required operating environment is necessary for all operating personnel. Many times the operating personnel do not get involved with different technical issues until a complete failure has occurred. Hence, the awareness of the most dominant failure causes has a significant impact on assisting operators to avoid catastrophic failures and tremendous economic losses due to VFD shutdown. Proper plant design, accurate monitoring and data logging, following manufacturer preventive maintenance schedule, and choosing qualified team of operators can be the key to an efficient operation and a long lifetime for any VFD system. In this paper, we have analyzed the electrical and non-electrical causes of VFD failures based on a case study of a typical medium voltage VFD pumping station. Finally, recommendations are given from field analysis and observations.**

Index Terms—**Variable Speed Drives (VFD), Adjustable Speed Drives (ASD), Electrical Fault Causes, Non-Electrical Fault Causes, SCADA, Twelve-Pulse Power Electric Inverter, Rectifier Flashover, Inverter Short-Circuit**

I. INTRODUCTION

A variable frequency drive (VFD) is an electrical adjustable speed drive (ASD). In general, an adjustable speed drive (ASD) should be able to control the torque and speed of the drive to satisfy the process requirements. It should also save energy and improve efficiency [1]. ASD's are classified into three main categories: mechanical, hydraulic, and electrical ASD's. A major advantage of electrical ASD's over the non-electrical types is that both the human interface and control equipment are placed in the electrical room away from the process area. This is considered an important safety feature. Figure 1 shows the setup for the different types of ASD's. In mechanical and hydraulic ASD's, the control equipment is located between the motor and the mechanical load (i.e. in the process area), which makes maintenance very difficult. However, in an electrical ASD only the driving motor is placed in the process area. This is just one benefit of electrical ASD's [2].

Electrical ASD's can be of the dc type or the ac type. Only the ac type may be referred to as a VFD. A VFD can indeed control both the speed and torque of a squirrel-cage induction motor by varying the motor input frequency (and voltage). The fundamental components of a VFD system are

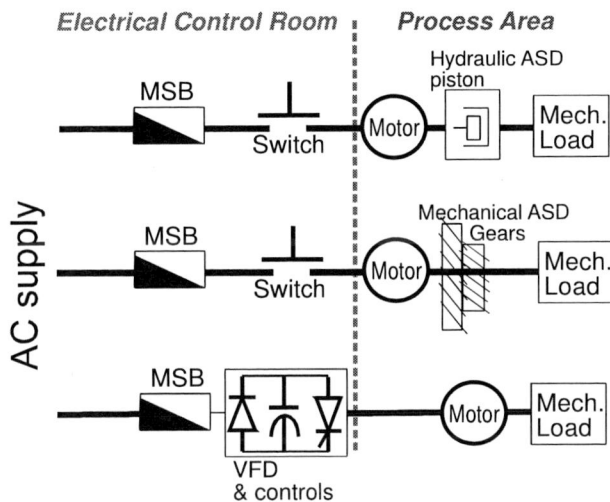

Fig. 1. Types of Adjustable Speed Drives showing the location of the control equipment for each.

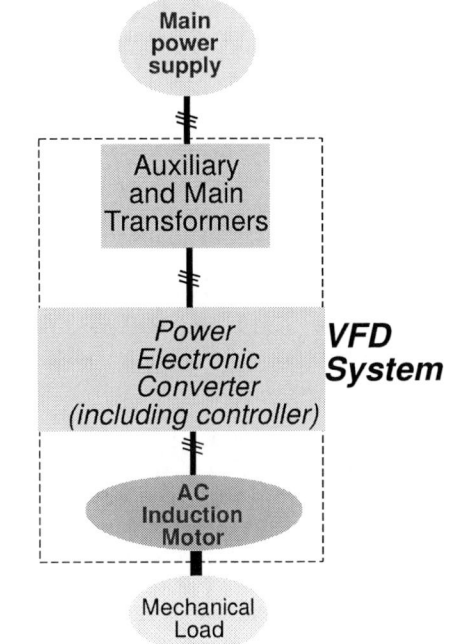

Fig. 2. Components of a VFD system.

978-1-4799-0024-4/13 $31.00 © 2013 IEEE

an electric motor, a power electronic converter (modulator), and a controller as shown in Fig. 2.

VFD's have been improved greatly with the development of semiconductor devices and the field of power electronics. Their applications are widespread due to their high efficiency and performance compared to mechanical and hydraulic ASD's. Some of the common applications of VFD's are milling stations, water pumping plants, and irrigation systems.

Compared to direct on-line (DOL) operation, VFD's have the advantage of soft-starting, soft-stopping, and speed and torque control of inductor motors. Hence, overcoming the problems of direct on-line motor starting. By controlling the speed of an ac motor, a VFD can vary the flow rate of a water pump. Speed regulation in a VFD system is significantly lower than that in a DOL system. On the other hand, DOL starting subjects an induction motor to high starting currents much greater than the rated current, while the soft-starting capability of VFD's allow for a gradual increase in starting current and torque. This lowers the inrush current, minimizes undesirable jerking during start, and reduces stress on the mechanical components which saves in maintenance cost. Indeed, VFD's offer smoother starting and controlled speed running. However, VFD's are relatively expensive systems. If speed control is not a requirement during continuous running, then, soft-starter systems are a better choice as they are less expensive [3].

Unlike the sine wave output from an ac generator, the electric ac output of a VFD contains harmonics. These harmonics may damage motor bearings as well as winding insulation. Reducing harmonics is typically done using a higher pulse VFD or by using active semiconductor devices in the rectifier stage. Furthermore, any remaining harmonics may then be mitigated by applying filters that are tuned to trap the desired harmonics. Nowadays, VFD's can synthesize near sine wave outputs.

II. LAYOUT OF A VFD-CONTROLLED SITE

The layout of a VFD-controlled site should include a separate electrical room where the VFD units and switchgear are to be installed as shown in Fig. 3. The ac motors are typically installed at another location in the site. Because of extreme vulnerability of the VFD electronic components and control boards, the VFD electrical room must be maintained clean, dry, dust-free, and at a cool temperature. Hence, it is advisable to locate any heat emitting components (*such as main and auxiliary VFD transformers*) in a separate room adjacent to the main VFD electrical room. Main transformers feed power to the VFD rectifier while auxiliary transformers feed power to the control and protection sensors. Typically, total heat losses in medium voltage drives are roughly 1.5% for VFD and 1.5% for transformers. Hence, if the transformers were to be placed outside the VFD electrical room, the heat load on the electrical room climate control and cooling system can be reduced by half.

The distance between the electrical room and the process area should be minimized in order to reduce losses and voltage drop. The transformers must include surge arresters to protect against internal and external voltage surges. Surge arresters must not be placed more than 3m away from the transformers.

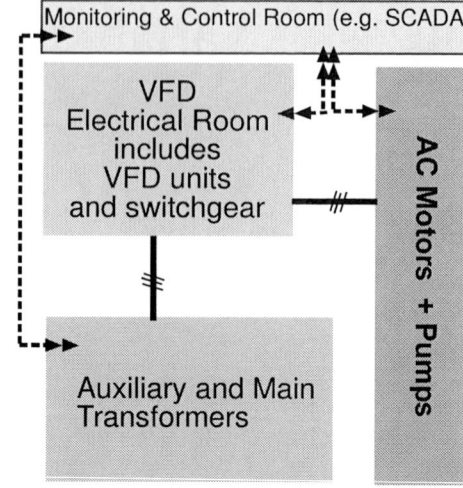

Fig. 3. Layout of VFD-controlled site.

In many applications (similar to sewage treatment and pumping stations), it is recommended to construct the electrical VFD room on ground level and not on (-) G level (i.e. not underground level). However, placing the electrical room underground subjects the electronic VFD components to the possibility of damage in case of accidental overflow beyond the normal water level or leakage of underground water due to poor civil work. When the electrical room is above ground, the VFD's, transformers, switchgear, (etc.) will all be protected from such accidents. This means that the investment in the electrical room could be easily salvaged, repaired, and returned back to service.

Generally, for harsh and dusty environments, variable frequency drives should be of the water-cooled type. However, water-cooled drives require a water circulating and cooling system (e.g. chiller).

Incorporating a monitoring system may provide an efficient early warning mechanism to protect, maintain, and optimize the operation of equipment and protect capital investment. Hence, using an accurate and continuously running supervisory control and data acquisition (SCADA) system is a must in industry.

III. DESCRIPTION OF VFD STRUCTURE

The basic structure of a VFD consists of a rectifier, a dc-link, and an inverter. A rectifier may be of the uncontrolled type if it includes diodes, whereas a controlled-type rectifier includes active semiconductors devices such as thyristors, IGBT's, IGCT's, etc.. On the other hand, inverters always include active semiconductor devices in order to be able to turn them on and off during the synthesis of ac output voltages. In between a rectifier and an inverter is a dc-link capacitor for voltage stabilization. Filters and protection circuitry should always be included to mitigate harmonics and protect the semiconductor devices [4].

A VFD may include a twelve-pulse uncontrolled rectifier feeding two series dc-link capacitors functioning as a dc voltage source for a three-level Neutral-Point-Clamped

978-1-4799-0024-4/13 $31.00 © 2013 IEEE

(NPC) three-phase inverter. Next is a description of the main sections of the VFD system analyzed in this paper.

A. 12-Pulse Uncontrolled Rectifier (Series Configuration)

Figure 4 shows a twelve-pulse uncontrolled rectifier circuit of the series configuration. Notice that it consists of two six-pulse rectifiers connected in series. The twelve-pulse rectifier charges two series dc-link capacitors. The node NP, between the two capacitors, is the neutral point of the subsequent NPC inverter. Input to the twelve-pulse rectifier is fed from a three-winding transformer with a delta primary, a delta secondary, and a wye tertiary (Dyn11d0 vector group). This means that one of the six-pulse rectifiers receives its input from a delta winding while the other six-pulse rectifier receives its input from a wye winding. Hence, there is 30° phase shift for the ac input voltages of the twelve-pulse rectifier.

Figure 5 shows a typical dc output voltage waveform of a series twelve-pulse rectifier. The plot in Fig. 5 was obtained using PSCAD simulation of the twelve-pulse rectifier used inside the medium voltage (3.3kV) VFD units of the pumping station studied in this paper. Specifications of the main three-winding main transformer feeding the series twelve-pulse rectifier are given in Table I. Notice that the output voltage shown in Fig.5 has twelve pulses over one complete ac cycle. The average dc output voltage of a series type twelve-pulse rectifier is given in (1) [5].

$$V_{dc\_avg} = (0.98862)\,(1.932)\,\sqrt{2}\ V_{secondary\_rms} \quad (1)$$

$V_{secondary\_rms}$ is the rms line voltage across the delta secondary of the three-winding main transformer. Hence, from the rectifier specifications, the average dc output voltage is calculated as given in (2).

$$V_{dc\_avg_{case\_study}} = (0.98862)\,(1.932)\,\sqrt{2}\ 1.9kV = 5.13kV \quad (2)$$

The diode conduction sequence over one complete ac cycle, for the series type twelve-pulse rectifier is given in Table II.

TABLE I
SPECIFICATIONS OF THREE-WINDING MAIN TRANSFORMER FEEDING THE SERIES TWELVE-PULSE RECTIFIER OF THE CASE STUDY VFD

Rated Voltage rms, kV HV / LV / LV	Vector Group	Rated Power, kVA HV / LV / LV	Rated Current , A HV / LV / LV
11 / 1.9 / 1.9	Dyn11d0	960/ 480 /480	50.4 / 145.6/ 145.6

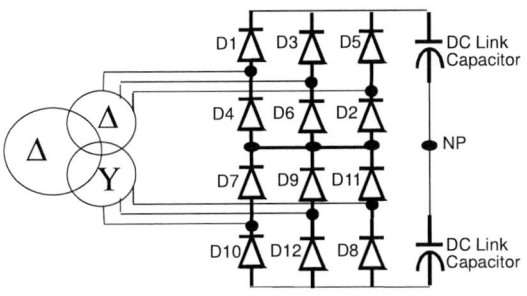

Fig. 4. Twelve-Pulse Uncontrolled Rectifier.

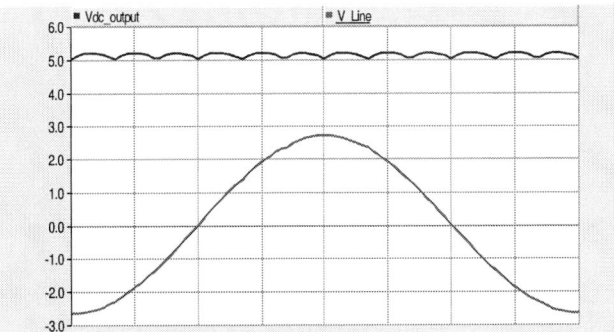

Fig. 5. Twelve-pulse dc output voltage compared to one line cycle

TABLE II
DIODE CONDUCTION SEQUENCE OF A SERIES TWELVE-PULSE RECTIFIER

Each cell below represents 30° showing the diodes conducting during each 30° interval.											
D6	D6	D1	D1	D2	D2	D3	D3	D4	D4	D5	D5
D1	D1	D2	D2	D3	D3	D4	D4	D5	D5	D6	D6
D11	D12	D12	D7	D7	D8	D8	D9	D9	D10	D10	D11
D12	D7	D7	D8	D8	D9	D9	D10	D10	D11	D11	D12

Each cell in Table II represents a 30° portion of one complete ac cycle.

B. Three-Level Neutral-Point-Clamped (NPC) Inverter

Figure 6 shows the second main section of this VFD structure which is the three-level NPC three-phase inverter [6]. The dc-link section will be considered as a common section to both the rectifier and the inverter. PSCAD simulation showing the three-level line-to-neutral ac output voltage of the NPC inverter is shown in Fig. 7. The line-to-line ac output voltage is shown in Fig. 8.

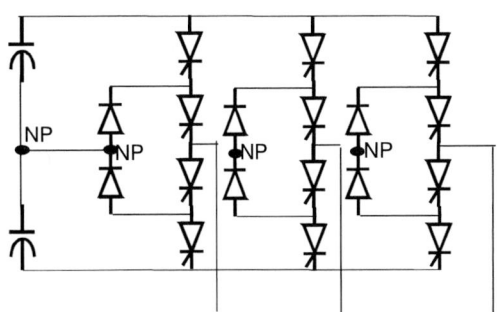

Fig. 6. Three-level NPC inverter.

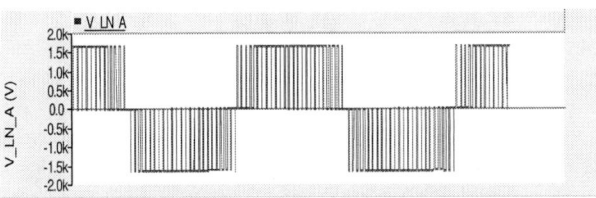

Fig. 7. 3-level line-to-neutral ac output voltage of a 3-level NPC inverter.

978-1-4799-0024-4/13 $31.00 © 2013 IEEE

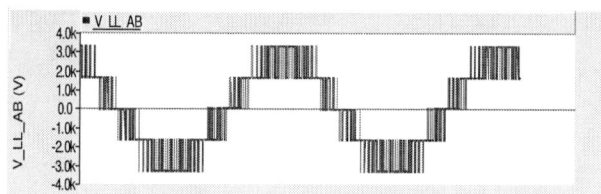

Fig. 8. 5-level line-to-line ac output voltage of a 3-level NPC inverter.

IV. CLASSIFICATION OF VFD FAILURE CAUSES BASED ON FIELD ANALYSIS OF A VFD PUMPING STATION

Field analysis of failures in the VFD system (described in Section III) of a major pumping station has been carried out. The failure lead to taking five out of nine VFD units out of service. Based on field analysis, the causes of VFD failure are classified as internal and external causes. Internal causes are those which occurred inside the VFD units while external causes are those which occurred outside the VFD units but lead to the failure of internal VFD parts. The internal causes may be subdivided into electrical and non-electrical causes. Similarly external causes may also be sub divided as electrical and non-electrical causes. Below is a list of the possible failure causes based on field analysis.

A. ELECTRICAL INTERNAL CAUSES
A1. Failure of VFD internal cooling (circulation) fans.
A2. Failure of VFD control electronics and gating units.
A3. Failure of VFD heater to reduce humidity inside the VFD.
A4. Failure of VFD auxiliary supply backup batteries.

B. NON-ELECTRICAL INTERNAL CAUSES
B1. Contamination due to dust, humidity, corrosive gases, which affects the lifetime of components and reduce the clearance distance and causes weaker cooling of the components.
B2. Thermal stress due to increased ambient temperature along with lack of internal VFD cooling (circulation).

C. ELECTRICAL EXTERNAL CAUSES
C1. Failure of Main transformers.
C2. Loss of Main Supply.
C3. Failure of Auxiliary transformers.
C4. Loss of Auxiliary Supply.
C5. Non-stability of utility ac power supply (Medium Voltage Input).
C6. Absence of surge arrestors.
C7. Faulty switchgear.

D. NON-ELECTRICAL EXTERNAL CAUSES
D1. Contamination due to dust, humidity, corrosive gases, which affects the lifetime of components and reduce the clearance distance and causes weaker cooling of the components.
D2. Failure of Electrical Room cooling system.
D3. Lack of spare parts.
D4. Lack of preventive maintenance.
D5. Low operator qualifications.
D6. Operating the VFD system under severe conditions against the manufacturer specifications.

V. IMPACT OF VFD FAILURE CAUSES ON VFD OPERATION

The pumping station analyzed in this paper includes nine twelve-pulse VFD units. The VFD units are supplied with main power through medium voltage 11/1.9/1.9kV main transformers (described in Section III). Power for the control and protection components of the VFD units is fed through 11/0.433kV auxiliary transformers. The layout of this VFD pumping station differs from that of Fig. 3 in that the VFD units and transformers are all placed in the same electrical room.

Table III, below, shows the actual failure types and their causes as concluded from field analysis of the nine VFD units. Unfortunately, failure of the five VFD units lead to overload on the four VFD units which remained in service. This, in turn, lead to a lengthy interruption in the operation of the pumping station.

TABLE III
VFD FAILURE TYPES AND THEIR CAUSES FROM FIELD ANALYSIS

	Failure Causes	Failure Type
VFD # 1		In service
VFD # 2		In service
VFD # 3	A1, A2, A3, B1, B2, C6, D1, D2, D3, D4, D5, D6	Flashover in the diode rectifier section
VFD # 4		In service
VFD # 5	A1, A2, A3, B1, B2, C6, D1, D2, D3, D4, D5, D6	Flashover in the diode rectifier section
VFD # 6	A1, A2, A3, B1, B2, D1, D2, D3, D4, D5, D6	Short-circuit in inverter section
VFD # 7	A2, B1, B2, C4, D1, D2, D3, D4, D6	Short-circuit in inverter section
VFD # 8		In service
VFD # 9	A1, A2, A3, B1, B2, C6, D1, D2, D3, D4, D5, D6	Flashover in the diode rectifier section

From Table III, it is clear that three failure cases out of five are of the rectifier flashover type. Two failure cases are of the inverter short-circuit type.

A. Causes Leading to Rectifier Flashover

It is clear that operation at extreme environmental conditions (high temperature, dust, humidity, and corrosive gases) highly participated in damaging the electronic components of VFD 3,5, or 9. This practice is against manufacturer's specifications. Indoor dust is often more harming than outdoor dust. Fine mode particles of indoor dust are characterized with large surface area which attracts moisture more easily due to capillary wetting [7-8]. Additionally, all transformers lacked surge arresters. Due to the absence of surge arrestors in the main transformers, higher than normal voltages may appear across the VFD main input terminals. These high voltages may initiate (trigger) and increase an arcing current bypassing a reverse-biased diode in the rectifier. This flashover path is formed by high temperature, dust, humidity, and corrosive gases. Hence, right before the failure of VFD 3,5,or 9, the resistance of the flashover path must have reached a minimum value which lead to exceeding the maximum current rating of the diode that is conducting in the short-circuit (shown in red) in Fig. 9. The short-circuit current leads to diode destruction (hot spots in areas of high current concentration). Diode D2, in Fig. 9, will fail due to the short-circuit current. No failure is expected to be seen in the diode

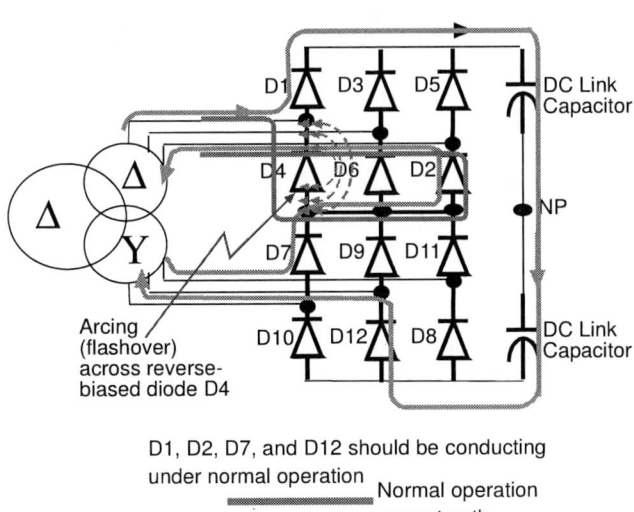

D1, D2, D7, and D12 should be conducting under normal operation

▬▬▬ Normal operation current path

◄- - - Arcing current flow

▬▬▬ Short circuit current path

Fig. 9. One possible arcing (flashover) across reverse-biased diode (D4) due to high temperature, dust, humidity, and corrosive gases.

that was bypassed by the arcing (flashover) current. Indeed, contamination due to dust, humidity, corrosive gases, which affects the lifetime of components, reduces the clearance distances and causes weaker cooling of the components.

PSCAD simulation has been done to approximate the effect of extreme environmental conditions on reducing the clearance distances of components and leading to conduction of high currents across a reverse-biased diode. Figure 10 is a PSCAD output waveform showing high short-circuit current flowing through diode D2 of Fig. 9 during flashover. The amount of flashover (short-circuit) current increases, reaching as high as 1.3 kA, as the resistance across diode D4 decreases. This high current will definitely damage diode D2 as in the case of VFD 3, 5, and 9.

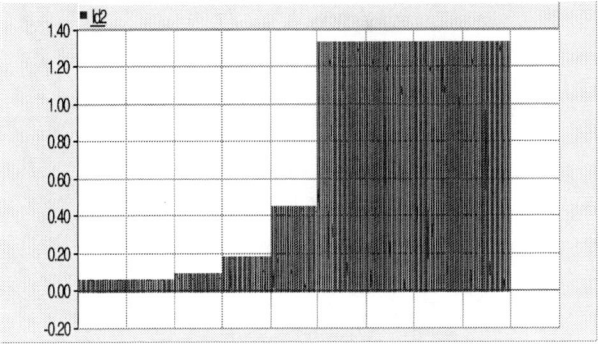

Fig. 10. PSCAD output waveform showing the high current flowing through diode D2 during rectifier flashover.

B. Causes Leading to Inverter Short-Circuit

Similar to the case of rectifier flashover, extreme environmental conditions will easily damage the control electronics which are the most vulnerable components of the VFD. Electronic control boards of a VFD are responsible for protection, monitoring, maintaining dc-link charge level, and correctly triggering the inverter semiconductor devices [9].

Continuously extreme environmental conditions lead to the failure of more control boards. Failure or erratic operation of electronic control circuitry may lead to operating an inverter in an unidentified state which may result in inverter short-circuit. Figure 11 shows some of the possible cases of normal inverter operation. Fig. 12 shows one case of semiconductor short-circuit due to false-triggering. One possible aftermath of false-triggering is failure of three thyristors and one diode as in the case of VFD 6 and 7.

In the case of the pumping station studied in this paper, forty percent of the failures were attributed to inverter side short-circuit due to false-triggering. The impact such failures on industry is disastrous and can be estimated in millions of dollars due to the high interruption cost and extremely high corrective maintenance cost.

In desert areas and environments of high temperature, it is of extreme necessity to provide and maintain a clean and cool operating environment that satisfies manufacturer requirements. Regular preventive maintenance also contributes to limiting the occurrence of such failures.

978-1-4799-0024-4/13 $31.00 © 2013 IEEE

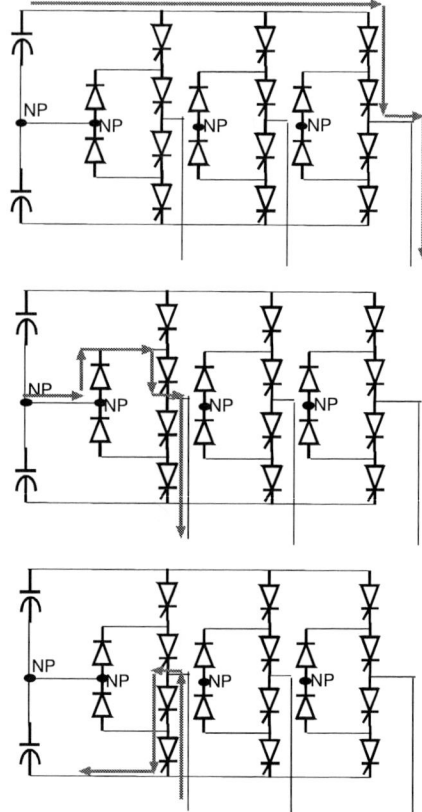

Fig. 11. Inverter cases of normal operation.

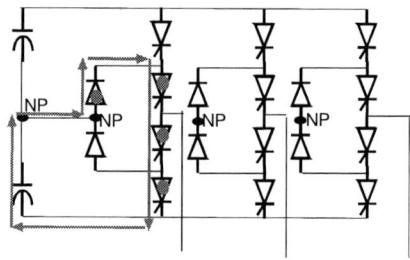

Fig. 12. Inverter short circuit due to false-triggering.

VI. CONCLUSIONS AND RECOMMENDATIONS

Based on field analysis of an existing major VFD pumping station, a study of possible causes of failure of a medium voltage VFD has been presented in this paper. The causes of failure were classified as electrical and non-electrical causes. From the failures analysis, five out of nine VFD units failed due to a combination of both electrical and non-electrical causes. However, the results of the failure analysis suggest that the most dominant causes are non-electrical. It is evident that the two most common failures were flashover in the rectifier side and short-circuit in the inverter side. Although these failures are of an electrical nature, their causes are evidently non-electrical, suggesting the crucial effect of maintaining clean environmental

conditions to prolong the lifetime of the VFD's. Awareness of such environmental factors cannot be further emphasized. Operators with limited experience may contribute to the failure. Operators should be able to perform the necessary maintenance and to accurately monitor the VFD system in order to prevent unnecessary failures, especially due to dust, humidity, and high temperature.

In order to protect the VFD against diode short-circuit in the rectifier side, an excellent choice would be to place fast semiconductor short-circuit protection devices in series with each diode in a six-pulse rectifier as shown in the Fig. 13.

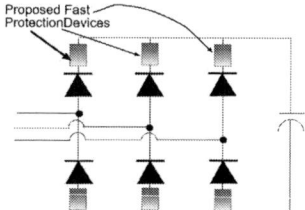

Fig. 13. Recommended protection against diode short-circuit due to flashover in the rectifier.

The electronic control circuitry of a VFD system is strongly affected by harsh environmental and operational conditions. Protection of inverter side semiconductor devices against false-triggering is mainly achieved through maintaining fully operational electronic control circuitry.

VII. REFERENCES

[1] P. Iwanciw, "Current Inverter Technology- Advantages, Problems, and Solutions," IEE, Savoy Place, London WC2R OBL, UK, 1994.

[2] B. Widmer (ABB-Presentation), "Overview MV Drives Medium Voltage Drives for Energy Savings and Lifecycle Improvements"

[3] Walter J Lukitsch, Senior Member IEEE , "Soft Start Vs AC Drives - Understand the Differences," Allen-Bradley Company, Milwaukee, WI

[4] M. Barnes, Practical Variable Speed Drives and Power Electronics, Elsevier, Great Britain, 2003.

[5] M. H. Rashid, Power Electronics Handbook, Second Edition, Academic Press, 2007

[6] A. Nabae, I. Takahashi, and H. Akagi, "A New Neutral-Point-Clamped PWM Inverter," in IEEE Transactions on Industry Application IA-17, no. 5, pp. 518–200, 1981.

[7] B. Song, et.al. "Impact of Dust on Printed Circuit Assembly Reliability by Bo Song", Center for Advanced Life Cycle Engineering (CALCE), University of Maryland, College Park, MD., 2012

[8] D. G. DeNure and E. S. Sproles, Jr. , "Dust test results on multicontact circuit board connectors", IEEE Transactions on Components, Hybrids, and Manufacturing Technology, Volume 14, No. 4, 1991

[9] D. Kasha., B.K. Bose, "Investigation of Fault Modes of Voltage-Fed Inverter Systems for Induction Motor Drives", IEEE Transactions on Industry Applications, Vol. 30, No.4, 1994, pp.1028-1038.

978-1-4799-0024-4/13 $31.00 © 2013 IEEE

Detecting High-Resistance Connection Asymmetries in Inverter Fed AC Drive Systems

G. Stojčić, T. M. Wolbank

Abstract – **The wiring system between inverter and the machine consists of different parts connected by metal-to-metal joints. High load currents lead to an increase temperature of the connections and higher material degradation. Basically, these effects result in an increased resistance in one phase. If such faults are not identified timely an unexpected outage of the whole system can be a consequence. However, the degree of reliability of the wiring system should not be less than that of the machine or inverter. Detecting such small resistance incensement provide the possibility to react timely on the occurred fault. In this paper a method is presented to detect open-circuit faults in the stator windings of an inverter fed drive. The method gets along with the hardware already present and doesn't need additional sensors etc. By applying voltage phasor steps to the machine terminals and measuring the resulting current the phase resistance can be estimated. By combination of the current direction with the phase resistance value resistance phasors can introduced. With these phasors a high sensitive fault indicator is generated able to detect connection fault severity and position. Measurements on a laboratory test stand for different fault configurations prove the applicability and accuracy of the proposed method.**

Index Terms--fault detection, wiring system, high resistance connections, induction machine, inverter fed,

I. INTRODUCTION

In industry applications and propulsion systems the electrical circuit between the inverter and the machine is composed of numerous conductors, terminal blocks, switches, fuses, circuit breakers etc. All these parts are connected through joints which are usually metal-to-metal connections and introduce a source of additional resistance in the electrical circuit.

Considering a metal-to-metal connection at a joint it is well known that real surfaces are not flat but perforated. Thus, only small regions of the connecting surface are connected. Basically, metals which are usually used in electrical circuits like aluminum, copper and brass are surrounded by a non-conductive oxide film. Only in areas where this oxide film is fractured by the contact pressure a real metal-to-metal connections is established and thus, this cluster of micro-spots is the conducting part [1]-[2].

Exposing the connection to high currents as this is common in industry and propulsion applications leads to a temperature increase at the metal-to-metal areas above the material bulk temperature. This temperature increase has an influence on the degradation process and thus oxide film growth. Especially overload or short circuit

currents lead to an intensive over-temperature and high degradation and resistance increase.

Another effect influencing the connection resistance is given by load cycling. This effect has no influence on the degradation process but on the mechanical pressure and force of the connecting parts like bolts and clamps. Here the thermal expansion of different materials is the leading cause. Considering a connection joint where one material has an expansion coefficient twice of the other one the pressure after heating up the connection will also double. This leads to a deformation of the geometrical properties of the connecting parts and finally to cracks [2].

Finally also the ambient effects in rough areas like moisture, dust, debris, ambient temperature changes, vibrations and external mechanical forces, etc. lead to A higher wear and corrosion of the regarding parts.

All these effects together, result in an increased resistance of a connection and thus, in reduced efficiency of the connecting parts and the whole system. However, the degree of reliability of the electrical circuit should not be less than that of the inverter and machine. Investigations have shown that high resistance connections are among the main causes for failures in electrical systems [5],[6],[7]. Furthermore not only the high resistance itself is the main problem but also the following propagation. For example a resistance increase in one phase leads to voltage asymmetry in the stator winding and finally to stator winding defects which are among the most frequent failures [3]-[10].

In the past years numerous investigations have been done and methods developed to detect high resistance connections [8]-[14]. All the methods show good accuracy when detecting high resistance connections. However, most of the methods have been developed for line fed machines and can thus be hardly applied to inverter fed drives. Another drawback is given by the usage of additional sensor as e.g. voltage sensors which are usually not applied to inverter fed drives and raise the system costs. Furthermore most of the investigations have focused on already occurred faults in the electrical circuit.

Therefore reliable condition monitoring methods must be developed applicable to inverter fed drives. The common visual inspections and manual resistance measurements are reliable but not always applicable. Considering the drive in a traction system, the connections and joints are usually not easy accessible. Thus, such methods are associated with the disassembling of the system and thus not usable for continuous monitoring.

G. Stojčić is with the Department of Energy Systems and Electrical drives at the Vienna University of Technology, Vienna, Austria (e-mail:goran.stojcic@tuwien.ac.at).

T. M. Wolbank is with the Department of Energy Systems and Electrical drives at the Vienna University of Technology, Vienna, Austria (e-mail: thomas.wolbank@tuwien.ac.at).

978-1-4799-0024-4/13 $31.00 © 2013 IEEE

The fault developing mechanism in the starting phase develops in a special manner. Considering the joint connection temperature equal to the ambient temperature at the beginning, the resistance increase will not be significant and negligible and thus also hardly detectable by conventional methods. By increasing the current through the connections due to demanded machine load the resistance will also rise. Assuming now a faulty connection, the resistance slope in one phase will be higher than in the remaining phases. Such fault cases will be denoted as hot contact point (HCP) in the following, due to the temperature dependent behavior and will be the point of main effort in this paper. Detecting HCP faults provides the possibility to react timely and to reduce the impact of subsequent fault.

In this article a method is presented to detect HCP faults. The method is based on the measurement of resulting current reaction to voltage phasor steps generated by the inverter. A benefit of the proposed method is given by the fact that only the inverter built-in standard current sensors and no additional hardware is needed. Measurements on a test stand with a voltage source inverter, an 11kW induction machine and a special designed terminal board will prove the method's applicability as well as the accuracy. However, for the sake of completeness, it must be mentioned that the presented method is only applicable while machine is at standstill and not running.

II. IDENTIFICATION OF PHASE RESISTANCE

Identification of the resistance value of a system is usually achieved by applying a voltage or current to the systems terminals and measuring the resulting current or voltage. The resistance value can then be easily obtained by Ohm's Law. This procedure is usually applied to ohmic' materials like resistors and wires as present in an electrical drive system.

Considering a drive system consisting of an inverter, wiring, connectors and the electrical machine, the mentioned procedure can be realized by applying a voltage phasor by the inverter to the machine terminals and measuring the current by the inverter built-in current sensors. Now, due to the knowledge of both values the resistance value of the system can be easily calculated. But it must be mentioned that the voltage applied to the machines terminals is disturbed by inverter non-idealities and usually very difficult to be accurately identified without additional inverter output voltage sensors. Such voltage sensor are related with higher costs and are not necessary for the control and thus not applied to common drive systems.

The inverter output voltage is generated by switching of the inverter, known as pulse width modulation (PWM). Additionally to the discrete pulses, there are also other phenomena present arising from the switching devices inherent characteristics, namely: voltage drop, output voltage transitions slope, turn off/on time and the inverter dead time. All this phenomena influence the resistance estimation through the distortion of the inverter output voltage. These effects must be eliminated or reduced to achieve and accurate resistance estimation.

The inverter dead time carries the majority of all disturbing effects. Due to the fact that the inverter dead time is defined by hardware its influence can be clearly reduced by a special voltage pattern procedure given in the following.

Basically the method presented is based on two applied voltage phasor steps and measurement of the current response within these voltages (1). These measurements results are combined to eliminate the inverter dead time influence. The applied voltage phasors have the same direction but different magnitudes. The measurement is applied at standstill with zero flux and no load. As only the resistance value is identified and only steady state values are considered the resulting current response has also the same direction.

$$
\begin{aligned}
v_{S,1} &\rightarrow i_{S,1} \\
v_{S,2} &\rightarrow i_{S,2}
\end{aligned}
\tag{1}
$$

The resistance can then be calculated by the differences from both obtained values as given in (2) and (3).

$$
\begin{aligned}
\Delta v &= v_{S,2} - v_{S,1} \\
\Delta i &= i_{S,2} - i_{S,1}
\end{aligned}
\tag{2}
$$

$$
r = \frac{\Delta v}{\Delta i} = \frac{v_{S,2} - v_{S,1}}{i_{S,2} - i_{S,1}}
\tag{3}
$$

This obtained resistor value now represents the phase resistance in the corresponding phase direction. The procedure is repeated for each phase direction so all phases are identified and compared and combined to one resulting value. Thus symmetrical influences (stator resistance) can be clearly reduced and a representation of the phase asymmetry is obtained.

III. SIGNAL PROCESSING AND FAULT INDICATOR

As shown above a special voltage pattern provide the possibility of phase resistance estimation for inverter-fed machines. A high-resistance connection at a joint leads to an increased resistance value in the corresponding phase and therefore a specific signal processing and fault indicator is developed to identify such asymmetries.

In a first step a voltage phasor is applied in one of the main phase directions (Phase U). Meanwhile the reacting current is measured. After that a second voltage phasor is applied to the same phase direction but with a different magnitude and the current is sampled again. All the parameters are forwarded to a resistance estimation block realized by (3). The voltage signal applied to phase U is shown in Fig. 1. The procedure is repeated subsequently to the remaining phases. Finally, resistance values of all three phases are obtained (r_U, r_V, r_W). In the following this values are denoted as 'resistance phasors'. Basically these phasors are only representative of the phase resistance values and are not comparable with other phasor values like voltage, current or flux in the space vector frame.

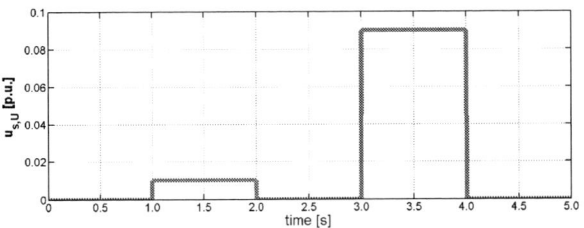

Fig. 1: Applied voltage signal to phase U.

Now, as the resistance values can be allocated to the phase direction the resistance is transformed from a scalar to a vector with the estimated value as magnitude and corresponding phase direction as angular position (4).

$$\underline{r}_U = r_U \cdot e^{j0}$$
$$\underline{r}_V = r_V \cdot e^{j2\pi/3} \qquad (4)$$
$$\underline{r}_W = r_W \cdot e^{j4\pi/3}$$

In the following step, the set of resistance phasors is combined by adding them up spatially. As a result a single space vector is obtained. In the healthy/symmetrical machine state all phase values have the same magnitude leading to a zero sequence component that is eliminated by the vector calculation. The healthy state result is thus a zero vector, independent of symmetrical resistance changes caused for example by temperature in the stator winding.

Assuming a high-resistance connection in a phase the single vector will point in this phase direction and the magnitude will correspond to the fault severity. Thus, this calculated single vector will be denoted fault indicator in the following. For a clearer presentation the described procedure is given as a block diagram in Fig. 2.

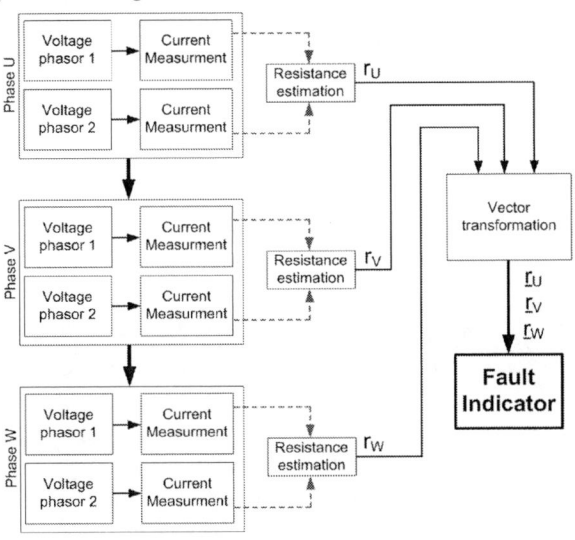

Fig. 2: Block diagram of proposed method

IV. EXPERIMENTAL SETUP AND MEASUREMENT RESULTS

A. Experimental Setup

To verify the applicability of the proposed method a laboratory test stand was installed with a 280V 11kW squirrel cage rotor induction machine with the parameters given in Table I and a voltage source inverter. The control and measurement system is realized on a computer system programmable under MATLAB/Simulink. Additionally a special terminal board was designed to realize a simple and non-destructive simulation of high-resistance connections in the inverter-to-machine wiring system. By changing the conductor between the input and output side of the terminal board a distinct resistance change can be adjusted. This is preferably done by copper wire or a bridge made of tin. A temperature measurement unit enables the verification of temperature changes in the stator winding system and the connectors. The schematic diagram and the terminal board are presented in and , respectively.

TABLE 1: PARAMETERS OF TEST MACHINE

Parameter	value
Nominal Voltage	280 V
Nominal frequency	75 Hz
Nominal current	30 A
Number of poles	4
Number of Stator slots	36
Stator resistance per phase	0.145 Ω

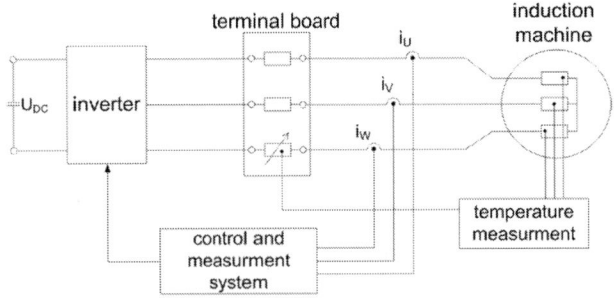

Fig. 3: Schematic diagram of the test stand

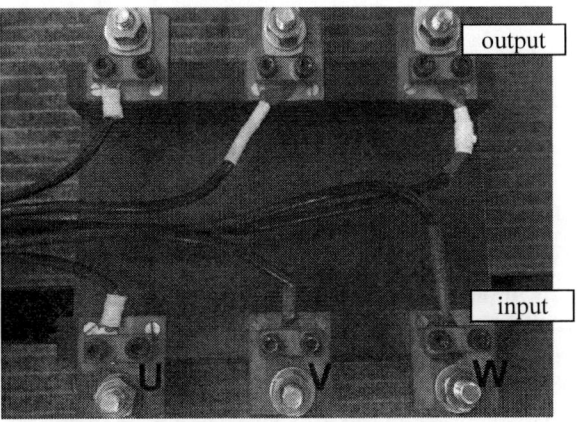

Fig. 4: Terminal board with for HCP fault emulation

As can be seen in the terminal board is realized by joint connections made of solid copper on the incoming and outgoing side. In the symmetrical case all three joints are connected by equal wires and/or couplers. By applying coupler of another material and/or dimension a distinct change in one phase can be realized without destruction of the setup. So, an easy and fast switching between faultless and faulty cases is realized for investigation of the proposed method.

B. Measurements to increased phase resistance

To prove the accuracy of the method several measurements and configurations were realized. In a first step a symmetrical configuration was investigated and all junctions were connected by copper wires. These copper conductors are emulating real connectors in a drive system. All the copper conductors have the same geometrical dimensions and thus equal resistance values. These measurements serve as reference and can be seen as an initial point of an excising system before a fault has occurred. As described in the previous section the method is based on the current reaction to a voltage phasor step. In Fig. 5 the fault indicator (green star) and the phase resistance values r_U, r_V and r_W (blue crosses) for the healthy case are presented. The axis scaling is given in arbitrary units [a.u.] corresponding to the signal processor's internal representation and the complex plane is representing the stator fixed frame. As already mentioned, the combination of the resistance phasors to one single fault indicator phasor leads to a zero component in the healthy case as can be very well seen in the Figure. However, this measurement is realized by a low magnitude of the voltage phasor steps (first step: 0.005 p.u. and second step: 0.02 and the step duration was set to 5 seconds.

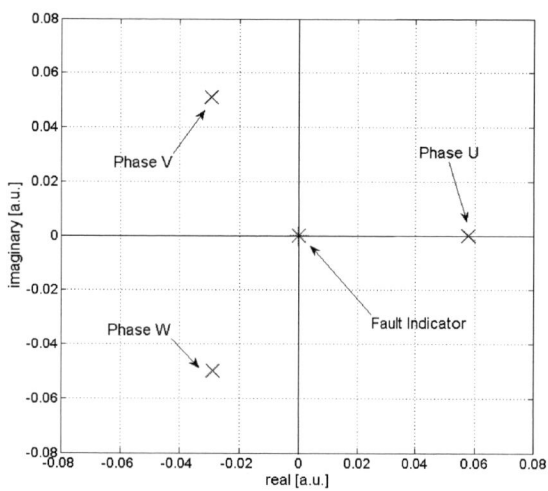

Fig. 5: Phase resistance values and fault indicator for healthy case in the complex plane.

In further consequence the copper conductor in phase W was exchanged with a longer conductor to prove the methods accuracy to already existing increased connection resistance. The resistance of healthy phase paths was measured to 10.8 mΩ while in the faulty path

to 18.0 mΩ, what leads to an increase of 60%. The fault indicator in the complex plane for this case together with the indicator for the faultless case is presented in Fig. 6. The indicator has a clear deviation from the origin and the faultless case, respectively. Furthermore, due to the position of the fault indicator in phase direction W the faulty phase can be identified.

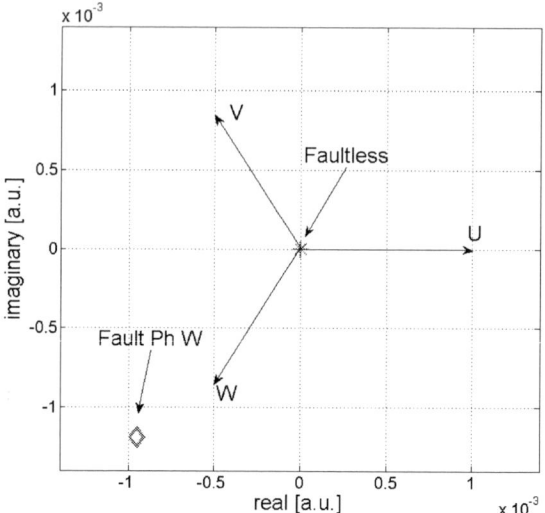

Fig. 6: Fault indicator in the complex plane for fault realized in Phase W by a resistance 60% higher than for healthy phases (U and V).

C. Measurements for HCP fault detection

As already mentioned, to keep the reliability at a high level it is necessary to detect faults in their early stage. Therefore, an experiment assembly was set up to emulate the behavior of a HCP in an early stage. To realize the temperature increase and thus the resistance increase of the faulty connection by current load, a bridge made of tin was applied to the terminal board in phase W as can be seen in Fig. 7. For identification of the resistance-to-temperature relation the temperature of the tin bridge is measured by a thermo couple. Due to the fault characteristics the resistance of all three connections is equal at ambient temperature. Thus, the tin bridge resistance at ambient temperature was set by the geometrical dimensions equal to the copper conductor resistance of 10.8mΩ.

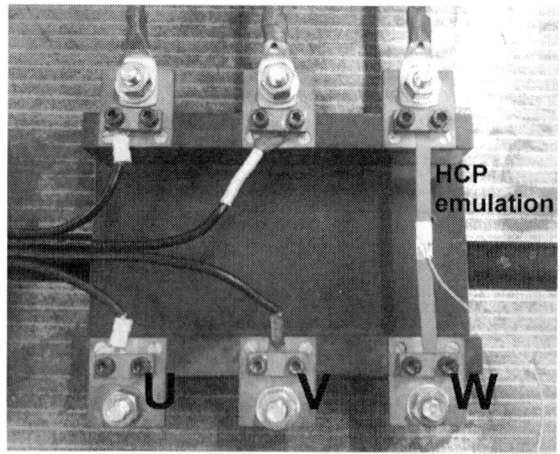

Fig. 7: Terminal board with HCP emulation in phase W by a tin bridge.

978-1-4799-0024-4/13 $31.00 © 2013 IEEE   230

To get a clear impression of the experimental setup behavior during current load, a voltage phasor step with duration of 120 s was applied to the phase direction W. The pulse magnitude was set to 0.05 p.u. for both cases. The temperature of the HCP and copper conductor, the current response as well as the current difference are presented in Fig. 8 for the pulse duration. The temperature increase of the copper conductor can be neglected compared to the tin bridge, emulating the HCP. The current value in phase W for both cases starts at 0.857 as the resistance of tin bridge and copper conductor is equal at ambient temperature. At the end of the voltage pulse a clear difference of the current values can be seen due the increased resistance by current load induced temperature rise of the HCP. The current load has not only an impact on the HCP but also on the stator winding temperature. Thus, also the stator winding temperature was measured (as indicated in) in all three stator winding phases. However, due to the small increase of the stator winding temperature the impact of the stator winding resistance on the proposed method can be neglected. These temperature measurement results are given in Table II.

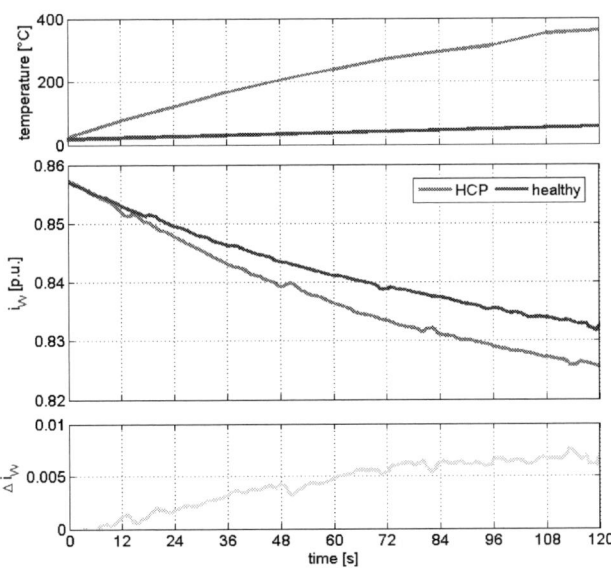

Fig. 8: Temperature, current [p.u.] and current difference progress of healthy and HCP case in Phase W due to a voltage phasor step in phase W (duration 120s).

TABLE II
STATOR WINDING TEMPERATURE DURING VOLTAGE PULSE IN PHASE W

Phase	U	V	W
start temp. [°C]	26	26	26
end temp. [°C]	33	33	38
environmental temp. [°C]	24		

In the next step the measurement procedure for fault indicator estimation, as presented in the previous section, was executed for the healthy case (all connections realized by copper conductor of equal length). The voltage phasor step duration was set to 30s each. The first voltage phasor was applied with a magnitude of 0.01 p.u. while the second with 0.05 p.u. The current measurement within the voltage phasor step duration was done with 5 kHz. To suppress the signal noise a running average calculation based on a digital filter was implemented in the control and measurement system. The used filter type is a direct form 2 FIR filter with a window length of 1000. The impact of the filtering process to the fault indicator is given in Fig. 9.

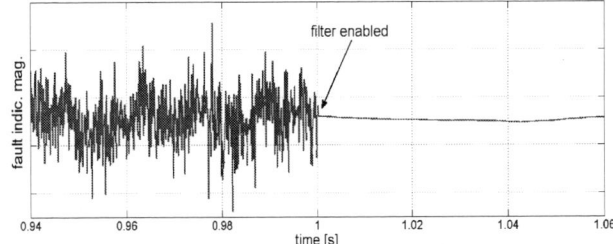

Fig. 9: Impact of a digital filter based running average calculation on the fault indicator signal (direct form 2 FIR filter with a window length of 1k).

Subsequently the copper conductor in phase W was exchanged by a tin bridge (HCP simulation) and the measurement procedure was repeated with the same settings as for the healthy case. The fault indicator signal of both cases is presented in . By the three dimensional frame a clear impression is given regarding the fault indicator trace within the voltage phasor step duration. The complex plane spans the base level as already shown in Fig. 5 and Fig. 6, respectively. The z-axis corresponds to the pulse duration (here 30 s).

The results prove that for the healthy case the fault indicator (blue trace) shows only negligible deviation from the origin and thus indicates a symmetrical configuration. On the other hand the fault indicator for a HCP configuration (red trace) shows a clear deviation from the origin with time. Furthermore, with the fault indicator direction the HCP can be assigned to a distinct phase, here phase W.

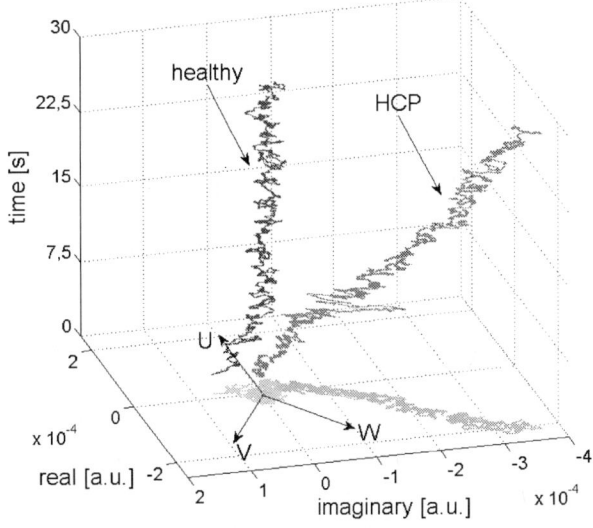

Fig. 10: 3D presentation of fault indicator vs. time.
Blue: healthy. Red: HCP.

Finally, to get an impression of the coherences between already occurred faults and the fault indicator a measurement for such a case was done. The connection of phase W on the terminal board was realized by a copper conductor with a resistance value of 52 mΩ. This is an increase of 481% with respect to the remaining phases. In Fig. 11 all the measurement results (healthy, HCP, occurred fault) are plotted together. As can be very well seen the fault indicator magnitude for the already occurred fault case (pink trace) remains almost constant versus time and can be clearly distinguished from the HCP case (red trace).

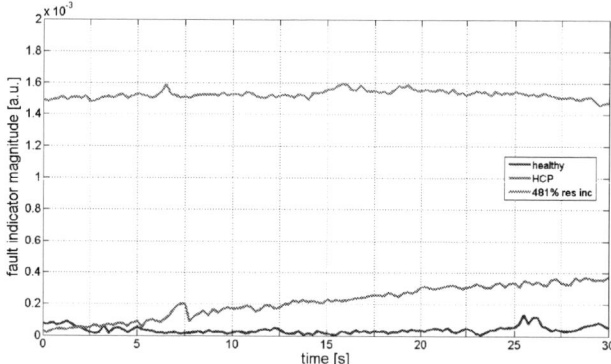

Fig. 11: Fault indicator magnitude vs. time for the healthy case and two faulty cases.

V. CONCLUSION

In the present paper a method was proposed and investigated to detect bad junction connections in the wiring system of inverter-fed drives. A big advantage of the method is given by the fact that not only already occurred faults but also incipient fault cases can be identified. The method is based on the identification of the phase resistance. The resistance estimation is realized by two voltage phasor steps and measurement of the current reaction. By transformation of the obtained phase resistance values to complex values and combining them to one phasor leads to a high sensitive fault indicator. By applying longer voltage phasor steps to the machine the time trace of the fault indicator can be used to identify not only already occurred connection faults but also incipient faults. A test stand was set up to prove the methods applicability. The test stand includes a special designed terminal board to realize non-destructive fault emulation. Measurements for different fault cases have proved the methods applicability and accuracy. It has been also shown that not only occurred but also incipient faults can be clearly identified by its magnitude and position with the proposed method.

VI. REFERENCES

[1] Naybour, R.D.; Farrell, T., "Degradation mechanisms of mechanical connectors on aluminium conductors," Electrical Engineers, Proceedings of the Institution of , vol.120, no.2, pp.273,280, February 1973

[2] Braunovic, M., "Effect of connection design on the contact resistance of high power overlapping bolted joints," Components and Packaging Technologies, IEEE Transactions on , vol.25, no.4, pp.642,650, Dec 2002

[3] MOTOR RELIABILITY WORKING GROUP, "Report of large motor reliability survey of industrial and commercial installations, Part II,"*IEEE Trans. Ind. Appl.*, vol. IA-21, no. 4, pp. 865–872, Jul. 1985

[4] MOTOR RELIABILITY WORKING GROUP, "Report of large motor reliability survey of industrial and commercial installations, Part I," *IEEE Trans. Ind. Appl.*, vol. IA-21, no. 4, pp. 853–864, Jul. 1985

[5] Bonneville Power Administration, Electrical distribution system tune-up, Jan. 1995.

[6] R.S. Colby, "Detection of high-resistance motor connections using symmetrical component analysis and neural networks," Proc. of IEEE SDEMPED, pp. 2-6, Atlanta, GA, 2003

[7] Washington State Energy Office, Keeping the spark in your electrical system: an industrial electrical maintenance guidebook, Oct. 1995.

[8] Jangho Yun; Kwanghwan Lee; Kwang-Woon Lee; Sang-Bin Lee; Ji-Yoon Yoo, "Detection and Classification of Stator Turn Faults and High-Resistance Electrical Connections for Induction Machines," Industry Applications, IEEE Transactions on , vol.45, no.2, pp.666,675, March-april 2009

[9] Keeping the Spark in Your Electrical System: An Industrial Electrical Maintenance Guidebook, 1995

[10] G. A. McCoy and J. G. Douglass Energy Management for Motor Driven Systems, 2000 :Office Ind. Technol., U.S. Dept. Energy Available:http://www1.eere.energy.gov/industry/bestpractices/ techpubs_motors.html

[11] J. Yun , J. Cho , S. B. Lee and J. Yoo "On-line detection of high-resistance connections in the incoming electrical circuit for induction machines", Proc. IEEE-IEMDC, pp.583 -589 2007

[12] D. Almand, "Fault zone analysis—Power circuit," in *Proc. PDMA Motor*

[13] *Rel. Techn. Conf.*, 2004. [Online]. Available: http://www.pdma.com/ oldart.html

[14] J. Bockstette, E. Stolz, and E. J.Wiedenbrug, "Upstream impedance diagnostics for three phase induction machines," in *Proc. IEEE-SDEMPED*, Cracow, Poland, 2007, pp. 411–414.

VII. BIOGRAPHIES

G. Stojčić received the B.Sc. degree in Electrical Engineering and the M.Sc. degree in Power Engineering from Vienna University of Technology, Vienna, Austria in 2009 and 2011, respectively. He is currently a Project Assistant at the Department of Energy Systems and Electrical Drives, Vienna University of Technology and working towards his PhD degree. His special fields of interest are fault detection and condition monitoring of inverter fed drives.

T. M. Wolbank received the doctoral degree and the Associate Prof. degree from Vienna University of Technology, Vienna, Austria, in 1996 and 2004, respectively. Currently, he is with the Department of Energy Systems and Electrical Drives, Vienna University of Technology, Vienna, Austria. He has coauthored more that 100 papers in refereed journals and international conferences. His research interests include saliency based sensorless control of ac drives, dynamic properties and condition monitoring of inverter-fed machines, transient electrical behavior of ac machines, and motor drives and their components and controlling them by the use of intelligent control algorithm

FPGA-based Smart-sensor for Fault Detection in VSD-fed Induction Motors

A.G. Garcia-Ramirez, R.A. Osornio-Rios, *Member, IEEE*, A. Garcia-Perez, *Member, IEEE*, R.J. Romero-Troncoso, *Senior Member, IEEE*.

Abstract – **Nowadays, different industrial processes use induction motors fed through variable speed drives (VSD). In order to improve these processes, the industry demands the use of smart sensors to detect the faults, reduce the cost of maintenance, and decrease power consumption. In this work, broken rotor bars, unbalance and misalignment are automatically detected in induction motors fed by a VSD using the three current phases online, with a smart sensor. The proposed smart sensor is implemented in a field programmable gate array offering a low computational load methodology, low-cost, and portable solution for fault detection in induction motors VSD-fed. Results show a high effectiveness detection of the treated faults.**

Index Terms--**Fault detection; Field programmable gate arrays; Induction motors; Variable speed drives.**

I. INTRODUCTION

I NDUCTION motors are key elements in industry due to their robustness, easy construction, low cost and versatility, representing 85% of power consumption worldwide. Consequently, early fault detection in induction motors is one of the most important subjects for industry [1], because these faults may produce unanticipated interruptions on product lines, with severe consequences in product quality, such as safety and cost. Therefore, the detection of incipient faults has attracted the interest of many researchers in recent years [2]. Around 40% to 50% of induction motor faults are bearing related, rotor faults represent 5% to 10% and unbalance and misalignment faults are about 12% [3]. Furthermore, connection of induction motors through variable speed drives (VSD), allows extending their useful life, and saving energy, but making the detection of faults more difficult due to the spurious harmonics induced by the VSD operation [4]. Moreover, the extensive use of VSD allows new possibilities for the on-line detection of faults [5]. Concerning to fault detection in induction motors, a

---

This work was supported in part by FOFIUAQ2012, under ACAEC188088 CONACyT-2012 and by CONACyT under Scholarship 229736.

R. J. Romero-Troncoso is with HSPdigital CA-Telematica at DICIS, University of Guanajuato, Salamanca, Gto. 36885 Mexico (corresponding author; e-mail: troncoso@hspdigital.org).

A.G. Garcia-Ramirez is with HSPdigital CA-Mecatronica at Facultad de Ingenieria, Campus San Juan del Rio, Universidad Autonoma de Queretaro, San Juan del Rio, Qro. 76807 Mexico (e-mail: aggarcia@hspdigital.org).

R.A. Osornio-Rios is with HSPdigital CA-Mecatronica at Facultad de Ingenieria, Campus San Juan del Rio, Universidad Autonoma de Queretaro, San Juan del Rio, Qro. 76807 Mexico (e-mail: raosornio@hspdigital.org).

A. Garcia-Perez is with HSPdigital CA-Procesamiento Digital de Señales at DICIS, University of Guanajuato, Salamanca, Gto. 36885 Mexico (e-mail: arturo@salamanca.ugto.mx).

number of techniques had been proposed. For instance, in [6] Shahin *et al.* investigate the recent advances on digital signal processing techniques for induction motors diagnosis. These techniques cover the frequency domain as the fast Fourier transform (FFT), the zoom-FFT, the chirp Z-transform, and the time-frequency domain such as multiple signal classification (MUSIC), short time Fourier transform, wavelet transform, etc. The best technique is chosen depending on the physical phenomena to observe. Garcia-Perez *et al.* [3] proposed a methodology which combines a filter bank of finite impulse response (FIR) filters with high resolution spectral analysis based on MUSIC for detecting multiple combined faults, with current and vibration signals. The results show the analytical predetermined fault frequency location for single, two or three combined faults (broken rotor bars, unbalance and bearings damage). In [7], it is proposed an artificial neural network (ANN) methodology to detect broken rotor bars (BRB) with statistical patterns of time-domain data, current spectrum and the combination of two ANN inputs to classify the motor condition. Riera-Guasp *et al.* [8] studies the relation between the amplitude of the components of the air-gap fault field produced by a double bar breakage and the relative position of the broken bars. Otherwise, [9] describes a method for the diagnosis and detection of BRB and bearing damage in induction motors by motor current signal analysis and multiple features extracted from transformations on current and voltage signals, using the hidden Markov model as classifier. Regarding the unbalance condition (UNB), Kral *et al.* in [10] proposed a technique to sense the specific modulation of the electric power of faults such as eccentricities as well as load torque perturbation without using the frequency spectrum. Otherwise, in [11] the misalignment condition (MAL) is computed by the FFT extracting the unique vibration features exhibited in the full spectrum. All the aforementioned works, need a heavy computational load and can be difficult to implement on hardware for online operation because most of these techniques requires offline processing and an expert technician for interpreting results. In [12], two techniques namely Park transform approach and Concordia transform are presented and compared for the detection of bearing damage in induction motors. The results of this work show that these techniques are valid to identify faulty patterns making suitable for hardware implementation, such as a smart sensor.

A smart sensor is a device that includes primary sensors,

signal processing, communication, and integration capabilities. The term "smart sensor" is employed according to the functionality classification. Nowadays the use of smart sensors improves the monitoring system demands due to their features in communication and data processing functionalities, and their versatility and ability to work in environments where the access for field workers is limited [13].

In this work the development of a smart sensor, based on a field-programmable gate array (FPGA), for automatic and online detection of faults in induction motor fed through VSD is presented, covering operating frequencies from as low as 3 Hz and up to 60 Hz. The methodology used in this work is based on the Park transform that converts the *ABC* current system to the *D-Q* current system, having the advantage of a lower computational load than the FFT and other spectrum-based methodologies; then the magnitude of the *DQ* current system and the mean are calculated to be the inputs of an artificial neural network, which gives the identification of the fault. Three different faults on induction motor: broken rotor bars, unbalance and misalignment are investigated, and the results show the potentiality of the smart sensor to classify the induction motor faults, automatically and online.

II. THEORETICAL BACKGROUND

A. Park Transform

The Park transform, permits to express the three current phases of an induction motor through a two-axis system in quadrature. The components of this system: direct and quadrature (i_D and i_Q), are given by (1) and (2), respectively:

$$i_D = \sqrt{\frac{2}{3}}i_A - \sqrt{\frac{1}{6}}i_B - \sqrt{\frac{1}{6}}i_C \tag{1}$$

$$i_Q = \sqrt{\frac{1}{6}}i_B - \sqrt{\frac{1}{6}}i_C \tag{2}$$

Where, i_A, i_B and i_C are the stator current phases [14]. Using (1) and (2) the transformation of the stator current phases *ABC* to the *D-Q* system is very simple. Fig. 1 shows the Lissajous figures of the current in *D-Q* system for a healthy motor (Fig. 1a) and a faulty motor (Fig. 1b).

B. Induction Motor Faults

This work focuses on three different faulty conditions: broken rotor bars (BRB), misalignment (MAL) and unbalance (UNB).

The BRB fault appears because of welding defects, high strength joints, expansion and mechanical stresses [15]. The presence of BRB in induction motors produces several problems, such as power quality degradation [16]. On the other hand, MAL is presented when the motor and the load pulleys are not aligned. The MAL fault can cause over 70% of the rotating machinery vibration problems, and it is the second most commonly fault in rotating machines [11].

Finally, the UNB condition is presented when the rotor weight is not uniformly distributed around its geometrical center, which means that the center of mass is not on the center of rotation. The UNB condition is the most observed fault in induction motors, and if is not attended, the results for the machinery can be catastrophic [10].

C. Artificial Neural Networks

An Artificial Neural Network (ANN) is a computational model that provides a method to characterize synthetic neurons to solve problems in the same way as the human brain [1]. The most popular architecture for ANN is the multilayer feed-forward networks (MFN), which has an input layer, one or more hidden layers and an output layer. In this architecture, the data goes in one direction, from the input layer through the hidden layer to the output layer, as shown in Fig. 2.

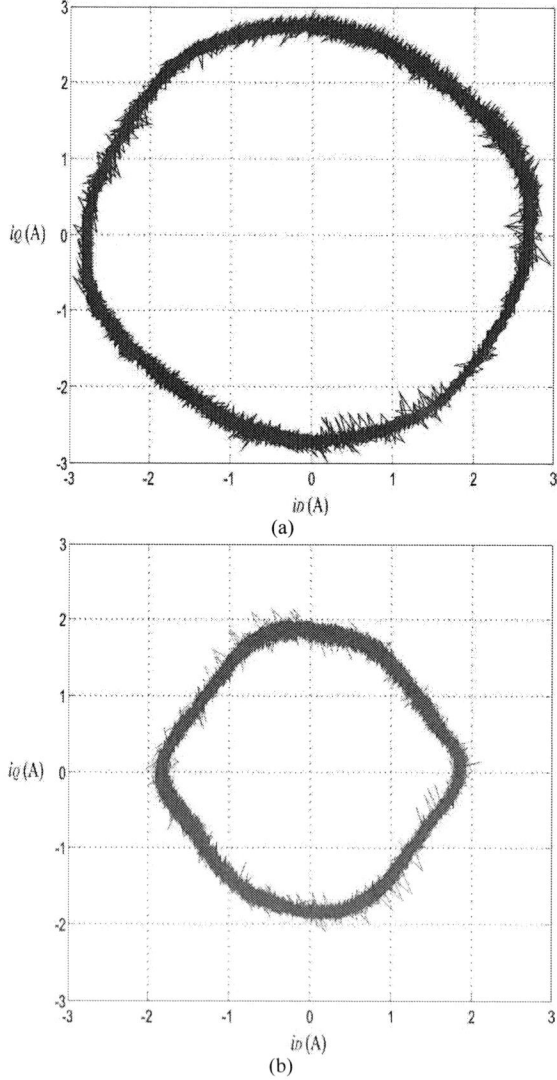

Fig. 1. Lissajous figures of the *D-Q* system current of (a) Healthy motor, and (b) Faulty motor.

Where, X_i ($i = 1,2,3,...,n$) are inputs and Y_i ($i = 1,2,3,...,m$) are outputs. The back-propagation algorithm (BPA), is the most conventional method to train an MFN, which is a supervised learning method, that consists on mapping the process inputs to the desired outputs by minimizing the error between the desired outputs and the calculated outputs [17]. The MFN is an excellent candidate to be implemented in FPGA, due to its simplicity, practicality and low computational load [18].

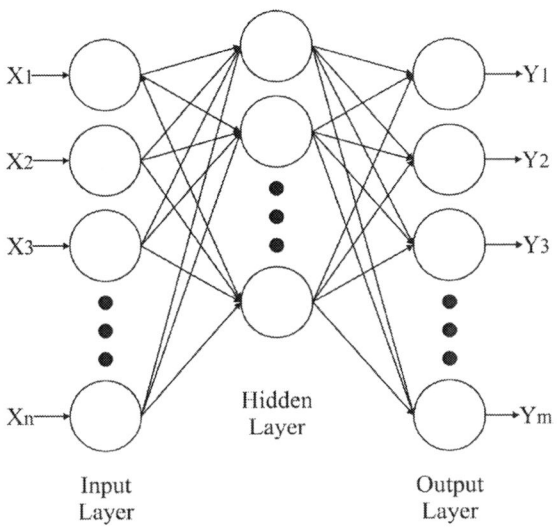

Fig. 2. Multilayer feed-forward network architecture.

III. METHODOLOGY

The methodology is based on the Park transform that allows representing the three current phases of an induction motor in a two-dimensional system. Fig. 3 shows the Lissajous figure obtained from the Park transform. Then, through a magnitude calculation (3) of each k sample, as shown also in Fig. 3, the D-Q system is translated to a vector $i_{DQ\text{-}k}$ that is the resultant representation of the Lissajous figure. In Fig. 4 the $i_{DQ\text{-}k}$ vector is presented. Afterward, by a mean computation (4) of the $i_{DQ\text{-}k}$ vector, the average radius of the Lissajous figure can be taken. This mean value of the $i_{DQ\text{-}k}$ vector is also presented in Fig. 4. Finally, the data from the mean computation is an input of the proposed ANN which delivers the motor condition.

$$i_{DQ\text{-}k} = \sqrt{i^2_{D\text{-}k} + i^2_{Q\text{-}k}} \tag{3}$$

$$Mean = \frac{1}{L}\sum_k i_{DQ\text{-}k} \tag{4}$$

Where, L is the length of the $i_{DQ\text{-}k}$ vector.

A. Proposed Artificial Neural Network

The proposed ANN implements an MFN with three inputs nodes that receive the mean calculation of the i_{DQ} vector in each treated frequency, thirty nodes in the hidden layer and four output nodes for the detection of: healthy condition (HLT), broken rotor bars (BRB), unbalance (UNB) and misalignment (MAL). The output nodes correspond to each fault condition.

IV. EXPERIMENT

A. Experimental Setup

The steady-state current signal provided by a VSD (model WEG CFW08) connected to the induction motor is used for detecting the faults and to classify them. The VSD has an operation range from 0 Hz to 100 Hz using a frequency resolution of 0.01 Hz. In Fig. 5(a) the experimental setup is shown, where one 1-hp three-phase induction motors (model WEG 00136APE48T) is used to test the performance of the proposed methodology to identify the fault conditions considered in this work. The rotational speed of the motor is controlled by a VSD at 3Hz, 30Hz and 60Hz. The tested motors have two poles, twenty eight bars and receive a power supply of 220 V AC. The applied mechanical load is from an ordinary alternator, which represents a quarter of load for the motor. The three-phase current signals are acquired using three hall-effect sensors model L08P050D15, from Tamura Corporation. A 16-bit 4-channerl serial-output sampling analog-to-digital converter ADS8341 from Texas Instrument Incorporated is used in the data acquisition system (DAS). The instrumentation system which was calibrated through the Fluke 435 that uses a sampling frequency f_s = 12 KHz obtains 120,000 samples of each current phase during 10 seconds of the induction motor steady-state. The motor start-up is controlled by a relay to automate the test run. Fig. 5(b) shows the proposed smart sensor for induction motor fault detection. The three current phases of the induction motor fed by VSD are acquired by the hall-effect sensors; then it is conditioned and analog-to-digital (A/D) converted in the DAS. Afterwards, in a smart unit implemented into a proprietary Spartan 3E XC3S1600 FPGA platform running at 48 MHz, the three digital current phases are transformed in i_D and i_Q phases by the Park transform. Then, the magnitude (3) and the mean of this magnitude (4) are calculated to be a single data for an input of an ANN which gives the induction motor condition. However, the current magnitude depends on the load and the motor power; then, to minimize the undesired effects of magnitude variation that could modify the diagnosis, the *ABC* current signals are normalized before the Park transform is applied. Fig. 6 shows the mean calculation of the i_{DQ} vector obtained by each fault condition in the three studied frequencies Table I summarizes the resource usage of the FPGA.

B. Investigated Faults

The BRB condition was artificially produced by drilling a 7.938mm diameter hole in a rotor bar without harming the shaft of the rotor. Fig. 7(a) shows the rotor with the BRB condition used in the test. The misalignment condition (MAL) was carried out by shifting forward the band in the alternator pulley, so that the transverse axes of rotation for the motor and its load were not aligned. Fig. 7(b) shows the misaligned motor. The UNB test was produced by attaching a bolt in an arm of the rotor pulley as shown in Fig. 7(c).

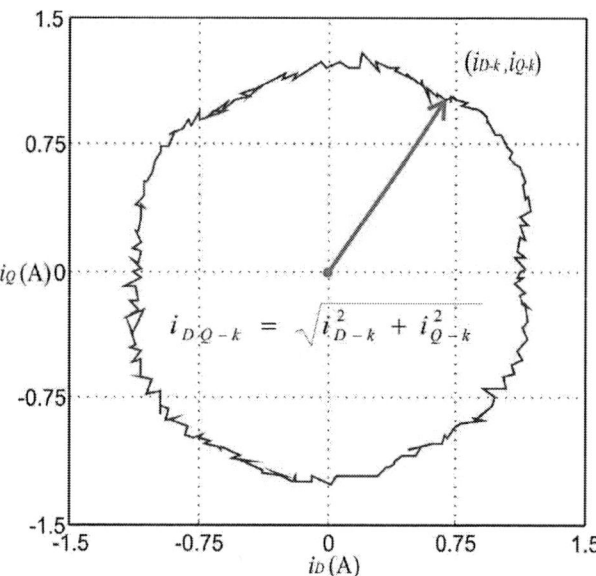

Fig. 3. Lissajous figure obtained from the Park transform.

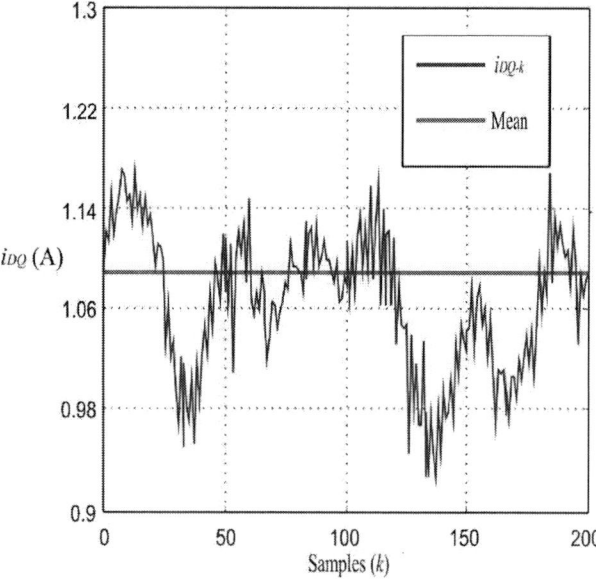

Fig. 4. Representation of the Lissajous figure by the $i_{DQ\text{-}k}$ vector.

C.   Network Training

The training set for the ANN training was obtained with 500 random synthetic values for every condition in each treated frequency within the range $[\mu - \sigma, \mu + \sigma]$, where μ is

the mean and σ the standard deviation of the mean values of the i_{DQ} vector on the first ten trials. Forty trials are carried out under each motor condition in every treated frequency; which were used as validation set for the diagnosis. Using Matlab neural network toolbox, the weights and the biases of each layer in the ANN were calculated for being implemented on the FPGA for the diagnosis. In some cases when the design of the motor is changed, a new training is required for adjusting the calibration of the smart sensor and improving the classification results. For future development, it is proposed to add an additional input to the ANN where the load value is given.

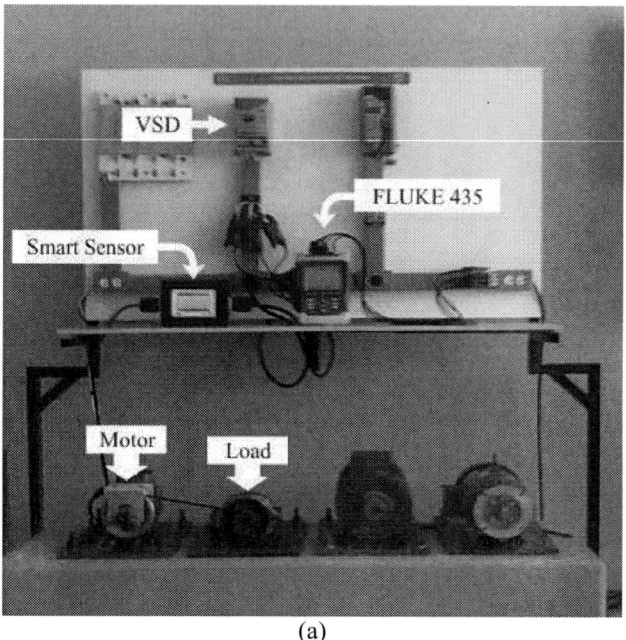

(a)

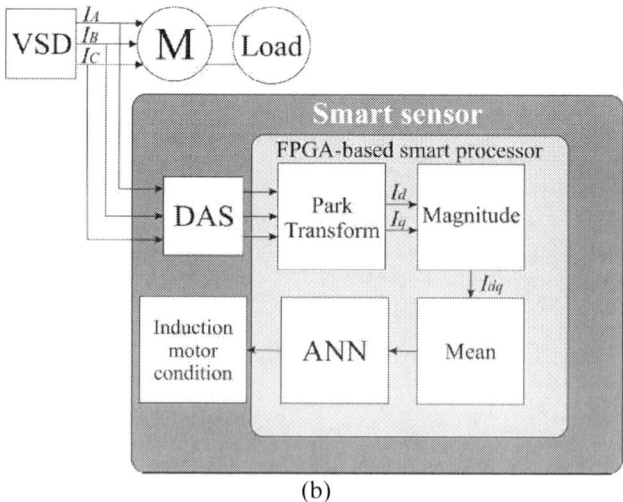

(b)

Fig. 5. (a) Experimental setup. (b) Block diagram of the proposed smart sensor for induction motor fault detection.

TABLE I
RESOURCE USAGE OF THE FPGA.

Resource utilization	Xilinx Spartan 3E XC3S1600E
Slices	1,495/14,752 (10%)
Flip-flops	979/29,504 (3%)
4-input LUTs	2,662/29,504 (9%)
Maximum operation frequency	58.28 MHz

V. RESULTS

A. Fault Identification Results

Table II shows the effectiveness by the proposed smart sensor during the induction motor condition classification of every frequency studied. The results include the identification of the healthy condition, broken rotor bar, unbalance and misalignment for each selected frequency. In Fig. 8 some examples of Lissajous figures are presented in order to observe the behavior of the faults in the D-Q system compared to the healthy condition. In order to obtain statistically significant results, forty tests were performed to acquire the three current phases from the induction motor in all treated cases for each selected frequency.

B. Discussion

Three different frequency cases of the VSD are studied in order to fulfill a range from low to high frequencies: 3Hz, 30Hz and 60Hz. Results of the smart sensor with motor running in healthy condition (HLT) show a detection effectiveness of 100% in 3Hz and 60Hz, while at 30Hz the detection effectiveness is 90%. Results at the broken rotor bars (BRB) condition present a detection effectiveness over 97% in 3Hz and 100% at 30Hz and 60Hz. In the unbalance (UNB) condition the 100% of effectiveness is present in each studied frequency. Finally, with the misalignment (MAL) condition the smart sensor delivers an effectiveness of 100% for 3Hz and 60Hz, and 95% at 30Hz. A characteristic of the proposed smart sensor is the automatic detection of multiple faults in VSD-fed induction motors using a simple methodology, different from other works that have a heavy computational load, requiring offline processing and an expert technician for interpreting results. For instance, as stated in Table I, the resources of the FPGA are around 10% with the proposed methodology. In [19], the implementation of a 1024-point FFT requires over 30% of FPGA resources, regardless the magnitude computation and an ANN. Otherwise, in [20] it is reported a 34% of resource usage of a reconfigurable FPGA-based system for wavelet analysis. On the other hand, [9-12] report offline methodologies for the detection of single isolated faults in induction motors connected through the power line supply. The FPGA implementation of the proposed smart sensor offers an online hardware implementation with a low computational load methodology, low-cost, and portable solution for fault detection in VSD-fed induction motors, different from other works that employ techniques with higher computational load, which are not suited for automatic online hardware/software implementation.

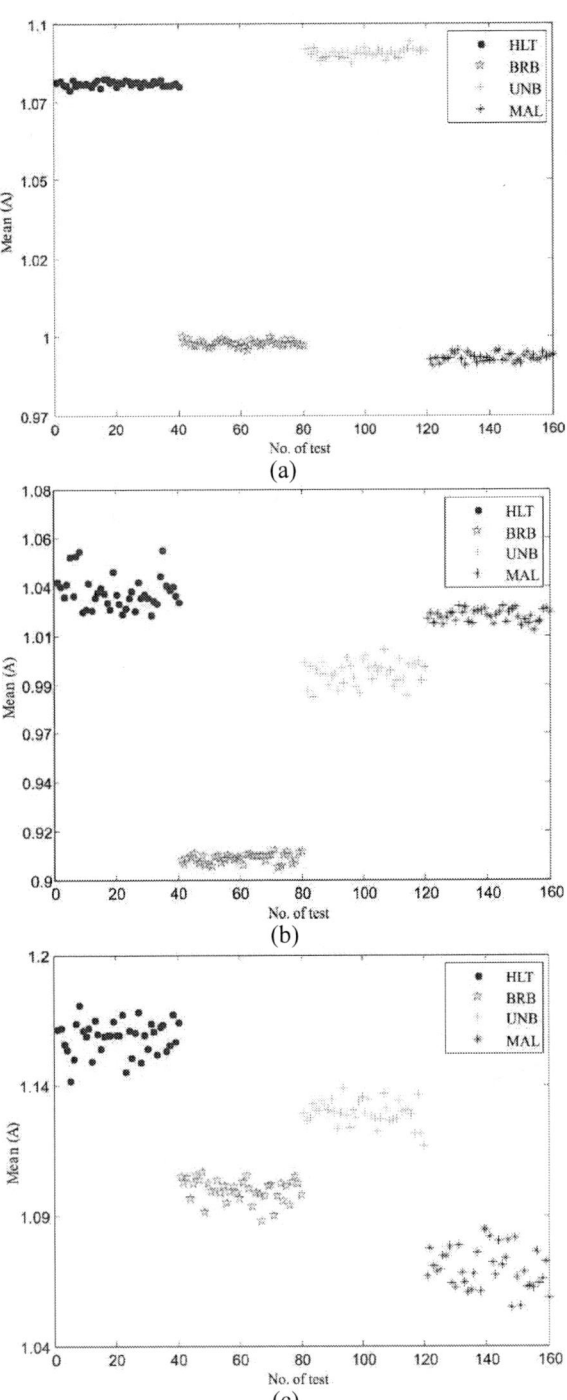

Fig. 6. Mean calculation of the i_{DQ} vector obtained of each fault condition in: (a) 3Hz, (b) 30 Hz and (c) 60 Hz.

VI. CONCLUSIONS

This work proposes a new smart sensor for detection of faults in VSD-fed induction motors using the three current phases. The proposed methodology is based on the Park transform and an MFN in order to determine the motor condition when is fed by a VSD, due to their simplicity and lower computational load different from other techniques, it is an excellent candidate to be implemented on hardware. The functionality of the smart sensor was successfully tested in forty tests of each category of the faults.

(a)

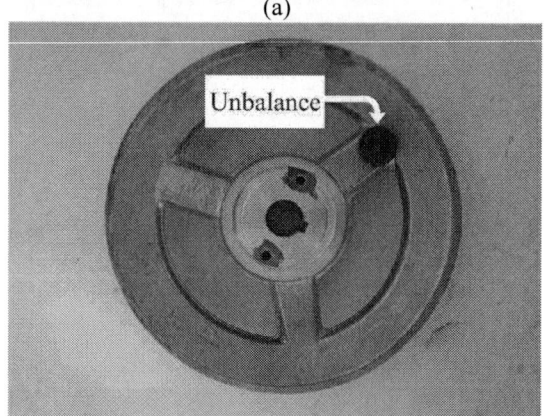

(b)

(c)

Fig. 7. (a) Broken rotor bar. (b) Unbalance. (c) Misalignment.

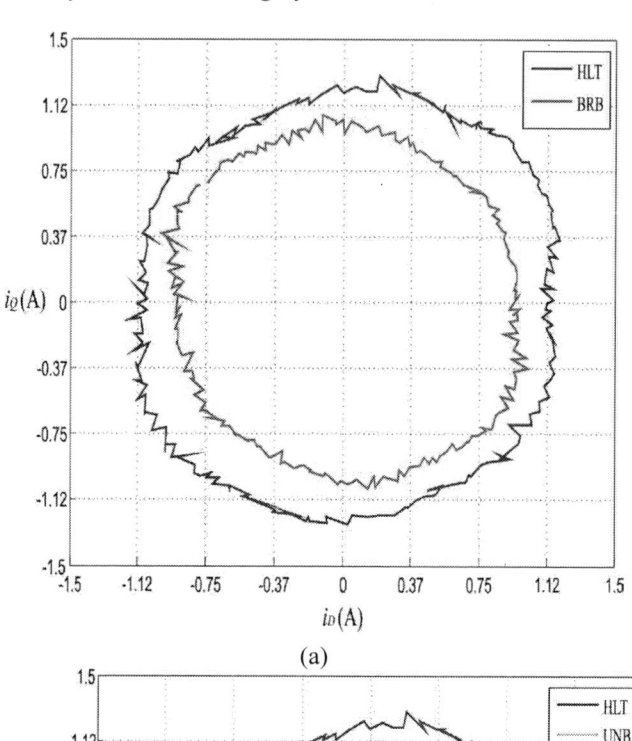

(a)

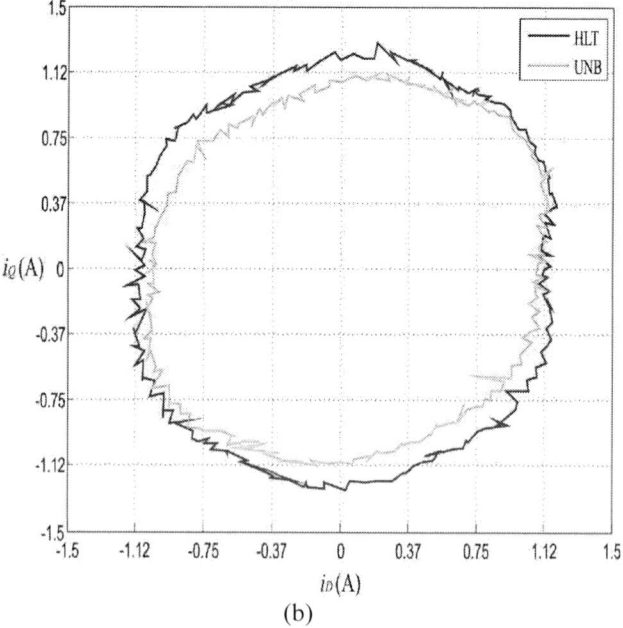

(b)

TABLE II
EFFECTIVENESS OF THE PROPOSED SMART SENSOR ON IDENTIFYING THE INDUCTION MOTOR CONDITION.

Induction motor condition	3Hz Effectiveness	30Hz Effectiveness	60Hz Effectiveness
HLT	100%	90%	100%
BRB	97%	100%	100%
UNB	100%	100%	100%
MAL	100%	95%	100%

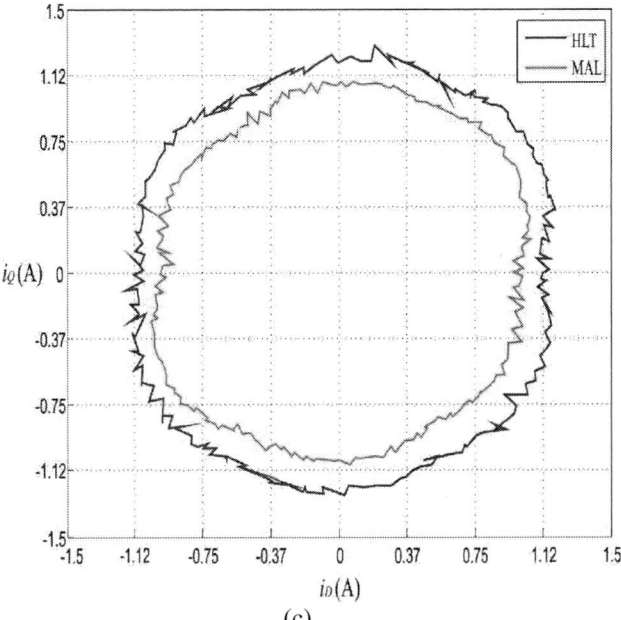

(c)

Fig. 8. Examples of Lissajous figures of the *D-Q* system current for comparing the healthy condition against (a) Broken rotor bar, (b) Unbalance and (c) Misalignment.

VII.   REFERENCES

[1]   A. Bellini, F. Fillipeti, C. Tassoni, G.A. Capolino, "Advances in diagnostic techniques for induction machines," *IEEE Trans. Industrial Elextronics,* vol. 55, pp. 4109-4126, Dec. 2008.

[2]   D.J. Siyambalapitiya, P.G. Mclaren, "Reliability improvement and economic benefits of on-line monitoring systems for large induction machines," *IEEE Trans. Industrial Applications,* vol. 26, pp. 1018-1025, Nov-Dec. 1990.

[3]   A. Garcia-Perez, R.J. Romero-Troncoso, E. Cabal-Yepez, R.A. Osornio-Rios, "The application of high-resolution spectral analysis for identifying multiple combined faults in induction motors," *IEEE Trans. Industrial Elextronics,* vol. 58, pp. 2002-2010, May 2011.

[4]   O. Duque-Perez, L.A. Garcia-Escudero, D. Moringo-Sotelo, P.E. Gardel, M. Perez-Alonso, "Condition monitoring of induction motors fed by voltage source inverters. Statistical Analysis of spectral data," in *Proc. Of the International Conference of Electric Machines and Drives (ICEM12)*, pp. 2479-2484, 2012.

[5]   S.H. Kia, H. Henao, G.A. Capolino, "Some digital signal processing techniques for induction machines diagnosis," *IEEE International Symposium on Diagnostics for Electric Machines, Power Electronics & Drives (SDEMPED).* pp. 322-329, Sept. 5-8 2011.

[6]   J. Urresty, J. Riba, H. Saavedra, J. Romeral, "Analysis of demagnetization faults in surface-mounted permanent magnet synchronous motors with symmetric windings," *IEEE International Symposium on Diagnostics for Electric Machines, Power Electronics & Drives (SDEMPED).* pp. 240-245, Sept. 5-8 2011.

[7]   P. Gardel, D. Moringo-Sotelo, O. Duque-Perez, M. Perez-Alonso, L.A. Garcia-Escudero, "Neural network broken bar detection using time domain and current spectrum data," in *Proc. Of the International Conference of Electric Machines and Drives (ICEM12)*, pp. 2492-2497, 2012.

[8]   M. Riera-Guasp, J.Pons-Llinares, F. Vedreno-Santos, J.A. Antonino-Daviu, M. Fernandez-Cabanas, "Evaluation of the amplitudes of high-order fault related components in double bar faults," *IEEE International Symposium on Diagnostics for Electric Machines, Power Electronics & Drives (SDEMPED).* pp. 307-315, Sept. 5-8 2011.

[9]   A. Soualhi, G. Clerc, H. Razik, A. Lebaroud, "Fault detection and diagnosis of induction motors based on hidden Markov model," in *Proc. Of the International Conference of Electric Machines and Drives (ICEM12),* pp. 1693-1699, 2012.

[10]   C. Kral, T.G. Habetler, R.G. Harley, "Detection of mechanical imbalances of induction machines without spectral analysis of time-domain signals," *IEEE Trans. Industrial Applications,* vol. 40, pp. 1101-1106, July-Aug. 2004.

[11]   T. H. Patel, A.K. Darpe, "Experimental investigations on vibration response of misaligned rotors," *Elseiver Mechanical systems and signal processing,* vol. 23, pp. 2239-2252, May. 2009.

[12]   I.Y. Onel, M.E.H. Benbouzid, "Induction motor bearing failure detection and diagnosis: Park and Concordia transform approaches comparative study," *IEEE Trans. Mechatronics,* vol. 13, pp. 257-262, April. 2008.

[13]   J. Rivera, G. Herrera, M. Chacon, P.Acosta, M. Carrillo, "Improved progressive polynomial algorithm for self-adjustment and optimal response in intelligent sensors" *MDPI Sensors,* vol. 9, pp. 3767-3789, Nov. 2008.

[14]   H. Nejjari, M.E.H. Benbouzid, "Monitoring and diagnosis of induction motors electrical faults using a current Park's vector pattern learning approach," *IEEE Trans. Industrial Applications,* vol. 36, pp. 730-735, May-June. 2000.

[15]   M.J. Picazo-Rodenas, R. Royo, J. Antonino-Daviu, J. Roger-Folch, "Use of infrared thermography for computation of heating curves and preliminary failure detection in induction motors," in *Proc. Of the International Conference of Electric Machines and Drives (ICEM12),* pp. 525-531, 2012.

[16]   M.E.H. Benbouzid, "A review of induction motors signature analysis as a medium for faults detection," *IEEE Trans. Industrial Elextronics,* vol. 47, pp. 984-993, Oct. 2000.

[17]   Y. Huang, "Advances in artificial networks – Methodological development and application" *MDPI Algorithms,* vol. 2, pp. 973-1007, Aug. 2009.

[18]   L. Capocchi, G.A. Capolino, "An efficient architecture of multi-stage neural network for rotor induction generator short-circuit fault classification," in *Proc. Of the International Conference of Electric Machines and Drives (ICEM12),* pp. 1565-1571, 2012.

[19]   J. A. Vite-Frias, R. J. Romero-Troncoso, A. Ordaz-Moreno, "VHDL core for 1024-point radix-4 FFT computation," in *Proc. Int. Conf. Reconfigurable Computing and FPGAs ReConFig 2005,* pp. 26/1-4. 2005.

[20]   R.J. Romero-Troncoso, M. Pena-Anaya, E. Cabal-Yepez, A. Garcia-Perez, R.A. Osornio-Rios, "Reconfigurable SoC-based smart sensor for wavelet and wavelet packet analysis," *IEEE Trans. Intrumentation and Measurement,* vol. 61, pp. 2458-2468, Sep. 2012.

VIII.   BIOGRAPHIES

Armando G. Garcia-Ramirez received the B. E. degree from the Mazatlan Institute of Technology, Mazatlan, Mexico, and the M.E. degree (with Honors) from the University of Guanajuato, Salamanca, Mexico in 2011, where he did research work at the HSPdigital group. Currently, he is a Ph.D. student at the University of Queretaro, Queretaro, Mexico. His research interest includes hardware signal processing and smart sensors on a field-programmable gate arrays for applications on mechatronics.

Roque A. Osornio-Rios (M'10) received the B.E. degree from the Instituto Tecnologico de Queretaro, Queretaro, Mexico, and the M.E. and the Ph.D. degrees from the University of Queretaro, Queretaro, Mexico, in 2007. He is a National Researcher with CONACYT. He is currently Professor whit the University of Queretaro. He was an Advisor of over 30 theses, and a coauthor of over 40 technical papers in international journals and conferences His fields of interest include hardware signal processing and mechatronics. Dr. Osornio-Rios received the "2004 ADIAT National Award on Innovation" for his works in applied mechatronics.

Arturo Garcia-Perez (M'10) received the B.E. and M.E. degrees in electronics from the University of Guanajuato, Salamanca, Mexico, in 1992 and 1994, respectively, and the Ph.D. degree in electrical engineering from the University of Texas at Dallas, Richardson, in 2005. He is currently a Titular Professor with the Department of Electronic Engineering, University of Guanajuato. He is a National Researcher with the Consejo Nacional de

Ciencia y Tecnología level 1. He was an Advisor of over 50 theses. His fields of interest include digital signal processing for applications in mechatronics

Rene de J. Romero-Troncoso (M'07–SM'12) received the Ph.D. degree in mechatronics from the Autonomous University of Queretaro, Queretaro, Mexico, in 2004. He is a National Researcher level 2 with the Mexican Council of Science and Technology, CONACYT. He is currently a Head Professor with the University of Guanajuato and an Invited Researcher with the Autonomous University of Queretaro, Mexico. He has been an advisor for more than 180 theses, an author of two books on digital systems (in Spanish), and a coauthor of more than 90 technical papers published in international journals and conferences. His fields of interest include hardware signal processing and mechatronics. Dr. Romero–Troncoso was a recipient of the 2004 Asociación Mexicana de Directivos de la Investigación Aplicada y el Desarrollo Tecnológico Nacional Award on Innovation for his work in applied mechatronics, and the 2005 IEEE ReConFig Award for his work in digital systems.

IGBT Fault Diagnosis using Adaptive Thresholds during the Turn-on Transient

M. A. Rodríguez-Blanco, A. Vázquez-Pérez, L. Hernández-González, A. Pech-Carbonell, M. May-Alarcón.

*Abstract -- **This paper presents the design of an electronic fault detection circuit in the insulated gate bipolar transistor (IGBT) based on the exclusive measurement of the gate signal during the turn-on transient. In order to increase the effectiveness of the detection and to tolerate the variations of input to system, adaptable thresholds have been added to the circuit. There are three important aspects in this research, specifically: 1 - Early detection, since the evaluation is realized during the turn-on transient; 2 – Reducing false alarms, because the variations of input to the system are considered; 3 – A realistic design, since the components used are commercially available, and the IGBT model used is the standard for PSpice software which has already been widely validated in the literature.***

*Index Terms-- **Adaptable Thresholds, Fault Detection, FDI System, Insulated Gate Bipolar Transistor (IGBT) and Motor Inverter System.***

I. NOMENCLATURE

A_{GD}	Gate-drain area
C_{GD}	Gate-drain capacitance
C_{GDJ}	Gate-drain depletion capacitance
C_{OXD}	Gate-drain oxide capacitance
N_B	n-layer doping concentration
q	Electron charge
R_G	Gate driver resistor
V_{DS}	Drain-Source voltage
V_{GD}	Gate-Drain voltage
V_{GE}	Gate-emitter voltage
ε_{si}	Electrical insulation

II. INTRODUCTION

TODAY there is a growing need and interest in developing fault-tolerant systems to make computers more reliable, given that a fault in a system not only can put human integrity at risk, but also cause great economic losses, both the stoppage of production, spending on repairs attributed to the spread of the fault, and the resulting poor quality of the product. One of the mechatronics subsystems that is highly demanded by the industry is the three-phase inverter-motor system, which is part of the final control element and has been extensively researched. [1]-[3] Most of the research in this area focuses on the design of new algorithms for diagnosing or reconfiguring the failure [4] and analysis of new tolerant schemes for the inverter-motor system under the material redundancy approach . [1], [2], [5] Other important areas that still require more research are the importance of early detection of the failure in the device´s power to prevent its spread, and the right timing for the replacement of the damaged element thus reducing the tracking in the control error. [6] During the diagnosis of a high-dynamic-response system which is tolerant to faults, one of the most important stages is detection, because it is the first contact with the fault so you have the greatest amount of time for a complete evaluation. As soon as possible it is important to generate and evaluate a signal containing information about the fault, commonly known as residual signal, and this becomes the key part of the tolerant system.

There are various fault-tolerant designs applied to the inverter-motor system. However, most do the failure detection during the steady-state and only one [6] in the transient-state during the turn-on transient of insulated gate bipolar transistor. It allows early detection and prevents the spread of the failure, which is of great interest in fault tolerant design, when the inverter is made up of several single IGBT modules as back-up units rather than a complete inverter leg (two devices in the same package). The manufacturing trends for high power devices are focused on this type of packaging. In this paper, the failure-free and under-failure behavior of the IGBT gate signal reported [6] is used and a new failure detection circuit for the IGBT with adaptive detection thresholds is proposed. The electronics circuit is designed and simulated with PSpice software using fully commercial components and the standard IGBT model, which has already been widely validated in the literature [7].

III. DETECTION OF FAILURE IN THE IGBT WITH ADAPTIVE THRESHOLDS

In general, a simple failure detection design is one that allows you to detect and identify the system's fault by evaluating the symptoms inherent in a signal referred to as residue and identified, in continuous time, as $r(t)$; this is obtained through a residual generation stage and evaluated through the decision mechanism stage. A simple failure-detection design problem is that the decision mechanism considers the healthy state of constant system without considering the inconsistencies of the temporary input signal $u(t)$ which is done in [8]. The problem can be solved by

---

The authors wish to acknowledge gratefully to CONACYT-CVU: 362028 and UNACAR by financial support.

M. A Rodríguez-Blanco, A. Vázquez-Pérez, A. Pech-Carbonel, M. May-Alarcón are with Autonomous University of Carmen City (UNACAR), Department of Electronics Engineering, Ciudad del Carmen Campeche, Zip code: 24115, México (e-mail: marblanco73@hotmail.com).

L. Hernández-González is with the National Polytechnic Institute, Electrical and Mechanical Engineering School, Campus Culhuacan, México City, México, (e-mail: bilbito_98@yahoo.com).

978-1-4799-0024-4/13 $31.00 © 2013 IEEE

adding an adaptive threshold to the detection design, as proposed in [4] and shown in Fig. 1.

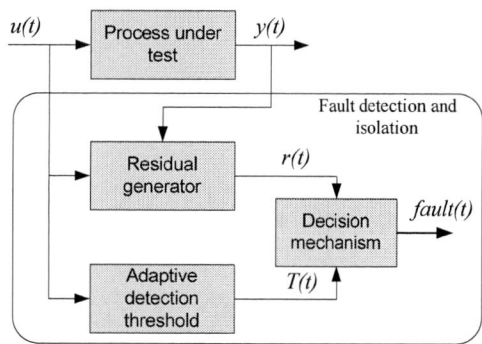

Fig. 1. General scheme for fault detection and isolation with adaptive thresholds.

For an industrial application in any power electronics system, where the IGBT acts as a process under failure, the failure detection problems lie essentially in the handling of large quantities of current and/or voltage as well as the high dynamic response. The IGBT failure detection design therefore sets the output signal $y(t)$ for small quantities with electrical insulation, without affecting the high response dynamics of the system's failure detection, by making sure the stages of the residual generator, adaptive threshold, and decision mechanism are installed on the side of IGBT's gate drive circuit, using analog and digital electronics to ensure early detection and to avoid the use of optical couplers, DAC/ADC converters and sensors of large magnitudes.

A. Process under test

In this research, the IGBT is the process being tested to detect failures. The IGBT, is the electronic device most widely used in power electronics and in applications with greater probability of failure due to thermal stress attributed to the loss of power that arises, both in the steady-state as well as in the transient-state. The IGBT is a power semiconductor device that combines the features of the bipolar junction transistor (BJT) and the metal oxide semiconductor field effect transistor (MOSFET), allowing low switching losses, due to MOSFET's high- input impedance. This provides low conduction losses because of BJT's reduced voltage collector-emitter, V_{CE}, during the conduction stage. Commonly the IGBT gate control voltage is $\pm$ 15 V, which offers the advantage of being able to control power systems with a weak input current.

An analysis of the measuring technique for the gate in the transient-state starts at the equivalent circuit of the gate signal; the gate-to- drain region is modeled by capacitance C_{GD}, which is the most vulnerable to failure because the conduction channel is highly mobile during IGBT switching. This region is made up of a variable capacitance C_{GDJ}, which models the deflection zone, and a fixed capacitance C_{OXD}, which models the oxide region. Fig. 2 shows the simplified

equivalent circuit seen from the emitter of the IGBT gate, not including the small R_B modulating voltage and the voltage of the internal diode of the bipolar transistor for the internal model when it is turned on. This is because the collector to emitter voltage applied is the conduction voltage, which is very small in comparison to the supply voltage. Therefore it is assumed that the drain terminal D is approximately equal to the collector terminal C. IGBT's V_{CE} source can be substituted by drain-source voltage V_{DS} as shown in Fig. 2.

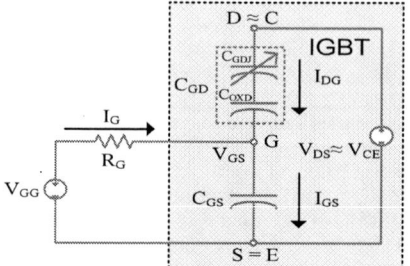

Fig. 2. Equivalent circuit seen from the gate of IGBT.

C_{GDJ} behavior is expressed in equation (1) showing that the variation of design parameters ε_{si} and material A_{GD} directly affect the C_{GDJ} deflection region. The variation of these parameters may appear as a result of the destructive effect of IGBT [9]. This modulation of C_{GDJ} also affects the descriptive equation of C_{GD} (2), which is defined by the transient behavior of V_{GS} and V_{DS} during the turn-on transient of IGBT ($V_{DS}> V_{GS}$).

$$C_{GDJ}\left(V_{DG}\right) = A_{GD}\sqrt{\frac{q.N_B.\varepsilon_{si}}{2\left(V_{DG}+V_{TD}\right)}} \qquad (1)$$

$$C_{GD} = \begin{vmatrix} C_{OXD} & \xrightarrow{if}(V_{GS}\text{-}V_{TD}) \geq V_{DS} \\[2mm] \dfrac{C_{OXD}.C_{GDJ}}{C_{OXD}+C_{GDJ}} & \xrightarrow{if}(V_{GS}\text{-}V_{TD}) < V_{DS} \end{vmatrix} \qquad (2)$$

The gate current $I_G(t)$ of the equivalent circuit is apparent in Fig. 2 and is given by the following expression.

$$I_G(t) = C_{GS}.\frac{dV_{GS}(t)}{dt} - C_{GD}.\frac{dV_{DS}(t)}{dt} + C_{GD}.\frac{dV_{GS}(t)}{dt} \qquad (3)$$

The expression above shows that $I_G(t)$ is directly affected by the variation of C_{GD}, causing a significant change in the gate voltage of the IGBT. Fig. 3 shows the IGBT gate charge characterized by three phases due to its internal capacitances; this measurement was taken of a chopper circuit proposed in [10] with a gate current constantly forced to move during the turn-on transient for better definition of each phase.

978-1-4799-0024-4/13 $31.00 © 2013 IEEE

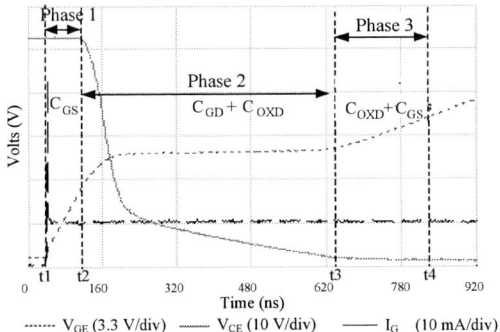

Fig. 3. Turn-on transient of IGBT with constant gate current I_G.

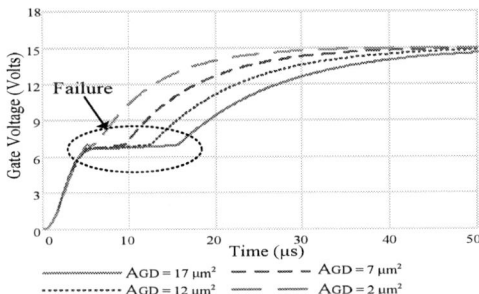

Fig. 4. Simulation of IGBT gate signal varying A_{GD}.

The behavior of the IGBT during the turn-on transient depends on V_{GS} and V_{DS} voltage levels which can be represented in three phases.

Phase 1 (t1 <t <t2): At this point $V_{DS}(t)$ is larger than $V_{GS}(t)$; the equivalent capacitance seen from the gate to the emitter of IGBT depends on C_{GS} because at this point C_{GDJ} is smaller and therefore C_{GD} is omitted for simplicity in the analysis.

Phase 2 (t2 <t <t3): This phase is the most complex behavior because the flat area of the gate signal is generated by the Miller effect, causing the first and third terms of IG(t) in equation (3) to be omitted for simplicity; then the gate current $I_G(t)$ is determined only by the negative slope of V_{DS} and C_{GD} when $C_{GD}=C_{OXD}+C_{GDJ}$.

Phase 3 (t3 <t <t4): In line with equation (2), the value of C_{GD} during this phase may be considered equal to C_{OXD} because $V_{DS}(t)$ has a value smaller than $V_{GS}(t)$ during the turn-on transient of IGBT; then the total equivalent capacitance seen from the gate to emitter is the electrical parallel of C_{OXD} and C_{GS}.

Fig. 4 shows a parametric analysis of C_{GDJ} through A_{GD} variation, showing a modulation of the flat area of the gate signal attributed to the modulation of the deflection region. Therefore this behavior can be used to detect failure by short-circuit and open-circuit devices. The reduction of the flat zone indicates a short-circuit device failure and an extension of the flat area indicates a failure by an open-circuit device. The purpose of conducting early detection in the IGBT is to prevent the failure from spreading to the complementary components of an inverter, by inhibiting the command signal, as well as allowing tolerance to failures through the online replacement of the damaged element.

Generally, the IGBT destruction sequence initially occurs due to over-current in the turn-on transient of the power device due to internal problems, such as aging of the device, or external problems by overload; it can also occur during Fault Under Load (FUL) and Hard Switch Fault (HSF) switching. Initially, when there is an over-current situation on the device, there is usually a short-circuit, and then the high temperature within the package destroys the connecting cables, causing an open-circuit failure. But before it appears as a short-circuit or open-circuit in the power device, the physical parameters of the IGBT are degraded causing significant variation in the turn-on transient signal of IGBT. A critical aspect to consider in the design of a failure detection circuit for the IGBT, or any application of power electronics, is to consider the parasitic inductance Le of the emitter which contributes to the noise induced in the gate signal [11]. The suitability of the detection system must consider additional circuits as differential amplifiers with a gain k, and one additional measurement signal for the IGBT gate, which requires the prior description of the circuit that will be diagnosed.

This has already been reviewed in [10] so the analysis can be omitted in this paper. In Fig. 5 (a) a chopper circuit is used as a power electronics application, showing the parasitic components inherent in the circuit, as well as the input voltages to the differential circuit; V_G and V_{Le2} are required for induced noise reduction attributed to Le. Fig. 5 (b) shows the complete circuit of the process being tested, highlighting the differential circuit. Fig. 5 (c) shows the circuit implemented in PSpice.

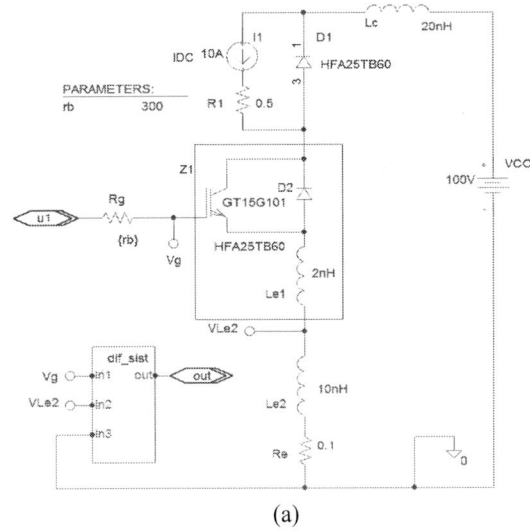

(a)

978-1-4799-0024-4/13 $31.00 © 2013 IEEE   243

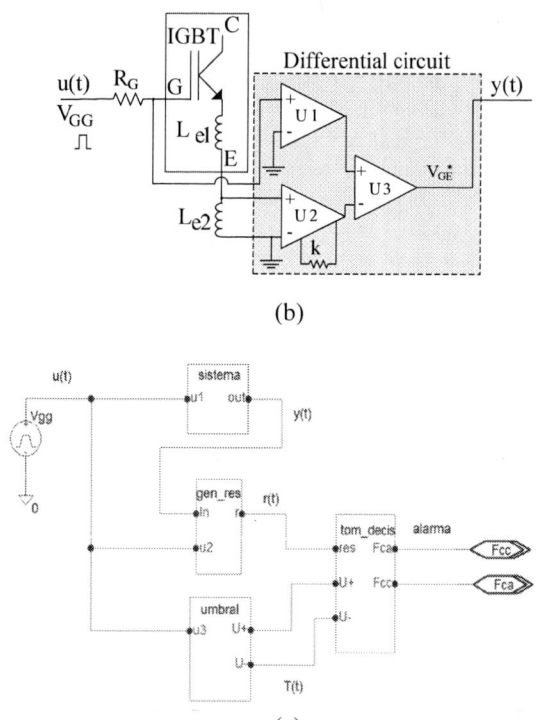

(b)

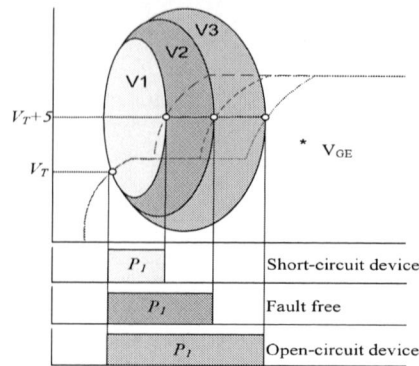

Fig. 6. Window detector for pulse width P1.

(c)

Fig. 5. Proposed circuit: a) chopper circuit, b) process being tested, c) Failure detection circuit implemented in PSpice.

B. Residual Generator

The residual generator in continuous-time is the stage where you can detect any abnormality in the process under failure using a signal called residue $r(t)$. However, it should be done as accurately as possible, because even when there are no failures in the system, the residue is never zero $r(t) \neq 0$ [4]; often the residual signal depends not only on the changes occurring in the plant but also the amplitude and frequency of volatility of the input $u(t)$ [12].

Since the Phase 2 is mainly affected by the aging of the IGBT model parameters [9] a wide margin between thresholds is proposed in this method. However, the appropriate benchmark to establish the fault-free condition has not been formally studied.

The residual generation technique proposed produces a pulse width P1 corresponding to the detection window only during the IGBT turn-on transient being tested, from a voltage V_T to $(V_T + 5V)$. This ensures the existence of phase 2 of the gate load in the detection window, as shown in Fig. 6. The failure-free condition can be set by knowing the IGBT parameters using the PSpice parameter extraction software or through a series of electrical measurements proposed in Rodriguez et. al [13]. The failure condition with fixed thresholds can be set by narrowing of the pulse width P1 by ± 50%. However, a variation in the input signal $u(t)$ can cause false alarms.

The proposed sequence for the residual signal $r(t)$ is to use P1 to activate a ramp S1 with a magnitude which is proportional to the pulse width and maintains the voltage level reached up until a few nanoseconds before a new evaluation; the discharge is triggered by a pulse P2, which can be obtained by evaluating $y(t)$ from $(V_T - 2V)$ to V_T.

A failure-free reference signal S2 adaptable to changes in the process input $u(t)$ ensures that, after the evaluation area, the difference between S1 and S2 is zero, expressed as rx(t1 <t <t3)=0, with a filter to minimize measurement error within the evaluation area such that $r(t) = 0$. Fig. 7 shows the timing diagram of the proposed circuit for the failure-free example and Fig. 8 shows the proposed electronic circuit.

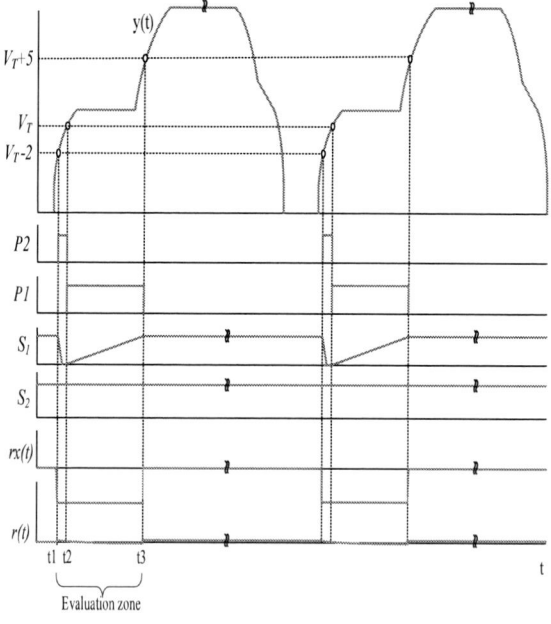

Fig. 7. Timing diagram of the proposed circuit for failure-free case.

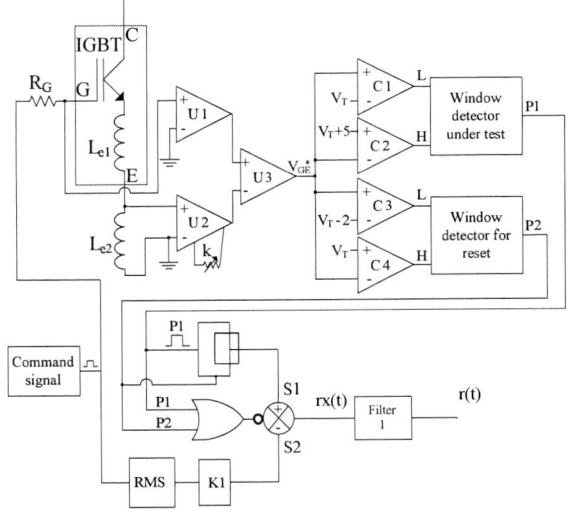

Fig. 8. Schematic diagram of the proposed circuit.

The signal S2 can be obtained using a parameter K1 to create a linear equation. Given the non-linearity of this process, only a range of *u(t)* of 12V to 15V was achieved. Therefore expanding these failure tolerance ranges using smart techniques and implementing them in real time is a still an open issue. Furthermore, to hide wrong residue during the transient time of evaluation, we used a design which will take a previous measure of time with a low-pass filter tuned to the switching frequency. Fig. 9 shows the drawing of the residual generator in PSpice, showing window-comparing sub-circuits, hysteresis detectors, a ramp generator and a subtract circuit. The latter is enabled by the absence of P1 and P2. It is worthy to note that none of these sub-circuits is considered to be an ideal element.

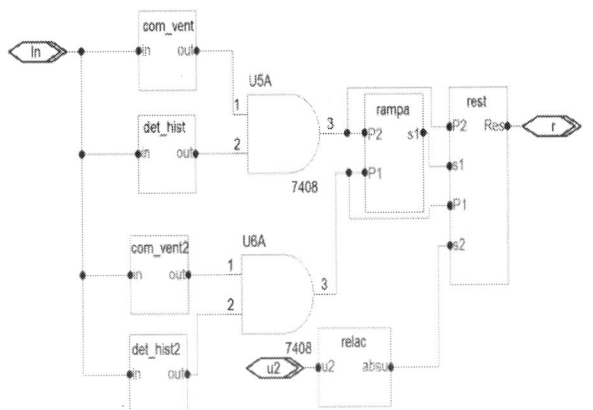

Fig. 9. Schematic for residual generator.

C. Adaptive Threshold

The decision mechanism stage normally involves a process for establishing thresholds when using fixed detection thresholds, sensitivity to failure can be reduced

even more if the chosen threshold is very high, however the false alarm rate is very high when the threshold chosen is very low. Using an adaptive detection threshold rather than a fixed one can significantly improve performance of a failure detection scheme as measured by the detection time and false alarm rate [14].

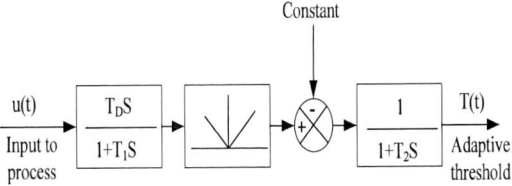

Fig. 10. Schematic for the adaptive detection threshold.

The threshold consists of a filter with lead-lag behavior driven by the input signal *u(t)*. This causes the output of the filter to be zero for steady-state inputs; otherwise it is a measure of the stimulation of dynamic input so as to calculate the absolute value to rectify the signal. A constant value is added due to the effects of noise in the measurement of the residual signal. Optionally a first-order low-pass filter (LPF) can be applied to soften the adaptive threshold. The time constants T1 and T2 are chosen according to the control over time of the process constant; in this case, T1 is related to the line frequency and T2 to the switching frequency, so the application of proposed method is restricted to systems with constant switching frequency, as in the most of applications that only work with the pulse width modulation as example in frequency converters.

Fig. 11 shows the adaptive threshold circuit where you can see the blocks used and two different proportionality constants for the positive and negative thresholds. This is due to the nonlinearity of *y(t)* in regards to parametric changes in the process being tested. However, both direct linearization in the case of increases as well as decreases, helps to broaden the range of failure-free operations. In this particular case, the location of the positive threshold was set at 5V and the negative threshold was -1.38V.

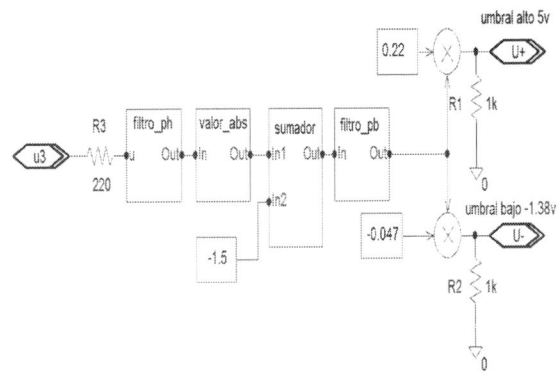

Fig. 11. Implementation in PSpice of the adaptive detection thresholds.

D. Decision Mechanism

The decision mechanism is useful to determine the kind

of error which occurred in the signal $y(t)$; if the residue exceeds the positive threshold, it is diagnosed to be open-circuit. Where the residue is below the threshold, it is diagnosed as a short-circuit.

IV. RESULTS

This section shows the good performance offered by the residual generator block when it is healthy and under fault. Furthermore, we observe the robustness that has the decision mechanism block to prevent false alarms when the input signal to the system change. Since the proposed circuit was developed in Pspice using only commercial components in a future implementation ensures good performance.

Fig. 12 shows the full circuit proposed, according to the design of detection and identification of faults and *Fig. 13* shows the results of simulation in PSpice.

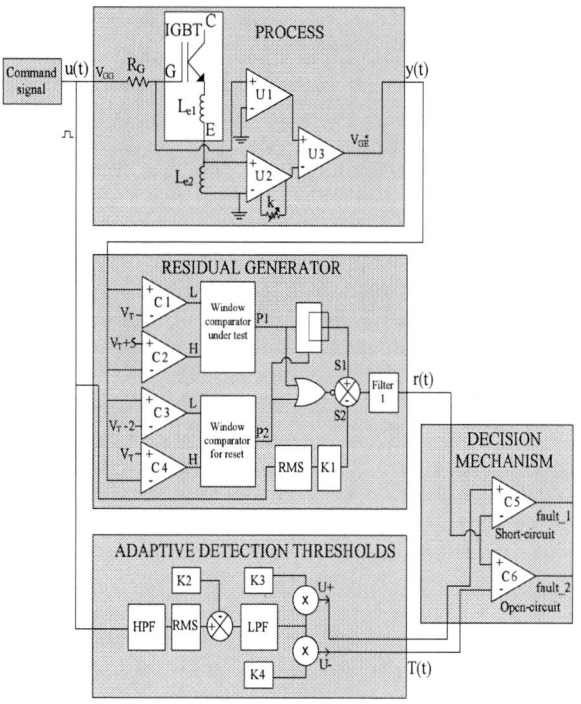

Fig. 12. Full circuit proposed to detect failures in the IGBT with adaptive detection threshold.

In *Fig. 13* (a) and 14 (b), respectively, the IGBT gate signal is shown without noise induced $y(t)$ and with the adaptive thresholds in a failure-free condition. The diagnosis is correct i.e. fault_1=0 and the residual signal $r(t)$ is within the permitted failure-free region. Furthermore, Figures 14 (c) and 14 (d) show the under-failure condition for short-circuit and open-circuit devices respectively; it indicates that at time t1, the residual signal $r(t)$ leaves the failure-free region, causing the signals for fault_1 and fault_2 to be different from zero, for a diagnosis of short-circuit and open-circuit respectively. Figures 14 (e) and 14 (f) show the

adaptation of the threshold U+ and U- concerning the magnitude of the residual $r(t)$ when the input signal is modified from 15V to 12V and from 12V to 15V respectively.

To emulate the behavior of phase 2 of the IGBT gate signal for failure by short-circuit, failure-free and open-circuit, the external gate resistance of R_G was modified to 100 Ω, 300 Ω and 500 Ω respectively.

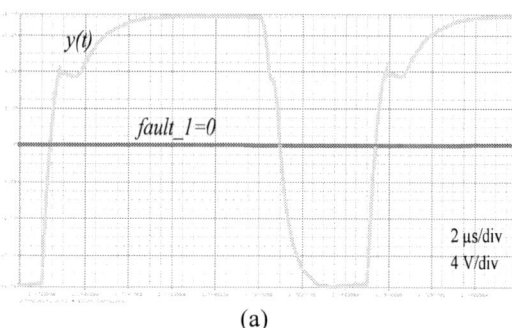

(a)

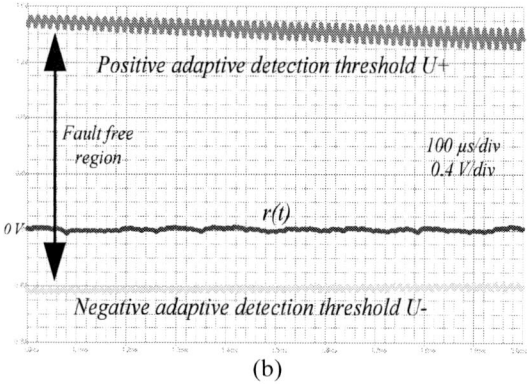

(b)

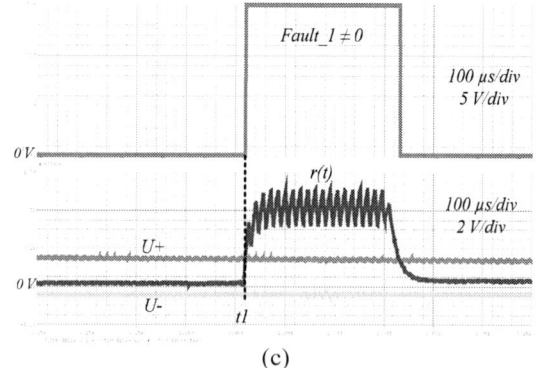

(c)

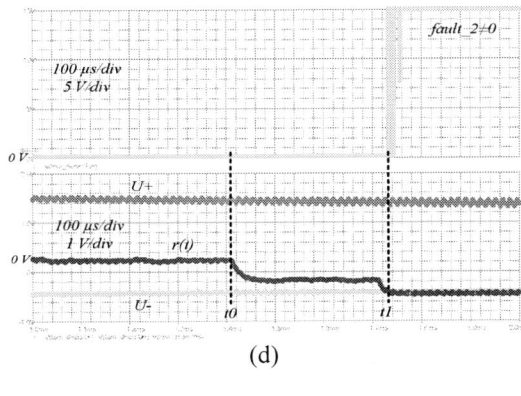

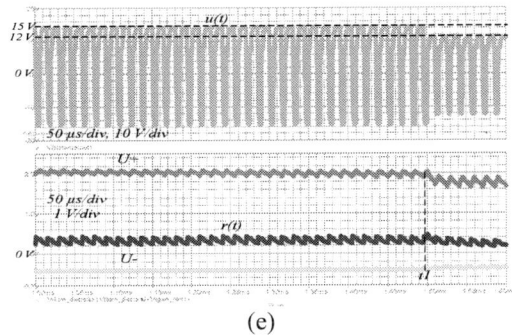

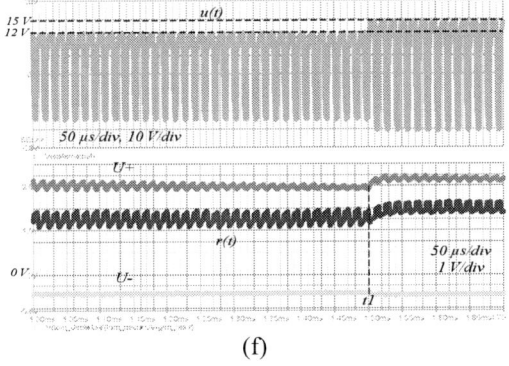

Fig. 13. Simulation results. (a) Signal *y(t)* and Fault_1(t) = 0 in failure-free condition. (b) Adaptive detection thresholds and residual signal *r(t)*. (c) Failure by short-circuit device, the residual signal is greater than the positive adaptive detection threshold. (d) Failure by open-circuit device, the residual signal is less than the negative adaptive detection threshold. (e) Adaptive thresholds and residual signal *r(t)* decreasing the input signal *u(t)* of 15V to 12V. (f) Adaptive thresholds and residual signal *r(t)* increasing the input signal *u(t)* of 12V to.

of the failure to the complementary device in the same affected leg, when a short-circuit or over-current occurs. Furthermore, the main advantage of adding a detection process with adaptive threshold is that the false alarm rate decreases because the system is not as vulnerable to variations in the power supply of the IGBT gate driver circuit. The proposed design is not suitable for power applications, where the switching frequency is above 20 kHz since the introduction of an external resistance of R_G limits the IGBT switching speed. However, the delay introduced helps to extend the ignition transient providing more time to evaluate the failure without the current manufacturing technology of the operational amplifiers affecting the early diagnosis.

V. CONCLUSION

In this research, early detection is obtained by the evaluation during the turn-on transient of IGBT, through continuous processing using analog and digital electronics; opto-couplers and digital-to-analog converters (DAC) are not used since they often introduce greater propagation time in any application including signal processing with simple algorithms.

The main goal of an early failure detection design in the IGBT is to obtain a corrective action to prevent propagation

VI. REFERENCES

[1] B. A. Welchko, T. A. Lipo, T. I. M. Jahns, and S. E. Schulz, "Fault tolerant three-phase AC motor drive topologies: a comparison of features, cost, and limitations," *IEEE Transactions on Power Electronics,* vol. 19, no. 4, pp. 1108-1116, 2004.

[2] D. Campos, T. Espinoza and E. Espinoza, "Fault-tolerant control in variable speed drives: a survey," *IET Electric Power Application letter,* vol. 2, pp. 121-134, 2008.

[3] V. Ayhan Maraba, A. Emin Kuzucuoglu, "Speed Control of an Asynchronous Motor Using PID Neural Network," *Studies in Informatics and Control,* vol. 20, no.3, pp. 199-208, 2011.

[4] J. Chen, and R. J. Patton, *Robust Model-Based Fault Diagnosis For Dynamic Systems,* Kluwer Academic Publishers, chapter 2, 1999, pp. 51-55.

[5] S. Abramik, W. Sleszynski, J. Nieznanski and H. Piquet. "A diagnostic method for on-line fault detection and localization in VSI-fed AC drives," *In Proc. 2003, 10th European Conference on Power Electronics and Application,* France. 2003.

[6] M. A. Rodríguez, A. Claudio, D. Theilliol, L.G. Vela, "A new fault detection technique for IGBT based on gate voltage monitoring," *In Proc. 2007 IEEE Power Electronics Specialists Conference PESC,* Orland Florida U.S., 2007.

[7] M. Cotorogea, "Physics-based SPICE-model for IGBTs with transparent emitter," *IEEE Transactions on Power Electronics,* vol. 24, no. 12, pp. 2821-2832, 2009.

[8] Miguel Hernandez, Basilio del Muro, Domingo Cortes, Juan Carlos Sanchez, "An Easy to Apply Methodology for Fault Detection and Isolation in Linear Systems," *Studies in Informatics and Control,* vol. 21 no. 3, pp. 275-282, 2012.

[9] A. H. Craing, Ronald H. Randall and Joe Yedinak, "IGBT ghost failures in boost topology circuits explained through third quadrant operation," *In Proc. 2000 IEEE Applied Power Electronics Conference and Exposition,* pp. 1103-1108.

[10] M. A. Rodríguez-Blanco, A. Claudio-Sánchez, D. Theilliol, L. G. Vela-Valdés, P. Sibaja Terán, L. Hernández-González and J. Aguayo-Alquisira, "A failure detection strategy for IGBT based on gate voltage behavior applied to a motor drive system," *IEEE Transactions on Industrial Electronics,* vol. 58, no. 5, pp. 1625-1633, 2011.

[11] M. A. Rodríguez, M. López, P. Sibaja, L. Hernández, J. L. Vázquez, "Aspectos críticos en el diseño de un circuito de detección de fallas en el IGBT basado en la medición de la señal de compuerta," *In Proc. IEEE Congreso Internacional sobre Innovación y Desarrollo Tecnológico,* Cuernavaca Morelos México, 2010

[12] R. Isermann, *Fault-Diagnosis Systems* Springer-Verlag, Berlin Heidelberg, 2006, p. 107.

[13] M. A. Rodríguez, A. Claudio, M. Cotorogea, L. H. González, J. Aguayo, "Reconfigurable special test circuit of physic-based IGBT models parameter extraction," *Solid State Electronics,* vol. 54, pp. 1246-1256, 2010.

[14] T. Höflig, and R. Isermann, "Fault detection based on adaptive parity equations and single-parameter tracking," *Control Engineering Practice – CEP,* vol.4, no.10, pp. 1361-1369, 1996.

VII. BIOGRAPHIES

Marco A. Rodríguez-Blanco received M.S. and Ph.D. degrees in electronics engineering from the National Center for Research and Technological Development (cenidet), Cuernavaca, Morelos, México in 2001 and 2009 respectively. Since 2001, he has been with the Autonomous University of Carmen City UNACAR, Campeche, México as a full time professor in the Department of Electronics Engineering. His research interests are focused on the area of electronics circuit systems, characterization and parameters extraction of power semiconductor devices, and fault detection in motor drives.

Amsi Vázquez-López received B.S. and M.S. degrees in electronics engineering and electronics science, specializing in the area of automatic controls, from the Autonomous University of Carmen City UNACAR, Campeche, México in 2010 and 2012 respectively.

Leobardo Hernández-González received an M.S. degree in Microelectronics from the National Polytechnic Institute of México City, México in 2001, and a Ph.D. degree in electronics engineering, specializing in the area of Power Electronics, from the National Center for Research and Technological Development CENIDET, Cuernavaca, Morelos, México in 2009. Since 1992, he has been a full time professor in the ESIME of the National Polytechnic Institute of México. His research interests are Modeling, Characterization of Power Devices and Failure Detection.

Abraham Pech-Carbonel received B.S. in electronics engineering in the area of automatic controls, from the Autonomous University of Carmen City UNACAR, Campeche, México in 2013.

Manuel May-Alarcón received a M. S. degree in electronics science from the Center for Scientific Research and Higher Education at Ensenada CICESE, Baja California, México in 1998, and a Ph.D. degree in Science specializing in the area of Optics, from National Institute Astrophysics, Optics and Electronics INAOE, Puebla, México in 2003. Since 2004, he has been with the Autonomous University of Carmen City UNACAR, Campeche, México as a full time professor in the Department of Electronics Engineering.

Fault-Tolerant Converter for AC Drives using Vector-Based Hysteresis Current Control

Nuno M. A. Freire, A. J. Marques Cardoso

Abstract -- **This paper presents a fault-tolerant converter intended for AC regenerative drives requiring improved reliability and availability. The standard six-switch three-phase converter is rearranged in order to obtain a four-switch tree-phase converter (FSTPC) during post-fault operation. Thus, its performance relies heavily on the fault diagnostic technique to trigger the remedial procedures, and on the control strategy to guarantee proper post-fault operation of the drive. With the aim to achieve these two goals effectively without increasing significantly the control system complexity, it is proposed to integrate, into the drive controller, a new and simple voltage-based approach for open-circuit fault diagnosis (without additional sensors), and vector-based hysteresis current control of the FSTPC. The performance of proposed fault-tolerant converter is evaluated by means of experimental results with a permanent magnet synchronous machine (PMSM) drive.**

Index Terms-- **Condition monitoring, fault diagnosis, fault tolerance, power semiconductor switches, pulse width modulation converters, semiconductor device reliability, variable speed drives.**

NOMENCLATURE

*	Reference value
$\alpha\beta$	Stationary reference frame axes
B_h	Hysteresis comparators bandwidth
d_n	Diagnostic variables
i_s	Measured stator currents
$I_{C,rms}$	Capacitors maximum rms current
k_d	Diagnostic threshold
L_s	Synchronous inductance
R_s	Stator resistance
S_a, S_b, S_c	Converter switching states
u_n^*	Reference phase voltages
u_α^*, u_β^*	Reference voltages in $\alpha\beta$ axes
$u_{c\alpha}, u_{c\beta}$	Converter voltages in $\alpha\beta$ axes
V_c	Converter voltage space vectors
V_{dc}	DC bus voltage
V_{EMF}	Back-emf voltage space vector
ω_n	Rated mechanical speed
ω_r	Rotor mechanical speed
Δu_{dcmax}	Capacitors maximum permissible oscillation
$\langle x \rangle$	Average value of the variable x

I. INTRODUCTION

Regenerative drives have seen their acceptance increased in numerous applications in order to harness the braking power of a motor drive or to improve the overall performance of the drive. Nevertheless, such electric drives are prone to suffer several kinds of faults. Among them, power converter faults are the most frequent, which are usually related to semiconductor or control circuit failures. This is reported in some statistical studies [1]-[2], where such faults are attributed to 60% of the power devices failures. Despite this, power switch open-circuit fault diagnosis is not included in motor drives as a standard feature and the industry still seeks reliable monitoring methods for power electronics [3], which justifies the recent noticeable number of published studies on this subject.

Converter self-diagnosis intends to handle unforeseen faults, by triggering remedial actions in fault-tolerant systems or simply by forcing a shutdown, thereby avoiding catastrophic consequences. Therefore, such procedure has to be suitable for integration into the drive controller, and ensure a reliable and effective diagnostic. Ideally, this should be accomplished without increasing the overall system cost and complexity, thus, additional hardware and great increase of the computational burden should be avoided. These considerations turn some recently proposed approaches [4]-[7] to be optimum choices for diagnosing open-circuit faults.

A fault-tolerant converter is intended to maintain its operation after an internal fault occurrence until a maintenance operation can be scheduled, with an acceptable performance and without endangering the overall system. Generally, hardware reconfiguration for circuit topology rearrangement is compulsory as well as software reconfiguration for an effective control of the obtained power converter.

To endow a standard three-phase converter with fault-tolerant capabilities, various topologies have been proposed. A redundant leg can be connected to all the converter phases through TRIACs [8]-[9], permitting the replacement of the faulty leg and keeping the same converter topology. The four-switch three-phase converter (FSTPC) has been addressed in various studies, by connecting the midpoint of the dc bus to the converter phases [10]-[11] or to the machine neutral point [12]-[13]. An extra leg connected to the machine neutral point is another alternative topology [14]-[15]. In a back-to-back converter, the phases of both

---

The authors gratefully acknowledge the financial support of the Portuguese Foundation for Science and Technology (FCT) under Project No. SFRH/BD/70868/2010.

Nuno M. A. Freire is with the Department of Electrical and Computer Engineering, University of Coimbra, P-3030-290 Coimbra, Portugal (nunofr@ieee.org).

A. J. Marques Cardoso is with the Department of Electromechanical Engineering, University of Beira Interior, P - 6201-001 Covilhã, Portugal (ajmcardoso@ieee.org).

The authors are also with the Instituto de Telecomunicacões, P-3030-290 Coimbra, Portugal.

sides (grid/machine) can also be connected to each other through TRIACs [16]-[17].

Taking into account the recent interest in fault-tolerant systems as well as in converter topologies with reduced number of power switches [18]-[22], this paper introduces the use of vector-based hysteresis current control (HCC) for FSTPCs. Therefore, by integrating the proposed control strategy into the drive controller, together with a suitable real-time fault diagnostic technique, an effective fault-tolerant converter is achieved. Switching tables are formulated for the three possible case scenarios, i.e., fault occurrences in each of the three phases. The performance of the proposed control system is analyzed by means of experiments with a back-to-back converter topology for PMSM drives. The aims of this paper are: (1) to propose the use of a simple and effective diagnostic technique for power switch open-circuit faults in fault-tolerant systems; (2) to propose vector-based hysteresis current control for FSTPCs; and (3) to evaluate the performance of the proposed fault-tolerant drive by means of experimental results, by analyzing its response to a fault occurrence in real-time, and by comparing vector-based HCC and conventional HCC.

II. OPEN-CIRCUIT FAULT DIAGNOSIS

Fault diagnosis is a vital task in a fault-tolerant system, since it has the duty to trigger the remedial procedures in order to ensure proper post-fault operation. By considering the need for a simple and reliable diagnostic technique for open-circuit faults, a voltage-based approach was proposed in [4], which does not require extra hardware. In this paper it is proposed the use of a new and simpler version, only requiring the reference voltages and the dc link voltage. When compared to [4], voltage observers and low pass filters are removed simplifying both implementation and tuning of the algorithm, and thus eliminating the inherent delay on the detection time. The block diagram of Fig. 1 depicts the proposed diagnostic technique, in which the average values of the normalized reference voltages are used as diagnostic variables:

$$d_n = \left\langle \frac{u_n^*}{V_{dc}} \right\rangle \tag{1}$$

where $n=a, b, c$. The reference voltages are directly available from the main control system when a voltage modulator (SVM, for instance) is applied, or they can be reconstructed from the switching pattern, when HCC is the employed control strategy, as follows:

$$\begin{cases} u_\alpha^* = \frac{2}{3} V_{dc} (S_a - \frac{1}{2} S_b - \frac{1}{2} S_c) \\ u_\beta^* = \frac{1}{\sqrt{3}} V_{dc} (S_b - S_c) \end{cases} \tag{2}$$

Under normal operating conditions the diagnostic variables, d_n, assume null values, because $\langle u_n^* \rangle = 0$. After the occurrence of a single open-circuit fault, the converter is not able to follow the imposed reference voltage, and the control targets of the closed-loop system cannot be fulfilled, resulting in high errors between the input reference signal and the feedback signal, which are reflected in the controller output signals (reference voltages), as an attempt to compensate the fault effects. Therefore, the reference voltages may contain the necessary information to perform the fault diagnosis. Defining a threshold value k_d and according to the diagram of a typical voltage source converter shown in Fig. 2a, the signatures for fault detection and localization are shown in Table I. Although all the diagnostic variables may exceed the threshold values (k_d / -k_d) as a consequence of an open-circuit fault, the faulty switch corresponds to the first verified condition in Table I. The k_d value is empirically established to be equal to 0.05.

III. FAULT-TOLERANT CONVERTER TOPOLOGY FOR AC REGENERATIVE DRIVES

The fault-tolerant converter is composed of two six-switch three-phase converters (SSTPCs) in a back-to-back topology, each one comprising six IGBTs with the respective antiparallel diodes, and six additional TRIACs (Fig. 2a), which remain open under normal operating conditions. The included TRIACs are intended to reconfigure the circuit topology, offering a path through the affected phase for the current flowing in both directions. Therefore, an FSTPC is suggested for the post-fault operation of the SSTPC, for instance, if open-circuit faults occur on both converter sides in phases A and a, the post-fault topology is the four-leg converter shown in Fig. 2b.

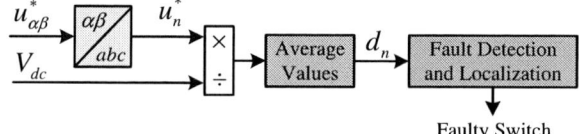

Fig. 1. Block diagram of the fault diagnostic technique.

TABLE I
OPEN-CIRCUIT FAULT DIAGNOSIS SIGNATURES

Faulty Switches	d_a	d_b	d_c
T1	$> k_d$		
T2	$< -k_d$		
T3		$> k_d$	
T4		$< -k_d$	
T5			$> k_d$
T6			$< -k_d$

978-1-4799-0024-4/13 $31.00 © 2013 IEEE   250

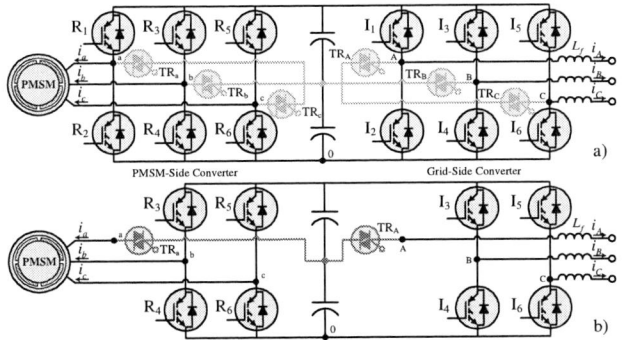

Fig. 2. Fault-tolerant converter topologies: a) under normal operating conditions; b) after fault occurrences in phase a (R1 or R2) and phase A (I1 or I2).

Compared with the standard SSTPC, the fault-tolerant converter requires power switches (IGBTs) and a capacitor bank with doubled voltage rating, whereas the TRIACs must have half of the IGBTs' voltage rating and equal current rating. The maximum current of the two dc link capacitors is increased by the phase currents of both converter sides, then its design should be carefully considered because a bulky capacitor bank might be required in order to fulfill the desired current rating as well as the maximum permissible voltage oscillation.

A cost-effective solution can be obtain by designing the dc-link bank in order to allow the converter to operate at reduced power levels under post-fault operation. Consequently, the maximum values assumed by the phase currents in each converter should be limited in order to not exceed the maximum current stress on the capacitors ($I_{C,rms}$):

$$\left(\frac{I_{1,rms}}{\sqrt{2}k_{PWM1}}\right)^2 + \left(\frac{I_{2,rms}}{2k_{PWM2}}\right)^2 + \left(\frac{\hat{I}_{1,rms}}{2k_{f1}}\right)^2 + \left(\frac{\hat{I}_{2,rms}}{2k_{f2}}\right)^2 \leq I_{C,rms}^2 \quad (3)$$

nor to exceed the maximum permissible voltage oscillation of each capacitor $\left(\Delta u_{dc\,max}\right)$:

$$\frac{1}{\sqrt{2}\,2\pi C}\left(\frac{\hat{I}_{1,rms}}{f_1} + \frac{\hat{I}_{2,rms}}{f_2}\right) \leq \Delta u_{dc\,max} \quad (4)$$

where subscripts 1 and 2 stand for the PMSM- and grid-side converters, respectively, $I_{1,rms}$ and $I_{2,rms}$ for the rated rms values, and $\hat{I}_{1,rms}$, $\hat{I}_{2,rms}$ for the currents maximum rms values when a given phase is connected to the dc bus midpoint. The constant values k_{PWM1}, k_{PWM2}, k_{f1} and k_{f2} result from the frequency dependence of the capacitors equivalent series resistance (R_{ESR}), permitting to obtain the R_{ESR} at the PWM frequencies and the currents fundamental frequencies (f_1, f_2), respectively. R_{ESR} at 100 Hz ($R_{ESR,100Hz}$) is usually the reference value, being k_{PWM} and k_f given by: $k_{PWM} = \sqrt{R_{ESR,100Hz}/R_{ESR,PWM}}$; $k_f = \sqrt{R_{ESR,100Hz}/R_{ESR,f}}$. In (3) the first two terms are obtained following the design guidelines in [23], and the last two appear from the fundamental currents flowing through the capacitors.

Therefore, the converter current limits under post-fault can be pre-calculated as a function of the PMSM speed and stored into a look-up table. A cost-effective option could be to design the capacitor bank to operate the converter up to the full power if a fault occurs in the grid-side converter, since its currents have a constant fundamental frequency (50Hz/60Hz), making easier the design. As a consequence, during post-fault operation of the PMSM-side converter, the drive is not able to develop high torque at low speeds (<50Hz), which remains as the main drawback of this topology. However, it is well suited for applications in which the torque increases proportionally to the speed (such as centrifugal pumps, fans, blowers, and compressors).

Regarding the fault-tolerant drive control system, further modifications are required. The remedial procedures are composed of up to three steps: fault isolation, hardware and software reconfigurations. The detection of an open-circuit fault is followed by the isolation of the faulty phase (inhibition of its control signals), then, the connection of that phase to the dc bus midpoint through the respective TRIAC (hardware reconfiguration), and software reconfiguration, by selecting a new switching table and imposing an adequate dc bus reference voltage. The only difference between the procedures for the two converter-sides is found on the selection of the dc bus reference voltage, which has to be twice as high when the grid-side converter is a FSTPC, because the grid voltage is constant and the output voltage capability of the converter is halved. On the other hand, for the machine-side converter the required output voltage highly depends on the machine speed (because the back electromotive force is proportional to the speed), thus, the reference dc-link voltage should be set accordingly to the reference speed:

$$\begin{cases} V_{dc}^{**} = V_{dc}^{*}, & if\ \omega_r^* \leq \omega_n/2 \\ V_{dc}^{**} = 2V_{dc}^{*}\dfrac{\omega_r^*}{\omega_n}, & if\ \omega_r^* > \omega_n/2 \end{cases} \quad (5)$$

where V_{dc}^{*} and V_{dc}^{**} stand for the reference dc-link voltages under normal and post-fault operation, respectively, ω_r^* for the reference rotor speed, and ω_n for the machine rated mechanical speed. Therefore, (5) results in the curve shown in Fig. 3.

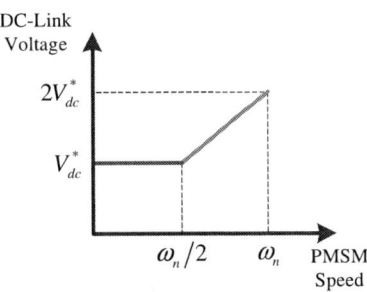

Fig. 3. Required DC-link voltage for post-fault operation of the PMSM-side converter as a function of the reference mechanical speed.

978-1-4799-0024-4/13 $31.00 © 2013 IEEE

IV. Vector-Based Hysteresis Current Control

Vector-based HCC [24]-[26] has been proposed to overcome some disadvantages of the conventional HCC – high switching frequency and band violation up to twice of the defined bandwidth due to interphases dependency – but retaining its main advantages – simplicity of implementation, fast transient response, and parameters independence.

Concerning HCC practical implementation for SSTPCs, the conventional HCC requires three two-level hysteresis comparators, so that each one controls a converter phase. Thus, the state change in one phase - that leads to a different voltage space vector - does not take into account its influence on the remaining phases (interphases dependency). On the other hand, the vector-based HCC in the stationary reference frame uses multilevel hysteresis comparators (three or four levels) and a switching table to control the currents in $\alpha\beta$ axes. Thus, a voltage space vector is selected in the switching accordingly to the current error information of all phases, eliminating interphases dependency and reducing significantly the switching frequency.

The converter voltage space vectors are defined by:

$$V_c = \frac{2}{3}\left(u_{AN} + u_{BN} e^{j2\pi/3} + u_{CN} e^{j4\pi/3}\right) = u_{c\alpha} + ju_{c\beta} \quad (6)$$

where u_{AN}, u_{BN} and u_{CN} stand for the phase-to-neutral voltages, and the basic principle of the vector-based HCC can be illustrated through the equivalent mathematical model of the PMSM in the stationary reference frame:

$$V_c = V_{EMF} + R_s i_s + L_s \frac{di_s}{dt} \quad (7)$$

Therefore, by neglecting the stator resistance, it is obtained that the sign of the current variation only depends on the applied voltage space vector and the back electromotive force voltage space vector (V_{EMF}):

$$\frac{di_s}{dt} = \frac{1}{L}\left(V_c - V_{EMF}\right) \quad (8)$$

Accordingly, with the knowledge of the available voltage vectors, the switching table can be easily formulated, as discussed later on in sub-sections IV-A and IV-B.

The block diagram of the proposed fault-tolerant vector-based HCC is shown in Fig. 4. In addition to the switching table required for normal operation, three new switching tables are included for post-fault operation, which are selected by using the information of the fault diagnosis method. Moreover, the faulty phase as to be isolated and the respective TRIAC turned on.

A. Six-Switch Three-Phase Converter

The standard two-level three-phase voltage source converter is composed of six power switches (IGBTs), resulting in eight possible states. It can be deduced that the two null and the six non-null voltage space vectors (Fig. 5a) are given by:

$$\begin{cases} u_{c\alpha} = \frac{2}{3}V_{dc}\left(S_a - \frac{1}{2}S_b - \frac{1}{2}S_c\right) \\ u_{c\beta} = \frac{1}{\sqrt{3}}V_{dc}\left(S_b - S_c\right) \end{cases} \quad (9)$$

where V_{dc} represents the dc bus voltage and S_a, S_b and S_c are the switching states (ON: 1, OFF: 0) of the upper IGBTs (I_1, I_3, I_5) of phases a, b and c, respectively.

Therefore, adopting three-level hysteresis comparators [24], the switching table (Table II) can be formulated by substituting (9) in (8).

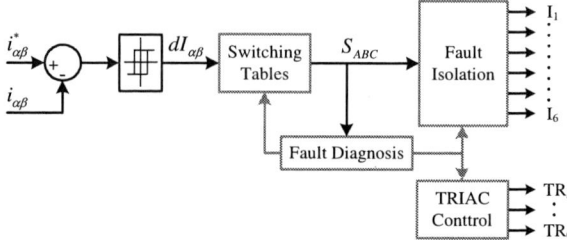

Fig. 4. Block diagram of the proposed fault-tolerant vector-based HCC.

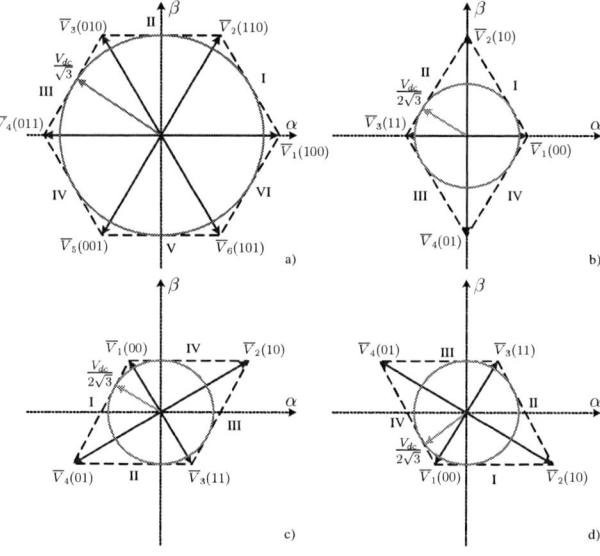

Fig. 5. Voltage space vectors synthesized by: a) a six-switch tree-phase converter (SSTPC); b) a four-switch tree-phase converter (FSTPC) with phase A connected to the dc bus midpoint; c) an FSTPC with phase B connected to the dc bus midpoint; d) an FSTPC with phase C connected to the dc bus midpoint.

TABLE II
Switching Table for the Six-Switch Three-Phase Converter

		dI_β		
		0	1	2
dI_α	0	V_5 (001)	V_4 (011)	V_3 (010)
	1	V_5/V_6	V_0/V_7 (000)/(111)	V_2/V_3
	2	V_6 (101)	V_1 (100)	V_2 (110)

B. Four-Switch Three-Phase Converter

By connecting a given phase to the dc bus midpoint the converter is composed of only four switches, and accordingly only four voltage vectors can be synthesized. Considering that phase A is connected to the dc bus midpoint, the voltage vectors in $\alpha\beta$ axes can be deduced as follows (Fig. 3b):

$$
\begin{cases}
u_{c\alpha} = \dfrac{V_{dc}}{3}(1 - S_B - S_C) \\
u_{c\beta} = \dfrac{V_{dc}}{\sqrt{3}}(S_B - S_C)
\end{cases}
\tag{10}
$$

In a similar way, the FSTPC voltage vectors can be obtained for the case in which phase B is connected to the dc link midpoint (Fig. 3c):

$$
\begin{cases}
u_{c\alpha} = \dfrac{V_{dc}}{3}\left(2S_A - \dfrac{1}{2} - S_C\right) \\
u_{c\beta} = \dfrac{V_{dc}}{\sqrt{3}}\left(\dfrac{1}{2} - S_C\right)
\end{cases}
\tag{11}
$$

as well as for the case where phase C is connected to the dc link midpoint (Fig. 3d):

$$
\begin{cases}
u_{c\alpha} = \dfrac{V_{dc}}{3}\left(2S_A - S_B - \dfrac{1}{2}\right) \\
u_{c\beta} = \dfrac{V_{dc}}{\sqrt{3}}\left(S_B - \dfrac{1}{2}\right)
\end{cases}
\tag{12}
$$

Thus, for the three case scenarios of post-fault operation, by adopting two-level hysteresis comparators:

$$
dI_{\alpha\beta} =
\begin{cases}
0, & if\ i^{*}_{\alpha\beta} - i_{\alpha\beta} < -B_h \\
1, & if\ i^{*}_{\alpha\beta} - i_{\alpha\beta} > B_h
\end{cases}
\tag{13}
$$

and substituting (11)-(12) in (8), three switching tables are formulated (Tables III-V). It is worth noting that the maximum output voltage is halved in comparison to the SSTPC.

TABLE III
SWITCHING TABLE FOR THE FOUR-SWITCH THREE-PHASE CONVERTER WITH PHASE A CONNECTED TO THE DC LINK MIDPOINT

		dI_β	
		0	1
dI_α	0	V_3 (11)	V_2 (10)
	1	V_4 (01)	V_1 (00)

TABLE IV
SWITCHING TABLE FOR THE FOUR-SWITCH THREE-PHASE CONVERTER WITH PHASE B CONNECTED TO THE DC LINK MIDPOINT

		dI_β	
		0	1
dI_α	0	V_4 (01)	V_1 (00)
	1	V_3 (11)	V_2 (10)

TABLE V
SWITCHING TABLE FOR THE FOUR-SWITCH THREE-PHASE CONVERTER WITH PHASE C CONNECTED TO THE DC LINK MIDPOINT

		dI_β	
		0	1
dI_α	0	V_1 (00)	V_4 (01)
	1	V_2 (10)	V_3 (11)

V. EXPERIMENTAL RESULTS

The experimental setup comprises a 2.2 kW PMSG coupled to a four-quadrant test bench, two Semikron SKiiP three-phase two-level voltage source converters in a back-to-back topology, two TRIACs, a dSPACE DS1103 digital controller, a Yokogawa WT3000 precision power analyzer, a dc bus capacitor bank of 1.1 mF and an output filter of 5 mH. The experimental setup is depicted in Fig. 6. The PMSM parameters are given in Table VI. Together with Matlab/Simulink and dSPACE ControlDesk software, the DS1103 controller provides real-time control and monitoring of the overall system with a sampling time of 50 µs. An open-circuit fault is introduced by removing the IGBT gate command signal. The proposed vector-based HCC is implemented together with rotor field oriented control (RFOC) and voltage oriented control (VOC) strategies for the PMSM- and grid-side converters, respectively.

All the experiments were carried out at reduced voltage levels, a grid phase-to-phase voltage of 75 V, and a reference dc link voltage of 150 V under normal operating conditions, allowing the machine to operate up to approximately half of its rated mechanical speed (≈900 rpm).

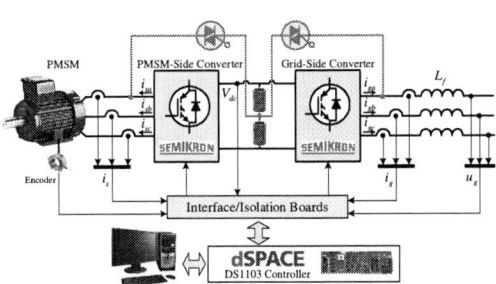

Fig. 6. Diagram of the experimental setup.

TABLE VI
PERMANENT MAGNET SYNCHRONOUS GENERATOR PARAMETERS

Power	P	2.2 kW
Speed	N	1750 rpm
Torque	T_n	12 Nm
Voltage	V	146 V
Current	I	10.4 A
Number of pole pairs	p	5
Armature resistance	R_s	0.415 Ω
Magnet flux linkage	ψ_{PM}	0.121 Wb
Synchronous inductance	L_s	5.13 mH

A. Fault-Tolerance in the Grid-Side Converter

Fig. 7 shows the system response to a fault in the IGBT I_1 of the grid-side converter at $t=0.22$ s. The fault detection is accomplished in 3 ms (15% of the current fundamental period), triggering the remedial procedures, which means that phase A gate command signals are turned off and the TRIAC is immediately after turned on, while the appropriate switching table is selected and a reference dc link voltage twice as high is imposed (300 V). The operation of the PMSM is not affected, and the steady-state operation is reached when the DC bus voltage becomes equal to its reference value.

B. Fault-Tolerance in the Machine-Side Converter

Fig. 8 shows the system response to a fault in the IGBT R_1 of the PMSM-side converter at $t=0.053$ s. After 6 ms (30% of the current fundamental period), the same remedial actions taken for the grid-side converter are triggered. However, the DC bus voltage does not need to be doubled, because the PMSM is not operating at the rated speed. Then, the reference DC link voltage is chosen as a function of the reference speed (in accordance to (5)), being increased to 200 V. As consequence of the DC bus voltage increase, a transient state is imposed to the grid-side converter currents. Finally, it can be seen that the taken measures allow the mechanical speed to be kept always close to its reference value, and then the drive stable operation is guaranteed.

C. Operation as Four-Leg Converter

In Fig. 9 both converters operate as an FSTPC (Fig. 2b), providing two balanced three-phase current systems. It is important to notice that despite having a constant DC bus voltage equal to 300V, the voltage of each capacitor oscillates as a consequence of both currents of phases a and A flowing through the capacitors. However, practically sinusoidal and balanced currents are achieved with the lack of need for compensation methods to reduce the currents distortion [18], which are usually required when a voltage modulator is employed (such as in S-PWM and SV-PWM techniques).

D. Comparison between Conventional and Vector-Based Hysteresis Current Control

In order to evaluate the advantages of the proposed vector-based HCC for the FSTPC, results at the same operating conditions with vector-based HCC (Fig. 10a) and conventional HCC (Fig. 10b) are compared, using the same bandwidth for the hysteresis comparators (B_h=0.1 A). By analyzing the PMSM phase currents, it can be seen that with the vector-based HCC the high frequency ripple slightly increases, which may indicate the reduction of the switching frequency. Then, calculating the average switching frequency for the considered operating condition, it is concluded that its value is approximately halved when using vector-based HCC (Table VII), resulting in the increase of the converter efficiency from 91.3% to 92.8% (measurements obtained by means of a Yokogawa WT3000 precision power analyzer). Concerning the grid-side converter operating as an FSTC, a

significant reduction of the average switching frequency is also verified when employing the vector-based HCC (Table VII).

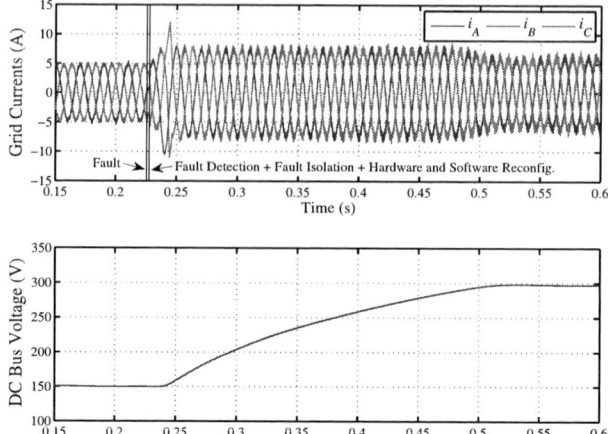

Fig. 7. Experimental results regarding the system response to a fault in phase A of the grid-side converter (Motoring operation: 600 rpm; 30%T_n).

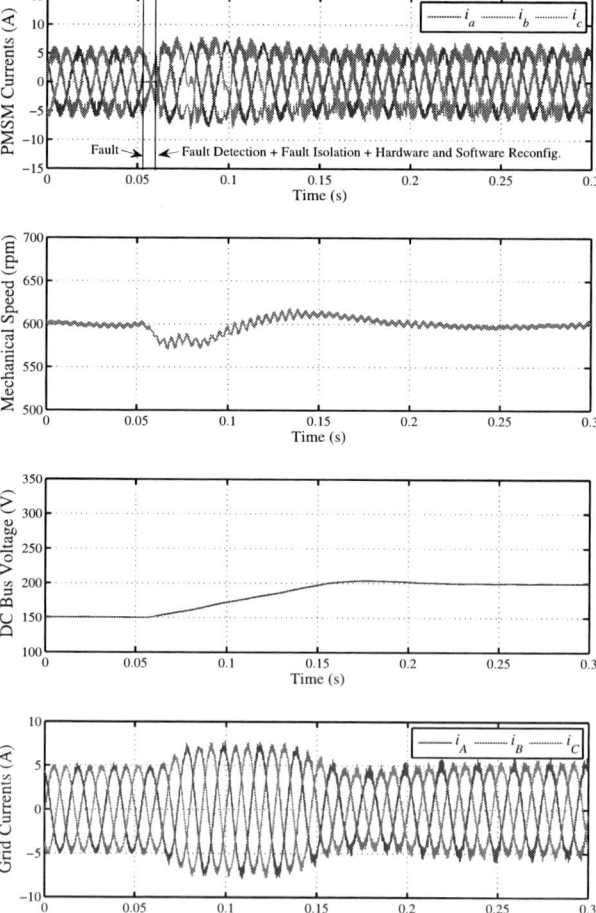

Fig. 8. Experimental results regarding the system response to a fault in phase a of the PMSM-side converter (Motoring operation: 600 rpm; 30%T_n).

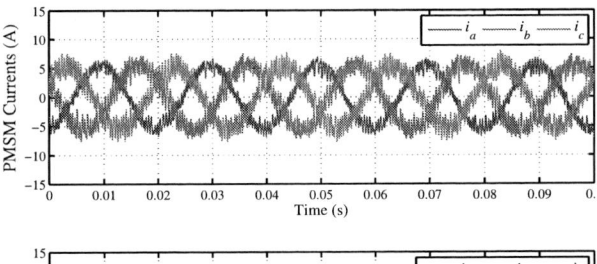

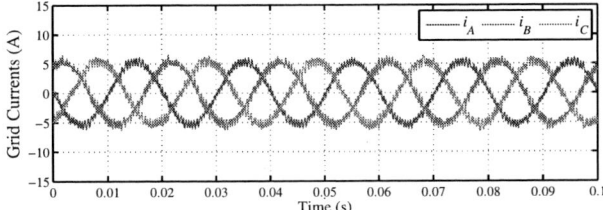

Fig. 9. Experimental results regarding the system operation with the four-leg converter topolgy (motoring operation: 600 rpm; 30%T_n).

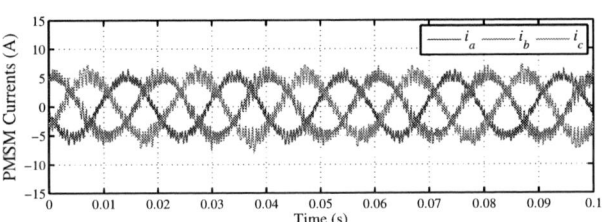

a)

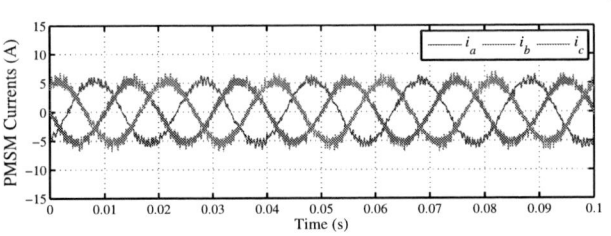

b)

Fig. 10. Experimental results regarding the PMSM-side converter operation as an FSTPC (braking operation: 600 rpm; 30%T_n): a) vector-based HCC; b) conventional HCC.

TABLE VII
EXPERIMENTAL EVALUATION OF THE AVERAGE SWITCHING FREQUENCY OF
THE FOUR-SWITCH THREE-PHASE CONVERTER

	Average Switching Frequency	
	Grid-side converter	PMSM-side converter
Vector-based HCC	2200 Hz	2100 Hz
Conventional HCC	3600 Hz	4000 Hz

VI. CONCLUSION

The control system for a fault-tolerant converter with application in AC regenerative drives has been proposed in this paper. The proposed fault diagnostic technique and control strategy contribute to achieve fault tolerance at the cost of a very low increase of the computational and implementation efforts.

The diagnostic technique is able to diagnose open-circuit faults in closed-loop controlled PWM AC voltage source converters, eliminating the typical requirement of additional hardware by using the information contained in the reference voltages available from the control system. Fault diagnosis is possible during both inverter and rectifier operating modes, and the detection speed is fast enough to allow continuous operation of the drive by switching to the post-fault control strategy.

The proposed vector-based hysteresis current control (HCC) permit to control a four-switch three-phase converter (FSTPC) through an extremely simple algorithm and with a satisfactory performance. Although it achieves an average switching frequency lower than the conventional HCC, the variable switching frequency remains as its main disadvantage. However, this can be perfectly acceptable during the period of time until a maintenance intervention.

Although this paper has focused the application of the FSTPC in fault-tolerant drives, it is worth pointing out that the proposed vector-based HCC is a valid control strategy for this converter topology regardless of the application.

VII. REFERENCES

[1] F. Spinato, P. J. Tavner, G. J. W. Bussel, E. Koutoulakos, "Reliability of wind turbine subassemblies", *IET Renewable Power Generation*, vol. 3, no. 4, pp. 387-401, Dec. 2009.

[2] S. Yang, D. Xiang, A. Bryant, P. Mawby, L. Ran and P. Tavner, "Condition monitoring for device reliability in power electronic converters: a review", *IEEE Transactions on Power Electronics*, vol. 25, no. 11, pp. 2734-2752, Nov. 2010.

[3] S. Yang, A. Bryant, P. Mawby, D. Xiang, L. Ran and P. Tavner, "An industry-based survey of reliability in power electronic converters", *IEEE Transactions on Industry Applications*, vol. 47, no. 3, pp. 1441-1451, May-June 2011.

[4] N. M. A. Freire, J. O. Estima and A. J. M. Cardoso, "A voltage-based approach for open-circuit fault diagnosis in voltage-fed SVM motor drives without extra hardware", *XX International Conference on Electrical Machines (ICEM)*, pp. 2378-2383, Sept. 2012.

[5] J. O. Estima and A. J. M. Cardoso, "A new approach for real-time multiple open-circuit fault diagnosis in voltage source inverters", *IEEE Transactions on Industry Applications*, vol. 47, no. 6, pp. 2487-2494, Nov./Dec. 2011.

[6] J. O. Estima and A. J. M. Cardoso, "A new algorithm for real-time multiple open-circuit fault diagnosis in voltage-fed PWM motor drives by the reference current errors", *IEEE Transactions on Industrial Electronics, IEEE Transactions on Industrial Electronics*, vol. 60, no. 8, pp. 3496-3505, Aug. 2013.

[7] N. M. A. Freire, J. O. Estima and A. J. M. Cardoso, "Open-circuit fault diagnosis in PMSG drives for wind turbine applications", *IEEE Transactions on Industrial Electronics*, vol. 60, no. 9, pp. 3957-3967, Sept. 2013.

[8] S. Karimi, A. Gaillard, P. Poure and S. Saadate, "FPGA-based real-time power converter failure diagnosis for wind energy conversion systems", *IEEE Transactions on Industrial Electronics*, vol. 55, no. 12, pp. 4299-4308, Dec. 2008.

[9] R. R. Errabelli and P. Mutschler, "Fault-tolerant voltage source inverter for permanent magnet drives", *IEEE Transactions on Power Electronics*, vol. 27, no. 2, pp. 500-508, Feb. 2012.

[10] Gi-Taek Kim, T. A. Lipo, "VSI-PWM rectifier/inverter system with a reduced switch count", *IEEE Transactions on Industry Applications*, vol. 32, no .6, pp. 1331-1337, Nov./Dec. 1996.

[11] M. B. R. Correa, C. B. Jacobina, E. R. C. da Silva and A. M. N. Lima, "A general PWM strategy for four-switch three-phase inverters", *IEEE Transactions on Power Electronics,* vol. 21, no. 6, pp. 1618-1627, Nov. 2006.

[12] T. H. Liu, J. R. Fu and T. A. Lipo, "A strategy for improving reliability of field-oriented controlled induction motor drives", *IEEE Transactions on Industry Applications,* vol. 29, no. 5, pp. 910-918, Sept./Oct. 1993.

[13] A. Gaeta, G. Scelba and A. Consoli, "Modeling and control of three-phase PMSMs under open-phase fault", *IEEE Transactions on Industry Applications,* vol. 49, no. 1, pp. 74-83, Jan.-Feb. 2013.

[14] S. Bolognani, M. Zordan and M. Zigliotto, "Experimental fault-tolerant control of a PMSM drive", *IEEE Transactions on Industrial Electronics,* vol. 47, no. 5, pp. 1134-1141, Oct. 2000.

[15] O. Wallmark, L. Harnefors and O. Carlson, "Control algorithms for a fault-tolerant PMSM drive," *IEEE Transactions on Industrial Electronics,* vol. 54, no. 4, pp. 1973-1980, Aug. 2007.

[16] A. Bouscayrol, B. Francois, P. Delarue and J. Niiranen, "Control implementation of a five-leg AC–AC converter to supply a three-phase induction machine", *IEEE Transactions on Power Electronics,* vol. 20, no. 1, pp. 107-115, Jan. 2005.

[17] C. B. Jacobina, I. S. de Freitas, E. R. C. da Silva, A. M. N. Lima and R. L. de A. Ribeiro, "Reduced switch count DC-link AC–AC five-leg converter", *IEEE Transactions on Power Electronics,* vol. 21, no. 5, pp. 1301-1310, Sep. 2006.

[18] J. Kim, J. Hong, and K. Nam, "A current distortion compensation scheme for four-switch inverters", *IEEE Transactions on Power Electronics,* vol. 24, no. 4, pp. 1032-1040, Apr. 2009.

[19] Q. T. An, L. Sun, K. Zhao and T. M. Jahns, "Scalar PWM algorithms for four-switch three-phase inverters", *Electronics Letters* , vol. 46, no. 13, pp. 900-902, June 2010.

[20] R. Wang, J. Zhao and Y. Liu, "Comprehensive investigation of four-switch three-phase voltage source inverter based on double fourier integral analysis", *IEEE Transactions on Power Electronics,* vol. 26, no. 10, pp. 2774-2787, Oct. 2011.

[21] S. Dasgupta, S. N. Mohan, S. K. Sahoo and S. K. Panda, "Application of four-switch-based three-phase grid-connected inverter to connect renewable energy source to a generalized unbalanced microgrid system", *IEEE Transactions on Industrial Electronics,* vol. 60, no. 3, pp. 1204-1215, March 2013.

[22] B. El Badsi, B. Bouzidi and A.Masmoudi, "DTC scheme for a four-switch inverter-fed induction motor emulating the six-switch inverter operation", *IEEE Transactions on Power Electronics*, vol. 28, no. 7, pp. 3528-3538, July 2013.

[23] J. W. Kolar and S. D. Round, "Analytical calculation of the RMS current stress on the DC-link capacitor of voltage-PWM converter systems", *IEE Proceedings - Electric Power Applications,* vol. 153, no. 4, pp. 535-543, July 2006.

[24] M. P. Kazmierkowski, M. A. Dzieniakowski and W. Sulkowski, "Novel space vector based current controllers for PWM-inverters", *IEEE Transactions on Power Electronics,* vol. 6, no. 1, pp. 158-166, Jan. 1991.

[25] M. Mohseni and S. M. Islam, "A new vector-based hysteresis current control scheme for three-phase PWM voltage-source inverters," *Power Electronics, IEEE Transactions on* , vol.25, no.9, pp.2299-2309, Sept. 2010

[26] M. Mohseni, S. M. Islam and M. A. S. Masoum, "Enhanced hysteresis-based current regulators in vector control of DFIG wind turbines", *IEEE Transactions on Power Electronics*, vol. 26, no. 1, pp. 223-234, Jan. 2011.

VIII. BIOGRAPHIES

Nuno M. A. Freire was born in Coimbra, Portugal in 1987. He received the MSc degree from the University of Coimbra, Portugal, in 2010. Currently, he is a PhD student at the Department of Electrical and Computer Engineering of the Faculty of Sciences and Technology of the University of Coimbra, Portugal. His research interests are focused on condition monitoring and diagnostics of power electronics, fault-tolerant systems, and wind energy conversion systems.

A. J. Marques Cardoso (S'89–A'95–SM'99) was born in Coimbra, Portugal, in 1962. He received the E. E. diploma, the Dr. Eng. Degree and the Habilitation Degree, all from the University of Coimbra, Coimbra, Portugal, in 1985, 1995 and 2008, respectively. From 1985 till 2011, he was with the University of Coimbra, where he was Director of the Electrical Machines Laboratory. Since 2011, he has been with the University of Beira Interior, where he is a Full Professor at the Department of Electromechanical Engineering and Director of the Electromecatronic Systems Laboratory. His teaching interests cover electrical rotating machines, transformers, and maintenance of electromechatronic systems, and his research interests are focused on condition monitoring and diagnostics of electrical machines and drives. He is the author of a book entitled *Fault Diagnosis in Three-Phase Induction Motors* (Coimbra Editora, Coimbra, 1991) and of more than 300 papers published in technical journals and conference proceedings.

Finite Element Investigation of the Short-Circuit Fault in the Stator Winding of Induction Motors and Harmonics of the Neighboring Magnetic Field

V. Fireteanu, A-I. Constantin, R. Romary, R. Pusca, S. Ait-Amar

Abstract **– Based on a time domain finite element analysis of the electromagnetic field, the paper studies effects of the short-circuit fault in the stator winding of an induction motor and the influence of this fault on the magnetic field outside the motor. The detection of the short-circuit fault through the magnetic field in the motor neighboring is based on the comparison of harmonics of the output voltage of coil sensors in the healthy and faulty motor states.**

The influence of the motor frame on the efficiency of fault detection is studied.

Index Terms—**Induction machine, short-circuit in the stator winding, fault detection, finite element analysis**

I. INTRODUCTION

THE interest for the evaluation of different faults in electrical machines through non-invasive techniques continues to grow the last period [1 – 10]. The magnetic field outside the machine is influenced by different faults. The investigation of this field with coil sensors represents a very simple and efficient method [1], [8].

The study and diagnose of different faults based on the finite element model of the motor becomes more and more attractive [5], [7], [9]. The execution of virtual faults and the study of different solutions for faults detection offers important advantages in comparison with the physical experiment.

Based on dedicated finite element models of the induction motor [5], respectively on the analysis of the electromagnetic field in time domain, this paper studies the short-circuit fault in the stator winding. The fault effects on the time variation of the current, of the electromagnetic torque and of the unbalanced force acting on the rotor are studied in the first part. In the second part, the magnetic field outside the motor is investigated, through the time variation of the output voltage of coil sensors, in case of healthy and faulty motor state, for no-load and loaded motor operation [3]. Analytical models related the influence of the short-circuit fault on the particular harmonics amplitude of the magnetic field with frequency over 500 (Hz) are presented in the references [7] and [8].

V. Fireteanu and A-I. Constantin are with POLITEHNICA University of Bucharest, EPM_NM-Laboratory, 313 Splaiul Independentei, 060042 Bucharest, Romania (email: virgiliu.fireteanu@upb.ro).

R. Romary R. Pusca and S. Ait-Amar are with University of ARTOIS, LSEE, 62400 Béthune, France (e-mails: raphael.romary@univ-artois.fr, sonia.aitamar@univ-artois.fr, remus.pusca@univ-artois.fr).

II. DESCRIPTION OF THE FINITE ELEMENT MODEL FOR THE INVESTIGATION OF THE SHORT-CIRCUIT FAULT

The geometry and mesh of the electromagnetic field computation domain, Fig. 1 (a), (b) and the circuit model, Fig. 1 (c), correspond to a four poles squirrel-cage induction motor of 11 (kW), with a rated supply of 3 x 380 (V), 50 (Hz). The computation domain contains the stator and rotor cores, which are magnetic and nonconductive regions, the 48 stator slots - nonconductive, nonmagnetic and source regions, the 32 bars of the squirrel-cage and the motor shaft - regions of solid conductor type, the motor airgap and the infinitely extended region outside the motor.

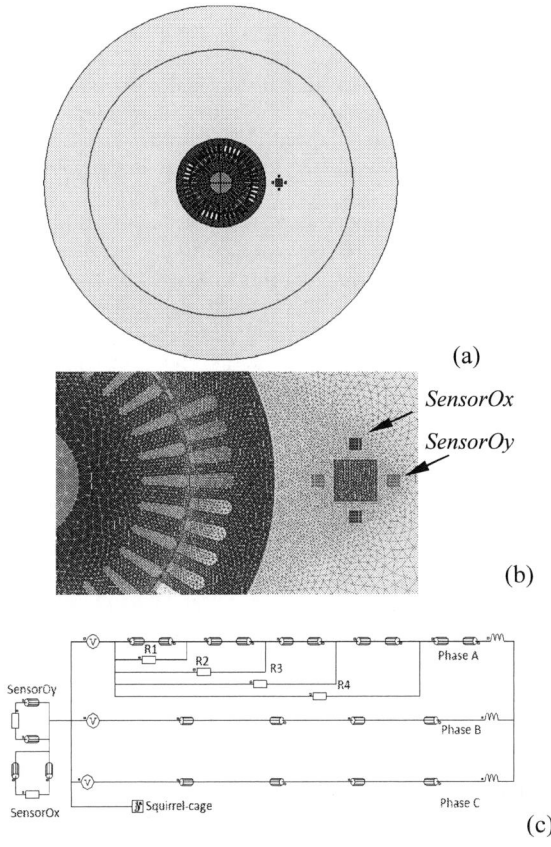

Fig. 1. Geometry of the electromagnetic field computation domain (a), mesh (b) and the circuit model (c)

978-1-4799-0024-4/13 $31.00 © 2013 IEEE

The phases B and C of the stator winding, Fig. 1 (c), are represented by four coil components that reflect the eight go-sides and the eight return-sides of the elementary coils. The three inductances in Fig. 1 (c) reflect the parts of the stator winding outside the magnetic core.

The short-circuit fault concerns the phase A of the stator winding. The part of this phase representing one stator pole, which regroups four elementary coils, is considered in the circuit model illustrated on Fig. 1 (c) through eight coil components. These are the go-sides and the return-sides of the four elementary coils. The resistors R1 - R4 in the circuit model, Fig. 1 (c), simulate the short-circuit of one, of two, of three or all four elementary coils of phase A.

The motor frame has an important influence on the magnitude and on the time variation of the magnetic field outside the motor. An annular frame of solid conductor type is considered in the last section of the paper: an aluminum alloy frame of 8 mm thickness and a magnetic non-linear steel frame of 5 mm thickness.

In order to investigate the magnetic field outside the motor, the computation domain, Fig. 1 (a), (b), includes two coil sensors on a common magnetic core. The *SensorOx* voltage is generated by the radial component of the magnetic field and the *SensorOy* voltage by the azimuth one. The circuit model, Fig. 1 (c), includes components for the evaluation of the two output voltages.

The investigation of the electromagnetic field and the evaluation of global quantities use transient magnetic models of the motor with constant rotor speed. The speed 1450 (rpm) corresponds to the rated load motor operation and the synchronous speed 1500 (rpm) to the no-load operation. The value of the time step in the step-by-step in time domain analysis is 0.05 (ms). Thus, all harmonic variations with frequency equal or less than 1000 (Hz) are evaluated in at least 20 steps on a period.

The investigations in this paper are interested in the steady state of the electromagnetic field, when the amplitudes of all electric and field quantities are time independent. Since the steady state of the transient numerical solution is reached in about 0.5 (s), the time interval from 0.52 (s) to 0.6 (s) is considered in the results analysis.

The lines of the magnetic field inside the healthy motor without frame and the maps of the magnetic flux density and relative permeability are presented in Fig. 2 for the last time step. The two maps reflect a high saturated motor.

III. EFFECTS OF THE SHORT-CIRCUIT FAULT IN THE STATOR WINDING ON THE MOTOR OPERATION PARAMETERS

The results presented in this paper correspond to the short-circuit of the first elementary coil of phase A, Fig. 1 (c). In case of faulty motor the resistor R1 has the resistance 3 (Ω). A value of 1 (MΩ) is associated in this case to the resistors R2, R3 and R4 or to all four resistors R1, R2, R3 and R4 in the case of healthy motor.

Four study cases were considered: HE0 - no-load healthy motor operation, HE1 - healthy motor loaded operation, FA0 - no-load faulty motor and FA1 - faulty motor loaded operation. The rms values of the three phase currents for the HE1 and FA1 cases are presented in Table 1 for loaded

motor operation. The unbalance of the three phase currents is not negligible in case of the faulty motor.

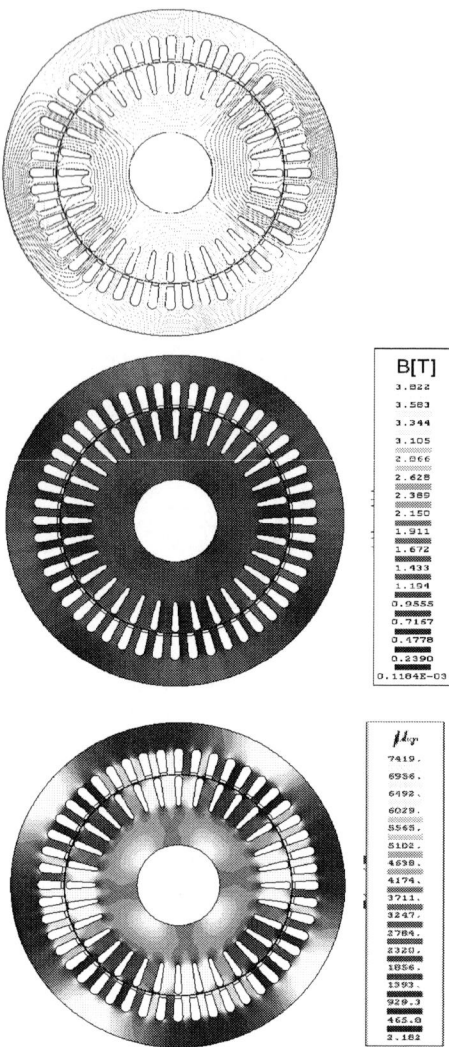

Fig. 2. Magnetic field lines, maps of the magnetic flux density and of the relative magnetic permeability

Table 1. Rms values of the stator currents in (A)

	Phase A	Phase B	Phase C	Coil in short-circuit
HE1	12.867	12.869	12.868	12.867
FA1	13.810	13.654	13.038	10.415

There are relatively slight differences between the time variations of the phase A current, Figs. 3 and 4. The rms values are 12.867 (A) in the HE1 case and 13.810 (A) in the FA1 case. There are important differences in the spectra of current harmonics in the range of 500 to 1000 (Hz). For example, the amplitude of the 725 (Hz) harmonic - number 58, has an important increase when it pass from the healthy to the faulty state, from 8.825 (mA) to 146.3 (mA).

The time variation of the current in the short-circuited elementary coil of phase A, Fig.5, is characterized by important amplitude of some harmonics in the range of

500 to 1000 (Hz). The most important, with amplitude over 2 (A), is the harmonic of 825 (Hz) (harmonic number 66).

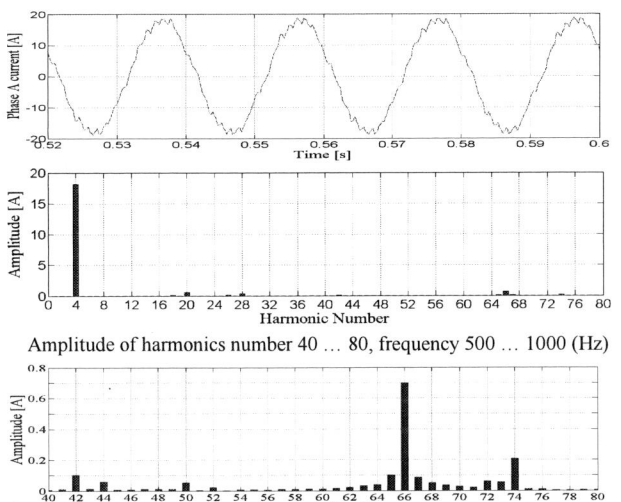

Amplitude of harmonics number 40 ... 80, frequency 500 ... 1000 (Hz)

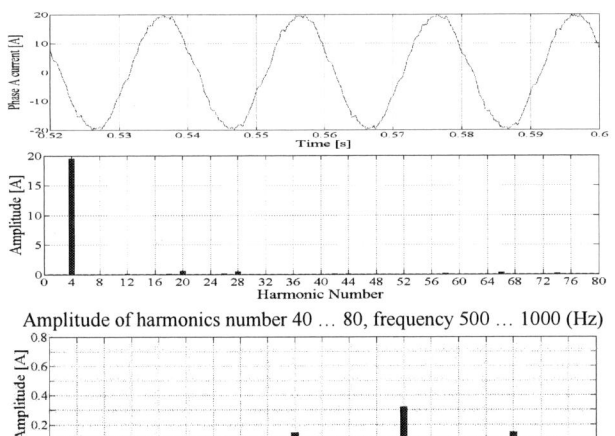

Fig. 3. Time variation and harmonics of the phase A current – HE1case

Fig. 4. Time variation and harmonics of the phase A current – FA1 case

The short-circuit fault determines a slight decrease of the mean value in time of the electromagnetic torque, from 76.79 (Nm) to 76.27 (Nm). The time variations of this quantity and the spectra of harmonics in Figs. 6, 7 show some changes associated with the fault presence. The most important is the appearance of the harmonic of 100 (Hz), with non-negligible amplitude, practically inexistent in the healthy case (HE1).

The most important effect of the short-circuit fault is reflected in the time variation of the module of unbalanced electromagnetic force acting on the rotor. If the mean value in time of this force, 0.102 (N), is completely negligible in case of the healthy motor, Fig. 8, this quantity has the value of 146.4 (N) in the FA1 case, Fig. 9. The 100 (Hz) harmonic (number 8) of this force has important amplitude.

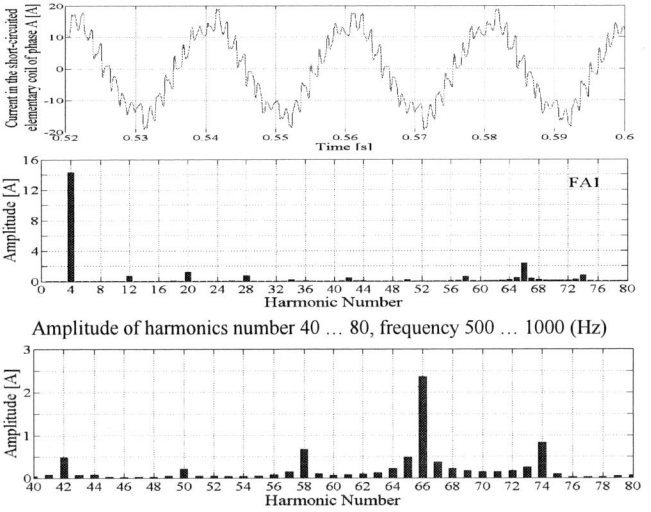

Amplitude of harmonics number 40 ... 80, frequency 500 ... 1000 (Hz)

Fig. 5. Time variation and harmonics of the current in the short-circuited elementary coil of phase A – FA1 case

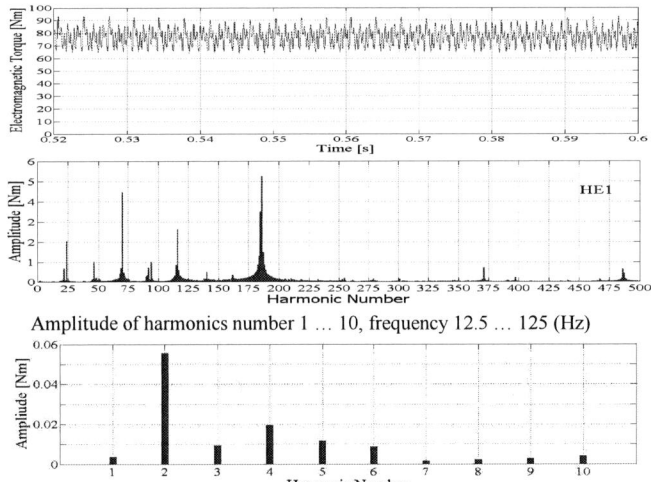

Amplitude of harmonics number 1 ... 10, frequency 12.5 ... 125 (Hz)

Fig. 6. Electromagnetic torque time variation and harmonics - HE1 case

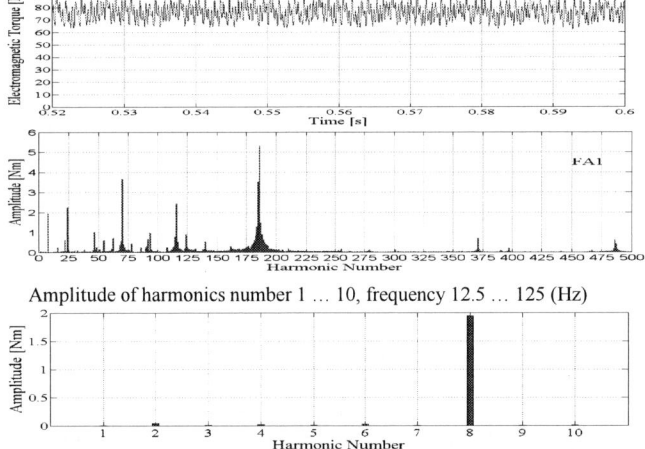

Amplitude of harmonics number 1 ... 10, frequency 12.5 ... 125 (Hz)

Fig. 7. Electromagnetic torque time variation and harmonics – FA1 case

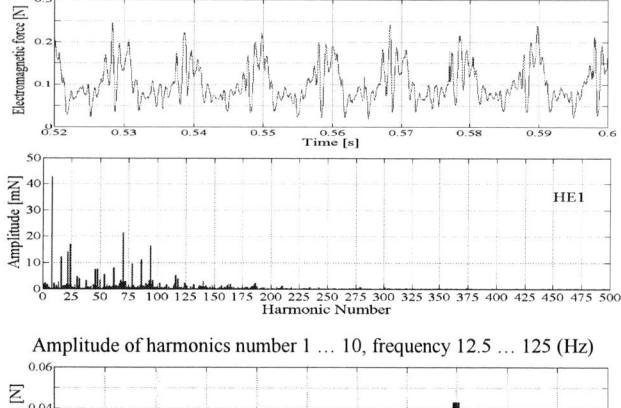

Amplitude of harmonics number 1 … 10, frequency 12.5 … 125 (Hz)

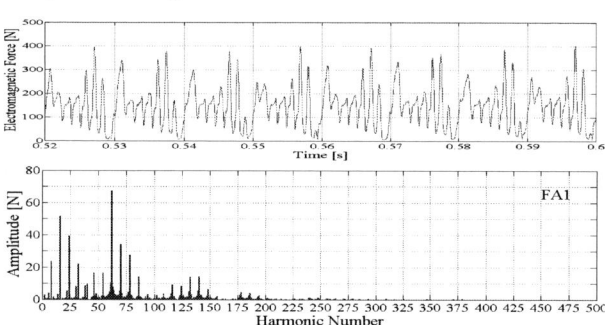

Fig. 8. Electromagnetic force time variation and harmonics- HE1case

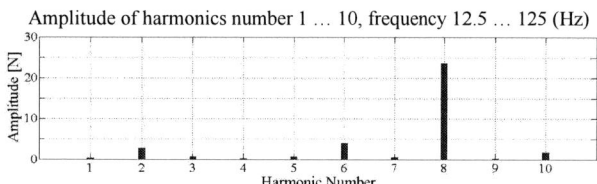

Amplitude of harmonics number 1 … 10, frequency 12.5 … 125 (Hz)

Fig. 9. Electromagnetic force time variation and harmonics- FA1case

IV. THE SHORT-CIRCUIT FAULT IN THE STATOR WINDING AND THE MAGNETIC FIELD IN THE MOTOR NEIGHBORING

This section studies the possibility to detect de short-circuit fault through the magnetic field in the motor neighboring. The time variation and the harmonic spectra of the output voltage of the two coil sensors, *SensorOx* and *SensorOy*, Fig. 1, are analyzed. These harmonics correspond to the slotting effect [3]. The results for healthy and faulty motor are compared for loaded motor operation (HE1 and FA1) - speed 1450 (rpm), and for the no-load motor operation (HE0 and FA0) – speed 1500 (rpm).

<u>Loaded motor operation</u>. There are differences between the *SensorOx* voltage and *SensorOy* voltage time variation, Figs. 10 and 11. There are no important changes between the healthy (HE1) and faulty (FA1) states.

The amplitudes of the most important harmonics in the range of 500 to 1000 (Hz) are presented in Tables 2 and 3. For both sensors, the ratio FA1 / HE1 has values

lower than one, the amplitude of these harmonics decreases when the short-circuit fault appears.

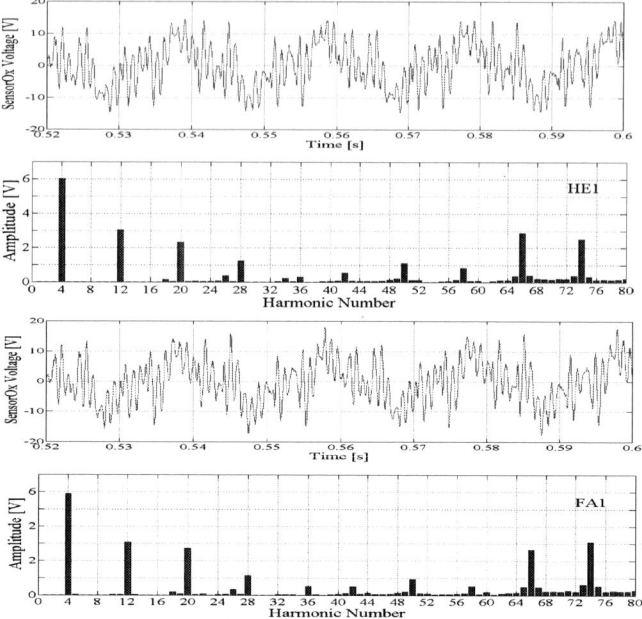

Fig. 10. Time variation and harmonics of *SensorOx* output voltage. Healthy (HE1) and faulty (FA1) states, loaded motor operation

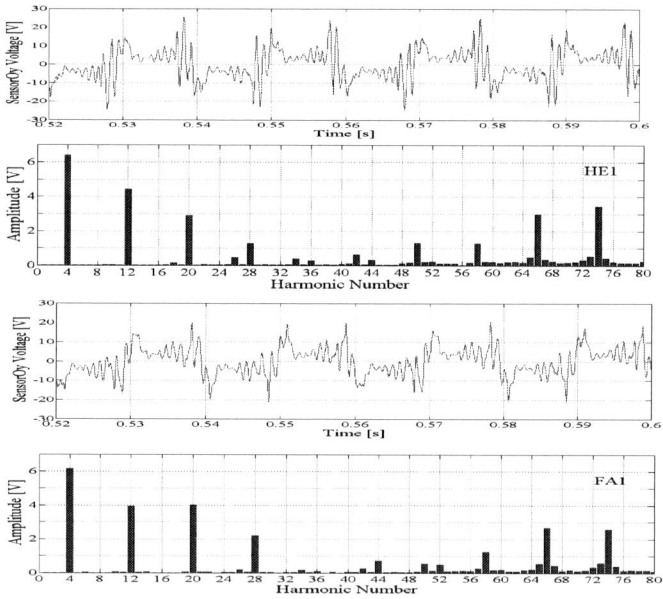

Fig. 11. Time variation and harmonics of *SensorOy* output voltage. Healthy (HE1) and faulty (FA1) states, loaded motor operation

Table 2. Amplitude of *SensorOx* harmonics (mV), loaded operation

Harmonic number/**Freq** (Hz)	42 / **525**	50 / **625**	58 / **725**	66 / **825**
HE1	553.6	1103	828.2	2868
FA1	519.8	945.5	540.0	2681
FA1 / HE1	**0.94**	**0.86**	**0.65**	**0.93**

Table 3. Amplitude of *SensorOy* harmonics (mV), loaded operation

Harmonic number/**Freq**(Hz)	42 / **525**	50 / **625**	66 / **825**	74 / **925**
HE1	622.6	1298	2983	3439
FA1	266.7	544.7	2705	2624
FA1 / HE1	**0.43**	**0.42**	**0.91**	**0.76**

Table 4. Amplitude of *SensorOx* harmonics (mV), no-load operation

Harmonic number/**Freq**(Hz)	44 / **550**	52 / **650**	60 / **750**	68 / **850**
HE0	64.4	85.2	68.2	359.4
FA0	76.7	748.7	314.2	1451
FA0 / HE0	**1.19**	**8.78**	**4.61**	**4.04**

Table 5. Amplitude of *SensorOy* harmonics (mV), no-load operation

Harmonic number/**Freq**(Hz)	44 / **550**	52 / **650**	60 / **750**	76 / **950**
HE0	31.1	67.2	101.4	322.5
FA0	130.3	659.1	721.5	679.3
FA0 / HE0	**4.19**	**9.81**	**7.11**	**2.11**

No-load motor operation. Related the voltage of the two coil sensors, there are differences between the healthy (HE0) and faulty (FA0) states as it is showed Figs. 12, 13. Values higher than one characterise the ratio FA0 / HE0 of the amplitude of some important harmonics, Tables 4 and 5.

The increase of the amplitude of some harmonics determined by the short-circuit fault, for both coil sensors, Tables 4 and 5, in case of no-load motor operation, is much more important than the decrease emphasized for loaded motor operation, Tables 2 and 3. Consequently, the no-load motor operation represents the better choice for the non-invasive detection of the short-circuit fault in the stator winding by using any of *SensorOx* or *SensorOy* sensors.

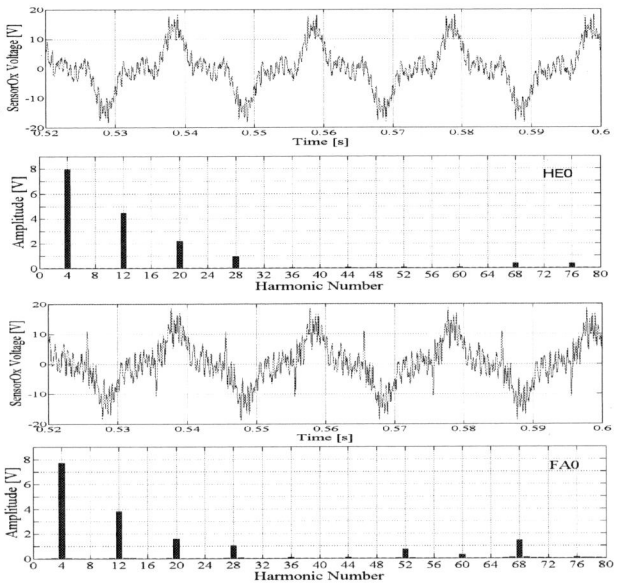

Fig. 12. Time variation and harmonics of *SensorOx* output voltage. Healthy (HE0) and faulty (FA0), no-load motor operation

V.   INFLUENCE OF THE MOTOR FRAME

The values of the amplitude of the voltage harmonics for motor without encasing are in the range of 64 to 1451 (mV). The decrease of the electromagnetic field from the inner to the outer face of the frame, Figs. 14 and 15, corresponds to an aluminum frame of 8 (mm) thickness and resistivity 0.045 (μΩm).

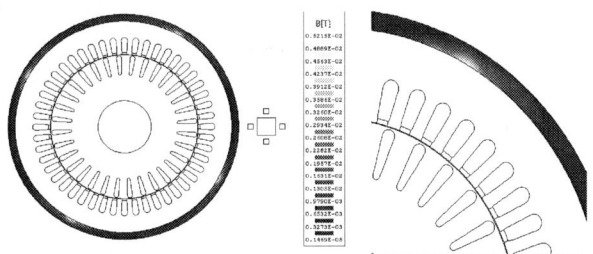

Fig. 14. Map of the magnetic flux density in the aluminium frame

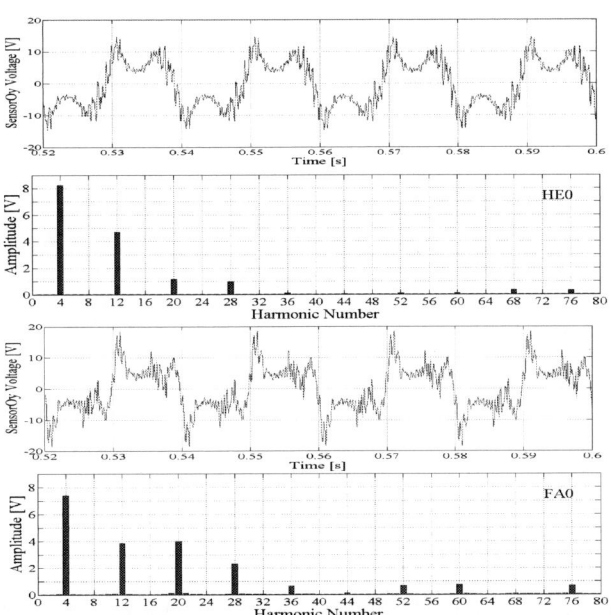

Fig. 13. Time variation and harmonics of *SensorOy* output voltage. Healthy (HE0) and faulty (FA0), no-load motor operation

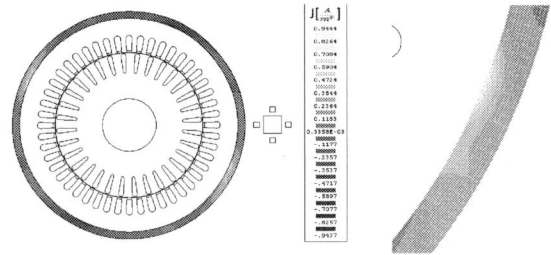

Fig. 15. Map of the current density induced in the aluminium frame

The frame presence determines the decrease of the main harmonics of *SensorOx* voltage, whose amplitudes are in the range of 1.163 to 20.75 (mV), Table 6. The ratio FA0 / HE0 characterizing the influence of the short-circuit fault has also important values in case of an aluminium frame and for the no-load motor operation. It is the case of the 650 (Hz) and 850 (Hz) harmonics, Table 6.

978-1-4799-0024-4/13 $31.00 © 2013 IEEE

Table 6. Amplitude of *SensorOx* harmonics (mV), aluminium frame

Harmonic number/**Freq**(Hz)	44 / **550**	52 / **650**	60 / **750**	68 / **850**
HE0	2.729	1.163	2.726	2.520
FA0	4.382	13.72	4.355	20.75
FA0 / HE0	**1.61**	**11.8**	**1.60**	**8.23**

The decrease of the magnetic field in a magnetic nonlinear steel frame of 5 (mm) thickness and resistivity 0.2 ($\mu\Omega$m), saturation magnetic flux density 1.8 (T) and relative initial permeability of 1000 is much more important, Figs. 16 and 17. The amplitude of the most important harmonics, Tables 7 and 8, are roughly 1000 times lower than in case of the aluminium frame.

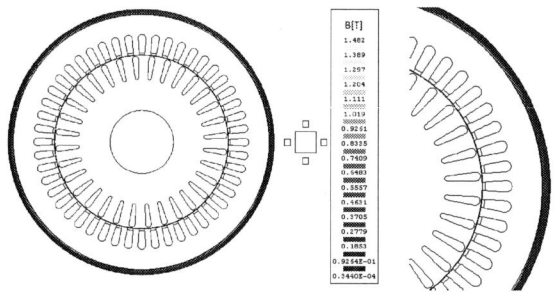

Fig. 16. Map of the magnetic flux density in the steel frame

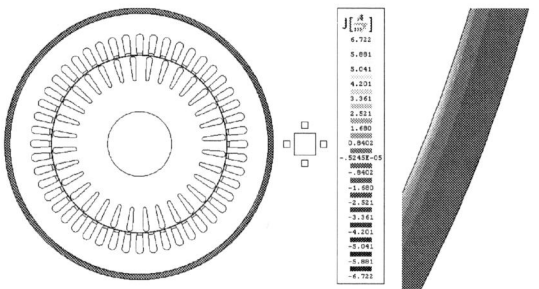

Fig. 17. Map of the current density induced in the steel frame

The decrease of the amplitudes of harmonics of *SensorOx* voltage for the loaded motor operation, Table 8, is much more important than the increase of the amplitudes in case of no-load motor operation, Table 7. Consequently, the loaded motor operation represents a better choice for the short-circuit fault detection in case of the steel frame.

Table 7. Amplitude of *SensorOx* harmonics (μV), no-load, steel frame

Harmonic number/**Freq**(Hz)	44 / **550**	52 / **650**	60 / **750**	68 / **850**
HE0	3.643	3.065	2.638	2.331
FA0	6.634	5.559	4.790	4.208
FA0 / HE0	**1.82**	**1.81**	**1.81**	**1.81**

Table 8. Amplitude of *SensorOx* harmonics (μV), loaded motor, steel frame

Harmonic number/**Freq**(Hz)	42 / **525**	50 / **625**	58 / **725**	66 / **825**
HE1	1.933	1.528	1.343	1.189
FA1	0.407	0.258	0.234	0.207
FA1 / HE1	**0.211**	**0.169**	**0.174**	**0.174**

VI. References

[1] H. Henao, C. Demian, and G.A. Capolino, "A frequency-domain detection of stator winding faults in induction machines using an external flux sensor", *IEEE Trans. Ind. Appl.*, vol. 39, pp. 1272–1279, Sept/Oct. 2003.

[2] T. Assaf, H. Henao, G.A. Capolino, "Simplified axial flux spectrum method to detect incipient stator inter-turn short-circuits in induction machine", in *Proc. ISIE. 2004*, Ajaccio, France, pp. 815–819, vol. 2.

[3] R. Romary, D. Roger, J.F. Brudny. "Analytical computation of an AC machine external magnetic field". European Physical Journal-Applied physics, EPJ-AP, EDP Sciences, September 2009, Vol.47, N° 3, pp. 31102.

[4] M. D. Negrea, "Electromagnetic Flux Monitoring for Detecting Faults in Electrical Machines", doctoral disertation at Helsinki University of Technology, Laboratory of Electromechanics, November, 2006.

[5] V. Fireteanu, P. Taras, "Teaching induction machine through finite element models", *ICEM Conf.*, Sept. 6-9, 2008, Vilamoura, Portugual.

[6] B. Vaseghi, N. Takorabet, and F. Meibody-Tabar, "Transient finite element analysis of induction machines with stator winding turn fault," *PIER Journals*, vol. 95, pp. 1-18, 2009.

[7] A. Ceban, "Methode globale de diagnostique des machines electriques" , PhD Thesis, Université d'Artois, France 2012.

[8] R. Pusca, R. Romary, A. Ceban, "Detection of Inter-turn Short Circuits in Induction Machines Without Knowledge of the Healthy State", *Proc. of ICEM Conf.*, September 2-5, 2012, Marseille, France.

[9] V. Fireteanu, P. Taras, "Influence of the magnetic steel encasing of induction motors on the efficiency of the rotor faults diagnosis based on the harmonics of the coil sensors output voltage", *Proc. of ICEM Conf.*, September 2-5, 2012, Marseille, France.

[10] A-I. Constantin, V. Fireteanu, V. Leconte, "Effects of the Short-Circuit Faults in the Stator Winding of Induction Motors and Fault Detection through the Magnetic Field Harmonics, *Proc. of ATEE Conf.*, May 23-25, 2013, Bucharest, Romania.

VII. Biographies

Virgiliu Fireţeanu graduated in 1970 the former Polytechnic Institute of Bucharest, Electrotechnical Faculty. From 1994 he is Full Professor of POLITEHNICA University of Bucharest, Electrical Engineering Faculty. His actual field of interest it is the finite element analysis of electro-mechanical and electro-thermal energy conversion systems. He animates the activity in higher education and research of the EPM_NM Laboratory (http://www.amotion.pub.ro/~epm).

Alexandru-Ionel Constantin received in 2012 the undergraduate diploma in electrical engineering – applied informatics from Electrical Engineering Faculty of POLITEHNICA University of Bucharest, Romania. He is now associated with EPM_NM Laboratory, prepares a master degree in the finite element evaluation and diagnosis of faulty operation states in electrical machines.

Raphaël Romary received the Ph.D. from Lille University in 1995 and the D. SC degree from Artois University in 2007. He is currently Full Professor in that University and researcher at the Laboratory of Electrical Systems and Environment (LSEE). His research interest concerns the analytical modeling of electrical machines with applications to noise and vibration, losses, electromagnetic emissions, diagnosis.

Remus Pusca received in 1995 the electrical engineering degree from Technical University of Cluj-Napoca, Romania. He obtained in 2002 Ph.D. degree in electrical engineering, from the University of Franche-Comté, France. Since 2003 he has joined the Laboratory of Electrical Systems and Environment (LSEE). He is Associate Professor at the Artois University. His research interest is control of electrical systems and diagnosis of electrical machines.

Sonia Ait-Amar received the diploma of Electrical Engineer and Doctor in Electrical Engineering from the University of Bejaia in 1994 and 1997, respectively. Thereafter, she taught at the University of Bejaia and pursued her research in electrical discharges. In 2003, she joined ABB France as a researcher. She received the Ph.D. degree in plasma physics from Paris Sud 11 University in 2006. She is Associate Professor at the University of Artois since September 2008. Her research interests are diagnosis of electrical machines and methods for predictive maintenance.

978-1-4799-0024-4/13 $31.00 © 2013 IEEE

Modeling and Simulation of Stator Turn Faults. Detection Based on Stator Circular Current and Neutral Voltage

Yassine MAOUCHE; *Student Member IEEE*, Abdelfettah BOUSSAID; *Student Member IEEE*,
Mohamed BOUCHERMA; *Member IEEE* and Abdelmalek KHEZZAR; *Member IEEE*
Laboratoire d'Électrotechnique de Constantine LEC
Département d'électrotechnique
Université Constantine 1
25000, Constantine, Algeria
Email: Yassine.MAOUCHE@lec-umc.org

Abstract—Most existing electrical faults in induction machines are stator turn faults; therefore the modeling of this defect appears very interesting. A number of authors have treated the modeling of this defect by completely changing the model from healthy to a defective one. This paper deals with the modeling of induction machines considering both healthy and faulty stator cases. The aim of this work is to make the electrical matrix parameters as a function of stator winding distribution. The consequence of both, fault degree and fault location are also considered. Stator faults detection are based on frequency analysis of the zero sequence for delta connected winding and on neutral voltage for wye connected winding. The simulation results show that the significant increase or decrease of certain harmonics can be used to detect short circuited turns in the machine stator.

Keywords—Condition monitoring, faut detection, faut diagnosis, Fourier transforms, induction motor, inter-turn short circuits, neutral voltage, spectrum analysis, stator circular current.

Nomenclature

I_{sijk}	Stator current (i phase, j pole and k coil).
R_{sijk}	Stator resistance (i phase, j pole and k coil).
L_{sijk}	Stator inductance (i phase, j pole and k coil).
$M^{sr}_{sijk\_m}$	Stator rotor mutual inductance between (i phase,j pole and k coil) and (rotor loop m).
R_s^d	Resistance of defective phase.
h	Order of harmonic.
M^{sr}_{sA}	Mutual inductance between the stator phase A and the rotor loops.
R_s	Stator phase resistance.
f_s	Main voltage supply frequency.
w_s	Main voltage supply pulsation.
R_b, R_e	Rotor bar and end ring resistances respectively.
N_e	Number of slots per phase per pole.
P	Number of poles pairs.
M^{srd}_{sA}	Mutual inductance between defective phase of stator and rotor.

I. Introduction

Stator turn faults in induction machines are related to the short-circuit of a small number of winding turns. For industrial processes, the IM fault monitoring and diagnosis is essential to identify motor failures to avoid severe damage to induction motors. According to published surveys, IM failures include bearing failures (which are responsible for 40%-50% of all faults), inter-turn short circuits in stator windings (which represent 30%-40% of the reported faults), and broken rotor bars and end ring faults (which represent 5%-10% of the IM faults) [1].

Paper [1] is centered on electrically measurable faults that take place in the stator windings and rotor cage, namely inter-turn stator shorts and broken rotor bars. Other techniques consist of using only the DC current, artificial intelligence, an external-flux-density sensor, multidimensional demodulation techniques [2][6]. In [7] an analysis of inter-turn faults symptoms for the induction motor drive based on the phase-shift calculation.

A good quantity of literature is accessible on modeling of induction machines. General models assuming sinusoidally distributed stator and rotor windings is described in [8]. In [9] models with multiple stator coil details and squirrel cage rotors, termed multiple coupled circuit models are presented. Model based on the winding function and multiple coupled circuit model approach including slot skewing effect [10]. The vector form of the multiple coupled circuit model is presented in [11]. Every part of these models is difficult and is not disposed to theoretical analysis. Morever, these models doesn't precise the location for the fault and allow only the number of short-circuited turns to vary.

The aim of the present paper is to develop a motor model based on the winding functions. We use a connection matrix to generate electrical parameters matrices for healthy and faulty motor and taking into account the fault degree and its position.

One can notice that the inter-turn short circuit gives rise to some harmonic components in the line current, stator circular current and neutral voltage.

II. Winding Topology

The stator is divided into three phases, every phase has P sections of coils connected in series; each section of coils has N_e coils. One coil has N_c turns or N_c conductors in any coil-side Fig.1.

Clearly, the stator faults have an influence on the sta-

978-1-4799-0024-4/13 $31.00 © 2013 IEEE

tor parameters (resistances, leakage inductances, and mutual inductances between phases) and on the mutual inductances between the stator and the rotor.

The Following matrix is used to transform electrical parameters of coils to the electrical parameters of phase:

$$[C] = \begin{bmatrix} x_1 \\ \vdots \\ \vdots \\ x_i \end{bmatrix}_{(P \cdot N_e) \times 1} \quad (1)$$

where $\quad x_i = \begin{cases} 1 - \dfrac{N_{cc}}{N_c} & defective\ coil \\ 1 & healthy\ coil \end{cases}$

N_{cc} is the number of turns of a short-circuit in coil i. The Association matrices $[C]$ allow us to develop electrical parameters matrix according to winding's topology:

The Resistance matrices of all phase coils are:

$$[R_{sijk}] = \begin{bmatrix} R_{si11} & 0 & \cdots & 0 \\ 0 & \ddots & \vdots & \vdots \\ \vdots & \vdots & \ddots & 0 \\ 0 & 0 & & R_{siPN_e} \end{bmatrix} \quad (2)$$

The equivalent phase resistance is:

$$R_s = [C]^t [R_{sijk}] \cdot [C] \quad (3)$$

The leakage inductances is:

$[L_{sijk}] =$

$$\begin{bmatrix} L_{si11} & \cdots & \cdots & M_{si11\_iPN_e} \\ M_{si12\_i11} & \ddots & \vdots & \vdots \\ \vdots & \vdots & \ddots & M_{si(P-1)N_e\_iPN_e} \\ M_{siPN_e\_i11} & \vdots & \vdots & L_{siPN_e} \end{bmatrix} \quad (4)$$

The equivalent leakage inductance is:

$$L_s = [C]^t [L_{sijk}] \cdot [C] \quad (5)$$

The mutual inductances between P coils of different phases are:

$[M_{sijk}] =$

$$\begin{bmatrix} M_{si11\_(i+1)11} & \cdots & \cdots & M_{si11\_(i+1)PN_e} \\ M_{si12\_(i+1)11} & \ddots & \vdots & \vdots \\ \vdots & \vdots & \ddots & M_{si(P-1)N_e\_(i+1)PN_e} \\ M_{siPN_e\_(i+1)11} & \vdots & \vdots & M_{siPN_e\_(i+1)PN_e} \end{bmatrix} \quad (6)$$

The equivalent mutual inductance is:

$$M_s = [C]^t [M_{sijk}] \cdot [C] \quad (7)$$

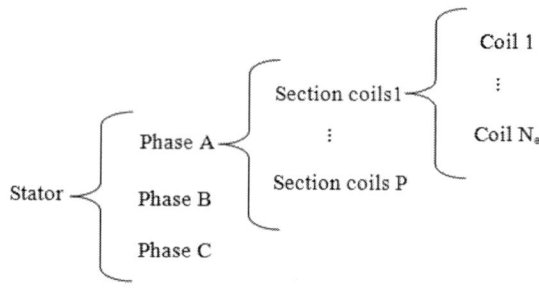

Fig. 1.   Stator arrangement of three-phase induction motor.

The mutual inductances between stator and rotor are

$$\left[M^{sr}_{sijk}\right] =$$

$$\begin{bmatrix} M^{sr}_{si11\_1} & \cdots & \cdots & M^{sr}_{s111\_n_b} & 0 \\ \vdots & \cdots & \cdots & \vdots & \vdots \\ \vdots & \cdots & \cdots & \vdots & \vdots \\ M^{sr}_{siPN_e\_1} & \cdots & \cdots & M^{sr}_{siPN_e\_n_b} & 0 \end{bmatrix} \quad (8)$$

The equivalent matrix of mutual inductance between stator phase A and rotor loops is:

$$[M^{sr}_{sA}] = [C]^t [M^{sr}_{sijk}] \cdot [C] \quad (9)$$

A. Delta-connected

Suppose that stator three-phase windings are Δ connected Fig.2. The primary equations of the induction machine can be written in vector-matrix form as follows:

$$[U_s] = [R_s][I] + \frac{d}{dt}[\psi_s] \quad (10)$$

$$\begin{bmatrix} [V_r] \\ V_e \end{bmatrix} = [0] = \begin{bmatrix} [R_r] & \frac{R_e}{n_b} \\ & \vdots \\ \frac{R_e}{n_b} & \cdots & R_e \end{bmatrix} \begin{bmatrix} [i_r] \\ i_e \end{bmatrix} \quad (11)$$

$$+ \frac{d}{dt}[\psi_r]$$

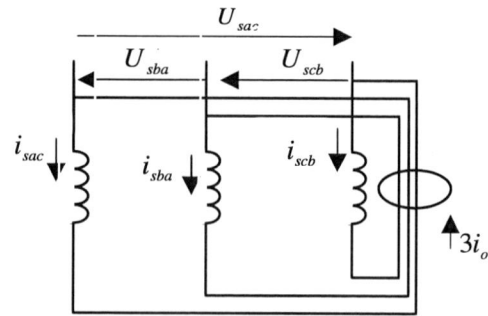

Fig. 2.   Delta connection of three-phase induction machine.

$$[\psi_s] = [L_s][i_s] + [M_{sr}][i_r] \tag{12}$$

$$\begin{bmatrix} [\psi_r] \\ \psi_e \end{bmatrix} = \begin{bmatrix} [M_{sr}] \cdot [i_s] \\ 0 \end{bmatrix} \tag{13}$$

$$+ \begin{bmatrix} [L_r] & & \frac{L_e}{n_b} \\ & & \vdots \\ \frac{L_e}{n_b} & \cdots & L_e \end{bmatrix} \cdot \begin{bmatrix} [i_r] \\ i_e \end{bmatrix}$$

where

$$[U_s] = \begin{bmatrix} [U_{sab}] & [U_{sbc}] & [U_{sca}] \end{bmatrix}^t$$

$$[I_s] = \begin{bmatrix} [I_{sab}] & [I_{sbc}] & [I_{sca}] \end{bmatrix}^t$$

$$[\psi_s] = \begin{bmatrix} [\psi_{sab}] & [\psi_{sbc}] & [\psi_{sca}] \end{bmatrix}^t$$

are the stator voltage, current and flux vectors respectively. $[V_r] = \begin{bmatrix} V_{r1} & \cdots & V_{rn_b} \end{bmatrix}^t$, $[i_r] = \begin{bmatrix} i_{r1} & \cdots & i_{rn_b} \end{bmatrix}^t$ and $[\psi_r] = \begin{bmatrix} \psi_{r1} & \cdots & \psi_{rn_b} \end{bmatrix}^t$ are the rotor voltages, current and flux linkage vectors, respectively with dimensions $1 \times n_b$. $[R_s]$ and $[R_r]$ are the stator and rotor resistance matrices respectively and are given by:

$$[R_s] = \begin{bmatrix} R_s & 0 & 0 \\ 0 & R_s & 0 \\ 0 & 0 & R_s \end{bmatrix} \tag{14}$$

$[R_r] =$

$$\begin{bmatrix} 2(R_b + R_e) & -R_b & 0 & \cdots & -R_b \\ -R_b & 2(R_b + R_e) & -R_b & \cdots & 0 \\ 0 & -R_b & 2(R_b + R_e) & & 0 \\ \vdots & & \ddots & \ddots & \vdots \\ -R_b & 0 & 0 & \cdots & 2(R_b + R_e) \end{bmatrix} \tag{15}$$

and $[L_r]$ are the stator and rotor inductance matrices and are described by the same manner as in [9].

The circular current is expressed by:

$$3i_o = I_{sac} + I_{sba} + I_{sca} \tag{16}$$

B. Wye-connected

In case of the Wye-connected stator windings Fig.3, relation (10) can be rewritten in vector-matrix form as follows:

$$[v_s] = [R_s][I_s] + \frac{d}{dt}[\psi_s] + [V_n] \tag{17}$$

where $[V_n]$ is the line neutral voltage and is given by:

$$3V_n = V_{sa} + V_{sb} + V_{sc} \tag{18}$$

as there is no method to infer it, we have to set the stator equations of the model in line-to-line voltage form.

$$\begin{bmatrix} V_{sab} \\ V_{sbc} \end{bmatrix} = \begin{bmatrix} R_s & -R_s \\ R_s & 2R_s \end{bmatrix} \cdot \begin{bmatrix} I_{sa} \\ I_{sb} \end{bmatrix} + \frac{d}{dt} \begin{bmatrix} \psi_{sa} \\ \psi_{sb} \end{bmatrix} \tag{19}$$

This global model allows us to simulate the behavior of both healthy and defective machines.

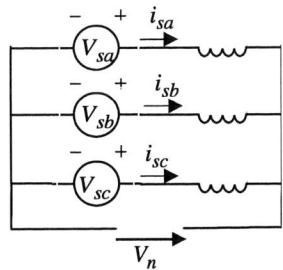

Fig. 3. Wye connection of three-phase induction machine.

III. DETERMINATION OF INDUCTANCES

The machine inductances can be developed according to the winding function theory, the mutual inductance between any two windings (i) and (j) in any electric machine can be computed by the following equation [12]:

$$L_{ij}(\phi) = \mu_0 Lr \int_0^{2\pi} \frac{n_i(\phi, \theta) N_j(\phi, \theta)}{e(\phi, \theta)} \tag{20}$$

where ϕ is the angular position of the rotor with respect to some stator reference, θ is a particular angular position along the stator inner surface, e is the air gap function, L is the length of stack and r is the average radius of the air gap. $n_i(\phi, \theta)$ is the winding distribution of coil i; it is introduced to describe the considered coil j ; and $N_j(\phi, \theta)$ is the winding function of coil j.

Using (20), the magnetizing inductance of coil i is:

$$L_{sijk} = \frac{4\mu_0 Lr}{e} \frac{N_c^2}{\pi} \sum_{h=1}^{\infty} \frac{1}{h^2} \left[\sin\left(h\frac{\alpha_s}{2}\right) \right]^2 \tag{21}$$

The mutual inductance between the coils of stator is:

$$M_{sijk\_ijk+1} = \frac{4\mu_0 Lr}{e} \frac{N_c^2}{\pi} \sum_{h=1}^{\infty} \frac{1}{h^2} \cos\left[h\left(k_1 \frac{2\pi}{N_s} + f_1 \frac{2\pi}{p}\right) \right] \times$$
$$\cos\left[h\left(k_1 \frac{2\pi}{N_s} + f_1 \frac{2\pi}{p} + \beta\right) \right] \tag{22}$$

The mutual inductance between the coils of stator and rotor is:

$$L_{ij} = \frac{4\mu_0 Lr}{e} \frac{N_c}{\pi} \sum_{h=1}^{\infty} \frac{1}{h^2} \sin\left(h\frac{\alpha_s}{2}\right) \times \sin\left(h\frac{\alpha_r}{2}\right) \times$$
$$\cos\left[h\left(\theta_r + (k - \frac{1}{2})\alpha_r - \frac{\alpha_s}{2} - \right.\right.$$
$$\left.\left. k_1 \frac{2\pi}{N_s} - f_1 \frac{2\pi}{p} - q\beta\right) \right] \tag{23}$$

Using the vetor of connection $[C]$ one can get:

The Magnetizing inductance of a phase which has $N_t = P \times N_e \times N_c$ turns.

$$L_{ii} = \frac{4\mu_0 Lr}{e} \frac{N_t^2}{p^2 \pi} \sum_{h=1}^{\infty} \frac{k_{bh}^2}{h^2} \tag{24}$$

The mutual inductance between phases of stator:

$$L_{ij} = \frac{4\mu_0 Lr}{e} \frac{N_t^2}{p^2\pi} \sum_{h=1}^{\infty} \frac{k_{bh}^2}{h^2} \cos(h\beta) \qquad (25)$$

The mutual inductance between phases of stator and rotor:

$$M_{sA}^{sr} = \frac{4\mu_0 Lr}{e} \frac{N_c}{\pi} \sum_{h=1}^{\infty} \frac{k_{bh}}{h^2} \sin\left(h\frac{\alpha_r}{2}\right) \times$$
$$\cos\left[h\left(\theta_r + \left(k-\frac{1}{2}\right)\alpha_r - \frac{\alpha_s}{2} - q\beta\right)\right] \quad (26)$$

IV. SIMULATION AND DISCUSSION

The stator winding arrangement of this motor is shown in Fig. 4. In order to well observe the effects of inter-turn short circuit on the stator/rotor mutual inductances, we showed the $M_{s1jk\_1}^{sr}$ inductances relating to a healthy state in the same graph Fig. 5. The association matrix of phase A in this first case is defined as:

$$[C] = \begin{bmatrix} 1 \\ 1 \\ 1 \\ 1 \end{bmatrix}_{(2\cdot2)\times1} \qquad (27)$$

Now, considering the case of a short-circuit which touch single coil 15% of the turns of the first slot of phase A, that is to say 11 turns of the coil placed in slots 1 and 7 correspondingly (it means that approximately 4% of the turns of phase A are short circuited). The resistance R_{cc} of the turns short-circuited will be, in this case $0.16 \times R_{s112}$, and the new resistance of phase A becomes $3.84 \times R_{s112}$, where R_{s112} is the resistance of the healthy coil. The association matrix of phase A in the second case is defined as:

$$[C] = \begin{bmatrix} 0.8429 \\ 1 \\ 1 \\ 1 \end{bmatrix}_{(2\cdot2)\times1} \qquad (28)$$

Fig. 6 shows the new mutual inductances between the slots of phase A and the rotor loop r1. One can notice that the inter-turn short-circuit affects only the inductances of the injured coil.

The deferential equations derived above can be solved by fourth-order Runge-Kutta method using MATLAB simulation software. The simulated motor has 24 stator slots 22 bars, 4 poles and a rated power of 1.1Kw. The dynamic simulation was carried out with a three-phase symmetrical sinusoid voltages and with a constant load point (50% of nominal load). The spectrum of the line current, obtained by simulation, is shown in left side of Fig. 7 ; one can see that the line current in healthy machine contains additional to the fundamental harmonic of stator current only the higher-harmonic component frequencies caused by the rotor field harmonics. The frequencies of these harmonics are given by [13]:

$$f_{sh} = f_s \left[\frac{\lambda n_b}{p}(1-s) \pm s \right] \qquad (29)$$

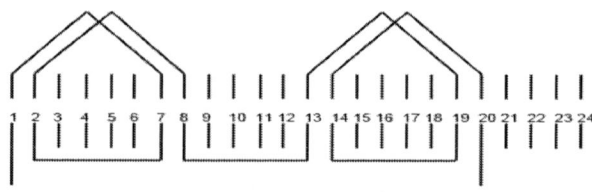

Fig. 4. The stator winding distribution of the simulated motor.

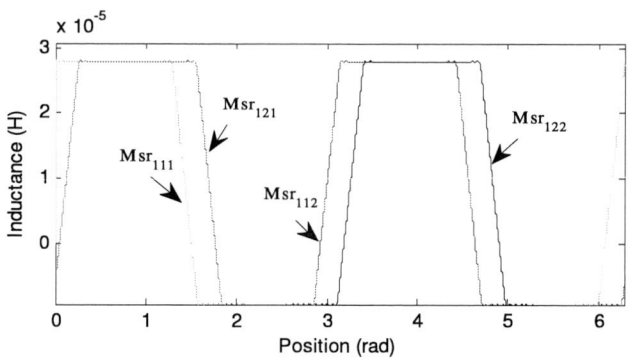

Fig. 5. Mutual inductances between slots of phase A and rotor loop r1 (healthy stator winding).

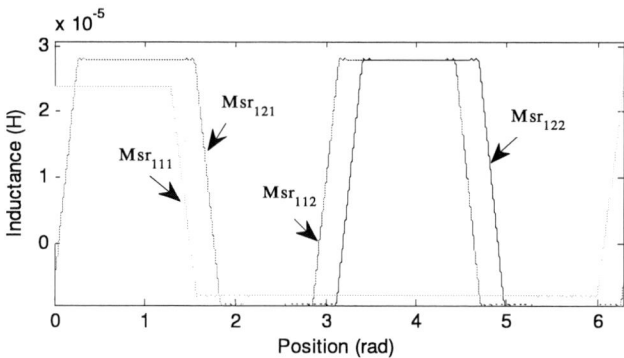

Fig. 6. Mutual inductances between slots of phase A and rotor loop r1 (motor with 10 turns of single coil short circuited).

Note that in this spectrum , for the first significant RSH harmonic is corresponding to $\lambda = 2$ and the second RSH to $\lambda = 4$. When $\lambda = 2$ and using (29) the first RSH appears at the frequency $f_{sh}^\lambda = 22.4f_s$Hz versus the harmonic order $(2(n_b/p) + 1) = 23$. The second rotor slot harmonic corresponding to $\lambda = 2$ versus harmonic order $(2(n_b/p) - 1) = 21$ did not show up in healthy machine.

The spectrum in the right side of Fig. 7 shows the additional harmonics of phase current in the case of 15% of all turns of phase A are short circuited. Short-circuit faults in a three-phase induction motor causes imbalance of the stator windings which induces negative-sequence currents (backward field). In addition to the RSH harmonics given by the positive sequence, an additional RSHs at frequencies $f_{sh}^\lambda = 20.68f_s$Hz and $f_{sh}^\lambda = 44.36f_s$Hz show up. A pulsating torque at multiples of the double supply frequency is produced due to interaction of the negative-sequence current

978-1-4799-0024-4/13 $31.00 © 2013 IEEE

with the fundamental-frequency rotor currents at frequencies [14]:

$$f_{sh}^{h,k} = f_s\left[h(1-s) \pm s \pm 2k\right] \quad (30)$$

The pulsating torque induces the following frequencies when $k = 1$: $f_{sh}^{\lambda,1} = 18.64 f_s Hz$ and $f_{sh}^{\lambda,1} = 24.68 f_s Hz$

One can notice the appearance of 150Hz, 250Hz, 350Hz, these harmonics are due to dissymmetry in stator .It should be pointed out that the faulty related harmonic frequencies corresponding to γf_s also exists in the healthy machine (time harmonics). It is apparent that the detection of short circuited turns by surveillance γf_s can be based only on significant increase of these harmonics.

The top left and right sides of Fig. 8. show the harmonic components of both the circular current and neutral voltage under healthy condition of the delta connection and and star connection respectively. Only the RSHs can be seen on the circular current spectrums or neutral voltage at the order $3h = (\lambda(^{n_b}/_p) - 1) = 21$ and $3h = (\lambda(^{n_b}/_p) - 1) = 45$ which are the larger components in the spectrum and they

are absent in the spectrum of the phase current in case of healthy condition. These RSHs are caused by only the space distribution of rotor bars. The RSHs orders of both the circular current and neutral voltage under healthy condition belong to [14]:

$$F = \quad (31)$$
$$\left\{ h = (2k+1)_{k=0,1,2...} \cap h = \frac{1}{3}\left(\lambda\frac{n_b}{p} \pm 1\right)_{\lambda=1,2} \right\}$$

According to the previous condition, it seems that the circular current and neutral voltage spectrums are the complement of the line current spectrum. The zero-sequence components in the voltage and current supply are:

$$\begin{cases} I_{so} = \sum_{h=1}^{\infty} I_{soh} \cos(h w_s t + \phi_{vh}) & \text{delta connection} \\ V_{so} = \sum_{h=1}^{\infty} V_{soh} \cos(h w_s t + \phi_{ih}) & \text{star connection} \end{cases}$$
$$(32)$$

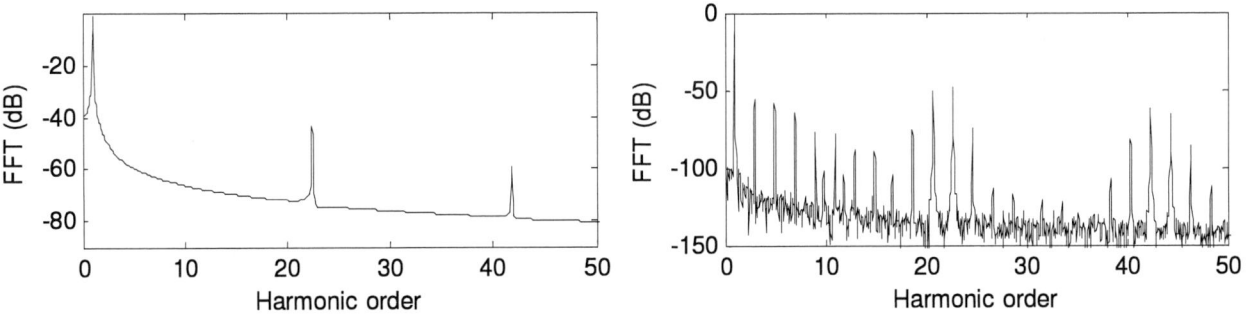

Fig. 7.   Simulated, normalized current spectrum of healthy motor (left) and faulty one, with 15% of the turns short circuited (right).

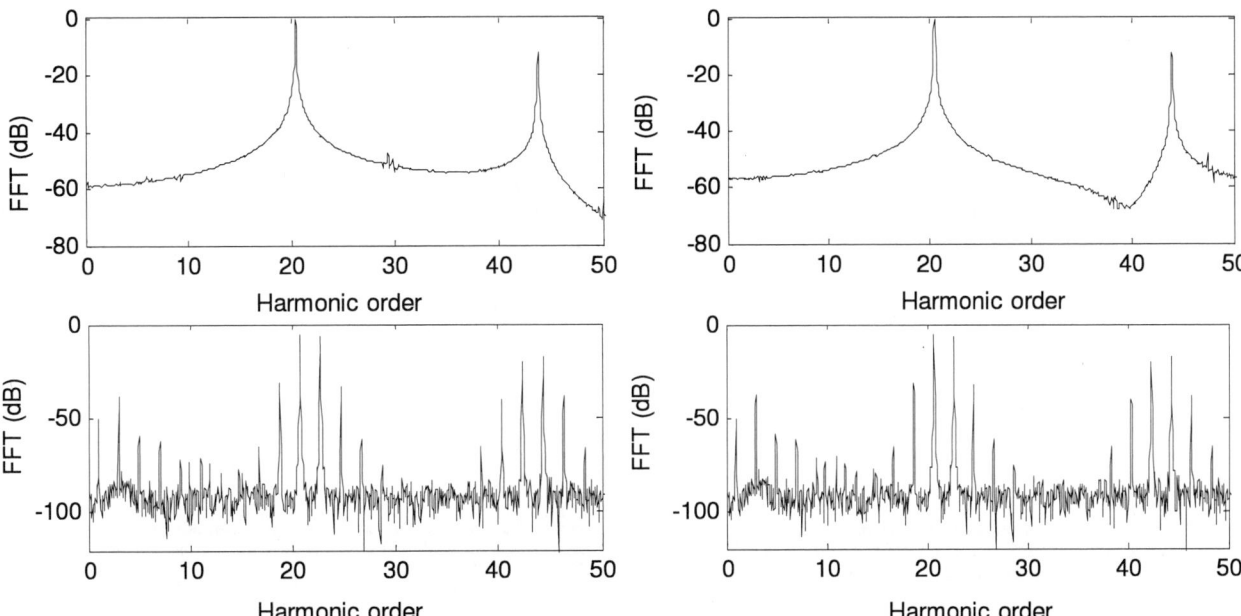

Fig. 8.   Simulated, normalized spectrum of neutral voltage o(top left), circular current (top right)of healthy motor. neutral voltage (bottom left), circular current (bottom right) of faulty motor with 15% of the turns of phase A short circuited.

TABLE I

HARMONIC COMPONENTS OF STATOR CURRENT, CIRCULAR CURRENT AND NEUTRAL VOLTAGE.

	Stator currents	
	Healthy conditions	Faulty conditions
RSH_s frequencies	$f_s\left[h\left(1-s\right)\pm s\right]$	$f_s\left[h\left(1-s\right)\pm s\pm 2k\right]$
Time harmonics frequencies	f_s	γf_s

	Circular current or Neutral voltage	
	Healthy conditions	Faulty conditions
RSH_s frequencies	$f_s\left[3h\left(1-s\right)\pm s\right]$	$f_s\left[3h\left(1-s\right)\pm s\pm 2k\right]$
Time harmonics frequencies	$-$	γf_s

The spectrums in the bottom (right and left side) of Fig. 8. show the component harmonics of circular current and neutral voltage in the case of a short circuit of 15% of all turns of phase A. It is important to note that both line voltage (Delta-connection) and circular current (Wye-connection) give a similar spectrum frequencies of stator current in case of turn faults. Table 1 summarizes harmonic components of stator current, line voltage and circular current.

V. CONCLUSION

In this work, a new mathematical model of the squirrel cage induction motor was presented in case of stator windings fault. It allows the simulation of the stator defects as the inter-turn short-circuits with both delta and wye connected stator windings. This model is established by taking into account all the space harmonics and stator winding connection. It is shown that the circular current and neutral voltage offer larger magnitudes of the RSHs components when compared to the RSHs components of the line stator current.

REFERENCES

[1] S. Nandi and H. A. Toliyat, "Condition monitoring and fault diagnosis of electrical machinesA review," in Conf. Rec. 34th IAS Annu. Meeting,1999, pp. 197204.J. Clerk Maxwell, A Treatise on Electricity and Magnetism, 3rd ed., vol. 2. Oxford: Clarendon, 1892, pp.6873.

[2] Siwei Cheng, Yilu Zhang, Nawrocki, S., Habetler, "using only the dc current information to detect stator turn faults in automotive claw-pole alternators", diagnostics for electric machines, power electronics and drives (sdemped), 2011 ieee international symposium on, on page(s): 119 125.

[3] Gandhi, Arun Corrigan, Timothy Parsa, Leila "recent advances in modeling and online detection of stator interturn faults in electrical motors", IEEE Transactions on Industrial Electronics, Volume.58, Issue.5, pp.1564, 2011, ISSN: 02780046,

[4] Romary, R. Jelassi, S. Brudny "Stator-Interlaminar-Fault Detection Using an External-Flux-Density Sensor", IEEE Transactions on Industrial Electronics, Volume.57, Issue.1, pp.237, 2010, ISSN: 02780046,

[5] Choqueuse, V., Benbouzid, M.E.H., Amirat, Y., Turri, S., "Diagnosis of Three-Phase Electrical Machines Using Multidimensional Demodulation Techniques", Industrial Electronics, IEEE Transactions on, On page(s): 2014 - 2023 Volume: 59, Issue: 4, April

[6] A. Bouzida, S. Hamdani, O. Touhami, R. Ibtiouen, M. Fadel and A. Rezzoug, "An experimental study on stator and rotor defects of squirrel cage induction machines", in Proc. IEEE 19th International Conference on Electrical Machines ICEM 2010, pp. 1-5.

[7] Kowalski, C.T., Orlowska-Kowalska, T., Wierzbicki, R., Wolkiewicz, M., "Analysis of inter-turn fault symptoms for the converter-fed induction motor based on the phase-shift calculation", IECON 2010 - 36th Annual Conference on IEEE Industrial Electronics Society, On page(s): 766 - 771

[8] P. C. Krause, O. Wasynczuk, and S. D. Sudhoff," Analysis of Electrical Machinery". New York: IEEE Press, 1996.

[9] A. R. Mu noz, and T. A. Lipo, "Complex vector model of the squirrel-cage induction machine including instantaneous rotor bar currents," IEEE Trans. Ind. Appl., vol. 35, no. 6, pp. 1332-1340, Nov./Dec. 1999.

[10] X. Luo, Y. Liao, H. A. Toliyat, A. El-Antably, T. A. Lipo, "Multiple coupled circuit modeling of induction machines," IEEE Trans. Ind. Appl.,vol. 31, no. 2, pp. 311-318, Mar./Apr. 1995.

[11] M. Sahraoui, A. Ghoggal, S. E. Zouzou A. Aboubou. Razik, "Modelling and Detection of Inter-Turn Short Circuits in Stator Windings of Induction Motor," Energy Convers., vol. 19, no. 1, pp.164-169, Mar.2006

[12] M. Arkan, D. Kostic-Perovic, and P. J. Unsworth, "Modelling and simulation of induction motors with inter-turn faults for diagnostics,"Electr. Power Syst. Res., vol. 75, no. 1, pp. 57-66, Jul. 2005. IEEE Trans. Ind. Applicat Vol. 38, pp. 101-109, Jan./Feb. 2002.

[13] J. Penman, H.G. Sedding, B.A. Lloyd, W.T. Fink, "Detection and location of inter turn short-circuits in the stator windings of operating motors",IEEE Transactions on Energy Conversion,Vol.9, pp. 652658, No.4, Dec. 1994

[14] M. E. K. Oumaamar, A. Khezzar, M. Boucherma, H. Razik,R. N. Andriamalala, and L. Baghli, " Neutral voltage analysis for broken rotor bars detection in asynchronous motors using Hilbert transformphase," in Conf. Rec. 42nd IEEE IAS Annu. Meeting, New Orleans, LA,Sep. 2327, 2007, pp. 19401947.

VI. BIOGRAPHIES

Yassine MAOUCHE (SM'12) was born in Chelghoum Laid, Algeria, in 1987. He received the Licence degree in electrical engineering from the University of Mentouri Constantine, in 2009 and the Master. degree from the University of Mentouri Constantine in 2011. He is currently working toward the Ph.D. degree on the diagnosis of faults of induction machines in the laboratory Laboratoire dElectrotechnique de Constantine. He is a member of the IEEE Industry Applications Society, the IEEE Industrial Electronics Society, and the IEEE Power Engineering Society.

Abdelfettah BOUSSAID (SM'12) was born in Ferdjioua MILA, Algeria, in 1988. He received the Licence degree in electrical engineering from the University of Mentouri Constantine, in 2009 and the Master. degree from the University of Mentouri Constantine in 2011. He is currently working toward the Ph.D. degree on the electrical engennering his interesting is on the active filtering of harmonics in electrical networks in the laboratory Laboratoire dElectrotechnique de Constantine. He is a member of the IEEE Industry Applications Society, the IEEE Industrial Electronics Society, and the IEEE Power Engineering Society.

Mohamed BOUCHERMA (M'12) received the B.Sc. degree in electrical engineering from Annaba University,Annaba, Algeria, in 1985, the M.Sc. degree in power system engineering from the University of Strathclyde, Glasgow, U.K., in 1989, and the Ph.D.degree in electrical engineering from The University of Shefeld, Shef field, U.K., in 1994. He is currently a Lecturer with the Department of Electrical Engineering, Mentouri University of Constantine, Constantine, Algeria. His main research interests are power systems and electrical machines.

Abdelmalek KHEZZAR (M'07) was born in Batna, Algeria, in 1969. He received the B.Eng. degree in electrical engineering from the University of Batna, Batna, in 1993 and the Ph.D. degree from Institut National Polytechnique de Lorraine, Nancy, France,in 1997.
In 2000, he joined the Mentouri University of Constantine, Constantine, Algeria, as a Lecturer with the Department of Electrical Engineering, where, since December 2008, he has been a Full Professor and, in April 2009, he was elected as Director of the laboratory "Laboratoire dElectrotechnique de Constantine." His main research interests are power electronics, drives, and analysis of electrical machine with special emphasis on fault diagnosis. Prof. Khezzar is a member of the IEEE Industry Applications Society, the IEEE Industrial Electronics Society, and the IEEE Power Engineering Society.

Electromagnetic and Temperature Field Analyses of Winding Short-circuits in DFIGs

Zheng Liu, Wenping Cao, Zheng Tan, Xueguan Song, Bing Ji, and Guiyun Tian

Abstract—The doubly fed induction generator (DFIG) plays an important role in the wind power generation. With the ever-increasing demand for improving the reliability and energy efficiency whilst reducing the maintenance and operational costs, there is a trend to develop the state-of-the-art condition monitoring technologies in wind applications especially for offshore applications. Winding short-circuits faults are among major electrical failures in DFIGs and can be caused by many reasons and can lead to undesirable heating to impact on the performance of the machine. They have not been fully understood due to the complexity of the problem which requires 3D electromagnetic and thermal fields to understand the fault mechanisms. This paper presents coupled electromagnetic and thermal field analyses of DFIGs with a focus on winding faults. Finite element tools are used for analyzing the characteristics of magnetic field, temperature distribution and heat flow during the healthy and faulty operations. This work can provide an insight into the DFIG's stator and rotor winding faults and suggestions for improvement in thermal design of the DFIG machines.

Index Terms—Doubly-fed induction generator (DFIG), finite element analysis (FEA), magnetic field, temperature field, winding short-circuits.

I. INTRODUCTION

The doubly fed induction generator (DFIG) is the dominant wind turbine machine technology in the wind energy industry, especially for medium- and large-scale installations. As a result, it is of critical importance to improve their reliability and energy efficiency as well as reducing their maintenance and operational costs.

Wind turbine generators are getting larger and more expensive, especially in offshore installations. More than a third of annual downtime of wind turbines is related to generator failures in the wind energy industry [1]. As a consequence, it is essential to fully understand the failure mechanisms of DFIGs and to improve their thermal performance subsequently.

In the literature, major machine faults are attributed to bearing faults, rotor winding failures, stator winding failures and cooling system faults [2]. Machines larger than 1MW tend to have more mechanical problems with the bearings and overheat issues [3] whilst smaller machines have higher likelihood of winding failures. Stator and rotor faults are generally caused by contamination or insulation breakdown

This work was supported by FP7 Marie Curie Exchanges Scheme (IRSES/318925).

The authors are with the School of Electrical and Electronic Engineering, Newcastle University, Newcastle upon Tyne, NE1 7RU, United Kingdom. (z.liu6@newcastle.ac.uk)

at various locations of connecting conductors and the banding, such as turn-turn, winding-winding, phase-phase, or winding-iron. In addition, the voltage spike arising from PWM control, end winding overheating during the dynamic wind speed variations, and the harsh offshore environment (corrosion and moisture) can also be contributing factors to cause insulation breakdowns or even machine failures [4]. A fault generally starts from a short circuit, creates local overheating and if left untreated then propagates onto adjacent windings. However, the deterioration of winding insulation and the hotspots may be detected at an early stage by appropriate condition monitoring measures. Two major condition monitoring technologies widely used are motor current signature analysis (MCSA) [5][6] and wavelet analysis [7]. They have been well developed and reported in the literature. This paper provides a study on their thermal performance with a focus on stator and rotor winding failures by using finite elements electromagnetic and temperature field analysis tools, namely, MagNet and ThermNet, respectively. A finite element model of the doubly fed induction generator is established for simulating electromagnetic and temperature fields impacted by winding faults. The characteristics of harmonic components in magnetic flux density of the air gap and the temperature distribution are compared between healthy and fault DIFGs.

II. MODELS OF INDUCTION GENERATORS

In this paper, a 3 phase 15 kW doubly fed induction generator has been modeled and built in the finite element software as an example to analyze electromagnetic and thermal performance. The machine simulation model and its specifications are shown in Fig. 1 and Table I, respectively.

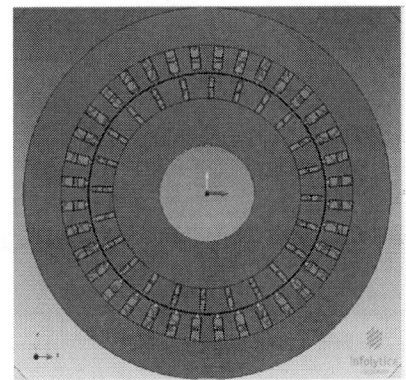

Fig. 1. 2D simulation model of DFIG.

978-1-4799-0024-4/13 $31.00 © 2013 IEEE

TABLE I
SPECIFICATIONS OF THE DFIG

Item	Value
Rate power	15 kW
Pole number	4
Phase number	3
Stator rated current	31 A
Stator rated voltage	380 V
Rated speed	1460 rpm
Stator slot	36
Rotor slot	24
Outer diameter of stator	260mm
Inner diameter of stator	170mm
Airgap length	0.8mm

III. ELECTROMAGNETIC ANALYSIS

Electromagnetic performance of the DFIG is evaluated at healthy and faulty conditions.

A. Healthy Conditions

Based on the actual parameters for this 15kW off-the-shelf machine, the numerical model is built and analysed through the simulation. Figs. 2 and 3 show the 2D finite element mesh and magnetic flux distribution of this DFIG, respectively. The grid has 252318 nodes to increase the accuracy especially in the slot and air gap.

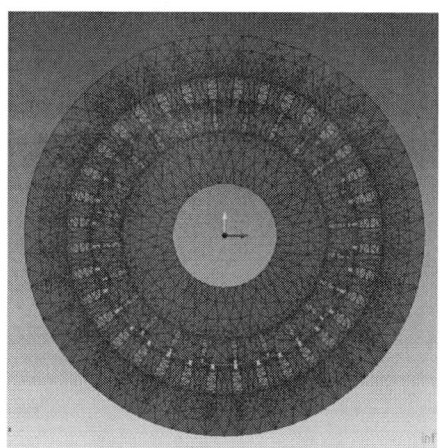

Fig. 2.   Finite element mesh of the DFIG model

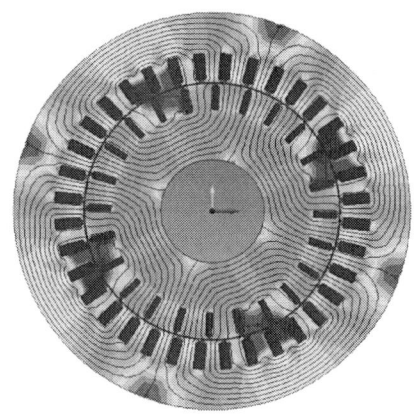

Fig. 3.   Magnet flux distribution of the DFIG model

Using the toolbox in the MagNet, the air gap magnetic flux density of the DFIG can be obtained where the radius is chosen from the middle position between the inside radius of the stator and the outside radius of the rotor. In this case, the air gap magnetic flux density of the healthy DIFG at no-load condition is observed in Fig. 4.

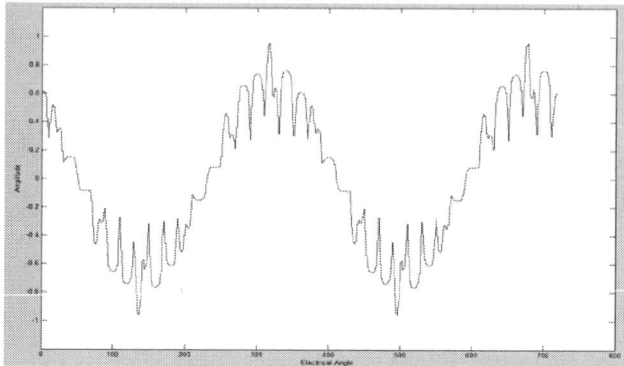

Fig. 4.   Air gap magnetic flux density of the DFIG on no-load

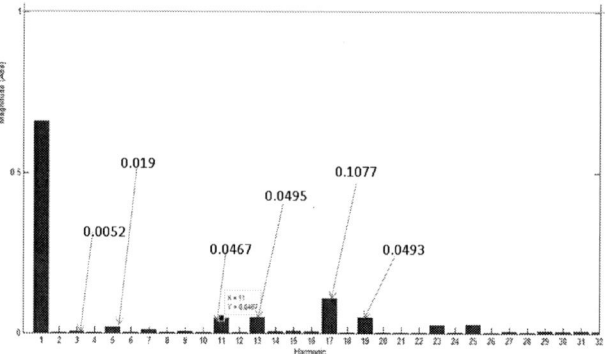

Fig. 5.   Harmonic components in the air gap flux density

By using fast Fourier transform (FFT) in MatLab, the harmonics of the air gap flux density in the DFIG model are attained, as shown in Fig. 5. From the results, it can be seen that the even harmonic components are very low whilst the odd harmonic components are clearly present, including 17th, 13th, 19th and 11th in the order of reducing magnitude. As a result, these odd spatial harmonic components should be minimized by optimizing the slot shape.

These simulation results are used to validate the DFIG model built in MagNet for the healthy operational condition of the DFIG prior to introducing any winding faults.

B. Faulty Conditions

In this part of simulation, some winding short-circuits are arbitrarily created and applied to the rotor of the DFIG. Since the turn-turn short circuits are the most common type of faults, they are simulated in the DFIG model with a 4% short-circuit in the rotor winding (23 turns).

The flux distribution in the air gap is a straightforward indicator of the DFIG's magnetic performance where all the changes are reflected on it directly. The air gap flux density and its harmonic components are depicted in Figs. 6 and 7, respectively.

978-1-4799-0024-4/13 $31.00 © 2013 IEEE   270

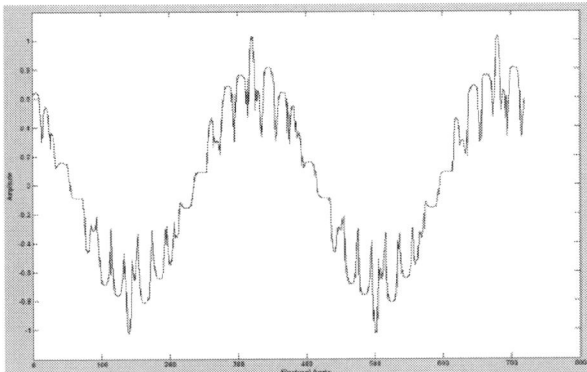

Fig. 6.   The airgap magnetic flux density with 4% shorted winding on the rotor

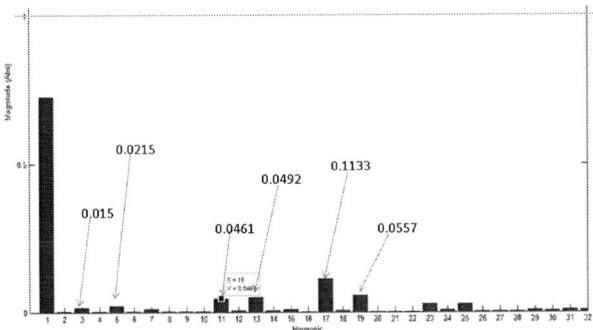

Fig. 7.   Harmonic components in the air gap flux density with 4% shorted winding on the rotor

The differences in the amplitudes of harmonics in the air gap flux density between healthy and fault DFIG conditions are shown in Table II. As the 11th and 13th harmonic components are closely related to the slot/teeth, their amplitudes do not change much. However, the 3rd, 5th 17th and 19th harmonic components change more notably with the faults between two situations.

TABLE II
HARMONIC CONTENTS OF THE AIR GAP FLUX DENSITY
(% OF THE FUNDAMENTAL COMPONENT)

Major harmonics	Healthy condition	Short-circuit condition	Difference in %
3rd	0.0052	0.015	188
5th	0.019	0.0215	13
11th	0.0467	0.0461	1.3
13th	0.0495	0.0492	0.6
17th	0.1077	0.1133	5
19th	0.0493	0.0557	13

IV. THERMAL SIMULATIONS

Any winding faults in DFIGs will impact on the thermal performance of the machine and the undesirable heating may cause catastrophic damage to the machine if left untreated.

A.   3D Temperature field of the DFIG

In the magnetic analysis, the DFIG machine can be analysed using 2D magnetic model with reasonable accuracy. But for the thermal problems, it is needed to use 3D models and also to consider the ventilation and heat transfer with the machine. The DFIG model in this part has been added some ducts in the rotor for cooling including totally enclosed air-cooling convection.

The following assumptions are made for temperature filed solving:

- All the slice faces of the DFIG model are perfectly insulated.
- The insulators in the slot are identical.
- The heat convection coefficient is identical at all the position of the air gap.

The reason for temperature increase in the machine is the heat losses. The total loss is given by:

$$P_{all} = P_{1cu} + P_{2cu} + P_{iron} + P_M + P_S \qquad (1)$$

where P_{all} is the total loss, P_{1cu} is the copper loss in the stator, P_{2cu} is the copper loss in the rotor, P_{iron} is the iron loss, P_M is the mechanical loss, and P_S is the stray load loss. By determining the individual loss in the generator, it is then possible to sum up to find the total loss. Then all the heat sources of different generator components can be set by referring corresponding losses.

In an air-cooled machine, the heat radiation is very small and thus neglected in this study for simplification. The main routes for heat transfer are through convection and conduction. The 3D steady state heat conduction is:

$$\frac{\partial}{\partial x}\left(k_x \frac{\partial T}{\partial x}\right) + \frac{\partial}{\partial y}\left(k_y \frac{\partial T}{\partial y}\right) + \frac{\partial}{\partial z}\left(k_z \frac{\partial T}{\partial z}\right) = -Q \qquad (2)$$

where k_x, k_y, k_z are the heat conduction coefficient in the x-, y- and z- directions, respectively and Q is the heat generation ratio.

Additionally, the heat convection coefficient h is also need to be calculated to set the boundary conditions for the outer surface of the stator core. This is calculated by [8]:

$$h = \rho\, cp\, D\, \frac{v}{4L * [1 - EXP(-M)]} \qquad (3)$$

$$M = 0.1448L^{0.96}/D^{1.16}\{k/(\rho\, cp\, v)\}^{0.214} \qquad (4)$$

Where D is the outside diameter, L is the characteristic length of the surface; v is the fluid velocity, ρ is the fluid density, c_p is the fluid specific heat capacity, k is the thermal conductivity.

B.   Thermal modeling of the Healthy DFIG

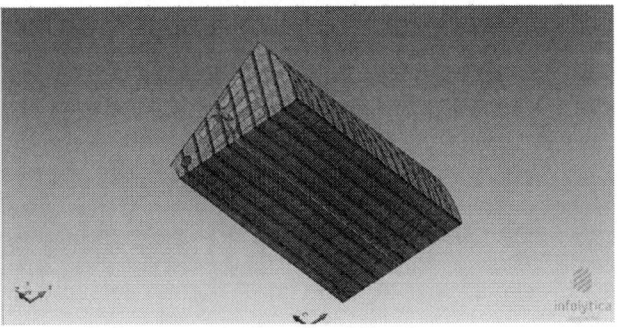

Fig.8.   3D model of the DFIG

As heat transfers through x, y, and z directions, the thermal simulation cannot be fully explained by the 2D model so that a time-consuming 3D model is needed. Fig. 8 shows the 3D model of the DFIG by using ThermNet (coupled with MagNet). To reduce the solving time of thermal simulations, the 3D DFIG model is built with a portion of the motor body for one phase with one pole. The types of convection heat transfer used in the DFIG are nature convection and forced convection.

By adding all the losses calculated from MagNet, the temperature field distribution of DFIG with nature convection is illustrated in Fig. 9. It can be observed that the hottest part is in the center of the rotor, with a recorded temperature of 125.48°C.

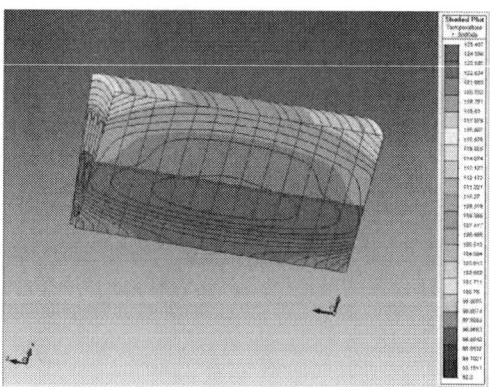

Fig.9.   3D temperature field with nature convection

On the basis of the nature convection model, the forced convection can be added in the FEA. Generally, the forced convection in DFIGs can be achieved by installing a fan on the shaft and rotating it with the rotor. In this case, the air is blown into the airgap and ventilation duct. In the FEA, it is assumed that the direction of forced convection is from the left of the model to the right in all diagrams. When air flows through the DFIG, the convection heat transfer coefficient h is different in the axial direction. Consequently, the simulation model has been sliced into 13 sections along the axial direction. According to Eqs (3) and (4), each part has been set a different convection heat transfer coefficient. Figs. 9 and 10 show the 3D temperature distribution of the healthy DFIG model.

After a 5.5-hour simulation, the temperature field of the DFIG mode has reached a steady state. Figs. 10 and 11 demonstrate that the hottest part of this whole model is the outlet of the air path which leads to a probability of winding insulation damage. But the highest temperature in the generator is only 107.5°C, which is already much cooler than that with the nature convection. Moreover, the temperature distributions of three different coils in the stator are the same; so are the two coils in the rotor.

By setting the losses calculated from the faulted DFIG model in MagNet (with winding short circuits), the temperature field of faulted DFIG model is obtained, as shown in Figs. 12 and 13. The short circuit occurs only in one coil of the rotor winding. From Fig. 13, the temperature

field distribution of the short-circuit coil is different to that of the healthy one. More specifically, the area of the highest temperature (117.7°C) in the rotor shorted coil is larger than the healthy case, which is easier to cause a winding failure.

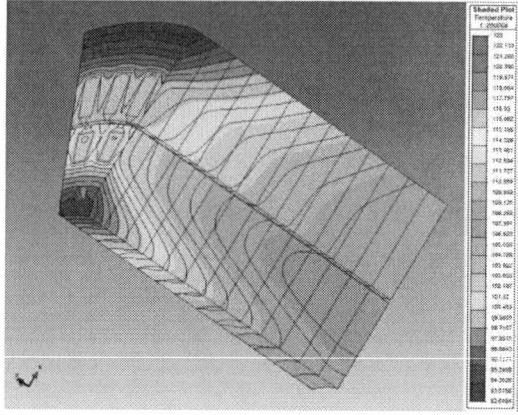

Fig.10.   3D temperature field with forced convection

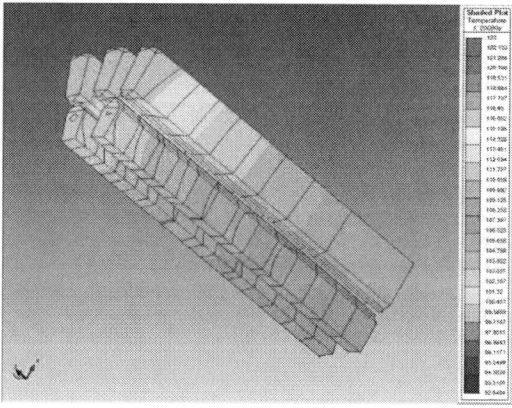

Fig.11.   3D temperature field under healthy conditions

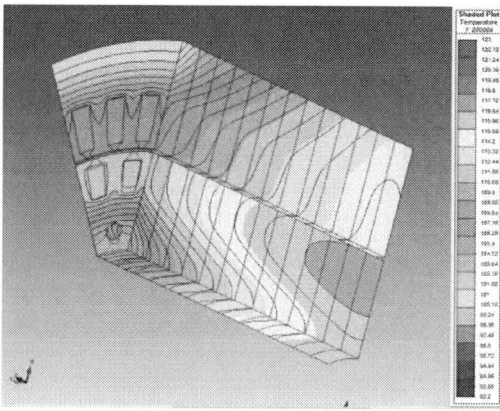

Fig.12.   3D temperature field with rotor short-circuits

Comparing the healthy and fault DFIG models, the major difference is the heat source which is induced by the current flowing through the copper winding. During a winding short circuit event, the current generates heat and creates hotspots at the beginning. When stabilized, the winding temperature in the fault model is much higher than that in the healthy model, giving rise to the likelihood of insulation failure and fault propagation.

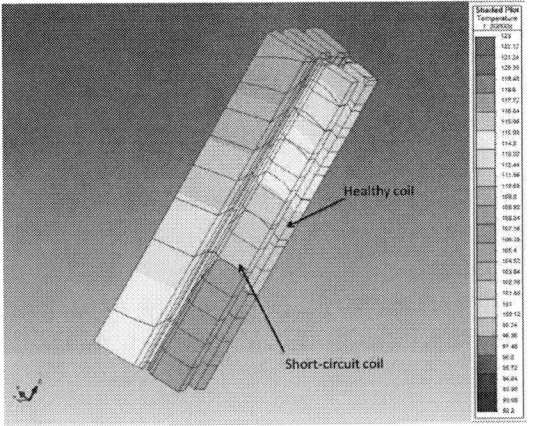

Fig.13.   3D temperature field with rotor short circuits

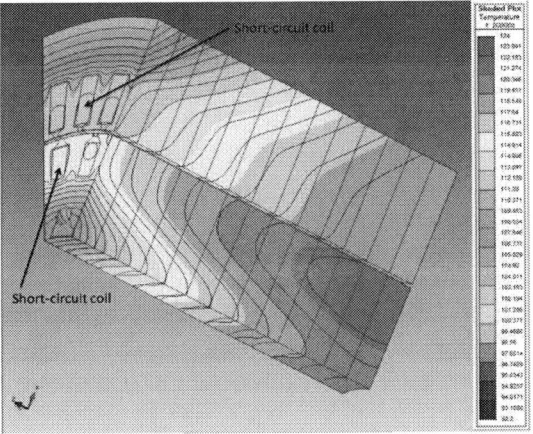

Fig.14.   3D temperature field with short circuits in both the rotor and stator windings

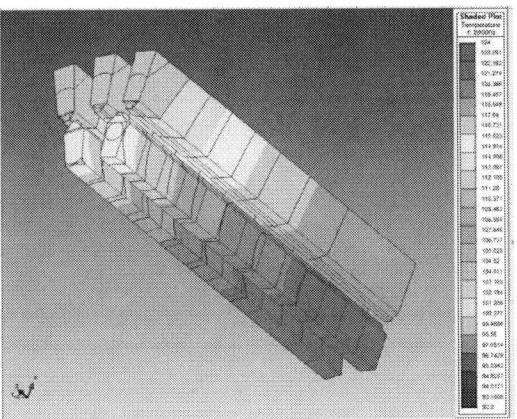

Fig.15.   3D temperature field of with short circuits in both the rotor and stator windings

With the turn-to-turn winding fault occurred in the stator or the rotor, the excessive heat generated from short-circuits will diffuse through the airgap so that the average temperature of the generator increases, leading to hotspots, insulation degradation or new short-circuits in the neighboring windings of the DFIG. When the winding faults present in both the stator and the rotor, damages can be more severe or even catastrophic to the generator. Figs. 14 and 15 show the temperature field distribution of the DIFG with short-circuits occurring at one rotor coil and one stator coil. From simulation results with the rotor and stator short circuits, the highest temperature (123.09°C) is similar to the case of natural convection machine but much higher than the previous fault simulation in air-cooled cases. Furthermore, the isotherm of the stator windings at the inlet of the air is quite different from the healthy model and fault rotor winding models. Comparing with the two healthy stator windings, the trend of the bulging isotherm in the faulty windings is more significant than the healthy ones.

V. CONCLUSION

This paper has presented a finite element analysis of electromagnetic and temperature fields for the DFIG with different winding faults. The simulation results show that the 3rd, 5th 17th and 19th harmonic components are affected by winding short circuits whilst 11th and 13th are less sensitive to short-circuit faults. By using the heat losses calculated from the magnetic analysis, the thermal analysis can be carried out. Simulation results show the temperature of hotspots in the fault DFIG model is much higher than the healthy ones. In this paper, analyzing electromagnetic and temperature fields are capable of providing an in-depth understanding of failure mechanisms of doubly-fed induction generators and also providing an guidance to develop condition monitoring technologies for detecting the DFIG's faults. In the future work, the hardware for winding fault detection will be developed and tested on a 30kW prototype machine which can also be used to validate the simulation models developed in this paper.

VI. References

[1] R. Johan and B. Lina Margareta, "Survey of failures in wind power systems with focus on Swedish wind power plants during 1997 & ndash; 2005," *IEEE Transactions on Energy Conversion,* vol. 22, pp. 167-173, 2007.

[2] K. Alewine and W. Chen, "A review of electrical winding failures in wind turbine generators," *IEEE Electrical Insulation Magazine,* vol. 28, pp. 8-13, 2012.

[3] F. Spinato, P. J. Tavner, G. J. W. van Bussel, and E. Koutoulakos, "Reliability of wind turbine subassemblies," *IET Renewable Power Generation,* vol. 3, pp. 387-401, 2009.

[4] G. Gao and W. Chen, "Design challenges of wind turbine generators," in *IEEE Electrical Insulation Conference (EIC),* pp. 146-152, 2009.

[5] D. Shah, S. Nandi, and P. Neti, "Stator-interturn-fault detection of doubly fed induction generators using rotor-current and search-coil-voltage signature analysis," *IEEE Transactions on Industry Applications,* vol. 45, pp. 1831-1842, 2009.

[6] M. N. Zaggout, P. J. Tavner, and L. Ran, "Wind turbine condition monitoring using generator control loop signals," *6th IET International Conference on Power Electronics, Machines and Drives (PEMD 2012),* pp. 1-6, 2012.

[7] Y. Gritli, A. Stefani, F. Filippetti, and A. Chatti, "Stator fault analysis based on wavelet technique for wind turbines equipped with DFIG," *2009 International Conference on Clean Electrical Power,* pp. 485-491, 2009.

[8] F. Heiles, "Design and arrangement of cooling fins," *Elecktrotecknik und Maschinenbay,* vol. 69, No. 14, July 1952.

Induction Motor Stator Faults Diagnosis by Using Parameter Estimation Algorithms

Fang Duan, Rastko Živanović
School of Electrical and Electronic Engineering
The University of Adelaide, South Australia, Australia
fduan@eleceng.adelaide.edu.au, rastko@eleceng.adelaide.edu.au

Abstract—**Parameter estimation is a cost-effective method for fault detection of induction motors. This method is based on detecting change of the characteristic parameters at presence of fault. However, the challenge of parameter estimation is nonlinearity of a machine model which results in multiple local minima involved during the computation process. This paper investigates the suitability of local and global search methods to be used in the estimation of characteristic parameters that are indicating stator short circuit faults. Results of practical case studies are presented where two search methods (local and global) are evaluated and compared. A further study in noisy environment proves the feasibility of diagnosing the fault based on stator currents with low signal to noise ratio.**

I. INTRODUCTION

Induction motors play a critical role in many industrial processes due to their efficient and cost-effective performance. They are widely utilized in industrial drives, providing core capabilities essential to industry success. Because of such great importance of induction machines, intensive efforts have been devoted to the investigation of reliable methods for the machine fault diagnosis.

Condition monitoring is the process of monitoring parameters describing condition of a machine while in operation. Through the parameters monitoring process, motor failures can be diagnosed before irreversible damage has occurred. The condition monitoring can be broadly categorized into two basic types: intrusive and non-intrusive techniques [1]. The intrusive method is complex and costly because particular sensors and measurement equipments have to be employed to monitor the deviation of air-gap torque [2], magnetic field [3], vibration [4] and so on. For the non-intrusive method, the condition monitoring can be achieved by processing of supply voltage and current signals. Therefore, the advantages of the non-intrusive method are reduced cost and simplicity.

Motor current signature analysis (MCSA) is one of the most widely applied non-intrusive methods. Motor operating conditions can be obtained by monitoring stator current [5] and analyzing it in either time-domain or frequency-domain. Frequency-domain analysis can provide more detailed information of machine's status [6]. Several of induction motor faults detection and identification methods are based on

Fast Fourier Transform (FFT) spectral signature analysis [7]–[9]. However, the harmonic frequency components are too small to be detected under the low intensity fault condition, which limits application of those methods. On the other hand, it was reported that the artificial intelligence, such as fuzzy logic [10], [11], genetic algorithms [12], [13], neural network [14], [15] as well as techniques based on the Bayesian classifier approach [16] can improve the diagnostic accuracy significantly. However, their heavy computational burden will increase both system cost and detection time.

The induction motors are symmetrical structures and any kind of fault will break their symmetrical property. The change of this balance will result in the drift of characteristic parameters, which offers a method to detect motor fault by monitoring these parameters. In [17], the stator short circuit fault has been successfully detected through the estimation of characteristic parameters from the recorded stator current. In that paper, the parameter estimation technique is based on a MATLAB/SIMULINK induction motor model presented in [18] and a global direct search algorithm, named Hyperbolic Cross Points (HCP) algorithm [19]. The algorithm has been employed previously to find global minima in the estimation of DC offset time constants on transmission line [20] and diagnosis of broken rotor bar fault in induction motor [21]. In continuation of our previous work, in this paper, we test two industry-proven optimization algorithms implemented in the *SIMULINK Parameter Estimation Toolbox* [22]. We applied those algorithms to estimate characteristic parameters of stator short circuit faults in an induction motor. The parameter estimation has been conducted in $\alpha\beta$ coordinate reference frame to simplify the analysis process.

After a brief review of mathematical model of a machine with stator short circuit fault in Section II, the parameter estimation algorithm is described and discussed in Section III. Apart from two characteristic parameters indicating stator short circuit faults, additional two parameters are defined for the calibration purposes. In Section IV, these four parameters are estimated by using both local and global search methods [23], implemented in the *SIMULINK Parameter Estimation Toolbox*. Section V investigates the parameter estimation in a noisy environment. It is demonstrated that stator short circuit faults have been correctly identified from

978-1-4799-0024-4/13 $31.00 © 2013 IEEE

the recorded voltage and current by choosing the global search method while the local method performance depends strongly on the provided initial search point. The results also indicate the feasibility of diagnosing the fault from the recorded stator currents in the noisy environment.

II. MATHEMATICAL MODEL OF A STATOR SHORT CIRCUIT FAULT IN AN INDUCTION MOTOR

The stator short circuit fault in an induction motor is normally characterized by two parameters, localization parameter θ_f and fault level μ_f. In the previous research, an induction motor model with stator short circuit fault has been built based on a mathematical model in [18], [24]. The proposed model has been validated by comparing simulated and measured stator current with different number of shorted turns and load levels. In this paper, the model has been adapted and utilized in the parameter estimation algorithm. For the sake of convenience, the mathematical model of induction motor with short circuit fault is briefly described in matrix form as follows.

$$\frac{d}{dt}\begin{bmatrix}\boldsymbol{\lambda}_s^{abc}\\\boldsymbol{\lambda}_r^{abc}\\\lambda_f\end{bmatrix} = \begin{bmatrix}\boldsymbol{u}_s^{abc}\\0\\0\end{bmatrix} - \boldsymbol{r}\begin{bmatrix}\boldsymbol{i}_s^{abc}\\\boldsymbol{i}_r^{abc}\\i_f\end{bmatrix}, \qquad (1)$$

$$\begin{bmatrix}\boldsymbol{\lambda}_s^{abc}\\\boldsymbol{\lambda}_r^{abc}\\\lambda_f\end{bmatrix} = \begin{bmatrix}\boldsymbol{L}_{ss}^{abc} & \boldsymbol{L}_{sr}^{abc} & \boldsymbol{L}_{sf}^{abc}\\\boldsymbol{L}_{rs}^{abc} & \boldsymbol{L}_{rr}^{abc} & \boldsymbol{L}_{rf}^{abc}\\\boldsymbol{L}_{fs} & \boldsymbol{L}_{fr} & L_f\end{bmatrix}\begin{bmatrix}\boldsymbol{i}_s^{abc}\\\boldsymbol{i}_r^{abc}\\i_f\end{bmatrix}, \qquad (2)$$

where

$\boldsymbol{\lambda}_s^{abc}$, $\mathbf{u}_s^{abc}$ and $\mathbf{i}_s^{abc}$	stator flux, voltage and current
$\boldsymbol{\lambda}_r^{abc}$, $\mathbf{u}_r^{abc}$ and $\mathbf{i}_r^{abc}$	rotor flux, voltage and current
λ_f and i_f	short circuit flux and current

$$\mathbf{r} = \begin{bmatrix}\boldsymbol{r}_s & 0 & 0\\0 & \boldsymbol{r}_r & 0\\0 & 0 & r_f\end{bmatrix} \quad \text{resistance matrix}$$

$\boldsymbol{r}_s = r_s \times \boldsymbol{I}$	stator resistance
$\boldsymbol{r}_r = r_r \times \boldsymbol{I}$	rotor resistance
$r_f = \mu_f \cdot r_s$	short circuit resistance

$$\boldsymbol{I} = \begin{bmatrix}1 & 0 & 0\\0 & 1 & 0\\0 & 0 & 1\end{bmatrix} \quad \text{Identity matrix}$$

The stator-to-stator winding inductance $\boldsymbol{L}_{ss}$, rotor-to-rotor winding inductance $\boldsymbol{L}_{rr}$ and short-circuit inductance L_f are given by [24],

$$\boldsymbol{L}_{ss} = \begin{bmatrix}L_g + L_p & -L_p/2 & -L_p/2\\-L_p/2 & L_g + L_p & -L_p/2\\-L_p/2 & -L_p/2 & L_g + L_p\end{bmatrix}, \qquad (3)$$

$$\boldsymbol{L}_{rr} = \begin{bmatrix}L_p & -L_p/2 & -L_p/2\\-L_p/2 & L_p & -L_p/2\\-L_p/2 & -L_p/2 & L_p\end{bmatrix}, \qquad (4)$$

$$L_f = \mu_f^2(L_p + L_g), \qquad (5)$$

where L_p and L_g are mutual inductance and self inductance referred to the stator, respectively.

The stator, rotor and short-circuit mutual inductances can be expressed as [24],

$$\boldsymbol{L}_{sr} = L_p\begin{bmatrix}cos(\theta_r) & cos(\theta_r + 2\pi/3) & cos(\theta_r - 2\pi/3)\\cos(\theta_r - 2\pi/3) & cos(\theta_r) & cos(\theta_r + 2\pi/3)\\cos(\theta_r + 2\pi/3) & cos(\theta_r - 2\pi/3) & cos(\theta_r)\end{bmatrix}, \qquad (6)$$

$$\boldsymbol{L}_{sf} = \mu_f L_p\begin{bmatrix}cos(\theta_f)\\cos(\theta_f - 2\pi/3)\\cos(\theta_f + 2\pi/3)\end{bmatrix}, \qquad (7)$$

$$\boldsymbol{L}_{rf} = \mu_f L_p\begin{bmatrix}cos(\theta_f - \theta_r)\\cos(\theta_f - \theta_r - 2\pi/3)\\cos(\theta_f - \theta_r + 2\pi/3)\end{bmatrix}, \qquad (8)$$

$$\boldsymbol{L}_{rs} = \boldsymbol{L}_{rs}^T, \quad \boldsymbol{L}_{fs} = \boldsymbol{L}_{sf}^T, \quad \boldsymbol{L}_{fr} = \boldsymbol{L}_{rf}^T, \qquad (9)$$

where θ_r is rotor angle.

In order to reduce the number of model variables, Concordia transformation is employed to convert from three phase quantities abc to two phase quantities $\alpha\beta$ coordinate. The transformation is defined as,

$$\begin{bmatrix}u_\alpha\\u_\beta\end{bmatrix} = \sqrt{\frac{2}{3}}\begin{bmatrix}1 & -1/2 & -1/2\\0 & \sqrt{3}/2 & -\sqrt{3}/2\end{bmatrix}\begin{bmatrix}u_a\\u_b\\u_c\end{bmatrix}. \qquad (10)$$

III. STATOR SHORT CIRCUIT FAULT DIAGNOSIS USING PARAMETER ESTIMATION

The mathematical model indicates that a stator short circuit fault can be described by the fault localization θ_f and fault level μ_f. Therefore, it is feasible to detect and locate a stator short circuit fault by using the parameter estimation method to identify θ_f and μ_f.

In this paper, the methods used to estimate these two characteristic parameters are based on the optimization techniques [23], [25] implemented in *SIMULINK Parameter Estimation Toolbox* [22]. The parameter estimation case study is devised by using the three phase 800 W induction motor. The parameters of the motor are tabulated in Table. I. Fig. 1 presents a schematic diagram of the parameters estimation procedure. The SIMULINK machine model is the same as the one developed in [18], and it will not be discussed here in detail. The recorded three phase voltage and current signals are used as input to the model. If the voltage signal is not available, it is also feasible to use simulated voltage signal, which will be addressed in Section V. By using Eq. (10), the voltage and current signals are converted from abc coordinate

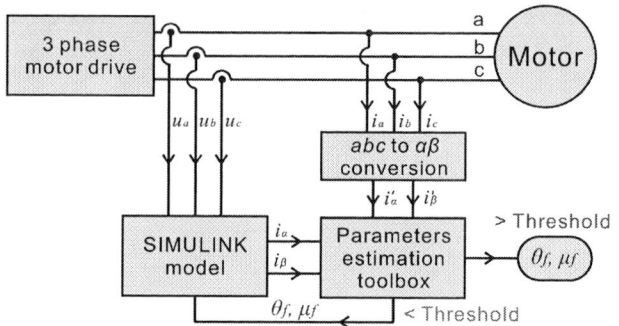

Fig. 1. Induction motor stator winding short circuit fault detection using *SIMULINK Parameter Estimation Toolbox*.

TABLE I
INDUCTION MOTOR PARAMETERS

Ouput power	800 W
Line voltage	380 V
Rotor inertia	0.0025 Kg m^2
Rated frequency	50 Hz
Number of poles	4
Rated power factor	0.74
Stator winding resistance	8.4 ohm
Stator winding reactance	10.3 ohm
Stator magnetizing reactance	137.5 ohm
Referred rotor winding resistance	8.2 ohm
Referred rotor winding reactance	10.3 ohm

to $\alpha\beta$ reference frame in the model to simplify the analysis.

In the initial stage, the fault level μ_f is set to zero, representing a healthy motor without any stator short circuit fault. Using the recorded voltage signals, the simulated stator current i_α and i_β in $\alpha\beta$ coordinates are shown in Fig. 2(a) and (b), respectively. The measured stator currents i_a, i_b and i_c are also converted to i'_α and i'_β, as shown in Fig. 2(c) and (d). It can be noted that the simulated currents are in good agreement with measured ones. The dashed lines at 0.5 s of α and β plots reveal only slight offset in time between simulated and measured currents. The offset time T_D will be estimated as one of the parameters. There are also small difference in the peak values between measured and simulated currents, resulting from environmental effect and measurement error. A parameter K is defined to solve this issue. To these implement delay and scaling functions, a new block is added to the model proposed in [18]. The block is shown in Fig. 3. The simulated $i_{\alpha,\beta}$ is delayed T_D and then is multiplied by a scaling parameter K. Therefore, there are four parameters to be estimated in the system. These variables and related ranges are listed in the Table. II.

The recorded voltage $u_{a,b,c}$ and post-processed measured current $i'_{\alpha,\beta}$ are imported into the *SIMULINK Parameter*

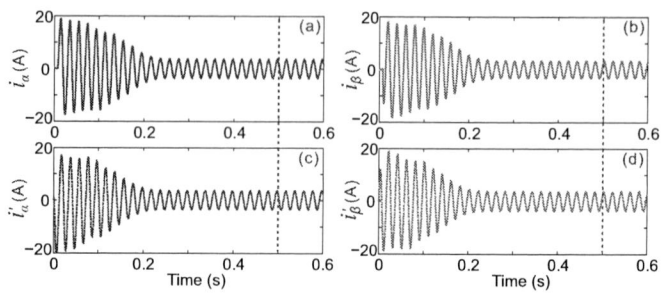

Fig. 2. The simulated and measured stator current of healthy motor in $\alpha\beta$ coordinate. (a) and (b) Simulated current; (c) and (d) Measured current.

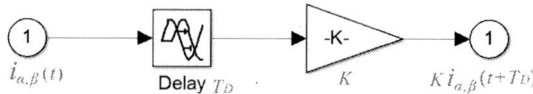

Fig. 3. The alignment of simulated currents with measurement in time series and amplitude.

Estimation Toolbox as Input and Output Data, respectively. Four parameters θ_f, μ_f, T_D, and K are defined as variables to be estimated. If the difference between $i'_{\alpha,\beta}$ and $i_{\alpha,\beta}$ is larger than the error threshold, the $i_{\alpha,\beta}$ will be generated again by using a set of new parameters given by the selected algorithm. The process iterates until the error threshold is satisfied. The stator short circuit fault can be revealed by the estimated characteristic parameters θ_f and μ_f.

The toolbox offers two local search methods (gradient descent and nonlinear least squares) and one global search method (pattern search) [23], [25]. The gradient descent method is a first-order local search algorithm, in which the local minimum. is detected by taking steps proportional to the negative of the gradient. The algorithm has been employed in the optimization of induction motor control [26]. The nonlinear least squares method is commonly utilized to identify a non-linear model with n unknown parameters base on a set of m observations ($m > n$). For example, Wang et al. [27] used this method to identify the parameters of an induction motor. As a global search method, pattern search method finds minimum by generating a sequence of points towards an optimal point [28]. The values of the objective function of the sequence points either decrease or keep constant. The parameter estimation methods of gradient descent, nonlinear least squares, pattern search methods have been applied in the parameter estimation.

TABLE II
UNKNOWN PARAMETERS AND THEIR RANGES

T_D	Offset time	[0 0.02]
K	Amplifier	[0.5 1.5]
θ_f	Fault localization	0, $2\pi/3$, $4\pi/3$
μ_f	Fault level	[0 50%]

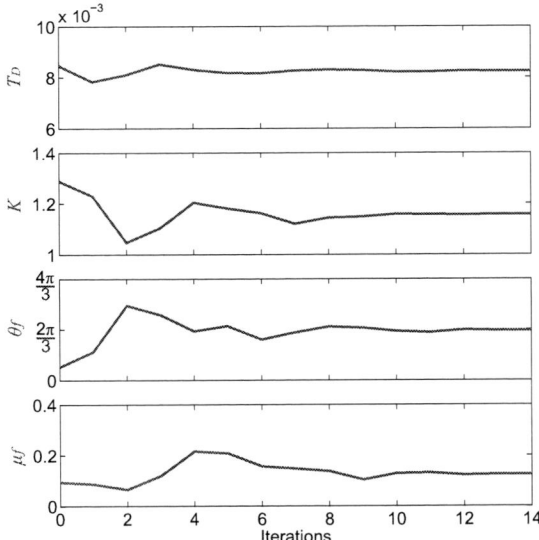

Fig. 4. Trajectories of estimated parameters with the start points close to the global minima.

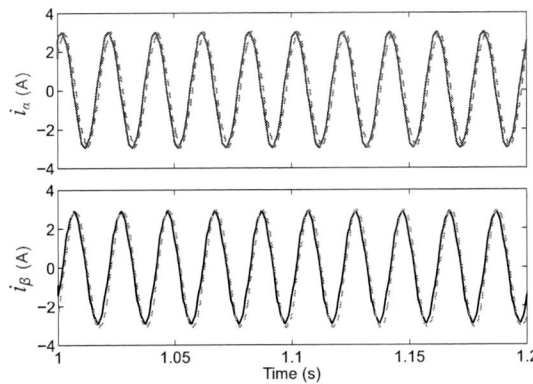

Fig. 5. Comparison between simulated (continue line) and measured (dash line) current at $\alpha\beta$ coordinate.

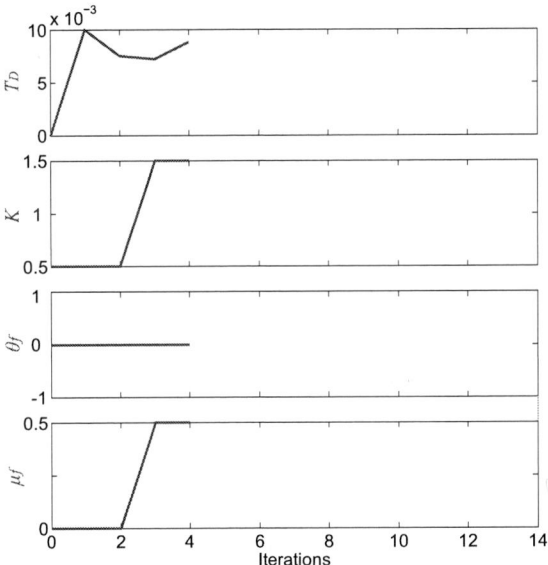

Fig. 6. Trajectories of estimated parameters with the start points far from the global minima.

IV. EXPERIMENT RESULTS

The experiment data of 10% ($\mu_f = 0.1$) stator short circuit fault in phase b ($\theta_f = 2\pi/3$) is employed to evaluate the proposed diagnostic technique based on parameter estimation. The estimation is done by using local and global search methods as discussed in section III. The starting point in the parameter space is very critical for local search method. Fig. 4 shows the trajectories of estimated parameters by using gradient descent method with active-set algorithm [25]. With the start points that are close to the global minima, the estimated parameters are comparable with the real values. Fig. 5 shows that the simulated currents are almost overlap with the measured currents. If the start points are far from the global minima, the algorithm may be trapped into local minima as shown in Fig. 6. The trajectories indicate that the algorithm only iterates in 4 steps to reach the threshold. Similar results have been obtained by using the Nonlinear Least Squares method [25]. In these parameter estimation processes, the data window length is equal to 1 s (i.e. from 1 to 2 s). To investigate the effect of the data window length, the length is increased to 2 s. The simulation results shown in Fig. 7 reveals that the increased length has negligible improvement of the estimation accuracy. However, it will increase the number of iteration steps and hence the computation time.

The global search method, the Pattern Search [23], is applied to estimate parameters θ_f, μ_f, T_D, and K. The start values for these parameters are set to be the same as for the gradient descent and least squares methods (see initial values in Fig. 6). In Fig. 8, the global minima is detected after 49 iterations of the pattern search method. The estimated parameters are very close to the real parameters. Although the global search method requires more computation time compared to a local search method, it guarantees robustness in parameter estimation.

V. PARAMETER ESTIMATION IN NOISY ENVIRONMENT

The recorded phase voltages and stator currents are with high ratio of signal to noise (S/N) because these signals have been measured locally in the experiment. However, the transmitted signals may be distorted by noise in the transmission in the remote monitoring and control system. The S/N could be further reduced during the process of data format conversion [29]. To study the feasibility of using the parameter estimation method in the remote monitoring and control system, the Gaussian white noise is artificially added to the recorded signals.

The symmetrical motor and 10% stator winding circuit short experimental setups are available in the current

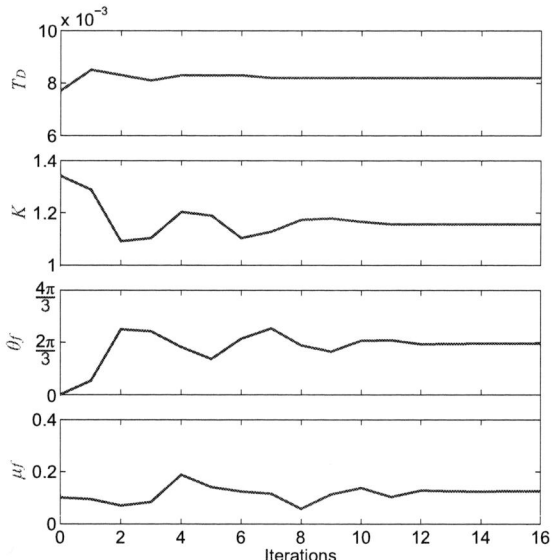

Fig. 7. Trajectories of estimated parameters with data length of 2 s.

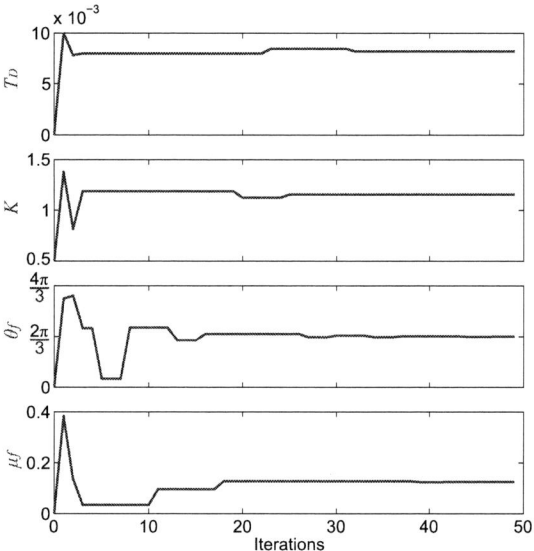

Fig. 8. Trajectories of estimated parameters by using the global search method.

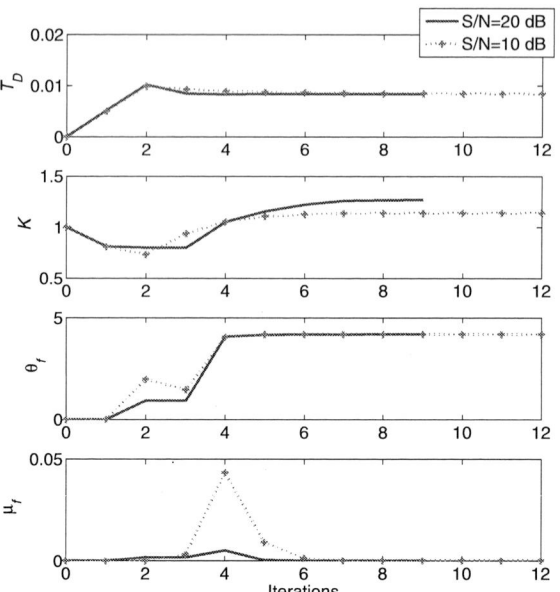

Fig. 9. Trajectories of the estimated parameters using recorded voltage and stator current with S/N of 20 and 10 dB.

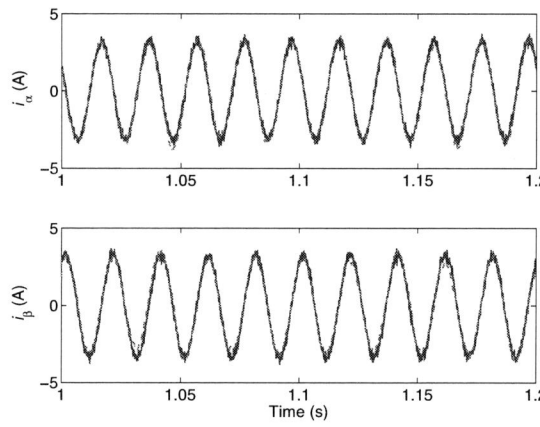

Fig. 10. Comparison between recorded stator current (S/N 20 dB) and simulated current using estimated parameters. The parameters are estimated based on recorded voltage and stator current with S/N of 20 dB.

measurement. The data of symmetrical motor and 10% stator winding circuit short are utilized in this paper to verify the proposed parameter estimation method. The further research will include low percentage fault level and remote motor monitoring and control system [29]. The recorded currents of the symmetrical motor have been utilized to study parameter estimation in the noisy environment. Fig. 9 shows the trajectories of the estimated parameters using recorded voltage and stator current with S/N of 20 and 10 dB. The toolbox needs more iterations to estimate parameters when the low S/N signals are applied. In both cases, the estimated fault level μ_f indicates the symmetrical motor. The noise polluted currents of S/N 20 and 10 dB as well as the corresponding simulated currents are shown

in Fig. 10 and 11, respectively. The good alignment of these currents in terms of amplitude and phase indicates the successfully estimation of the offset time T_D and amplifier K.

The measurement of the phase voltages is not always available for short circuit diagnosis in some plants. Therefore, the parameter estimation is also conducted only based on the recorded stator currents. The input three phase voltages V_a, V_b, and V_c of the model are calculated based on the motor parameters in Table I using the following equations,

$$
\begin{aligned}
V_a &= V_m \cos(2\pi f t + \phi) \\
V_b &= V_m \cos(2\pi f t - 2\pi/3 + \phi) \\
V_c &= V_m \cos(2\pi f t + 2\pi/3 + \phi)
\end{aligned}
\tag{11}
$$

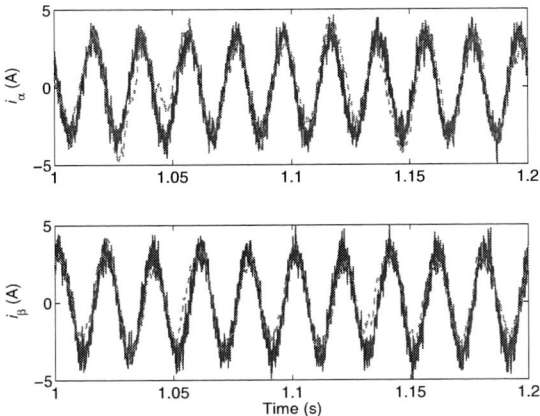

Fig. 11. Comparison between recorded stator current (S/N 10 dB) and simulated current using estimated parameters. The parameters are estimated based on recorded voltage and stator current with S/N of 10 dB.

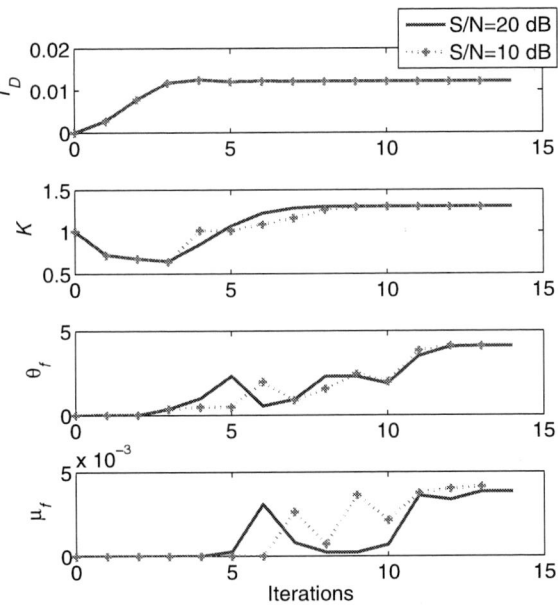

Fig. 12. Trajectories of estimated parameters using recorded stator current only with S/N of 20 and 10 dB.

where V_m is the magnitude of phase voltage, f is the rated frequency in Hz, and ϕ is the initial phase. The phase difference between the measured and simulated currents is controlled by the estimated parameter off time T_D. Thus, the initial phase ϕ is set to 0. The estimated parameters in Fig. 12 are very close to these values in Fig. 9. The simulated stator currents signals are free from noise since the calculated three phase voltages are employed, as shown in Fig. 13 and 14. The results in these figures reveal that it is feasible to diagnose stator winding circuit fault by using stator current only in the noisy environment.

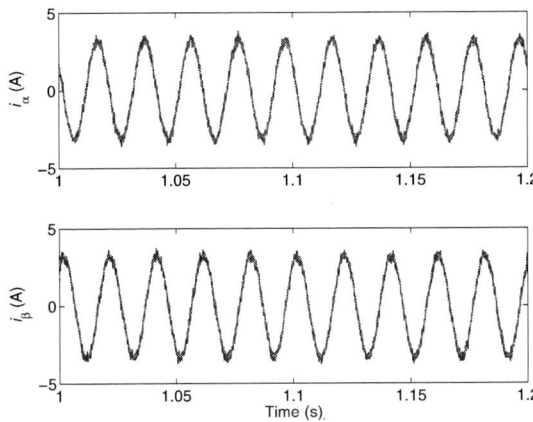

Fig. 13. Comparison between recorded stator current (S/N 20 dB) and simulated current using estimated parameters. The parameters are estimated based on recorded stator current with S/N of 20 dB.

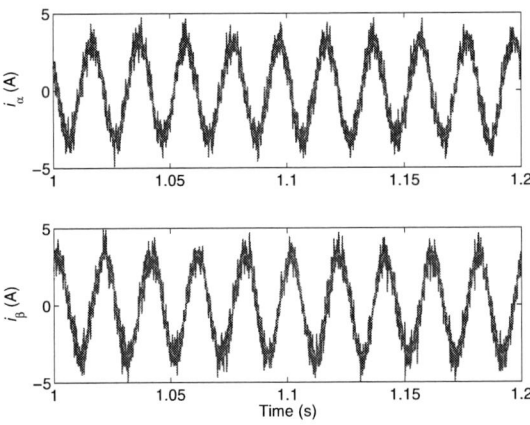

Fig. 14. Comparison between recorded stator current (S/N 10 dB) and simulated current using estimated parameters. The parameters are estimated based on recorded stator current with S/N of 10 dB.

VI. CONCLUSION

Induction machine stator fault diagnosis is achieved by estimating the selected set of parameters from recorded voltages and currents at machine power supply terminal. Both local (gradient descent and nonlinear least squares) and global (pattern search) search methods are applied in the parameter estimation method. As expected, the experimental results reveal that the accuracy of local search methods highly depend on the algorithm initial point in the parameter space. The global pattern search method is not sensitive to initial search point and achieves accurate estimation results by guaranteed detection of global minima with a slightly increased computation time. The results indicate that it is feasible to diagnose stator winding circuit faults by using stator currents only. The parameter estimation method can be conducted based on the stator currents with low S/N. The proposed model-based approach and the MATLAB/SIMULINK implementation can be easily extended to the similar tasks of detecting other types of faults in induction motors.

REFERENCES

[1] P. Vas, *Parameter Estimation, Condition Monitoring, and Diagnosis of Electrical Machines.* Clarendron Press Oxfords, 1993, ch. 3.

[2] D. Dorrell, W. Thomson, and S. Roach, "Analysis of airgap flux, current, and vibration signals as a function of the combination of static and dynamic airgap eccentricity in 3-phase induction motors," *IEEE Transactions on Industry Applications*, vol. 33, no. 1, pp. 24 –34, Jan. 1997.

[3] M. G. Melero, M. F. Cabanas, F. R. Faya, C. H. Rojas, and J. Solares, "Electromagnetic torque harmonics for on-line inter turn short circuits detection in squirrel cage induction motors," in *8th European Conf. on Power Electronics and Appl. EPE, Lausanne, Switzerland*, Sep. 1999, p. 9.

[4] A. Sadoughi, M. Ebrahimi, and E. Razaei, "A new approach for induction motor broken bar diagnosis by using vibration spectrum," in *International Joint Conference SICE-ICASE, 2006.*, Oct. 2006, pp. 4715 –4720.

[5] J.-H. Jung, L. Jong-Jae, and B.-H. Kwon, "Online diagnosis of induction motors using MCSA," *IEEE Transactions on Industrial Electronics*, vol. 53, no. 6, pp. 1842–1852, 2006.

[6] H. Douglas, P. Pillay, and A. Ziarani, "A new algorithm for transient motor current signature analysis using wavelets," *Industry Applications, IEEE Transactions on*, vol. 40, no. 5, pp. 1361–1368, 2004.

[7] M. Benbouzid, "A review of induction motors signature analysis as a medium for faults detection," in *Proceedings of the 24th Annual Conference of the IEEE Industrial Electronics Society, 1998. IECON '98*, vol. 4, 1998, pp. 1950–1955.

[8] G. Kliman, R. Koegl, J. Stein, R. Endicott, and M. Madden, "Noninvasive detection of broken rotor bars in operating induction motors," *IEEE Transactions on Energy Conversion*, vol. 3, no. 4, pp. 873 –879, Dec. 1988.

[9] A. Bellini, F. Filippetti, G. Franceschini, C. Tassoni, and G. Kliman, "Quantitative evaluation of induction motor broken bars by means of electrical signature analysis," *IEEE Transactions on Industry Applications*, vol. 37, pp. 1248 –1255, Sep. 2001.

[10] M. Benbouzid and H. Nejjari, "A simple fuzzy logic approach for induction motors stator condition monitoring," in *IEEE International Electric Machines and Drives Conference, IEMDC*, 2001, pp. 634–639.

[11] B. Ayhan, M.-Y. Chow, and M.-H. Song, "Multiple discriminant analysis and neural-network-based monolith and partition fault-detection schemes for broken rotor bar in induction motors," *IEEE Transactions on Industrial Electronics*, vol. 53, no. 4, pp. 1298 –1308, Jun. 2006.

[12] A. Siddique, G. Yadava, and B. Singh, "A review of stator fault monitoring techniques of induction motors," *IEEE Transactions on Energy Conversion*, vol. 20, no. 1, pp. 106–114, Mar. 2005.

[13] L. Cristaldi, M. Lazzaroni, A. Monti, F. Ponci, and F. Zocchi, "A genetic algorithm for fault identification in electrical drives: a comparison with neuro-fuzzy computation," in *Proceedings of the 21st IEEE Instrumentation and Measurement Technology Conference, 2004. IMTC 04.*, vol. 2, May 2004, pp. 1454–1459.

[14] F. Filippetti, G. Franceschini, and T. Carla, "Neural networks aided on-line diagnostics of induction motor rotor faults," in *Conference Record - IAS Annual Meeting (IEEE Industry Applications Society)*, vol. 1, 1993, pp. 316 – 323.

[15] H. Su and K.-T. Chong, "Induction machine condition monitoring using neural network modeling," *IEEE Transactions on Industrial Electronics*, vol. 54, no. 1, pp. 241–249, 2007.

[16] M. Haji and H. Toliyat, "Pattern recognition - a technique for induction machines rotor broken bar detection," *IEEE Transactions on Energy Conversion*, vol. 16, no. 4, pp. 312 –317, Dec. 2001.

[17] F. Duan and R. Živanović, "Diagnosis of induction machine stator faults by parameter estimation techniques based on direct search on sparse grid," in *The 9th IET International Conference on Advances in Power System Control, Operation and Management*, 2012, pp. 1–6.

[18] ——, "A model for induction motor with stator faults," in *AUPEC12 - 22th Australasian Universities Power Engineering Conference*, Bali, Indonesia, Sep. 2012, pp. 1–4.

[19] E. Novak and K. Ritter, "Global optimization using hyperbolic cross points," in *State of the Art in Global Optimization*, C. Floudas and P. Pardalos, Eds. Kluwer Academic Publishers, 1972, pp. 19–33.

[20] F. Duan and R. Živanović, "Estimation of dc offset parameters based on global optimization," in *AUPEC08 - 18th Australasian Universities Power Engineering Conference*, Sydney, Australia, Dec. 2008, pp. 1–4.

[21] ——, "Induction motor fault diagnostics using global optimization algorithm," in *AUPEC09 - 19th Australasian Universities Power Engineering Conference*, Adelaide, Australia, Sep. 2009, pp. 1–4.

[22] Mathworks. [Online]. Available: http://www.mathworks.com/help/sldo/

[23] M. Avriel, *Nonlinear Programming: Analysis and Methods.* Courier Dover Publications, 2003.

[24] S. Bachir, S. Tnani, J. Trigeassou, and G. Champenois, "Diagnosis by parameter estimation of stator and rotor faults occurring in induction machines," *IEEE Transactions on Industrial Electronics*, vol. 53, no. 3, pp. 963–973, 2006.

[25] J. Nocedal and S. J. Wright, *Numerical optimization.* Springer Science+ Business Media, 2006.

[26] N. Sadati, S. Kaboli, H. Adeli, E. Hajipour, and M. Ferdowsi, "Online optimal neuro-fuzzy flux controller for DTC based induction motor drives," in *Twenty-Fourth Annual IEEE Applied Power Electronics Conference and Exposition, APEC 2009.*, 2009, pp. 210–215.

[27] K. Wang, J. Chiasson, M. Bodson, and L. Tolbert, "A nonlinear least-squares approach for identification of the induction motor parameters," *IEEE Transactions on Automatic Control*, vol. 50, no. 10, pp. 1622–1628, 2005.

[28] R. M. Lewis and V. Torczon, "Pattern search algorithms for bound constrained minimization," *SIAM J. on Optimization*, vol. 9, no. 4, pp. 1082–1099, Apr. 1999.

[29] F. Duan and R. Živanović, "Automated multi-motor condition monitoring based on IEC 61850," in *ECCE Asia DownUnder 2013 - 5th Annual International Energy Conversion Congress and Exhibition*, Melbourne, Australia, Jun. 2013, pp. 1–5.

Diagnosis of Stator Winding Inter-turn Short Circuit in Three-Phase Induction Motors by Using Artificial Neural Networks

P. J. Broniera, W. S. Gongora, A. Goedtel, W. F. Godoy

Abstract-- **The application of induction motors in industry is widespread. Thus, several studies have presented strategies for the diagnosis and prediction of failures in these motors. One technique used is based on the recent utilization of intelligent systems for detecting faults in electric motors. Thus, this paper proposes an alternative tool to traditional techniques for fault detection of a short circuit between the inter-turns of the stator winding using artificial neural networks to analyze stator current signals in the time domain. Experimental results are presented to validate the proposed approach.**

Index Terms-- **Artificial Neural Networks, Stator faults, Three phase induction motors.**

I. INTRODUCTION

The application of induction motors is extensive in the industry context, as it is the primary means of electromechanical energy conversion, due to its consolidated features such as low cost, versatility and robustness [1].

Despite the fact that such motors are usually well rugged and constructed the possibility of failures is inherent to the device. Incipient faults inside the machine usually affect its performance even before significant failures occur, thus bringing significant damage to the industrial process [1-5].

According to [6] the major faults found in electric motors are related with electrical or mechanical problems. Among the electrical faults highlights stator faults which are of the order of 38% [7].

The diagnosis of this type of failure can be performed by means of traditional non-invasive techniques. These strategies are based on the analysis of quantities such as vibration, voltage, current, torque and speed [4-9].

The work presented in reference [7] defines a detection system stator faults for induction motors wich analyze the symmetrical components by spectral harmonics of current signals. Reference [8] is an overview of fault detection methods in the literature using inteligent systems. The work of [10] describes the use of signatures of thestator current signals to diagnose short circuit faults among the turns of the stator winding of a three phase induction motor.

Among the techniques used for the fault diagnosis in electric motors it can be highlighted the use of intelligent systems, especially Artificial Neural Networks (ANN). This methodology provides robustness for the treatment of uncertainties in the signal processing.

Furthermore, it significantly reduces the number of sensor elements that monitor machine operating conditions. Such sensors contribute directly to the increased costs of implementing traditional techniques [2].

The work presented in paper [11] defines a standard neural classifier, which uses a MLP network whose inputs are statistics data taken from the sampled current signals. The current signals are collected from the induction motor phases in order to detect inter-turn short circuit faults in the stator winding.

In [12] it is defined a pattern classifier which uses artificial neural networks to classify the stator current signals by means of principal component of analysis (PCA) to diagnose short circuit faults in the stator winding of induction motors.

The purpose of this paper is to present a strategy to predict stator winding faults in induction motors based on artificial neural networks by monitoring currents signals at time domain, monitoring the electrical current in the stator power supply of an induction motor line connected. More specifically, the network must be capable of learning patterns of stator current signals and identify short circuit faults among the turns.

This article is organized as follows: Section 2 presents a description of the major faults in electric motors. Section 3 presents aspects of artificial neural networks. Section 4 the methodology used for signal treatment is presented with experimental results. Finally, in Section 5, the conclusions of the study are presented.

This work was supported by Fundação Araucária de Apoio ao Desenvolvimento Científico e Tecnológico do Paraná (Process Nr. 06/56093-3), Conselho Nacional do Desenvolvimento Científico e Tecnológico - CNPq (Process Nr. (Process Nr 474290/2008-5, Nr 473576/2011-2 and Nr 552269/2011-5) and Capes-DS.

P. J. Broniera is with the Department of Electrical Engineering, UTFPR University, Cornélio Procópio, PR 86300-000 Brazil (e-mail: paulobrj@hotmail.com).

W. S. Gongora is with the Department of Electrical Technician, IFPR Institute, Assis Chateaubriand, PR 85935-000 Brazil (wylliam.gongora@ifpr.edu.br).

A. Goedtel is with the Department of Electrical Engineering, UTFPR University, Cornélio Procópio, PR 86300-000 Brazil (e-mail: agoedtel@utfpr.edu.br).

W. F. Godoy is with the Department of Electrical Engineering, UTFPR University, Cornélio Procópio, PR 86300-000 Brazil (e-mail: wagnergodoy@utfpr.edu.br).

II. INDUCTION MOTORS FAULTS CLASSIFICATION

Currently most machines are operated by using electric motors which may, with the usage, present some kind of failure. These faults can be classified into two groups: i) electrical and ii) mechanical faults.

Figure 1 shows a block diagram of the main types of failures. The electrical faults are highlighted due to the problems related to stator winding, rotor winding, rotor broken bars and rings and also rotor broken connections. Moreover, the mechanical failures may be derived from bearings problems, eccentricity, wear coupling and misalignment as reported by [13].

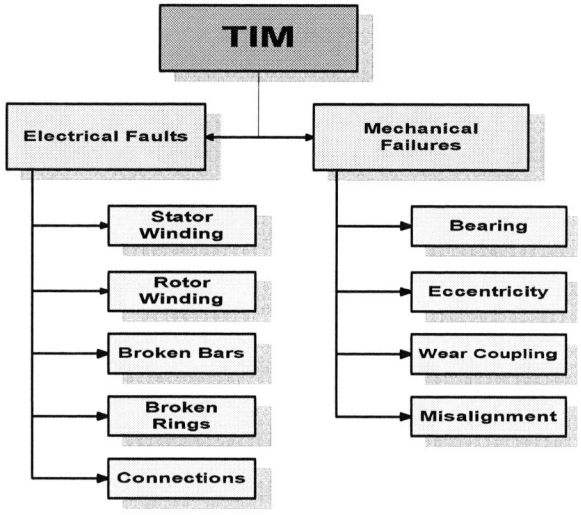

Fig. 1. Failure classification.

The stator windings are subjected to undesirable situations by several factors such as temperature rise due to mechanical loads, vibrations and mechanical problems related with power quality. According to [14], the stator winding failures may represent up to 38% of undesired stops of electric motors, as presented in Figure 2.

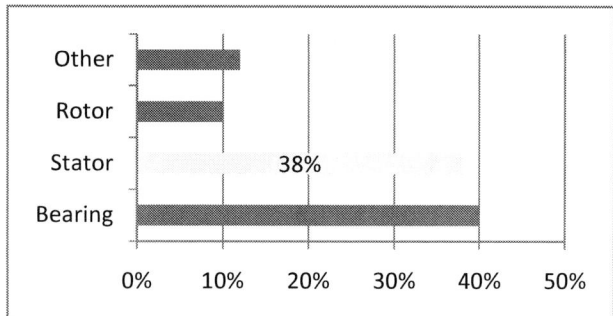

Fig. 2. Main failures in induction motors.

The stator faults can be divided into the following categories: short-circuit among turns, short circuit among windings and short-circuit among winding and machine carcase.

The deterioration of the stator isolation typically begins with a short-circuit involving few turns. According to [15] the short-circuit current is approximately twice of the blocked rotor current that causes localized heating which quickly extends to other winding sectors. The duration of the failure depends on the motor operating conditions and it is hard to estimate. What is known is that the evolution is rapid, thus becomes essential to continually monitor the motor in order to detect such failure [16].

III. ARTIFICIAL NEURAL NETWORKS

The basic element of a neural network is the artificial neuron (Figure 3), which is also known by node or processing element.

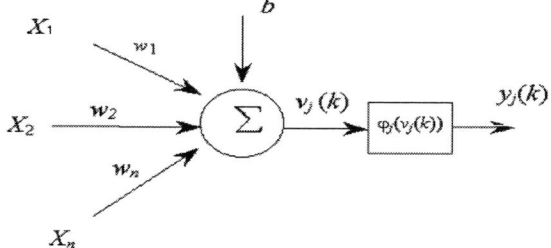

Fig. 3. Representation of the artificial neuron.

The model of the artificial neuron illustrated in Fig. 1 can be mathematically modeled by:

$$v_j(k) = \sum_{i=1}^{n} X_i . w_i + b \qquad (1)$$

$$y_j(k) = \varphi_j(v_j(k)) \qquad (2)$$

where:

n is the number of input signals of the neuron.

X_i is the i-th input signal of the neuron.

w_i is the weight associated with the i-th input signal.

b is the threshold associated with the neuron.

$v_j(k)$ is weighted response (summing junction) of the j-th neuron with respect to the instant k.

$\varphi_j(.)$ is the activation function of the j-th neuron.

$y_j(k)$ is the output signal of the j-th neuron with respect to the instant k.

Each artificial neuron is able to compute from input signals the respective output signal. The activation functions used to compute the output signal are typically nonlinear. The neural networks that process analog data have often used activation functions like sigmoid or hyperbolic tangent type. The adjustment process of the network weights (w_j) associated with the j-th output neuron is done from computation of the error signal $e_j(k)$ linked to the k-th iteration or k-th input vector. This error signal is provided by:

$$e_j(k) = d_j(k) - y_j(k) \qquad (3)$$

978-1-4799-0024-4/13 $31.00 © 2013 IEEE

where $d_j(k)$ is the desired response to the j-th output neuron. Adding all squared errors produced by the output neurons of the network with respect to k-th iteration, we have:

$$E(k) = \frac{1}{2} \sum_{j=1}^{p} e_j^2(k) \qquad (4)$$

where p is the number of output neurons. Considering N the number of training patterns and $E(k)$ the error energy function of the neural network, it is possible to calculate the mean squared error energy function E_m of the network as follows:

$$E_m = \frac{1}{N} \sum_{k=1}^{N} E(k) \qquad (5)$$

The objective of the backpropagation algorithm is minimizing $E(k)$ and E_m through the adjustment of w_i and b.

To achieve the objective of reducing the error, the training algorithm presents the input data set to the neural network and the output is then computed as described in (1) and (2). The error calculated in each iteration is used as a parameter to the weight adjustment. After presentation of all training data set to the network, the mean error can be calculated using (5). The parameter E_m estimates the convergence of the algorithm and determines if the algorithm should stop when it reaches the mean desired error. Equation (6) describes the weight adjustment in the multilayer structure.

$$w_{ji}(k) \leftarrow w_{ji}(k) - \eta \frac{\partial E(k)}{\partial w_{ji}(k)} \qquad (6)$$

where w_{ji} is the weight connecting the j-th neuron of the output layer to i-th neuron of the previous layer, and η is a constant that determines the learning rate of the backpropagation algorithm.

The backpropagation algorithm is based on the Least Mean Square (LMS) method and it applies a correction in the synaptic weights, called $\Delta w_{ji}(k)$, to the synaptic weight $w_{ji}(k)$. This correction is proportional to the partial derivative $\partial E(k)/\partial w_{ji}(k)$ as described in [15]. Using the chain rule, it is possible to express this gradient in the following form:

$$\frac{\partial E(k)}{\partial w_{ji}(k)} = \frac{\partial E(k)}{\partial e_j(k)} \cdot \frac{\partial e_j(k)}{\partial y_j(k)} \cdot \frac{\partial y_j(k)}{\partial v_j(k)} \cdot \frac{\partial v_j(k)}{\partial w_{ji}(k)} \qquad (7)$$

This partial derivative is called "*sensitivity factor*" and indicates the search direction with respect to weight $w_{ji}(k)$ [15]. The terms in equation (7) are given by:

$$\frac{\partial E(k)}{\partial e_j(k)} = e_j(k) \qquad (8)$$

$$\frac{\partial e_j(k)}{\partial y_j(k)} = -1 \qquad (9)$$

$$\frac{\partial y_j(k)}{\partial v_j(k)} = \varphi_j'(v_j(k)) \qquad (10)$$

$$\frac{\partial v_j(k)}{\partial w_{ji}(k)} = y_i(k) \qquad (11)$$

Equation (8) is calculated through the derivative of (4) with respect to $e_j(k)$. The derivative of the error function in

(3) with respect to the j-th output, i.e. $y_j(k)$, results in (9). The derivative of (2) with respect to $v_j(k)$ results in (10). Equation (11) is the result of the derivative of (1) considering w_{ji} the weight connecting the j-th neuron of the output layer to i-th neuron of the previous layer.

Equation (12) is the result of the grouping of (8)-(11) and it is described as follows:

$$\frac{\partial E(k)}{\partial w_{ji}(k)} = -e_j(k)\varphi_j'(v_j(n))y_i(n) \qquad (12)$$

The synaptic correction $\Delta w_{ji}(k)$ with respect to the weight $w_{ji}(k)$ is described through the delta rule as described in (6), i.e.:

$$\Delta w_{ji}(k) = -\eta \frac{\partial E(k)}{\partial w_{ji}(k)} \qquad (13)$$

The use of the negative signal in (13) indicates the descendent gradient in relation to the search of synaptic weights to reduce $E(k)$. The substitution of (12) in (13) results in the following equation:

$$\Delta w_{ji}(k) = -\eta \delta_j(k)y_i(k) \qquad (14)$$

where $\delta_j(k)$ is the local gradient defined by:

$$\delta_j(k) = -\frac{\partial E(k)}{\partial v_j(k)} = -\frac{\partial E(k)}{\partial e_j(k)}\frac{\partial e_j(k)}{\partial y_j(k)}\frac{\partial y_j(k)}{\partial v_j(k)} = e_j(k)\varphi_j'(v_j(k)) \qquad (15)$$

The local gradient shows the direction of the synaptic weights in order to reduce $E(k)$.

This work aims to diagnosis stator faults by using ANN. The approach of this proposal consider the neural network as pattern classifier. According to [17], both Multilayer Perceptron and Radial Basis Function networks (RBF) can be used for this purpose.

The Multilayer Perceptron networks (MLP) have a feedforward architecture, which training is performed in a supervised form. Thus, the network properly adjusts its synaptic weights for a given application.

The Radial Base Function (RBF) topology is comprised by an input layer, just one neural intermediate layer and an output layer. The neurons of intermediate layer consider as activation rule the Gaussian function and as for the neurons of the output layer the activation function is of linear type [17].The RBF is also characterized by containing two different training stages, the first step is associated with the adjustment of the weights of the intermediate layer through an unsupervised learning method which is exclusively dependent on the characteristics of the input data.

The second training stage is associated with adjustments of the neural weights of the output layer neurons, which uses the delta generalized rule as learning technique [17].

IV. FAILURE IDENTIFICATION USING NEURAL NETWORKS

This work consists in using the stator currents readings in the time domain for an induction motor operating with stator

978-1-4799-0024-4/13 $31.00 © 2013 IEEE

283

faults and also with no stator faults. These signals are presented to an ANN with the aim of identifying short-circuit faults between the stator windings. Figure 3a shows an overview of the structure of the laboratory test bench.

The test bench is composed primarily by the Direct Current Machine (DCM) with tachogenerator, which is coupled to the TIM connected by rotating torque sensor. Figure 3(a) shows the picture of the test bench. Although, in this figure are shown the sets of signal conditioning: i) current and voltage of the DC machine, ii) the torquemeter signal conditioning and iii) data acquisition board. The DCM is configured to operate as a Direct Current Generator (DCG), whose goal is to impose load torque to TIM. The voltage generated in the armature of the DCG is applied to the resistive load. Thus, it was possible to vary the resistive torque imposed by the TIM with DCG acting on the field coil voltage supply via a DC power source. The current signals were individually measured by three Hall sensors as showed in Figure 3b, which were connected to a signal acquisition board and linked to a microcomputer.

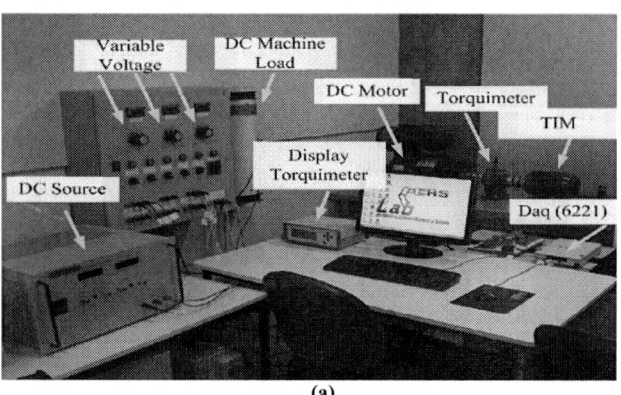

(a)

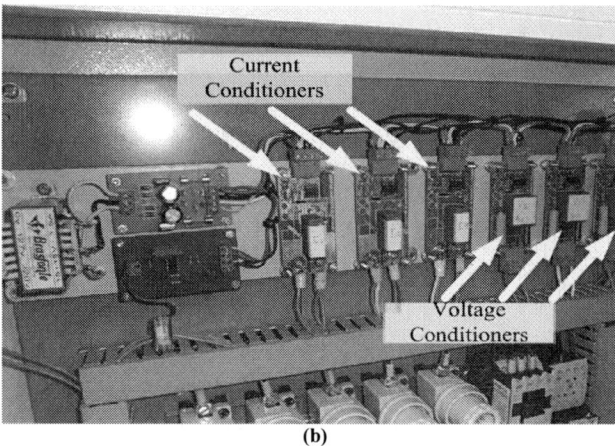

(b)

Fig. 3. Data acquisition workbench: a) Data acquisition workbench in an overview b) Conditioning of current and voltage.

Signal processing was performed according to the block diagram illustrated in the Figure 4.

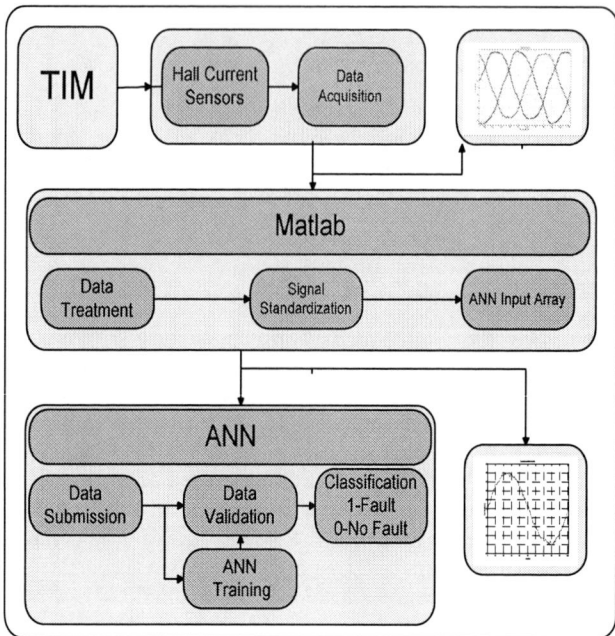

Fig. 4. Data treatment routine.

The sampling rate of the current signals was 25k samples/s in an acquisition time of 5s. Thus, it was carried out 134 tests and it was obtained storage of 125,000 sampling points for each phase. From this set of data 5000 points were randomly selected and stored in vectors for treatment, as showed in Figure 5 (a).

After the storage of the each phase currents, the respective signals were conditioned resulting in a vector of 5000 points. This amounted 10 sampling wave cycles, each with 500 points. To simplify and reduce the number of network inputs it is used a data processing routine to identify a full wave cycle, data was normalized and subsampled resulting in an array of 100 points as showed in Figure 5 (b).

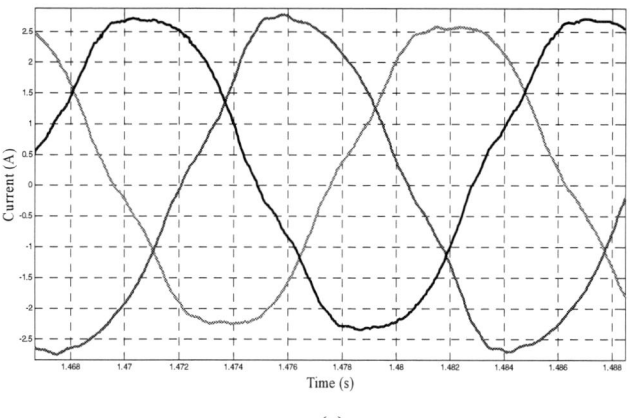

(a)

978-1-4799-0024-4/13 $31.00 © 2013 IEEE   284

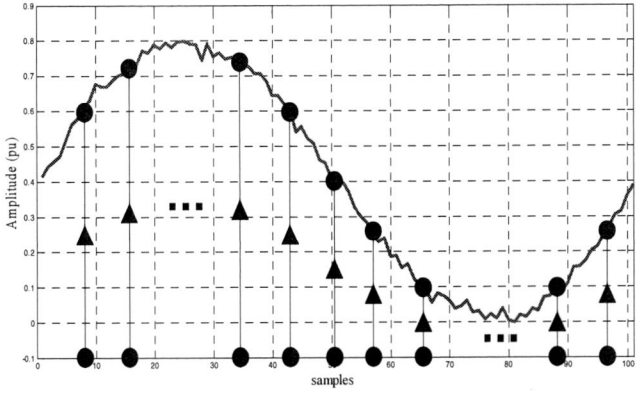

(b)

Fig. 5. Stator current signals: a) Stator Currents (phase a, b and c) stored each one in a vector of 5.000 points; b) Full wave cycle subsampled in 100 points resulting the conditioned stator currents (network inputs).

From this procedure it was generated a table containing 134 samples with and without short-circuit faults between the turns of the stator winding. Among these samples 89 were used for training and 55 for validation. The respective samples with stator faulty signals received a tag with desired output equals 1, and samples of signals without fail received tag with desired output equal to 0.

In order to evaluate the ANN in front of possible interference with presented current signals, it was randomly inserted at the same frequency of the sampled signal a white noise. This caused changes throughout the sampled signal, as showed in Figure 6.

For this it was developed a routine that establishes random values from -1 to 1, which are limited to 10% of the current signal normalized from 0 to 1. Thus, the sampled signal containing noise were presented to the network in the validation step, in order to verify the robustness of the ANN, in front of the interference contained in the current signal presented in its inputs.

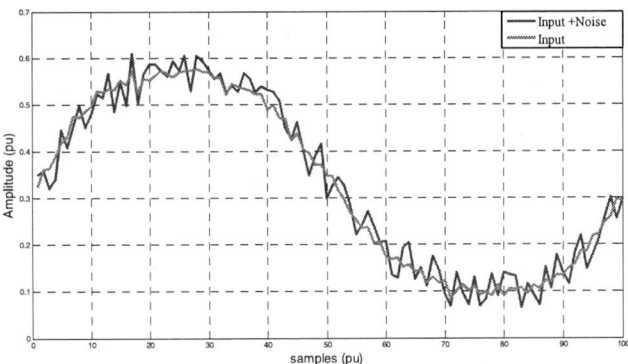

Fig. 6. Input signal with noise.

After the step of collecting, treatment and robustness test, data were submitted to three network architectures with two

different topologies, as per information showed in the Table 1, in order to classify such structures based on their respective tags.

TABLE 1 - ANN parameters

Type	Network 1	Network 2	Network 3
Architecture	RBF	MLP	MLP
Training	PS	PS	PS
Nº of layers	2	2	2
Neurons in the 1º layer	2	3	8
Neurons in the 2º layer	1	1	2
Training Algorithm	SO/DR	BP	BP+LM
Activation function 1º layer	G	HT	HT
Activation function 2º layer	Linear	Linear	Linear

(LM) Levenberg Maquardt; (G) Gaussian (HT) Hyperbolic Tangent (SO) Self-organized (DR) Delta rule (BP) Backpropagation

Proposed networks were subjected to the training considering same input signal and with a learning rate as per information showed in Table 2. As stopping criteria it was established the Mean Squared Error (MSE), which is defined according to the best performance of each network.

TABLE 2 - Experimental Results

Type	Network 1	Network 2	Network 3
Training samples	89	89	89
Validation samples	55	55	55
Objective error	1.10^{-13}	1.10^{-11}	1.10^{-11}
Learning coefficient	0.0001	0.001	0.1
Epochs	287.832	349	45
False positive	5	13	3
False negative	3	4	0
Classification error	8/55	17/55	3/55
Accuracy percentage	85.4%	69.1%	94.54%

Network 1 converged with 287.832 epochs and with 85.4% of accuracy, while Network 2 has reached the stopping criterion with 349 epochs reaching only 69.1% of accuracy. In case of the Network 3, the number of epochs was lower; it was reached 45 epochs and 94.54% of accuracy.

V. CONCLUSIONS

This paper has proposed an alternative method to the traditional techniques for fault detection of a short circuit among the stator winding turns. This method used artificial neural networks to analyze the stator current signals in the time domain with 100 amplitude points of each signal.

It was collected and treated 134 samples of stator current signals, which were trained and validated by three different networks architectures. Based on the obtained results, it can be stated from the three proposed architectures, the network 3 presented the best performance for that application, thus allowing the validation of the proposed method, with accuracy of 94.54% of test set.

VI. REFERENCES

[1] J. N. Brito, *Development of a Hybrid Intelligent System for Fault Diagnosis in Three-Phase Induction Motors,* PhD Thesis (*in Portuguese*), Campinas, SP: Universidade Estadual de Campinas-PDGEM/UNICAMP, 2002.

[2] M. Suetake, *Intelligent Systems for Monitoring and Fault Diagnostics in Three-Phase Induction Motors,* PhD Thesis (*in Portuguese*), São Carlos: Universidadede São Paulo, 2012.

[3] A. Bellini, F. Filippett and C. Tassoni, "Advances in Diagnostic Techniques for Induction Machines," *IEEE Transactions on Industrial Electronics 55(12)* , vol. 55, pp. 4109-4126, 2008.

[4] P. Filho, *Coupling Predictive Three-Phase Induction Motors through analysis of Magnetic Flux,* PhD Thesis (*in Portuguese*), Campinas: Universidade Estadual de Campinas, 2003.

[5] S. M. A. Cruz e A. J. M. Cardoso, " Diagnosis of Stator Inter-Turn Short Circuits in DTC Induction Motor Drives," *IEEE Transactions on Indutry Applications,* pp. 1332-1339, 2003.

[6] R. M. Tallam, S. B. S. G. C. Lee, G. B. Kliman e J. Yoo, "A survey of methods for detection of stator-related faults in induction machines," *IEEE Transactions on Indutry Applications 43(4),* pp. 920-933, 2007.

[7] M. B. K. Bouzid e G. Champenois, "New Expressions of Symmetrical Components of the Induction Motor Under Stator Faults," *IEEE Transactions on Industrial Electronics,* vol. 99, pp. 1-8, 2012.

[8] F. M. d. C. Santos, I. N. d. Silva e M. Suetake, "On the Application of Intelligent Systems for Fault Diagnosis in Induction Machines - An Overview," (*in Portuguese*), *Controle & Automação,* vol. 23, n. 5, pp. 553-569, 2012.

[9] M. F. S. V. D'Angelo, R. M. Palhares, R. M. Takahash, R. H. Loschi, L. M. R. Baccarini e W. M. Caminhas, "Incipient fault detection in induction machine stator winding using a fuzzy-bayesian change point detection approach," *Applied Soft Computing Journal 11(1),* pp. 179-192.

[10] A. Ukil, S. Chen e A. Andenna, "Detection of stator short circuit faults in three-phase induction motors using motor current zero crossing instants," *Electric Power Systems Research,* pp. 1036-1044, 2011.

[11] V. N. Ghate e S. V. Dudul, "Optimal MLP neural network classifier for fault detection of three phase induction motor," *Expert Systems Whit Applications 37(4),* pp. 3468-3481, 2010.

[12] J. Martins, V. F. Pires e A. J. Pires, "Unsupervised neural-network-based algorithm for an on-line diagnosis of tree-phase induction motor stator fault," *IEEE Transactions on Industrial Electronics,* pp. 259-264, 2007.

[13] G. Singh e S. A. Kazzaz, "Induction machine drive condition monitoring and diagnostic research a survey," *Electric Power Systems Research 64 (2),* pp. 145-158, 2010.

[14] T. Han, B. yang e Z. Yin, "Feature-based fault diagnosis system of induction motors using vibration signal," *Journal of Quality in Maintenance Engineering,* pp. 163-175, 2007.

[15] L. M. R. Baccarini, B. R. d. Menezes e W. M. Caminhas, "Fault Induction dynamic model, suitable for computer simulation: Simulation results and experimental validation," *Mechanical Systems and Signal Processing,* pp. 300-311, 2010 .

[16] T. M. Wolbank, K. A. Loparo e R. Wohrnschimmel, " Inverter statistics for online detection of stator asymmetries in inverter-fed induction motors," *IEEE Transactions on Industry Applications,* pp.1102-1108, 2003.

[17] Haykin, Simon. Neural networks and learning machines. – 3rd ed. Pearson Prentice Hall, 2008.

VII. BIOGRAPHIES

Broniera, P. J. Technologist in Industrial Automation from Federal Technological University of Paraná. Currently, he is student of Master´s degree in electrical engineering at the Federal Technological University of Paraná, Cornélio Procópio, Brazil. His research interests are within the fields of electrical machinery, intelligent systems, and power electronics.

Gongora, W. S. Bachelor's at Engenharia de Controle e Automação from Faculdade Assis Gurgacz (2007) and master's at Engenharia Elétrica from Universidade Tecnológica Federal do Paraná (2013). He is a professor at the Federal Institute of Paraná (IFPR), campus Assis Chateaubriand. Working mainly on the themes: artificial neural networks, fault prediction, three-phase induction motor.

Godoy, W. F. was born in Cornélio Procópio, Brazil, in 1977. He received the B.S. degree in electrical engineering from the University Norte do Paraná, Londrina, Brazil, in

2003, the M.Sc. degree in eletrical engineering from the University of Londrina, Londrina, Brazil, in 2010. Currently, he is an Assistant Professor with the Federal Technological University of Paraná, Cornélio Procópio, Brazil. His research interests are within the fields of electrical machinery, intelligent systems, and electrical maintenance.

Goedtel, A. was born in Arroio do Meio, Brazil, in 1972. He received the B.S. degree in electrical engineering from the Federal University of Rio Grande do Sul, Porto Alegre, Brazil, in 1997, the M.Sc. degree in industrial engineering from the São Paulo State University (UNESP), São Paulo, Brazil, in 2003, and the Ph.D. degree in electrical engineering from the University of São Paulo (USP), São Paulo, in 2007. Currently, he is an Assistant Professor with the Federal Technological University of Paraná, Cornélio Procópio, Brazil. His research interests are within the fields of electrical machinery, intelligent systems, and power electronics.

Naïve Bayes classifier for Temporary short circuit fault detection in Stator winding

D. A. Asfani, M. H. Purnomo, D. R. Sawitri

Abstract – This paper is proposing Naïve Bayes classifier detection system to identify the symptom of stator winding deterioration. The proposed system is based on probabilistic classifier with strong independence assumption of each fault case. The temporary short circuit case is defined as non permanent short circuit fault with high impedance. This fault case is representing the early stage of stator insulation break down. The laboratory experiment is performed to simulate the fault cases consist of induction motor with stator modification and current measurement system. The detection system is trained to identify the temporary short circuit occurrence consist of transient starting, steady state and ending of temporary short circuit. The system is also tested using non trained data to clarify the detection performance.

Index Terms--Fault detection, induction motor, stators, Wavelet transforms, bayesian methods, kernel.

I. INTRODUCTION

STATOR winding is one of the weakest components of electric machine. This part is contributes around 40% of machine failure [1]. Combination several stresses that categorized into electrical, thermal, mechanical and environmental during operating condition are involved the ageing process. Investigation on combination of several stresses show that temperature and voltage have higher decreasing effect on insulation lifespan than frequency [2].

Based on this fact, the condition and early fault detection system is needed to increase reliability and reduce maintenance cost. The industrial survey shows that the downtime cost due to unpredictable fault can reach up to $200k/hour in automotive industries [3]. Since past few decades, development of diagnostic techniques had received intense interest of researcher. The innovation of signal processing and decision techniques are become main improvements of proposed system [4]. Mainly, the signal processing techniques can be classified into three categories related to frequency domain analysis, time-frequency/time-scale domain analysis and time domain analysis [5]. A classical technique of frequency domain analysis is widely known based on FFT. In recent years, the improvement of

this method named power spectrum density (PSD) is successfully applied in machine diagnostic [6]. A well-known of time-frequency/time-scale analysis is short-time Fourier transform (STFT). Using this method, a frequency spectrum of time varied signal is obtained based on local spectrum of the signals that change over the time. The alternative is time-scale analysis based on wavelet transform. The wavelet analysis is based on function scale dilation and time translation. These two functions enable providing a certain resolution in time and frequency that suitable with system requirement. In some cases, fault is easily identified using amplitude and phase modulation of stator current. In order to extract the fault information, the time domain analysis is needed. Hilbert transform is one of the method can be used to demodulation both of amplitude and phase current signal. The other method of time domain analysis are noise cancellation techniques, speed and torque estimation [5].

Beside signal processing techniques, the decision system is one of the main concerns for developing diagnostic system of induction machine. This system mainly can be categorized into two groups that are deterministic model or model-based system and artificial intelligent (AI) system. Deterministic model is more complex and difficult to apply a single application but provide more flexible system that can consider different fault cases. Furthermore, AI technique such as expert system, fuzzy logic, and neural network has advantages of reducing human intervention which increase the system efficiency. However, they need an initial training process that affects operation performance. By providing sufficient training data, system will be efficient, simple and successfully applied as diagnostic electric machines [4]. Induction motor diagnostic based on hybrid fuzzy neural network combined with classification and regression tree (CART) is proposed [7]. In this study, signal noise is considered to clarify the effectiveness of the proposed methods. The result shows that the effectiveness of the proposed method is decreasing with higher noise level.

This paper is propose an expert system based on simple probabilistic classifier that known as naïve Bayes classifier to detect the presence of short circuit fault in stator winding. Since the assumption of independent between input variable, Naïve Bayesian classifier is known as a simple non parametric classifier method. However, it provide compete result with other complex algorithm for classification and detection study such as classifying text document [8]. In the case of motor diagnostic, Naïve Bayesian classifier is used to

---

This work was supported in part by Directorate General of Higher Education, Department of Education and Culture (DIKTI), Indonesia no. 013674.107/IT2.7/PN.08.01/2013.

D. A. Asfani and M. H. Purnomo are with Electrical engineering department of Institut Teknologi Sepuluh Nopember, Surabaya, Indonesia, 60111(e-mail: anton@ee.its.ac.id, hery@ee.its.ac.id).

D.R. Sawitri is with Electrical Engineering Departmen, Dian Nuswantoro University, Indonesia (e-mail: drsawitri@gmail.com).

detect the existence of fault based on the simulation data [9]. In this system, the input variables are assumed independently to the probability of fault cases. Since the training data set has unequal distribution, the kernel density estimation is applied as smoothing method. The temporary short circuit fault is investigated in this paper representing the initial stage of insulation winding deterioration. The fault has nonpermanent occurring and high value of fault impedance. This fault is divided into three stage based on the occurrence. The first stage is starting transient of fault, the second is steady state of fault and the last stage is the end of fault. The three stages together with normal operation are defined as operating states of motor. During operation, motor will belong to one of this state.

Stator signal is processed using wavelet transform to obtain variable input. Wavelet analysis is known as the suitable method to capture the fast transient phenomena and provide better result than frequency analysis such as Fourier transforms [10]. Haar wavelet third level is selected as the digital filter because it is known as the most appropriate filter for this short circuit case [11]. The designed system is tested using no trained data set and shows that all cases is detected properly. The proposed method also compared with other classifier method that is linear discriminant analysis. It shows the proposed method gives the better result. The next section of this paper is organized into three part consist of temporary short circuit fault; propose system for temporary short circuit detection; detection result and discussion.

II. TEMPORARY SHORT CIRCUIT FAULT

In this section the fault cases that investigated in this paper is presented. The background of the fault cases and the laboratory based simulation of the fault are described in the subsection.

A. Background of Temporary short circuit

The incipient of short circuit fault in stator winding is represented by temporary short circuit fault in this paper. This fault has low current short circuit and non permanent occurring. In the early stage of stator insulation deterioration, the insulation withstand voltage is decreasing to certain level that can isolate the voltage in the normal operating voltage. However, when an over voltage occur due

to some normal or fault operation such as electric drive operation, circuit breaker operation or other voltage surge disturbances, the insulation will breakdown for a while. The other disturbance is temporary mechanical or temperature stress during operation lead to insulation breakdown for certain time period and back to normal operation when the disturbance is cleared. A long with time, the short circuit fault become worst to be permanent fault and generating higher level of short circuit fault. The investigated short circuit fault is shown in Fig. 1 [12].

B. Short Circuit Occurrence and Detection

The main objective of this paper is detecting the temporary short circuit even though it occurred in very short time for example in one cycle. The transient current during fault occurred is used to identify the fault. In order to detect this fault, the motor current spectrum is categorized into four states that is normal, starting short circuit, steady state short circuit and end of short circuit. Based on those states, short time occurrence short circuit fault can be detected. Moreover, the duration of short circuit can be provided from the information when the fault is starting occur and ending. Also how often this fault occurred in certain time period can be calculated to determine the severity of incipient short circuit fault.

The laboratory experiment is performed to support this research. Single phase induction motor, 0.25 hp, 110/220 Volt 4.8/2.4 Amp, 1400 rpm, 50 Hz, is used in this experiment. Since the detection system is mainly based on the transient phenomena during short circuit occurred, the phenomena in single phase motor is equal to three phase motor. When short circuit starting occurred, the motor current is increasing rapidly for view millisecond period until a certain level and noticed as spectrum discontinuity of stator current as shown in Fig.1. This period is defined as transient of starting short circuit. Conversely, during short circuit is cleared, the motor current is decreased. This transient is quietly faster than increasing or decreasing current during sudden load change.

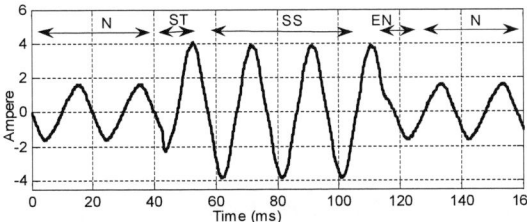

Fig. 1. Motor current spectrum under temporary short circuit (N: Normal; ST:Starting S.C; SS: Steady State S.C; EN:Ending S.C) [5]

TABLE I
TEMPORARY SHORT CIRCUIT DATA SET [10]

Operation Case	Training and Validation		Testing	
	Number of Cases	Variation (Ampere)	Number of Cases	Variation (Ampere)
Normal	40	1 ; 1.3	10	1.2 ; 1.5
Steady state S.C	100	2.75 ; 3.2; 3.6; 3.8 ; 4	10	3.5; 4.5
Starting S.C	50	2.3;3.1; 3.25; 3.75; 4	15	2.4; 3.2; 4.5
Ending S.C	50	2.3; 3.1; 3.6; 3.8; 4	15	3.1; 3.6; 3.86
Total	240		50	

978-1-4799-0024-4/13 $31.00 © 2013 IEEE

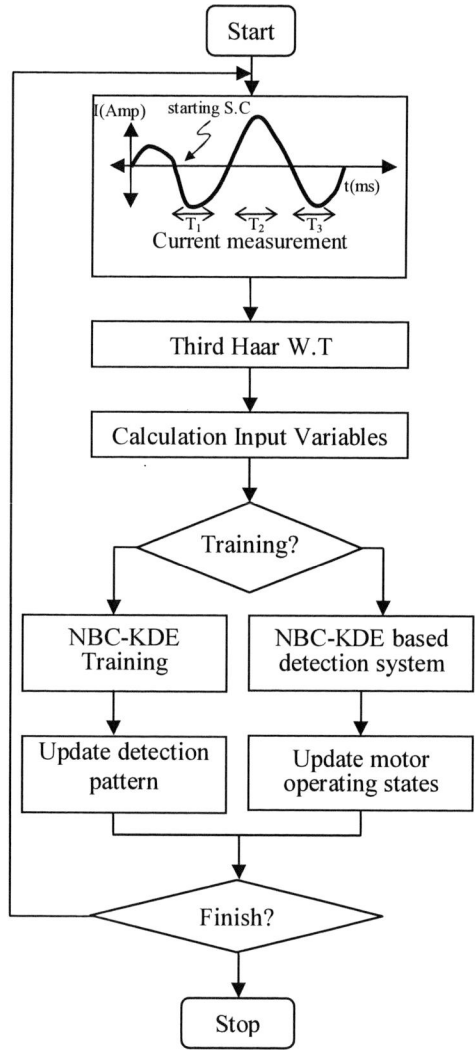

Fig. 2 Flowchart of the proposed

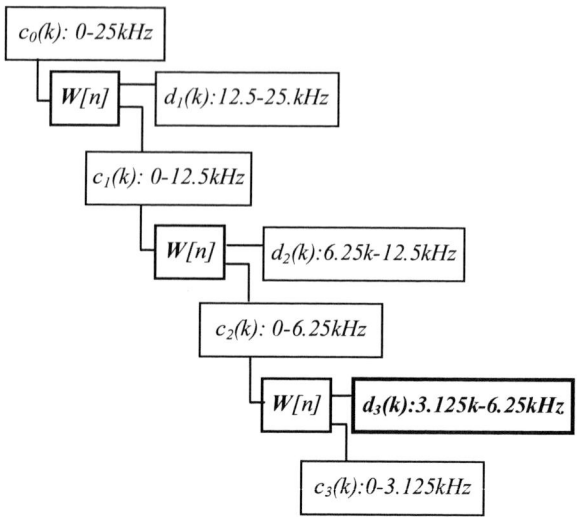

Fig. 3 Wavelet transformation tree

estimation (KDE) is used to improve the detection performance of NBC because the each data set based on the motor operating states has difference data distribution.

III. Proposed Method for Temporary short circuit Detection

Generally, the designing of proposed method can be summarized into two main steps that is training and testing process. However, both of two processes have similar data processing. From the measurement until detection of the motor state, there are three processes to handling data. The first is data metering, the second is wavelet filtering, the third is energy calculation and the last is motor state detection. All of the data processes and handling is shown in Fig. 2. The following subsection is discussed the designing process of the proposed method consist of determining third level Haar wavelet filter, calculation variable input and NBC-KDE detection system.

A. Third Level Haar Wavelet Filter

Discrete wavelet Haar transform can simply calculated by multiplying the discrete signal $x[n]$ using wavelet transform matrix $W[n]$, when n is number of data and j is transformation level.

$$c_j(k), d_j(k) = W[n]x[n] \qquad (1)$$

High frequency $d_j(k)$ and low frequency $c_j(k)$ of j-level wavelet transform can be calculated from signal of previous level based on wavelet tree as shown in Fig. 3. This figure also shows the frequency band of each wavelet coefficient based on half-band wavelet filter. Haar wavelet is known as

The three variables are used to distinguish the four defined state. The three variables are obtained from motor current. Motor current is measured and sampled into digital data. There are three consecutive period is recorded. Each period has half cycle time period or 10 ms of 50 Hz current frequency. Then, the wavelet filter is applied to digital current spectrum to obtain the high frequency signal. In this paper the Haar wavelet third level is used because its simplicity and appropriate for investigated cases. The energy of high frequency signal is then calculated to obtain the variable input [11]. Fault cases are simulated using laboratory experiment and varied by fault current magnitude. Table I shows the data set which used in this paper. The data test is selected randomly varied by fault current magnitude. In this proposed method, the naïve Bayes classifier (NBC) is used as identification system that detecting the motor states from the three input variables. Moreover, the kernel density

the most simple wavelet family. The wavelet matrix of discrete Haar wavelet transform can be expressed using equation as follows

$$W[n] = \begin{bmatrix} \dfrac{1}{\sqrt{2}} & \dfrac{1}{\sqrt{2}} \\ \dfrac{1}{\sqrt{2}} & -\dfrac{1}{\sqrt{2}} \end{bmatrix} \tag{2}$$

Substituting equation (2) into (1) we can obtain high frequency and low frequency of discrete wavelet transform as following equation.

$$\begin{bmatrix} c_j(k) \\ d_j(k) \end{bmatrix} = \frac{1}{\sqrt{2}} \begin{bmatrix} 1 & 1 \\ 1 & -1 \end{bmatrix} \begin{bmatrix} c_{j-1}(2k) \\ d_{j-1}(2k+1) \end{bmatrix} \tag{3}$$

Rearrange equation (3), we can get equation (4) and (5) as follows

$$c_j(k) = \frac{1}{\sqrt{2}} \left(\left(c_{j-1}(2k) \right) + \left(c_{j-1}(2k+1) \right) \right) \tag{4}$$

$$d_j(k) = \frac{1}{\sqrt{2}} \left(\left(c_{j-1}(2k) \right) - \left(c_{j-1}(2k+1) \right) \right) \tag{5}$$

Substituting $j=3$ for third level transformation into equation (4) and (5) and rearranging the equation we can get the high frequency of third level transformation as follows .

$$d_3(k) = \frac{1}{2\sqrt{2}} \begin{pmatrix} c_0(8k) + c_0(8k+1) + c_0(8k+2) \\ + c_0(8k+3) - c_0(8k+4) - \\ c_0(8k+5) - c_0(8k+6) - \\ c_0(8k+7) \end{pmatrix} \tag{6}$$

B. Energy of high frequency signal of third level Haar wavelet

The energy of three consecutive current signals each 10ms periods are used as input variable detection system. The high frequency signal of third level Haar wavelet is denoted by $d3_i(t)$, where variable $i=1,2,$ and $3,$ is represent three consecutive periods signal. Input variable in_i is then calculated using the following formula.

$$in_i = \sum_{T_i} d3_i(t)^2 \tag{7}$$

C. Naïve Bayes Classifier

Naïve Bayes classifier is used as prediction of data x which belong to class C pattern using Bayes' theorem.

$$P(C|X = x) = \frac{P(X = x|C) \bullet P(C)}{P(X = x)} \tag{8}$$

Where $P(C)$ is probability of C class, $P(X=x|C)$ is class conditional probability density or likelihood for x to be generated from C class and $P(X=x)$ is normalized by unconditional density [12]. The restrictive assumption is used in NBC that between the domain features in the given class is assumed independent. Based on the assumption, the computation and decomposition of likelihood employing local probability densities can be calculated as follows.

$$P(X|C) = \prod_{i=1}^{n} P(X_i|C) \tag{9}$$

When applied into continues data sample, the BNC has major disadvantage that causing the difficulty of implementation. The probability of continues variable in the certain interval cannot be estimated similarly. Based on this disadvantage, the other methodology is needed solve this problem. The common method can be used is parametric density estimation and non parametric density estimation.

In parametric density estimation, we need a model to describe the data density. For example, Gaussian model can be used to estimate the data variance and mean. A single Gaussian estimation can be implemented because the simplicity and easy and computational application. However, when the degree of deviation of data from normality is decreased, the accuracy will drop off. And this phenomenon is very common in real application. The second method to solve the disadvantage of NBC is non parametric density estimation. In this method, the model of data density is not needed. That's mean there is no model to generate data. The density of the data is calculated based on the data set itself. One of the common non-parametric methods is kernel density estimation (KDE). This method is used in this paper.

D. Kernel Density Estimation

Kernel density estimation (KDE) is non-parametric method of estimating the probability density function of population data [13]. Conditional probability densities of each class can be calculated as follows.

$$p_s(x) = \frac{1}{S.h} \sum_{t=1}^{S} K\left(\frac{x - x_t}{h} \right) \tag{10}$$

Where K is kernel function with its bandwidth or smoothing parameter is h, and number of data belonging of its class is S. For training purposes, the data set is defined as x_t. A kernel that commonly used in KDE analysis is standard Gaussian. In this case, the width of h is defined equal to

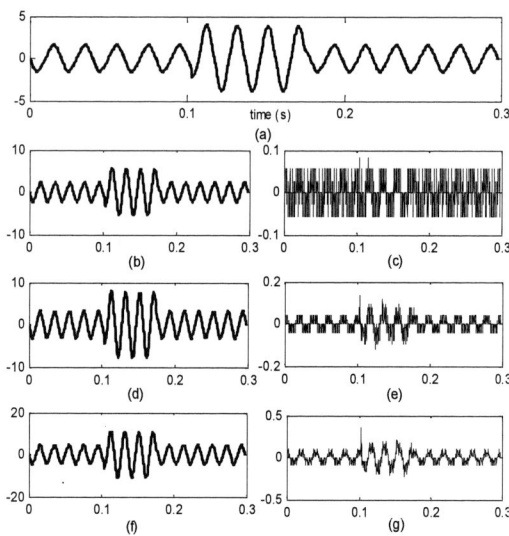

Fig.4. Wavelet decomposition of motor current signal. (a) Original current signal. (b) Low frequency (L.F) signal first level. (c) High frequency (H.F) signal first level. (d) L.F signal 2nd level. (e) H.F signal 2nd level. (f) L.F signal 3rd level. (g) H.F signal 3rd level.

$1/\sqrt{S}$. When applied data testing to the trained system, a class-conditional density is calculated as follows.

$$p\left(X_i = x_{im}^{tst} \big| C = k\right) = \frac{1}{N_{trk}} \sum_{t=1}^{N_{trk}} \frac{1}{\sqrt{2\pi}\sigma} e^{-\frac{\left(x_{im}^{tst} - x_{it}^{tr}\right)^2}{2\sigma^2}}$$

(11)

Where X_i is i-th variable, k is determined class, x_{im}^{tst} is m-th tested data and x_{it}^{tr} is training data that used. The Gaussian kernel has width σ around each of the N_{trk} training pattern of class k.

IV. RESULT AND DISCUSSION

The wavelet decomposition of the motor current containing temporary short circuit is shown in Fig. 4. The motor is initiated by normal operation and after 0.1s, short circuit fault is simulated by closing the short circuit switch. The current is increasing faster than normal current sinusoidal form and noticed as discontinues form. This phenomenon is causing higher high frequency magnitude as shown in Fig. 4 (c) and more significant noticed in Fig. 4 (g). The fault period also easily identified using magnitude of high frequency signal. The magnitude is higher than the normal operation and the density of the high frequency value is increased during starting of the fault. This phenomenon is represented by the energy of high frequency signal. When the signal has higher magnitude and density then the energy is increased.

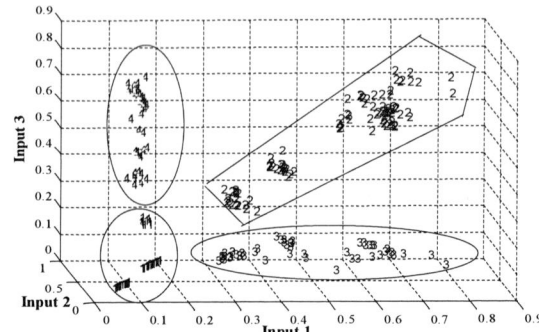

Figure 5. Training data classification illustration using 3rd level Haar wavelet transforms. (1) Normal, (2) Steady state, (3) Starting S.C, (4) Ending S.C. [5].

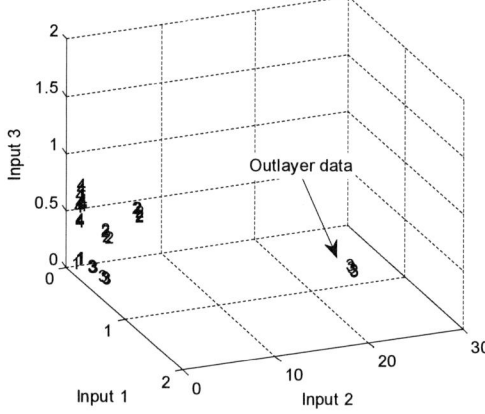

Figure 6. Testing data illustration (1) Normal, (2) Steady state, (3) Starting S.C, (4) Ending S.C.

Figure 5 shows the plot of data set according the motor state to identify the temporary fault case. There are three input variable to identify the temporary fault. The first input (Input 1) is calculated from energy of high frequency signal of first of three consecutive period current sampling, while Input 2 and Input three are from the second and third period consecutively. This data is then used as training data set of NBC detection system. Totally, 240 cases are used as data training and 50 cases are used as testing data varied by the magnitude of fault current.

Moreover, in order to confirm effectiveness of proposed method is compared with other classifier technique that is linear discriminant analysis (LDA) [11]. The different of this method with the proposed method is that LDA use the linear function to separate the group or classes of data while the proposed method based on the probability of the data belong to the group or class. The result of training system and testing is provided in confusion matrix as shown in Table II.

The result of training process shows that the proposed method is perfectly recognized all of cases. In this process, the proposed method is giving the better performance than

TABLE II
THE DETECTION PERFORMANCE

Detection Method			Actual training cases				Actual testing cases			
			N	SS	ST	EN	N	SS	ST	EN
Naïve Bayes classifier (Proposed method)	Detection cases	N	40	-	-	-	10	-	-	-
		SS	-	60	-	-	-	10	-	-
		ST	-	-	50	-	-	-	10	-
		EN	-	-	-	50	-	-	-	15
		Unknown	-	-	-	-	-	-	5	-
Wavelet-LDA		N	40	-	-	10	10	-	-	-
		SS	-	60	-	-	-	10	-	-
		ST	-	-	50	-	-	-	15	-
		EN	-	-	-	40	-	-	-	15
		Unknown	-	-	-	-	-	-	-	-

the LDA which has 10 cases false detection. Moreover, for the testing with non-trained data, the proposed method is give the 5 (five) cases error detection. The proposed method cannot identify the starting of short circuit and also not recognized to be other operating state. These cases occurred because those cases have the out layer of Input-2 as shown in Figure 6. While using the NBC-KDE methods, all of data training is tries to fit properly instead of generalizing. This case will cause a problem call overfitting. When the system is tested using data outside the training set, it can be resulting "unknown" classification [14]. As shown in Table 2, the proposed method is give perfect result in training data set, however there are five data tests are recognized as "unknown" class. The "unknown" data is shown in Fig.6 as out layer data. The value of Input-2 is around 20 of 1 as usual cases. In this range the density of the trained data is very low or no cases trained in this range. Practically, these cases are causing by the fault cases which have transient current very high due to the high percentage of shorted winding up to 75% and low impedance fault. However, the better performance is given by LDA with 100% detected fault cases. Since the LDA use the linear equation to separate the group of the data, it is possible to recognizing these cases if the input variable covered by the border function.

V. CONCLUSION AND FUTURE WORK

The probability-based classifier system is proposed in this paper to detect the temporary short circuit fault in stator winding of induction motor. The naïve Bayesian classifier with kernel density estimation is used to distinguish the operating state of the motor based on the current spectrum. The operating state of the motor is categorized based on the occurrence of temporary short circuit fault and divided into normal, starting fault, steady state and end of fault. By detecting the state of the motor, the symptom of stator deterioration can be detected, even though the current is very small and the duration very short. The third level of Haar wavelet filter is used as signal processing. The variable input is obtained from energy of high frequency signal of wavelet filter. During training process, the proposed method is successfully classifying the motor state. However, five cases of 50 testing cases are failed detecting because the second input variable is too high. This cases is shows the disadvantage of the proposed method that fail to recognize out layer data.

The advantage of the proposed method is successfully solving the problem of LDA system that failed to recognize all of data set. However, several strategies are needed to improve the proposed method such as providing additional training data to cover all of possible fault cases. Several extreme cases should be included to cover all of possible cases. Analysis of training data distribution is needed to avoid overfitting problem. The other improvements of this method are considering inverter fed drive, dynamics fault and supply voltage asymmetries. It will be more complicated since its can generates high frequency current.

VI. REFERENCES

[1] M.B.K. Bouzid, G. Champenois, "New Expressions of Symmetrical Components of the Induction Motor Under Stator Faults," IEEE Transactions on Industrial Electronics, vol.60, no.9, pp.4093-4102, Sept. 2013.
[2] N. Lahoud, J. Faucher, D. Malec, P. Maussion, "Electrical Aging of the Insulation of Low Voltage Machines: Model definition and test with the Design of Experiments," Industrial Electronics, IEEE Transactions on, vol.PP, no.99, pp.1-9.2013.

[3] S. Grubic, J. Aller, B. Lu, and T. Habetler, "A survey on testing and monitoring methods for stator insulation systems of low-voltage induction machines focusing on turn insulation problems," Industrial Electronics, IEEE Transactions on, vol. 55, no. 12, pp. 4127 –4136, dec. 2008

[4] A. Bellini, F. Filippetti, C. Tassoni, G.-A. Capolino, "Advances in Diagnostic Techniques for Induction Machines," Industrial Electronics, IEEE Transactions on , vol.55, no.12, pp.4109-4126, Dec. 2008.

[5] S.H. Kia, H. Henao, G.-A. Capolino, "Some digital signal processing techniques for induction machines diagnosis," Diagnostics for Electric Machines, Power Electronics & Drives (SDEMPED), 2011 IEEE International Symposium on , vol., no., pp.322,329, 5-8 Sept. 2011.

[6] J . Cusido, L. Romeral, J.A Ortega, J.A Rosero, A.G Espinosa, "Fault Detection in Induction Machines Using Power Spectral Density in Wavelet Decomposition," IEEE Transactions on Industrial Electronics, vol.55, no.2, pp.633,643, Feb. 2008.

[7] M. Seera, C.P. Lim, D. Ishak, H. Singh, "Fault Detection and Diagnosis of Induction Motors Using Motor Current Signature Analysis and a Hybrid FMM–CART Model," IEEE Transactions on Neural Networks and Learning Systems, vol.23, no.1, pp.97-108, Jan. 2012.

[8] A. Kelemen, Hong Zhou, P. Lawhead, Yulan Liang, "Naive Bayesian classifier for microarray data ", Proceedings of the International Joint Conference on Neural Networks 2003, Vol. 3, pp. 1769 - 1773, July 2003.

[9] S.P. Santos, J.A.F. Costa, "A Comparison between Hybrid and Non-hybrid Classifiers in Diagnosis of Induction Motor Faults," The 11th IEEE International Conference on Computational Science and Engineering, 2008. pp.301-306, 16-18 July 2008.

[10] D.A Asfani, Syafaruddin, M.H Purnomo, T. Hiyama,"Temporary Short Circuit Detection in Induction Motor Winding Using Second Level Haar-Wavelet Transform", IEEJ Transactions on Industry Applications, Volume 131, Issue 9, pp. 1093-1102, 2011.

[11] D.A. Asfani, Syafaruddin, M.H. Purnomo, T. Hiyama, "Wavelet-LDA-neural network based short circuit occurrence detection in induction motor winding," Diagnostics for Electric Machines, Power Electronics & Drives (SDEMPED), 2011 IEEE International Symposium on , vol., no., pp.330,336, 5-8 Sept. 2011.

[12] Y. Gurwicz, B. Lerner, "Bayesian network classification using spline-approximated kernel density estimation", Pattern Recognition Letters 26 (2005) 1761–1771.

[13] Y. Murakami, K. Mizuguchi, "Applying the Naïve Bayes classifier with kernel density estimation to the prediction of protein–protein interaction sites", Bioinformatics, Vol. 26 no. 15 , pages 1841–1848, 2010.

[14] J. Cheng, R. Greiner, Comparing Bayesian network classifiers, in: Proceedings of the 15th Conference on Uncertainty in Artificial Intelligence (UAI-1999), Morgan Kaufmann Publishers. Inc., San Francisco, CA, USA, 1999, pp. 101–108.

VII. BIOGRAPHIES

Dimas Anton Asfani (S'10) received his B.Eng, M.T degrees in Electrical Engineering from Institut Teknologi Sepuluh Nopember Surabaya, Indonesia, in 2004 and 2006. Ph.D. degree in the Power system Laboratory Kumamoto University in 2012..He was joined Institut Teknologi Sepuluh Nopember, Surabaya, Indonesia in 2005 as lecture and research asistent. His research concentrates mainly on fault detection in Induction machine and Power system protection.

Mauridhi Heri Purnomo (Member) received the B.S degree in power system engineering from Institut Teknologi Sepuluh Nopember Surabaya, Indonesia, in 1985. M.Eng and Ph.D degrees in control engineering and Intelligence System from Osaka City Univ. Japan, in 1995 and 1998. He is currently a Professor of Departement of Electrical Enginering. His research interest include fuzzy control, intelligence control and control applications.

Dian R. Sawitri is doctoral student in Institut Teknologi Sepuluh Nopember (ITS), Surabaya, Indonesia. She received M.Eng degree in electrical engineering from Gadjah Mada University, Yogyakarta, Indonesia in 2002 and Bachelor Degree from Diponegoro University in 1993. She is currently working in electrical engineering department, Dian Nuswantoro University in Semarang, Indonesia. Her research interest include intelligent system application in power system, power electronic and power quality.

Analytical Study of Pulsating Torque and Harmonic Components in Rotor Current of Six-Phase Induction Motor under Healthy and Faulty Conditions

Yassine MAOUCHE; *Student Member IEEE*, Abdelfettah BOUSSAID; *Student Member IEEE*,
Mohamed BOUCHERMA; *Member IEEE* and Abdelmalek KHEZZAR; *Member IEEE*
Laboratoire d'Électrotechnique de Constantine LEC
Département d'électrotechnique
Université Constantine 1
25000, Constantine, Algeria
Email: Yassine.MAOUCHE@lec-umc.org

Abstract—In literature, open phase faults in multi-phase induction motor have been investigated. However rotor faults and their impact on the performance of multi-phase induction motor aren't widely discussed. An analytical study of the pulsating torque and harmonic components in rotor current of six-phase induction machine as a function of displacement between sub-windings in healthy and faulty cases (the rotor broken bar) is developed. The novel formula of pulsation torques of six-phase induction motor presented in this paper can be used as a fault indicator for broken rotor bars. The used model is based on multiple coupled circuits and takes into account the geometry and winding layout, including the calculation of pulsating torques. The simulation results show the disappearance of some higher rotor slot harmonics in torque spectrums as well as a significant improvement in the rotor current harmonics for a specific displacement between sub-windings.

Keywords—*Broken bar fault, condition monitoring, faut detection, faut diagnosis, Fourier transforms, six-phase induction motor, spectrum analysis, torque.*

I. INTRODUCTION

IN the last years, the interest in multiphase induction motor has been growing in industrial; it is adopted for their greater fault tolerance in open phases. The increasing of phase number reduces also the current of each power electronic device and provides suitable solutions in high-power. Multiphase induction machines have been getting much attention in the literature, as summarized in [1]. Their advantages have been identified in many works. Multiphase machine offer higher efficiency [2], [3], reduced torque pulsations [4], and gives greater fault tolerance in open phases [5].

On the other hand, the specific number extensively discussed is the six-phase induction machine [6], [8]. In [7, 9, 10] , cope with control aspects of six-phase induction machine there has been significant importance in fault tolerance. Even most of these papers suppose ideal sinusoidal windings distribution which is more suitable for control design, neglecting harmonic losses in rotor current and pulsating torques can be considered as a real drawback.

The six-phase is similar to three-phase case; it can be submitted to external and internal stresses of various natures, degradation can occur in their performances. The broken bar fault can damage motor performance because it can cause the torque and speed ripple, decrease average torque, noise and imbalance of stator current. For these reasons, amount of paper has been published dealing with broken bar fault detection. Most of them are based on use of motor current signature analysis (MCSA) techniques [15]-[18]. These techniques investigate the fault indicative frequencies around fundamental supply frequency or by surveillance the rotor space harmonic components as a fault indicator but in this case a high sampling frequency value is need. The main advantage of these techniques is that cheap equipment, only one line current is needed to detect broken bar fault. These techniques are used also in modern speed detection [11] and [12] in modern vector control technique [13]. The short-pitched stator winding is not able to attenuate the space harmonic components which are much interest in squirrel cage induction motor [24]. On the other hand, it is possible that the induction motor does not generate space harmonic components when the number of rotor bars is odd.

Other researchers have been investigated the influence of the broken bar fault on the instantaneous power, instantaneous power factor, current Park compenents, inverter input current analysis and supply voltage modulation in case the machine is supplied power electronic converters [15]-[22]. The benefit of these techniques is that the frequency components due to broken bar fault are next to zero frequency and close to the dc component witch facilitate the detection of lower and upper sidebands currents [23]. Broken bar fault detection through torque monitoring is also investigated [25]–[29], for instance in [29] shows the influence of the broken bar fault only on the frequency range around 300 Hz of the spectrum of the electromagnetic torque instead of monitoring the traditional low frequencies components [25]–[28].

The aim of this paper is to develop a motor model of six-phase as a function of displacement between winding-sets. The model is based on the winding functions and takes into account the real winding distributions. Firstly the harmonic components of the rotor current and torque for a healthy machine are analytically analysed and confirmed by simulation results. Then, the proposed study is used to analyze the impact of the broken rotor bars on harmonic components in both rotor current and torque, the displacement between winding-sets is also considered.

978-1-4799-0024-4/13 $31.00 © 2013 IEEE

II. ANALYTICAL MODEL FOR INDUCTION MACHINE

Compared with a three-phase machine, the six-phase induction machine has the same rotor and the same magnetic core, but the stator windings are split into two three-phase sets shifted by α electrical degrees Fig.1.

As a consequence, the mathematical modeling approach of symmetrical and asymmetrical six-phase induction machines is also similar to its three-phase counterpart. To have a clear nomenclature, the terms ASP and SSP are adopted for the machine to indicate the two separate sets of three phase windings shifted respectively by 30 and 60 electrical degrees, and TP for the ordinary three-phase machine. The original three-phase machine can rewounded as shown in Fig.2.

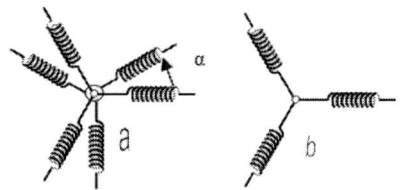

Fig. 1. (a) ASP and SSP (b) TP winding configurations.

Traditional three-phase winding

Symmetrical six-phase (SSP) winding

Asymmetrical six –phase (ASP) winding

Fig. 2. Stator windings.

A. Electrical equations

Assume that the two three-phase windings shifted by α electrical degrees are Y connected with isolated neutral and the rotor cage having n_b bars is viewed as n_b identical spaced loops and the current distribution can be specified in terms of $n_b + 1$ independent rotor currents. These currents are formed of n_b rotor loop currents $[i_r]$ plus a current i_e circulating in one of the end rings.

The primary equations of the induction machine can be written in vector-matrix form as follows:

$$[v_s] = [R_s][i_s] + \frac{d}{dt}[\psi_s] + [v_n] \tag{1}$$

$$\begin{bmatrix} [v_r] \\ v_e \end{bmatrix} = [0] = \begin{bmatrix} [R_r] & \frac{R_e}{n_b} \\ & \vdots \\ \frac{R_e}{n_b} & \cdots & R_e \end{bmatrix} \begin{bmatrix} [i_r] \\ i_e \end{bmatrix} + \frac{d}{dt}[\psi_r] \tag{2}$$

$$[\psi_s] = [L_s] \cdot [i_s] + [M_{sr}] \cdot [i_r] \tag{3}$$

where: $[v_s] = [v_{sa} \quad v_{sd} \quad v_{sb} \quad v_{se} \quad v_{sc} \quad v_{sf}]^t$, $[i_s] = [i_{sa} \quad i_{sd} \quad i_{sb} \quad i_{se} \quad i_{sc} \quad i_{sf}]^t$ and $[\psi_s] = [\psi_{sa} \quad \psi_{sd} \quad \psi_{sb} \quad \psi_{se} \quad \psi_{sc} \quad \psi_{sf}]^t$ are the stator voltage, current and flux vectors, respectively. $[v_n]$ is the neutral to neutral vector voltage.

$[v_r] = [v_{r1} \quad \cdots \quad v_{rn_b}]^t$, $[i_r] = [i_{r1} \quad \cdots \quad i_{rn_b}]^t$ and $[\psi_r] = [\psi_{r1} \quad \cdots \quad \psi_{rn_b}]^t$ are the rotor voltage, current and flux linkage vectors, respectively. $[R_s]$ and $[R_r]$ are the stator and rotor resistance matrices which are given by:

$$[R_s] = R_s \cdot [1]_{6 \times 6} \tag{4}$$

$$[R_r] = \begin{bmatrix} 2(R_b + R_e) & -R_b & 0 & 0 & \cdots & -R_b \\ -R_b & 2(R_b + R_e) & -R_b & 0 & \cdots & 0 \\ 0 & -R_b & 2(R_b + R_e) & -R_b & \cdots & 0 \\ 0 & 0 & -R_b & 2(R_b + R_e) & \cdots & 0 \\ \vdots & \vdots & \vdots & \ddots & \ddots & \vdots \\ -R_b & 0 & 0 & \cdots & -R_b & 2(R_b + R_e) \end{bmatrix} \tag{5}$$

The mutual inductance matrix can be expressed as:

$$[M_{sr}] = \sum_{h=1}^{\infty} M_{sr}^h \times \begin{bmatrix} \cos(h(\theta + \phi_h)) & \cdots & \cos(h(\theta + \phi_h + (k-1)a)) & \cdots \\ \cos(h(\theta + \phi_h - \alpha)) & \cdots & \cos(h(\theta + \phi_h + (k-1)a - \alpha)) & \cdots \\ \cos(h(\theta + \phi_h - \frac{2\pi}{3})) & \cdots & \cos(h(\theta + \phi_h + (k-1)a - \frac{2\pi}{3})) & \cdots \\ \cos(h(\theta + \phi_h - \frac{2\pi}{3} - \alpha)) & \cdots & \cos(h(\theta + \phi_h + (k-1)a - \frac{2\pi}{3} - \alpha)) & \cdots \\ \cos(h(\theta + \phi_h - \frac{4\pi}{3})) & \cdots & \cos(h(\theta + \phi_h + (k-1)a - \frac{4\pi}{3})) & \cdots \\ \cos(h(\theta + \phi_h - \frac{4\pi}{3} - \alpha)) & \cdots & \cos(h(\theta + \phi_h + (k-1)a - \frac{4\pi}{3} - \alpha)) & \cdots \end{bmatrix} \tag{6}$$

where: $a = p\frac{2\pi}{n_b}$ is the electrical angle of a rotor loop, ϕ_h is the initial phase angle,
$h = 1$ and $h = (6\nu \pm 1)_{\nu=1,2,\ldots}$, and $k = 1 \cdots n_b$ is a counter of a rotor loops.

B. Mechanical equations

The mechanical equations can be expressed as:

$$\frac{d\omega}{dt} = \frac{P}{J}(\Gamma_e - \Gamma_r) \tag{7}$$

$$\frac{d\theta}{dt} = \omega \tag{8}$$

where Γ_e is the electromagnetic torque, Γ_r is the load torque and ω is the electrical rotor speed. Using the basic principle of energy conversion, the torque developed by the machine Γ_e can be obtained by considering the change in co-energy W_{co} of the system produced by a small change in rotor position when the currents are held constant.

$$\Gamma_e = \left[\frac{\partial W_{co}}{\partial \theta_{mec}}\right]_{(i_s, i_{rn}.const)} \tag{9}$$

where $\theta_{mec} = P \times \theta$ is the stator rotor mechanical angular. This gives:

$$\Gamma_e = [i_s]^t \frac{\partial [M_{sr}]}{\partial \theta_{mec}} \cdot [i_r] \tag{10}$$

III. ANALYTICAL EXPRESSIONS FOR THE ROTOR CURRENT AND THE ELECTROMAGNETIC TORQUE

Only the rotor flux is needed to describe the different harmonic components in the rotor currents.

$$[\psi_{rk}] = [L_r] \cdot [i_{rk}] + [M_{sr1}] \cdot [i_{s1abc}] + [M_{sr2}] \cdot [i_{s2abc}] \tag{11}$$

where

$$[M_{sr1}] = \sum_{h=1}^{\infty} M_{sr}^h \cos\left[h\left(\theta + \phi_h + ka - (i-1)\frac{2\pi}{3}\right)\right] \tag{12}$$

$$[M_{sr2}] = \sum_{h=1}^{\infty} M_{sr}^h \times$$
$$\cos\left[h\left(\theta + \phi_h + ka - (i-1)\frac{2\pi}{3} - \alpha\right)\right] \tag{13}$$

$[i_{s1abc}]$ and $[i_{s2abc}]$ are the stator line current vectors given for each i phase and h harmonic orders by:

$$i_{s1_{h_s}} = I_{sh_s} \cos\left[\theta_{sh_s} - h_s(i-1)\frac{2\pi}{3}\right] \tag{14}$$

$$i_{s2_{h_s}} = I_{sh_s} \cos\left[\theta_{sh_s} - h_s(i-1)\frac{2\pi}{3} - h_s\alpha\right] \tag{15}$$

θ_{sh_s} is the stator angle of the h_s harmonic component related to frequency f_{sh_s} defined by:

$$\theta_{sh_s} = f_{sh_s}t + \theta_{0hs} \tag{16}$$

$$[\psi_{rk}] = [L_r] \cdot [I_{rk}] + \frac{1}{2}\sum_{i=1}^{3}\sum_{h}^{\infty}\sum_{h_s}^{\infty} M_{sr}^h I_{sh_s} \times$$

$$\left[\cos\left(\theta_{sh_s} - h\theta - h(k-1)a - (h_s-h)(i-1)\frac{2\pi}{3}\right) +\right.$$

$$\cos\left(\theta_{sh_s} + h\theta + h(k-1)a - (h_s+h)(i-1)\frac{2\pi}{3}\right) + \tag{17}$$

$$\cos\left(\theta_{sh_s} - h\theta - h(k-1)a - (h_s-h)(i-1)\frac{2\pi}{3} - (h_s-h)\alpha\right) +$$

$$\left.\cos\left(\theta_{sh_s} + h\theta + h(k-1)a - (h_s+h)(i-1)\frac{2\pi}{3} - (h_s+h)\alpha\right)\right]$$

The electromagnetic torque can be developed using (10), $[i_r]$ is the rotor line current vector, with the kth elements for the h_r harmonic component given by:

$$i_{rh_r,k} = I_{rhr}\cos\left(\theta_{rh_r} - h_r ka\right) \tag{18}$$

θ_{rh_r} is the stator angle of the h_r harmonic component related to frequency f_{rhr}

$$\theta_{rh_r} = f_{rh_r}t + \theta_{0hr} \tag{19}$$

For the electromagnetic expression Γ_e see the top of the next page.

IV. THE ROTOR CURRENT FREQUENCIES AND TORQUE PULSATION COMPONENTS IN HEALTHY CONDITION

We suppose that the two sub-windings of induction motor shifted by α are supplied by a symmetrical sinusoidal voltage system and with a constant load point.

Under this condition the stator current has a series of high-frequency components belong to the following set [14]

$$h_s = \left\{h = \left(\lambda\frac{n_b}{p} \pm 1\right)_{\lambda=1,2...} \cap\atop h = (6\gamma \pm 1)_{\gamma=1,2,3....}\right\} \tag{21}$$

The simulated machine has 24 stator slots 22 bars, 2 poles and a rated power of 1.5 kW. Calculations were made using MATLAB simulation software. Fig.3 shows the stator current harmonic components of TP, SSP and ASP induction motors in healthy conditions. Additional to the fundamental harmonic of stator current a serie of higher-harmonic component frequencies are given by the following expression:

$$f_{sh_s}(h_s) = (h_s(1-s) \pm s) \cdot f_s \tag{22}$$

Using expression (17) the rotor harmonic components can be exprssed as:

$$f_{rh_r}(h_s, h) = f_{sh_s}(h_s) \pm h(1-s)f_s$$
$$= (h_s(1-s) \pm s)f_s \pm h(1-s)f_s \tag{23}$$

Substituting (18), (14), (15) and (6) in (10) leads to:

$$\Gamma_e = -p \sum_{h=1}^{\infty} \sum_{k=1}^{n_b} \sum_{i=1}^{3} \sum_{h_s}^{\infty} \sum_{h_r}^{\infty} I_{rh_r} h I_{sh_s} M_{sr}^h \times$$

$$\left[\sin\left(h\theta + \theta_{rh_r} + \theta_{sh_s} + (h-1)ka - (h+1)(i-1)\frac{2\pi}{3} - h\phi_h\right) + \right.$$

$$\sin\left(h\theta + \theta_{rh_r} - \theta_{sh_s} + (h-1)ka - (h-1)(i-1)\frac{2\pi}{3} - h\phi_h\right) +$$

$$\sin\left(h\theta - \theta_{rhr} + \theta_{sh_s} + (h+1)ka - (h+1)(i-1)\frac{2\pi}{3} - h\phi_h\right) +$$

$$\sin\left(h\theta - \theta_{rh_r} - \theta_{sh_s} + (h+1)ka - (h-1)(i-1)\frac{2\pi}{3} - h\phi_h\right) + \qquad (20)$$

$$\sin\left(h\theta + \theta_{rh_r} + \theta_{sh_s} + (h-1)ka - (h+1)(i-1)\frac{2\pi}{3} - h\phi_h + (h_s+h)\alpha\right) +$$

$$\sin\left(h\theta + \theta_{rh_r} - \theta_{sh_s} + (h-1)ka - (h-1)(i-1)\frac{2\pi}{3} - h\phi_h - (h_s-h)\alpha\right) +$$

$$\sin\left(h\theta - \theta_{rh_r} + \theta_{sh_s} + (h+1)ka - (h+1)(i-1)\frac{2\pi}{3} - h\phi_h + (h_s+h)\alpha\right) +$$

$$\left. \sin\left(h\theta - \theta_{rh_r} - \theta_{sh_s} + (h+1)ka - (h-1)(i-1)\frac{2\pi}{3} - h\phi_h - (h_s-h)\alpha\right) \right]$$

The existence of these components in the rotor current is the combination of the harmonic components in the stator currents and the harmonic components of the mutual inductances.

From equation (17) we deduce the following condition:

$$\left\{ h \pm h_s = 3\gamma|_{\gamma=0,1,2,\cdots} \right\} \qquad (24)$$

where $h = 1$ and $h = 6k \pm 1|_{k=0,1,2,\cdots}$ and h_s is given by (21).

The analysis of this condition gives two different cases:

The first one is that TP and SSP of induction machines where the harmonic ranks of rotor currents are given by:

$$h_r = 1 \text{ and } 6k \pm 1|_{k=1,2,\cdots} \qquad (25)$$

and the harmonic frequencies by:

$$\begin{aligned} f_{rh_r}(h_r) &= h_r \cdot f_s \\ &= s f_s \text{ and } (6k(1-s) \pm s) f_s|_{k=1,2,\cdots} \end{aligned} \qquad (26)$$

The second one is that of the ASP machine where one can notice in (17) that

$$(h_s \pm h)\alpha = \pi \qquad (27)$$

this leads to the cancellation of odd rows of frequencies in relation (17) and the ranks of the harmonic components in the rotor currents become:

$$h_r = 1 \text{ and } 12k \pm 1|_{k=1,2,\cdots} \qquad (28)$$

and the harmonic frequencies as:

$$\begin{aligned} f_{rh_r}(h_r) &= h_r \cdot f_s \\ &= s f_s \text{ and } (12k(1-s) \pm s) f_s|_{k=1,2,\cdots} \end{aligned} \qquad (29)$$

Fig.4 shows the rotor loop current spectrum in the case of TP or SSP. One can notice the appearance of harmonic components in concordance with (26) where the harmonic frequencies are:

$$\left\{ \begin{aligned} & s f_s, (6-7s) f_s, (6-5s) f_s, (12-13s) f_s, \\ & (12-11s) f_s, (18-19s) f_s, (18-17s) f_s, \\ & (24-25s) f_s, (24-23s) f_s, \cdots \end{aligned} \right\}$$

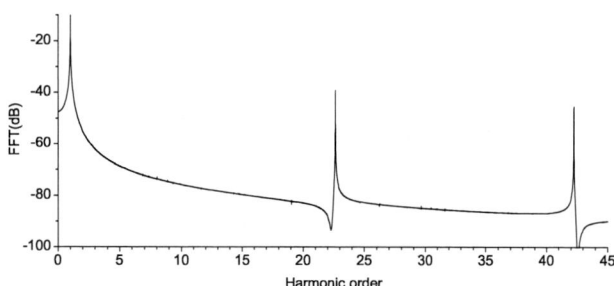

Fig. 3. FFT spectrum of stator current of a healthy induction machine.

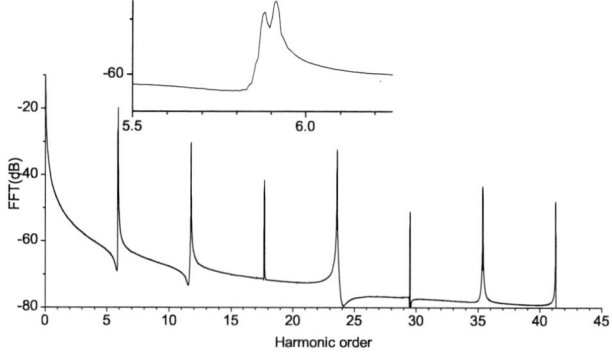

Fig. 4. FFT spectrum of rotor loop current of healthy SSP or TP machine.

978-1-4799-0024-4/13 $31.00 © 2013 IEEE

Fig.5 shows the rotor loop current spectrum in healthy condition of ASP induction motor. The sumilation results confirm the anlytical relation given by (29) which gives the harmonic frequencies at:

$$\left\{ sf_s, (12-13s)f_s, (12-11s)f_s, \right.$$
$$\left. (24-25s)f_s, (24-23s)f_s, \cdots \right\}$$

By the same way and by analyzing the relation giving the electromagnetic torque (20), the harmonic stator current orders contributing to the creation of the pulsating torque in the case of TP or SSP machines are:

$$h_s = \left\{ \left(h = 6k \pm 1 \big|_{k=1,3,5,\cdots} \cap \right. \right.$$
$$\left. \left. h = \frac{kn_b}{p} \pm 1 \big|_{\lambda=1,2,3,\cdots} \right) \right\} \quad (30)$$

at the frequencies :

$$f_{Th}(h_s, h_r, h) = h(1-s)f_s \pm f_{rh_r} - \zeta f_{sh_s} \quad (31)$$

where:

$$\zeta = \left\{ \begin{array}{ll} +1 & \text{if } h_s \in F \\ -1 & \text{if } h_s \in G \end{array} \right.$$

$F = \{1, 7, 13, 19...\}$ *is the set of forward components.*

$G = \{5, 11, 17...\}$ *is the set of backward components.*

which are confirmed by the sumilation results reported on Fig.6 where the stator current harmonics are related to the orders:

$$\{1, 23, 43, ...\}$$

therefore the torque harmonics frequencies are:

$$\{0, (24-22s)f_s, (42-44s)f_s, ...\}$$

In the case of ASP machine and by analyzing equations (20) and (27) , we get the following set of stator current harmonic components orders:

$$h_s = \left\{ \left(h = 12k \pm 1 \big|_{k=1,2,3,\cdots} \cap \right. \right.$$
$$\left. \left. h = \frac{kn_b}{p} \pm 1 \big|_{\lambda=1,2,3,\cdots} \right) \right\} \quad (32)$$

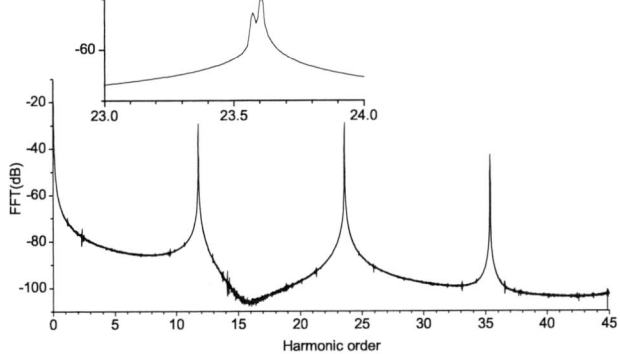

Fig. 5. FFT spectrum of rotor loop current of healthy ASP induction machine.

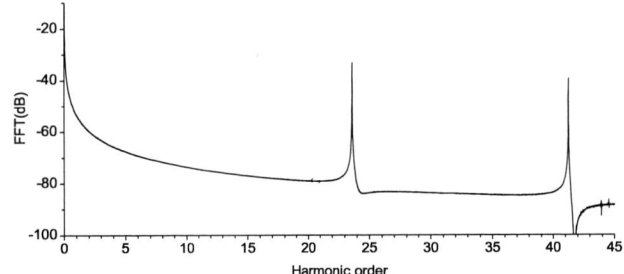

Fig. 6. FFT spectrum of torque of a healthy TP and SSP induction machine.

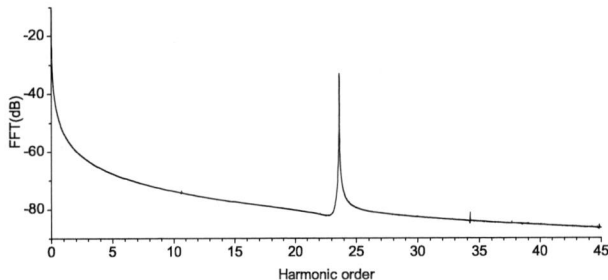

Fig. 7. FFT spectrum of torque of a healthy ASP induction machine.

The simultaion results as gives on Fig. 7 show the harmonic torque components at the frequencies:

$$\{0, (24-22s)f_s, ...\}$$

One can notice that there is less harmonic components in the electromagnetic torque in the case of ASP machine than in the cases of TP or SSP machines. In addition to this, the existence of rotor slot harmonics in case of six-phase winding in torque spectrum is depended of the displacement between winding sets, number of pole pairs and number of rotor bars.

V. THE ROTOR CURRENT FREQUENCIES AND TORQUE PULSATION COMPONENTS IN FAULTY CONDITION (BROKEN ROTOR BARS)

A fault on the rotor, such as broken rotor bars or constructional dissymmetry, causes asymmetrical working condition within the rotor. This dissymmetry is simulated as a variation in the rotor bar resistance. The maximum values of the rotor loop currents are not equal. In this case, the rotor faults induce additional RSHs components in the stator current frequencies [14]:

$$f_{sh_s}(h_s, k) = (h_s(1-s) \pm s \pm 2ks)f_s \quad (33)$$

where: $h_s = \left\{ h = (6\gamma \pm 1)_{\gamma=1,2,3....} \right\}$

Using (23) and (33) the rotor current harmonic components caused by the RSHs harmonics of rotor asymmetry are:

$$f_{rh}(h_s, h, k) = (h_s \pm 2ks)f_s \pm h(1-s)f_s$$
$$= f_{rh}(h_s, h, k)\big|_{healthy \; conditions} \pm 2ks \quad (34)$$

Figs.8 and 9 present the spectrum of the rotor current, we can notice that the broken bar fault causes lower and upper side bands close to the existent harmonic components in healthy condition at equal distance $\pm 2ks$.

In the same manner appear new additional components of torque induced by the RSHs components of stator current. Using (20) and (33) the torque harmonic frequencies become:

$$f_{Th}(h_s, h_r, h) = f_{Th}(h_s, h_r, h)|_{healthy\ conditions} \pm 2ks \quad (35)$$

where $k = 1, 2, 3, \cdots$

In the case of TP or SSP condition (30) becomes :

$$h_s = \left\{ \left(h = 6k \pm 1|_{k=1,3,5,\cdots} \right) \right\} \quad (36)$$

In the case of ASP machine condition (32) becomes :

$$h_s = \left\{ \left(h = 12k \pm 1|_{k=1,2,3,\cdots} \right) \right\} \quad (37)$$

Fig.10 presents the spectrum of torque in the case of TP and SSP, and their corresponding frequencies as described by (35) are:

$$\left\{ \begin{array}{l} (0 \pm 2ks)f_s, (6(1-s) \pm 2ks)f_s, (12(1-s) \pm 2ks)f_s, \\ (18(1-s) \pm 2ks)f_s, (24(1-s) \pm 2ks)f_s, (30(1-s) \pm \\ 2ks)f_s, (36(1-s) \pm 2ks)f_s, (42(1-s) \pm 2ks)f_s, \cdots \end{array} \right\}$$

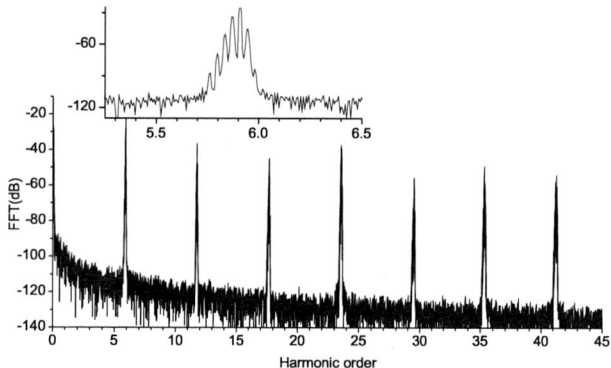

Fig. 8. FFT spectrum of rotor loop current of SSP and TP induction machine with one broken bar.

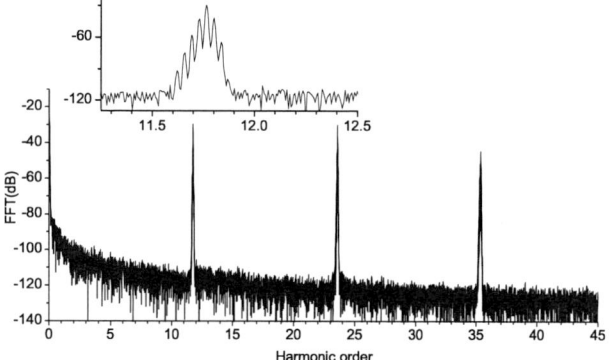

Fig. 9. FFT spectrum of rotor loop current of ASP induction machine with one broken bar.

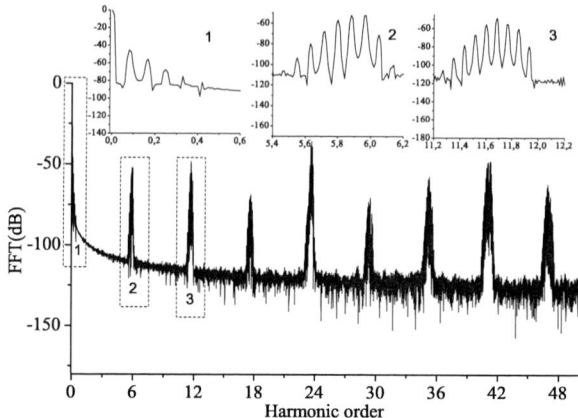

Fig. 10. FFT spectrum of torque of SSP and TP induction machine with one broken bar.

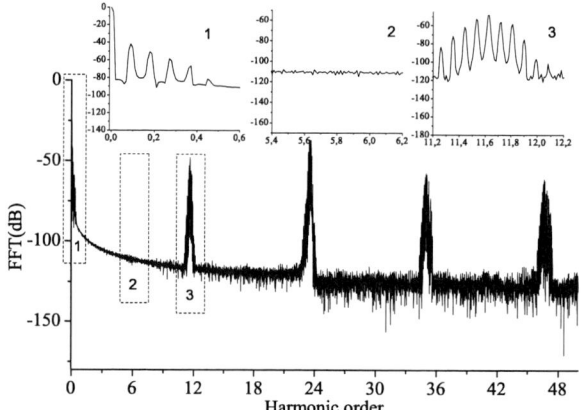

Fig. 11. FFT spectrum of torque of ASP induction machine with one broken bar.

Fig.11 presents the spectrum of torque in the case of ASP, and their additional harmonics described by (35) and their frequencies are:

$$\left\{ \begin{array}{l} (0 \pm 2ks)f_s, (12(1-s) \pm 2ks)f_s, \\ (24(1-s) \pm 2ks)f_s, (36(1-s) \pm 2ks)f_s \cdots \end{array} \right\}$$

It is remarkable that the broken bar fault in both torque and rotor current spectrums generates less singature in case of ASP winding compared to the SSP and TP windings and the singature of broken bar fault is depended of the displacement between winding sets.

VI. CONCLUSION

An analytical study of pulsating torque and harmonic components in rotor current of six-phase induction machine as a function of displacement between winding sets in healthy and faulty conditions (the rotor electrical fault) was presented. The developed expressions for both pulsating torque and rotor current are useful for the choice of the multi-phase induction motors congurations and also for the diagnosis purposes.

978-1-4799-0024-4/13 $31.00 © 2013 IEEE

REFERENCES

[1] Singh, G.K.: "Multi-phase induction motor drive research a survey", Electr. Power Syst. Res., 2002, 61, pp. 139147

[2] Toliyat, H.A., Lipo, T.A., andWhite, J.C.: "Analysis of a concentrated winding induction machine for adjustable speed drive applications Part 2": motor design and performance, IEEE Trans. Energy Convers., 1991, 6, (4), pp. 684692

[3] J. M. Apsley , S. Williamson , A. C. Smith and M. Barnes "Induction motor performance as a function of phase number", Proc. Inst. Elect. Eng.-Elect. Power Appl., vol. 153, no. 6, pp.898 -904 2006.

[4] Jahns, T.M.: "Improved reliability in solid state AC drives by means of multiple independent phase drive units", IEEE Trans. Ind. Appl., 1980,16, (3), pp. 321331.

[5] L. Alberti and N. Bianchi "Experimental Tests of Dual Three-Phase Induction Motor Under Faulty Operating Condition", IEEE Transactions On Industrial Electronics, VOL. 59, NO. 5,MAY 2012

[6] A. Boglietti , R. Bojoi , A. Cavagnino and A. Tenconi "Efficiency analysis of PWM inverter fed three-phase and dual three-phase high frequency induction machines for low/medium power applications", IEEE Trans. Ind. Electron., vol. 55, no. 5, pp.2015 -2023 2008

[7] Lyra, R.O.C., and Lipo, T.A.: "Torque density improvement in a six-phase induction machine with third harmonic current injection", IEEE Trans. Ind. Appl., 2002, 38, (5), pp. 13511360

[8] Bojoi, R., Lazzari, M., Profumo, F., and Tenconi, A.: "Digital field oriented control for dual three-phase induction motor drives", IEEE Trans. Ind. Appl., 2003, 39, (3), pp. 752760

[9] E. Levi , R. Bojoi , F. Profumo , H. A. Toliyat and S. Williamson "Multiphase induction motor drives A technology status review", IET Electr. Power Appl., vol. 1, no. 4, pp.489 -516 2007

[10] Toliyat, H.A.: "Analysis and simulation of five-phase variable-speed induction motor drives under asymmetrical connections", IEEE Trans. Power Electron., 1998, 13, (4), pp. 748756

[11] J. Jiang and J. Holtz, "High dynamic speed sensorless AC drive with online model parameter tuning for steady-state accuracy," IEEE Trans. Ind. Electron., vol. 44, no. 2, pp. 240246, Apr. 1997.

[12] P. Vaclavek and P. Blaha, "Speed estimation scheme for small AC induction machine sensorless control," in Proc. 33rd IEEE IECON, Nov. 58,2007,pp.986991.

[13] C. S. Staines, C. Caruana, G. M. Asher, and M. Sumner, "Sensorlesscontrol of induction machines at zero and low frequency using zero sequence currents," IEEE Trans. Ind. Electron., vol. 53, no. 1, pp. 195206,Dec. 2005.

[14] A. Khezzar, M.Y. Kaikaa, , M. E. Oumaamar, M. Boucherma and H .Razik " On the Use of Slot Harmonics as a Potential Indicator of Rotor Bar Breakage in the Induction Machine"; , vol. 56 Issue: 11, 2009, pp. 4592 - 4605

[15] J.-H. Jung, J.-J. Lee, and B.-H. Kwon, "Online diagnosis of induction motors using MCSA," IEEE Trans. Ind. Electron., vol. 53, no. 6, pp. 18421852, Dec. 2006.

[16] S. H. Kia, H. Henao, and G. A. Capolino, "A high-resolution frequency estimation method for three-phase induction machine fault detection," IEEE Trans. Ind. Electron., vol. 54, no. 4, pp. 23052314, Aug. 2007.

[17] S. Nandi, H. A. Toliyat, and X. Li, "Condition monitoring and fault diagnosis of electrical motorsA review," IEEE Trans. Energy Convers., vol. 20, no. 4, pp. 719729, Sep. 2005.

[18] H. Henao, H. Razik, and G.-A. Capolino, "Analytical approach of the stator current frequency harmonics computation for detection of induction machine rotor faults," IEEE Trans. Ind. Appl., vol. 41, no. 3, pp. 801807, May/Jun. 2005.

[19] M. Eltabach, A. Charara, I. Zein, and M. Sidahmed, "Detection of broken rotor bar of induction motors by spectral analysis of the electromagnetic torque using luenberger observer," in Proc. IEEE 27th Annu. Conf. Ind.Eletron. Soc., Denver, CO, USA, Nov. 29Dec. 02 2001, pp. 658663.

[20] V. V. Thomas, K. Vasudevan, and V. J. Kumar, "Online cage rotor fault detection using air-gap torque spectra," IEEE Trans. Energy Convers., vol. 18, no. 2, pp. 265270, Jun. 2003.

[21] M. Drif and A. Cardoso "Discriminating the simultaneous occurrence of three-phase induction motor rotor faults and mechanical load oscil-lations by the instantaneous active and reactive power media signature analyses", IEEE Trans. Ind. Electron., vol. 59, no. 3, pp.1630 -1639 2012

[22] M. Nemec , K. Drobnic , D. Nedeljkovic , R. Fiser and V. Ambrozic "Detection of broken bars in induction motor through the analysis of supply voltage modulation", IEEE Trans. Ind. Electron., vol. 57, no. 8, pp.2879 -2888 2010

[23] C. Kral, T. G. Habetler, R. G. Harley, F. Pirker, G. Pascoli, H. Oberguggenberger, and C. J. M. Fenz, "A comparison of rotor fault detection techniques with respect to the assessment of fault severity," in Proc. SDEMPED, Atlanta, GA, Aug. 2426, 2003, pp. 265270.

[24] T. Jokinen and V. Hrabovcova, "Design of Rotating Electrical Machines".Hoboken, NJ: Wiley, 2009.

[25] J. S. Hsu, "Monitoring of defects in induction motors through air-gap torque observation," IEEE Trans. Ind. Appl., vol. 31, no. 5, pp. 1016 1021, Sep./Oct. 1995.

[26] V. V. Thomas, K. Vasudevan, and V. J. Kumar, "Online cage rotor fault detection using air-gap torque spectra," IEEE Trans. Energy Convers., vol. 18, no. 2, pp. 265270, Jun. 2003.

[27] S. J. Manolas and J. A. Tegopoulos, "Analysis of squirrel cageinduction motors with broken bars and rings," IEEE Trans. Energy Convers., vol. 14, no. 4, pp. 13001305, Dec. 1999.

[28] M. Eltabach, A. Charara, I. Zein, and M. Sidahmed, "Detection of broken rotor bar of induction motors by spectral analysis of the electromagnetic torque using luenberger observer," in Proc. IEEE 27th Annu. Conf. Ind. Eletron. Soc., Denver, CO, USA, Nov. 29Dec. 02 2001, pp. 658663.

[29] K. N.Gyftakis, V. S. Dionysios, Joya C. Kappatou and Epaminondas D. Mitronikas "A Novel Approach for Broken Bar Fault Diagnosis in Induction Motors Through Torque Monitoring,"in Energy Conversion, IEEE Transactions on Vol. 28 , no. 2, pp.267 - 277 2013

VII. BIOGRAPHIES

Yassine MAOUCHE (SM'12) was born in Chelghoum Laid, Algeria, in 1987. He received the Licence degree in electrical engineering from the University of Mentouri Constantine, in 2009 and the Master. degree from the University of Mentouri Constantine in 2011. He is currently working toward the Ph.D. degree on the diagnosis of faults of induction machines in the laboratory Laboratoire dElectrotechnique de Constantine. He is a member of the IEEE Industry Applications Society.

Abdelfettah BOUSSAID (SM'12) was born in Ferdjioua MILA, Algeria, in 1988. He received the Licence degree in electrical engineering from the University of Mentouri Constantine, in 2009 and the Master. degree from the University of Mentouri Constantine in 2011. He is currently working toward the Ph.D. degree on the electrical engennering his interesting is on the active filtering of harmonics in electrical networks in the laboratory Laboratoire dElectrotechnique de Constantine. He is a member of the IEEE Industry Applications Society, the IEEE Industrial Electronics Society, and the IEEE Power Engineering Society.

Mohamed BOUCHERMA (M'12) received the B.Sc. degree in electrical engineering from Annaba University,Annaba, Algeria, in 1985, the M.Sc. degree in power system engineering from the University of Strathclyde, Glasgow, U.K., in 1989, and the Ph.D.degree in electrical engineering from The University of Shefeld, Shef field, U.K., in 1994. He is currently a Lecturer with the Department of Electrical Engineering, Mentouri University of Constantine, Constantine, Algeria. His main research interests are power systems and electrical machines.

Abdelmalek KHEZZAR (M'07) was born in Batna, Algeria, in 1969. He received the B.Eng. degree in electrical engineering from the University of Batna, Batna, in 1993 and the Ph.D. degree from Institut National Polytechnique de Lorraine, Nancy, France,in 1997.
In 2000, he joined the Mentouri University of Constantine, Constantine, Algeria, as a Lecturer with the Department of Electrical Engineering, where, since December 2008, he has been a Full Professor and, in April 2009, he was elected as Director of the laboratory "Laboratoire dElectrotechnique de Constantine." His main research interests are power electronics, drives, and analysis of electrical machine with special emphasis on fault diagnosis.
Prof. Khezzar is a member of the IEEE Industry Applications Society, the IEEE Industrial Electronics Society, and the IEEE Power Engineering Society.

The Zero-Sequence Current Spectrum as an On-Line Static Eccentricity Diagnostic Mean in Δ-Connected PSH-Induction Motors

K. N. Gyftakis, *Member IEEE* and J. C. Kappatou, *Member IEEE*

Abstract -- It has been reported that the only-static and only-dynamic eccentricity cannot be reliably detected in induction motors which produce Principal Slot Harmonics (PSH) with the use of the traditional MCSA method. Most work was focused in the detection of dynamic and mixed, rather than static eccentricity. In this paper, the detection of the only-static eccentricity fault in Δ-connected induction motors with the use of the zero-sequence current spectrum is thoroughly investigated. The frequency spectrum of the zero-sequence current is studied with the application of the Fast Fourier Transform in a PSH-induction motor operating under different constant speed cases and for different fault levels. The work is carried out with transient FEM simulations. The results will show that the proposed method succeeds in the only static eccentricity diagnosis and its severity, while the MCSA proves to be unreliable.

Index Terms-- Fault diagnosis, FEM, induction motor, static eccentricity.

I. INTRODUCTION

THE air-gap eccentricity is a condition when the electrical machine's air-gap between the rotor and the stator is unequal [1]-[2]. Two eccentricity types there exist: static and dynamic. When the two types coexist, it is referred to as a mixed eccentricity condition.

The static eccentricity is strongly connected to the manufacturing process of the machine [3]. It may be caused by the ovality of the stator core or, most commonly, by the incorrect relative positioning of the rotor and stator at the commissioning stage [2]. In the case of static eccentricity, the air-gap asymmetry is fixed in space and time. The static eccentricity, if not detected at an early stage, could evolve into mixed eccentricity. Unless detected at an early stage, the eccentricity fault can lead to the rubbing between the rotor and the stator and finally to the total breakdown of the machine [4].

Many methods have been proposed during the years for the eccentricity fault detection. Most papers focus on the MCSA method [5]-[11]. The following formula [12], gives the static and dynamic eccentricity related frequencies in a 3-phase induction motor line current, but also the Principal Slot Harmonics (PSH):

$$f_h = \left[\left(kR \pm n_d \right) \frac{(1-s)}{p} \pm v \right] f \qquad (1)$$

where:

$n_d = 0$ in the case of static eccentricity;

$n_d = 1, 2, 3, \ldots$ in case of dynamic eccentricity;

f fundamental supply frequency;

R rotor slot number;

s slip;

p fundamental pole pairs number;

k any positive integer;

v order of the stator time harmonics, which are present in the power supply driving the motor.

However, it has been shown in [5] and [13] that, only certain combinations between the machine's magnetic poles and the rotor slot number will reveal only-static or only-dynamic eccentricity fault signatures with the use of formula (1). In these cases, the rotor slot number R obeys to:

$$R = 2p \left[3(m \pm q) \pm r \right] \pm k \qquad (2)$$

where: $m \pm q = 0, 1, 2, 3, \ldots$, $r = 0$ or 1 and $k = 1$ or 2.

When k=1, which means odd rotor slot number, the proposed formula (1) offered satisfying results. On the other hand, results from motors with k=2 offered very weak fault signatures, which were noticeable at speed very close to the synchronous one.

If setting k=0 in formula (2), then the rotor slot number for PSH-induction motors occurs and it is given by (3):

$$R = 2p \left[3(m \pm q) \pm r \right] \qquad (3)$$

In the case of PSH-induction motors, the formula (1) does not offer any reliable results, if there is only-static or only-dynamic eccentricity [5]. This happens because the PSH

This work was supported by the Research Program: "K. Karatheodori 2010" of the University of Patras, Research Committee.

K. N. Gyftakis is with the Laboratory of Electromechanical Energy Conversion, Electrical and Computer Engineering Dept., University of Patras, Greece. (phone: 0030-2610-996413; fax: 0030-2610-997362; e-mail: kosgyftak@upatras.gr).

J. C. Kappatou is with the Laboratory of Electromechanical Energy Conversion, Electrical and Computer Engineering Dept., University of Patras, Greece. (e-mail: joya@ece.upatras.gr).

978-1-4799-0024-4/13 $31.00 © 2013 IEEE

signatures are located at the same frequencies which are indicative of the eccentricity fault.

Other methods, proposed in the literature, address the eccentricity fault diagnosis as well. In [14], the use of the current's Park vector was proposed. In [15] the use of the instantaneous power and in [16] the instantaneous reactive power spectrum has been used. Furthermore, the application of an additional excitation in induction motor drives for static, dynamic and mixed eccentricity diagnosis was shown, while the air-gap eccentricity is a kind of saliency, which affects the zero-sequence voltage and allows the use of the additional excitation for eccentricity diagnosis [17]. Also, the SPRT (Single Phase Rotation Test) has been applied lately to the eccentricity fault diagnosis [18]. Finally, the only-static eccentricity in PSH-induction motors has been proposed to be detected with the use of the SFD-MCSA method [19]. Nevertheless, the method shown in [19] is an off-line diagnostic method.

This paper constitutes a study of an on-line diagnostic method for the only-static eccentricity diagnosis in PSH Δ-connected induction motors. The zero-sequence current spectrum is used as the diagnostic mean. The work has been carried out with transient FEM analysis. The FFT (Fast Fourier Transform) will be applied with the use of Matlab. The simulated motor has the characteristics of a real 3-phase, 4-pole, 380V, 50Hz, 4kW squirrel cage induction motor with 36 stator and 28 rotor slots. The 3 phases are delta connected. The number of rotor slots obeys to formula (3), thus the studied motor is a PSH-induction motor.

Firstly, the healthy and faulty motors will be simulated to operate under constant speed. Three different speed cases are considered: 1400rpm, 1460rpm and 1490rpm. The eccentricity level has been deliberately low at 20%. For each case, the traditional MCSA methods results will be also illustrated. Secondly, the motor will be simulated to operate at 1460rpm under 40% and 60% level of the static eccentricity fault, in order to investigate the proposed method's effectiveness on the fault severity identification. The results clearly reveal, that this paper's proposed method for the detection of only-static eccentricity in 3-phase PSH-induction motors, is effective and reliable, non-invasive and offers better results than the traditional MCSA method. Moreover, the method can be performed during motor operation and consequently it is an on-line diagnostic method. Finally, the proposed diagnostic mean is capable of revealing the static eccentricity fault severity.

II. FEM SIMULATIONS

A. The Simulated Induction Motor

The motor is simulated with the FEM software OPERA Electromagnetic Design. The motor operates with constant speed for three different cases. The selected speed values are 1400rpm, 1460rpm and 1490rpm. For every case, two models have been created: the healthy and the faulty one with only static eccentricity. The level of the static eccentricity fault is for all cases 20%. The FEM model of the simulated healthy motor is illustrated in Fig. 1. The numbers of nodes and elements are about 81000 and 161000 respectively.

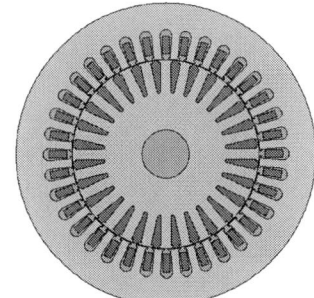

Fig. 1. The simulated model of the real induction motor.

B. Traditional MCSA Method's Results

The phase currents for the healthy motor and the faulty one are presented for every speed in Fig. 1. The occurring frequency spectra of the current waveforms are presented in Fig. 2, with the application of the FFT.

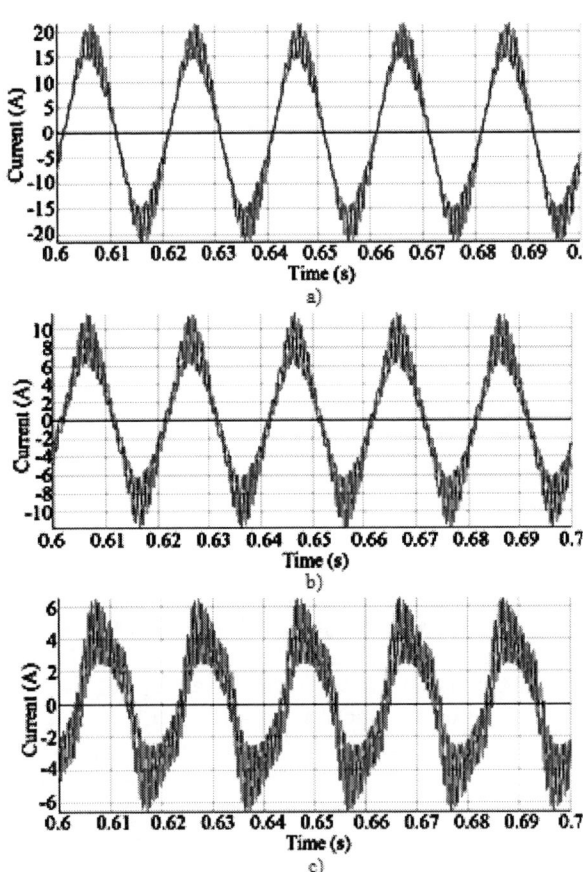

Fig. 2. The phase current waveforms of the healthy motor (blue) and the faulty one (red) for speed: a) 1400rpm, b) 1460rpm and c) 1490rpm.

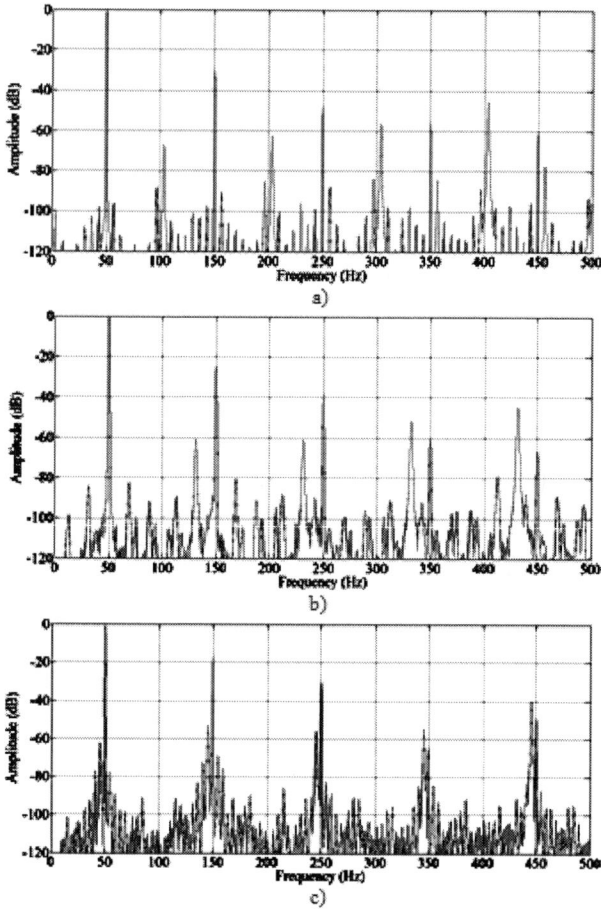

Fig. 3. The current frequency spectra of the healthy motor (blue) and the faulty one (red) for speed: a) 1400rpm, b) 1460rpm and c) 1490rpm.

It is quite obvious at this point that, the frequency spectrum of a single current cannot reveal the only-static eccentricity fault in this induction motor (Fig. 3), since the frequency spectra of the healthy and faulty motors coincide almost perfectly.

C.   Zero-Sequence Current Spectrums

In this section, the zero-sequence current waveforms and their spectra will be presented for the healthy and faulty motors operating under the previously mentioned constant speed cases: 1400rpm, 1460rpm and 1490rpm.

The occurring zero-sequence current waveforms are shown in Fig. 4. Several remarks can be deducted. It is clear that, the zero-sequence current is strongly dependant on the non-linearity of the magnetic B-H iron core characteristic and as a consequence on the saturation effect. This comes from the fact that, the zero-sequence current waveform has frequency 150Hz as one may observe in Fig. 4. Moreover, since the motor is simulated un-skewed, the presence of the slotting effect introduces a significant harmonic index in the zero-sequence current waveforms for every speed. Furthermore, while the speed increases, the stator phase current decreases as expected (Fig.2). In Fig. 4-a, one may

observe that the zero-sequence current reaches 8A amplitude for speed 1400rpm. In Fig. 4-b and Fig. 4-c the zero-sequence current amplitude remains the same and equal to 6A. This can be explained only by the fact that, for both the cases of 1460rpm and 1490rpm, the motor's iron core is greatly saturated. This is shown through Fig. 5 and Fig. 6, in which the cross section of the motor is shown, at time 0.6s, for healthy and faulty operation respectively, for all speed cases. The color distribution in Fig. 5 and Fig. 6 concerns the amplitude of the magnetic flux density. Moreover, in Fig. 7, the stator and rotor iron core B-H magnetic characteristic is presented, extracted from the manufacturer's datasheets. From Fig. 5 and Fig.6, one may derive that, although the maximum of the magnetic flux density decreases with the increase of the speed, a greater part of the machine's iron core rests in saturation levels.

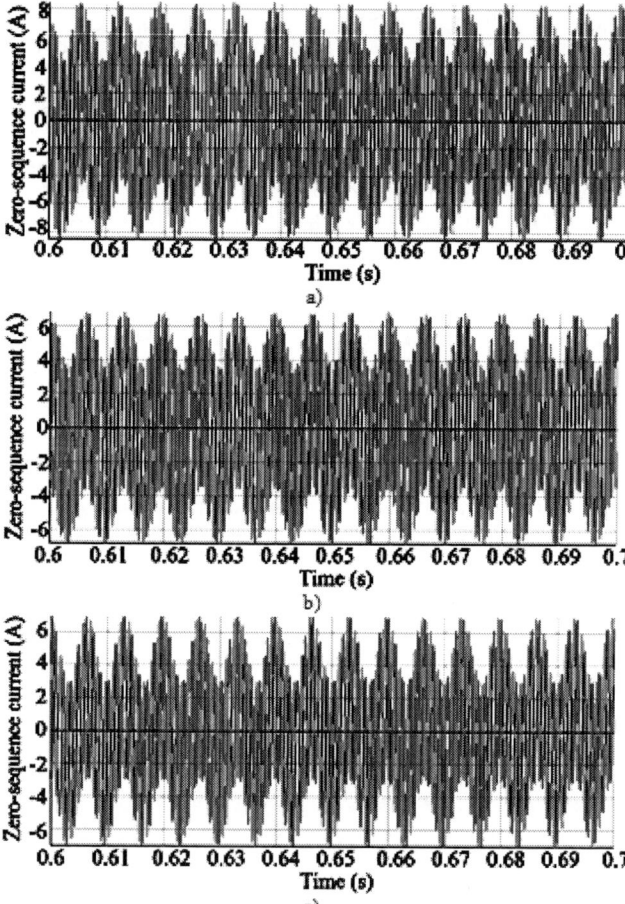

Fig. 4. The zero-sequence current waveforms of the healthy motor (blue) and the faulty one (red) for speed: a) 1400rpm, b) 1460rpm and c) 1490rpm.

978-1-4799-0024-4/13 $31.00 © 2013 IEEE

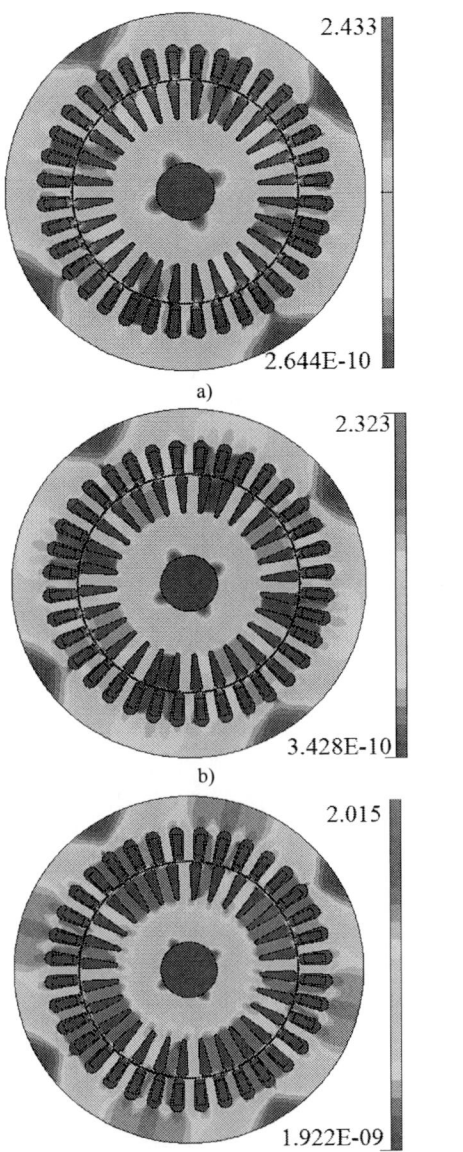

Fig. 5. The healthy motor's magnetic flux density amplitude in Tesla for speed: a) 1400rpm, b) 1460rpm and c) 1490rpm.

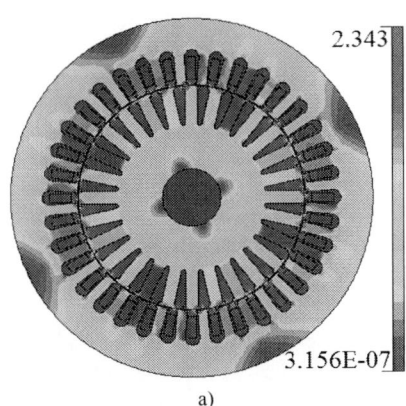

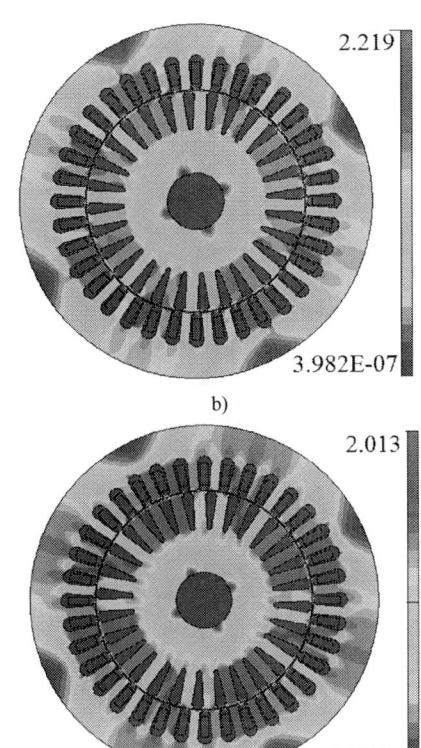

Fig. 6. The faulty motor's magnetic flux density amplitude in Tesla for speed: a) 1400rpm, b) 1460rpm and c) 1490rpm.

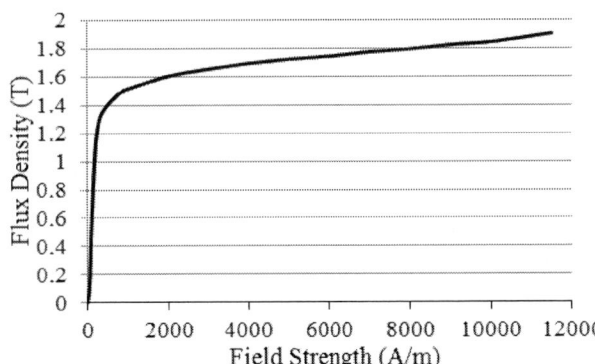

Fig. 7. The rotor and stator iron core's B-H magnetic characteristic.

The frequency spectra of the zero-sequence current waveforms of Fig. 4 are illustrated in Fig. 8 and the fault signatures' amplitudes are presented in Table I. In Fig. 8-a, the 50Hz component has increased in the faulty motor more than 16dB, presenting an amplitude -72.63dB. The component at 250Hz is practically the same in the healthy and faulty motors. Furthermore, One can observe in Fig. 8-b that, for speed 1460rpm both the two components at 50Hz and 250Hz have increased in the motor with the static eccentricity fault. Those components have amplitudes -70.39dB and -77.97dB respectively , while for the healthy motor they are -104.2dB and -123.5dB. Finally, in Fig. 8-c the 50Hz and 250Hz components have amplitudes equal to -

69dB and -79.38dB respectively in the faulty motor, while the same components are -111.6dB and -116.6dB in the healthy motor.

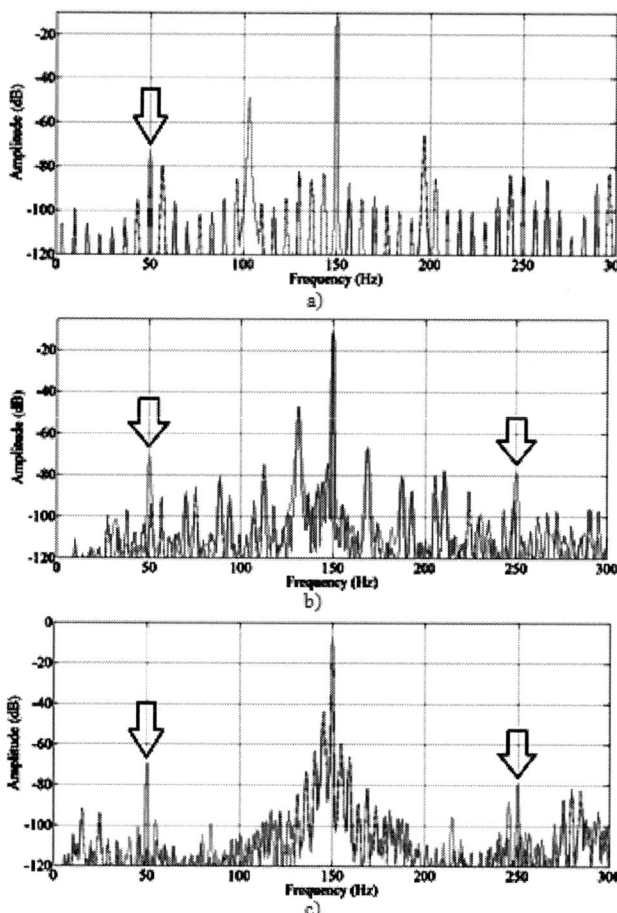

Fig. 7. The zero-sequence current spectra of the healthy motor (blue) and the faulty one (red) for speed: a) 1400rpm, b) 1460rpm and c) 1490rpm

TABLE I
FAULT RELATED HARMONICS AMPLITUDES FOR THE HEALTHY AND FAULTY MOTORS

Speed	Motor cases	Fault Signatures' Amplitudes	
		50Hz	250Hz
1400rpm	Healthy	-89.25dB	-83.52dB
	Faulty	-72.63dB	-87.38dB
1460rpm	Healthy	-104.2dB	-123.5dB
	Faulty	-70.39dB	-77.97dB
1490rpm	Healthy	-111.6dB	-116.6dB
	Faulty	-69dB	-79.38dB

D. Increased Levels of the Static Eccentricity Fault

It is crucial for a diagnostic method to be able to detect not only the fault's existence, but its severity at the same time. For this purpose, two more cases are studied, where the level of the fault increases. Two more models were simulated. A motor with 40% and one more with 60% static eccentricity level operating at speed 1460rpm.

The zero-sequence current spectra for both cases are presented in Fig. 8 compared to the healthy case. The 50Hz and 250Hz fault signatures' amplitudes are presented in Table II for all cases.

From Table II it occurs that, for the motor with 40% static eccentricity, the two fault signatures present similar amplitudes. This does not happen in the motor with 60% static eccentricity, where the signature at 50Hz is significantly stronger (more than 8dB), than the one at 250Hz. Despite that, it seems that the only-static eccentricity fault severity can be reliably predicted through the 250Hz signatures, which increase in amplitude according to the fault level's increment.

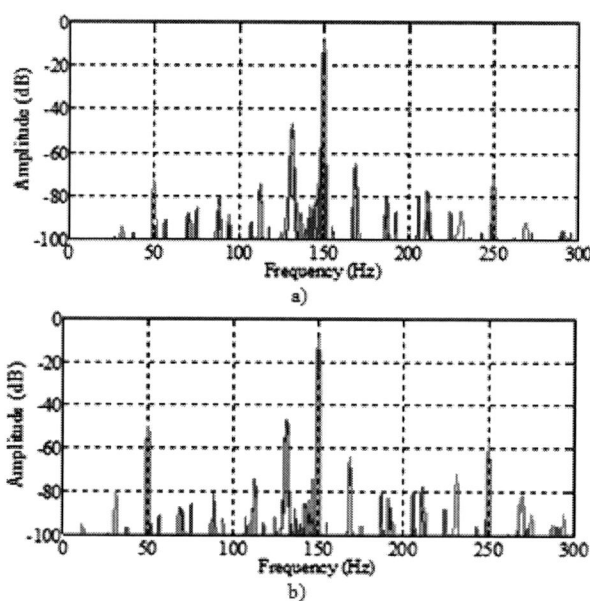

Fig. 7. The zero-sequence current spectra of the motor with: a) 40% and b) 60% static eccentricity (red), compared to the healthy (blue) for speed 1460rpm.

TABLE II
FAULT RELATED HARMONICS AMPLITUDES FOR THE HEALTHY AND FAULTY MOTORS OF DIFFERENT FAULT LEVELS AT 1460RPM

Fault severity	Fault Signatures' Amplitudes	
	50Hz	250Hz
Healthy	-104.2dB	-123.5dB
20%	-70.39dB	-77.97dB
40%	-72.71dB	-72.09dB
60%	-49.76dB	-58.18dB

E. Discussion

The above presented results lead to the following remarks:

- A single current's frequency waveform is incapable of revealing the only-static eccentricity fault in a PSH-induction motor. The traditional MCSA method fails to detect the fault for every studied speed (Fig. 3).
- The zero-sequence current is strongly affected by the saturation level of the induction motor. As a consequence, the zero-sequence current presents an important amplitude, while the speed increases.
- The only-static eccentricity fault can be detected with the zero-sequence current spectrum in delta connected PSH-induction motors, even when the fault's severity is low (in this case 20%). The fault signatures are located at 50Hz and 250Hz.
- The only-static eccentricity fault signatures' amplitudes increase simultaneously with the increase of the induction motor's operating speed. Consequently, this diagnostic method's reliability is high when it is performed at low-load operation.
- The proposed method is capable of detecting the severity of the only-static eccentricity fault, through the observation of the 250Hz fault signature's amplitude.

III. CONCLUSIONS

In this paper, the use of the zero-sequence current spectrum for the only-static eccentricity diagnosis in delta connected PSH-induction motors is proposed. The results illustrate the method's effectiveness and reliability. The proposed method can be applied during motor operation and so it is an on-line and non intrusive diagnostic method. Since the zero-sequence current is strongly connected to the saturation of the induction motor, it is recommended that the proposed method should be applied at low-load operation. Moreover, the results illustrate the capability of the proposed method to detect the fault's severity, through the observation of the 250Hz harmonic component in the zero-sequence current spectrum. Future work should concern the study of the method's effectiveness in non-PSH induction motors, as well as other electrical machine types.

IV. REFERENCES

[1] P. Zhang, Y. Du, T. G. Habetler and B. Lu, "A Survey of Condition Monitoring and Protection Methods for Medium-Voltage Induction Motors", *IEEE Trans. Ind. Appl.*, Vol. 47, No.1, pp. 34-46, Jan/Feb 2011.

[2] S. Nandi, H. A. Toliyat and X. Li, "Condition Monitoring and Fault Diagnosis of Electrical Motors-A Review", *IEEE Trans. Ener. Conv.*, Vol. 20, No.4, pp. 719-729, Dec 2005.

[3] D. G. Dorrell, W. T. Thomson and S. Roach, "Analysis of airgap flux, current, vibration signals as a function of the combination of static and dynamic airgap eccentricity in 3-phase induction motors", *IEEE Trans. Ind. Appl.*, Vol. 33, No.1, pp. 24-34, Jan/Feb 1997.

[4] A. Barbour and W. T. Thomson, "Finite element study of rotor slot designs with respect to current monitoring for detecting static airgap eccentricity in squirrel-cage induction motor", Proc. *IEEE Ind. Appl. Soc. Annual Meeting Conf.*, New Orleans, LA, Oct. 5-8, 1997, pp. 112-119.

[5] S. Nandi, S. Ahmed and H. Toliyat, "Detection of Rotor Slot and Other Eccentricity Related Harmonics in a Three Phase Induction Motor with Different Rotor Cages", *IEEE Trans. Ener. Conv.*, VOL.16, NO. 3, pp. 253-260, Sep. 2001.

[6] S. Nandi, R. M. Bharadwaj and H. Toliyat, "Performance Analysis of a Three-Phase Induction Motor Under Mixed Eccentricity Condition", *IEEE Trans. Ener. Conv.*, VOL.17, NO. 3, pp. 392-399, Sep. 2002.

[7] R. N. Andriamalala, H. Razik, L. Baghli anf F. M. Sargos, "Eccentricity Fault Diagnosis of a Dual-Stator Winding Induction Machine Drive Considering the Slotting Effects", *IEEE Trans. Ind. Electronics*, VOL. 55, NO. 12, pp. 4238-4251, Dec. 2008.

[8] M. Blodt, P. Granjon, B. Raison and G. Rostaing, "Models for Bearing Damage Detection in Induction Motors Using Stator Current Monitoring", *IEEE Trans. Ind. Electronics*, VOL. 55, NO. 4, pp. 1813-1822, Apr. 2008.

[9] J. Faiz, B. M. Ebrahimi, B. Akin and H. A. Toliyat, "Finite-Element Transient Analysis of Induction Motors Under Mixed Eccentricity Fault", *IEEE Trans. Magn.*, VOL. 44, NO. 1, pp. 66-74, Jan. 2008.

[10] J. Faiz, B. M. Ebrahimi, B. Akin and H. A. Toliyat, "Comprehensive Eccentricity Fault Diagnosis in Induction Motors Using Finite Element Method", *IEEE Trans. Magn.*, VOL. 45, NO. 3, pp. 1764-1767, Mar. 2009.

[11] J. Faiz, B. M. Ebrahimi and H. A. Toliyat, "Effect of Magnetic Saturation on Static and Mixed Eccentricity Fault Diagnosis in Induction Motor", *IEEE Trans. Magn.*, VOL. 45, NO. 8, pp. 3137-3144, Aug. 2009.

[12] P. Vas, Parameter Estimation, Condition Monitoring, and Diagnosis of Electrical Machines. Oxford: Clarendron Press, 1993.

[13] A. Ferreah, P. J. Hogben-Liang, K. J. Bradley, G. M. Asher, and M. S. Woolfson, "The effect of rotor design of sensorless speed estimation using rotor slot harmonics identified by adaptive digital filtering using the maximum likelihood approach," Proc. *IEEE Ind. Appl. Soc. Ann. Meeting Conf.*, New Orleans, LA, Oct. 5-8, 1997, pp. 128-135.

[14] A. J. Cardoso and E. S. Saraiva, "Computer-aided detection of air-gap eccentricity in operating three-phase induction motors by Park's vector approach," *IEEE Trans. Ind. Appl.*, Vol. 29, No. 5, pp. 897-901, Sep/Oct 1993.

[15] Z. Liu, X. Yin, Z. Zhang, D. Chen and W. Chen, "Online rotor mixed fault diagnosis way based on spectrum analysis of instantaneous power in squirrel cage induction motors," *IEEE Trans. Ener. Conv.*, Vol. 19, No. 3, pp. 485-490, Sep. 2004.

[16] M. Drif and A. J. M. Cardoso, "Airgap-Eccentricity Fault Diagnosis, in Three-Phase Induction Motors, by the Complex Apparent Power Signature Analysis", *IEEE Trans. Ind. Electronics*, VOL. 55, NO. 3, pp. 1404-1410, Mar. 2008.

[17] G. Bossio, C. D. Angelo, J. Solsona, G. O. Garcia and M. I. Valla, "Application of an Additional Excitation in Inverter-Fed Induction Motors for Air-Gap Eccentricity Diagnosis", *IEEE Trans. Ener. Conv.*, VOL. 21, NO. 4, pp. 839-847, Dec. 2006.

[18] S. Lee, J. Hong, S. Bin Lee and S. Nandi, "Detection of Airgap Eccentricity for Induction Motors Using the Single-Phase Rotation Test," *IEEE Trans. Ener. Conv.* Vol. 27, No. 3, pp. 689-696, Sep. 2012.

[19] K.N. Gyftakis and J.C. Kappatou, " A Novel and Effective Method of Static Eccentricity Diagnosis in 3-Phase, PSH-Induction Motors," *IEEE Trans. Ener. Conv.*, Vol. 28, No. 2, pp. 405-412, Jun. 2013.

V. BIOGRAPHIES

Konstantinos N. Gyftakis was born in Patras, Greece, in May 1984. He received the diploma in Electrical and Computer Engineering from the University of Patras, Patras, Greece in 2010. He is a PhD Candidate in the Department of Electrical and Computer Engineering, University of Patras. His research activities are in FEM design, fault diagnosis and optimization of electrical machines. He is an IEEE member, member of IEEE PES, IAS, IES and Magnetics Society, member of the HELIEV (Hellenic Institute of

Electric Vehicles) and finally member of the Technical Chamber of Greece. (E-mail: **kosgyftak@upatras.gr**)

Joya C. Kappatou was born in Argostoli, Greece. She received the diploma in Electrical Engineering from the University of Patras, Patras, Greece and the PhD from the same University in 1991 in the field of Electrical machines and Power Electronics. She is Assistant Professor in the Electrical and Computer Engineering Department of the University of Patras. Her teaching and research activities are in electrical machines, power electronics, modeling and design using FEM, faults diagnosis in electrical machines. Dr. Kappatou is a member of IEEE and the Technical Chamber of Greece. (University of Patras, Electrical and Computer Engineering Department, 26500 Rion-Patras, Greece, Tel: +30 2610/996413, Fax: +30 2610/997362, E-mail: **joya@ece.upatras.gr**)

Discriminating time-varying loads and rotor cage fault in Induction motors

A. E. Mabrouk, S. E. Zouzou, M. Sahraoui and S. Khelif

Abstract -- Diagnosis of electrical machines are becoming more and more important issues in the field of electrical machines as new data processing technique. Motor Current Stator Analysis (MCSA) are usually used to detect the broken bars. In several industrial applications, the motor is subjected to load torque variations of low frequencies, which have effects similar to rotor faults in the current spectrum and result of diagnostic procedure may be ambigues. Discriminating rotor cage fault from oscillating load effects in Induction motors must be considered.

In this paper, we present a study based on the application of the active and reactive power signature analyses for discriminating broken rotor bars from mechanical load oscillation effects in operating three-phase squirrel cage induction motors. This method is attractive because it does not need to interrupt the operating system. Finite element method was used to perform dynamical simulation, which leads to more precise results than other models, as the reel geometry and winding layout of the machine are used. The computer simulations and laboratory experiments results show the interest and the efficiency of this technique for the correct distinction between broken rotor bars and load oscillations.

Index Terms-- Squirrel cage induction motor, instantaneous active power, instantaneous reactive power, broken bar, finite element analysis, load oscillation.

I. INTRODUCTION

Nowadays three-phase induction motors are used in a wide variety of industrial applications. They are the most used kind of electrical machines. However, electrical and mechanical faults pose a particular challenge to the industry and end users which often interrupt the productivity and require maintenance. In literature, rotor faults have been shown to account for a large portion of induction motor failures, sometimes they are the single biggest cause of failure in the field. Under the assumption of a constant load torque, Motor Current Stator Analysis (MCSA) are usually used to detect the broken bars [1]-[2]. The corresponding side-band characteristic harmonics are located at:

$$f_b = (1 \pm 2ks)f_s \qquad (1)$$

Where f_s is the electrical supply frequency and s is the rotor

slip.

In several industrial applications, like cement industry, the motor is subjected to load torque variations of low frequencies, due to cyclic variations of the industrial process. Considering only the fundamental component in the Fourier series development, the load torque is described by:

$$T_{load}(t) = T_{avg} + T_{osc} \cos(2\pi f_0 t) \qquad (2)$$

Where T_{load} is the torque load, T_{avg} is the average load, T_{osc} is oscillated part of torque, f_0 is the load torque frequency. If f_0 is similar to $2sf_s$, the same spectral components mentioned in (1) can be found in the motor current spectrum even when the motor is in healthy conditions [3]. In this case, the commonly used MCSA does not allow discriminating between a motor with broken rotor bars from a motor driven by an oscillating load. Several research works addressing this problem have been published, extended Park's vector approach [4], detecting negative sequence harmonics information [5], model reference estimation [3], time-frequency approach [6], Vienna monitoring methods [7], neural networks [8], current space vector [9], Instantaneous Active and Reactive Currents [10] and in recently publications, different powers and their derived quantities [11]-[13].

In this paper, the instantaneous active and reactive powers spectra are used to distinguish between broken rotor bars and square-wave torque effect. This analysis is validated through the use of a FEM model and laboratory experiments.

.

II. ACTIVE AND REACTIVE POWERS ANALYSES

The instantaneous active power $p_0(t)$ and reactive power $q_0(t)$ in the healthy case are [11]:

$$p_0(t) = v_a i_a + v_b i_b + v_c i_c = 3VI \cos(\varphi) \qquad (3)$$

$$q_0(t) = \sqrt{3}(v_a i_b + v_b i_a) = 3VI \sin(\varphi) \qquad (4)$$

The spectra of the instantaneous active and reactive powers contain only a dc component corresponding to their mean values. The rotor cage fault or mechanical abnormality is characterized by the appearance of a sequence of sideband components around the fundamental [11]-[14]:

A. E. Mabrouk, S. E. Zouzou, M. Sahraoui and S. Khelif are with Laboratoire LGEB, Université de Biskra, 07000 Biskra, Algérie (e-mail: h_mabrouk@rocketmail.com). (e-mail: zouzou_s@hotmail.com). (e-mail: s_moh78@yahoo.fr).(khelifsamia@gmail.com)

$$i_a(t) = i_f \cos(\omega_s t - \alpha_0) + \sum_{k=1}^{\infty}\left\{ I_{dl,k} \cos\left[(1-2ks)\omega_s t - \beta_{l,k}\right]\right.$$
$$\left. + \sum_{k=1}^{\infty} I_{dr,k} \cos\left[(1+2ks)\omega_s t - \beta_{r,k}\right]\right\} \qquad (5)$$

$$i_b(t) = i_f \cos(\omega_s t - \alpha_0 - 2\pi/3) + \sum_{k=1}^{\infty}\left\{ I_{dl,k} \cos\left[(1-2ks)\omega_s t - \beta_{l,k} - 2\pi/3\right]\right.$$
$$\left. + \sum_{k=1}^{\infty} I_{dr,k} \cos\left[(1+2s)\omega_s t - \beta_{r,k} - 2\pi/3\right]\right\} \qquad (6)$$

$$i_c(t) = i_f \cos(\omega_s t - \alpha_0 + 2\pi/3) + \sum_{k=1}^{\infty}\left\{ I_{dl,k} \cos\left[(1-2ks)\omega_s t - \beta_{l,k} + 2\pi/3\right]\right.$$
$$\left. + \sum_{k=1}^{\infty} I_{dr,k} \cos\left[(1+2s)\omega_s t - \beta_{r,k} + 2\pi/3\right]\right\} \qquad (7)$$

Where:

i_f : maximum value of the fundamental of phase current (A);
α_0 : initial phase angle of the fundamental current;
$i_{dl,k}$: maximum value of the current lower sideband component, at a frequency of $(1-2ks)f_s$ (A);
$\beta_{l,k}$: initial phase angle of the current lower sideband component;
$i_{dr,k}$: maximum value of the current upper sideband component, at a frequency of $(1+2ks)f_s$ (A);
$\beta_{r,k}$: initial phase angle of the current upper sideband component.

Similarly, the instantaneous power can be described as:

$$p(t) = 3VI \cos(\varphi) + 3V\left\{ \sum_{k=1}^{\infty}\left[I_{dl,k} \cos(2ks\omega_s t + \beta_{l,k})\right.\right.$$
$$\left.\left. - I_{dr,k} \cos(2ks\omega_s t - \beta_{r,k})\right]\right\} \qquad (8)$$

$$q(t) = 3VI \sin(\varphi) + 3V\left\{ \sum_{k=1}^{\infty}\left[I_{dl,k} \sin(2ks\omega_s t + \beta_{l,k})\right.\right.$$
$$\left.\left. - I_{dr,k} \sin(2ks\omega_s t - \beta_{r,k})\right]\right\} \qquad (9)$$

It can be seen that the instantaneous total active and reactive powers become different from the healthy case ones. The additional component at the disturbance frequency $f = 2ksf_s$ in their spectra, provide extra diagnostic information about the machine condition in the presence of broken rotor bars or load torque oscillations.

III. SIMULATION RESULTS

In recent years, Finite Element Method (FEM) was widely used in the design and analysis of electric machines. It was proved that the FEM based analysis is an effective and inexpensive method for studying the influence of the faults on the electrical machine behavior [15]. Several program packages for magnetic field computations have been developed (Flux, Maxwell, Ansys, femm …).
The magnetic circuit of the considered induction motor is shown in Fig. 1. The mesh made on the magnetic circuit is denser in the vicinity of the air gap, where the electromagnetic exchanges between the stator and the rotor are carried out.

The 2D electromagnetic field computation model in (x, y) Cartesian coordinates is based on the magnetic vector potential formulation characterized by the following partial differential equation:

$$\nabla \times \left(\frac{1}{\mu}\nabla \times \vec{A}\right) + \sigma\left(\frac{\partial \vec{A}}{\partial t} - \vec{v} \times (\nabla \times \vec{A})\right) = \vec{J_s} \qquad (10)$$

Where:

μ : Magnetic permeability;
A : Magnetic vector potential;
σ : Electrical conductivity;
v : Velocity;
J_s : Current density.

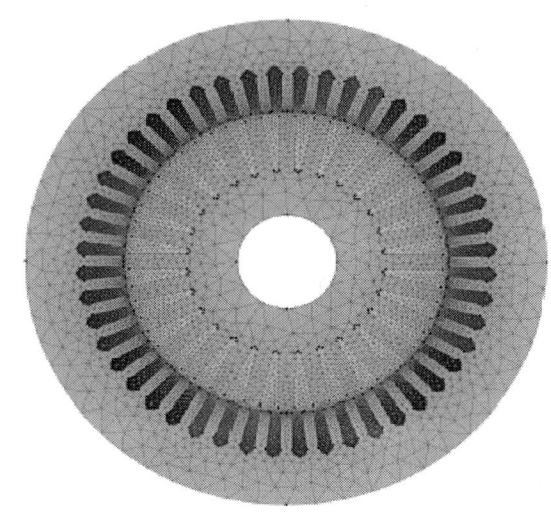

Fig. 1. Geometry and mesh of computation domain.

The use of the rotating air gap, allows considering the rotation of the rotor in the magneto-evolutionary study without making a new mesh of the geometry at each position of the rotor and keeps the same number of unknowns, hence, eliminate the velocity term in (10) [16]. Since the studied induction motor is voltage supplied, the motor is excited to its rated voltage and frequency using a three-phase voltage source, the two unknowns quantities A and J_s in (10) are determined by a field-circuit coupling model of the machine. Once the magnetic field is determined by the time-stepping finite-element method, the magnetic torque is calculated using the Maxwell stress tensor. The mechanical equation determines a new angular and radial position of the. Rotor motion is governed by the mechanical equation:

$$J\dot{w} = T_m - T_l - B\,w \qquad (11)$$

Where:

J : is the moment of inertia;

$\ddot{w}$: is the angular acceleration;
w : is the angular velocity;
T_m : is the electromagnetic torque;
T_l : is the torque load;
B : is the friction coefficient.

Hence having the mechanical equations coupled with magnetic equations, the dynamic behavior of the motor can be simulated in any moment.

The characteristics of the induction motor are shown in Table I in appendix.

In our simulation, two cases are simulated: two broken bars, healthy motor with oscillating torque.

Fig. 2 shows the waveform of the torque imposed to the motor. The torque oscillates between two values according to time. In this case, the load torque presents a harmonic components at the frequency f_o=1 Hz similar to $2sf_s$, used to detect the broken bars.

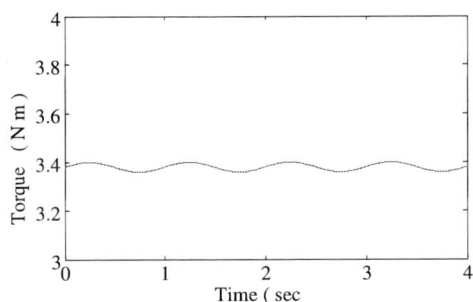

Fig. 2. The shaft torque

Fig. 3 shows the magnetic flux distribution at the transient state for two broken bars and healthy rotor with oscillating torque. When the slip is large, the eddy current shows a large value. This high value of slip is necessary to illustrate the effects of the broken bars on the magnetic field.

The concentration of magnetic flux is observed around the broken bar and creates asymmetric magnetic flux distribution. This is due to the fact that in the broken bar region there is no localized conductor demagnetization effect since these bars carry no currents.

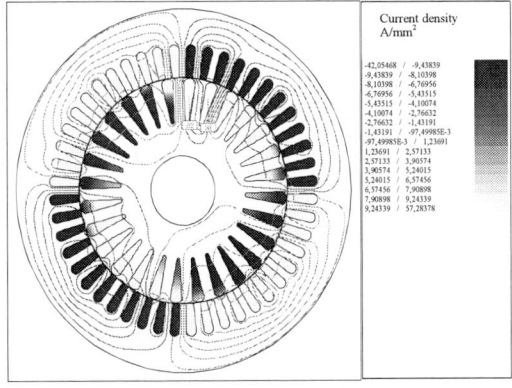

a) Two broken rotor bars

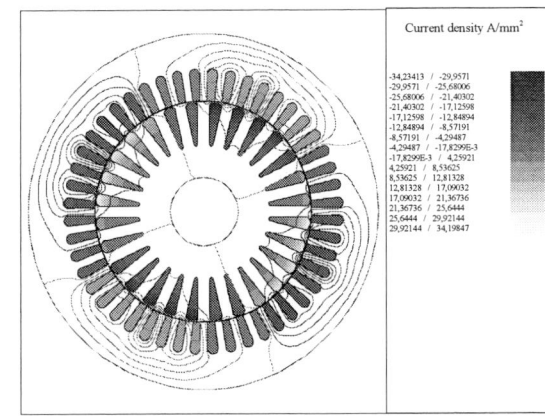

b) Healthy Motor with oscillating torque.
Fig. 3. Magnetic flux and current density at start up.

On Fig. 4, it is obvious that the region around the broken bar of the rotor has a higher degree of saturation compared to the same region in healthy state with oscillating torque. The current densities in the neighbored rotor bars to those broken are significantly increased.

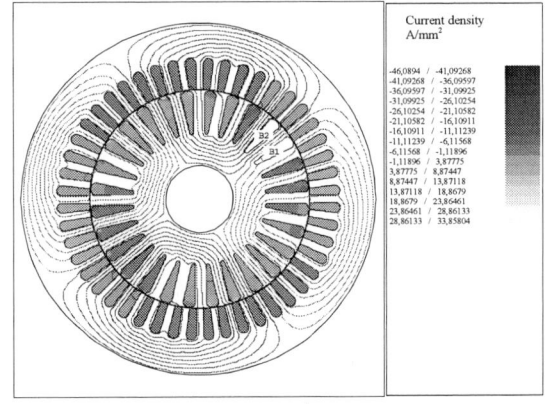

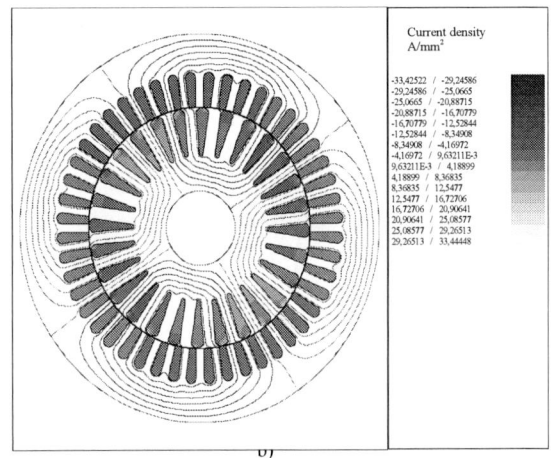

Fig. 4. Magnetic flux and current density at t= 2.297 s.
a) Two broken rotor bars b) Healthy Motor with oscillating torque.

Fig. 5 shows the waveform of the air gap flux density along a circular contour in the air-gap. The flux densities have a symmetrical distribution in healthy state with an oscillating torque. The perturbation in the magnetic field produced by two broken bars results in a non-symmetrical field.

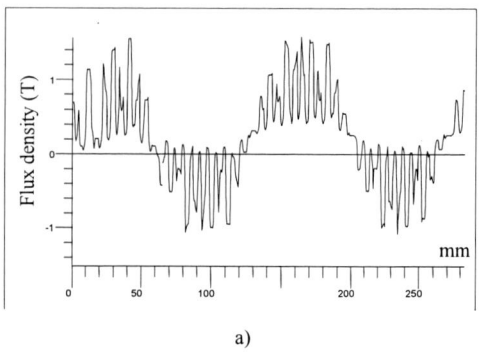

a)

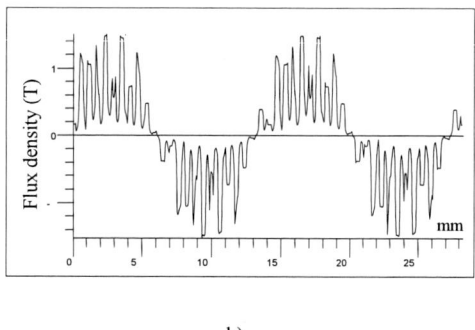

b)

Fig. 5. Waveform of the Air gap flux density at t= 2.297 s.
  a) Two broken rotor bars b) Healthy Motor with oscillating torque.

The evolution of the electromagnetic torque and speed, during the first seconds after the connection, for the case of two broken bars and for healthy rotor with load torque oscillation is shown in fig. 6 and fig. 7, respectively.

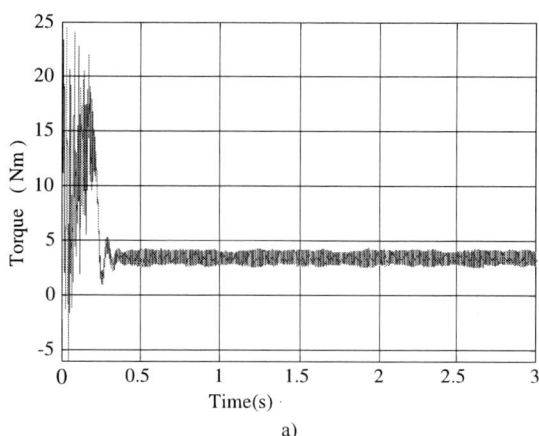

a)

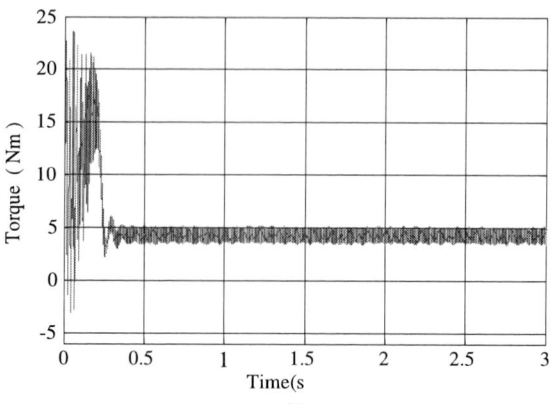

b)

Fig.6. Electromagnetic torque (40 % rated load load).
a) Two broken rotor bar b) Healthy Motor with oscillating torque.

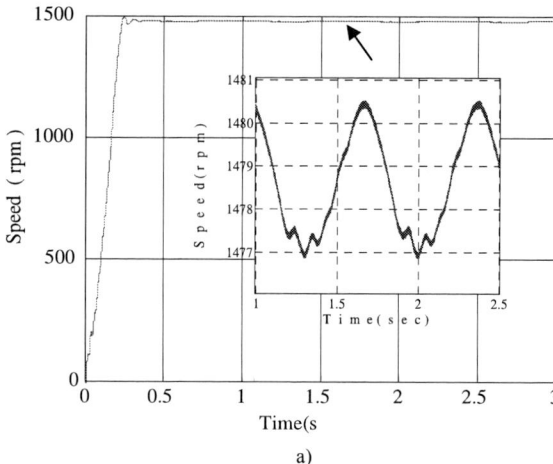

a)

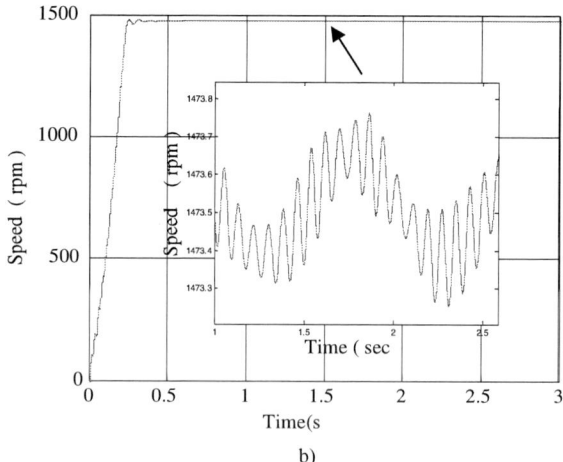

b)

Fig.7. Rotational Speed (40 % of rated load).
  a) Two broken rotor bars b) Healthy Motor with oscillating torque.

978-1-4799-0024-4/13 $31.00 © 2013 IEEE

IV. STATOR CURRENT ANALYSIS

The torque oscillation at a frequency (Fig. 8) generates a speed oscillation at this frequency whose amplitude depends on the inertia of the motor-load (Fig. 9).

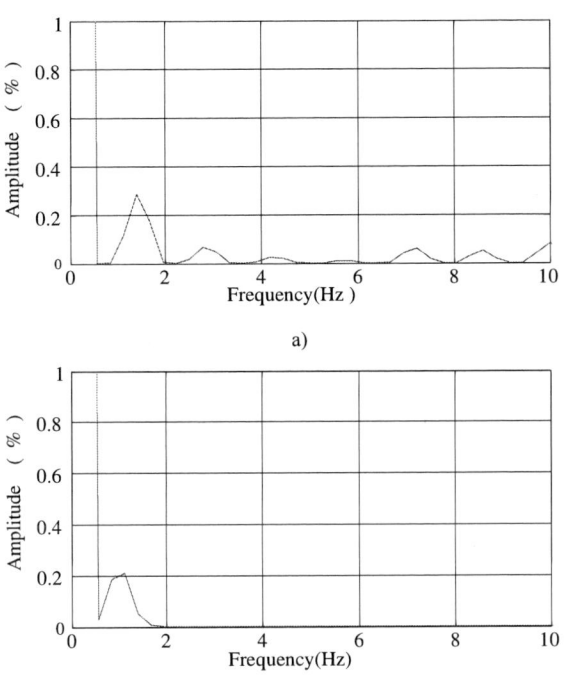

a)

b)

Fig.8. Torque spectrum (40 % of rated load)
   a) Two broken rotor bars b) Healthy Motor with oscillating torque.

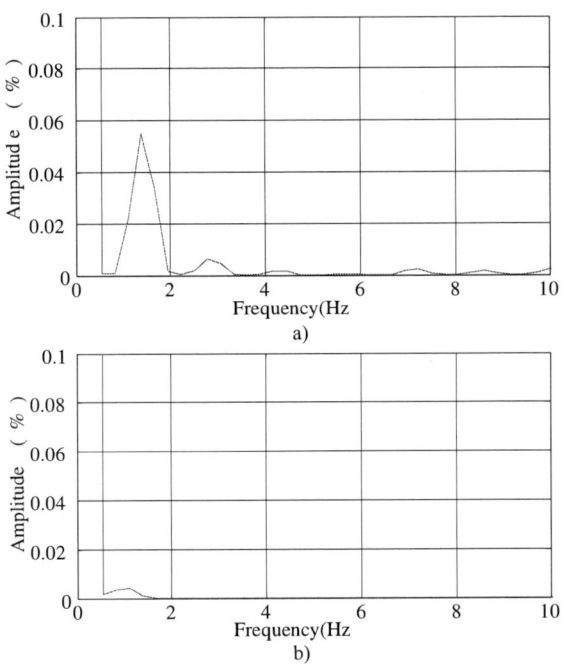

a)

b)

Fig.9. speed spectrum (40 % of rated load)
   a) Two broken rotor bars b) Healthy Motor with oscillating torque.

Fig. 10.a, shows the current spectrum, sidebands at $+2sf$ are produced by broken bars, fig. 10.b shows the stator current spectrum with sidebands at f_0=1 Hz produced by oscillating load which can be confused with broken bars. the broken rotor bars can't be distinguished from the oscillating load with MCSA.

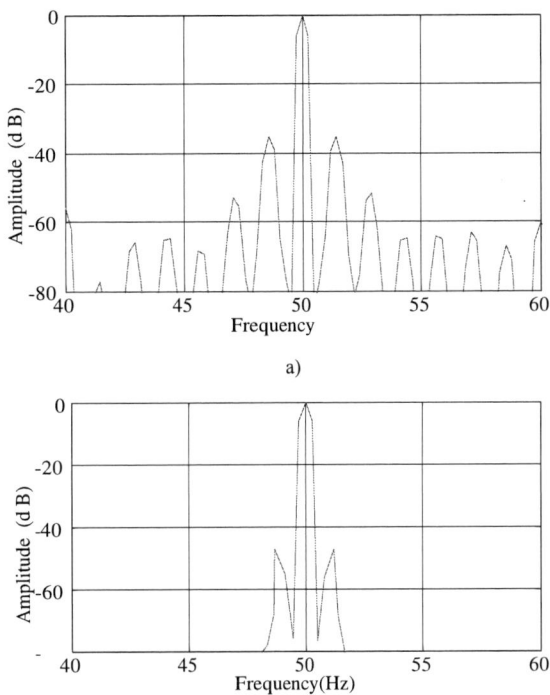

a)

b)

Fig.10. Stator current spectrum (40 % of rated load).
   a) Two broken rotor bars b) Healthy Motor with oscillating torque.

V. ACTIVE AND REACTIVE POWER ANALYSIS

The normalized spectra of the instantaneous active and reactive power for the case of two broken bars (Figs. 11.a and b) show the appearance of a characteristic component directly at the frequency of speed oscillation, $f_0 = 2sf_s$, aside from the dc component. This characteristic component presents higher amplitude in the instantaneous reactive power spectrum with respect to the active power one. In the case of the load torque oscillation is exactly the opposite behavior as compared to the broken rotor bar case (Figs. 12.a and b).

The oscillating active power clearly reflect the effect of the oscillating load, the reactive power is practically negligible.

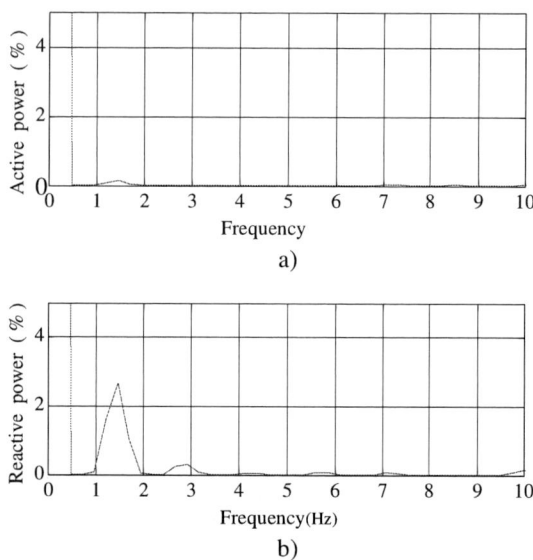

Fig. 11. Motor with two broken rotor bars (40 % of rated load)
(a) p(t) (b) q(t) spectrums.

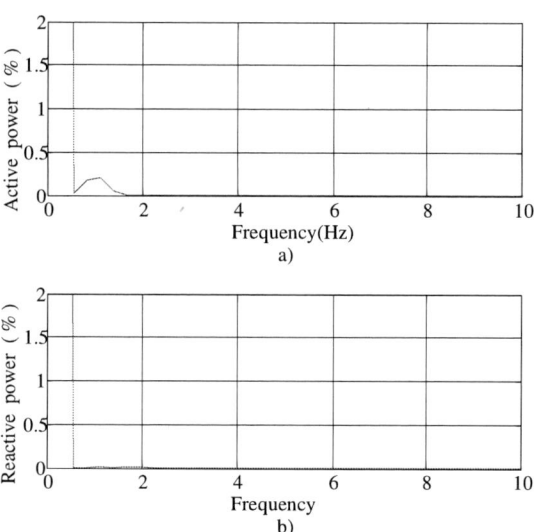

Fig. 12. Motor with oscillating torque (40 % of rated load)
(a) p(t) (b) q(t) Spectrums.

VI. EXPERIMENTAL RESULTS

In order to validate the simulation results, a special test bench was used. The test bench used in the exerimental investegation is avilable at the LGEB Biskra-Algeria (Fig. 13). The experimental test were carried out on three-phase induction machines (3 KW, 50-Hz, 4pole, 28 bars Y connected). The data acquisition was fixed to 10 seconds at a sampling frequency of 10 kHz through a D-Space 1104 board. A simulator torque which oscillates between two values according to time. The cycle frequency is adjustable from 0 to 200s and the cyclic ratio from 0% to 100%. Fig.

14.a shows the waveform of the torque used in the experiment. In this case, the load torque presents a sequence of harmonic components with a fundamental one at the frequency $f_o=1$ Hz (Fig 14.b). Their total effect can be considered as a superposition of a series of single frequency oscillating loads.

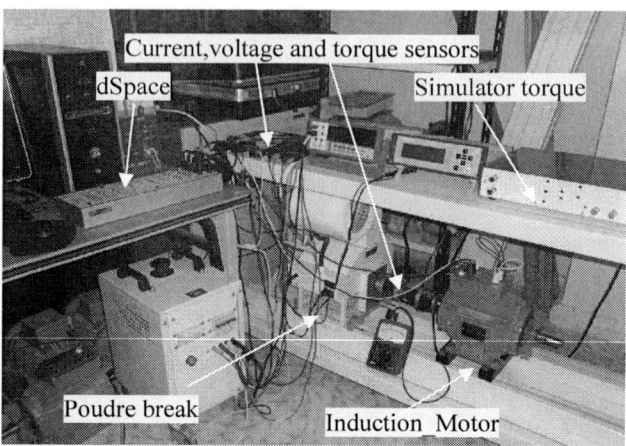

Fig. 13. Test-bed used for experimental analysis.

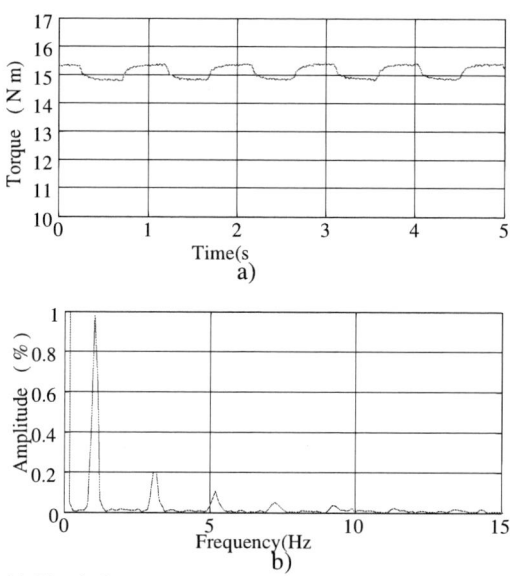

Fig. 14. The shaft torque
a) The waveforme b) Spectrum.

Fig. 15.a shows the current spectrum of the machine with two broken bars, Fig. 15.b shows the current spectrum of the healthy machine with load torque oscillation. Prominent sideband components appear at 51 Hz and 49 Hz, i.e., at 50 Hz (fundamental frequency) plus and minus 1 Hz (frequency of speed oscillation). A frequency component at 53 Hz, i.e., 50 Hz $+2 \times 1$ Hz, was generated by the second harmonic of the pulsating torque, It is evident that the FFT analysis alone cannot distinguish between broken rotor bars and load oscillations.

978-1-4799-0024-4/13 $31.00 © 2013 IEEE

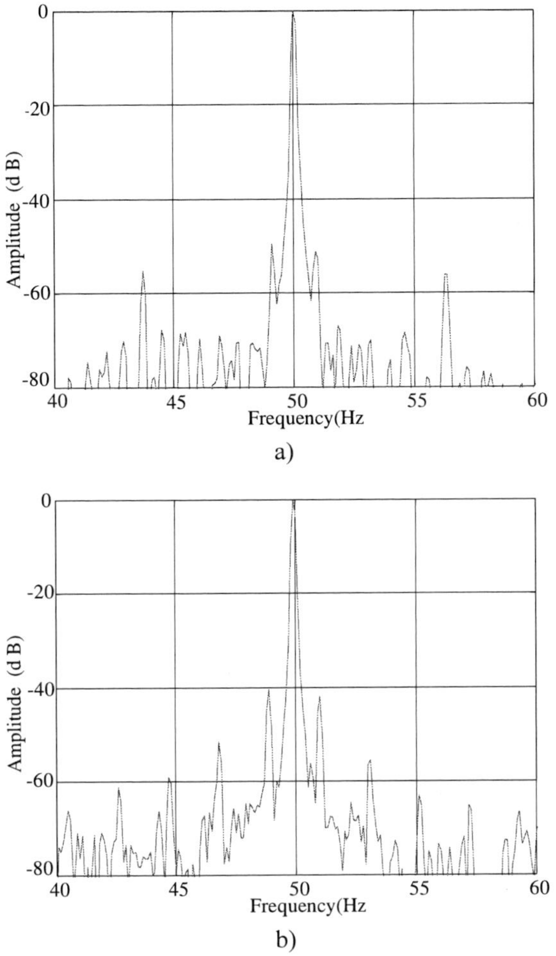

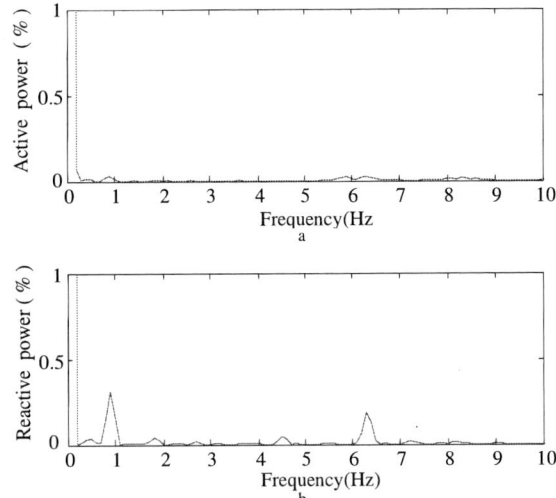

Fig. 16. Motor with two broken rotor bars (40 % of rated load).
(a) p(t) (b) q(t) spectrums.

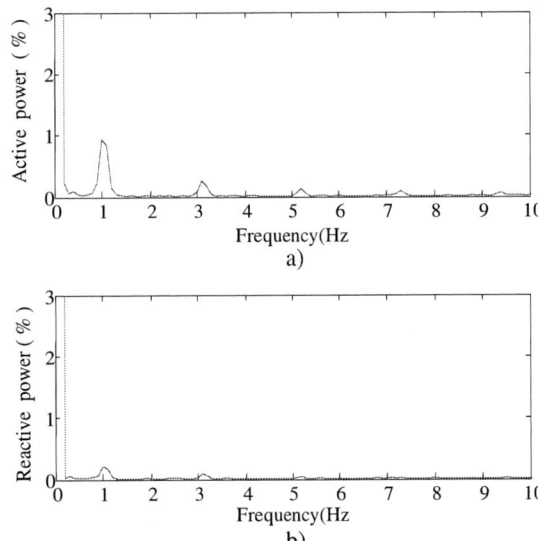

Fig. 17. Motor with an oscillating torque (40 % of rated load).
(a) p(t) (b) q(t) spectrums.

Fig.15. Stator current spectrum (40 % of rated load).
a) Two broken rotor bar b) Healthy Motor with oscillating torque.

The spectrum of the instantaneous active and reactive power in the case of motor with two broken bars is shown in Fig. 16.a and b, respectively. It can be appreciated that the instantaneous active power spectrum amplitude at $2sf$ is less than of the reactive power. In the case of the load torque oscillation, the normalized spectra of the instantaneous active and reactive power spectrum shows the appearance of sidebands at $n \times f_o$ (n=1,3,5,.....) due to the oscillating load. The magnitude of the instantaneous active power spectrum at f_o =1Hz which can be confused with the broken bar condition $2sf_s$ (s=0.01), is much greater than the reactive power one as shown in Figs.17.a and b, respectively.

VII. CONCLUSION

This paper presents the circuit coupled finite element method used to modeling the Three-Phase Squirrel Cage Induction Motor. For this purpose, the time-stepping finite element method (TSFE) was proposed. The determination of magnetic flux density waveform, magnetic flux distribution was obtained. The perturbation in the air-gap magnetic field produced by broken bars results in a non-symetrical field. The flux densities have a symmetrical distribution in healthy state driven by oscillating load.

Discriminating rotor cage fault from oscillating load effects in Induction motors are considered. The computer simulations results shows that rotor fault conditions can be separated from the abnormal mechanical load conditions by the combination of the active and reactive power signature analyses at frequencies of twice the slip frequency. The experimental results were confirmed that the proposed approach constitutes a useful tool for the study and diagnostics of induction motors.

VIII. APPENDIX

TABLE I
ELECTRICAL AND GEOMETRICAL CHARACTERISTICS OF SIMULATED INDUCTION MOTOR

characteristic	value
Rated power	1.1 Kw
Rated source voltage	380 V
Rated source frequency	50 Hz
Number of stator slots	48
Number of rotor slots	28
Outer diameter of stator core	145 mm
Inner diameter of stator core	90.4 mm
Outer diameter of rotor core	89.8 mm
Inner diameter of rotor core	27.94 mm
Air gap thickness	0.3 mm
Length of the stator and rotor core	55 mm

IX. ACKNOWLEDGMENT

The authors would like to thank Professor Champenois at the LAII laboratory, Poitiers, France, for his help.

X. REFERENCES

[1] G.B. Kliman, l. Stein, "Method of Current Signature Analysis", *Electric Machines and Power Systems*, vol. 20, pp. 463-474,1992.

[2] A. Bellini, F. Filippetti, G. Franceschini, C. Tassoni, and G. Kliman, "Quantitative evaluation of induction motor broken bars by means of electrical signature analysis ", *IEEE Trans. Industry Application*, vol. 37, pp. 1248–1255, Sept./Oct. 2001.

[3] R. R. Schoen, T. G. Habetler, "Effects of time-varying loads on rotor fault detection in induction machines", *IEEE Trans. Industry Application*, vol. 31, pp. 900–906, July/Aug. 1995.

[4] S. M. A. Cruz and A. J. M. Cardoso, "Rotor cage fault diagnosis in operating three-phase induction motors, under the presence of time-varying loads", presented at the Eur. Conf. Power Electronics Application, Graz, Austria, 2001.

[5] W. Long, T. G. Habetler, and R. G. Harley, "Separating load torque oscillation and rotor fault effects in stator current-based motor condition monitoring ", presented at the IEEE Int. Conf. Electrical Machines and Drives, San Antonio, USA, May 2005.

[6] K. Bacha, M. Gossa, H. Henao, and G. A. Capolino, "A time-frequency method for multiple fault detection in three-phase induction machines", presented at the IEEE Int. Symp. Diagnostics Electrical Machines, Power Electronics and Drives, Vienna, Austria, 2005.

[7] C. Kral, H. Kappeler, F. Pinker, G. Pascoli, "Discrimination of rotor faults and low frequency load torque modulations of squirrel cage I.M.by means of the Vienna monitoring method ", presented at the IEEE Power Electronics Specialists Conference, Recife, Brazil, June, 2005.

[8] G. Salles, F. Filippetti, C. Tassoni, G. Crellet, and G. Franceschini, "Monitoring of induction motor load by neural network techniques", *IEEE Trans. Power Electronics*, vol. 1, pp. 762–768, Jul. 2000.

[9] C. Concari, G. Franceschini and C. Tassoni, " Induction machine current space vector features to effectively discern and quantify rotor faults and external torque ripple ", *IET Electric Power Applications*, Vol. 6, pp. 310-32, 2012.

[10] G. R. Bossio, C.H. De Angelo, J. M. Bossio, C. M. Pezzani, and G. O. Garcia, " Separating Broken Rotor Bars and Load Oscillations on IM Fault Diagnosis Through the Instantaneous Active and Reactive Currents ", *IEEE Trans. Industrial Electronics*, vol.56, pp.4571-4580, 2009.

[11] M. Drif, A. J. Marques Cardoso, "Discriminating the Simultaneous Occurrence of Three-Phase Induction Motor Rotor Faults and Mechanical Load Oscillations by the Instantaneous Active and Reactive Power Media Signature ", *IEEE Trans. Industrial Electronics*, vol. 59, pp. 1630-1639, 2012.

[12] C.H. De Angelo, G.R. Bossio and G.O. Garcia, " Discriminating broken rotor bar from oscillating load effects using the instantaneous active and reactive powers", *IET Electric Power Applications*, Vol. 4, pp. 281-290, Jan. 2010.

[13] S. M. A. Cruz, "An Active–Reactive Power Method for the Diagnosis of Rotor Faults in Three-Phase Induction Motors Operating Under Time-Varying Load Conditions", *IEEE Trans. Energy Conversion*, vol. 27, pp. 71-84, 2012

[14] S. M. A. Cruz and A. J. M. Cardoso, "Rotor cage fault diagnosis in three-phase induction motors by the total instantaneous power spectral analysis," presented at the Industry Applications Society Annu. Meeting, Phoenix, USA, 1999.

[15] S. E. Zouzou, S. Khelif, N. Halem, M. Sahraoui, "Analysis of induction motor with broken rotor bars using finite element method ", presented at the 2nd international conference on electric power and energy conversion systems, Sharjah, UAE, 2011.

[16] T. H. Pham, P. F. Wendling, S. J. Salon, and H. Acikgoz, "Transient finite element analysis of an induction motor with external circuit connections and electromechanical coupling ", *IEEE Trans. Energy Conversion*, vol. 14, pp. 1407–1412, Dec. 1999.

XI. BIOGRAPHIES

Mabrouk Abd Elhamid was born in Biskra (Algeria) on 1985. He received the Engineer and "Magistère" in electrical engineering from university of Med Khider of Biskra, Algeria in 2009 and 2012 respectively. He is currently pursuing Ph.D degrees at Med Khider University, Algeria. His research interests include modeling and simulation of faulty electrical machines and intelligent systems application.

Salah Eddine Zouzou was born in Biskra, Algeria on 1963. He received the B.S degree from the "Ecole Nationale Polytechnique d'Alger", Algeria in 1987 and the M.S and Ph.D degrees from the "École Nationale Polytechnique de Grenoble" France, in 1988 and 1991 respectively. His fields of research interests deal with the design and condition monitoring of electrical machines. He has authored or co-authored more than 50 scientific papers in national and international conferences and journals. Prof. Zouzou is a Professor at the University of Biskra, Algeria and he is the director of the "Laboratoire de Génie Electrique de Biskra" since 2003.

Mohamed Sahraoui was born in Biskra, Algeria in 1978. He received the Engineer, "Magistère" and the Ph degrees in electrical engineering from the University of Med Khider of Biskra, Algeria in 2001, 2003 and 2010 respectively. He is interested in the modelling, condition monitoring and faults diagnosis of electrical machines.

Samia KHELIF was born in ourgla (Algeria) on 1983. She received the Engineer and "Magistère" in electrical engineering from university of Med Khider of Biskra, Algeria in 2008 and 2012 respectively. She is currently pursuing Ph.D degrees at Med Khider University, Algeria. Her research interests include monitoring and faults diagnosis of electrical machines.

Mathematical Modeling of Eccentricities in Induction Machines by the Mono-harmonic Model

A. J. Fernández Gómez, A. Dziechciarz, T. J. Sobczyk

Abstract -- The use of mathematical models of faulty induction machine has proved a useful tool for prediction of main effects due to faults. Researches develop mathematical models to use them as base of analysis of practical cases. The induction machine has become one of the most important elements in the industry, which has resulted in increased interest in controlling its operation. Among the usual faults in induction machines mechanical faults represent the 50-60% of the total failures. Those faults have associated in most cases some degree of eccentricity. In this paper a mathematical model of mixed eccentricity effects on induction machine for industrial drives applications is presented. The purpose of the paper is going one step further than previous approach of static and dynamic eccentricities showing that only one model is enough to reproduce successfully all types of eccentricities. Effects of main magneto-motive forces and steady-state performance of the motor supplied by a symmetrical sinusoidal source have been studied making it easy to understand the physical phenomena inside the machine and simplify the mathematical representation of those eccentricities. Based on this approach Matlab/Simulink models can be created for fault diagnosis purposes.

Index Terms--AC machines, Condition Monitoring, Dynamic Eccentricity, Mixed Eccentricity, Modeling, Static Eccentricity.

I. INTRODUCTION

Induction machine is the mostly used electrical machine on the market. Practically they are installed in every production system due to their characteristics. Since these machines are often cruitial in industry, their reliability is very important issue [1]. Eccentricities in electric machines are often a consequence of other common faults such as damaged bearings or caused by incorrect positioning of the rotor inside the stator during assembly.

Eccentricity of an induction machine is a type defined as unbalanced air gap between the rotor and the stator. Unbalanced air gap affects self and mutual inductances of both rotor and stator [5]-[6]-[11]-[12]-[15]. These changes

---

This work has the financial support from the Marie Curie FP7-ITN project "Energy savings from smart operation of electrical, process and technical equipment – ENERGY-SMARTOPS", Contract No: PITN-GA-2010-264940 is gratefully acknowledged.

T. J. Sobczyk is with Inst. on Electromechanical Energy Conversion, Cracow University of Technology, Kraków, ul. Warszawska 24, 31-155 Poland (e-mail: pesobczy@cyf-kr.edu.pl).

A. J.Fernández Gómez is with Inst. on Electromechanical Energy Conversion, Cracow University of Technology, Kraków, ul. Warszawska 24, 31-155 Poland (e-mail: afernandezpk@gmail.com).

A. Dziechciarz is with Inst. on Electromechanical Energy Conversion, Cracow University of Technology, Kraków, ul. Warszawska 24, 31-155 Poland (e-mail:arkadiusz.dziechciarz@gmail.com).

occurring in the values of inductances, cause additional harmonics in stator current spectrum. In [16] is shown that those harmonics could even be detected analysing the d-q components of stator currents. It is very important to distinguish characteristic symptoms of eccentricity from typical harmonics occurring in healthy state in machine's current spectrum.

In [14] autor present an study of static and dynamic eccentricties effects around fundamental harmonic and principle slot harmonic (PSH) for different severities and load conditions based on the amplitude of characteristics frequencies. Authors report that harmonics around main component at 50Hz cannot be used to detect the type of fault but could be used as severity fault index. However the use of PSH can detect the type of eccentricity. An extensive study is presented in [7] where authors define "Eccentricity Severity Factor" to evaluate the presence of mixed eccentricity.

Mathematical modeling of faulty state of the machine can be used to represent the main effects of eccentricities on the frequency spectra of phase stator currents. Those models contain the additional frequency components associated with the eccentricity type. Applying MCSA techniques those frequencies could be detected and studied. In [3] authors presents mathematical models which represent the dynamic state of the indution machine (IM) under static and dynamic eccentricity for monoharmonic effect. Nevertheless, models presented cannot be used to represent mixed eccentricity effects.

In this paper an extended study of the work presented in [3] is carried out. A simple dynamic model of induction machine under mixed eccentricity has been developed for industrial drives applications. Usually drives use healthy models of IM with parameters estimators to ensure the dynamic behavior of the drive. With the porpuse of improving the performance of the drives under fualt condition of the machine, the model has to be adecuate to reproduce main effects on stator currents.

The permeance function of the unbalanced air gap is too complicated to be written in analytical form hence it has been approached by Fourier series. To simplify the model of the machine and limit the required calculations, the effects of eccentricity caused by the basic harmonic of the air gap field and linear magnetic circuit have been studied.

The paper proves that only one model is needed to simulate all types of eccentricities regarding the assumption made. Mathematical procedure followed to reach the final equation of mixed eccentricity model is shown.

II. Calculating Winding Inductances

The inductances of stator and rotor, both self and mutual, have been calculated basing on circuit modeling (or modified winding function approach). Assuming infinite permeability of the iron core of the stator and rotor, the total field of the machine is situated in the air gap. The magneto-motive forces of stator and rotor for the basic frequency are as follows:

$$\theta_{sk}(x,t) = i_{sk}(t)\frac{4}{\pi}\frac{w_{sk}k_c}{2p}cos\ (px - (k-1)\frac{2\pi}{3}) \tag{1}$$

$$\theta_{rk}(x,t) = i_{rk}(t)\frac{4}{\pi}\frac{w_{rk}k_c}{2p}cos\ (p(x-\varphi) - (k-1)\frac{2\pi}{3}) \tag{2}$$

where k is the number of windings in stator and rotor.

The magnetic flux density in the air gap is given by formula:

$$B_k(x,t) = \mu_0\lambda(x,\varphi)\left(\theta_k(x,t) - \frac{\int_x^{x+2\pi}\lambda(x',\varphi)\theta_k(x',t)dx'}{\int_x^{x+2\pi}\lambda(x',\varphi)dx'}\right) \tag{3}$$

In this model smooth air gap has been assumed and presence of the slots is represented by Carter's coefficients k_c for stator and rotor. When distribution of the windings of stator and rotor is known, self and mutual inductances can be computed using formula:

$$L_{n,n} = \frac{\Psi_n}{i_n}\bigg|_{\substack{i_m \neq i_n \\ i_m = 0}} \tag{4}$$

$$L_{n,k} = \frac{\Psi_n}{i_k}\bigg|_{\substack{i_m \neq i_k \\ i_m = 0}} \tag{5}$$

Fluxes are given by formula:

$$\Psi_{n,k} = rl\sum_{p=1}^{\infty}\left(\int_{\alpha_{n,p}-\frac{\pi}{p}}^{\alpha_{n,p}}\left(\sum_{m=1}^{p}\left(-\frac{2}{\pi}w_n k_{n,p}sin p(x-\alpha_{n,p})\right)\int_{\alpha_{p,m}}^{\alpha_{k,m}}B_k(\xi,\varphi,t)d\xi\right)dx\right) \tag{6}$$

In real machine the magnetic flux density in the air gap is not purely sinusoidal hence it can be approached by Fourier series. The Fourier series contains infinite number of harmonics whose amplitudes become lower. The model presented in this paper is based on the first basic harmonic of the magnetic flux density in the air gap. Its frequency is proportional to the pair pole number of the machine. The inductances of the machine contain only the first harmonic of the field.

III. Influence of Air-gap Asymmetry on Permeance Function

In a healthy induction machine the size of the airgap is approximately constant. A certain degree of static eccentricity is an inherent part of every induction machine. On the other hand, due to the slots in stator and rotor, the average size of the air gap is slightly bigger. This fact can be included in machine's equations by Carter's coefficients. When an eccentricity occurs, the air gap is unbalance, its length is not constant around the inner circumference of the stator. Angular positions of the minimal and maximum length of the air-gap can be constant or can change with rotation of the rotor. The unbalance air-gap causes asymmetry of mutual and self-inductances of both stator and rotor because the magnetic field distribution on the air-gap is not uniform. Furthermore self-inductances of the rotor and stator might not be constant depending on the type of the eccentricity.

Changes of the magnetic field force lines in the air-gap can be represented through the permeance function λ.

$$\lambda(x,\varphi) = \frac{1}{\delta_e} \tag{7}$$

where δ_e is the length of the airgap.

This function is periodic and it can be approximated by Fourier series. In case of healthy machine the value is constant; in turn, for eccentricities it depends on the position of the stator and rotor windings respect to stator reference axis, Fig. 1.

Summarizing, the permeance function describes the changes of the minimal air-gap regarding the type of eccentricity.

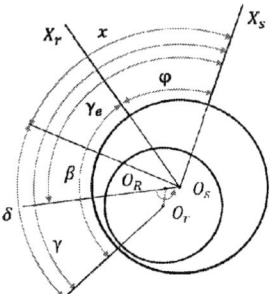

Fig. 1. Reference angles for permeance function calculation

For static eccentricity the position of the minimal air-gap γ_e is fixed, the rotation axis of the rotor O_R coincides with the mass center of the rotor O_r but not with the stator axis O_s. The permeance function depends on the position x of the stator windings respects to the stator reference axis X_s.

$$\lambda(x) = \sum_{r=-\infty}^{\infty}{}_r e^{jrx} \tag{8}$$

where $_r$ are the Fourier coefficients for static eccentricity.

For dynamic eccentricity the position of the minimal air-gap rotates with the angle φ but the length is still constant, the rotation axis of the rotor O_R coincides with the stator axis O_s. The permeance function depends on the rotation angle φ.

$$\lambda(\varphi) = \sum_{s=-\infty}^{\infty}{}_s e^{js\varphi} \tag{9}$$

where $_s$ are the Fourier coefficients for dynamic eccentricity.

Finally, for mixed eccentricity the position and the length of the air-gap are not constant along the circumference. The permeance function depends on the position of each winding of the stator and the rotation angle φ (Rotation direction shown in red color in Fig. 1.).

$$\lambda(x,\varphi) = \sum_{r=-\infty}^{\infty} \sum_{s=-\infty}^{\infty} {}_{r,s} e^{jrx} e^{js\varphi} \qquad (10)$$

where $_{r,s}$ are the Fourier coefficients for mixed eccentricity.

Following figures show the influence of eccentricities in the values of the 2D Fourier series of the permeance function including symmetrical case [5]. All calculations have been done using as example an induction motor with 36 open slots on the stator, 28 bars on the rotor, internal stator radius $R = 0.0594\,m$, external rotor radius $r = 0.059\,m$, equivalent length of a machine core $l = 0.115\,m$ and 2 pair of poles. Results of permeance function calculations were suitable for a dimension set of harmonics: $r = 501$ and $s = 501$.

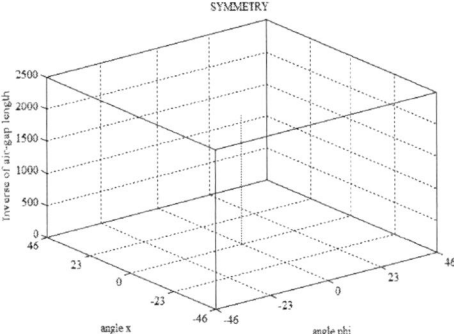

Fig. 2. Fourier coefficient for Symmetrical machine

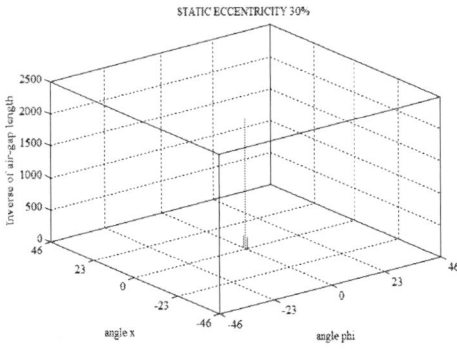

Fig. 3. Fourier coefficients $_r$ for Static eccentricity

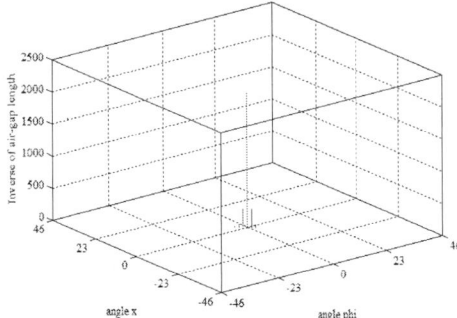

Fig. 4. Fourier coefficients $_s$ for Dynamic eccentricity

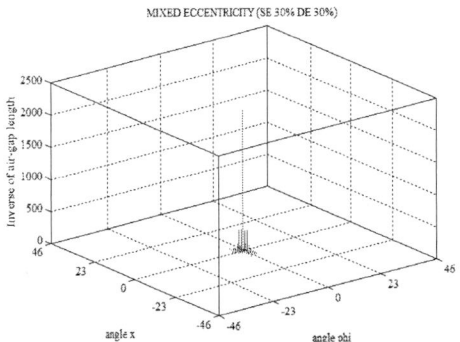

Fig. 3. Fourier coefficients $_{r,s}$ for Mixed eccentricity

The 2-D Fourier coefficients are distributed in a particular way, which is a characteristic of each type of eccentricity. For symmetrical machine only the main coefficient is not equal to zero, and as a consequence there is only one bar in the center of the graph; for static eccentricity the coefficients higher than zero are distributed in one column; for dynamic eccentricity they are distributed in only one diagonal; and for mixed eccentricity they occupy the area around the main coefficient.

Due to the inclusion of the slot effect and magnetic saturation new Fourier coefficients appear in the permeance function. Following the distribution characteristic of each eccentricity, Fourier coefficients will appear around the slot harmonic in case of slot effect assumption and in random positions of the spectra for saturation effect. The resulting mathematical models need more computing time, hence, they are not a suitable tool to be used in drives in which the computation time is a critical variable.

Permeance function can be expressed as a sum of sines and cosines or in complex exponentials form.

$$\lambda(x,\varphi) \approx \sum_{r,s=0}^{\infty} \left\{ \begin{array}{l} a_{r,s} \cos(rx)\cos(s\varphi) + b_{r,s}\cos(rx)\sin(s\varphi) \\ + c_{r,s}\sin(rx)\cos(s\varphi) + d_{r,s}\sin(rx)\sin(s\varphi) \end{array} \right\} \qquad (11)$$

$$\lambda(x,\varphi) \approx \sum_{r,s=-\infty}^{\infty} \left\{ \begin{array}{l} {}_{r,s}\, e^{jrx}e^{js\varphi} + \\ +{}_{r,-s}\, e^{jrx}e^{-js\varphi} + \\ +{}_{-r,s}\, e^{-jrx}e^{js\varphi} + \\ +{}_{-r,-s}\, e^{-jrx}e^{-js\varphi} \end{array} \right\} \qquad (12)$$

where $a_{r,s}$, $b_{r,s}$, $c_{r,s}$, $d_{r,s}$ are the Fourier coefficients.

Relation between Fourier coefficients expressed in both frames can be easily found applying:

$$\begin{bmatrix} Re({}_{r,s}) \\ Imag({}_{r,s}) \\ Re({}_{r,-s}) \\ Imag({}_{r,-s}) \end{bmatrix} = \frac{1}{4} \cdot \begin{bmatrix} 1 & 0 & 0 & -1 \\ 0 & 1 & 1 & 0 \\ 1 & 0 & 0 & 1 \\ 0 & -1 & 1 & 0 \end{bmatrix} \cdot \begin{bmatrix} a_{r,s} \\ b_{r,s} \\ c_{r,s} \\ d_{r,s} \end{bmatrix} \qquad (13)$$

$$\begin{array}{l} {}_{-r,-s} = conj({}_{r,s}) \\ {}_{-r,s} = conj({}_{r,-s}) \end{array} \qquad (14)$$

IV. MIXED ECCENTRICITY MATRIX

Using circuit modeling for the case of study of three windings in stator and rotor, the obtained inductance matrices in natural components are the following:

$$L_{s,s} = L_{\sigma s}\begin{bmatrix} 1 & & \\ & 1 & \\ & & 1 \end{bmatrix} + L_{0,m}^{ss}\begin{bmatrix} 1 & -\frac{1}{2} & -\frac{1}{2} \\ -\frac{1}{2} & 1 & -\frac{1}{2} \\ -\frac{1}{2} & -\frac{1}{2} & 1 \end{bmatrix} +$$

$$+L_{2p,m}^{ss}\begin{bmatrix} \cos(2p\gamma) & \cos\left(2p\gamma-\frac{2\pi}{3}\right) & \cos\left(2p\gamma-\frac{4\pi}{3}\right) \\ \cos\left(2p\gamma-\frac{2\pi}{3}\right) & \cos\left(2p\gamma-\frac{4\pi}{3}\right) & \cos(2p\gamma) \\ \cos\left(2p\gamma-\frac{4\pi}{3}\right) & \cos(2p\gamma) & \cos\left(2p\gamma-\frac{2\pi}{3}\right) \end{bmatrix} +$$

$$-L_{p,m}^{ss}\begin{bmatrix} 1 & -\frac{1}{2} & -\frac{1}{2} \\ -\frac{1}{2} & 1 & -\frac{1}{2} \\ -\frac{1}{2} & -\frac{1}{2} & 1 \end{bmatrix} - L_{p,m}^{ss}\begin{bmatrix} \cos(2p\gamma) & \cos\left(2p\gamma-\frac{2\pi}{3}\right) & \cos\left(2p\gamma-\frac{4\pi}{3}\right) \\ \cos\left(2p\gamma-\frac{2\pi}{3}\right) & \cos\left(2p\gamma-\frac{4\pi}{3}\right) & \cos(2p\gamma) \\ \cos\left(2p\gamma-\frac{4\pi}{3}\right) & \cos(2p\gamma) & \cos\left(2p\gamma-\frac{2\pi}{3}\right) \end{bmatrix} \quad (15)$$

$$L_{r,r} = L_{\sigma r}\begin{bmatrix} 1 & & \\ & 1 & \\ & & 1 \end{bmatrix} + L_{0,m}^{rr}\begin{bmatrix} 1 & -\frac{1}{2} & -\frac{1}{2} \\ -\frac{1}{2} & 1 & -\frac{1}{2} \\ -\frac{1}{2} & -\frac{1}{2} & 1 \end{bmatrix} +$$

$$+L_{2p,m}^{rr}\begin{bmatrix} \cos(2p\gamma-2p\varphi) & \cos\left(2p\gamma-2p\varphi-\frac{2\pi}{3}\right) & \cos\left(2p\gamma-2p\varphi-\frac{4\pi}{3}\right) \\ \cos\left(2p\gamma-2p\varphi-\frac{2\pi}{3}\right) & \cos\left(2p\gamma-2p\varphi-\frac{4\pi}{3}\right) & \cos(2p\gamma-2p\varphi) \\ \cos\left(2p\gamma-2p\varphi-\frac{4\pi}{3}\right) & \cos(2p\gamma-2p\varphi) & \cos\left(2p\gamma-2p\varphi-\frac{2\pi}{3}\right) \end{bmatrix} +$$

$$-L_{p,m}^{rr}\begin{bmatrix} 1 & -\frac{1}{2} & -\frac{1}{2} \\ -\frac{1}{2} & 1 & -\frac{1}{2} \\ -\frac{1}{2} & -\frac{1}{2} & 1 \end{bmatrix} +$$

$$-L_{p,m}^{rr}\begin{bmatrix} \cos(2p\gamma-2p\varphi) & \cos\left(2p\gamma-2p\varphi-\frac{2\pi}{3}\right) & \cos\left(2p\gamma-2p\varphi-\frac{4\pi}{3}\right) \\ \cos\left(2p\gamma-2p\varphi-\frac{2\pi}{3}\right) & \cos\left(2p\gamma-2p\varphi-\frac{4\pi}{3}\right) & \cos(2p\gamma-2p\varphi) \\ \cos\left(2p\gamma-2p\varphi-\frac{4\pi}{3}\right) & \cos(2p\gamma-2p\varphi) & \cos\left(2p\gamma-2p\varphi-\frac{2\pi}{3}\right) \end{bmatrix} \quad (16)$$

$$M_{sr} = L_{0,m}^{sr}\begin{bmatrix} \cos(p\varphi) & \cos\left(p\varphi+\frac{2\pi}{3}\right) & \cos\left(p\varphi+\frac{4\pi}{3}\right) \\ \cos\left(p\varphi-\frac{2\pi}{3}\right) & \cos(p\varphi) & \cos\left(p\varphi+\frac{2\pi}{3}\right) \\ \cos\left(p\varphi-\frac{4\pi}{3}\right) & \cos\left(p\varphi-\frac{2\pi}{3}\right) & \cos(p\varphi) \end{bmatrix} +$$

$$+L_{2p,m}^{sr}\begin{bmatrix} \cos(p\varphi+2p\gamma) & \cos\left(p\varphi+2p\gamma+\frac{2\pi}{3}\right) & \cos\left(p\varphi+2p\gamma+\frac{4\pi}{3}\right) \\ \cos\left(p\varphi+2p\gamma+\frac{2\pi}{3}\right) & \cos\left(p\varphi+2p\gamma+\frac{4\pi}{3}\right) & \cos(p\varphi+2p\gamma) \\ \cos\left(p\varphi+2p\gamma+\frac{4\pi}{3}\right) & \cos(p\varphi+2p\gamma) & \cos\left(p\varphi+2p\gamma+\frac{2\pi}{3}\right) \end{bmatrix} +$$

$$-L_{p,m}^{sr}\begin{bmatrix} \cos(p\varphi) & \cos\left(p\varphi+\frac{2\pi}{3}\right) & \cos\left(p\varphi+\frac{4\pi}{3}\right) \\ \cos\left(p\varphi-\frac{2\pi}{3}\right) & \cos(p\varphi) & \cos\left(p\varphi+\frac{2\pi}{3}\right) \\ \cos\left(p\varphi-\frac{4\pi}{3}\right) & \cos\left(p\varphi-\frac{2\pi}{3}\right) & \cos(p\varphi) \end{bmatrix} +$$

$$-L_{p,m}^{sr}\begin{bmatrix} \cos(p\varphi+2p\gamma) & \cos\left(p\varphi+2p\gamma+\frac{2\pi}{3}\right) & \cos\left(p\varphi+2p\gamma+\frac{4\pi}{3}\right) \\ \cos\left(p\varphi+2p\gamma+\frac{2\pi}{3}\right) & \cos\left(p\varphi+2p\gamma+\frac{4\pi}{3}\right) & \cos(p\varphi+2p\gamma) \\ \cos\left(p\varphi+2p\gamma+\frac{4\pi}{3}\right) & \cos(p\varphi+2p\gamma) & \cos\left(p\varphi+2p\gamma+\frac{2\pi}{3}\right) \end{bmatrix} \quad (17)$$

The values of inductances for main magneto-motives force in the machine can be obtained as:

$$L_{0,m}^{ss} = L_{0,m}^{rr}v^2 = L_{0,m}^{sr}v = r \cdot l \cdot \mu_0\frac{4}{\pi}\left(\frac{w_s k_s}{p}\right)^2 f_1(\varphi) \quad (18)$$

$$L_{2p,m}^{ss} = L_{2p,m}^{rr}v^2 = L_{2p,m}^{sr}v = r \cdot l \cdot \mu_0\frac{4}{\pi}\left(\frac{w_s k_s}{p}\right)^2 f_2(\varphi) \quad (19)$$

$$L_{p,m}^{ss} = L_{p,m}^{rr}v^2 = L_{p,m}^{sr}v = r \cdot l \cdot \mu_0\frac{4}{\pi}\left(\frac{w_s k_s}{p}\right)^2 f_3(\varphi) \quad (20)$$

$$f_1(\varphi) = \sum_{m=0}^{\infty}\Lambda_{0,m}\cos m(\varphi+\beta) \quad (21)$$

$$f_2(\varphi) = \frac{1}{2}\sum_{m=0}^{\infty}\Lambda_{2p,m}\cos m(\varphi+\beta) \quad (22)$$

$$f_3(\varphi) = \frac{1}{4}\left(\sum_{m=0}^{\infty}\Lambda_{p,m}\cos m(\varphi+\beta)\right) \cdot \frac{\sum_{m=0}^{\infty}\Lambda_{p,m}\cos m(\varphi+\beta)}{\sum_{m=0}^{\infty}\Lambda_{0,m}\cos m(\varphi+\beta)} \quad (23)$$

$$v = \frac{w_s k_s}{w_r k_r} \quad (24)$$

In these equations p is the number of pole pairs; φ is the rotation angle; β is the position of the minimal air-gap respect to reference axis of the rotor X_r; w_s, w_r are turn numbers of stator and rotor winding respectively and; k_s, k_r are winding coefficients; γ is the position of the minimal air-gap respect to the stator. Parameters r and l are machine's radius and length. In equations (15) and (16), $L_{\sigma s}$ and $L_{\sigma r}$ are leakage inductances of stator and rotor respectively.

Equations containing matrices presented above are difficult to solve since the matrices of mutual inductances are angle dependent. To simplify these equations and therefore limit the calculations made by computer, some special transformations are used. Symmetrical components transformation reduces equations of a three phase system to three independent circuits, which reduces the number of operations made by the computer while solving the equations. Furthermore matrices of mutual inductances become constant. Transformation matrices to symmetrical components system are given below:

$$S = \frac{1}{\sqrt{3}}\begin{bmatrix} 1 & 1 & 1 \\ 1 & a & a^2 \\ 1 & a^2 & a \end{bmatrix} \quad (25)$$

$$T_r = \begin{bmatrix} 1 & & \\ & e^{j\vartheta_r} & \\ & & e^{-j\vartheta_r} \end{bmatrix} \quad (26)$$

where $a = e^{j\frac{2\pi}{3}}$ and the angle ϑ_r is a transformation angle which equals to $\vartheta_r = p\varphi$, where p is pair pole number and φ is rotational angle of the rotor.

Currents of stator and rotor in new coordinate system are given by the formula:

$$\begin{bmatrix} i_s^0 \\ i_s^1 \\ i_s^2 \end{bmatrix} = S\begin{bmatrix} i_{s1} \\ i_{s2} \\ i_{s3} \end{bmatrix} \quad (27)$$

$$\begin{bmatrix} i_r'^0 \\ i_r'^+ \\ i_r'^- \end{bmatrix} = T_r S\begin{bmatrix} i_{r1}' \\ i_{r2}' \\ i_{r3}' \end{bmatrix} \quad (28)$$

Inductance matrices in symmetrical components have the following form:

$$L_{s,s} = \begin{bmatrix} L_{\sigma s}{}^0 & 0 & 0 \\ 0 & L_{\sigma s}{}^1 & 0 \\ 0 & 0 & L_{\sigma s}{}^2 \end{bmatrix} + L_{0,m}^{ss} \begin{bmatrix} 0 & 0 & 0 \\ 0 & \frac{3}{2} & 0 \\ 0 & 0 & \frac{3}{2} \end{bmatrix} + L_{2p,m}^{ss} \cdot \left(\frac{3}{2}\right) \cdot \begin{bmatrix} 0 & 0 & 0 \\ 0 & 0 & e^{j2p\gamma} \\ 0 & e^{-j2p\gamma} & 0 \end{bmatrix} + \qquad (29)$$

$$+ L_{p,m}^{ss} \begin{bmatrix} 0 & 0 & 0 \\ 0 & \frac{3}{2} & 0 \\ 0 & 0 & \frac{3}{2} \end{bmatrix} + L_{p,m}^{ss} \cdot \left(\frac{3}{2}\right) \cdot \begin{bmatrix} 0 & 0 & 0 \\ 0 & 0 & e^{j2p\gamma} \\ 0 & e^{-j2p\gamma} & 0 \end{bmatrix}$$

$$L_{r,r} = \begin{bmatrix} L_{\sigma r}{}^0 & 0 & 0 \\ 0 & L_{\sigma r}{}^1 & 0 \\ 0 & 0 & L_{\sigma r}{}^2 \end{bmatrix} + L_{0,m}^{rr} \begin{bmatrix} 0 & 0 & 0 \\ 0 & \frac{3}{2} & 0 \\ 0 & 0 & \frac{3}{2} \end{bmatrix} + L_{2p,m}^{rr} \cdot \left(\frac{3}{2}\right) \cdot \begin{bmatrix} 0 & 0 & 0 \\ 0 & 0 & e^{j2p\gamma} \\ 0 & e^{-j2p\gamma} & 0 \end{bmatrix} + \qquad (30)$$

$$+ L_{p,m}^{rr} \begin{bmatrix} 0 & 0 & 0 \\ 0 & \frac{3}{2} & 0 \\ 0 & 0 & \frac{3}{2} \end{bmatrix} + L_{p,m}^{rr} \cdot \left(\frac{3}{2}\right) \cdot \begin{bmatrix} 0 & 0 & 0 \\ 0 & 0 & e^{j2p\gamma} \\ 0 & e^{-j2p\gamma} & 0 \end{bmatrix}$$

$$M_{sr} = L_{0,m}^{sr} \begin{bmatrix} 0 & 0 & 0 \\ 0 & \frac{3}{2} & 0 \\ 0 & 0 & \frac{3}{2} \end{bmatrix} + L_{2p,m}^{sr} \cdot \left(\frac{3}{2}\right) \cdot \begin{bmatrix} 0 & 0 & 0 \\ 0 & 0 & e^{-j2p\gamma} \\ 0 & e^{j2p\gamma} & 0 \end{bmatrix} + \qquad (31)$$

$$+ L_{p,m}^{sr} \begin{bmatrix} 0 & 0 & 0 \\ 0 & \frac{3}{2} & 0 \\ 0 & 0 & \frac{3}{2} \end{bmatrix} + L_{p,m}^{sr} \cdot \left(\frac{3}{2}\right) \cdot \begin{bmatrix} 0 & 0 & 0 \\ 0 & 0 & e^{-j2p\gamma} \\ 0 & e^{j2p\gamma} & 0 \end{bmatrix}$$

V. Obtaining static eccentricity from Mixed Eccentricity Model

Understanding the physical phenomena it is possible to deduce the model of static eccentricity from the model of mixed eccentricity.

For mixed eccentricity it has been shown that the length of the minimal air-gap is not constant and its position changes along the circumference. For static eccentricity the position of the minimal air-gap is fixed in space and the length is constant. For static eccentricity λ is independent of the rotation angle "φ" because the position of minimal air-gap is fixed in space, and it only depends on angle x. For each winding of the stator the length of the air-gap is constant. The equivalent Fourier coefficients of the permeance function $_r$ for the case of static eccentricity can be found from permeance function of mixed eccentricity considering only main 'm' harmonic in (17) (18) (19), hence:

$$f_1(\varphi) = \Lambda_0 \qquad (32)$$

$$f_2(\varphi) = \frac{1}{2}\Lambda_{2p} \qquad (33)$$

$$f_3(\varphi) = \frac{1}{4}\Lambda_p \cdot \frac{\Lambda_p}{\Lambda_0} \qquad (34)$$

Please note that the inductance matrix $L_{s,s}$ is constant due to the fact that the position of the minimal air-gap does not change in time meanwhile for rotor quantities $L_{r,r}$ the dependence on the rotation angle still remains due to the length of the air-gap for one slot of the rotor change in time.

Using above simplifications, the model for static eccentricity deduced from the mixed eccentricity applying

symmetrical components transformation coincides with the model proposed in [3].

VI. Obtaining Dynamic eccentricity from Mixed Eccentricity Model

For dynamic eccentricity the position of the minimal air-gap is rotating with the angle "φ" (rotational angle of the rotor) but the length is still constant in rotor's coordinate system. For a fixed point on stator, the air-gap changes.

The equivalent Fourier coefficients for the case of dynamic eccentricity can be obtained considering only main m harmonic in (17) (18) (19).

$$f_1(\varphi) = \Lambda_0 \qquad (35)$$

$$f_2(\varphi) = \frac{1}{2}\Lambda_{2p} \qquad (36)$$

$$f_3(\varphi) = \frac{1}{4}\Lambda_p \cdot \frac{\Lambda_p}{\Lambda_0} \qquad (37)$$

For dynamic eccentricity the angle γ in equations of the machine in natural components is equal to $\gamma = \varphi + \delta$ where δ is the position of the minimal air gap with reference to the rotor and φ is the rotational angle of the rotor. Transforming equations (11) - (13) to symmetrical components, model of induction machine with dynamic eccentricity has been obtained, which is similar to the model shown in [3].

VII. Conclusions

In this paper a mathematical model for mixed eccentricity in induction machines for drives applications has been presented under assumptions of main magneto-motive harmonics effect in natural components. Equations presented in the paper are suitable to be implemented in Matlab/Simulink for diagnosis purposes. Simplifications applied to the models have eliminated the angle dependence in inductances matrices.

Description of the physical phenomena inside the electrical machine under eccentricities faults has been used as a reference to explain the relation between eccentricities. Permeance function has been defined by Fourier coefficients and differences between faults has been shown.

It has been proved that static eccentricity and dynamic eccentricity can be deduced from the model of mixed eccentricity.

VIII. References

[1] "Report of Large Motor Reliability Survey of Industrial and Commercial Installations", *IEEE Transactions on Industry Applications*, vol. IA-23, no. 1, pp.153-158, Jan. 1987.

[2] T. J. Sobczyk, P. Vas, C. Tassoni,"A comparative study of effects due to eccentricity and external stator and rotor asymmetries by monoharmonic models", ICEM In International conference on electrical machines, pp. 946-950, Aug. 2000.

[3] T. J. Sobczyk, P. Vas, C. Tassoni, "Models for Induction motors with air-gap asymmetry for diagnosis purposes", *ICEM Proceedings International Conference on Electrical Machines*, vol. II, pp. 79-78, Sep. 1996.

[4] J. Rosero, J. Cusido, A. García Espinosa, J.A. Ortega, L. Romeral, "Broken Bearings Fault Detection for a Permanent Magnet Synchronous motor under non-constant working conditions by means of a joint time frequency analysis", IEEE International Symposium on Industrial Electronics, pp. 3415-3419, Jun. 2007.

[5] T. Wegiel, K. Weinreb, M. Sulowicz, "Main inductances of induction motor for diagnostically specialized mathematical models", Archives of Electrical Engineering, vol. 59 (1-2), pp. 51-66, 2010.

[6] T. J. Sobczyk, K. Weinreb, T. Wegiel, M. Sulowicz, "Theoretical study of effects due to rotor eccentricities in induction motors", *in Proc. IEEE Symp. Electrical Machines, Power Electronics and Drives, the SDEMPED 1999*, pp.289 -295,Sep. 1999.

[7] R. Samaga, K.P. Vittal,"Air gap mixed eccentricity severity detection in an induction motor", IEEE Recent Advances in Intelligent Computational Systems (RAICS), pp. 115-119, Sep. 2011.

[8] T. J. Sobczyk, P. Drozdowski, "Inductances of electrical machine winding with a non-uniform air-gap", Archiv fur Elektrotechnik, vol. 76, pp. 213-218, 1993.

[9] A. Bellini, F. Filippetti, F. Franceschini, T.J. Sobczyk, C. Tassoni, "Diagnosis of induction machines by d-q and i.s.c rotor models", *in Proc. 5th IEEE Symp. on Diagnostics of Electric Machines, Power Electronics and Drives, the SDEMPED 2005*, pp.41-46, Sep. 2005.

[10] T.J. Sobczyk, "Direct determination of two periodic solutions for non-linear dynamic systems", COMPEL International Journal for Computation and Mathematics in Electrical and Electronic Engineering, vol. 13, No. 3, 1994, pp.509–529.

[11] J. Faiz, B. M. Ebrahimi, M. Valavi, H. A. Toliyat, "Mixed eccentricity fault diagnosis in salient-pole synchronous generator using modified winding function method", Progress In Electromagnetics Research B, vol 11, pp. 155-172, 2009.

[12] G. M. Joksimović, "Dynamic simulation of cage induction machine with air gap eccentricity", IEE Proceedings-Electric Power Applications, vol. 152, no 4, pp. 803-811, Jul. 2005.

[13] S. Nandi, H. A. Toliyat, X. Li, "Condition monitoring and fault diagnosis of electrical motors-a review". Energy Conversion, IEEE Transactions on, vol. 20, no 4, pp. 719-72, Dec. 2005.

[14] J. Faiz, "Comprehensive eccentricity fault diagnosis in induction motors using finite element method", IEEE Transactions on Magnetics, vol. 45, no 3, p.p 1764-1767, March 2009.

[15] J. Faiz, I. T. Ardekanei, H. A. Toliyat, "An evaluation of inductances of a squirrel-cage induction motor under mixed eccentric conditions'. Energy conversion, ieee transactions on, vol. 18, no 2, pp. 252-258, Jun. 2003.

[16] R. Samaga, K. P. Vittal, "Investigation into effect of mixed air gap eccentricity on dq components of currents in induction motor", 6th IEEE International Conference on In Industrial and Information Systems (ICIIS), pp. 271-276, Aug. 2011.

[17] J. Fraile Mora, *Electrical Machines*, edn.VI. Madrid: McGraw-Hill, 2003, pp. 259-378.

[18] G. M. Fitchtenholz, *Differential and integral calculus*, tom. 3, Warsaw: PWN (in Polish), 1980, pp. 390-408.

[19] T.J. Sobczyk, *Metodyczne aspekty modelowania matematycznego maszyn indukcyjnych*, Warsaw: WNT (in Polish), 2004.

IX. BIOGRAPHIES

Tadeusz J. Sobczyk is a professor at the Faculty of Electrical & Computer Engineering of the Cracow University of Technology. In the years 1993-1999, he was the Dean of this Faculty. Presently he is the Director of the Institute on Electromechanical Energy Conversion and the Head of Department of Electrical Machines in this Institute. Since 1991 is a member of the Committee of Electrical Engineering of the Polish Academy of Science. In 2000 he was awarded the honorary title Doctor Honorees Causa of the Russian Academy of Sciences. His main research fields are: electrical machines and drives, electromechanical systems, electrical energy conversion and transformation by power electronic systems.

Alejandro J. Fernández Gómez was born (1984) and educated in Spain. In 2010 he received M.Sc degrees in electrical engineering from the Faculty of Industrial Engineering at the University of Vigo. Since 2011 he has been working as early stage researcher in the Institute on Electromechanical Energy Conversion at the Faculty of Electrical & Computer Engineering at the Cracow University of Technology enrolled in ITN Energy Smartops project of Marie Curie Actions. His research field is electromechanical energy conversion and fault diagnosis of electrical machines.

Arkadiusz Dziechciarz Received his M.Sc degree from the Faculty of Electrical & Computer Engineering at the Cracow University of Technology. Currently he is a science teaching assistant in the Institute on Electromechanical Energy conversion at the Faculty of Electrical & Computer Engineering at the Cracow University of Technology. His main research field is fault diagnosis of electrical machines.

Winding Function Approach for Induction Machine Fault Detection

Pu. Shi, Zheng. Chen, Yuriy. Vagapov, Zoubir. Zouaoui

Abstract -- This paper introduces a novel method for motor fault detection under varying load conditions. Model based induction machine broken rotor-bar fault performances are discussed. Winding function approach (WFA) based mathematical model is developed to provide indication information regarding motor parameter evolution under the influence of load level and fault severity. Through further analysis the motor parameters evolution, broken rotor-bar fault detection indexes are developed. Moreover, these indexes can assess rotor fault severity based on stator current and rotor speed. Stator current and rotor speed are used to demonstrate correlations between these parameters and broken rotor bar severity. Simulations and experimental results confirm the validity and effectiveness of the proposed approach.

Index Terms--Winding Function Approach, Induction Machine, Broken Rotor Bars

I. INTRODUCTION

INDUCTION machines have been widely employed in industrial production processes due to its reliability, robustness and cost-effective. Although the design of induction machines is constantly improving, induction machines are still prone to failure sooner or later [1].

Besides, motors are often exposed to hostile environments and conditions which can age the motor and make it subject to incipient fault. These faults, failures and gradual deterioration, will lead to motor break down if left undetected and resulting costly unplanned downtime. Therefore, condition monitoring of electrical machines has received considerable attention for many years [2] [3].

Walliser have shown that when a bar is broken near an end ring, then significant inter-bar currents may flow between the broken and adjacent bars [1]. Ellison verified from tests carried out in an anechoiv chamber that slot harmonics in the acoustic noise spectra from a small power induction motor [2]. Verma and Morita have studied the changes in the air-gap field as a function of static eccentricity using search coils in the stator core [3] [4].

Most researches are highly relying on experimental data which is not always available due to hardware or software problems. Model based approaches can easily overcome this issue by simulating fault cases on computer programs.
In this report, a novel approach for predicting the dynamic performance of induction machines which can be used in model-based diagnosis is presented. The performances of rotor broken bars on the machine stator current and rotor

P. Shi, Z. Chen, Y. Vagapov, and Z. Zouaoui are with Engineering Department, Glyndwr University, Wrexham, UK (e-mail: s07001957@mail.glyndwr.ac.uk).

speed signature under varying load conditions are discussed.

II. COUPLED CIRCUIT MODEL OF INDUCTION MACHINE

Winding function approach (WFA) stems from Ampere's law which accounts for all the space harmonics in induction machine. The motor model based on differential equations simulates the performance of m-phase induction machine with n rotor bar [5] [6] [7] [8].

WFA employs the coupled magnetic theory which treats the current in each rotor bar as an independent variable. In this report, WFA based induction motor model is developed initially with the following assumptions [5]:

(1) The machine is symmetric and air-gap is uniform and smooth;

(2) The rotor bars are insulated from the rotor and there is no inter-bar current flows through the laminations;

(3) The permeability of machine armatures is assumed infinite.

A. Stator Voltage Equations

With an induction machine stator constructed of m-phase concentric windings, the stator equations for an induction machine can be written in vector matrix form as [5] [6]:

$$[V_s] = [R_s][I_s] + \frac{d}{dt}[\Phi_s] \quad (1)$$

Where,

$$[\Phi_s] = [L_{ss}][I_s] + [L_{sr}][I_r] \quad (2)$$

The matrix $[R_s]$ and $[L_{ss}]$ are symmetric 3×3 matrix. The mutual inductance matrix $[L_{sr}]$ is a $3 \times n$ matrix comprised of the mutual inductances between the stator coils and the rotor loops.

$$[L_{sr}] = [L_{rs}]^T = \begin{bmatrix} L_{sr11} & L_{sr12} & \cdots & L_{sr1n} & L_{sr1e} \\ L_{sr21} & L_{sr22} & \cdots & L_{sr2n} & L_{sr2e} \\ L_{sr31} & L_{sr32} & \cdots & L_{sr3n} & L_{sr3e} \end{bmatrix} \quad (3)$$

B. Rotor Voltage Equations

A rotor cage consists of n bars can be treated as *n* identical and equally spaced rotor loop. As illustrated in Fig.1, each loop is formed by two adjacent rotor bars and the connecting portions of the end-ring [6] [8].

As shown in Fig.2, the voltage equations for the rotor loops can be written in vector matrix form as [5] [6]:

$$[V_r] = [R_r][I_r] + \frac{d}{dt}[\Phi_r] \quad (4)$$

Where,

$$[V_r] = [V_{r1} \quad V_{r2} \quad \cdots \quad V_{rn} \quad V_{re}]^T \quad (5)$$

In case of a cage rotor, the rotor end-ring voltage is $V_{re}=0$, and the rotor loop voltages are $V_{rk}=0$, $(k=1,2...n)$.

The resistance matrix $[R_r]$ is a symmetric $(n+1)$ by $(n+1)$ matrix given by:

$$[R_r] = \begin{bmatrix} 2(R_b+R_e) & -R_b & 0 & \cdots & 0 & -R_b & -R_e \\ -R_b & 2(R_b+R_e) & -R_b & \cdots & 0 & 0 & -R_e \\ \vdots & \vdots & \vdots & \vdots & \vdots & \vdots & \vdots \\ 0 & 0 & 0 & \cdots & 2(R_b+R_e) & -R_b & -R_e \\ -R_b & 0 & 0 & \cdots & -R_b & 2(R_b+R_e) & -R_e \\ -R_e & -R_e & -R_e & \cdots & -R_e & -R_e & nR_e \end{bmatrix} \quad (6)$$

In relation (4), the rotor flux can be written as:

$$[\Phi_r] = [L_{rs}][I_s] + [L_{rr}][I_r] \quad (7)$$

Where, the matrix $[L_{rs}]$ is the transpose of the matrix $[L_{sr}]$. Due to the structural symmetry of the rotor, $[L_{rr}]$ can be written in matrix form (8), where L_{kk} is the self-inductance of the kth rotor loop, L_b is the rotor bar leakage inductance, L_e is the rotor end-ring leakage inductance and L_{ki} is the mutual inductance between two rotor loop.

$$[L_{rr}] = \begin{bmatrix} L_{11}+2(L_b+L_e) & L_{12}-L_b & L_{13} & \cdots & L_{1(n-1)} & L_{1n}-L_b & -L_e \\ L_{21}-L_b & L_{21}+2(L_b+L_e) & L_{23}-L_b & \cdots & L_{2(n-1)} & L_{2(n-1)} & -L_e \\ \vdots & \vdots & \vdots & \cdots & \vdots & \vdots & \vdots \\ L_{(n-1)1} & L_{(n-1)2} & L_{(n-1)3} & \cdots & L_{(n-1)(n-1)}+2(L_b+L_e) & L_{(n-1)n}-L_b & -L_e \\ L_{n1}-L_b & L_{n2} & L_{n3} & \cdots & L_{n(n-1)}-L_b & L_{nn}+2(L_b+L_e) & -L_e \\ -L_e & -L_e & -L_e & \cdots & -L_e & -L_e & nL_e \end{bmatrix} \quad (8)$$

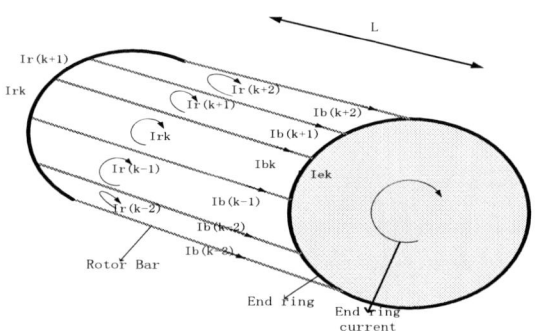

Fig.1. Elementary rotor loops and current definitions

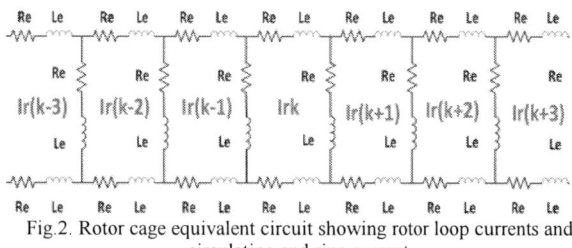

Fig.2. Rotor cage equivalent circuit showing rotor loop currents and circulating end ring current.

C. Calculation of Torque

The mechanical equation of the machine is [5] [6]:

$$T_{em} - T_L = J\frac{d\Omega_m}{dt} \quad (8)$$

Where, T_{em} is the electromagnetic torque produced by the machine, T_L is the load torque, J is the inertia of the rotor and Ω_m is the mechanical speed.

$$\Omega_m = \frac{1}{P}\frac{d\theta}{dt} \quad (9)$$

The electrical torque can be found from the magnetic coenergy W_{co} as

$$T_e = \left(\frac{\partial W_{co}}{\partial \theta_{rm}}\right)_{(I_s, I_r, \text{constant})} \quad (10)$$

In a linear magnetic system the coenery is equal to the stored magnetic energy so that,

$$W_{co} = \frac{1}{2}I_s^t L_{ss} I_s + \frac{1}{2}I_s^t L_{sr} I_r + \frac{1}{2}I_r^t L_{rs} I_s + \frac{1}{2}I_r^t L_{rr} I_r \quad (11)$$

It is obvious that L_{ss} and L_{rr} contain only constant elements and T_e is a scalar quantity. Therefore the electromagnetic torque is finally given by the following equation [5] [6]:

$$T_{em} = PI_s^t\left\{\frac{d}{d\theta}[L_{sr}]\right\}I_r \quad (12)$$

where θ is the angular position of the rotor and P denotes the number of motor pole pairs.

D. Calculation of Inductances

According to winding function theory, the mutual inductance between two windings i and j in any electric machine can be computed by the following equation [9] [10] [11]:

$$L_{ij}(\theta_r) = \frac{\mu_0 lr}{g}\int_0^{2\pi} N_i(\theta,\Phi)N_j(\theta,\Phi)d\Phi \quad (13)$$

Where, $\mu_0 = 4\pi.10^{-7}$ H/m, g is the air gap length, θ_r is the rotor angular position, r is the average radius of the air gap, l is the active stack length of the motor, Φ is angular position, and $N_i(\theta,\varphi)$, $N_j(\theta,\varphi)$ is called the winding function of circuit i and j and represents the magneto motive force (MMF) distribution along the air gap.

The specifications of the induction machine simulated in this paper listed in Table.I. Fig.3(a) shows the winding distribution of the stator phase. The MMF distribution of the stator phase is represented by the winding function in Fig.3(b) [12].

Table.I. Specifications of an Induction Machine

Specifications	Value
Rated Power	3.7 kW
Horse Power	5 HP
Input Voltages	220/380 V
Input Currents	13.8/8.0 A
Pole pairs	1
Frequency	50 Hz
Speed	3000 rpm
Number of Stator Slots	36 EA
Number of Rotor Bars	28 EA

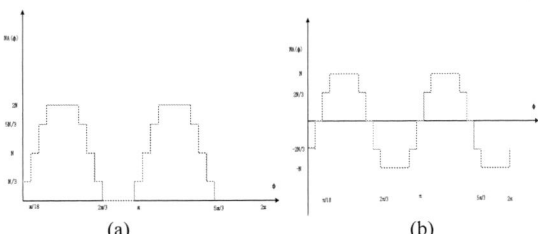

Fig.3. Stator-winding graph. (a) Winding distribution. (b) Winding function.

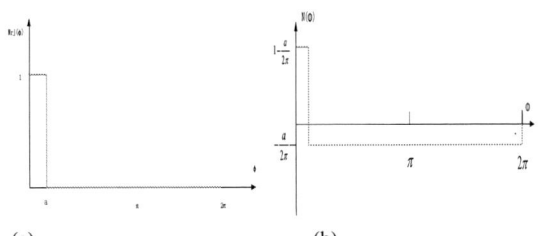

Fig. 4. Rotor-winding graph. (a) Winding distribution. (b) Winding function.

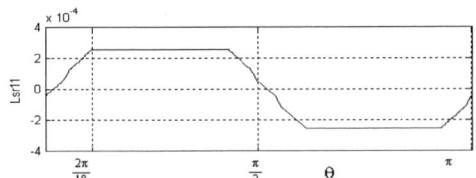

Fig.5. Stator-rotor mutual inductance between the stator phase "1"and rotor loop "1"

To obtain the winding distribution and winding function of the next rotor loop, the graphs in Fig. 4(a) and (b) should be shifted to the positive angle a as

$$a = \frac{2\pi}{N} \quad (15)$$

Where a is the radian angle between two adjacent rotor bars and N is the number of rotor bars.

The mutual inductance between stator and rotor branches will be a function of the rotor position angle, θ as shown in Fig.5.

III. SIMULATION AND ANALYSIS

When induction machine operate under ideal, healthy circumstances, stator current is sinusoidal signal. However, in practical unexpected distortions and noises always exist in stator current, which make it difficult to accurate detect and diagnosis faults.

Fig.6 illustrates the stator phase currents of the induction machine during steady under balanced sinusoidal voltage supply. 0%, 20%, 40%, 60% and 100% load level cases were simulated. Stator current amplitudes increasing are clearly observed due to load levels rising.

Fig.7 shows the steady state speed of the machine. Rotor speeds change from the 2998 at no load to 2962, 2947, 2933, 2912 and 2888 at 20%, 40%, 60%, 80% and 100% load level.

Fig.6 and Fig.7 depict load level raising causes stator current increase and rotor speed decrease. Table.II lists the detail values of the stator current and speeds values under different load levels.

When broken bar fault occurs in rotor, negative component at frequency (1-2s)f is generated in stator current. This component produces a torque ripple and causes rotor speed variation. This speed variation cause mechanical angular variation and lead to motor produces a phase modulation in the stator flux.

Fig.8 (a) shows a case of one rotor bar totally disconnected, which means no current will flow through it. Fig.9 and Fig 10 show the steady state phase current envelops and speed of the machine for one to three broken bars under no load level. Comparing to Fig.6 and Fig.7, ripples exist in the machine phase currents and speeds. These symptoms are not obvious enough to diagnosis by observing the figures especially under low load level. As shown in Fig.10, rotor speed fluctuate around 2988±2 rpm, these cannot be sensed by ordinary sensor.

Fig.11 and Fig.12 depict the steady state phase currents and speeds of the machine of 80% load under different broken bars. Comparing to Fig.6 and Fig.7 and Fig.9 and Fig.10, stator phase currents and rotor speeds keep decreasing under same load level due to more broken rotor bars fault occurs. The riffles caused by broken rotor bar fluctuate more and more obviously.

Fig.8(c) shows the winding circuit of three broken bars. Fig.13 and Fig.14 illustrate the phase currents and speeds of the machine for the case of three broken rotor bars during steady state with a balanced sinusoidal voltage supply.

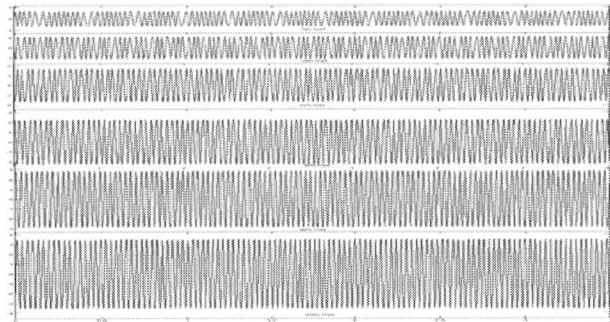

Fig.6. Healthy induction motor stator current under different load level

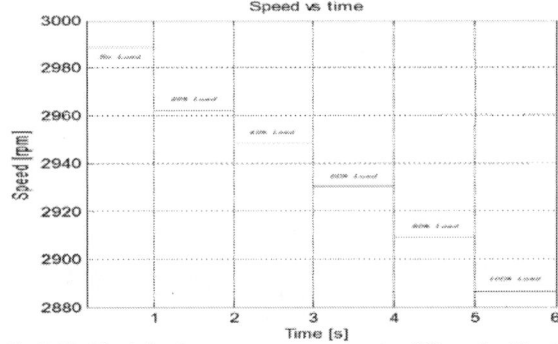

Fig.7. Healthy induction motor rotor speed under different load level.

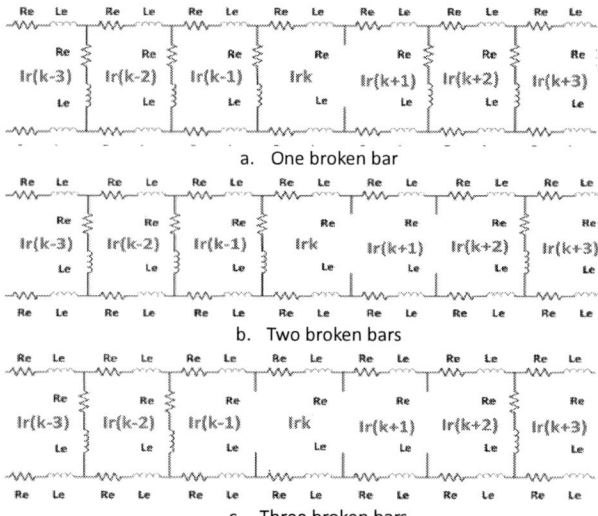

a. One broken bar

b. Two broken bars

c. Three broken bars

Fig.8. Winding circuits representation of one broken rotor bar.

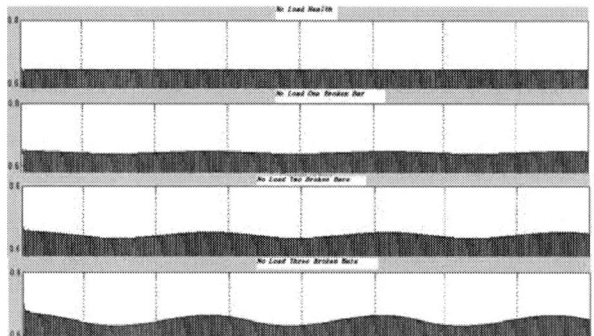

Fig.9. Broken under no load level.

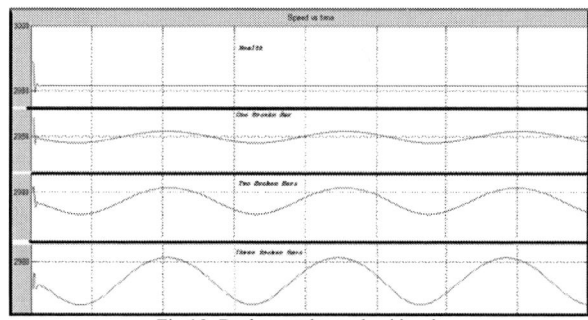

Fig.10. Broken under no load level.

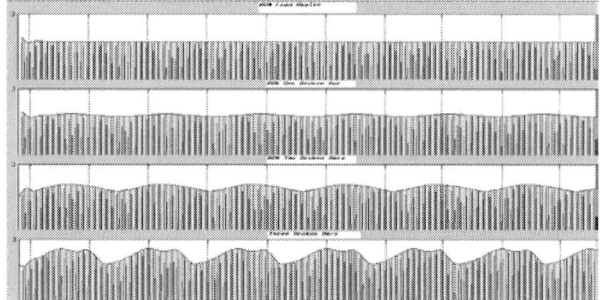

Fig.11. Two broken rotor bars induction motor stator current under different load level.

Table.II. Healthy induction machine current and speed under different load level

Load Level (Percentage)	Stator Currents (A)	Rotor Speed (RPM)
No Load	0.64	2998
20%	1.17	2962
40%	1.8	2947
60%	2.4	2933
80%	2.9	2912
100%	3.6	2888

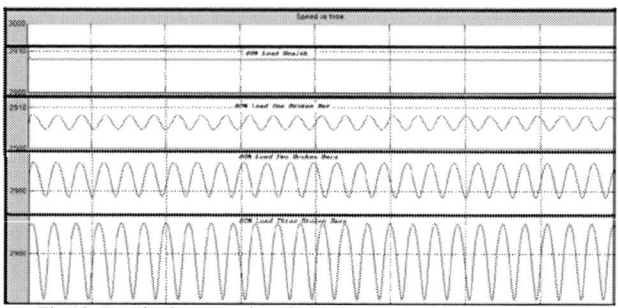

Fig.12. Two broken rotor bars induction motor rotor speed under different load level.

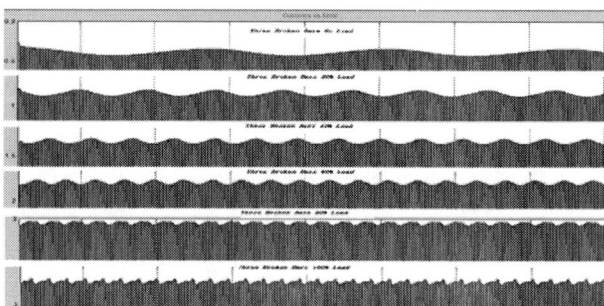

Fig.13. Three broken rotor bars induction motor stator current under different load level.

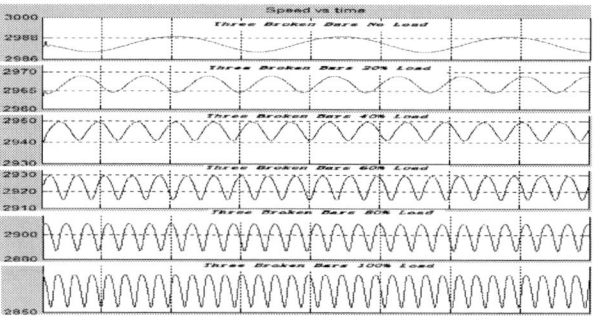

Fig.14. Three broken rotor bars induction motor rotor speed under different load level.

IV. FAULT DETECTION APPROACH

A special correlation between number of broken bars, rotor speed, load level and stator current can be developed from above analysis. Stator current and rotor speed decrease with load level increase regardless operate condition. But these wave fluctuations come with fault severity and load level, as broken bar number increase high magnitude and period fluctuations will show up.

Table.III shows the mean abs values and standard deviation values of induction under different operation condition. From Table.III mean *abs* and *SD* both increase with load. It is hard to make any decision based on these two values. To solve this problem, a specified fault index equation (16) is introduced

$$Index = \frac{SD}{Meanabs} \quad (16)$$

Where SD represents standard deviation, Meanabs represents mean absolute values.

By calculating the fault index equation in (16), it is much easier to judge the induction machines condition. As shown in Table.III(a), under healthy situation this index is around 1.67. Under one broken rotor bar occurs in this induction machine, this index will drop to 1.40 as shown in Table. III(b). As more rotor bars breakage happen, the index will continue drop to 1.36 and 1.31 under two and three broken bars.

From Table.III, a special relationship between load level and stator current has developed. Through index equation (16), accurate induction machine operation condition can be evaluated regarding the load level.

V. EXPERIMENTAL TEST AND RESULTS

The experiment was composed of three phase induction machines, current transducer, A/D converter, and computer. Transient current signals were collected and wavelet transform (WT) was used to decompose the acquired time domain signal into time-frequency domain. Then, fault features frequency *(1±2s)f* was extracted from WT transform. Finally, the individual diagnosis results were used to validate the developed model.

The characteristics of the 3 phase induction machines used in this experiment are listed in Table.I. Among these tests, one is normal (healthy), which was set as a benchmark condition compared with another faulty motor with three broken rotor bars. A current Hall Effect sensor was placed in one of the line current cables. The stator current was sampled with a *1* kHz rate and interfaced to a PC by an ADC-11 acquisition board.

Fig.16 and Fig.17 show the time and frequency analysis of stator current under healthy and three broken rotor bars. From Fig.16(a) and Fig.17(a), which show stator current wave under time domain, there is no obvious difference between these two. In contrast to Fig.16(a) and Fig.17(a), Fig.16(a) and Fig.17(a) are more obvious to judge the motor condition. According to the fault features frequency *(1±2s)f* equation, six frequencies appear around the fundamental frequency in Fig.17 (b).

Although frequency analysis has the ability diagnosis motor condition according to Fig.17(b), the fault feature magnitudes are too small to detect and measure, sometimes even buried by the noise. This causes lot difficulties during industrial induction machine fault detection.

To avoid this problem and improve the detection result,

the proposed approach is used in this experiment to decide the motor situation as shown in Table .IV. The fault indices shown in Table .IV shows a very accurate result coincide with the real situation. Comparing to Fig.17(b), quantification number have much more demonstration influence.

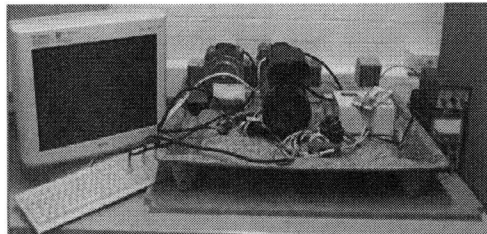

Fig.15. View of the experimental setup and broken rotor bars

Table.III. Simulation result analysis based on mean and SD values

Healthy						
	No Load	20%	40%	60%	80%	100%
Mean abs	0.272	0.4625	0.7023	1.0093	1.3063	1.5923
Standard Deviation	0.4543	0.7813	1.177	1.699	2.191	2.677
SD/Mean abs	1.670220588	1.689297297	1.675921971	1.683344892	1.677256373	1.681215851

(a).Healthy

One broken bar						
	No Load	20%	40%	60%	80%	100%
Mean abs	0.3235	0.5563	0.8368	1.136	1.4415	1.7578
Standard Deviation	0.4546	0.7818	1.179	1.601	2.03	2.475
SD/Mean abs	1.405255023	1.405356822	1.408938815	1.409330986	1.40825529	1.408010013

(b).One broken bar

Two broken bar						
	No Load	20%	40%	60%	80%	100%
Mean abs	0.339	0.5728	0.8598	1.177	1.4956	1.8288
Standard Deviation	0.4641	0.7818	1.171	1.603	2.04	2.501
SD/Mean abs	1.369026549	1.364874302	1.361944638	1.361937128	1.36400107	1.36756343

(c).Two broken bars

Three broken bar						
	No Load	20%	40%	60%	80%	100%
Mean abs	0.347	0.5978	0.9055	1.229	1.5395	1.9135
Standard Deviation	0.4566	0.7864	1.188	1.618	2.027	2.513
SD/Mean abs	1.315850144	1.31549013	1.31198233	1.316517494	1.316661254	1.313300235

(d).Three broken bars

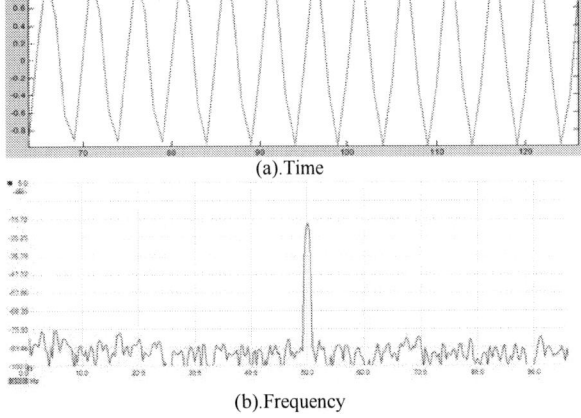

(a).Time

(b).Frequency

Fig.16. Experimental plots of stator current spectrum around fundamental of the healthy machine

Table IV. Mean and standard deviation values of WT under healthy and three broken rotor bars

Experimental Results		
	Healthy	Three broken bars
Mean abs	0.2132	0.3572
Standard Deviation	0.3517	0.4652
SD/Mean abs	1.649624765	1.302351624

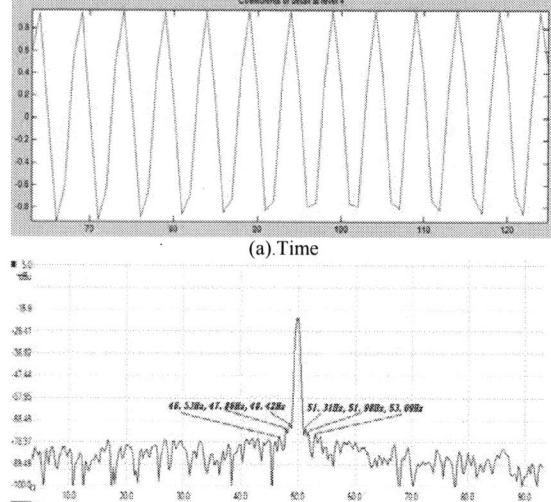

(a).Time

(b).Frequency

Fig.17. Experimental plots of stator current spectrum around fundamental of the faulty machine with three broken bars

VI. CONCLUSION

A detailed model of a squirrel-cage induction machine has been developed. In order to simulate broken rotor bars defects, the machine was modeled as a group of coupled magnetic circuits by considering the current in each rotor bar as an independent variable. The model has the ability to simulate the performance of induction machines during transient as well as at steady state, including the effect of rotor faults. Simulation and experimental data were analyzed to study rotor faults cause-effect relationship in the stator current and the current spectrum signature. A novel criterion was derived to assess rotor fault severity based on stator current. Simulations and experimental results confirm the validity of the proposed approach.

VII. REFERENCES

[1] Walliser, R F and Landy, C F, "Determination of interbar current effects in the detection of broken rotor bars in squirrel cage induction motors," *IEEE Trans.Energy Conversion,* vol. 9, no. 1, pp. 152-158, 1994.

[2] Ellison, A J and Yang, S J, "Effects of rotor eccentricity on acoustic noise from induction machines," *Proc.Inst.Electr.Eng,* vol. 118, no. 1, pp. 174-184, 1974.

[3] Verma, S P and Natarajan, R, "Effects of eccentricity in induction motors," *in Proc.Int.Conf.Electrical Machines,Budapest,Hungary,* pp. 930-933, 1982.

[4] Morita, I, "Air-gap flux analysis for cage rotor diagnosis," *Elec.Eng.In Japan,* vol. 112, no. 3, pp. 171-181, 1992.

[5] Toliyat, H A, Lipo, T A and White, J C, "Analysis of a concentrated winding induction machine for adjustable speed drive application part 1 (Motor analysis)," *IEEE Transaction on Energy Conversion,* vol. 6, no. 4, pp. 679-684, 1991.

[6] Toliyat, H A and Lipo, T A, "Transsient analysis of cage induction machines under stator rotor bar and end-ring faults," *IEEE Transaction on Energy Conversion,* vol. 10, no. 2, pp. 241-247, 1995.

[7] Ah-jaco, A, "Modélisation des moteurs asynchrones triphasés en régime Transitoire avec saturation et harmoniques d'espace. Application au diagnostic," *PhD thesis, Université de Lyon,* juillet 1997.

[8] Munoz, A R and Lipo, T A, "Complex vector model of the squirrel-cage induction machine including instantaneous rotor bar currents," *IEEE-IAP,* vol. 35, no. 6, 1999.

[9] Lipo, T A and Toliyat, H A, "Feasibility study of a converter optimized induction motor," *Palo Alto,CA,Electric Power Research Institute,EPRI Final Rep,* pp. 2624-02, 1989.

[10] Luo, X, et al., "Multiple coupled circuit modeling of induction machines," *IEEE Transaction on Industry apllications,* vol. 31, no. 2, pp. 311-318, March/April 1995.

[11] AI-Nuaim, N A and Toliyat, H A, "A novel method for modeling dynamic air-gap eccentricity in synchronous machines based on modified winding function theory," *IEEE Transaction on Energy Conversion,* vol. 13, no. 2, pp. 156-162, June 1998.

[12] Houdouin, G, et al., "Coupled Magnetic Circuit Modeling of the Stator Windings Faults of Induction Machines Including Saturation Effect," *In proceedings of the IEEE International Conference on industrial Technology(ICIT),* pp. 148-153, 2004.

[13] Chow, M Y, *Methodologies of using neural network and fuzzy logic technologies for motor incipient fault detection,* s.l. : World Scientific Publishing Co.Pte.Ltd,Singapore, 1997.

[14] Patton, R J and Chen, J, "On-line residual compensation in robust fault diagnosis of dynamic systems," *in IFAC Symp Artificial Intelligence in Real-time Control, Delft, The Netherlands,* no. 17, pp. 221-227, 1992.

[15] Bertenshaw, D R, et al., "Detection of stator core faults in large electrical machines," *IET Electric Power Applications,* vol. 6, no. 6, pp. 295-301, 2012.

VIII. BIOGRAPHIES

Pu Shi was born in Henan, China, in 1982. He received the M.Eng. degree from Glyndwr University, Wrexham, UK, in 2009 and the B.Eng. degree from Southwest University for Nationalities, Chengdu, China in 2007.

In September 2009, he started his PhD degree in Glyndwr University, UK. He currently is focusing his research on induction machine fault detection, applying artificial intelligence technology such as neural network, expert system and fuzzy logic.

On-line Inter-Turn Short-Circuit detection in Permanent Magnet Synchronous Generators

B. Aubert, J. Regnier, S.Caux, D. Alejo

Abstract -- **This paper focus on inter-turn short-circuit detection in Permanent Magnet Synchronous Generators (PMSG). Inter-turn short-circuit current are among the most critical in the PMSG. For safety considerations, a fast detection is required when a fault occurs. In this work, a fault indicator for on-line diagnosis is proposed. Based on a faulty PMSG model expressed in Park's reference frame, the number of short-circuited turns is estimated using Extended Kalman Filter (EKF). Simulation and experimental results are provided to demonstrate the quickness and the robustness of the fault indicator with regards to various operating points on an electrical network.**

Index Terms—**Extended Kalman filter, fault diagnosis, on-line parameter estimation, permanent magnet synchronous generator, inter-turn short-circuit**

I. INTRODUCTION

PERMANENT Magnet Synchronous Generators (PMSG) are increasingly used in various industrial fields such as aerospace, railway and automotive sectors or renewable energy [1]-[2]. With regard to safety considerations, inter-turn short-circuits are the root of high fault current values in stator windings and may cause dramatic damage in PSMG as long as the machine is rotating [3]. Thus, an early on-line detection is required to perform the appropriate safety request and to avoid serious damages.

Three main approaches could be distinguish to detect inter-turn short-circuits in stator windings. The first approach is based on signal analysis, such as stator current analysis in three-phase's [4], Park's [3] or Concordia's [5] frame, axial flux leakage measurements [6], wavelet transforms [7] or symmetrical components analysis [3],[8]. These methods often use spectral tools to underline the occurrence of specific frequency components related to inter-turn short-circuits. These approaches require waiting several periods of the analyzed signal to highlight a stator winding fault. Moreover, these methods are not suitable for variable-speed applications and are not robust with regard to various operating point such as power, harmonic or unbalanced load variations. The second way is related to knowledge-based methods. These methods imply an *a priori* knowledge of fault signature to allow an *a posteriori* classification during the operation of the monitored system. Artificial intelligence tools as fuzzy logic system [9], artificial neural networks [10] and pattern recognition methods [11] can be used but time response and computation considerations are often not matching with on-line detection requirements. The third approach is based on state or parameters estimation and implies the use of mathematical models of the studied system. The identification process can be done off-line [12] but such algorithms are not suitable for fast detection. The identification process can also be done on-line using recursive techniques [13]. Few papers deal with detection scheme suitable with the required low time response imposed by the high dynamic of short-circuit current. Moreover, for an advanced detection scheme, an indicator must be built from the estimated parameters to obtain an automated monitoring process. Additionally, the constraints related to on-line monitoring considerations must be taken into account and imply a detection algorithm matching the performances of on-board computers.

To match these requirements, this paper proposes an identification of the number of short-circuited turns in a faulty PMSG model using an Extended Kalman Filter (EKF). This paper is organized as follow. In section II, the faulty PMSG model and its application on the EKF filter are described. In section III, the experimental test bench, including a PMSG specifically wounded to allow inter-turn short-circuit fault reproduction, and its associated PMSG simulation model are presented. Experimental results are compared to simulation ones to validate the PMSG model. Section IV presents simulation results of the EKF algorithm in both healthy and faulty conditions. Using the simulation data, the ability of the EKF to detect inter-turn short circuit is demonstrated through different scenarii and operating conditions. The proposed automated fault indicator build from the short-circuited turns' estimation is also described. Finally, the proposed algorithm is embedded on a real-time computation target and on-line experimental results are presented.

II. PARAMETER IDENTIFICATION PROCEDURE

In this section, the state space representation of the faulty PMSG model and its application on the EKF algorithm is presented.

B. Aubert, J. Regnier and S. Caux are with the Laboratoire Plasma et Conversion d'Energie, CNRS/ENSEEIHT-INPT/UPS, Université de Toulouse, 31071 Toulouse, France (e-mail: brice.aubert@laplace.univ-tlse.fr;jeremi.regnier@laplace.univ-tlse.fr; stephane.caux@laplace.univ-tlse.fr)

B. Aubert and D. Alejo are with Aeroconseil, AKKA Technologies Group, 31703 Blagnac, France (e-mail: brice.aubert@aeroconseil.com; Dominique.alejo@aeroconseil.com)

A. The Faulty PMSG model

The faulty PMSG model used for inter-turn short-circuit detection is based on a former study [12] with less modeling assumptions (voltage drops due to short-circuit reduction is taken into account) to make the PMSG model more sensitive to windings faults. The model enables the fault localization with the use of angle $\theta_{s/c}$ (equal to 0, $2\pi/3$ or $4\pi/3$ for a short-circuit respectively on phase A, phase B or phase C) and the determination of the number of short-circuited turns $n_{s/c}$ (ratio between the number of short-circuited turns and the entire turns number on a stator winding).

Fig. 1 shows the basic model for fault diagnosis with inter-turn short-circuit on phase C ($\theta_{s/c} = 4\pi/3$).

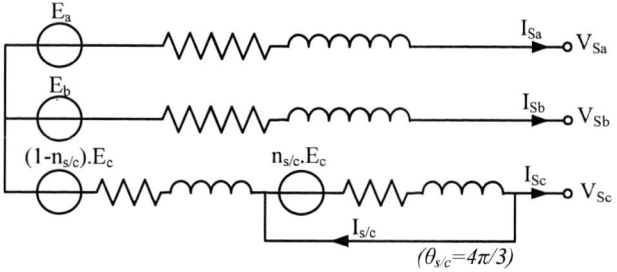

Fig. 1. Inter-turn short-circuit model used for fault diagnosis

According to this faulty model, the stator voltages and the short-circuit loop equations are expressed in (1):

$$
\begin{cases}
[V_S] = -R_S.[I_S] - n_{s/c}.R_S.T_{s/c}.I_{s/c} - [\dot\Phi_S] \\
0 = n_{s/c}.R_S.T_{s/c}{}^T.[I_S] + n_{s/c}.R_S.I_{s/c} + \dot\Phi_{s/c} \\
[\dot\Phi_S] = [L].[\dot I_S] + \sqrt{\dfrac{3}{2}}.n_{s/c}.L_{ps}.T_{32}.\begin{bmatrix}\cos\theta_{s/c}\\ \sin\theta_{s/c}\end{bmatrix}.\dot i_{s/c} \\
\qquad + n_{s/c}.L_{ls}.T_{s/c}.\dot i_{s/c} - [E] \\
\dot\Phi_{s/c} = \sqrt{\dfrac{3}{2}}.n_{s/c}.L_{ps}.\left(T_{32}.\begin{bmatrix}\cos\theta_{s/c}\\ \sin\theta_{s/c}\end{bmatrix}\right)^T.[\dot I_S] + n_{s/c}.L_{ls}.T_{s/c}^T.[\dot I_S] \\
\qquad + n_{s/c}^2.(L_{ps}+L_{ls}).\dot i_{s/c} - n_{s/c}.T_{s/c}^T.[E]
\end{cases}
\tag{1}
$$

Where:

$[V_S]$	: Stator voltages vector
$[I_S]$	: Stator current vector
$I_{s/c}$	: Short-circuit current
$[E]$	: Electromotive forces vector
R_S	: Stator resistance

$$[L]=\begin{bmatrix}L_{ls}+L_{ps} & -L_{ps}/2 & -L_{ps}/2 \\ -L_{ps}/2 & L_{ls}+L_{ps} & -L_{ps}/2 \\ -L_{ps}/2 & -L_{ps}/2 & L_{ls}+L_{ps}\end{bmatrix}$$: Inductance matrix

$$[T_{32}]=\frac{1}{\sqrt{3}}.\begin{bmatrix}\sqrt{2} & 0 \\ -1/\sqrt{2} & \sqrt{3/2} \\ -1/\sqrt{2} & -\sqrt{3/2}\end{bmatrix}$$: Concordia transformation matrix

$$T_{s/c}=\frac{1}{3}.\begin{bmatrix}1+2.\cos(\theta_{s/c}) \\ 1+2.\cos\left(\theta_{s/c}-\dfrac{2\pi}{3}\right) \\ 1+2.\cos\left(\theta_{s/c}-\dfrac{4\pi}{3}\right)\end{bmatrix}$$: Short-circuit matrix

$$\left(=\begin{bmatrix}1\\0\\0\end{bmatrix}\text{ if }\theta_{s/c}=0;\ =\begin{bmatrix}0\\1\\0\end{bmatrix}\text{ if }\theta_{s/c}=\frac{2\pi}{3};\ =\begin{bmatrix}0\\0\\1\end{bmatrix}\text{ if }\theta_{s/c}=\frac{4\pi}{3}\right)$$

After a Park transformation from (1), the faulty model can be expressed as:

$$
\begin{cases}
[V_S]_{dq} = -R_S.[I'_S]_{dq} - \omega.L_S.\begin{bmatrix}0 & -1\\1 & 0\end{bmatrix}.[I'_S]_{dq} - L_S.[\dot I'_S]_{dq} + [E]_{dq} \\
[I_S]_{dq} = [I'_S]_{dq} - [\tilde I_{s/c}]_{dq} = [I'_S]_{dq} - \dfrac{1}{[Z_{s/c}]_{dq}}.[V_S]_{dq}
\end{cases}
\tag{2}
$$

Where:

$L_S = 3/2.L_{ps}+L_{ls}$: Stator synchronous inductance

$P(\theta)=\begin{bmatrix}\cos\theta & -\sin\theta \\ \sin\theta & \cos\theta\end{bmatrix}$: Park transformation matrix

$\dfrac{1}{[Z_{s/c}]_{dq}}=\dfrac{2.n_{s/c}}{(3-2.n_{s/c}).R_S}.P(\theta)^T.Q(\theta_{s/c}).P(\theta)$: Equivalent short-circuit fault impedance.

$Q(\theta_{s/c})=\begin{bmatrix}\cos^2(\theta_{s/c}) & \cos(\theta_{s/c}).\sin(\theta_{s/c}) \\ \cos(\theta_{s/c}).\sin(\theta_{s/c}) & \sin^2(\theta_{s/c})\end{bmatrix}$
: Fault localization matrix

It can be noticed that (2) is the same equation as a classical PMSG taking account $[I_S']$ as the stator current with the addition of a faulty equation. The short-circuit current is represented by the equivalent fault impedance $[Z_{s/c}]$ at the output of the machine which deflects a part of the stator current. Expending this model to each phase, three short-circuit impedances ($Z_{s/c1}$, $Z_{s/c2}$, $Z_{s/c3}$ for $\theta_{s/c}=0$, $2\pi/3$, $4\pi/3$ respectively) are added to the PMSG model as shown on Fig. 2.

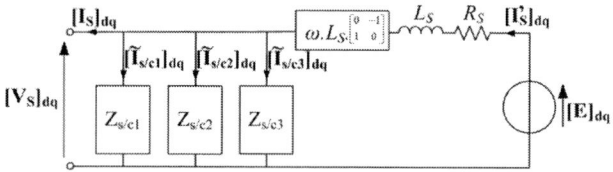

Fig. 2. Global inter-turn faulty PMSG model in Park's frame

This model can be used to make an on-line estimation of the short-circuited turns ratio $n_{s/ci}$ through the EKF algorithm to detect a stator winding fault.

B. The Extended Kalman Filter

The Kalman filter is a recursive optimum state estimator for a linear system represented by a state model disturbed by non-correlated, zero-mean white Gaussian noises:

$$
\begin{cases}
\dot X(t) = A.X(t)+B.U(t)+W(t) \\
Y(t) = C.X(t)+D.U(t)+V(t)
\end{cases}
\tag{3}
$$

Where:

$X(t)$	: state vector
$U(t)$	: input vector
$Y(t)$	: output vector
A,B,C,D	: state, input, output and feedforward matrix
$W(t)$	: state noises vector with covariance matrix $Q(t)$
$V(t)$	: measurement noises vector with covariance matrix $R(t)$

To estimate one or several parameters (λ) of the model, additional state variables must be added, resulting in the following non-linear system:

$$\begin{cases} \dot{X}_e(t) = \begin{bmatrix} \dot{X}(t) \\ \dot{\lambda}(t) \end{bmatrix} = \begin{bmatrix} A(\lambda(t)) & 0 \\ 0 & 0 \end{bmatrix} \cdot \begin{bmatrix} X(t) \\ \lambda(t) \end{bmatrix} + \begin{bmatrix} B(\lambda(t)) \\ 0 \end{bmatrix} \cdot U(t) + \begin{bmatrix} W_X(t) \\ W_\lambda(t) \end{bmatrix} \\ Y(t) = C(\lambda(t)) \cdot X(t) + D(\lambda(t)) \cdot U(t) + V(t) \end{cases} \quad (4)$$

Where X_e is the extended state vector. The additional equation $\dot{\lambda} = 0$ in (2) reflects that the dynamic response of the estimated parameter is assumed unknown.

For PMSG application, (2) can be written as a non-linear state space representation with $[I_S']_{dq}$ as the state vector and $(n_{s/c})_{1,2,3}$ as the additional state variables. The inter-turn fault is expressed through the feedforward matrix D, depending on the number of short-circuited turns $n_{s/c1}$, $n_{s/c2}$, $n_{s/c3}$. This non-linear state space model must be linearized for the EKF application. After the linearization step around the current state (with the use of the Jacobian matrices) and the sampling step using a first order approximation ($e^{A.Te} = I + A.Te$), the discrete linearized system with constant sample time Te is:

$$\begin{cases} \tilde{X}_{e_{k+1}} = \tilde{F}_k \cdot \tilde{X}_{e_k} + W_k \\ \tilde{Y}_k = \tilde{H}_k \cdot \tilde{X}_{e_k} + V_k \end{cases} \quad (5)$$

Where:

$$\begin{cases} \tilde{F}_k = \begin{bmatrix} I + Te.A(\lambda_k) & Te.\left(\dfrac{\partial A(\lambda_k)}{\partial \lambda_k} \cdot X_k + \dfrac{\partial B(\lambda_k)}{\partial \lambda_k} \cdot U_k \right) \\ 0 & I \end{bmatrix} \\ \tilde{H}_k = \begin{bmatrix} C(\lambda_k) & \dfrac{\partial C(\lambda_k)}{\partial \lambda_k} X_k + \dfrac{\partial D(\lambda_k)}{\partial \lambda_k} U_k \end{bmatrix} \end{cases} \quad (6)$$

Thus, the discrete Extended Kalman Filter algorithm is written and summarized in Fig. 3. In EKF algorithm, a special attention should be paid to covariances matrices, particularly for the state noises covariance matrix Q to let the estimated parameter dynamic response high and also take into account model variations [14]. Q can be separated into two covariance matrix Q_X(for measured states) and Q_λ(for the extended parameter). It is assumed that Q depends on noises measurements and Q_λ values are weighted according to λ theoretical value in order to allow a fast variation of the estimated parameter. These considerations enable an early inter-turn short-circuit detection. The covariance matrices used for the estimation of $(\hat{n}_{s/c})_i$ are expressed in (7):

$$\begin{aligned} P_0 &= 10^{-1} \cdot \begin{bmatrix} I_{2\times2} & 0 \\ 0 & 10^{-3}.I_{3\times3} \end{bmatrix}; \ R = 10^{-2}.I_{2\times2} \\ Q &= 10^{-5} \cdot \begin{bmatrix} I_{2\times2} & 0 \\ 0 & 3.10^{-5}.I_{3\times3} \end{bmatrix} \end{aligned} \quad (7)$$

III. EXPERIMENTAL AND SIMULATION TEST BENCHES

A. Experimental test bench

The PMSG studied in this paper to evaluate fault indicators is a three-phase 3.6 kW four-pole Permanent Magnet Synchronous Machine mechanically connected with a DC brushless machine to operate in generator mode. Stator windings of PMSG are modified with additional connection points in the stator coils on Phase A, allowing the

introduction of several inter-turn short-circuit levels (4%, 8%, 12% and 16% of short-circuit turns in the stator winding, respectively corresponding to 3, 6, 9 and 12 turns on 72 whole turns of a stator winding).

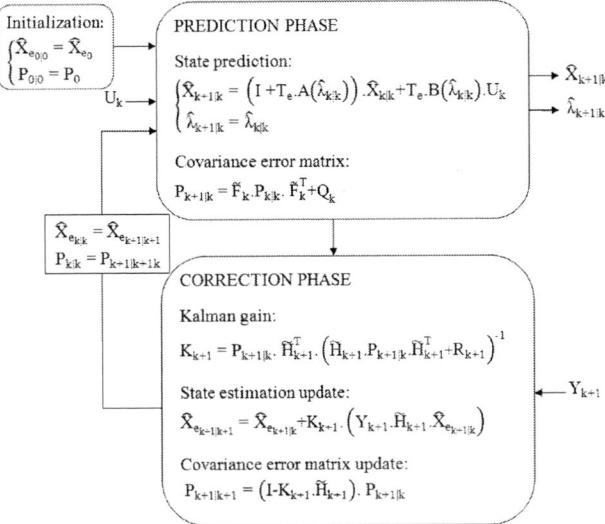

Fig. 3. Discrete Extended Kalman Filter algorithm

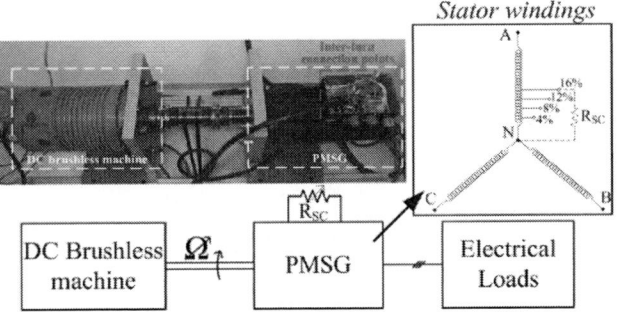

Fig. 4. Experimental test bench scheme

B. Simulation test bench

The PMSG faulty model used for simulation is based on Electrically Coupled Magnetic Circuit (ECMC) [15]. It consists in a semi-analytical computation of the PMSG inductances from the stator winding layout including additional connections for inter-turn short-circuit. Thus, several faulty PMSG simulation models have been developed on SABER software with 4%, 8%, 12% and 16% of short-circuit turns in the stator winding.

A comparison between an experimental and a simulated short-circuit current with 16% short-circuited turns is shown on Fig. 5. A resistance R_{SC} is introduced in the short-circuit loop to contain the short-circuit current to $15A_{RMS}$. The experimental waveform is very close to the simulation one, implying a good behavior of the faulty model. Moreover, Table I compares experimental and simulation data for voltage and current RMS values in phase A and current in the short-circuit loop for several faults severity. The relative error is always under 2.5%, which is accurate enough to consider that simulation tests match the experimental results with a good confident rate.

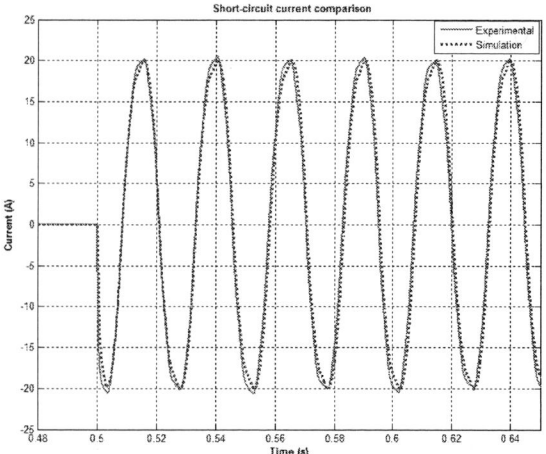

Fig. 5. Comparison between experimental and simulation 15A fault current with 16% short-circuited turns at 40Hz.

TABLE I
COMPARISON BETWEEN RMS EXPERIMENTAL AND SIMULATION DATA

Short-circuited turns	Stator voltage			Stator current			Short-circuit current		
	Exp	Sim	Err	Exp	Sim	Err	Exp	Sim	Err
4%	47.5 V	47.7 V	0.34%	5.12 A	5.24 A	2.28%	3.92 A	3.87 A	1.29%
8%	47.3 V	47.7 V	0.85%	5.10 A	5.21 A	2.15%	7.59 A	7.61 A	0.22%
12%	47.0 V	47.1 V	0.34%	5.08 A	5.17 A	1.79%	11.1 A	11.1 A	0.30%
16%	45.8 V	46.3 V	1.05%	5.01 A	5.08 A	1.56%	14.8 A	14.6 A	0.94%

IV. SIMULATION RESULTS

For all simulation results, the sampling period Te is set to 100µs.

A. Estimated parameter response

As shown in Fig. 6b where a 16% inter-turn short-circuit with low resistance ($R_{s/c} = 0.1m\Omega$) is generated at t=0.5s on phase A, estimated parameters $n_{s/c}$ are impacted. Indeed, a modification of mean value and large oscillations at two times the electrical frequency appear as soon as the winding fault occurs. It is also noticeable that $n_{s/c1}$ is not the only estimated parameter sensitive to the short-circuit: $n_{s/c2}$ and $n_{s/c3}$ are also modified in a lesser extent due to some simplifying assumption about PMSG symmetry used for the faulty model. However, the fault can clearly be located on phase A in this test. It is also noticeable that the short-circuit current (Fig. 6a) reaches high value ($80A_{RMS}$ in this case) compared to the nominal current of the PMSG ($15A_{RMS}$) which confirms that early detection of inter-turn stator short-circuit is required to avoid deterioration.

B. Inter-turn short-circuit indicator calculation

Noticing that the estimated parameters are impacted with oscillations at two times the electrical frequency when a fault occurs, the chosen indicator to detect inter-turn short-circuits expressed in (8) is based on the sum of these three short-circuited turns ratio. Each sampling period, a sliding mean value with overlap is applied over one electrical half period.

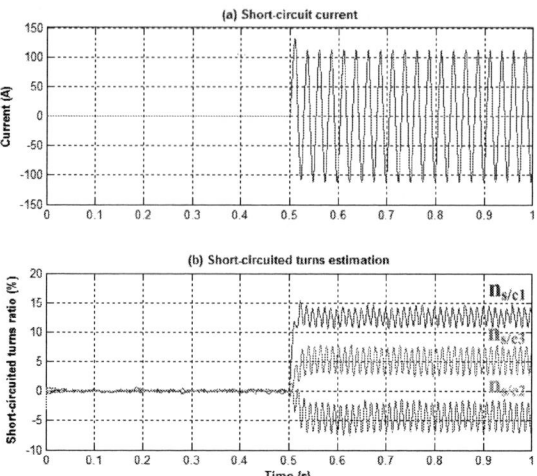

Fig. 6. Estimated parameter ($n_{s/c}$) response to a 16% inter-turn short-circuit (I_{LOAD} = 10A per phase, 40Hz)

$$\text{Fault\_indicator} = \langle \sum_{i=1}^{3} |n_{s/c}|_i| \rangle_{T/2} \qquad (8)$$

This sliding mean calculation principle, depicted on Fig. 7, will be used as a fault indicator to take benefits of its filtering characteristic. Indeed, in case of a threshold use for fault detection, reducing the indicator's oscillations will improve its robustness. Moreover, time response of the indicator remains quick (less than one electrical period) in spite of the filtering characteristic of the sliding mean calculation.

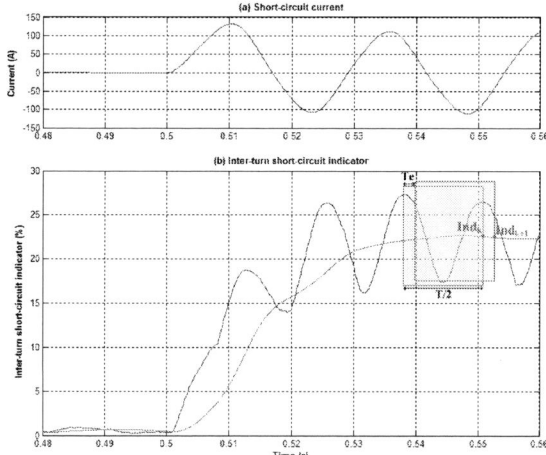

Fig. 7. Inter-turn short-circuit indicator building

Fig. 8 shows the evolution of the indicator in healthy case (for t < 0.5s) and in faulty case (for t > 0.5s) with several values of short-circuited turns. In healthy case, the indicator remains close to 0, reflecting a good estimation of the theoretical electrical parameters. In faulty cases, the indicator increases and raises with the number of short-circuited turns presented, which confirms that the fault detection is trickiest for a few number of turns short-circuited.

978-1-4799-0024-4/13 $31.00 © 2013 IEEE   332

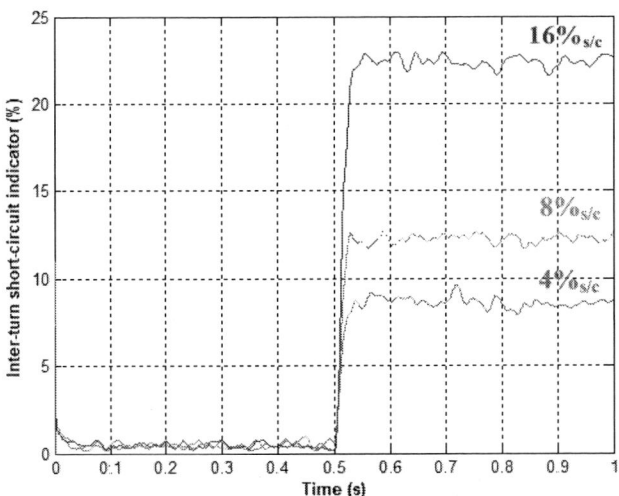

Fig. 8. Indicator response to various inter-turn short-circuits (I_{LOAD} = 10A per phase, 40Hz)

C. Robustness Analysis

In order to evaluate the robustness of the chosen fault indicator to the inter-turn short-circuit diagnosis for several operating points, the following two simulation tests have been performed:

- **Test 1 - Frequency variation** from 20Hz to 60 Hz with 10Hz step and with 10A load current per phase.
- **Test 2 - Power Variation** from 10A to 0A load current per phase with 2.5A step and with 40Hz electrical frequency.

Within each frequency or load current step, an inter-turn short-circuit is generated during 200ms to assess the fault indicator behavior.

Fig. 9 shows the evolution of the fault indicator for various operating electrical frequencies. Even if the proposed indicator is not completely constants, its robustness with regard to frequency variations is demonstrated. The indicator values for the faulty cases is always between 16% and 23% whatever the considered frequency for a 16%$_{s/c}$ short-circuit. In healthy case, the indicator remains close to 0%. Distinguish healthy and faulty operations is then always possible. Notice that in the healthy cases, the indicator is closer to zero for high electrical frequencies. Indeed, the signal to noise ratio for voltage measurements are more significant at low frequency due to lower electromotive forces which can slightly disturb the EKF.

The power variation test (Fig. 10) shows that load current values has no influence on the fault indicator which is still proportional to the number of short-circuited turns. These two tests confirm that a threshold on this indicator can be used to make a quick (about one electrical period) and robust (against frequency and load variations) fault decision. The following section will corroborate these simulation results with additional experimental tests.

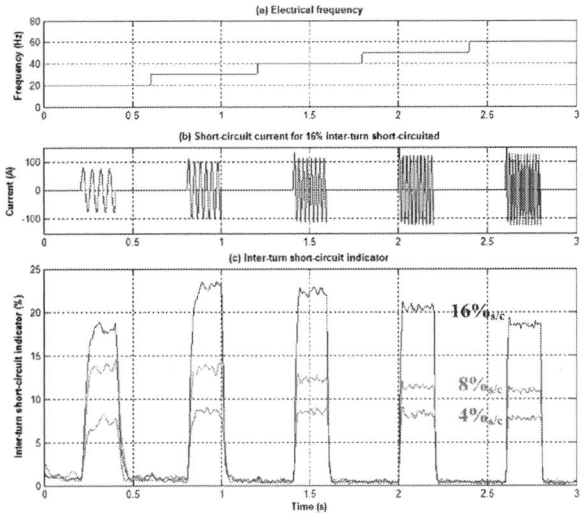

Fig. 9. Indicator response to various electrical frequencies (I_{LOAD} = 10A)

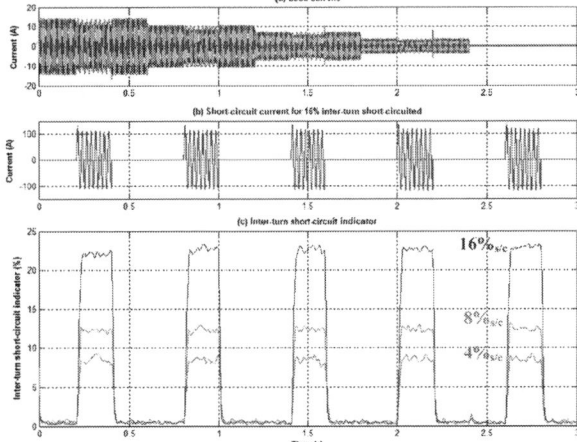

Fig. 10. Indicator response to various electrical loads (f = 40 Hz)

V. EXPERIMENTAL TESTS

A. Experimental Procedure

For safety conditions, high short-circuit current generated in simulation tests cannot be reached in the laboratory test bench. However, it is possible to validate this detection method by adding a short-circuit resistance R_{SC} to limit the short-circuit current value to 20A$_{RMS}$. As for simulation tests, the robustness of the chosen fault indicator is studied depending on frequency and power load variations for a 16% inter-turn short-circuit. The EKF algorithm is implemented on a DSK6713 DSP board in association with an Analog to Digital daughter card for data acquisition. Simulation and experimental results are then compared.

B. Experimental Results

Fig. 11 and Fig. 12 shows the evolution of the short-circuit indicator for several operating points depending on frequency and power load in healthy and faulty conditions.

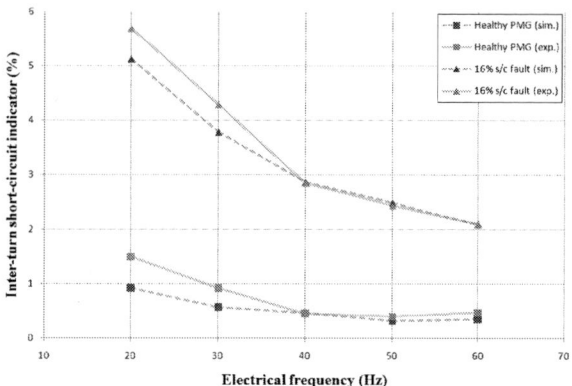

Fig. 11. Short-circuit indicator versus electrical frequency (I_{LOAD} = 5A)

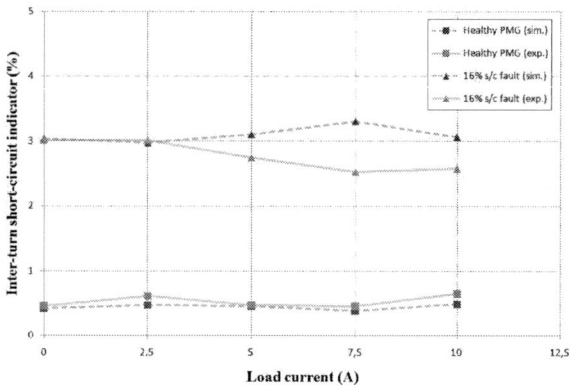

Fig. 12. Short-circuit indicator versus power load (f = 40Hz)

The experimental results are quite similar to simulation ones in both cases which confirms the good representativeness of the simulation model even for a resistive inter-turn short-circuit. Moreover, as in simulation tests (Fig. 9), Fig. 11 confirms that the fault indicator value depends on frequency variation when a short-circuit occurs. It can also be noticed that the indicator value slightly increases in healthy case for low frequency because the signal to noise ratio for voltage measurements is more significant. However, healthy and faulty case can easily be differentiated in this test. Fig. 12 confirms that power variation has no influence on the fault indicator as it was previously pointed out on simulation (Fig. 10).

These two tests shows that if a fault threshold is set for example to 2% on this indicator to avoid false alarms caused by model and measurement noises, this indicator can also detect resistive short-circuits which are not critical for PMSG windings. This early detection could prevent serious damage caused by a potential higher short-circuit current.

CONCLUSION

This paper has described a method for on-line inter-turn short-circuit detection in stator windings based on the short-circuited turn ratio estimation on a faulty PMSG model using an EKF algorithm. Simulation and experimental tests have shown that this fault indicator enables to locate the winding fault with a fast time response for several operating

points depending on frequency and load current. The prospects of this work are to carry on the robustness analysis with the addition of other operation points including inductive, unbalanced and harmonics loads.

REFERENCES

[1] J. E. Hill, S. Mountain, "Control of a Variable Speed, Fault Tolerant Permanent Magnet Generator", IEE International Conference on Power Electronics, Machines and Drives (PEMD 2002), pp. 492-497, June 2002.

[2] D. M. Saban, C. Bailey, D. Gonzalez-Lopez and L. Luca, "Experimental evaluation of a high-speed permanent-magnet machine", *Proc. 55th IEEE PCIC*, pp.1 -9 2008.

[3] S. M. A. Cruz and A. J. M. Cardoso, "Multiple reference frames theory: A new method for the diagnosis of stator faults in three-phase induction motors", *IEEE Trans. Energy Convers.*, vol. 20, no. 3, pp.611 -619 2005.

[4] M. Sahraoui, A. Ghoggal, S. E. Zouzou, A. Aboubou and H. Razik, "Modelling and detection of inter-turn short circuits in stator windings of induction motor", *Proc. Int. IECON*, pp.4981 -4986 2006.

[5] D. Diallo, M. E. H. Benbouzid, D. Hamad and X. Pierre, "Fault detection and diagnosis in an induction machine drive: A pattern recognition approach based on Concordia stator mean current vector", *IEEE Trans. Energy Convers.*, vol. 20, no. 3, pp.512 -519 2005.

[6] J. Penman, H. G. Sedding, B. A. Lloyd and W. T. Fink, "Detection and Location of Interturn Short Circuits in the Stator Windings of Operating Motors", IEEE Transactions on Energy Conversion, Vol. 9, pp. 652-658, Dec. 1994.

[7] Z. Chen, R. Qi and H. Lin, "Inter-turn short circuit fault diagnosis for PMSM based on complex gauss wavelet", ICWAPR 2007, Vol. 4, pp.1915-1920, 2007.

[8] S. Cheng, P. Zhang and T.G. Habetler, "An Impedance Identification Approach to Sensitive Detection and Location of Stator Turn-to-Turn Faults in a Closed-Loop Multiple-Motor Drive", *Industrial Electronics, IEEE Transactions on*, vol.58, no.5, pp.1545-1554, May 2011.

[9] M. A. Awadallah, M. M. Morcos, S. Gopalakrishnan and T. W. Nehl, "A neuro-fuzzy approach to automatic diagnosis and location of stator inter-turn faults in CSI-fed PM brushless DC motors", IEEE Trans. Energy Conversion, vol. 20, no. 2, pp. 253-259, June 2005.

[10] M. B. K. Bouzid, G. Champenois, N. M. Bellaaj, L. Signac and K. Jelassi, "An effective neural approach for the automatic location of stator interturn faults in induction motor", *IEEE Trans. Ind. Electron.*, vol. 55, no. 12, pp.4277 -4289 2008.

[11] A.Soualhi, G. Clerc and H. Razik, "Faults classification of induction machine using an improved ant clustering technique", *Diagnostics for Electric Machines, Power Electronics & Drives (SDEMPED), 2011 IEEE International Symposium on*, vol., no., pp.316-321, 5-8 Sept. 2011.

[12] S. Bachir, S. Tnani, J.-C. Trigeassou and G. Champenois, "Diagnosis by parameter estimation of stator and rotor faults occurring in induction machines", *IEEE Trans. Ind. Electron.*, vol. 53, no. 3, pp.963 -973 2006.

[13] M. Khov, J. Regnier, J. Faucher, "Detection of turn short-circuit faults in stator of PMSM by on-line parameter estimation", *Power Electronics, Electrical Drives, Automation and Motion, 2008. SPEEDAM 2008. International Symposium on*, vol., no., pp.161-166, 11-13 June 2008.

[14] B. de Fornel and J.-P. Louis, "Electrical Actuators: Identification and Observation", ISTE Ltd, 2010.

[15] A. Ali Abdallah, J. Regnier, J. Faucher, "Simulation of Internal Faults in Permanent magnet Synchronous Machines", 6th International Conference on Power Electronics and Drive Systems, Kuala Lumpur, Malaysia, 2005.

BIOGRAPHIES

Brice Aubert was born in 1985. He received the M.S. degree in electrical engineering from the Institut National Polytechnique de Toulouse, Toulouse, France, in 2008. After working for two years on aircraft electrical system research at Aeroconseil, he is currently working toward the Ph.D. degree at the Laboratoire Plasma et Conversion d'Energie, CNRS/ENSEEIHT-INPT/UPS, Université de Toulouse, Toulouse. His research interests include fault diagnosis for electrical systems.

Jeremi Regnier was born in 1975. He received the Ph.D. degree in electrical engineering from the Institut National Polytechnique de Toulouse (INP Toulouse), Toulouse, France, in 2003. Since 2004, he has been an Assistant Professor with the Electrical Engineering and Control Systems Department, INP Toulouse. He is also a Researcher with the Laboratoire Plasma et Conversion d'Energie, Université Paul Sabatier–Institut National Polytechnique de Toulouse, Toulouse. His research interests include modeling and simulation of faulty electrical machines and drives as well as the development of monitoring techniques using signal-processing methods.

Stéphane Caux was born in 1970. He received the Ph.D. degree in robotics from the University of Montpellier 1, Montpellier, France, in 1997. Since 1998, he has been an Assistant Professor at the Laboratoire d'Electrotechnique et d'Electronique Industrielle, ENSEEIHT, INPT, Toulouse, France, and at the Laboratoire Plasma et Conversion d'Energie, CNRS/ENSEEIHT-INPT/UPS, Université de Toulouse, Toulouse. His main interests include robust control and observers for electrical systems (electrical motors, inverters, and fuel cell systems) and real-time energy management for multisource electrical systems (optimization and fuzzy management).

Dominique Alejo received the Ph.D. degree in electrical engineering from the Institut National Polytechnique de Toulouse (INP Toulouse), Toulouse, France, in 2003. He is currently the head of Research, Power Electronics and Simulation department at Aeroconseil. His research interests include electrical systems in future aircrafts.

978-1-4799-0024-4/13 $31.00 © 2013 IEEE

Saturation Independent Detection of Dynamic Eccentricity Fault in Salient-Pole Synchronous Machines

T. Ilamparithi, Subhasis Nandi

Abstract -- This paper proposes a novel scheme based on current signature analysis to estimate the severity of dynamic eccentricity fault independent of saturation in salient-pole synchronous machines (SPSM). The detection technique uses short circuit test data and exploits the presence of inherent static eccentricity in a machine. Implementation of the scheme involves two steps. First, the percentage increases in both mixed eccentricity component and static eccentricity component are computed. Then the difference between them is used as a factor to estimate the severity of dynamic eccentricity. The method is also found capable of distinguishing between static and dynamic eccentricity faults. Results obtained from an experimental 3 phase, 2.2 kW machine have been used to validate the proposed eccentricity severity estimation scheme.

Index Terms— condition monitoring, eccentricity, frequency spectrum, finite element, saturation, synchronous machine.

I. INTRODUCTION

In an ideal machine, the axis of the stator and that of the rotor coincide resulting in a perfectly symmetrical distribution of air gap around the periphery of the rotor. When the two axes do not coincide, the perfect symmetry is lost and the condition is known as eccentricity [1], [2]. Eccentricity is predominantly classified into two types: static eccentricity (SE) and dynamic eccentricity (DE). In case of SE the point of minimum air gap length remains fixed in space. DE occurs when the rotor, whose axis is displaced from that of the stator, is rotating about the stator's geometric axis. Under such a circumstance, the point of minimum air gap length rotates along with the rotor. Usually both SE and DE tend to exist simultaneously leading to what is known as mixed eccentricity (ME) [3].

Non-invasive eccentricity fault diagnosis in SPSM has been gaining significance of late [4-6]. Current signature analysis has been mostly preferred for the above mentioned purpose. In this method, the fault specific frequency components of the machine current are first identified and then monitored for variation in magnitude to estimate the severity of the fault [7]. The line current harmonics arising

---

This work was supported by the Natural Sciences and Engineering Research Council of Canada (NSERC), the Canadian Foundation for Innovation (CFI) and the University of Victoria.

T. Ilamparithi is with OPAL – RT Technologies, Montreal, Canada and S. Nandi is with the Department of Electrical and Computer Engineering, University of Victoria, Victoria, BC, Canada

(e-mail: t.ilamparithi@opal-rt.com; snandi@ece.uvic.ca)

due to different eccentricity conditions have been identified in [8] by following the magneto-motive force – specific permeance approach and are given by (1).

$$f_{ste} = \left(n \pm 6h \pm \frac{km}{p}\right)f \qquad (1)$$

where f_{ste} is the frequency of the harmonic component present in the stator current under different eccentric conditions, f is the stator line frequency, $n = 1,5,7,...$; $h = 1,2,3...$; $k = 0,1,2,3...$; $m = 0$ for healthy (HE) and dynamic eccentricity condition, $m = 0, 2p, 4p, ...$ for static eccentricity condition and $m = 0,1,2,3...$ for mixed eccentricity condition, p = fundamental pole pair number of the machine.

From (1) it has been found that DE modulates the magnitudes of odd harmonics other than triplen (multiples of third) harmonics that are present under healthy condition. With SE, triplen harmonics are introduced in addition to other odd harmonics. ME has been found to result in all harmonics, both odd and even along, with sideband components. Table I shows the first few harmonic components under different eccentricity conditions. The space vector rather than the line current is used because the space vector can be used for double sided FFT, thereby showing the components produced due to both forward and reverse rotating MMF.

TABLE I
HARMONIC COMPONENTS IN STATOR CURRENT UNDER ECCENTRIC CONDITIONS

HE	+1f	-5f	+7f	-11f	+13f	-17f
SE	±1f	±3f	±5f	±7f	±9f	±11f
DE	+1f	-5f	+7f	-11f	+13f	-17f
ME	±0.5f	±1f	±1.5f	±2f	±2.5f	±3f

The signatures in the induced voltage of a synchronous machine operating at no-load have been used to detect SE in [9], [10]. However, these works do not report on estimating the severity of eccentricity fault. Eccentricity fault identification and classification using Artificial Neural Network (ANN) for low power permanent magnet synchronous motor (PMSM) has been reported in [11], [12]. The major drawback of the reported detection scheme is that it requires a large training data set. Moreover, the accuracy of DE severity estimation has been found affected by noise level. Thus, a technique that is simple and capable of identifying unambiguously the severity of eccentricity fault, especially for large power machines, is desired. In this paper

---

978-1-4799-0024-4/13 $31.00 © 2013 IEEE

the objective is achieved by formulating a scheme that estimates the severity of DE in SPSM independent of saturation.

II. PROPOSED TECHNIQUE FOR SEVERITY ESTIMATION

Initially, online DE severity estimation was tested by

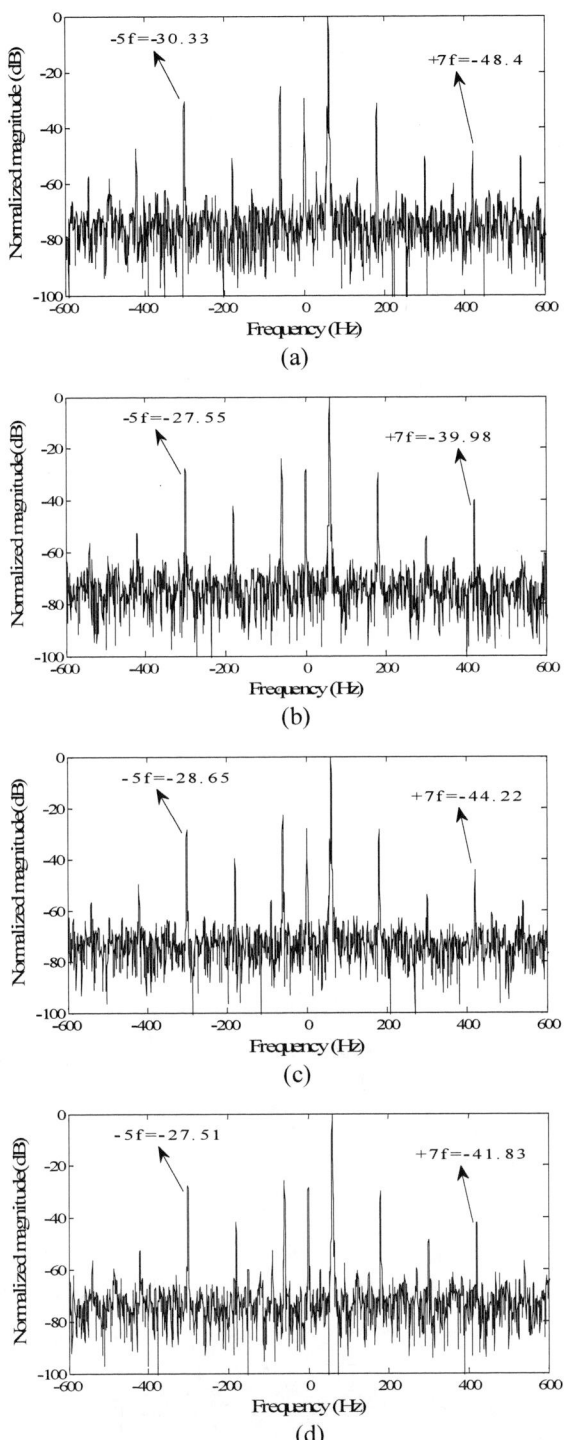

Fig.1. Stator line current space phasor harmonic spectrum under different DE conditions at full load; (a) Healthy (b) 17% DE (c) 33% DE (d) 50% DE

running the SPSM as a motor. Line currents of the motor were acquired and current phasor was calculated for each condition. Harmonic spectrums of line current phasors under HE, 17%, 33% and 50% DE conditions are shown in Fig. 1.

From Fig. 1, it is clear that the fault specific frequency components do not follow a trend with variation in the severity of DE. Therefore, it was decided that residual estimates be computed for the purpose of severity estimation. Also, from Fig. 1, it was observed that 7th harmonic component has higher sensitivity compared to 5th harmonic. Therefore, it was selected as the frequency of interest. Residual estimates of the motor's line current were computed following the method mentioned in [3], [8]. The residual estimates of the positive sequence seventh harmonic component of the line current data acquired under rated excitation was used to estimate the severity of DE. Different power factors of operation were also attempted by changing

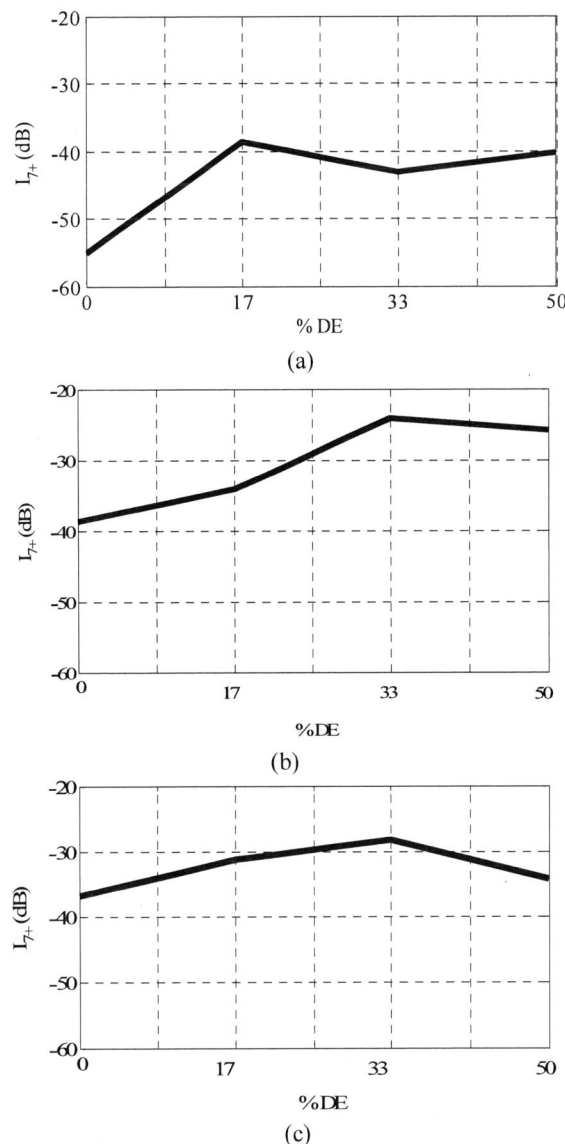

978-1-4799-0024-4/13 $31.00 © 2013 IEEE

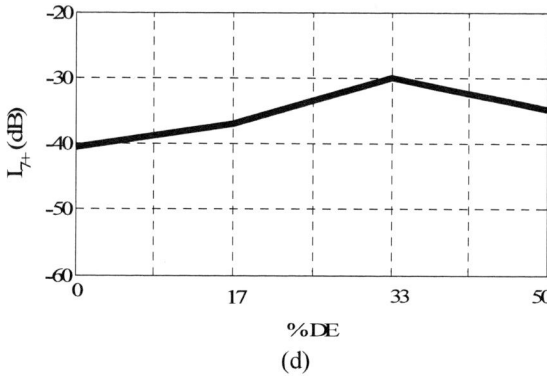

(d)

Fig.2. Variation of +420 Hz component's residual estimate of the stator line current space phasor under different DE conditions at full load; (a) at rated excitation (b) 0.9 lag, (c) unity power factor, (d) 0.9 lead

the rotor field excitation. Nevertheless, the results were not satisfactory as shown in Fig. 2. Thus, online detection of DE severity was not possible. Therefore, off-line methods were explored.

The machine was made to function as a generator. Initially, the stator was open circuited and the field circuit was supplied with rated current. Open circuit test was conducted under HE as well as increasing levels of DE. The field currents under all these conditions were acquired. The harmonic spectrum of field currents as well as their residual estimates were generated and analyzed. As in the previous case, open circuit test based current signature analysis failed to detect the severity of DE. In order to find out the reason, finite element (FE) simulations were carried out in Maxwell-2D, a commercial FE software package. A 2D model of the experimental machine was first simulated under rated excitation to obtain the flux density plots. The details of the machine model and the B-H characteristics of the core material used for stator and rotor (Fig.5) are shown in appendix. Based on the FE simulations, it was found that both the healthy machine and the DE machine were operating under considerable saturation at rated excitation. Fig. 3 shows the flux density plots of both the healthy and the 50% DE SPSM obtained under open circuit test. It is clear that there is only a marginal increase in the magnitude of flux density under 50% DE condition as compared to the healthy condition.

To minimize the effect of saturation, the FE models were then simulated under short circuit test condition. The stator windings of the generator were short circuited. The field current was adjusted to produce rated current in the stator windings. The flux density plots obtained under short circuit condition are shown in Fig. 4. From Fig.4 it is clear that both healthy and 50% DE SPSM operate in the linear magnetic region (please refer to Fig. 5 for B-H curve) under short circuit test condition. Also, the change in the magnitude of flux density with eccentricity is significant and clearly distinguishable.

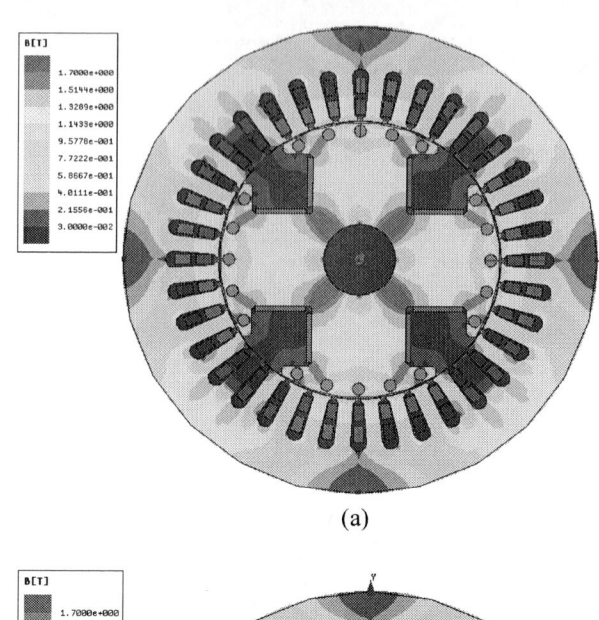

(a)

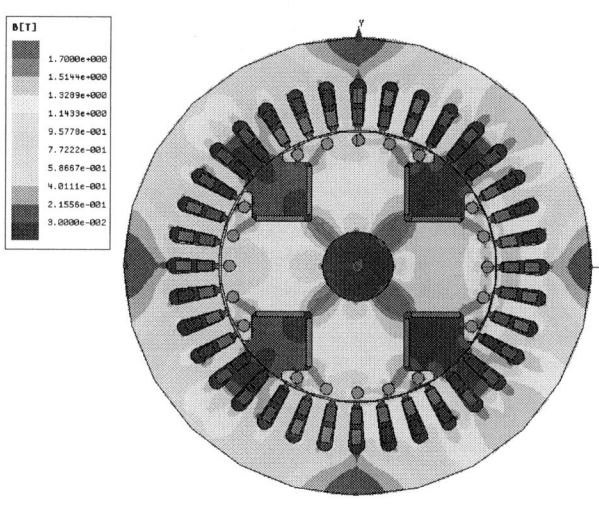

(b)

Fig.3. Flux density plots obtained under open circuit test (a) healthy SPSM (b) 50% DE SPSM

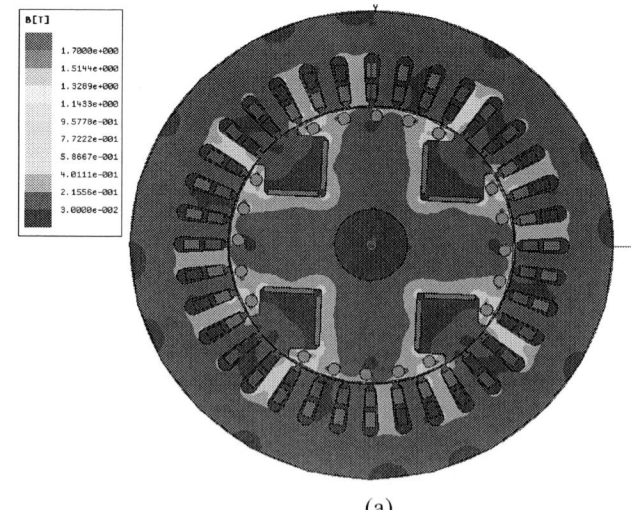

(a)

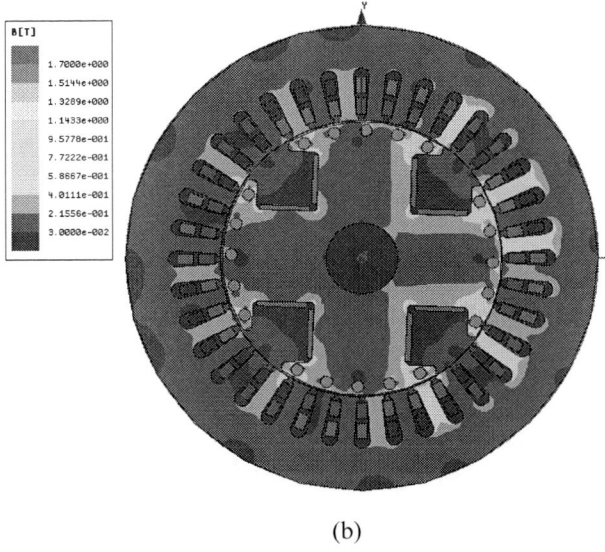

(b)

Fig.4. Flux density plots obtained under short circuit test (a) healthy SPSM (b) 50% DE SPSM

As pointed out earlier, with rated excitation under open circuit condition, the machine is already in saturation even under healthy condition. Therefore, any change in the air gap does not result in a significant increase in the magnitude of flux density hindering the detection of DE severity. On the contrary, under short circuit condition, the machine operates in the linear magnetic region and consequently a change in the air gap results in a significant change in the magnitude of flux density around the point of minimum air gap.

Thus, to ensure saturation independent DE severity detection, line currents acquired under short circuit test were used to estimate the severity of the fault. The proposed scheme computes a novel severity estimation factor by exploiting the inherent SE present in the machine. Due to inherent SE, any increase in the severity of DE will also result in an increase in the magnitude of ME component. However, the magnitudes of SE associated triplen harmonics should ideally remain unaffected. Thus, by finding the difference between the percentage increase in ME related component and the percentage increase in SE associated triplen harmonics, increasing severity of DE can be estimated. Also, the logic is independent of any inherent DE as that will not affect the SE related third harmonic components. The percentage change in a frequency component can be computed using (2).

$$\% \; increase \; in \; a \; frequency \; component = \left(1 - \frac{|frequency \; component \; at \; eccentricity \; condition|}{|frequency \; component \; at \; healthy \; condition|}\right) \times 100 \tag{2}$$

For example

$$\% \; increase \; in \; positive \; sequence \; 30 \; Hz \; for \; 50\%DE = \left(1 - \frac{|-50.84|}{|-64.44|}\right) \times 100 = 21.105 \; \%$$

It is to be noted that the harmonic components are

normalized with respect to fundamental and therefore, their values will be negative. Thus, in (2) only the absolute values are used.

III. EXPERIMENTAL SETUP

A 3 phase, 2.2 kW, 208V, 4 pole, 60 Hz, star connected salient pole synchronous machine has been used as the laboratory prototype for validating the proposed diagnostic scheme. The enclosure of the machine has been modified so as to facilitate air gap measurements using feeler gauge. The bearings and end bells have also been changed to introduce eccentricity. Eccentric bushings, offset by 16.67%, 33.33%, and 50% of the airgap, have been used to obtain DE of varying severity [12]. Fig. 4 shows the modified enclosure and a new bearing along with a DE bushing.

(a)

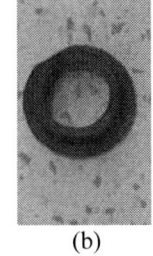

(b)

(c)

Fig. 4. (a) Experimental machine in the modified enclosure; (b) DE bushing (c) New bearing [8]

Short circuit test has been performed on the test machine under increasing levels of DE conditions. For each eccentricity setting 10 seconds of steady state data of machine's line currents have been acquired using a data acquisition system at 3600 Hz sampling frequency.

IV. RESULTS AND DISCUSSION

Double sided FFT on the stator current space phasor has been obtained and the magnitudes of the desired frequency components have been identified. The percentage increases of +30 Hz, -180 Hz and +540 Hz have been computed. The harmonics are selected based on higher energy content and higher sensitivity to eccentricity faults. The differences between the ME related +30 Hz component and the SE related triplen harmonic components (-180 Hz and +540 Hz) have then been obtained. The average of these two components has been used to estimate the severity of DE as shown in Table II.

TABLE II
DE SEVERITY ESTIMATION IN SPSM

Fault Level		17% DE	33% DE	50% DE
% increase in positive sequence 30 Hz	(A)	0.997	10.022	21.105
% increase in positive sequence 540 Hz	(B)	-3.471	-2.745	-0.135
% increase in negative sequence 180 Hz	(C)	0.723	1.533	2.480
Difference between (A) and (B)	(A)-(B) = (D)	4.468	12.767	21.240
Difference between (A) and (C)	(A)-(C) = (E)	0.274	8.489	18.625
Average of (D) and (E)		2.371	10.628	19.933

To ensure unambiguous detection of DE fault, the proposed detection scheme has also been tested using data obtained under increasing levels of SE. Unlike the results shown in Table II, with increasing levels of SE, the proposed scheme does not show any trend as can be seen from Table III. Thus, the presented DE severity estimation scheme can also be used to distinguish between the two types of eccentricity faults.

Thus the advantage of the proposed scheme is that it can not only identify the type of eccentricity but also the severity level of DE. Also the measurement technique being off-line is independent of load level. However, a low power auxiliary motor will be required to run the synchronous motor under short circuit condition.

TABLE III
SE SEVERITY ESTIMATION IN SPSM

Fault Level		17% SE	33% SE	50% SE
% increase in positive sequence 30 Hz	(A)	-6.127	7.251	11.574
% increase in positive sequence 540 Hz	(B)	-1.695	4.198	10.091
% increase in negative sequence 180 Hz	(C)	-0.155	1.085	5.373
Difference between (A) and (B)	(A) - (B) = (D)	-4.432	3.053	1.483
Difference between (A) and (C)	(A) - (C) = (E)	-5.972	6.166	6.201
Average of (D) and (E)		-5.202	4.610	3.842

V. CONCLUSION

A non-invasive technique to estimate the severity of dynamic eccentricity independent of saturation in salient-pole synchronous machines has been proposed in this paper. The proposed method has made use of current signature analysis of short circuit test data to achieve its objective. The difference between the percentage increases in mixed eccentricity and static eccentricity related components are monitored to predict the severity of dynamic eccentricity fault. Moreover, the scheme has been able to distinguish between static and dynamic eccentricity. Experiments conducted on a three hp salient pole synchronous machine have been used to validate the proposed technique.

VI. APPENDIX

SPSM details:

Rating: 3ϕ, 208 V, 60 Hz, 2.2 kW, 4 pole machine
Stator: 36 slots, double layer lap winding, 144 turns/ph
Rotor: 20 damper bars, standard 4 pole field winding, 1260 turns
Stator inner diameter = 148 mm
Rotor outer diameter = 146.8 mm
Field winding resistance = 81 Ω
Field winding inductance = 6 H
Moment of inertia = 0.05 kg-m^2
Polar arc = 1.1652 rad
Inter polar arc = 0.4056 rad

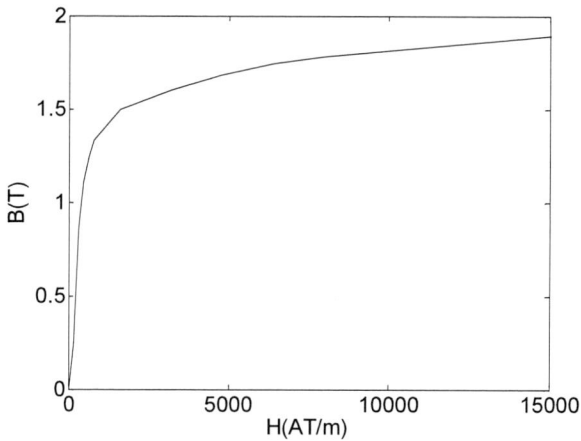

Fig. 5 B-H characteristics of stator and rotor core material used in simulation.

VII. ACKNOWLEDGMENT

The authors express their sincere thanks to Mr. Rodney Katz for his valuable suggestions and to Dr. Mike Milroy for the fabrication of the experimental set-up.

VIII. REFERENCES

[1] P. Zhang, Y. Du, T. G. Habetler, and B. Lu, "A Survey of Condition Monitoring and Protection Methods for Medium-Voltage Induction Motors," *IEEE Trans. Industry Applications*, vol. 47, no. 1, pp. 34-46, Jan./Feb. 2011.

[2] P. J. Tanver, "Review of condition monitoring of rotating electrical machines," *IET Trans. Electric Power Applications*, vol. 2, no. 4, pp. 215-247, 2008.

[3] B. M. Ebrahimi, J. Faiz and M. J. Roshtkari, "Static-, dynamic-, and mixed-eccentricity fault diagnoses in permanent-magnet synchronous motors," *IEEE Trans. Industrial Electronics*, vol. 56, no. 11, pp. 4727-4739, Nov. 2009.

[4] M. Babaei, J. Faiz, B. M. Ebrahimi, S. Amini and J. Nazarzadeh, "A detailed analytical model of a salient-pole synchronous generator under dynamic eccentricity fault," *IEEE Trans. Magnetics*, vol. 47, no. 4, pp. 764-771, Apr. 2011.

[5] J. Hong, S. B. Lee, C. Kral and A. Haumer, "Detection of airgap eccentricity for permanent magnet synchronous motors based on the d-axis inductance," in *Proc. of 2011 IEEE International Symposium on Diagnostics for Electric Machines, Power Electronics and Drives*, pp. 378-384, Sep. 2011.

[6] Bruzzese and G. Joksimovic, "Harmonic signatures of static eccentricities in the stator voltages and in the rotor current of no-load salient-pole synchronous generators," *IEEE Trans. Industrial Electronics*, vol. 58, no. 5, pp. 1606-1624, May 2011.

[7] M. Odavic, M. Sumner, P. Wheeler and Jing Li, "Real-time fault diagnostics for a permanent magnet synchronous motor drive for aerospace applications," in *Proc. of IEEE Energy Conversion Congress and Exposition*, 2010, pp. 3044-3049.

[8] T. Ilamparithi and S. Nandi, "Current residue based load independent eccentricity detection in salient pole synchronous machines," *in Proc. of IEEE International Conference on Electric Machines*, 2012, pp. 1625-1630.

[9] G. Joksimovic, C. Bruzzese, and E. Santini, "Static eccentricity detection in synchronous generators by field current and stator voltage signature analysis – Part I: Theory," in *Proc. of International Conf. on Electrical Machines*, 2010, pp. 1-6.

[10] G. Joksimovic, C. Bruzzese, and E. Santini, "Static eccentricity detection in synchronous generators by field current and stator voltage signature analysis – Part II: Measurements," in *Proc. of International Conf. on Electrical Machines*, 2010, pp. 1-5.

[11] B. M. Ebrahimi, J. Faiz, and B. N. Araabi, "Pattern identification for eccentricity fault diagnosis in permanent magnet synchronous motors using stator current monitoring," *Trans. IET Electric Power Applications*, vol. 4, no. 6, pp. 418-430, 2010.

[12] T. Ilamparithi, Non-Invasive Detection of Air Gap Eccentricity in Synchronous Machines Using Current Signature Analysis, Ph.D. dissertation, University of Victoria, Victoria. 2012.

IX. BIOGRAPHIES

T. Ilamparithi was born in Neyveli, India in 1983. He received the B.E. degree from Anna University, Chennai, India in 2005 and the M.Tech. degree from Indian Institute of Technology, Delhi, New Delhi, India in 2007. He is currently working as Simulation Engineer in OPAL-RT Technologies, Montreal, Canada. His main research interest is fault diagnosis of electrical machines.

Subhasis Nandi received the B.E. degree from Jadavpur University, Calcutta, India, in 1985, the M.E. degree from Indian Institute of Science, Bangalore, India, in 1988 and the Ph.D. degree from Texas A&M University, College Station, USA in 2000 all in electrical engineering. He joined the Department of Electrical and Computer Engineering, University of Victoria, Victoria, Canada where he is currently employed as an Associate Professor.
Between 1988 and 1996 he was with TVS Electronics and Central Power Research Institute, Bangalore, India, working in the areas of power electronics and drives. Between September 2008 and August 2009, he was a contract professor with the Department of Electrical Engineering, Korea University, Seoul, South Korea. His main research interests are power electronics and drives and analysis and design of electrical machines, with special emphasis on fault diagnosis.

978-1-4799-0024-4/13 $31.00 © 2013 IEEE

Magnetically Coupled Circuit Based Magnetic Vibrations Modeling of PMSM

Guillaume Verez, Ouadie Bennouna, Yacine Amara, Ghaleb Hoblos, Georges Barakat

Abstract -- **This paper presents an analytical model for the study of magnetic vibrations in a permanent magnet synchronous motor. The analytical model is based, for the magnetic aspect, on the magnetically coupled circuits approach widely used for modeling electrical machines under faults. This approach associated with the winding functions technique yields to a system of differential equations governing the machine. Its solution helps to compute the Maxwell stress on the stator core. An analytical mechanical model describing the structure displacement and the relied vibrations has also been developed. It is based on the resolution of the equation of motion with Flügge-Byrne-Lur'ye theory. A good agreement has been obtained between results issued from the developed analytical model with those issued from finite element analyses. Analytical models have the advantage to be more convenient for the diagnosis purpose because of the consequent reduction of the calculation time as compared to numerical methods.**

Index Terms--**Analytical models, Force, Finite element methods, Harmonic analysis, Permanent magnet machines, Vibrations**

I. NOMENCLATURE

θ Angular position of the rotor relatively to a fixed stator reference

θ_S Angular position in the stator frame

z Axial position

p Number of poles

μ_0 Magnetic permeability constant

II. INTRODUCTION

PERMANENT magnet synchronous machines are good candidates for industrial applications requiring low torque ripple and high torque density like automotive traction. However, vibrations and noise are generally critical aspects for on-board use. It is thus necessary to study them in order to be able to reduce them. There are basically two methods to reduce vibrations and noise. Improving the stator's stiffness and reducing the electromagnetic exciting force. Vibrations of electromagnetic origin are produced by the currents inside the machine. Currents produce magnetic forces between the stator and the rotor that make the stator core vibrate. These vibrations are then transmitted to

surrounding air through the crankcase. The three main types of forces that cause vibrations of electromagnetic origin are Maxwell stress, the Laplace force and magnetostriction. In this paper, only the radial component of Maxwell stress is studied while other forces are being neglected. The benefits of magnetically coupled circuits (MCC) applied to vibrations will be exposed here. Vibrations are characterized by static displacements. The latter are obtained with the Maxwell stress tensor and natural frequencies of the stator that are given by the resolution of the equation of motion using Flügge-Byrne-Lur'ye theory. Finite element analysis (FEA) is widely used nowadays to investigate vibrations but analytical models are needed for fast computation during early design stages or optimization methods where a large number of machines with modified parameters are simulated. Thus, even if FEA is precise, it is time consuming and analytical approach can lead to a better performance over computation time ratio.

The layout of the article is as follows. A review of the main analytical methods used for the determination of vibrations of electromagnetic origin is presented in section III. Section IV describes both electromagnetic and mechanical analytical models which will be tested using an 8-poles 3-phases permanent magnet synchronous motor operating at constant speed. The following section investigates FEA on the same machine and a comparison of both results is given in section VI. Relevance of the work done for fault diagnosis purposes is explained in VII. Finally, concluding remarks are presented.

III. REVIEW OF MAIN ANALYTICAL METHODS

Modern methods employed to find out Maxwell stress and resulting vibrations are often conducted by FEA [1]-[3]. However, they can be estimated with various analytical approaches. As a result, a review of main methods to determine the air gap flux density in order to solve the Maxwell stress tensor, as well as analytical mechanical methods to obtain vibrations from this source, is conducted.

The air gap flux density is usually calculated using the product of air gap permeance by the magnetomotive force and these operands are often conducted by Fourier series. First works on Fourier series applied to permeance by Jordan [4] and Alger [5] were successively improved by considering contributions of the stator and the rotor [6]-[10]. Effects of simplified slots can be taken into account by introducing permeance functions using another transformation which is conformal mapping [11][12], like Schwarz-Christoffel mapping, which allows a more precise representation of spike phenomena occurring at poles boundaries [13]. The

---

G. Verez, Y. Amara and G. Barakat are with the Department of Electronic and Electrical Engineering, GREAH, University of Le Havre, 76057 Le Havre, France (e-mails: guillaume.verez@etu.univ-lehavre.fr, yacine.amara@univ-lehavre.fr, georges.barakat@univ-lehavre.fr).

O. Bennouna and G. Hoblos are with the Department of Energy and Electrical Engineering, IRSEEM, ESIGELEC, 76801 Saint-Etienne du Rouvray, France (e-mails: ouadie.bennouna@esigelec.fr, ghaleb.hoblos@esigelec.fr)

permeance function needs to be two-dimensional (2-D) since one-dimensional is not suitable for permanent magnet machines where the effective air gap length is larger than the induction machine one [14]. Conformal mapping can calculate one [11] or every point [15] of the slot opening but it neglects magnets deformations and the path to predict flux density [14]. However, with this type of 2-D permeance, the magnetic field generated by permanent magnets can be derived on the basis of a magnetic vector potential [16]. Thus, it is possible to obtain tangential and radial fields. Permeance calculations can also be obtained with a permeance network modeling the magnetic circuit of the machine [17]. Regarding the flux density due to permanent magnets, Fourier series can be used [8][18]. In this approach, the flux path permeance function taking into account stator slotting can be determined using a magnetic reluctance network [9]. It is also possible to use a statistical model based on data obtained with FEA in order to deduce a precise expression of permeance [19]. The magnetomotive force generated by the rotor and the stator can be obtained with Fourier series [20] or a model based on the cotangent [21].

Estimations of the air gap flux density can also be obtained without the above product. Even if it is possible to use separation of variables [22], a typical analytical solution consists in using 2-D field theory in Cartesian or polar coordinate system. The method presented in [23] suggests the study of magnets floating between two smooth cylinders representing a rotor and a stator. However, the magnetic field produced by a magnet is equivalent to one created by a current-carrying coil and placed at the surface of a magnet. The sum of contributing currents is thus made. This method does not allow to efficiently take into account slotting effects on flux density. Moreover, permanent magnets relative permeability is approximated to unity which can lead to errors [24]. The method proposed in [24] is based on a 2-D model in polar coordinate system and implies the resolution of Laplace and Poisson equations in both air and magnets regions while assuming that magnets relative permeability is constant but not necessarily unity. As a result, both components of the flux density can also be obtained thanks to boundary value problems in polar coordinates [25] where the relative permeance function [11] is used. Also, an exact analytical subdomain model for the open-circuit extends the scalar potential distribution along the slot opening of every stator slot into Fourier series processed over the pole pitch period [26]. As the aim of this flux density characterization is the magnetic force, it can be directly calculated with the equivalent magnetizing current method [27] when affecting the surface of the structure.

Magnetically coupled electric circuits theory identifies the main flux paths and a magnetic circuit relates stator currents to flux in all parts of the machine. This circuit has been extendedly used to describe winded or bar rotors of induction motors [28] and especially faults inside them [29]. MCC theory is able to adequately capture space harmonics and their interaction with time harmonics [30] which is of a great importance in the determination of vibrations.

For modal analysis applied to the stator, a system can be solved using a transfer function between displacement and force. Stiffness, mass and damping depend on the stator geometry [27][31]. An analytical model based on stress and strain equations combined with a static equilibrium condition can be used [2]. It then leads to the resolution of a differential equation on the estimated radial displacement. However, instead of considering the stator as a whole, it can be represented as a double ring consisting of an outer frame and an inner stator core [8]. As a result, an equivalent formula can be used to deduce natural frequencies.

IV. ANALYTICAL MODELS

A. Electromagnetic model

In order to obtain the air gap flux density normal component which is required to calculate Maxwell stress, the MCC is used. The air gap flux density is obtained thanks to the product of magnetomotive force by the air gap permeance. First, general governing equations are expressed. Then, the calculation of inductances through distribution and winding functions is provided. Next, the expression of permeance is obtained. Finally, the air gap flux density and Maxwell stress are computed.

Many hypotheses are emitted and particularly it is supposed that iron laminations have an infinite relative permeability and that the magnetic circuit is non-conductive. In order to present the model, a permanent magnet synchronous motor with a sole winding per phase on a magnetic period is considered, along with a 3-phase wye configuration. A double-layer full-pitch distributed winding is taken into account. Electrical equations of stator phases can be written by the mean of the matrix equation (1).

$$[V_S] = [R_S][I_S] + \frac{d[\Phi_S]}{dt} \qquad (1)$$

$[V_S]$ and $[I_S]$ are respectively stator phases voltage and current vectors. $[R_S]$ is the stator phases resistance diagonal matrix and $[\Phi_S]$ is the vector of the flux crossing stator phases. Moreover, the relationship between $[\Phi_S]$ and the vector of the magnet flux crossing stator phases, $[\Phi_{SR}]$, is given in (2).

$$[\Phi_S] = [L_{SS}][I_S] + [\Phi_{SR}] \qquad (2)$$

$[L_{SS}]$ is the stator self and mutual inductance matrix. As a result, the system of matrix differential equations governing the machine operation is given by (3).

$$\begin{cases} \frac{d[I_S]}{dt} = [L_{SS}]^{-1}\left\{-\left([R_S] + \Omega\frac{d[L_{SS}]}{d\theta_m}\right)[I_S] - \Omega\frac{d[\Phi_{SR}]}{d\theta_m} + [V_S]\right\} \\ \frac{d\Omega}{dt} = \frac{1}{2J}[I_S]^t\left\{\frac{d[L_{SS}]}{d\theta_m}[I_S] + \frac{d[\Phi_{SR}]}{d\theta_m}\right\} - \frac{v_r}{J}\Omega - \frac{1}{J}\Gamma_L \\ \frac{d\theta_m}{dt} = \Omega \end{cases} \qquad (3)$$

Ω is the rotor's mechanical speed, Γ_L is the load torque, J is the moment of inertia of every rotating mass and v_r is the viscosity resistance. Stator mutual inductance matrix components between windings i and j can be written as (4) suggests.

978-1-4799-0024-4/13 $31.00 © 2013 IEEE

$$L_{ij}(\theta) = \int_0^{Lax} \int_0^{2\pi} \left\{ \begin{array}{c} F_{Di}(\theta, \theta_S, z) F_{Wj}(\theta, \theta_S, z) \Lambda(\theta, \theta_S, z) \\ \times r_{av}(\theta, \theta_S, z) d\theta_S dz \end{array} \right\} \quad (4)$$

This expression originates from the air gap main flux path. L_{ax} is the effective air gap axial length and r_{av} is the average air gap radius. The distribution function of a coil, F_D, is described in [32]. For more than one coil in series per phase, contribution of each coil adds to one another. The winding function of a coil, F_W, is also described in [32].

The permeance function, Λ, has been derived from a model in the linear case and considerers a doubly slotted air gap. A statistical study was made with FEA for typical slot shapes and gives a good representation of the air gap flux density drop in front of a slot. As local saturation in the machine is hard to take into account with MCC, the fundamental component of saturation is used to represent non-linearity by including it in stator magnetizing and mutual inductances calculations. Saturated permeance is expressed in [33] where a saturation factor describes relative deviation of magnetizing inductance value due to saturation. Linear permeance is obtained with design of experiments applied on FEA solutions. Permeance is then constructed on each air gap arc segment as shown in [19].

The magnets flux crossing stator winding i is expressed in (5) where F_{PM} is the magnets magnetomotive force given in [21].

$$\Phi_i(\theta) = \int_0^{Lax} \int_0^{2\pi} \left\{ \begin{array}{c} F_{Di}(\theta, \theta_S, z) F_{PM}(\theta, \theta_S, z) \Lambda(\theta, \theta_S, z) \\ \times r_{av}(\theta, \theta_S, z) d\theta_S dz \end{array} \right\} \quad (5)$$

Now that the currents are derived thanks to the above equations, the radial component of the air gap flux density is given by the product of air gap permeance, Λ, by the total magnetomotive force, F_{tot}, as described by (6).

$$B^{rad} = \Lambda \times F_{tot} = \Lambda \times (F_{stator} + F_{rotor}) \quad (6)$$

The stator magnetomotive force is obtained thanks to the calculated currents. Using the Maxwell stress tensor, the magnetic pressure in the air gap is given by (7).

$$P_{Maxwell}^{rad} = \frac{1}{2\mu_0}\left(B^{rad^2} - B^{tan^2}\right) \quad (7)$$

The tangential component of the flux density, B^{tan}, is often neglected in the computation of radial forces since its normal component is generally larger than its tangential one [27]. Indeed, as the ferromagnetic core's magnetic permeability is higher than the air gap one, the magnetic flux lines are almost perpendicular to the stator and rotor cores [34]. On top of that, harmonic orders of both flux density components are equal and neglecting the tangential component only impacts the harmonic amplitude [35]. With this assumption, the tangential Maxwell force density is also neglected and thus the normal component of the magnetic pressure is approximated by (8) since $\mu_0 \approx \mu_{air} \ll \mu_{iron}$ [36].

$$P_{Maxwell}^{rad} \approx \frac{B^{rad^2}}{2\mu_0} \quad (8)$$

The effect of switching could add current harmonics and thus force harmonics but it is not taken into account [37].

B. Mechanical model

The resulting vibration of Maxwell stresses is obtained with an analytical mechanical model. First, natural frequencies are derived. Then, amplitude of deformations is deduced. As stator vibrations are of importance, the machine can be decomposed into three parts which are the stator core, teeth and windings. End windings and frame are not taken into account. Stator core and teeth with windings are modeled separately and are viewed as two coaxial cylinders of infinite length. Assuming infinite length allows to only consider circumferential modes. Moreover, if two cylinders are coaxial, the resultant moment of inertia is equal to the sum of each of their moment. Thus (9) can be used to obtain an equivalent expression of the natural frequency of the mth circumferential vibrational mode of the stator-teeth-winding assembly [34][3].

$$f_m = \frac{1}{2\pi}\sqrt{\frac{K_m^c + K_m^w}{M_c + M_w}} \quad (9)$$

K_m^c and M_c are respectively the lumped stiffness and the lumped mass of the stator core. K_m^w and M_w are ones of the teeth-winding system. The stator core is modeled as a thin circular cylindrical shell of infinite length. Its natural frequencies are given by (10) [34].

$$f_m^c = \frac{\Omega_{c,m}}{2\pi R_c}\sqrt{\frac{E_c}{\rho_c(1 - v_c^2)}} \quad (10)$$

R_c is the mean radius of the stator core, E_c is the elasticity modulus of laminations, ρ_c is the mass density of the stator core and v_c its Poisson ratio. $\Omega_{c,m}$ is a frequency parameter that, when squared, is a root of the uncoupled second order determinant arising from the equation of motion. Many eight order shell systems of equations and thus theories exist and here the Flügge-Byrne-Lur'ye theory is applied since it gives good results for circumferential modes among other theories [38]. As a result, for $m > 0$, $\Omega_{c,m}^2$ is expressed by the mean of (11) [39].

$$\Omega_{c,m}^2 = \frac{1}{2}\left\{(1 + m^2 + km^4) \pm [(1 + m^2)^2 - 2km^6]^{\frac{1}{2}}\right\} \quad (11)$$

k is a dimensionless thickness parameter defined by (12) [39] in which h_c is the thickness of the stator core.

$$k = \frac{h_c^2}{12 R_c^2} \quad (12)$$

As (9) will be used to derived the equivalent frequency of the stator system, a lumped stator core stiffness has to be obtained thanks to (13) [34] with consideration of the axial stator core length L_c.

$$K_m^c = \frac{2\Omega_{c,m}^2}{R_c}\frac{\pi L_c h_c E_c}{(1 - v_c^2)} \quad (13)$$

A lumped teeth-with-windings stiffness is given in (14) [34] where $\Omega_{w,m}^2$ is also obtained by (11) but in which k is defined in (15) [39].

978-1-4799-0024-4/13 $31.00 © 2013 IEEE

$$K_m^w = \frac{2\Omega_{w,m}^2}{R_w} \frac{V_w E_w}{(1-v_w^2)} \quad (14)$$

$$k = \frac{h_w^2}{12R_w^2} \quad (15)$$

R_w is the mean radius of the stator winding, V_w is the volume of teeth with windings, h_w teeth height. v_w and E_w are equivalent Poisson ratio and elasticity modulus for winding and insulation.

Thanks to the equivalent natural frequency derived in (9) and amplitude $\widehat{P_{M,m}}$ of the radial component of Maxwell stresses derived in (8), static displacements of the stator assembly can be obtained. Each force wave of mode number m gives rise to a set of vibrations of order j and same frequency f_f. Static deformation occurs when the deforming force does not depend on time and is thus obtained for $m = j$ and $f_f = 0$. Assuming the machine to behave like a thick ring, the average displacement is obtained using (16) for $m = 0$, and (17) for $m \geq 2$ [6]. The first mode is particular since it tends to displace the rotor off his center [20] and it is thus not investigated.

$$Y_0^{stat} = \widehat{P_{M,0}} \frac{R_{av} R_{in}}{E_c h_c} \quad (16)$$

$$Y_{m \geq 2}^{stat} = \widehat{P_{M,m \geq 2}} \frac{12 R_{av}^3 R_{in}}{E_c h_c^3 (m^2 - 1)^2} \quad (17)$$

In these expressions, R_{av} and R_{in} are the average and inner radii.

V. FEA MODELS

Two finite element analysis models have been developed in order to validate the proposed electromagnetic and mechanical analytical models. As radial magnetic pressure and circumferential modes have been computed, 2-D FEA has been used. Even though the importance of structural 3-D FEA has been explained in [3], analytical models are 2-D and a strict validation with FEA models can only be made on the same basis. A 3-phase permanent magnet synchronous motor running at constant 1200 rpm has been used. Its parameters are described in Table I. This type of machine is employed for electrical vehicle traction.

A. Electromagnetic model

The electromagnetic model has been modeled with Flux 2D by CEDRAT. Fig. 1 shows a cross-sectional view of half of the stator-rotor assembly that has been modeled. Surface permanent magnets are used and teeth have constant width. An equivalent electric circuit has been used to supply stator conductors. The voltage source is a sine wave of 200 V_{rms}. Coil conductors link the electric circuit with the finite element model. As this model is 2-D, end windings are not shown but they have to be taken into account. Flux in the end windings is considered linear and as a result, a leakage inductance can be modeled. The end windings resistance is 0.1 $m\Omega$. There are 13 turns per conductors, and those conductors are connected in series. There are four windings in parallel per phase and two coils in series per winding.

TABLE I
MACHINE PARAMETERS

Parameters	Value
Number of stator slots	48
Pole pair number	4
Rotor radius	$9.2 \times 10^{-2}\ m$
Magnet thickness	$5.0 \times 10^{-3}\ m$
Air gap length	$6.0 \times 10^{-4}\ m$
Stack length	$7.5 \times 10^{-2}\ m$
Magnet pole arc	140°
Slot tooth depth	$1.0 \times 10^{-3}\ m$
Total stator slot depth	$3.1 \times 10^{-2}\ m$
Slot opening	1.237°
Stator phase resistance	88 $m\Omega$
Stator phase leakage inductance	0.159 mH

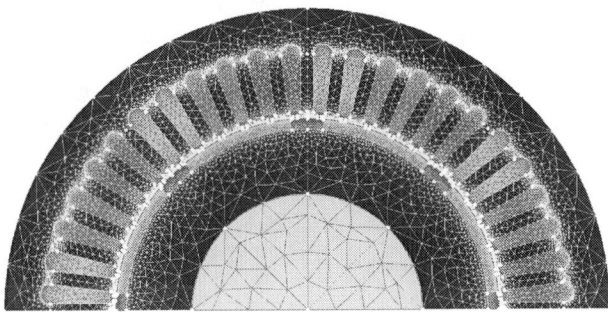

Fig. 1. Cross-sectional view of half of the meshed stator-rotor assembly with Flux 2D.

Over one eighth of the entire machine, there are 21522 nodes which represents 10705 surface elements.

B. Mechanical model

The FEA mechanical model for vibrations has been processed with ANSYS. As the analytical model does not take into account axial modes, the FEA geometry was modeled in 2-D. The exact same stator construction used in Flux 2D has been considered. Teeth and core are laminated iron with an equivalent modulus of elasticity taken as $E_c = 200 \times 10^9\ Pa$. Copper, insulation and encapsulation give an equivalent modulus of elasticity for slot regions of $E_w = 9,4 \times 10^9\ Pa$ [34]. Fig. 2 shows the displaced structure due to firsts circumferential modes. One eighth of the stator comprises 18940 nodes and 6087 surface elements.

VI. COMPARISONS

The electromagnetic model is simulated with Matlab and compared to results from FEA achieved with Flux 2D. Fig. 3 shows the air gap flux density issued from the magnetically coupled circuit model and FEA. Based on the above comparison, one can conclude that the MCC model is able to locally compute Maxwell stress. Indeed, results for the air gap flux density are satisfactory and the Maxwell stress is proportional to the square of the air gap flux density.

Mechanical natural frequencies of the stator obtained with FEA computation on ANSYS and the analytical model programmed in Matlab are shown in Table II. The analytical theory of thin infinite cylindrical shells is applied on a thick

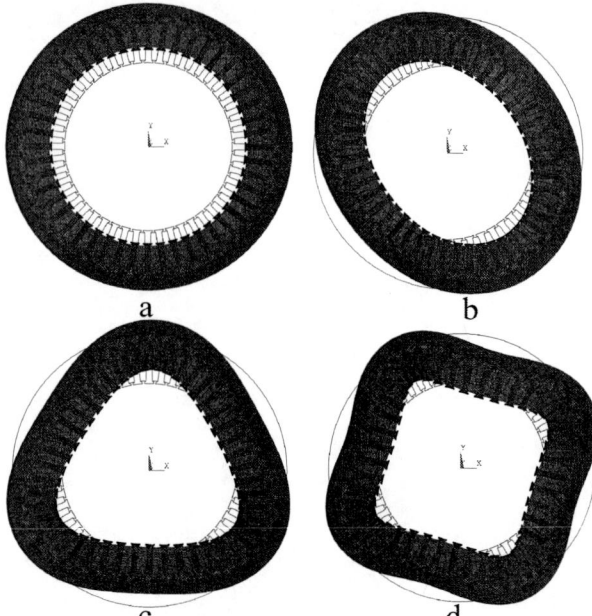

Fig. 2. Displaced structure due to circumferential modes:
(a) 0, (b) 2, (c) 3 and (d) 4.

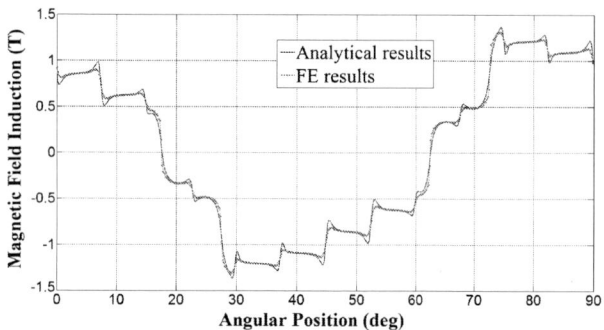

Fig. 3. Radial component of the air gap flux density issued from analytical model and FEA.

ring and explains part of the difference. Shape of teeth and slots also contribute to this difference since they cannot be taken into account analytically.

In order to efficiently compare the analytical and FEA models, a single harmonic of the force wave at a particular time is taken into account. It corresponds to a sine wave given by (18). This force is applied on nodes of the FEA mesh for a given amplitude of F_0.

$$F(\theta_S) = F_0 \sin(m\theta_S) \qquad (18)$$

Mean displacements of the structure obtained with ANSYS and the analytical model for various values of m is given in Table III. In this example, $F_0 = 200\ N$. The difference can be explained for the same reasons as for natural frequencies but the material used analytically can only be unique and thus, considering steel for the entire stator structure leads to inaccuracy. This fact is illustrated in Fig. 4 for circumferential mode 2 and different materials for

TABLE II
NATURAL FREQUENCIES OF THE STATOR SYSTEM

Circumferential mode	Natural frequency found analytically (Hz)	Natural frequency found with FEA (Hz)	Difference (%)
0	4025.3	3957.2	1.7
2	781.9	812.7	−9.5
3	1811.4	2064.7	−15
4	3266.5	3577.3	−10.8
5	5156.9	5267.4	−4.4

TABLE III
MEAN DISPLACEMENTS OF THE STRUCTURE

Circumferential mode	Mean displacements found analytically (m)	Mean displacements found with FEA (m)	Difference (%)
2	7.687×10^{-8}	6.484×10^{-8}	15.65
3	1.081×10^{-8}	1.213×10^{-8}	−12.19
4	3.075×10^{-9}	3.496×10^{-9}	−13.69

slots on one side and teeth and core on the other side. 'Ana' refers to mean displacements being computed analytically. In this case 'Both' is assuming another analytical formula given in [34] where copper and steel can both be taken into account. As it can be seen, good agreements are obtained between analytical theory considering steel material as suggested in [6] and FEA considering steel for core and teeth and copper for slots.

The mechanical model gives satisfactory results even if slots have a detailed geometry. The time required to obtain such results is less than a second. Thus, in front of such a fast computation, the performance over computation time ratio is substantial.

VII. FAULT DIAGNOSIS

The important reduction of calculation time as compared to numerical methods serves diagnosis purpose. A vibrational spectrum can thus be given as an input for diagnosis models. Also, noise spectra can also be obtained. As a result, the link between particular displacement or noise components and their origin can be investigated rapidly. For example, main types of harmonics are those due to winding distribution, permeance, input currents, and so on.

A diagnosis approach combining wavelet transform and artificial neural networks was described in [40]. In order to test such diagnosis models, a fast analytical computation can provide many healthy and faulty signals by changing some parameters. Thus, the focus can be made on data classification and parametrization of the neural network back propagation algorithm.

VIII. CONCLUSION

This paper deals with the development of an analytical model for the study of magnetic vibrations in a permanent magnet synchronous motor. For the magnetic sub-model, the analytical model is derived thanks to the magnetically coupled circuits (MCC) approach with the use of the

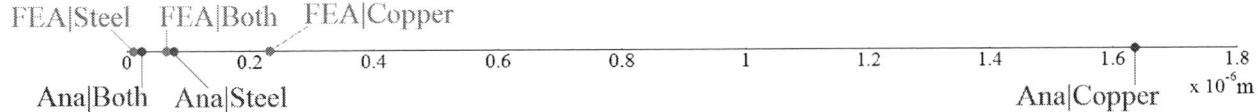

Fig. 4. Mean displacements obtained analytically and with FEA for different material configurations.

winding functions technique widely used for modeling electrical machines under faults. The main steps leading to the differential equations system governing the machine are exposed. Maxwell stress on the stator core is then computed thanks to the air gap flux density issued from the MCC model. The development of the mechanical sub-model describing the structure displacement and the relied vibrations is also explained. The effectiveness of the suggested analytical model is finally studied in the case of a 48 slots, 4 poles PM synchronous motor for which a good agreement is obtained between results issued from the developed analytical model with those issued from finite element analyses. The acquisition of the full vibrational spectrum for a large frequency band can rapidly be achieved. This could serve fault recognition evaluation using diagnosis tools such as artificial neural networks algorithms. Indeed, an electric fault can easily be modeled with the electromagnetic method and thus the influence on the magnetic pressure and displacement can be identified. In addition to the diagnosis purpose, the developed analytical model can also be used in the early design stages in order to prevent excessive magnetic vibrations.

IX. REFERENCES

[1] J.-W. Jung, S.-H. Lee, J.-P. Hong, D.-H. Lee, and K.-N. Kim, "Reduction design of vibration and noise in IPMSM type integrated starter and generator for HEV," *IEEE Trans. Magnetics*, vol. 46, no. 6, pp. 2454-2457, June 2010.

[2] R. Islam, and I. Husain, "Analytical model for predicting noise and vibration in permanent-magnet synchronous motors," *IEEE Trans. Industry Applications*, vol. 46, no. 6, pp. 2346-2354, Nov.-Dec. 2010.

[3] D. Torregrossa, F. Peyraut, B. Fahimi, J. M'Boua, and A. Miraoui, "Multiphysics finite-element modeling for vibration and acoustic analysis of permanent magnet synchronous machine," *IEEE Trans. Energy Conversion*, vol. 26, no. 2, pp. 490-500, June 2011.

[4] H. Jordan, *Geräuscharme Elektromotoren*, Essen: W. Girardet, 1950.

[5] P. L. Alger, *The nature of polyphase induction machines*, General Electric series, New-York: Wiley, 1951.

[6] P. L. Tímár, A. Fazekas, J. Kiss, A. Miklós, and S. J. Yang, *Noise and vibration of electrical machines*, Studies in Electrical and Electronic Engineering 34, Amsterdam: Elsevier, 1989.

[7] J.-F. Brudny, "Modélisation de la denture des machines asynchrones. Phénomène de résonance," *Journal de Physique III*, vol. 7, no. 5, pp. 1009-1023, May 1997.

[8] S. Huang, M. Aydin, and T. A. Lipo, "Electromagnetic vibration and noise assessment for surface mounted PM machines," *in Proc. 2001 Power Engineering Society Summer Meeting*, no. 3, pp. 1417-1426.

[9] G. Dajaku, and D. Gerling, "Stator slotting effect on the magnetic field distribution of salient pole synchronous permanent-magnet machines," *IEEE Trans. Magnetics*, vol. 46, no. 9, pp. 3676-3683, Sept. 2010.

[10] J.-F. Brudny, and J.-P. Lecointe, "Rotor design for reducing the switching magnetic noise of AC electrical machine variable-speed drives," *IEEE Trans. Industrial Electronics*, vol. 58, no. 11, pp. 5112-5120, Nov. 2011.

[11] Z.-Q. Zhu, and D. Howe, "Instantaneous magnetic field distribution in brushless permanent magnet DC motors. III. Effect of stator slotting," *IEEE Trans. Magnetics*, vol. 29, no. 1, pp. 143-151, Jan. 1993.

[12] U. Kim, and D. K. Lieu, "Magnetic field calculation in permanent magnet motors with rotor eccentricity: with slotting effect considered," *IEEE Trans. Magnetics*, vol. 34, no. 4, pp. 2253-2266, July 1998.

[13] Y. J. Zhang, S. L. Ho, H. C. Wong, and G. D. Xie, "Analytical prediction of armature-reaction field in disc-type permanent magnet generators," *IEEE Trans. Energy Conversion*, vol. 14, no. 4, pp. 1385-1390, Dec. 1999.

[14] J. Fu, and C. Zhu, "Subdomain model for predicting magnetic field in slotted surface mounted permanent-magnet machines with rotor eccentricity," *IEEE Trans. Magnetics*, vol. 48, no. 5, pp. 1906-1917, May 2012.

[15] D. Zarko, D. Ban, and T. A. Lipo, "Analytical calculation of magnetic field distribution in the slotted air gap of a surface permanent-magnet motor using complex relative air-gap permeance," *IEEE Trans. Magnetics*, vol. 42, no. 7, pp. 1828-1837, July 2006.

[16] H.-J. Shin, J.-Y. Choi, H.-I. Park, and S.-M. Jang, "Vibration analysis and measurements through prediction of electromagnetic vibration sources of permanent magnet synchronous motor bases on analytical magnetic field calculations," *IEEE Trans. Magnetics*, vol. 48, no. 11, pp. 4216-4219, Nov. 2012.

[17] N. Bracikowski, M. Hecquet, P. Brochet, and S. V. Shirinskii, "Multiphysics modeling of a permanent magnet synchronous machine by using lumped models," *IEEE Trans. Industrial Electronics*, vol. 59, no. 6, pp. 2426-2437, June 2012.

[18] G. He, Z. Huang, and D. Chen, "Two-dimensional field analysis on electromagnetic vibration-and-noise sources in permanent-magnet direct current commutator motors," *IEEE Trans. Magnetics*, vol. 47, no. 4, pp. 787-794, April 2011.

[19] G. Houdouin, "Contribution à la modélisation de la machine asynchrone en présence de défauts rotoriques," Ph.D. dissertation, Dept. Electro. Elec. Eng., Univ. Le Havre, France, 2004.

[20] A. Aït-Hammouda, "Pré-dimensionnement et étude de sensibilité vibro-acoustique de machines à courant alternatif et à vitesse variable," Ph.D. dissertation, Dept. Electrotech. Pow. Electro., Ecole Centrale Lille and Univ. Lille, France, 2005.

[21] G. Barakat, T. El-Meslouhi, and B. Dakyo, "Analysis of the cogging torque behavior of a two-phase axial flux permanent magnet synchronous machine," *IEEE Trans. Magnetics*, vol. 37, no. 4, pp. 2803-2805, July 2001.

[22] Q. Gu, and H. Gao, "Air gap field for PM electric machines," Electric Machines and Power Systems, vol. 10, no. 5-6, pp. 459-470, May 1985.

[23] N. Boules, "Prediction of no-load flux density distribution in permanent magnet machines," *IEEE Trans. Industry Applications*, vol. IA-21, no. 3, pp. 633-643, May 1985.

[24] Z.-Q. Zhu, D. Howe, E. Bolte, and B. Ackermann, "Instantaneous magnetic field distribution in brushless permanent magnet DC motors. I. Open-circuit field," *IEEE Trans. Magnetics*, vol. 29, no. 1, pp. 124-135, Jan. 1993.

[25] U. Kim, and D. K. Lieu, "Effects of magnetically induced vibration force in brushless permanent-magnet motors," *IEEE Trans. Magnetics*, vol. 41, no. 6, pp. 2164-2172, June 2005.

[26] L. J. Wu, Z. Q. Zhu, D. Staton, M. Popescu, and D. Hawkins, "An improved subdomain model for predicting magnetic field of surface-mounted permanent magnet machines accounting for tooth-tips," *IEEE Trans. Magnetics*, vol. 47, no. 6, pp. 1693-1704, June 2011.

[27] S.-H. Lee, J.-P. Hong, S.-M. Hwang, W.-T. Lee, J.-Y. Lee, and Y.-K. Kim, "Optimal design for noise reduction in interior permanent-

magnet motor," *IEEE Trans. Industry Applications*, vol. 45, no. 6, pp. 1954-1960, Nov.-Dec. 2009.

[28] X. Luo, Y. Liao, H. A. Toliyat, A. El-Antably, and T. A. Lipo, "Multiple coupled circuit modeling of induction machines," *IEEE Trans. Industry applications*, vol. 31, no. 2, pp. 311-318, March-April 1995.

[29] H. A. Toliyat and T. A. Lipo, "Transient analysis of cage induction machines under stator, rotor bar and end-ring faults," *IEEE Trans. Energy Conversion*, vol. 10, no. 2, pp 241-247, June 1995.

[30] J.F. Bangura, "Directly coupled electromagnetic field-electric circuit model for analysis of a vector-controlled wound field brushless starter generator," *IEEE Trans. Energy Conversion*, vol. 26, no. 4, pp. 1033-1040, Dec. 2011.

[31] S. Yu, and R. Tang, "Electromagnetic and mechanical characterizations of noise and vibration in permanent magnet synchronous machines," *IEEE Trans. Magnetics*, vol. 42, no. 4, pp. 1335-1338, April 2006.

[32] G. Houdouin, G. Barakat, B. Dakyo, and E. Destobbeleer, "A winding function theory based global method for the simulation of faulty induction machines," *in Proc. 2003 IEEE International Conference on Electric Machines and Drives*, no. 1, pp. 297-303.

[33] G. Barakat, B. Dakyo, H. Henao, and G. A. Capolino, "Coupled magnetic circuit modeling of the stator windings faults of induction machines including saturation effect," *in Proc. 2004 IEEE International Conference on Industrial Technology*, no. 1, pp. 148-153.

[34] J. F. Gieras, C. Wang, and J. C. Lai, *Noise of polyphase electric motors*, CRC Press, Boca Raton: Taylor & Francis Group, 2006.

[35] M. Valavi, A. Nysveen, and R. Nilssen, "Characterization of radial magnetic forces in low-speed permanent magnet wind generator with non-overlapping concentrated windings," *in Proc. 2012 XXth International Conference on Electrical Machines*, no. 1, pp. 2943-2948.

[36] K.-T. Kim, K.-S. Kim, S.-M. Hwang, T.-J. Kim, and Y.-H. Jung, "Comparison of magnetic forces for IPM and SPM motor with rotor eccentricity," *IEEE Trans. Magnetics*, vol. 37, no. 5, pp. 3448-3451, Sept. 2001.

[37] Z. Makni, M. Besbes, and C. Marchand, "Multiphysics design methodology of permanent-magnet synchronous motors," *IEEE Trans. Vehicular Technology*, vol. 56, no. 4, pp. 1524-1530, July 2007.

[38] A. Farshidianfar, and P. Oliazadeh, "Free vibration analysis of circular cylindrical shells: comparison of different shell theories," *International Journal of Mechanics and Applications*, vol. 2, no. 5, pp. 74-80, 2012.

[39] A. W. Leissa, *Vibration of shells*, Melville: Acoustical Society of America, 1993.

[40] O. Bennouna, H. Chafouk, O. Robin, J. P. Roux, "A diagnosis approach combining wavelet transform and artificial neural networks," *9th International Conference on Sciences and Techniques of Automatic Control & Computer Engineering*, Sousse, Tunisia, Dec. 2008.

X. BIOGRAPHIES

Guillaume Verez was born in Lille, France, in 1987. He received the B.Sc. and M. Sc. in Electrical Engineering from Ecole d'Ingénieurs en Energie Eau et Environnement, Grenoble, France, in 2008 and 2010 respectively. He also received the M. Sc. in Energy and Environmental Engineering of the Norwegian University of Science and Technology, Trondheim, Norway, in 2011. He is currently working towards the Ph. D. degree at University of Le Havre, Le Havre, France. His research interests are vibration and noise analysis in electric machines.

Ouadie Benouna received the Ph. D. degree from the University of Corsica, France, in 2006, and the Dipl. Ing. degree in mechanical engineering from the ENSAM (Ecole Nationale Supérieure d'Arts & Métiers), in 2003. He joined the IRSEEM (Institut de Recherche en Systèmes Electroniques EMbarqués) in February 2008. Since September 2009, he is an associate professor at the ESIGELEC, Rouen, France. Currently, he is the head of Mechatronics major at the same school. His research interests include diagnosis and fault detection, signal processing, and fault tolerant control. The main applications concern aircraft control, and wind energy conversion systems.

Yacine Arama received the Ph.D degree from the University of Paris Sud XI (Orsay), in 2001. He is currently a lecturer and works with Electrotechnic and Automatic Research Team of Le Havre (GREAH), University of Le Havre. His research interests include design, modeling and permanent magnet machines for automotive applications.

Ghaleb Hoblos received the PhD and HDR (Habilitation to Supervise Research) degrees in Automatic Control and Signal Processing from the University of Lille, France in 2001 and the University of Rouen, France in 2008 respectively. In 2002, he joined the Electrical Engineering and Energies Department, ESIGELEC, Rouen, France where he currently teaches in the areas of fault diagnosis systems, signal processing and automatic control. His research interests include fault tolerant control, diagnosis methods based on a model or not, hybrid systems. He is also concerned with the implementation of real-time diagnosis algorithms in embedded electronic units. His application domains are: aeronautic, aerospace and technological risk management.

Georges Barakat received the M.Sc and Ph.D degree from the Institut National Polytechnique de Grenoble, Grenoble, France, in 1992 and 1995, respectively. Currently he is a Professor of electrical Engineering Department, Electrotechnic and Automatic Research Team of Le Havre (GREAH), University of Le Havre. His research interests include electrical machines modeling, design and diagnosis.

978-1-4799-0024-4/13 $31.00 © 2013 IEEE

2-Pole Turbo-Generator Eccentricity Diagnosis By Split-Phase Current Signature Analysis

Claudio Bruzzese

Abstract – **2-pole turbogenerators are usually very high power round-rotor synchronous machines used in generating plants. Turbogenerator monitoring is crucial, especially in nuclear plants. Very long turbogenerator rotors experience excessive run-out due to thermal, mechanical, and magnetic unbalances, resonances, etc. The rotor eccentricity causes vibrations and winding and frame loosening. Since the stators of turbogenerators are often parallel-connected for maximum power output rating, the rotor eccentricities also make rise unbalances in the split-phase currents. This paper resorts to the split-phase current monitoring for detection and estimation of rotor eccentricities. A method is proposed, based on a special theoretical analysis of 2-pole round-rotor machines, through symmetrical components. Time-stepping finite-elements are used for confirmatory simulations of a 500MVA machine.**

Index Terms—**Analysis, current, diagnosis, eccentricity, FFT, monitoring, rotor, split-phase, turbogenerator, winding.**

I. Nomenclature

B_{he}	maximum flux density in the healthy machine
e	column vector of voltages induced by the rotor
f	fundamental frequency ($=\omega/2\pi$)
F_6	Fortescue's 6x6 generalized transformation matrix
g_0	air gap length in the healthy machine
K_d, K_p	stator distribution and pitch factors
$K_{d,h}, K_{p,h}$	harmonic distribution and pitch factors
l_S	stator iron stack length
$L_{k\delta}$	inductance of PPGs displaced of the angle $k\delta$
$L^{(q)}$	q-th sequence inductance, q=0, 1, …, 5.
M	magneto-motive force
n	number of elementary stator PPGs ($=6$)
n_E	rotor winding turn function
N_{slot}	number of stator slots
N_{turn}	number of turns in a stator coil side (half slot)
P	air gap permeance function
r_m	mean air gap radius
R_{SS}, L_{SS}	stator resistance and inductance 6x6 matrices
R_S	resistance of a stator pole-phase-group (PPG)
SCR	machine short-circuit ratio
v_S, i_S	stator PPG voltage and current column vectors
$v^{(q)}, i^{(q)}$	complex q-th sequence voltages and currents
V_{LL}	line-line rms voltage at the machine terminals
X_s, X_l	machine synchronous and leakage reactances
w_E	rotor winding function

z	stator slots per-pole and per-phase ($=N_{slot}/6$)
α	complex unit vector ($=e^{j\delta}$)
β	coil pitch angle ($\leq 180°$)
δ	displacement angle of consecutive PPGs ($=60°$)
$\Delta\Psi_{s/d}$	flux increment due to eccentricity
ζ	air gap field angle on the direct axis (power angle)
θ	rotor rotation mechanical angle ($=\omega t$)
θ_S, θ_R	angular abscissas in the stator and rotor frames
ρ_s, ρ_d	static and dynamic p.u. eccentricities, $0\leq\rho_s+\rho_d<1$
ξ_s, ξ_d	angles of static and dynamic eccentricities
$\psi^{(q)}_{(\omega)}$	$2q$-pole, $\omega/2\pi$-frequency flux linkage space vector
Ψ_{he}	max PPG flux linkage in the healthy machine
σ	slot pitch angle ($=360°/N_{slot}$)
τ	stator winding pole pitch length
τ_h	stator winding pole pitch for the h-th harmonics
$*$	complex conjugate operator

II. Introduction

TWO-POLE turbo-generators (TGs) are usually steam or gas turbine-driven round-rotor machines ranging from few MVAs up to about 2000 MVA [1]. These machines are generally used in nuclear and fossil-fueled power plants, cogeneration plants, and combustion turbine units. Although expected to last about 30 years, large TG lives are often extended far beyond that, if well maintained and operated. In fact, replacing an operating unit is very capital intensive and, thus, done only when a catastrophic failure has occurred [2].

TG rotors are high-speed rotating members which undergo severe dynamic mechanical, electromagnetic and thermal loading. Rotor rotation-related vibrations are one of the main causes of mechanical and electrical problems [3]. Mechanical problems affect bearings, retaining rings, balance weights and bolts, collector rings, and grounding brushes [1], [4]. Vibrations are also detrimental for the stator stack, which is usually made up of many thousands of lamination segments firmly aligned and held together by keybars and end-plates. In hydrogen-cooled machines, vibrations may also endanger the containment frame (wrapper plate or pressure vessel) and seals. Electrical problems concern the windings which is a very costly item. Loose windings may fail due to insulation fretting against the slot walls [1], [5], [6].

The fault diagnosis and condition monitoring of electrical machines has grown very much in the last decades [1]-[22]. Also the interest for power synchronous machines and

---

□C. Bruzzese is with the Dept. of Astronautical, Electrical, and Energy Eng., University of Rome – Sapienza, Via Eudossiana 18, 00184 Rome, Italy (e-mail: claudio.bruzzese@uniroma1.it).

978-1-4799-0024-4/13 $31.00 © 2013 IEEE

generators rotor eccentricity (RE) diagnosis is recently arising [2]-[4], [7]-[9]. However few papers investigated REs in round-rotor high-power machines [2], and even less in 2-pole TGs. This paper studies the effect of a generic mixed RE made up of a static RE (SRE) and a dynamic RE (DRE) in a 2-pole round-rotor TG endowed with parallel connections in the armature winding.

Previous papers analyzed the split-phase currents in machines with four or more poles [10], [11], [14]. Two-pole machines however require a more refined analysis for the adequate calculation of the air gap flux density in case of eccentric rotor. In a $2p$-pole machine with $p>1$, usually new $2(p\pm1)$-pole air gap flux density waves arise in case of REs. However, only $2(p+1)$-pole waves possibly appear in a two-pole machine, i.e. four-pole waves. This is proved by using the winding function theory in this paper.

This paper proposes a symmetrical-component based model of a 2-pole winding with parallel-connections, and taking in account the REs. The split-phase currents are formally calculated, and then the formulas are reversed to obtain the RE degrees from current measurements. The split-phase current signature analysis (SPCSA) is proved for a 2-pole 500MVA TG by using time-stepping FEM (TSFEM).

Section III of this paper shows the SPCSA theory, leading to practical closed-form formulas for RE diagnosis. Section IV shows the TSFEM simulations for the 500MVA TG.

III. THEORY OF SPLIT-PHASE CURRENTS IN 2-POLE TGS WITH REs

A. Two-Pole Split-Phase Winding Model

Several types of armature windings are used in TGs, such as concentric, lap, or wave windings, with Roebel bars inserted in open slots. However modern TGs typically are wound with double-layer whole-coiled lap windings [1], since they permit greater freedom of choice about the connection (series or parallel) to achieve the desired terminal voltage and current ratings, Fig. 1. Single-turn bars and parallel connections are very often used for maximum machine power output, since the terminal voltages must be limited between 13.8kV and 27kV. The parallels are connected on the machine front-ends.

The three-phase armature winding of a 2-pole TG is basically made up of six pole-phase-groups (PPGs), i.e. groups of consecutive coils series-connected over a pole in a phase. The PPGs are identical and symmetrically distributed as shown in Fig. 2. A practical winding scheme is shown in Appendix A. The $n=6$ PPGs can be modeled as unconnected mutually coupled circuits with independent currents:

$$v_S = R_{SS}\, i_S + L_{SS}\frac{d}{dt}i_S + e \quad . \tag{1}$$

In (1), $i_S=(i_1,\ i_2,...,\ i_n)^t$ represents the stator internal currents, $v_S=(v_1,\ v_2,...,\ v_n)^t$ are the PPG voltages, R_{SS} is a diagonal matrix, and L_{SS} contains the stator inductances. The voltages e induced by the rotor are influenced by REs, due to the modulation of the flux density wave traveling in the abnormal air gap. After calculation of the voltages e, the internal currents i_S can be carried out from (1) through model transformation and circuit connection [10].

B. 2-Pole TG Model Transformation

The symmetrical components (SCs) are a powerful tool used for analysis of both healthy and faulty machines [12], [13]. Here a SC-based transformation is used for the eccentric-rotor 2-pole TG model. The basic generalized current space-transformation is carried out through a 6x6 Fortescue's matrix $F_6{}^*$ with elements $f_{xy}=a^{(x-1)(y-1)}/\sqrt{6}$ [12]:

$$
\begin{vmatrix} i^{(0)} \\ i^{(1)} \\ i^{(2)} \\ i^{(3)} \\ i^{(4)} \\ i^{(5)} \end{vmatrix}
= \frac{1}{\sqrt{6}}
\begin{vmatrix}
1 & 1 & 1 & 1 & 1 & 1 \\
1 & \alpha & \alpha^2 & \alpha^3 & \alpha^4 & \alpha^5 \\
1 & \alpha^2 & \alpha^4 & \alpha^6 & \alpha^8 & \alpha^{10} \\
1 & \alpha^3 & \alpha^6 & \alpha^9 & \alpha^{12} & \alpha^{15} \\
1 & \alpha^4 & \alpha^8 & \alpha^{12} & \alpha^{16} & \alpha^{20} \\
1 & \alpha^5 & \alpha^{10} & \alpha^{15} & \alpha^{20} & \alpha^{25}
\end{vmatrix}
\cdot
\begin{vmatrix} i_1 \\ i_2 \\ i_3 \\ i_4 \\ i_5 \\ i_6 \end{vmatrix} \quad . \tag{2}
$$

In (2), the complex terms $i^{(q)}$ are space-sequence currents (also termed current space-vectors). The TG stator model (1) is transformed in case of centered (healthy) rotor machine, i.e. with L_{SS} which is constant and circulant (any row is obtained from the precedent by a circular right-shift, so only the first row is shown, see Fig. 2 for symbols):

$$\textit{first row of } L_{SS} = \begin{vmatrix} L_0 & L_{60} & L_{120} & L_{180} & L_{120} & L_{60} \end{vmatrix} \quad . \tag{3}$$

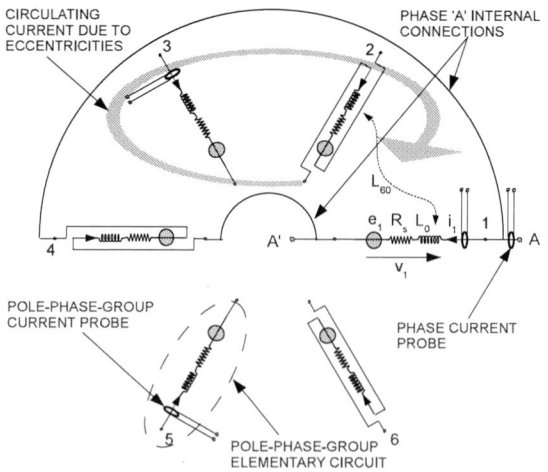

Fig. 2. Two-pole three-phase winding with parallel-connected PPGs. Only the phase A is shown connected.

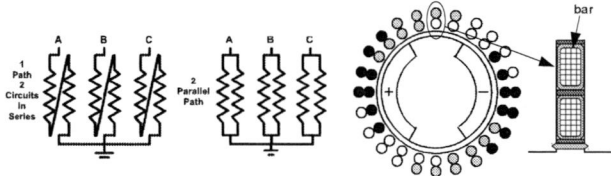

Fig. 1. Two-pole three-phase winding and connection options [1].

978-1-4799-0024-4/13 $31.00 © 2013 IEEE

Transformation of (1) leaves unchanged R_{SS}, and reverts L_{SS} in a diagonal constant real matrix (details can be found in [10]) so obtaining 6 decoupled sequence circuits, Fig. 3:

$$v^{(q)} = R_s i^{(q)} + L^{(q)} \frac{d\,i^{(q)}}{dt} + e^{(q)} \;,\; q=0,...,5. \qquad (4)$$

The q-th circuit in (4) is complex conjugate of the $(6-q)$-th one, so the latter is redundant. The circuit for $q=1$ is responsible of the main electromechanical energy conversion, and it is actually a per-pole per-phase machine equivalent circuit. In fact, the sequence inductance $L^{(1)}$ is twice the machine synchronous inductance, i.e. $L^{(1)}=2X_s/\omega$. The generic q-th circuit is sensitive to eventual $2q$-pole air gap flux density waves. The following steps calculate the sequence voltages $e^{(q)}$ in (4), for both healthy and eccentric rotor case. Note finally that L_{SS} in (1) is influenced by REs due to a first-order space-harmonic in the air-gap distribution. The additional first harmonic in L_{SS} can be neglected in practice.

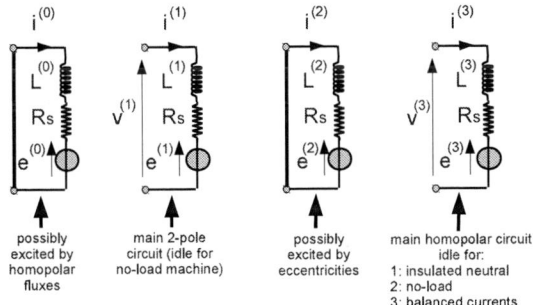

Fig. 3. Stator space-sequence circuits for a 2-pole machine.

C. 2-Pole TG with Centered Rotor

For healthy machine, and neglecting higher order space harmonics, the air gap permeance function of a TG can be assumed constant: $P=P_0=\mu_0/g_0$. Also in a 2-pole machine, the air gap MMF wave due to stator and rotor currents is written in the rotor frame as:

$$M = M_0 \cos(\theta_R + \zeta) \;. \qquad (5)$$

where $\zeta > 0$ for generator operation. The MMF wave (5) rotates in the stator frame due to the coordinate change $\theta_R = \theta_S - \omega t$. The resulting flux density wave, from $B=PM$, is:

$$B = B_{he} \cos(\omega t - \theta_S - \zeta) \qquad (6)$$

where $B_{he}=P_0 M_0$. The fluxes linked by (6) with the 6 PPGs make up a single-frequency 2-pole symmetric system, Fig. 4:

$$\psi_k = \Psi_{he} \cos(\omega t - \delta(k-1) - \zeta) \;,\; k=1,...,6 \qquad (7)$$

where Ψ_{he} is proportional to B_{he} in the unsaturated machine:

$$\Psi_{he} = \frac{2}{\pi} B_{he}\, \tau\, l_S\, N_{turn}\, z\, K_d\, K_p \;. \qquad (8)$$

In case of parallel-connected PPGs, the amplitude Ψ_{he} is fixed by the line-line voltage V_{LL} through $\Psi_{he}=\sqrt{2}V_{LL}/\sqrt{3}\omega$.

Fluxes in (7) are posed in complex form by applying the same transformation as in (2), and two rotating flux space vectors are obtained, $\psi_{(\omega)}^{(1)}$ and $\psi_{(\omega)}^{(5)}=\psi_{(\omega)}^{(1)*}$:

$$\psi_S' = F_6^* \psi_S = \begin{pmatrix} 0 & \psi_{(\omega)}^{(1)} & 0 & 0 & 0 & \psi_{(\omega)}^{(5)} \end{pmatrix}' . \qquad (9)$$

In (9), $\psi_S=(\psi_1, \psi_2,..., \psi_6)'$ is the column vector of the real PPG flux linkages (7), and ψ_S' contains the correspondent sequence fluxes; the 2-pole complex vector $\psi_{(\omega)}^{(1)}$ in (9) is:

$$\psi_{(\omega)}^{(1)} = \frac{\sqrt{6}}{2} \Psi_{he}\, e^{j(\omega t - \zeta)} \;. \qquad (10)$$

The correspondent 2-pole induced voltage space vector to put in (4) in case of centered-rotor machine is:

$$e_{(\omega)}^{(1)} = \frac{d}{dt} \psi_{(\omega)}^{(1)} \;. \qquad (11)$$

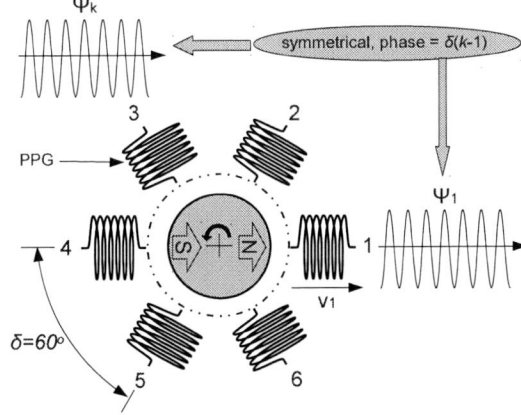

Fig. 4. Centered-rotor machine.

D. 2-Pole TG with Eccentric Rotor

A general mixed eccentricity of the rotor is made up of a static and a dynamic component. The air gap lenght function is written as (the symbols are explained in Fig. 5):

$$g = g_0 \big(1 - \rho_s \cos(\theta_S - \xi_s) - \rho_d \cos(\theta_R - \xi_d)\big) \qquad (12)$$

and the permeance function can be approximated for little eccentricities (as in the case of TGs) as:

$$P = \mu_0/g \simeq P_0 \big(1 + \rho_s \cos(\theta_S - \xi_s) + \rho_d \cos(\theta_R - \xi_d)\big) \;. \qquad (13)$$

978-1-4799-0024-4/13 $31.00 © 2013 IEEE

At this point, usually a multiplication of permeance (13) and MMF (5) gives the actual air gap flux density distribution in the eccentric rotor machine, as in [10]. Papers [11], [14] carried out the analysis of eccentric-rotor machines with four or more poles, showing a good match with experimental results. It is proven that additional $2(p\pm1)$-pole flux density waves appear in the air gap due to REs. However, the approach of [10], [11], [14] is correct only for machines with $p>1$. In fact for a 2-pole machine we obtain:

$$
\begin{aligned}
B = &\, B_{he}\cos\left(\omega t - \theta_S - \zeta\right) \\
&+ 0.5\,\rho_s B_{he}\cos\left(\omega t - \xi_s - \zeta\right) \\
&+ 0.5\,\rho_s B_{he}\cos\left(\omega t - 2\theta_s + \xi_s - \zeta\right) \\
&+ 0.5\,\rho_d B_{he}\cos\left(-\xi_d - \zeta\right) \\
&+ 0.5\,\rho_d B_{he}\cos\left(2\omega t - 2\theta_s + \xi_d - \zeta\right)
\end{aligned}
\tag{14}
$$

and the $2(p-1)$-pole waves degenerate in two uniform flux density distributions in the air gap, with constant amplitude in case of dynamic RE and with oscillating amplitude in case of static RE (see the fourth and second term at second member of (14), respectively). Figs. 6, 7 show a representation of fluxes (14) as produced by 'virtual' rotors. The homopolar flux distributions imply a net axial flux, which is not contemplated in a 2-dimension model since a reluctive return path for the axial flux is not specified. So in a 2-pole machine, the MMF (5) actually leads to a violation of the Gauss's law in case of eccentric rotor, i.e. the condition of zero net flux through a surface which encloses the rotor, as the contour C in Fig. 5. A coherent 2-D model can be obtained by using the winding function approach (WFA), as done in the following.

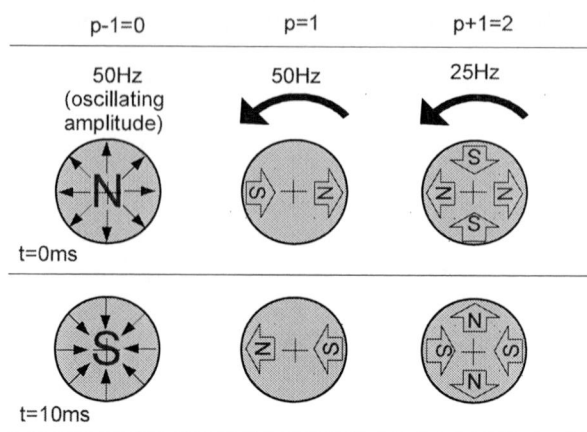

Fig. 6. Virtual rotors, case of static eccentricity.

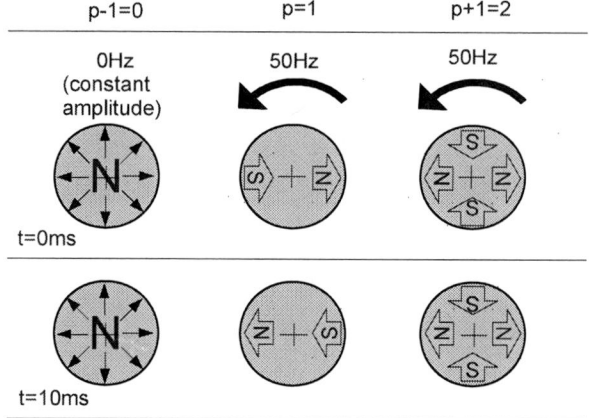

Fig. 7. Virtual rotors, case of dynamic eccentricity.

E. WFA Applied to a 2-Pole TG with Eccentric Rotor

The WFA is based on the Ampere's law and Gauss's law applied to a planar machine geometry, supposing no drop of MMF in the iron [15], [16]. So axial components of flux are not allowed. This condition may be associated to machines with very high reluctance paths for flux reclosing on the ends, or to very long-rotor TGs. In such cases, the homopolar fluxes in (14) are clearly wrong. In fact, even if any reclosing magnetic path does exist through the machine frame, the front covers, the sleeve bearings, and the solid shaft (axial path), this path has a reluctance much more higher than any path in the air gap. Usually the stator core is elastically suspended with keybars fixed to circumferencial ribs attached to the frame, as in Fig. 8, for vibration damping. So the axial fluxes should be much smaller than those given by (14). Obviously, a machine model should include the axial reclosing paths for an accurate calculation of the small axial fluxes in case of eccentricity, with great increase of model complexity. These fluxes are neglected in the following.

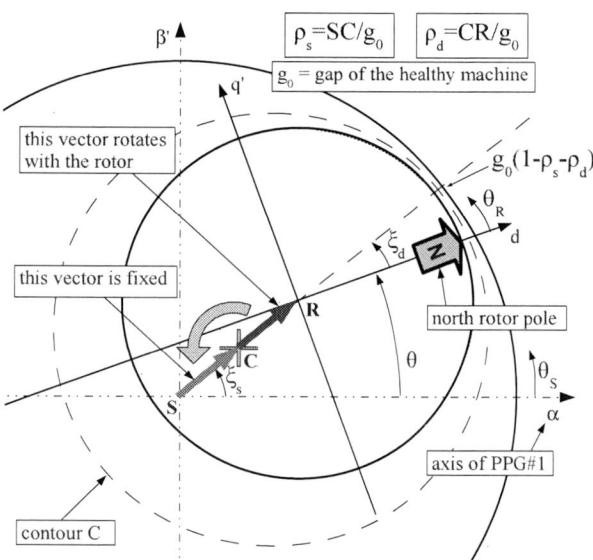

Fig. 5. Rotor eccentricity scheme and symbols used in the formulas.

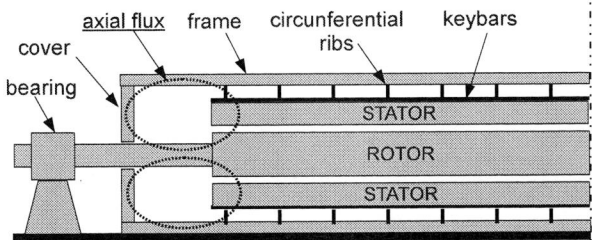

Fig. 8. Flux reclosing paths at a TG end.

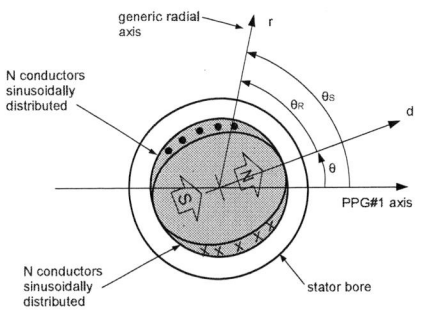

Fig. 9. Rotor winding sinusoidally distributed for WFA application.

To apply the WFA, we consider the no-load case with the air gap MMF being produced by the excitation winding, Fig. 9. An ideal sinusoidally-distributed rotor winding is considered, with turn function defined in the rotor frame as:

$$n_E(\theta_R) = -0.5\,N\cos\theta_R . \qquad (15)$$

The excitation MMF M_E is defined as:

$$M_E = w_E i_E \qquad (16)$$

where the winding function w_E appears [19]:

$$w_E = \frac{\langle n_E g^{-1}\rangle}{\langle g^{-1}\rangle} - n_E . \qquad (17)$$

By using (15) and (12) in (17), from (16) we obtain:

$$M_E = M_0\left(\cos\theta_R - \frac{\rho_s}{2}\cos(\omega t - \xi_s) - \frac{\rho_d}{2}\cos(\xi_d)\right) \qquad (18)$$

where $M_0 = N\,i_E/2$ is posed. The MMF (18) differs from (5) due to two subtractive terms, which actually lead to cancellation of the homopolar fluxes. In fact, the air-gap flux density distribution is obtained by multiplying (13) and (18), and, also neglecting higher-order small terms, we obtain:

$$\begin{aligned}B = &\,B_{he}\cos(\omega t - \theta_S)\\ &+ 0.5\,\rho_s B_{he}\cos(\omega t - 2\theta_S + \xi_s)\\ &+ 0.5\,\rho_d B_{he}\cos(2\omega t - 2\theta_S + \xi_d)\end{aligned} \qquad (19)$$

where no homopolar flux appears. So the homopolar virtual rotors in Figs. 6, 7 can be neglected in the model adopted here, only retaining the four-pole virtual rotors for further study. The reasoning above can be easily adapted to the loaded operation, only considering an equivalent conductor distribution on the rotor displaced of the angle ζ which takes in account the stator reaction. The result is again (14) without the homopolar terms:

$$\begin{aligned}B = &\,B_{he}\cos(\omega t - \theta_S - \zeta)\\ &+ 0.5\,\rho_s B_{he}\cos(\omega t - 2\theta_S + \xi_s - \zeta)\\ &+ 0.5\,\rho_d B_{he}\cos(2\omega t - 2\theta_S + \xi_d - \zeta)\end{aligned} \qquad .(20)$$

F. Flux Linkages in a 2-Pole TG with Eccentric Rotor

The four-pole waves in (20) are not able to link fluxes with an integer-step 2-pole winding. However, usually the armatures of TGs have a coil-step reduction, and a linkage with four-pole waves turns out to be possible. The stator-linked fluxes are obtained from (20) by integration as:

$$\begin{aligned}\psi_k = &\,\Psi_{he}\cos(\omega t - p\,\delta(k-1) - \zeta)\\ &+ \Delta\Psi_s\cos(\omega t - 2\delta(k-1) + \xi_s - \zeta)\\ &+ \Delta\Psi_d\cos(2\omega t - 2\delta(k-1) + \xi_d - \zeta)\end{aligned} \qquad .(21)$$

The amplitudes of additional fluxes in (21) cannot be derived from the formulas in [10], [14] valid in general for $p>1$. The 2-pole machine requires a different calculation. The additional flux systems have amplitudes:

$$\Delta\Psi_{s/d} = \frac{2}{\pi}\frac{\rho_{s/d}\,B_{he}}{2}\tau_2\,l_S\,N_{turn}\,z\,K_{d,2}\,K_{p,2} \qquad (22)$$

where τ_2, $K_{d,2}$, and $K_{p,2}$ are pole pitch, distribution factor, and pitch factor for the second harmonic (see Appendix B). By using (8) in (22), we obtain:

$$\Delta\Psi_{s/d} = \frac{\rho_{s/d}\Psi_{he}}{2}K_2 \qquad (23)$$

where the synthetic coefficient K_2 is given by:

$$K_2 = \frac{K_{d,2}\,K_{p,2}}{2\,K_d\,K_p} \approx 0.866\sqrt{1 - K_p^2} . \qquad (24)$$

The transformation of the additional fluxes in (21) produces two 2-pole rotating space vectors, with amplitudes proportional to ρ_s and ρ_d, respectively:

$$\psi_{(\omega)}^{(2)} = \frac{\sqrt{6}}{2}\Delta\Psi_s\,e^{j(\omega t + \xi_s - \zeta)} \qquad (25)$$

$$\psi_{(2\omega)}^{(2)} = \frac{\sqrt{6}}{2}\Delta\Psi_d\,e^{j(2\omega t + \xi_d - \zeta)} . \qquad (26)$$

Finally, the time-derivative of the flux vectors (25) and (26) gives the overall 4-pole sequence voltage which excites the correspondent circuit in Fig. 3:

$$e^{(2)} = \frac{d}{dt} \left(\psi_{(\omega)}^{(2)} + \psi_{(2\omega)}^{(2)} \right) . \tag{27}$$

The voltage (27) is not able to induce any component in the machine phase current due to the pole pair mismatch with the 2-pole sequence circuit shown in Fig. 3. However, the 4-pole sequence circuit is properly stimulated. The eccentric rotor TG can be finally thought as the superimposition of three virtual machines with centered rotor: the healthy 2-pole original machine and other two healthy 4-pole machines running with mechanical frequencies 25Hz and 50Hz, respectively, Fig. 10.

G. Calculation of the Fault-Related Sequence Currents

When a RE appears, currents arise in the 4-pole circuits of Fig. 10, which can be used for fault diagnosis. By using (4), (23), and (25)-(27), and by neglecting the resistance R_S, the steady-state 4-pole current can be carried out as:

$$
i^{(2)} = i_{(\omega)}^{(2)} + i_{(2\omega)}^{(2)} = \\
\frac{-\sqrt{6}\,\Psi_{he}/4}{L^{(2)}/K_2} \left[\rho_s e^{j(\omega t + \xi_s - \zeta)} + \rho_d e^{j(2\omega t + \xi_d - \zeta)} \right] \tag{28}
$$

In (28) there are two components proportional to the SRE and DRE respectively. The bi-periodic vector (28) traces a closed curve on the complex plane, in form of a centered trochoid or centered Pascal's limaçon, as shown in Section IV. The use of (28) for fault diagnosis is straightforward, passing through the FFT analysis of the complex current carried out for a given machine by (2). However, the parameter $L^{(2)}$ is hardly furnished in the manufacturer's data sheets, which is an obstacle to the handy use of (28) for an absolute estimation of REs. This problem is solved in Appendic C, where $L^{(2)}$ is calculated for a generic 2-pole TG.

H. Definition of Practical Fault Indicators

The complex current (28) is first rewritten in a more practical way as follows:

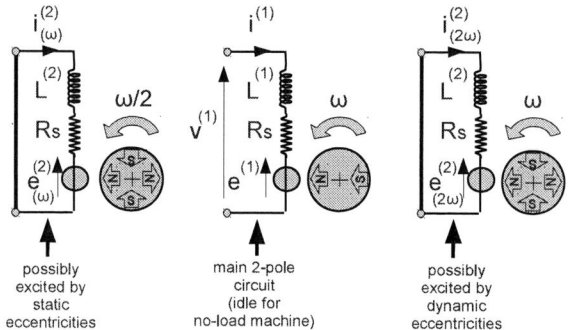

Fig. 10. Virtual healthy machines representing an eccentric-rotor machine.

$$i^{(2)} = S_2 e^{j(\omega t + \xi_s - \zeta)} + D_2 e^{j(2\omega t + \xi_d - \zeta)} . \tag{29}$$

The symbols S_2, D_2 represent the amplitudes of the frequency components obtained through complex FFT of the vector carried out from the measured currents. The SRE and the DRE can be estimated from (28) and (29) as follows:

$$\rho_s = \frac{2 L^{(2)}/K_2}{V_{LL}/\omega} S_2 \tag{30}$$

$$\rho_d = \frac{2 L^{(2)}/K_2}{V_{LL}/\omega} D_2 \tag{31}$$

where $\Psi_{he} = \sqrt{2}V_{LL}/\sqrt{3}\omega$ has been also used.

Equations (30), (31) are the basis of the proposed eccentricity diagnosis method in 2-pole round-rotor machines. The needed parameters are the actual line-line voltage, the coefficient K_2 (24), and the inductance $L^{(2)}$ (C17).

IV. SIMULATIONS OF A 500MVA TURBO-GENERATOR

Time-stepping FEM simulations of a 500MVA 2-pole TG (machine data are in Appendix A) were carried out with time-step=0.5ms, in no-load (Fig. 11) and full-load (Fig. 12) conditions. The parameters S_2, D_2 to put in (30), (31) were obtained by FFT in Figs. 11-f and 12-f. The other parameters were as follows: V_{LL}=18kV, ω=314rad/s, K_2=0.296 from(24), and $L^{(2)}$=1.12mH from (C17). The REs estimated were $\rho_s \approx \rho_d \approx 5.5\%$ in no-load, and $\rho_s \approx \rho_d \approx 6.3\%$ at full-load, which is quite a good approximation of the REs actually simulated in the model (5% of both SRE and DRE). The RE angles simulated were ξ_s=0°, ξ_d=180°. (the SRE points toward the PPG#1 axis, and the DRE points toward the South pole of the rotor). The axis of the trochoid in Fig. 11-d is not perfectly horizontal due to the time-step lag (9° at 3000rpm).

V. CONCLUSION

A practical method (SPCSA) has been carried out for assessment of rotor misalignments in round-rotor 2-pole synchronous machines (turbogenerators) with parallel circuits in the stator. The f and $2f$ frequency signatures in the spectrum of the four-pole space-vector obtained from the internal currents have to be used in the formulas provided in this paper, to obtain the diagnosis. The following machine parameters are needed: line-line rms voltage, frequency, synchronous and leakage reactance, number of stator slots, and winding pitch factor. Time-stepping FEM simulations of a 500MVA TG confirm the SPCSA applicability.

Only three split-phase currents must be measured in practical machines (one per phase), since the other three are obtained for subtraction from the phase currents. Rogowski coils could be used, embedded in the front-ends of the TG. However some TG is equipped with split-phase current protections, and so the measurement is already available. Experimental research is ongoing on this promising topic.

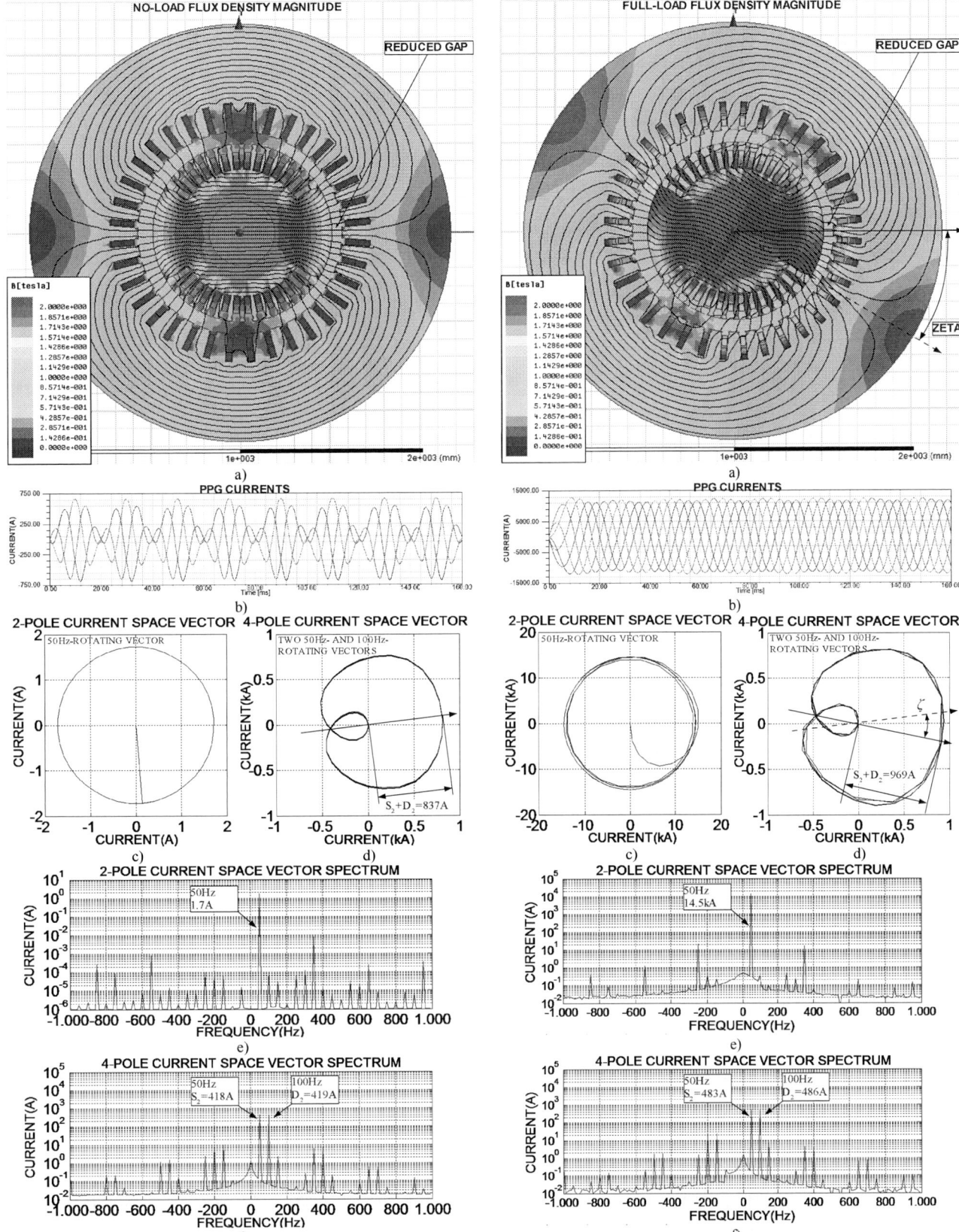

Fig. 11. a)-f): 500MVA TG in no-load condition. 5% SRE and 5% DRE.

Fig. 12. a)-f): 500MVA TG in full-load condition. 5% SRE and 5% DRE.

VI. APPENDIX A

Machine data as used in the paper are reported in Table I. The winding scheme for one phase is shown in Fig. 13.

TABLE I
500MVA TURBO-GENERATOR RATINGS

Apparent power	500MVA	Voltage/current	18kV/16kA
Active power	450MW	Frequency	50Hz
Speed	3000rpm	cosphi	0.9
SCR	0.6	Rotor current	1425A
X_s (unsat.)	1.19Ω (1.83p.u.)	X_l	59.5mΩ (0.09p.u.)
air gap	7mm	core lenght	7.21m

Armature winding: three-phase, double layer, lap-wound, two-pole, whole coiled winding with 36 slots, 6 slots per-pole and per-phase, step reduction=4 slots, and two parallel pole-phase-groups per phase.

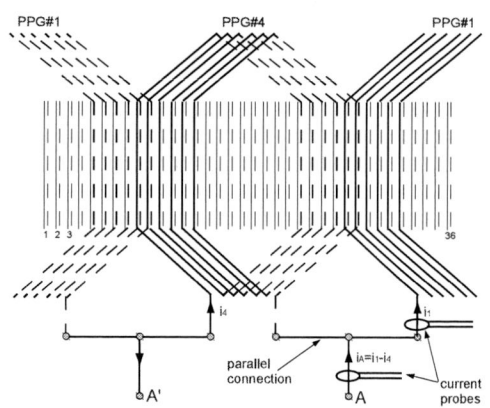

Fig. 13. Phase A winding scheme. The phases B and C are identical, only displaced of 120 and 240 degrees.

VII. APPENDIX B

The distribution and step-reduction (pitch) factors and the pole pitch for the h-th harmonic are recalled here below:

$$K_{d,h}=\frac{\sin(zh\sigma/2)}{z\sin(h\sigma/2)} \qquad (B1)$$

$$K_{p,h}=\sin(h\beta/2) \qquad (B2)$$

$$\tau_h=\tau/h . \qquad (B3)$$

VIII. APPENDIX C

The calculation of $L^{(2)}$ is carried out here for a generic 2-pole TG. The stator inductances are transformed by [10]:

$$L_{SS}'=F_n^* L_{SS} F_n=diag\{L^{(q)}\} \qquad (C1)$$

and the sequence inductances can be directly obtained by:

$$L^{(q)}=L^{(n-q)}=\sum_{u=0}^{n-1} L_{u\delta}\cos(qu\delta) . \qquad (C2)$$

In particular, from (C2) and (3) we obtain $L^{(1)}$ and $L^{(2)}$:

$$L^{(1)}=L_0+L_{60}-L_{120}-L_{180} \qquad (C3)$$

$$L^{(2)}=L_0-L_{60}-L_{120}+L_{180} . \qquad (C4)$$

Note that $L^{(1)}$ is known from the synchronous reactance:

$$L^{(1)}=2X_s/\omega \qquad (C5)$$

and that the self-inductance L_0 embeds the leakage inductance X_l/ω (which is also assumed known):

$$L_0=2X_l/\omega+L_{1,1} \qquad (C6)$$

where $L_{1,1}$ is the PPG self inductance due to the air gap flux linkage. The problem of calculating $L^{(2)}$ starting from (C3)-(C6) is solved as follows. From the WFA theory, the winding function of a single coil with N_{turn} turns spanning β radians in the stator bore is (Fig. 14):

$$w_{coil}(\gamma)=\sum_{u=1}^{\infty}\frac{2N_{turn}}{\pi u}\sin\frac{u\beta}{2}\cos u\gamma . \qquad (C7)$$

The winding function w_k of the k-th PPG (made up of z coils in series) is the summation of z functions as in (C7), displaced of a slot pitch one from the other:

$$w_k=N_{turn}w_k' \qquad (C8)$$

where w_k' is the winding function of a single-turn PPG:

$$w_k'=\sum_{h=0}^{z-1}\sum_{u=1}^{\infty}\frac{2}{\pi u}\sin\frac{u\beta}{2}\cos u\left(\gamma-h\sigma-(k-1)\frac{\pi}{3}\right). \qquad (C9)$$

Note that w_k' can be numerically evaluated for a given machine, only requiring the parameters σ, β, and $z=60°/\sigma$.

Finally, the mutual inductance between the first and k-th PPG is in general:

$$L_{1,k}=K_{WFA}L_{1,k}' \qquad (C10)$$

where K_{WFA} regroups all the unknown parameters:

$$K_{WFA}=\frac{2\pi\mu_0 r_m l_s N_{turn}^2}{g_0} \qquad (C11)$$

whereas $L_{1,k}'$ ($k=1,...,4$) are 'non-dimensional inductances' which can be numerically pre-calculated, known σ, β, and z:

$$L_{1,k}'=\langle w_1' w_k'\rangle . \qquad (C12)$$

So the inductances appearing in (C3), (C4) are:

978-1-4799-0024-4/13 $31.00 © 2013 IEEE   356

$$L_0 = 2\,X_l/\omega + L_{1,1} = 2\,X_l/\omega + K_{WFA}\,L_{1,1}' \qquad (C13)$$

$$L_{60} = L_{1,2} = K_{WFA}\,L_{1,2}' \qquad (C14)$$

$$L_{120} = L_{1,3} = K_{WFA}\,L_{1,3}' \qquad (C15)$$

$$L_{180} = L_{1,4} = K_{WFA}\,L_{1,4}' \quad . \qquad (C16)$$

By using (C13)-(C16) in (C3), (C4), and solving the system for $L^{(2)}$, it is easy to eliminate K_{WFA} so obtaining the following formula with only known parameters:

$$L^{(2)} = \frac{2\,X_l}{\omega} + \frac{2(X_s - X_l)}{\omega} \cdot \frac{L_{1,1}' - L_{1,2}' - L_{1,3}' + L_{1,4}'}{L_{1,1}' + L_{1,2}' - L_{1,3}' - L_{1,4}'} \; . \; (C17)$$

Equation (C17) together with (C9), (C12), solves the problem of calculating $L^{(2)}$, by few lines of computer code. Only X_s, X_l, β, $\sigma = 360°/N_{slot}$, and $z = N_{slot}/6$ are needed. If X_l is unknown, (C17) gives fair results by assuming $X_l \approx 0.1 \div 0.2$ p.u., as usual in large TGs with SCR $\approx 0.4 \div 0.6$ [1]. N_{slot} is not essential, since it is usually large and (C17) converges very quickly for $\sigma \to 0$ and $z \to \infty$.

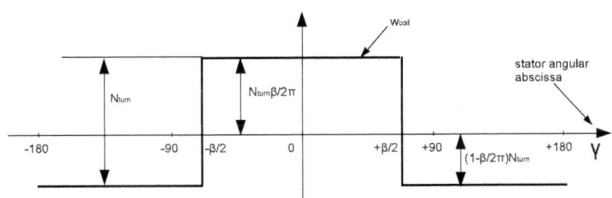

Fig. 14. Winding function w_{coil} of a single coil with N_{turn} turns, centered with respect to a proper stator angular abscissa γ.

IX. REFERENCES

[1] G. Klempner and I. Kerszenbaum, *Handbook of Large Turbo-Generator Operation and Maintenance*, IEEE Power Engineering Series, Wiley Inc., New York, 2008.

[2] M. Biet, "Rotor faults diagnosis using features selection and nearest neighbors rule: Application to a turbogenerator," *IEEE Trans. Ind. Electron.*, IEEE early access.

[3] D. Zarko, D. Ban, I. Vazdar, and V. Jaric, "Calculation of unbalanced magnetic pull in a salient-pole synchronous generator using finite-element method and measured shaft orbit," *IEEE Trans. Ind. Electron.*, vol. 59, no. 6, pp. 2536-2549, Jun. 2012.

[4] J. J. Simond, M. Tu Xuan, and R. Wetter, "An innovative inductive air-gap monitoring for large low speed hydro-generators," in *Proc. ICEM Conf.*, Vilamoura, Portugal, Sep. 6-9, 2008, paper ID 760.

[5] L. Romeral, J. C. Urresty, J. R. R. Ruiz, and A. G. Espinosa, "Modeling of surface-mounted permanent magnet synchronous motors with stator winding interturn faults," *IEEE Trans. Ind. Electron.*, vol. 58, no. 5, pp. 1576-1585, May 2011.

[6] P. Neti and S. Nandi, "Stator interturn fault detection of synchronous machines using field current and rotor search-coil voltage signature analysis," *IEEE Trans. Ind. Appl.*, vol. 45, no. 3, pp. 911-920, May/Jun. 2009.

[7] T. C. Ilamparithi, and S. Nandi, "Detection of eccentricity faults in three-phase reluctance synchronous motor," *IEEE Trans. Ind. Appl.*, vol. 48, no. 4, pp. 1307-1317, Jul./Aug. 2012.

[8] C. Bruzzese and G. Joksimovic, "Harmonic signatures of static eccentricities in the stator voltages and in the rotor current of no-load salient pole synchronous generators," *IEEE Trans. Ind. Electron.*, vol. 58, no. 5, pp. 1606-1624, May 2011.

[9] C. Bruzzese and T. Mazzuca, "DIEM project's outcomes: an automated air-gap monitoring approach for Italian Navy's on-board low-voltage generators," in *Proc. IEEE ESARS Conf.*, Bologna, Italy, Oct. 16-18, 2012.

[10] C. Bruzzese, "Study of cardioid-shaped loop current space vector trajectories for rotor eccentricity detection in power synchronous machines," in *Proc. IEEE SDEMPED Conf.*, Bologna, Italy, Sep. 5-8, 2011.

[11] C. Bruzzese, "A virtual instrument for on-line evaluation of alternator's shaft misalignments through ICSVA (Internal Current Space-Vector Analysis)," in *Proc. IEEE SDEMPED Conf.*, Bologna, Italy, Sep. 5-8, 2011.

[12] C. Bruzzese, "Analysis and application of particular current signatures (symptoms) for cage monitoring in non-sinusoidally fed motors with high rejection to drive load, inertia, and frequency variations", *IEEE Trans. Ind. Electron.*, vol. 55, no. 12, pp. 4137-4155, Dec. 2008.

[13] C. Bruzzese, O. Honorati, and E. Santini, "Evaluation of classic and innovative sideband-based broken bar indicators by using an experimental cage and a transformed (n,m) complex model", in *Proc. IEEE ISIE Conf.*, Vigo, Spain, Jun. 4-7, 2007.

[14] C. Bruzzese, "Field experience with the split-phase current signature analysis (SPCSA): Eccentricity assessment for a stand-alone alternator in time-varying and unbalanced load conditions," in *Proc. IEEE WEMDCD Conf.*, Paris, France, Mar. 11-12, 2013, pp. 255-268.

[15] C. Bruzzese, E. Santini, V. Benucci, and A. Millerani, "Model-based eccentricity diagnosis for a ship brushless generator exploiting the machine voltage signature analysis (MVSA)," in *Proc. IEEE SDEMPED Conf.*, Cargese, France, Aug. 31 – Sep. 3, 2009.

[16] S. Nandi, R. M. Bharadwaj, and H. A. Toliyat, "Performance analysis of a three-phase induction motor under incipient mixed eccentricity condition," *IEEE Trans. Energy Convers.*, vol. 17, no. 3, pp.392-399, Sep. 2002.

[17] R. Perers, U. Lundin, and M. Leijon, "Saturation effects on unbalanced magnetic pull in a hydroelectric generator with an eccentric rotor," *IEEE Trans. Magn.*, vol. 43, no. 10, pp. 3884-3890, Oct. 2007.

[18] H. Henao, S. H. Kia, and G.-A. Capolino, "Torsional-vibration assessment and gear-fault diagnosis in railway traction system," *IEEE Trans. Ind. Electron.*, vol. 58, no. 5, pp. 1707-1717, May 2011.

[19] C. Bruzzese, A. Giordani, and E. Santini, "Static and dynamic rotor eccentricity on-line detection and discrimination in synchronous generators by no-load E.M.F. space vector analysis," in *Proc. SPEEDAM Conf.*, Ischia, Italy, Jun. 18-20, 2008.

[20] Z. Daneshi-Far, G. A. Capolino, and H. Henao, "Review of failures and condition monitoring in wind turbine generators," in *Proc. ICEM Conf.*, Rome, Italy, Sep. 6-8, 2010.

[21] G. Traxler-Samek, R. Zickermann, and A. Schwery, "Cooling airflow, losses, and temperatures in large air-cooled synchronous machines," *IEEE Trans. Ind. Electron.*, vol. 57, no. 1, pp. 172-180, Jan. 2010.

[22] M. J. DeBortoli, S. J. Salon, D. W. Burow, and C. J. Slavik, "Effects of rotor eccentricity and parallel windings on induction machine behavior: A study using finite element analysis," *IEEE Trans. Magn.*, vol. 29, no. 2, pp. 1676-1682, Mar. 1993.

X. BIOGRAPHY

Claudio Bruzzese (S'05-M'08) received the M.Sc. (*cum laude*) and Ph.D. degrees from the University of Rome "Sapienza," Rome, Italy, in 2002 and 2008, respectively.

He worked as designer of electric plants between 1998 and 2002. After graduation, he was with the National Power System Management Company. Since September 2002, he has been with the Dept. of Electrical Eng., University of Rome "Sapienza," as Researcher Associate, and from 2011 as Assistant Professor. He is consultant for the Italian Ministry of Defence, and has developed projects in the framework of the Military Research National Program. He was visiting researcher with the University of Victoria, Victoria, Canada, in 2012. His interests cover fault diagnosis of power induction and synchronous machines, railway and naval power systems, linear drives, and electromechanical design and advanced modeling. He is author or coauthor of about 50 technical papers, and holds four patents.

Dr. Bruzzese is a Registered Professional Engineer in Italy. He is member of the IEEE Industrial Electronics Society.

Gear Tooth Surface Damage Fault Detection Using Induction Machine Electrical Signature Analysis

Shahin Hedayati Kia, Humberto Henao, *Senior Member IEEE*, Gérard-André Capolino, *Fellow IEEE*

Abstract - **The aim of the present work is the diagnosis of tooth surface damage fault in gears using the induction machine electrical signature analysis. The condition monitoring of gears is a crucial task due to its importance in the mechanical power transmission in industrial, aerospace and automotive applications. The vibration analysis has been commonly used as an effective tool for gear fault diagnosis in several studies. The gear torsional vibration effect in the stator current and the estimated electromagnetic torque has been previously studied based on the observation of gear mechanical characteristic frequencies in the spectrum of the load torque. This paper investigates the profile generated by a gear tooth surface damage fault in the load torque. It will be shown that the periodic behavior of this particular profile produces fault-related frequencies in the stator current and hence harmonics namely integer multiple of rotation frequency in the instantaneous frequency of the stator current space vector and the estimated electromagnetic torque. The obtained results show a possible non-invasive gear tooth surface damage fault detection with a fault sensitivity comparable to the one obtained with invasive methods. A set-up based on a 250W three-phase squirrel-cage induction machine shaft-connected to a single-stage gear has been used for this purpose.**

Index Terms-- **AC motor protection, Fault diagnosis, Gears, Induction motor, Monitoring, Signal processing, Space vector, Stator current analysis, Vibration measurement.**

I. INTRODUCTION

THE condition monitoring of electromechanical systems is considered crucial for several industrial processes. This task is commonly realized by using vibration analysis. The vibration signal is the image of periodic events in the mechanical system which may be representative of its free and natural dynamic behavior being even excited by external sources [1]. This behavior will be changed in case of any kind of mechanical abnormality in the electromechanical system. Gears, which are extensively used in such systems to transmit the mechanical power from one shaft to an other one by adapting both the speed and the torque, have received considerable attention in the field of condition-based maintenance for years [2]. The main sources of vibration, for a healthy gear, are principally due to the time-varying gear mesh stiffness and the gear transmission errors. The evolution of the contact between teeth of a gear is the main reason of mesh stiffness time variation [3]. The gear transmission errors are related to the tooth-profile error,

the eccentricity of pinion and wheel, the different imperfections and any other transmission irregularities. The common gear faults are related to gear tooth irregularities namely the tooth breakage, the tooth root crack, the chipped tooth, the spalling damage and the tooth surface damage which are typical localized faults [4]. When such faults occur, an impact will be generated by the gear fault at the rotational frequency which results in a wide frequency distribution in the vibration spectrum [5].

In spite of the extensive usage of the vibration measurement for diagnosis of different types of gear faults, it has several drawbacks such as a signal background noise due to external excitation motion, the inaccessibility in mounting the vibration transducer and the sensitivity to the installation position [6]. The induction machine electrical signal analysis (IMESA), namely the stator current and the estimated electromagnetic torque analysis, are two non-invasive methods which can represent good alternatives to the vibration analysis with minimum changes in the system installation. The effect of gear torsional vibrations in the stator current has been previously studied on the basis of observations of the torque spectrum for a healthy gear [7]. It has been detected that the mesh frequency cannot be easily measured due to its weak magnitude in the stator current spectrum. Moreover, the presence of mesh-related frequencies in the stator current is mainly due to the modulation phenomenon around the mesh frequency in the vibration and consequently in the torque [5], [8]. Particularly [8], the effect of the pinion tooth surface damage fault on mesh-related frequency magnitudes in the stator current spectrum has been investigated. These frequencies can be also observed in the instantaneous frequency of the stator current space vector [9]. It was shown that the amplitudes of some of them are sensitive to gear localized fault in the stator current spectrum. Nevertheless, the reasons for such increases or decreases in amplitudes are still ambiguous. The principal motivation of this work is to study the effect of gear tooth localized fault in the stator current according to its signature in the torque. Indeed, the gear localized fault may not necessary influence the amplitudes of mesh and mesh-related frequencies in the stator current which was the main assumption for non-invasive gear tooth fault detection in previous works.

The estimation of mechanical variables, particularly the rotor speed and the electromagnetic torque by using the measured electrical quantities such as stator currents and voltages, are other interesting non-invasive methods which avoid installation changes. The observers can be used as well in order to minimize static estimation errors but they are not

---

This work is supported by the Regional Council of Picardie, Amiens, France. Shahin Hedayati Kia, Humberto Henao and Gérard-André Capolino are with the Department of Electrical Engineering, University of Picardie "Jules Verne," 80039 Amiens Cedex 1, France (e-mails: Shahin.Hedayati.Kia@u-picardie.fr; Humberto.Henao@ieee.org; Gerard.Capolino@ieee.org).

978-1-4799-0024-4/13 $31.00 © 2013 IEEE

vital in condition monitoring applications, because in most of cases the fault-related frequencies should be evaluated in the ac part of estimated variables. Both estimators and observers have been used for mechanical faults detection in induction machines [10]-[11]. The electromagnetic torque estimation has shown its performances in case of loss of lubrication fault detection for a gear-based motor drive [12] and chipped tooth and gear surface wear faults in a reduced-scale railway traction system [10]. The tooth breakage fault in high-ratio gear in cement kiln drives has been detected by means of this analysis [13]. It has been shown how the torque periodical impulse excitations from tooth breakage fault influence the stator current and consequently creates additive frequencies in the electromagnetic torque.

This paper introduces the concept of fault profile analysis for a gear fault detection. The fault profile appears in the torque due to the occurrence of any tooth localized fault in gears. This signature is periodic with a specific time interval related to the rotation frequency corresponding to the fault location (pinion side or wheel side for a one-stage gear). Since any periodic function can be represented by its Fourier series, it is possible to show that each particular signature produces fault-related frequencies with magnitudes corresponding to the Fourier series coefficients in the stator current spectrum and hence harmonics explicitly integer multiple of rotation frequencies in the electromagnetic torque spectrum. This property leads to an early stage non-invasive gear fault detection which was not previously investigated. This fact is verified for non-invasive gear tooth surface damage fault diagnosis. In this paper, a set-up based on a 250W three-phase squirrel-cage induction machine connected to a single-stage gearbox has been used for experiments.

II. INDUCTION MACHINE ELECTRICAL SIGNATURE ANALYSIS (IMESA)

The tooth localized fault (Fig. 1) on the contrary to tooth distributed faults in gears, produces large impacts with the rotation frequency periodicity corresponding to the fault location in the vibration signal which is principally due to the transmission of the mechanical torque. This effect can be well identified in the torque signal. Fig. 2 shows the chipped tooth fault effect at the output stage of a reduced-scale railway traction simulator including a 1.2kW 220V/380V 50Hz, two-pair-pole induction machine and a two-stage gearbox with 3.77 gear transmission ratio and numbers of

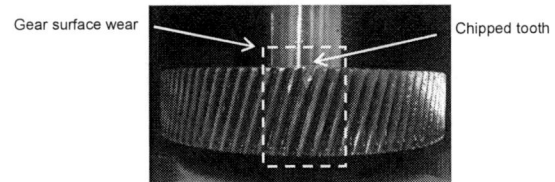

Fig.1. Chipped tooth and gear surface wear faults at the output wheel of a gearbox.

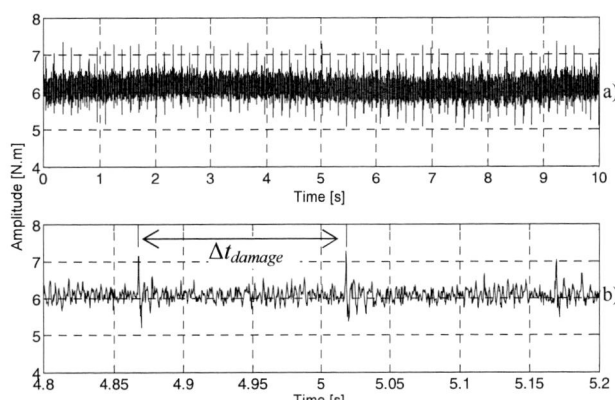

Fig. 2. Steady-state experimental result at rated-load for chipped tooth and gear surface wear faults: a) Torque signal - b) Zoom on some periods.

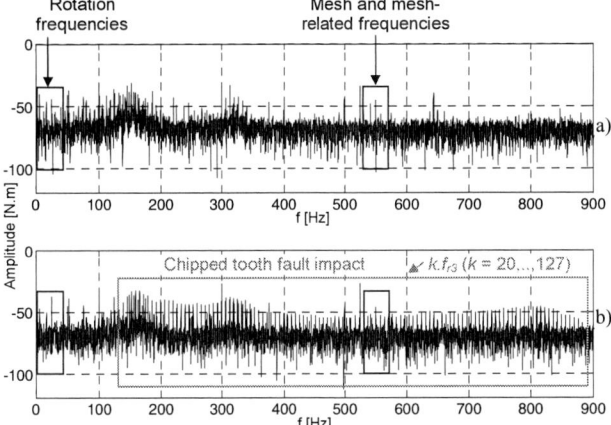

Fig. 3. Spectrum of torque at rated-load in the [0Hz, 900Hz] frequency bandwidth:
a) Healthy gear - b) Gear with chipped tooth and gear surface wear faults

wheel teeth N_1=21, N_2=31, N_3=38, N_4=97 [10]. The chipped tooth fault is clearly observed in form of pulses in the torque signal. The occurrence of these pulses depends on the output wheel rotating speed (f_{r3}=6.63Hz) and it can be computed with Δt_{damage}=1/($f_{r3}N_4$)=1.56ms for the induction machine rotation speed at 1500rpm. Fig. 3 shows the torque spectrum for healthy and chipped tooth fault conditions. The rotation frequencies in addition to mesh and mesh-related frequencies are detected in the torque spectrum of both healthy and chipped tooth gear fault. Moreover, the fault impact produces harmonics at integer multiple of output wheel rotation frequencies kf_{r3} which are concentrated in the [120Hz, 900Hz] frequency bandwidth with k=20,...,127. This effect is directly related to the periodic characteristic of the fault profile and its shape in each time period (Δt_{damage}). This observation leads to a proposal for a new formulation of the mechanical load torque experimented by the induction machine in order to predict fault-related frequencies in the stator current for the detection of a tooth gear localized fault. Based on this last observation (Fig. 3.a), the load torque observed by an induction machine coupled to a healthy gear

$T_{LH}(t)$, consists in a mean value and a time-varying part and it can be written as [7]:

$$T_{LH}(t) = T_0 + T_{osc}(t) \tag{1}$$

where T_0 is the average load torque, $T_{osc}(t)$ is the load torque oscillation related to the healthy gear namely both pinion and wheel rotations and mesh frequencies as it has been previously detailed [7]. The fault impact generates an extra periodic component (Fig. 3.b) in addition to the healthy condition. Therefore, the load torque in faulty condition $T_{LF}(t)$ can be formulated as:

$$T_{LF}(t) = T_{LH}(t) + T_{fp}(t) \tag{2}$$

where $T_{fp}(t)$ is a periodic component produced by the gear tooth localized fault. Since any periodical signal can be represented by its Fourier series, the $T_{fp}(t)$ term can be rewritten as:

$$T_{fp}(t) = \sum_{k=-\infty}^{+\infty} C_k e^{jk\omega_{fp}t} \tag{3}$$

$$C_k = \frac{1}{\tau_{fp}} \int_{-\tau_{fp}/2}^{\tau_{fp}/2} T_{fp}(t) e^{-jk\omega_{fp}t} dt$$

$$\omega_{fp} = 2\pi f_{fp}$$

where f_{fp} is the frequency of the fault profile, τ_{fp} is the period of the fault profile and C_k are the Fourier series coefficients.

The stator current can be represented by using only the main frequency components associated to the torsional vibrations induced by a healthy gear specifically rotation, mesh and mesh-related frequencies. Thus, these last frequency components, which are mainly the gear characteristic frequencies, can be detected around the supply frequency in the stator current spectrum [8].

In case of a gear tooth localized fault, $T_{fp}(t)$ appears in the load torque. Considering the same concept for fault-related frequencies since the fault profile can be replaced by its Fourier series, the stator current of an induction machine in gear tooth localized fault condition can be represented by:

$$I_{faulty}(t) = I_{healthy}(t) + \sum_{n=1}^{\infty} I_{l,n} \cos\left(\left(\omega_s - n\omega_{fp}\right)t - \varphi_{l,n}\right)$$

$$+ \sum_{n=1}^{\infty} I_{r,n} \cos\left(\left(\omega_s + n\omega_{fp}\right)t - \varphi_{r,n}\right) \tag{4}$$

where $I_{healthy}(t)$ is the stator current in healthy condition, $I_{l,n}$, $I_{r,n}$ and $\varphi_{l,n}$, $\varphi_{r,n}$ are magnitudes and phases of left and right side-band fault frequencies respectively with $n=1,2,3,\ldots$. The magnitude of each component is proportional to the coefficient C_k based on the modulation index previously mentioned [7]. It should be emphasized that the shape of the fault profile determines coefficients of Fourier series C_k. Thus, fault frequency components in the stator current can be formulated as:

$$f_{faulty} = \left| f_s \pm n f_{fp} \right| \tag{5}$$

where f_s is the supply frequency.

III. FAULT PROFILE RECONSTRUCTION

The fault-related frequencies in the stator current spread in a wide frequency range which makes their tracking difficult. It has been demonstrated that the frequency components of the electromagnetic torque, the instantaneous frequency of the stator current space vector and the load torque are similar [9], [10]. Thus, these last methods can be used to simplify the fault diagnosis process. Moreover, it is possible to reconstruct the periodic fault profile in time domain through a simple analysis on the spectrum (Fig. 4). This technique computes initially the discrete Fourier transform (DFT) of a signal. The obtained spectrum is multiplied by a vector with zeros at all elements except for kf_{fp} frequencies (elements equal to 1) which are the frequencies of interest. The computation of inverse Fourier transform of this multiplication reconstructs the fault profile in the time domain. It is obvious that the input should be a stationary signal in order to obtain correct results. The energy of the reconstructed signal E_{fp} can be used as a criterion for fault detection:

$$E_{fp} = \sqrt{\frac{1}{N_w} \sum_{i=1}^{N_w} x^2[i]} \tag{6}$$

where N_w is the number of data points in a window and $x[\cdot]$ is the reconstructed fault profile. This last profile can be achieved from the measured load torque, the instantaneous frequency of the stator current space vector or the estimated electromagnetic torque. It should be noted that the fault detection based on the energy computation of reconstructed fault profile needs all the three stator-phase currents for stator current space vector instantaneous frequency computation and needs all the three stator-phase currents and voltages for electromagnetic torque estimation respectively.

IV. EXPERIMENTAL RESULTS

A. Experimental set-up

A 250W, 50Hz, 400V, star-connected, 0.77A, 4-pole, 1380rpm, 24 rotor bars, three-phase squirrel-cage induction machine is connected to a digital controllable brake through a one-stage gear with numbers of teeth at the input $N_{r1}=25$ and at the output $N_{r2}=75$ (Fig. 5). f_{r1} and f_{r2} represent the rotation frequencies at the input and output stages of the gearbox, respectively. The digital controllable brake system can simulate the load by keeping the rotation speed constant at the output stage of the gearbox through a pulley-belt transmission system. The instrumentation consists in three current sensors with the same 0.1V/A sensitivity and three voltage sensors with the same 1/200 transformation ratio. In

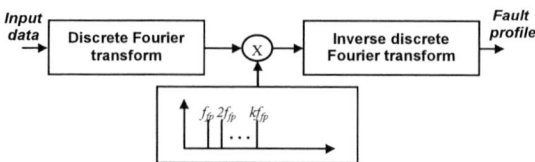

Fig. 4. Reconstruction of the fault profile in time domain.

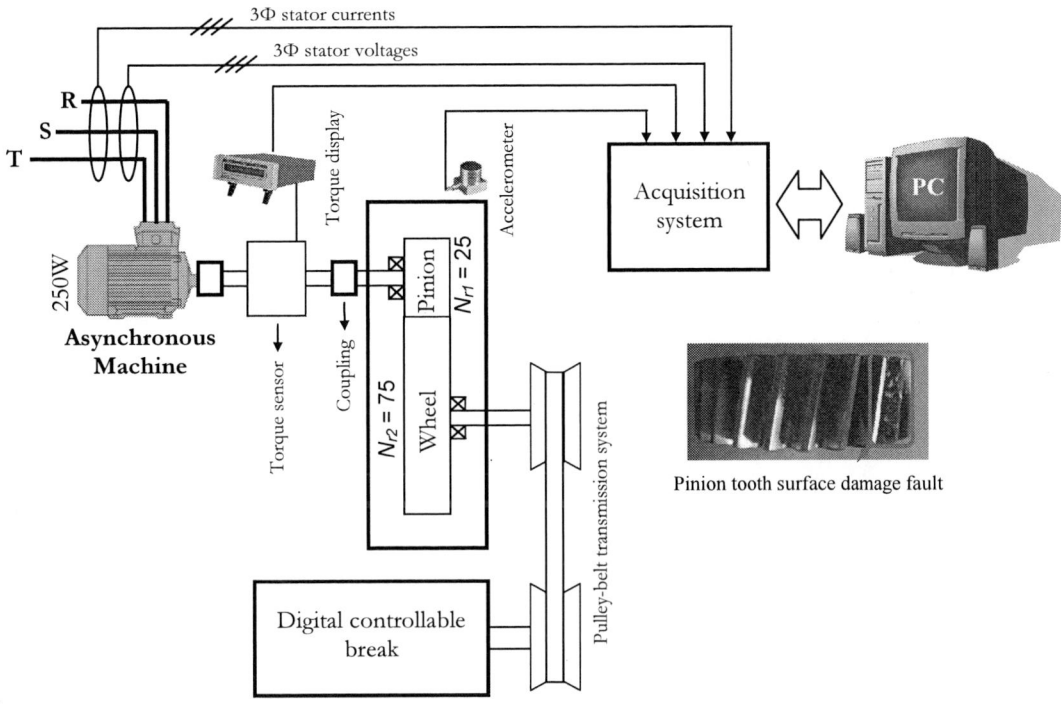

Fig. 5. Schematic of the proposed set-up.

addition, an accelerometer with a 500mV/g sensitivity is installed near the gearbox input stage. Besides, a torque sensor with 5kHz frequency bandwidth has been implemented between the induction machine shaft and input stage of the gearbox. For data collection, a 24-bit resolution modular data acquisition system with built-in signal conditioning filter has been used. The acquisition time has been chosen with T_{acq}=60s. However, in order to get a stationary signal only the first 10s of acquisition time has been used for the fault profile reconstruction. The power spectrum density (PSD) using the Welch method has been also used to minimize the PSD variance. The averaged PSD of the stator current has been normalized to the magnitude of the fundamental frequency component. Finally, the squared value of the averaged PSD (SPSD) has been computed in order to fit the magnitude scale between -120dB and 0dB. The Hanning window has been used as a window function for the Welch method. The electromagnetic torque estimation is proposed in the stationary reference frame with only stator voltage and current measurements and the stator resistance estimation (R_s=43.11Ω) [10]. The instantaneous frequency of the stator current space vector is also computed using three-phase current measurements [9]. The faulty gear mentioned later has been related to the pinion tooth surface damage fault in the proposed set-up.

B. Results analysis

The results of experimental tests for the induction machine working at rated-load have been described in this section. The rotation frequencies in this last case are f_{r1}=23.1Hz and f_{r2}=7.7Hz. The frequency of the fault profile is f_{rp}=f_{r1} since the pinion tooth surface damage fault has been

studied. In steady-state condition, the vibration has been shown (Fig. 6). The fault signature on the vibration is clearly identified with $1/f_{r1}$ period due to the pinion tooth damage fault because the accelerometer is close to the fault location. This last signature can be also observed in the measured torque with some time delay (Fig. 7). The measured torque of the healthy gear includes several harmonics and mostly dominated by the torsional vibration induced by the gear characteristic frequencies and mechanical system structure. In addition to these last frequencies, a periodic profile appears in the measured torque in case of a pinion tooth surface damage fault (Fig. 7.b). It has been verified but not shown that the spectrum of the measured torque includes kf_{r1} frequency components with significant magnitudes at

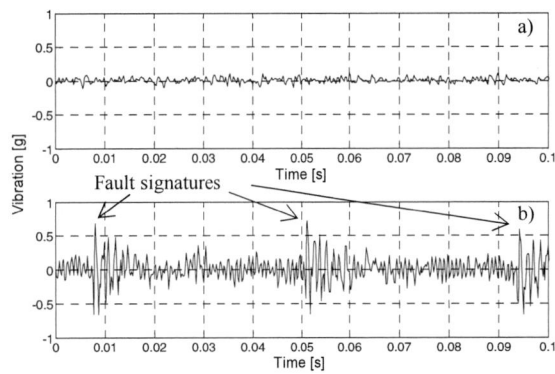

Fig. 6. Vibration in the steady-state working condition at rated-load:
a) Healthy gear - b) Faulty gear.

k=5,...,34. Thus, the reconstruction algorithm can be implemented in order to extract the fault profile from the measured torque (Fig. 8). It should be noted that this algorithm determines the average variations of the fault signature and it draws a stationary result. The comparison between two cases (Fig. 8.a and b) shows the presence of the fault profile due to the pinion tooth surface damage fault. The fault-related frequencies can be observed in the PSD of the stator current based on (5). Within all frequency components which are used for fault profile reconstruction (kf_{r1} with k=5,...,34) some of them are well identified in the stator current spectrum particularly 88.6Hz and 188.6Hz ($|f_s \pm 6f_{r1}|$), 111.7Hz and 211.7Hz ($|f_s \pm 7f_{r1}|$), 134.8Hz and 234.8Hz ($|f_s \pm 8f_{r1}|$), 157.9Hz and 257.9Hz ($|f_s \pm 9f_{r1}|$), 181Hz and 281Hz ($|f_s \pm 10f_{r1}|$) which are all concentrated in [80Hz, 300Hz] frequency bandwidth ($|f_s \pm kf_{r1}|$ with k=6,...,10) and which are illustrated in Figs. 9 and 10. The magnitude sensitivities of fault-related frequencies in the stator current spectrum are also presented in Table I. The reconstruction method can be applied to quantities such as the instantaneous frequency of the stator current space vector and the estimated electromagnetic torque which have similar frequency components. Figs. 11 and 12 demonstrate the extracted fault profile from these two last quantities. The energy criterion (6) has been used as a fault index and the results are presented in Table II. It can be concluded that the instantaneous frequency of the stator current space vector

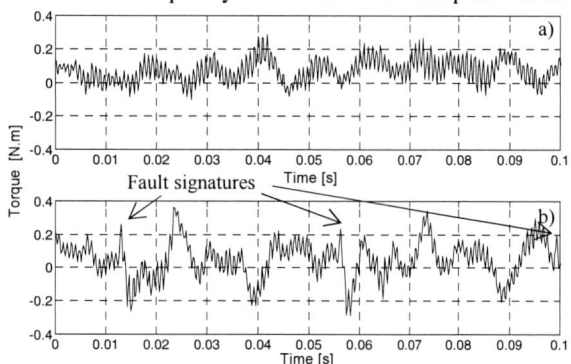

Fig. 7. AC part of the measured torque at rated-load:
a) Healthy gear - b) Faulty gear.

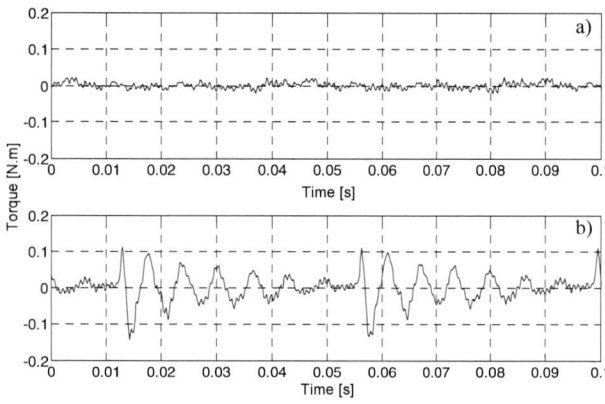

Fig. 8. Reconstruction of fault profile from the measured torque at rated-load:
a) Healthy gear - b) Faulty gear.

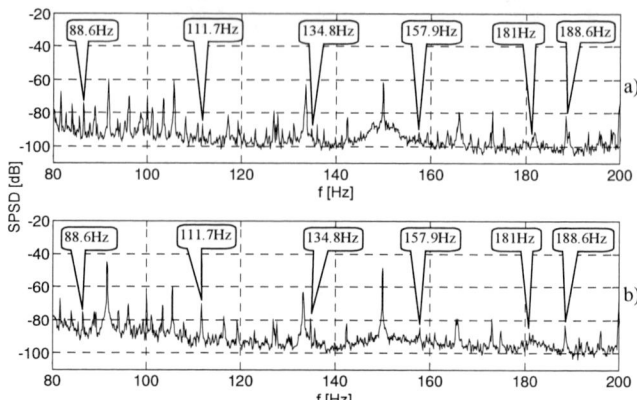

Fig. 9. SPSD of the normalized stator current at rated-load in the frequency bandwidth [80Hz, 200Hz]:
a) Healthy gear - b) Faulty gear.

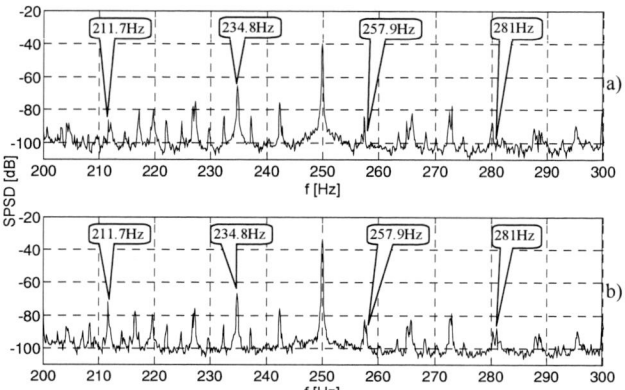

Fig. 10. SPSD of the normalized stator current at rated-load in the frequency bandwidth [200Hz, 300Hz]:
a) Healthy gear - b) Faulty gear.

gives a comparable fault sensitivity to the measured torque for pinion tooth surface damage fault diagnosis. The same energy value has been computed for the measured torque and the instantaneous frequency of the stator current space vector. Hence, the fault energy is concentrated in frequencies kf_{r1} with k=6,...,10 of these two last signatures.

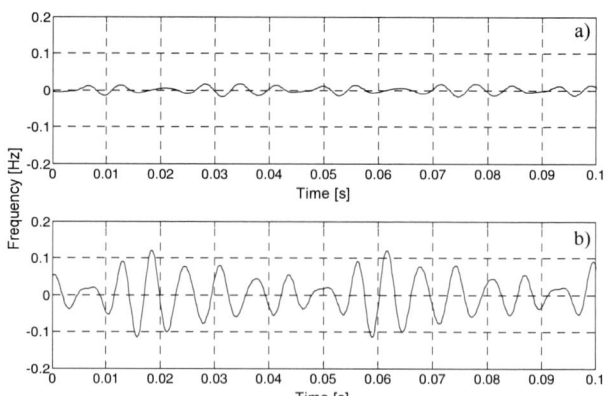

Fig. 11. Reconstruction of fault profile from the instantaneous frequency of the stator current space vector at rated-load:
a) Healthy gear - b) Faulty gear.

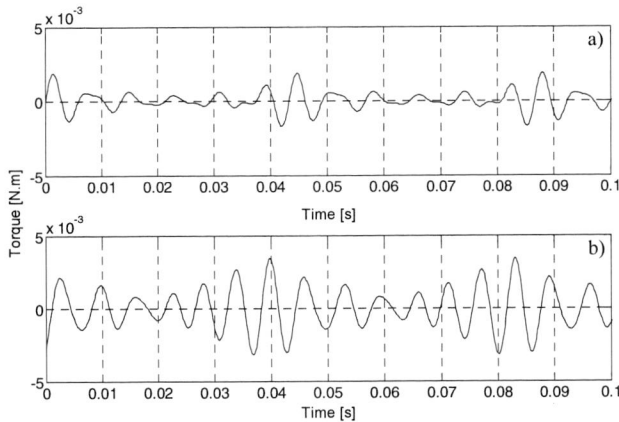

Fig. 12. Reconstruction of fault profile from the estimated electromagnetic torque at rated-load:
a) Healthy gear - b) Faulty gear.

TABLE I

MAGNITUDE SENSITIVITY OF FAULT-RELATED FREQUENCIES IN THE STATOR CURRENT SPSD FOR PINION TOOTH SURFACE DAMAGE FAULT (dB).

	$|f_s \pm 6f_{rI}|$	$|f_s \pm 7f_{rI}|$	$|f_s \pm 8f_{rI}|$	$|f_s \pm 9f_{rI}|$	$|f_s \pm 10f_{rI}|$
Stator current	5	19	3	12	9

TABLE II

ENERGY COMPUTATION OF RECONSTRUCTED FAULT PROFILE.

	E_{fp}	
	Healthy gear	Faulty gear
Instantaneous frequency of stator current space vector	0.0089	0.0524
Estimated electromagnetic torque	0.0006	0.0015
Measured torque	0.0052	0.0393

V. CONCLUSION

In this paper, a non-invasive technique for fault diagnosis of a pinion tooth surface damage fault has been presented. The main idea is based on the fact that in faulty condition an extra frequency component appears in the load torque due to the presence of fault in addition to initial healthy condition frequencies. This periodic signature can be represented by its Fourier series and hence spread in frequency domain in the spectrum of the stator current spectrum. It has been also shown that, by using a simple algorithm, it is always possible to reconstruct the fault profile from the measured torque, the instantaneous frequency of the stator current space vector and the electromagnetic torque. The proposed energy criterion proves its effectiveness for a non-invasive fault diagnosis of a pinion tooth surface damage fault detection in a one-stage gear system driven by a three-phase induction machine.

VI. REFERENCES

[1] C.W. de Silva, Vibration: Fundamentals and Practice, Boca Raton, FL: CRC Press, 2000.

[2] J. Ma and C.J. Li, "Gear defect detection through model-based wideband demodulation of vibration," *Journal of Mechanical Systems and Signal Processing*, vol. 10, no. 5, pp. 653-665, 1996.

[3] J.H. Kuang and A.D. Lin, "Theoretical aspects of torque responses in spur gearing due to mesh stiffness variation," *Mechanical Systems and Signal Processing*, vol. 17, no. 2, pp. 255–271, March 2003.

[4] S. Jia, and I. Howard, "Comparison of localised spalling and crack damage from dynamic modelling of spur gear vibrations," *Mechanical*

Systems and Signal Processing, vol. 20, no. 2, pp. 332–349, Feb. 2006.

[5] W. Wang, "Early detection of gear tooth cracking using the resonance demodulation technique," *Journal of Mechanical Systems and Signal Processing*, vol. 15, no. 5, pp. 887-903, 2001.

[6] A.R. Mohanty, C. Kar, "Monitoring gear vibrations through motor current signature analysis and wavelet transform," *Mechanical Systems and Signal Processing*, vol. 20, no. 1, pp. 158-187, January 2006.

[7] S.H. Kia, H. Henao, G.-A. Capolino, "Analytical and experimental study of gearbox mechanical effect on the induction machine stator current signature," *IEEE Transactions on Industry Applications*, vol. 45, no. 4, pp. 1405-1415, July/Aug. 2009.

[8] S.H. Kia, H. Henao, G.-A. Capolino, "A comparative study of acoustic, vibration and stator current signatures for gear tooth fault diagnosis," in *Proc. of ICEM'2012*, Marseille (France), 2-5 Sept. 2012, pp. 1512-1517.

[9] B. Trajin, M. Chabert, J. Regnier, J. Faucher, "Hilbert versus Concordia transform for three-phase machine stator current time-frequency monitoring," *Mechanical Systems and Signal Processing*, vol. 23, pp. 2648-2657, 2009.

[10] H. Henao, S.H. Kia, G.-A. Capolino, "Torsional vibration assessment and gear fault diagnosis in railway traction system," *IEEE Transactions on Industrial Electronics*, vol. 58, no. 5, May 2011.

[11] J. Guzinski, M. Diguet, Z. Krzeminski, A. Lewicki, H. Abu-Rub, "Application of speed and load torque observers in high speed train drive for diagnostic purpose," *IEEE Transactions on Industrial Electronics*, vol. 56, no. 1, pp. 248–256, Jan. 2009.

[12] S.H. Kia, H. Henao, G.-A. Capolino, "Torsional vibration effects on induction machine current and torque signatures in gearbox-based electromechanical system," *IEEE Transactions on Industrial Electronics*, vol. 56, no. 11, pp. 4689-4699, Nov. 2009.

[13] I. Bogiatzidis, A. Safacas, E. Mitronikas, "Detection of backlash phenomena appearing in a single cement kiln drive using the current and the electromagnetic torque signature," to be published *IEEE Transactions on Industrial Electronics*, 2013.

VII. BIOGRAPHIES

Shahin Hedayati Kia received the M.Sc. in electrical engineering from the *Iran University of Science and Technology* (IUST), Tehran, Iran, in 1998 and the M.Sc. and the Ph.D. in power electrical engineering from the *University of Picardie "Jules Verne"*, Amiens, France, in 2005 and 2009, respectively. From 2008 to 2009, he was a lecturer at *INSSET de Saint-Quentin*, France. From September 2009 to September 2011 he was a post-doctoral associate at the School of Electronic and Electrical Engineering of Amiens (ESIEE Amiens). In September 2011, he joined the *University of Picardie "Jules Verne"* as an Associate Professor in the Department of Electrical Engineering. His research interests include application of modern digital signal processing in electrical power systems and diagnosis of electrical machines.

Humberto Henao (M'95, SM'05) received the M.Sc. in electrical engineering from *Universidad Tecnologica de Pereira*, Colombia in 1983, the MSc in power system planning from *Universidad de los Andes*, Bogota, Colombia in 1986, the Ph.D. in electrical engineering from *Institut National Polytechnique de Grenoble*, Grenoble, France in 1990. From 1987 to 1994, he was consultant for companies as *Schneider Industries* and *GEC Alstom* in the Modeling and Control Systems Laboratory (UMCS), *Mediterranean Institute of Technology*, Marseille, France. In 1994, he joined the *Ecole Supérieure d'Ingénieurs en Electrotechnique et Electronique*, Amiens, France as Associate Professor. In 1995, he joined the *University of Picardie "Jules Verne"*, Amiens, France as an Associate Professor in the Department of Electrical Engineering. He was promoted Professor of Electrical Engineering within the same University in 2010.

He is currently the Department representative for the international programs and exchanges (SOCRATES). He is also the power group leader within the Laboratory of Innovative Technologies (LTI) of the *University of Picardie "Jules Verne"*. Dr. Henao main research interests are modeling, simulation, monitoring and diagnosis of electrical machines and electrical drives.

Gérard-André Capolino (A'77, M'82, SM'89, F'02) was born in Marseille, France. He received the B.Sc. in electrical engineering from *Ecole Centrale de Marseille*, Marseille, France in 1974, the M.Sc. from *Ecole Supérieure d'Electricité*, Paris, France in 1975, the Ph.D. from University Aix-Marseille I, Marseille, France in 1978 and the D.Sc. from

Institut Polytechnique de Grenoble, Grenoble, France in 1987. In 1994, he joined the *University of Picardie "Jules Verne"* in Amiens, France as a Full Professor and he is now Director of the European Master in Advanced Power Electrical Engineering (MAPEE).

His recent research interests have been focussed on fault tolerant control of multiphase induction machines and on condition monitoring and fault detection of AC electrical machinery. He has published more than 450 papers in scientific journals and conference proceedings since 1975. He is Associate Editor of *IEEE Transactions on Power Electronics* and of *IEEE Transactions on Industrial Electronics*. He is also the acting chair for the Steering Committee of the International Conference on Electrical Machines (ICEM). He is also the President of IEEE Industrial Electronics Society (IES) for 2012-20013. Dr. Capolino has been the recipient of the 2008 IEEE-IES Dr.-Ing. Eugene Mittelmann Achievement Award, the 2010 ICEM Arthur Ellison Achievement Award, and the 2011 IEEE-PELS Diagnostics Achievement Award.

Sensorless Speed Estimation and Diagnosis of Induction Motors Based on Purified Space Vectors

Dongfeng Shi
Optimized Systems and Solutions
Rolls-Royce Group
Derby, UK
Email: dongfeng.shi@o-sys.com

Abstract— A new scheme based on the demodulation technique and interpolated fast Fourier transform (IFFT) is proposed to construct purified space vector for the purpose of sensorless speed measurement and fault diagnosis for induction motors. The phase and amplitude fluctuation associated with the eccentricity harmonics is demodulated first. Interpolated fast Fourier transform is then performed to estimate the speed of induction motors running under constant speed and transient conditions. Experimental studies have further confirmed good agreement between the estimated motor speed using the proposed scheme and the measured speed using an encoder. The effectiveness of proposed purified space vector has been further demonstrated in the detection of coupling misalignments and stator faults.

I. Introduction

Three phase induction motors are the most popular equipment types for prime movers globally and are widely deployed in all industrial sectors. Global competitions have stimulated industrial plants to reduce production losses and breakdown cost caused by the failures within prime mover systems. An industrial motor, beyond its prime function as an actuator, may also function effectively as an intelligent sensor by embedding data acquisition and signal processor. The motor current signature analysis (MCSA), as part of the preventive maintenance (PM) program has been widely employed to enhance safety, efficiency, reliability, availability, and longevity in manufacturing processes and proved highly cost effective. The common faults, i.e. bearing, stator, rotor and eccentricity related conditions, and corresponding diagnosis technologies have been reviewed [1]. Additionally, to realize robust and high-precision control of induction motors, sensorless speed estimation has received more and more attention in recent years [2].

Although several signal processing techniques, i.e. fast Fourier transform (FFT), envelope analysis, space vectors and wavelet transform, have been proposed to conduct sensorless speed measurement and diagnosis, the deployment of such monitoring systems in a shop floor environment is still rare due to some inherent drawbacks within traditional motor current signature analysis procedures. Firstly, the amplitude spectrum and phase spectrum can be obtained through FFT of the motor current signal. Unfortunately, the latter is always overlooked in application. If the phase information in motor current analysis is ignored, a considerable degree of information about motor behaviour will be lost. Secondly, the traditional spectral analysis method is unable to express the relationship between the motor current signals acquired from different phases. A multi-sensor fusion strategy may be explored to combine motor current signals from each of the three phases to pinpoint the actual faults within induction motors. Finally, in addition to the specific features of malfunction, the original space vector [3] contains plenty of interference components and noise. It is not feasible to diagnose the malfunctions by using the original space vector directly [2]. An efficient tool is required to extract definitive features of malfunction from the original space vector. This paper aims at overcoming the shortcomings in traditional signal processing and feature extraction techniques, and provides a novel scheme to purify and extract features from the original space vector for sensorless speed measurement to enhance diagnosis for induction motors. The paper is organized as follows. The motor current signal in frequency domain will be characterized in Section 2. Section 3 introduces a new space vector purification technique based on a high-resolution spectrum combined with demodulation procedure to accurately calculate frequency, phase and amplitude of all frequency components in the motor current signal. In Section 4, the performance of proposed purified space vectors will be validated by several simulation results. The application of proposed scheme in sensorless speed measurement and fault diagnosis for induction motors are investigated in Section 5. Conclusions and recommendations are presented in the last section.

II. Eccentricity and Slot Harmonics

In general, the motor current signals are composed of supply frequency (50 or 60Hz) related component and other frequency components caused by imperfections in the electrical supply. The saliency harmonic is a typical speed–related component in the induction motor current caused by variations in air-gap permeance due to the rotor slotting or eccentricity. The rotor slotting or eccentricity-related saliency harmonics are modulated by the supply frequency of the stator current when the induction motor is running. Owing to the presence of the eccentricity and rotor slots, saliency harmonics appear at frequency f_h in the frequency spectrum of motor current, and f_h can be expressed as [2]:

978-1-4799-0024-4/13 $31.00 © 2013 IEEE

$$f_h = f_s\left(\left(kR + n_d\right)\left(\frac{1-s}{p}\right) + n_w\right) \quad (1)$$

$n_d=0$	for static eccentricity;
$n_d=1,2,3$	for dynamic eccentricity;
f_s	fundamental supply frequency(Hz);
R	number of rotor slots;
s	slip;
p	number of pole pairs;
k	any positive integer;
$n_w=1, 3, 5$	the air-gap magnetomotive force

(MMF) harmonic order.

It is obvious that the parameters that determine saliency harmonics are independent on changes in operational parameters, e.g. load and temperature. Since multiple slot harmonics are present in the motor current as illustrated in Eq. (1), several no-load tests have to be conducted to specify which harmonic component represents the real primary harmonic, which is characterized by its amplitude that is consistently the strongest among all the harmonics. Consequently, it is difficult to estimate the motor speed accurately in practice. However, eccentricity harmonics exist at any nonzero shaft speed and are independent of the number of slots. There are two types of air-gap eccentricity in induction motors: static air-gap eccentricity and dynamic air-gap eccentricity. Static eccentricity can occur due to incorrect positioning of the stator or rotor at the commissioning stage. Dynamic eccentricity can be generated from a bent rotor, worn bearings, or coupling misalignment. Taking only eccentricity into account, side-band eccentricity related components f_h will appear around the power supply frequency in the motor current spectrum, and Eq. (1) can be rewritten as:

$$f_h = f_s\left(1 \pm m\left(\frac{1-s}{p}\right)\right) \quad (2)$$

where m is the order of the eccentricity-related harmonic, which can be any positive integer.

In general, the harmonic that is consistently the strongest when $m=1$ can be selected as the primary eccentricity harmonic for motor speed estimation. Apparently, there exists a modulated relationship between the power supply component and the eccentricity harmonic. Since the amplitude of the power supply component is much higher than that of the primary eccentricity harmonic component, the latter is mostly masked in the motor current spectrum. Therefore, a reliable demodulation approach must be developed to accurately extract the eccentricity harmonic information from the motor current signals for the purpose of sensorless motor speed estimation.

III. PURIFIED SPACE VECTOR

A. Demodulation

The space vector [3] is an effective format to describe three-phase induction motor phenomena in the two-dimensional representation. As a function of three-phase motor currents (i_a, i_b, i_c), the space vector (i_d, i_q) can be expressed as

$$i_d = \sqrt{\frac{2}{3}}i_a - \frac{1}{\sqrt{6}}i_b - \frac{1}{\sqrt{6}}i_c$$

$$i_q = \frac{1}{\sqrt{2}}i_b - \frac{1}{\sqrt{2}}i_c \quad (3)$$

The space vector of motor currents in d-q domain is close to a circle if only balanced supply frequency component is considered. Due to the presence of noise interference, imperfections and speed-related harmonics, some fluctuations may occur in the space vector. Since the amplitude of speed-related harmonics is much smaller than supply frequency, the fluctuations due to modulation between them are not obvious. The original space vector can be expressed as the sum of several frequency components and noise component in d-q domain in Eq. (4)

$$\vec{I} = \sum A_n \sin\left(2\pi f_n t + \alpha_n\right) + j\sum B_n \sin\left(2\pi f_n t + \beta_n\right) + \vec{I}_{noise} \quad (4)$$

Where $j = \sqrt{-1}$, Furthermore, this original space vector can be purified and decomposed into positive and negative sequence components and expressed as:

$$\vec{I} = \sum\left(P_n e^{j2\pi f_n t} + P_{-n} e^{-j2\pi f_n t}\right) \quad (5)$$

where P_n and P_{-n} indicate the amplitude of positive and negative sequence components, respectively and can be calculated by

$$P_n = \sqrt{A_n^2 + B_n^2 + 2\left|A_n B_n \sin(\alpha_n - \beta_n)\right|} \quad , \quad (6)$$

$$P_{-n} = \sqrt{A_n^2 + B_n^2 - 2\left|A_n B_n \sin(\alpha_n - \beta_n)\right|} . \quad (7)$$

The amplitude, frequency and phase of supply frequency components can be calculated by the high-resolution spectrum technique as stated in Section 3.B. Additionally, since the amplitude of speed-related harmonics is much smaller than supply frequency component, it is not wise to calculate amplitude, frequency and phase of speed related harmonic components directly. Consequently, an efficient procedure must be introduced to demodulate the weak speed-related harmonic from the dominant supply frequency component. The straightforward demodulation approach is based on the amplitude fluctuations and can be expressed as

$$B = \sqrt{i_d^2 + i_q^2} \quad (8)$$

Hence, supply frequency component can be removed as a direct component (DC) and residual signal contains the speed-related harmonic component. Moreover, the phase envelope can be presented to extract speed related harmonic component through calculating the instantaneous phase angle in d-q domain as well. Obviously, instantaneous phase angle of space vector can be defined as

$$\phi(t) = \arctan(\frac{i_q}{i_d}) \qquad (9)$$

The time interval between adjacent points in the space vector will not be constant due to the presence of eccentricity harmonic component and can be expressed as

$$\Delta t = 2\pi \frac{f_c}{f_s} + \tau \qquad (10)$$

where f_s and f_c is the supply frequency and sampling frequency respectively, and τ denotes the phase fluctuation due to the presence of eccentricity harmonics.

The phase fluctuation τ_i at time instant i can be defined as the difference between instantaneous phase angle of motor current space vector and reference space vector, which rotates at uniform supply frequency, and can be expressed as

$$\tau_i = \phi_i - 2\pi \frac{f_c}{f_s} rem\left(i, \frac{f_c}{f_s} \right) \qquad (11)$$

where ϕ_i denotes the instantaneous phase angles and can be obtained by Equation (10), and $rem(i, f_c/f_s)$ denotes the remainder from the division of i by f_c/f_s. In order to reduce estimation error, the sample frequency is specified as an integral multiple of the supply frequency, e.g. 3600Hz.

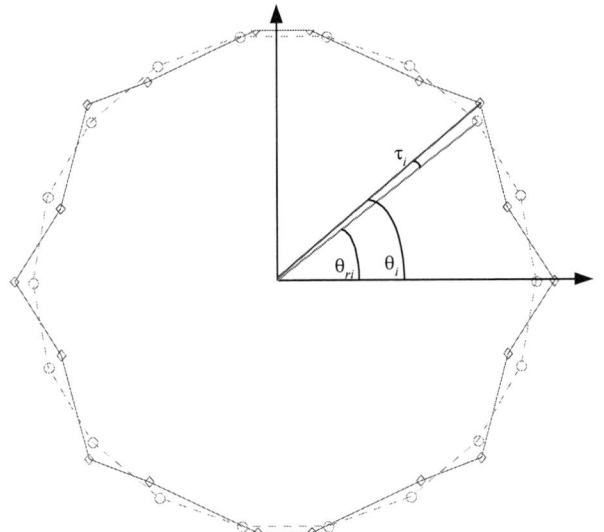

Figure 1 Amplitude and phase fluctuations in motor current in the presence of eccentricity harmonics (dashed line: reference signal only contains supply frequency; solid line: real signal with eccentricity harmonics)

Figure 1 shows the samples taken from the space vector in the presence of eccentricity harmonics, and samples taken from an ideal reference signal containing only the supply frequency. Phase angles of actual ϕ_i, reference ϕ_{ri}, and the fluctuation between them τ_i are marked clearly in Figure 1. The fluctuation of the space vector of three-phase motor current can be calculated using Equation (11) and a corresponding time

series can be obtained in time domain. The motor speed can be detected through Fourier transform of this time series into the frequency domain.

B. High-resolution Spectrum

The Fourier transform is a powerful signal-processing tool to analyse the composition of signals in frequency domain. However, due to signal truncation in time domain, leakage effects will appear in discrete Fourier spectra despite the use of windowing functions. In addition, an FFT spectrum is the result of a continuous spectrum sampled with a frequency interval (Δf). In general, the real spectrum line may not be located in the centre of the main-lobe, and the estimated frequency, amplitude and phase of the signal component are not accurate. This is the so-called comb-effect as illustrated in Figure 2. Only if the sampling frequency is an exact multiple times of the frequency of sinusoidal signal, will the estimated frequency be located at the centre of main-lobe and equal to the real frequency. Otherwise, this comb-effect leads to serious error in estimating the frequency, amplitude, and phase of a sinusoidal signal. If only a single sinusoidal signal is considered, the corresponding error of frequency, amplitude and phase of single sinusoidal signal can be estimated by [4]

$$f_e = \min \left(\left| k\Delta f - f \right|, \left| (k+1)\Delta f - f \right| \right) \qquad (12)$$

$$A_e = A_1 \cdot (W(f_e) - 1) \qquad (13)$$

$$\alpha_e = \frac{\pi f_e}{\Delta f} \qquad (14)$$

where, W(f) is the Fourier transform of any window function used. It is obvious that the errors in calculated frequency, amplitude and phase can be large. The frequency error can be up to Δf, the resolution of the spectrum. Therefore, a new high-resolution spectrum based on interpolation is introduced to estimate the precise frequency, amplitude and phase of the sinusoidal signal in this section. Given N sample values of the single sinusoidal signal, $x(0), x(1) \cdots x(N-1)$, the discrete Fourier spectrum can be calculated by

$$X(k) = \frac{1}{N} \sum_{n=0}^{N-1} x(n) e^{-j2\pi nk/N} \qquad (15)$$

Adjacent peaks can be detected at y_k and y_{k+1} as stated in Figure 2, and the corresponding frequency can be denoted as $k\Delta f$ and $(k+1)$ Δf respectively. Due to the symmetry of window function, the following equation can be obtained

$$\frac{y_k}{y_{k+1}} = \frac{W(\delta)}{W(\Delta f - \delta)} \qquad (16)$$

Where δ is the distance between the right frequency and the correct signal frequency. The corrected frequency of sinusoidal signal can be obtained by

$$f_0 = (k+1)\Delta f - \delta \qquad (17)$$

The corresponding amplitude and phase of sinusoidal signal can be calculated respectively by

$$A = \frac{y_{k+1}}{W(\delta)} \qquad (18)$$

$$\alpha = \tan^{-1}\left(\frac{I_{k+1}}{R_{k+1}}\right) + \delta\pi \qquad (19)$$

Finally, if the value of δ can be obtained, the corrected frequency, amplitude and phase information can be calculated using Equation (17)-(19). The value of δ depends on the type of window function. In this work, since the Hanning window is employed to calculate the FFT of signal, δ can be estimated from

$$\delta = \frac{2y_{k-}y_{k+1}}{y_{k+1} + y_k} \qquad (20)$$

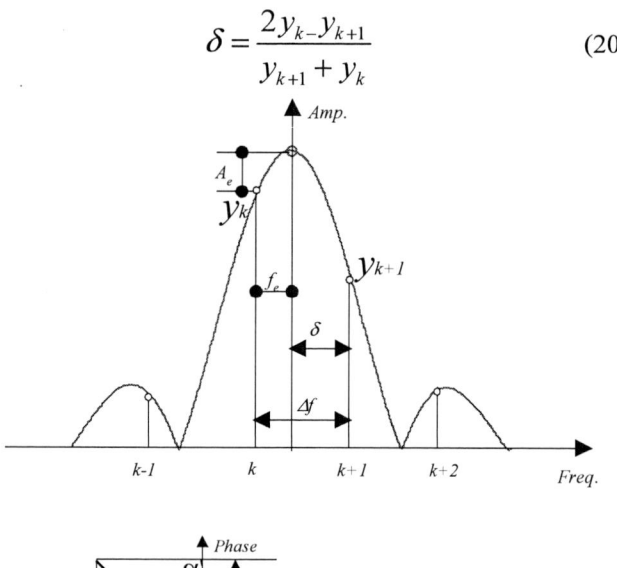

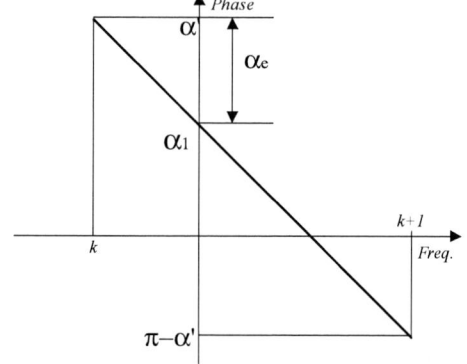

Figure 2 Amplitude, frequency and phase errors in the normal FFT

IV. SIMULATION RESULTS

To prove this novel scheme is effective in the detection of motor speed when the induction motor is operating under transient conditions, such as starting acceleration or shut-down processes, the simulated motor current was generated using the induction motor model reported in [5]. Figure 3 shows waveform of motor current during starting acceleration. Since the amplitude of motor current increases during acceleration, the amplitude fluctuations were almost obscured. Therefore, it is not feasible to estimate motor speed using amplitude fluctuations. Fortunately, the phase fluctuations are unaffected,

and can be extracted by demodulation procedure as shown in Figure 4.

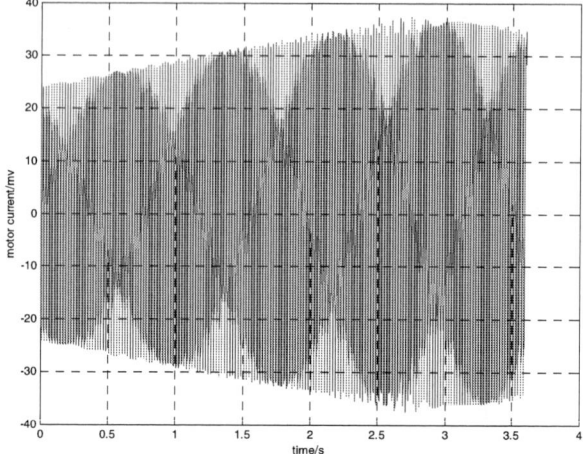

Figure 3 Waveform of motor current in run-up stage

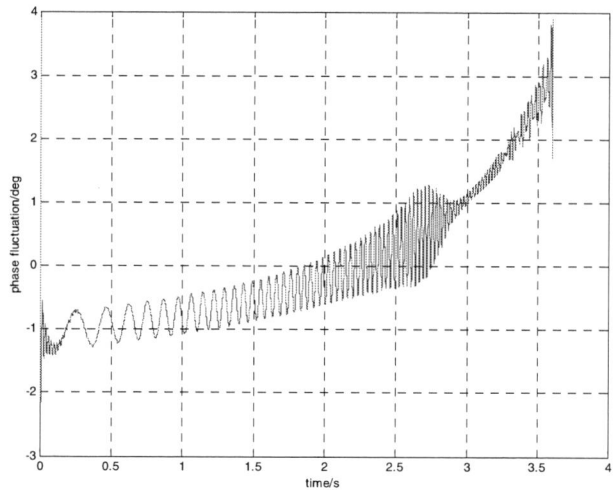

Figure 4 Phase fluctuations demodulated by in run-up stage

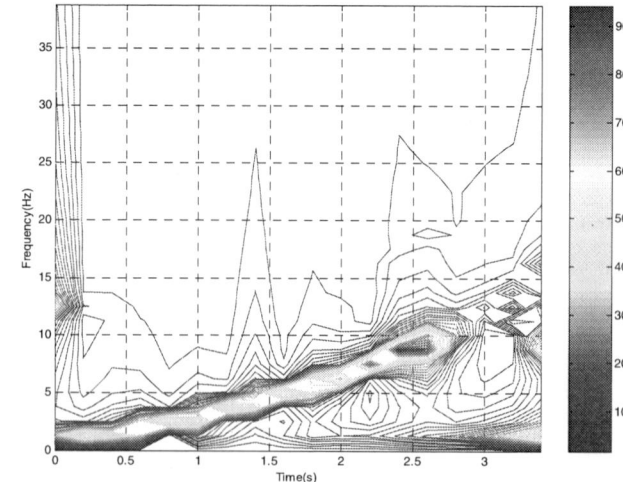

Figure 5 Short time Fourier transform of Phase fluctuations in run-up stage

978-1-4799-0024-4/13 $31.00 © 2013 IEEE

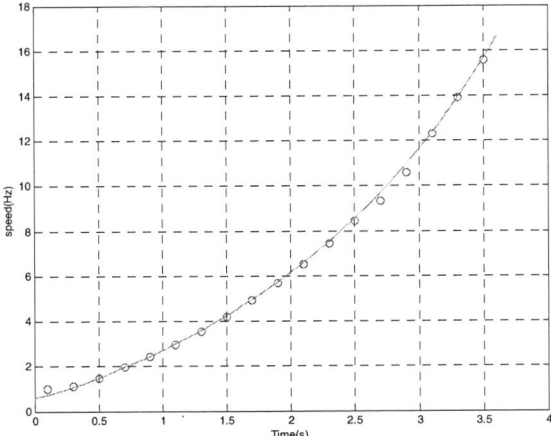

Figure 6 Comparison between speeds estimated by novel scheme (o) and calculated by model (solid line)

Obviously, the phase fluctuations contain the speed information, which can be proved further by taking the short time Fourier transform (STFT) of the data as shown in Figure 5. It can be seen that given motor accelerated from 0 to 15Hz. However, since the frequency resolution is extremely low in STFT, the high-resolution spectrum has to be further introduced to estimate the accuracy speed of induction motor. The speeds estimated after high-resolution enhancement are shown in Figure 6. The estimated speed agrees very well with the real speed from the motor simulation model.

V. EXPERIMENTAL RESULTS

To evaluate the performance of the proposed scheme, several experiments were conducted using an ABB motor, under constant speed and during acceleration. An optical incremental encoder with 1,024 pulses per revolution was coupled to the motor shaft for direct, comparative measurement of the motor speed. The experiments were conducted with a three-phase Variac transformer to vary the supply voltage and control the run-up processes of a pumping system. The data acquisition system was based on a Pentium 266MHz PC, fitted with an Amplicon PC30G 12-bit 100kHz plug-in card. Two sets of data were collected from the motor by HP VEE software package and processed by the Matlab program. A sixth-order analog Butterworth anti-aliasing filter with a cut-off frequency at 100Hz was employed to pre-process the current signal and remove interference frequency component, such as harmonics of the supply frequency. The three-phase motor current signal and instantaneous angular speed signal were A/D converted and sampled at a rate of 6,600 Hz.

Table 1 illustrates the speed estimated by the demodulation approach based on the space vector and IFFT technique, compared with the speed measured by the encoder when the motor was running at constant speeds. It is seen that the speed obtained by the developed scheme is in good agreement with the speed measured by the shaft-encoder, for a speed as low as 1Hz. Phase fluctuations demodulated from motor current signals and further processed by STFT are shown in Figure 7.

Figure 8 shows the speed estimated by the sensorless schemes and measured by the encoder during the acceleration phase (from zero to operational speed). These results confirm that the proposed scheme was able to efficiently estimate the motor speed during the transient phase.

Table 1 Comparison of sensorless motor speed estimates with encoder (Motor controlled by a Variac transformer at constant speeds)

	1	2	3	4	5	6	7
Encoder(Hz)	1.07	3.32	8.10	12.07	17.13	20.76	22.54
Estimate(Hz)	1.07	3.29	8.11	12.10	17.12	20.76	22.46

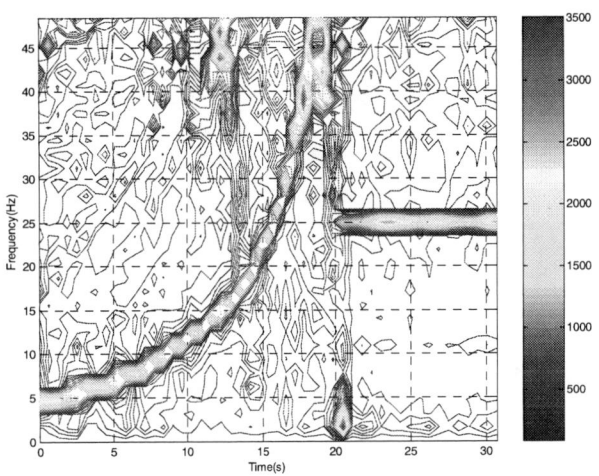

Figure 7 Short time Fourier transform of Phase fluctuations in run-up stage

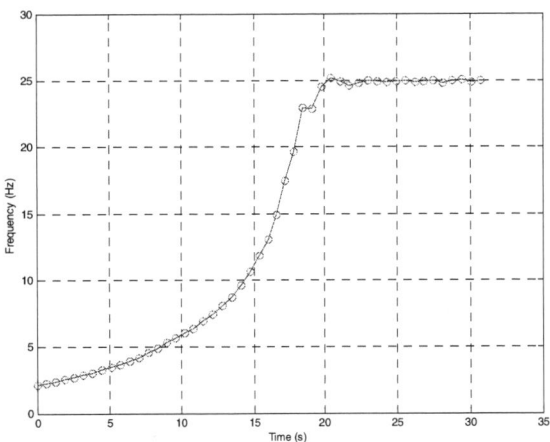

Figure 8 Comparison between speeds estimated by proposed scheme (o) and measured by encoder (solid line)

Coupling Misalignment is a very common failure in machinery and earlier detection of misalignment is helpful to maintain the performance of machinery. In order to prove the effectiveness of purified space vector scheme, further experiments have been conducted with the same motor. The three phase motor current signals have been acquired under conditions of zero misalignment, and of different level of angular and parallel misalignment. The accurate amplitude, frequency and phase information of eccentricity component caused by coupling misalignment were calculated by high-resolution spectrum. Then, the purified space vector was formed as an ellipse using these accurate amplitude, frequency and phase information of relative component. It is obvious that energy of corresponding purified space vector increased along with the worsening of misalignment as shown in figure 9 and 10.

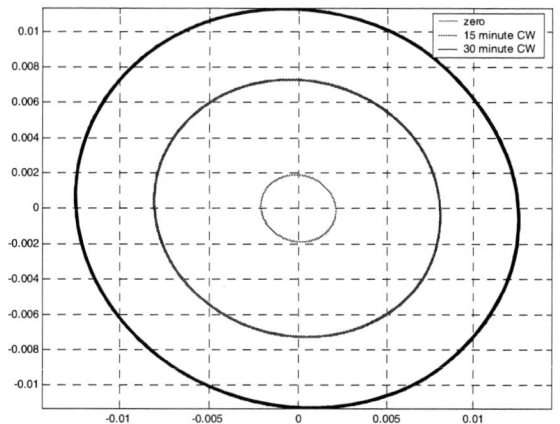

Figure 9 Purified space vectors of eccentricity components under different angular misalignments.

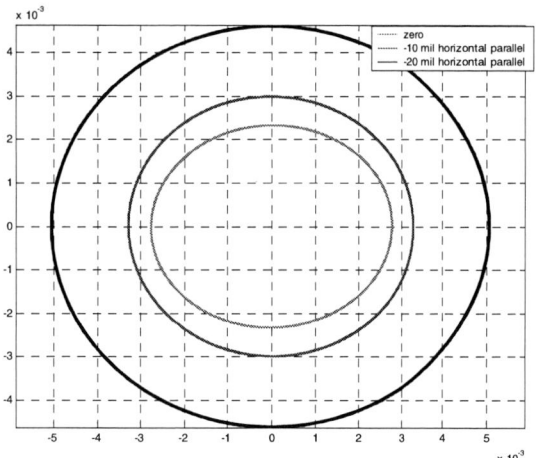

Figure 10 Purified space vectors of eccentricity components under different parallel misalignments.

Finally, another 2HP 4-hole Reliance PreAlert motor has been specially wound for testing smaller inter-turn faults. Stator faults were simulated through taps on the windings, which could be connected to short circuit two, three and four neighboring turns. The taps have been brought to a switch box where the number of shorted turns and phase can be selected. The accurate amplitude, frequency and phase of supply frequency component were calculated by high-resolution spectrum and corresponding purified space vectors are shown in figure 11. It can be seen that the length of major axis of purified space vector is changed with the onset of stator fault. It can be explained that the stator short circuit resulted in the presence of negative sequence component.

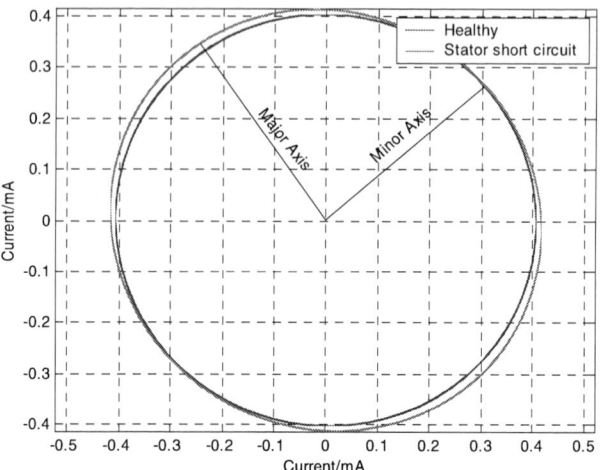

Figure 11 Purified space vectors of supply components with/without stator short circuit.

CONCLUSIONS

A new sensorless scheme, purified space vector, was proposed to estimate the speed and diagnose faults within induction motors using a demodulation approach combined with interpolated FFT. The proposed purified space vector has been proved effective in the estimation of speed of induction motors under low speed as well as during the speed transient periods. The performance of purified space vector has been further demonstrated in the diagnosis of faults within induction motors, e.g. coupling misalignment and stator short circuits. Furthermore, given that the demodulation and IFFT procedure can be efficiently implemented through PC or DSP programming, the proposed scheme has a low requirement on calculation time, e.g. only 0.58 ms is required to conduct a 1024-point real FFT on a Motorola floating-point DSP (model 96002). Application of the developed scheme to real-time speed estimation and fault diagnosis in motor drives will be investigated in future studies.

REFERENCES

[1] S Nandi, HA Toliyat and XD Li "Condition monitoring and fault diagnosis of electrical motors - A review" IEEE TRANSACTIONS ON ENERGY CONVERSION 20 (4): 719-729 DEC 2005
[2] P. Vas, Sensorless vector and direct torque control, Oxford University Press, UK, 1998
[3] SMA Cruz, AJM Cardoso "Stator winding fault diagnosis in three-phase synchronous and asynchronous motors, by the Extended Park's Vector Approach" IEEE TRANSACTIONS ON INDUSTRY APPLICATIONS 37 (5): 1227-1233 SEP-OCT 2001
[4] DF Shi, WJ Wang, L. S. Qu, "Purification and feature extraction of shaft orbits for diagnosing large rotating machinery, " Journal of Sound and Vibration, Vol. 279, pp. 581-600, 2005
[5] M Arkan, DK Perovic, and P Unsworth "Online stator fault diagnosis in induction motors," IEE Proc. Electrical Power Application. Vol. 148, pp537-547, No. 6, November, 2001

Bearing Faults Detection in Induction Machines Based on Statistical Processing of the Stray Fluxes Measurements

Ciprian Harlişca, Loránd Szabó, Lucia Frosini, Andrea Albini

Abstract -- **Frequent defects of induction machines are due to diverse bearing faults. The detection of such faults in their incipient phase can decisively contribute to the prevention of unplanned breakdowns in industrial plants. In this paper the detection of three types of bearing faults by means of statistical processing of the stray fluxes measurements is detailed. The developed noninvasive method requires only both simple probes and easy computations. Numerous measurements had been performed for all the combinations of bearing faults, loads and stray flux probes taken into study. All the results emphasized the effectiveness of the applied simple fault diagnosis method.**

Index Terms--**ac machines, ball bearings, electric machines, fault detection, fault diagnosis, induction motors, rotating machines.**

I. INTRODUCTION

Electrical machines with small faults can still work, but these faults will evolve in time and they may cause a complete breakdown. Preventive measures should be taken in order to protect the machines.

Faults can occur in any part of the machine. These can be of electrical or mechanical origin. The main electrical faults can be in the stator and rotor windings (or cage) [1], [2]. Mechanical faults include bearing faults, air-gap eccentricity, gearbox faults, misalignments, etc. The most of the failures (about 40%) are related to the bearings [3]. The bearing faults do not cause immediate breakdown, but they evolve in time until they produce a critical failure of the machine. Unfortunately these failures finally results both in costly repair costs and long downtimes.

The bearings faults can be caused by material fatigue, overheating, harsh environments, inadequate storage, contamination, corrosion, wrong handling and installation, unbalanced loads, bearing currents, etc. [4]. However the main cause of their failure is due to their poor lubrication, which can be easily avoided by a correct maintenance plan.

This paper was supported by the project "*Improvement of the doctoral studies quality in engineering science for development of the knowledge based society-QDOC*" contract no. POSDRU/107/1.5/S/78534, project co-funded by the European Social Fund through the Sectorial Operational Program Human Resources 2007-2013.

C. Harlişca and L. Szabó are with Department of Electrical Machines and Drives, Technical University of Cluj-Napoca, Romania (e-mails: Ciprian.Harlisca@mae.utcluj.ro, Lorand.Szabo@emd.utcluj.ro).

L. Frosini and A. Albini are with Department of Electrical, Computer and Biomedical Engineering, University of Pavia, Italy (e-mails: lucia@unipv.it, andrea.albini@unipv.it).

In the literature the bearing faults are classified according to:

- the location of the fault: inner race, outer race, balls, and cage;
- the fault signature: single-point defects and generalized roughness.

Generalized roughness faults are the most frequent causes of bearing failures. They usually occur in industrial environment due to various mechanical causes which lead to a faster wear of the components of the bearing, especially of the raceways and balls. Such faults can be easily determined because the bearing spins roughly or difficultly.

The detection of single-point bearing defects is more difficult. If a moving component passes over a defected surface in the bearing, it creates a succession of oscillations which repeat with each pass over the damaged area [5], [6]. The repetition frequency of the impact depends on the position of the fault within the bearing and has an indirect impact on the current, magnetic flux, noise and vibration of the machine.

In this paper the detection of three types of single-point bearing faults by means of statistical processing of the stray fluxes measurements is presented. The developed noninvasive method requires only simple, even hand-made, magnetic flux probes and some easy computation steps.

II. BEARING FAULT DETECTION TECHNIQUES

Upon performing a literature survey on the bearing fault detection techniques it can be stated that a huge number of methods are proposed.

A significant part of the papers are dealing with the rolling bearings fault detection based on analyzing the stator current and the vibration of the electrical machines [7]. These methods essentially are based on finding some well-defined specific fault frequency components in the spectrum of the current or vibration signals [8], [9].

Unfortunately vibration based monitoring techniques require expensive precise vibration sensors and special equipment. They also need direct access to the machine under testing, which is not always possible in industrial environment. On the other hand current monitoring requires only simple current sensors [10]. The current monitoring based techniques can be used to detect a large number of other faults, too: broken rotor bars, shorted windings, air-gap eccentricity [11], load faults, etc.

Several research teams studied the detection of bearing faults in electrical machines by using the stray flux around the motor [12], [13]. The stray flux of an electric machine is the magnetic flux that radiates outside the housing of the machine. It is residual and undesirable, since it is not participating in the torque generation [14].

The stray flux detection used in electrical machines fault diagnosis is applied since about 30 years as an effective technique for noninvasive diagnosis since the sensor can be outside the machine without being necessary to measure voltage, current or other electric or non-electric quantities [15].

This method can be applied for detecting also other electrical machine faults, as stator winding faults [16], rotor defects [17] or voltage source dissymmetry [18].

Also another diagnosis method, the Park's Vector Approach (PVA) is frequently used in detecting bearing faults [19], [20].

In the last years several artificial intelligence (AI) based bearing fault detection methods were developed [21]. The most significant results were obtained by using artificial neural networks (ANNs) [22], fuzzy logic [23], Support Vector Machine (SVM) approach [24], particle swarm optimization (PSO) [25], Hilbert-based bispectral analysis approach [26], etc.

Several other fault detection methods are based on processing the measured signals by means of the wavelet transform [27], [28].

III. THE EXPERIMENTAL SETUP

The test bench for performing the experimental study was built up in the Laboratory of Electric Drives of University of Pavia, Italy (see Fig. 1).

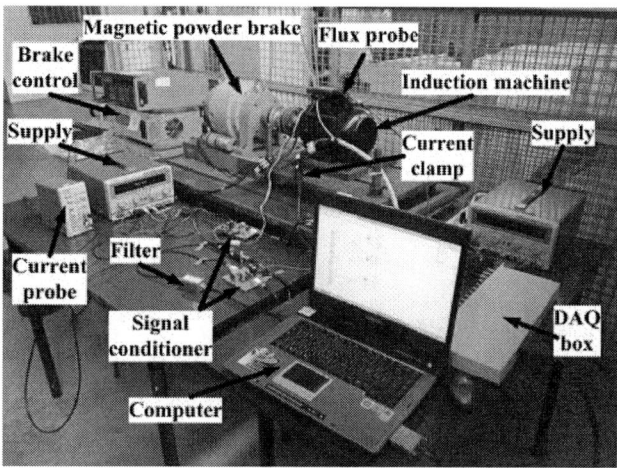

Fig. 1. The experimental setup

The grid connected three-phase 2445T4050 type (FIR Elettromeccanica S.R.L) induction machine has the following rated data: power 2.2 kW, current 8.7 / 5 A, speed 2800 r/min.

The induction machine has two NSK 6205Z type rolling ball bearings with nine balls, which are lubricated with grease.

The induction machine is joined with a magnetic powder brake of 5 kW and 100 N·m through an elastic couple. The load can be set and measured by using the control unit of the brake.

The standard measurements (of the RMS values of the currents, voltages and active power) were performed via a three phase power meter. The speed was measured with a mechanical tachometer. For the phase current measurements a simple Hall-effect based current probe was used.

For the specific stray flux measurements two magnetic flux sensors were applied (see Fig. 3):

- a hand-made flux probe consisting of a semicircular ferrite core with a 44 mm outer and a 40 mm inner diameter. The ferrite core is wound with 300 turns of enameled copper of 0.112 mm diameter [13].
- an industrial one, of M-343F-1204 type, produced by Emerson. It has a circular form and consists of several turns wound around an air-core. This probe can be used to measure the leakage axial flux and it is similar to most of the flux sensors cited in the literature, e.g. [29].

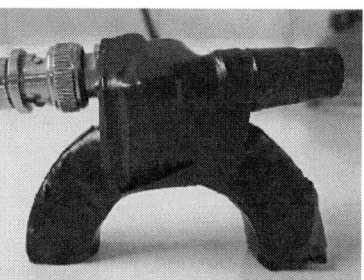

a) hand made

b) commercial (Emerson)

Fig. 2. The applied flux probes

For a greater effectiveness of the measurements a first order low pass RC filter (R = 1 kΩ, C = 82 F and a cut-off frequency of 1942 Hz) was connected to each of the probes [30].

The flux and current probes were connected to a portable National Instruments data acquisition board and to a personal computer with NI LabVIEW software. The used data acquisition board is a NI USB 6212 type with 16 analog inputs, 16 bit resolution, an input range of ±10 V and a bus-

978-1-4799-0024-4/13 $31.00 © 2013 IEEE

powered USB for high mobility. The data acquisition was controlled through a special created virtual instrument, which allowed the simultaneous data acquisition from all the connected probes.

IV. THE PERFORMED MEASUREMENTS

The current probe was placed on one of the phases and the flux probes were positioned around the induction machine in various places. The hand-made flux probe was placed in three different locations around the housing of the machine in order to measure the axial body flux, the radial flux on the end winding and the radial body flux, respectively, as it is shown in Fig. 3a ,b and c. The Emerson flux probe was positioned outside the fan end of the induction machine, with its axis coincident with the shaft axis, as shown in Fig. 3d.

a) axial body flux   b) radial flux on the end-winding

c) radial body flux   d) axial flux

Fig. 3. Positions of the flux probes
for measuring different stray fluxes

Three bearing faults were experimentally simulated and studied [9]:

- crack in the outer race, similar to a fault caused by excessive wear (Fig. 4);
- hole in the outer race (Fig. 5a);
- deformation of the seal (Fig. 5b).

Fig. 4. The faulted bearing
crack in the outer race

a) hole in the outer race

b) deformation of the seal
Fig. 5. The faulted bearings

The induction machine was tested in its healthy condition and having one of its bearings substituted by a faulty one. For each machine condition the motor was tested at no-load, at 50% of the rated load and at the rated load. Exceptionally when the machine was tested with a cracked bearing, the full load condition was not possible to be achieved due to very strong vibrations of the machine which could cause the destruction of the machine.

During all the measurements performed 500,000 samples were acquired at 10 kHz sampling frequency. For each set of measurements ten consecutive acquisitions were collected, with an acquisition time of 50 seconds each.

V. THE STATISTICAL DATA PROCESSING

After finishing the measurements all the saved data acquisitions were processed via FFT in order to obtain the harmonic spectrum of the signal. Here, in Fig. 6, only a single result set of the spectrum analysis is given.

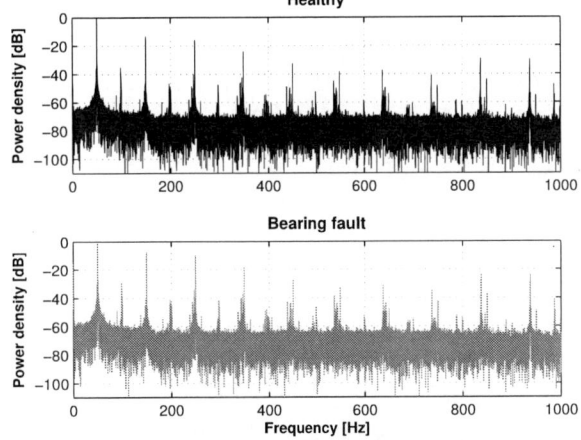

Fig. 6. The power density of the radial flux on the end-winding
for the healthy machine and that with a hole in the outer race
of one of its bearing at 50% of the rated load

As it can be seen, several frequency components (mainly very close to the integer multiples of the fundamental) indicating a bearing fault can be clearly distinguished in the figure.

The implementation of the proposed statistical method follows the subsequent steps [30]:

- each integer multiple of the fundamental, between 100 and 1000 Hz, is normalized with respect to the 50 Hz fundamental, as to be set at 0 dB. Only these harmonic components were taken into study, since they do not depend on the specific parameters of the machines (e.g. number of rotor slots);
- for every harmonic component taken into account a mean value of the ten acquisitions is computed, for both the healthy and the faulty condition of the machine (m_h, m_f, respectively);
- the difference of the mean values mentioned above ($d = m_h - m_f$) is calculated;
- in order to evaluate the diagnostic content of the processed data, the absolute value of the difference ($\Delta = |d|$) is compared with the standard deviation (s) of the harmonics in the case of the healthy motor.

Based on this comparison, three levels of fault significance were considered [30]:

- if $\Delta > s$, the level of significance is low (⚠);
- if $\Delta > 2s$ it is medium (⚠⚠);
- if $\Delta > 5s$ it is high (⚠⚠⚠).

If these conditions are not fulfilled an incipient bearing failure is impossible to be detected by means of this method.

The proposed bearing faults detection method based on statistical computations hopefully is really applicable in industrial environment as a truly accurate and robust fault indicator.

The effectiveness of the proposed statistical computations based method was studied for 20 cases, for all the combinations of three bearing faults (crack in the outer race, hole in the outer race and deformation of the seal), three loads (no-load, 50% of the rated load and the rated load) and the two stray flux probes in the positions taken into study (given in Fig. 3). From the huge amount of result only four can be given here, those considered the most significant ones.

In the following tables, for each experimental case, the mean values of the logarithmic power spectral density for the healthy and faulty machine, their difference, the standard deviation, respectively the significance level of the diagnostic index for all the frequency components taken into study between 100 and 1000 Hz at a 50 Hz step are all given.

TABLE I

RESULTS FOR DAMAGED BEARING SEAL CONDITION AT 50% OF THE RATED LOAD OBTAINED BY MEANS OF MEASURING THE RADIAL FLUX ON THE END-WINDING

f [Hz]	m_h [dB]	m_f [dB]	Δ	s	$\Delta > s$	$\Delta > 2s$	$\Delta > 5s$
100	-40.74	-47.61	6.87	2.32	⚠	⚠⚠	
150	-15.09	-14.03	-1.06	2.39			
200	-48.89	-49.68	0.79	1.98			
250	-13.49	-16.11	2.61	1.63	⚠		
300	-48.33	-52.11	3.78	2.13	⚠		
350	-24.90	-28.72	3.81	1.38	⚠	⚠⚠	
400	-62.99	-69.85	6.86	1.98	⚠	⚠⚠	
450	-35.63	-42.69	7.07	1.99	⚠	⚠⚠	
500	-59.40	-72.00	12.60	1.93	⚠	⚠⚠	⚠⚠⚠
550	-40.21	-51.32	11.11	2.22	⚠	⚠⚠	⚠⚠⚠
600	-67.00	-74.77	7.77	1.87	⚠	⚠⚠	
650	-47.30	-65.12	17.82	1.91	⚠	⚠⚠	⚠⚠⚠
700	-66.26	-73.82	7.55	2.05	⚠	⚠⚠	
750	-66.25	-66.61	0.35	1.43			
800	-67.79	-71.77	3.99	1.22	⚠	⚠⚠	
850	-46.63	-64.74	18.11	1.94	⚠	⚠⚠	⚠⚠⚠
900	-68.02	-73.19	5.17	1.13	⚠	⚠⚠	
950	-60.44	-72.81	12.37	2.34	⚠	⚠⚠	⚠⚠⚠
1000	-67.45	-74.54	7.09	1.62	⚠	⚠⚠	

As it can be seen in the table several harmonic components indicate clearly the damaged bearing of the induction machine in study.

TABLE II

RESULTS FOR DAMAGED BEARING SEAL CONDITION AT THE RATED LOAD OBTAINED BY MEANS OF MEASURING THE RADIAL FLUX ON THE END-WINDING

f [Hz]	m_h [dB]	m_f [dB]	Δ	s	$\Delta > s$	$\Delta > 2s$	$\Delta > 5s$
100	-46.47	-52.45	5.98	1.45	⚠	⚠⚠	
150	-20.64	-18.23	-2.41	1.47	⚠		
200	-52.49	-57.87	5.37	1.17	⚠	⚠⚠	
250	-19.98	-21.88	1.90	1.66	⚠		
300	-52.81	-61.97	9.16	1.14	⚠	⚠⚠	⚠⚠⚠
350	-29.62	-31.90	2.29	1.49	⚠		
400	-67.79	-68.89	1.10	1.51			
450	-41.99	-44.28	2.29	2.42			
500	-62.89	-76.00	13.10	2.25	⚠	⚠⚠	⚠⚠⚠
550	-44.26	-50.03	5.78	6.57			
600	-72.70	-77.78	5.07	2.67	⚠		
650	-54.61	-57.21	2.60	5.83			
700	-70.77	-78.57	7.81	2.10	⚠	⚠⚠	
750	-63.48	-61.66	-1.82	6.12			
800	-73.34	-77.19	3.86	2.21	⚠		
850	-59.28	-61.37	2.10	1.94	⚠		
900	-72.77	-79.16	6.39	2.11	⚠	⚠⚠	
950	-64.67	-64.36	-0.31	3.09			
1000	-73.39	-78.61	5.22	1.26	⚠	⚠⚠	

TABLE III
RESULTS FOR DAMAGED BEARING SEAL CONDITION AT THE RATED LOAD OBTAINED BY MEANS OF MEASURING THE AXIAL BODY FLUX

f [Hz]	m_h [dB]	m_f [dB]	Δ	s	$\Delta > s$	$\Delta > 2s$	$\Delta > 5s$
100	-45.12	-49.86	4.75	1.96	⚠	⚠⚠	
150	-25.24	-24.81	-0.42	1.69			
200	-57.03	-58.42	1.39	2.38			
250	-16.48	-18.74	2.26	1.97	⚠		
300	-74.77	-72.39	-2.38	2.11	⚠		
350	-33.21	-37.68	4.47	1.69	⚠	⚠⚠	
400	-63.85	-73.85	9.99	2.80	⚠	⚠⚠	
450	-58.47	-64.03	5.56	2.06	⚠	⚠⚠	
500	-65.91	-74.80	8.90	2.26	⚠	⚠⚠	
550	-44.08	-55.76	11.68	2.12	⚠	⚠⚠	⚠⚠⚠
600	-69.87	-79.11	9.23	1.41	⚠	⚠⚠	⚠⚠⚠
650	-56.01	-63.51	7.49	2.06	⚠	⚠⚠	
700	-68.46	-79.50	11.04	3.30	⚠	⚠⚠	
750	-58.28	-67.99	9.71	2.95	⚠	⚠⚠	
800	-75.36	-77.74	2.39	2.08	⚠		
850	-53.45	-68.27	14.81	3.59	⚠	⚠⚠	
900	-70.38	-80.68	10.30	3.26	⚠	⚠⚠	
950	-66.19	-79.17	12.98	3.99	⚠	⚠⚠	
1000	-72.97	-78.65	5.68	2.08	⚠	⚠⚠	

Upon the results given in Table III it can be stated that also the axial body flux signals of the induction machine can be used in fault detection, in a similar way as the radial flux on the end-windings.

TABLE IV
RESULTS FOR DAMAGED OUTER BEARING RACE CONDITION AT THE RATED LOAD OBTAINED BY MEANS OF MEASURING THE RADIAL BODY FLUX

f [Hz]	m_h [dB]	m_f [dB]	Δ	s	$\Delta > s$	$\Delta > 2s$	$\Delta > 5s$
100	-36.71	-28.11	-8.60	1.45	⚠	⚠⚠	⚠⚠⚠
150	-14.83	-12.29	-2.54	1.86	⚠		
200	-49.71	-39.83	-9.88	1.40	⚠	⚠⚠	⚠⚠⚠
250	-24.37	-20.91	-3.45	1.89	⚠		
300	-65.94	-55.92	-10.02	2.29	⚠	⚠⚠	
350	-30.99	-28.38	-2.61	2.01	⚠		
400	-59.06	-49.95	-9.11	2.47	⚠	⚠⚠	
450	-54.64	-46.82	-7.82	1.81	⚠	⚠⚠	
500	-68.96	-62.87	-6.09	1.57	⚠	⚠⚠	
550	-47.81	-48.19	0.37	1.91			
600	-66.37	-60.18	-6.19	1.45	⚠	⚠⚠	
650	-50.48	-50.78	0.30	1.92			
700	-66.99	-67.31	0.32	2.06			
750	-56.73	-57.38	0.64	1.77			
800	-67.66	-66.66	-1.00	1.39			
850	-65.47	-65.19	-0.28	2.54			
900	-66.71	-65.71	-0.99	1.75			
950	-67.90	-66.53	-1.38	1.77			
1000	-68.68	-65.96	-2.72	1.83	⚠		

The detection method based on measuring the radial flux body flux at the end windings can be also used to detect outer bearing faults, as a crack in the outer race (see Table IV).

VI. CONCLUSIONS

Bearing faults are one of the most frequent defects of induction machines. Therefore the detection of such faults already in their incipient phase is quite important in the industrial environment.

In the paper a noninvasive bearing fault detection methodology is presented, which involves simple measurements of the stray flux around the machine by means of different flux probes in different positions. It was proven that by applying the method three basic bearing faults types can be detected by simple measurements and computations.

The most effective detection was performed by measuring the radial stray flux of the induction machine's end-winding by using the hand-made flux probe. Insignificant results were obtained by measuring the axial stray flux of the machine via the Emerson flux probe.

Also the stator currents were measured and processed in the frame of the study [30]. However the diagnostic content given by this parameter was less significant with respect to the information given by the stray flux, therefore they were not detailed in the paper.

The tests performed at different loads have shown noteworthy similarities, which means that the developed fault detection method can be a useful tool regardless of the tested machine loading.

The fault detection method detailed in the paper seems to be quite effective. Moreover, the applied hand-made magnetic flux probe used to measure the stray flux round the induction machine is very simple and cheap, as compared with other expensive ones presented in several papers dealing with this diagnosis method, e.g. [12].

In industrial environment, the sensitivity of the flux coil could be affected by the existence of other possible magnetic fields, especially in case of motors supplied by power converters and installed near each other. In this case, it will be necessary to measure the magnetic field around each motor in different positions in order to define the place in which the flux measurement is less influenced by other nearby electrical drives.

The increase of the harmonics multiple of the fundamental in presence of bearing defects has not been yet theoretically justified. However, generally, any asymmetry in an induction motor could excite these harmonics. The problem consists in distinguishing if this increase is due to a bearing fault, or to another kind of fault. Future theoretical and experimental works will be focused to solve this problem.

All the results are encouraging future works concerning the use of other sensors (for measuring currents, stray fluxes and accelerations) also for detecting other faults of the induction machine.

VII. REFERENCES

[1] W.T. Thomson, "A review of on-line condition monitoring techniques for three-phase squirrel-cage induction motors – Past present and future," in *Proceedings of the IEEE Symposium on Diagnostics for Electrical Machines, Power Electronics and Drives (SDEMPED '99)*, Gijon (Spain), 1999, pp. 3-18.

[2] M.E.H. Benbouzid, "A review of induction motors signature analysis as a medium for faults detection," *IEEE Transactions on Industrial Electronics*, vol. 47, pp. 984-993, 2000.

[3] Motor Reliability Working Group, "Report of large motor reliability survey of industrial and commercial installations Part I and II," *IEEE Transactions on Industry Applications*, vol. IA21, pp. 853-872, 1985.

[4] W. Saadaoui and K. Jelassi, "Induction motor bearing damage detection using stator current analysis," in *Proceedings of the IEEE International Conference on Power Engineering, Energy and Electrical Drives (POWERENG '2011)*, Malaga (Spain), 2011.

[5] A.A. Elfeky, M.I. Masoud, and I.F. El-Arabawy, "Fault signature production for rolling element bearings in induction motor," in *Proceedings of the IEEE Conference on Compatibility in Power Electronics (CPE' 2007)*, Gdansk (Poland), 2007, pp. 1-5.

[6] J. Antoni and R.B. Randall, "On the use of the cyclic power spectrum in rolling element bearings diagnostics," *Journal of Sound and Vibration*, vol. 281, pp. 463-468, 2005.

[7] F. Immovilli, A. Bellini, R. Rubini, and C. Tassoni, "Diagnosis of bearing faults in induction machines by vibration or current signals: a critical comparison," *IEEE Transactions on Industry Applications* vol. 46, pp. 1350-1359, 2010.

[8] R.R. Schoen, T.G. Habetler, F. Kamran, and R.G. Bartheld, "Motor bearing damage detection using stator current monitoring," *IEEE Transactions on Industry Applications*, vol. 31, pp. 1274-1279, 1995.

[9] L. Frosini and E. Bassi, "Stator current and motor efficiency as indicators for different types of bearing faults in induction motors," *IEEE Transactions on Industrial Electronics*, vol. 57, pp. 244-251, 2010.

[10] R.B. Randall and J. Antoni, "Rolling element bearing diagnostics - A tutorial," *Mechanical Systems and Signal Processing*, vol. 25, pp. 485-520, 2011.

[11] D.G. Dorrell and W.T. Thomson, "Analysis of airgap flux, current, and vibration signals as a function of the combination of static and dynamic airgap eccentricity in 3-phase induction motors," *IEEE Transactions on Industry Applications*, vol. 33, pp. 24-34, 1997.

[12] O. Vitek, M. Janda, V. Hajek, and P. Bauer, "Detection of eccentricity and bearing faults using stray flux monitoring," in *Proceedings of the IEEE International Symposium on Diagnosis for Electrical Machines, Power Electronics & Drives (SDEMPED '2011)*, Bologna (Italy), 2011, pp. 456-461.

[13] L. Frosini, A. Borin, L. Girometta, and G. Venchi, "Development of a leakage flux measurement system for condition monitoring of electrical drives," in *Proceedings of the IEEE International Symposium on Diagnostics for Electric Machines, Power Electronics & Drives (SDEMPED '2011)*, Bologna (Italy), 2011, pp. 356-363.

[14] H. Henao, C. Demian, and G.A. Capolino, "A frequency-domain detection of stator winding faults in induction machines using an external flux sensor," *IEEE Transactions on Industry Applications*, vol. 39, pp. 1272-1279, 2003.

[15] J. Penman, H.G. Sedding, B.A. Lloyd, and W.T. Fink, "Detection and location of interturn short circuits in the stator windings of operating motors," *IEEE Transactions on Energy Conversion*, vol. 9, pp. 652-658, 1994.

[16] S.-B. Han, D.-H. Hwang, S.-H. Yi, and D.-S. Kang, "Development of diagnosis algorithm for induction motor using flux sensor," in *Proceedings of the International Conference on Condition Monitoring and Diagnosis (CMD '2008)*, Beijing (China),, 2008, pp. 140-142.

[17] A. Bellini, S. Concari, G. Franceschini, C. Tassoni, and A. Toscani, "Vibrations, currents and stray flux signals to asses induction motor rotor condition," in *Proceedings of the 32nd IEEE Annual Conference on Industrial Electronics (IECON '2006)*, Paris (France), 2006, pp. 4963-4968.

[18] H. Henao, T. Assaf, and G.A. Capolino, "Detection of voltage source dissymmetry in an induction motor using the measurement of axial leakage flux," in *Conference Record of the International Conference on Electrical Machines (ICEM '2000)* Espoo (Finland), 2000, pp. 1110-1114.

[19] C. Harlişca and L. Szabó, "Wavelet analysis and Park's Vector based condition monitoring of induction machines," *Journal of Computer Science and Control Systems*, vol. 4, pp. 35-38, 2011.

[20] N. Mehala and R. Dahiya, "Detection of bearing faults of induction motor using Park's Vector Approach," *International Journal of Engineering and Technology*, vol. 2, pp. 263-266, 2010.

[21] W.-Y. Chen, J.-X. Xu, and S.K. Panda, "Application of artificial intelligence techniques to the study of machine signatures," in *Proceedings of the XX IEEE International Conference on Electrical Machines (ICEM' 2012)*, Marseille (France), 2012, pp. 2390-2396.

[22] B.K.N. Rao, P. Pai Srinivasa, and T.N. Nagabhushana, "Failure diagnosis and prognosis of rolling-element bearings using Artificial Neural Networks: A critical overview," in *Journal of Physics: Conference Series*, 2012.

[23] M.S. Ballal, Z.J. Khan, H.M. Suryawanshi, and R.L. Sonolikar, "Induction motor: fuzzy system for the detection of winding insulation condition and bearing wear," *Electric Power Components and Systems*, vol. 34, pp. 159-171, 2006.

[24] K.C. Gryllias and I.A. Antoniadis, "A Support Vector Machine approach based on physical model training for rolling element bearing fault detection in industrial environments," *Engineering Applications of Artificial Intelligence*, vol. 25, pp. 326–344, 2011.

[25] B. Samanta and C. Nataraj, "Use of particle swarm optimization for machinery fault detection," *Engineering Applications of Artificial Intelligence*, vol. 22, pp. 308-316, 2009.

[26] D.-M. Yang, "The application of artificial neural networks to the diagnosis of induction motor bearing condition using Hilbert-based bispectral analysis," in *Proceedings of the 5th IEEE Conference on Industrial Electronics and Applications (ICIEA '2010)*, Taichung (Taiwan), 2010, pp. 1730-1735.

[27] K.S. Gaeid, H.W. Ping, M.K. Masood, and L. Szabó, "Survey of wavelet fault diagnosis and tolerant of induction machines with case study," *International Review of Electrical Engineering (IREE)*, vol. 7, pp. 4437-4457, 2012.

[28] E. Ayaz, A. Ozturk, and S. Seker, "Continuous Wavelet Transform for bearing damage detection in electric motors," in *Proceedings of the IEEE Mediterranean Electrotechnical Conference (MELECON '2006)*, Malaga (Spain), 2006, pp. 1130-1133.

[29] H. Henao, G.A. Capolino, and C.S. Marţiş, "On the stray flux analysis for the detection of the three-phase induction machine faults," *Conference Record of the IEEE Industry Applications Conference IAS '2003 (38th IAS Annual Meeting)*, vol. 2, pp. 1368-1373, 2003.

[30] L. Frosini, A. Borin, L. Girometta, and G. Venchi, "A novel approach to detect short circuits in low voltage induction motor by stray flux measurement," in *Proceedings of the XX IEEE International Conference on Electrical Machines (ICEM '2012)*, Marseille (France), 2012, pp. 1538-1544.

Identification of Variable Mechanical Parameters using Extended Kalman Filters

M. Perdomo, M.Pacas, T. Eutebach (Lenze Automation GmbH), J. Immel (Lenze Automation GmbH)

Abstract -- **The automatic operation of processes requires accurate and up-to-date information about the current state of the system parameters, which frequently cannot be measured during operation. Furthermore, this parameters can change in time due to several factors such as the own dynamics of the system. For electrically powered systems a correct description of the mechanical part and its dynamics is a requirement for a good control performance. The present work describes a Kalman Filter approach to the identification of mechanical parameters. The online identification of time variable mechanical parameters is a task of prime importance for the tuning of self-adaptive controls. In this paper a method for the identification of constant and variable mechanical parameters in industrial drives, introduced in the past by other authors, is analyzed and experimentally tested.**

Index Terms-- **Nonlinear dynamical systems, identification, extended Kalman filter, mechanical system.**

I. NOMENCLATURE

E_k	Total kinetic energy
$i_q *$	Reference value of i_q (proportional to the reference value of the electromagnetic torque)
J	Moment of inertia
J_Σ	Total moment of inertia
J_e	Reduced moment of inertia
K	Kalman gain matrix
M_M	Electromagnetic torque
$m_{element}$	Mass of the mechanism element (crank, connecting rod, slider)
M_L	Load torque
p	Mechanism element position
$\hat{P}$	Error covariance matrix
Q	Process noise covariance matrix
R	Measurement noise covariance matrix
s	Actuator position
$\underline{u}$	Input vector
$\underline{x}$	State vector
$\underline{\hat{x}}$	Estimated state vector
$\underline{y}$	Output vector
$\underline{\hat{y}}$	Estimated output vector
Φ	Transition matrix

ϕ	Angular position connecting rod
θ	Angular position crank
ω_M	Motor angular speed
$W_{element}$	Weight of the mechanism element (crank, connecting rod, slider)

EKF   Extended Kalman Filter
PRBS Pseudo Random Binary Signals

II. INTRODUCTION

THE dynamic response of a system is influenced by its mechanical parameters. An optimal control design for mechatronic systems can only be accomplished if accurate informations about these parameters are provided. If the parameters are unknown taking apart the system to size the components would be an inconvenient task and in some cases not even possible. The identification of the mechanical parameters in drive systems has a significant connotation for the automatic commissioning of industrial machines. In many industrial applications the mechanical parameters present a change during the operation; this can be due to the inherent characteristics of its mechanisms or the proper nature of the process. On the other hand, this behavior could also indicate an emerging failure. For both scenarios on-line identification of these parameters is desirable. In the first case, the controllers must be automatically tuned to the changing mechanical parameters in order to meet the desired dynamic performance. In the second case, it allows the development of online condition monitoring strategies.

Several identification methods are described in the literature, which include some techniques in the frequency domain and others in the time domain as well as deterministic or stochastic identification methods [1]. One of these non-deterministic methods is the Kalman Filter in several varieties. In the present work the characteristics, advantages and drawbacks of this technique are analyzed by simulation and through physical experimentation.

In the last years several approaches for identification have been proposed whose suitability depends on the purpose of the application [1] [2]. The selection of an identification scheme depends principally on the available information of the system. In the literature numerous identification techniques are presented but only few of them have been successfully implemented: Fourier analysis, PLL utilization

---

The present work is supported by the Mexican National Council for Science and Technology (CONACyT) and the German Academic Exchange Service (DAAD) with a scholarship granted to María Perdomo who is currently pursuing her PhD at the University of Siegen.

978-1-4799-0024-4/13 $31,00 © 2013 IEEE

for frequency identification, step response evaluation and other analytical methods.

Müller and Mutschler [2] propose two methods for the mechanical parameter identification: the first one estimates the discrete transfer function minimizing the quadratic prediction error sum. The second method evaluates the fast Fourier transform of the angular acceleration by terms of a graphical procedure; the parameters are analytically calculated based on the estimated poles and zeros and the plant inertia (which are assumed to be known).

Schütte [3] and Wertz [4] use PRBS in order to excite the system in order to calculate the correlation functions, the power spectral density and obtain the frequency response of the mechanical system from them. PRBS has been used in different identification methods to produce transients which frequently make the identification possible. In our research group identification methods in the frequency domain were developed and experimentally validated [5] [6]

A. Parametric model of the system

The mechanical part of the electrically drive systems can be modeled mostly as a system of one or two elastically coupled masses which may include nonlinearities such as friction or backslash. Fig. 1 shows a model with current and speed control in cascade where the current control is represented by a first order lag and the speed control by a PI-controller.

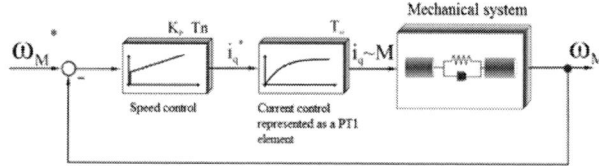

Fig. 1: Electromechanical system with speed and current control

In [7] and [8] Beineke introduces a technique for mechanical parameter identification in time domain using the Extended Kalman Filter as estimator, allowing identification also for time variable parameters.

B. Extended Kalman Filter as non-linear parameter estimator

The extended Kalman Filter is a recursive procedure used to obtain the best possible estimation of the state vector of a linear model in base of non-parametric data [9]. With the utilization of a model based on the available information on the system, real time measurements can be processed to calculate an estimate of the state vector. This is achieved by minimizing the error between the estimated states and the obtained measurements. The Kalman Filter can be extended for nonlinear systems by linearization using a Taylor approximation. The utilization of this filter has reported satisfactory results [7]. The flowchart of a continuous-discrete EKF is illustrated in Figure 2.

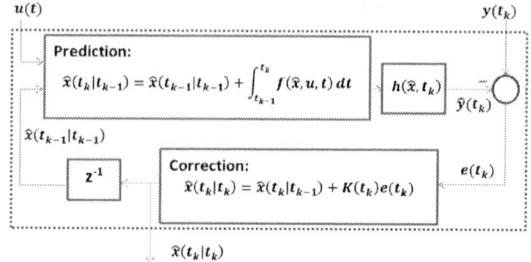

Fig. 2: Extended Kalman Filter structure

The system model is described by nonlinear equations:

$$\frac{d}{dx}\underline{x} = f(\underline{x},\underline{u},t) + \underline{q}(t) \tag{1}$$

$$\underline{y}(t_k) = h(\underline{x},\underline{u},t_k) + \underline{r}(t_k)$$

Where $q(t)$ and $r(t)$ represent the process and the measurement noises. The Kalman Filter assumes that $q(t)$ and $r(t)$ are Gaussian noises (with zero mean and covariance Q and R respectively). The EKF gives an estimation of states $\hat{\underline{x}}$ and outputs $\hat{y}$ by minimizing the error between the measured signals and the observed states with the Kalman gain. The following terms are to be mentioned:

- The *a-priori* values of $\hat{\underline{x}}(t_k|t_{k-1})$ and $\widehat{P}(t_k|t_{k-1})$ are the extrapolation of the observation up to time t_{k-1}
- The *a-posteriori* values of $\hat{\underline{x}}(t_k|t_k)$ and $\widehat{P}(t_k|t_{k-1})$ are the corrected prediction after the measurement in time t_k.

The identification utilizing the EKF executes the next recursive steps:

1) **Prediction:** The a-priori estimation can be obtained from the numerical integration of the system model.

$$\hat{\underline{x}}(t_{k+1}|t_k) = \hat{\underline{x}}(t_k|t_k) + \int_{t_k}^{t_{k+1}} f(\hat{\underline{x}}(t|t_k),\underline{u},t)dt \tag{2}$$

$$\widehat{P}(t_{k+1}|t_k) = \Phi(t_k)\widehat{P}(t_k|t_k)\Phi^T(t_k) + \widehat{Q}(t_k) \tag{3}$$

The transition matrix $\Phi(t_k)$, linearized in the current estimate is defined by:

$$\Phi(t_k) \approx \Phi(T,\hat{\underline{x}}(t_k|t_k)) = \sum_{m=0}^{\infty} \frac{\left(F(\hat{\underline{x}}(t_k|t_k),t_k)\right)^m \cdot T^m}{m!} \tag{4}$$

where the Jacobi Matrix $F(\hat{\underline{x}}(t_k|t_k),t_k)$ evaluated in the current estimate step is:

$$F(\hat{\underline{x}}(t_k|t_k),t_k) = \frac{\partial}{\partial \underline{x}}f(\underline{x},\underline{u},t)\Big|_{\underline{x}=\hat{\underline{x}}(t_k|t_k),t=t_k} \tag{5}$$

2) **Correction:** The a-priori estimation is corrected using the Kalman gain.

$$\hat{\underline{x}}(t_k|t_k) = \hat{\underline{x}}(t_k|t_{k-1}) + K(t_k)\left[y(t_k) - h(\hat{\underline{x}},\underline{u},t_k)\right] \tag{6}$$

$$\widehat{P}(t_k|t_k) = [\mathbb{1} - K(t_k)H(t_k)]\widehat{P}(t_k|t_{k-1})] \tag{7}$$

The linearized observation matrix depends on the prediction:

$$H(t_k) = \frac{\partial}{\partial \hat{\underline{x}}}h(\hat{\underline{x}},\underline{u},t_k))\Big|_{\hat{\underline{x}}=\hat{\underline{x}}(t_k|t_{k-1})} \tag{8}$$

The Kalman gain can be calculated in terms of:

$$K(t_k) = \widehat{P}(t_k|t_{k-1})H^T(t_k)\left[H(t_k)\widehat{P}(t_k|t_{k-1})H^T(t_k) + R(t_k)\right]^{-1} \tag{9}$$

C. EKF identification using a single mass model system

A block diagram of a single mass system with a cascade control structure is show in the Fig. 3. PRBS as additional excitation is also depicted in the figure. The PI-control parameters are adjusted according to the identified parameters of the mechanical system.

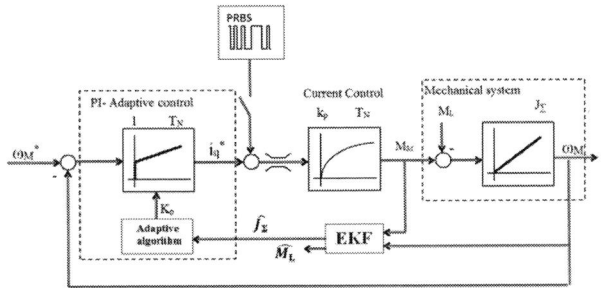

Fig. 3: Structure for a single mass system

The mechanical equation of a single mass system is described by the motion equation:

$$M_M - M_L = J_\Sigma \cdot \frac{d\omega_M}{dt} \tag{10}$$

The state vector is defined and is extended to include the inertia and the load torque as states to be estimated:

$$[\underline{x}]^T = \left[\omega_M(t) \quad M_L(t) \quad \left(\frac{1}{J_\Sigma}\right)(t)\right] \tag{11}$$

Thus the nonlinear model is defined:

$$\frac{d}{dt}\underline{x} = f(\underline{x},\underline{u},t) + \underline{q}(t)$$
$$= \begin{bmatrix} \left(\frac{1}{J_\Sigma}\right)(t) \cdot (M_M(t) - M_L(t)) \\ 0 \\ 0 \end{bmatrix} + \begin{bmatrix} q_{\omega_M(t)} \\ q_{M_L(t)} \\ q_{J_\Sigma(t)} \end{bmatrix} \tag{12}$$

$$\underline{y}(t_k) = h(\underline{x},\underline{u},t_k) + \underline{r}(t_k) = \omega_M(t_k) + r(t_k) \tag{13}$$

To obtain the transition matrix, the Jacobi Matrix of the system model is calculated as:

$$\mathbf{F}\left(\hat{\underline{x}}(t_k|t_k)\right) = \begin{bmatrix} 0 & \left(-\frac{1}{J_\Sigma}\right)(t_k|t_k) & M_M(t_k) - \widehat{M}_L(t_k|t_k) \\ 0 & 0 & 0 \\ 0 & 0 & 0 \end{bmatrix} \tag{14}$$

Therefore the transition matrix turns to be:

$$\Phi(t_k) = \begin{bmatrix} 1 & \left(-\frac{1}{J_\Sigma}\right)(t_k|t_k) \cdot T & [M_M(t_k) - M_L(t_k|t_k)] \cdot T \\ 0 & 1 & 0 \\ 0 & 0 & 1 \end{bmatrix} \tag{15}$$

The process noise is represented by its covariance matrix:

$$Q(t_k) = diag[q_{11} \quad q_{22} \quad q_{33}] \approx Q_c(t_k) \cdot T \tag{16}$$

The variances q_{11} and q_{22} represent the noise due to $\omega_M(t)$ and $M_L(t)$; q_{33} determines the dynamics of the parameter estimation. To achieve a nearly constant signal-to-noise ratio Beineke [7] proposes to recalculate q_{33} for each new estimation according to:

$$q_{33} = \left(\frac{1}{J_{\Sigma_{off}}} - \frac{1}{J_{\Sigma_{off}} + \Delta J_\Sigma}\right)^2$$
$$\approx \tilde{q}_{33} \cdot \frac{\hat{x}(t_k)}{1/J_{\Sigma_{off}}^2} = \tilde{q}_{33} \cdot \frac{\widehat{1/J_\Sigma^2}}{1/J_{\Sigma_{off}}^2} \tag{17}$$

In this expression, $\tilde{q}_{33}$ represents the value variation of q_{33} with respect to the previous estimation.

III. SLIDER-CRANK MECHANISM AS A VARIABLE INERTIA SYSTEM

The experimental performance of the EKF algorithm identification for variable inertia and load torque is intended to be tested using a lab set-up designed and built especially for this task. The entire mechanical system depicted in Fig. 4 consists of two separated mechanisms: a vertical slider-crank which shifts a load up and down and a horizontal slider-crank which moves a pair of jaws of a clamping fixture. The experimentation is intended to be performed on the vertical slider-crank mechanism.

Fig. 4: Front view of the mechanical system
1) Connecting rod of the vertical slider-crank
2) Connecting rods of the horizontal slider-crank

Each mechanism is driven by a synchronous servo motor with its own independent controller. The experimentation was performed on the vertical slider-crank mechanism.

As explained in [10] it is possible to characterize the dynamics of composed mechanism, whose elements move around parallel axes as a system of a single equivalent inertia or reduced moment of inertia. Utilizing the analysis of the kinetic energy of the whole mechanism the slider-crank mechanism can be modeled as a single inertia system where the inertia is varying in function of the position of the crank. Because of the vertical configuration of the slide the load torque changes between positive and negative values depending on the position of the angle of rotation obeying a sinusoidal characteristic. Modeling the vertical slider-crank mechanism as a single variable inertia and the variable load torque the behavior of the EKF estimation can be evaluated.

IV. EQUIVALENT MODEL OF THE VERTICAL SLIDER-CRANK AND LOAD TORQUE

A. Equivalent single inertia system

1) Multibody system kinetics

A generalized coordinate frame is sufficient to describe the position of a single body. Rigid bodies have characteristic mass parameters, center of mass, mass and moment of inertia. Usually the moment of inertia of a body and its position is given with respect to a frame on the center of mass or a local frame. For mechanisms composed by a number of rigid bodies the geometrical relation between the position of the actuator s(t) which is portrayed in the generalized coordinate frame and the position of each component p(t) which is given in local coordinates is determined by the structure of the mechanism and the dimensions of its components. Therefore, the position in a local frame can be expressed as function of the position of the actuator. Where the position of the actuator is a function of time:

$$p = p(s(t)) \tag{18}$$

The speed is calculated from the differentiation of position with respect to time:

$$\dot{p} = \dot{p}\big(s(t)\big) = \frac{dp}{ds}\frac{ds}{dt} = p'\,\dot{s} \tag{19}$$

The Kinetic energy of the whole mechanism is the sum of the kinetic energy of each component of the mechanism:

$$E_k = \frac{1}{2}\sum_i^l (m \cdot \dot{p}^2) = \frac{1}{2}\sum_i^l (m \cdot p'^2)\dot{s}^2 = \frac{1}{2}J_e(s)\dot{s}^2 \tag{20}$$

$J_e(s)$ is the equivalent inertia which is a function of the actuator position and represents all the inertia changes of every single element of the mechanism with respect to the generalized coordinate frame due to the elements motion. This kinematic analysis makes it possible to represent a composed mechanism as a single variable inertia system.

2) Equivalent inertia of the slider-crank mechanism.

An overview of the vertical slider-crank is depicted in Fig. 5. The slider-crank mechanism is composed by 4 elements (slider, crank, connecting rode and guide frame). Each element is a rigid body and the moving elements can be approximated as a massless rod connecting two mass particles. The mass particles concentrate the mass of the elements proportionally to the distance between the center of mass and the assumed position of the mass particle. Utilizing this procedure a model of the slider-crank as a system of mass particles can be obtained in order to calculate the equivalent inertia.

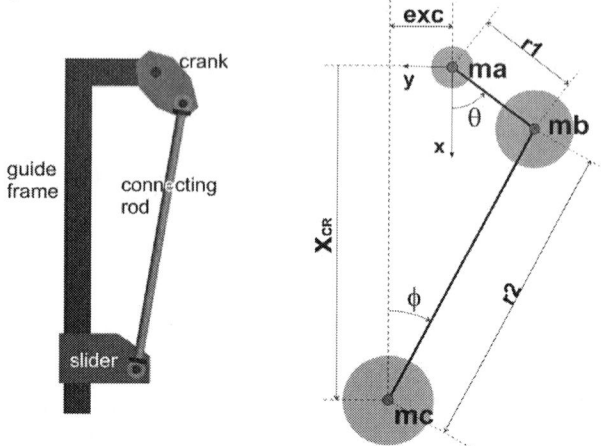

Fig. 5: Slider-crank overview and dynamically equivalent model of mass particles

The equivalent model can be defined following the method described in [11]. The center of mass of each element depends on its geometry and must be calculated to evaluate the proportional distribution of mass on each particle. Assuming the mass particles are located at the articulated joints and proportionally to the distance of each element's center of mass. As depicted in Fig. 5, the mass particles are:

$$ma = \frac{4}{5}m_{crank} \tag{21}$$

$$mb = \frac{1}{5}m_{crank} + \frac{1}{2}m_{CR} \tag{22}$$

$$mc = \frac{1}{2}m_{CR} + m_{slider} \tag{23}$$

The total kinetic energy is:

$$E_k = \frac{1}{2}\,mb(r_1\dot{\theta})^2 + \frac{1}{2}mc(\dot{x}_{CR}{}^2 + \dot{y}_{CR}{}^2) \tag{24}$$

From the geometry of the system, depicted in Fig. 5, it follows:

$$x_{CR} = r_1cos\theta + r_2cos\phi \tag{25}$$

$$exc + r_1sin\theta = r_2sin\phi \tag{26}$$

An expression for ϕ as function of θ can be obtained solving equation (26):

$$\phi = arcsin\left(\frac{exc + r_1sin\theta}{r2}\right) \tag{27}$$

Derivation the equations (25) and (27) yields:

$$\dot{x}_{CR} = r_1\dot{\theta}(-sin\theta) + r_2\dot{\phi}(-sin\phi) \tag{28}$$

$$\dot{\phi} = \frac{1}{\sqrt{1 - \left(\frac{exc + r_1sin\theta}{r_2}\right)^2}} \cdot \frac{r_1}{r_2}\dot{\theta}\,cos\theta \tag{29}$$

978-1-4799-0024-4/13 $31.00 © 2013 IEEE

Substituting (29) and (26) in (28) yields an expression of $\dot{x}_{CR}$ in terms of θ :

$$\dot{x}_{CR} = -r_1\dot{\theta}sin\theta - r_2\frac{\frac{r_1}{r_2}\dot{\theta}cos\theta}{\sqrt{1-\left(\frac{exc+r_1sin\theta}{r_2}\right)^2}} \cdot \frac{exc+r_1sin\theta}{r2} \qquad (30)$$

Only vertical movement is possible for the slider, therefore:

$$\dot{y}_{CR} = 0 \qquad (31)$$

With (30) and (31) the total kinetic energy results in:

$$E_k = \tfrac{1}{2}mb(r_1\dot{\theta})^2 + \tfrac{1}{2}mc\left[-r_1\dot{\theta}sin\theta - r_2\frac{\frac{r_1}{r_2}\dot{\theta}cos\theta}{\sqrt{1-\left(\frac{exc+r_1sin\theta}{r_2}\right)^2}} \cdot \frac{exc+r_1sin\theta}{r2}\right]^2 \qquad (32)$$

With

$$E_k = \frac{1}{2}J_e\dot{\theta} \qquad (33)$$

the equivalent inertia for the vertical slider-crank follows with respect to the crank rotation axis as:

$$J_e = mb(r_1)^2 + mc\left[r_1sin\theta + \frac{\frac{r_1}{r_2}cos\theta}{\sqrt{1-\left(\frac{exc+r_1sin\theta}{r_2}\right)^2}} \cdot exc + r_1sin\theta\right]^2 \quad (34)$$

B. Load Torque

To find the load torque acting on the crank axis of rotation, a static force analysis is carried out. The static force equilibrium equations are obtained from the free-body diagrams of each element and from them the interacting forces. As no external forces are present the only forces acting on the mechanism are the weight of the elements. The static force analysis was done considering that the weight of each element acts on its center of mass.

Free Body Diagrams:

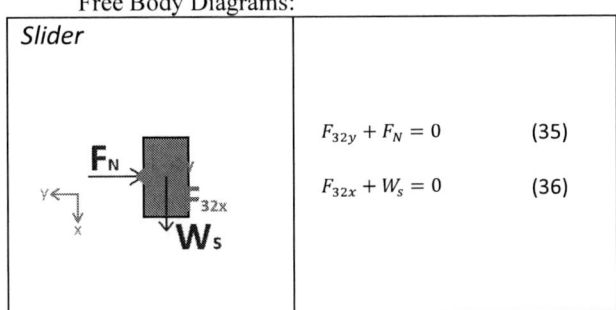

Slider

$$F_{32y} + F_N = 0 \qquad (35)$$

$$F_{32x} + W_s = 0 \qquad (36)$$

Connecting Rod

$$F_{23y} + F_{21y} = 0 \qquad (37)$$

$$F_{23x} + W_{CR} + F_{21x} = 0 \qquad (38)$$

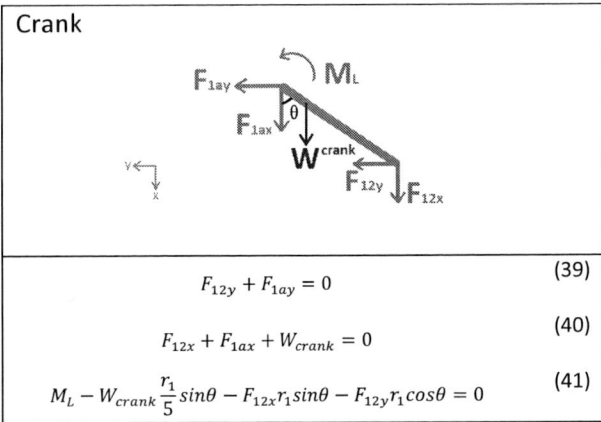

Crank

$$F_{12y} + F_{1ay} = 0 \qquad (39)$$

$$F_{12x} + F_{1ax} + W_{crank} = 0 \qquad (40)$$

$$M_L - W_{crank}\frac{r_1}{5}sin\theta - F_{12x}r_1sin\theta - F_{12y}r_1cos\theta = 0 \qquad (41)$$

Because of the action-reaction Newton's Law, the forces in the joints are:

$$-F_{32x} = F_{23x}; \ -F_{21x} = F_{12x} \qquad (42)$$

The horizontal components are only related to the normal force F_N which depends on the slider weight and the connecting rod position:

$$F_N = -W_s\frac{sin\phi}{cos\phi} \qquad (43)$$

The position of the connecting rod can be expressed in terms of the crank position:

$$F_N = -W_s \tan\left(arcsin\left(\frac{exc+r_1sin\theta}{r2}\right)\right) \qquad (44)$$

Substituting the equations (36), (38), (42) and (44) in (41) an expression for M_L can be found:

$$M_L = \left(\frac{1}{5}W_{crank} + W_{CR} + W_s\right)r_1sin\theta + F_Nr_1cos\theta \qquad (45)$$

According to the kinematic analysis and the static forces evaluation both the equivalent inertia and the load torque can be expressed as functions of the crank position:

$$J_e = mb(r_1)^2 + mc\left[r_1sin\theta + \frac{\frac{r_1}{r_2}cos\theta}{\sqrt{1-\left(\frac{exc+r_1sin\theta}{r_2}\right)^2}} \cdot exc + r_1sin\theta\right]^2 \quad (34)$$

$$M_L = \left(\frac{1}{5}W_{crank} + W_{CR} + W_s\right)r_1sin\theta + F_Nr_1sin\theta \qquad (45)$$

The total moment of inertia is the summary of the motor, coupling and equivalent inertia:

$$J_\Sigma = J_{motor} + J_{coupling} + J_e \qquad (46)$$

A graphic representation of the load torque and total moment of inertia is depicted in Fig. 6. The equations (34) and (45) were evaluated utilizing the dimensions and masses of the vertical slider-crank presented in Table I.

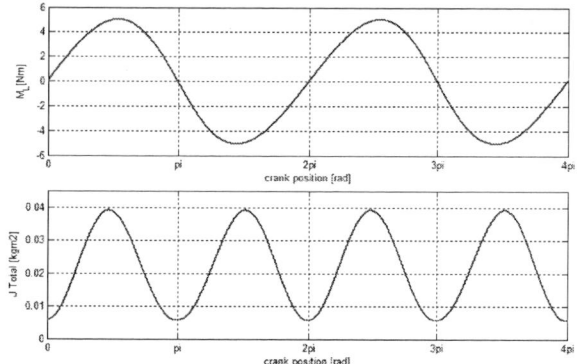

Fig. 6: Load torque and equivalent inertia as function of the crank position

V. RESULTS OF EXPERIMENTS AND SIMULATION

In base of the previously described mechanical system model a simulation was executed in SIMULINK using the structure depicted in Fig. 3. A PI-controller is used for the speed while the closed control loop of the current control is represented as a first order lag. The identification is performed by an EKF programed in a MATLAB embedded code block which requires two signals as input: the electromagnetic torque generated by the motor and the motor angular speed.

Several simulations were performed by assuming a constant inertia, time varying inertia and the simulated slider-crank device. Fig. 7 shows how the EKF-algorithm can track the changing inertia and converges rapidly to the real value.

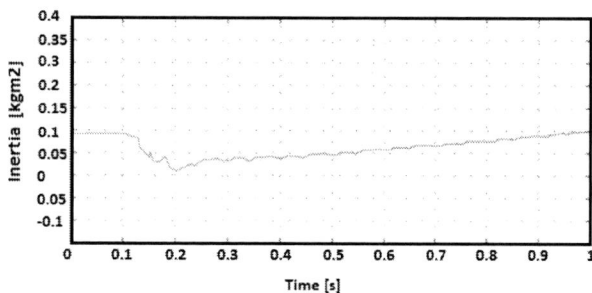

Fig. 7: Simulation Results for a variable inertia with constant rate of change (yellow-actual; pink-EKF estimation)

The fast convergence to the actual values is only possible if the initial values and noise parameters are correctly indicated, such as initial states estimation $\hat{x}_0$, initial error estimation covariance $\hat{P}_0$, measurement noise covariance R and process noise covariance Q which are determined from the known information about the physical system; the measurement noise covariance R is calculated from the sensor resolution and the accuracy of the system model is represented by the process noise covariance Q.

In this case the variable inertia was simulated with a ramp function with a rate of change of 0.1 Kg·m² per second. The change of the mechanical parameters in time can affect the control performance. To avoid instability an adaptive control was utilized [12]. The correct estimation of the mechanical

parameters required additional excitation to the system which was achieved by adding PRBS to the reference value i_q* which is proportional to the electromagnetic torque reference value.

In a further step the simulated mechanical system was modified to include the variable load torque and the variable inertia. To simulate the speed measurement error due to the encoder resolution the output of the simulated mechanical system was quantized.

The simulated system response to a 10rpm step in the speed reference value ω_M^* at t=1s, compared to the EKF identification is shown in Fig. 8. It can be noticed that the estimation of speed and load torque quickly converges to values close to the actual ones. The estimation of the inertia converges after 1s, and from then on, the estimation error is less than 10% yielding an acceptable estimation result.

Physical experimentation on the vertical slider crank was carried out after the simulation. In a similar way to the simulation the speed reference value was set to 10rpm. For the physical experimentation the platform dSPACE DS1104 was programed to execute the EKF algorithm and the measure processing. This platform was connected to the motor controller through an encoder interface to obtain the measured speed and through Serial Peripheral Interface to get the $iq*$ reference value which is proportional to the electromagnetic torque of the motor. These two signals are the inputs to the EKF algorithm.

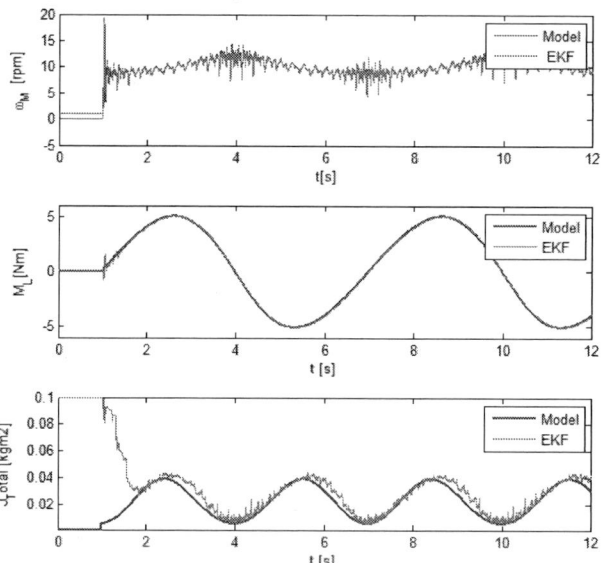

Fig. 8: Simulation Results for the EKF estimation of variable load torque and variable moment of inertia.

Initially some attempts of EKF identification were executed without additional excitation producing extremely poor results. To bring the required additional excitation to the system, PRBS of different amplitudes can be generated by the dSPACE platform and added to the i_q* reference value through an analog interface. Fig. 9 shows the experimental results of the EKF estimation compared to the model adapted to the vertical slider-crank, considering the proportionality of i_q to the electromagnetic torque and the effect of dry friction on the load torque.

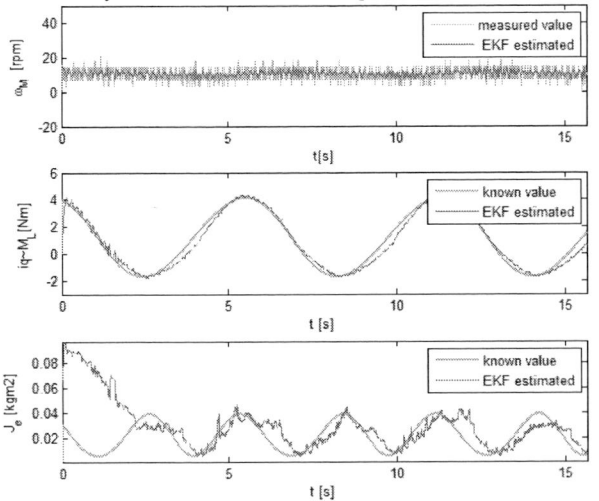

Fig. 9: Experimental results for the EKF estimation of variable load torque and variable moment of inertia.

It can be noticed that the EKF takes up to 2 seconds longer to converge to a value close to the model. A better selection on the initial values and process noise covariance matrix, as well as PRBS amplitude and frequency, could lead to better results. An identification of the variable mechanical parameters is also desired for all the ranges of different speed reference values of the motor, these points are aims of future work.

VI. APPENDIX

Table I

Masses	Dimensions	Weight	Inertia
mb=0.9 Kg	r1=0.07 m	Wb=8.8 N	J_{motor}=0.00012 Kgm^2
mc=6.5 Kg	r2=0.31 m	Wc=63.7 N	$J_{coupling}$=0.001 Kgm^2

VII. CONCLUSIONS AND FUTURE WORK

- The EKF identification technique has been tested in simulation to identify both, constant and variable mechanical parameters in electromechanical machines.
- The initial values and noise covariance are of prime importance to the accurate estimation and fast convergence to the actual values.
- The configuration of the mechanical set-up is a good example for the identification of mechanical parameters in repetitive production cycles that are often utilized in several industrial applications.
- A model for the load torque and equivalent moment of inertia of the vertical slider-crank can be found from the kinetic energy and the static force analysis of the whole mechanism.
- In this case both, the load torque and total moment of inertia are modeled as functions of the actuator`s position. As position changes in time depending on the reference angular speed, they are also functions of time that can be used to evaluate the estimation using the single inertia model in the EKF algorithm.

- Future works will analyze of the influence of different factors such as inertia rate of change, frequency and amplitude of additional excitation.
- The development of condition monitor strategies based on the EKF identification is a further topic of investigation. Comparing the identified mechanical parameters with those obtained at the commissioning of the EKF, can offer an instance for fault detection.

VIII. REFERENCES

[1] R. Iserman, *Identifikation Dynamischer Systeme*. Vol. I and II., Berlin: Springer-Verlag, 1988.

[2] I. Mutschler und P. Müller, "Two Reliable Methods for Estimating the Mechanical Parameters of a Rotating Three-Inertia-System," Dubrovnik, Croatia: Proc. EPEPEMC-Conf., 2002, pp. CD-ROM..

[3] Schütte, F.; Beineke, S.; Rolfsmeier, A.; Grotstollen, H.; , "Online identification of mechanical parameters using extended Kalman filters," *Industry Applications Conference, 1997. Thirty-Second IAS Annual Meeting, IAS '97., Conference Record of the 1997 IEEE* , vol.1, no., pp.501-508 vol.1, 5-9 Oct 1997.

[4] Wertz, H.; Schütte, F.; , "Self-tuning speed control for servo drives with imperfect mechanical load," *Industry Applications Conference, 2000. Conference Record of the 2000 IEEE* , vol.3, no., pp.1497-1504 vol.3, 2000

[5] Pacas, M.; Villwock, S.; , "Development of an expert system for identification, commissioning and monitoring of drives," *Power Electronics and Motion Control Conference, 2008. EPE-PEMC 2008. 13th* , vol., no., pp.2248-2253, 1-3 Sept. 2008

[6] Zoubek, H.; Pacas, M.; , "A method for speed-sensorless identification of two-mass-systems," *Energy Conversion Congress and Exposition (ECCE), 2010 IEEE* , vol., no., pp.4461-4468, 12-16 Sept. 2010

[7] S. Beineke, *Online-Schätzung von Mechanischen Parametern, Kennlinien und Zustandsgrößen geregelter Elektrischer Antriebe*, Düsseldorf: VDI Verlag, 2000.

[8] Beineke, S.; Schütte, F.; Grotstollen, H.; , "Online identification of nonlinear mechanics using extended Kalman filters with basis function networks," *Industrial Electronics, Control and Instrumentation, 1997. IECON 97. 23rd International Conference on* , vol.1, no., pp.316-321 vol.1, 9-14 Nov 1997

[9] S. Maybeck, *Stochastic Model, Estimation and Control*, Bd. Vol. I, New York: Academic Press, 1979.

[10] H. a. H. F. Dresig, *Maschinendynamik*, Berlin: Springer-Verlag, 2005.

[11] A. Ghosh und A. Kumar Mallik, *Theory of Mechanisms and Machines*, New Delhi, India: Affiliated East-West Press , 1988.

[12] P. F. Orlowski, *Practische Regeltechnik*, Berlin: Springer-Verlag.

[13] S. Villwock, "Identifikationsmethoden für die automatisierte Inbetriebnahme und Zustandsüberwachung elektrischer Antriebe," Ph.D. dissertation, Institute for Power Electronics and Electrical Machines: University of Siegen, 2007.

978-1-4799-0024-4/13 $31.00 © 2013 IEEE

Partial Discharge measurements in Electrical Machines controlled by Variable Speed Drives: from Design Validation to permanent PD Monitoring

Luca Fornasari, Andrea Caprara

Techimp H.Q. SpA
Zola Predosa, Bologna, Italy

Gian Carlo Montanari

Department of Electrical, Electronic and Information
Engineering - DEI
University of Bologna
Bologna, Italy

Abstract – **The new IEC 60034-18-41 Standard deals with Partial Discharge (PD) measurements on random wound electrical rotating machines controlled by Variable Speed Drives (VSD). It settles down guidelines for PD measurements during both off-line and on-line conditions.**

The research work described in the current paper applies IEC standard requirements regarding the measurement setup, showing working cases where this approach was successful. The main purpose was not only to check for the presence of PD phenomena during design validation under specified stress conditions, but also to detect and monitor PD behavior versus time during service conditions.

Index Terms – *Electrical Insulation System, Partial Discharge, Random Winding, Rotating Machines, Variable Speed Drives, Monitoring.*

I. INTRODUCTION

During the last decade the use of inverters is increased significantly worldwide. The main benefits are better control and improved performance of AC motors. However, it has been proved that use of inverters can also affect motor service reliability, due to the supply waveform characteristics (as voltage impulses with fast rise time, high repetition rate and voltage overshoots) when compared to the traditional 50/60 Hz supply. It is important, therefore, to understand deeply the impact of voltage waveform characteristics on insulation system performance, in order to minimize their adverse effect on motor reliability [1,5].

Partial Discharge Analysis (PDA), being an excellent diagnostic indicator, has been adopted worldwide. With respect to the PDA carried out at 50/60 Hz power frequency, that performed under pulsed voltage conditions is by far more challenging. This is mainly because the overlapping noise caused by the voltage source (commutation noise of the power electronic). One additional factor is the reproducibility, and thus the comparability of the PD measurement under repetitive pulse voltage generated by different drives, as voltage rise and fall time, repetition rate, overshoots change with the driver technology and characteristics.

The IEC 60034-18-41 Standard [4] proposes solutions for some of the described challenges, defining test voltage parameters, measurement setups and expected results.

In this paper, PD measurements performed according to such IEC Standard are presented and discussed. Features conceived to help the PDA will be shown, together with a method aimed at discriminating PD activity with respect to noise.

Results coming from a permanent PD monitoring on a VSD motor are also reported.

II. PD MEASUREMENT SETUP USED FOR QUALITY CONTROL

PD tests were carried out on different samples by means of a Ultra Wide Band (UWB) PD acquisition unit. The Equipment Under Test (EUT) were fed through two different pulse generators with different repetition frequency of the pulses: 60 Hz (from now on referred to as Pulse Generator #1) and 2,5 Hz (from now on referred to as Pulse Generator #2). A UHF antenna placed directed on the EUT was used to get PD signals, and a 60 dB frequency shifter was used to match the PD detector characteristics and improve the Signal-to-Noise-Ratio (SNR). Another antenna placed near the surge generator, without frequency shifter, provided the synchronization, through the irradiated commutation noise of the pulse generator. The signal was brought to the synchronization conditioning block (ITSM) and then to the PD detection unit [6,7]. Figure 1 describes the test setup used for the PD tests.

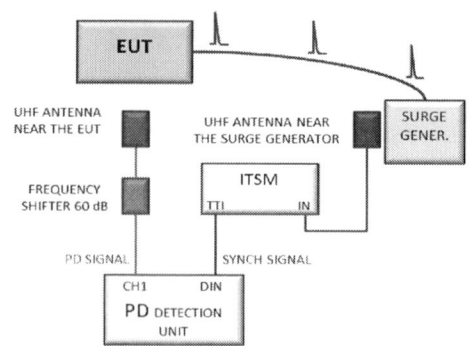

Figure 1: Circuit and connections layout during PD measurements.

A. Instruments and accessories

The measurement circuit is made of the following parts:

- **PD detection unit:** A PD detector able to measure PD under both sinusoidal and impulsive voltages was used. The detection bandwidth is 16 kHz up to 35 MHz. The unit can acquire the entire pulse shapes of a large number of PD signals, allowing a deep analysis of them to be performed. It can also monitor PD continuously, so that PD time behavior (generally PD magnitude and repetition rate) is made available. The software is able to assess whether the insulation system at a defined voltage level is PD free or not, reporting PDIV (Partial Discharge Inception Voltage) and RPDIV (Repetitive Partial Discharge Inception Voltage).

- **Sensors for PD detection and synchronization:** the UHF antenna is a PD sensor designed for the purpose of acquiring electromagnetic emissions from PD occurring in the tested object. It is a broadband antenna with a flat response in the UHF range, which can be used also to get the synchronization signal. Furthermore, because antenna does not require electrical contact with the motor terminals, detection can be easier and safer. The UHF antenna bandwidth is 100 MHz – 3 GHz as a stand-alone sensor.

- **Frequency Shifter:** The PD antenna sensor is connected to a Frequency Shifter that adapts the UHF spectrum of the acquired pulses to the lower frequency band of the PD detector. It is based on a peak envelop modulator (it uses a square law) and it is able to detect low signals without phase errors. Additional high-pass filters are embedded in the frequency shifter to increase the rejection of the electrical disturbance from the inverter, thus improving the SNR.

- **Synchronization block (ITSM):** The Impulsive Test Synchronization Module (ITSM) is a device aimed at generating a digital synchronization signal, which makes the PD detection unit able to perform impulsive voltage PD tests. The digital signal generated by the ITSM is synchronized to the voltage impulse of the impulsive source.

- **Analysis and communication system:** The acquired data can be analyzed in real time by connecting the PC to the detection unit by means of fiber optic link and using appropriate software for separation, identification and data storage.

B. Test objects

Test objects consisted of:

- VPI Resin Impregnated samples
- Formed wound wires without resin
- Twisted pairs with different resins
- Random wound motors

III. PD TEST PROCEDURES AND DATA ANALYSIS

During PD measurements under impulsive voltage, it is not possible to achieve a Phase Resolved Partial Discharge (PRPD) pattern which is used under AC voltage for the identification of PD phenomena. In order to assess whether PD are present or not, the only way is to look at the real time pulse waveforms coming from the PD sensor. Dedicated software able to show and store a configurable number of complete pulse waveforms from the acquisition unit, shall be designed for such purpose.

PD tests were aimed at finding RPDIV for each sample under test. By raising voltage level step by step, it was possible to obtain PD inception voltage from the observation and separation of the recorded signals.

An example of two signals coming from the UHF antenna sensor right before and right after RPDIV is shown in Figure 2. As it can be seen, in this case it is quite easy to understand when PD is incepted, because PD signals are clearly visible after the surge pulse coming from the generator.

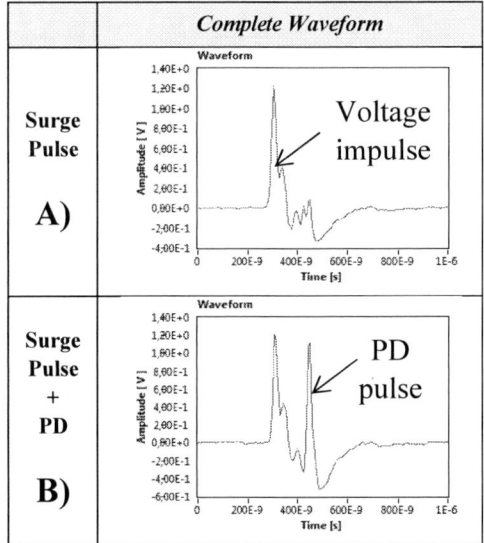

Figure 2: PD detection: A) only voltage impulse; B) voltage impulse + PD.

A key characteristic of the PD detector is its capability to separate PD pulses from voltage impulses. An effective separation tool is the so-called T-F map [8]. It provides two parameters for each recorded pulse: the equivalent time, T, and equivalent frequency [9], F (bandwidth), calculated as shown in eq. (1), being s(t) the recorded signal, t time, L the time length of the observation window and $\widetilde{S}(f)$ the Fourier transform of the normalized PD pulse (Euclidean norm = 1). Quantities τ and f are integration constants in the time and frequency domain.

$$\widetilde{s}(t) = s(t) / \sqrt{\left(\int_0^L s(\tau)^2 \, d\tau\right)}$$

$$
\begin{cases}
\begin{cases}
t_0 = \int_0^L t \widetilde{s}(t)^2 \, dt \\
T = \sqrt{\int_0^L (t - t_0)^2 \, \widetilde{s}(t)^2 \, dt} \\
\quad F = \sqrt{\int_0^\infty f^2 \left| \widetilde{S}(f) \right|^2 \, df}
\end{cases}
\end{cases}
\tag{1}
$$

The F term provides indications about the frequency content of a pulse, being larger for pulses with higher frequency content. Therefore pulses generated by different sources, including disturbances and noise, will have different signatures in the T-F map. This allows separation between PD and noise/disturbances. Each cluster on the T-F map includes all pulses having the same waveform characteristics. Therefore, recorded signals with and without PD should be located in different regions of the T-F Map.

An example of such separation is shown in Figure 3.

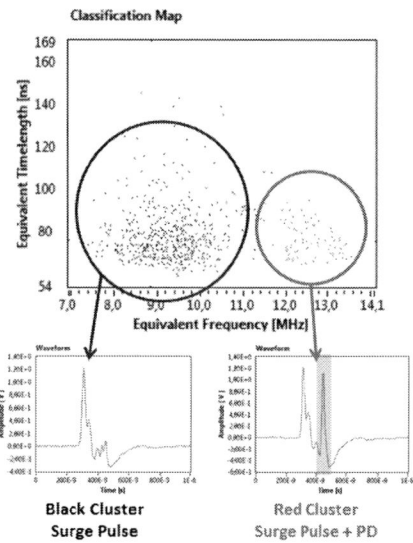

Figure 3: Example of separation of PD pulse from voltage impulse (see Figure 2) through the T-F Map.

IV. PD TEST RESULTS

Many samples were tested through the above-described technique in order to support validity of the circuit layout proposed. A few examples are reported in the following.

A. Test #1: VPI Resin Impregnated samples

Surge Generator: Pulse Generator #1 (60 Hz)

Five samples were tested using the layout shown in Figure 4. The UHF antenna for PD detection was used, placed in front of the sample under test.

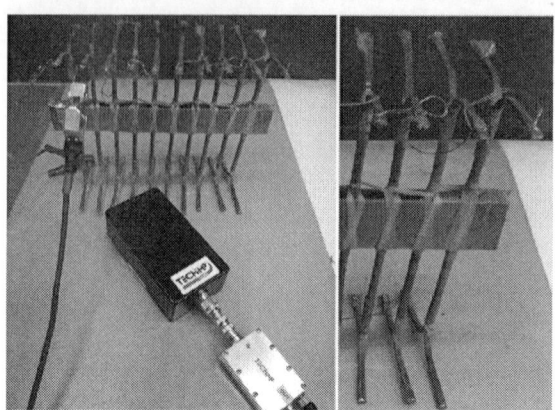

Figure 4: VPI samples PD Test.

Examples of PD measurement results are reported in Table 1.

Table 1: Example of PD measurement results. Test #1

Sample	RPDIV	PD waveform
1	3,1kV	
2	3,0kV	
4	3,1kV	
6	3,5kV	
8	3,3kV	

B. Test #2: Formed wound wires without resin

Surge Generator: Pulse Generator #1 (60 Hz)

Four samples were tested using the layout shown in Figure 4. The UHF antenna for PD detection was used, placed in front of the sample under test.

Examples of PD measurement results are reported in Table 2.

Table 2: Example of PD measurement results. Test #2.

Sample	RPDIV	PD waveform
1 TYPE 1	1,4kV	
2 TYPE 1	1,4kV	
3 TYPE 2	1,5kV	
4 TYPE 2	1,7kV	

C. Test #3: Formed wound wires without resin

Surge Generator: Pulse Generator #2 (2,5 Hz)

Four samples were tested using the layout of Figure 4. A UHF antenna for PD detection was used, placed in front of the sample under test.

Examples of PD measurement results are reported in Table 3.

Table 3: Example of PD measurement result. Test #3.

Sample	RPDIV	PD waveform
1 TYPE 1	1,4kV	
2 TYPE 1	1,5kV	
3 TYPE 2	1,5kV	
4 TYPE 2	1,8kV	

It is possible to notice that RPDIV results here are very close to those obtained by the Pulse Generator #1, Test #2.

D. Test #4: Twisted pairs with different resins

Surge Generator: Pulse Generator #1 (60 Hz)

Nine samples were tested with the layout of Figure 5. A TEM antenna for PD detection was used, placed in front of the sample under test.

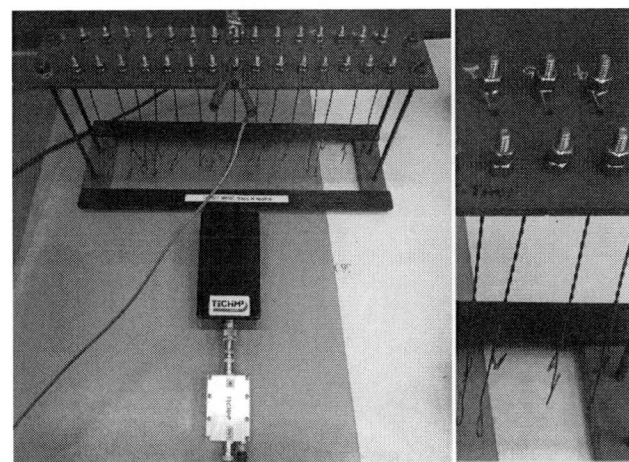

Figure 5: Twisted pairs samples PD Test.

Examples of PD measurement results are reported in Table 4.

Table 4: Example of PD measurement results. Test #4.

Sample	RPDIV	PD waveform
1 RESIN 1	2,4kV	
2 RESIN 1	4,5kV	
3 RESIN 1	4,8kV	
6 RESIN 2	1,7kV	
7 RESIN 2	2,8kV	
8 RESIN 2	1,9kV	
11 RESIN 3	2,8kV	
12 RESIN 3	2,8kV	
13 RESIN 3	2,3kV	

E.   Test #5: Random wound test motor 1

<u>Surge Generator:</u> Pulse Generator #1 (60 Hz)

Six PD measurements were performed using the layout shown in Figure 6. The UHF antenna for PD detection was used, placed in front of the terminal box.

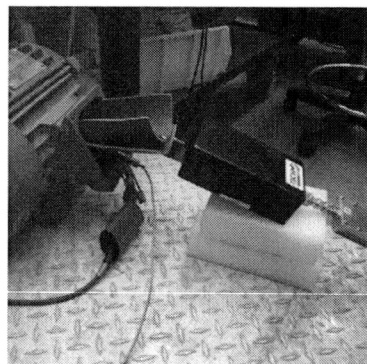

Figure 6: Random wound motor PD Test.

Examples of PD measurement results are reported in Table 5.

Table 5: Example of PD measurement results. Test #5.

Configur.	RPDIV	PD waveform
Phase T1	3,5kV	
Phase T2	3,1kV	
Phase T3	3,0kV	
Phase T1 Vs (T2+T3)	3,3kV	
Phase T2 Vs (T1+T3)	3,9kV	
Phase T3 Vs (T1+T2)	4,8kV	

The first three tests were performed feeding one phase at a time with the motor body grounded, leaving floating the other two phases. In this way the phase to ground insulation of each winding is tested.

To test the phase to phase insulation, other three tests were performed by feeding one phase at a time with the other two phases grounded and motor body floating.

F.   Test #6: Random wound test motor 1

<u>Surge Generator:</u> Pulse Generator #2 (2,5 Hz)

Test procedure was the same as Test #5, but only phase to ground insulation was considered, to verify possible difference in RPDIV value. Examples of PD measurement results are reported in Table 6.

Table 6: Example of PD measurement results. Test #6.

Configur.	RPDIV	PD waveform
Phase T1	3,5kV	
Phase T2	3,1kV	
Phase T3	3,0kV	

G.   Test #7: Random wound test motor 2

<u>Surge Generator:</u> Pulse Generator #1 (60 Hz)

A motor insulation using a different material was tested using the same procedure as in Test #5, in order to evaluate the impact of insulating material and processes: see Table 7.

Table 7: Example of PD measurement results. Test #7.

Configur.	RPDIV	PD waveform
Phase T1	3,2kV	
Phase T2	3,6kV	
Phase T3	3,8kV	
Phase T1 Vs (T2+T3)	4,3kV	
Phase T2 Vs (T1+T3)	3,9kV	
Phase T3 Vs (T1+T2)	4,6kV	

The PD measurement results obtained from Test #5 to #7 highlight the expected outline of the approach described in IEC 60034-18-41 [4]. According to this standard, type tests on complete random wound machine windings should be carried out at different voltage levels and impulse rise-times depending on the defined stress category. Motor stress category depends on expected working conditions of the machine under test and can be divided into A-Benign, B-Moderate, C-Severe and D-Extreme. If we consider the tested 460V winding, assuming a stress category equal to C (Severe), the maximum peak/peak operating voltages,

978-1-4799-0024-4/13 $31.00 © 2013 IEEE

currently under review by the IEC TC2 MT 10, are 1912 V for phase to ground and 2732 V for phase to phase. The standard as coming out from the MT activity does not recommend a fixed safety factor for testing (or enhancement factor in that standard). Instead, it provides a reference table based on which manufacturers and users can establish their own acceptance criteria. Table 8 is an example where a safety factor of 1.3 is used.

Table 8: Test results on motor 1 and 2, assuming safety factor of 1.3.

Configur.	RPDIV Measured	RPDIV Limit	Result
Ph-Gr T1	3,5kV	2,49kV	PASSED
Ph-Gr T2	3,1kV	2,49kV	PASSED
Ph-Gr T3	3,0kV	2,49kV	PASSED
Ph-Ph T1	3,3kV	3,56kV	NOT PASSED
Ph-Ph T2	3,9kV	3,56kV	PASSED
Ph-Ph T3	4,8kV	3,56kV	PASSED
Motor 1	NOT PASSED		

Configur.	RPDIV Measured	RPDIV Limit	Result
Ph-Gr T1	3,2kV	2,49kV	PASSED
Ph-Gr T2	3,6kV	2,49KV	PASSED
Ph-Gr T3	3,8kV	2,49kV	PASSED
Ph-Ph T1	4,3kV	3,56kV	PASSED
Ph-Ph T2	3,9kV	3,56kV	PASSED
Ph-Ph T3	4,6kV	3,56kV	PASSED
Motor 2	PASSED		

V. ON LINE PD MONITORING

The same PD measurement setup was used on a motor connected to a PWM converter. The only difference in the setup was that having a VSD power supply, the reference voltage for synchronization was related to the modulating frequency. This allows the correlation of the acquired PD pulses to the applied voltage and, consequently, PRPD Pattern generation. In this configuration, the capability of the PD detection unit to acquire periodically full waveforms and store the data belonging to of PRPD pattern and T-F map into the internal Compact Flash was exploited. Data were stored in the PD detection unit and downloaded afterward connecting the PC to the unit by means of a standard Ethernet connection.

Any possible PD events occurring inside the machine is synchronized and correlated with the supply voltage fundamental component obtained through low pass filtering. In this way it is a PRPD pattern under PWM waveform can be obtained, discriminating PD activities inside the motor from those uncorrelated i.e., noise. In particular, after noise suppression performed by means of the frequency shifters and after synchronizing the detector with the low frequency component of applied voltage, PD will appear in the PRPD pattern as a series of columns [10-12], while external

disturbances and noise will be horizontally spread out in the pattern phase domain, as shown Figure 7.

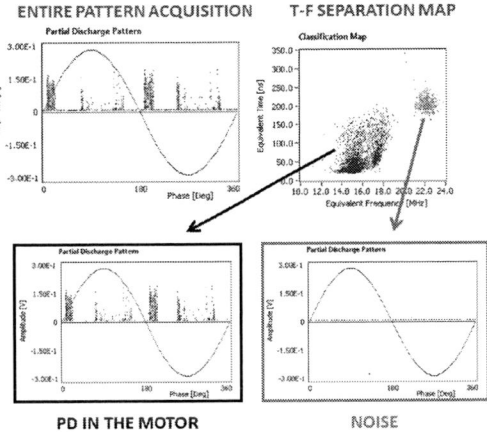

Figure 7: PRPD Patten and T-F map related to PD phenomenon and noise under VSD.

Once connected to the PWM converter, PD occurred immediately in the motor, even if it had passed the off-line Type Test according to the above mentioned standard. Furthermore, both PD amplitude and repetition rate were quite high compared to several other tested motors previously tested and affected by PD. In particular, the repetition rate, expressed in pulse per period of the modulating signal, was significantly high, reaching values up to 2.5 pulse per period. This is summarized in Figure 8, where the $Q_{MAX98\%}$ (98th percentile of amplitude distribution) and repetition rate (N_w) trend is reported over the monitoring time. As can be seen, the PD activity was quite constant in terms of N_w over the monitoring period, while the $Q_{MAX98\%}$ parameter slightly increased during the monitoring time, until reaching the maximum value just few minutes before experiencing a ground-wall failure, i.e. after 133 h.

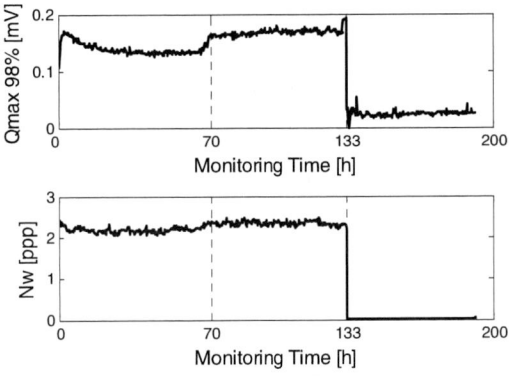

Figure 8: $Q_{MAX98\%}$ and Nw trend of PD detected during monitoring time, from the time the motor was energized to breakdown, occurring after 133 hours.

In order to better describe and compare the on-line results between different motors, a diagnostic Severity Index Ix, was devised in [13], by simply multiplying $Q_{MAX98\%}$ and N_w parameters as shown in eq. (2). As an example, Figure 8 shows the I_x trend during the monitoring period of 133 hours. Note the significant variation of the trend during the last few hours, with increasing Severity Index, approaching failure time.

$$Ix = Q_{max\,98\%} \cdot N_w \qquad (2)$$

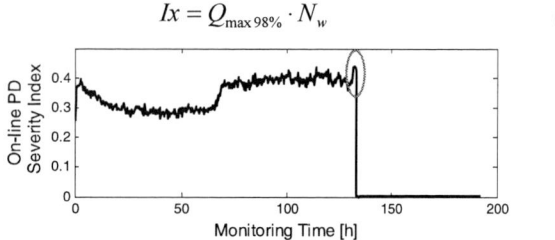

Figure 9: On-line PD monitoring Severity Index relevant to the measurements of Fig. 8 and till breakdown after 133 h.

VI. CONCLUSIONS

The testing procedure described in this paper highlight feasibility and effectiveness of both off-line and on-line tests under impulsive supply voltage on winding specimens and full motors.

Off-line test results indicated that only small differences can be expected when comparing PDIV values obtained with the two different pulse generators used for during testing. The T-F map is helpful to improve SNR, especially when there is significant spectrum interference between voltage impulse and PD pulse (which happens more and more with the new generations of solid state switches). To improve further SNR, specially designed UHF filters can be connected between the UHF antenna and frequency shifter. With the proper software interface, the operator can filter out the voltage surge detected by the PD sensor, focusing the attention only on PD phenomena happening during or after impulsive voltage application. Synchronization signal can be easily taken from another UHF antenna, which makes the PD measurement easy and not intrusive.

The on-line tests provided clear and straightforward information on PD trend and allowed indications of insulation system degradation to be extracted, supporting the validity of PD on-line measurements as the better diagnostic tool. In conclusion, it is proved that the test circuit used is effective in assessing insulation conditions of an electrical machine fed by PWM converters.

VII. REFERENCES

[1] G. C. Stone, S. Campbell, and S. Tetreault, "Inverter-fed drives: Which motor stators are at risk?" IEEE Ind. Appl. Mag., vol. 6, no. 5, pp. 17-22, Oct. 2000.

[2] J.C.G. Wheeler. "Effects of converter pulses on the electrical insulation in low and medium voltage motors," IEEE-DEIS Electr. Insul. Mag., vol. 21, no. 2, pp. 22-29, Mar.-Apr. 2005.

[3] M. Kaufhold, G. Borner, M. Eberhardt, and J. Speck, "Failure mechanism of low voltage electric machines fed by pulse-controlled inverters," IEEE Electr. Insul. Mag., vol. 12, no. 5, pp. 9-15, 1996.

[4] Rotating electrical machines – Part 18-41: Qualification and type tests for Type I electrical insulation systems used in rotating electrical machines fed from voltage converters, IEC 60034-18-41 TS Ed. 1.0, 2006.

[5] Rotating electrical machines – Part 18-42: Qualification and acceptance tests for partial discharge resistant electrical insulation systems (Type II)

used in rotating electrical machines fed from voltage converters, IEC 60034-18-42 TS Ed 1.0, 2008.

[6] A. Cavallini, G.C. Montanari, L. Fornasari, "The Evolution of IEC 60034-18-41 from Technical Specification to Standard: Perspectives for Manufacturers and End Users", IEEE IPMHVC, San Diego, USA, June 2012.

[7] M. Tozzi, G. C. Montanari, D. Fabiani, A. Cavallini, G. Gao, "Offline and on-line PD measurements on induction motors fed by power electronic impulses," in Proceedings of IEEE EIC, Montreal, Quebec, pp. 420-424, May 2009.

[8] A. Contin, A. Cavallini, G.C. Montanari, G. Pasini, F. Puletti, "Digital detection and fuzzy classification of partial discharge signals", IEEE Trans. Dielectr. Electr. Insul., Vol. 9, No. 3, pp. 335-348, 2002.

[9] Franks, L. E., Signal Theory, Prentice-Hall, 1975.

[10] D. Fabiani and G. C. Montanari, "The effect of voltage distortion on aging acceleration of insulation systems under partial discharge activity", IEEE Electrical Insulation Magazine, vol. 17, no. 3, pp. 24–33, Jun. 2001.

[11] M. Tozzi, A. Cavallini, G.C. Montanari, "Monitoring Off-line and On-line PD under Impulsive Voltage on Induction Motors. Part I: Standard Procedure", IEEE Electr. Insul. Mag, Vol.26, No.4, July/August 2010.

[12] M. Tozzi, G. C. Montanari, D. Fabiani, A. Cavallini and G. Gao, "Off-Line and On-Line PD Measurements on Induction Motors Fed by Power Electronic Impulses", IEEE EIC, Montreal, Quebec, Canada, 1-3 June, 2009.

[13] A. Cavallini, G. C. Montanari and M. Tozzi, "Electrical aging of inverter-fed wire-wound induction motors: from quality control to end of life", Conference Record of the 2010 IEEE International Symposium on Electrical Insulation (ISEI).

VIII. BIOGRAPHIES

Luca Fornasari (M'12) was born in Forli, Italy, on 17 July 1986. He received the Degree in Electrical Engineering in 2008 and the Master Degree in Electrical Engineering in 2012 both from the University of Bologna. Since 2007 he has had an active involvement with LIMAT, the University of Bologna High Voltage research laboratory. He has worked since 2008 as a Senior Service Engineer in Techimp HQ, performing PD tests on many types of electrical assets i.e. High Voltage cables, Generators, Motors, PWM fed Motors, Transformers, GIS systems, MV networks and becoming Service Manager in 2011. From mid-2012 he is Research Director inside Techimp HQ where he is leading research activities on global diagnostics of insulating systems.

Andrea Caprara was born in Bologna, Italy on March 25th, 1978. He received the Master Degree in Electrical Engineering in 2004 from the University of Bologna. Since then he joined Techimp HQ research and development department, focusing mainly on the design of products and algorithms related to the PD measurement, in particular on rotating machines and PD monitoring systems. He was also committed to several training tutorials and commissioning to customers about Techimp HQ systems. Since 2007 he was software manager in Techimp HQ, taking care of the design and development of the control software interface of products. From mid-2012 he is product manager of the PD related products focusing on the development, standardization and innovation of the new products.

Gian Carlo Montanari (M'87-SM'90-F'00) is currently Full Professor of Electrical Technology in the Department of Electrical Engineering of the University of Bologna, and teaches courses on Technology, Reliability and Asset Management. He has worked since 1979 in the field of aging and endurance of insulating materials and systems, diagnostics of electrical systems, and innovative electrical materials (magnetics, electrets, super-conductors, nanomaterials). He has been engaged also in the fields of power quality and energy market, power electronics, reliability and statistics of electrical systems, and smart grid. He has been recognized with the IEEE Ziu-Yeda, Dakin and Whitehead awards. He is a member of AEI and Institute of Physics. Since 1996 he has been President of the Italian Chapter of the IEEE DEIS. He is an Associate Editor of IEEE Transactions on Dielectrics and Electrical Insulation, and founder and President of the spin-off Techimp, established on 1999. He is author or co-author of about 700 scientific papers.

An Applied Laboratory Characterisation Approach for Electric Machine Insulation

D. F. Kavanagh, D. A. Howey and M. D. McCulloch

Abstract—Insulation is a fundamental part of electric machines as it provides suitable electrical isolation between the different parts and sub-components. Developing techniques for analysing properties of insulation is extremely useful for the assessment of operating conditions, insulation quality and degradation (state of health). This paper investigates a novel non-invasive experimental approach to characterise phase-to-phase and phase-to-ground insulation by measuring electrical impedance and derived parameters (capacitance, dissipation factor) over a wide frequency range (100 to 2 MHz) at various different temperatures. Due to its simplicity the method lends itself better to an electrostatics interpretation compared with the typical twisted pair approach. Analysis is presented for the method which show its effectiveness under different operating conditions. This work has important applications in the area of condition monitoring of insulation using impedance based measurements.

I. INTRODUCTION

Insulation materials perform an essential role in electric machines, providing electrical isolation between conductors, components and subsystems in the form of a high resistance medium (TΩ/mm). Diverse materials are used including polyester, polyamide-imide, mica, asphalt, nylon and epoxy [1], [2], [3] and these may be used for strand (or sub-conductor) insulation, turn-to-turn insulation or ground-wall and other insulation in a machine stator and rotor [3].

Over a machine's lifetime these insulation materials are subjected to thermal, electrical, environmental (ambient) and mechanical stresses, together sometimes abbreviated 'TEAM' [3]. These stresses contribute to insulation breakdown, initially causing localised electric discharge, which may result in material degradation, leading eventually to catastrophic failures caused by short-circuits. This mainly occurs in the stator coils [4] where types of short circuit include: within a single phase (closed loop), between the phase and the stator frame (ground) or between two or more phases as shown in Fig 1.

Condition monitoring research to date in this area has focused on detection and diagnosis of short-circuits in electric machines [5], [6], [7], [8], [9]. This is limited since it only considers degradation in a binary manner, that is to say either there is no short circuit or there is a short circuit. However, insulation materials degrade gradually with time, hence suitable techniques that give an estimation of insulation 'state of health' at a point in time would be more useful for preventative maintenance and control purposes (e.g. current derating to avoid failure). In order to estimate this, measurement of

changes in insulation material properties such as: capacitance, dissipation factor and impedance is required, since these are related directly to changes caused by factors such as ageing, deformation and delamination, as well as temperature.

Insulation material properties may be estimated using electrical impedance measurements of the insulation system [10]. Various small equivalent circuit impedance models [11], [12], [13] of electric machines have been proposed in the literature for the purposes of design, fault diagnosis and control. We attempt here to show the impact of temperature on these model parameters to create *temperature*-dependent models which has important applications in control systems and machine cooling as well as providing a valuable precursor towards developing *age*-dependent models to characterise state of health.

Polyamide-imide (PAIs) are typically used in electric machines as an insulation material for conductor wires (stator coils) as they have strong high temperature performance [14], [15]. In comparison to other materials e.g. Polyimide (PI) less is known about the dielectric properties of (PAIs) in particular as they approach the glass transition temperature (T_g) [15]. In addition, there has been even less investigation on materials coated on copper wires [14]. This paper outlines an experimental characterisation approach for polymeric materials (polyester and PAI) coated on copper wires and Nomex paper (ground wall insulation) using impedance measurements, focusing on variations from sample to sample and under elevated temperature conditions up to 250 °C.

II. BACKGROUND THEORY

The proposed methodology is to measure impedance as well as parameterised capacitance and dissipation factor to characterise insulation materials. These are each measured over a frequency range from 100 Hz to 2 MHz.

1) Impedance: The impedance $Z(\omega)$, given by the complex ratio of the voltage to current phasors in the frequency domain, enables complete characterisation of any arbitrary equivalent circuit model of an insulation system. However in practice the

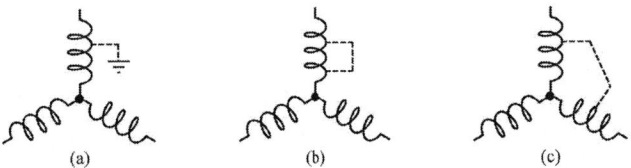

Fig. 1. Star (wye) connection of three-phase stator windings for an electric machine with (a) phase-to-ground, (b) turn-to-turn, (c) phase-to-phase short circuits.

Manuscript received April 30, 2013; revised June 12, 2013. All authors are with the Department of Engineering Science at the University of Oxford, Oxford, OX1 3PJ. Corresponding author: D. F. Kavanagh (email: darren.kavanagh@eng.ox.ac.uk).

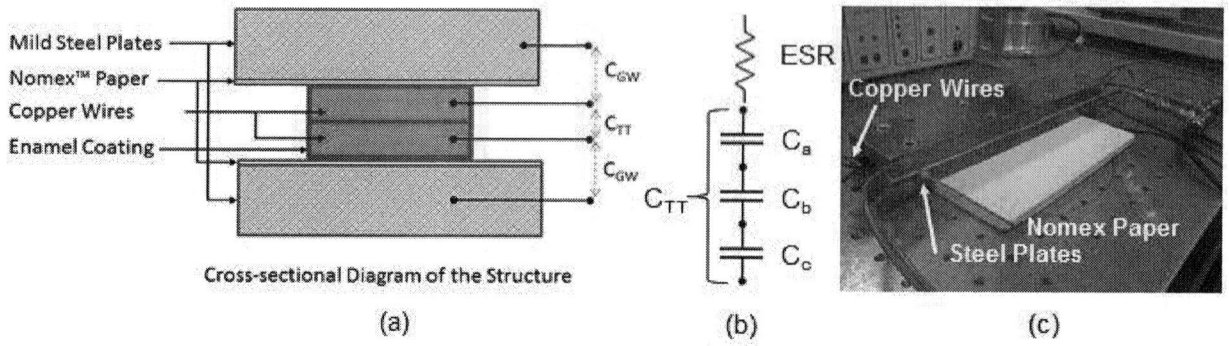

Fig. 2. (a) Diagram of the measurement set-up, (b) equivalent circuit model for C_{TT}, and (c) the physical structure.

assumption of a simple capacitor or capacitor plus equivalent series resistor is a valid representation of an insulation system, as discussed in the next two sections.

2) Capacitance: The capacitance for a simple parallel plate capacitor is defined as:

$$C = \frac{\epsilon_r \epsilon_0 A}{d} \qquad (1)$$

where C is the capacitance, ϵ_r the dielectric constant, ϵ_0 the permittivity of free space, A the plate area and d the distance between the plates. Usually a physical structure such as adjacent copper turns with insulation in-between is geometrically more complex and may contain many different capacitive elements. In this research the experimental specimens under test comprised two sections of insulated copper strip wire held together using a clamping device, Fig. 2. Thus the effective dielectric medium between the plates contained three capacitive elements in series, comprising two insulation layers (C_a and C_b) at the top and bottom, and microscopic air pockets in-between. The total capacitance is therefore:

$$\frac{1}{C_{TT}} = \frac{1}{C_a} + \frac{1}{C_b} + \frac{1}{C_c} \qquad (2)$$

$$\frac{1}{C_{TT}} = \frac{2d_t}{\epsilon_r \epsilon_0 A} + \frac{d_a}{\epsilon_0 A} \qquad (3)$$

$$C_{TT} = \frac{\epsilon_r \epsilon_0 A}{2d_t + d_a \epsilon_r} \qquad (4)$$

where $C_a = C_b$ and the permittivity of air is ϵ_0. The distribution of the air pockets is not uniform between the two surfaces, however for the purpose of modelling this is grouped into one capacitive element using an average air gap distance d_a.

3) Dissipation Factor DF: The complex permittivity $\hat{\epsilon}$ is used to describe the interaction of a material with an electric E field:

$$\hat{\epsilon} = \epsilon_r' - j\epsilon_r'' \qquad (5)$$

where the real part of the permittivity ϵ_r' describes energy stored in the material and the imaginary part ϵ_r'' describes the

relative dissipative properties of the material. The dissipation factor (DF) is defined as the ratio of these two quantities:

$$DF = \frac{\epsilon_r''}{\epsilon_r'} \equiv \tan \delta \qquad (6)$$

which is equivalent to the tangent of the angle δ between ϵ_r' and ϵ_r'' on a complex plane. This can be related to an Equivalent Series Resistance (ESR) representing losses in a non-ideal capacitor. Dissipation factor is used to indicate the quality of a capacitors but has also found an important role in assessing the health of insulation/dielectric materials [16], [17].

III. EXPERIMENTAL PROCEDURE

The experimental set-up is shown in Figure 3. A novel bespoke clamping device was constructed consisting of two plates of mild steel with symmetric slots cut in them to hold samples of insulated copper strip wire aligned together, Fig. 2. The clamp was designed to apply uniform pressure across the samples for repeatable measurements by tightening the bolts using a digital precision torque wrench to 5 Nm each. The electrical connections are also given in the diagram, along with a representative equivalent circuit showing C_{TT} with an ESR.

Samples of rectangular copper wire with class H insulation (polymer coated) for a large electrical traction machine were tested. The wire dimensions were 15.24 mm wide by 2.54 mm thick with an insulation coating thickness of approximately 0.12-0.17 mm, comprising of a base coat of polyester-imide and an outer coat of polyamide-imide. The rationale behind using large rectangular (strip) wire was to maximise the plate area of the capacitor. Also rectangular wire is more suitable for constraining it within clamping arrangement with uniform pressure on the plates. We note from manufacturers product specifications that the nominal insulation thickness does not vary significantly over different wire sizes and therefore the results presented here may be applicable to a range of wire sizes.

Impedance, capacitance and dissipation factor were measured using a N4L precision impedance/LCR analyser (PSM1735 with Impedance Analysis Interface), at room temperature and at various elevated temperatures up to 250 °C.

978-1-4799-0024-4/13 $31.00 © 2013 IEEE

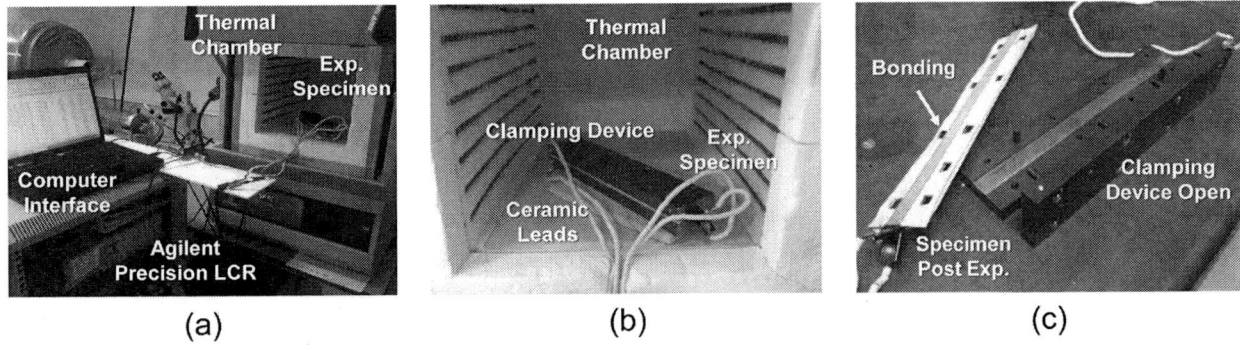

Fig. 3. (a) Experimental rig including Agilent Impedance/LCR meter, furnace, temperature probes, computer Interface; (b) Close-up of test piece in furnace with high temperature connections; (c) test sample after an experiment.

IV. RESULTS AND DISCUSSION

The results are presented for room temperature measurements and at higher temperatures up to 250 °C. For each of the tests, two samples of strip wire of length 250 mm were used to form the capacitance structure as given in Fig. 2. Using 6 samples in total, $^6C_2 = 15$ unique combination pairs were studied (i.e. 15 test cases).

A. Room Temperature Measurements

The impedance measurements are presented in Fig. 4 which show a steady roll-off per decade for $|Z|$ from $1M\Omega$ downwards with increases in frequency. The phase of impedance as expected is close to -90 degrees over the majority of the frequency range, characteristic of a capacitor. At higher frequencies greater than 800 kHz, transmission line effects or parasitic inductance cause our model assumptions to break down.

The capacitance measurements are given in Fig. 5 at room temperature. These show a relatively constant value of capacitance over the frequency range with average at around 1 nF, which compares well with the calculated capacitance using Eq. 4 of 1.16 nF assuming a relative permittivity of 4.3 for the dielectric medium, given by the manufacturer (F.D. Sims). The dissipation factor is shown in Fig. 6. Similar to the capacitance this is relatively constant over the frequency range with average value around 0.01, comparing well with the manufacturer's value of 0.0125.

B. Elevated Temperature Measurements

The impedance measurements at different temperatures are given in Fig. 7 for three different thermal cycles. The magnitude of $|Z|$ at 150 °C and 200 °C shows only small changes occur compared to the room temperature measurements in Fig. 4. However the magnitude at 250 °C does decrease significantly at lower frequencies due to the increase in capacitance, since the impedance of a capacitor is $Z = \frac{1}{j\omega C}$. The impedance phase plots, Fig. 7 show changes do occur 200°C approximately an 8 degree phase shift is observed at low frequency, but at 250 °C a dramatic shift in phase at lower frequencies occurs, up to 50 degrees increase, again due to an increase in ESR.

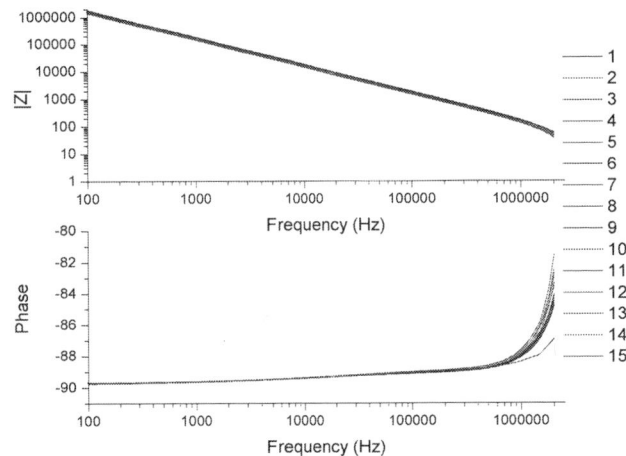

Fig. 4. Measurements of impedance, magnitude Z and phase θ at 25°C. This shows the sample to sample variation for 15 test cases.

The capacitance measurements for elevated temperatures are given in Fig. 8 for three different thermal cycles. The capacitance at 150 °C shows marginal changes compared with the room temperature measurements, perhaps due to the increase in the surface area of the plates and the distance between the plates reducing due to the thermal expansion of the copper wire, refer to Eq. 1. It may be noticed that the capacitance becomes heavily frequency-dependent at higher temperatures (200 and 250 °C). The temperature of 250 °C is approaching glass transition temperature T_g which for PAI is 275 °C. Around this temperature the material becomes more rubber-like and has much larger thermal expansion, closing the air gaps between the two surfaces as described in Sec. II. This temperature Tg is known to have a significant effect on the rate of molecular interactions which increases $\hat{\epsilon}$ due to the dipolar groups becoming more mobile, allowing dipole polarisation to contribute to $\hat{\epsilon}$ [18]. This is attributed by Diaham et al. [15] to micro-brownian segmental motions of the molecular chains. This phenomena is observed more for measurements at lower frequency which gradually rolls-off with increasing frequency which is perhaps due to the relative relaxation time of the

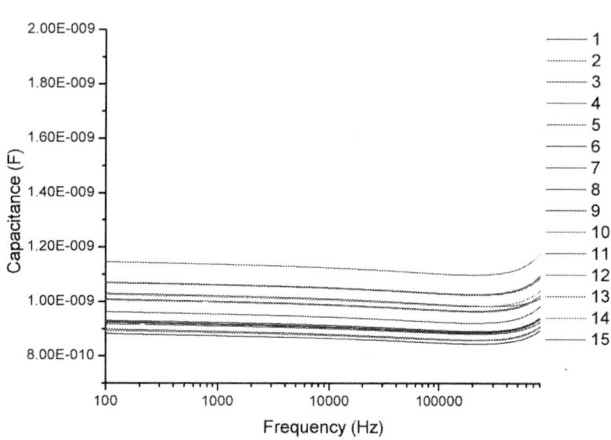

Fig. 5. Measurements of capacitance, C_{TT} at 25°C. This shows the sample to sample variation for 15 test cases.

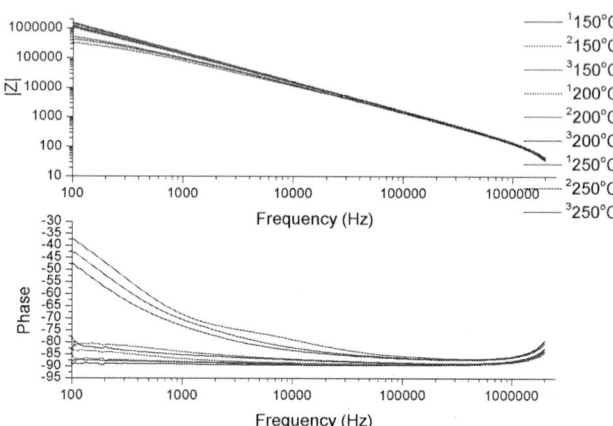

Fig. 7. Measurements of impedance, magnitude Z and phase θ performed at temperatures of $150, 200, 250$°C for three different thermal cycles.

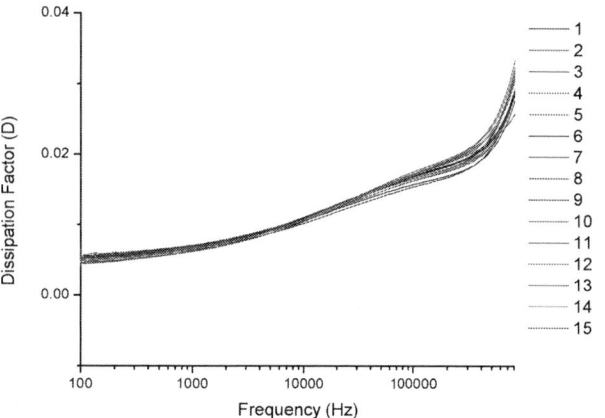

Fig. 6. Measurements of dissipation factor, D at 25°C. This shows the sample to sample variation for 15 test cases.

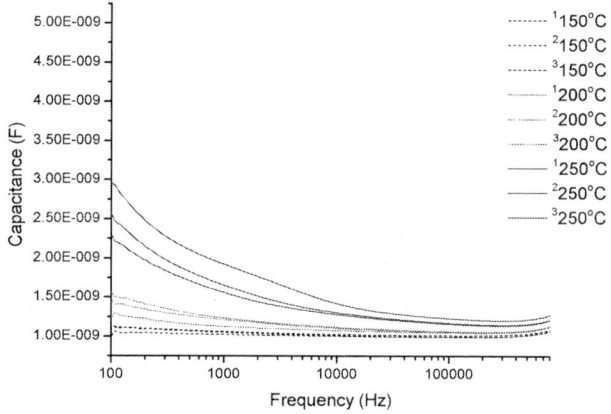

Fig. 8. Measurements of capacitance, C_{TT} performed at temperatures of $150, 200, 250$°C for three different thermal cycles.

material properties at this elevated temperature. The dissipation factor plots in Fig. 9 show that the structure is becoming more lossy or dissipative under higher temperatures.

This proposed approach could be developed further to include adding epoxy resin to bond the two wire specimens together within the clamp structure. This is common in many stator coil arrangements and is usually achieved through a Vacuum Pressure Impregnation (VPI) process, hence this would be a beneficial addition as it would make the methodology even more representative of a stator slot in an electric machine.

V. CONCLUSION

This paper has demonstrated an experimental methodology derived from first principles to characterise insulation used in electric machines in a controlled manner with a novel experimental arrangement using actual insulated rectangular copper wire. The approach applies to surface insulation on conductor wires, and ground-wall insulation. Using the method three types of measurements have been investigated: capacitance, dissipation factor and impedance. Of these three measurements, dissipation factor and impedance phase $\angle Z(\omega)$ give detailed information about changes in insulation properties with temperature, and were relatively insensitive to sample-to-sample variations. Ultimately, this characterisation framework can now be applied to insulation health assessment using impedance measurements. Hence future work will apply this approach to characterise trends surrounding insulation degradation and state of health over prolonged ageing studies using fixed and cyclic temperature stressing. There are also possibilities to investigate humidity and atmosphere pressure conditions using the framework.

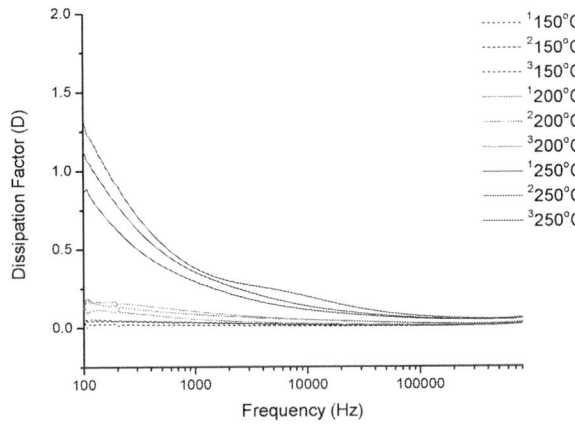

Fig. 9. Measurements of dissipation factor, D performed at temperatures of $150, 200, 250°C$ for three different thermal cycles.

ACKNOWLEDGMENT

This work was funded by the UK Engineering and Physical Sciences Research Council Grant No. EP/I038586/1. Valuable technical support was provided by Graham Haynes and Maurice Keeble-Smith, whose support is gratefully acknowledged.

REFERENCES

[1] L. Rux, "The physical phenomena associated with stator winding insulation condition as detected by the ramped direct high-voltage method," *Ph.D. Thesis Mississipi State University*, pp. 1–168, 2004.

[2] V. Sihvo, "Insulation system in an integrated motor compressor," *Ph.D. Thesis - Lappeenranta University of Technology*, 2010.

[3] G. C. Stone, E. A. Boulter, I. Culbert, and H. Dhirani, *Electrical Insulation for Rotating Machines - Design, Evaluation, Aging, Testing and Repair*. IEEE Press Series on Power Engineering, 2004.

[4] M. Arkan, D. K. Perovic, and P. Unsworth, "Online stator fault diagnosis in induction motors," *IEE Proceedings Electric Power Applications*, vol. 148, pp. 537–547, 2001.

[5] L. Frosini, E. Bassi, and L. Girometta, "Detection of stator short circuits in inverter-fed induction motors," *Proceedings of the 38th Annual Conference on IEEE Industrial Electronics Society*, pp. 5102–5107, 2012.

[6] D. Daz, M. C. Amaya, and A. Paz, "Inter-turn short-circuit analysis in an induction machine by finite elements method and field tests," *Proceedings of the XXth International Conference Electrical Machines (ICEM)*, pp. 1757–1763, 2012.

[7] J. Urresty, J. Riba, H. Saavedra, and L. Romeral, "Detection of inter-turns short circuits in permanent magnet synchronous motors operating under transient conditions by means of the zero sequence voltage," *Proceedings of the 14th European Conference on Power Electronics and Applications*, pp. 1–9, 2011.

[8] M. A. Awadallah, M. M. Morcos, S. Gopalakrishnan, and T. W. Nehl, "Detection of stator short circuits in vsi-fed brushless dc motors using wavelet transform," *IEEE Transactions on Energy Conversion*, vol. 21, no. 1, pp. 1–8, 2006.

[9] S. M. A. Cruz and A. J. M. Cardoso, "Diagnosis of stator inter-turn short circuits in dtc induction motor drives," *IEEE Transactions on Industry Applications*, vol. 40, no. 5, pp. 1349–1360, 2004.

[10] S. Diaham, M. Locatelli, and T. Lebey, "Improvement of polyimide electrical properties during short-term thermal aging," *Annual Report Conference on Electrical Insulation Dielectric Phenomena*, pp. 79–82, 2008.

[11] B. Mirafzal, G. Skibinski, R. Tallam, D. Schlegel, and R. Lukaszewski, "Universal induction motor model with low-to-high frequency-response characteristics," *IEEE Transactions on Industry Applications*, vol. 43, no. 5, pp. 1233–1246, 2007.

[12] B. A. Potter and S. A. S. M. D. McCulloch, "Study of the variation of the input impedance of induction machines with frequency," *IET Electric Power Applications*, vol. 1, no. 1, pp. 36–42, 2007.

[13] E. Zhong and T. Lipo, "Improvements in emc performance of inverter-fed motor drives," *IEEE Transactions on Industry Applications*, vol. 21, no. 6, pp. 1247–1256, 1995.

[14] B. Petitgas, G. Seytre, O. Gain, G. Boiteux, I. Royaud, A. Serghei, A. Gimenez, and A. Anton, "High temperature aging of enameled copper wire-relationships between chemical structure and electrical behaviour," *Annual Report Conference on Electrical Insulation and Dielectric Phenomena*, 2011.

[15] S. Diaham, M. L. Locatelli, T. Lebey, and S. Dinculescu, "Dielectric and thermal properties of polyamid-imide (pai) films," *Annual Report Conference on Electrical Insulation and Dielectric Phenomena*, pp. 482–485, 2009.

[16] K. Younsi, P. Neti, M. Shah, J. Y. Zhou, J. Krahn, and K. Weeber, "On-line capacitance and dissipation factor monitoring of ac stator insulation," *IEEE Transactions on Dielectrics and Electrical Insulation*, vol. 17, no. 5, pp. 1441–1452, 2010.

[17] T. K. Saha, "Review of modern diagnostic techniques for assessing insulation condition in aged transformers," *IEEE Transactions on Dielectrics and Electrical Insulation*, vol. 10, no. 5, pp. 903–917, 2003.

[18] Wiley-Interscience, *Characterization and Analysis of Polymers*. John Wiley and Sons, New Jersey, 2008.

Darren F. Kavanagh is a Postdoctoral Researcher in the Department of Engineering Science at the University of Oxford. His postdoctoral research is in the area of degradation and failure analysis of electric machines with specic applications in electric vehicles. Prior to this he completed his Ph.D. research at Trinity College Dublin in 2011. His doctoral research was in the area of advanced signal processing and pattern recognition for acoustic signals. In 2006, he was awarded an EMBARK scholarship by the Irish Research Council to pursue a Ph.D. on the topic of acoustic signal processing. He was awarded the prestigious Ministers Silver Medal for Science by the Minister for Education (Ireland) in 2004. He has gained valuable academic experience at educational institutions such as Trinity College Dublin, Institute of Technology Tallaght, ITT Dublin, and at the National University of Ireland, NUI Maynooth. He has also beneted greatly from industrial experience at Alcatel Lucent-Bell Laboratories, Intel, and Xilinx.

David A. Howey PhD (Imperial 2010) MA MEng (Cantab. 2002) MIEEE MIET is University Lecturer in Engineering Science at the University of Oxford. His research interests are in condition monitoring and management of electric vehicle components including batteries. He currently leads projects on fast electrochemical modelling, model-based battery management systems, battery thermal management, and motor degradation.

Malcolm D. McCulloch is a Lecturer at the University of Oxford and is a Senior Tutor in Engineering at Christ Church, Oxford. Malcolm moved to Oxford in 1993 and is the Head of the Energy and Power Group at the University of Oxford. The Groups focus is to develop and commercialise sustainable energy technologies in the three sectors of domestic energy use, transport and renewable generation. Malcolm is non-executive director and founder of Navetas Energy Management, YASA Motors and Kepler Energy. He is Director of the Institute for Carbon and Energy Reduction in Transport, based at the Oxford Martin School, which is researching the techno-economic diffusion on new mobility modes. He helped develop Tipping Point, an annual event to bring some of the top artists and climate scientists together. Malcolm acted as an advisor to Ian McKewan for his book Solar. Malcolm has over 60 publications and 15 patent applications.

978-1-4799-0024-4/13 $31.00 © 2013 IEEE

A Wideband Partial Discharge Meter using FPGA

Radek Sedláček, Josef Vedral, Ján Tomlain

Abstract -- **This paper describes a hardware design of a fully digital wideband PD meter based on application of FPGA as well as design of coupling device required for PD measurements. The designed coupling device has frequency bandwidth of 1 kHz - 10 MHz. The PD signal is digitalized by a fast 14-bit A/D convertor sampling at frequency of 50 MSa/s. The digital samples of PD signal are read by the FPGA, subsequently filtered by a number of digital FIR filter banks and stored in a 32 MB DDR memory. On request from PC software, the FPGA send samples in reduced form through Ethernet interface for the next signal processing and evaluation all important parameters of PD analysis. The paper also describes a design of smart charge calibrator especially developed for the PD meter testing and calibration.**

Index Terms -- **partial discharges; partial discharge measurement; coupling impedance; nondestructive diagnostics; electrical insulation measurement; charge calibrator**

I. INTRODUCTION

Partial discharge (PD) measurement is one of nondestructive diagnostic method which allows discovering defects in high-voltage (HV) insulation systems of power transformers, electric power generators or HV cables. Long-term monitoring of a PD level and its statistical evaluation can predict a residual lifetime of all these electrical machines and helps us to increase the reliability of electrical power transmission and distribution. Predictions of electrical machines residual lifetime enable better scheduling of economical investments in the area of maintenances and renovations. Due to these facts, PD measurement is widely used diagnostic method in energy industry and accepted diagnostic tool all over the world.

The PD can be measured directly by detection of induced charge or indirectly by detection of physical quantities like discharge light, mechanical vibration, ultrasonic waves or chemical changes of SF_6 gas caused by discharge arc. Overview of all measurement methods used for PD is given in [1]. This paper is particularly focused on galvanic method of PD measurement. The principle of the method described here is based on voltage pulses measurement between terminals of the electrical machine under test. However a suitable HV coupling capacitance standard together with coupling device must be used for PD signal separation from AC HV supplying tested machine. A great advantage of that method consists in possibility to use it for on-line and off-line testing. The physical processes occurring during PD in voids as well as mechanisms responsible for their development has been widely described in [2]. According to experimental results there is also shown that pulse width is fundamentally independent of overvoltage but increases rapidly width increasing gap or thickness of dielectrics. Typical pulse width is in range a few nanoseconds till a few microseconds. Main guarded parameter of the PD analysis is apparent charge and its phase angle regarding period of the testing high voltage. A meter based on analogue signal processing of PD signal ware typical based on application of peak detector system [3]. Main disadvantage of analog solution consists in dead time during the period when voltage of PD is sampled and hold capacitor is discharged.

Actually, owing to the availability of fast digital electronics, there is possible to realized hardware that provides fully digital signal processing width PD signal. Calculation of apparent charge is done just in time domain by summation of discrete samples forming each PD event. According to [4] wideband PD meters should work with the frequency bandwidth 30 kHz - 500 kHz, in some case the frequency bandwidth should be extended up to 2 MHz [5].

With respect of all above mentioned facts, we have designed and constructed a PD meter enabling to sample the PD signal at frequency of 50 MSa/s. If PD signal is processed at so high frequency, obtained data have a huge volume (rough sample data 100 MB/s) therefore from that point of view there is a need to have a very powerful computational hardware, particularly in case of real-time data processing. This is one of main reason, why have decided to use FPGA as the main control component of the PD meter. The advantage of using the FPGA in these case is presented in [6][7]. The main aim of our effort was design and development PD measurement system with fully digital signal processing of the PD events proposed for off-line testing rotary and non-rotary machines.

The full description of the wideband PD meter is given in the following chapter. An integral part of the PD meter is own sophisticated coupling device. Conventional commercial available coupling devices have two inputs, one for connection the PD signal, second input serves for measuring actual value of RMS voltage over the device

---

This work has been financial supported by Technology Agency of the Czech Republic in framework of No. TA02010311.

R. Sedláček is with Czech Technical University in Prague, Faculty of Electrical Engineering, Department of measurement, Prague, 166 27 Czech Republic (e-mail: sedlacr@fel.cvut.cz)

J. Vedral is with Czech Technical University in Prague, Faculty of Electrical Engineering, Department of measurement, Prague, 166 27 Czech Republic (e-mail: vedral@fel.cvut.cz)

J. Tomlain is with Czech Technical University in Prague, Faculty of Electrical Engineering, Department of measurement, Prague,166 27 Czech Republic (e-mail: johny@tind.sk)

978-1-4799-0024-4/13 $31.00 © 2013 IEEE

under test (DUT). In our case, we have complex coupling device that allows to gain the PD signal and to measure RMS value of testing voltage over DUT simultaneously. The coupling device works up to $100\ V_{RMS}$ over input terminals and includes also zero-cross circuit that can be used for internal triggering of used AD convertor. The level of the PD signal on the input of the CD should not exceed value of $10\ V_{RMS}$.

Before the PD meter is used for diagnostics, there is a need to calibrate it because amplitude of the PD signal is affected by capacitance of a coupling device as well as cable connected between DUT and coupling device input. This paper contains also description of designed smart charge calibrator for the calibration of the PD meter.

II. HARDWARE OF PD METER

The block structure of the developed wideband PD meter and its connection to device under test (DUT) and high-voltage AC source is shown in Fig. 1. The PD meter consists of two parts – complex coupling device (CD) and control electronics utilizing the FPGA. Input of the PD meter is connected to the DUT by means of high voltage coupling capacitor C_C with nominal value of 100 pF or 1 nF.

The complex coupling device serves as separator of the PD signal from testing 50 Hz voltage. In addition main function (filtering testing voltage), the complex coupling device allows measuring RMS (RSM block) of input voltage in whole range up to $100\ V_{RMS}$. Detailed description of the coupling device is given in chapter IV.

Input of the PD meter is AC coupled and galvanically isolated (GI) by two impulse transformers. The transformer's output is led to wideband programmable-gain amplifier (PGA), type PGA870. The wide gain range of -11.5 dB to +20 dB can be adjusted in 0.5 dB gain step. On the output of amplifier, anti-aliasing filter (AAF) with cutoff frequency of 25 MHz is put. The PD signal is sampled by 14-bit ADC, type ADS6143 at 50 MHz frequency. The ADC control and reading is done by FPGA circuit. Due to fast prototyping PD meter, we have utilized a development kit DE0-NANO. The kit includes the FPGA (Altera Cyclone IV EP4CE22F17C6), auxiliary circuits like configure memory EPS16 debug interface and also 32 MB SDRAM. A great advantage of application FPGA consists in sufficiently high computational power. The FPGA drives ADC, SDRAM, communication circuits over ethernet and provides also signal processing with the PD signal. Inside FPGA, the PD signal after digitalization is fed through the FIR digital filters (256th-order, multiplexed 4 banks of coefficients, full parallel implementation). The user has a change to restrict the frequency bandwidth of the PD signal (25 MHz, 10 MHz, 2 MHz and 1 MHz). This is advantage in case of some high frequency disturbances. After filtering, data must be stored in memory and on the request from control software running on PC would be send over Ethernet 100BASE-TX. Due to security reason, the ethernet link

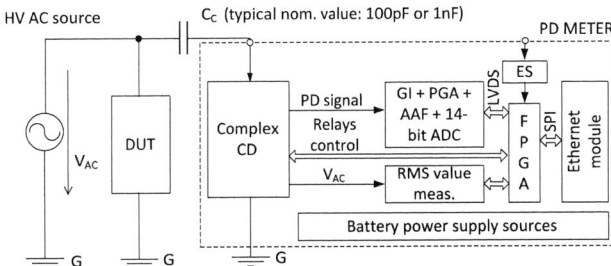

Figure 1. Block structure of the PD meter and its connection to the DUT and HV AC source

connection between the PD meter and control computer is realized by optical fiber. We are using two commercial-available media convertors equipped by SC connectors and multimode optical fiber.

III. DATA SIGNAL PROCESSING

If PD signal is samples at frequency 50 MHz, there is continual data stream of size 100 MB/s. So we had to find way how to reduce or compress a data stream. Therefore we have implemented quite simple algorithm which defines so-called comparison level. By means of control software, we may set this parameter and if the amplitude of PD signal is bellow that comparison level, it mean that this data sample is thrown away and it does not stored into external memory on FPGA kit. This idea is based on the fact that if the PD signal is sampled at frequency 50 MHz, during time of 20 ms (one period of testing voltage), 10^6 samples is acquired but only less than 10 % of samples carries essential information about PD event. The rest of samples contains only a noise. By proper setting of comparison level it can be reach reduction more than 100 times. The next volume data reduction is done by partial integration of several data samples going consecutive. Of course, these data samples must fulfill the comparison condition. After signal data processing, samples are stored in external memory. There is used regular circular buffer. For each record in memory is used pair amplitude of sample and time index (order of samples in range 0 up to 999 999). When data should be sent through ethernet, the FPGA reads data stored in memory and sends via ethernet module in the same format. A socket has fixed length of 8 bytes, first byte is preamble followed by time index (3 bytes) and sample (4 bytes). We have implemented also full expert mode which enables to take all samples per net period and send it into PC for full time analysis.

For evaluation of the PD event is very important their localization regarding period of testing voltage. In this case there are two different ways. One possibility of synchronization is via external optically coupled input (ES). Second one is derived by means of complex coupling devices, which includes zero-cross circuit working in wide range $1\ V_{RMS}$ up to $100\ V_{RMS}$. The PD meter is battery-operated device, does not have any controls and is fully

Figure 2. Main assemled PCB of the PD meter based on utilizing FPGA circuit (Altera Cyclone IV)

Figure 3. PD meter realization (coupling device on left side, main FPGA board on right side)

controlled by a PC. Assembled PCB of the PD meter is shown in Fig. 2. Final encapsulation of the PD meter can be seen in Fig. 3.

IV. COMPLEX COUPLING DEVICE

Very important integral part of each partial discharge meter is auxiliary input circuit so-called coupling device. The main purpose of the coupling device is a separation of the PD signal from exciting testing 50 Hz voltage. In general form, the coupling device behaves like high-pass filter having cut-off frequency around a few kHz so that net frequency of 50 Hz would be suppressed. Actual cut-off frequency is given by value capacitors which are switched on. According to Fig. 1, the testing voltage generated by independent HV AC source is applied over DUT. A coupling HV capacitor C_C of nominal value of 100 pF or 1 nF in conjunction with coupling device forms a voltage divider having ratio of 1000:1 or 100:1. In our case, we have developed more complex, more sophisticated coupling device. A block structure of the developed complex coupling device (CCD) is shown in Fig. 4. The CCD has only one input for measuring both PD signal as well as RMS level of

input 50 Hz voltage. The CCD contains coupling impedance formed by three high-quality low loss capacitors C_1, C_2, C_3, inductor L_S and resistor R_S. Used capacitors WIMA MKP-X2 R series have high capacitance stability over time and typically is applied in capacitor voltage dividers. They have metallized polypropylene dielectric, dissipation factor better than 25×10^{-4} at the frequency of 100 kHz and are rated for voltage of 400 VAC.

A choice of capacitor in coupling device has influence on cut-off frequency of course and on the total divider ratio. For examples, if coupling capacitor of 1 nF is used and it is select 1μF capacitor in coupling device, ratio of 1000:1 is realized. In this case, the cut-off frequency is about 2.92 kHz. Measured amplitude frequency characteristic of developed complex coupling device is shown in Fig. 5. According to Fig. 5, the coupling device has a frequency bandwidth from a few kHz up to 10 MHz. The frequency characteristic of the CCD has been measured only by means oscilloscope with analog bandwidth of 350 MHz. The drop at the end over the frequency 10 MHz is probably caused by parasitic capacitances of used inductor L_S. In the next design of inductor L_S we would like to minimized this unwanted effect.

Excepting coupling impedance, the CCD contains two auxiliary circuits. The first one provides scaling of input AC signal by means of compensated voltage divider with ratio 10:1 or 1:1. Switching of ratios is automatically driven by secondary circuit containing a voltage divider, a rectifier, a peak detector, a comparator and an optocoupler. The AC voltage from divider is fed on the input of isolation (optically coupled) amplifier, type ISO 124. The application of the isolation amplifier ensures full galvanic isolation of whole coupling device from the PD meter. The state of divider ratio is indicated by divider status output in the FPGA.

The maximal input voltage of coupling device is 100 V_{RMS}. The maximal level of the PD signal is 10 V_{RMS}.

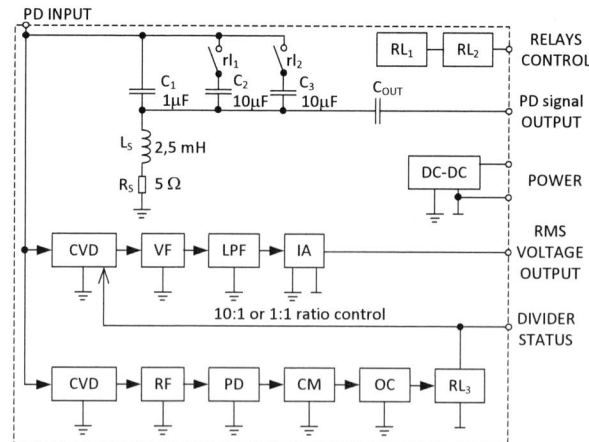

Figure 4. A block structure of developed couplig device (CVD – compensated voltage divider, VF – voltage follower, LPF – lowpass filter, IA – isolating amplifier, RL – relay, DC-DC –isolated power supply, RF – active half -wave rectifier, PD – peak detector, CM – compator, OC – optocoupler, RL1, RL2, RL3 - relays)

978-1-4799-0024-4/13 $31.00 © 2013 IEEE

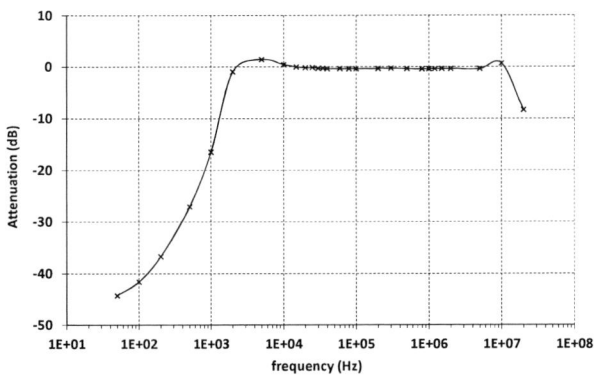

Figure 5. Measured transfer function of coupling device when capacitor $C_l=1$ μF is select

V. PD METER CALIBRATION

Before starting measurement with the PD meter, there is a need to perform calibration procedure. Therefore we have designed own smart charge calibrator that is controlled by microcontroller.

The calibrator can generate charge pulse in the range from 100 pC to 100 nC. Beside charge size, user can set a number of pulses generated per each half period in range from 1 to 16. And it is possible to change width of generated pulse. The rising edge of charge pulse is less than 100 ns, falling edge can be set in four different values: 1μs, 5 μs, 10 μs and 50 μs. The principle of calibrator is quite simply. There is used a reference voltage for charging reference capacitors. After reference capacitor is charged in proper time this capacitor is discharged by means of given reference resistor through analogue switch. On the output stage, there is implemented convertor voltage/current. Due to this fact the calibrator have a current output. A typical waveform from charge calibrator is shown in Fig. 6. Fig. 7 and Fig. 8

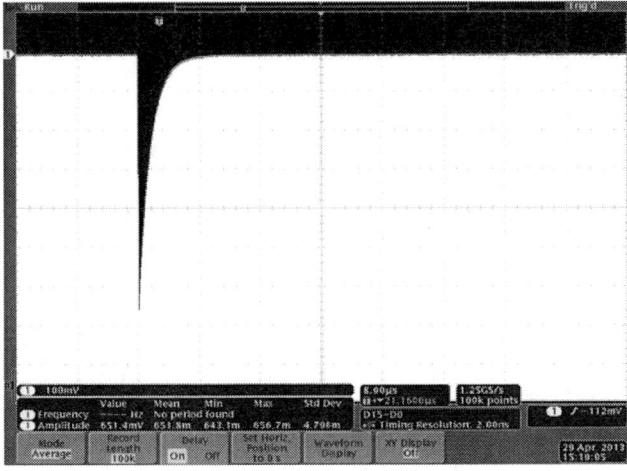

Figure 6. Waveform of calibration pulse genereated from samrt charge calibrator

Figure 7. Enclosure of smart charge calibrator

Figure 8. Assembed PCB of smart charge calibrator

present the final enclosure of the charge calibrator and assembled PCB of the charge calibrator. This unique solution of the charge calibrator enables very easy to verify the linearity (transfer function) of the PD meter due to fact that user can select number of calibration pulses which are generated during one half period 10 ms (1 up to 16 pulses).

VI. EXPERIMENTAL DATA

The first laboratory experiment was done in a laboratory of the ORGREZ company. As DUT was used a reference capacitor of 290 nF that was supplied from regulated AC power source (max. output voltage 30 kV). There was used coupling capacitor of value 1 nF. The results of this experimental measurement are presented in Fig. 9. During experiment it has been verified that complex coupling device and the PD meter works correctly. Due to fact that actually we are still working on development of main software for control and displaying results, axis Y express in both figures number of LSB of the used ADC. This value must be already converted on apparent charge. In all cases the PD meter was in full expert mode (all samples are send into PC for analysis) and trigger was derived from zero-cross detector in complex coupling device.

978-1-4799-0024-4/13 $31.00 © 2013 IEEE

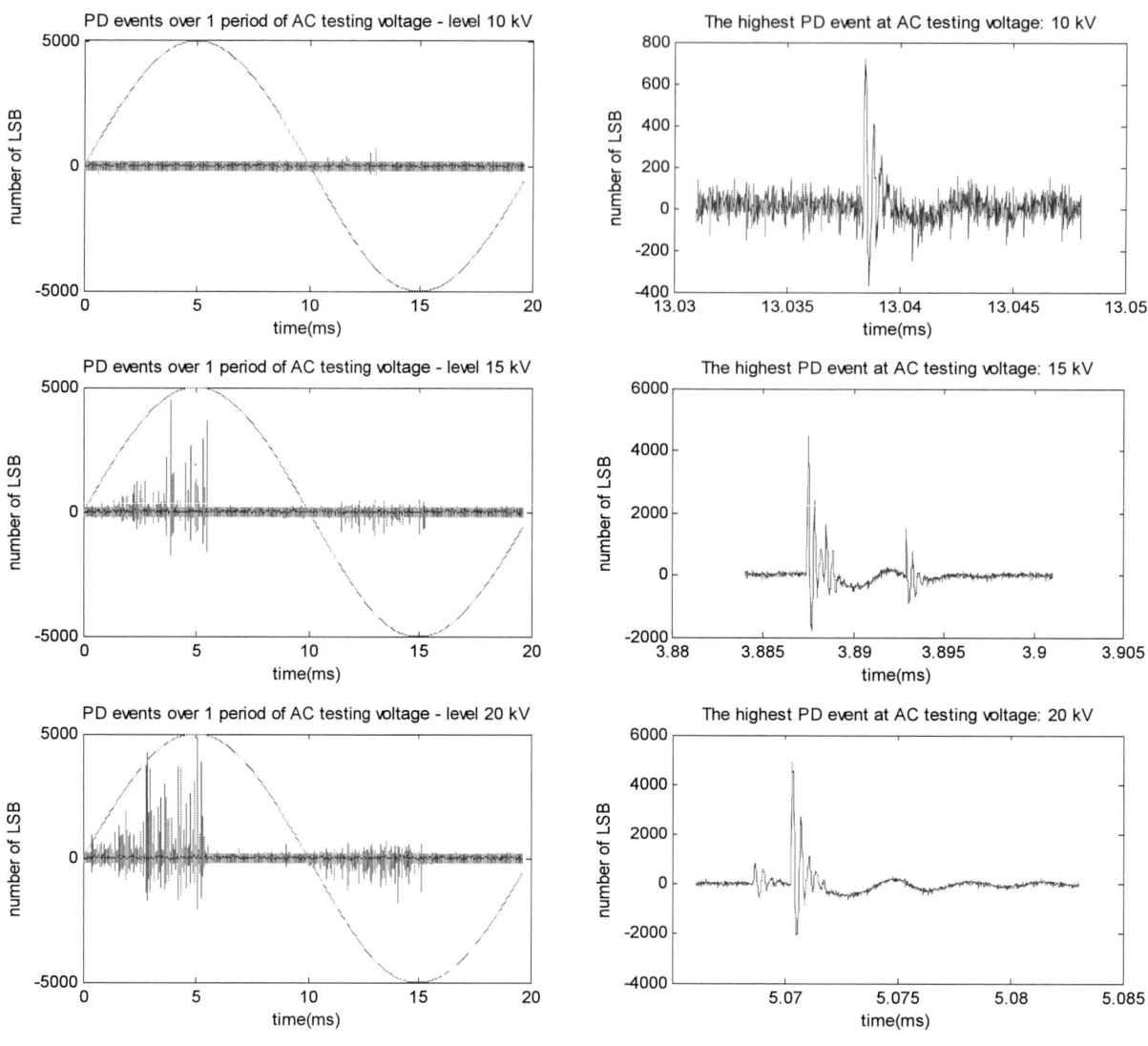

Figure 9. Measured PD signals for three different voltage level (DUT – air capacitor of 290 nF)

VII. CONCLUSION

In this paper authors describe the design and realization of the wideband PD meter based on application of the FPGA. The PD meter works with the frequency bandwidth of 30 kHz – 10 MHz. Actually the bandwidth is limited especially by the coupling device. Beside the PD meter description there is also presented our sophisticated coupling device and smart charge calibrator that should be used before each PD measurement.

The main aim of authors was development of the PD meter for off-line diagnostic of rotary and non-rotary machines. The presented solution is very cost-effective in comparison with similar commercially available PD measurement systems (less than 1500 EURO without the coupling capacitors). Actually we are still working on control software for PC. The first testing was done in accredited laboratory of ORGREZ company. In the future work we would like to increase frequency bandwidth of coupling device over 60 MHz, implemented algorithms for the PD analyses directly in the FPGA. With the highest performance of the used ADC (ADS6143) we are able to reach sampling frequency of 125 MHz without any changes on other part of the PD meter hardware. We also plan to add one more ADC channel for simultaneous measuring PD signal and immediate voltage of testing 50 Hz net voltage for estimation discharges power. Actually this statistic is computed from RMS value and immediate phase angle of testing voltage.

VIII. ACKNOWLEDGMENT

Authors would like to acknowledge J. Brázdil and P. Justiz from OGREZ company for their helpful discussions by development of the wideband PD meter and their help by testing in accredited laboratory of the high voltage.

IX. REFERENCES

[1] M. Hikita, S. Okabe, H. Murase, H. Okubo, "Cross-equipment Evaluation of Partial Discharge Measurement and Diagnosis Techniques in Electric Power Apparatus for Transmission and Distribution," *Dielectrics and Electrical Insulation, IEEE Transactions on* , vol.15, no.2, pp.505-518, April 2008.

[2] J. C. Devins, "The 1984 J. B. Whitehead Memorial Lecture the Physics of Partial Discharges in Solid Dielectrics, " *Electrical Insulation, IEEE Transactions on* , vol. EI-19, no.5, pp.475-495, Oct. 1984

[3] J. Vedral, M. Kříž, "Signal Processing in Partial Discharge Measurement," Metrology and Measurement Systems, vol. XVII, no.1, pp.55-63, 2010.

[4] "IEEE Trial-Use Guide to the Measurement of Partial Discharges in Rotating Machinery," IEEE Std 1434-2000, 2000.

[5] IEC60270, High-voltage test techniques - Partial discharge measurements, CEI/IEC 60270:2000.

[6] Y. Cheng, X. Hu, X. Chen, P. Li, "Partial Discharge Online Monitoring System Based on FPGA," Proceedings of 2005 International Symposium on Electrical Insulating Materials, Kitakyushu, Japan, pp. 486 – 489, 2005.

[7] Emilliano, C.K. Chakrabarty, A. K. Ramasamy, A. B. A. Ghani, "Partial discharge detection system for counting PD signals in high voltage underground cable by using FPGA technology," Modern Electric Power Systems (MEPS), 2010 Proceedings of the International Symposium, pp.1-6, Sept. 2010.

X. BIOGRAPHIES

Radek Sedláček born on June 27, 1977 in Humpolec, Czech Republic. He received the Ph.D. degree in Measurement and Instrumentation from the Czech Technical University in Prague - Faculty of Electrical Engineering (CTU – FEE) in 2007. His Ph.D. work was related to calibration methods of self-inductance standards in the low frequency range. Since 2005 he acts as assistant professor at the CTU - FEE, Department of Measurement. He also engaged in the field of digital signal processing and implementation of DSP algorithms on FPGAs and digital signal processors. He is an author or co-author of more than 11 conference and journal papers in the area of precise impedance measurements. He is actually collaborating with ORGREZ company on developing the intelligent diagnostic system of high-voltage electrical equipment

Josef Vedral born May 18, 1947 in Tachov. In 1971 he graduated from the Faculty of Electrical Engineering of the Czech Technical University in Prague. In 1975 he defended his doctoral thesis in the field of measuring technology. In 1990 he qualified in the same field at the same university. Long-term he deals with the processing of signals in measurement, testing Digitizer continuous signals. He is currently collaborating with ORGREZ company on the developing intelligent diagnostic system of high-voltage electrical equipment. He is the author of 52 papers at international conferences, 26 articles and 12 patents.

Ján Tomlain was born in Myjava in Slovakia, on October 31, 1990. He graduated high school L. Novomeskeho, Senica and actually studying at the Czech technical university, Faculty of electrical engineering. So far he received bachelors degree.

Broken Bar Fault Diagnosis in Single and Double Cage Induction Motors Fed by Asymmetrical Voltage Supply

K. N. Gyftakis, *Member IEEE*, D. K. Athanasopoulos and J. C. Kappatou, *Member IEEE*

Abstract -- The diagnosis of faults in induction motors is a crucial scientific subject since it affects the industrial production functionality and reliability. Lately, a new scientific area is born: the electrical machines prognostics. So, aiming to this direction the complete knowledge of all abnormal parameters during the induction motor operation should be identified. In this paper, a methodology is described in order to identify not only if there is a broken bar fault, but also if the induction motor voltage supply is asymmetrical at the same time. The investigation is carried out with transient FEM simulations. The proposed methodology comes to supplement the traditional MCSA method with desired satisfaction.

Index Terms--Asymmetrical voltage supply, Broken bar, Fault diagnosis, FEM, FFT, Induction motor.

I. INTRODUCTION

THE induction machine constitutes the primary and most common selection in industrial environments operating as motor. Despite its robustness, faults of different origins may appear and violate the production process or also lead to a breakdown. This is the reason why, during the last years extensive studies and research have led to a variety of induction motor diagnostic methods.

The broken bar fault belongs to the induction motor rotor faults area. It appears in about 10% of all induction motor faults [1]. The broken bar fault is a fault which progresses during time. This means that, if a bar cracks it will eventually break and the neighboring bars will follow to break, one after the other, if the fault is not diagnosed early enough. If the fault is not diagnosed at all, there will be a total breakdown of the induction motor itself and the damage could be irreparable.

Many methods have addressed the broken bar fault detection so far. Means such as the current [2]-[6], torque [7]-[12] and power waveforms [12]-[14], etc have been studied in the past offering diagnostic potential. For this specific fault, the dominant method for most operating cases is the traditional MCSA (Motor Current Signature Analysis) method. It is based on the identification of the fault's signatures in the current's frequency spectrum. Generally,

---

This work was supported by the research program: "K. Karatheodori 2010", of the Research Committee, University of Patras, Greece.
K. N. Gyftakis, D. K. Athanasopoulos and J. C. Kappatou are with the Laboratory of Electromechanical Energy Conversion, Electrical and Computer Engineering Dept. , University of Patras, Greece. (e-mail: kosgyftak@upatras.gr, joya@ece.upatras.gr).

---

the current is a valuable diagnostic mean because it is measured easily and at low cost. The MCSA method according to the literature seems to have several weaknesses, despite its simplicity. It is affected by the saturation level of the induction motor [15], has low reliability when the motor operates under low or no-load [16] and finally it is not reliable concerning the case of the double cage induction motors [17].

In an installation, it is possible that a motor operates under unbalanced voltage supply for many reasons. The asymmetry could be caused by the grid or the installation itself (asymmetrical wiring, connections etc). The voltage imbalance causes an asymmetry in the magnetic field distribution which could lead to an eccentricity level of the rotor, while the motor windings are charged in a different level.

In this paper, a methodology is presented in order to identify the broken bar fault while the induction motor operates under asymmetrical voltage supply. The studied motors are delta-connected. The proposed method is based on the monitoring of the zero-sequence current waveform and the analysis of its frequency spectrum. It will be shown that this diagnostic mean reveals both the two faults at the same time. The method itself can be of great value if one thinks its characteristics: It is a non-intrusive and on-line method. It is based on current measurements and the Fast Fourier Transform so it is easily implemented. Generally, it is very similar to the traditional MCSA method with the difference of the number of the current sensors which are three (one for each phase) instead of one. Finally, with this method one can detect two faults at the same time, with very strong and reliable indicators.

The work will be divided and presented into two paragraphs. The first and second paragraphs concern the FEM simulations and results of a single and a double cage induction motor respectively. Both the two simulated motors, have identically the same stator from a real 3-phase induction motor which is characterized by 4kW, 4 poles, 400V. Also, they have 28 rotor slots. For all simulations the non-linear B-H magnetic characteristic of the rotor and stator iron core is taken into consideration. The performed FEM simulations are transient under constant speed 1460rpm. All simulated motors are considered un-skewed because of the high computational time. The FEM software used is OPERA while the application of the FFT is implemented with MATLAB. The simulated single cage induction motor FEM

978-1-4799-0024-4/13 $31.00 © 2013 IEEE

model has been verified experimentally in previously published works, under healthy and faulty conditions and under different operating conditions [11],[18].

II. SINGLE CAGE INDUCTION MOTOR FEM SIMULATION

In this paragraph, a NEMA's class A, aluminum cage induction motor is studied. 4 different cases are investigated: a) The motor is healthy and the voltage supply is symmetrical, b) The motor is healthy and the voltage supply is asymmetrical, c) The motor has a broken bar and the voltage supply is symmetrical and d) The motor has a broken bar and the voltage supply is asymmetrical. When the voltage supply is symmetrical the applied voltage is 380V for all phases. In the asymmetrical supply condition the first and second phases have 380V, while the third has 353.5V. The simulated, healthy, single cage induction motor is presented in Fig. 1.

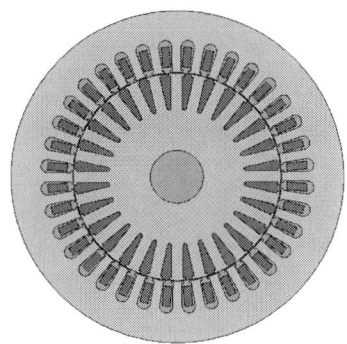

Fig. 1. The healthy single cage induction motor.

A. Simulation Results with Symmetrical Voltage Supply

In this section two cases are studied. The single cage induction motor under healthy and with one broken bar operation at 1460rpm. In Fig. 2-a, the FFT of the line current is presented for both cases. The broken bar fault diagnosis offers reliable signatures close to the stator MMF harmonics at 50Hz, 250Hz and 350Hz, totally agreeing with previous knowledge. Best fault signature can be considered the one at 47.6Hz with -31.4dB amplitude (Fig.2-b).

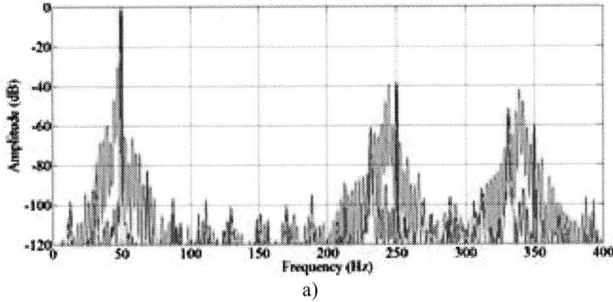

a)

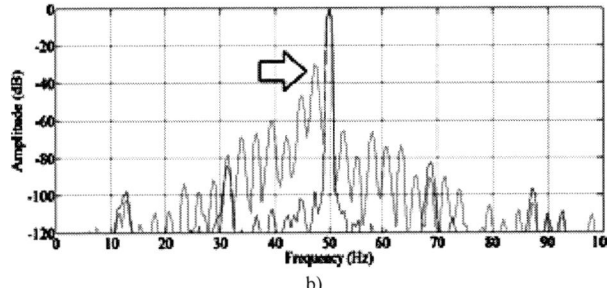

b)

Fig. 2. Line current frequency spectrum for the healthy (blue) and faulty (red) single cage induction motors under symmetrical voltage supply for: a) the 0-400Hz frequency range and b) the 50Hz frequency area.

In Fig. 3-a, the zero-sequence current spectrum for the two cases is presented. The fault can easily be identified close to the 3rd and 9th harmonic which are saturation related. Best fault signature can be considered the one at 147.6Hz with -18dB amplitude (Fig. 3-b). At this point one could claim that, one can gain stronger broken bar fault signatures with the zero-sequence current spectrum than with the traditional MCSA, at the cost of 2 extra current sensors.

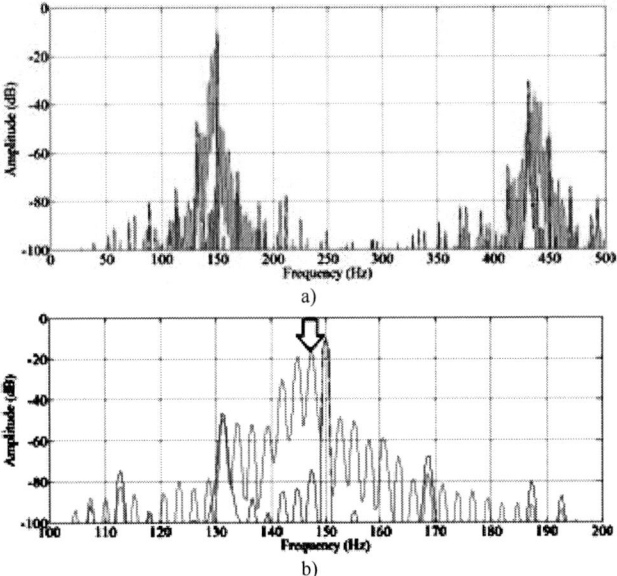

Fig. 3. The zero-sequence current spectrum for the healthy (blue) and faulty (red) single cage induction motors under symmetrical voltage supply for: a) the 0-500Hz frequency range and b) the 150Hz frequency area.

B. Simulation Results with Asymmetrical Voltage Supply

In this section, the single cage healthy motor and the one with the broken bar are simulated to operate under asymmetrical voltage supply for speed 1460rpm. In Fig. 4-a, one can see the frequency spectrum of the line current for both cases. The spectrum is very similar to the spectrum of Fig. 2-a. The main difference is that now, due to the voltage asymmetry, the 3rd current harmonic makes its presence and close to it one can find broken bar fault signatures also. But the problem here is that the third current harmonic is saturation related. So, there is no certain indication of the voltage asymmetry in a general situation. Best signature in the current spectrum is the one at 47.5Hz with -31.3dB amplitude (Fig. 4-b).

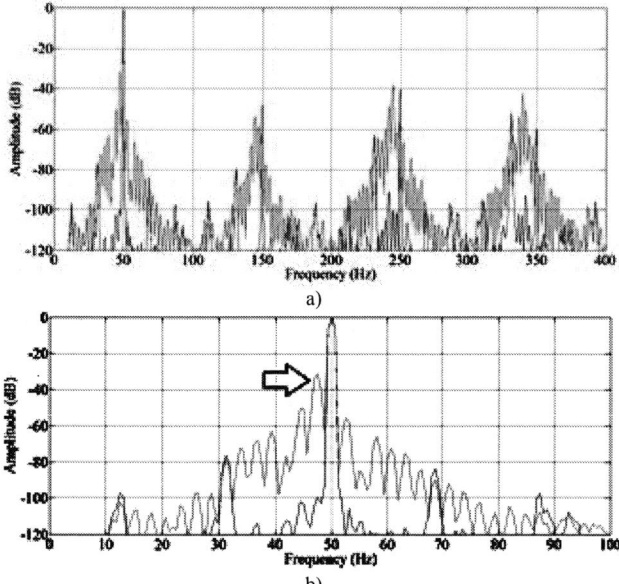

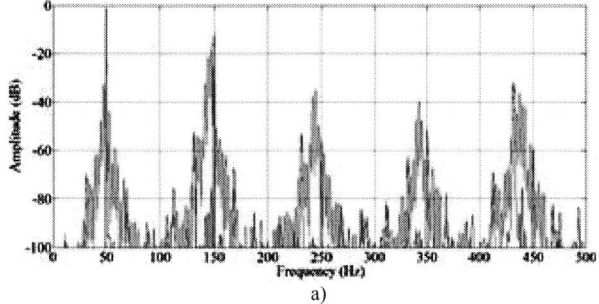

Fig. 4. Line current frequency spectrum for the healthy (blue) and faulty (red) single cage induction motors under asymmetrical voltage supply for: a) the 0-400Hz frequency range and b) the 50Hz frequency area.

In Fig. 5-a, one can observe the zero-sequence current spectrum for the two motor cases. It is revealed that, the presence of the voltage asymmetry induces strong stator MMF related harmonics (50Hz, 250Hz and 350Hz) in this signal's spectrum for the healthy and faulty cases. A comparison with Fig. 3-a shows that those harmonics did not exist at all when the voltage supply is symmetrical. Concerning the identification of the broken bar fault, strongest signature is the one at 147.5Hz with -17.5dB amplitude (Fig. 5-b).

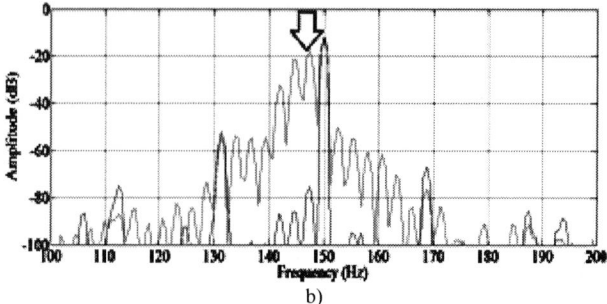

Fig. 5. The zero-sequence current spectrum for the healthy (blue) and faulty (red) single cage induction motors under asymmetrical voltage supply for: a) the 0-500Hz frequency range and b) the 150Hz frequency area.

III. DOUBLE CAGE INDUCTION MOTOR FEM SIMULATION

In this paragraph, a NEMA's class C, double cage induction motor is studied [19]. This motor has exactly the same stator with the previously studied single cage induction motor. Also, the rotor slots are 28. The upper cage is from aluminum while the inner cage from copper. The middle slot area between the 2 different bars is considered to be iron. A bar which belongs to the upper cage is considered to be broken. Same like in the previous paragraph II, 4 different cases are investigated: a) The motor is healthy and the voltage supply is symmetrical, b) The motor is healthy and the voltage supply is asymmetrical, c) The motor has a broken bar and the voltage supply is symmetrical and d) The motor has a broken bar and the voltage supply is asymmetrical. When the voltage supply is symmetrical the applied voltage is 380V for all phases. In the asymmetrical supply condition the first and second phases have 380V, while the third has 353.5V. The simulated, healthy, double cage induction motor is presented in Fig. 6.

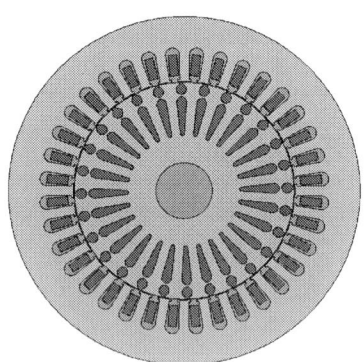

Fig. 6. The healthy double cage induction motor.

A. Simulation Results with Symmetrical Voltage Supply

In this section, the double cage induction motor operates for both healthy and faulty conditions at 1460rpm under symmetrical voltage supply. In Fig. 7-a, the line current frequency spectrum, for both cases, is presented. The results totally agree with previous work [17], since it is obvious that the fault signatures present much weaker amplitudes than the ones in the single cage induction motor (Fig. 2). The strongest signature seems to be the one at 47.5Hz with amplitude -45.6dB (Fig. 7-b).

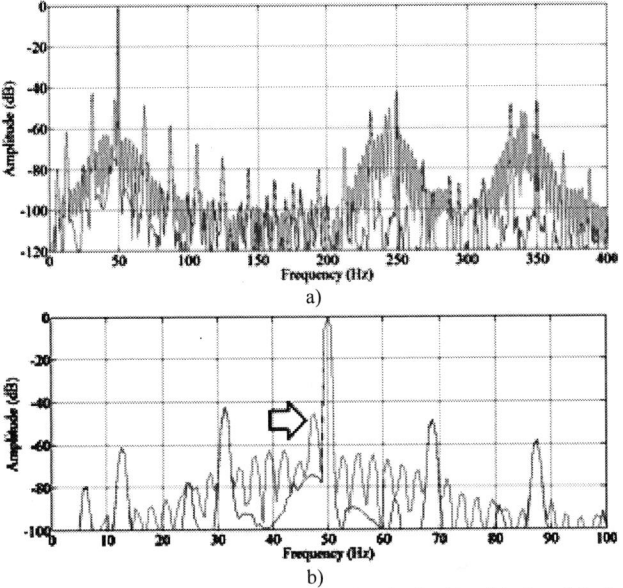
a)

b)

Fig. 7. Line current frequency spectrum for the healthy (blue) and faulty (red) double cage induction motors under symmetrical voltage supply for: a) the 0-400Hz frequency range and b) the 50Hz frequency area.

In Fig. 8-a the zero-sequence current spectrum is presented, for the healthy double-cage induction motor and the one with the broken bar fault. The strongest fault signature is located at 147.5Hz with amplitude -28.3dB (Fig. 8-b).

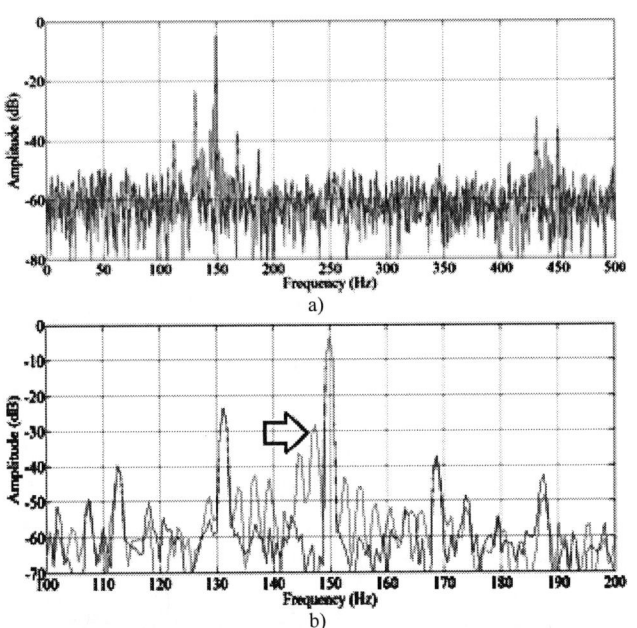
a)

b)

Fig. 8. The zero-sequence current spectrum for the healthy (blue) and faulty (red) double cage induction motors under symmetrical voltage supply for: a) the 0-500Hz frequency range and b) the 150Hz frequency area.

B. Simulation Results with Asymmetrical Voltage Supply

In this section the healthy and faulty, double cage induction motor will be simulated to operate under unbalanced voltage supply, at 1460rpm. In Fig. 9-a, the line current frequency spectrums for both cases, are presented. Same like in Fig. 4-a, one can easily observe the presence of the saturation related harmonic at 150Hz. Close to this harmonic the fault signatures are very weak compared to the ones located close to 50Hz, 250Hz and 350Hz. In this spectrum, the greatest fault signature is located at 47.5Hz with amplitude -46.7dB (Fig. 9-b).

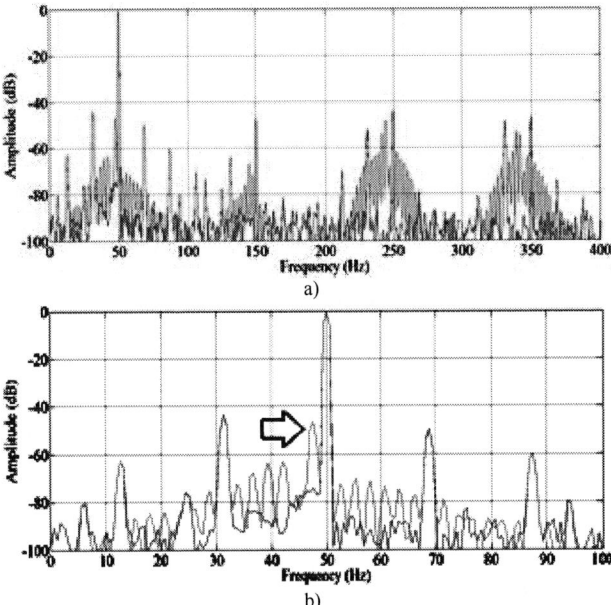
a)

b)

Fig. 9. Line current frequency spectrum for the healthy (blue) and faulty (red) double cage induction motors under asymmetrical voltage supply for: a) the 0-400Hz frequency range and b) the 50Hz frequency area.

The zero-sequence current spectrum of the healthy and faulty double cage induction motors is presented in Fig. 10-a. One can easily observe the existence of the stator MMF related harmonics at 50Hz, 250Hz and 350Hz which were missing in Fig. 8-a, for symmetrical voltage supply. The greatest broken bar fault signature is located at 147.5Hz with amplitude -28.43dB (Fig. 10-b).

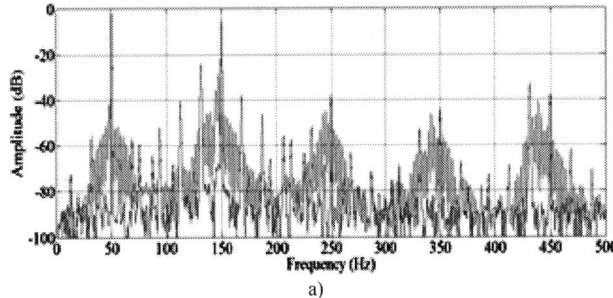
a)

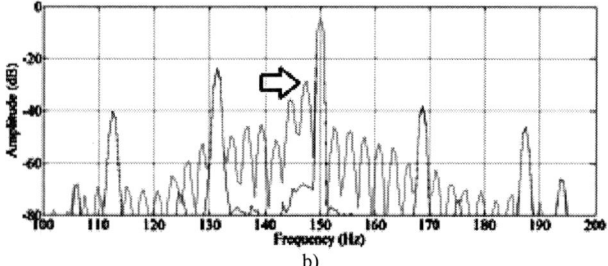

b)

Fig. 10. The zero-sequence current spectrum for the healthy (blue) and faulty (red) double cage induction motors under asymmetrical voltage supply for: a) the 0-500Hz frequency range and b) the 150Hz frequency area.

IV. CONCLUSIONS

This work's results lead to several remarks. Firstly, in all cases, the strongest broken bar fault signatures exist in the zero-sequence current spectrum close to the 150Hz harmonic. They prove to be about 12dB and 18dB greater in the single and double cage induction motor respectively, compared to the traditional MCSA signatures close to 50Hz. Secondly, the presence of the voltage imbalance affects the third line current harmonic. Unfortunately, one cannot be sure for this harmonic's origin since it is strongly related to the saturation level of the induction motor and does not predict directly the voltage supply asymmetry. Finally, the best indication of the co-existence of the two faults, can be retrieved through the zero-sequence current frequency spectrum and by reliably great fault signatures.

V. REFERENCES

[1] S. Nandi, H. Toliyat and X. Li, "Condition Monitoring and Fault Diagnosis of Electrical Motors-A Review," *IEEE Trans.Ener.Conv.,* Vol. 20, No. 4, pp. 719-729, Dec 2005.

[2] J. Milimonfared, H. M. Kelk, S. Nandi, A. D. Minassians and H. A. Toliyat, "A Novel Approach for Broken-Rotor-Bar Detection in Cage Induction Motors," *IEEE Trans.Ind.Applicat.,* Vol. 35, pp. 1000-1006, Sep/Oct 1999.

[3] Pinjia Zhang, ,YiDu , Thomas G. Habetler, Bin Lu, "A Survey of Condition Monitoring and Protection Methods for Medium-Voltage Induction Motors", IEEE Trans. Ind. Applicat. Vol.47, No.1, pp. 34-46, Jan/Feb. 2011.

[4] Humberto Henao, Hubert Razik, Gerard-Andre Capolino, "Analytical Approach of the Stator Current Frequency Harmonics Computation for Detection of Induction Machine Rotor Faults", IEEE Trans. Ind. Applicat. Vol.41,No 3, pp. 801-807, May/June 2005.

[5] J. Cusido, J. Rosero and E. Aldabas, "New Fault Detection Techniques For Induction Motors," *Electrical Power Quality and Utilisation, Magazine* Vol. II, No. 1, 2006.

[6] Andrzej M. Trzynadlowski, Ewen Ritchie," Comparative Investigation of Diagnostic Media for Induction Motors: A Case of Rotor Cage Faults" IEEE Trans. Indust. Electronics, Vol.47, No. 5, pp. 1092-1099, Oct. 2000.

[7] J. S. Hsu, "Monitoring of defects in induction motors through air-gap torque observation," *IEEE Trans. Ind. Appl.,* Vol. 31, No. 5, pp. 1016–1021, Sep./Oct. 1995.

[8] V. V. Thomas, K. Vasudevan, and V. J. Kumar, "Online cage rotor fault detection using air-gap torque spectra," *IEEE Trans. Energy Convers.,* vol. 18, no. 2, pp. 265–270, Jun. 2003.

[9] ST. J. Manolas and J. A. Tegopoulos, "Analysis of Squirrel CageInduction Motors with Broken Bars and Rings", *IEEE Trans. Energy Conv.,* vol. 14, no. 4, pp. 1300–1305, Dec. 1999.

[10] M. Eltabach, A. Charara, I. Zein and M. Sidahmed, "Detection of broken rotor bar of induction motors by spectral analysis of the electromagnetic torque using Luenberger observer", IECON'01, 27th Annual Conference of the IEEE IES, Denver, CO, pp.658-663, 29 Nov.-02 Dec. 2001.

[11] K. N. Gyftakis, D. V. Spyropoulos, J. Kappatou and E. D. Mitronikas, " A Novel Approach for Broken Bar Fault Diagnosis in Induction Motors through Torque Monitoring," IEEE Trans. Ener. Conv., Vol. 28, No. 2, pp. 267-277, Jun. 2013.

[12] Mario Eltabach, Ali Charara,, Isamil Zein, "A Comparison of External and Internal Methods of Signal Spectral Analysis for Broken Rotor Bars Detection in Induction Motors", IEEE Trans. Indust. Electronics, vol. 51, No. 1, pp. 107-121, Feb.2004.

[13] Andrzej M. Trzynadlowski, Ewen Ritchie," Comparative Investigation of Diagnostic Media for Induction Motors: A Case of Rotor Cage Faults" IEEE Trans. Indust. Electronics, Vol.47, No. 5, pp. 1092-1099, Oct. 2000.

[14] Zhenxing Liu, Xianggen Yin, Zhe Zhang, Deshu Chen, Wei Chen, "Online Rotor Mixed Fault Diagnosis Way Based on Spectrum Analysis of Instantaneous Power in Squirrel Cage Induction Motors", IEEE Trans. Energy Convers., Vol.19, No. 3, pp. 485-490, Sep.2004.

[15] Jonathan Sprooten, Jean-Claude Maun, "Influence of Saturation Level on the Effect of Broken Bars in Induction Motors Using Fundamental Electromagnetic Laws and Finite Element Simulations", IEEE Trans. Energy Convers. Vol. 24,No. 3, pp. 557-564, Sep.2009.

[16] R. Puche-Panadero, M. Pineda-Sanchez, M. Riera-Guasp, J. Roger-Folch, E. Hurtado-Perez and J. Perez-Cruz, "Improved Resolution of the MCSA Method Via Hilbert Transform, Enabling the Diagnosis of Rotor Asymmetries at Very Low Slip," IEEE Trans. Ener. Conv., Vol. 24, No. 1, pp. 52-59, Mar. 2009.

[17] Jongbin Park, Byunghwan Kim, Jinkyu Yang, Sang Bin Lee, Ernesto J. Wiedenbrug, Mike Teska, Seungoh Han, "Evaluation of the Detectability of Broken Rotor Bars for Double Squirrel Cage Rotor Induction Motors", Energy Conversion Congress and Exposition(ECCE), pp 2493 – 2500, IEEE 2010.

[18] K.N. Gyftakis and J.C. Kappatou, " A Novel and Effective Method of Static Eccentricity Diagnosis in 3-Phase, PSH-Induction Motors," *IEEE Trans. Ener. Conv.,* Vol. 28, No. 2, pp. 405-412, Jun. 2013.

[19] K. N. Gyftakis, D. Athanasopoulos and J. Kappatou, "Study of Double Cage Induction Motors with Different Rotor Bar Materials", IEEE International Conference on Electrical Machines XXth ICEM 2012, Marseille, France, 2-5 Sep. 2012.

VI. BIOGRAPHIES

Konstantinos N. Gyftakis was born in Patras, Greece, in May 1984. He received the diploma in Electrical and Computer Engineering from the University of Patras, Patras, Greece in 2010. He is a PhD Candidate in the Department of Electrical and Computer Engineering, University of Patras. His research activities are in FEM design, fault diagnosis and optimization of electrical machines. He is an IEEE member, member of IEEE PES, IAS, IES and Magnetics Society, member of the HELIEV (Hellenic Institute of Electric Vehicles) and finally member of the Technical Chamber of Greece. (E-mail: **kosgyftak@upatras.gr**).

Dimitrios K. Athanasopoulos was born in Patras, Greece, in November 1989. He received the diploma in Electrical and Computer Engineering from the University of Patras, Patras, Greece in 2012. He is a PhD Candidate in the Department of Electrical and Computer Engineering, University of Patras. His research activities are in optimization of electrical machines' design with FEM, power electronics and fault diagnosis. (E-mail: **athanasd@upatras.gr**).

Joya C. Kappatou was born in Argostoli, Greece. She received the diploma in Electrical Engineering from the University of Patras, Patras, Greece and the PhD from the same University in 1991 in the field of Electrical machines and Power Electronics. She is Assistant Professor in the Electrical and Computer Engineering Department of the University of Patras. Her teaching and research activities are in electrical machines, power electronics, modeling and design using FEM, faults diagnosis in electrical machines. Dr. Kappatou is a member of IEEE and the Technical Chamber of Greece. (University of Patras, Electrical and Computer Engineering Department, 26500 Rion-Patras, Greece, Tel: +30 2610/996413, Fax: +30 2610/997362, E-mail: **joya@ece.upatras.gr**).

978-1-4799-0024-4/13 $31.00 © 2013 IEEE

Broken Bar Detection Using Current Analysis - A Case Study

Dragan Matić, Željko Kanović, Dejan Reljić, Filip Kulić, Đura Oros, Veran Vasić

Abstract -- **This paper covers a case study of broken bar detection for 3.15 MW motor in a thermal power plant application. The motor current is measured in one phase. Feature extraction is based on transient and steady state analysis. Hilbert and Wavelet transforms are used to extract broken bar features. To discuss rotor condition in time domain skewness and kurtosis of current envelope are also considered. Low shaft-load conditions are present. In case of high-voltage, high-power induction motor reliable broken bar detection is possible when contemporary digital signal processing techniques are used .**

Index Terms – **rotor fault, broken bar, induction motor, Hilbert transform, Wavelet transform**

I. Nomenclature

The following nomenclature is used:

f_n - nominal supply frequency,
f_{bb} – broken bar feature frequency,
P_n – nominal active power,
U_n – nominal terminal voltage,
I_n – nominal motor current,
$i_a(t)$ – motor phase current,
$i_b(t)$ – motor phase current when broken bar is present,
I_H – modulus of analytical signal,
s – slip,
x_i – sampled data,
N – total number of data,
n_b – number of broken bars,
N_b – number of rotor bars.

II. Introduction

Induction motors cover over 80% of overall electro-mechanical conversion with installed power of 3 kW per person [1]. They are widely used in domestic and industrial applications. Malfunction of induction motor can potentially cause harm to environment, human being or produce a significant financial loss depending on an application they are involved. To prevent unwanted situations early fault detection techniques are used. Generally, they are financially more acceptable than periodic maintenance procedures [2], [3].

Common induction motor faults are related to stator, rotor and bearings malfunction, Fig.1. Rotor type of faults occupy around 10% and has a significant interest at academic and industrial level [4].

Motivation for this work is found in the broken bar detection for high-power induction motors in real industrial applications. This recearch is a case study wich will lead to practical use of known methods.

The paper is organized into the nine sections. The first section is nomenclature, the second section is intruduction; the third section describes industrial application, the fourth section gives overview of signal processing techniques used for extraction of broken bar features; in the fift section broken bar verification is shown, into six section conclusion is made. The rest of sections are appendix, references and biographies.

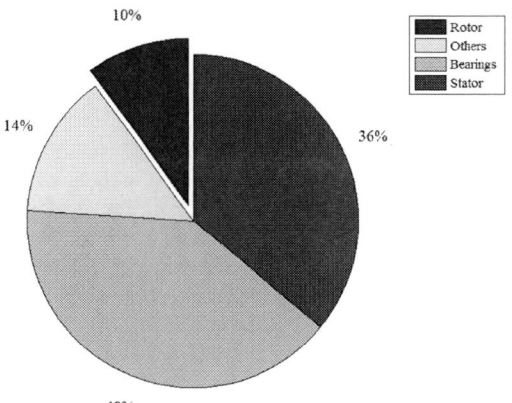

Fig.1. Survey of common induction motor faults

III. Thermal Power Plant Application

Thermal Power Plant – Heating Plant (TPP-HP) was constructed for a combined production cycle. The plant produces energy in a modern, co-generation process. TPP-HP consists of the main production unit (MPU) and the supporting one, which is used to actuate the MPU. MPU has two boilers, capacity of 330 t/h; 11.77 MPa each.

The high-pressure pump, which feeds the boiler, is a heart of the thermal power station as it supplies feed-water to the boiler continuously. The feed-water from the reservoir is pumped to the suction of the high-pressure boiler feed pump by low-pressure pump. The amount of feed-water is controlled with geared variable-speed hydrodynamic coupling. Both pumps are driven by high voltage squirrel-cage induction motor, 3.15MW each. A part of the motor

---

This work was supported by the Ministry of Education, Science and Technological Development of the Republic of Serbia under projects TR032018, TR033013, III42004 and by the Provincial Secretariat for Science and Technological Development of the Autonomous Province of Vojvodina under project 114-451-3508/2013-04.

D. Matić, Ž. Kanović and F. Kulić are with Sub-department for Automatic Control and Systems Engineering, Faculty of Technical Sciences, University of Novi Sad, Serbia, (dmatic@uns.ac.rs).

D. Reljić, Đ. Oros and V. Vasić are with Chair of Power Electronics and Converters, Faculty of Technical Sciences, University of Novi Sad, Serbia, (reljic@uns.ac.rs).

978-1-4799-0024-4/13 $31.00 © 2013 IEEE

drive is shown in Fig.2.

It is observed that one motor operates with the level of a mechanical vibration above normal, and the other operates properly. There has been a suspicion of some sort of a mechanical fault. The common types of mechanical faults in an induction machine with this kind of the symptoms are rotor broken bar and eccentricity faults.

Broken bar fault causes increase of vibration, noise, star-up sparking, motor losses and the detriment of torque [5]. To determine a cause of the motor malfunction data acquisition is required.

Fig.2. Thermal power plant application

IV. CURRENT ACQUISITION AND FEATURE EXTRACTION

For the current acquisition are used: National Instruments USB-6251 digital acquisition card, standard PC, and measuring clamps with ratio 400/5 A. One phase current signal is acquired from the standstill to the nominal speed (Fig.3 and Fig.6); motor data are given in the appendix. The current signals are acquired from healthy and malfunction motor. Sampling frequency is set to 5 kHz.

A broken bar in a rotor can be successfully detected by analyzing the current or vibration signal of the motor. Motor current signature analysis (MCSA) is widely spread method for a broken bar detection [6], [7]. In this paper broken bar detection is based on motor current analysis. The motor current is measured only in one phase.

A. Steady state analysis

Steady state analysis is based on MCSA method, which searches for the abnormalities in the amplitude spectra of the current signal. The amplitude spectra is estimated by Fast Fourier Transform (FFT), and normalized to 0 dB. If the broken bar is present and the shaft load is nominal, the characteristic features appear at frequencies [8]:

$$f_{bb} = [1 \pm 2ks] f_n, \quad k = 1, 2, 3 \dots \quad (1)$$

Discussing the magnitudes at the given frequencies (1) the rotor condition is established. A common case is to observe the magnitudes of the first characteristic features

around the supply frequency, left side band (LSB) and right side band (RSB).

Broken bar detection based on FFT analysis has several disadvantages: spectral leakage due to finite time window, need for high frequency resolution, varying load conditions and confusing mechanical frequencies [9].

The low load working conditions are present in the case study due to MCSA approach suffers from spectral leakage that buries LSB feature under the supply frequency of 50 Hz, Fig.4. Consequently, the rotor condition cannot be determent and an alternative approach is required.

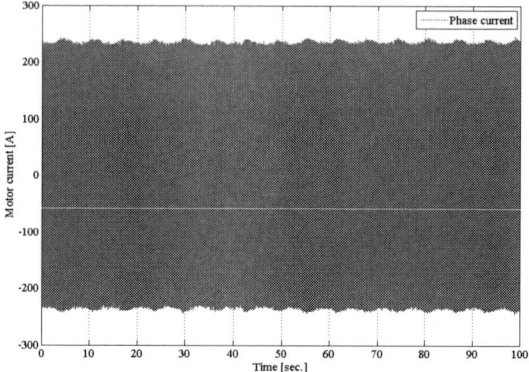

Fig.3. Steady state motor current signal

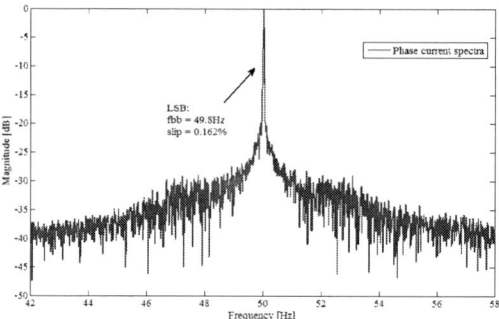

Fig.4. Features buried under supply frequency

B. Spectral analysis of analytical signal modulus

Due to disadvantages of MCSA alternative approach is used. Method is based on the amplitude spectra analysis of the modulus of the analytical motor current signal. The modulus of the analytic signal shows a pulsation with the characteristic frequency of the machine fault [9]:

$$\vec{i}_b(t) = \left[1 + \frac{n_b}{N_b} \cos(2\pi(2sf_s)t) \right] I_m e^{j\omega t}. \quad (2)$$

If the characteristic spike exists in the low frequency range, a broken bar is present. Analytical signal is obtained by Hilbert transform (HT). Direct component (DC) of the observed signal is removed, which provides reliable broken bar feature that covers full load range. Observed variable is shown as a pseudo code [9]:

$$I_H = abs(hilbert(i_a)) - mean(abs(hilbert(i_a))). \quad (3)$$

In the case steady of 3.15 MW motor, the spectrum of the modulus of analytical signal is observed, Fig. 5. Existence of the spike at frequency of 0.162 Hz clearly indicates presents of the broken bar fault. To verify the clam additional signal processing techniques is used.

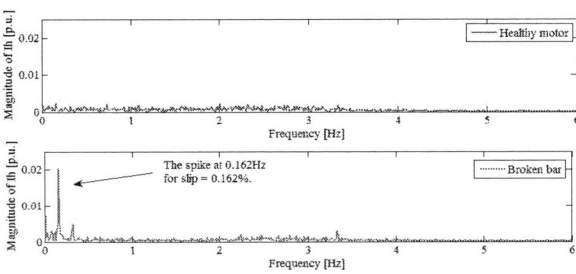

Fig.5. Broken bar detection

C. Digital Wavelet Decomposition

By analyzing the star-up current of an induction motor, it is possible to determine the existence of a broken bar. In this paper, Digital Wavelet Transform (DWT) is used to perform transient signal decomposition. The original signal is decomposed into details and approximations. The level of decomposition depends on a sampling frequency and need to be chosen in the way to cover full frequency range of the broken bar fault. If LSB is observed frequency range is 0 - 50 Hz [10], [11]. When Daubechies 44 mother wavelet is used, for given sampling frequency of 5 kHz, characteristic features of broken bar appears at 8th level of details [10]. In Fig.7 and Fig.8 comparison of healthy and broken bar case is shown. In Fig7 two characteristic segments on lower plot can be spotted. Presents of higher oscillations indicate broken bar presents. In Fig.8 comparison of broken bar and healthy features is done. It is clear that the magnitudes of the characteristic features are higher when the broken bar is present, Fig.8.

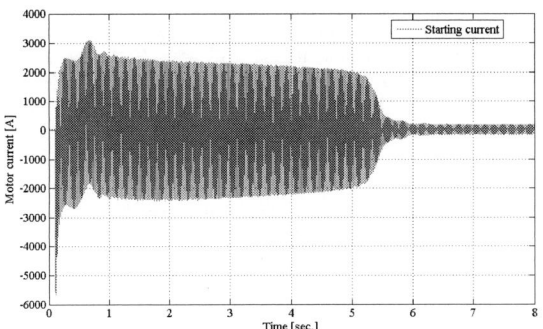

Fig.6. Starting motor current signal

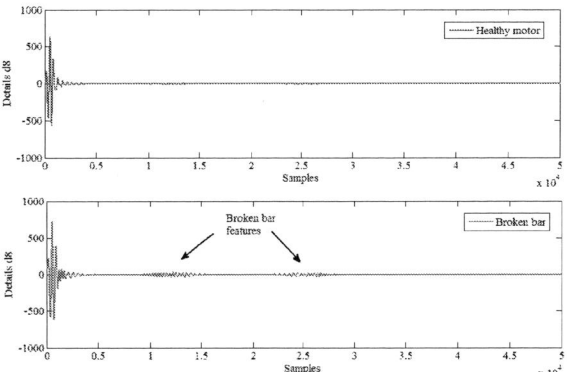

Fig.7. Wavelet decomposition at level details 8th

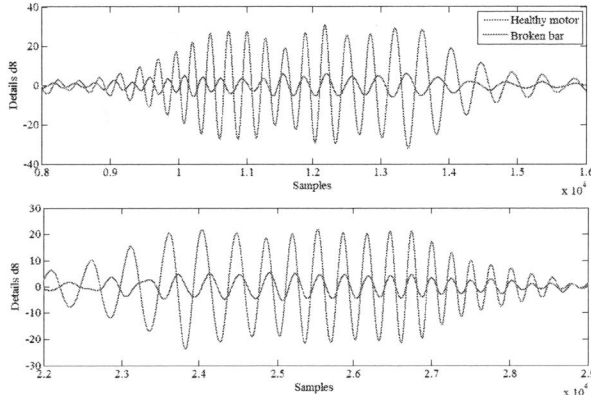

Fig.8. Comparison of broken bar features

D. Current Envelope Analysis

To eliminate presents of supply frequency of 50 Hz whichs dominates current spectrum in search for broken bar fault, analisys of current evelop is adviced [12], Fig.9.

In this case it is challeging to determen reliable broken bar features. As a measure, statistical moments like skewnes and kurtosis can be used [13]:

$$\mu = \frac{1}{N} \sum_{i=1}^{N} x_i, \quad (4)$$

$$\sigma = \sqrt{\frac{1}{N} \sum_{i=1}^{N} (x_i - \mu)^2}, \quad (5)$$

$$c = \frac{1}{N\sigma^3} \sum_{i=1}^{N} (x_i - \mu)^3, \quad (6)$$

$$k = \frac{1}{3N\sigma^4} \sum_{i=1}^{N} (x_i - \mu)^4. \quad (7)$$

In Tab.1 values of statistical moments for envelop signals are presented. It is obvious that skewnes (6) has higher value when broken bar is present, same conclusion can be made when kurtosis (7) is observed.

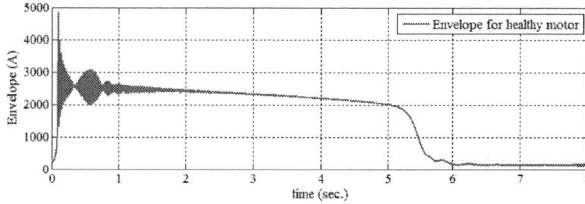

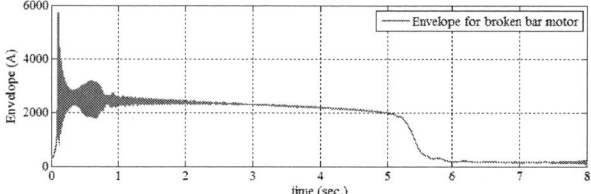

Fig.9. Envelope of starting motor current

Table 1. Statistical moments for envelope signals

	Skewnes	Kurtosis
Healthy	-0.51	1.55
Bro.bar	-0.41	1.7

V. BROKEN BAR VERIFICATION

Based on the results presented in previous section can be concluded the presents of broken bar fault. The tested rotor is dismounted. By conducting the visual inspection, it is verified presents of broken bar fault, Fig.10.

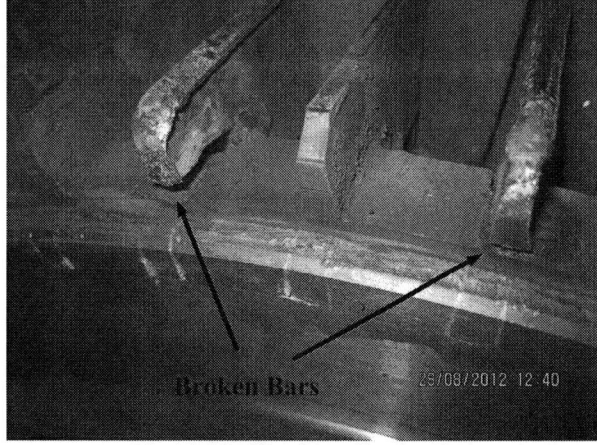

Fig.10. Verification of broken bar

VI. CONCLUSION

In this paper it is shown process of broken bar detection for high-power high-voltage induction motor. Low load working conditions force use of digital signal processing techniques alternative to MCSA approach. MCSA cannot be used due to spectral leakage, which made feature extraction difficult.

Shown procedures are applicable in real working conditions, relatively easy to implement and deploy.

Diagnostic procedures have advantage to periodic maintain, they are less costly and more reliable.

VII. APPENDIX

Induction motor data: type: 1 ZKV6 630 M-2, P_n = 3150 KW, U_n = 6 kV, I_n = 373 A, f_n = 50 Hz, ω_n = 2982 rpm, $\cos\varphi$ = 0.92, connection: star, rotor type: one cage, 56 bars.

VIII. REFERENCES

[1] R. Crowder, *Electric drives and electromechanical systems.* Oxford: Elsevier, 2006.
[2] H. A. Taliyat and T.A. Lipo, "Transient analysis of cage induction machines under stator, rotor bar and end ring faults," *IEEE Transactions on Energy Conversion*, vol.10 no.2,pp.241-247, 1995.
[3] W.T. Thomson and M. Fenger, "Current Signature Analysisto Detect Induction Motor Faults," *IEEE Industry Application Magazine*,vol.7, no.4, pp.26-34, 2001.
[4] D. Matić, F. Kulić, and V. Bugarski, "Survey of the methods for online broken bar induction motor fault detection," *Journal on Processing and Energy in Agriculture*, vol.14, no.2, pp.90-93, 2010.
[5] A. Bellini, F. Fillippeti, G. Franceschini, and C. Tassoni,"Quantitative Evaluation of Induction Motor Broken Bars by Means of Electrical Signature Analysis," *IEEE Transactions on Industry Applications*, vol.37, no.5, pp.1248-1255, 2001.
[6] M.E.H. Benbouzid, "Bibliography on induction motors faults detection and diagnosis," *IEEE Transactions on Energy Conversion*, vol.14, pp.1065–1074, 1999.
[7] M.E.H. Benbouzid,"A review of induction motors signature analysis as a medium for faults detection," *IEEE Transactions on Industrial Electronics*, vol.47, pp.984–993, 2000.
[8] G.B. Kilman and J. Stein, "Methods of motor current signature analysis," *Electric Machines and Power Systems*, vol.20, pp.463–474, 1992.
[9] R. Puche-Panadero, M. Pineda-Sanchez, M. Riera-Guasp, J. Roger-Folc, E. Hurtado-Perez and J. Perez-Cruz, "Improved resolution of the MSCA method via Hilbert transform, enabling the diagnosis of rotor asymmetries at very low slip," *IEEE Transactions on Energy Conversion*, vol.24, no.1, pp.52–59, 2009.
[10] J.A. Antonino-Daviu, M. Riera-Guasp, J.R. Folch and M.P.M. Palomares, "Validation of a new method for the diagnosis of rotor bar failures via wavelet transform in industrial induction machines," *IEEE Transactions on Industry Applications*, vol.42, no.4, pp.990-996, 2006.
[11] J.A. Antonino-Daviu, V. Climente-Alarco, J. Pons-Llinares, R. Puche, and M. Pineda-Sanchez, "Transient-based analysis for the detection of broken damper bars in synchronous motors," *Mechanical Systems and Signal Processing*, vol.34, pp.367-377, 2013.
[12] N. Ertugrul; W.L. Soong, D.A. Gray, C. Hansen and J. Grieger, "Detection of broken rotor bars in induction motor using starting-current analysis and effects of loading," *Electric Power Applications, IEE Proceedings*, vol.153, no.6, pp.848-855, 2006.
[13] P. Stepanić, I.V. Latinović and Ž. Đurović, "A new approach to detection of defects in rolling element bearings based on statistical pattern recognition," *Journal of Advanced Manufacturing Technology*, vol.45,pp.91-100, 2009.

IX. BIOGRAPHIES

Dragan Matić was born in Novi Sad, Serbia in 1978. He received PhD from Faculty of technical science, University of Novi Sad in 2012. He is involved in lecturing as an assistant professor and currently works on several projects. Fields of interests are: artificial intelligence, system control and fault detection. He is a member of IEEE.

Željko Kanović was born on July 18th, 1976, in Sombor, Serbia. He received the B.Sc. degree in Mechanical Engineering in 2000, and the M.Sc. and Ph.D. degree in Electrical Engineering in 2007 and 2012, all from the University of Novi Sad, Serbia. Currently, he is a Teaching Assistant at the Computing and Control Department at the same University. He is a member of IEEE.

Dejan Reljić was born in Prijepolje, Republic of Serbia, in 1977. He received the Dipl. Ing. and M.Sc. degrees in electrical engineering from the

Faculty of Technical Sciences, University of Novi Sad, Serbia, in 2002 and 2006, respectively. In 2002 he joined the Department for Power, Electronics and Telecommunications Engineering, Faculty of Technical Sciences, University of Novi Sad, where he is currently a Teaching and Research Assistant. His main research and teaching interests include control and application of electrical drives, modeling and simulation of electrical machines. He is a member of IEEE.

Filip Kulić was born in Novi Sad, Serbia in 1968. He received PhD from Faculty of technical science, University of Novi Sad in 2003. He is involved in lecturing as an associate professor and currently works on several projects. Fields of interests are: artificial intelligence, system control and fault detection. He is a member of IEEE.

Đura V. Oros was born in Ruski Krstur, Serbia, 1957. He received the MS, Diploma degree from the Electrical Engineering Faculty, University of Belgrade and PhD degree from the University of Novi Sad, 1997 and 2008, respectively, all in Electrical Engineering. He is at the University of Novi Sad, teaching course of electrical machine drives. His main research interest is in the area of electric drives and electrical machines parameter estimation. He is a member of IEEE.

Veran V. Vasic was born in Sabac, Serbia on December 8, 1970. He received the BS Diploma degree from the University of Novi Sad, the MS and PhD degrees from the Electrical Engineering Faculty, University of Belgrade, in 1994, 1996 and 2001, respectively, all in Electrical Engineering. Since September 1994, he is at the University of Novi Sad, teaching course of electrical machine and drives. His main research interest is in the area of high-performance electric drives, modeling and simulation of electric machines. He is a member of IEEE.

978-1-4799-0024-4/13 $31.00 © 2013 IEEE

A Novel Broken Rotor Bar Fault Detection Method Using Park's Transform and Wavelet Decomposition

Ramin Salehi Arashloo, José Luis Romeral Martinez, Mehdi Salehifar,

Department of Industrial Engineering, Universitat Politècnica de Catalunya. BarcelonaTech, Spain

Center of Motor Control and Industrial Applications (MCIA)

ramin.salehi@mcia.upc.edu   luis.romeral@upc.edu   mehdi.salehifar@mcia.upc.edu

Abstract

Detection of broken rotor bars has been an important but difficult work in fault diagnosis area of induction motors. The characteristic frequency components of faulted rotor are very close to the power frequency component but by far less in amplitude, which brings about great difficulty for accurate detection.

In the present study, a new method is proposed in order to remove the main frequency component, resulting in more efficient detection of the rotor fault characteristics in the frequency spectrum of stator currents. The method is based on Park's transformation in combination with discrete wavelet decomposition to eliminate the effect of main frequency and zoom on the energy of objective fault related frequency components. In addition, the method efficiency is evaluated using Simulations in Matlab.

Keywords: Induction Motor, Fault Detection, Broken Rotor Bars, Discrete Wavelet Transform, Park's Transform.

I. Introduction

Induction motors are among the most significant electromechanical systems, which are widely spread in almost every industry. These machines often provide the core capabilities for industrial success because they are mechanically robust and cost effective. In order to keep the machine at its best performance level, techniques such as condition monitoring and fault detection have become increasingly essential [1].

Induction machine faults are generally classified as either mechanical or insulation faults. Common mechanical faults include rotor bar breakage, rotor end ring cracking, bearing or gearbox failures, static and/or dynamic air-gap irregularities, stator winding faults, and bent shaft. Statistical data demonstrate that the mechanical faults are responsible for more than 90% of all failures [2]. Among these many mechanical faults, ongoing research work is in particular being focused on broken rotor bars faults and on the development of fault diagnostic techniques [3].

There are different methods for the detection of mechanical signals of an incipient fault. Typical examples as related to this may include detection of bearing faults or air-gap eccentricity by monitoring the vibration signal or by analyzing the lubricating oil debris in the case of former fault [4], [5]. Other diagnostic methods include temperature measurement, acoustic noise analysis, infrared analysis, partial discharge measurement and radio frequency emission monitoring [6], [7]. In addition, recently some important issues on motor fault diagnosis methods have been published [8], [9].

Even though mechanical sensing techniques based on thermal and vibration monitoring have been utilized widely, these methods are in second-order if we compare them to stator current analysis. For example, in the case of vibration monitoring, the analyses are affected by the base excitation motion due to presence of a number of machineries in the production site. Moreover, in a number of cases, the severity of the fault has to be high enough to make it detectable at all by the mentioned-specified diagnostic methods [10].

Motor-current-signature analysis (MCSA) is a condition monitoring technique which has been widely implemented to diagnose incipient faults in electrical motors. MCSA focuses its efforts on the frequency spectrum analysis of the stator current and has been successfully applied to detect rotor failures, shorted turns in stator windings and abnormal levels of air-gap eccentricity among other mechanical faults [11], [12]. In addition, MCSA has been successfully implemented in many industrial cases since the 1980s. Furthermore, Induction motor is modeled and rotor broken bars fault has been detected experimentally using Park's transform [13].

Rotor fault characteristic frequencies are close to the stator current main frequency. Especially, for low slip values, where the fault harmonic characteristics are very close to the supply frequency, the spectral leakage phenomenon can obscure the fault characteristic frequencies [14].

On the other hand, the task of distinguishing faulty rotor conditions based on simple signal processing methods like FFT can only be done accurately as long as the signals are stationary, the terminal voltages are sinusoidal, and the induction motors are used around full-load condition. But the stator currents are usually nonstationary signals whose properties change with

978-1-4799-0024-4/13 $31.00 © 2013 IEEE

the time-variant operating conditions of the motor such as fluctuations in mechanical torque and power supply.

Wavelet transform is a mathematical tool that has recently emerged for applications such as time-frequency analysis, detection of irregularities, and feature extraction. The efficiency of wavelets is due to properties such as translation property and dilation property that can be used to automatically zoom in and out in order to locate positions of low and high frequency changes [2]. In 1995, wavelet decomposition analysis is used to detect broken rotor bars of induction motors [15], and variance of wavelet coefficients energy in different areas is used to detect broken rotor bars [16].

In the present study, a new method is proposed in order to remove the main frequency component, resulting in more efficient detection of the rotor fault characteristics in the frequency spectrum of stator currents. The method is based on Park's transformation in combination with discrete wavelet decomposition to eliminate the effect of main frequency and zoom on the energy of objective fault related frequency components. The remainder of this paper is organized as follows:

Section II describes the most common mechanical faults in induction motors and their diagnosis using classical MCSA; a brief review of wavelet transform is presented in section III; in sections IV and V Park's transform and proposed detection method are discussed; simulation results obtained by proposed algorithm are presented and compared for different conditions of motor in section VI, and finally, the summary and conclusions drawn from this study are presented in Section VII.

II. Spectrum Analysis of Stator Current and Problem Statement

The rotating magnetic field induces rotor voltages and rotor currents at slip frequency, and this produces an effective three phase magnetic field rotating at slip frequency with regard to the rotor. Two different cases can be appeared due to this rotating magnetic field:

1) Symmetrical cage winding → Only forward rotating field is produced.
2) Asymmetric rotor → A backward rotating field will result at slip frequency with respect to the rotor.

This backward-rotating field induces balanced three phase voltages in stator windings at the corresponding frequency, and a related current which modifies the stator-current frequency spectrum also appears. Different rotating fields appear with different faults in the induction machine, such as air-gap eccentricity, broken rotor bars and bearing damage. In such faults, the current frequencies associated with the rotating fields will be expressed by (1), (2) and (3).

1) Broken rotor bars:

$$f_{brb} = f_s \left[l \left(\frac{1-s}{p} \right) \pm s \right] \tag{1}$$

where $\dfrac{l}{p}$ = 1, 5, 7, 11, 13, . . . are the characteristic values of the motor.

2) Air-gap-eccentricity:

$$f_{ecc} = f_s \left[1 + m \left(\frac{1-s}{p} \right) \right] \tag{2}$$

where m = 1, 2, 3, . . . is a positive integer number, p is the number of pole pairs, s is the per-unit slip, and f_s is the electrical supply frequency.

3) Bearing damage:

$$f_{bng} = \left| f_s \pm f_{i,o} \right|$$
$$f_{i,o} = \frac{n_b}{2} f_r \left[1 \pm \frac{b_d}{p_d} \cos \beta \right] \tag{3}$$

where n_b is the number of bearing balls, $f_{i,o}$ are the characteristic vibration frequencies, f_r is the mechanical rotor speed in hertz, b_d is the ball diameter, p_d is the bearing pitch diameter, and β is the contact angle of the balls on the races [17].

III. Wavelet Transform

The wavelet transform [WT] is a powerful tool in the field of power systems signal processing [18]. It has the advantage of flexibility in describing nonstationary signals which is an important advantage for variable load applications and power quality problems. Similarly, wavelet transform is the breaking up of a signal into shifted and scaled versions of the source (or mother) wavelet.

The wavelet transform is divided into two main categories, continuous wavelet transform and discrete wavelet transform [19]. In wavelet transformation, the analysing wavelet functions will adjust their time-widths to their frequencies in such a way that lower frequency wavelets will be very broad and higher frequency ones will be narrower. Therefore, transient components of the signal in the upper frequency witch is isolated in a shorter part of power frequency cycle can be detected. The ability of WT to concentrate on long time intervals for low frequency components and short time intervals for high-frequency components leads to better evaluation of the signals with localized transients [18].

A. Continuous Wavelet Transform
Wavelet transform has the ability of performing through a Multi-Resolution Analysis (MRA). For a

signal $f(t)$, the generating function of the MRA can be expressed as:

$$\varphi_k^j = 2^{-j/2} \varphi\left(2^{-j} t - k\right) \quad (4)$$

where, φ is the so-called mother wavelet; k is the time shift factor, and j indicates the decomposition level. The wavelet coefficients obtained by applying an orthogonal wavelet can be expressed as:

$$d_k^j = \int_{-\infty}^{\infty} f(t) \psi_k^j(t) dt \quad (5)$$

Where ψ_k^j is the wavelet analyzing functions, for instance, Shannon, Morlet, Debauche and Haar could be used. During the transformation process, the mother wavelet is shifted and scaled contiguously and the correlation of the examined signal and the mother wavelet produces the wavelet coefficients.

B. Discrete Wavelet Transform

In discrete wavelet transform, the mother wavelet is scaled in the power of 2. Therefore, it is a good choice for implementation in digital computers. Assume S is a discrete-time signal to be decomposed using the discrete wavelet analysis into its approximate and detailed versions. In the first decomposition level, the coefficients are cA_1 and cD_1; where, cD_1 is the detailed representation of S, and cA_1 is the approximate version of S. cA_1 and cD_1 are expressed as:

$$cA_1 = \sum_k^n L(k - 2n)S(k) \quad (6)$$

$$cD_1 = \sum_k^n H(k - 2n)S(k) \quad (7)$$

where, L and H are the decomposition filters of $S(n)$ in cA_1 and cD_1 respectively. The base of the second decomposition level is cA_1 and the coefficients can be expressed as:

$$cA_2 = \sum_k^n L(k - 2n)cA_1(k) \quad (8)$$

$$cD_2 = \sum_k^n H(k - 2n)cA_1(k) \quad (9)$$

Computation of the higher-level decompositions has a similar way. Hence, it can be seen that cA_1 is the approximate version of the original signal S. L behaves as a low-pass filter. If cD_1 contains only high frequency components of signal S, then H will behave as a high-pass filter. After the decomposition process, the original signal can be reconstructed again.
This recursive process can be expressed as:

$$S' = A_n + D_n + D_{(n-1)} + \ldots \ldots D_2 + D_1 \quad (10)$$

and if the detail and approximation coefficients are not modified before the reconstruction process then S'

will be equal to S. Fig. 1 shows frequency distribution up to 3 levels of decomposition [19].
In Fig. 2 the implementation procedure of a discrete wavelet transform is depicted. In this picture s is the original signal, $2\downarrow$ denotes a down sampling by a factor of 2, and LPF and HPF are the low-pass and high-pass frequency filters respectively. The original signal is divided into two halves of the frequency bandwidth at the first stage and sent to both LPF and HPF. The same procedure will be repeated and the output of the LPF will be further cut into half of the frequency bandwidth and will form the input of the second level.

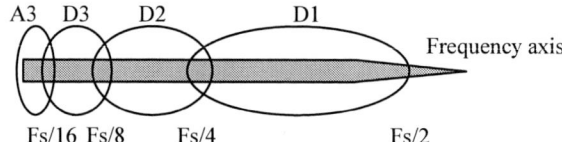

Fig.1 Frequency distribution of 3 levels wavelet decomposition

This procedure is repeated until the signal is decomposed to the desired level.

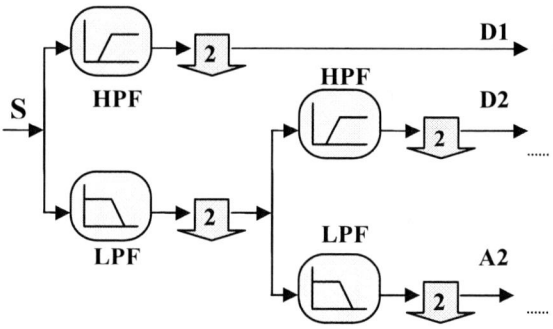

Fig. 2 Implementation procedure of discrete wavelet transforms

IV. Park's Transform

Park's transform is the most widely used method for torque and flux control in Vector Control applications of induction motors. This transform is in addition a useful way for analyzing stator current harmonics, especially those of less amplitude. It is a simple and powerful mathematical tool that is based on an arithmetical base rotation and is able to convert a symmetrical system with three variables (a, b, c) to another system with two perpendicular variables (d, q). Fig. 3 shows the diagram of abc frame and dqo frame in Park's transformation.

978-1-4799-0024-4/13 $31.00 © 2013 IEEE   414

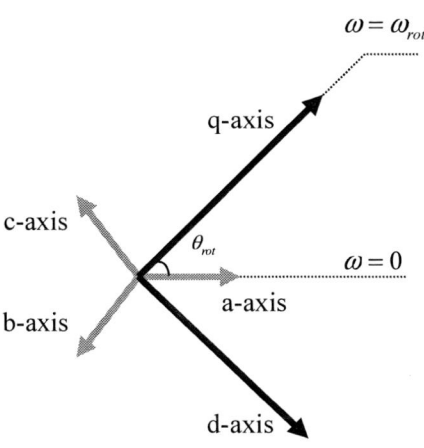

Fig. 3 Park's transformation rotating frame, (d-axis and q-axis)

This transformation is detailed on the following equations. Where transformed variables can be voltages, currents or flux, and θ_{rot} is the Park's rotating frame angle [20].

$$\theta_{rot} = \omega_{rot} t \tag{11}$$

$$\left[F_{qdo} \right] = \left[T_{qdo}\left(\theta_{rot} \right) \right]\left[F_{abc} \right] \tag{12}$$

$$\left[T_{qdo}\left(\theta_{rot} \right) \right] = \frac{2}{3}\begin{bmatrix} \cos\theta_{rot} & \cos\left(\theta_{rot} - \dfrac{2\pi}{3} \right) & \cos\left(\theta_{rot} + \dfrac{2\pi}{3} \right) \\ \sin\theta_{rot} & \sin\left(\theta_{rot} - \dfrac{2\pi}{3} \right) & \sin\left(\theta_{rot} + \dfrac{2\pi}{3} \right) \\ \dfrac{1}{2} & \dfrac{1}{2} & \dfrac{1}{2} \end{bmatrix} \tag{13}$$

$$\left[I_{abc} \right] = \begin{bmatrix} I_m \cos\omega t \\ I_m \cos\left(\omega t - \dfrac{2\pi}{3} \right) \\ I_m \cos\left(\omega t + \dfrac{2\pi}{3} \right) \end{bmatrix} \tag{14}$$

$$\left[I_{qdo} \right] = \left[T_{qdo} \right]\left[I_{abc} \right] \tag{15}$$

$$\left[I_{qdo} \right] = \frac{2}{3}\begin{bmatrix} \cos\theta_{rot} & \cos\left(\theta_{rot} - \dfrac{2\pi}{3} \right) & \cos\left(\theta_{rot} + \dfrac{2\pi}{3} \right) \\ \sin\theta_{rot} & \sin\left(\theta_{rot} - \dfrac{2\pi}{3} \right) & \sin\left(\theta_{rot} + \dfrac{2\pi}{3} \right) \\ \dfrac{1}{2} & \dfrac{1}{2} & \dfrac{1}{2} \end{bmatrix}$$

$$\times \begin{bmatrix} I_m \cos\omega t \\ I_m \cos\left(\omega t - \dfrac{2\pi}{3} \right) \\ I_m \cos\left(\omega t + \dfrac{2\pi}{3} \right) \end{bmatrix} \tag{16}$$

$$\left[I_{qdo} \right] = I_m \begin{bmatrix} \cos\left(\theta_{rot} - \omega t \right) \\ \sin\left(\theta_{rot} - \omega t \right) \\ 0 \end{bmatrix} \tag{17}$$

V. Proposed Approach

By analyzing stator current harmonic spectrum, incipient state of rotor faults could be detected, especially as abnormalities in the energy of fault harmonic characteristics. However, most of those harmonics amplitudes are not high enough for good detection and correct diagnosis of the fault.

The effect of broken rotor bar in three phase induction motors can be considered as an additional rotating magnetic field caused by rotor in the air gap which induces appropriate fault harmonic characteristics in stator currents. As stator three phase windings are placed symmetrical in motor, it can be inferred that the induced fault characteristic harmonics are balanced in stator three phase windings with same magnitude and frequency, but with $2\pi/3$ phase deference in angle. In the following equations, the resultant dqo currents in faulty condition with only one fault characteristic component ($i_{m_{brb}} \cos\left(\omega_{brb} t \right)$) is computed:

$$\left[I_{abc} \right] = \begin{bmatrix} I_m \cos\left(\omega t \right) + i_{m_{brb}} \cos\left(\omega_{brb} t \right) \\ I_m \cos\left(\omega t - \dfrac{2\pi}{3} \right) + i_{m_{brb}} \cos\left(\omega_{brb} t - \dfrac{2\pi}{3} \right) \\ I_m \cos\left(\omega t + \dfrac{2\pi}{3} \right) + i_{m_{brb}} \cos\left(\omega_{brb} t + \dfrac{2\pi}{3} \right) \end{bmatrix} \tag{18}$$

$$\left[F_{qdo} \right] = \left[T_{qdo}\left(\theta_{rot} \right) \right]\left[F_{abc} \right] \tag{19}$$

$$\left[I_{qdo} \right] = \left[T_{qdo} \right]\left[I_{abc} \right] \tag{20}$$

$$
\left[I_{qdo} \right] = \frac{2}{3}
\begin{bmatrix}
\cos\theta_{rot} & \cos\left(\theta_{rot} - \dfrac{2\pi}{3}\right) & \cos\left(\theta_{rot} + \dfrac{2\pi}{3}\right) \\[2mm]
\sin\theta_{rot} & \sin\left(\theta_{rot} - \dfrac{2\pi}{3}\right) & \sin\left(\theta_{rot} + \dfrac{2\pi}{3}\right) \\[2mm]
\dfrac{1}{2} & \dfrac{1}{2} & \dfrac{1}{2}
\end{bmatrix}
$$

$$
\times
\begin{bmatrix}
I_m \cos\left(\omega t\right) + i_{br} \cos\left(\omega_{br} t\right) \\[2mm]
I_m \cos\left(\omega t - \dfrac{2\pi}{3}\right) + i_{br} \cos\left(\omega_{br} t - \dfrac{2\pi}{3}\right) \\[2mm]
I_m \cos\left(\omega t + \dfrac{2\pi}{3}\right) + i_{br} \cos\left(\omega_{br} t + \dfrac{2\pi}{3}\right)
\end{bmatrix}
\tag{21}
$$

$$
\left[I_{qdo} \right] = \frac{2}{3}
\begin{bmatrix}
\dfrac{3}{2}\left[I_m \cos\left(\theta_{rot} - \omega t\right) + i_{br} \cos\left(\theta_{rot} - \omega_{br} t\right) \right] \\[2mm]
\dfrac{-3}{2}\left[I_m \sin\left(\omega t - \theta_{rot}\right) + i_{br} \sin\left(\omega t - \theta_{rot}\right) \right] \\[2mm]
0
\end{bmatrix}
\tag{22}
$$

$$
\left[I_{qdo} \right] =
\begin{bmatrix}
I_m \cos\left(\theta_{rot} - \omega t\right) + i_{br} \cos\left(\theta_{rot} - \omega_{br} t\right) \\[2mm]
I_m \sin\left(\theta_{rot} - \omega t\right) + i_{br} \sin\left(\theta_{rot} - \omega t\right) \\[2mm]
0
\end{bmatrix}
\tag{23}
$$

As it can be seen, each frequency will be shifted through Park's transformation. Equations 24 and 25 express the relationship between the frequency of current harmonics before and after Park's transformation, and Fig. 4 shows the relationship of different harmonics before and after Park's transformation.

$$
f_{qdo} = f_{abc} - f_s \qquad ; f_s \le f_{abc} \tag{24}
$$

$$
f_{qdo} = f_s - f_{abc} \qquad ; f_s \ge f_{abc} \tag{25}
$$

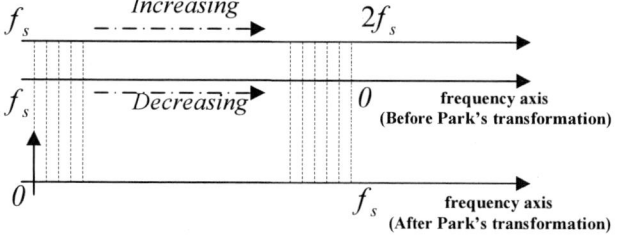

Fig. 4 Frequency spectrum of stator current before and after Park's transformation with synchronous rotating frame ($\omega_{rot} = \omega_s$).

As demonstrated in Fig. 4, in the case of synchronous rotating frame, DC component of dqo currents is now representative of abc currents main frequency. In addition, fault related characteristic components, in three phase stator currents, are now appeared near zero in dqo currents frequency spectrum.

Eliminating DC component, main frequency effect can be easily neglected in dqo currents. On the other hand, as demonstrated in Fig. 1, the frequency band widths of wavelet details are narrower in lower frequencies which results in more specific focus on fault related frequency components of dqo currents.

In this paper, energy evaluation of fault characteristic harmonics in dqo currents with neglected DC component is suggested to detect faulty rotor conditions. More specifically, relative energy value of the correspondent frequency band in q current to the total energy is used as diagnosis criterion in this study. Fig. 5 shows an overview of fault diagnosis algorithm.

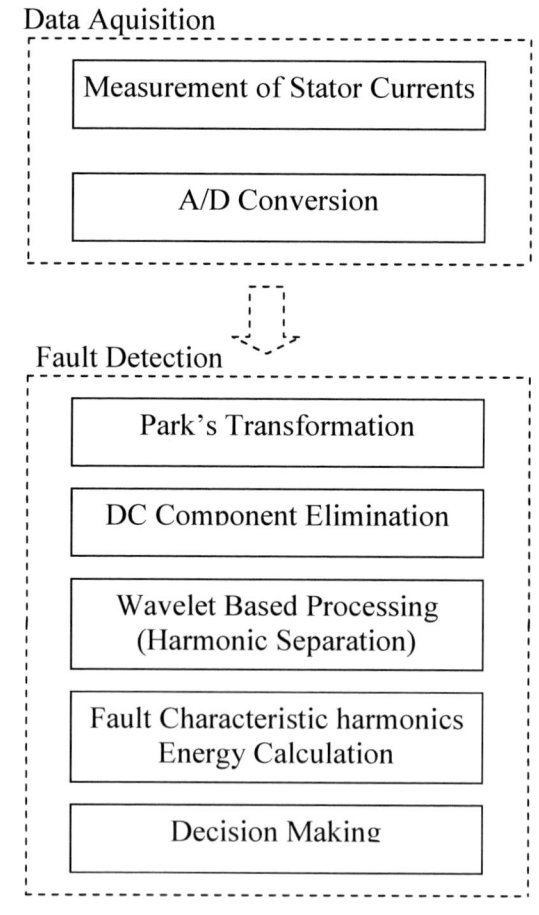

Fig. 5 Overview of the induction motor fault detection algorithm.

VI. Implementation and Analysis of the Results

As it can be seen from (1), goal harmonic frequencies for detection of broken rotor bar faults are $f_s[(1-s)\pm s]$, $f_s[5(1-s)\pm s]$ etc.

Supply frequency of the test set is 50 Hz, so the frequency band of 30 Hz to 50 Hz in stator currents is a good choice for detection of broken rotor bar faults. The sampling frequency is adjusted to 10 kHz. Table I shows the frequency levels of an eighth level wavelet decomposition for dqo currents and their corresponding frequency bands in stator three phase currents.

TABLE I. frequency limits of different wavelet details and approximation for an eighth level wavelet decomposition on dqo currents and their corresponding frequency bands in stator current (abc frame).

	Frequency Band Limits of Wavelet Details and Wavelet Approximation (Hz)		
	After Park's Transformation	Before Park's Transformation	Before Park's Transformation
D1	2500-5000	2550-5050	-
D2	1250-2500	1200-2550	-
D3	625-1250	675-1200	-
D4	312-625	365-675	-
D5	156-312	206-365	-
D6	78.1-156	120.1-200	-
D7	39-78.1	89-120.1	0-11
D8	19.5-39	69.5-89	11-30.5
A8	0-19.5	50-69.5	30.5-50

Daubechies-8 wavelet is used in this paper as mother wavelet. The efficiency of Daubechies wavelets based on the accurate reconstruction of power system transient signals and the suitability of this family for the analysis of power system transients is the basis for choosing Daubechies-8 [18].

Simulated motor is a three-phase 1.5-hp 380-V 50-Hz induction motor in normal and faulty conditions (two, three, four and five broken rotor bars). The rated speed of the induction motor is 1445 r/min, and the machine is run around full-load condition while the terminal voltages are sinusoidal [13].

Stator current and its frequency spectrum for healthy and damaged conditions are shown in Fig. 6 to 9. Frequency amplitudes of the selected frequency band (30 Hz to 50 Hz) are greater in the case of faulty motor but the frequency components are not clearly visible in the spectrum because of the main frequency large amplitude. Fig. 10 and 11 show the frequency spectrum of q currents with neglected DC component for two different conditions.

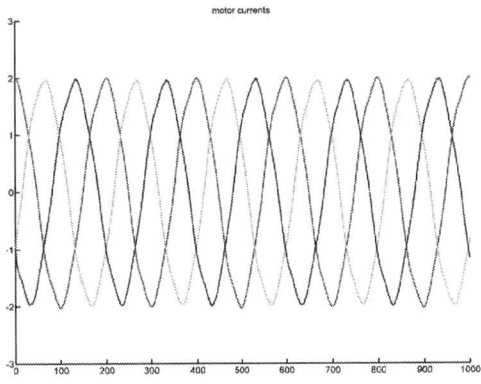

Fig.6 three phase stator currents, healthy condition

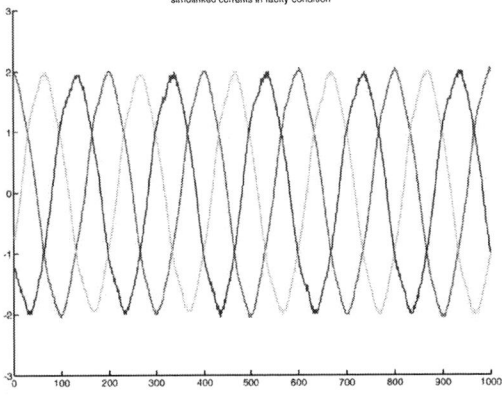

Fig.7 three phase stator currents, two broken rotor bars

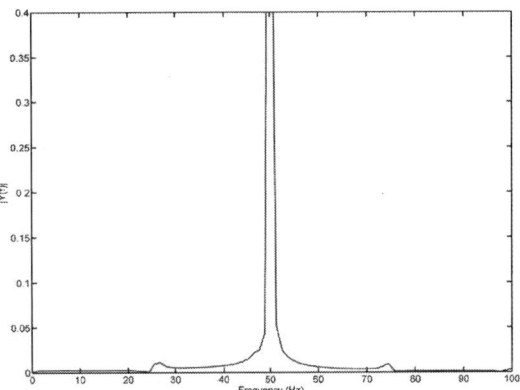

Fig. 8 Frequency spectrum of healthy motor

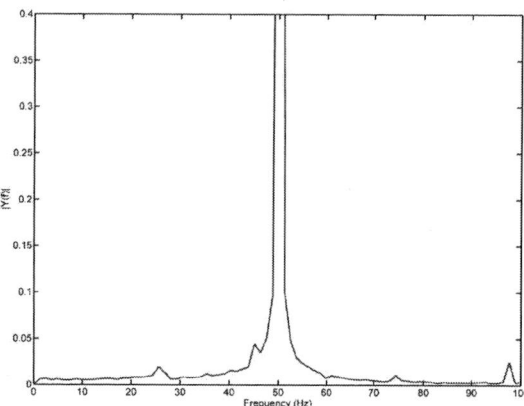

Fig. 9 Frequency spectrum of a motor with two broken rotor bars

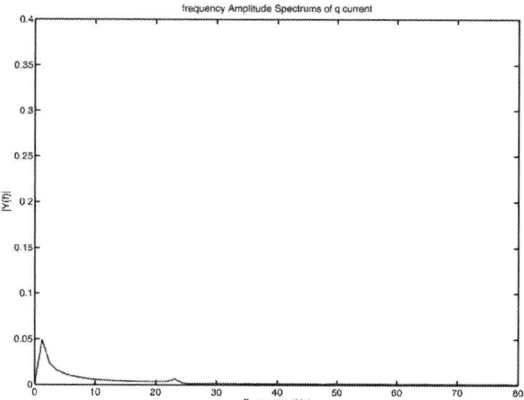

Fig. 10 frequency spectrum of q current after elimination of zero component, healthy condition

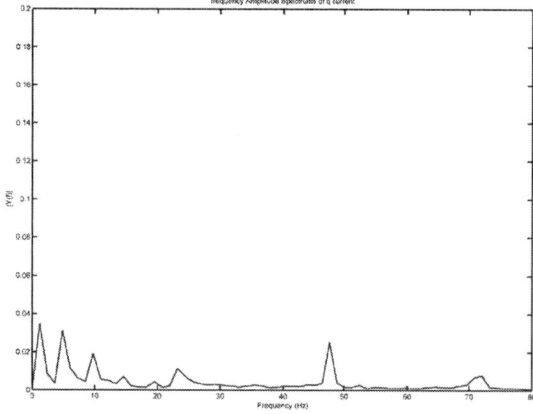

Fig. 11 frequency spectrum of q current after elimination of zero component; two broken rotor bars

Variation of the proposed diagnostic criteria for different conditions of motor is demonstrated in table 2 and the ratio of the proposed criterion value in different conditions to its value in healthy condition is demonstrated in table 3. As it can be seen, relative energy value of the appropriate frequency band in q current with neglected DC component is a good choice for the detection of broken rotor bar fault in induction motor and simulation results show the validity of the proposed approach to successfully distinguish healthy condition from other faulty conditions.

TABLE 2. Variation of the proposed criteria in different conditions of motor using db-8 wavelet

Motor condition	A8 relative energy value
Healthy	0.14469
Two broken bars	0.45566
Three broken bars	0.53427
Four broken bars	0.48412
Five broken bars	0.53688

TABLE 3. Ratio of the proposed criterion value in different conditions to its healthy condition value.

Motor condition	Ratio
Healthy	1
Two broken bars	3.15
Three broken bars	3.69
Four broken bars	3.34
Five broken bars	3.71

VII. Conclusions

In this paper, the development and testing of a signal processing-based fault detection methodology for the detection of induction-motor broken rotor bar fault, based on wavelet analysis of the dqo currents is presented. To demonstrate its efficiency, the fault detection and diagnosis method is applied on simulation results of a faulty motor in MATLAB environment. Obtained results show that proposed algorithm is able to detect fault characteristic harmonics very well and distinguish healthy condition from other faulty conditions with high accuracy. Furthermore, the method allows an easy implementation for further expert system implementation.

VIII.   References

[1] K. Kim, A. G. Parlos, "Induction Motor Fault Diagnosis Based on Neuropredictors and Wavelet Signal Processing", IEEE/ASME Transactions on Mechatronics, vol. 7, no. 2, June 2002.

[2] Z. Ye, B. Wu, A. Sadeghian, "Current Signature Analysis of Induction Motor Mechanical Faults by Wavelet Packet Decomposition", IEEE Transactions on Industrial Electronics, vol. 50, no. 6, December 2003.

[3] Z. Zhang, Z. Ren, W. Huang, "A Novel Detection Method of Motor Broken Rotor Bars Based on Wavelet Ridge", IEEE Transactions on Energy Conversion, vol. 18, no. 3, September 2003.

[4] D. Dorrell, W. Thomson, and S. Roach, "Analysis of Air Gap Flux, Current and Vibration Signals as Function of the Combination of Static and Dynamic Air Gap Eccentricity in 3-phase Induction Motors", IEEE Transactions on Industrial Applications, vol. 33, pp. 24–34, January/February 1996.

[5] G. Kliman et al., "Noninvasive Detection of Broken Bars in Operating Induction Motors", IEEE Transactions on Energy Conversion, vol. 3, pp. 874–879, December 1988.

[6] M. E. Benbouzid et al., "Induction Motor Asymmetrical Faults Detection using Advanced Signal Processing Techniques", IEEE Transactions on Energy Conversion, vol. 14, pp. 147–152, June 1999.

[7] T. Breen et al., "New Developments in Noninvasive Online Motor Diagnostics", in Proc. IEEE PCIC, 1996, PP. 231–236.

[8] M. Benbouzid, "A Review of Induction Motors Signature Analysis as a Medium for Faults Detection," IEEE Transactions on Industrial Electronics, vol. 47, pp. 984–993, October 2000.

[9] F. Filippetti, G. Franceschini, C. Tassoni, and P. Vas, "Recent Development of Induction Motor Drives Fault Diagnosis using AI Techniques", IEEE Transactions on Industrial Electronics, vol. 47, pp. 994–1004, October 2000.

[10] P. Vas, "Parameter Estimation, Condition Monitoring, and Diagnosis of Electrical Machines", Oxford, U.K.: Clarendon Press, 1993.

[11] J. Milimonfared, H. M. Kelk, S. Nandi, A. D. Minassians, and H. A. Toliyat, "A Novel Approach for Broken-Rotor-Bar Detection in Cage Induction Motors", IEEE Transactions on Industrial Applications, vol. 35, no. 5, pp. 1000–1006, September/October 1999.

[12] S. Bachir, S. Tnani, G. Champenois, and J. C. Trigeassou, "Induction Motor Modeling of Broken Rotor Bars and Fault Detection by Parameter Estimation," in Proc. IEEE SDEMPED, Gorizia, Italy, September 2001, pp. 145–149.

[13] J. Urresty, J. Riba Ruiz, L. Romeral, "Diagnosis of Interturn Faults in PMSMs Operating Under Nonstationary Conditions by Applying Order Tracking Filtering", IEEE Trans. on Power Elect, Vol. 28 , no. 1, pp. 507-515, 2013

[14] F. Harris, "On the Use of Windows for Harmonic Analysis with the Discrete Fourier Transform", Proc. IEEE, vol. 66, no. 1, pp. 51–83, January 1978.

[15] R. Burnett, J. F. Watson, and S. Elder, "The Application of Modern Signal Processing Techniques to Rotor Fault Detection and Location within Three Phase Induction Motors", Europe. Signal Processing J., vol. 49, pp. 426–431, 1996.

[16] O. Poncelas, J. A. Rosero, J. A. Ortega, J. Cusido, L. Romeral, "Motor Fault Detection Using a Rogowski Sensor Without an Integrator", IEEE Trans. on Ind. Elec., Vol. 56, no. 10, pp. 4062-4070, 2009

[17] J. Cusidó, L. Romeral, J. A. Ortega, J. A. Rosero, A. G. Espinosa, "Fault Detection in Induction Machines Using Power Spectral Density in Wavelet Decomposition", IEEE Transactions on Industrial Electronics, vol. 55, no. 2, February 2008.

[18] J. Faiz, B. M. Ebrahimi, A. R. Rajabioun and H. A. Toliyat, "A Criterion Function for Broken Bar Fault Diagnosis in Induction Motor under Load Variation using Wavelet Transform", Proceeding of International Conference on Electrical Machines and Systems 2007.

[19] G. Didier, E. Temisien, O. Caspa, H. Razik, H. Henao, A. Yazidi, and G. A. Capolino, " Rotor Fault Detection Using the Instantaneous Power Signature", IEEE International Conference on Industrial Technology, vol. 1, pp. 170-174, December 2004.

[20] C. M. Ong, "Dynamic Simulation of Electric Machinary", Printice Hall Company, 1998.

Analytical Evaluation of inductances for induction machine with dynamic eccentricity using MWFA and FE methods

S. Hamdani, O. Touhami, R. Ibtiouen and M. Hasni

Abstract – **An analytical procedure for computing the inductances of an induction machine in presence of dynamic eccentricity fault is presented. The proposed method is based on the modified winding function approach and takes into consideration harmonics in the winding and the inverse air-gap functions. Results obtained by the analytical method are compared to these calculated by finite elements (FE) method and satisfactory match was found between them. Also, the developed equations allow a considerable saving of computing time compared to the finite element method, in particular when they are considered in the dynamic simulation of the induction machine.**

Index Terms-- **induction motor, dynamic eccentricity, modified winding approach, air-gap, harmonics.**

I. INTRODUCTION

INDUCTION motors are widely used in manufacturing, transportation, mining, power system and so on due to reliability, simplicity of construction, high overload capability and high efficiency. Compared with a DC machine, induction motors are more rugged, less expensive and require less maintenance. However, operational environment, duty, and installation issues may combine to accelerate induction motor failure far sooner than the designed motor lifetime. Faults can occur in the stator, rotor, bearing, or the external systems connected to the induction motor. The Statistical analysis of induction motor faults show that 37% of the failures are due to the stator faults caused by deterioration of winding insulation, 41% are bearing related and 10% are due to the rotor faults [1, 2].

Rotor faults can be induced by electrical failures such as a bar defect or mechanical failures such as rotor eccentricity. The first fault occurs from thermal stresses, hot spots, or fatigue stresses during transient operations such as start-up, especially in large motors. A broken bar changes torque significantly and is dangerous to the safe and consistent operation of electric machines [3].

The second type of rotor fault is related to air gap eccentricity. This fault is a common effect related to a range of mechanical problems in induction motors such as load unbalance or shaft misalignment. Long-term load unbalance

can damage the bearings and the bearing housing and influence air gap symmetry. Shaft misalignment means horizontal, vertical or radial misalignment between a shaft and its coupled load. With shaft misalignment, the rotor will be displaced from its normal position because of a constant radial force [4].

The induction motor modelling with dynamic eccentricity using the winding function approach is examined in many papers [5-8]. The frequency components generated by dynamic eccentricity are explained analytically in [5]. Using the rotating field approach, authors confirm the existence of specific frequency components around the fundamental, caused by the dynamic air gap eccentricity. The interactions between the dynamic eccentricity and the inherent static eccentricity are also illustrated using an adequate mathematical model of the induction machine. For this model, the calculation of inductances is performed by the modified winding function approach.

Authors in [6] use the coupled magnetic circuit approaches for modelling the induction machine under dynamic eccentricity. Different inductances are calculated from the geometry and layout of the machine by the turn and winding functions approach. Simulation results are also illustrated to show the behaviour of the induction machine under this fault.

In [7], the authors use the extension of winding function approach to calculate and evaluate machine inductances for a non-uniform air-gap. In this approach, the modified winding function is defined as the difference between the turn function for uniform air gap and its average in one period. Using this approach, they confirm that the mutual inductances calculated from the stator or from the rotor are not the same. The modified winding function approach (MWFA) is also used in [8]. Comparing to [7], authors introduce the axial variable (z) in addition to the polar variable (φ, r) for inductances calculation.

One can note that the winding function approach is used in all the papers discussed above, but the analytical expressions of the self and mutual inductances of the induction machine are never presented in these papers. For that reason, this paper attempts to introduce the analytical formulation of these inductances using the MWFA and taking into consideration space harmonics. To validate our analytical expressions, a comparison with results obtained by finite elements using a free software method will be performed.

S. Hamdani is with Electrical and industrial systems laboratory , USTHB Houari Boumedien University of Science and Technology, BP 32 El Alia 16111 Bab Ezzouar Algiers Algeria, (e-mail shamdani@usthb.dz)
O. Touhami and R. Ibtiouen are with the Research Laboratory of Electrotechnics, 10, Av Pasteur El Harrach, BP182, 16200, Algeria.
M. Hasni is with Electrical and industrial systems laboratory , USTHB Houari Boumedien University of Science and Technology, BP 32 El Alia 16111 Bab Ezzouar Algiers Algeria

978-1-4799-0024-4/13 $31.00 © 2013 IEEE

II. INDUCTANCES FOR AN ECCENTRIC ROTOR

Let us consider two windings A and B. The MMF produced by current i_A of winding A, can simply be found by the product of the modified winding function and the current following in the winding:

$$F_A(\varphi_s,\theta) = N_A(\varphi_s,\theta).i_A \qquad (1)$$

φ_s is the stator circumferential position, θ_r is the rotor position and $N_A(\varphi_s,\theta)$ is the Modified Winding Function defined as:

$$N(\varphi_s,\theta) = n(\varphi_s,\theta) - <n(\varphi_s,\theta)> \qquad (2)$$

with :

$$<n(\varphi_s,\theta)> = \frac{1}{2\pi.<g^{-1}(\varphi_s,\theta)>}.\int_0^{2\pi} g^{-1}(\varphi_s,\theta).n(\varphi_s,\theta).d\varphi_s \quad (3)$$

where $n(\varphi,\theta)$ is the turn function, $g(\varphi_s,\theta)$ is the air-gap function and $<g^{-1}(\varphi_s,\theta)>$ is the average value of the inverse air-gap function. In general, the differential flux across the gap through a differential area $l.r.d\varphi$ can be written as:

$$d\phi = F_A(\varphi_s,\theta).\frac{\mu_0 rl d\varphi_s}{g(\varphi_s,\theta)} \qquad (4)$$

As illustrate by figure 1 and using (4), it is possible now to calculate the flux linkage in the region $\varphi_{sj}<\varphi_s<\varphi_{sj'}$ by:

$$\phi_{Bj\_j'} = \mu_0 rl \int_{\varphi_{sj}}^{\varphi_{sj'}} n_{Bj\_j'}(\varphi_s,\theta).F_A(\varphi_s,\theta).g^{-1}(\varphi_s,\theta)d\varphi_s \quad (5)$$

Where $n_{Bj\_j'}(\varphi_s,\theta)$ is the turns function located between the reference angles φ_{sj} and $\varphi_{sj'}$. This function is equal to zero when φ_s takes on value outside the span of the integral. So, (5) can also be written as:

$$\phi_{Bj\_j'} = \mu_0 rl \int_0^{2\pi} n_{Bj\_j'}(\varphi_s,\theta).F_A(\varphi_s,\theta).g^{-1}(\varphi_s,\theta)d\varphi_s \quad (6)$$

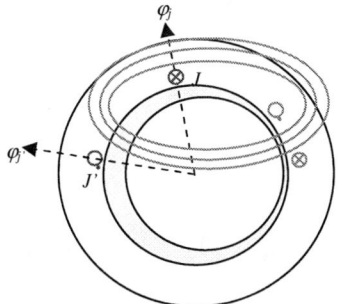

Fig. 1. Cross section of an elementary induction machine with two arbitrary windings

The total flux linking the winding B produced by the winding A is found by summing all elementary fluxes link defined as above, or

$$\phi_{BA} = \sum_{j=1}^{N_B} \phi_{Bj\_j'} = \mu_0 rl \int_0^{2\pi}\left(\sum_{j=1}^{N_B} n_{Bj\_j'}(\varphi_s,\theta)\right).F_A(\varphi_s,\theta).g^{-1}(\varphi_s,\theta)d\varphi_s \quad (7)$$

The term in the brackets, however, is simply the turns function for the B winding defined by:

$$n_B(\varphi_s,\theta) = \sum_{j=1}^{N_B} n_{Bj\_j'}(\varphi_s,\theta) \qquad (8)$$

The flux linkage of winding B due to a current winding A becomes:

$$\phi_{BA} = \mu_0 rl \int_0^{2\pi} n_B(\varphi_s,\theta).F_A(\varphi_s,\theta).g^{-1}(\varphi_s,\theta)d\varphi \quad (9)$$

The mutual inductance L_{BA} is defined as the flux linkage of winding B divided by the current flowing in winding A. Substituting (1) into (9) and dividing by i_A yield:

$$L_{BA} = \mu_0 rl \int_0^{2\pi} n_B(\varphi_s,\theta).N_A(\varphi_s,\theta).g^{-1}(\varphi_s,\theta)d\varphi_s \quad (10)$$

It should be noted that the results that have been derived are clearly valid for cases where windings A and B are one and the same. Hence, the inductance of winding A associated with flux crossing the air gap (magnetizing inductance) is given by the integral [9,10].

$$L_A = \mu_0 rl \int_0^{2\pi} n_A(\varphi_s,\theta).N_A(\varphi_s,\theta).g^{-1}(\varphi_s,\theta)d\varphi_s \quad (11)$$

III. AIR-GAP FUNCTION

Dynamic eccentricity in induction motor occurs when the center of the rotor is not at the center of rotation and the minimum air-gap revolves with rotor. This fault can exist as a result of problems associated with rotor misalignment, manufacturing tolerances, rotor shaft bending and weak bearings. In addition, such fault causes unbalanced magnetic pull (UMP) on the rotor, which brings up mechanical stress on some part of the shaft and bearing. After a prolong operation, these factors can snowball into broken mechanical part or even stator or rotor rub, causing major breakdown of the motor [5]. The occurrence of dynamic eccentricity introduces unequal air-gap and therefore modifies the expression of the air-gap function. Figure 2 illustrates the geometric configuration for a dynamic rotor displacement and the corresponding nomenclature. For simplicity, we assume that the displacement occurs perpendicularly to the shaft of the machine. The axial displacement will not be taken into account. Under these conditions, the calculation is performed in a two-dimensional plane [11,12].

The rotor center C_r is located by its Cartesian coordinates (a,b). The equation of the rotor circle of radius R_{dr} can be given by:

$$(x-a)^2 + (y-b)^2 = R_{dr}^2 \qquad (12)$$

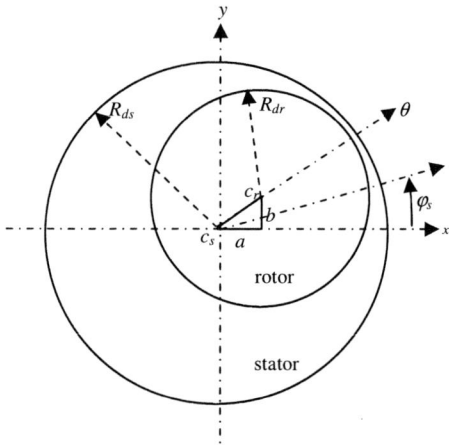

Fig. 2. Air-gap geometric configuration for a dynamic rotor displacement

Using $x=r.\cos(\varphi_s)$ and $y=r.\sin(\varphi_s)$, it possible to convert (12) into polar coordinates. A quadratic equation is obtained for the radius $r(\varphi_s)$ of the rotor outer surface. The solution of this equation is:

$$r(\varphi_s) = a.\cos(\varphi_s) + b.\sin(\varphi_s) + \sqrt{R_{dr}^2 - (a.\sin(\varphi_s) + b.\cos(\varphi_s))^2} \quad (13)$$

The air-gap length is the difference between the stator radius R_{ds} and $r(\varphi_s)$.

$$\varepsilon(\varphi_s) = R_{ds} - r(\varphi_s)$$
$$= R_{ds} - a.\cos(\varphi_s) - b.\sin(\varphi_s) - R_{dr}.\sqrt{1 - \frac{1}{R_{dr}^2}(a.\sin(\varphi_s) + b.\cos(\varphi_s))^2}$$

$$(14)$$

Induction machine air-gap is very small compared to the rotor radius R_{dr}. Therefore the displacement the rotor designated by the coordinates (a, b) is also small compared to R_{dr}. Eq. (14) can be simplified as follows:

$$g(\varphi_s) = g_0 - a.\cos(\varphi_s) - b.\sin(\varphi_s) \qquad (15)$$

Where $g_0 = R_{ds} - R_{dr}$ is air-gap length without eccentricity.
Let us introduce the dynamic eccentricity degree designed by the ration between the displacement length C_sC_r and g_0:

$$\delta_d = \frac{C_s C_r}{g_0} \qquad (16)$$

Equation (16) is used to calculate the coordinates a and b as follow:

$$a = g_0.\delta_d.\cos(\theta) \ ; \quad b = g_0.\delta_d.\sin(\theta) \qquad (17)$$

Substituting expressions of a and b given by (17) into (15) and rearranging it using trigonometric relations.

$$g(\varphi_s, \theta) = g_0.(1 - \delta_d.\cos(\varphi_s - \theta)) \qquad (18)$$

The inverse of the air-gap function can be given by the Fourier series development as [13]:

$$g^{-1}(\varphi_s, \theta_r) = \lambda_0 + \sum_{k=1}^{\infty} \lambda_k.\cos(k.(\varphi_s - \theta)) \qquad (19)$$

with:
$$\begin{cases} \lambda_0 = \dfrac{1}{g_0.\sqrt{1-\delta_d^2}} \\ \lambda_k = 2.\lambda_0.\left(\dfrac{1-\sqrt{1-\delta_d^2}}{\delta_d}\right)^k \end{cases} \qquad (20)$$

IV. TURN AND WINDING FUNCTION CALCULATION

The approach described in the second section is used to calculate the turn and the winding function of a three phases four poles induction machine with 36 stator slots single layer winding and 28 rotor bars. The winding distribution of the stators phases is illustrated by figure 3.
In order to develop the analytical formulation of the turn function for one stator phase, let us analyse at first, the turn function generated by only one stator coil which has n_e turns per slot. As illustrated by figure 4, the corresponding turn function has a rectangular shape and can be expressed by the following equation:

$$n_{se}(\varphi_s) = \begin{cases} n_e & \text{if} \quad -\dfrac{\tau}{2} \le \varphi_s < \dfrac{\tau}{2} \\ 0 & \text{otherwise} \end{cases} \qquad (21)$$

where τ is the turn pitch. Taking the coil axis as the origin, this rectangular waveform can be decomposed into Fourier series as follows:

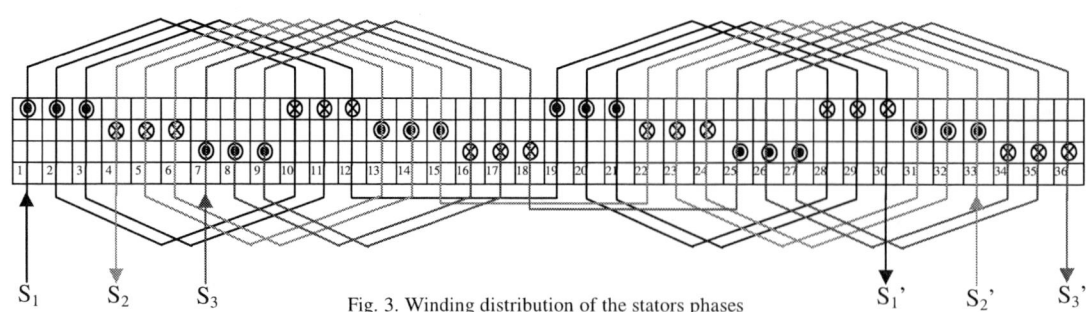

Fig. 3. Winding distribution of the stators phases

978-1-4799-0024-4/13 $31.00 © 2013 IEEE   422

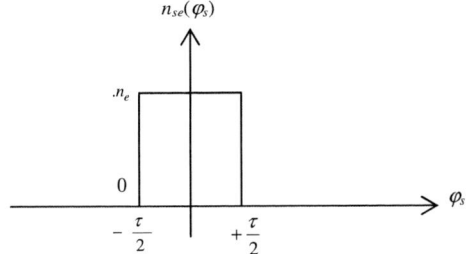

Fig. 4. Stator coil turn function

$$n_{se}(\varphi_s) = \frac{n_e.\tau}{2\pi} + \sum_{n=1}^{\infty} \frac{2.n_e}{\pi.n} \sin\left(\frac{n.\tau}{2}\right) \cos(n.\varphi_s) \qquad (22)$$

Note that $n_{se}(\varphi_s)$ contains all space harmonics with rank equal to n. Only if the turn had a full pitch ($\tau = \pi/p$), the even harmonics would disappear and only odd harmonics will be present in the turn function expression.

The stator phase contains N_{es} slots per pole. Therefore, the shape represented in figure 4 will be duplicated N_{es} times and each duplication is shifted of $2\pi/N_s$ where N_s is the number of stator slots. In the same manner, for a p poles pair induction machine, the same figure will be duplicated p times and each duplication is shifted of $2\pi/p$. So, the stator phase turn function can be obtained by the addition of all coils turns function contribution:

$$n_s(\varphi_s) = \sum_{m1=0}^{Nes-1} \sum_{m2=0}^{p-1} \sum_{n=1}^{\infty} n_{se}\left(\varphi_s - m_1.\frac{2\pi}{N_s} - m_2.\frac{2\pi}{p}\right) \qquad (23)$$

By considering N_{sp} the number of series coils per phase which is equal to $n_e.p.N_{es}$ and translating the origin with $(N_{es}-1).\pi/N_s$ angle, after development and simplification of equation (23), the analytical expression of the stator turn function can be written as:

$$n_s(\varphi_s) = \frac{N_{sp}.\tau}{2\pi} + \frac{2.N_{sp}}{\pi.p} \sum_{n=1}^{\infty} \frac{K_b(k)}{k}.\cos(n.p.\varphi_s) \qquad (24)$$

where $K_b(n)$ is the winding factor of the space harmonic of rank n. The turn function of the first stator phase is illustrated by figure 5. Using (2), (3) and (19), the modified winding function (MWF) expression of the ith stator phase is given by (25). This expression shows that the MWF has a mean value different from zero and depends on the rotor position. However, this component exist only if the harmonic order k of the inverse air-gap function is a multiple of the pair poles ($k=n.p$)

$$N_{si}(\varphi_s,\theta) = \begin{bmatrix} \frac{2.N_{sp}}{\pi}.\sum_{n=1}^{\infty} \frac{K_b(n)}{n.p}.\cos(n.p(\varphi_s - \varphi_{si})) \\ -\frac{N_{sp}}{\pi}.\sum_{n=1}^{\infty} \frac{K_b(n)}{n.p}.\frac{\lambda_{(n.p)}}{\lambda_0} \cos(n.p(\theta - \varphi_{si})) \end{bmatrix} \qquad (25)$$

The rotor can be considered equivalent to an m-phase two layer winding. A turn (loop) is formed by the conductors in the top layer of one slot and the bottom layer of the adjacent slot. The number of phases in such a winding would be $m=Nr/p$ where Nr expresses the number of rotor bars [14]. The phase shift between the currents in two adjacent turns is $2\pi/m = 2\pi p/Nr$, since the rotor bars are spaced by the angle $\alpha_r = 2.\pi/N_r$. Using this approach, the turn function n_{rj} of the jth loop represented in figure 6 can be written as:

$$n_{rj}(\varphi_r) = \begin{cases} 1 & \text{if} \quad (j-1)\alpha_r \leq \varphi_r < j.\alpha_r \\ 0 & \text{otherwhise} \end{cases} \qquad (26)$$

By considering the loop axis as origin, this function can be decomposed into Fourier series as follows:

$$n_{rj}(\varphi_r) = \frac{1}{N_r} + \sum_{n=1}^{\infty} \frac{2}{\pi.n} \sin\left(n.\frac{\alpha_r}{2}\right) \cos(n.\varphi_r) \qquad (27)$$

with the same manner, one can found the MWF of the jth rotor loop for $k=n$ as:

$$N_{rj}(\varphi_r,\theta_r) = \begin{bmatrix} \frac{2}{\pi}.\sum_{n=1}^{\infty} \sin\left(\frac{n.\alpha_r}{2}\right).\cos(n.(\varphi_r - \varphi_{rj})) \\ -\frac{1}{\pi}.\sum_{n=1}^{\infty} \sin\left(\frac{n.\alpha_r}{2}\right).\frac{\lambda_n}{\lambda_0}\cos(n.\varphi_{rj}) \end{bmatrix} \qquad (28)$$

As we can see, the rotor winding function has mean value independent of the rotor position, since the rotor rotate with the inverse gap function at the same rotational speed. Also, this component exist only if the harmonic order (k) of the inverse air-gap function is equal to the harmonic order of the rotor turn function ($k=n$).

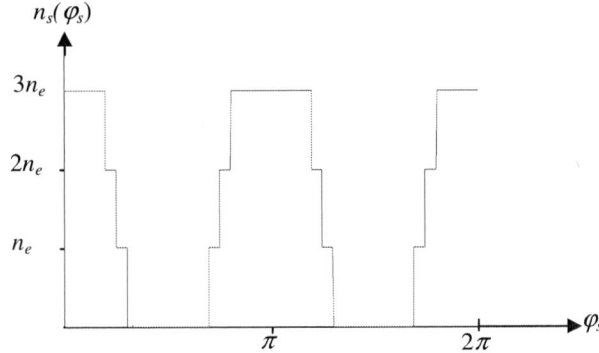

Fig. 5. Turn function of the first stator phase

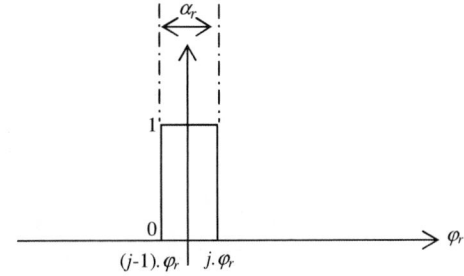

Fig. 6. Rotor loop turn function

V. INDUCTANCE CALCULATION USING MWFA

A. Stator phases inductances

The inductance between any stator phases i and j can be given by:

$$L_{sij} = \mu_0 rl \int_0^{2\pi} n_{si}(\varphi_s - \varphi_{si}, \theta).N_{sj}(\varphi_s - \varphi_{sj}, \theta).g^{-1}(\varphi_s, \theta)d\varphi_s \quad (29)$$

For a full pitch winding and substituting n_{si}, N_{sj} and g^{-1} by their expressions given respectively by (24), (25) and (19), the mutual inductance between two stator windings can be obtained after calculation and simplification by:

$$L_{sij} = (L_{sij})_h + (L_{sij})_{de1} - (L_{sij})_{de2} \quad (30)$$

Detailed expressions of the terms in (30) are given in appendix. One can note that the first term is the inductance of the healthy machine multiplied by λ_0, the second term results from the product of the "cosine" that exists in the three functions n_{si}, N_{sj} and g^{-1}. This term exist only if the harmonic rank (k) of the inverse air-gap function is equal to : $(n+m).p$, $(n-m).p$ or $(m-n).p$. However, the last term result from the contribution of the mean value in the modified winding function and the inverse air-gap function. For the machine whose construction parameters and winding details are given in the appendix, figure 7 shows the variation of the self inductance of the first stator phase and the mutual inductance between the first and the second stator phase for δ_d=60% of dynamic eccentricity. The calculation was performed by considering the first nine terms (harmonics) of the turn and winding functions. This figure clearly shows that in presence of dynamic eccentricity, the self and mutual inductances of the stator phases increase with fault and became function of the rotor position. This fact should be taken into account during the calculation of the electromagnetic torque.

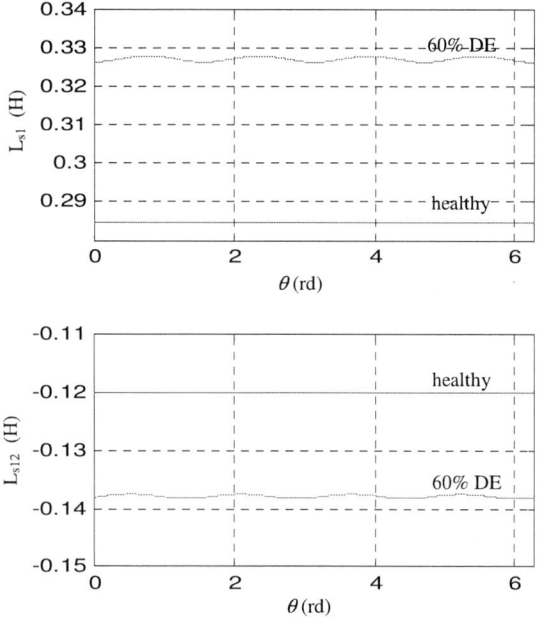

Fig. 7. Self and mutual inductances of the stator phase

B. Rotor loops inductances

Similarly, the inductance L_{rij} between any two rotor loops i and j can be calculated by:

$$L_{rij} = \mu_0 rl \int_0^{2\pi} n_{ri}(\varphi_r - \varphi_{ri}).N_{rj}(\varphi_r - \varphi_{rj}).g^{-1}(\varphi_r)d\varphi_r \quad (31)$$

After development and simplification, the analytical expression of the rotor inductances is :

$$L_{rij} = (L_{rij})_h + (L_{rij})_{de1} - (L_{rij})_{de2} \quad (32)$$

As we can see in the detailed expression of terms given in the appendix, the rotor inductances are independent on the rotor position because the air-gap picture doesn't change with rotor movement. Hence, these inductances should be calculated only once outside of the iterative process of calculation.

C. Mutual inductances between stator and rotor

The mutual inductance between any stator phase i and any rotor loop j is the ratio between the flux linked to the rotor loop which is generated by the stator phase and the current following this phase. Hence, the expression of this mutual inductance can be calculated by the flowing integral:

$$L_{sirj} = \mu_0 rl \int_0^{2\pi} n_{rj}(\varphi_r - \varphi_{rj}).N_{si}(\varphi_s - \varphi_{si}).g^{-1}(\varphi_r)d\varphi_r \quad (33)$$

To write the expression of the stator modified winding function in the rotor reference, we must substitute φ_s by $\varphi_r + \theta$. After calculation of this integral and simplification using trigonometric relations, the mutual inductance between the stator and the rotor can be expressed by:

$$L_{sirj} = (L_{sirj})_h + (L_{sirj})_{de1} - (L_{sirj})_{de2} \quad (34)$$

We can see from the detailed expressions of all the terms in equation (34) given in the appendix, that the mutual inductance between the stator phases and the rotor loops are dependent on the rotor position. A comparison have been shown in figure 8 between the mutual inductance of the first stator phase and the first rotor loop for a healthy machine and for a machine with 60% of dynamic eccentricity. This figure shows clearly that the mutual inductances increase in presence of dynamic eccentricity. Figure 9 shows the curves of mutual inductances between the first stator phase and the first three rotor loops. It is obvious that the value of the mutual inductance declines with the rise of the air gap length.

VI. INDUCTANCE CALCULATION USING FE

For the purpose of validation, the results obtained by the analytical method based on the MWFA are compared to these calculated by finite elements (FE) method using the free software FEMM 4.2. In order to be in concordance with the analytical approach, the magnetic saturation was neglected.

978-1-4799-0024-4/13 $31.00 © 2013 IEEE   424

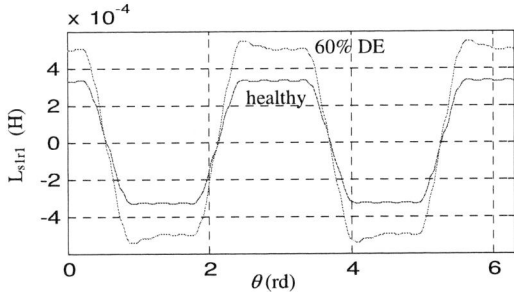

Fig. 8. Mutual inductance between the first stator phase and the first rotor

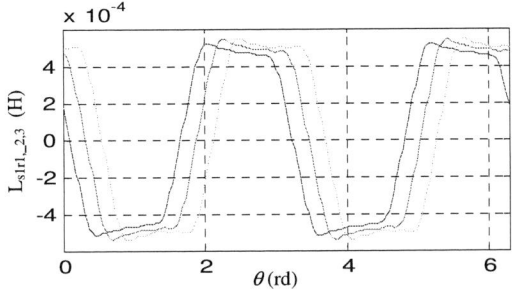

Fig. 9. Mutual inductance between the first stator phase and the first three rotor loops with δ_d =60%

In this condition, the inductance is calculated by the ratio between the flux linked with the winding and the current into the winding. For the stator winding (i), the magnitude of current is fixed to 1A and for the other windings, the current is zero. Figure 10 shows the field solution for the case when a current of 1A follows in the first stator phase. The linked flux can be evaluated by:

$$\phi_{si} = \frac{n_e l}{S} . \sum_{k=1}^{N_s} q_i(k) . \int_S A.dS \qquad (35)$$

Where A is the vector potential, S is the slot surface and $q_i(k)$ is vector of dimension equal to the number of stator slots and its elements are equal to 1 if the slot belongs to the phase (i) with a current flowing in the forward direction, equal to -1 if the slot belongs to the phase (i) with a current flowing in the reverse direction, equal to 0 if the slot does not belong to the phase (i). The plots of self inductance of the first stator phase and the mutual inductance between the first and the second stator phases are shown in figure 11. The mutual inductance between the first stator phase and the first rotor loop is illustrated by figure 12. The comparison of results obtained by FE indicates a good agreement with these of MWFA, especially their shape and their variation with the rotor position. For the amplitudes, a difference exists, because the exact material details about the laminations and some geometrical dimensions of the machine are not available. Compared to the MWFA, FE method presents some advantages like the consideration of the slots effect which is clearly shown in the inductance profile and the saturation. Using the MWFA, these factors can be introduced but the evaluations of inductance expression becomes more difficult and leads to a very longer formulas. However, the FE method was never considered as an alternative to the MWFA-based methods, simply because

of the computational time involved [14]. For example, by FE method, computation of the first stator phase self inductance and the mutual inductance between this phase and the two other phases, for only one rotor position takes around 85 seconds. With MWFA, the same calculation taking into account nine terms in the turn and winding function takes about 4 seconds. Therefore, the time for FE calculation is much large (more than 20 times) compared with MWFA, which is the main reason to use MFWA rather than FE for the dynamic simulation of the whole machine [15].

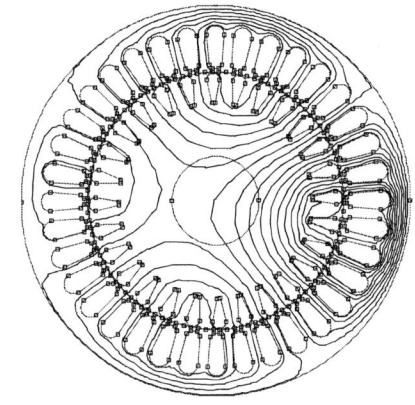

Fig. 10. Flux plot when only the first stator phase is supplied

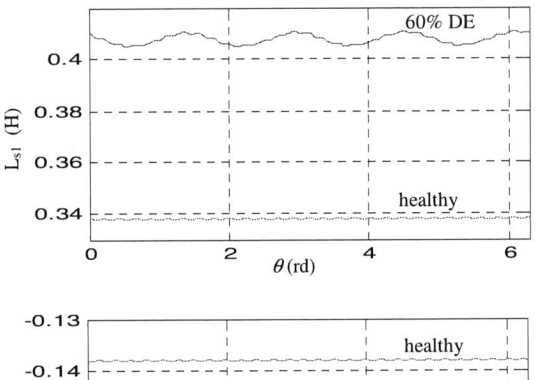

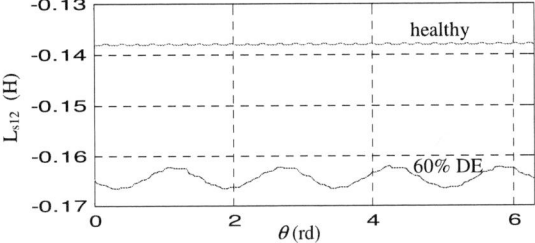

Fig. 11. Self and mutual inductances of the stator phase by FE method

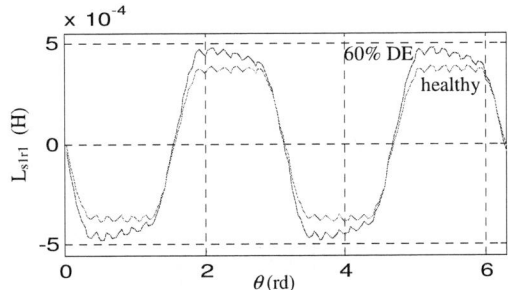

Fig. 12. Mutual inductance between the first stator phase and the first rotor loop by FE method

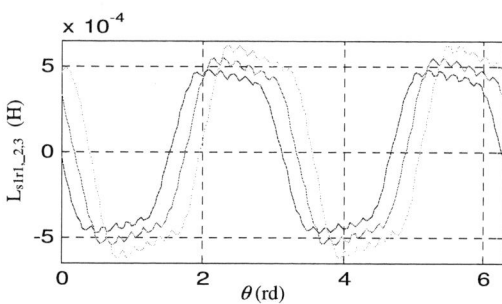

Fig. 13. Mutual inductance between the first stator phase and the first three rotor loops by FE method

VII. CONCLUSION

Analytical expressions of the self and mutual inductances of the induction machine in presence of dynamic eccentricity are presented. These inductances are obtained by the MWFA. For a machine with eccentricity defect the amplitude of inductances increases with dynamic eccentricity degree. In addition, the self inductance and mutual inductance of the stator become variable with the rotor position.

The comparison between the results obtained by the proposed method and those obtained by FE technique shows a good agreement between the two methods. However, the main advantage of the MWFA computation time is low, which allows us to say that this method is very suitable for the simulation of the induction machine.

VIII. APPENDIX

Squirrel cage induction machine parameters
Pn = 4kW; Un =220/380 V; (Δ/Y); In=15.2/8.8 A; Nn=1435 tr/mn ; p=2; f=50Hz; cosφ=0.83; Number of rotor bars: 28; air-gap: 0.28mm; slots number: 36; number of turns in series per phase : 156; stator outer diameter: 140mm; axial stack length: 120mm

Detailed expressions of inductances

A. Stator inductances

$$\left(L_{sij}\right)_h = \frac{4.\mu_0.r.l.N_{sp}^2.\lambda_0}{\pi.p^2}\left[\sum_{n=1}^{\infty}\left(\frac{K_b(n)}{n}\right)^2.cos\left(n.p\left(\varphi_{sj}-\varphi_{si}\right)\right)\right]$$

$$\left(L_{sij}\right)_{de1} = \frac{2.\mu_0.r.l.N_{sp}^2}{\pi.p^2}\left[\begin{array}{l}\sum_{n=1}^{\infty}\frac{K_b(n)}{n}\sum_{m=1}^{\infty}\frac{K_b(m)}{m}.\lambda_{(n+m).p}\,cos\left((n+m).p.\theta-p.\left(m.\varphi_{sj}+n.\varphi_{si}\right)\right)\\+\sum_{n=2}^{\infty}\frac{K_b(n)}{n}\sum_{m=1}^{n-1}\frac{K_b(m)}{m}.\lambda_{(n-m).p}\,cos\left((n-m).p.\theta+p.\left(m.\varphi_{sj}-n.\varphi_{si}\right)\right)\\+\sum_{m=2}^{\infty}\frac{K_b(m)}{m}\sum_{n=2}^{m-1}\frac{K_b(n)}{n}.\lambda_{(m-n).p}\,cos\left((m-n).p.\theta+p.\left(m.\varphi_{sj}-n.\varphi_{si}\right)\right)\end{array}\right]$$

$$\left(L_{sij}\right)_{de2} = \frac{2.\mu_0.r.l.N_{sp}^2}{\pi.p^2.\lambda_0}\left[\sum_{n=1}^{\infty}\frac{K_b(n)}{n}.\lambda_{(n.p)}\,cos\left(n.p.\left(\theta-\varphi_{sj}\right)\right).\sum_{m=1}^{\infty}\frac{K_b(m)}{m}\lambda_{(m.p)}\,cos\left(m.p.\left(\theta-\varphi_{si}\right)\right)\right]$$

B. Rotor inductances

$$\left(L_{rij}\right)_h = \frac{4.\mu_0.r.l.\lambda_0}{\pi}\left[\sum_{n=1}^{\infty}\left(\frac{sin\left(\frac{n.\alpha_r}{2}\right)}{n}\right)^2.cos\left(n.\left(\varphi_{rj}-\varphi_{ri}\right)\right)\right]$$

$$\left(L_{rij}\right)_{de1} = \frac{2.\mu_0.r.l}{\pi}\left[\begin{array}{l}\sum_{n=1}^{\infty}\frac{sin\left(\frac{n.\alpha_r}{2}\right)}{n}.\sum_{m=1}^{\infty}\frac{sin\left(\frac{m.\alpha_r}{2}\right)}{m}.\lambda_{(n+m)}cos\left(m.\varphi_{rj}+n.\varphi_{ri}\right)+\sum_{n=2}^{\infty}\frac{sin\left(\frac{n.\alpha_r}{2}\right)}{n}.\sum_{m=1}^{n-1}\frac{sin\left(\frac{m.\alpha_r}{2}\right)}{m}.\lambda_{(n-m)}cos\left(m.\varphi_{rj}-n.\varphi_{ri}\right)\\+\sum_{m=1}^{\infty}\frac{sin\left(\frac{m.\alpha_r}{2}\right)}{m}.\sum_{n=1}^{\infty}\frac{sin\left(\frac{n.\alpha_r}{2}\right)}{n}.\lambda_{(m-n)}cos\left(m.\varphi_{rj}-.\varphi_{ri}\right)\end{array}\right]$$

$$\left(L_{rij}\right)_{de2} = \frac{2.\mu_0.r.l}{\pi.\lambda_0}\left[\sum_{n=1}^{\infty}\frac{sin\left(\frac{n.\alpha_r}{2}\right)}{n}.\lambda_n.cos(n.\varphi_{ri})\sum_{m=1}^{\infty}\frac{sin\left(\frac{m.\alpha_r}{2}\right)}{m}.\lambda_m\,cos\left(m.\varphi_{rj}\right)\right]$$

C. Mutual inductance between the stator and the rotor

$$\left(L_{sirj}\right)_h = \frac{4.\mu_0.r.l.N_{sp}.\lambda_0}{\pi.p^2}\left[\sum_{n=1}^{\infty}\frac{K_b(n)}{n^2}.sin\left(n.p\frac{\alpha_r}{2}\right)cos\left(n.p\left(\varphi_{si}-\varphi_{rj}-\theta\right)\right)\right]$$

$$\left(L_{sirj}\right)_{del} = \frac{\mu_0.r.l.N_{sp}}{\pi.p}\left[\sum_{n=1}^{\infty}\frac{sin\left(\frac{n.\alpha_r}{2}\right)}{n}\sum_{m=1}^{\infty}\frac{K_b(m)}{m}.\lambda_{(n+m.p)}\,cos\left(m.p.(\theta-\varphi_{si})-n.\varphi_{rj}\right)+\sum_{m=1}^{s}\frac{K_b(m)}{m}\sum_{n=m.p+1}^{\infty}\frac{sin\left(\frac{n.\alpha_r}{2}\right)}{n}.\lambda_{(n-m.p)}\,cos\left(m.p(\theta-\varphi_{si})+n\varphi_{rj}\right)\right.$$
$$\left.+\sum_{m=1}^{\infty}\frac{K_b(m)}{m}\sum_{n=1}^{m.p-1}\frac{sin\left(\frac{n.\alpha_r}{2}\right)}{n}.\lambda_{(m.p-n)}\,cos\left(m.p(\theta-\varphi_{si})+n\varphi_{rj}\right)\right]$$

$$\left(L_{sirj}\right)_{de2} = \frac{2.\mu_0.r.l.N_{sp}.\lambda_0}{\pi.p\lambda_0}\left[\sum_{m=1}^{\infty}\frac{K_b(m)}{m}.\lambda_{m.p}.cos\left(m.p(\theta-\varphi_{si})\right).\sum_{n=1}^{\infty}\frac{sin\left(n.\frac{\alpha_r}{2}\right)}{n}.\lambda_n.cos\left(n.\varphi_{rj}\right)\right]$$

IX. REFERENCES

[1] A. H. Bonnett 'Root Cause AC Motor Failure Analysis with a Focus on Shaft Failures' IEEE transactions on industry applications, vol. 36, no. 5, september/october 2000.

[2] W. T. Thomson, D. Rankin, and D. G. Dorrell, "On-line current monitoring to diagnose air-gap eccentricity in large three-phase induction motors-industrial case histories verify the predictions," IEEE Transactions on Energy Conversion, Vol. 14, No. 4, Dec. 1999, pp1372-1378.

[3] W. T. Thomson, D. Rankin, and D. G. Dorrell, "On-line current monitoring to diagnose air-gap eccentricity-an industrial case history of a large high-voltage three-phase induction motors," Electric Machines and Drives Conference Record, 1997, ppMA2/4.1-MA2/4.3.

[4] W. T. Thomson and A. Barbour, "The on-line prediction of air-gap eccentricity levels in large (MW range) 3-phase induction motors," IEEE International Electric Machines and Drives Conference, 1999, pp383-385.

[5] M. Sahraoui, A. Ghoggal, S.E. Zouzou and M.E. Benbouzid, "Dynamic eccentricity in squirrel cage induction motors – Simulation and analytical study of its spectral signatures on stator currents," Simulation Modelling Practice and Theory, Elsevier, pp. 1-11, August 2008.

[6] H. A. Toliyat, M. S. Arefeen and A. G. Parlos, "A method for dynamic simulation of air-gap eccentricity in induction machines," IEEE Trans. Ind. Appl., Vol. 32, N°4, pp. 910-918, July/August 1996.

[7] J. Faiz and I. Tabatabaei, "Extension of Winding Function Theory for Nonuniform Air Gap in Electric Machinery," IEEE Transaction on Magnetics, Vol. 38, N° 6, pp. 3654- 3657, Novembre 2002

[8] G. Bossion, C. D. Angelo. J. Solsona and M. I. Valla,"A 2D-model of the induction motor: an extension of the modified winding function approch," 144 IEEE Trans. On Energy Conv., Vol. 19, N° 1, pp. 144-150, March 2004

[9] N. Al-Nuaim and H. Toliyat, "A novel method for modeling dynamic air-gap eccentricity in synchronous machines based on modified winding function theory," *IEEE Transactions on* Energy Conversion, Vol. 13, N° 2, pp. 156-162, Jun 1998.

[10] J. Faiz and I. Tabatabaei, "Extension of Winding Function Theory for Nonuniform Air Gap in Electric Machinery," *IEEE Trans. On Magnetics,* Vol. 38, N°. 6, pp. 3654-3657, Nov. 2002.

[11] H. Guldemir, "Detection of air-gap eccentricity using line current spectrum of induction motors", *Electric Power Systems Research,* Vol. 64, N° 2, pp. 109–117, Feb. 2003.

[12] M. Blodt, J. Regnier and J. Faucher, "Distinguishing load torque oscillations and eccentricity faults in induction motors using stator current Wigner distributions," pp. 1549-1556, 2006

[13] G. M. Joksimovic, M. D. Durovic, J. Penman and N. Arthur, "Dynamic simulation of dynamic eccentricity in induction machines-winding function approach," IEEE Transactions on Energy Conversion, Vol. 15, No. 2, June 2000, pp 143-148.

[14]) H. Henao, C. Martis and G. A. Capolino, "An Equivalent Internal Circuit of the Induction Machine for Advanced Spectral Analysis," *IEEE Trans. Ind. Appl.*, Vol. 40, N°3, pp. 726-734, May/June 2004.

[15] S. Nandi, " Induction machines including stator and rotor slot effects," IEEE Trans. On Ind. Appl. , Vol 40, N° 4, pp. 1058-1065, July/August 2004

X. BIOGRAPHIES

S. Hamdani received his engineering, master and doctorate degrees in electrical engineering from Ecole Nationale Polytechnique (ENP) of Algiers in 1995, 1999 and 2012 respectively. He is currently a Professor at the Electrical Engineering department of Université des Sciences et de la Technologie (USTHB) Houari Boumediene Algiers. His current research interests and experience include fault diagnosis of electric machinery, analysis and design of electrical machines, Control and identification of electrical machines.

O. Touhami was born in Tizi-Ouzou, Algeria. He received the Engineer, Master and Doctorate degrees in Electrical Engineering in 1981, 1986 and 1994 respectively from Ecole Nationale Polytechnique of Algiers. His areas of scientific interest are Energy Conversion, Modeling of Complex Systems, Power Quality, Diagnostic and Monitoring of Dynamic Systems. He was Scientific Adviser to the Ministry of the higher education in Algeria and Expert Member of the CNEPRU commission since 1997-2012. He is actually reviewer in IEEE Transaction on Energy Conversion, IEEE Transaction on Industrial Electronics. He was also Director of Research Laboratory in Ecole Nationale Polytechnique of Algiers since 2000 on 2005.

R. Ibtiouen received the Ph.D. degree in electrical engineering from the Ecole Nationale Polytechnique ENP Algiers and from the Institut National Polytechnique de Lorraine (INPL), France, in 1993. From 1988 to 1993, he integrated the Groupe de Recherche en Electrotechnique et Electronique de Nancy (GREEN). Between 2009 and 2010, he was Invited Professor at Université Henri Poincaré at Nancy (France). He is currently a Professor and Associated Director of Research with the Ecole Nationale Polytechnique. Since January 2005, he was the Director of the Research Laboratory of Electrical Engineering (LRE) in ENP. His fields of scientific interest include the modeling electric systems and drives, particularly electrical machines.

Analysis of radiated EMI for power converters switching in MHz frequency range

A. Majid[1], J. Saleem[2], F. Alam[3], K. Bertilsson[4]

[1,2,4]*Department of Electronics Design, Mid Sweden University,*
Holmgatan 10, 851 70 Sundsvall, Sweden
[3]*SEPS Technologies AB,*
Holmgatan 10, 851 70 Sundsvall, Sweden
abdul.majid@miun.se

Abstract -- The higher switching frequency in combination with di/dt loops and dv/dt nodes in the power stages of high frequency power converters generates higher order harmonics which cause Electro Magnetic Interference (EMI). It is commonly perceived that high frequency converters are more vulnerable to radiated EMI, thus, it is very important to analyze the radiated emission of these converters. According to the author's knowledge, the analysis of the radiated emission of power converters switching in the MHz frequency region has not been presented until now. Therefore, in this paper, the measurements, and analysis of the near field radiated emissions of these emerging power converters is presented. These measurements are beneficial in the early design stage of power converters. Both E-field and H-field and captured and analyzed for a half bridge converter switching at 3 MHz and at the output power level of 25 W. The effects of the magnetic and electric field emissions with the addition of a Y-capacitor and secondary side common mode choke are analyzed.

Index Terms -- Common Mode, Differential Mode, Electro Magnetic Interference, Electro Magnetic Compatibility

I. INTRODUCTION

The switching frequency of power converters is increased in order to increase their power density and improve their dynamic performance. With the utilization of emerging power semiconductor devices such as super-junction, GaN and SiC switching devices, the power converters are designed and developed in the MHz frequency region. The development of high frequency multi-layered PCB and hybrid (POT+I) core power transformers is another revolutionary step for the design and implementation of these converters [1]–[2]. These power converters are compact in size and are highly energy efficient. In the design of these high frequency power converters surface mount devices (SMD) are generally used which are better than through-hole (leaded) devices from the EMI point of view. The SMDs are smaller in size, have reduced parasitic components and a closer placement of components is possible. As a result of this, the length of the return current path is reduced. It is a common belief that high frequency power converters are more vulnerable to EMI. Therefore, it is important to analyze the EMI spectrum of these converters.

The analysis of conducted EMI for half bridge converters, switching in frequency range of 1–3 MHz, is presented in the research work [3]. The radiated emissions based on the harmonics components of two and three layers multilayered coreless PCB transformers are estimated in [4]. Since the PCB and hybrid core power transformers are used in high frequency power converters, it is important to analyze their radiated EMI in switching frequency range of 1–4 MHz. These converters are in an early design phase and the analysis of radiated noise emission is extremely important in order to develop them according to the Electromagnetic Compatibility (EMC) regulations, such as the International Special Committee on Radio Interference (CISPR) and the Federal Communication Commission (FCC) etc. It is also very important to suggest measures in order to suppress or eliminate the EMI generated by these converters. In the early design stages of power converters, it is not necessary to characterize the radiated EMI at a special EMC laboratory. Therefore, near field measurements are helpful to pin point the sources of radiated EMI and, based on these results, certain measures can be taken to suppress or eliminate it.

The analysis of near field measurements is valuable because it confirms the nature of offending emissions and the high frequency response of the circuit. These measurement techniques are of assistance in increasing throughput at a lower cost. The near field probes used for these measurements have finite dimensions, therefore, the orientation of probes or device under test can affect the measurements and thus these measurements only provide relative results. The repeatability of measurements does not provide the same results if the orientation of probe or device is slightly changed. By performing these measurements, the types and strengths of the electric and magnetic fields in different areas of the circuits can be analyzed. Therefore, based on these measurements, the sources of radiated EMI can be detected and remedial measures can be performed in order to solve the problem.

In this paper, near field measurements of a half bridge converter, switching at 3 MHz are analyzed. The HAMEG HZ540 EMI probe kit is used for the measurements. The Electric Field (E-field) and Magnetic Field (H-field) strengths are measured. The harmonics amplitudes are plotted and particular measures are taken to suppress the harmonics amplitudes.

978-1-4799-0024-4/13 $31.00 © 2013 IEEE

II. HIGH FREQUENCY HALF BRIDGE CONVERTER USED FOR NEAR FIELD MEASUREMENTS

The near field radiation measurements are performed for an unshielded AC-DC half bridge converter switching at 3 MHz and at an output power level of 25 W. The converter has an input line filter and it is not shielded. The hybrid (POT +I) core center tapped 4:1 power transformer and GaN power MOSFETs are used in the converter. These hybrid transformers have lower values of inter-winding capacitance, i.e, 22.2 pF, and leakage inductance i.e, 28 nH [2], which are very important from an EMI point of view. The leakage inductance of the power transformer resonates with its inter-winding capacitance and the junction capacitances and secondary side diode. As a result of this, high CM noise peaks occur [5]. Theoretically, the resonance of the leakage inductance with inter-winding capacitance of these power transformers occurs at around 200 MHz. Therefore the near field radiated spectrum of the converter is analyzed in order to investigate this resonance effect on the radiated emission. The analysis of both the H-field and E-field radiated emissions are presented in sections III and IV respectively.

III. ANALYSIS AND SUPPRESSION OF RADIATED MAGNETIC FIELD FOR POWER CONVERTER AND CABLES

The radiated signals of a half bridge power converter are measured by using a HAMEG H-Field probe HZ552. These probes provide a better reliability than the conventional field probes. The probe is fixed by means of a nonconductive fixture and connected to a HAMEG HMS300 Spectrum Analyzer. The measurements are performed in the very near field area (at approximately 1-2 cm from the converter and cables). With the assistance of these probes, the effectiveness of the filters can also be judged by measuring the radio frequency interference (RFI), which is conducted along the cables [6]. The measurement results are analyzed by the variation of certain circuit parameters.

A. Power converter H-field measurement

The magnetic components such as the inductors and power transformer are the main sources of magnetic field radiations. Because of non-ideal magnetic cores, fringing fluxes exist, which can cause a magnetic field [7]. The major causes of magnetic fields are high currents and low voltages [8], therefore, in order to investigate the sources of radiated magnetic fields, the probe is placed at a distance of 1 cm from the power transformer and secondary side rectifier diodes. The results are plotted in Fig. 2 (a). It is observed that the radiated H-field spectrum contains numerous clusters of higher harmonics. The first cluster contains the highest harmonics peak (approximately 42 dB μV) which occurs at around 200 MHz. The second peak (approximately 35 dB μV) occurs between 250−300 MHz. These harmonics are mainly due to the resonance effect of transformer leakage inductance and inter-winding capacitance.

In order to suppress CM EMI, the C_Y capacitor is connected between the primary and the secondary grounds as shown in Fig. 1. The addition of the C_Y capacitor has shown a considerable reduction in the amplitude of conducted common mode EMI in these converters [2]. Therefore, the measurements are performed to analyze its effect on radiated emission. From the results of Fig. 2 (b) it can observed that with the addition a 1.5 nF C_Y capacitor, the radiated emission from the converters is reduced. Almost all the harmonics are suppressed especially the peaks which are prominent in Fig. 2 (a) are significantly reduced.

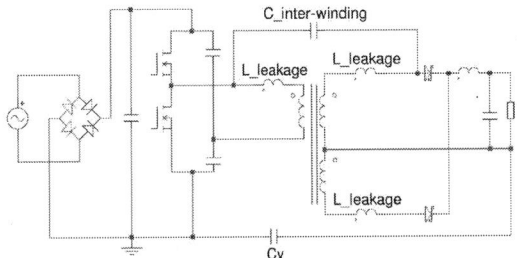

Fig. 1. Half bridge converter with Cy

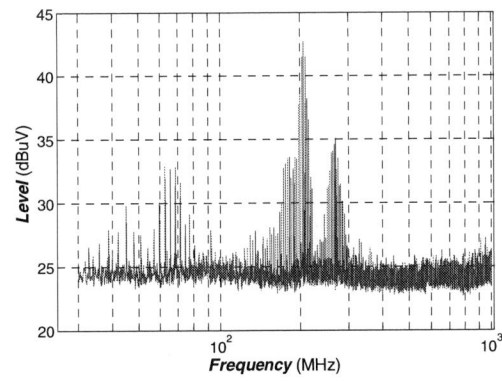

(a)

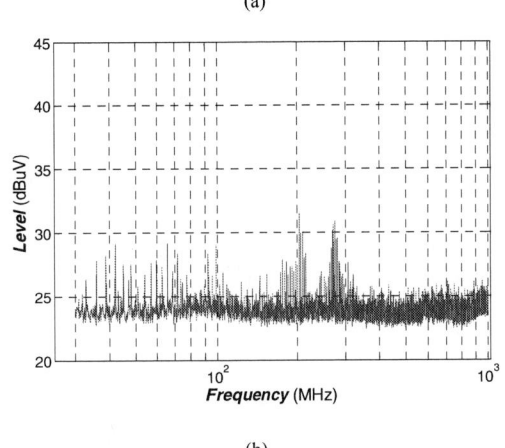

(b)

Fig. 2. Secondary side H-Field spectrum (a) without C_Y (b) with C_Y

B. Analysis and Suppression of radiated H-Field from secondary side cables

In order to analyze the radiations from cables, the measurements are performed by placing the H-Field probe at a distance of 1 cm from the cables. The EMI generated from these converters may spread while diffusing on the conductor surface of the power transmission lines [9]. The CM EMI because of propagates along the secondary side cables and through the secondary side ground [8], and as a result, the radiations from power lines can also occur. Therefore at higher switching frequencies the dominant propagation mode of cables becomes radiation instead of conduction [10].

The H-field measurements of the secondary side cables are shown in Fig. 4 (a). It is observed that the radiations from the cables occur between a frequency range 30−400 MHz. The two peaks, which were prominent in the measurement results of *section-II A*, are also present in these measurements. In addition to these two peaks, there are more harmonics in the frequency range of 30−200 MHz.

The measurement results of *section-II A* reveal that the addition of a 1.5 nF C_Y capacitor in converter circuit proves to be of assistance in reducing the radiated H-field from the converter. In order to analyze the effect of this C_Y addition on the radiations from the secondary cables, the measurements are performed and the H-field spectrum is shown in Fig. 4 (b). It is observed that the harmonics peaks around 200 MHz are completely suppressed. There are still some harmonics between 90−160 MHz at a level of 35 BμV and 250−300 MHz at a level of 32 dBμV. In order to reduce the interference due to the cables, it is desired that these harmonics should be suppressed.

In order to further reduce the emission from the secondary side cables, a common mode choke is placed on the secondary side of the power converter as shown in Fig. 3. It is obvious from the analysis of the H-field measurements on the cables (Fig. 4 (a) and (b)) that the harmonic amplitudes are higher between the 30−300 MHz frequency range. Therefore it is desired that the CM choke should have maximum attenuation within this frequency range. The attenuation of CM choke used for the reduction of radiated emission from the cables is shown in Fig. 5. The measurements performed with the addition of the Y-capacitor and the secondary side common mode choke are shown in Figure 6. It is obvious from the results that there is considerable suppression of the harmonic amplitude and the noise spectrum is at the noise floor of the spectrum analyzer.

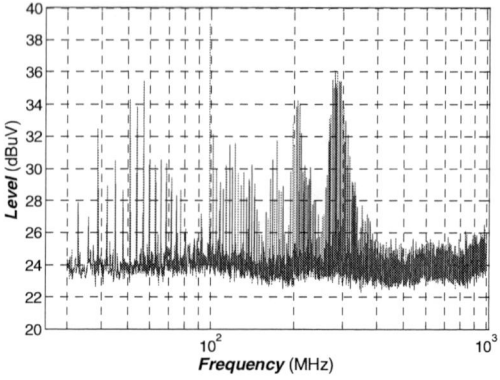

(a)

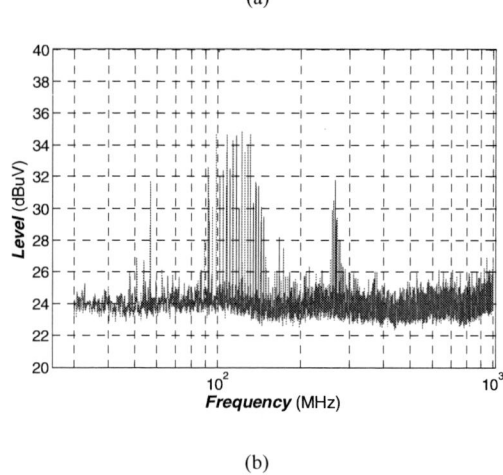

(b)

Fig. 4. Secondary side cables H-Field (a) without C_Y (b) with C_Y

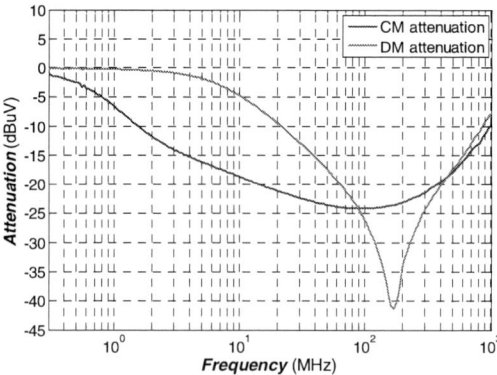

Fig. 5. Secondary side choke attenuation of

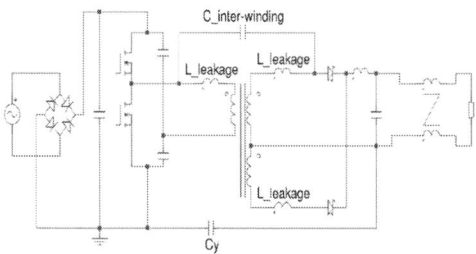

Fig. 3. Half bridge converter with secondary side choke

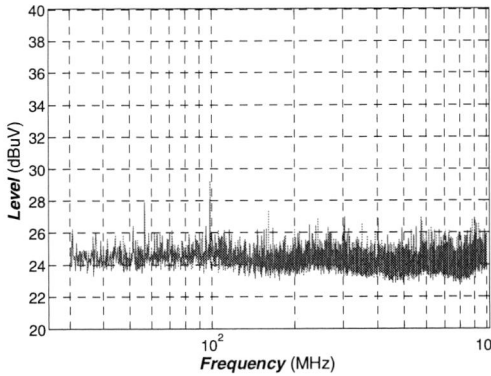

Figure 6 Secondary side cables H-Field with choke

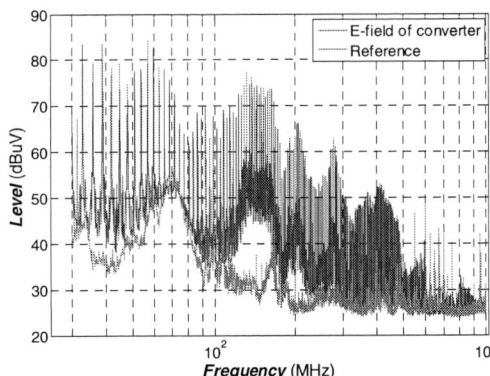

Fig. 7. Comparison of E-Field measurements of the converter (without adding C_Y and choke) with ambient field

IV. ANALYSIS AND SUPPRESSION OF RADIATED ELECTRIC FIELD (E-FIELD)

In order to analyze the radiated E-field, a highly sensitive HAMEG E-Field probe HZ551 is used. It has the capability to measure the entire radiations from the circuit. This probe can be used to perform relative measurements for pre-compliance tests [6]. The E-field probe is usually placed at a distance of 0.5 to 1.5 meters from the RFI source. The near field measurements are not absolute therefor, in order to know the ambient environment level, the E-Field is measured when the converter is off and the frequency range (30 MHz to 1 GHz) of interest is swept. This is considered as the reference for E-Field measurements. When the converter is on, the measured signals consist of both the ambient and the converter emissions.

All the E-field measurements discussed in this paper are performed by placing an E-field probe at a distance of 0.7 meters from the power converter. First of all, the measurements are taken for the half bridge converter without using the C_Y capacitor and a secondary side CM choke. The comparison of the power converter radiated E-field spectrum with the ambient or reference field is shown in Fig. 7. It is observed that, in the frequency range of 30-600 MHz, the amplitude of the harmonics is higher than the reference level. From the analysis of the H-field in section-III, it was observed that the radiations occur from the converter as well as from the cables. The same phenomenon is observed in E-field measurements. There are higher peaks of harmonics between 30-200 MHz and between 250-300 MHz in the E-field spectrum (Fig. 7). These peaks were also visible in the spectrum of the H-field measurements of the transformer (Fig. 2 (a)) and the secondary side cables (Fig. 4 (a)). However, in the E-field emission spectrum, there are additional peaks at around 400 MHz, which are not present in the H-field measurements. From this analysis, it can be inferred that the transformer, as well as the secondary side cables, have both E and H field radiations.

A. Analysis of E-field with addition of C_Y capacitor

The analysis of the H-field measurements presented in *section-II B* reveal the strength of radiated field with the addition of C_Y. Therefore, E-field measurements are performed for the half bridge converter after the placement of a 1.5 nF Y-capacitor. The comparison of the E-Field strength with and without C_Y is shown in Fig. 8. From the comparison results it is observed that the harmonics amplitudes around 200 MHz are suppressed. However some additional harmonics components appear within the 90-150 MHz frequency range. The same phenomenon was observed in the measurements of the H-field radiations from the secondary side cables (Fig. 4(b)). Therefore, it can be concluded that the addition of a Y-capacitor has a similar effect for both the E and H field radiations from the cables.

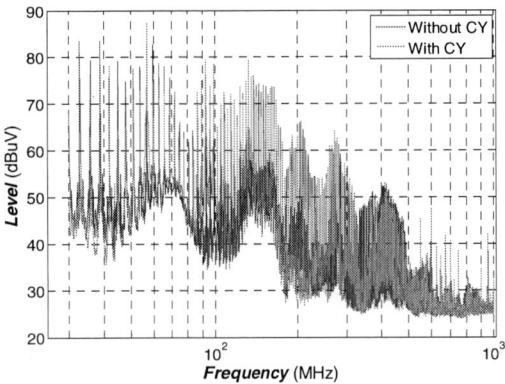

Fig. 8. Comparison of E-Field strength with and without C_Y

B. Analysis of E-field with addition of secondary side common mode choke

The common mode choke is used on the secondary side of the power converter and this resulted in a significant reduction of the radiated H-field strength from the output cables. Therefore, in order to analyze the effect of the CM choke placement on the E-field strength, the measurements are performed for the converter having both a 1.5 nF C_Y capacitor and a secondary side choke.

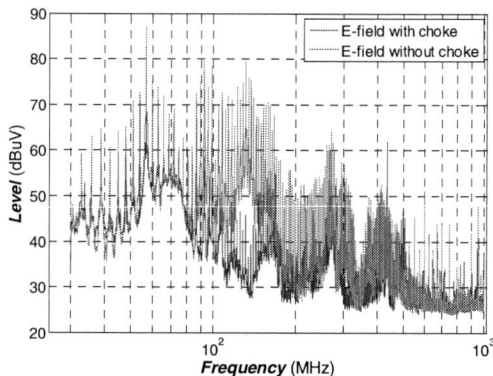

Fig. 9. E-field spectrums comparison with and without CM Choke

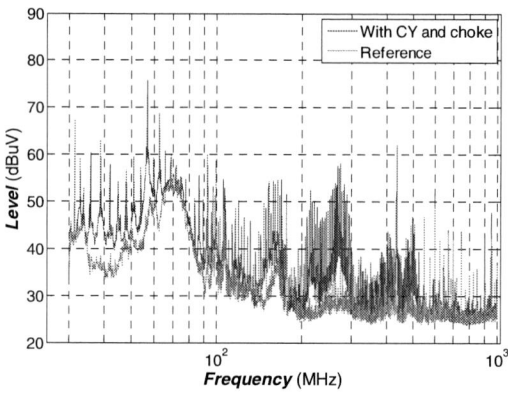

Fig. 10. Comparison of E-Field strength with reference

The E-field spectrum for the converter, with and without CM choke, is shown in Fig. 9. It is observed that with the addition of the CM choke on the secondary side, that the harmonics amplitudes in range of 70-150 MHz are suppressed. The cluster of harmonics around 200 MHz is also suppressed. The E-field measurement results of the converter (having both C_Y and CM choke) are compared with the reference spectrum (noise floor of the spectrum analyzer) as shown in Fig. 10. It is obvious that majority of the harmonics are suppressed, however there are still some harmonics between 30-70 MHz and 200-300 MHz frequency range.

From the analysis of near field measurements, it is observed that the placement of a 1.5 nF C_Y capacitor and secondary side common mode choke are necessary for a reduction of the radiated EMI. The RF shielding of the converter will further reduce the emission.

V. CONCLUSION

The focus of this paper is to measure and analyze the near field radiated EMI of the half bridge converter, switching at the 3 MHz frequency range. Both the E-field and H-field measurements are performed by using HAMEG near field probes. From the measurement results, it is observed that these high frequency power converters and, additionally their output cables, generate radiated EMI. It is observed that the addition of a 1.5 nF C_Y capacitor is helpful for the reduction of both E and H fields due to power converter circuit as well as secondary side cable. In order to eliminate the radiated emission from the output cables, a common mode choke is used on the secondary side of the power converter. It is observed that, with the addition of the CM choke, both E and H-field strengths are significantly reduced.

REFERENCES

[1] Kotte, H. B, Ambatipudi, R, Bertilsson, K, "High-Speed (MHz) Series Resonant Converter (SRC) Using Multilayered Coreless Printed Circuit Board (PCB) Step-Down Power Transformer," *Power Electronics, IEEE Transactions on* , vol.28, no.3, pp.1253,1264, March 2013.

[2] Kotte H, Ambatipudi R, Haller S, Bertilsson K, "A ZVS Half Bridge DC-DC Converter in MHz Frequency Region using Novel Hybrid Power Transformer", *Proceedings of International Exhibition and Conference for Power Electronics, Intelligent Motion, Power Quality (PCIM)* 2012, 8-10 May 2012, Nuremberg, Germany.. 2012;:399-406.

[3] Majid. A, Saleem. J, Alam. F, Bertilsson, K, "EMI Suppression in High Frequency (MHz) Half Bridge Converter" Paper submitted for Elektronika ir Elektrotechnika ISSN 1392-1215 (Lithuania)

[4] Ambatipudi, R.; Kotte, H.B.; Bertilsson, K., "Radiated emissions of multilayered coreless printed circuit board step-down power transformers in switch mode power supplies," *Power Electronics and ECCE Asia (ICPE & ECCE), 2011 IEEE 8th International Conference on* , vol., no., pp.960,965, May 30 2011-June 3 2011

[5] Kong. P, Lee. F.C, "Transformer structure and its effects on common mode EMI noise in isolated power converters," *Applied Power Electronics Conference and Exposition (APEC), 2010 Twenty-Fifth Annual IEEE* , vol., no., pp.1424-1429, 21-25 Feb. 2010.

[6] Hameg Near Field catatog http://www.hameg.com/downloads/flyer/HAMEG_FLYER_E_HZ540 _HZ550.pdf (Last accessed June 10, 2013)

[7] Yu Chen; Xuejun Pei; Songsong Nie; Yong Kang, "Monitoring and Diagnosis for the DC–DC Converter Using the Magnetic Near Field Waveform," *Industrial Electronics, IEEE Transactions on* , vol.58, no.5, pp.1634,1647, May 2011

[8] Rashid. M. H, "Electronics Handbook: Devices, Circuits, and Applications" 3rd *Edition ISBN 978-0-12-382036-5* pp 1103-1104

[9] Mutoh, N.; Nakashima, J.; Kanesaki, M.; , "Multilayer power printed structures suitable for controlling EMI noises generated in power converters," *Industrial Electronics, IEEE Transactions on* , vol.50, no.6, pp. 1085- 1094, Dec. 2003

[10] Tim. W, "EMC for Product Designers" 4th edition ISBN–13: 978-0-75-068170-4.

Design of Current Source DC/DC Converter and Inverter for 2kW Fuel Cell Application

A. Andreiciks, I. Steiks, O. Krievs, F. Blaabjerg

Abstract -- In order to use hydrogen fuel cell in domestic applications either as main power supply or backup power source, the low DC output voltage of the fuel cell has to be matched to the voltage level and frequency of the utility grid AC voltage. The interfacing power converter systems usually consist of a DC/DC converter and an inverter. In this paper a detailed simulation study of such interfacing converter system comprising a double inductor push-pull step-up DC/DC converter and a cascaded H-bridge inverter has been carried out and further confirmed with experimental results. The power converter system is designed for interfacing a 2kW proton exchange membrane (PEM) fuel cell.

Index Terms-- DC/DC converter, H-bridge inverter, Fuel cell system, High frequency power converters.

I. NOMENCLATURE

P_{cond} - conduction losses,

P_{SW} - switching losses,

P_l - power to be dissipated in the clamp resistor,

I_{SW} - RMS drain current,

I_{FC} - fuel cell current,

V_{clamp} - maximum voltage of the clamp capacitor,

V_{norm} - normal voltage of the clamp capacitor,

R_{on} - on state resistance,

t_r - turn-on time of the transistor,

t_f - turn-off time of the transistor,

f_s - switching frequency,

D - duty cycle of the converter;

V_p - primary transformer voltage

T_S - compensated closed loop transfer function

T_{M1-M1a} - the maximum time delay between turning off the main switch and turning on the auxiliary switch,

C_{clamp} - capacitance of the clamp capacitor,

C_l - effective drain-source capacitance of the Power MOSFET,

L_{lk} - leakage inductance of the transformer.

Development of this article is co-financed by the European Regional Development Fund within the project „Wind and Hydrogen Based Autonomous energy Supply System", Agreement No. 2010/0188/2DP/2.1.1.1.0/10/APIA/VIAA/ 031.

A. Andreiciks is with Department of Power and Electrical Engineering, Riga Technical University, Kronvalda bulv 1a, LV1010, Riga, Latvia (e-mail: aleksandrs.andreiciks@rtu.lv).

I. Steiks is with Department of Power and Electrical Engineering, Riga Technical University, Kronvalda bulv 1a, LV1010, Riga, Latvia (e-mail: ingars.steiks@rtu.lv).

O. Krievs is with Department of Power and Electrical Engineering, Riga Technical University, Kronvalda bulv 1a, LV1010, Riga, Latvia (e-mail: oskars@eef.rtu.lv).

F. Blaabjerg is with Institute of Energy Technology, Aalborg University, Pontoppidanstrade 101, 9220, Aalborg, Denmark (e-mail: fbl@et.aau.dk).

II. INTRODUCTION

The hydrogen fuel cells used as main power supply or backup power source in domestic applications usually consist of a DC/DC converter and an inverter. Due to the comparatively high input and output voltage difference, most frequently used converters with high frequency transformer are acknowledged as the optimal solution for the DC/DC stage [1], [7], [11]. There are many well known transformer isolated DC/DC converter topologies, which are able to perform the necessary voltage boosting to match the inverter's DC-link voltage level: full-bridge, half-bridge, flyback, forward, basic push-pull topologies, as well as a number of their derived topologies [8], [9]. These topologies can be classified into two groups – voltage source converters and current fed converters. In this paper a current fed topology is selected, since it is characterized by a low input current ripple, which is more appropriate for proton exchange membrane fuel cells [1],[12].

For the purpose of increasing the overall efficiency of the system, the DC/DC converter is split into two modules, connecting the inputs in parallel and outputs – in series (Fig.1.). The first advantage of such approach is the reduced power handling of individual transformers, which results in smaller transformer size and simpler construction. Secondly, the primary switch conduction losses are significantly reduced, due to smaller switch currents. Thirdly, such configuration facilitates the application of a multilevel inverter at the output of the DC/DC converter stage, which results in a decreased step-up transformer ratios, as well as lighter switching conditions (dv/dt).

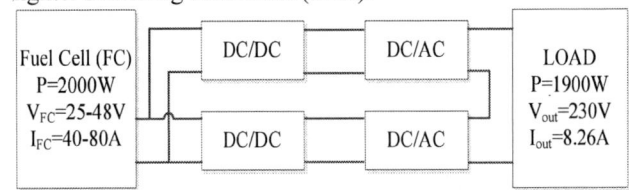

Fig 1. The structure of the interfacing converter of the fuel cell

III. THE SELECTED DC/DC CONVERTER TOLOLOGY

Considering the necessity of high voltage boosting with low input current ripple, the most appropriate converters are current fed full-bridge and push-pull configurations. Since the DC/DC converter efficiency can be considerably improved by reducing the count of the primary switches and implementing a simple transformer structure (without split windings), a Double Inductor push-pull Converter topology (DIC) [16] is selected and analysed further in the paper.

Each of the two DIC converter modules is designed for increasing the low input voltage of the fuel cell (25Vdc-48Vdc) to output voltage of 130Vdc, with estimated maximum efficiency $\eta = 95\%$.

Considering the high switching frequency and low voltage, the MOSFET devices are the best choice for the primary switches in terms of cost and performance. Besides the peak voltage, that the switch should be able to block, and the RMS current that will flow through it, the selection of MOSFET devices is confined also by the lowest possible on-state resistance in order to minimize the conduction losses.

As it is well known, the total switch losses comprise the conduction losses and the switching losses. The primary switch conduction losses of the DIC converter depend on the on-state resistance and the RMS drain current:

$$P_{cond} = 2 \cdot I_{SW}^2 \cdot R_{on} = 2 \cdot (I_{FC}\sqrt{(2-D/2)})^2 \cdot R_{on} \qquad (1)$$

The switching losses depend also on the switching frequency and the switching times of the transistor:

$$P_{SW} = 2(I_{FC}/2) \cdot U_p \cdot f_s \cdot (t_r + t_f). \qquad (2)$$

The hard-switching converters have a drawback with voltage overshoots at turn-off due to the energy stored in the parasitic inductances. These voltage spikes are not only dangerous to power transistors - they substantially increase the switching losses as well. There are two basic ways to protect the transistor switches from being damaged by overvoltage. The first is using transistors with blocking voltage ratings that exceed the overvoltage stresses. This, however, results in poor utilization of the transistor, since the on state resistance of the MOSFET transistors increases dramatically with increasing the blocking voltage. The other way is to limit the stresses within safe levels using snubber circuits.

The highest voltage spikes across the switches of the DIC converter appear at turn-off, when the overvoltage occurs due to the transformer leakage inductance [8], [13], [14]. A simple passive voltage clamping circuit for the topology under consideration is described in [17]. Nevertheless, since active clamping circuits have several very important advantages: the ability to operate at zero voltage switching (ZVS) both - turn-on and turn-off transients for all switches and also the possibility to recover the energy stored in the leakage inductance, an active clamping circuit has been chosen in the current case.

IV. OVERVOLTAGE PROTECTION OF THE DC/DC CONVERTER SWITCHES

The DIC converter with active clamping circuit comprises two controlled clamp transistors and a capacitor (Fig. 2).

Such clamping circuit can improve the overall efficiency of the converter, since it does not waste the energy of the leakage inductor, instead feeding it to the converter's output.

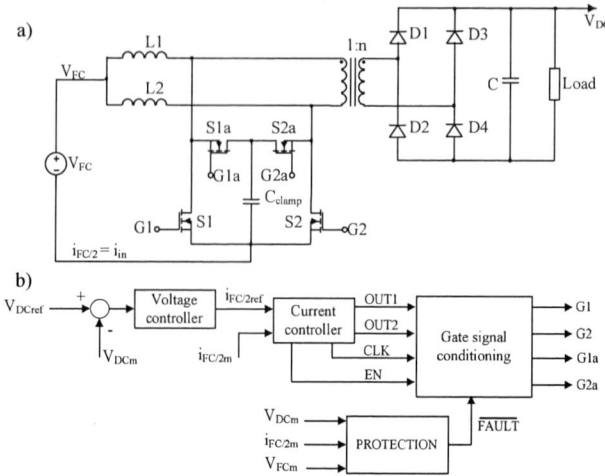

Fig. 2. (a) the implemented DIC converter topology; (b) the control block diagram

The basics of operation of the presented clamping circuit is using the resonant phenomena between the leakage inductance of the transformer and the clamping capacitor.

The value of the clamping capacitor should be set so that one half of the resonant period formed by clamping capacitor and leakage inductance of the transformer exceeds the maximum turn-off time of the main switches [13]. The clamping capacitance must satisfy the following condition:

$$C_{clamp} > T_s^2 (1-D)^2 / \pi^2 \cdot L_{lk} \qquad (3)$$

To achieve ZVS for the main switch, it must be turned on after turning-off the auxiliary switch. This delay should be less than one quarter of the resonant period of the transformer's leakage inductance and effective drain-source capacitance of the power switch [7]:

$$T_{M1-M1a} = \sqrt{L_{lk} \cdot C_i} \cdot \pi / 2 \qquad (4)$$

The maximum time delay between turning off the main switch and turning on the auxiliary, using the calculated minimum value for the clamping capacitor, can be estimated as follows:

$$T_{M1-M1a} = \sqrt{L_{lk} \cdot C_{clamp}} \cdot \pi / 2 \qquad (5)$$

V. CONTROL OF THE DC/DC CONVERTER

Two different control methods are compared for the designed converter - the duty-cycle control and the current control.

In the duty-cycle control method the output voltage is measured and then compared to the reference voltage. The error signal is used as the input in the PI controller, which will compute the duty-cycle reference for the pulse-width modulator. PI controller is chosen for the duty-cycle regulation. Tuning of PI gains can be realized using various methods. The method is illustrated in Fig. 3a.

978-1-4799-0024-4/13 $31.00 © 2013 IEEE

In the current control method (Fig. 3b) the converter's output is controlled by limiting the transistor's peak current. The control signal is current and a simple hysteresis control network switches on and off the transistor so that their peak current follows the reference.

The sources [9] and [10] state, that the current control, in the case of an isolated boost push-pull converter has some advantages over the duty-cycle control. First, it has a simpler dynamics (removes one pole from the control-to output transfer function). Second, it makes use of the current sensor information in normal operation mode – transistor failures due to excessive currents can be prevented by limiting the reference switch current.

A DC bias current may be introduced in the transformer by small voltage imbalances due to differences in boost inductors or switches. The current control method will alter the switch duty cycles in a way that these imbalances tend to disappear and the transformer volt-second balance is maintained. The disadvantage is that the current control has a susceptibility to noise in the reference and the measured switch current signals. In general, a small amount of filtering is necessary for the measured current. The current control becomes unstable whenever the duty-cycle becomes larger than 0.5. However, this drawback can be overcome by adding an artificial ramp to the reference current signal. Considering the above arguments, the current control is more attractive for the present application.

A. The DIC converter with current control

The model of the current controller implies that the calculation of the relation between the inductor's current and the control signal (Fig. 3b) must be done. Since there are two inductors, the sum of the two inductor currents is being used in the relation.

As mentioned before, the current control becomes unstable when the duty cycle is less than 0.5, which can be solved by adding an adequate ramp to the sensed signal. Since in the present case the maximum duty-cycle is 0.6, only a small ramp is required, to improve the noise immunity

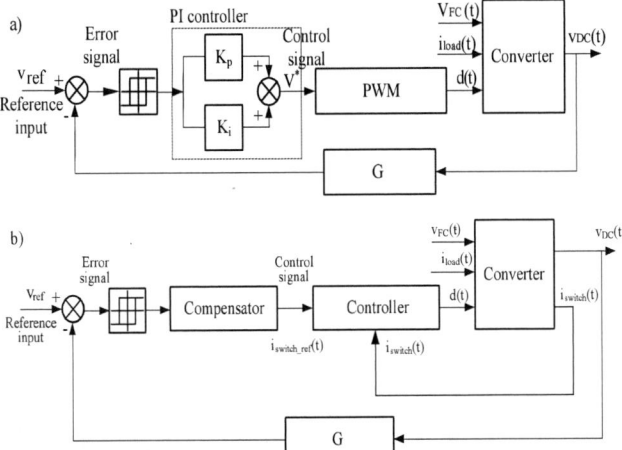

Fig. 3. Control of the DC/DC converter: (a) direct duty-cycle control method; (b) current control method

of the system. Furthermore, this small ramp does not influence the converter's transfer functions and the design of the controller.

The most important properties of the current control method are the following: it keeps the input current at the desired value and realizes the current balance between positive and negative half-waves in the transformer, preventing the saturation.

The measured signals are the fuel cell voltage (V_{FC}) and current (I_{FC}) and the output voltage (V_{DC}). The fuel cell current and output voltage are used as feedback signals in the control system. The protection is realized by using all the measured signals.

VI. THE MULTILEVEL INVERTER (DC/AC)

In general there exists three main topologies of multilevel inverters applicable in the present case.

Diode clamped converter (DCC) is commonly used in high-power applications [7], [11]. The output voltage levels are formed from series connected DC-bus capacitors and clamping diodes. In most cases this topology is used with up to 3-levels, because of the balancing DC-bus capacitors and voltage overshoots on diodes [2], [15]. The advantages of DCC are the high efficiency due to devices being switched at the fundamental frequency, as well as the possibility to control the reactive power flow. The disadvantage are ,on the other hand, the excessive number of clamping diodes, if the number of levels is too high, and the complexity to realize real power flow control for the individual converters.

The second topology is the Flying Capacitor Converter (FCC) consisting of series connected capacitors [8] and having no need for clamping diodes compared to DCC. The advantages of FCC are as follows:
- a redundancy of switch combinations is available for balancing the output voltage levels;
- the real and reactive power flow can be controlled.
Disadvantages of FCC are the following:
- high number of required storage capacitors (in case of many levels);
- high level systems are more difficult to package and capacitors are also expensive;
- the switching frequency it switching losses are high for real power transmission and therefore complicates the control of the converter.

Cascaded H-bridge Converter (CHC) is based on series connected H-bridges as shown in (Fig.4.) and was first introduced in [16]. Each module of CHC must be isolated, which is often named as a disadvantage of this topology. However, for the fuel cell interfacing converter system discussed here, this is not an obstacle. One of the most important advantages of CHC is the possibility to use switches of relatively low voltage [3], which, however, increases the required amount of switches and correspondingly – the complexity of the driver circuitry [6].

Analyzing the three multilevel inverter topologies, the CHC appears to be the most suitable solution, since the

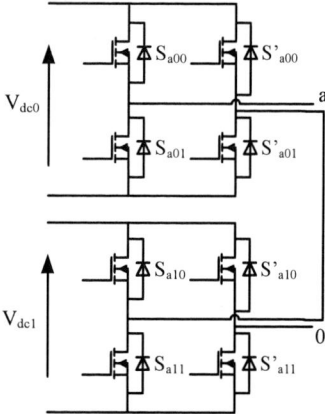

Fig. 4. Topology of a 5-level Cascaded H-bridge inverter

requirement for the isolated DC sources is fulfilled and it requires the lowest number of components.

VII. SIMULATION OF THE INTERFACING CONVERTER SYSTEM

The overall system consists of one fuel cell stack module, two current-fed DIC converter modules with active clamping, two cascaded H-bridge converters (CHC) and the AC load. The simulation was done in MATLAB/SIMULINK environment according to the structure shown in Fig.5.

The operating conditions of the simulation are adjusted to the rated parameters of the converter. Active load with a small filter capacitor of 10 µF is considered. The value of leakage inductance ($L_{lk} = 4.8$ µH) was obtained experimentally, by measuring the transformers mutual and self-inductances of the experimental prototype.

Simulation waveforms of a steady state operation are shown in Fig.6. From waveforms it is evident that the transistor voltage does not exceed the maximum allowed voltage of transistor 150V. Also the transistor current is well within the allowable limits. The fuel cell output current ripple is ± 2.5A (less than 5% of the average value of the current), which is satisfactory for fuel cell operation. It can also be noted, that the Current-fed DIC DC/DC converter with active clamping circuit operates as desired, providing 130 V output voltage at rated input and load conditions. The cascaded h-bridge converter operates as desired also.

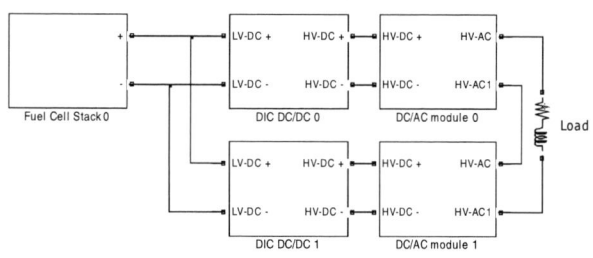

Fig. 5. Fuel Cell, Current-fed DIC Converters with Active Clamping Circuit and Cascaded H-bridge Converters simulation model in MATLAB Simulink.

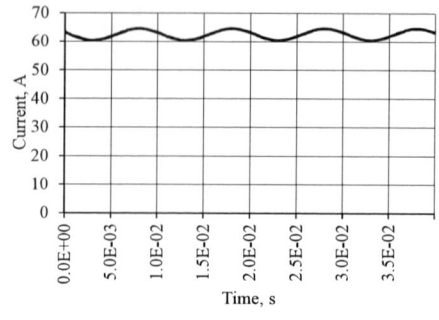

a)

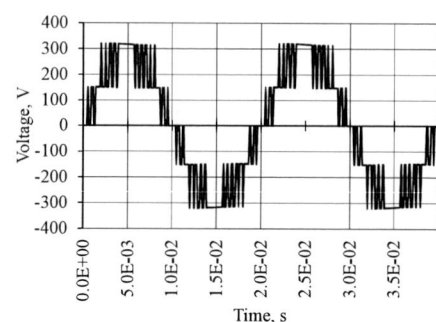

b)

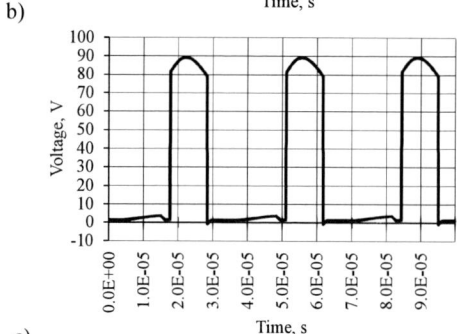

c)

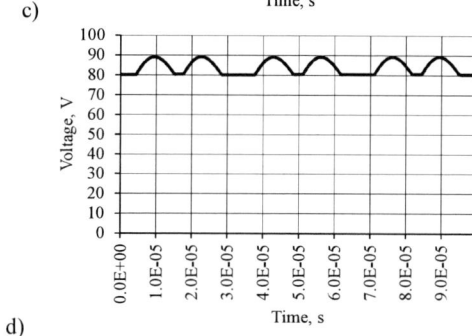

d)

Fig. 6. Simulation results: the output current of the fuel cell (a); the output voltage of multilevel inverter (b); the voltage of one of the primary switches of DC/DC converter module (c); the voltage of the clamp capacitor of DC/DC converter module (d)

VIII. EXPERIMENTAL RESULTS

Experimental testing of the fuel cell interfacing converter system was carried out as well. The experimental setup of the converter system is shown on Fig.6. The testing was performed using HyPM HD8 PEM [18] fuel cell module with a rated electrical power of 8 kW, connecting the output of the interfacing converter system to a resistive load.

978-1-4799-0024-4/13 $31.00 © 2013 IEEE   436

Fig. 6. The prototype of CHC and DIC converter.

The experimental setup of control systems of the DIC converter modules, as well as the five level CHC was implemented on Xilinx Spartan 3E FPGA hardware and Xilinx ISE WebPACK 14.2 software.

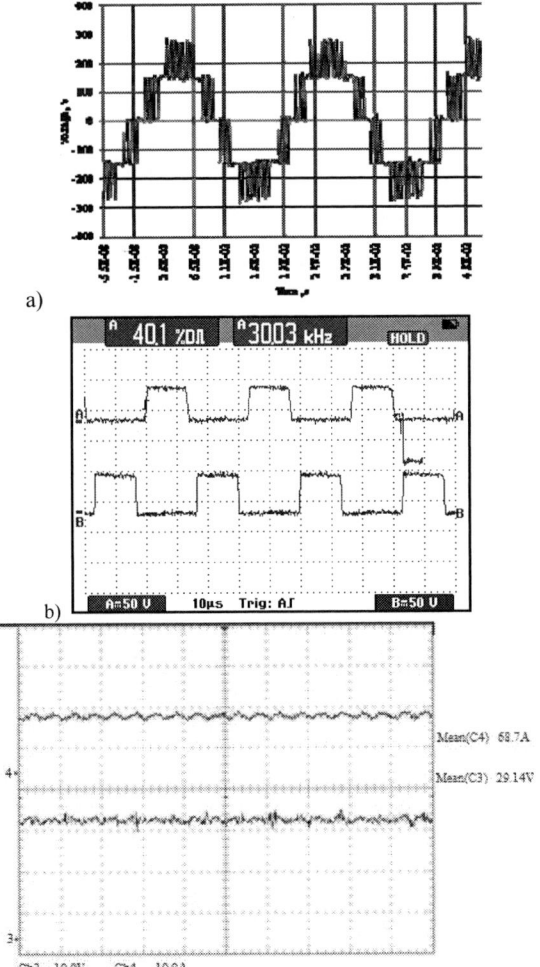

Fig. 7. Experimental waveforms: output voltage of the CHC inverter (a); voltage across primary power transistors of the DIC converter module (b); input current, input voltage of DIC converter (c)

Measurements of the input current, input voltage and load voltage and current were carried out, calculating the efficiency of the overal system. Generally, the experimental results (Fig. 7.) have shown appropriate performance of the interfacing converter system - the input current ripple of the DC/DC converter stage is ± 3A (less than 5% of the average value of the current). The fundamental frequency of the output voltage of the CHC inverter is 50.1Hz and the RMS voltage – 230V. The efficiency of the two current-fed DIC converter modules with active clamping and two CHC inverters was relatively low at 10% of load, reaching 93% at the rated load conditions.

Fig 7b. shows a successful reduction of turn-off overvoltages across the primary power switches of the DC/DC converter stage. The main and the active clamping switches are operated under the ZVS conditions at turn-on and turn-off transients.

The input current and voltage of the DIC converter prototype are shown on Fig.7c. As evident, the input current ripple is ± 3A (less than 5% of the average value of the current), which is satisfactory for fuel cell operation. Experimental results of the converter system confirm simulation data.

IX. CONCLUSIONS

Analysis and development of interfacing power converter system for hydrogen fuel cell has been done, comprising a current-fed DC/DC step-up stage with active clamping and a multilevel inverter with a cascaded H-bridge converter topology.

The experimentally acquired input current ripple of the DC/DC step-up stage is below ± 5% of the mean value, which is within acceptable limits for the safe operation of the fuel cell. The results show successful reduction of turn-off overvoltages across the primary power switches of the DC/DC stage. The main and the active clamping switches of the DC/DC stage are operated under ZVS conditions at turn-on and turn-off transients.

The CHC inverter topology has been chosen as the most suitable for the system under consideration, requiring fewer components and low switch voltages.

Generally, the simulation and experimental results of the interfacing converter system agree with the design specifications, ensuring a rated efficiency of 93% at nominal operation and with an output voltage of 230V.

X. REFERENCES

Periodicals:

[1] W. Choi, P.N. Enjeti, J. W. Howze, G. Joung, "An experimental evaluation of the effects of ripple current generated by the power conditioning stage on a proton exchange membrane fuel cell stack", Journal of Materials Engineering and Performance, New York, USA, Vol. 13. pp.3257-264, 2004.

[2] Ranganathan, V.T., Ziogas, P.D., Stefanovic, V.R. "A DC-AC power conversion technique using twin resonant high frequency links", Proc. of IAS, vol.IA-19, issue:3, May, 1983, pp. 393-400.

[3] Rodriguez, J., Lai, J. S., ZhengPeng, F. "Multilevel inverters: a survey of topologies, control and applications", IEEE Trans. on Industrial. Application, vol. 49. August, 2002, pp. 724-738

[4] Sedghisigarchi, K., Feliachi, "Dynamic and transient analysis of power distribution systems with fuel Cells-part I: fuel-cell dynamic model" IEEE Transactions on energy conversion, vol.19, no. 2, June, 2004, pp. 423-428.

[5] Marchesoni M. "High-performance current control techniques for applications to multilevel high-power voltage source inverters" IEEE Transactions on Power Electronics, vol. 7, no.1, January, 1992, pp. 189 -204.

[6] Song, S.G., Kang, F.S., Park, S.J. "Cascaded multilevel inverter employing three-phase transformers and single dc input", IEEE Trans. Ind. Electron., vol. 56, no. 6, June, 2009, pp. 2005-2014.

[7] Lembeye, Y., Bang, V. D., Lefevre, G., Ferrieux, J.P., "Novel half-bridge inductive dc-dc isolated converters for fuel cell application", IEEE Trans. Energy Convers., vol. 24, no.1 Mar. 2009, pp. 203-210.

Books:

[8] EGG Services Parsons Inc., "Fuel cell handbook (6th edition)", United States Department of Energy, USA, November 2002.

[9] N. Mohan, T. Undeland, W.P. Robbins, "Power Electronics. Converters, Applications and Design", John Wiley & Sons, ISBN:0-471-22693-9, 2003.

[10] R.W. Erickson, D. Maksimovic, "Fundamentals of Power Electronics", Chapman & Hall, ISBN 0-7923-7270-0, 2001.

[11] Charles L. Phillips, Royce D. Harbor: Feedback Control Systems– Prentice Hall, ISBN: 0-13-949-090-6, 2000

Papers from Conference Proceedings (Published):

[12] S. Meo, A. Perfetto, L. Piegari, F. Esposito "A ZVS current fed DC/DC converter oriented for applications fuel-cell-based", Proc. of IECON 2004, Busan, South Korea, November 2004. pp. 932-937.

[13] S. De Caro, A. Testa, D. Triolo, M. Cacciato, A. Consoli, "Low Input Current Ripple Converters for Fuel Cell Power Units// in Proc. Hard Switching Converters and Control EPE 2005. – Germany, 2005.

[14] J.T. Kim, B.K. Lee, T.W. Lee, S.J. Jang, S.S. Kim, C.Y. Won, "An Active Clamping Current-Fed Half-Bridge Converter for Fuel-Cell Generation Systems", 35th Annual IEEE Power Electronics Specialists Conference, Aachen, Germany, 2004.

[15] F.J. Nome, I. Barbi, "A ZVS Clamping Mode - Current-Fed Push-pull DC-DC Converter", Proc. of ISIE`98, Pertoria, South Africa, July 1998, pp. 617-621.

[16] Rajagopalan, V., Al Haddad, K., Ayer, J. (1985). Innovative Utility-Interactive D.C. to A.C. Power Conditioning System. *Proc. of IEEE IECON '85*, vol. 2.

[17] Ojo, O., Konduru, S., (2005). A discontinuous carrier-based PWM modulation method for the control of neutral point voltage of three phase three-level diode clamped converters. Proc. of PESC`05, Recife, Brazil, June, 2005, pp 1652-1658.

[18] Andreičiks, A., Steiks, I., Krievs, O. "Design of Efficient Current Fed DC/DC Converter for Fuel Cell Applications", Proc. of ISIE2011, Gdansk, Poland, 27-30 June, 2011, pp. 206-210.

Web pages:

[19] *IEEE Guide for Application of Power Apparatus Bushings*, IEEE Standard C57.19.100-1995, Aug. 1995.

XI. BIOGRAPHIES

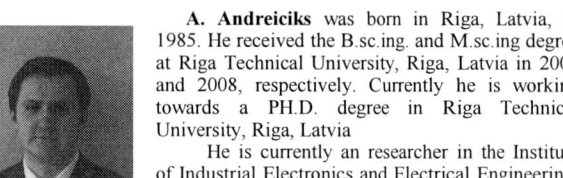

A. Andreiciks was born in Riga, Latvia, in 1985. He received the B.sc.ing. and M.sc.ing degree at Riga Technical University, Riga, Latvia in 2006 and 2008, respectively. Currently he is working towards a PH.D. degree in Riga Technical University, Riga, Latvia

He is currently an researcher in the Institute of Industrial Electronics and Electrical Engineering, Riga Technical University. His main field of interest is the design and optimization of power electronic circuits for renewable energy systems.

I. Steiks received the B.sc.ing., M.sc.ing. and D.sc.ing. degree from the Faculty of Power and Electrical Engineering, Riga Technical University, Riga, Latvia, in 2003, in 2005 and in 2011 accordingly.

Since 2006, he has been a researcher and since 2012 - a leading researcher in the Institute of Industrial Electronics and Electrical Engineering, Riga Technical University. His main research interests include DC/AC converters and DC/DC converters for fuel cells.

O. Krievs has received Bachelor's (2001), Master's (2003) and Doctor's (2007) degrees in the field of electrical engineering at the Faculty of Power and Electrical Engineering of Riga Technical University.

O. Krievs has been working in Riga Technical University since 2001 and currently is in the positions of dean and assistant professor at the Department of Power and Electrical Engineering of Riga Technical University. His main research fields include active power filters, frequency converters and DC/DC converters.

F. Blaabjerg He was employed at ABB-Scandia, Randers, from 1987-1988. During 1988-1992 he was PhD. student at Aalborg University, Denmark, became Assistant Professor in 1992, Associate Professor in 1996 and full professor in power electronics and drives in 1998. He has been part-time research leader at Research Center Risoe in wind turbines. In 2006-2010 he was dean of the faculty of Engineering, Science and Medicine and became visiting professor at Zhejiang University, China in 2009. His research areas are in power electronics and its applications like in wind turbines, PV systems and adjustable speed drives. He has been Editor in Chief of the IEEE Transactions on Power Electronics 2006-2012. He was Distingueshed lecturer for the IEEE Power Electronics Society 2005-2007 and for IEEE Industry Applications Society from 2010-2011. He has been Chairman of of EPE'2007 and PEDG'2012 – both held in Aalborg

He received the 1995 Angelos Award for his contribution in modulation technique and the Annual Teacher prize at Aalborg University. In 1998 he received the Outstanding Young Power Electronics Engineer Award from the IEEE Power Electronics Society. He has received thirteen IEEE Prize paper awards and another prize paper award at PELINCEC Poland 2005. He received the IEEE PELS Distingueshed Service Award in 2009 and the EPE-PEMC 2010 Council award. Finally he has received a number of major research awards in Denmark

An Adaptive Robust Position Control for Induction Machines using a Sliding Mode Flux Observer

Oscar Barambones

Abstract— An adaptive sliding-mode position control for induction motors using the field oriented control theory is presented. The proposed sliding-mode control law incorporates an adaptive switching gain to avoid calculating an upper limit for the system uncertainties. The design also incorporates a sliding-mode flux estimator that operates on the principle of flux and current observer. The proposed observer uses a plant model and also includes a sliding mode terms to overcome the model uncertainties. The stability analysis of the proposed observer and controller under parameter uncertainties and load disturbances is provided using the Lyapunov stability theory. Finally experimental results show that the proposed controller with the proposed observer provides high-performance dynamic characteristics and that this control scheme is robust with respect to plant parameter variations and external load disturbances.

Index Terms- Position Control. Sliding Mode Control. Robust Control. Induction Machines. Lyapunov Stability. Nonlinear Control.

NOMECLATURE

i_{ds}, i_{qs}	Stator currents in stationary reference frame
i_{ds}^e, i_{qs}^e	Stator currents in synchronously ref. frame
V_{ds}, V_{qs}	Stator voltages in stationary reference frame
ψ_{dr}, ψ_{qr}	Rotor flux in stationary reference frame
ψ_{dr}^e, ψ_{qr}^e	Rotor flux in synchronously reference frame
R	Resistance
L	Inductance
w_r	Rotor electrical speed
w_m	Rotor mechanical speed
T_r	Rotor time constant
J	Inertia moment
B	Viscous friction coefficient
θ_m	Rotor mechanical position
θ_r	Rotor electrical position
p	Pole numbers
T_L	Load torque
S	Sliding variable
β	Switching gain

I. INTRODUCTION

Position control is often used in some applications of electrical drives like robotic systems, conveyor belts, etc. In these applications uncertainty and external disturbances are present and therefore a robust control system that maintain the desired control performance under this situations are

Oscar Barambones is with Dpto. Ingeniería de Sistemas y Automática. EUI de Vitoria. Nieves Cano 12. 01006 Vitoria, Spain. oscar.barambones@ehu.es

frequently required [1],[2]. The sliding-mode control strategy has been focussed on many studies and research for the position control of the induction motors [3]-[8].

However the traditional sliding control schemes requires the prior knowledge of an upper bound for the system uncertainties, since this bound is employed in the switching gain calculation. This upper bound should be determined as precisely as possible, because as higher is the upper bound higher value should be considered for the sliding gain, and therefore the control effort will also be high, which is undesirable in a practice. In order to surmount this drawback, in the present paper it is proposed an adaptive law to calculate the sliding gain.

On the other hand in the last decade remarkable efforts have been made to reduce the number of sensors in the control systems. The sensors increases the cost and also reduces the reliability of the control system because this elements are generally expensive, delicate and difficult to instal [9]-[11].

This paper presents a sliding mode control scheme consisting, on the one hand of a flux estimation algorithm in order to avoid the flux sensors, and on the other hand, of a variable structure control algorithm that overcome the system uncertainties and load torque disturbances. Moreover, the proposed control scheme do not present a high computational cost and therefore can be implemented easily in a real time applications over a low-cost DSP processor.

This manuscript is organized as follows. The sliding mode flux observer is introduced in Section II. Then, the proposed adaptive variable structure robust position control is presented in Section III. In the Section IV, some simulation results are presented. Finally, concluding remarks are stated in Section V.

II. SLIDING MODE ROTOR FLUX ESTIMATOR

Many schemes based on simplified motor models have been devised to estimate some internal variables of the induction motor from measured terminal quantities. This procedure is frequently used in order to avoid the presence of some sensors in the control scheme. In order to obtain an accurate dynamic representation of the motor, it is necessary to base the calculation on the coupled circuit equations of the motor.

Since the motor voltages and currents are measured in a stationary frame of reference, it is also convenient to express the induction motor dynamical equations in this stationary frame.

The system state space equations in the stationary reference frame can be written in the form [12]:

$$
\begin{aligned}
\dot{i}_{ds} &= \frac{-1}{\sigma L_s}\left(R_s + \frac{L_m^2}{L_r^2}R_r\right)i_{ds} + \frac{L_m}{\sigma L_s L_r}\frac{1}{T_r}\psi_{dr} \\
&\quad + \frac{L_m}{\sigma L_s L_r}w_r\psi_{qr} + \frac{1}{\sigma L_s}V_{ds} \\
\dot{i}_{qs} &= \frac{-1}{\sigma L_s}\left(R_s + \frac{L_m^2}{L_r^2}R_r\right)i_{qs} - \frac{L_m}{\sigma L_s L_r}w_r\psi_{dr} \\
&\quad + \frac{L_m}{\sigma L_s L_r}\frac{1}{T_r}\psi_{qr} + \frac{1}{\sigma L_s}V_{qs} \\
\dot{\psi}_{dr} &= \frac{L_m}{T_r}i_{ds} - \frac{1}{T_r}\psi_{dr} - w_r\psi_{qr} \\
\dot{\psi}_{qr} &= \frac{L_m}{T_r}i_{qs} + w_r\psi_{dr} - \frac{1}{T_r}\psi_{qr}
\end{aligned}
\tag{1}
$$

where V_{ds}, V_{qs} are stator voltages; i_{ds}, i_{qs} are stator currents; ψ_{dr}, ψ_{qr} are rotor fluxes; w_r is motor speed; R_s, R_r are stator and rotor resistances; L_s, L_r are stator and rotor inductances; L_m, is mutual inductance; $\sigma = 1 - \dfrac{L_m^2}{L_s L_r}$ is leakage coefficient; $T_r = \dfrac{L_r}{R_r}$ is rotor-time constant.

From singular perturbation theory [14], and based on the well-known induction motor model dynamics [12], the slow variables of the system are ψ_{dr}, ψ_{qr} and the fast variables are i_{ds}, i_{qs}. Therefore, the corresponding singularly perturbed model of (1) is:

$$
\begin{aligned}
\varepsilon\dot{i}_{ds} &= -L_m\alpha_r i_{ds} + \alpha_r\psi_{dr} + w_r\psi_{qr} + \frac{L_r}{L_m}(V_{ds} - R_s i_{ds}) \\
\varepsilon\dot{i}_{qs} &= -L_m\alpha_r i_{qs} - w_r\psi_{dr} + \alpha_r\psi_{qr} + \frac{L_r}{L_m}(V_{qs} - R_s i_{qs}) \\
\dot{\psi}_{dr} &= L_m\alpha_r i_{ds} - \alpha_r\psi_{dr} - w_r\psi_{qr} \\
\dot{\psi}_{qr} &= L_m\alpha_r i_{qs} + w_r\psi_{dr} - \alpha_r\psi_{qr}
\end{aligned}
\tag{2}
$$

where $\varepsilon = \dfrac{\sigma L_s L_r}{L_m}$ and $\alpha_r = \dfrac{1}{T_r}$.

The proposed sliding mode observer is a copy of the original system model, which has corrector terms with switching functions based on the system outputs. Therefore, considering the measured stator currents as the system outputs, the corresponding sliding-mode-observer can be constructed as follows:

$$
\begin{aligned}
\varepsilon\dot{\hat{i}}_{ds} &= -L_m\alpha_r i_{ds} + \alpha_r\hat{\psi}_{dr} + w_r\hat{\psi}_{qr} + \frac{L_r}{L_m}(V_{ds} - R_s i_{ds}) \\
&\quad - k_1 e_{id} + g_{i_d}\,\mathrm{sgn}(e_{id}) \\
\varepsilon\dot{\hat{i}}_{qs} &= -L_m\alpha_r i_{qs} - w_r\hat{\psi}_{dr} + \alpha_r\hat{\psi}_{qr} + \frac{L_r}{L_m}(V_{qs} - R_s i_{qs}) \\
&\quad - k_2 e_{iq} + g_{i_q}\,\mathrm{sgn}(e_{iq}) \\
\dot{\hat{\psi}}_{dr} &= L_m\alpha_r i_{ds} - \alpha_r\hat{\psi}_{dr} - w_r\hat{\psi}_{qr} + g_{\psi_d}\,\mathrm{sgn}(e_{id}) \\
\dot{\hat{\psi}}_{qr} &= L_m\alpha_r i_{qs} + w_r\hat{\psi}_{dr} - \alpha_r\hat{\psi}_{qr} + g_{\psi_q}\,\mathrm{sgn}(e_{iq})
\end{aligned}
\tag{3}
$$

where $\hat{i}$ and $\hat{\psi}$ are the estimations of i and ψ; k_1 and k_2 are positive constant gains; g_{i_d}, g_{i_q}, g_{ψ_d} and g_{ψ_q} are the observer

gain matrix; $e_{i_d} = \hat{i}_{ds} - i_{ds}$ and $e_{i_q} = \hat{i}_{qs} - i_{qs}$ are de current errors, and $\mathrm{sgn}()$ is the sign function.

Subtracting (2) from (3), the estimation error dynamics are:

$$
\begin{aligned}
\varepsilon\dot{e}_{i_d} &= \alpha_r e_{\psi_d} + w_r e_{\psi_q} - k_1 e_{id} + g_{i_d}\,\mathrm{sgn}(e_{id}) \\
\varepsilon\dot{e}_{i_q} &= -w_r e_{\psi_d} + \alpha_r e_{\psi_q} - k_2 e_{iq} + g_{i_q}\,\mathrm{sgn}(e_{iq}) \\
\dot{e}_{\psi_d} &= -\alpha_r e_{\psi_d} - w_r e_{\psi_q} + g_{\psi_d}\,\mathrm{sgn}(e_{id}) \\
\dot{e}_{\psi_q} &= w_r e_{\psi_d} - \alpha_r e_{\psi_q} + g_{\psi_q}\,\mathrm{sgn}(e_{iq})
\end{aligned}
\tag{4}
$$

where $e_{\psi_d} = \hat{\psi}_{dr} - \psi_{dr}$, $e_{\psi_q} = \hat{\psi}_{qr} - \psi_{qr}$

The previous equations can be expressed in matrix form as:

$$
\begin{aligned}
\varepsilon\dot{e}_i &= +Ae_\psi + K_i e_i + G_i\Upsilon_e \\
\dot{e}_\psi &= -Ae_\psi + G_\psi\Upsilon_e
\end{aligned}
\tag{5}
$$

where

$A = \alpha_r I_2 - w_r J_2$, $e_i = [e_{i_d}\ e_{i_q}]^T$, $e_\psi = [e_{\psi_d}\ e_{\psi_q}]^T$,

$\Upsilon_e = [\mathrm{sgn}(e_{id})\ \mathrm{sgn}(e_{iq})]^T$,

$$
G_i = \begin{bmatrix} g_{i_d} & 0 \\ 0 & g_{i_q} \end{bmatrix}, \qquad
G_\psi = \begin{bmatrix} g_{\psi_d} & 0 \\ 0 & g_{\psi_q} \end{bmatrix}
$$

$$
I_2 = \begin{bmatrix} 1 & 0 \\ 0 & 1 \end{bmatrix}, \qquad
J_2 = \begin{bmatrix} 0 & -1 \\ 1 & 0 \end{bmatrix},
$$

$$
K_i = \begin{bmatrix} -k_1 & 0 \\ 0 & -k_2 \end{bmatrix}
$$

Following the two-time-scale approach, the stability analysis of the above system can be considered determining the observer gains G_i and K_i of the fast subsystem or measured state variables (i_{ds}, i_{qs}), to ensure the attractiveness of the sliding surface $e_i = 0$ in the fast time scale. Thereafter, the observer gain G_ψ of the slow subsystem or inaccessible state variables (ψ_{dr}, ψ_{qr}), are determined, such that the reduced-order system obtained when $e_i \cong \dot{e}_i \cong 0$ is locally stable [14].

From singular perturbation theory, the fast reduced-order system of the observation errors can be obtained by introducing the new time variable $\tau = (t - t_0)/\varepsilon$ and thereafter setting $\varepsilon \to 0$ [14]. In the new time scale τ, taking into account that $d\tau = dt/\varepsilon$, (5) becomes:

$$
\begin{aligned}
\frac{d}{d\tau}e_i &= Ae_\psi + K_i e_i + G_i\Upsilon_e \\
\frac{d}{d\tau}e_\psi &= 0
\end{aligned}
\tag{6}
$$

Therefore, if the observer gains G_i and K_i are adequately chosen, the sliding mode occurs in (6) along the manifold $e_i = [e_{i_d}\ e_{i_q}]^T = 0$.

The attractivity condition of the sliding surface $e_i = 0$ given by:

$$
e_i^T\frac{de_i}{d\tau} < 0
\tag{7}
$$

is fulfilled with the following inequalities,

$$
\begin{aligned}
g_{i_d} &< -\left|\alpha_r e_{\psi_d} + w_r e_{\psi_q}\right| + k_1\left|e_{id}\right| \tag{8} \\
g_{i_q} &< -\left|-w_r e_{\psi_d} + \alpha_r e_{\psi_q}\right| + k_2\left|e_{iq}\right| \tag{9}
\end{aligned}
$$

978-1-4799-0024-4/13 $31.00 © 2013 IEEE

Proof:

Let us define the following Lyapunov function candidate,

$$V = \frac{1}{2} e_i^T e_i$$

whose time derivative is,

$$
\begin{aligned}
\frac{dV}{d\tau} &= e_i^T \frac{de_i}{d\tau} \\
&= e_i^T [A e_\psi + K_i e_i + G_i \Upsilon_e] \qquad (10) \\
&= \begin{bmatrix} e_{id} \left\{ g_{i_d} \operatorname{sgn}(e_{id}) + \alpha_r e_{\psi_d} + w e_{\psi_q} - k_1 e_{id} \right\} \\ e_{iq} \left\{ g_{i_q} \operatorname{sgn}(e_{iq}) - w e_{\psi_d} + \alpha_r e_{\psi_q} - k_2 e_{iq} \right\} \end{bmatrix}
\end{aligned}
$$

Taking into account that all states and parameters of induction motor are bounded, then there exist sufficiently large negative numbers g_{i_d}, g_{i_q}, and positive numbers k_1 and k_2 so that the inequalities defined in (9) are verified and then the attractivity condition defined in (7) is fulfilled.

Then, once the currents trajectory reaches the sliding surface $e_i = 0$, the observer error dynamics given by (6) behaves, in the sliding mode, as a reduced-order subsystem governed only by the rotor-flux error e_ψ, assuming that $e_i = \dot{e}_i = 0$.

The slow error dynamics (when $e_i = 0$ and $\dot{e}_i = 0$), can be obtained setting $\varepsilon = 0$ in the system equation presented in (5):

$$
\begin{aligned}
0 &= +A e_\psi + G_i \Upsilon_e \\
\dot{e}_\psi &= -A e_\psi + G_\psi \Upsilon_e \qquad (11)
\end{aligned}
$$

In order to demonstrate de stability of the previous system, the following Lyapunov function candidate is proposed:

$$V = \frac{1}{2} e_\psi^T e_\psi \qquad (12)$$

The time derivative of the Lyapunov function candidate is:

$$\frac{dV}{dt} = \dot{e}_\psi^T e_\psi \qquad (13)$$

From (11) it is deduced that:

$$
\begin{aligned}
e_\psi &= -A^{-1} G_i \Upsilon_e \qquad (14) \\
\dot{e}_\psi &= (G_i + G_\psi) \Upsilon_e \qquad (15)
\end{aligned}
$$

Then from (13), (14) and (15)

$$
\begin{aligned}
\frac{dV}{dt} &= -\Upsilon_e^T (G_i + G_\psi)^T A^{-1} G_i \Upsilon_e \\
&= -\Upsilon_e^T (G_i + G_\psi)^T A^{-1} G_i \Upsilon_e \\
&= -\Upsilon_e^T \left[(I_2 + G_\psi G_i^{-1}) G_i \right]^T A^{-1} G_i \Upsilon_e \\
&= -\Upsilon_e^T G_i^T (I_2 + G_\psi G_i^{-1})^T A^{-1} G_i \Upsilon_e \\
&= -\Upsilon_e^T G_i^T (A^{-1})^T A^T (I_2 + G_\psi G_i^{-1})^T A^{-1} G_i \Upsilon_e \\
&= -(A^{-1} G_i \Upsilon_e)^T A^T (I_2 + G_\psi G_i^{-1})^T A^{-1} G_i \Upsilon_e \\
&= -e_\psi^T A^T (I_2 + G_\psi G_i^{-1})^T e_\psi \\
&= -e_\psi^T (I_2 + G_\psi G_i^{-1}) A e_\psi \qquad (16)
\end{aligned}
$$

To ensure that $\dot{V}$ is negative definite the following sufficient condition can be requested:

$$(I_2 + G_\psi G_i^{-1}) A \geq \varrho I_2 \qquad (17)$$

where ϱ is a positive constant

Solving the gain matrix G_ψ in (17) yields:

$$
\begin{aligned}
(I_2 + G_\psi G_i^{-1}) &\geq \varrho I_2 A^{-1} \qquad (18) \\
(I_2 + G_\psi G_i^{-1}) &\geq \varrho A^{-1} \qquad (19) \\
G_\psi G_i^{-1} &\geq \varrho A^{-1} - I_2 \qquad (20) \\
G_\psi &\leq (\varrho A^{-1} - I_2) G_i \qquad (21)
\end{aligned}
$$

Therefore, the time derivative of the Lyapunov function will be negative definite if the observer gain G_ψ is chosen taking into account (21). As a result from (16) it is concluded that the equilibrium point ($e_\psi = 0$) of the flux observer error dynamic given by (11) is exponentially stable; that is, the flux observer error converges to zero with exponential rate of convergence.

III. ADAPTIVE VARIABLE STRUCTURE POSITION CONTROL

The mechanical equation of an induction motor can be written as:

$$J \ddot{\theta}_m + B \dot{\theta}_m + T_L = T_e \qquad (22)$$

where J and B are the inertia constant and the viscous friction coefficient of the induction motor respectively; T_L is the external load; θ_m is the rotor mechanical position, which is related to the rotor electrical position, θ_r, by $\theta_m = 2\,\theta_r/p$ where p is the pole numbers and T_e denotes the generated torque of an induction motor, defined as [12]:

$$T_e = \frac{3p}{4} \frac{L_m}{L_r} (\psi_{dr}^e i_{qs}^e - \psi_{qr}^e i_{ds}^e) \qquad (23)$$

where ψ_{dr}^e and ψ_{qr}^e are the rotor-flux linkages, with the subscript 'e' denoting that the quantity is refereed to the synchronously rotating reference frame; i_{qs}^e and i_{ds}^e are the stator currents, and p is the pole numbers.

The relation between the synchronously rotating reference frame and the stationary reference frame is performed by the so-called reverse Park's transformation:

$$
\begin{bmatrix} x_a \\ x_b \\ x_c \end{bmatrix} = \begin{bmatrix} \cos(\theta_e) & -\sin(\theta_e) \\ \cos(\theta_e - 2\pi/3) & -\sin(\theta_e - 2\pi/3) \\ \cos(\theta_e + 2\pi/3) & -\sin(\theta_e + 2\pi/3) \end{bmatrix} \begin{bmatrix} x_d^e \\ x_q^e \end{bmatrix}
$$
(24)

where θ_e is the angle position between the d-axis of the synchronously rotating reference frame and the a-axis of the stationary reference frame, and it is assumed that the quantities are balanced.

The estimated angular position of the rotor flux vector ($\bar{\psi}_r$) related to the d-axis of the stationary reference frame may be calculated by means of the rotor flux components in this reference frame ($\hat{\psi}_{dr}$, $\hat{\psi}_{qr}$) as follows:

$$\hat{\theta}_e = \arctan 2 \left(\hat{\psi}_{qr}, \hat{\psi}_{dr} \right) \qquad (25)$$

where $\hat{\theta}_e$ is the estimated angular position of the rotor flux vector.

Using the field-orientation control principle [12], the current component i_{ds}^e is aligned in the direction of the rotor flux vector $\bar{\psi}_r$, and the current component i_{qs}^e is aligned in the perpendicular direction to it. At this condition, it is satisfied that:

$$\psi_{qr}^e = 0, \qquad \psi_{dr}^e = |\bar{\psi}_r| \qquad (26)$$

Taking into account the results presented in equation (26), the equation of induction motor torque (23) is simplified to:

$$T_e = \frac{3p}{4}\frac{L_m}{L_r}\psi_{dr}^e i_{qs}^e = K_T\, i_{qs}^e \qquad (27)$$

where K_T is the torque constant, defined as follows:

$$K_T = \frac{3p}{4}\frac{L_m}{L_r}\psi_{dr}^{e*} \qquad (28)$$

where ψ_{dr}^{e*} denotes the command rotor flux.

With the above mentioned proper field orientation, the dynamics of the rotor flux is given by [12]:

$$\frac{d\psi_{dr}^e}{dt} + \frac{\psi_{dr}^e}{T_r} = \frac{L_m}{T_r}i_{ds}^e \qquad (29)$$

Then, the mechanical equation (22) becomes:

$$\ddot{\theta}_m + a\,\dot{\theta}_m + f = b\,i_{qs}^e \qquad (30)$$

where the parameters are defined as:

$$a = \frac{B}{J}, \quad b = \frac{K_T}{J}, \quad f = \frac{T_L}{J}; \qquad (31)$$

Now, we are going to consider the previous mechanical equation (30) with uncertainties as follows:

$$\ddot{\theta}_m = -(a + \triangle a)\dot{\theta}_m - (f + \triangle f) + (b + \triangle b)i_{qs}^e \qquad (32)$$

where the terms $\triangle a$, $\triangle b$ and $\triangle f$ represents the uncertainties of the terms a, b and f respectively.

Let us define the position tracking error as follows:

$$e(t) = \theta_m(t) - \theta_m^*(t) \qquad (33)$$

where θ_m^* is the rotor position command.

Taking the second derivative of the previous equation with respect to time yields:

$$\ddot{e}(t) = \ddot{\theta}_m - \ddot{\theta}_m^* = u(t) + d(t) \qquad (34)$$

where the following terms have been collected in the signal $u(t)$,

$$u(t) = b\,i_{qs}^e(t) - a\,\dot{\theta}_m(t) - f(t) - \ddot{\theta}_m^*(t) \qquad (35)$$

and the uncertainty terms have been collected in the signal $d(t)$,

$$d(t) = -\triangle a\, w_m(t) - \triangle f(t) + \triangle b\, i_{qs}^e(t) \qquad (36)$$

Now, we are going to define the sliding variable $S(t)$ as:

$$S(t) = \dot{e}(t) + k\,e(t) + k_i \int e(t)\,dt \qquad (37)$$

where k and k_i are positive constant gains.

Then, the sliding surface is defined as:

$$S(t) = \dot{e}(t) + k\,e(t) + k_i \int e(t)\,dt = 0 \qquad (38)$$

The variable structure position controller is designed as:

$$u(t) = -k\dot{e}(t) - k_i e(t) - \hat{\beta}\gamma\,\mathrm{sgn}(S) \qquad (39)$$

where the k is the previously defined gain, $\hat{\beta}$ is the switching gain, S is the sliding variable defined in (37) and $\mathrm{sgn}(\cdot)$ is the sign function.

Finally, the switching gain $\hat{\beta}$ is adapted according to the following updating law:

$$\dot{\hat{\beta}} = \gamma\,|S| \qquad \hat{\beta}(0) = 0 \qquad (40)$$

where γ is a positive constant that let us choose the adaptation speed for the sliding gain.

It should be noted that in same cases the adaptation law (40) should be equipped with a dead-zone to avoid an unbounded growth of the gain caused by the measurement noise. In this cases the sliding gain adaptation law can be modified to:

$$\dot{\hat{\beta}} = \gamma\,|S_o| \qquad \hat{\beta}(0) = 0 \qquad (41)$$

where S_o is defined by:

$$S_o = \begin{cases} S - \xi & if \quad |S| > \xi \\ 0 & otherwise. \end{cases} \qquad (42)$$

where ξ is a small positive constant.

It is interesting to point out that S_o is a measure of the distance from the sliding surface S to the interval $[-\xi,\ \xi]$:

From the previous equation it is concluded that $\dot{S}_o = \dot{S}$ when S is outside the interval $[-\xi,\ \xi]$, while $\dot{S}_o = 0$ otherwise.

In order to obtain the position trajectory tracking, the following assumption should be formulated:

($\mathcal{A}1$) There exists an unknown finite and positive switching gain β such that

$$\beta > d_{max} + \eta \qquad \eta > 0$$

where $d_{max} \geq |d(t)| \quad \forall\ t$ and η is a positive constant.

Note that this condition only implies that the system uncertainties are bounded magnitudes.

Theorem 1: Consider the induction motor given by (32). Then, if the assumption ($\mathcal{A}1$) are verified, the control law (39) leads the rotor mechanical position $\theta_m(t)$ so that the position tracking error $e(t) = \theta_m(t) - \theta_m^*(t)$ tends to zero as the time tends to infinity.

The proof of this theorem will be carried out using the Lyapunov stability theory.

Proof : Define the Lyapunov function candidate:

$$V(t) = \frac{1}{2}S(t)S(t) + \frac{1}{2}\tilde{\beta}(t)\tilde{\beta}(t) \qquad (43)$$

where $S(t)$ is the sliding variable defined previously and $\tilde{\beta}(t) = \hat{\beta}(t) - \beta$.

Its time derivative is calculated as:

$$
\begin{aligned}
\dot{V}(t) &= S(t)\dot{S}(t) + \tilde{\beta}(t)\dot{\tilde{\beta}} \\
&= S \cdot [\ddot{e} + k\,\dot{e} + k_i e] + \tilde{\beta}(t)\dot{\hat{\beta}} \\
&= S \cdot [u + d + k\,\dot{e} + k_i e] + \tilde{\beta}\,\gamma|S| \\
&= S \cdot \left[-k\,\dot{e} - k_i e - \hat{\beta}\gamma\,\mathrm{sgn}(S) + d + k\,\dot{e} + k_i e\right] \\
&\quad + (\hat{\beta} - \beta)\gamma|S| \\
&= d\,S - \hat{\beta}\gamma|S| + \hat{\beta}\gamma|S| - \beta\gamma|S| \qquad (44) \\
&\leq |d|\,|S| - \beta\gamma|S| \\
&\leq |d|\,|S| - (d_{max} + \eta)\gamma|S| \\
&= |d|\,|S| - d_{max}\,\gamma|S| - \eta\,\gamma|S| \\
&\leq -\eta\,\gamma|S| \qquad (45)
\end{aligned}
$$

then

$$
\dot{V}(t) \leq 0 \qquad (46)
$$

It should be noted that the eqns. (34), (37), (39) and (40), and the assumption $(\mathcal{A}\,1)$ have been used in the proof.

Using the Lyapunov's direct method, since $V(t)$ is clearly positive-definite, $\dot{V}(t)$ is negative semidefinite and $V(t)$ tends to infinity as $S(t)$ and $\tilde{\beta}(t)$ tends to infinity, then the equilibrium at the origin $[S(t), \tilde{\beta}(t)] = [0, 0]$ is globally stable, and therefore the variables $S(t)$ and $\tilde{\beta}(t)$ are bounded. Since $S(t)$ is bounded then it is deduced that $e(t)$ and $\dot{e}(t)$ are bounded.

On the other hand, making the derivative of equation (37) it is obtained,

$$
\dot{S}(t) = \ddot{e}(t) + k\dot{e}(t) + k_i e(t) \qquad (47)
$$

then, substituting the equation (34) and (39) in the above equation,

$$
\begin{aligned}
\dot{S}(t) &= u(t) + d(t) + k\dot{e}(t) + k_i e(t) \\
&= -k\,\dot{e} - k_i e - \hat{\beta}\gamma\,\mathrm{sgn}(S) + d(t) + k\,\dot{e} + k_i e \\
&= d(t) - \hat{\beta}\gamma\,\mathrm{sgn}(S) \qquad (48)
\end{aligned}
$$

From equation (48) we can conclude that $\dot{S}(t)$ is bounded because $d(t)$, γ and $\hat{\beta}$ are bounded.

Now, from equation (44) it is deduced that

$$
\ddot{V}(t) = d\,\dot{S}(t) - \beta\,\gamma\frac{d}{dt}|S(t)| \qquad (49)
$$

which is a bounded quantity because $\dot{S}(t)$ is bounded.

Under these conditions, since $\ddot{V}$ is bounded, $\dot{V}$ is a uniformly continuous function, so Barbalat's lemma let us conclude that $\dot{V} \to 0$ as $t \to \infty$, which implies that $S(t) \to 0$ as $t \to \infty$.

Therefore $S(t)$ tends to zero as the time t tends to infinity. Moreover, all trajectories starting off the sliding surface $S = 0$ must reach it asymptotically and then will remain on this surface. This system's behavior once on the sliding surface is usually called *sliding mode* [16].

When the sliding mode occurs on the sliding surface (38), then $S(t) = 0$, and therefore the dynamic behavior of the tracking problem (34) is equivalently governed by the following equation:

$$
\dot{S}(t) = 0 \quad \Rightarrow \quad \dot{e}(t) = -k\,e(t) \qquad (50)
$$

Then, since k is a positive constant, the tracking error $e(t)$ and its derivative $\dot{e}(t)$ converges to zero exponentially.

It should be noted that, a typical motion under sliding mode control consists of a *reaching phase* during which trajectories starting off the sliding surface $S = 0$ move toward it and reach it in finite time, followed by *sliding phase* during which the motion will be confined to this surface and the system tracking error will be represented by the reduced-order model (50), where the tracking error tends to zero.

Finally, the torque current command, $i_{qs}^{e*}(t)$, can be obtained directly substituting (39) in (35):

$$
i_{qs}^{e*}(t) = \frac{1}{b}\left[-k\,\dot{e} - k_i e - \hat{\beta}\gamma\,\mathrm{sgn}(S) + a\,\dot{\theta}_m + \ddot{\theta}_m^* + f(t)\right] \qquad (51)
$$

Therefore, the proposed variable structure control resolves the position tracking problem for the induction motor in presence of some uncertainties in mechanical parameters and load torque.

The semi-global asymptotic stability of the closed-loop system with the proposed sliding mode observers, is provided by the separation principle, which requires the asymptotic stability of the observer fast enough, such that it brings the state estimate close enough to its real value in a short time and restores the stabilizing powers of the controller as a necessary and sufficient condition [15].

IV. SIMULATION RESULTS

In this section we will study the position regulation performance of the proposed sliding-mode field oriented control versus reference and load torque variations by means of simulation examples.

The induction motor used in this case study is a 50 HP, 460 V, four pole, 60 Hz motor having the following parameters: $R_s = 0.087\,\Omega$, $R_r = 0.228\,\Omega$, $L_s = 35.5\,mH$, $L_r = 35.5\,mH$, and $L_m = 34.7\,mH$.

The system has the following mechanical parameters: $J = 1.662\,kg.m^2$ and $B = 0.1\,N.m.s$. It is assumed that there are an uncertainty around 20 % in the system parameters, that will be overcome by the proposed sliding control.

In addition the following values have been chosen for the controller parameters: $k = 50$, $k_i = 10$, $\gamma = 20$ and $\hat{\beta}(0) = 0$.

In this example the motor starts from a standstill state and we want that the rotor position follows a ramp command, that starts from 0 rad and finish at 2.5 rad. The system starts with an initial load torque $T_L = 25\,N.m$, and at time $t = 1\,s$, the load torque steps from $T_L = 25\,N.m$ to $T_L = 200\,N.m$. Then at time $t = 1.5\,s$, the load torque steps from $T_L =$

$200\,N.m$ to $T_L = 400\,N.m$. In this case, as before, it is assumed that there is an uncertainty around 20 % in the load torque.

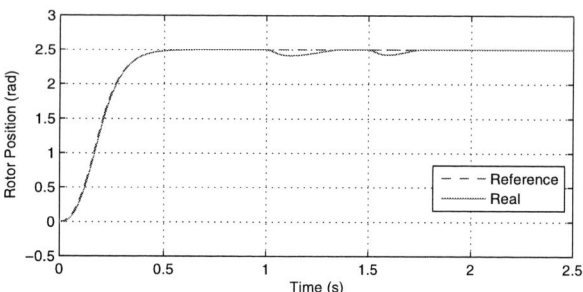

Fig. 1.   Reference and real rotor position signals (rad)

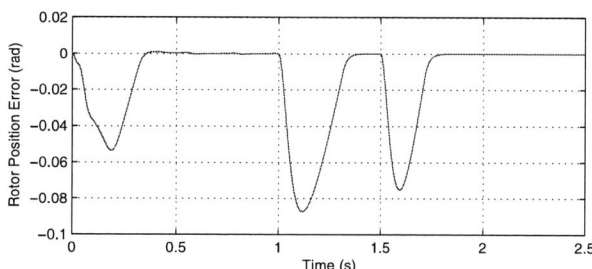

Fig. 2.   Rotor position error (rad)

Fig. 1 shows the desired rotor position (dashed line) and the real rotor position (solid line), and Fig. 2 shows the rotor position error. As it may be observed, after a transitory time in which the sliding gain is adapted, the rotor position tracks the desired position in spite of system uncertainties. Nevertheless, at time $t = 1\,s$ and $t = 1.5\,s$ a little position error can be observed. This error appears because there is a torque increment at this time, and then the controlled system lost the so called 'sliding mode' because the actual sliding gain is too small for the new uncertainty introduced in the system due to the load torque increment. But, after a small time, the sliding gain is adapted so that this gain can compensate for the new system uncertainties and then the rotor position error is eliminated.

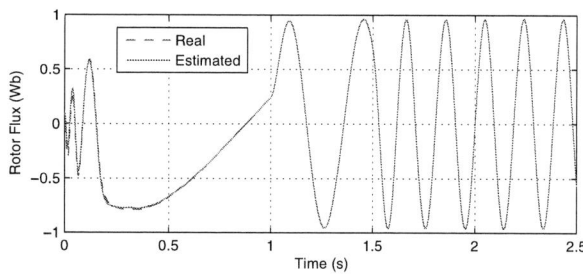

Fig. 3.   Estimated and real rotor Flux signals (Wb)

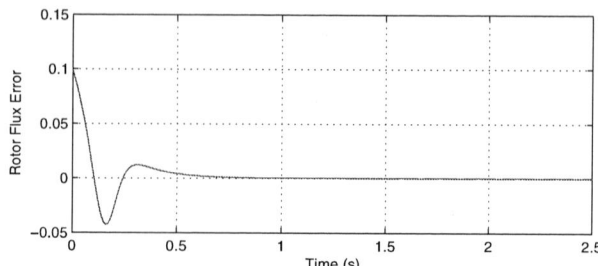

Fig. 4.   Rotor Flux error (Wb)

Fig. 3 shows the estimated rotor flux (solid line) and the real rotor flux (dashed line), and Fig. 4 shows the rotor flux error. As it may be observed, after a transitory time, the estimated rotor flux converges to the real rotor flux value in spite of system uncertainties.

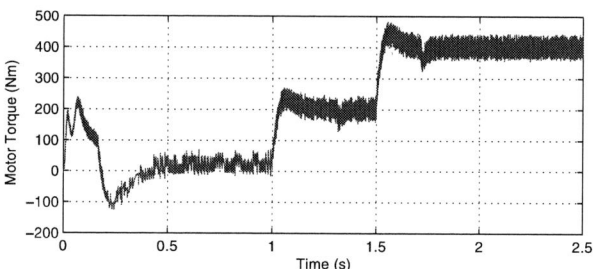

Fig. 5.   Motor torque (N.m)

Fig. 5 shows the motor torque. The motor torque has a high initial value in the speed acceleration zone because it is necessary a high torque to increment the rotor speed and then the value decreases in a deceleration region. Later at time $t = 1\,s$ and $t = 1.5\,s$ the torque increases due to the load torque increment. This figure shows that the so-called chattering phenomenon appears in the motor torque. Although this high frequency changes in the torque will be reduced by the mechanical system inertia, they could cause undesirable vibrations in the rotor, which may be a problem for certain systems. However, for the systems that do not support this chattering, it may be eliminated substituting the sign function by the saturation function in the control signal.

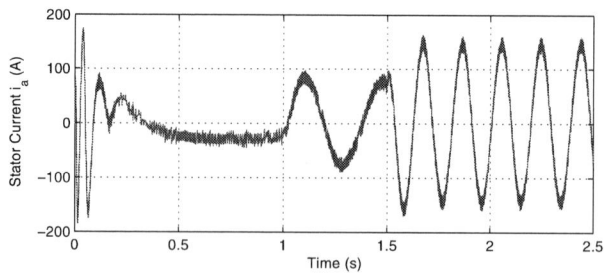

Fig. 6.   Stator Current i_{sa} (A)

Fig. 6 shows the stator current i_{sa}. As in the case of the motor torque, the current signal presents a high value in the initial state. Next, in the constant position region the current is lower because the motor torque only has to compensate the load torque. Then, at time $t = 1\,s$ and $t = 1.5\,s$ the current increases due to the load torque increment.

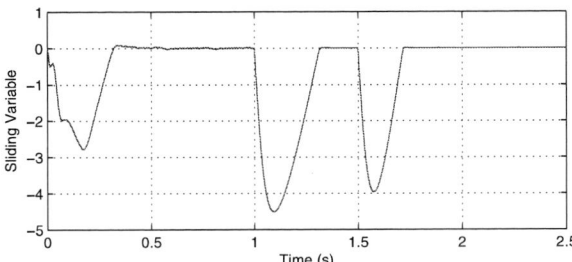

Fig. 7. Sliding Variable

Fig. 7 shows the time evolution of the sliding variable. In this figure it can be seen that the system reach the sliding condition ($S(t) = 0$) at time $t = 0.35\,s$, but the system lost this condition at time $t = 1\,s$ and $t = 1.5\,s$ due to the load torque increment, which produces an increment in the system uncertainties that could not be compensated by the actual value of the sliding gain.

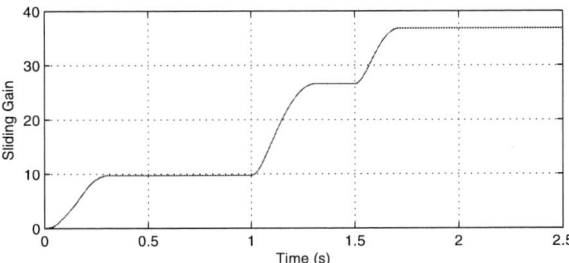

Fig. 8. Sliding Gain

Fig. 8 presents the time evolution of the adaptive sliding gain. The sliding gain starts from zero and then it is increased until its value is high enough in order to compensate for the system uncertainties. Then, after $t = 0.35\,s$, the sliding gain is remained constant because the system uncertainties remain constant as well. Later at time $t = 1\,s$ and $t = 1.5\,s$, there is an increment in the system uncertainties caused by the rise in the load torque. Therefore the sliding gain is adapted once again in order to overcome the new system uncertainties. As it can be observed in the figure after this adaptation the sliding gain remains constant again, since the system uncertainties remains constant as well.

It should be noted that the adaptive sliding gain allows to employ a smaller sliding gain, because the sliding gain does not have to be chosen high enough to compensate for all the possible uncertainties that can be appear in the system. In this way in the proposed adaptive scheme the sliding gain will be adapted (if necessary) when a new uncertainty will appear in the system in order to surmount this uncertainty.

V. CONCLUSIONS

In this paper a robust position regulation for an induction motors using an adaptive sliding mode control has been presented. The proposed controller incorporates an adaptive algorithm to calculate the sliding gain value. The sliding gain adaptation, on the one hand avoids the necessity of calculate the upper bound for the system uncertainties, and on the other hand allows to employ a smaller sliding gain in order to overcome the system uncertainties. Therefore, the control signal of the proposed sliding mode control scheme will be smaller that the control signals of the traditional variable structure control schemes, because in the last one the sliding gain value should be chosen high enough to overcome all the possible uncertainties that could appear in the system along the time.

Additionally, in order to avoid the flux sensors, that increases the cost and also reduces the reliability, a sliding mode based flux observer is proposed. The observer uses the measured stator voltages and currents to estimate the rotor flux.

Furthermore, the proposed control scheme do not present a high computational cost and therefore this control scheme can be implemented in a low cost DSP-processor.

Finally, by means of simulation examples, it has been shown that the proposed position control scheme performs reasonably well in practice, and that the position tracking objective is achieved under uncertainties in the system parameters and under load torque variations.

VI. APPENDIX

Induction motor data sheet.

- V, voltage, 460 V
- f, frequency, 60 Hz
- P_N, nominal power, 50 HP
- R_s, stator resistance, 0.087 Ω
- R_r, rotor resistance, 0.228 Ω
- L_m, magnetizing inductance, 34.7 mH
- L_s, stator inductance, 35.7 mH
- L_r, rotor inductance, 35.7 mH
- p, pair of poles, 2
- J, moment of inertia, 1.662 $kg\,m^2$
- B, viscous friction coefficient, 0.12 $N\,m/(rad/s)$

ACKNOWLEDGMENTS

The authors are very grateful to the Basque Government by the support of this work through the project S-PE12UN015 and to the UPV/EHU by its support through the projects GIU10/01 and UFI11/07.

REFERENCES

[1] C. Cecati "Position control of the induction motor using a passivity-based controller". *IEEE Trans. on Industry Applications*, vol. 36, 1277-1284. Sep/Oct 2000.

[2] F. Betin and G.-A. Capolino, "Shaft Positioning for Six-Phase Induction Machines With Open Phases Using Variable Structure Control", *IEEE Trans. Ind. Electron.*, vol. 59, no. 6, pp. 2612-2620, June. 2012

[3] W.J. Wang, and J.Y. Chen , "Passivity-based sliding mode position control for induction motor drives" *IEEE Trans. on Energy conversion*, vol. 20, 316-321. 2005

[4] B. Veselić, B. Peruničić-Draženović, and Č. Milosavljević, "High-Performance Position Control of Induction Motor Using Discrete-Time Sliding-Mode Control", *IEEE Trans. Ind. Electron.*, vol. 55, no. 11, pp. 3809-3817, Nov. 2008.

[5] B. Castillo-Toledo, S. Di Gennaro, A.G. Loukianov, J. Rivera. (2008) Discrete time sliding mode control with application to induction motors, *Automatica*, vol. 44, pp. 3036-3045.

[6] O. Barambones, J.M. Gonzalez de Durana and L.M. Camarero, A Robust Position Control for Induction Machines, in *XIX IEEE Int. Conf. on Electrical Machines. ICEM 2010*. Roma, Italy. Sep. 6-8, 2010.

[7] J.B. Oliveira, A.D. Araujo, S.M. Dias. (2010) Controlling the speed of a three-phase induction motor using a simplified indirect adaptive sliding mode scheme, *Control Engineering Practice*, vol. 18, pp. 557-584.

[8] M. A. Fnaiech, F. Betin, G.A. Capolino, and F. Fnaiech, "Fuzzy Logic and Sliding-Mode Controls Applied to Six-Phase Induction Machine With Open Phases", *IEEE Trans. Ind. Electron.*, vol. 57, no. 1, pp. 354-364, Jan. 2010.

[9] K. Ohyama, G. M. Asher, and M. Sumner,, "Comparative analysis of experimental performance and stability of sensorless induction motor drives", *IEEE Trans. Ind. Electron.*, vol. 53, no. 1, pp. 178-186, Feb. 2006.

[10] M. Ghanes and G. Zheng, "On Sensorless Induction Motor Drives: Sliding-Mode Observer and Output Feedback Controller", *IEEE Trans. Ind. Electron.*, vol. 56, no. 9, pp. 3404-3413, Sep. 2009.

[11] M. Comanescu, "An Induction-Motor Speed Estimator Based on Integral Sliding-Mode Current Control", *IEEE Trans. Ind. Electron.*, vol. 56, no. 9, pp. 3414-3423, Sep. 2009.

[12] B.K. Bose, B.K., 2001, *Modern Power Electronics and AC Drives.*, Prentice Hall, New Jersey.

[13] O. Barambones, A.J. Garrido and F.J. maseda, Integral sliding mode controller for induction motor based on field oriented control theory, *IET Control Theory & Applications*, vol. 1, no. 3, pp. 786-794, May. 2007

[14] P.V. Kokotovic, H. Khalil, J. OReilly. (1996) Singular Perturbation Methods in Control: Analysis and Design *Academic Press*, New York.

[15] A. N. Atassi, H. K. Khalil. (2000) Separation results for the stabilization of nonlinear systems using different high-gain observer designs, *Systems & Control Letters*, Vol. 39, pp.183-191.

[16] Utkin V.I., 1993, Sliding mode control design principles and applications to electric drives, *IEEE Trans. Ind. Electron.*, vol. 40, no. 1, pp. 2335, Feb. 1993.

VII. BIOGRAPHIES

Oscar Barambones was born in Vitoria, Spain in 1973.He received the M.Sc. degree in applied physics, the Ph.D. degree in control systems and automation, and the M.Sc. degree in electronic engineering, from the University of the Basque Country in 1996, 2000 and 2001, respectively. Since 1999 he has held several teaching positions at the Systems Engineering and Automatic Control Department in the Basque Country University, where he is currently a Professor of systems and control engineering. He has also been the Vice Dean of Research and master in the University College of Engineering of Vitoria from 2009 to 2013. He has more than 100 papers published in the main international conferences of the automatic control area, book chapters, and Journal Citation Report indexed journals. He has served as a Reviewer in several international indexed journals and conferences, and has supervised several Ph.D. theses. His current main research interests include the applied control of dynamic systems, particularly induction machines.

Calculation of Stator Winding Parameters to Predict the Voltage Distributions in Inverter Fed AC Machines

Oliver Magdun, Sébastien Blatt and Andreas Binder, *Senior Member IEEE*

Abstract – **Transmission line models are commonly used to calculate and predict the voltage distributions along the inverter-fed electrical machine windings. In this paper, frequency-dependent parameters of the stator windings are calculated via finite element and analytical methods. They are implemented in a transmission line model to predict the voltage distribution in a round wire stator winding induction machine. As a difference from the existent models, the laminated iron core effects are taken into account by an eddy current loss resistance. A metal-oxide varistor model is implemented in the cable-machine winding simulation model, and the voltage distribution is calculated. Comparisons with the measurements are given.**

Index Terms—**variable speed drives, transmission line modeling, parameter estimation, finite element methods, eddy currents, transient overvoltage, varistors, surge protection.**

I. INTRODUCTION

MODERN inverter-fed drives are equipped with fast switching IGBT-converters (IGBT: Insulated Gate Bipolar Transistor). The fast switching transitions of the inverter output voltage, within tens of nanoseconds range, cause traveling electromagnetic waves along the feeding cable between the inverter and machine, which entail a significant voltage increase at the machine terminals due to the voltage reflections at the motor clamps [1]. The occurring step voltage variations with the inverter switching frequency and high *dv/dt* due to the reflections at the motor terminals, almost twice the inverter DC link voltage in the worst case, are directly responsible for the winding insulation failures [2], [3]. Special converter topologies with modern modulation techniques or output voltage filters can reduce the overvoltage at the machine terminals, hence decreasing the insulation winding stress. Generally, they are expensive, and thus they have to be chosen selectively for each type of drive. A less expensive and promising solution, compared to filters and other mitigation methods, might be the metal-oxide varistors that limit the overvoltage at the

machine winding entry at least to $1.5 \cdot V_{DC}$ (V_{DC} – DC link inverter voltage) [4]. Yet, to be efficient for the machine insulation protection the metal-oxide varistors need a special design and optimization [5]. They can be optimized if the voltage distribution along the machine windings is calculated beforehand. The common mode inverter voltage causes, in addition to the voltage reflection effect, harmful bearing currents, which cannot be accomplished by the varistors. In this context, several induction machine modifications are necessary in order to protect the machine bearings against the harmful parasitic currents [6].

In this paper, for a 4-pole, 7.5 kW cage induction motor, the stator winding parameters are calculated via the analytical and finite element (FE) methods, and the voltage distribution along the stator winding is calculated with a transmission line model, which as a difference from the available models in the literature [7]-[9] takes into account the laminated iron core effects by an eddy current loss resistance. Further, a model of a metal-oxide varistor is added to the drive simulation model, and the voltage distribution along the machine winding is calculated. Comparisons with the measurements are presented.

II. STATOR WINDING TRANSMISSION LINE MODEL

A. Equivalent Circuit Representation

The transmission line models are commonly used for modeling the wave propagation effects in inverter-fed machine windings and the long motor-feeding cables [7] – [9]. In a simple way, a transmission line can be represented by lumped parameter equivalent circuits, modelled as a series of equivalent Γ- or π-sections. The high frequency components, which need to be processed by the lumped parameter circuit models, are considered up to a frequency f_{max}, usually larger than the cut-off frequency: $f_{co} = 1/(\pi t_r)$ of the feeding voltage signal FFT spectrum (t_r - the rise time of the impulse stator voltage due to the reflections). With $f_{max} \cong 3 \cdot f_{co} \approx 1/t_r$, the smallest wavelength of the reflected waveform along the transmission line can be approximated: $\lambda_{min} = v_c/f_{max} \approx v_c \cdot t_r$, where v_c is the velocity of wave propagation. For cables, this velocity is almost half of the velocity of light in vacuum $v_c \cong 150$ m/µs [9], and for the machine windings, due to the laminated iron core effects, it is $v_c < 75$ m/µs [10]. For a rise time of the impulse voltage at the motor winding entry, typically of $t_r = 200 \dots 300$ ns, and for

---

This work was supported in part by Deutsche Forschungsgemeinschaft DFG, FOR575 "High-frequency parasitic effects in inverter-fed electrical drives". The financial support of Hitachi Corporation is gratefully acknowledged by the authors.

O. Magdun and A. Binder are with the Institute for Electrical Energy Conversion, Darmstadt University of Technology, 64283 Darmstadt, Germany (e-mail: omagdun@ew.tu-darmstadt.de, abinder@ew.tu-darmstadt.de).

S. Blatt is with the Institute for High Voltage Technology, Darmstadt University of Technology, 64283 Darmstadt, Germany (e-mail: blatt@hst.tu-darmstadt.de).

978-1-4799-0024-4/13 $31.00 © 2013 IEEE

a propagation speed of 75 m/µs, the smallest wavelength of the reflected electromagnetic wave is calculated as $\lambda_{\min} \approx v_c \cdot t_r = 75\,\mathrm{m/\mu}\cdot 200\,\mathrm{ns} = 15\,\mathrm{m}$. Transmission lines are properly modeled, when at least 5 … 10 Γ- or π-sections are taken per minimum wavelength. Thus we get a section length of $\lambda_{\min}/10 \approx 1.5\,\mathrm{m}$, which for the standard induction machines of up to 100 kW is larger than the length of one turn. For bigger machines (>100 kW), a few Γ- or π-sections per turn are more appropriate. Here, for the representation of the stator winding of a random-wound round wire winding, in a 4-pole 7.5 kW cage induction machine with an iron length of $l_{\mathrm{Fe}} = 0.137$ m, a transmission line model with one section per turn is used. In Fig. 1, the n_T = 17 turns of one coil of the single layer distribute three-phase winding 7.5 kW machine are represented separately for each of the four regions of the machine, passed by the winding turns: the two slots and the two end-winding regions. The reference ground conductor "0" of the transmission line model is the stator iron core, which is considered a perfect conductor.

Inside the slots, the self-inductances of each turn j are represented in Fig. 1b by the inductances $L_{j,j}$, and the mutual inductances between the turns j and k are represented by the inductances $L_{j,k}$, $j,k = 1, …, n_T$. Considering the slot walls as a flux barrier [7], the magnetic field inside the slot is only produced by those currents, flowing in the considered slot conductors. If each conductor within the slot carries the same current $i_j = i$, with $j = 1 … n_{Ts}$, the inductance per turn j inside the slot is calculated as:

$$L_{\mathrm{T,slot}\,j} = \sum_{k=1}^{n_T} L_{j,k}\,. \qquad (1)$$

In the overhang, the leakage flux can be considered by a total inductance L_{ov}, which at high frequencies is distributed per turn in equal inductances $L_{\mathrm{T,slot}\,j}$ [11]. Thus, for each turn j, we get:

$$L_{\mathrm{T}j} = L_{\mathrm{T,ov}\,j} + L_{\mathrm{T,slot}\,j} = L_{\mathrm{T,ov}\,j} + \sum_{k=1}^{n_T} L_{j,k}\,. \qquad (2)$$

In the slot regions, the capacitances between the turns and stator core are represented by the turn-to-stator frame (ground) capacitances $C_{i,0} = C_{\mathrm{slot}\,i,0}$, $i = 1 … n_T$, and the capacitances between the winding turns are represented by the turn-to-turn capacitances $C_{i,j} = C_{\mathrm{slot}\,i,j}$ $i \neq j$, $j = 1 … n_T$ (Fig. 1a). In the overhang, only the turn-to-turn capacitances $C_{\mathrm{ov}\,i,j}$ are considered. For simplicity, the capacitances $C_{\mathrm{g}\,i,0}$ and $C_{\mathrm{T}i,j}$ of each turn of the transmission line model of Fig. 1c are calculated here as:

$$\begin{aligned} C_{\mathrm{g}\,i,0} &\approx 2 \cdot C_{\mathrm{slot}\,i,0} \cong 2 \cdot C_{i,0} \\ C_{\mathrm{T}\,i,j} &= 2 \cdot \left(C_{\mathrm{slot}\,i,j} + C_{\mathrm{ov}\,i,j} \right) \cong 2 \cdot \left(C_{i,j} + \frac{l_{\mathrm{ov}}}{l_{\mathrm{slot}}} \cdot C_{i,j} \right), \end{aligned} \qquad (3)$$

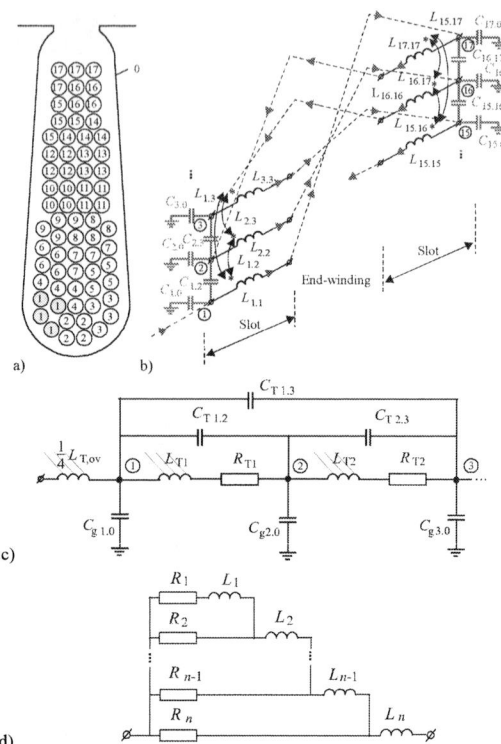

Figure 1. Representation of one coil with 17 turns of a single-layer winding of a 7.5 kW induction machine: a) - the geometrical arrangement of the 17 turns (4 round conductors per turn), b) – the self-inductances $L_{1,1}$, $L_{2,2}$, …, $L_{17,17}$, the mutual inductances $L_{1,2}$, $L_{1,3}$, $L_{2,3}$, … $L_{16,17}$, the stator turn-to-ground capacitances $C_{1,0}$, $C_{2,0}$, … $C_{12,0}$ and the turn-to-turn capacitances $C_{1,2}$, $C_{2,3}$, … $C_{11,12}$ of one coil within the slots, c) – a transmission line model derived from the model of b), with RL-ladder circuits for all the pairs $R_{\mathrm{T}1}$, $L_{\mathrm{T}1}$, $R_{\mathrm{T}2}$, $L_{\mathrm{T}2}$, … , d) – a ladder circuit with n branches for each pair $R_{\mathrm{T}1}$, $L_{\mathrm{T}1}$, $R_{\mathrm{T}2}$, $L_{\mathrm{T}2}$, …, of c).

where l_{ov} is the average conductor length in the overhang and $l_{\mathrm{slot}} = l_{\mathrm{Fe}}$ is the conductor length within the stator slot. A detailed calculation of the model capacitances can consider the coil distribution with twisted turns of Fig. 1b.

The copper losses are included in the model of Fig. 1c by the resistances $R_{\mathrm{T}i}$, which take into account the skin and proximity effects. At frequencies larger than 10 kHz, the increase of the stator winding resistance within the slots is much larger than the increase of the winding resistance in the overhang due to the smaller overhang stray field. Thus, only the resistances within the slots are considered, and the overhang resistances are neglected:

$$R_{\mathrm{T}j} \cong R_{\mathrm{T,slot}\,j} = \sum_{k=1}^{n_T} R_{j,k} \qquad (4)$$

The inductances and resistances per turn are not constant for the considered large frequency spectrum, i.e. 10 kHz … 1 MHz [11], [12], but they are frequency-dependent because of the skin and proximity effects. For modeling them, therefore, instead of each pair: $R_{\mathrm{T}j}$, $L_{\mathrm{T}j}$, the RL-ladder circuits of Fig. 1d are used. The capacitances are considered constant and the dielectric losses in the wire and slot insulation are neglected.

978-1-4799-0024-4/13 $31.00 © 2013 IEEE   448

B. Calculation of Stator Winding Capacitances

The electric energy method is used to calculate the capacitances of a system of conductors located in a linear electric medium. For n conductors and the reference ground conductor "0", the electric energy is written:

$$W_{el} = \frac{1}{2} \cdot \sum_{i=1}^{n} C_{i,0} \cdot u_i^2 + \frac{1}{2} \cdot \sum_{i=1}^{n-1} \sum_{\substack{j=1 \\ j>i}}^{n} C_{i,j} \cdot (u_i - u_j)^2 \; . \quad (5)$$

where $C_{i,0}$ and $C_{i,j}$ are the system capacitances, and u_i and u_j are the conductor potentials of the i-th and j-th conductor. In order to determine the capacitances of a multi-conductor system with n conductors and the reference conductor "0", a number of $m = \dfrac{n \cdot (n+1)}{2}$ independent combinations of 0 V and 1 V of the conductor potentials u_i and u_j are needed. These combinations can be selected in such a way that the energy expression (5) is converted to an equation system:

$$W_{el} = \frac{1}{2} \cdot k_u \cdot C, \quad (6)$$

where W_{el} represents the energy vector (m lines x 1 column), C represents the capacitance vector (m lines x 1 column) and the k_u represents the connection matrix (m lines x m columns) between the two vectors W_{el} and C. Further, the capacitances are extracted from the equation system:

$$C = \left(\frac{1}{2} \cdot k_u \right)^{-1} \cdot W_{el} \quad (7)$$

As an example, consider the multi-conductor distribution of Fig. 1a, but with only the first 6 turns (turns 1, 2, 3, 4, 5, 6) inserted inside the slot of the 7.5 kW machine. The four round conductors of one turn have the same potential, and thus, for the extraction of the system capacitances, the six turns are represented as "6 conductors" $n = 1, \ldots, 6$.

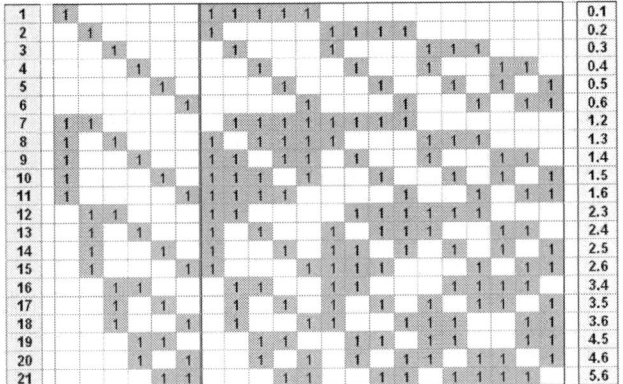

Figure 2. Matrix equation (6) obtained from the electrical energy expression (5) for six conductors. The "green" vector (left side: 21 lines x 1 column) represents the energy vector W_{el}, the "yellow" vector (right side: 21 lines x 1 column) represents the capacitance vector C, and the matrix k_u (middle: 21 lines x 21 columns) represents the connection between the vector W_{el} and the vector C, determined by the energy expression (5).

The stator yoke is considered as the grounded conductor "0". In this case, the multi-conductor system is described by 21 capacitances (turn-to-turn and turn-to-stator frame capacitances): $C_{1,0}, \ldots, C_{5,6}$ (Fig. 2, right side matrix, yellow colour). The voltage potentials u_i and u_j are considered in (5) as combinations of normalized electric potentials 0 V and 1 V (the first 6 columns of the matrix k_u) in order to calculate consecutively 21 values of the electrical energies $W_{el,k}$, $k = 1 \ldots 21$. For calculating the energy $W_{el,1}$ of (5), a potential of $u_1 = 1$ V is applied to the conductor "1" and 0 V to all the other conductors. The following differences of potential are calculated: $u_{1,0} = u_1 - u_0 = u_1 = 1$, $u_{2,0} = 0$, $u_{3,0} = 0$, $u_{4,0} = 0$, $u_{5,0} = 0$, $u_{6,0} = 0$, $u_{1,2} = u_1 - u_2 = 1$, $u_{1,3} = 1$, $u_{1,4} = 1$, $u_{1,5} = 1$, $u_{1,6} = 1$, $u_{2,3} = 0$, $u_{2,4} = 0$, $u_{2,5} = 0$, $u_{2,6} = 0$, $u_{3,4} = 0$, $u_{3,5} = 0$, $u_{3,6} = 0$, $u_{4,5} = 0$, $u_{4,6} = 0$, $u_{5,6} = 0$, and the first line of the equation system (6) is obtained:

$$W_{el,1} = \frac{1}{2} \cdot \big(C_{1,0} \cdot u_{10}^2 + C_{1,2} \cdot u_{1,2}^2 + C_{1,3} \cdot u_{1,3}^2 + C_{1,4} \cdot u_{1,4}^2 +$$
$$+ C_{1,5} \cdot u_{1,5}^2 + C_{1,6} \cdot u_{1,6}^2 \big).$$

The process is repeated 21 times, while all the 21 lines of the matrix k_u are determined. The electrical energies $W_{el,k}$, $k = 1 \ldots 21$ of the vector W_{el} can be obtained via a FEM software, considering the potential distributions of Fig. 2 (the first six columns of the matrix k_u).

With the presented method, the capacitances $C_{i,0}$ and $C_{i,j}$ of the stator coil model of Fig. 1a with 17 turns are calculated via FEM. Thus, the system of (7) is solved for $m = \dfrac{n_T \cdot (n_T + 1)}{2} = \dfrac{17 \cdot (17+1)}{2} = 153$ equations. The turn-to-stator ground capacitance $C_{g i,0}$ and the turn-to-turn capacitances $C_{T i,j}$ of the model of Fig. 1c are calculated with (1). Only the significant values of the turn-to-turn capacitances are kept for the implementation in the model of Fig 1c (Table I). The turn-to-turn capacitances, which are smaller than 1 pF, are neglected, because they are ca. 100 times smaller than the capacitances between the adjacent turns (>100 pF).

C. Calculation of Stator Winding Inductances and Resistances

In [7], the self and mutual inductances $L_{j,k}$ of the multi-conductor transmission line models of single-layer or double-layer windings are calculated within the slot, using the magnetic field energy:

$$W_m = \frac{1}{2} \cdot \sum_{j=1}^{n_{Ts}} L_{j,j} \cdot i_j^2 + \sum_{j=1}^{n_{Ts}-1} \sum_{\substack{k=1 \\ k>j}}^{n_{Ts}} L_{j,k} \cdot i_j i_k = \frac{1}{2} \cdot \sum_{j=1}^{n_T} \sum_{k=1}^{n_T} L_{j,k} \cdot i_j i_k , \quad (8)$$

where i_j and i_k, with $j = 1, \ldots, n_T$ and $k = 1, \ldots, n_T$, are the sinusoidal current sources of conductors j and k in the slot with the coil turns $1, \ldots, n_T$, having combinations of normalized conductor currents 0 and 1 A, which are introduced as initial conditions in the FEM problem. The resistances result from the dissipated power:

$$P_{el} = \sum_{j=1}^{n_{Ts}} R_{j,j} \cdot i_j^2 + 2 \cdot \sum_{j=1}^{n_{Ts}-1} \sum_{\substack{k=1 \\ k>j}}^{n_{Ts}} R_{j,k} \cdot i_j i_k = \sum_{j=1}^{n_{Ts}} \sum_{k=1}^{n_{Ts}} R_{j,k} \cdot i_j i_k \ . \quad (9)$$

The energy and copper loss calculation method can provide the self and mutual inductances and resistances per turn within the slot, but it is very time consuming. At high frequencies (100 kHz … 10 MHz), the 2D mesh within the slot needs a very high number of finite elements for giving an accurate solution of the FEM computation. Additionally, the magnetic energy W_m and the copper losses P_{el} need to be calculated consecutively via FEM for a large number of combinations of 0 and 1A (153 combinations for the turns of the coil of Fig. 1a). Moreover, the low voltage induction machines with single-layer round-wire windings, which are automatically wound, have an arbitrary distribution of the conductors in the slot. Consequently, even if the inductances and resistances are calculated exactly for an assumed conductor arrangement of the individual turns of one coil within the slot, they could be quite different for another arrangement. To overcome this problem, instead of an explicit representation of the conductors within the slot, like in Fig. 3a, a continuum representation [13] of the entire coil is used (Fig. 3b). Following the equivalent foil approach of [13], the proximity and skin effect are included in the 2D FEM calculation by use of the analytical expressions of the equivalent conductivity and permeability in the regions, filled with packed round wire windings.

TABLE I

CALCULATED CAPACITANCES IN NF, VIA FEM (FEMM PROGRAM), FOR ONE COIL OF A FOUR-POLE 7.5 KW CAGE INDUCTION MACHINE WITH THE ASSUMED TURN DISTRIBUTIONS OF FIG. 1

Turn-to-stator frame (ground) capacitances $C_{g\,io}$/nF		Turn-to-turn capacitances (> 1pF) $C_{T\,i,j}$/nF			
$C_{g\,1.0}$	0.057	$C_{T\,2.1}$	0.251	$C_{T\,10.9}$	0.244
$C_{g\,2.0}$	0.041	$C_{T\,4.1}$	0.337	$C_{T\,11.9}$	0.072
$C_{g\,3.0}$	0.057	$C_{T\,3.2}$	0.251	$C_{T\,11.10}$	0.207
$C_{g\,4.0}$	0.016	$C_{T\,4.2}$	0.144	$C_{T\,12.10}$	0.221
$C_{g\,5.0}$	0.031	$C_{T\,4.3}$	0.100	$C_{T\,13.10}$	0.004
$C_{g\,6.0}$	0.031	$C_{T\,5.3}$	0.237	$C_{T\,12.11}$	0.004
$C_{g\,7.0}$	0.016	$C_{T\,5.4}$	0.111	$C_{T\,13.11}$	0.220
$C_{g\,8.0}$	0.021	$C_{T\,6.4}$	0.235	$C_{T\,13.12}$	0.213
$C_{g\,9.0}$	0.021	$C_{T\,7.4}$	0.107	$C_{T\,14.12}$	0.111
$C_{g\,10.0}$	0.023	$C_{T\,7.5}$	0.353	$C_{T\,15.12}$	0.112
$C_{g\,11.0}$	0.023	$C_{T\,7.6}$	0.213	$C_{T\,14.13}$	0.224
$C_{g\,12.0}$	0.027	$C_{T\,8.6}$	0.004	$C_{T\,15.14}$	0.408
$C_{g\,13.0}$	0.027	$C_{T\,9.6}$	0.201	$C_{T\,16.14}$	0.116
$C_{g\,14.0}$	0.025	$C_{T\,8.7}$	0.313	$C_{T\,16.15}$	0.219
$C_{g\,15.0}$	0.035	$C_{T\,9.7}$	0.004	$C_{T\,17.15}$	0.111
$C_{g\,16.0}$	0.020	$C_{T\,9.8}$	0.309	$C_{T\,17.16}$	0.333
$C_{g\,17.0}$	0.053	$C_{T\,11.8}$	0.172		

Note: For calculating the capacitances, a homogenous slot insulation with a relative permittivity of $\varepsilon_r = 4$ has been considered in Fig. 1a.

When a current i flows in a conductor within the slot, it will produce a flux ϕ around the conductor that passes the stator teeth and yoke (Fig. 4). At high frequencies, the flux ϕ is limited and constrained by the eddy currents i_{eddy} in the laminated core to a small penetration depth δ_{Fe} at the lamination surfaces. Depending on the pulsation frequency of the flux and the geometry and material properties of the

laminated core, the eddy currents flow inside the laminations and produce iron losses, which need to be further considered in the model. The hysteresis losses at high frequencies are much smaller than the eddy current losses and are therefore neglected. The eddy current losses can be represented as a resistance R_e, connected in parallel to the frequency-dependent stator winding resistance and inductance within the slots (Fig. 5). Thus, the total winding resistance will include a resistance due to the skin and proximity effects in the winding and an equivalent iron core resistance that is related to the power loss due to the eddy currents, flowing in the laminated core [14]. An exact method to calculate the eddy current loss resistance R_e at high frequencies (10 kHz … 10 MHz) is not available yet in the literature. It can be extracted from the experimental curves, for example: Fig. 6a [15] or Fig. 6b [16]. For the 7.5 kW machine, with a stator outer diameter of $d_{se} \cong 0.2$ m, the total eddy current loss resistance is extracted from Fig. 6a: $R_e \cong 3$ kΩ. For one turn of the 7.5 kW machine with $N_s = 136$ turns per phase, we get an average resistance per turn of $R_{e,T} = R_e/N_s = 3$ kΩ/136 = 22.06 Ω.

In earlier papers [7]-[10], for the calculation of the transmission line model inductances and resistances, the slot boundaries were enforced by zero magnetic vector potential under the assumption of an infinitely small penetration depth of the magnetic field in the laminated iron core. But the FEM simulations of [17] showed that a magnetic field may exist around the coil conductors, in the laminated iron core, not only at lower frequencies, but also at higher frequencies (< 1 MHz). With FEM (continuum approach of Fig. 3b), the resistances and inductances per coil are calculated here by imposing the restrictive conditions of the analytical calculation at low frequency: parallel flux lines crossing the slot, as slot stray flux (Fig. 3b), and no eddy currents in the iron core ($\sigma_{Fe} = 0$). The RL-ladder circuit parameters $R_{s,n}$, $L_{s,n}$, $n = 1 \ldots 6$ (Table II) of Fig. 5 are extracted via FEM calculation. Fig. 6a & b show the calculated FEM inductances and resistances, in the conditions of Fig. 3b. They are not fitting to the measured values with an LCR-meter. The measured values of resistances and inductances per turn are obtained from the average values of the measured resistances and inductances of the all 8 coils of the 7.5 kW machine, by dividing the average values by the number of turns per coil $n_T = 17$. Measured with the LCR-meter, the frequency-dependent resistance includes the winding overhang, the resistance due to the skin and proximity effects in the winding, the equivalent eddy current loss resistance and the mutual inductance between the conductors of neighboring slots, due to the magnetic field penetration in the laminated stator and rotor iron core. In Fig. 5, the eddy current loss resistance per turn $R_{e,T}$ is added in parallel to the RL-ladder circuit $R_{s,n}$, $L_{s,n}$. At frequencies larger than 10 kHz, it contributes to the strong increase of the equivalent resistance per turn and decrease of the equivalent inductance per turn (Fig. 7, curve 3).

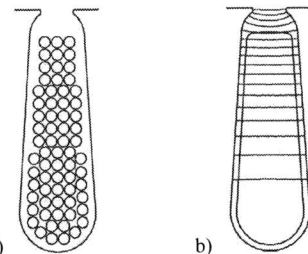

Figure 3. Exact (a) and continuum (b) representation of a coil within the slot, imposing parallel flux lines crossing the slot.

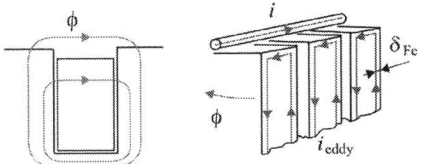

Figure 4. Patterns of the high frequency flux ϕ and eddy currents i_{eddy} in the stator laminated yoke.

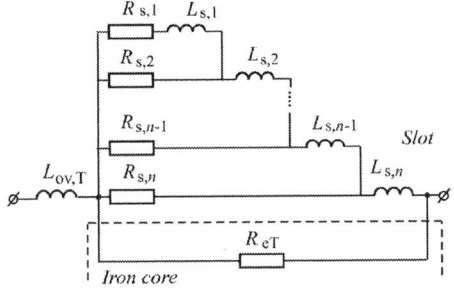

Figure 5. Equivalent circuit for modeling one turn at high frequency.

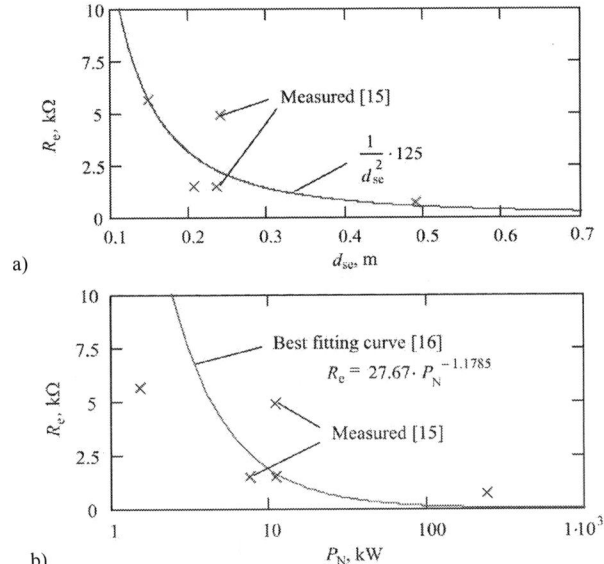

Figure 6. The eddy-current loss resistance R_e, obtained from the experimental measurements of [15], [16]: a) – represented as function of the stator outer diameter d_{se}, b) – represented as function of the nominal power P_N.

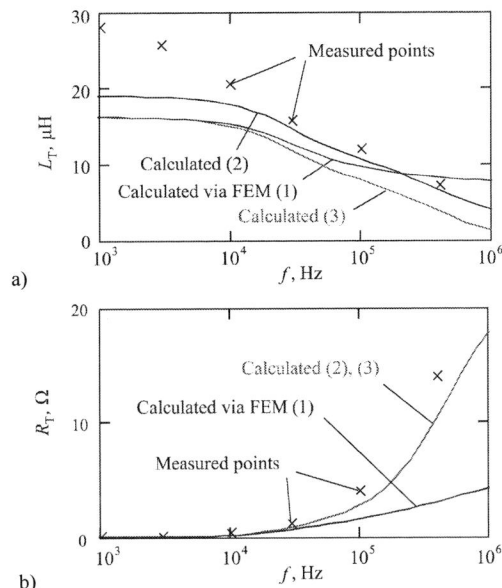

Figure 7. Inductance (a) and resistance (b) per turn: 1) – calculated via FEM, continuum and RL-ladder circuit approach, 2) – calculated via the circuit of Fig. 5, and 3) – calculated via the circuit of Fig. 5, neglecting the overhang inductance per turn $L_{T,ov} = 2.8\ \mu H$.

TABLE II
CALCULATED RL-LADDER CIRCUIT PARAMETERS FOR A TURN, VIA FEM

n	1	2	3	4	5	6
R_{sn} Ω	0.008	0.018	0.791	1.849	3.207	5.024
L_{sn} μH	0.652	0.140	2.228	2.679	2.151	7.567

The inductance per turn in the overhang (Fig. 5) is assumed constant and equal distributed per turn (Sec IIA):

$$L_{ov,T} \cong k_r \cdot \frac{1}{N_s} \cdot \left(\mu_0 N_s^2 \cdot (2/p) \cdot \lambda_b \cdot l_{ov} \right), \quad (10)$$

where $k_r \approx 0.3$ is a factor, which considers the reduction of the winding overhang inductance due to the skin effect in the stator iron sheets, N_s is the number of turns per phase, λ_b is the geometry coefficient and l_{ov} is the length of a turn in the overhang. For the 7.5 kW machine we get $L_{ov,T} = 2.8\ \mu H$. With the overhang inductance $L_{ov,T}$ (Fig. 5), the calculated equivalent inductance per turn (Fig. 7a, curve 2) is fitting quite well with the measured inductance per turn at frequencies larger than 10 kHz. At lower frequencies, they are not fitting. The model of Fig. 5 neglects the inductance due to the magnetic coupling between the conductors of neighboring slots. According to [17], the mutual inductance between the conductors of different slots, due to the magnetic field penetration in the stator and rotor core, might be almost 10% of the self-inductance. But the obtained differences of Fig. 7 could be also attributed to a lower estimated winding overhang inductance, at low frequency, by (10).

TABLE III
CALCULATED *RL*-LADDER CIRCUIT PARAMETERS FOR A TURN, VIA THE
EQUIVALENT CIRCUIT OF FIG. 5

n	1	2	3	4	5	6
R_n Ω	0.018	0.024	0.316	3.945	7.244	20.535
L_n μH	1.177	6.777	7.987	3.102	6.047	3.426

The calculated equivalent resistance per turn is nearly the measured one. The differences are attributed to the eddy current loss resistance, which has been estimated from the experimental curves. From the calculated values of the equivalent inductance and resistance per turn (Fig. 7, curve 2) the *RL*-ladder circuit parameters of the circuit of Fig. 1d are extracted (Table III).

III. PREDICTION OF VOLTAGE DISTRIBUTIONS IN THE MACHINE WINDINGS

The model of Fig. 5 can be implemented in the transmission line model of Fig. 1c, as "it is" or as the *RL*-ladder circuit of Fig. 1d with the parameters of Table III. In order to calculate the voltage distributions along the machine winding, a cable model [18] is connected together with the stator winding model of Fig. 1c. The voltage at the inverter output is measured and used as an input for the cable-machine winding model.

The 4-pole 7.5 kW cage induction machine (Fig. 1) was used for the measurements of the transient voltages along the machine winding coils [12]. It was fed by a non-shielded 100 m long cable [18] from an inverter with the DC link voltage of $U_{DC} = 560$ V and the switching frequency $f_{sw} = 3$ kHz. Fig. 8a shows the measured voltages between the machine coil terminals and ground in comparison to the calculated ones. The voltage (1) in Fig. 8a represents the measured line-to-ground voltage at the machine terminal, and the voltages (2), ..., (9) represent the measured voltages between the input terminal of each coil and ground. With the model of Fig. 5, which includes the eddy current loss resistance, the voltage transients are predicted well for the first four coils of the stator winding. Due to the winding asymmetries, which have not been considered in the calculation model, the line-to-ground voltages are not well predicted for the last four coils. However, even for the last coils the maximum of the voltage transients is fitting well with the measured one.

For comparison, the voltage distribution along the winding is calculated in Fig. 8b with a model that neglects the eddy current loss resistance R_{eT}. In this case, the prediction overestimates the voltage transients along the machine windings. This is due to the lacking eddy current loss resistance, which has a damping effect on the voltage transients.

A metal-oxide varistor with the measured capacitance of 13.54 nF and the measured nonlinear voltage-current characteristic of Fig. 9 is added at the machine entry terminals for reducing the coil overvoltages. Due to the large capacitance of the varistor, the voltage difference between two

consecutive measured traveling waves is reduced (Fig. 10). For predicting the voltage transients, a varistor model, consisting of a constant capacitance of 13.54 nF for the insulator state, connected in parallel to the nonlinear resistance of Fig. 9 for the conductive state, has been implemented in the cable-stator winding model. Fig. 10 shows that the voltage distributions along the machine windings are well predicted, if the eddy current loss resistance is considered in the winding simulation model.

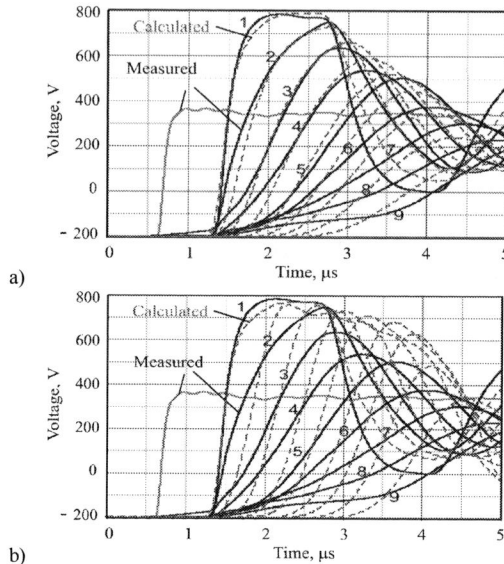

Figure 8. Measured and calculated line-to-ground voltages of an inverter fed cage induction machine 7.5 kW: a) - considering the eddy current loss resistance R_e in the simulation model of Fig. 5, b) – neglecting the iron core resistance R_e. For calculation, the cable model of [18] was connected to the winding simulation model of Fig. 1c. The measured voltage at the invertor output has been used as an input voltage source.

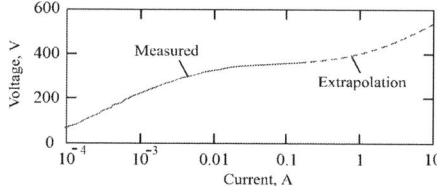

Figure 9. The measured non-linear characteristic of a varistor in the conductive state. Above 0.1 A, the voltages are taken from the extrapolation of typical varistor characteristics.

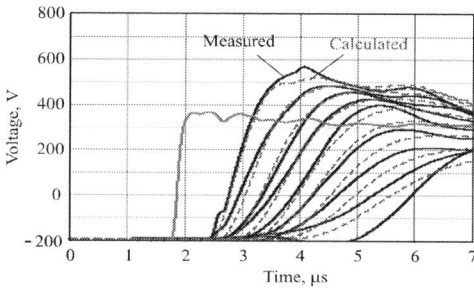

Figure 10. Measured and calculated line-to-ground voltages of an inverter-fed cage induction machine 7.5 kW, protected by a varistor with the capacitance of 13.54 nF and its non-linear voltage-characteristic of Fig. 9.

Considering the varistor behavior of Fig. 9 in the conductive state, at 350 V, a current impulse of 0.1 A flows to the ground via the varistor resistance. At a line-to-ground voltage of 600 V at the machine terminals, due to the voltage reflections along the machine feeding cable, a current impulse, bigger than 10 A may flow to the ground. This current is repetitive and will permanently stress the varistor, leading to early varistor damages. The voltage transients have to be therefore reduced below ± 350 V $\cong \pm 1.25 \cdot U_{DC}/2$ by increasing the varistor capacitance.

IV. Conclusions

The voltage distribution in the round wire randomly wound winding induction machines can be predicted, when the transmission line models with frequency-dependent resistance and inductance parameters are used. For an accurate prediction, the eddy current loss resistance must not be neglected in the transmission line models. Apart of other machine parameters, it has a strong damping effect on the voltage transients along the machine winding. The overhang inductance can be considered as a constant inductance, equally distributed per turn. In order to reduce the overvoltage at the inverter-fed machine entry and to equalize the voltage differences between the winding coils, metal oxide varistors can be used. They have to be selected according to the individual properties of each drive, e.g. the invertor voltage rise time, the inverter output voltage and the cable and machine winding parameters. Without a proper selection, the varistors may be early endangered due to a large impulse current stress.

V. References

[1] R. Kerkman, D. Leggate, J. Pankau, D. Schlegel, "Reflected Wave Modeling Techniques for PWM AC Motor Drives", Proc. of APEC'98, vol. 2, pp. 1021-1029, Anaheim, USA, 15-19 February 1998.

[2] L. Saunders, G. Skibinski, S. Evon, D. Kempkes, "Riding the reflected wave - IGBT Drive Technology Demands New Motor and Cable Considerations" IEEE IAS & Chemical Ind. Conf., pp. 75-84, Philadelphia, USA, 23-25 September 1996.

[3] M. Kaufhold, K. Schaefer, K. Bauer, A. Bethge, J. Risse, "Interface phenomena in stator winding insulation - challenges in design, diagnosis, and service experience", IEEE Electr. Ins. Mag., pp. 27-36, March/April, 2002.

[4] A. Rocks, V. Hinrichsen, "Application of varistors for overvoltage protection of machine windings in inverter-fed drives", Proc. of IEEE-SDEMPED'07, pp. 335-340, 6-8 September, Krakow, Poland, 2007.

[5] A. Rocks, V. Hinrichsen, "Overvoltage protection of inverter-fed drives with the help of energy varistors - Dimensioning rules for consideration of different cable types", Proc. of APEC'09, pp. 330-335, 15-19 February, Washington DC, USA, 2009.

[6] A. Muetze, "Thousands of hits: on inverter-induced bearing currents, related work, and the literature", Elektrotechnik und Informationstechnik, vol. 28, no. 11-12, pp 382-388, December 2011.

[7] G. Suresh, H.A. Toliyat, D.A. Rendussara, P.N. Enjeti, "Predicting the transient effects of PWM voltage waveform on the stator windings of random wound induction motors", Proc. of APEC'09, pp. 135-141, 23-27 February, Atlanta GA, USA, 1997.

[8] B. S. Oyegoke, "Voltage distribution in the stator winding of an induction motor following a voltage source", Electr. Eng., vol. 82, pp. 199 – 205, 2000.

[9] M. T. Wright, S. J. Yang, K. McLeay, "General Theory of Fast-Fronted Interturn Voltage Distribution in Electrical Machine Windings", IEE Proc., vol.130, Pt. B, no. 4, pp. 245-256, July 1983.

[10] P. J. Tavner, R.J. Jackson, "Coupling of discharge curents between conductors of electrical machines owing to laminated steel core", IEE Proc., Pt. B, vol. 135, no. 6, pp. 84-90, November 1988.

[11] T. Humiston, P. Pillay, "Parameter Measurements to Study Surge Propagation in Induction Machines", IEEE Trans. Ind. Appl., vol. 40, no. 5, pp. 1341-1348, September/October 2004.

[12] O. Magdun, A. Binder, C. Purcarea, A. Rocks, "High-frequency induction machine models for calculation and prediction of common mode stator ground currents in electric drives", Proc. of EPE'09, CD-ROM, 8 pages, 8-10 September, Barcelona, Spain, 2009.

[13] D. Meeker, "Continuum Representation of Wound Coils via an Equivalent Foil Approach" [online]. Available on 23.01.2013 at http://www.femm.info/examples/prox/notes.pdf.

[14] G. Grandi, M. K. Kazmierczuk, A. Massarini, U. Reggiani, G. Sancineto, "Model of Laminated Iron-Core Inductors for High Frequencies", IEEE Trans. on Magn., vol. 40, no. 4, pp. 1839-1845, July 2004.

[15] O. Magdun, A. Binder, "The high-frequency induction machine parameters and their influence on the common mode stator ground current", Proc. of ICEM'12, pp. 503-509, 2-5 September, Marseille, France, 2012.

[16] A. Boglietti, A. Cavagnino, M. Lazari, "Experimental High-Frequency Parameter Identification of AC Electrical Motors", IEEE Trans. on Ind. Appl., vol. 43, no. 1, pp. 23-29, January/February 2007.

[17] H. Jorks, E. Gjonaj, T. Weiland, O. Magdun, "Three-dimensional simulations of an induction motor including eddy current effects in core laminations", IET Sc., Meas. & Techn., vol. 6, no. 5, pp. 344-349, 2012.

[18] O. Magdun, A. Binder, C. Purcarea, A. Rocks, B. Funieru, "Modeling of asymmetrical cables for an accurate calculation of common mode ground currents", Proc. of IEEE - ECCE'09, pp. 1075-1082, 20-24 September, San Jose, USA, 2009.

VI. Biographies

Oliver Magdun received the Dipl. Eng. degree in electrical engineering from the Polytechnic University of Bucharest, Romania, and the Ph.D. degree in electrical engineering from Darmstadt University of Technology, Germany. From 1998 to 2006, he was an assistant and lecturer with Valahia University of Targoviste. Since 2006, he has been joined as a research associate to the Institute for Electrical Energy Conversion, Darmstadt University of Technology. His research interests include the calculation of high frequency parasitic effects in inverter-fed electrical machines.

Sébastien Blatt was born in Frankfurt am Main, Germany in 1981. He received the Dipl. Eng. degree in electrical engineering in 2009 from Darmstadt University of Technology, Germany, where he is currently working toward the Ph.D. degree in electrical engineering. He is currently a Research Assistant at the High Voltage Laboratories of Darmstadt University of Technology. His research interests include microvaristor filled insulation materials in inverter-fed drives.

Andreas Binder (M'97–SM'04) received the Dipl. Eng. and Dr. Tech. (Ph.D.) degrees in electrical engineering from Vienna University of Technology, Vienna, Austria, in 1981 and 1988, respectively. From 1981 to 1983, he was with ELIN-Union AG, Vienna, where he worked on synchronous generator design. From 1989 to 1997, he was with Siemens AG, Bad Neustadt and Erlangen, Germany, where he led a group that developed DC and inverter-fed AC motors and drives. Since 1994, he is a Lecturer at Vienna University of Technology (venia docendi). Since October 1997, he has been the Head of the Institute for Electrical Energy Conversion, Darmstadt University of Technology, Darmstadt, Germany, as a Full Professor, where he is responsible for teaching and research on electrical machines, drives, and railway systems. He is the author or co-author of more than 260 scientific publications and two books. He is the holder of several patents in this field. He was the recipient of the ETG-Literature Award of the VDE in 1997. Prof. Binder has been the Head of the Section "Electrical Machines, Drives, and Mechatronics" within the German Association of Electrical Engineers (VDE) between 1999 and 2006. He was guest professor at Graz University of Technology from 2001 to 2008. He received the title of Dr. Honoris Causa from the "Politehnica" University of Bucharest Romania in 2007. Prof. Binder is recipient of the VDE Medal of Honor in 2009.

Circulating Current Minimization and Current Sharing Control of Parallel Boost Converters Based on Droop Index

Sijo Augustine, *Student Member, IEEE*, Mahesh K. Mishra, *Senior Member, IEEE*, N. Lakshminarasamma

Abstract—**This paper investigates the relationship between current sharing difference and circulating current for two parallel connected dc-dc converters. In the proposed algorithm a new figure-of-merit called *Droop Index* is introduced, which is a function of normalized current sharing difference and losses in the output side of the converters. This algorithm minimizes the circulating current and current sharing difference between the converters. Although there may exist a trade-off between current sharing difference and voltage regulation, the proposed droop index algorithm gives better performance and low voltage regulation. The detailed analysis and design procedure are explained for two dc-dc boost converters connected in parallel. The effectiveness of proposed method is verified using MATLAB simulation.**

Index Terms—**Circulating current, current sharing, droop index, microgrid, parallel dc-dc converter.**

I. INTRODUCTION

Distributed energy systems with sources like solar, wind, fuel cell, micro-turbine etc. connected together becomes important due to two main reasons: (i) Conventional energy sources can be preserved. (ii) Non conventional sources of energy are renewable [1]. It is not necessary that the energy sources exist in the same site, but can be scattered depending upon the ease of energy harness. Integrating renewable energy sources by using power electronic interfaces gives flexibility in conversion and power level. In this, dc microgrid plays a major role as it is highly efficient, reliable, controllable and economic [2], [3]. In dc microgrid system, dc-dc and ac-dc converters are used to interconnect renewable sources and loads. Operation and control of parallel connected converters [4]- [6] are important and some advantages are (i) expandability of output power (ii) reliability (iii) efficiency (iv) ease of maintenance.

Several paralleling methods are reported in literature, the main schemes are droop control [7], [8] and active current sharing methods [9], [10]. The major problems associated with paralleling are voltage regulation, load sharing and circulating current [11]. If the output voltage of all parallel connected converters are constant and same then load current must be divided equally among the converters. The issue of circulating current will arise when converters are connected in parallel and there is a mismatch in the converter output voltages. Circulating current leads to increased flow of current through the switches which in turn increases the power electronic switch ratings and losses. Circulating current also gives rise to a difference in current sharing which causes an overload on the converters. The converter with higher output voltage has to supply more load current and proper load sharing cannot be possible.

A novel droop algorithm is reported in [7] for the converter parallel operation. In this method peak output current is compared with output current set value which controls the reference voltage of each converter. In this algorithm, the droop gain values are calculated based on rated power. So it will give better performance only when the source supples the rated power. A decentralized circulating current control method is proposed in [12], which is based on no-load circulating current values. This will reduce the error in current sharing without deteriorating voltage regulation. This algorithm is inefficient for boost converter because no load operation is not possible.

The major reasons for variations from the constant output voltage are changes in the input power, changes in load, parametric variations and error in voltage and current feedback [12]. A small mismatch in the output voltages ($\pm$ 1%) is enough to initiate a circulating current, which leads to current sharing difference. These two effects will together deteriorate the system performance.

In this paper a figure-of-merit called droop index (D.I.) is proposed to control the circulating current and load sharing for converters in parallel operation. This gives better control over circulating current and load sharing in both transient and steady state conditions.

This paper is organized as follows. In Section II the circulating current and current sharing issues are discussed based on converter equivalent circuits. Circulating current control using a series resistor, R_{droop} is analyzed in Section III. Proposed R_{droop} calculation is explained in Section IV. Simulation results are reported in Section V. Finally, conclusions are summarized in Section VI.

II. CIRCULATING CURRENT AND CURRENT SHARING ISSUES

In this section, circulating current and current sharing issues for parallel connected dc-dc converters are discussed in detail.

This work is supported by the Ministry of Science and Technology, DST, India (Project No.: DST/TM/SERI/2k10/47).

The authors are with the Department of Electrical Engineering, Indian Institute of Technology Madras, Chennai, India. (e-mail: sijomundackal@outlook.com; mahesh@ee.iitm.ac.in; lakshmin@ee.iitm.ac.in).

978-1-4799-0024-4/13 $31.00 © 2013 IEEE

Fig. 1 shows parallel dc-dc converters which interface PV panel and a constant voltage load, where V_{DC1}, V_{DC2}, R_1 and R_2 represent output voltages and cable resistance of converter-1 and converter-2 respectively. The output side of the converter

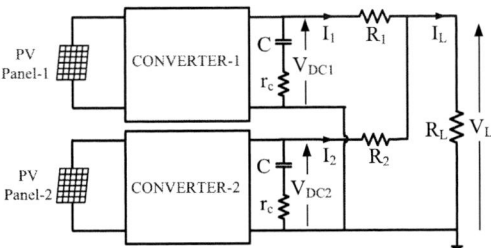

Fig. 1. Parallel dc-dc converters with different output voltages and a constant voltage load.

can be represented as a voltage source in series with the cable resistances and its equivalent circuit [13] which is shown Fig. 2. In Fig. 2, I_1 and I_2 are the converter-1 and converter-2 output currents respectively. I_{C12} is the circulating current component from converter-1 to converter-2 and $I_{1'}$ is the load component from converter-1. Case studies for current sharing

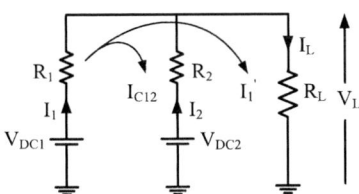

Fig. 2. Steady state equivalent circuit for the dc output side.

difference and circulating current based on the converter output voltage and cable resistance are listed in Table I.

TABLE I
CASE STUDIES FOR CURRENT SHARING AND CIRCULATING CURRENT BASED ON CONVERTER OUTPUT VOLTAGES, CABLE RESISTANCE

Case	V_{DC1}, V_{DC2}	R_1, R_2	I_1, I_2	$|I_{C12}|$, $|I_{C21}|$
1	equal	equal	equal	zero
2	different	equal	different	not zero
3	different	different	different	not zero

a) Mathematical analysis of circulating current for two converter system

By applying KVL in Fig. 2

$$V_{DC1} - I_1 R_1 - I_L R_L = 0 \tag{1}$$

$$V_{DC2} - I_2 R_2 - I_L R_L = 0. \tag{2}$$

The expression for converter currents I_1 and I_2 can be derived from (1) and (2) and is given as

$$I_1 = \frac{(R_2 + R_L)V_{DC1} - R_L V_{DC2}}{R_1 R_2 + R_1 R_L + R_2 R_L} \tag{3}$$

$$I_2 = \frac{(R_1 + R_L)V_{DC2} - R_L V_{DC1}}{R_1 R_2 + R_1 R_L + R_2 R_L}. \tag{4}$$

The circulating current can be expressed as

$$I_{C12} = -I_{C21} = \frac{V_{DC1} - V_{DC2}}{R_1 + R_2} \tag{5}$$

$$= \frac{I_1 - I_2}{2} \qquad (\text{if } R_1 = R_2)$$

$$= \frac{I_1 R_1 - I_2 R_2}{R_1 + R_2}. \quad (\text{if } R_1 \neq R_2)$$

As shown in Fig. 2, converter current I_1, is the sum of load component and circulating current component. If the cable resistances are small in comparison with load resistance then the product $R_1 R_2$ can be neglected. By substituting (5) in (3) and (4)

$$I_1 = \underbrace{\frac{R_2 V_{DC1}}{R_1 R_L + R_2 R_L}}_{load~current~(I_{1'})} + \underbrace{\frac{V_{DC1} - V_{DC2}}{R_1 + R_2}}_{circulating~current~(I_{C12})} \tag{6}$$

$$I_2 = \underbrace{\frac{R_1 V_{DC2}}{R_1 R_L + R_2 R_L}}_{load~current~(I_{2'})} - \underbrace{\frac{V_{DC2} - V_{DC1}}{R_1 + R_2}}_{circulating~current~(I_{C21})}. \tag{7}$$

b) Mathematical analysis of circulating current for 'n' converter system

The same analysis can be extended to n converter system. Fig. 3 shows n parallel connected converters with a constant voltage load. This can be analyzed based on two cases: (i) if the cable resistance values are same ($R_1 = R_2 = \ldots = R_n$) and (ii) if the cable resistance values are different ($R_1 \neq R_2 \neq \ldots \neq R_n$). A general expression for circulating current of

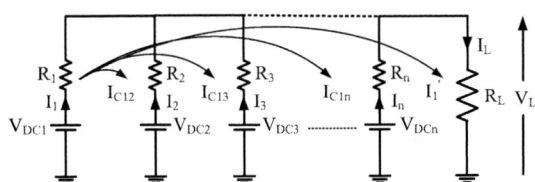

Fig. 3. Equivalent circuit for the dc output side for n parallel converters.

converter-1 for case (i) is derived as

$$I_{C1} = \frac{I_1 - I_1}{n} + \frac{I_1 - I_2}{n} + \ldots + \frac{I_1 - I_n}{n}$$

$$= \sum_{j=2}^{n} \frac{I_1 - I_j}{n}. \tag{8}$$

and due to all converters is given by

$$\begin{bmatrix} I_{C1} \\ \vdots \\ I_{Cj} \end{bmatrix} = \frac{1}{nR} \begin{bmatrix} (n-1) & \ldots & -1 \\ \vdots & \ddots & \vdots \\ -1 & \ldots & (n-1) \end{bmatrix} \begin{bmatrix} V_{DC1} \\ \vdots \\ V_{DCj} \end{bmatrix}. \tag{9}$$

where, $j = 1, 2, \ldots n$.

In case (ii), expression for circulating current of converter-1 is given as

$$I_{C1} = \frac{I_1 R_1 - I_1 R_1}{R_1 + R_1} + \ldots + \frac{I_1 R_1 - I_n R_n}{R_1 + R_n}. \tag{10}$$

978-1-4799-0024-4/13 $31.00 © 2013 IEEE

and a general expression is derived as

$$
\begin{bmatrix} I_{C1} \\ \vdots \\ I_{Cj} \end{bmatrix} = \begin{bmatrix} \sum_{m \neq 1}^{n}\left(\frac{1}{R_1+R_m}\right) & \cdots & \left(\frac{-1}{R_1+R_n}\right) \\ \vdots & \ddots & \vdots \\ \left(\frac{-1}{R_n+R_1}\right) & \cdots & \sum_{m \neq n}^{n-1}\left(\frac{1}{R_n+R_m}\right) \end{bmatrix} \begin{bmatrix} V_{DC1} \\ \vdots \\ V_{DCj} \end{bmatrix} \quad (11)
$$

where, $j = 1, 2, \ldots n$

From the above analysis, it can be observed that in case of voltage deviation converters have to supply circulating current component in addition with load current component. Thus the current sharing difference is determined by the output voltages of each converter and its cable resistances.

III. CIRCULATING CURRENT CONTROL BY ADDING R_{droop}

This section explains sharing of converter currents and minimization of circulating current by adding a series resistor, R_{droop} to each converter output as shown in Fig. 4. By placing

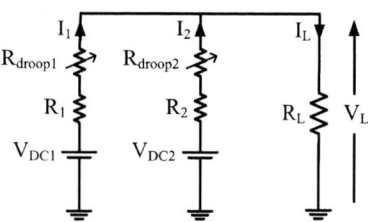

Fig. 4. Steady state equivalent circuit for the dc output side of two parallel converters with R_{droop}.

R_{droop1} and R_{droop2} the current sharing can be controlled and thus circulating currents can be minimized. This can be applied for parallel dc-dc boost converter [14] as shown in Fig. 5 where, S_1 and S_2 are the main switches, r_{DS} is the main switch ON state resistance, V_f is the cut-in voltage of diode, r_f is the ON state resistance of diode, L is the input inductor, r_L is the ESR (Equivalent Series Resistance) of inductor, C is the output capacitor, r_C is the ESR of filter capacitor and V_{i1} and V_{i2} are the input voltages. The virtual impedance method [15],

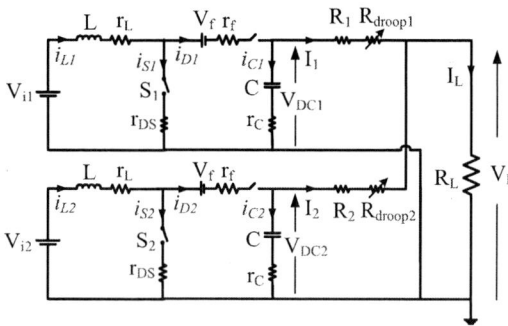

Fig. 5. Parallel boost converters with R_{droop}.

[16] is used to implement R_{droop1} and R_{droop2} and is shown in Fig. 6. This can be achieved by taking output currents feedback from converter-1, converter-2 and multiplied with calculated

R_{droop1} and R_{droop2} respectively. This resultant signals are subtracted from reference voltages (V_{DC}^{*}) of each converter and the new reference signals are now (V_{DC}^{**}) and is given as

$$
v_{DC}^{**} = v_{DC}^{*} - i_{DC}R_{droop}. \quad (12)
$$

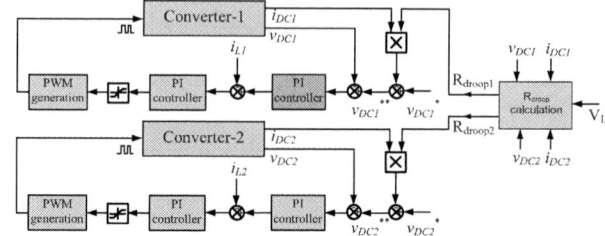

Fig. 6. Block diagram of parallel boost converters with R_{droop} calculation.

IV. PROPOSED DROOP INDEX CALCULATION

Calculation of R_{droop} values are based on the proposed figure-of-merit called droop index. The droop index is function of normalized current sharing difference and output power losses and is given as

$$
\text{Droop Index} = min\left[\frac{1}{2}\left[|I_1 - I_2|_{\text{N}} + (P_{loss})_{\text{N}}\right]\right]. \quad (13)
$$

In parallel connected system, as given in (5) circulating current directly proportional to the current sharing difference. If the current sharing is equal then the resultant circulating current become zero. Simultaneously, insertion of the series resistor will cause additional power loss in the system and it will leads to reduction in the load voltage. The output power loss based on Fig. 4 can be expressed as,

$$
P_{loss} = I_1^2(R_1 + R_{droop}) + I_2^2(R_2 + R_{droop}). \quad (14)
$$

In (13), 'N' stands for normalization. Normalization of current sharing is based on the rated load current and output power is based on maximum allowable losses in terms of converter rated power. For example, the relationship between normalized current sharing difference and losses are shown in Fig. 7(a), clearly shows the current sharing difference decreases and output power loss increases with increase in R_{droop}. So finding an optimum point will give better performance in terms of both current sharing and circulating current is given in Fig. 7(b).

Calculation of minimum droop index is by varying R_{droop} and its maximum value will depends on the voltage regulation. i.e., the product of converter output current and R_{droop} should not increase the maximum allowable voltage deviation. For example, if the rated voltage of dc bus is $48 \pm 5\%$ V and each converter rated current is 5 A then the product of converter rated current and R_{droop} should be less than the maximum deviation in dc bus voltage (2.4 V). The current sharing

978-1-4799-0024-4/13 $31.00 © 2013 IEEE

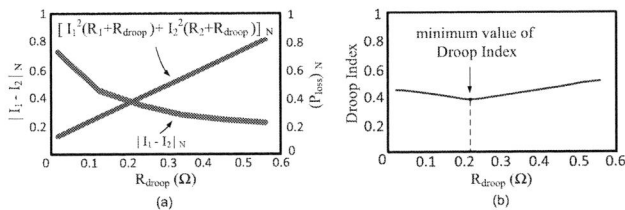

Fig. 7. Relationship between (a) normalized current sharing difference, output power loss and R_{droop} (b) droop index and R_{droop}

TABLE II
NOMINAL PARAMETERS OF BOOST CONVERTER

Parameters name	Symbol	Value
Input voltage	V_i	24V
Output voltage	V_L	48V
Cut-in voltage of MOSEFT	V_{on}	0.6V
ON state resistance of MOSEFT	r_{DS}	1mΩ
ON state resistance Diode	r_f	1mΩ
Filter inductor	L	100μH
ESR of inductor	r_L	0.03Ω
Filter capacitor	C	220μF
ESR of filter capacitor	r_C	0.05Ω
Load resistance	R_L	9.6Ω
Nominal switching frequency	f_{sw}	100kHz

difference can be calculated from (1) and (2) by introducing new variables x, y and m and given as

$$x = \frac{V_{DC1}}{V_{DC2}}, \quad y = \frac{R_1}{R_2}, \quad and \quad m = R_2 + R_{droop2}$$

$$|I_1 - I_2| = \frac{mxV_{DC2} - myV_{DC2} + 2(x-1)V_{DC2}R_L}{m^2y + mR_L(y+1)} \quad (15)$$

Similarly, the output power loss can be expressed as

$$P_{loss} = V_{DC2}^2 \left[\frac{[a - b + c]}{(m^2y + mR_L(y+1))^2} \right] \quad (16)$$

where,

$$a = myx^2(R_L^2 + (m + R_L))^2$$
$$b = 2xR_L(m + R_L) + mR_L^2x^2$$
$$c = my + R_L^2 - 2xR_L(my + R_L).$$

Substituting (15) and (16) in (13) and varying R_{droop2} from zero to its maximum value, the minimum value of droop index can be found out. So the value of R_{droop1} can be expressed as

$$R_{droop1} = yR_{droop2} \quad (17)$$

If both the converters output connected to same cable, i.e., R_1 is equal to R_2, then the magnitude of 'y' becomes unity and calculated values of R_{droop1} and R_{droop2} are same. If R_1 and R_2 are different, values of R_{droop1} and R_{droop2} are different and depends upon the 'y' magnitude.

V. SIMULATION STUDY

The proposed calculation of R_{droop1} and R_{droop2} for two parallel dc-dc boost converters given in Fig. 4 is simulated by using MATLAB / SIMULINK. The proposed control Algorithm was verified for the following cases: (i) a step change in output voltage of any one converter with both converters having same cable resistance (a) without any controller, (b) with proposed R_{droop} calculation (ii) a step change in output voltage of any one converter with with both converters having different cable resistance (a) without any controller, (b) with proposed R_{droop} calculation. The rated power of each boost converter is $240W$ and nominal parameters of boost converter are given in Table II.

The gains for the voltage and current PI controllers are 0.002, 100, 0.0009 and 400 respectively. The maximum variation in the converter output voltage is limited by $\pm 1\%$ of its nominal output voltage.

A. Case 1: A Step change in output voltage of any one converter with both converters having same cable resistance

a) *Without R_{droop} calculation:* In this case, the output voltage reference V_{DC}^* is changed in steps of $\pm 1\%$ of its nominal value (48 V) with a cable resistance of 100 $m\Omega$ in each converter. Fig. 8(a) shows the output voltages of converter-1 and converter-2 and load voltage, Fig. 8(b) shows the converter-1 and converter-2 output currents and load current and Fig. 8(c) shows the circulating current among the two converters. Initially up to 0.1 s the circuit is simulated at its nominal values, i.e, both the converter output voltages is 48 V. During this interval the load voltage is 47.7 V, the drop is due to the cable resistances. The load current current is 5 A and it is equally shared (2.5 A each) between the converters and the circulating current is also zero. At 0.1 s, converter-2 output voltage is increased by 1% of its nominal value (now V_{DC2} = 48.48 V). In this interval, load voltage is 47.9 V, converter-2 output current is 4.1 A and converter-1 output current is 0.9 A. The current sharing difference $|I_1 - I_2|$ is 3.2 A with an error of 64% (error is calculated with rated load current as base value). The magnitude of circulating current between the converters is 1.6 A. At 0.3 s, converter-2 output voltage is brought back to its nominal value 48 V . The values are same as in the first interval. At 0.5 s, converter-2 output voltage is decreased by 1% of its nominal value (now V_{DC2} = 47.52 V). During this interval, load voltage is 47.5 V, converter-1 output current is 4.1 A and converter-2 output current is 0.9 A. The current sharing difference, $|I_1 - I_2|$ is 3.2 A with an error of 64%. The magnitude of circulating current between the converters is 1.6 A.

b) *With proposed R_{droop} calculation:* In this simulation, the proposed method is implemented and the output voltage reference V_{DC}^* is changed in steps of $\pm 1\%$ of its nominal value 48 V with a cable resistance of 100 $m\Omega$ in each converter. Fig. 9(a) shows the output voltages of converter-1 and converter-2 and load voltage, Fig. 9(b) shows the converter-1 and converter-2 currents and load current, Fig. 9(c) shows the circulating current among the two converters and Fig. 9(d) shows the calculated R_{droop1} and R_{droop2} values. Initially up to 0.1 s the circuit is simulated at its nominal values, i.e, both the converter output voltages are 48 V. During this time interval load voltage is 47.7 V, the drop is due to the cable

978-1-4799-0024-4/13 $31.00 © 2013 IEEE

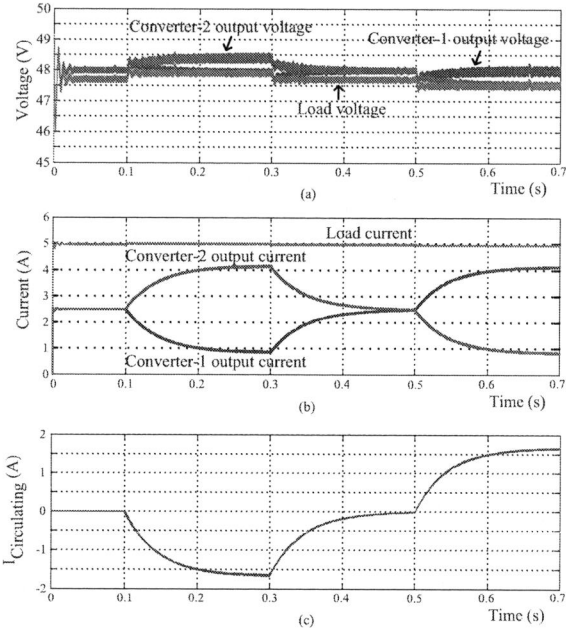

Fig. 8. Simulation results for step change in output voltage without R_{droop} calculation. (a) converter and load voltages (b) output currents (c) circulating current

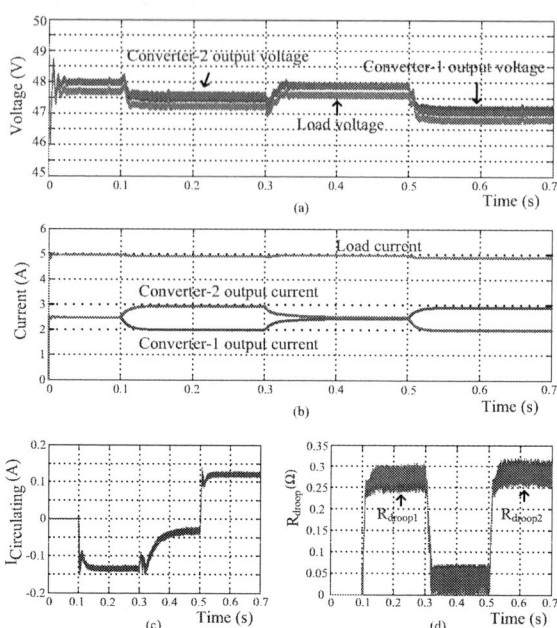

Fig. 9. Simulation results for step change in output voltage with proposed R_{droop} calculation. (a) converter and load voltages (b) output currents (c) circulating current (d) calculated R_{droop1} and R_{droop2} values

resistances. The load current current is 5 A and it is equally shared (2.5 A each) between the converters and the circulating current is zero. At 0.1 s, converter-2 output voltage is increased by 1% of its nominal value (now V_{DC2} = 48.48 V). In this interval, load voltage is 47.3 V, converter-2 output current is 2.9 A and converter-1 output current is 2.0 A. The current sharing difference, $|I_1 - I_2|$ is 0.9 A with an error of 18%. The magnitude of circulating current between the converters is 140 mA and calculated R_{droop1} and R_{droop2} values are same and is 0.26 Ω. At 0.3 s, converter-2 output voltage is brought back to its nominal value 48 V. The values are same as in the first interval. At 0.5 s, converter-2 output voltage is decreased by 1% of its nominal value (now V_{DC2} = 47.52 V). during this interval, load voltage is 46.9 V, converter-1 output current is 2.8 A and converter-2 output current is 2.0 A. The current sharing difference, $|I_1 - I_2|$ is 0.8 A with an error of 16%. The magnitude of circulating current between the converters is 130 mA and calculated R_{droop1} and R_{droop2} values are same and is 0.28 Ω. So by using proposed droop index, optimum value current sharing difference and circulating current can be achieved.

B. Case 2: A step change in output voltage of any one converter having different cable resistance

a) Without R_{droop} calculation: In this case, the output voltage reference V_{DC}^* is changed in steps of ±1% of its nominal value 48 V with an output cable resistance of 100 mΩ in converter-1 and 150 mΩ in converter-2. Fig. 10(a) shows the output voltages of converter-1 and converter-2 and load voltage, Fig. 10(b) shows the converter-1 and converter-2

currents and load current and Fig. 10(c) shows the circulating current among the two converters. Initially up to 0.1 s the circuit is simulated at its nominal values, i.e, both the converter output voltages is 48 V. During this time interval load voltage is 47.5 V, is due to drop in the cable resistances. Load current shared by each converter is not same (2.5 A) due to the difference in output cable resistance. The circulating current between the converters is also zero. At 0.1 s, converter-2 output voltage is increased by 1% of its nominal value (now V_{DC2} = 48.48 V). In this interval, load voltage is 47.3 V, converter-2 output current is 3.6 A and converter-1 output current is 1.4 A. The current sharing difference, $|I_1 - I_2|$ is 2.2 A with an error of 44%. The magnitude of circulating current between the converters is 1.4 A. At 0.3 s, converter-2 output voltage is brought back to its nominal value 48 V. The values are same as in the first interval. At 0.5 s, converter-2 output voltage is decreased by 1% of its nominal value (now V_{DC2} = 47.52 V). During this interval, load voltage is 47.0 V, converter-1 output current is 4.25 A and converter-2 output current is 0.75 A. The current sharing difference, $|I_1 - I_2|$ is 3.5 A with an error of 70%. The magnitude of circulating current between the converters is 1.5 A.

b) With proposed R_{droop} calculation: In this simulation, the proposed method is implemented and the output voltage reference V_{DC}^* is changed in steps of ±1% of its nominal value 48 V with an output cable resistance of 100 mΩ in converter-1 and 150 mΩ in converter-2. Fig. 11(a) shows the output voltages of converter-1 and converter-2 and load voltage, Fig. 11(b) shows the converter-1 and converter-2 currents and load current, Fig. 11(c) shows the circulating current among the two

978-1-4799-0024-4/13 $31.00 © 2013 IEEE   458

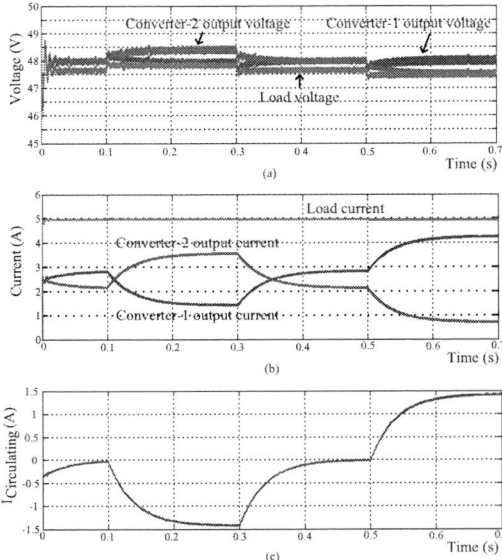

Fig. 10. Simulation results for step change in output voltage without R_{droop} calculation. (a) converter and load voltages (b) output currents (c) circulating current

converters and Fig. 11(d) shows the calculated R_{droop} values. Initially up to 0.1 s the circuit is simulated at its nominal

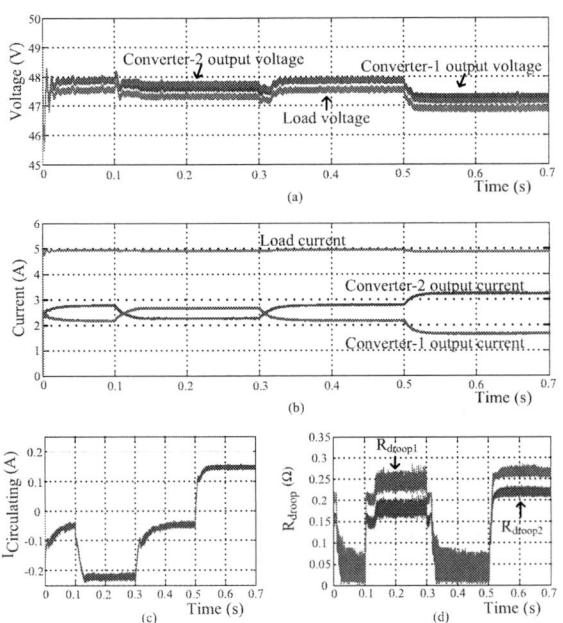

Fig. 11. Simulation results for step change in output voltage with proposed R_{droop} calculation. (a) converter and load voltages (b) output currents (c) circulating current

values, i.e, both the converter output voltages is 48 V. During this time interval load voltage is 47.5 V, is due to drop in the cable resistance. Load current shared by each converter is not same (2.5 A) due to the difference in output cable resistance. The circulating current among the converters is 30 mA is due

to slight mismatch in the calculation of R_{droop1} and R_{droop2} values. At 0.1 s, converter-2 output voltage is increased by 1% of its nominal value (now V_{DC2} = 48.48 V). In this interval, load voltage is 47.3 V, converter-2 output current is 2.7 A and converter-1 output current is 2.3 A. The current sharing difference, $|I_1 - I_2|$ is 0.5 A with an error of 10%. The magnitude of circulating current among the converters is 230 mA and calculated R_{droop1} and R_{droop2} values are 0.25 Ω and 0.18 Ω respectively. At 0.3 s, converter-2 output voltage is brought back to its nominal value 48 V. The values are same as in the first interval. At 0.5 s, converter-2 output voltage is decreased by 1% of its nominal value (now V_{DC2} = 47.52 V). During this interval, load voltage is 46.9 V, converter-1 output current is 3.25 A and converter-2 output current is 1.65 A. The current sharing difference, $|I_1 - I_2|$ is 1.6 A with an error of 32%. The magnitude of circulating current between the converters is 150 mA and calculated R_{droop} values are 0.23 Ω and 0.27 Ω respectively. Thus, by using proposed droop index, optimum value current sharing difference and circulating current can be achieved.

VI. CONCLUSION

In this paper, a figure-of-merit called Droop Index is proposed for dc-dc converter parallel operation. The analysis and design of dc-dc boost converter using proposed method has been successfully demonstrated through the MAT-LAB/SIMULINK. The proposed method calculates virtual resistance values instantaneously based on the converter output voltage mismatch. This improves the current sharing difference and decreases circulating current between the converters, in transient as well as in steady state condition.The effects of converter cable resistances are taken into account to verify the performance of the proposed method. Since this algorithm deals with parallel operation of dc-dc converters, this is a better choice for dc microgrid applications.

REFERENCES

[1] S. Bull, "Renewable energy today and tomorrow," *Proceedings of the IEEE*, vol. 89, no. 8, pp. 1216 –1226, Aug 2001.

[2] D. Salomonsson, L. Soder, and A. Sannino, "An adaptive control system for a dc microgrid for data centers," *Industry Applications, IEEE Transactions on*, vol. 44, no. 6, pp. 1910 –1917, Nov.-Dec. 2008.

[3] H. Kakigano, Y. Miura, and T. Ise, "Low-voltage bipolar-type dc microgrid for super high quality distribution," *Power Electronics, IEEE Transactions on*, vol. 25, no. 12, pp. 3066 –3075, Dec. 2010.

[4] I. Kondratiev, E. Santi, and R. Dougal, "Robust nonlinear synergetic control for m-parallel-connected dc-dc boost converters," in *Power Electronics Specialists Conference, 2008. PESC 2008. IEEE*, June 2008, pp. 2222 –2228.

[5] B. Choi, "Comparative study on paralleling schemes of converter modules for distributed power applications," *Industrial Electronics, IEEE Transactions on*, vol. 45, no. 2, pp. 194 –199, Apr 1998.

[6] S. Luo, Z. Ye, R.-L. Lin, and F. Lee, "A classification and evaluation of paralleling methods for power supply modules," in *Power Electronics Specialists Conference, 1999. PESC 99. 30th Annual IEEE*, vol. 2, 1999, pp. 901 –908 vol.2.

[7] J.-W. Kim, H.-S. Choi, and B. H. Cho, "A novel droop method for converter parallel operation," *Power Electronics, IEEE Transactions on*, vol. 17, no. 1, pp. 25 –32, Jan 2002.

[8] B. Irving and M. Jovanovic, "Analysis, design, and performance evaluation of droop current-sharing method," in *Applied Power Electronics Conference and Exposition, 2000. APEC 2000. Fifteenth Annual IEEE*, vol. 1, 2000, pp. 235 –241 vol.1.

978-1-4799-0024-4/13 $31.00 © 2013 IEEE

[9] J. Rajagopalan, K. Xing, Y. Guo, F. Lee, and B. Manners, "Modeling and dynamic analysis of paralleled dc/dc converters with master-slave current sharing control," in *Applied Power Electronics Conference and Exposition, 1996. APEC '96. Conference Proceedings 1996., Eleventh Annual*, vol. 2, Mar 1996, pp. 678 –684 vol.2.

[10] R.-H. Wu, T. Kohama, Y. Kodera, T. Ninomiya, and F. Ihara, "Load-current-sharing control for parallel operation of dc-to-dc converters," in *Power Electronics Specialists Conference, 1993. PESC '93 Record., 24th Annual IEEE*, Jun 1993, pp. 101 –107.

[11] J. Sun, Y. Qiu, B. Lu, M. Xu, F. Lee, and W. Tipton, "Dynamic performance analysis of outer-loop current sharing control for paralleled dc-dc converters," in *Applied Power Electronics Conference and Exposition, 2005. APEC 2005. Twentieth Annual IEEE*, vol. 2, March 2005, pp. 1346 –1352 Vol. 2.

[12] S. Anand and B. Fernandes, "Modified droop controller for paralleling of dc-dc converters in standalone dc system," *Power Electronics, IET*, vol. 5, no. 6, pp. 782 –789, July 2012.

[13] I. Batarseh, K. Siri, and H. Lee, "Investigation of the output droop characteristics of parallel-connnected dc-dc converters," in *Power Electronics Specialists Conference, PESC '94 Record., 25th Annual IEEE*, Jun 1994, pp. 1342 –1351 vol.2.

[14] M. Kazimierczuk, *Pulse-width Modulated DC-DC Power Converters*. John Wiley & Sons, 2008.

[15] C. Jamerson, T. Long, and C. Mullett, "Seven ways to parallel a magamp," in *Applied Power Electronics Conference and Exposition, 1993. APEC '93. Conference Proceedings 1993., Eighth Annual*, Mar 1993, pp. 469 –474.

[16] Y. Ito, Y. Zhongqing, and H. Akagi, "Dc microgrid based distribution power generation system," in *Power Electronics and Motion Control Conference, 2004. IPEMC 2004. The 4th International*, vol. 3, Aug. 2004, pp. 1740 –1745 Vol.3.

VII. BIOGRAPHIES

Sijo Augustine received his AMIE degree in Electrical Engineering from Institution of Engineers [India] in 2008 and M. Tech in Power Electronic and Drives from VIT University, Vellore, Tamil Nadu in 2010. Presently he is pursuing the Ph.D. degree in the Department of Electrical Engineering, Indian Institute of Technology Madras, Chennai, India. His research interests are in the fields of power electronics, power quality and microgrids.

Mahesh K. Mishra (S'2000-M'02-SM'10) received his Bachelor of Technology from College of Technology, Pantnagar, India and M.E. from University of Roorkee, India in 1991 and 1993 respectively. In Feb. 2002, he received the Ph.D. in Electrical Engineering from Indian Institute of Technology, Kanpur, India.

He has teaching and research experience of about 20 years. For about 10 year, he was a faculty in Electrical Engineering Department, Visvesvaraya National Institute of Technology, Nagpur, India. Currently, he is a Professor in Electrical Engineering Department, Indian Institute of Technology Madras, India. His interests are in the areas of Power Distribution Systems, Power Electronics and Control Systems.

Dr. Mahesh is life member of Indian Society of Technical Education (ISTE).

Prof. Lakshminarasamma obtained her Ph.D. degree in Electrical Engineering from the Indian Institute of Science and joined the faculty of Electrical Engineering at the Indian Institute of Technology, Madras as an Assistant Professor in the year 2009. She has coauthored four journal papers in peer-reviewed journals, including the IEEE Transactions on Power Electronics and several premier conferences. Her research interests are in the areas of Power Electronics, drives and renewable energy applications.

978-1-4799-0024-4/13 $31.00 © 2013 IEEE

A Simple and Robust Method for Open Switch Fault Detection in Power Converters

Mehdi Salehifar, Ramin Salehi Arashloo, Manuel Moreno-Eguilaz, Vicent Sala, L. Romeral
Department of Electronic Engineering, Universitat Politecnica de Catalunya, BarcelonaTech
Center of Motion Control and Industrial Applications (MCIA)
Barcelona, Spain
mehdi.salehifar@mcia.upc.edu

Abstract – **in this paper, a new fault detection method based on signal normalization using a simple trigonometric function is presented and applied to a five phase converter for fault tolerant application under nonsinusoidal unbalanced current waveforms. Generality, simplicity, ability to localize faulty switch, multiple switch fault detection and robustness are achieved using this approach. Once theory is explained, simulation results with Matlab/Simulink and experimental waveforms are described to show the effectiveness of the proposed detection method. Experimenal results corroborate these simulation results.**

Keywords – multiphase power converters, fault detection, fault tolerant converters, PMSM drive, and open switch fault

I. INTRODUCTION

There are some applications where continuous operation is vital. In such applications, a fault tolerant concept is mandatory to achieve safety and reliability. In order to operate a fault tolerant system, fast fault detection, localization and isolation of faulty components are of paramount importance [1].

Some review on fault types and detection methods have been presented in literature such as [2] and [3]. Generally, faulty types in power converters can be divided into open switch and short circuit faults. Short circuit faults are often so destructive that the system should be shut down immediately after fault detection. However, if it is detected fast enough, typically less than10 *μs* in case of an IGBT, it is possible to avoid system shutdown. Such fast detection is necessary to operate fault tolerant converters with an extra leg [2]. It should be noted that nowadays, short circuit detection and protection methods have been successfully included in high performance custom designed commercial drivers.

On the other side, open circuit faults are less destructive. Primary effect of such faults is a reduction in system performance. However, if it is not detected, secondary faults may happen [2]. Three different research lines have been developed in literature to detect open switch faults: signal (i.e. current or voltage of converter) based methods, model based detection methods and reference band methods [4].

Regarding model based method, a simple and fast open switch and open phase fault detection method was presented in [5], based on a model reference adaptive system for PMSM drive application. Main advantage of this method is that it is cheap since it does not need extra sensors.

However, it is well known that prediction of load response is sensitive to load parameters and it cannot work well in all operation work points.

It is a simple, fast, efficient and robust detection method, if difference between load current and reference value is measured [6]. To make diagnostic variable independent of load parameters it has been normalized with respect to load current average absolute value [7]. However, this method has a drawback which makes its application limited to a system with a closed loop control. It cannot be applied to open loop systems.

Signal based fault detection is another simple detection method based on converter output voltage or current. Both methods have been investigated in literature. Voltage based methods are fast, simple and efficient [8-13]. Detection times less than 10 *μs* have been presented [8-9]. However main disadvantage of this method is that it is expensive since it needs extra sensors to be included in power converter.

Last detection method is based on converter signal. Since converter output current measurement is usually necessary for control purposes, it is not necessary to add additional sensors in order to apply this detection method. Different methods have been presented to detect this kind of fault: Park's vector approach, normalized dc current method, slope method, frequency method, wavelet transform and so on [2]. However, all these methods have some disadvantages, such as slow detection, single switch detection, robustness during load transients and complexity. Recently, several high performance methods have been shown in literature. The proposed methods in [4], [14] and [15] are the best ones among others. Diagnostic variables in [4] and [14] are based on normalized average current with respect to absolute average current method. In [15], diagnostic variable is based on Park vector phase angle. In all of these studies, input current is a three phase sinusoidal balanced current. However, in case of multi phase fault tolerant converters, input current can be a nonsinusoidal unbalanced current. These characteristics make presented methods show a poor performance in some diagnostic scenarios, so in this paper this concerns are for first time presented and an effective detection is proposed.

Two contributions are shown in this paper. First, a new fault detection method has been proposed; second, it has been applied to a multiphase power converter.

Rest of this paper is presented as follows. In section II, case study, faulty scenarios and proposed detection method

978-1-4799-0024-4/13 $31.00 © 2013 IEEE

are explained. Simulation results are shown in section III to investigate performance of the presented method and a comparison is made with presented methods in literature. The proposed method is included in a fault tolerant algorithm of a five phase PMSM in section IV. Experimental results are shown in section V to validate theory. At section VI, a conclusion is made and finally references are shown.

II. PROPOSED DETECTION METHOD AND APPLICATION

Due to their fault tolerant capabilities, five phase permanent magnet synchronous motors (PMSM) have received a considerable attraction in applications where safety is of paramount importance such as electric and hybrid electric vehicle, aerospace applications, chemical power plants, traction, space craft and more electric aircrafts [16-17]. So, to be fault tolerant, such systems should detect and isolate the fault and continue operation with minimal derating. To achieve these targets, a multi phase fault tolerant converter with fault detection and isolation ability is considered here as shown in Fig. 1. The converter output current can be sinusoidal, non sinusoidal, balanced or unbalanced under different faulty modes. Triac TR_j shown in Fig. 1 is used to isolate faulty phase after fault detection.

Considering this case study, the following faulty modes are contemplated: single switch open circuit, single phase, two phase or three phase open circuit, double switch fault with combination of two upper, two lower and one upper one lower, triple switch fault with combination for three upper, three lower, one upper two lower, two lower one upper, four switch faults with combination for three upper one lower, one upper three lower, or two upper two lower. As mentioned, faulty mode combinations for multi phase converter can be much more than three phase converter. However, considering practical cases, a five phase PMSM drive can tolerate until two faulty phases.

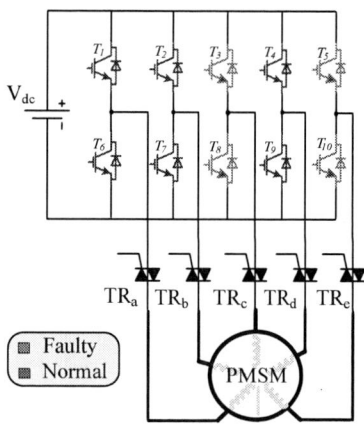

Fig. 1. Fault tolerant power converter configuration

A. Proposed Fault Detection Method

To evaluate a fault detection method, some performance criteria should be considered. First, it should be fast, what means that the detection method should detect fault with minimum time after fault. After detection, to realize fault tolerant operation it is necessary to localize and isolate faulty switch, so localization is next performance criteria. Third important characteristic is robustness, it means that detection method should avoid false alarm during load transients. As next important criteria, it should be general; this means that detection strategy should be able for application at any converter configuration such as three phase, four phase, five phase or any multiphase converter. Finally, simple implementation can bring an important advantage for detection strategy. Consequently, regarding all of aforementioned advantages, a novel detection method is proposed as follows.

To detect a fault, a diagnostic variable is first calculated. Because in practical application, converter output current is usually a noisy signal, prior to do a signal processing, it is necessary to apply a low pass filter which is done in this paper. A cut off frequency at least 10 times the fundamental frequency is recommended. If the cut off frequency is chosen too high, it cannot eliminate distortion components effectively while if a low value is selected, filter imposes a high delay on signal. This delay can slow down detection speed which is a critical criterion at fault detection method design.

In order to improve detection method robustness in case of fast transients and also being independent from load operating conditions, some methods has been presented in literature. Normalizing with respect to Park's vector modulus has been presented in [14]; however, it is less effective in case of unbalanced nonsinusoidal multiphase converters. Moreover, since there is more than one equivalent perpendicular page, Park's vector modulus calculation becomes even more complicated. A simple trigonometric function is proposed in this paper. Due its special waveform, converter output current is normalized to values between $-/2$ and $/2$, while signal shape is still suitable for fault detection. It is known that in faulty mode, current signal becomes zero for some time; presented function is not sensitive to this period.

Following filtering signal, its delayed current sample equal to one quarter of fundamental cycle T and current signal value are calculated and then divided. This value is labeled d_j. Then, inverse tangent of signal d_j is calculated. Therefore, a first variable d is calculated as:

$$d_j = \tan^{-1}\left(\frac{i_j(t)}{i_j(t-T/4)} \times K\right), \quad j\varepsilon\{a,b,c,d,e\} \qquad (1)$$

Where K is a sensitivity factor. Higher values for K provides better transient response. In order to obtain a suitable diagnostic signal D, average absolute value of d signal at two times of fundamental frequency is calculated as:

$$D_j = \frac{1}{T/2} \int_0^{T/2} |d|\,dt, \quad j\varepsilon\{a,b,c,d,e\} \qquad (2)$$

Value of diagnostic variable D is equal to $/2$ under healthy mode conditions. Generally two fault types can happen in one converter leg. In case of both faults, D value decreases. First fault is single switch open circuit and second type is double switch fault. Under single switch fault, diagnostic variable reduces below $/2$, and an oscillatory component with fundamental frequency appears at

978-1-4799-0024-4/13 $31.00 © 2013 IEEE   462

diagnostic variable. This is due to the fact that cut off frequency of filter applied at signal d is at two times the fundamental component. On the other side, at double switch faulty mode, D approaches to zero, since current value becomes zero immediately after fault happens. To detect fault, it is necessary to compare diagnostic signal with a threshold value. This value can be determined by doing extensive simulations and analysis while regarding different faulty modes for power converters.

In case of both single switch and open phase fault, D value is compared with a threshold value; here 1.1 is considered. Using this threshold, it is possible to detect both single switch and open phase faults. However, to determine phase fault type, D is compared with a second threshold value. Since under open phase fault, D value reduces to zero; this advantage is utilized to detect open phase faults in this paper. So, D average at fundamental frequency is calculated and if it is lower than the second threshold, faulty type mode will be open phase.

In order to maintain fault tolerant concept in a multiphase power converter, it is necessary to locate faulty switch, so it can be replaced with an extra leg or isolated totally from the converter. Here a simple approach is applied which is based on current polarity. According to this method, input phase current is passed through a weight function. This function estimates input current i to values between -1 and 1 as follows:

$$y_j(i) = \begin{cases} 1 & i \geq 0.1 \\ 0 & -0.1 < i < 0.1 \quad j\varepsilon\{a,b,c,d,e\} \\ -1 & i \leq -0.1 \end{cases} \quad (3)$$

This block samples and calculates average of converter current simplified by equation (3) for one fundamental cycle as follows:

$$\overline{I}_j = \frac{1}{M}\sum_{j=1}^{M} y_j, \quad j\varepsilon\{a,b,c,d,e\} \quad (4)$$

Where M is number of samples in one fundamental period. Average of converter curent I_j is compared with a positive and a negative threshold. If average value is higher than positive threshold, then lower switch is the faulty component at corresponding phase. A value lower than the negative threshold is related to the upper faulty switch. For simplicity, a code is generated based on current value; code 1 is considered for upper switch fault, code -1 is related to lower switch fault and code 2 corresponds to open phase fault.

Fault detection, diagnostic variable waveform at different faulty modes and fault localization blocks are shown in Fig. 2. Fundamental frequency is denoted by F_l in this figure. Considering proposed fault detection method, its main advantage is simple implementation. Since presented method is mainly sensitive to current values close to zero, it is a robust method to high transients which makes it superior to presented methods in literature. On the other side, it is fast, since average period is half of what is normally used in literature. Another important advantage for this method is its generality is demonstrated, since no special transformation related to converter topology is used; this method can be used in a two level converter with any arbitrary number of

phases. Finally, this method only needs the converter output current measurement, which means that it can be used in a system with open loop or closed loop control while some methods in literature need reference current which is not available in an open loop control system. Moreover, it is possible to detect open phase fault without using an auxiliary variable in contrast to presented methods in literature.

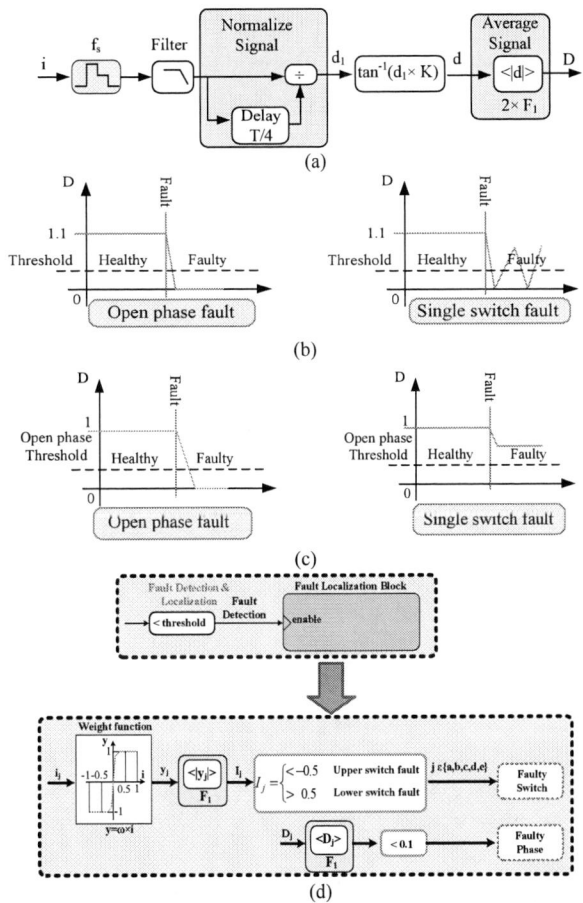

Fig. 2. Proposed fault detection method (a) calculation of diagnostic variable (b) D average at half of fundamental period (c) D average at one fundamental period (d) fault localization block

III. SIMULATION RESULTS

In order to show the effectiveness of the proposed method, simulation results are included. Moreover, to show the robustness of the presented method, a step change from full load to 10 percent of nominal load is applied in all simulation results. On the other side, current waveforms are simulated with first and third harmonic(10%), forcing a nonsinusoidal current waveform. In addition, current waveforms are unbalanced. Different faulty modes are simulated. Moreover, in order to make simulation as close to the real system, 1 μW white noise is added to all measured currents. Converter parameters for all the simulations are: switching frequency is 2 kHz, DC link voltage is 310 V, fundamental frequency is 50 Hz, load equivalent resistance and inductance are 2 Ω and 7 mH, respectively.

To show the performance of the diagnostic method under high transients, a 90 % step change is applied to input voltage at 0.1 s. Load waveforms and diagnostic variables for all five phases are shown in Fig. 3. Current waveforms and detection index of phase currents are shown by *a*, *b*, *c*, *d*, and *e*, respectively. Besides, thresholds for open switch and open phases faults are indicated by Th_1 and Th_2, respectively. As seen, minimum value for D is 1.4, which is much bigger than diagnostic threshold; this value was determined equal to 1.1 after doing simulation for all possible faulty modes.

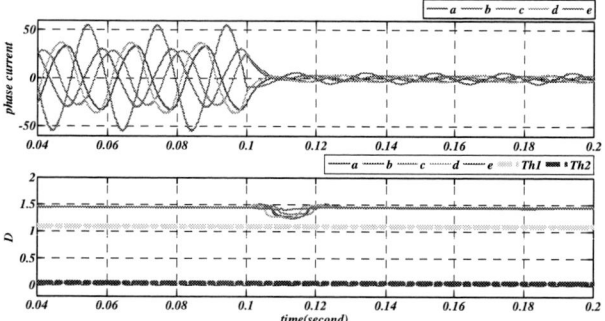

Fig. 3. Current waveforms and diagnostic variables under healthy mode

One faulty switch was simulated at lower switch in phase A at 0.25 s. Resultant simulation waveforms of converter current, diagnostic variable, fault signal and fault code are shown in Fig. 4. As shown, the fault is detected and localized effectively in less than a quarter of one cycle.

Similarly two faulty switches are simulated and results are shown in Fig. 5. The fault is detected and localized at phases A and B. At fault time, B phase current is positive. Due to this reason, around 0.255 s, the current value is changed to zero.

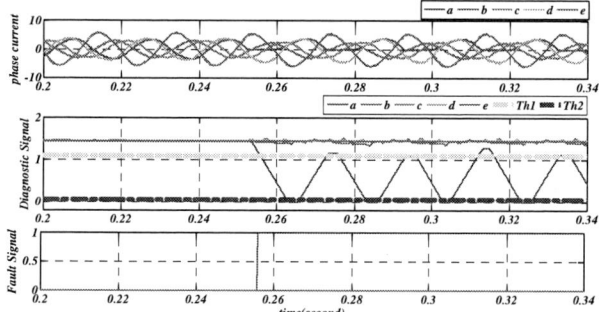

Fig. 4. Simulation results under one faulty switch T_6 at time 0.25 s

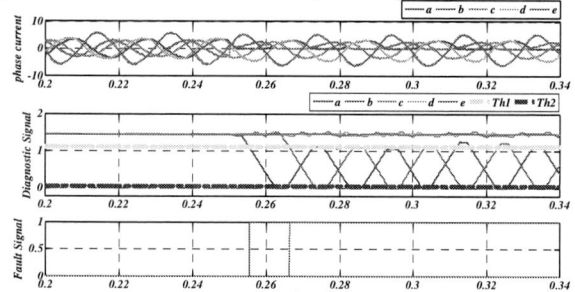

Fig. 5. Simulation results under two faulty switch T_6, T_7 at time 0.25 s

Considering this fact, detection time is around 6 ms, which is fast enough.

To further emphasize multiple switch fault detection and localization, a fault is simulated at two lower switches in phases A and B at time 0.24 s. Two upper switch faults are also considered at upper switches in phase D and E at time 0.28 s. Final waveforms are shown in Fig. 6. According to results diagnostic method detects and localizes faulty switches effectively. Since current at phases A and D is close to zero when fault is initiated, detection time is longer for this case.

Since, another well known faulty mode in power converters is single phase open circuit fault. This case is also simulated. A fault is started in phase A at 0.24 s, an intermediate fault is implemented at phase E at 0.28. Simulated current waveforms and diagnostic variables are shown in Fig. 7; as seen presented method can effectively detect in such cases around 3 ms which is quite low, also diagnostic variable D under such fault reduces to zero in contrast to single switch faults where an oscillatory diagnostic signal is produced. It should note that fault detection is fast (i.e. quarter of a cycle). However, in this case, fault type is localized at half a cycle.

Combination of faulty phase and faulty switch is simulated at next step. Here a fault is considered in phases A and E at 0.24 s, and 0.28 s, respectively. An upper switch fault in phase B at time 0.26 s is also simulated. Simulation waveforms are shown in Fig. 8. Diagnostic variable is able to detect and localize fault successfully.

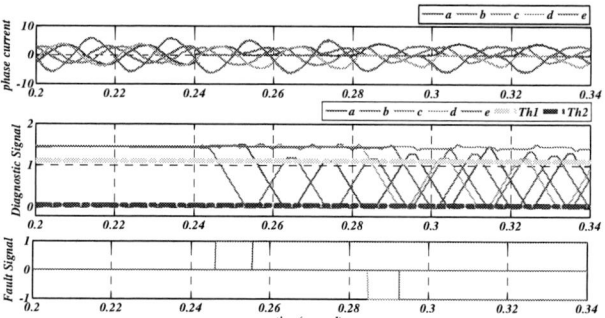

Fig. 6. Simulation results under two faulty switch T_6, T_7 at time 0.24 s and T_2, T_5 at time 0.28 s

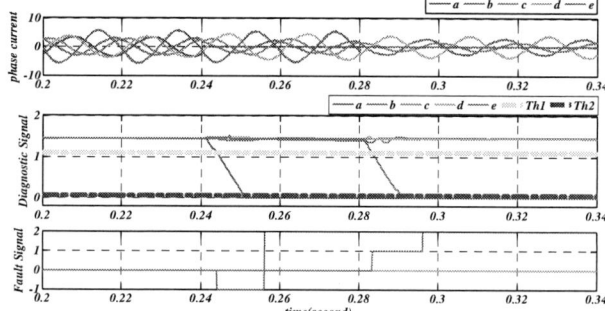

Fig. 7. Simulation results under two faulty phases *A* at 0.24 s and *E* at 0.28 s

In order to avoid false alarms, detection threshold in all simulation was 1.1. So, all of them were eliminated. However, it is possible to increase it in order to improve detection speed. In this case, multiple switch fault detection will be more difficult in some cases. Another point considering detection speed is that it depends on time that fault occurs, since diagnostic method is sensitive to zero period of current waveform, it has high effect on detection speed.

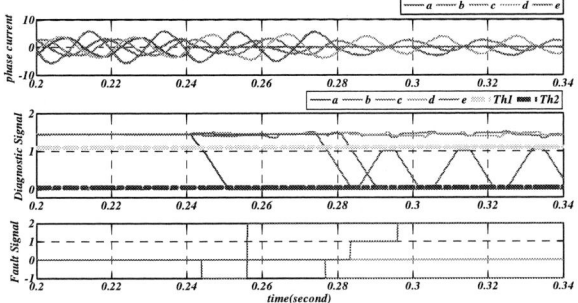

Fig. 8. Simulation results under two faulty phases *A* at 0.24 s, *E* at 0.28 s and one fault switch T_2 at 0.26 s

IV. FAULT TOLERANT CONTROL OF A FIVE PHASE PMSM USING PROPOSED DETECTION METHOD

In this part, presented fault detection method is used to maintain fault tolerant concept in a five phase PMSM with a closed loop control while motor neutral is available. Besides, detection capability is studied. Motor parameters are shown in table I. Switching frequency is 10 kHz and DC link voltage is 600 V. The threshold is shown by *Th* in all figures.

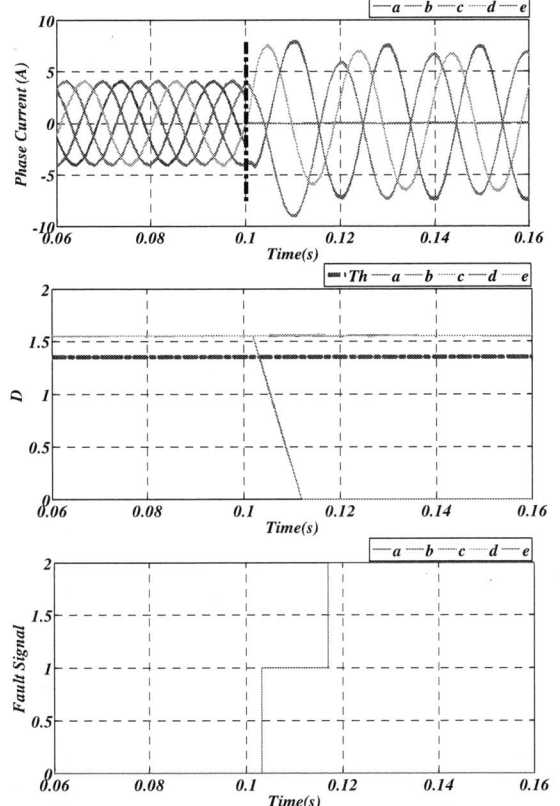

Fig. 9. Simulation results of fault detection for two open phase faults at phases A and B.

First, two case studies are simulated to shown fault detection performance of the proposed method. Two open phase faults are initiated in phases A and B at time 0.1 s. Simulated waveforms are shown in Fig. 9. According to results, fault is detected after 3 ms. It should be noted that fundamental period of phase current is 20 ms.

So, fault detection is done in less than a quarter of one cycle. As second case study, two open switch faults are initiated in upper switch of phase A and lower switch of phase B at time 0.1 s. According to simulation waveforms shown in Fig. 10, in both cases the fault is detected and localized at less than 3 ms.

TABLE I: MOTOR PARAMETERS

Number of Pole Pairs		2
Stator Resistance		0.55 Ω
Stator Inductance	Laa	35 mH
	Lab	10 mH
	Lac	7 mH
Nominal Torque		12 Nm
Nominal Speed		1527 rpm
Permanent Magnet Flux		0.48 Wb
Moment of Inertia		235 μkgm²

Now, proposed method is included in a fault tolerant control algorithm of a five phase PMSM with accessible neutral point. Control method, modulation strategy, and fault tolerant algorithm are discussed in [18] and [19]. Here, only final results are shown.

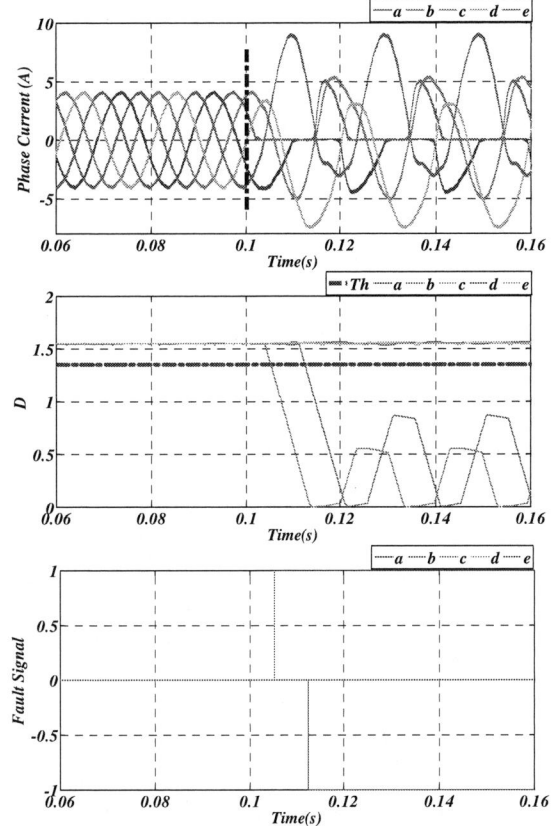

Fig. 10. Simulation results of fault detection for two open switch faults at phases A and B.

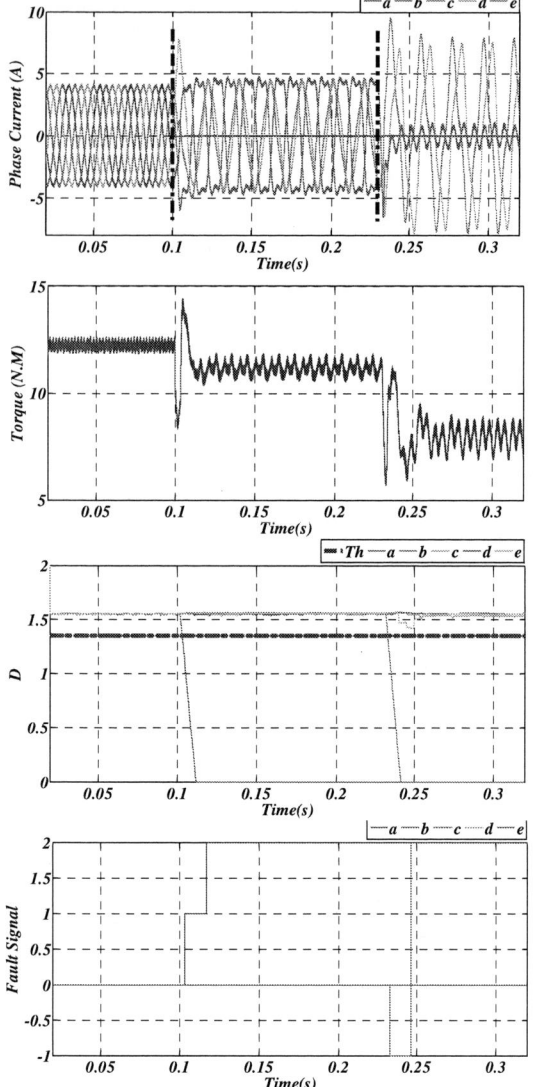

Fig. 11. Fault tolerant simulation waveforms for two adjacent open phase faults at phases A and B.

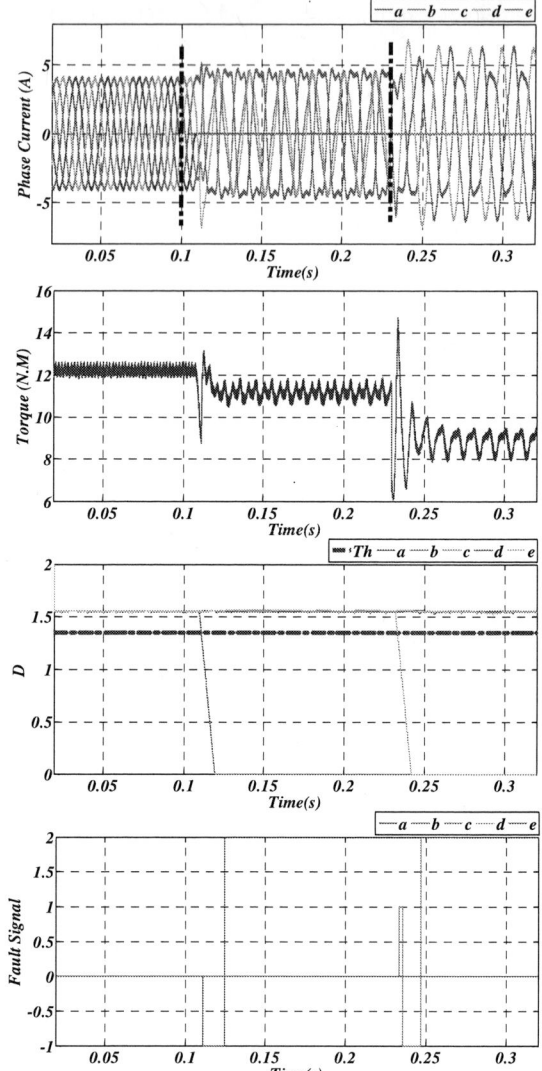

Fig. 12. Simulation results for two nonadjacent faults at phases A and C.

Motor drive is able to continue operation with one faulty phase, two adjacent faulty phases and two nonadjacent faulty phases. First a fault is initiated in phase A at time 0.1 s., then a second fault is initiated in phase B at time 0.23 s. which is adjacent to phase A. The simulation waveforms are shown in Fig. 11. As it can be seen, fault is detected at both cases successfully, at the same time, and the fault tolerant control is adapted according to new faulty mode.

Finally, two nonadjacent faulty mode is studied. Therefore one fault is enabled at upper switch in phase A at time 0.1 s and another fault is included in phase C at time 0.23 s, which is nonadjacent to phase A. Simulation waveforms of phase current, motor torque, diagnostic signal, and fault localization are shown in Fig. 12. Since current is negative in phase A at 0.1 s, it takes more time to detect fault at phase A in contrast to previous method.

According to present results, proposed method has a high performance for application in fault tolerant algorithm of a multi phase drive.

V. EXPERIMENTAL RESULTS

In order to validate simulation, experimental results are provided. Fault detection method is included in a fault tolerant algorithm of a five phase PMSM motor. Under one and two faulty phases, fault is detected by presented method, and then, according to faulty phase, corresponding control block is initiated. Here all detection and control method is implemented in a DSPACE model 1005; experimental setup is shown in Fig. 13; current and torque scales are shown below the figures.

As first case study, motor was operated under healthy mode. Then, a fault was randomly included in phase A after another fault was started in phase B which is adjacent to phase A. Experimental waveforms are shown in Fig. 14; as seen, fault tolerant algorithm is done effectively. Also, a second case

978-1-4799-0024-4/13 $31.00 © 2013 IEEE

study from healthy mode to one faulty phase A and faulty phase C, which is nonadjacent to phase A was considered. Final results are shown in Fig. 15. According to presented results, proposed method can be effectively applied in a system with fault tolerant capability.

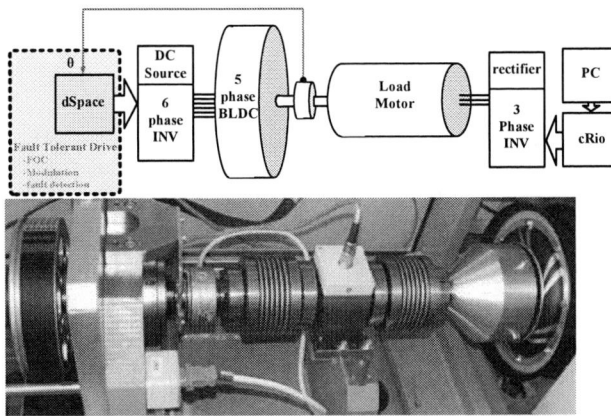

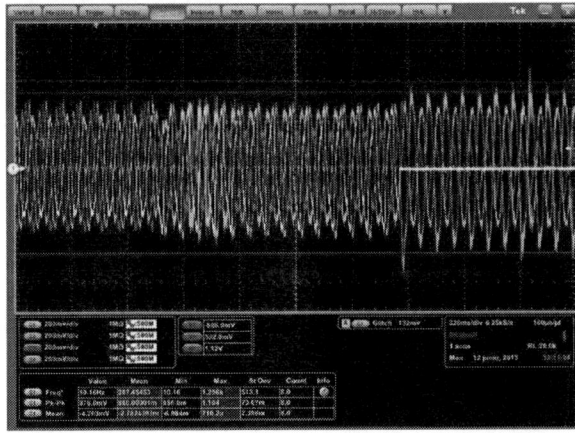

Fig. 13. Experimental setup

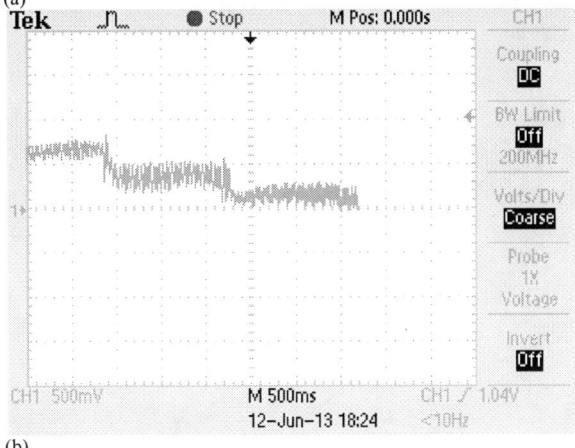

Fig. 14. Two adjacent faulty phases (a) phase currents (scale: 1V is equivalent to 16.67 A) (b) torque (scale: 1V is equivalent to 5.56 N.M)

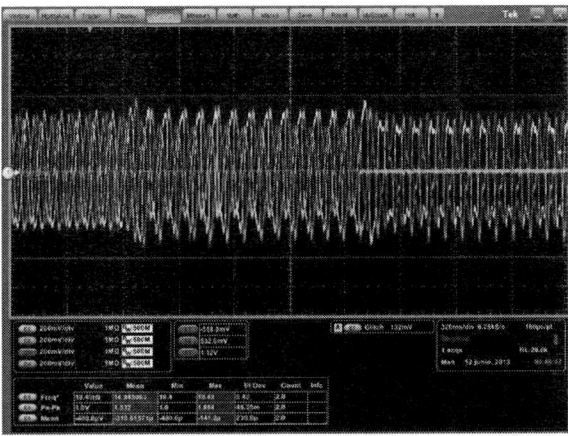

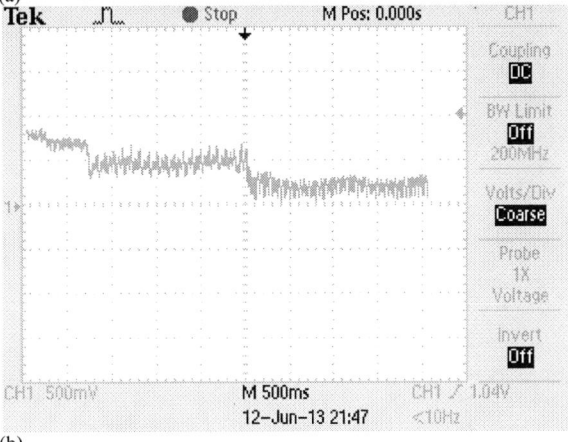

Fig. 15. Two nonadjacent faulty phases (a) phase currents (b) torque

VI. CONCLUSION

A new open circuit fault detection method was proposed in this paper, where current phase was estimated by a trigonometric function and used to define a detection variable. Simulation results and experimental waveforms were presented for a five phase converter under different faulty modes and nonsinusoidal unbalanced current. According to results, proposed method is fast, robust and reliable. Furthermore, it is a general, simple and efficient method. It is possible to detect and localize multiple switch faults as well. Detection speed less than a quarter of a fundamental cycle can be claimed. However, it depends on load current at fault initiation point. A high performance was achieved under fault tolerant control of a five phase PMSM drive with closed loop control.

ACKNOWLEDGEMENT

This work was supported by Spanish Ministry of Economic Affairs and Competiveness under the Research Project TRA 2010-21598-C02-01 and CD2009-00046 Consolidor Project.

REFERENCES

[1] Yantao Song and Bingsen Wang, "Survey on Reliability of Power Electronic Systems," *IEEE Trans. Power Electron.*, vol. 28, no. 1, pp. 591 – 604, Jan. 2013.

[2] Bin Lu, and Santosh K. Sharma, "A Literature Review of IGBT Fault Diagnostic and Protection Methods for Power Inverters," *IEEE Trans. Ind. Appl.*, Vol. 45, No. 5, pp. 1770-1777, Oct. 2009.

[3] K. Rothenhagen and F. W. Fuchs, "Performance of diagnosis methods for IGBT open circuit faults in three phase voltage source inverters for ac variable speed drives," in *Proc. Eur. Power Electron. Appl. Conf.*, 2005, pp. 1–10.

[4] W. Sleszynski, J. Nieznanski, and A. Cichowski, "Open-Transistor Fault Diagnostics in Voltage-Source Inverters by Analyzing the Load Currents," *IEEE Trans. Ind. Electron.*, vol. 56, no. 11, pp. 4681–4688, Nov. 2009.

[5] Shin-Myung Jung, Jin-Sik Park, Hag-Wone Kim, Kwan-Yuhl Cho, and Myung-Joong Youn, "An MRAS-Based Diagnosis of Open-Circuit Fault in PWM Voltage-Source Inverters for PM Synchronous Motor Drive Systems," *IEEE Trans. Power Electron.*, vol. 28, no. 5, pp. 2514-2526, May 2013.

[6] C. Kral and K. Kafka, "Power electronics monitoring for a controlled voltage source inverter drive with induction machines," *IEEE Power Electron. Specialists Conf.*, vol. 1, pp. 213-217, 2000.

[7] J. O. Estima and A. J. M. Cardoso, "A New Algorithm for Real-Time Multiple Open-Circuit Fault Diagnosis in Voltage-Fed PWM Motor Drives by the Reference Current Errors," *IEEE Trans. Ind. Electron.*, vol. 60, no. 8, pp. 3496-3505, Aug. 2013.

[8] M. Shahbazi, P. Poure, S. Saadate, M. R. Zolghadri, "FPGA-based Fast Detection with Reduced Sensor Count for a Fault-Tolerant Three-Phase Converter," *IEEE Trans. Ind. Informatics*, Volume: PP , Issue: 99, 2012.

[9] M. Shahbazi, E. Jamshidpour, P. Poure, S. Saadate, M. Zolghadri, "Open and Short-Circuit Switch Fault Diagnosis for Non-Isolated DC-DC Converters Using Field Programmable Gate Array," *IEEE Trans. Ind. Electron.*, vol. 60, no. 9, pp. 4136-4146, Sep. 2013.

[10] Q.-T. An, L.-Z. Sun, K. Zhao and L. Sun, "Switching function model based fast-diagnostic method of open-switch faults in inverters without sensors," *IEEE Trans. Power Electrons*, vol. 26, no. 1, pp.119-126, Jan. 2011.

[11] S. Karimi, P. Poure, and S. Saadate, "Fast power switch failure detection for fault tolerant voltage source inverters using FPGA," *IET Power Electron.*, vol. 2, no. 4, pp. 346–354, Jul. 2009.

[12] S. Karimi, A. Gaillard, P. Poure, and S. Saadate, "FPGA-Based Real-Time Power Converter Failure Diagnosis for Wind Energy Conversion Systems," *IEEE Trans. on Ind. Electron.*, vol. 55, no. 12, pp: 4299-4308, Dec. 2008.

[13] S. Karimi, P. Poure, S. Saadate, "FPGA-based fully digital fast power switch fault detection and compensation for three-phase shunt active filters," Electric Power Systems Research 78 (2008) 1933–1940.

[14] Jorge O. Estima and Antonio J. Marques Cardoso, "A New Approach for Real-Time Multiple Open-Circuit Fault Diagnosis in Voltage-Source Inverters," *IEEE Trans. Ind. Appl.*, vol. 47, no. 6, pp. 2487-2494, Dec. 2011.

[15] Nuno M. A. Feire, J. O. Estima, and A. J. M. Cardoso, "Open-Circuit Fault Diagnosis in PMSG Drives for Wind Turbine Applications," *IEEE Trans. Ind. Electron.*, vol. 60, no. 9, pp. 3957-3967, Sep. 2013.

[16] L. Parsa and H. A. Toliyat, "Fault-tolerant five-phase permanent-magnet motor drives," *IEEE Trans. Ind. Appl.*, vol. 41, no. 1, pp. 30-37, Feb. 2005.

[17] P. Zheng, Y. Sui, J. Zhao, C. Tong, T.A. Lipo, A. Wang, "Investigation of a Novel Five-Phase Modular Permanent-Magnet In-Wheel Motor," *IEEE Trans. Magnetics*, vol. 47, no. 10, pp. 4084-4087, Oct. 2011.

[18] M. Salehifar, R. Salehi Arashloo, M. Moreno-Eguilaz, V. Sala, L. Romeral, "Fault Tolerant Operation of a Five Phase Converter for PMSM Drives," *Applied Power Electronics Conference and Exposition (APEC), 2013 Twenty-Eighth Annual IEEE* , pp. 1177 – 1184, 2013.

[19] R. Salehi Arashloo, M. Salehifar, L. Romeral, "On the Effect of Accessible Neutral Point in Fault Tolerant Five Phase PMSM Drives," *IECON 2012 - 38th Annual Conference on IEEE Industrial Electronics Society*, pp. 1934 – 1939, 2012.

Efficiency Optimization on Vector Controlled Six-Phase Induction Motor in Healthy and Faulted Mode

M. Moghadasian, A. Sivert, A. Yazidi, F. Betin, G.A. Capolino

Abstract – This paper introduces a new technique for efficiency optimization of 6-phase induction motor (6PIM) drive, based on fuzzy online search control (SC) method. The modified SC algorithm does not require extra hardware and is insensitive to motor parameters. The results are compared to classic loss model control (LMC) efficiency optimization algorithm in healthy and faulted operational mode due to up to three stator open phases. The experimental results obtained on a dedicated test-bed validate the efficiency of the proposed method.

Index Terms—AC motor drives, multiphase induction machine, adaptive control, fuzzy control, loss model control, optimization methods, search control.

I. NOMENCLATURE

$\|X\|$	Euclidian norm of vector X
X^{T}	transpose of vector X
V	instantaneous phase voltage vector
I	instantaneous phase current vector
v	instantaneous phase voltage
i	instantaneous phase current
I_m	rated *rms* phase current
φ	linkage flux
r	phase resistance
l	phase inductance
M	mutual inductance
P	number of pole pairs
J	total shaft inertia
F	friction coefficient
K	stiffness coefficient
Ω	rotor mechanical shaft speed
ω_s	stator *mmf* angular speed
ω_r	rotor electrical angular speed ($P.\Omega$)
θ_r	mechanical rotor shaft position
T_m	electromagnetic torque
T_L	load torque
T_r	machine equivalent rotor time constant ($T_r=l_r/r_r$)

II. INTRODUCTION

INDUCTION motors (IMs) are widely used in the industry, serving as one of the most important roles during the energy conversion between electrical power and mechanical power. Especially the 3 phase induction machines (3PIM) with squirrel cage rotors are the dominant types due to the simple structures, easy connections, robust to severe operating conditions, low costs, and maintenance free features. They are used in drive fans, pumps, compressors, power tools, mills, elevators, cranes, electrical vehicles, ship applications, etc.

Nowadays, multi-phase induction machines are more and more substituting 3PIM for industrial applications such as automotive, aerospace, military and nuclear where high reliability and safety are required [1]. In the multiphase drive systems, the electrical machine presents more than three phases on the stator side and the same number of branches is required for the inverter. Therefore, the main advantage of multiphase machines compared to classical three-phase structures is that they are still operating even in a particular faulted mode resulting of the loss of excitation of one or more stator phases. One particular multiphase machine configuration that is often used for reliable positioning tasks or variable speed in industrial applications is the symmetrical six-phase induction machine (6PIM) [2]. As said just before, a 6PIM is still operating when up to three phases are lost.

Now that safety and reliability of the drive have been solved by using more than three phase, a new challenge for scientists consists in reducing the consumption of the machine. Indeed, induction machines, whatever the number of phases, consume most of the world's electrical energy every year. Improving efficiency of electrical drives is important not only for energy saving, but also for environmental protection. It is well-known that induction motors are high efficiency electrical machines when working closed to rated torque and speed; however, for IMs working below the full load value (rated value), rated flux causes excessive core loss and thereby reduces the efficiency of the system [3].

Since it is not possible to optimize the motor efficiency for every operating point by optimizing machine design, energy savings by reducing operating loss can be obtained with optimal control strategies [4]. The operating loss in an induction motor includes: stator and rotor copper losses, core

---

M. Moghadasian, A. Sivert, A. Yazidi and F. Betin are with the Department of Electrical Engineering - University of Picardie "Jules Verne", Amiens, France, (e-mails: {mahmood.moghadasian, arnaud.sivert, amine.yazidi, franck.betin, gerard.capolino}@u-picardie.fr).

losses, and mechanical losses. Under light loads, motor efficiency decreases due to an imbalance between the copper and the core losses. Hence, energy savings can be achieved by proper selection of the flux level in the motor.

In optimal control, there are two main approaches to improve the induction motor efficiency at light loads, namely loss model control (LMC) [5], and search control (SC) [3]. LMC determines the optimal air gap flux through the motor loss model while SC uses the input power of the drive and searches for optimum flux or excitation current.

The goal of this work is to compare both mentioned optimal control techniques to reach higher efficiency for 6PIM in healthy and faulted mode. After presentation of LMC method, an integrated fuzzy-logic-based control method is developed to perform SC algorithm. Since measurement of 6PIM input power needs additional hardware, the conventional fuzzy-SC method [3] could not be experimentally performed in our laboratory set-up. Therefore, new variables are selected to perform the proposed SC algorithm.

Indeed, the modified SC algorithm does not require extra hardware and is furthermore insensitive to motor parameters variations. The experimental results of both efficiency optimization algorithms are presented for 6PIM healthy and faulted operational mode.

III. HEALTHY AND FAULTED MODELS OF THE 6PIM

The stator windings of our 6PIM are uniformly distributed with a symmetrical phase-shift of $\gamma = 60°$ in between two consecutive phases as shown in Fig. 1. The different circuit elements associated to the stator and the rotor can be classified in term of their dependence or independence on the rotor position. The conductor resistances, self inductances of stator turns and rotor meshes, mutual inductances between stator turns and between rotor meshes can be considered as constant. It can be seen that mutual inductances between stator turns and rotor meshes are depending on the relative position between stator and rotor. For the time-domain analysis, the variation of conductor resistances due to the temperature changes is not taken into account [2].

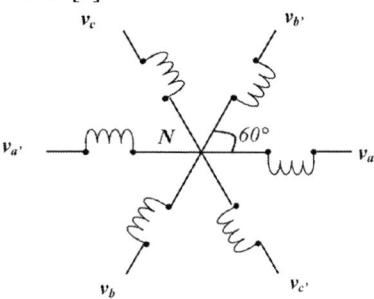

Fig. 1. Stator windings distribution of 6PIM

A. Healthy 6PIM Model

Model of 6-phase induction motor is derived based on the following assumptions: the machine windings are distributed

sinusoidaly; the mutual leakage inductances are negligible; the magnetic saturation and the core losses are neglected.

With these assumptions, the voltage equations for stator and rotor of the 6PIM can be written as :

$$[V_s] = [R_s].[i_s] + s([L_{ss}].[i_s] + [L_{sr}].[i_r]) \tag{1}$$

$$[V_r] = [R_r][i_r] + s([L_{rr}].[i_r] + [L_{rs}].[i_s]) \tag{2}$$

By applying the following transformation matrix $[T_6]$ as defined in [2], the 6PIM can be decomposed into three two-dimensional orthogonal subspaces: (α,β), (z_1,z_2) and (z_3, z_4) [2]:

$$[T_6] = \frac{1}{\sqrt{3}} \begin{bmatrix} \cos(0) & \cos\left(\frac{2\pi}{3}\right) & \cos\left(\frac{4\pi}{3}\right) & \cos(\pi) & \cos\left(\frac{5\pi}{3}\right) & \cos\left(\frac{\pi}{3}\right) \\ \sin(0) & \cos\left(\frac{2\pi}{3}\right) & \cos\left(\frac{4\pi}{3}\right) & \cos(\pi) & \cos\left(\frac{5\pi}{3}\right) & \cos\left(\frac{\pi}{3}\right) \\ \cos(0) & \cos\left(\frac{4\pi}{3}\right) & \cos\left(\frac{2\pi}{3}\right) & \cos(0) & \cos\left(\frac{4\pi}{3}\right) & \cos\left(\frac{2\pi}{3}\right) \\ \sin(0) & \sin\left(\frac{4\pi}{3}\right) & \sin\left(\frac{2\pi}{3}\right) & \sin\left(\frac{4\pi}{3}\right) & \sin\left(\frac{4\pi}{3}\right) & \sin\left(\frac{2\pi}{3}\right) \\ \frac{1}{\sqrt{2}} & \frac{1}{\sqrt{2}} & \frac{1}{\sqrt{2}} & \frac{1}{\sqrt{2}} & \frac{1}{\sqrt{2}} & \frac{1}{\sqrt{2}} \\ \frac{1}{\sqrt{2}} & \frac{1}{\sqrt{2}} & \frac{1}{\sqrt{2}} & \frac{-1}{\sqrt{2}} & \frac{-1}{\sqrt{2}} & \frac{-1}{\sqrt{2}} \end{bmatrix} \tag{3}$$

Then, with this transformation, the 6PIM field oriented control (FOC) is similar to the classical 3PIM one:

$$[T_6][V_s] = [T_6][R_s][T_6]^{-1}[T_6][I_s] + p([T_6][L_{ss}][T_6]^{-1}[T_6][I_s] + [T_6][L_{sr}][T_6]^{-1}[T_6][I_r]) \tag{4}$$

$$[0] = [T_6][R_r][T_6]^{-1}[T_6][I_r] + p([T_6][L_{rr}][T_6]^{-1}[T_6][I_r] + [T_6][L_{rs}][T_6]^{-1}[T_6][I_s]) \tag{5}$$

The components of the stator current in the rotating reference frame $dq0$ are obtained by applying the following transformation matrix $[T_2]$ ($\theta_s = \omega_s t$):

$$[T_2] = \begin{bmatrix} \cos(\theta_s) & \sin(\theta_s) \\ -\sin(\theta_s) & \cos(\theta_s) \end{bmatrix} \tag{6}$$

By assuming that the quadratic component is equal to zero on the contrary of the direct component, it leads to the basic field oriented control (FOC) equation $\varphi = \varphi_r = \varphi_{dr}$. Consequently, after these transformations, the 6PIM equations in a rotating reference frame can be written as [2]:

$$\begin{cases} v_{sd} = \sigma l_s s i_{sd} + \left(r_s + r_r \frac{M^2}{l_r^2}\right) i_{sd} - \sigma l_s \omega i_{sq} - \frac{M}{T_r l_r}\phi \\ v_{sq} = \sigma l_s s i_{sq} + \left(r_s + r_r \frac{M^2}{l_r^2}\right) i_{sq} + \sigma l_s \omega i_{sd} - \frac{M}{T_r l_r}\phi \\ J s \omega_r = \frac{M}{l_r} i_{sq} \phi - F \omega_r - T_L \\ s\phi = -\frac{1}{T_r}\phi + \frac{M}{T_r} i_{ds} \end{cases} \tag{7}$$

where

$$\begin{cases} \sigma = 1 - \frac{M^2}{l_s l_r} \\ T_r = \frac{l_r}{r_r} \end{cases} \tag{8}$$

B. Faulted 6PIM Model in the $\alpha\beta0$ Reference Frame

In this part, the 6PIM models with up to three open phases are given. In order to simplify the study, the case of one phase s_a open is under focus and the other cases will be built starting from this simple formulation. Whatever the open phases are, the stator and rotor voltage equations can be written as (1) and (2). When the phase s_a is opened, the current and voltage vectors are:

$$
\begin{cases}
[V_s] = [v_{sc}, v_{sb}, v_{sa'}, v_{sc'}, v_{sb'}]^T \\
[I_s] = [i_{sc}, i_{sb}, i_{sa'}, i_{sc'}, i_{sb'}]^T \\
[V_r] = [0 \quad 0 \quad 0 \quad 0 \quad 0 \quad 0]^T \\
[I_r] = [i_{rc}, i_{rb}, i_{ra'}, i_{rc'}, i_{rb'}]^T
\end{cases}
\tag{9}
$$

The drawback of this model lies in the coupled nonlinear expressions of the inductance matrices. In order to obtain a decoupled model, it is shown in [6] that two transformation matrices are required. These matrices, so-called $[T_5]$ and $[T_6]$, split the original model into two decoupled subspaces: ($\alpha\beta$) subspace and (z) subspace. The electromechanical energy conversion takes place only in ($\alpha\beta$) subspace which means that the EMF produced by five stator phases is equivalent to the EMF produced by two windings on the axes α and β with the currents $i_{s\alpha}$ and $i_{s\beta}$ which are defined by:

$$
\begin{bmatrix} i_{s\alpha} \\ i_{s\beta} \end{bmatrix} = [T_c][I_s] \qquad where \ [T_c] = \begin{bmatrix} [\alpha]/\|\alpha\| \\ [\beta]/\|\beta\| \end{bmatrix}
\tag{10}
$$

$[\alpha]$ and $[\beta]$ are obtained by suppressing the components from $[\alpha_0]$ and $[\beta_0]$, corresponding to the open phase(s). In healthy mode (no cut-off stator phases), $[\alpha_0]$ and $[\beta_0]$ are defined as (11), which is shown at the bottom of the page [2], where ϑ_i, $i = 1$ to 6 contains the stator current phase angles, defined as $\vartheta = [0, \ 2\pi/3, \ 4\pi/3, \ \pi, \ 5\pi/3, \ \pi/3]$. ϑ_0 is fixed to obtain two orthogonal vectors $[\alpha]$ and $[\beta]$ which is defined as:

$$
\vartheta_0 = \frac{1}{2} tan^{-1} \left(\frac{\sum_j sin(2\vartheta_i)}{\sum_j cos(2\vartheta_i)} \right)
\tag{12}
$$

for all j belonging to the set of active phases. For example,

when s_a is opened ($j = 2$ to 6), (12) gives $\vartheta_0 = 0$ and then, $[\alpha]$ and $[\beta]$ become as (13) which is presented below.

C. Decoupled Model

The simplified 6PIM model (7) contains cross-coupling terms between d and q components. In order to obtain a decoupled model, it is mandatory to remove these cross-coupling terms. A straightforward way is to use the following emf_d and emf_q quantities (equivalent to voltages):

$$
\begin{cases}
emf_d = \sigma l_s \omega i_{sq} + \frac{M}{T_r l_r} \phi \\
emf_q = -\sigma l_s \omega i_{sd} + \frac{M}{T_r l_r} \phi
\end{cases}
\tag{14}
$$

By adding these quantities to the nonlinear system (7), the decoupled model can be obtained as:

$$
\begin{cases}
v_{sd_1} = \sigma l_s s i_{sd} + \left(r_s + r_r \frac{M^2}{l_r^2} \right) i_{sd} \\
v_{sq_1} = \sigma l_s s i_{sq} + \left(r_s + r_r \frac{M^2}{l_r^2} \right) i_{sq}
\end{cases}
\tag{15}
$$

IV. CONVENTIONAL OPTIMIZATION CONTROL TECHNIQUES

Optimum control of IMs is an essential issue, because it is not possible to optimize the motor efficiency for every operating point by optimizing machine design. In many applications of constant speed operation, induction motor operate under partial load for prolong periods, such as spinning drive in textile industry [3], mine hoist load, drill presses and wood saw. In these applications, IM should operate at reduced flux that causes a balance between iron losses and copper losses and improves efficiency.

Many minimum-loss control schemes have been reported in the literature. These schemes can be separated into two categories: *Loss Model Control* (LMC) and *Search Control* (SC). The LMC method based on the IM loss minimization presents the advantage that it is simple and fast. However, this method requires accurate knowledge of motor parameters, which may change over the motor's operating temperature range and also vary from motor to motor. The SC method measures input power or DC bus power to search the flux where the motor runs at maximum efficiency. This approach is insensitive to motor parameters, but it does require extra hardware to measure DC bus current and voltage and cannot be used in the vector control system where additional sensor is not available.

A. Loss Model Control

The total loss of an IM consists of stator and rotor copper

$$
\begin{cases}
[\alpha_0] = [cos(\vartheta_0 + \vartheta_1) \quad cos(\vartheta_0 + \vartheta_2) \quad cos(\vartheta_0 + \vartheta_3) \quad cos(\vartheta_0 + \vartheta_4) \quad cos(\vartheta_0 + \vartheta_5) \quad cos(\vartheta_0 + \vartheta_6)]^T \\
[\beta_0] = [sin(\vartheta_0 + \vartheta_1) \quad sin(\vartheta_0 + \vartheta_2) \quad sin(\vartheta_0 + \vartheta_3) \quad sin(\vartheta_0 + \vartheta_4) \quad sin(\vartheta_0 + \vartheta_5) \quad sin(\vartheta_0 + \vartheta_6)]^T
\end{cases}
\tag{11}
$$

$$
\begin{cases}
[\alpha] = [cos(2\pi/3) \quad cos(4\pi/3) \quad cos(\pi) \quad cos(5\pi/3) \quad cos(\pi/3)] \\
[\beta] = [sin(2\pi/3) \quad sin(4\pi/3) \quad sin(\pi) \quad sin(5\pi/3) \quad sin(\pi/3)]
\end{cases}
\tag{13}
$$

losses P_{cu}, core losses P_{fe} and mechanical losses P_m as expressed in (16):

$$P_{loss} = P_{in} - P_{out} = P_{cu} + P_{fe} + P_m \qquad (16)$$

In the steady state, the stator and rotor copper losses are given by [7]:

$$P_{cu} = r_s\left(i_{sd}^2 + i_{sq}^2\right) + r_r\left(i_{rd}^2 + i_{rq}^2\right) \qquad (17)$$

The core losses include eddy current losses and hysteresis losses:

$$P_{fe} = k_h \omega_s \varphi_m^2 + k_e \omega_s^2 \varphi_m^2 \qquad (18)$$

where φ_m is the air-gap flux linkage, k_h and k_e are the hysteresis and eddy current loss coefficient. The air-gap flux linkage expression in terms of currents can be written as [7]:

$$\begin{cases} \varphi_{md} = M\left(i_{sd} + i_{rd}\right) \\ \varphi_{mq} = M\left(i_{sq} + i_{rq}\right) \end{cases} \qquad (19)$$

When the motor is running under the rotor flux field orientation, we have [7]:

$$\begin{cases} i_{rd} = 0 \\ i_{rq} = -\dfrac{M}{l_r} i_{sq} \end{cases} \qquad (20)$$

Therefore, the air-gap flux linkage equations (19) become:

$$\begin{cases} \varphi_{md} = M\, i_{sd} \\ \varphi_{mq} = M\left(1 - \dfrac{M}{l_r}\right) i_{sq} \end{cases} \qquad (21)$$

As a reasonable approximation, the mechanical losses are dependent on the rotor speed and can be expressed as (22):

$$P_m = k_m \omega_r^2 \qquad (22)$$

where k_m is mechanical loss coefficient. The electromagnetic torque of the 6PIM can be expressed by:

$$T_m = \frac{PM}{l_r}\left(\phi_{rd} i_{sq} - \phi_{rq} i_{sd}\right) = \frac{PM^2}{l_r} i_{sd} i_{sq} = K i_{sd} i_{sq} \qquad (23)$$

where $K = \dfrac{PM^2}{l_r}$.

In steady state, from (16) to (22), we can get the expression for total motor losses:

$$P_{loss} = r_s\left(i_{sd}^2 + i_{sq}^2\right) + r_r\left(i_{rd}^2 + i_{rq}^2\right) + \left(k_h \omega_e + k_e \omega_e^2\right)\times$$
$$\left[M^2 i_{sd}^2 + M^2\left(1 - \frac{M}{l_r}\right)^2 i_{sq}^2\right] + k_m \omega_r^2 \qquad (24)$$

Substituting (23) into (24), the total losses are:

$$P_{loss} = A\, i_{sd}^2 + \frac{B\, C_{em}^2}{K^2}\frac{1}{i_{sd}^2} + k_m \omega_r^2 \xrightarrow{P_m \approx 0} P_{loss} = f\left(i_{sd}\right) \quad (25)$$

where

$$\begin{cases} A = r_s + M^2\left(k_h \omega_e + k_e \omega_e^2\right) \\ B = r_s + r_r \dfrac{M^2}{l_r^2} + M^2\left(1 - \dfrac{M}{l_r}\right)^2\left(k_h \omega_e + k_e \omega_e^2\right) \end{cases} \qquad (26)$$

Equation (25) gives an expression of 6PIM losses in vector control system. It shows the relationship of motor losses with rotor flux (i_{sd}), motor torque and speed. It is obvious that if speed and torque are constant at each operation point, the total machine loss is just a function of i_{sd}. By taking the derivative of motor losses expression (25) with respect to i_{sd} and setting it to zero, the reference magnetizing current i_{sd}^* corresponding to the point of the minimum loss can be obtained as:

$$i_{sd}^* = \sqrt[4]{\frac{B T_m^2}{A K^2}} \qquad (27)$$

The above equation gives an expression for the exact minimum-loss rotor flux as a function of machine torque (T_m). The LMC control scheme for 6PIM is presented in Fig. 2.

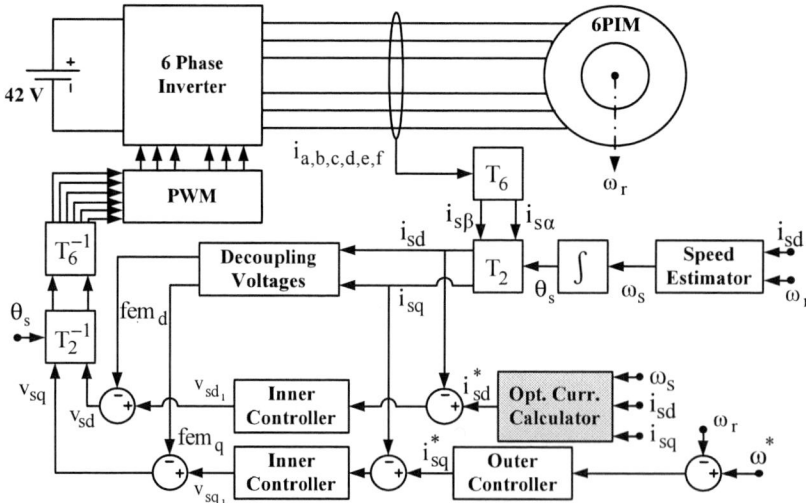

Fig. 2. Indirect vector control scheme of 6PIM with loss model control optimization algorithm

B. Search Control

Search controls (SCs) [3] use measured IM input power or dc-link power in the optimization process. For a given load torque and speed, at steady state, the flux is iteratively adjusted (normally reduced) until the point of minimum input power is reached. The controller searches for the operating point where the input power is at a minimum while keeping the motor output power constant.

Compared to LMC, the SCs are simpler to implement, do not depend on the knowledge of machine parameters, and consequently are insensitive to parameter variations. They are also guaranteed to yield the true optimal efficiency operation [3].

As the losses are parabolic functions of the flux, any nonlinear optimization search technique can be adapted to yield true optimum efficiency operation. The technique is particularly suitable to vector drives, where a natural decoupling of torque and flux control enables a simpler implementation of the search process. Initially, fixed-step sizes for magnetizing current were employed, and undesirable torque pulsations were observed, which were associated with the step changes in magnetizing currents. Fuzzy logic controllers were proposed to prevent torque pulsation [3]. Their main limitation is the relatively slow convergence speed, followed by the limited applicability, since the drive system must operate at a steady-state condition for most of the load cycle.

The conventional fuzzy SC method presented in [3] uses the dc-link power as a control variable which ensures that not only the IM losses, but also the overall drive system losses will be minimized. The input DC power is sampled and is compared with the previous value to determine the increment ΔP_{DC}. In addition, the last excitation reference current decrement (Δi_{sd}^*) is reviewed. On these basis, the decrement step of Δi_{sd}^* is generated from fuzzy rules through fuzzy inference system as shown in Fig. 3 [3]. The main problem to implement this algorithm is to add extra hardware to measure DC bus voltage and current and computing P_{DC}. To overcome this problem, P_{DC} should be replaced by (d,q) stator currents as described in the next section.

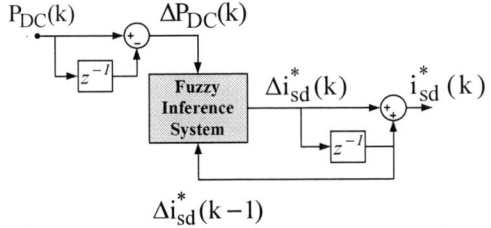

Fig. 3. Conventional fuzzy controller for d-axis reference stator current [3].

V. PROPOSED SEARCH CONTROL ALGORITHM

In this part, a fuzzy-logic-based energy optimizer operates on the basis of on-line optimization control, where the motor's air gap flux is decremented in steps until the measured input power reaches the lowest possible value. This type of control does not require knowledge of motor parameters; it is completely insensitive to parameter value changes; it is universally applicable to any arbitrary motor; no mathematical model is required; it can provide adaptive step sizes leading to fast convergence; and it can accept inaccurate and noisy signals.

A. Modified SC Scheme

The principle of efficiency optimization control with flux decrementation is illustrated in Fig. 4. The rotor air gap flux is decreased by reducing the magnetizing component of the stator current. This ultimately results in a corresponding increase in the torque component of the stator current by the action of the speed corrector, so that the developed torque and speed remain constant at the desired values. At the optimized operating point, as the air gap flux decreases in steps, the iron core loss decreases as well. However, due to the rise in the torque component of the stator current, the copper (resistive) loss increases and the minimum input power level is reached. The search is operating until the system reaches this minimum input power point.

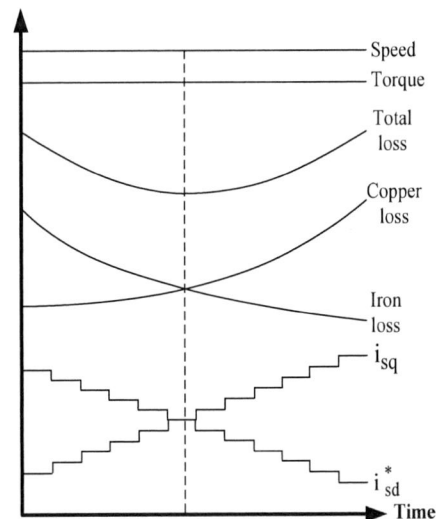

Fig. 4. Principle of efficiency optimization control by stator current

Based on the principle above, the fuzzy controller is designed as in Fig. 5. The fuzzy controller uses stator current change Δi_s and d-axis current change Δi_{sd} as its inputs:

$$\Delta i_s(k) = i_s(k) - i_s(k-1), \quad where \quad i_s = \sqrt{i_{sd}^2 + i_{sq}^2} \quad (28)$$

$$\Delta i_{sd}(k) = i_{sd}(k) - i_{sd}(k-1)$$

978-1-4799-0024-4/13 $31.00 © 2013 IEEE   473

The output is d-axis stator reference current increment Δi_{sd}^* and the d-axis stator reference current is calculated from:

$$i_{sd}^*(k) = \Delta i_{sd}^*(k) + i_{sd}^*(k-1) \qquad (29)$$

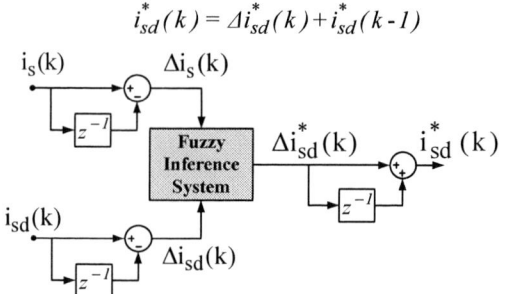

Fig. 5. Modified fuzzy controller for d-axis reference stator current

B. Fuzzy Controller

Over the past few years, the use of fuzzy set theory, in process control systems has found wide popularity. Today, fuzzy logic-based control systems, or simply fuzzy logic controllers (FLCs), can be found in a growing number of applications.

In the implementation of fuzzy control, standard procedures were utilized: input variables were fuzzified by triangular membership functions; the fuzzy control rules were evaluated by a rule base table; the outputs were combined by fuzzy composition, and then the result is defuzzified. The membership functions for fuzzy inputs and output are shown in Figs. 6 to 8. All membership functions are triangular for simplicity. The fuzzy output is calculated using COG (center of gravity) defuzzification approach. In order to make the control system shorter and faster than a program that reinterprets the rules at each control cycle, fixed lookup table has been used [8].

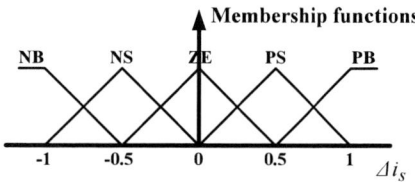

Fig. 6. Fuzzy set membership functions for input Δi_s

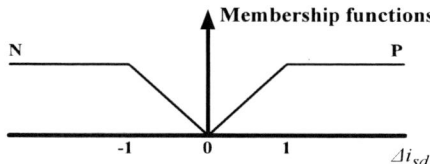

Fig. 7. Fuzzy set membership functions for input Δi_{sd}

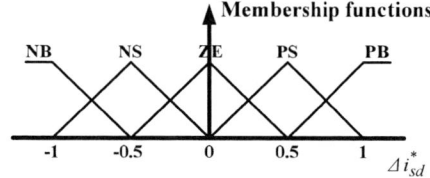

Fig. 8. Fuzzy set membership functions for output Δi_{sd}^*

The rule base for fuzzy control is given in table I. The basic idea is that if the last control action indicated a decrease of stator current, the search proceeds in the same direction. In case the last control action resulted in an increase of stator current, the search direction is reversed. For example, IF is $\Delta i_s = NS$ AND $\Delta i_{sd} = N$, THEN $\Delta i_{sd}^* = NS$. This rule means that IF the stator current increment Δi_s is negative small (NS) and the last d-axis current Δi_{sd} is negative (N), THEN new excitation current increment Δi_{sd}^* is negative small (NS).

TABLE I

FUZZY RULE TABLE

Δi_{sd} / Δi_s	N	P
PB	PB	NB
PS	PS	NS
ZE	ZE	ZE
NS	NS	PS
NB	NB	PB

VI. EXPERIMENTAL RESULTS

To validate the efficiency of the proposed method, an experimental setup has been developed. It is composed of a 6PIM supplied by two three-phase voltage source inverters with MOSFET as power switches whose dc link voltage is 42 V. The switching frequency is set at 10 kHz using the classical sampled natural PWM technique generated by an FPGA. The digital control board has been built around an Intel-Pentium 4 processor allowing easily an actual sampling period of 100μs. This board receives the stator current data through six 12-bit A/D converters with a maximum frequency of 20 kHz and the dc bus voltage with a frequency of 12.8 kHz. The speed is measured through a 4096-points encoder. The experimental setup is shown in Fig. 9 and the 6PIM drive main parameters are given in Table II. To open phase s_a, the corresponding VSI winding connection is removed.

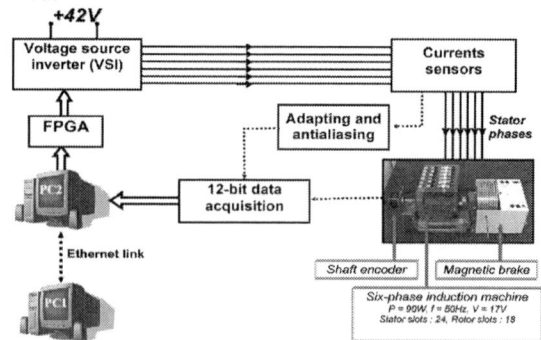

Fig. 9. Experimental setup

978-1-4799-0024-4/13 $31.00 © 2013 IEEE

A. LMC Experimental Results

In order to evaluate the performance of the LMC algorithm, the 6PIM system was set up. The test was implemented at speed $\omega_r = 200\ rad/sec$ and load torque $T_l = 0.2\ N.m$. Figs. 10 and 11 show the experimental results with LMC when zero to three stator phases are lost. The optimization process in all the cases is similar as Fig. 4; when the LMC begins at $t = 3(s)$, i_{sd}, rotor flux and core losses decrease, and i_{sq} and copper loss increase. Therefore, total 6PIM loss and input power decrease. Due to proper act of vector control system, the output power remains constant during all applications. Consequently, the efficiency of the 6PIM increases.

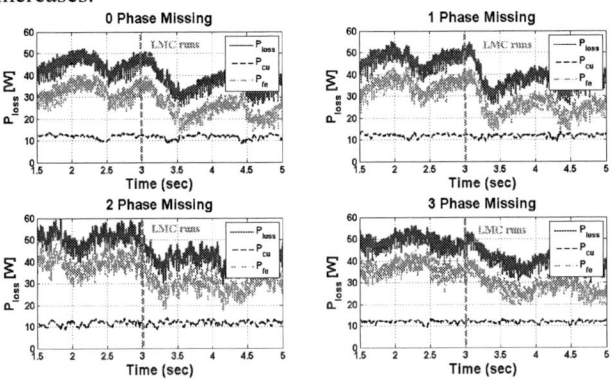

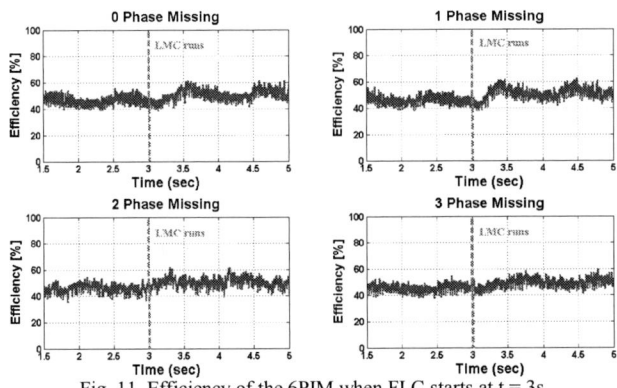

Fig. 10. Copper loss (P_{cu}), core loss (P_{fe}) and total loss (P_{loss}) of the 6PIM when FLC starts at t = 3s

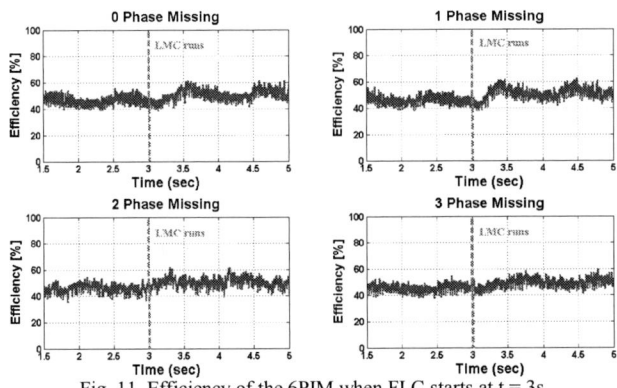

Fig. 11. Efficiency of the 6PIM when FLC starts at t = 3s

B. Modified SC Experimental Results

Figures 12 and 13 show the experimental results for applying fuzzy SC in healthy and three faulted modes same conditions as the previous section).

Comparing Figs. 11 and 13, it could be concluded that the modified SC method presents better performance rather than the LMC in increasing the efficiency of the 6PIM in various operational modes. The effects of the both implemented control algorithms on 6PIM drive parameters are summarized in table II. The table shows the approximated values of the 6PIM parameters. Using LMC, the efficiency of the 6PIM increases by 6%, while modified SC algorithm improves it by 22% in the best case (healthy mode).

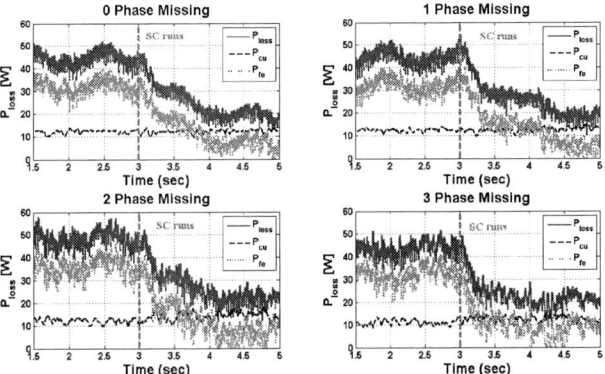

Fig.12. Copper loss (P_{cu}), core loss (P_{fe}) and total loss (P_{loss}) of the 6PIM when Fuzzy SC starts at t = 3s

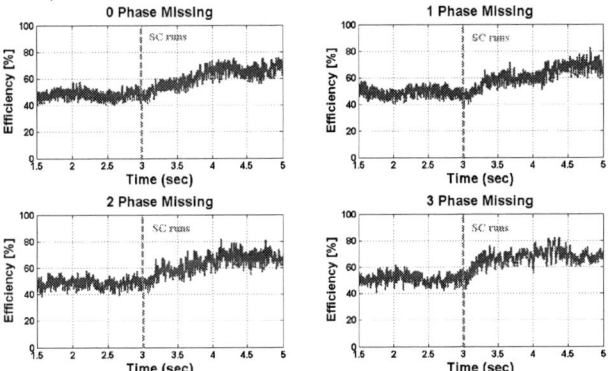

Fig. 13. Efficiency of the 6PIM when Fuzzy SC starts at t = 3s

VII. CONCLUSION

Two efficiency optimization methods for 6PIM vector control drive are presented in this chapter. The loss model control (LMC) is based on the IM loss minimization and it is simple and fast. However, this method requires accurate knowledge of motor parameters, which may change over the motor operating temperature range and also vary from motor to motor. The conventional search control (SC) method measures input power or DC bus power to find the flux that implies the motor to run at maximum efficiency. This approach is insensitive to motor parameters, but it does require extra hardware to measure DC bus current and voltage and cannot be used in vector control system, where additional sensor is not available. To overcome this problem, a new fuzzy logic system is suggested, which does not require extra hardware and is insensitive to motor parameters change. The fuzzy controller adjusts magnetizing current based on the dq stator currents, thus yielding true optimum efficiency operation with fast convergence. The relationship between magnetizing current and motor losses minimization in the 6PIM vector control system is investigated for all 6PIM control modes. The experimental results show that the modified SC algorithm is more effective and higher efficiency can be achieved using the fuzzy-SC method comparing with the LMC algorithm.

VIII. REFERENCES

[1] E. Levi, R. Bojoi, F. Profumo, H.A. Toliyat and S. Williamson, Multiphase induction motor drives – a technology status review, IET Electr. Power Appl., 2007, 1, (4), pp. 489–516.

[2] R. Kiani-Nezhad, B. Nahidmobarakeh, L. Baghli, F. Betin, and G. A. Capolino, Modeling and control of six-phase symmetrical induction machines under fault condition due to open phases, IEEE Trans. Ind. Electron., vol. 55, no. 5, pp. 1966–1977, May 2008.

[3] G. C. D. Sousa, B. K. Bose, and J. G. Cleland, Fuzzy Logic Based On-Line Efficiency Optimization Control of an Indirect Vector-Controlled Induction Motor Drive, IEEE Trans. Indus. Elec., 1995.

[4] S.S. Sivaraju, F.J.T.E. Ferreira, N. Devarajan, Genetic algorithm based design optimization of a three-phase multiflux Induction Motor, XXth International Conference on Electrical Machines (ICEM) ,2012.

[5] E. Poirier, M. Ghribi, and A. Kaddouri, Loss Minimization Control of Induction Motor Drives Based on Genetic Algorithms, Electric Machines and Drives Conference, MA, U.S.A., 2001, pp. 475-478.

[6] Y. Zhao, T.A. Lipo, Space vector PWM control of dual three-phase induction machine using vector space decomposition, IEEE Transactions on Industry Applications., Vol. 31, No.5, 1995.

[7] P. Vas, Sensorless Vector and Direct Torque Control, Oxford, U.K.: Oxford Univ. Press, 1998.

[8] D. A. Rutherford and G. C. Bloore, The implementation of fuzzy algorithms for control, Proc. IEEE, vol. 64, no. 4, , Apr. 1976.

TABLE II

6PIM PARAMETERS

Rated power	90 W
Rated torque	0.3 N.m
VSI DC source voltage	42 V
No. of poles	2
Mutual inductance	30.9 mH
Stator resistance	1.04 Ω
Stator leakage inductance	0.3 mH
Rotor resistance	0.64 Ω
Rotor leakage inductance	0.65 mH
Friction coefficient	4×10^{-4} N.m/rd/s
Inertia coefficient (J_{nom})	9.5×10^{-5} Kg.m^2

TABLE III

EFFECTS OF OPTIMIZATION ALGORITHMS ON 6PIM DRIVE PARAMETERS (APPROXIMATED VALUES)

Parameter	Conventional FOC				Loss minimization control				Search control			
	0 P. miss.	1 P. miss.	2 P. miss.	3 P. miss	0 P. miss.	1 P. miss.	2 P. miss.	3 P. miss	0 P. miss.	1 P. miss.	2 P. miss.	3 P. miss
P_{in} (W)	85	85	90	90	75	76	80	80	58	59	65	65
P_{out} (W)	40	40	40	40	40	40	40	40	40	40	40	40
P_{loss} (W)	45	45	50	50	35	36	40	40	18	19	25	25
P_{fe} (W)	33	33	38	38	22	23	27	27	4	5	9	9
P_{cu} (W)	12	12	12	12	13	13	13	13	14	14	16	16
i_{sd} (A)	2.7	2.7	2.7	2.7	2.6	2.5	2.4	2.4	1.9	1.9	1.9	1.9
i_{sq} (A)	2.9	2.9	3.3	3.3	3	3	3.4	3.4	3.5	3.5	4.1	4.1
η $(\%)$	47	47	44	44	53	53	50	50	69	68	61	61
$\Delta\eta$ $(\%)$	---	---	---	---	6	6	6	6	22	21	17	17

Magnetic Optimization of a Fault-Tolerant Linear Permanent Magnet Modular Actuator for Shipboard Applications

M. Bortolozzi, C. Bruzzese, F. Ferro, T. Mazzuca, M. Mezzarobba, G. Scala, A. Tessarolo, and D. Zito

Abstract – **The reliability is a key requirement in electric actuators to be used for moving ship control surfaces. This paper addresses a fault-tolerant design for an innovative naval actuator based on an inverter-fed permanent magnet linear synchronous motor. Due to the highly non-conventional actuator concept, a detailed FEM modeling approach is presented in this paper for comparison of different design solutions. A fault tolerant modular stator structure is proposed. The modular winding can be designed based on an integer-slots or fractional-slots concept. The two solutions are compared in this paper, considering encumbrance and thrusting force performances. The linear motion can be reverted to rotary motion through a prismatic-rotoidal joint for coupling to on board ship steering gears used to drive control surfaces (rudders, stabilizing fins). A drive prototype has been built and is presently under testing to assess and refine the results of the FEM modeling presented in the paper.**

Index Terms—**Comparative design, fault tolerance, FEM, force density, fractional slots, integer slots, linear actuator, modularity, optimization, permanent magnets, steering gears.**

I. INTRODUCTION

PRESENTLY, hydraulic drives are widely used on both civil and military large vessels to actuate such loads as rudders, stabilizing fins, capstans, bow thrusters and other loads requiring low actuating speed and high torque. The problems affecting hydraulic actuators have been largely highlighted in [1], [2]. Oil leaks, encumbrance, and poor efficiencies and control are the weakest points [3]. Reference has been also made in [4] to the convenience of shifting from hydraulic to electric drive systems, which lead to significant space savings and efficiency improvements, along with large operation and maintenance benefits. As an alternative to conventional gearbox-based rotating motors [5], a gearless electric drive solution, employing an inverter fed permanent-magnet linear synchronous actuator (PMLSA) has been

presented in [2], [6], [7]. The proposed drive is based on a direct force-torque conversion through a rotary-prismatic joint, for direct actuation of steering gears (rudder stocks, ship stabilizing fins) [6]. The PMLSA provides large actuating force and torque, however the drive affordability and redundancy against faults must be also granted. The fault tolerance is an important enabling feature in critical-mission drives, as also required in the Navy's classification rules [8]. A highly modular structure is proposed in this paper to achieve fault-tolerance to stator faults and to inverter faults. Two winding design solutions with fractional-slots and integer-slots are compared for the modular stator, by using FEM analysis, on the basis of machine encumbrance and thrusting force features.

Section II describes the modular PMLSA concept. Section III describes the fractional- and integer-slot winding alternatives. Section IV shows a comparative magnetic optimization, carried out by an original procedure. Final remarks and ongoing and future work on a laboratory PMLSA prototype are addressed in the Conclusion.

II. GENERAL DESCRIPTION OF THE FAULT-TOLERANT DRIVE SYSTEM

A scheme of the proposed drive system is shown in Fig. 1. The drive is based on a modular linear motor whose stator is split into several (N) independent units, each supplied by a power converter. All converters may either share the same input DC bus or be supplied from independent sources. The permanent-magnet mover actuates the rotary load (e.g. a steering gear for ship control surface actuation) by means of a rotary-prismatic joint that accomplishes the required torque to force conversion [2].

The arrangement of the drive makes it extremely resilient with respect to faults. In fact, in presence of a faulty stator unit, this can be disconnected and even removed (Fig. 2) for subsequent maintenance or repair with no service interruption. Thanks to the fractional-slot 10-pole 12-slot design employed for each linear motor unit, the removal of a module implies a loss of exactly $1/N$ of drive force capability [9]. The electromagnetic design assures the decoupling between stator modules, so that a possible short circuit occurring in a module does not affect the others.

As a further reliability feature, the linear motor is designed with a steel and aluminum structure that is extremely simple and robust, in order to minimize the

---

☐Financial support for the research covered in the paper has been provided by the Italian Ministry of Defence.

M. Bortolozzi, M. Mezzarobba, and A. Tessarolo are with the Dept. of Engineering and Architecture, University of Trieste, Italy. E-mail: atessarolo@units.it.

C. Bruzzese, F. Ferro, and D. Zito are with the Dept. of Astronautical, Electrical, and Energy Engineering, University of Rome–Sapienza, Italy. E-mail:claudio.bruzzese@uniroma1.it.

T. Mazzuca is with the Italian Navy General Staff (MARISTAT), Ministry of Defence, Rome, Italy. E-mail: teresa_mazzuca@marina.difesa.it.

G. Scala is with the General Directorate of Naval Armaments (NAVARM), Ministry of Defence, Rome, Italy. E-mail: giorgio_scala@marina.difesa.it.

978-1-4799-0024-4/13 $31.00 © 2013 IEEE

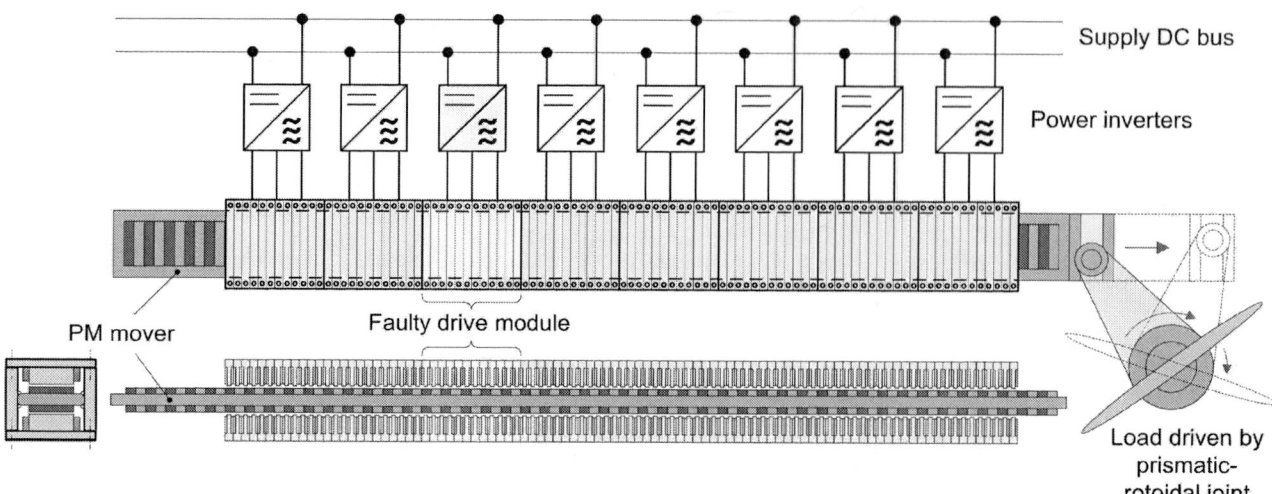

Fig. 1. Scheme of the modular linear motor drive (PMLSA).

probability of faults or malfunctioning relating to mechanical issues (fatigue, wear). The direct drive (gearless) arrangement, on the other side, removes all maintenance and reliability issues connected with gear-boxes [10]. Finally, compared to presently-used hydraulic actuators [11], the proposed drive solution exhibits the essential benefit of being totally oil-free and dramatically reducing the maintenance needs [1], [4].

However, very high force densities are needed for the proposed direct-drive application, since very high thrusting forces (tens of tons) must be generated in a limited space and weight. Due to the very low actuation speeds, the magnetic materials (ferritic steels) can be deeply saturated and exploited above ordinary limits. This requires a careful design and an ad-hoc magnetic optimization, for maximization of the machine force density. The stator back-iron has also structural function in the machine, so saving material and weight. Different stator solutions are considered in the following, also taking in account features such as structural simplicity and easy of machine repair.

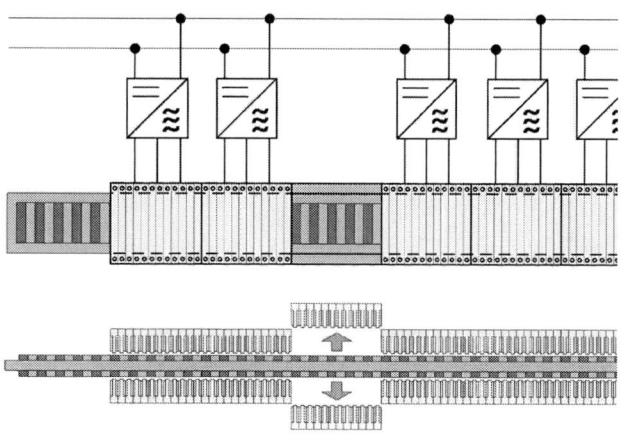

Fig. 2. Removal of a faulty module.

III. Alternatives Module Designs

A. Integer-Slot Winding Option

The maximum thrust per unit of air gap area is usually obtained through the simplest configuration of three-phase distributed winding with one slot per-pole and per-phase[12], [6]. An elementary module can contain as few as six teeth and two magnets as in Fig. 3 (design D_{6-2}). The MMF of a phase is a square wave, and both distribution and pitch factors are unit. However, the end-winding connections require a special design to obtain the desired modularity (the winding end-connections of neighboring modules cannot overlap), so the end connections must be put at least on three orders and they require more space and copper, with relevant joule losses.

B. Fractional-Slot Winding Option

A different winding configuration with fractional-slots is shown in Fig. 4. Due to the fractional slots, the magnetic field is periodic on the length covered by ten magnets, so a "module" is magnetically defined by twelve teeth, which must all be fed simultaneously. This design is referred to as D_{12-10}. However, the winding is built up of coils wound each on a tooth, so a tooth-module can be mechanically defined

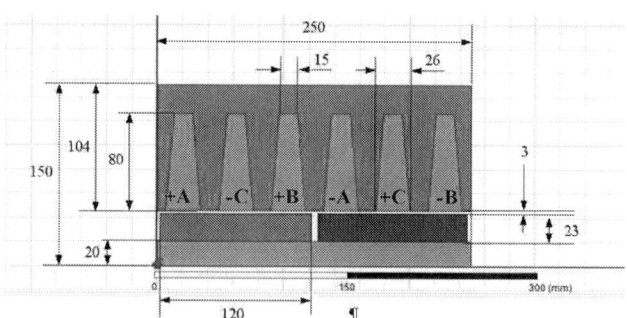

Fig. 3. Single stator module with six slots over two magnets (design D_{6-2}).

978-1-4799-0024-4/13 $31.00 © 2013 IEEE   478

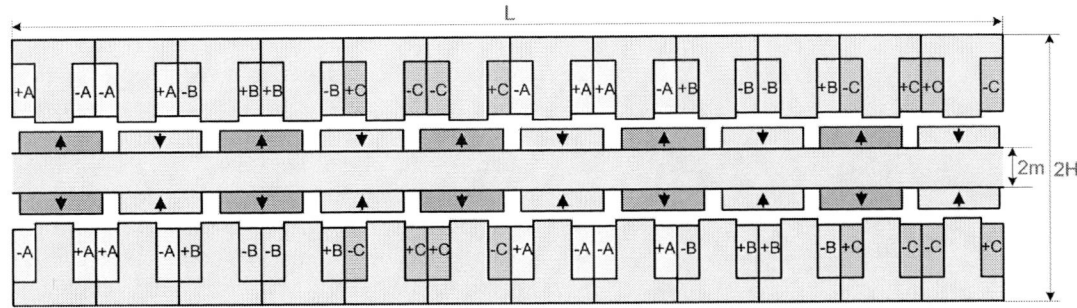

Fig. 4. Two paired stator modules (upper and lower module) with twelve slots over ten magnets (design 12-10, or D$_{12\text{-}10}$), and fractional-slot winding. Each stator module is made up of twelve removable tooth-modules.

as a single removable tooth with its coil. The front-end connections are much shorter than for the D$_{6\text{-}2}$ design, so copper, space, and joule losses are saved on the front-ends. This fact compensates for a smaller air gap force density, in fact the space saved can be used to design a larger module. Finally, the simplicity of construction of the fractional-slot design, the greater modularity and the easier module dismounting, substitution and reparation are the decisive features leading to the ultimate choice.

IV. MAGNETIC OPTIMIZATION

A. Optimization of the Fractional-Slot Design

The fractional-slot design in Fig. 4 has been optimized to obtain the maximum thrusting force per unit of active machine volume. The machine geometry has been completely parameterized as in Fig. 5. Fig. 6 shows the actual geometrical model in Cartesian coordinates. Table I reports the parameter starting values, preliminarily defined by using simplified sizing formulas [12]. Two different optimization algorithms have been used, i.e. the quasi-Newton (QN) algorithm and the sequential non-linear programming (SNLP) algorithm [6]. The mover eight m and the air gap g has been fixed due to mechanical safety reasons [8]. Both stator core and mover are designed as solid-steel pieces, i.e. no lamination is used in the machine. This is allowed due to the very low feeding frequencies [6], so that the iron losses (hysteresis and eddy losses) are neglectable. A special ferromagnetic material is used, i.e. high-strength

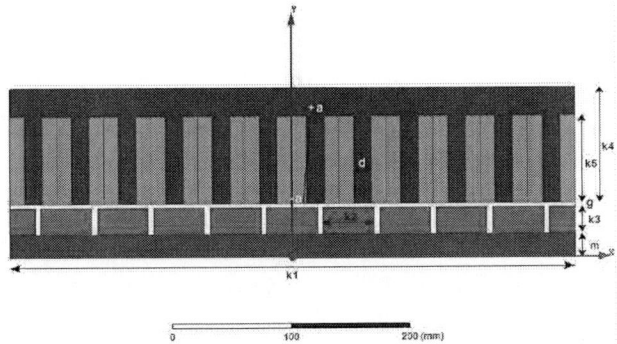

Fig. 5. Starting geometry for optimization of the fractional-slot design D$_{12\text{-}10}$.

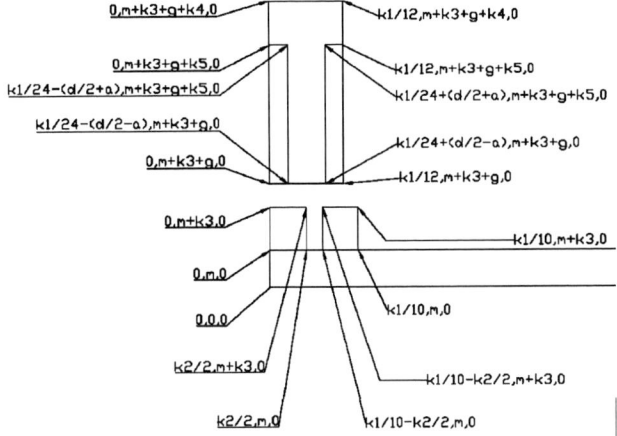

Fig. 6. Geometry definition through parameterized Cartesian coordinates.

TABLE I
STARTING PARAMETER VALUES FOR THE GEOMETRY D$_{12\text{-}10}$ (FIG. 5)

item	parameter	starting value
mover height	m (fixed)	20mm
module width	k_1	480mm
magnet width	k_2	43mm
magnet eight	k_3	23mm
air gap length	g (fixed)	3mm
stator eight	k_4	99mm
slot eight	k_5	75mm
tooth average width	d	$k_1/24$=20mm
stator yoke eight	k_4-k_5	24mm
tooth vertical semi-skew	a	0mm
rms current density	J	5A/mm^2
current angle	φ	0.6982rad
phase A curr. density	J_A	1.414Jcosφ
phase B curr. density	J_B	1.414Jcos(φ-2π/3)
phase C curr. density	J_C	1.414Jcos(φ-4π/3)
OVERALL LENGTH	k_1	**480mm**
OVERALL HEIGHT	m+k_3+g+k_4	**145mm**
Force$_x$/Volume	F_v	**632kN/m^3**

ferritic stainless steel. The relative magnetic permeability is very high, around 1500 in the average. The slot current density in Table I is fixed to a fairly high value, i.e. 5A/mm^2 for a supposed over-load operation with highly saturated iron. The phase angle φ is fixed to obtain the quadrature condition between the magnet field and the fifth harmonic of stator MMF. Seven parameters are freely varied during the automatic optimizations: a, d, k_1, k_2, k_3, k_4, k_5. The optimal geometries obtained are shown in Figs. 7, 8.

978-1-4799-0024-4/13 $31.00 © 2013 IEEE

Fig. 7. Flux density in the optimal geometry obtained by QN algorithm.

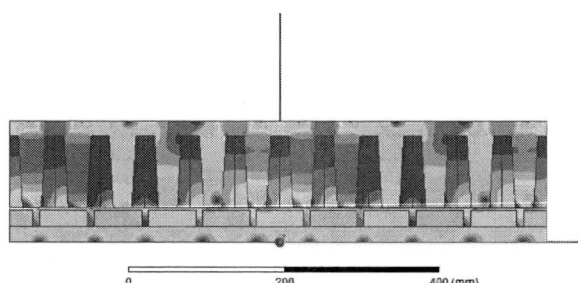

Fig. 8. Flux density in the optimal geometry obtained by SNLP algorithm (the color map is the same of Fig. 7. Also the geometric scale is the same, so that the geometrical proportions can be directly compared).

TABLE II
SIZES OF THE OPTIMAL GEOMETRIES IN FIGS. 7, 8 (DESIGN D_{12-10})

OPTIMAL SIZES (QN)				OPTIMAL SIZES (SNLP)		
	VALUE	UNIT			VALUE	UNIT
k1	539,72	mm		k1	696,23	mm
k2	46,47	mm		k2	60,61	mm
k3	22,29	mm		k3	21,35	mm
k4	86,59	mm		k4	111,6	mm
k5	73,86	mm		k5	91,92	mm
a	2	mm		a	2,6	mm
d	k1/24	mm		d	k1/24	mm
m	20	mm		m	20	mm
FORCE/VOLUME	751	kN/m³		FORCE/VOLUME	748	kN/m³

The two geometries in Figs. 7, 8 have very different sizes, althought they have similar shape. The geometry in Fig. 8 is 30% longer and 18% taller than that in Fig. 7. However, the two geometries feature very similar force densities, as shown in Table II. Many attempts were done, using different starting geometries, other optmization tools (e.g., Mode-Frontier), and also using different FEM softwares (Maxwell, FEMM). However, all the attempts finally led to different optimal geometries with similar thrust force densities. An interpretation for this fact is given below.

B. Geometrical Stretching Method (GSM)

As well known, local minima are a trouble for many optimization problems. Usually a cost function is used for the optimization, which must be minimized (here the cost function is the specific volume per unit of thrusting force). Gradient-based methods in particular suffer of local minima, so other methods are often tried, such as genetic algorithms. However, the problem arisen in the previous point is of different nature, and is referred to as numerical noise. This numerical noise as shown in Fig. 9 is produced in general by the limited numerical precision of the FEM solvers.

Gradient methods work well with smooth cost functions, since they try to define a minimization path based on local values of the cost function derivatives, estimated from the values of the same cost function in neighboring points. However, all gradient methods naturally evaluate points progressively closer and closer as the derivative values becomes more higher. This fact explains why the gradient methods fall trapped in points such as B, C in Fig. 9, when the cost function becomes flat near a minimum. The problem is more annoying with very flat minima. Different algorithms may get jammed in very distant points of the parameter space, as in the cases shown in Figs. 7, 8. To overcome the problem, the precision of the FEM software should be increased very much (i.e. the tolerance error usually set in the software should be lowered), with excessive increase of simulation time. Anyway, without other information on the shape of the cost function, the problem remains hard to be solved.

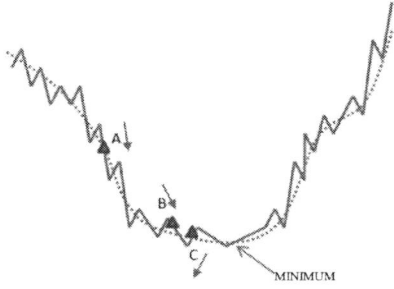

Fig. 9. Numerical noise due to FEM numerical imprecisions.

To overcome the problem of the numerical noise, a different approach has been used in this paper to explore the parameter space in a convenient, global, and intuitive way, also trying avoiding eventual local minima of the cost function.

Given an initial geometry for the machine to be optimized (e.g. the one in Fig. 5), the horizontal sizes are multiplied by a factor α_x, and the vertical sizes by an independent factor α_y, only keeping constant those sizes which have been fixed apriori (i.e. in our case the mover thickness m and the air gap lenght g). So the transformation applied is:

for horizontal sizes:

$$
\begin{aligned}
k_1 &= \alpha_x k_{1,start} \\
k_2 &= \alpha_x k_{2,start} \\
a &= \alpha_x a_{start} \\
d &= \alpha_x d_{start}
\end{aligned}
\tag{1}
$$

for vertical sizes:

$$
\begin{aligned}
k_3 &= \alpha_y k_{3,start} \\
k_4 &= \alpha_y k_{4,start} \\
k_5 &= \alpha_y k_{5,start}.
\end{aligned}
\tag{2}
$$

The machine overall length l and height h are so given by the following linear transformation:

978-1-4799-0024-4/13 $31.00 © 2013 IEEE

$$lenght = l = k_1 = \alpha_x k_{1,start}$$
$$height = h = k_3 + k_4 + m + g = \alpha_y(k_{3,start} + k_{4,start}) + m + g$$

so that l and h can be also assumed as independent variables instead of α_x, α_y:

$$\alpha_x = \frac{l}{k_{1,start}} \qquad (3)$$

$$\alpha_y = \frac{h - m - g}{k_{3,start} + k_{4,start}} \qquad . \qquad (4)$$

The linear transformation (3), (4) is applied by using (1), (2) to the machine geometry, for many points regularly spaced on the (l, h) plane. This operation is similar to an overall *stretching* of the horizontal and vertical sizes of the machine, although there are some fixed parameters. For any point a magnetostatic FEM simulation is carried out, so obtaining a surface which shows the overall trend of the thrusting force for different sizes of the machine, Fig. 10.

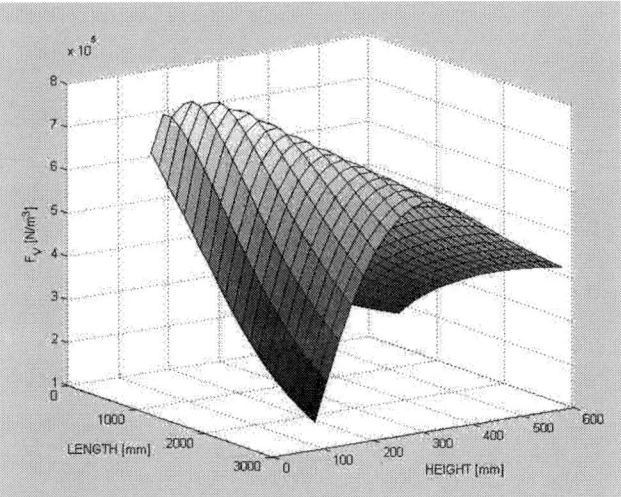

Fig. 10. Force per unit of machine volume (F_v) on the length-height plane (GSM applied to the starting geometry in Fig. 5).

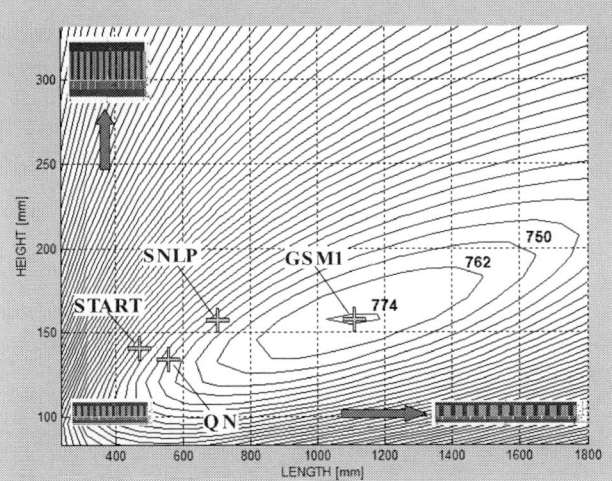

Fig. 11. Contours of the surface F_v on the length-height plane (values in kN/m³) (GSM of the starting geometry in Fig. 5).

A maximum of force is evident in Fig. 10. However, the contour-plot of the force surface in Fig. 11 reveals that the surface is very flat near the top. The two optimal geometries previously shown in Figs. 7, 8 and Table II (with $F_v \approx 750$kN/m³) are reported in Fig. 11, and it is evident that they are quite close to the contour for $F_v = 750$kN/m³ (obviously they *are not* on this contour, since the transformation (1), (2) cannot lead *exactly* to the geometries in Figs. 7, 8). Also the starting geometry (featuring 632kN/m³) in Fig. 5 and Table I is reported in Fig. 11, revealing the very short walks actually done by the two algorithms (QN and SNLP) on the way for the maximum force point. The best point obtained in Fig. 11 (marked GSM1) features 775kN/m³, Table III.

TABLE III
SIZES OF THE OPTIMAL GEOMETRY GSM1 IN FIG. 11 (DESIGN D$_{12-10}$)

item	parameter	value(mm)
mover height	m (fixed)	20
module width	k_1	1098
magnet width	k_2	98.8
magnet eight	k_3	25.2
air gap length	g (fixed)	3
stator eight	k_4	109.6
slot eight	k_5	82.2
tooth average width	d	45.8
stator yoke eight	k_4-k_5	27.4
tooth vertical semi-skew	a	0
OVERALL LENGTH	k_1	**1098mm**
OVERALL HEIGHT	$m+k_3+g+k_4$	**157.8mm**
Force$_x$/Volume	F_v	**775kN/m³**

Obviously the geometry GSM1 is not an absolute optimum, since only two degrees of freedom out of seven have been exploited. However, now the geometry GSM1 can be used as starting point of a new global optimization using, for example, the QN algorithm and leaving all the seven parameters free to vary in a wide range (e.g., 50% to 150% of the values of GSM1). The path of this new optimization procedure in the (l, h, F_v) space is shown in Fig. 12. The path searches new points above the surface of Fig. 10, showing that higher values of force density are reached. A new optimum point is finally gained, called QN2 in Fig. 12. The force density rises to 823kN/m³.

Finally, the GSM procedure can be applied to the new geometry QN2, to verify the global property of the optimum. Fig. 13 shows the contour plot of the force density surface obtained by stretching QN2. The top of the surface (geometry GSM2) is very close to the starting geometry QN2, and this proves that a convergence has been reached. The sizes and the shape of the final geometry GSM2 are shown in Table IV and Fig. 14, respectively.

The iterative GSM procedure is resumed as follows:

1) Design of any starting geometry with variable and fixed parameters, and definition of a proper cost function;
2) Definition of a linear transformation as (1), (2) along the x and y axes;
3) Parametric exploration of the cost function values on a given range of machine length and heights;
4) When a best point is chosen on the cost surface,

start a new optimization by using a classical gradient-based method, leaving all the parameters free to vary;

5) The best point lastly obtained can be used for a new GSM optimization going back to point 3).

However the space occupied by the winding end-connections is not taken in account. Fig. 16 shows a clear comparison between the two designs, with the contour lines of force density obtained from the GSM optimization. The contour lines of the two designs are very similar and concentric. The contour lines are useful for sizing machines featuring high force density but with different force ratings.

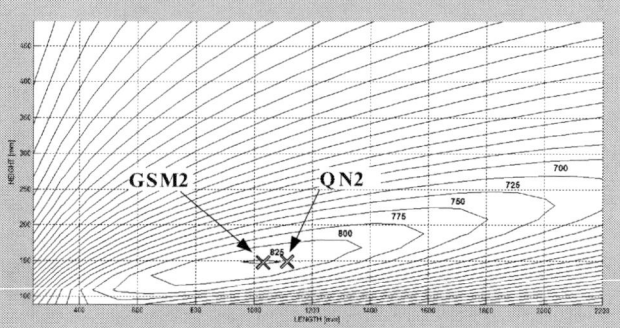

Fig. 13. Contours of the surface F_v on the length-height plane (values in kN/m³) (GSM of the geometry QN2).

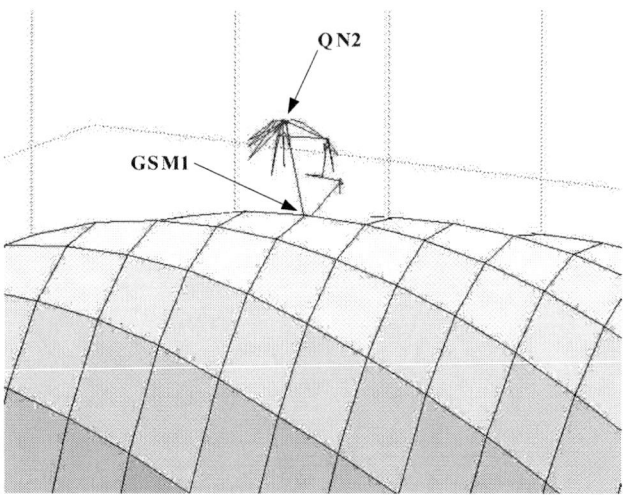

Fig. 12. Optimization path of the Quasi-Newton algorithm in the length-height-force density space, and optimum point found QN2. The surface is the same as in Fig. 10.

C. GSM applied to an integer-slot design

The GSM procedure has been applied also to an integer-slot machine geometry, for comparison with the fractional slot machine. The optimal geometry is shown in Fig. 15, with sizes as in Table V. By comparing Tables IV and V, it appears that the two optimized machines have almost the same overall sizes. The integer-slot machine has about 25% of force density more than the fractional slot machine.

TABLE IV
SIZES OF THE OPTIMAL GEOMETRY GSM2 IN FIG. 14 (DESIGN $D_{12\text{-}10}$)

item	parameter	value(mm)
mover height	m (fixed)	20
module width	k_1	1013.2
magnet width	k_2	104.1
magnet eight	k_3	30.7
air gap length	g (fixed)	3
stator eight	k_4	97.5
slot eight	k_5	78.8
tooth average width	d	35.1
stator yoke eight	$k_4\text{-}k_5$	18.7
tooth vertical semi-skew	a	1.48
OVERALL LENGTH	k_1	**1013.2mm**
OVERALL HEIGHT	$m+k_3+g+k_4$	**151.2mm**
Force$_x$/Volume	F_v	**825.6kN/m³**

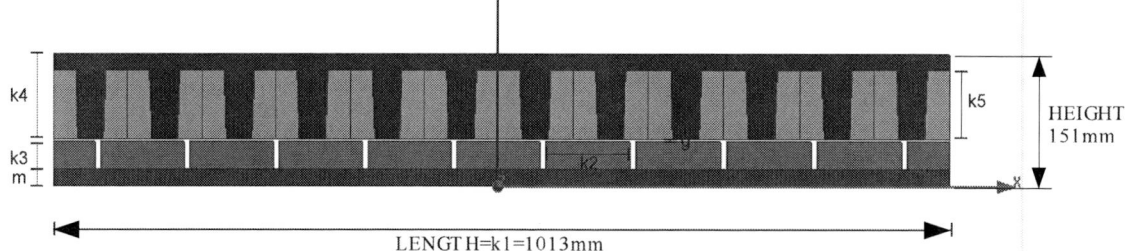

Fig. 14. Optimal geometry GSM2 (design $D_{12\text{-}10}$, fractional slots).

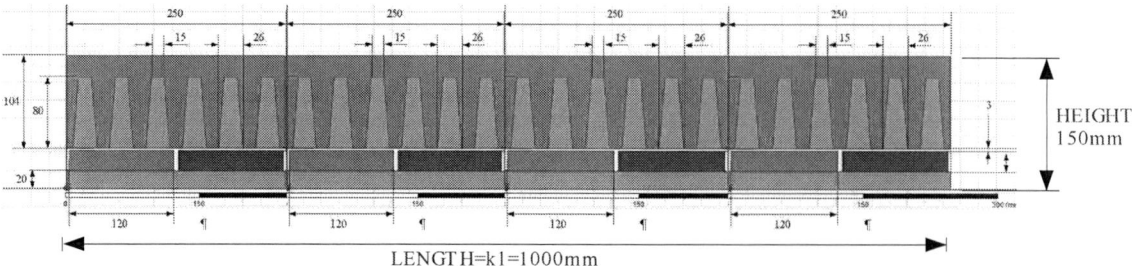

Fig. 15. Optimal geometry with integer slots (4 modules $D_{6\text{-}2}$, i.e. design $D_{24\text{-}8}$).

TABLE V
SIZES OF THE OPTIMAL GEOMETRY IN FIG. 15 (DESIGN D₂₄₋₈)

item	parameter	value(mm)
mover height	m(fixed)	20
module width	k_1	250
magnet width	k_2	120
magnet eight	k_3	23
air gap length	g(fixed)	3
stator eight	k_4	104
slot eight	k_5	80
tooth average width	d	20.5
stator yoke eight	$k_4 - k_5$	24
tooth vertical semi-skew	a	2.75
OV. LENGTH (4 modules)	$4k_1$	**1000mm**
OVERALL HEIGHT	$m+k_3+g+k_4$	**150mm**
Force$_x$/Volume	F_v	**1047kN/m³**

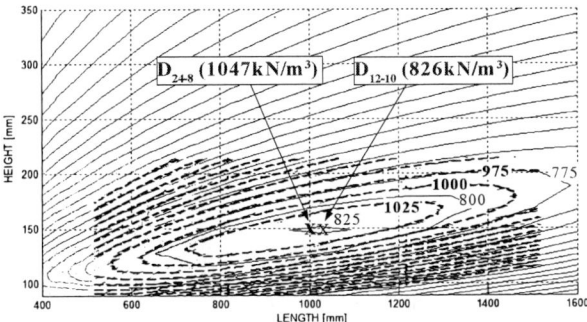

Fig. 16. Contours of the force density F_v on the length-height plane (values in kN/m³). Continuous lines: GSM for the geometry of Table IV (design D₁₂₋₁₀, fractional slots, same plot of Fig. 13). Dashed bold lines: GSM of the geometry in Table V (design D₂₄₋₈, integer slots).

V. CONCLUSION

The two solid models in Figs. 17, 18 show realistic implementations of the two optimal geometries in Figs. 15, 14, respectively. Since the two machines have similar sizes, the thrust axis, the suspension system (bearings), and the lateral frame axis are the same, and the possibility of building a bivalent machine has been considered for experimental purposes, only rearranging stator modules and magnets. Fig. 19 finally shows a laboratory prototype realized for the fractional-slot design of Fig. 18. Experiments are planned for thermal and mechanical testing, and for definition of static and dynamic safe operative areas of the drive. Future work also concerns the study of proper drive control strategies for application to typical on board ship tasks such as stabilizing fin and rudder control.

VI. REFERENCES

[1] C. Bruzzese, "Innovative Solutions for Onboard electromechanical actuators (ISO)–Phases 1-4," Research reports, Contract Rep. N. 19712 for the Italian Ministry of Defence, General Directorate of Naval Armaments (NAVARM), Jan. 23, 2012.

[2] T. Mazzuca and C. Bruzzese, "Project "ISO": innovative solutions for Italian Navy's onboard full-electric actuators", in *Proc. IEEE ESARS Conf.*, Bologna, Italy, Oct. 16-18, 2012.

[3] A. Akers, M. Gassman, and R. Smith, *Hydraulic Power System Analysis*. CRC Press, Taylor and Francis, 2006.

[4] C. Bruzzese, A. Tessarolo, T. Mazzuca, and G. Scala,"A Closer look to conventional hydraulic ship actuator systems and the convenience of shifting to (possibly) all-electric drives," in *Proc. IEEE ESTS Conf.*, Arlington, Virginia (USA), Apr. 22-24, 2013.

[5] CMC Marine, *"Stabilis Electra – Electrical Fin Stabilizer"* [Online] Available: http://www.cmcmarine.com/docs/download/ cmc-marine-stabiliselectra-brochure.pdf.

[6] C. Bruzzese, "A high absolute thrust permanent magnet linear actuator for direct drive of ship's steering gears: Concept and FEM analysis", in *Proc. ICEM Conf.*, Marseille, France, Sep. 2-5, 2012, pp. 556-562.

[7] A. Tessarolo, C. Bruzzese, T. Mazzuca, and G. Scala, "A novel fault-tolerant high-thrust inverter-controlled permanent magnet linear actuator as a direct-drive for shipboard loads," in *Proc. IEEE ESTS Conf.*, Arlington, Virginia (USA), Apr. 22-24, 2013.

[8] Rules for The Classification of Naval Ships-Part C–Machinery, Systems and Fire Protection – Steering Gear, Ch. 1, Sec. 11, RINAMIL 2007.

[9] F. Luise, S. Pieri, M. Mezzarobba, A. Tessarolo, "Regenerative testing of a concentrated-winding permanent-magnet synchronous machine for offshore wind generation—Part I: Test concept and analysis", *IEEE Trans. Ind. Appl.*, vol. 48, no. 6, pp. 1779-1790, Nov.-Dec. 2012.

[10] Y. Fujimoto, T. Kominami, and H. Hamada, "Development and analysis of a high thrust force direct-drive linear actuator", *IEEE Trans. Ind. Electron.*, vol. 56, no. 5, pp. 1383-1392, May 2009.

[11] Hatlapa Marine Equipment, *"Steering Gears – RAM Type Poseidon ST"*. [Online]. Available: http://www.hatlapa.de/products/steering-gear/223/steering-gear-downloads/EN/file.php?ID=893.

[12] I. Boldea and S. A. Nasar, *Linear Electric Actuators and Generators*. Cambridge University Press, 2005.

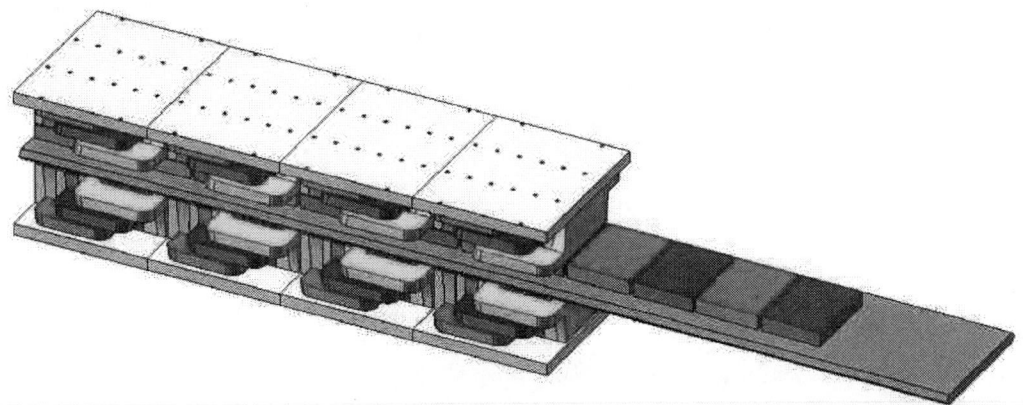

Fig. 17. PMLSA design with 24 slots per 8 magnets (D₂₄₋₈), integer slot winding (one slot per pole per phase). The three-phase modules can be dismounted.

978-1-4799-0024-4/13 $31.00 © 2013 IEEE

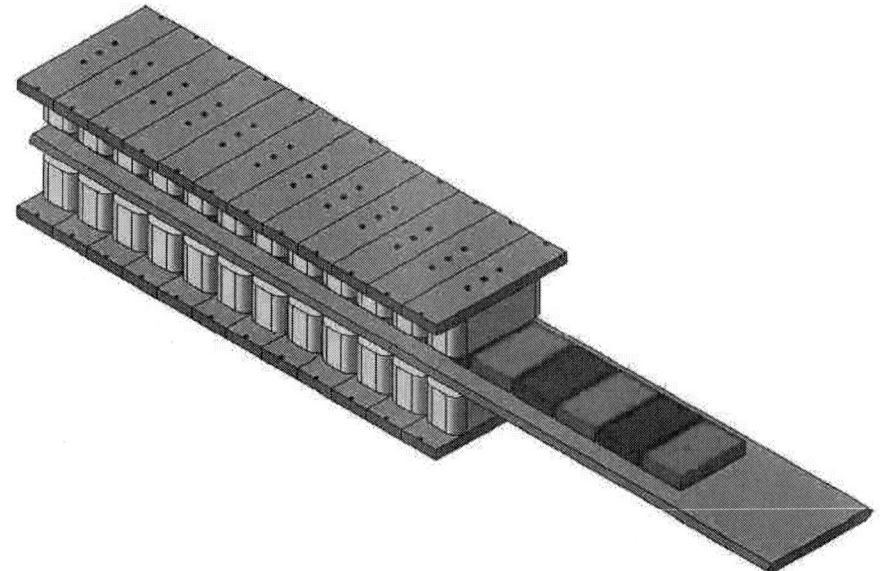

Fig. 18. PMLSA design with 12 slots per 10 magnets (D_{12-10}), fractional slot winding. The single tooth-modules can be dismounted.

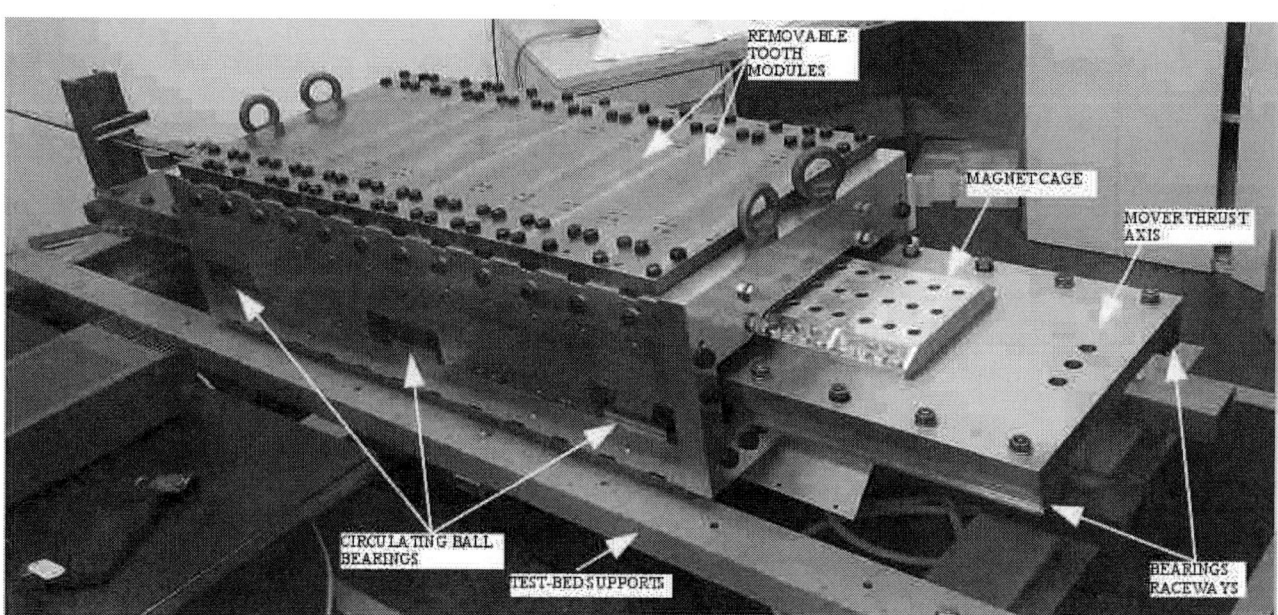

Fig. 19. Laboratory prototype of PMLSA based on the design in Fig. 18.

Fault Tolerant High Voltage Resonant Power Converter Application

A. Hultgren, S. Bui, J. Linnér, P. Ranstad, M. Lenells

Abstract – **A high voltage and high power fault tolerant application of a resonant converter is presented. The resonant converter load shows frequent short circuits making the load voltage drop to zero within some μs, implying high stress on the main circuit components. A load voltage short circuit fault tolerant system is suggested by including a novel way of augmenting the controller with a load voltage estimator. The estimator can detect load voltage short circuit and will then be a part of a fault tolerant high voltage resonant converter.**

Index Terms—**Fault tolerant, resonant converter, load voltage short circuit estimator.**

I. INTRODUCTION

The control of resonant converters in different applications has been of interest for several decades and several suggestions of controllers have been given. Often the control is taking place in a full bridge topology, as shown in Fig. 1, where the resonant circuit is situated between nodes A and B. The voltage across the resonant circuit is controlled by the four switching transistors, Z_1, Z_2, Z_3, and Z_4. The addressed resonant circuit, is used in a high voltage application which includes also a transformer, TR. To make the magnetic parts smaller, the resonance frequency is made high, which in its turn demands a fast controller. The addressed application is also a high power application, implying that faults can have a severe impact on the system. This paper suggests a novel augmentation of the control algorithm in order to shield the resonant converter components for one type of sudden appearing fault. In section two the converter application is described together with an overview of common control strategies of resonant converters. In section three the suggested control structure is presented together with the augmentation giving a fault tolerant behavior for a common appearing fault. In section four results from high voltage and high power tests of the proposed solution are shown. The comparison is performed

This work was supported in part by the Knowledge Foundation, Sweden, under Grant 2006/0265, Load adaptive control.

Anders G. Hultgren is with the Department of Electrical Engineering, School of Engineering, Blekinge Institute of Technology, 371 79 Karlskrona, Sweden. (e-mail: anders.hultgren@bth.se).

Sonny Bui is a student at Linnaeus university, 351 95 Växjö Sweden.

Jörgen Linnér is with Alstom Power AB, 352 41 Växjö, Sweden. (e-mail:jorgen.linner@power.alstom.com).

Per Ranstad is with Alstom Power AB, 352 41 Växjö, Sweden. (e-mail: per.ranstad@power.alstom.com).

MatzLenells is with the Department of Electrical Engineering at Linnaeus University, 351 95 Växjö, Sweden. (e-mail: matz.lenells@lnu.se).

with control strategies without the fault tolerant behavior. In section five a discussion can be found.

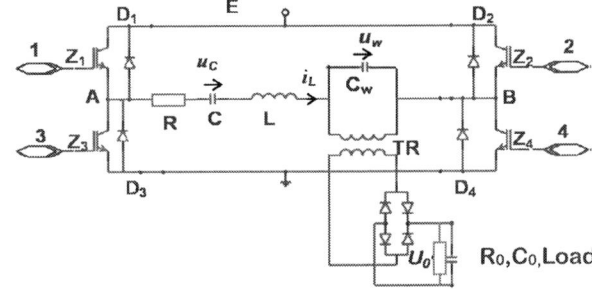

Fig. 1. Circuit diagram for the resonant converter, controlled by the four transistors Z_1, Z_2, Z_3, and Z_4.

II. POWER CONVERTER APPLICATION

The resonant circuit connected between node A and node B in Fig. 1 is of series parallel type. The series capacitance, C, and the series inductance, L, together with the parallel capacitance, C_w, models the resonant converter. The C_w parameter models parasitic effects in the high voltage transformer, TR, and has the capacitance value of about 15% of the series capacitance, C. The resistance R is neglected in the analysis. The controlled output of the converter is the current, I_0, to the load. The load main property is capacitive, and can be modeled as a capacitance, C_0, on the secondary side of the transformer. The capacitive load and high transformer ratio will give a comparatively large effective load capacitance, making the load voltage, U_0, only slowly varying. Hence, analyzing short time behavior of the converter system it is possible to model the load as a constant voltage source, with the value U_0', on the primary side of the transformer.

The load of the converter can show quick changes, that imply that the resonant circuit is exposed for high voltage and high current stress, with possible severe effects. The nominal load voltage can drop from about *70 kV* down to almost zero in some few micro seconds. When this happens the amplitudes of current and voltage in the resonant circuit will increase significantly.

The converter is supplied by the voltage, E, and the transistor bridge is controlling the voltage across the resonant circuit between the nodes A and B, u_{AB}. Assuming ideal transistor switches, the control signal can be taken out of a set of two levels, $u_{AB} \in \{-E, E\}$. The natural frequency and the characteristic impedance of the resonant tank are 20 kHz and 2,3 Ω, respectively. The controller is

978-1-4799-0024-4/13 $31.00 © 2013 IEEE   485

running the converter in a limit cycle above the resonance frequency, generating low switch-on losses in the transistors.

The output signal of the process is the load current, I_0. The feedback control of the resonant converter can be based on the averaged value of the output signal, I_0, or also on the resonant circuit voltage, u_C and u_w, and current i_L. In the addressed resonant converter the series resonant circuit, the LC-circuit, u_C and i_L, explains most of the behavior of the system.

Several suggestions for phase plane control of series resonant, parallel resonant, and combined, can be find in the literature. Oruganti and Lee [1] and [2] presented a set of switched state feedback controllers for the series loaded resonant converter and since then several authors have addressed design and analysis issues of resonant converter controllers. See for instance Rossetto [3], Chen et al [4], Melin et al [6] and [7], and Hultgren et al [5], [8] and [9]. Some of the suggested control laws generate piecewise systems, possibly piecewise linear systems and some of others generate a more general subclass of hybrid systems.

Suggestions for methods for over load protection of resonant converters can be found e.g. in King and Stuart [10] and in Jovcic and Wu [11].

This paper treats the problem of how to choose a control algorithm which is tolerant for short circuits in the load. The chosen control algorithm is of phase plane type with an augmentation for short circuit fault tolerance.

III. CONTROL ALGORITHMS AND FAULT TOLERANT AUGMENTATION

The controlled variable, in the addressed converter application, is the load current, I_0. As the parallel capacitance, C_w, is 15% of the series capacitance, C, the main property of the resonant converter is series loaded. It can be seen in Fig. 1 that the applied voltage magnitude , u_s, across the series resonant circuit, L and C, is the superposition of the control voltage, E, and the load voltage, U_0' on the primary side, $u_s = E - U_0'$. As the high voltage load frequently generates short circuit disturbances, which implies that U_0' drops within some few micro seconds to almost zero voltage and the applied voltage on the series resonant converter rises to $u_s = E$, there is a risk for current rushing in the resonant circuit. The resonant circuit components will then be exposed to high voltage and high current stress. It is of most importance that the control system of the application is fault tolerant for short circuit failure in the load.

The resonant converter can be controlled by different types of controllers. In the described application two types of controllers has been tested, one based on a time continuous model and one based on a piecewise model of the converter application.

A. Time continuous converter model and a linear controller design

Arranging the control voltage $u_{AB} \in \{-E, E\}$ as a square wave, it is possible to let the frequency of the square wave be the control signal, f_c. Controlling the resonant converter above the resonance frequency generates a time continuous behavior of the controlled system and enables zero voltage switching of the converter full bridge transistors. Increasing the frequency will decrease the output voltage across and current in the load. Lowering the frequency towards the resonance frequency will give high output voltage and current. As the resonance frequency is *20 kHz* the control signal frequency is *21 k < f_c < 45 kHz*. Such a load current controller can be designed using linear control theory.

Because the load shows frequent short circuits the controller should quickly react in order to prevent generation of high load current, high current or high voltage in the resonant converter. The design of the resonant series capacitor and the inductor must take into account that a short circuit in the load will generate high current and high voltage in the resonant circuit.

B. Piecewise converter model and a switched controller design

The converter can be described by a piecewise system. Choosing the two most important states as the normalized series capacitance voltage, u_{CN}, and inductance current, i_{LN}, the resonant converter is given by the second order system

$$
\begin{pmatrix} \dfrac{d}{dt} u_{CN} \\ \dfrac{d}{dt} i_{LN} \end{pmatrix} = \begin{pmatrix} 0 & \dfrac{1}{\sqrt{LC}} \\ \dfrac{-1}{\sqrt{LC}} & 0 \end{pmatrix} \begin{pmatrix} u_{CN} \\ i_{LN} \end{pmatrix} + \begin{pmatrix} 0 \\ \dfrac{s - U_{0N} \cdot sign(i_{LN})}{\sqrt{LC}} \end{pmatrix},
$$

(1)

where $\begin{pmatrix} u_{CN} & i_{LN} \end{pmatrix}^T = \begin{pmatrix} \dfrac{u_C}{E} & \dfrac{i_L}{E}\sqrt{\dfrac{L}{C}} \end{pmatrix}^T$, (2)

$$ U_{0N} = \dfrac{U_0'}{E} = \dfrac{U_0}{nE}, \quad s = \dfrac{u_{AB}}{E} \in \{-1, 1\}, $$

u_C is the series resonant capacitanc voltage,

i_L is the series resonant inductance current,

E is the applied voltage across the full bridge, and

n is the transform er ratio.

The normalized load current is given by

$$ I_{0N} = \dfrac{I_0}{E}\sqrt{\dfrac{L}{C}} $$

and the working range for the normalized load voltage and normalized load current is given by $0 < U_{0N} < 1$ and $0 < I_{0N} < 1.2$.

The system may be run in a limit cycle by choosing $s = 1$, each time the current i_L changes to positive sign, and by choosing $s = -1$ each time the current i_L changes to negative sign. This will give maximum output. By means of choosing

$s = -1$ at a certain state position (u_{CN}, i_{LN}) on the upper half plane and $s = 1$ at a certain state position (u_{CN}, i_{LN}) on the lower half plane, the size of the limit cycle can be adjusted to stationary fit the chosen desired load current I_{0N} for a certain load voltage U_{0N}.

The chosen control algorithm is based on Oruganti's optimal trajectory control, see [1] and [2]. The state position (u_{CN}, i_{LN}) on the upper half plane is chosen to be any position on the stationary trajectory for the given load voltage, set point current and the control signal value $s = -1$. The state position (u_{CN}, i_{LN}) on the lower half plane is chosen to be any position on the stationary trajectory for the given load voltage, set point current and the control signal value $s = 1$.

This choice implies that the switch curves for the controller is the stationary trajectories. An example of limit cycle and switch positions are shown in Fig. 2. It can be seen in the figure that as soon the trajectory reaches the circular arc shaped switch curve on the upper and lower half plane, the trajectory follows the switch curve.

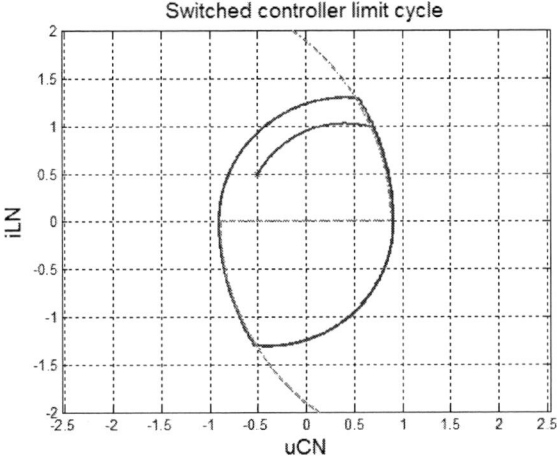

Fig. 2. A limit cycle of the resonant LC circuit is shown. The limit cycle runs counter clock wise. The trajectory switch-curves are shown as dash-dotted circular arcs on the upper half plane and dashed circular arcs on the lower half plane. The initial value is $(0.5, 0.5)$ and the simulation is performed for $U_{0N} = 0.6$ and $I_{0N} = 0.8$.

The stationary trajectories can be derived by the stationary solution to (1) and is given by

$$u_{CN} - u_N = i_{LN0} \sin(\frac{1}{\sqrt{LC}}(t - t_0)) + (u_{CN0} - u_N) \cos(\frac{1}{\sqrt{LC}}(t - t_0))$$

(3)
and

$$i_{LN} = i_{LN0} \cos(\frac{1}{\sqrt{LC}}(t - t_0)) - (u_{CN0} - u_N) \sin(\frac{1}{\sqrt{LC}}(t - t_0)) \quad (4)$$

where $u_N = \dfrac{s - U_{0N} \cdot sign(i_{LN})}{\sqrt{LC}}$.

Squaring and adding the solution (3) and (4) give

$$(u_{CN} - u_N)^2 + i_{LN}^2 = (u_{CN0} - u_N)^2 + i_{LN0}^2,$$

which describes circle arcs in the u_{CN}, i_{LN} space as shown in Fig. 3. Because u_N is a function of the switching value s, the centers of circles will shift when s is shifted. The radii of the circles are given by

$$r_s = (u_{CN0} - \frac{s - U_{0N} \cdot sign(i_{LN})}{\sqrt{LC}})^2 + i_{LN0}^2. \quad (5)$$

Fig. 3 shows how the stationary trajectory can be found for the upper half plane, $i_{LN} > 0$. The trajectories for the lower half plane is found by mirroring in the origin, see the

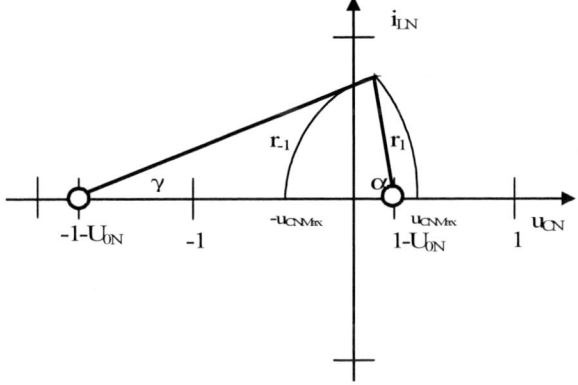

limit cycle example in Fig. 2.

As mentioned in section II, I_0 is the output from the system. The average I_{0N}, generated by the upper half plane limit cycle is proportional to the expression

$$\frac{1}{t_2 - t_1} \int_{t_1}^{t_2} i_{LN}(t) dt$$

where t_1 is the time instant when the trajectory passes into the upper half plane across the negative u_{CN} axis and t_2, when it passes out, across the positive u_{CN} axis.

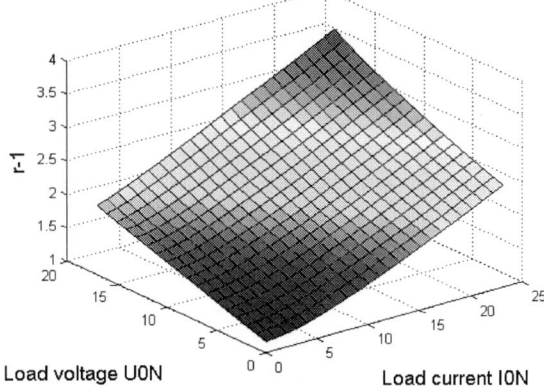

Fig. 4. Switch curve radii as a function of load voltage and load current.

978-1-4799-0024-4/13 $31.00 © 2013 IEEE

The symmetry of the limit cycle implies that the encircled area is related to the generated average load current. The larger the radii r_{-1} and r_1 are, the larger I_{0N} is generated. This means that, if the load voltage U_0 is given, then there is an one to one relationship between the radii r_{-1} and the load current I_0. Fig. 4 shows this relation. See [5], for a more detailed derivation of a switched controller of similar type.

This paper considers the case when the load current is controlled by choosing r_{-1} such that the integral expression is equal to the desired averaged value of I_0. By adding a slow outer loop control, where the measured load current is used as the output signal, one obtains a cascade control. In this paper we will only discuss the inner loop of such a controller.

In order to generate the correct switch curve for the controller, also the U_{0N} has to be known, as shown in Fig. 4 and expression (5). As frequently short circuits occur in the load, the controller should have the capability to quickly measure the load voltage and change the radius. The high voltage and the electrically disturbances makes it impossible to measure the load voltage with high bandwidth, implying that a short circuit, also with the use of a switched controller, will generate some over current and over voltage in the resonant converter.

The switched controller can be augmented with a load voltage estimator, as shown in the following section, and can then be part of a short circuit fault tolerant resonant converter.

C. Fault tolerant augmentation of the switched controller

The chosen state transformation (2) implies that the stored energy in the series resonant circuit, L and C, is proportional to the phase plane distance of the state value $\left(u_{CN}(t), i_{LN}(t)\right)$ to the origin. The distance can be supervised in order to detect if the resonant converter is increasing its stored energy above what is needed for the given load voltage and desired load current. A set of certain circular radius is chosen for a certain working point switch curves corresponding to higher energy stored in the resonant converter than needed for the actual working point. The circles are used as a load voltage estimator, with the functionality of detecting load voltage short circuit. Fig. 5 shows an example, where two circles are used to estimate the load voltage. The most inner limit cycle is the actual working point limit cycle. The two dashed circles constitute the estimator. Each circle corresponds to an energy level in the resonant converter corresponding to a certain estimated load voltage. The inner one to $U_{0N} = 0.4$ and the outer one to $U_{0N} = 0.15$. If the limit cycle, in the given example, passes the inner circle, the estimated load voltage is $U_{0N} < 0.4$ if it passes the outer one the estimated load voltage is $U_{0N} < 0.15$, i.e. a short circuit. The number of circles can be increased for higher resolution in the estimation.

The estimated load voltage is then used for generating a new controller switch curve suitable for the new working point. The estimation process is fast, it is performed during less than half of the limit cycle period time, i.e. within shorter time than *20 µs*.

The circular shaped estimator is not used as a switch curve for the controller, only as an estimator of if the controller's switch curve should be changed to a new working point. For origin centered circular switch curves, the stability of the limit cycle is not secured. A discussion about controlling a resonant converter with switch curves as origin centered circles can be found in [9]. Such switch curves can generate not stable limit cycles. In the suggested fault tolerant system, the optimal trajectory controller, with secured limit cycle stability, controls the limit cycles all the time, the set of circles is only used as an estimator.

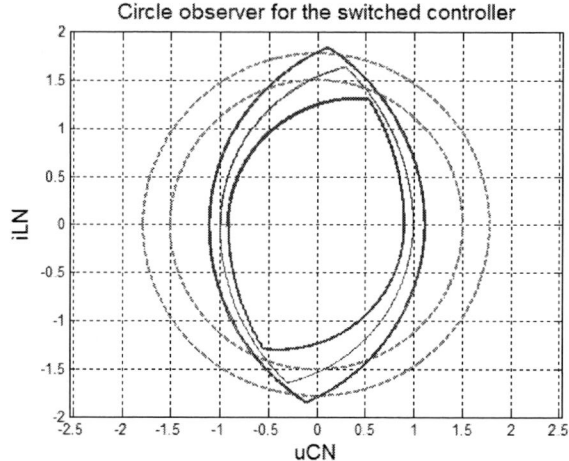

Fig. 5. The solid curves show limit cycles for I0N= 0.8 and three different load voltages. The dashed circles constitute a load voltage estimator. In the case the trajectory crosses the inner circle the load voltage is estimated to less than 0.4 and if it passes the outer circle the load voltage is estimated to less than 0.15.

IV. EXPERIMENTAL RESULTS

Experiments with the two different controllers, the frequency controller described in section III. A and the switched controller described in sections II B and C, have been performed in order to evaluate their abilities to limit the transient disturbances in the system as a consequence of an external fault. The experimental results are shown in two figures and in one table.

The external fault considered in these experiments is a sudden short circuit of the load caused by a spark. The experiments are performed using a converter, as described in Fig. 1, rated *70 kV/120 kW*. In addition to the circuit in Fig.1 a spark-gap has been connected in parallel to the load circuit, R, C. Once the spark-gap is triggered, the resulting spark-over will cause a short-circuit of the load and the resulting transients on the inductor current, i_L, and on the capacitor voltage, u_C, are studied. Prior to the occurrence of the short circuit the converter is operated at rated output voltage and power values.

978-1-4799-0024-4/13 $31.00 © 2013 IEEE

The first experiment, Fig. 6, shows the results when using the frequency controller. Fig. 6 shows the resonant current, i_L, and resonant voltage, u_C, as function of time, when a short circuit in the load appears and the frequency controller is used. From Fig. 6 it can be observed that the short circuit fault will generate high rise in the peak values of the current and voltage in the resonant circuit. It should be noted that the output voltage, U_0 (blue trace), is represented as a negative voltage. The slow response of the measured signal of U_0 is due to a limited bandwidth of the sensor in use.

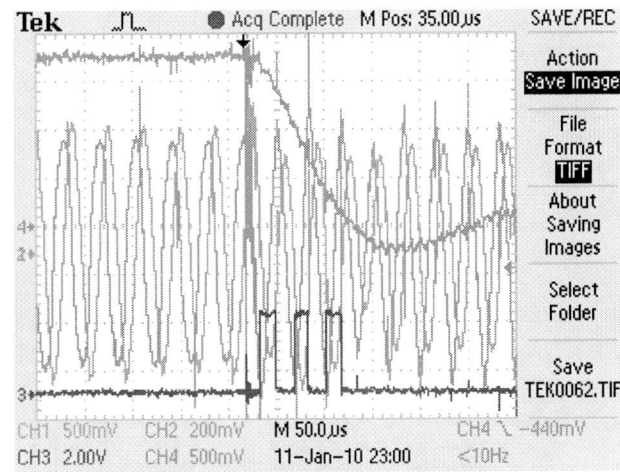

Fig. 7. Physical experiment. Short circuit response, frequency controller. i_L (blue lagging signal), u_C (yellow leading signal), U_0 (green), σ (purple square shaped).

TABLE 1 lists the relative peak values, $\hat{\imath}_{LN}$ and $\hat{u}_{CN}$ of i_L and u_C, respectively. The peak values, $\hat{\imath}_{LN}$ and $\hat{u}_{CN}$, are normalized to the peak values prior to the occurrence of the fault. The peak values are important inputs when selecting the design specification for the different components in the main circuit.

TABLE 1.
DESIGN PARAMETERS FOR THE RESONANT COMPONENTS

	Frequency controller	Switched controller
$\hat{\imath}_{LN}$ [%]	370	115
$\hat{u}_{CN}$ [%]	350	100

The elimination of the current transient is also expected to increase the reliability of the power semi-conductors, i.e. IGBTs and HV-diodes.

V. DISCUSSION AND CONCLUSIONS

In order to reduce the current and voltage transients, which occurs in the main circuit in a high power resonant converter as a consequence of an external fault, a novel switched controller including a fault observer has been developed, Section III.

The system performance in the context of response to an external fault, short circuit of the load, has been experimentally evaluated, Section IV.

The transients following the external fault are compared for the system operated with a frequency controller and with the proposed controller. It is evident, when comparing Fig. 6 and Fig. 7, that the switched controller including a fault observer significantly reduces the transients. In Table 1 the comparative parameter values are listed. It is clearly shown that, in the case of the frequency controller the transients peak at more than three times (370%) the steady state values while in the case of the proposed controller the transients are limited to very low levels (115%).

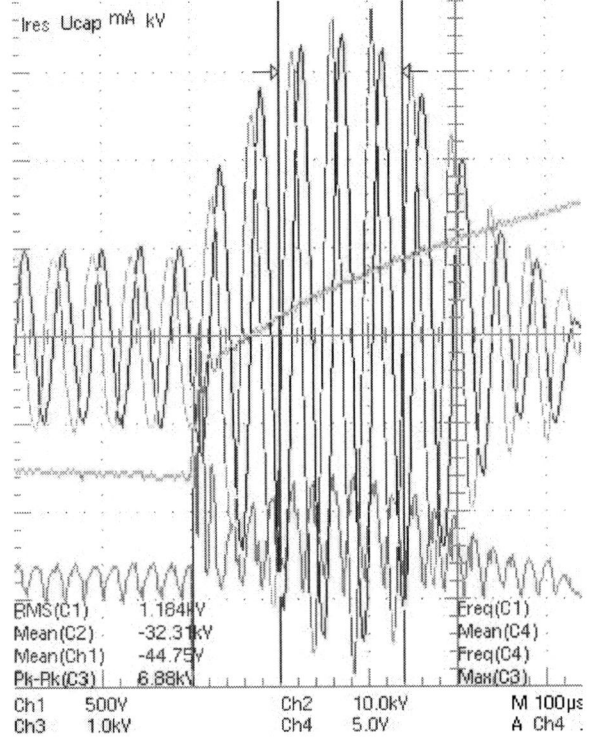

Fig. 6. Physical experiment. Short circuit response, frequency controller. i_L (yellow lagging signal), u_C (purple leading signal), U_0 (blue).

The result of the second experiment is shown in Fig. 7. In this experiment the switched controller including a fault detection estimator, as described in Section V, is used. Fig. 7 shows the resonant current, i_L and resonant voltage, u_C, as function of time, when a short circuit of the load occurs. From Fig. 7 it can be observed that the short circuit fault will generate only a minor increase of the peak value of i_L (yellow) and that the peak value of u_C is actually slightly reduced. In the Fig. 7 high level of the logical signal, σ, indicates that the fault detector is active. As can be observed, the fault detector is active during the first three periods succeeding the occurrence of the short circuit.

Hence, based on the experiments it is conclude that the implementation of the proposed controller, switched controller including a fault observer, can significantly reduce the voltage and current transients resulting from a sudden short circuit of the load. In the presented experiments these transients are practically eliminated.

From a system point of view a reduction of main circuit transients will impact positively on EMC, reliability, and the design of the main circuit components.

VI. REFERENCES

[1] Oruganti, R., Lee, F. C., "Resonant Power Processors Part 1 - State Plane Analysis, Part II - Methods of Control", *IEEE-IAS*, annual meeting 1984.

[2] Oruganti, R., Yang, J.J.; Lee, F.C.,"Implementation of optimal trajectory control of series resonant converter", *IEEE Transactions on Power Electronics*, v 3, n 3, p 318-27, July 1988.

[3] Rossetto L. A simple Control Technique for Series Resonant Converter. IEEE Transactions on Power Electronics, vol. 11, no. 4, pp 554-560, 1996.

[4] Chen H., Sng E., Tseng K.-J. Generalized Optimal Trajectory Control for Closed Loop Control of Series-Parallel Resonant Converter. IEEE Transactions on Power Electronics, vol. 21, no. 5, Sept. 2006.

[5] Hultgren A., Ingelbrandt P., Ranstad P., Nilsson M., Lenells M.," Switched Controllers Applied to an LCC Converter - Control Law Evaluation*"*, in *Proc. NORpie 2002*, Stockholm, 2002.

[6] Melin J. and Hultgren A. A limit cycle of a resonant converter. Proceedings of IFAC ADHS conference June 2003, St. Malo, France, 2003.

[7] [Melin J., Hultgren A. and Lindström T. Two Types of Limit Cycles of a Resonant Converter Modelled by a Three-dimensional system. Journal of Nonlinear Analysis: Hybrid Systems and Applications, 2 (2008) 1275-1286 Elsevier Inc. 2008.

[8] Hultgren A., Melin J. and Ranstad P. Asymmetric Limit Cycles in an Industrially Applied Controlled Resonant Converter, 3rd IFAC Conference on Analysis and Design of Hybrid systems, September 16-18, 2009, Zaragoza, Spain.

[9] Hultgren A., Melin J., Ranstad P., "Limit cycle control of an industrially applied resonant converter modelled as a hybrid system", in *Proc. - IEEE International Symposium on Circuits and Systems, p 1916-1919, 2011*, ISCAS 2011, Rio de Janeiro, Brazil.

[10] King R. J.and Stuart T. A., "Inherent Overload Protection for the Series Resonant Converter", in *IEEE Transactions on Aerospace and electronic Systems*, Vol. AES-19, Issue:6, pp 820-830, 1983.

[11] Jovic D. and Wu B., "Fast fault current interruption on high power DC networks", In *Proc. 2010 IEEE Power and energy Society general Meeting*, pp 1-6, 2010.

VII. BIOGRAPHIES

Per Ranstad (M'03) was born in Högby, Sweden, 1958. He received the M.Sc. degree in electrical engineering from Lunds Institute of Technology, Lund, Sweden, in 1987, the Licenciate and PhD degrees in electrical engineering from the Royal Institute of Technology, Stockholm, Sweden, in 2004, and 2010, respectively.

He is a Senior Researcher with Alstom Power and has been leading R&D activities in the area of power converters for industrial applications since more than 20 years. His main research interests are: high-voltage transformers, power semiconductors, resonant topologies, and converter control.

Dr. Ranstad was awarded the Masuda Award by the International Society for Electrostatic Precipitation (ISESP) at the ICESP XII in Nuremberg, May 2011.

Anders Hultgren was born in Norrköping, Sweden, 1953. He received the M.Sc. degree in Electrical Engineering from Chalmers Institute of Technology, Gothenburg, Sweden, in 1982. He is a Senior Lecturer at Blekinge Institute of Technology. He has been working in R&D projects in collaboration with different industries the last 20 years. His main research interests are: modeling of dynamic and hybrid physical systems and application of observers and controllers in industrial systems.

Jörgen Linnér was born in Växjö, Sweden, 1974. He received the B.Eng. in Electrical Engineering 1996 from Växjö University and is a development engineer with Alstom Power.

Sonny Bui was born in Hai Phong, Vietnam, 1968. He received the M.Sc. degree in applied mathematics 2004 from Vaxjö University. He has been a development engineer with Alstom Power and is now with Volvo Construction Equipment.

Matz Lenells was born in Hallsberg, Sweden, 1948. He received the B.Sc. degree in theoretical philosophy and mathematics 1972 from Lund University, Master of Engineering (Civilingenjörsexamen) in technical physics at Lund Institute of Technology in 1976, Ph D. (Teknologie Doktor) in Automatic Control at Lund Institute of Technology in 1982. He is a Senior Lecturer at Linnaeus University, Växjö. He has been working in R&D projects in collaboration with different industries the last 20 years. His main research interests are: modeling of dynamic and hybrid physical systems and application of observers and controllers in industrial systems.

978-1-4799-0024-4/13 $31.00 © 2013 IEEE

Experimental Evaluation of Combined Reference Frames Transformation for Stator Fault Detection in Multi-Phase Machines

C. Bianchini[1], F. Immovilli[1], E. Lorenzani, A. Bellini[1], E. Fornasiero[2],

[1]DISMI - University of Modena and Reggio Emilia, Italy

[2]DIE - University of Padova, Italy

Abstract—**This paper focuses on the modeling and experimental validation of a diagnostic index for fault detection in multiphase machines. Experiments are carried out on of a five-phase permanent-magnet machine designed for fault tolerant applications. The diagnostic index is aimed at stator faults detection and is based upon the combination of information from two different reference frames. The diagnostic index effectiveness and robustness are assessed by Finite Element analysis and experiments.**

I. INTRODUCTION

Multiphase machines can introduce an improvement in the area of medium to high power drives. Compared to traditional three–phase drives, multiphase machines exhibit higher efficiency, torque–to–weight ratio and torque–to–volume ratio [1]. Recently, multiphase machines are being investigated for high power drives and power generation [2]. For a given voltage and power, increasing the number of phases implies a reduction in the current per-phase, and therefore the power rating of the switches. Their use is increasing in those applications where fault tolerance and continuous operation is a mandatory request, such as in traction and aerospace applications [3], as well as a few industrial applications, where fault tolerant operation is appreciated [4].

Multiphase machines are inherently fault tolerant, as they can be designed to reduce the fault probability as well as to maintain operation in the presence of faults [5], [6]. Provided that the machine windings are independent (i.e. no star point connection) and a suitable converter architecture is chosen, a number of phases higher than three introduces a sort of redundancy in the system, in the electrical circuit as well as in the control system. In fact, a multiphase machine offers several degrees of freedom that can

be used to enhance the reliability of the drive [7], [8], provided that electrical and mechanical fault diagnosis techniques are available. The multiphase drive can continue to operate smoothly in the event of certain faults fault, such as the loss of one or more phases, maintaining control capabilities at the expense of a reduced torque [7], [9].

Fault diagnosis techniques for three phase electric machines is an extensively investigated topic: the main purpose being fault detection at an early state in order to compensate its negative effects and conduct predictive maintenance, to avoid loss of production or hazardous situations [10] – [12].

A fault causes an asymmetry in the machine, producing an electrical signature that can be detected by non–invasive techniques based on time domain analysis, frequency analysis, or time–frequency analysis. The most common technique, Motor Current Signature Analysis (MCSA), is based on the analysis of the harmonic content of the stator currents. The spectrum of the currents contains information (i.e. signatures) univocally related to the presence of electrical and/or mechanical faults and can even permit a quantitative analysis of damage progression [13].

Faults on multiphase machines are similar to those of three–phase machines, and they can be roughly classified in electrical (e.g., short–circuit fault) and mechanical (e.g., bearing damage) faults. A general approach for the diagnosis of stator asymmetries in a multiphase drive was presented in [14]. Other papers in literature investigated the effects of faults in multiphase motor drives, including short–circuit and open–circuit fault, and proposing methods to bypass the faulted components and techniques to increase the reliability of the drive [15] –

978-1-4799-0024-4/13 $31.00 © 2013 IEEE

[20]. The fault detection index proposed in [21] is here analyzed in greater depth and is experimentally assessed. The index is based upon the combination of information from two different reference frames and gives a DC component ideally equal to zero in healthy condition and different from zero during a fault (with a value proportional to damage severity). The present paper contribution is a more in-depth modeling of the fault and its application to a 5-phase machine together with experimental validation to assess the feasibility of multiple reference frame diagnostic techniques applied to multiphase machines. The idea is to develop a reliable index capable of an early detection of short–circuit faults. The applied technique is an on–line and non–invasive method, which uses the measured voltage of each phase to construct the fault detection index. It can be integrated in an existing electric drive as it does not require any dedicated/additional transducer for the fault detection.

The main drawback is that a measurement of the current (for a Voltage source inverter drive) or the voltage (for a Current source inverter drive) of each phase is required. Moreover a dedicated power converter having suitable architecture and multiphase output must be developed [22] – [25].

The diagnostic index and proposed method is employed for the detection of turn–to–turn short–circuit faults. The detection of this kind of fault is challenging, since the distortion of the flux is of modest entity and the asymmetry is very small (especially in case of a low number of short–circuited turns, i.e. the worst case scenario), even though the short–circuit current is high [26]. Section II details the diagnostic index construction. Section III presents the simulation results for the machine investigated Section IV reports the experimental results.

II. THE DIAGNOSTIC INDEX

A. Five–phase reference frame

In order to be effective the index should recognize the asymmetry of the phases, without giving false indications on the healthy components.

The effect of an inter–turn short–circuit is analyzed using the symmetrical components transformation [4], [27]. The transformation from phase components to orthogonal components is obtained as:

$$
\begin{bmatrix} v_\alpha \\ v_\beta \\ v_{\alpha_2} \\ v_{\beta_2} \\ v_0 \end{bmatrix} = K \begin{bmatrix} \mathfrak{Re}\left(\dot{a}^0 \quad \dot{a}^1 \quad \dot{a}^2 \quad \dot{a}^3 \quad \dot{a}^4\right) \\ \mathfrak{Im}\left(\dot{a}^0 \quad \dot{a}^1 \quad \dot{a}^2 \quad \dot{a}^3 \quad \dot{a}^4\right) \\ \mathfrak{Re}\left(\dot{a}^0 \quad \dot{a}^2 \quad \dot{a}^4 \quad \dot{a}^6 \quad \dot{a}^8\right) \\ \mathfrak{Im}\left(\dot{a}^0 \quad \dot{a}^2 \quad \dot{a}^4 \quad \dot{a}^6 \quad \dot{a}^8\right) \\ \frac{1}{\sqrt{2}} \quad \frac{1}{\sqrt{2}} \quad \frac{1}{\sqrt{2}} \quad \frac{1}{\sqrt{2}} \quad \frac{1}{\sqrt{2}} \end{bmatrix} \begin{bmatrix} v_a \\ v_b \\ v_c \\ v_d \\ v_e \end{bmatrix} \quad (1)
$$

where $\dot{a} = e^{j2\pi/5}$ and $K = 2/5$ for transformation with constant amplitude. The first two components define the space vector $\vec{v}_{\alpha\beta} = v_\alpha + jv_\beta$, while the third and fourth components define the space vector $\vec{v}_{\alpha_2\beta_2} = v_{\alpha_2} + jv_{\beta_2}$.

In literature, different variations exist for the transformation, especially for the space vector $\vec{v}_{\alpha_2\beta_2}$. In the present work, it is obtained by a double incrementation of the power of $\dot{a}$, as in [27].

B. Harmonic contents in the space vectors

The voltage, or any other electrical quantity, of the h–th phase of the five–phase machine can be expressed as the sum of its harmonic components as the real part of:

$$
v_h = \sum_{\nu=-\infty}^{\infty} V_\nu e^{j(\nu\omega t + \varphi_\nu - \nu h \frac{2\pi}{5})} \quad (2)
$$

where $h = 0, 1, 2, 3, 4$ for phase a, b, c, d and e, respectively. Therefore, omitting the coefficient K the space vector in the $\alpha\beta$ reference frame becomes

$$
\vec{v}_{\alpha\beta} = \sum_{h=0}^{4} \sum_{\nu=-\infty}^{\infty} V_\nu e^{j(\nu\omega t + \varphi_\nu + (1-\nu)h\frac{2\pi}{5})} \quad (3)
$$

By analyzing (3), one finds out that only the harmonics of order $\nu = \pm 1 \pm 5k$, with $k = 0, \pm 1, \pm 2, ...$ appear in the space vector $\vec{v}_{\alpha\beta}$. Similarly, the space vector in the $\alpha_2\beta_2$ reference frame is:

$$
\vec{v}_{\alpha_2\beta_2} = \sum_{h=0}^{4} \sum_{\nu=-\infty}^{\infty} V_\nu e^{j(\nu\omega t + \varphi_\nu + (2-\nu)h\frac{2\pi}{5})} \quad (4)
$$

In this case, only the harmonics of order $\nu = \pm 2 \pm 5k$, with $k = 0, \pm 1, \pm 2, ...$ appear in the space vector $\vec{v}_{\alpha_2\beta_2}$.

The diagnostic index here employed is constructed starting from the combined space vector:

$$
\vec{D} = \vec{v}_{\alpha\beta} \cdot \vec{v'}_{\alpha_2\beta_2} \quad (5)
$$

where $\vec{v}_{\alpha\beta}$ is obtained in (3) and $\vec{v}'_{\alpha_2\beta_2}$ is the complex conjugate of $\vec{v}_{\alpha_2\beta_2}$, obtained in (4).

C. Asymmetry of one phase

Let us assume that a fault occurs in one phase coil and that it affects only the amplitudes of voltage harmonics. As reported in [18], in the event of a fault each voltage component is decreased. By defining κ the voltage reduction factor and a the faulty phase, the harmonic of ν–th order of the space vector $\vec{v}_{\alpha\beta}$ becomes:

$$\vec{v}_{\nu\alpha\beta} = \kappa V_\nu e^{j(\nu\omega t + \varphi_\nu)} + \sum_{h=1}^{4} V_\nu e^{j(\nu\omega t + \varphi_\nu + (1-\nu)h\frac{2\pi}{5})} \tag{6}$$

The amplitude of the first term can be split as:

$$\kappa V_\nu = V_\nu + V_\nu(\kappa - 1) \tag{7}$$

In healthy condition, the spectrum contains only the harmonics of order ν equal to $(\pm 1 \pm 5k)$ in the space vector $\vec{v}_{\alpha\beta}$ and the harmonics of order $\nu = (\pm 2 \pm 5k)$ in the space vector $\vec{v}_{\alpha_2\beta_2}$. Other harmonics form a symmetric five phase system, with sum equal to zero. Under fault conditions, for each harmonic system, appears the additional term:

$$V_\nu(\kappa - 1)e^{j(\nu\omega t + \varphi_\nu)} \tag{8}$$

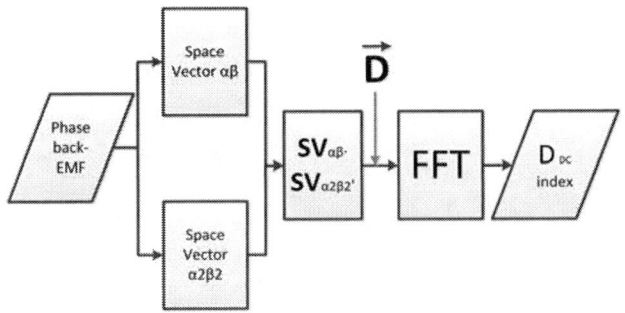

Fig. 1.   Diagnostic index construction

In the event of a fault, because of the machine unbalance, all the harmonic components appear in both the spectrum of $\vec{v}_{\alpha\beta}$ and $\vec{v}_{\alpha_2\beta_2}$. During fault condition, the combined space vector $\vec{D}$ is modified by the terms that arise in both $\vec{v}_{\alpha\beta}$ and $\vec{v}'_{\alpha_2\beta_2}$, as shown in Fig. 2.

If we consider a signal formed by several harmonics components (i.e. 1th, 3th, 5th, etc...); in

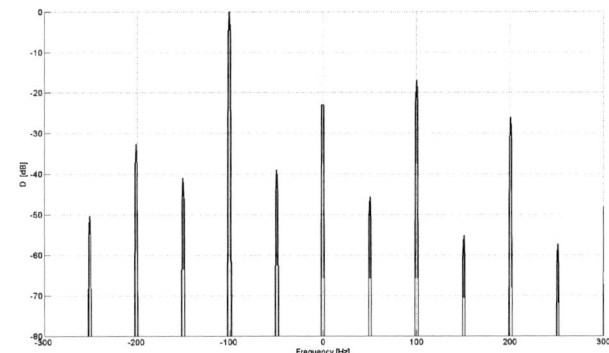

Fig. 2.   Combined space vector $\vec{D}$ computed from FEM simulations in case of partial stator short circuit.

case of healthy (ideal) condition only harmonics of frequency $(1 \pm 5k)f$ appear in the $\vec{v}_{\alpha\beta}$ reference frame and only the frequencies $(2\pm 5k)$ appear in the $\vec{v}'_{\alpha_2\beta_2}$ reference frame. In the event of a fault, all the frequencies that appear (with different amplitude) in both space vectors, in the combined space vector $\vec{D}$ will cause index terms in the form:

$$... + V_\nu e^{j(\nu\omega t + \varphi_\nu)} \cdot V_\nu e^{-j(\nu\omega t + \varphi_\nu)} + ... \tag{9}$$

The result from the multiplication of terms with opposite frequencies (i.e. $+f$ and $-f$) is the increase of the DC component (D_{DC}) in the combined space vector $\vec{D}$. The DC component of the combined space vector $\vec{D}$ is here used as the diagnostic index, as it will be different from zero in the event of a fault. Figure 1 shows the schematic diagram for the diagnostic index construction D_{DC}. In the present work, only the DC component is monitored to detect the occurrence of a stator short circuit fault.

III. SIMULATION RESULTS

A five–phase permanent–magnet machine with 20 stator slots and 18 poles was used in FEM modeling and experimental tests. Each stator coil is wound around a single tooth, achieving a single–layer winding with non-overlapped coils [9]. Each phase has a total number of turns equal to $N = 692$. Table I summarizes the characteristic parameters of the machine used in the tests.

The modeling of the machine under normal and faulty conditions were computed by means of Finite Element (FE) analysis, using FEMM software and custom scripts, Fig. 3. The linkage flux and back-EMF during healthy and faulty condition were computed by starting from FEA results, Fig. 4.

978-1-4799-0024-4/13 $31.00 © 2013 IEEE

TABLE I
Motor Parameters.

Name	Description	Value	Unit of Measurement
$2p$	number of poles	18	
R_f	Phase Resistance	23.6	Ω
L_f	Phase Inductance	82	mH
m	Number of Phases	5	
Q_s	Number of Slots	20	
D_e	External Diameter	120	mm
D_g	Gap Diameter	70	mm
L	Stack Length	50	mm
I_n	Rated Current	0.6	A_{rms}
n_n	Nominal Speed	333	rpm
T_n	Nominal Torque	5.5	Nm
P_n	Rated Power	200	W

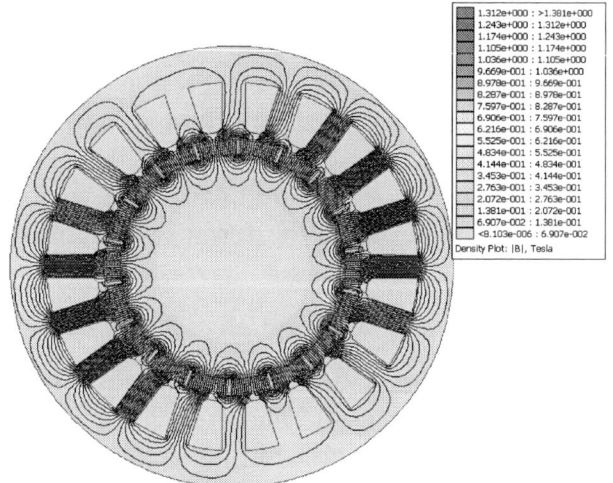

Fig. 3. FEMM simulation of the 5-phase machine.

The index $\vec{D}$ was computed at no-load condition and nominal speed (333 rpm), for the healthy machine (blue line) and for a partial short circuit in phase a in the case of 3% (red line), 6% (green line), 9% (black line) of turns short circuited, Fig. 5, according to the schematic outlined in Fig.1.

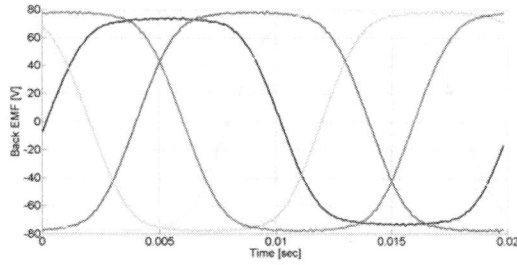

Fig. 4. Back emf computed from the FEM simulation

Figure 5 shows the results of the proposed index, for the machine modeled at different fault levels.

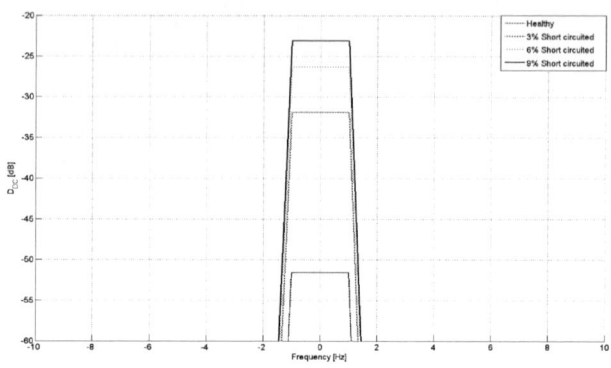

Fig. 5. Diagnostic index D_{DC} computed from FEM simulations under different levels of short circuit: blue - healthy machine; red - 3% of short circuited turns; green - 6% of short circuited turns; black - 9% of short circuited turns.)

The simulations show that the fault index is a good indicator of the stator short circuit, even in case of a very low number of short-circuited turns.

IV. Experimental Results

A prototype of the five–phase permanent–magnet machine with 20 slots and 9 pole pairs was built for the experimental tests. New windings were installed and one phase winding was center tapped with multiple taps to vary the amount of short circuited turns during fault simulation, Fig. 6.

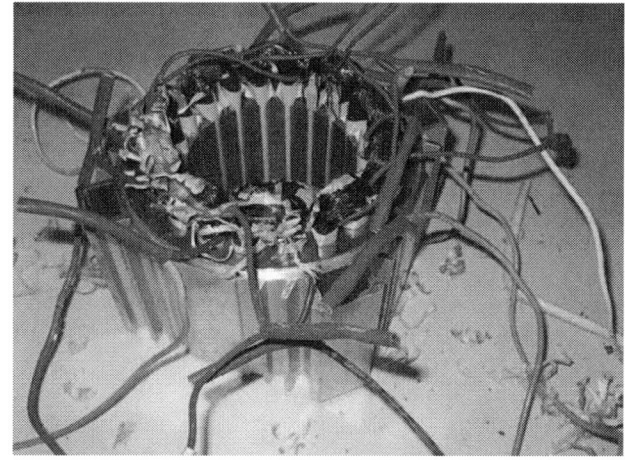

Fig. 6. Stator of the 5-phase machine during the re-winding process.

The machine was mounted on a test bench and was operated as a generator, Fig. 7. The back-emf were acquired at different speed and short circuit windings using two synchronized LeCroy WaveRunner 604Zi 12bit oscilloscopes . The sample frequency was equal to 100 kHz, the acquisition time was set to 5 seconds.

978-1-4799-0024-4/13 $31.00 © 2013 IEEE

Fig. 7. Experimental test setup.

The machine was run at a constant speed of 333 rpm and an increasing numbers of coils were short-circuited: 20, 40 and 60 turns, corresponding respectively to 3% 6% and 9% of the total winding turns.

The back-EMF acquired by the oscilloscopes was then elaborated in Matlab environment in order to reconstruct the proposed index. The voltage phase signal were combined to create the two space vector in the $\alpha\beta$ and $\alpha2\beta2$ reference frames. Then the combined vector $\vec{D}$ was obtained by multiplying the space vector in $\alpha\beta$ with the complex conjugate of the space vector in $\alpha2\beta2$, finally, FFT of the signal was computed and then the amplitude of the DC component is extracted to obtain the fault index D_{DC}. Figure 8 shows the results of the proposed index, for the machine running with different fault levels: healthy machine (blue), 3% of short circuited turns (red), 6% of short circuited turns (green), 9% of short circuited turns (black).

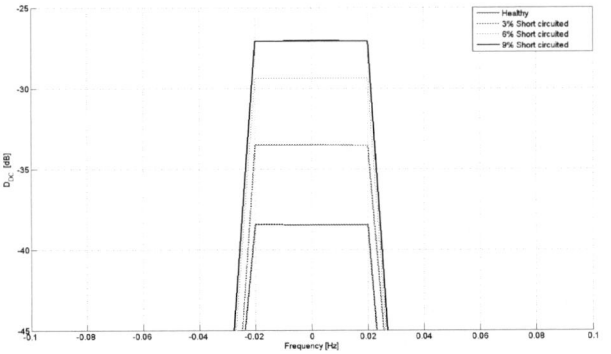

Fig. 8. Diagnostic index D_{DC} computed from experimental acquisitions under different levels of short circuit: blue - healthy machine; red - 3% of short circuited turns; green - 6% of short circuited turns; black - 9% of short circuited turns.

V. Conclusions

The detection of the short–circuit fault has been investigated relying on the analysis of a diagnostic index that combine signals from two different reference frames ($\alpha\beta$ and $\alpha_2\beta_2$) of the five phase machine. It is proved that the partial short–circuit fault can be efficiently detected by monitoring the amplitude of the proposed diagnostic index which gives a DC output signal equal to zero during normal operation and different from zero during a fault. The experimental results completely agree with the ones obtained from simulations. Moreover, as the experiments pointed out, the diagnostic index proved to be robust against the inherent unbalances in the construction of the real machine. The results in stator short circuit fault detection are promising. The effect of other faults (e.g. high resistance connection, magnet demagnetization, bearing/drivetrain faults) and fault detection and identification in the combined space vector $\vec{D}$ are currently being investigated and will be the subject of future work. A five–phase permanent–magnet machine designed for fault tolerant applications was employed in the experiments. It is characterized by a mutual inductance equal to zero and a high self inductance, in order to limit the short–circuit current. However, when short–circuit occurs on a limited number of turns, only the resistance limits the short–circuit current: therefore a quick detection of the fault is highly appreciated to avoid damage progression and catastrophic failure.

Acknowledgment

This work has been developed in the frame of PRIN2009 project High Reliability Multi-Phase Electric Drives for the More Electric Aircraft co-funded by the Italian Ministry of Research (MIUR).

The authors gratefully acknowledge the department of Electrical Engineering of the University of Padua, and in particular Professor Nicola Bianchi, for providing the prototype of the machine that was re-wound and employed in the experiments.

References

[1] L. Parsa, "On advantages of multi-phase machines," in *Industrial Electronics Society, 2005. IECON 2005. 31st Annual Conference of IEEE*, 2005, pp. 6 pp.–.

[2] E. Levi, "Multiphase electric machines for variable-speed applications," *Industrial Electronics, IEEE Transactions on*, vol. 55, no. 5, pp. 1893–1909, 2008.

[3] A. Cavagnino, Z. Li, A. Tenconi, and S. Vaschetto, "Integrated generator for more electric engine: Design and testing of a scaled size prototype," *Industry Applications, IEEE Transactions on*, vol. PP, no. 99, pp. 1–1, 2013.

[4] L. Parsa and H. Toliyat, "Five-phase permanent-magnet motor drives," *Industry Applications, IEEE Transactions on*, vol. 41, no. 1, pp. 30–37, 2005.

[5] M. Villani, M. Tursini, G. Fabri, and L. Castellini, "High reliability permanent magnet brushless motor drive for aircraft application," *Industrial Electronics, IEEE Transactions on*, vol. 59, no. 5, pp. 2073–2081, 2012.

[6] N. Bianchi, E. Fornasiero, and S. Bolognani, "Thermal analysis of a five-phase motor under faulty operations," *Industry Applications, IEEE Transactions on*, vol. PP, no. 99, pp. 1–1, 2013.

[7] M. Kang, J. Huang, J. Yang, D. Liu, and H. Jiang, "Strategies for the fault-tolerant current control of a multiphase machine under open phase conditions," in *Electrical Machines and Systems, 2009. ICEMS 2009. International Conference on*, 2009, pp. 1–6.

[8] N. Bianchi, S. Bolognani, and M. Pre, "Strategies for the fault-tolerant current control of a five-phase permanent-magnet motor," *Industry Applications, IEEE Transactions on*, vol. 43, no. 4, pp. 960–970, 2007.

[9] N. Bianchi, S. Bolognani, and M. Pre, "Design and tests of a fault-tolerant five-phase permanent magnet motor," in *Power Electronics Specialists Conference, 2006. PESC '06. 37th IEEE*, 2006, pp. 1–8.

[10] M. El Hachemi Benbouzid, "A review of induction motors signature analysis as a medium for faults detection," *Industrial Electronics, IEEE Transactions on*, vol. 47, no. 5, pp. 984–993, 2000.

[11] S. Nandi, H. Toliyat, and X. Li, "Condition monitoring and fault diagnosis of electrical motors-a review," *Energy Conversion, IEEE Transactions on*, vol. 20, no. 4, pp. 719–729, 2005.

[12] F. Immovilli, A. Bellini, R. Rubini, and C. Tassoni, "Diagnosis of bearing faults in induction machines by vibration or current signals: A critical comparison," *Industry Applications, IEEE Transactions on*, vol. 46, no. 4, pp. 1350–1359, 2010.

[13] A. Bellini, F. Filippetti, G. Franceschini, C. Tassoni, and G. Kliman, "Quantitative evaluation of induction motor broken bars by means of electrical signature analysis," *Industry Applications, IEEE Transactions on*, vol. 37, no. 5, pp. 1248–1255, 2001.

[14] L. Zarri, M. Mengoni, Y. Gritli, A. Tani, F. Filippetti, G. Serra, and D. Casadei, "Detection and localization of stator resistance dissymmetry based on multiple reference frame controllers in multiphase induction motor drives," *Industrial Electronics, IEEE Transactions on*, vol. 60, no. 8, pp. 3506–3518, 2013.

[15] L. Parsa and H. Toliyat, "Fault-tolerant five-phase permanent magnet motor drives," in *Industry Applications Conference, 2004. 39th IAS Annual Meeting. Conference Record of the 2004 IEEE*, vol. 2, 2004, pp. 1048–1054 vol.2.

[16] C. Jacobina, I. Freitas, T. Oliveira, E. da Silva, and A. M. N. Lima, "Fault tolerant control of five-phase ac motor drive," in *Power Electronics Specialists Conference, 2004. PESC 04. 2004 IEEE 35th Annual*, vol. 5, 2004, pp. 3486–3492 Vol.5.

[17] J. Apsley and S. Williamson, "Analysis of multiphase induction machines with winding faults," *Industry Applications, IEEE Transactions on*, vol. 42, no. 2, pp. 465–472, 2006.

[18] C. Bianchini, E. Fornasiero, T. Matzen, N. Bianchi, and A. Bellini, "Fault detection of a five-phase permanent-magnet machine," in *Industrial Electronics, 2008. IECON 2008. 34th Annual Conference of IEEE*, 2008, pp. 1200–1205.

[19] H.-M. Ryu, J.-W. Kim, and S.-K. Sul, "Synchronous-frame current control of multiphase synchronous motor under asymmetric fault condition due to open phases," *Industry Applications, IEEE Transactions on*, vol. 42, no. 4, pp. 1062–1070, 2006.

[20] N. Bianchi, S. Bolognani, and M. Pre, "Impact of stator winding of a five-phase permanent-magnet motor on postfault operations," *Industrial Electronics, IEEE Transactions on*, vol. 55, no. 5, pp. 1978–1987, 2008.

[21] C. Bianchini, E. Fornasiero, T. Matzen, N. Bianchi, and A. Bellini, "Stator fault detection for multi-phase machines with multiple reference frames transformation," in *Industrial Electronics, 2009. IECON '09. 35th Annual Conference of IEEE*, 2009, pp. 3467–3470.

[22] R. Bojoi, M. G. Neacsu, and A. Tenconi, "Analysis and survey of multi-phase power electronic converter topologies for the more electric aircraft applications," in *Power Electronics, Electrical Drives, Automation and Motion (SPEEDAM), 2012 International Symposium on*, 2012, pp. 440–445.

[23] O. Lopez, J. Alvarez, J. Doval-Gandoy, and F. Freijedo, "Multilevel multiphase space vector pwm algorithm," *Industrial Electronics, IEEE Transactions on*, vol. 55, no. 5, pp. 1933–1942, 2008.

[24] D. Casadei, D. Dujic, E. Levi, G. Serra, A. Tani, and L. Zarri, "General modulation strategy for seven-phase inverters with independent control of multiple voltage space vectors," *Industrial Electronics, IEEE Transactions on*, vol. 55, no. 5, pp. 1921–1932, 2008.

[25] L. De Lillo, L. Empringham, P. Wheeler, S. Khwan-On, C. Gerada, M. Othman, and X. Huang, "Multiphase power converter drive for fault-tolerant machine development in aerospace applications," *Industrial Electronics, IEEE Transactions on*, vol. 57, no. 2, pp. 575–583, 2010.

[26] A. Gandhi, T. Corrigan, and L. Parsa, "Recent advances in modeling and online detection of stator interturn faults in electrical motors," *Industrial Electronics, IEEE Transactions on*, vol. 58, no. 5, pp. 1564–1575, 2011.

[27] E. Levi, M. Jones, S. Vukosavic, A. Iqbal, and H. Toliyat, "Modeling, control, and experimental investigation of a five-phase series-connected two-motor drive with single inverter supply," *Industrial Electronics, IEEE Transactions on*, vol. 54, no. 3, pp. 1504–1516, 2007.

978-1-4799-0024-4/13 $31.00 © 2013 IEEE

The Performance of a Three-Phase Induction Motor fed by a Three-Level NPC Converter with Fault Tolerant Control Strategies

B. R. O. Baptista, M.B. Abadi, A.M. S. Mendes, S. M. A. Cruz

Abstract – **This paper presents the results of an experimental investigation regarding the performance of a three-phase induction motor fed by a three-level neutral-point-clamped converter with fault tolerant control strategies. A discussion about the fault diagnostic method used in this work, together with the used fault tolerant control strategy is presented. Experimental results regarding the thermal, electrical and mechanical behavior of a three-phase induction machine as well as the global efficiency of the ac drive under inverter normal and post-fault reconfiguration operations are discussed.**

Index Terms-- AC machines, fault diagnosis, fault tolerance, induction motor, multilevel converter, NPC converter, power quality, reliability, temperature.

I. INTRODUCTION

In recent years, multilevel converters (MC) have received increasing interest as shown by the large number of publications that can be found in the literature and their use in power conversion is also growing [1], [2]. The MC brings several advantages over a conventional two-level converter, e.g. better output voltage waveform quality, low common-mode input voltage, current with low distortion and they can operate at both high switching frequency and low switching frequency [3]-[5].

In many industrial applications, multilevel converters can feed motors which are the basis of the driving force in critical processes. In order to maintain service continuity and to prevent substantial losses, several authors have proposed hardware and control strategies in such a way that the driver continues to operate even in a failure situation of its components.

Some of the existing solutions for fault tolerance include: connection to the neutral point of the stator windings, inverters with redundant number of switches and systems with larger number of phases. The use of such converters

---

The authors wish to acknowledge the financial support of the Portuguese Science Foundation (FCT - Fundação para a Ciência e Tecnologia) under project number PTDC/EEA-EEL/100156/2008, titled "Fault Diagnosis in High Power Drives Based on Multilevel Converters".

Bruno Baptista [(1)], Mohsen Bandar Abadi[(2)], A. M. S. Mendes[(3)] and S. M. A. Cruz[(4)] are with the University of Coimbra/Instituto de Telecomunicações; Department of Electrical and Computer Engineering, Pólo II - Pinhal de Marrocos, P - 3030-290 Coimbra, Portugal; [(1)]brunobapt@co.it.pt,   [(2)]m.b.abadi@ieee.org,   [(3)]amsmendes@ieee.org; [(4)]smacruz@ieee.org.

leads to increased system reliability [1], [6], [7].

Unfortunately, as attractive as the fault tolerant topology is, it has some drawbacks. In many cases the continuity of service (under fault conditions) is guaranteed by a type of supply that is distant from that which was taken as reference in the design of the motor, since these fault tolerant strategies typically cause a reduction in the quality of the supply voltage waveform [6]. This supply condition can result in a different electromagnetic-thermal behavior of the motor, leading to a reduction of the efficiency and an increase in its temperature. This gives rise to an accelerated motor windings ageing, eventually leads to the breakdown of their insulation. The thermal behavior of the motor deserves great importance, since the temperature is a determining factor in the life of the machine. According to Montsinger the life of the insulation system is halved for each additional 10º C in the temperature at which it is exposed [8].

In a recent past, an experimental study regarding the thermal behavior of an induction motor fed by a two-level inverter with a fault-tolerant operating strategy was conducted [9]. However, in the context of multilevel inverters there is a paucity of published literature directed towards the thermal behavior of the motor.

In this paper, a comparative study of the thermal behavior and efficiency of an ac drive comprising a motor and a three-level neutral-point-clamped (3LNPC) inverter with fault tolerant control strategies was carried out.

The organization of the paper is as follows. After the introduction in section I, the 3LNPC control strategy and modulation technique is discussed for normal and faulty operation in section II. After that, the experimental prototype is presented in section III. In Section IV the obtained experimental results are presented and discussed. Finally, Section V presents the main conclusions of this paper.

II. THREE-LEVEL NPC CONTROL STRATEGY

Fig. 1 shows the topology of a three-level neutral-point-clamped (3LNPC) [10], it consists in three legs, each one with four active switches, labeled S_{x1}, S_{x2}, S_{x3}, S_{x4}, and two voltage clamping diodes labeled D_{x1} D_{x2} (x means leg A, B or C). In this work, the active switches are IGBTs. The DC bus has 2 capacitors, thus providing the middle point O.

Each inverter leg is usually characterized by three possible switching states, in this work designated by P, O, and N, according to the information provided in Table I for

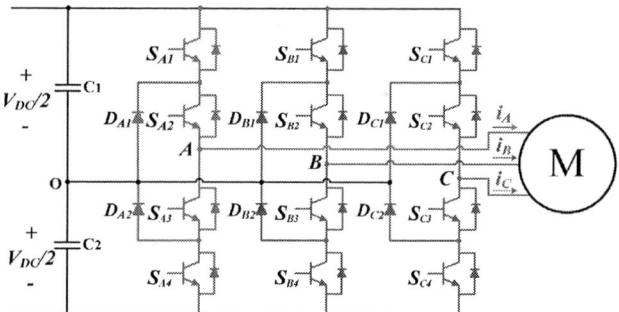

Fig. 1. Schematic representation of a 3LNPC AC motor drive.

the switches of leg x.

As can be observed in this table, if a P state is imposed by the controller to leg A, the switches S_{x1} and S_{x2} are both on. The states of IGBTs S_{x3} and S_{x4} are always the complementary of S_{x1} and S_{x2}, respectively. Under this control situation, the leg output pole voltage is $+V_{DC}/2$. If a converter global switching state is "PON" that means the output terminal of leg A is connected to the positive terminal of the DC bus, the terminal of leg B is connected to the DC bus middle point O and the terminal of leg C is connected to the negative terminal of the DC bus. Therefore, the leg output pole voltages v_{AO}, v_{BO}, v_{CO} are $+V_{DC}/2$, 0, and $-V_{DC}/2$, respectively. The converter switching states are dependent of the controller in order to reply to three-phase induction motor supply needs, by selecting the appropriate reference vector $\vec{V}_{ref} = \left| \vec{V}_{ref} \right| e^{j\theta}$.

To modulate the reference voltage vector $\vec{V}_{ref}$, several options are available for 3LNPC inverters. Space vector pulse width modulation (SVPWM) is one of the preferred real-time modulation techniques and is widely used for digital control of voltage source inverters [11], therefore, it was adopted in this work. Taking all three phases into account, the inverter has a total of 27 possible combinations of switching states. The 27 states correspond to 19 voltage vectors. Based on their magnitude (length), the voltage space vectors can be divided into four groups, designated by Zero Vectors and Small Vectors (TABLE II), Medium Vectors (TABLE III) and Large Vectors (TABLE IV). The Small Vectors have a length of $1/3V_{DC}$, while the Medium Vectors and Large Vectors have a length of $\sqrt{3}/3V_{DC}$ and $2/3\ V_{DC}$, respectively. The applied SVPWM technique is based on even-order harmonic elimination [12]. To implement the SVPWM technique, the complex plane is divided in 6 sectors (S_1 to S_6), each one with 6 regions (R_1 to R_6), as can be observed in Fig. 2 where the vector $\vec{V}_{ref}$ is positioned on the region 4 of sector I.

TABLE I
SWITCHING STATES OF LEG X

Leg state	Output voltage (v_{xo})	IGBTs state			
		S_{x1}	S_{x2}	S_{x3}	S_{x4}
P	$+V_{DC}/2$	on	on	off	off
O	0	off	on	on	off
N	$-V_{DC}/2$	off	off	on	on

TABLE II
ZERO AND SMALL VECTORS CHARACTERISTICS

Voltage space vector type	Zero vectors	Small vectors					
Length	0	$1/3\ V_{DC}$					
Label	V_0	V_1	V_2	V_3	V_4	V_5	V_6
Legs states	PPP OOO NNN	POO ONN	PPO OON	OPO NON	OPP NNO	OOP NNO	POP ONO

TABLE III
MEDIUM VECTORS CHARACTERISTICS

Length	$\sqrt{3}/3\ V_{DC}$					
Label	V_7	V_8	V_9	V_{10}	V_{11}	V_{12}
Legs States	PON	OPN	NPO	NOP	ONP	PNO

TABLE IV
LARGE VECTORS CHARACTERISTICS

Length	$2/3\ V_{DC}$					
Vector label	V_{13}	V_{14}	V_{15}	V_{16}	V_{17}	V_{18}
Leg States	PNN	PPN	NPN	NPP	NNP	PNP

Thus, this vector can be synthesized by three nearby stationary vectors of this region that are $\vec{V}_1$, $\vec{V}_2$ and $\vec{V}_7$. Assuming a small sampling period ST, the volt-second balancing equation for this example is given by (1) where T_a, T_b and T_c are the dwell times for the corresponding vectors $\vec{V}_1$, $\vec{V}_2$ and $\vec{V}_7$, respectively, given by (2), where m_a is modulation index and β is the angle of $\vec{V}_{ref}$ inside each 60° sector [12].

$$\vec{V}_{ref}.ST = \vec{V}_1.T_a + \vec{V}_7.T_b + \vec{V}_2.T_c$$
$$ST = T_a + T_b + T_c \tag{1}$$

$$T_a = ST(1 - 2.m_a.\sin\beta)$$
$$T_b = ST(2.m_a.\sin(\frac{\pi}{3}+\beta) - 1) \tag{2}$$
$$T_c = ST - (T_a + T_b)$$

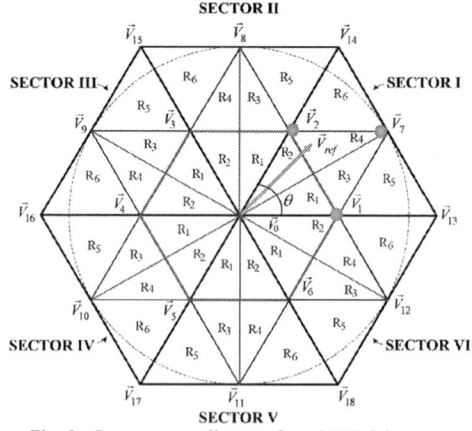

Fig. 2. Space vector diagram for a 3LNPC inverter.

In order to even-order harmonic elimination, the switching sequence in each *ST* should be divided in 7 parts (P_1 to P_7). Table V shows the switching sequence and dwell time of each seven-part switching sequence for the 3LNPC inverter when $\vec{V}_{ref}$ is located in S_1 and R_4.

Similar equations can be used to synthesize a reference voltage vector positioned in any other region number of any other sector as presented in [12].

A global schematic of a SVPWM for a 3LNPC inverter is shown in Fig. 3. As presented in this figure, to obtain the required IGBT pulses corresponding to a specific $\vec{V}_{ref}$ is necessary to know S_n, R_n and P_n. After that, by using a lookup table where all possible S_n, R_n are considered according with [12], the corresponding leg states are obtained.

TABLE V

DWELL TIME OF EACH PART WHEN $\vec{V}_{ref}$ IS LOCATED IN S_1 AND R_4

Part number	1	2	3	4	5	6	7
Voltage vector	$\vec{V}_2$	$\vec{V}_7$	$\vec{V}_1$	$\vec{V}_2$	$\vec{V}_1$	$\vec{V}_7$	$\vec{V}_2$
Dwell time	$\dfrac{T_c}{4}$	$\dfrac{T_b}{2}$	$\dfrac{T_a}{2}$	$\dfrac{T_c}{2}$	$\dfrac{T_a}{2}$	$\dfrac{T_b}{2}$	$\dfrac{T_c}{4}$

A. Behavior of the inverter under IGBT open-circuit Fault

When an IGBT open-circuit (OC) fault occurs, the faulty IGBT is unable to commutate in agreement with the output pulses generated by the modulator and remains in off-state permanently [13]. In this case, some inverter output voltage space vectors are unavailable due the impossibility to use some leg states.

Table VI shows the available and unavailable switching sates due to an OC fault in one semiconductor of leg x [13]. This table shows that when an OC fault occurs in one of the leg outer switches, all that leg output voltages that make use of the switching states P, (for S_{x1} fault) and N (for S_{x4} fault) are not available. If an OC fault occurs in an inner leg IGBT, the number of unavailable states is higher when compared with an outer IGBT fault.

Table VI shows also all possible output pole voltages of leg x (v_{xO}) for the inverter under an IGBT OC fault located in leg x.

B. Fault tolerant topology

With regard to the inverter fault tolerance capability, the main objective is to ensure the continuous operation of the system under faulty condition until the next maintenance service opportunity arrives.

In this work, a fault tolerant strategy with three legs, each one connected to the neutral-point (NP) of the inverter through a Triac (or equivalent) is used [13]. The purpose of these extra semiconductors is to connect the faulty leg to the NP when any of their switches fail. In this case, all the switching signals of the faulty leg are turned off. Under faulty situation only two converter legs are controllable but the motor continues to be powered by three balanced currents. Table VII lists all possible switching states after inverter post-fault reconfiguration, following an OC fault in one leg. As can be seen, this table shows that after connection the faulty leg to the NP, all the Large Vectors and some of the Medium Vectors are not available. Therefore, the maximum value for $\vec{V}_{ref}$ that can be synthetized by the converter under this hardware reconfiguration is halved. In order to compensate this output voltage reduction, the DC voltage should increase to double.

Moreover, to keep a balanced rotating voltage space vector, the Small Vectors should be the ones that must be used. In this inverter operation mode, the lengths of all Small Vectors are increased to the length of the Large Vectors used before inverter reconfiguration. Thus, the size of the small hexagon represented in Fig. 2 (orange color) should be increase to double. After that, the inverter works as 2-level topology. Subsequently, the fault tolerant operation can be achieved by using a modified SVPWM. Fig. 4 shows the block diagram of a SVPWM implementation for a post-fault reconfiguration operation. The IGBTs command pulses for a $\vec{V}_{ref}$ are now obtained from a new lookup table that uses only the Sector Number, Part Number and Faulty Phase Number as new input signals.

TABLE VI

INVERTER SITUATION UNDER AN OC FAULT IN LEG X

Condition of IGBTs located in leg x				Available state(s)	Unavailable state(s)	Possible output leg voltage v_{xO}
S_{x1}	S_{x2}	S_{x3}	S_{x4}			
OC	ok	ok	ok	O, N	P	$-V_{DC}/2$, 0
ok	OC	ok	ok	N	P, O	$-V_{DC}/2$
ok	ok	OC	ok	P	O, N	$V_{DC}/2$
ok	ok	ok	OC	P, O	N	0, $V_{DC}/2$

TABLE VII

AVAILABLE SWITCHING STATES AFTER SYSTEM RECONFIGURATION

Faulty leg	Available switching states			
	Zero vectors	Small vectors	Medium vectors	Large vectors
A	OOO	ONN, OON, OPO, OPP, OOP, ONO	ONP, OPN	---
B	OOO	POO, OON, NON, NOO, OOP, POP	PON, NOP	---
C	OOO	POO, PPO, OPO, NOO, NNO, ONO	NPO, PNO	---

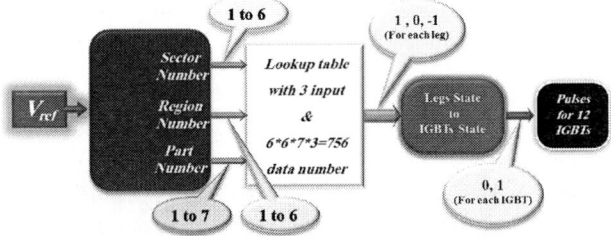

Fig. 3. Block diagram of a SVPWM implementation under inverter normal operation.

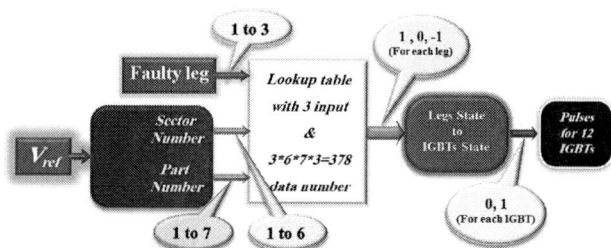

Fig. 4. Block diagram of a SVPWM implementation under post-fault reconfiguration operation.

C. Fault diagnosis system

After fault occurrence, in order to apply the correct fault tolerant control strategy, a fault diagnosis system is necessary. According with the purpose of this work, to implement the fault tolerant control strategy, the identification of the faulty leg is enough. Therefore, a fault diagnosis approach based on the average current Park's vector (ACPV) [14], [15] was implemented. With this diagnostic technique, the detection of an OC fault and the localization of the faulty leg are accomplished by examining the magnitude and position of the Park's vector of the average value of each motor line current. As the induction motor is fed by three wires, only two output currents are measured and acquired for diagnostic purposes, being the third current i_C estimated by $i_C = -i_A - i_B$. To implement this diagnostic method, the average value of each induction motor line current $(I_{A_{av}}, I_{B_{av}}, I_{C_{av}})$ is calculated by (3), where N is the total number of samples, k is the index of a sample and x is the leg index. Afterwards, the Park's vector transformation is applied to those values, in order to obtain the average motor line current Park's vector $\vec{I}_{s_{av}}$, characterized by a magnitude $\left|\vec{I}_{s_{av}}\right|$, and a phase angle $\theta_{s_{av}}$, given by (4).

$$I_{x_{av}} = \frac{1}{N} \sum_{k=1}^{N} i_{x,k}, \quad x \in \{A,B,C\} \tag{3}$$

$$\begin{cases} \vec{I}_{s_{av}} = I_{d_{av}} + j.I_{q_{av}} = \left|\vec{I}_{s_{av}}\right| \angle \theta_{s_{av}} \\[2mm] I_{d_{av}} = \frac{2}{3} I_{a_{av}} - \frac{1}{3}(I_{b_{av}} + I_{c_{av}}) \\[2mm] I_{q_{av}} = \frac{1}{\sqrt{3}}(I_{b_{av}} - I_{c_{av}}) \end{cases} \tag{4}$$

In order to obtain a diagnostic method independent of the load level and load rated power, $\vec{I}_{s_{av}}$ needs to be normalized. For that purpose, the motor line current Park's Vector magnitude $\left|\vec{I}_s\right|$ is calculated from the motor line

currents i_A, i_B, i_C as illustrated by (5). The normalized ACPV magnitude $\left|\vec{I}_{s_{av}}n\right|$ is then obtained by (6).

$$\begin{cases} \vec{I}_s = i_{s_d} + j.i_{s_q} = \left|\vec{I}_s\right| \angle \theta_s \\[2mm] i_{s_d} = \frac{2}{3} i_A - \frac{1}{3}(i_B + i_C) \\[2mm] i_{s_q} = \frac{1}{\sqrt{3}}(i_B - i_C) \end{cases} \tag{5}$$

$$\left|\vec{I}_{s_{av}}n\right| = \frac{\left|\vec{I}_{s_{av}}\right|}{\left|\vec{I}_s\right|} \tag{6}$$

The value of $\left|\vec{I}_{s_{av}}n\right|$ is used to detect an IGBT OC fault that under normal operation, is almost zero. However, if a fault occurs, $\left|\vec{I}_{s_{av}}n\right|$ will increase and assume a value higher than a predefined threshold value (near zero) and the OC fault is therefore detected.

To identify the faulty IGBT pair, the angle $\theta_{s_{av}}$ is used. Under normal operating conditions this angle is changing continuously between 0° and 360°. If a switch OC fault occurs, $\theta_{s_{av}}$ will have an almost constant value. Knowing the interval of Table VIII that better fits $\theta_{s_{av}}$ it is possible to identify the faulty IGBT pair and, consequently, the faulty leg. After this faulty leg identification, is possible to apply the corresponding fault tolerant control strategy.

TABLE VIII
LOOKUP TABLE TO FIND THE FAULTY IGBT PAIR

$\theta_{s_{av}}$	Faulty IGBTs	Faulty pair	Faulty leg
$150° < \theta_{s_{av}} < 210°$	S_{A1} or S_{A2}	1	A
$-30° < \theta_{s_{av}} < 30°$	S_{A3} or S_{A4}	2	
$270° < \theta_{s_{av}} < 330°$	S_{B1} or S_{B2}	3	B
$90° < \theta_{s_{av}} < 150°$	S_{B3} or S_{B4}	4	
$30° < \theta_{s_{av}} < 90°$	S_{C1} or S_{C2}	5	C
$210° < \theta_{s_{av}} < 270°$	S_{C3} or S_{C4}	6	

III. EXPERIMENTAL SETUP DETAILS

The experimental setup of this work consists in a WEG W22, F-class, TEFC, 4 kW, 400 V, 50 Hz, 8.0 A, 1450 rpm, three-phase squirrel-cage induction motor connected to a 3LNPC with fault tolerant capability. The stator windings are delta-connected. The motor under test is mechanically coupled to a WEG 4 kW permanent magnet synchronous generator that feeds a variable resistor load, which allows the user to impose any load torque profile.

The control strategies of the inverter were first simulated in the Matlab/Simulink environment and then converted to a

978-1-4799-0024-4/13 $31.00 © 2013 IEEE   500

dSPACE platform. This platform, allows a real-time monitoring of voltage, current and other parameters selected by the user. A global drive control schematic is shown in Fig. 5.

The 3LNPC inverter prototype was developed, based on Semix IGBT modules, with SKYPER 32 PRO drivers that constitute an interface between the IGBT modules and the controller. The inverter was supplied by a constant DC voltage of 165 V for normal operation mode and 330 V for a post-fault reconfiguration condition.

Tests were conducted for different values of the induction motor load torque and different speed references. Nevertheless, the results presented in this paper were obtained for a speed reference of 500 rpm and three load torque values: no-load, 5 N.m and 7 N.m.

During experimental test was used a power analyzer from Yokogawa that allows the acquisition of electrical and mechanical data. To perform the tests on the thermal behavior of the induction motor several sensors of the PT100 type were placed inside the machine as illustrated in Fig. 6. Fig. 6(a) shows the location of the sensors in the stator side distributed by the three windings as follows: S_1 placed in windings of phase C, S_2 and S_3 placed in windings of phase B and S_4 and S_5 placed in the windings of phase A. In addition, the stator core, the motor housing and room temperatures were also measured. Fig. 6(b) shows the rotor sensors positions placed 30 mm deep inside the rotor, labeled R_1, R_2, R_3 and R_4.

The signals from all sensors are conditioned by means of two independent data acquisition circuits. The first circuit includes the stator, core and room sensors while the second circuit includes the rotor sensors only. The system for conditioning the signals from the temperature sensors of the rotor was placed at one end of the motor shaft. All thermal sensor signals are sent to a PC through a wireless data acquisition system.

All thermal, electrical and mechanical data acquisition process was controlled and stored by software developed in *LabView* platform. A global view of the experimental setup is shown in Fig. 7.

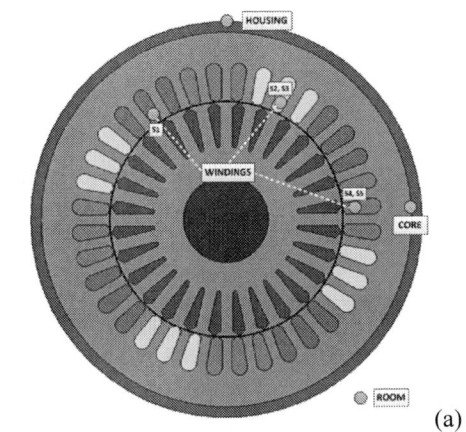

(a)

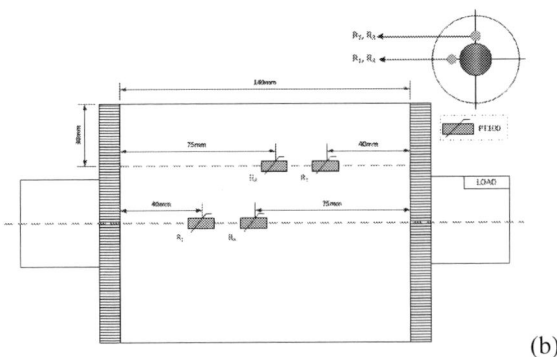

(b)

Fig. 6. Induction motor scheme with PT100 sensors location.

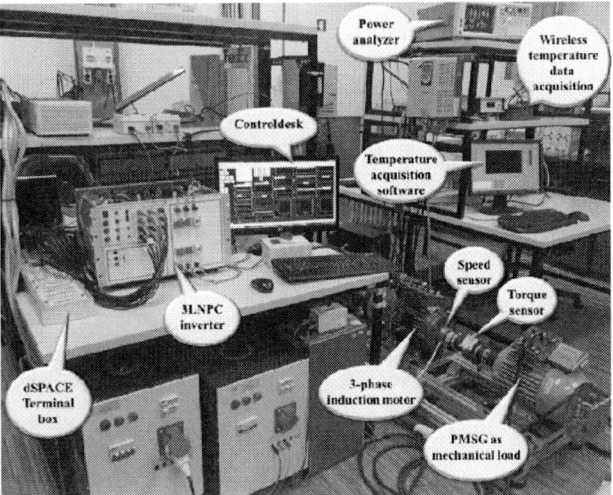

Fig. 7. Experimental setup overview.

IV. EXPERIMENTAL RESULTS

Each induction motor test ran for at least 3 hours to reach the steady state temperature, in accordance with the IEEE Std 112.

All thermal, electrical and mechanical parameters were acquired with a sample time of 1 minute. Motor line voltage and line currents were acquired with a sample time of 0.5 ms.

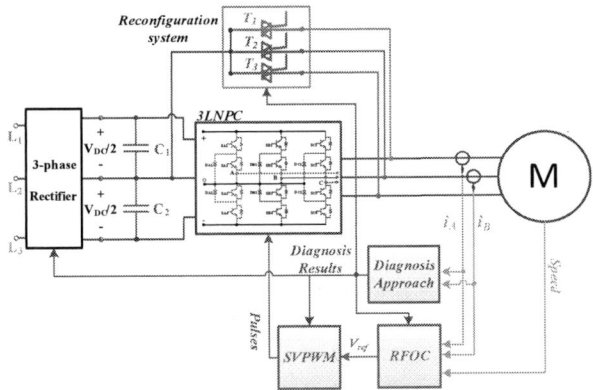

Fig. 5. Global drive control schematic.

Under 3LNPC normal operation the obtained motor line currents and line voltage V_{AB} are represented in Fig. 8 and Fig. 9, respectively, for a reference speed of 500 rpm and a load torque of 5 N.m. The others two motor line voltages waveforms are similar to this one. The corresponding mechanical data acquired during this test is shown in Fig. 10. As can be seen the electromagnetic torque, speed and consequently mechanical power were kept constant during the entire test which lasted for three-hours. Keeping constant the same mechanical output conditions, the post-fault reconfiguration strategy was introduced in the 3LNPC inverter. For this operation mode, the motor line currents and motor line voltages obtained for the inverter post-fault reconfiguration corresponding to an IGBT OC fault of inverter leg *A* are represented in Fig. 11 and Fig. 12, respectively.

As can be observed from the comparison between the line current waveforms, obtained under normal and faulty conditions, the amplitudes have similar values. However, regarding the V_{AB} voltage waveforms, they present different shapes. Under inverter normal operation, the V_{AB} voltage shows a typical three-level inverter output voltage while for inverter post-fault reconfiguration this voltage shows a typical two-level inverter output voltage. Thus, the total harmonic distortion of this voltage increases.

The electrical data that includes the total, active and reactive powers, power factor and efficiency for inverter normal and post-fault reconfiguration operations are shown in Fig. 13. The shadowed lines correspond to inverter post-fault operation. As can be observed all powers were increased in fault condition of the inverter.

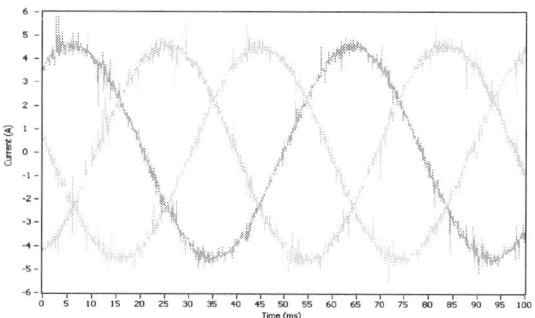

Fig. 8. Motor line currents under inverter normal operation.

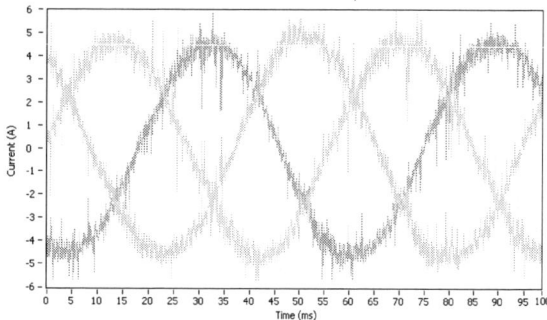

Fig. 11. Motor line currents under inverter post-fault reconfiguration.

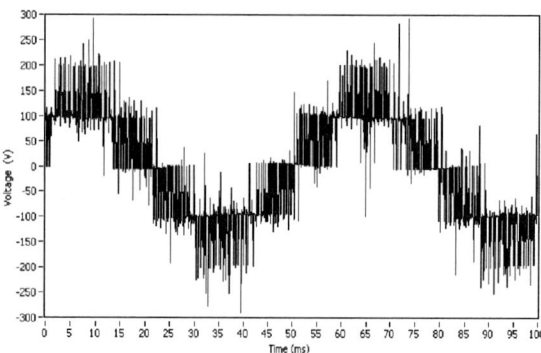

Fig. 9. Motor line voltages V_{AB} under inverter normal operation.

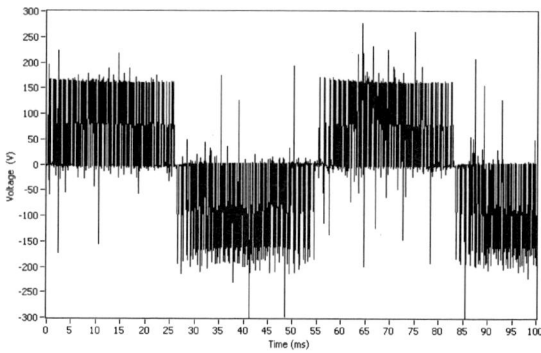

Fig. 12. Motor line voltage V_{AB} under inverter post-fault reconfiguration.

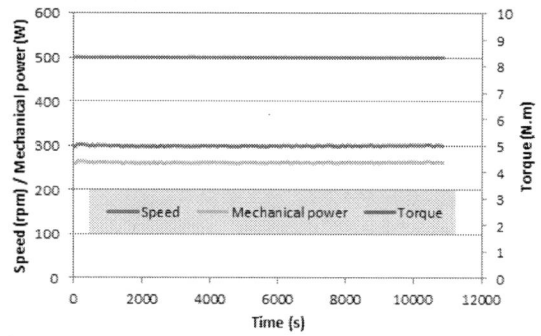

Fig. 10. Motor mechanical data under inverter normal operation.

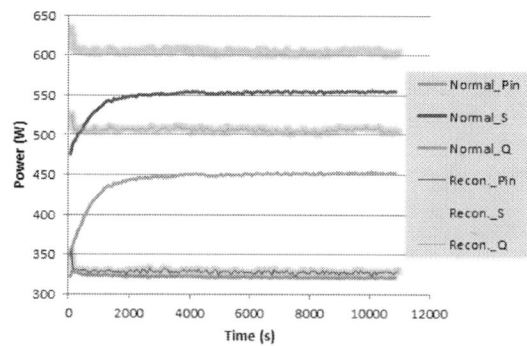

Fig. 13. Motor electrical data under inverter normal and post-fault reconfiguration operations.

978-1-4799-0024-4/13 $31.00 © 2013 IEEE

The motor temperature under normal condition of the inverter is presented in Fig. 14. This figure shows the evolution of the temperature obtained from all temperature sensors in the motor, and also the room temperature to be taken as reference.

The temperature increase in the stator and the rotor with respect to room temperature is shown in the bar graph of Fig. 15. It is particularly important to analyze the temperature rise in the stator winding, since the lifetime of winding insulation depends on the temperature at which the motor will operates. As stated earlier by the Montsinger rule, an increase of a few degrees in temperature can lead to a decrease of several years of useful life of the insulation system. For all cases, a temperature increase in the motor when the inverter is working in reconfigured mode was observed. In the specific case of 7 N.m there was an increase around ~10 °C.

The motor line current before and after the inverter reconfiguration is presented in Table IX. In addition, the percentage increase of the iron losses coefficients, η and χ, as defined in [16], are presented. The η is the coefficient which corresponds to the hysteresis losses and χ to the eddy-current losses. As can be seen in Table IX, in the case of fault condition there was an increase in total rms current and coefficients of iron losses. This has naturally led to increased temperature as shown in Fig. 15.

The inverter, motor and drive efficiencies are summarized in the bar chart in Fig. 16. As can be seen from this figure, the efficiency decreases in direction of lower loads. Furthermore, the reconfiguration mode always presents lower efficiency values for each individual operating point depicted.

TABLE IX
AVERAGE RECTIFIED MOTOR CURRENTS AND COEFFICIENTS OF IRON LOSSES

Torque (N.m)	I_{ABC} before fault (A)	I_{ABC} after fault (A)	η ↑(%)	χ↑ (%)
5	3.23	3.3	0.6	11.9
7	3.83	4.62	1.3	12.1

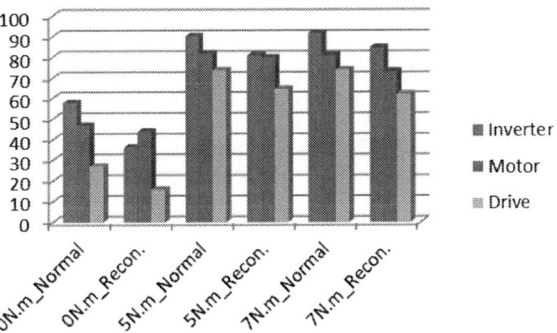

Fig. 16. Bar Chart – Efficiency.

V. CONCLUSION

The 3LNPC inverter has proven to be able to feed the motor under normal conditions as well as after a post-fault reconfiguration for the operating points analyzed. When the inverter operates in reconfigured mode, it starts working as a two levels inverter, resulting in quality loss of the waveform voltage. The results showed that this loss manifests itself in an increase of total rms current and iron losses coefficients. This leads to higher motor losses and consequently an increase in motor temperature. The thermal results showed that in all cases there was an increase in temperature for reconfigured mode. For instance in the case of 7 N.m, there was an increase (~10 °C), even though this is not a heavy load. Since the insulation lifespan of the windings is directly related to the temperature, the reconfiguration mode, if active for long periods of time, can shorten the lifetime of the motor.

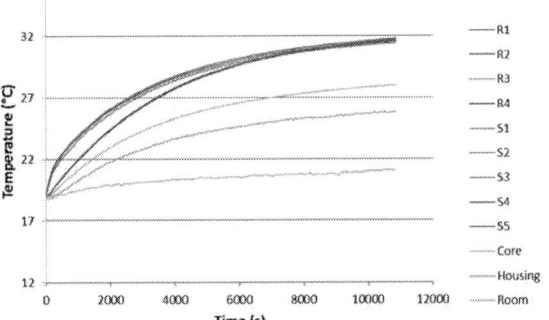

Fig. 14. Motor temperature evolution – All sensors.

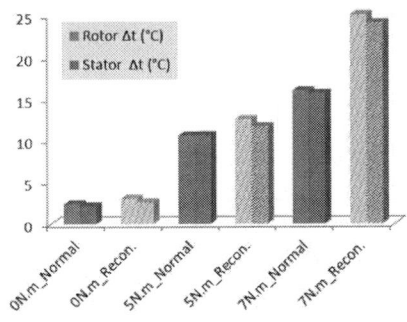

Fig. 15. Maximum temperature increase in the stator and the rotor over the room temperature for three differrent load torque.

VI. REFERENCES

[1] S. Kouro, M. Malinowski, K. Gopakumar, J. Pou, L. G. Franquelo, B. Wu, J. Rodriguez, M. A. Pérez, J. I. Leon,."Recent Advances and Industrial Applications of Multilevel Converters", *IEEE Trans. Industrial Electronics*, vol.57, no.8, pp. 2553-2580, August 2010.

[2] L. G. Franquelo, J. Rodriguez, J. I. Leon, S. Kouro,.R. Portillo, M. A. M. Prats, "The age of multilevel converters arrives", *IEEE Industrial Electronics Magazine*, vol.2, no.2, pp. 28,39, June 2008.

[3] J. Rodriguez, J. S. Lai, F. Z. Peng, "Multilevel inverters: a survey of topologies, controls, and applications", *IEEE Trans. Industrial Electronics*, vol.49, no.4, pp. 724- 738, Aug 2002.

[4] J. S. Lai, F. Z. Peng, "Multilevel converters-a new breed of power converters", *IEEE Trans. Industry Applications*, vol.32, no.3, pp. 509-517, May/Jun 1996.

[5] E. Najafi, A. H. M. Yatim, "Design and Implementation of a New Multilevel Inverter Topology", *IEEE Trans. Industrial Electronics*, vol.59, no.11, pp. 4148-4154, Nov. 2012.

[6] S. Ceballos, J. Pou, E. Robles, J. Zaragoza, J. L. Martín, "Performance Evaluation of Fault-Tolerant Neutral-Point-Clamped Converters", *IEEE Trans. Industrial Electronics*, vol.57, no.8, pp. 2709-2718, August 2010.

[7] P. Lezana, J. Pou, T.A. Meynard, J. Rodriguez, S. Ceballos, F. Richardeau, "Survey on Fault Operation on Multilevel Inverters" *IEEE Trans. Industrial Electronics*, vol.57, no.7, pp.2207-2218, July 2010.

[8] B. Baptista, A. M. S. Mendes, S. M. A. Cruz, A. J. M. Cardoso, "Temperature Distribution Inside a Three-Phase Induction Motor Running with Eccentric Airgap", Przeglad Elektrotechniczny, Vol. 88, No. 1a, pp. 96 - 99, January 2012.

[9] A. M. S. Mendes, X. M. Lopez-Fernandez, A. J. M. Cardoso, "Thermal Behavior of a Three-Phase Induction Motor Fed by a Fault-Tolerant Voltage Source Inverter", *IEEE Trans. Industry Applications*, vol.43, no.3, pp. 724-730, May-June 2007.

[10] A. Nabae, I. Takahashi, H. Akagi, "A New Neutral-Point-Clamped PWM Inverter", *IEEE Trans. Industry Applications*, vol. IA-17, NO. 5. pp. 518-523, September 1981.

[11] D.W. Feng, B. Wu, S. Wei, D. Xu, "Space vector modulation for neutral point clamped multilevel inverter with even order harmonic elimination", *in Proc. CCECE. 2004*, pp. 1471- 1475.

[12] B. Wu, "High-Power Converters and AC Drives", New Jersey: John Wiley & Sons, 2006. pp. 143-177.

[13] J. J., Park, T. J. Kim, D. S. Hyun "Study of neutral point potential variation for three-level NPC inverter under fault condition", *in Proc. IEEE IECON. 2008*. pp. 983-988.

[14] A. M. S. Mendes, A. J. M. Cardoso, "Voltage source inverter fault diagnosis in variable speed AC drives, by the average current Park's vector approach", *in Proc. IEMD. 1999*. pp. 704-706.

[15] M. B. Abadi, A. M. S. Mendes, S. M. A. Cruz, "Three-Level NPC Inverter Fault Diagnosis by the Average Current Park's Vector Approach", *in Proc. IEEE ICEM 2012*. pp. 1893-1898.

[16] Boglietti, A.; Cavagnino, A.; Ionel, D.M.; Popescu, M.; Staton, D.A.; Vaschetto, S., "A General Model to Predict the Iron Losses in PWM Inverter-Fed Induction Motors", IEEE Transactions on Industry Applications, vol.46, no.5, pp.1882-1890, Sept.-Oct. 2010.

VII. BIOGRAPHIES

B. R. O. Baptista was born in Coimbra, Portugal, in 1982. He received the B.S. degree in electrical engineering from Coimbra Institute of Engineering (ISEC), Coimbra, Portugal, in 2006, the M.S. degree in electrical engineering and the Diploma of Advanced Studies (Doctoral Studies) in electrical engineering from the University of Coimbra, Coimbra, in 2009 and 2012, respectively. He is currently working toward the Ph.D. degree at the University of Coimbra.

Since 2008, he has been a Fellow at the Instituto de Telecomunicações. His research interests include finite element analysis of rotating electrical machines, fault-tolerant adjustable-speed ac drive systems, high-frequency electronic ballasts and discharge lamp modeling.

Mohsen Bandar Abadi (S'13) was born in 24 March, 1983. He received his Diploma in Electrical Engineering at the University of Mashhad, Mashhad, Iran. He received his Master degree from university of Semnan, Semnan, Iran in the field of Power. Since January 2011, he has been PhD candidate on the Faculty of Electrical Engineering, University of Coimbra, Coimbra Portugal. His research areas are in power electronic such as Electrical machines, Fault Diagnosis Method.

André M. S. Mendes (S'95–M'04) was born in Castelo Branco, Portugal. He received Electrical Engineering. Diploma, the M.Sc. degree, and the Dr. Eng. degree from the University of Coimbra, Coimbra, Portugal, in 1993, 1998 and 2005, respectively.

Since 1991, he has been with the Department of Electrical and Computer Engineering, University of Coimbra-Portugal, where he is currently an Assistant Professor and the Director of the Power Electronics Laboratory.

His teaching interests cover electrical machines and power electronics and his research interests are focused on condition monitoring and diagnostics of ac motor drives and fault tolerant adjustable-speed ac drive systems.

Sérgio M. A. Cruz (S'96–M'04) was born in Coimbra, Portugal, in 1971. He received the E.E. diploma, and the M.Sc. and Dr. Eng. degrees in electrical engineering from the University of Coimbra, Coimbra, Portugal, in 1994, 1999, and 2004, respectively.

Since 1994, he has been with the Department of and Computer Engineering, University of Coimbra, where he is currently an Assistant Professor and the Director of the Electric Machines Laboratory.

He is the author of more than 60 journal and conference papers in his areas of interests. His research interests include electric machines, electric drives, and power electronic converters, with special emphasis on fault diagnosis, fault tolerance, and digital control.

Full Detection of High-Resistance Connections in Multiphase Induction Machines

L. Zarri, M. Mengoni, A. Tani, Y. Gritli, G. Serra, F. Filippetti, D. Casadei

Abstract -- High-resistance connections in electrical machines cause unbalances in the stator resistances, reduce the efficiency and increase the fire hazard. In this paper the problem of detection of high-resistance connections is investigated for multiphase induction machines with an odd number of phases. The main contribution of the paper is a control scheme that can determine the stator resistance unbalance of all phases, in transient and in steady-state operating conditions. The theoretical analysis and the feasibility of the control scheme are confirmed by experimental tests.

Index Terms-- Condition monitoring, fault detection, fault diagnosis, induction motors, variable-speed drives.

I. INTRODUCTION

Multiphase ac machines fed by multiphase inverters can be used in high power applications to obtain high power ratings without increasing the stator current per phase [1]. In addition, in multiphase machines, it is possible to control some of the spatial harmonics of the air-gap magnetic field independently of each other. This degree of freedom can be used to improve the behavior of the motor drive in case of faults.

During the last ten years the research activity on multiphase drives has considered a large variety of multiphase motors.

Six-phase induction machines are probably the most studied kind of multiphase induction machines. Some tecniques for fault-tolerant control in case of opened phases have already been developed for a symmetrical six-phase machine in [2]. The behavior of multiphase induction machines with an odd number of phases has been investigated under fault conditions specifically in [3]-[5].

The vast majority of the papers concerning the faul-tolerant control of multiphase machines does not consider the problem of diagnosis, which is usually tackled by using the same methods adopted for three-phase systems.

The present paper analyzes the behavior of multiphase induction machines with an odd number of phases under an unbalance of the resistances of the stator phases due to the damage of the connections. High resistance connections can be caused by a combination of poor workmanship, thermal cycling and vibration, or damage of the contact surfaces due

to pitting, corrosion or contamination. The contact resistance can increase to an unacceptable level, leading to thermal overloading, melting of copper conductors, short-circuit failures between conductors or to the ground. In addition the asymmetries of the stator voltage may cause a negative sequence current to circulate in the motor windings, thus reducing the motor output power and efficiency [6]-[8].

Consequently, if the evolution of this type of electrical fault is not detected at an early stage, its propagation can lead to serious failures. The traditional techniques for the detection of high resistance connections include methods such as the offline resistive imbalance test, visual inspection, the voltage drop survey and infrared thermography. To avoid specialized equipments, recently, sensorless on-line techniques based on the negative-sequence current and zero-sequence voltage have been proposed for three phase machines [9]-[10].

The present paper is specifically focused on multiphase induction machines and extends the results reported in [11]-[13], where an analytical model of a multiphase induction motor with unbalanced stator winding was developed.

The control system is able to compensate the unbalance due to high-resistance connections by using multi-reference frame controllers to cancel the inverse components of the stator currents. At the same time, the information generated by these regulators is used to detect the deviations from the average resistance of the phases.

This analysis is in agreement with results of other authors that have investigated the use of the negative sequence components with pioneering works [14] -[17].

The main contribution of the present paper is to show how the resistance unbalance can be determined simultaneously for all phases.

The validity of the theoretical analysis and the feasibility of the control scheme are confirmed by experimental tests.

II. MOTOR MODEL

Let us consider an electrical ac machine having M star connected stator phase windings, symmetrically distributed within the stator slots, where M is an odd number greater than three. The windings are supposed shifted by $2\pi/M$ electrical radians and with a single neutral point. The phases are equal to each other, except for the stator resistances. The voltage equation of the kth phase can be written as follows:

$$ v_{Sk} = R_{Sk} i_{Sk} + \frac{d\varphi_{Sk}}{dt} \quad (k=1,\dots,M) \qquad (1) $$

This work was in part supported by the Italian Ministry of Education, University and Research (MIUR) under the project PRIN 2009 "High-reliability multiphase electric drives for more electric aircraft".

The authors are with the Department of Electrical, Electronic and Information Engineering of the University of Bologna, Italy (e-mail: luca.zarri@ieee.org, michele.mengoni3@unibo.it, angelo.tani@unibo.it, yasser.gritli@unibo.it, giovanni.serra@unibo.it, fiorenzo.filippetti @unibo.it, domenico.casadei@unibo.it).

978-1-4799-0024-4/13 $31.00 © 2013 IEEE

where v_{Sk} is the voltage applied to the stator phase, R_{Sk} is its resistance and φ_{Sk} is the flux linkage. Each phase resistance can be expressed as the sum of the average value $R_{S,avr}$ and a deviation ΔR_k from the mean value.

$$R_{Sk} = R_{S,avr} + \Delta R_k \qquad (2)$$

As a consequence, the sum of all deviations is null, i.e.,

$$\sum_{k=1}^{M} \Delta R_k = 0 \qquad (3)$$

Equation (1), combined with (2)-(3), can be rewritten in terms of multiple space vectors (see Appendix), which are commonly adopted for the analysis of multiphase systems. The mathematical approach explained in [13] can be used to get the following voltage equations in the stator reference frame:

$$\bar{v}_{S\rho} = R_{S,avr}\bar{i}_{S\rho} + \frac{2}{M} \sum_{\sigma=1,3,...,M-2} \sum_{k=1}^{M} \Delta R_k \left(\bar{i}_{S\sigma} \cdot \overline{\alpha}_k^{\sigma} \right) \overline{\alpha}_k^{\rho} + \frac{d\overline{\varphi}_{S\rho}}{dt} \qquad (4)$$

where $\rho = 1, 3, ..., M-2$ and "$\cdot$" is the scalar product, defined as the real part of the product between the first operand and the complex conjugate of the second. The vectors $\bar{i}_{S\sigma}$, $\bar{v}_{S\sigma}$ and $\overline{\varphi}_{S\sigma}$ are the space vectors of the stator currents, stator voltages and flux linkages corresponding to the σth spatial harmonic component of the magnetic field.

The stator fluxes in (4) can be written as a function of the stator currents and of the rotor fluxes, as follows:

$$\overline{\varphi}_{S\rho} = \sigma_\rho L_{S\rho} \bar{i}_{S\rho} + \frac{M_\rho}{L_{R\rho}} \overline{\varphi}_{R\rho} \qquad (5)$$

where $\sigma_\rho L_{S\rho}$, M_ρ and $L_{R\rho}$ are respectively the total leakage inductance, the mutual inductance and the rotor self inductance related to the ρth spatial harmonic of the magnetic field. Finally, $\overline{\varphi}_{R\rho}$ is the space vector of the rotor flux in the plane α_ρ-β_ρ.

It is well-known that the rotor flux $\overline{\varphi}_{R\rho}$ can be related to the stator current vector $\bar{i}_{S\rho}$ by the following first-order differential equation, written in the stator reference frame:

$$\tau_{R\rho} \frac{d\overline{\varphi}_{R\rho}}{dt} + \left(1 - j\rho\omega_m \tau_{R\rho}\right) \overline{\varphi}_{R\rho} = M_\rho \bar{i}_{S\rho} \qquad (6)$$

where j is the imaginary unity, ω_m is the rotor speed in electrical radians, and $\tau_{R\rho}$ is a time constant defined as the ratio between the rotor self-inductance $L_{R\rho}$ and the rotor resistance $R_{R\rho}$.

III. CALCULATION OF THE RESISTANCE UNBALANCE

Let us suppose that the machine is controlled in such a way that $\bar{i}_{S1}$ is the only non-null current space vector, rotating at constant angular frequency ω with constant magnitude.

$$\bar{i}_{S\rho}(t) = \begin{cases} \bar{i}_{S1,ref}(t) = \bar{I}_{ref} e^{j\omega t} & \text{if } \rho = 1 \\ 0 & \text{otherwise} \end{cases} \qquad (7)$$

All the electrical quantities in (4) can be re-written as linear combinations of vectors rotating in counter-clockwise direction ("direct direction" with angular frequency ω) or in clockwise direction ("inverse direction" with angular frequency $-\omega$). The letters "d" and "i" are used to distinguish between the two directions. For example, the stator voltage vectors can be decomposed as follows:

$$\bar{v}_{S\rho} = \bar{v}_{S\rho}^{(d)} + \bar{v}_{S\rho}^{(i)} . \qquad (8)$$

Combining (4)-(7), one obtains the following expression of the direct components of the stator voltage vectors when $\rho \geq 3$ (the component when $\rho = 1$ is not calculated, since it is not interesting for the fault analysis):

$$\bar{v}_{S\rho}^{(d)} = \frac{1}{M} \left(\sum_{K=1}^{M} \Delta R_k \overline{\alpha}_k^{\rho-1} \right) \bar{i}_{S1,ref} \quad (\rho \geq 3), \qquad (9)$$

whereas the expression for the inverse components of the stator voltage vectors when $\rho \geq 1$ is shown in (10).

$$\bar{v}_{S\rho}^{(i)} = \frac{1}{M} \left(\sum_{K=1}^{M} \Delta R_k \overline{\alpha}_k^{\rho+1} \right) \bar{i}_{S1,ref}^* \quad (\rho \geq 1). \qquad (10)$$

It is convenient to define new variables starting from (9) and (10) to emphasize the effect of ΔR_k.

$$\overline{\eta}_\rho^{(d)} = v_{S\rho}^{(d)} \frac{\bar{i}_{S1}^*}{|\bar{i}_{S1}|^2} = \frac{1}{M} \sum_{K=1}^{M} \Delta R_k \overline{\alpha}_k^{\rho-1} \quad (\rho = 3, 5, ..., M-2) \quad (11)$$

$$\overline{\eta}_\rho^{(i)} = v_{S\rho}^{(i)} \frac{\bar{i}_{S1}}{|\bar{i}_{S1}|^2} = \frac{1}{M} \sum_{K=1}^{M} \Delta R_k \overline{\alpha}_k^{\rho+1} \quad (\rho = 1, 3, ..., M-2) \quad (12)$$

It is now straightforward to understand that the variables $\eta_\rho^{(d)}$ and $\eta_\rho^{(i)}$ are not independent of each other, since $\eta_\rho^{(d)}$ can be deduced from $\eta_\rho^{(i)}$ through the following equalities:

$$\overline{\eta}_\rho^{(d)} = \overline{\eta}_{\rho-2}^{(i)} \quad (3 \leq \rho \leq M - 2) \qquad (13)$$

The $\frac{M-1}{2}$ equations (12) can be decomposed along the real and imaginary axes, thus leading to a set of $M-1$ equations with real coefficients, where the unknowns are the M resistance unbalances. Since the number of unknowns is greater than the number of equations, to find a unique solution, it is necessary to consider also the constrain (3). Under this assumption, it is straightforward to verify that the resistance unbalances can be univocally expressed by the following relationships:

$$\Delta R_k = \frac{2}{M} \sum_{\rho=1,3,...,M-2} \overline{\eta}_\rho^{(i)} \cdot \overline{\alpha}_k^{\rho+1} \quad (k=1, 2, ..., M) \qquad (14)$$

If the negative sequence components are not due to a resistance unbalance, the estimated variation of the stator resistance is certainly not accurate. In this case, the proposed algorithm can just provide a generic malfunction alarm, and other more specific algorithms (not considered in this paper) are necessary to indentify the fault type.

a) Fault of one phase

Let us suppose that all stator resistances are equal to R_{S0}, except the one of phase 1, that is equal to $R_{S0}+r_1$. It is straightforward to verify that the average resistance is

$$R_{S,avr} = R_{S0} + \frac{r_1}{M} \qquad (15)$$

and the deviations are

$$\Delta R_k = R_{Sk} - R_{S,avr} = \begin{cases} r_1 \dfrac{M-1}{M} & if\ k = 1 \\[2mm] -\dfrac{r_1}{M} & otherwise \end{cases} \qquad (16)$$

b) Fault of two phases

If the resistances of phases 1 and 2 are affected respectively by the variations r_1 and r_2, whereas the other resistances are unchanged, than the average resistance become

$$R_{S,avr} = R_{S0} + \frac{r_1 + r_2}{M} \qquad (17)$$

and the deviations are

$$\Delta R_k = R_{Sk} - R_{S,avr} = \begin{cases} r_1 \dfrac{M-1}{M} - \dfrac{r_2}{M} & if\ k = 1 \\[2mm] r_2 \dfrac{M-1}{M} - \dfrac{r_1}{M} & if\ k = 2 \\[2mm] -\dfrac{r_1 + r_2}{M} & otherwise \end{cases} \qquad (18)$$

The results (17) and (18) are particularly useful to establish a connection between fault resistances and deviations.

IV. CONTROL SYSTEM WITH UNBALANCE DETECTION

The rotor field-oriented control is often adopted to control multiphase induction drives. It requires two PI regulators, implemented in a reference frame synchronous with the rotor flux vector, to track the references of the stator currents.

As shown in Fig. 1, this basic control scheme can be improved in such a way that the machine operation is not affected by stator unbalances (fault-tolerance property) and to allow the calculation of the resistance unbalances (fault-detection property).

The currents i_{S1d}^{sync} and i_{S1q}^{sync} are controlled by the PI regulators (a) and (b) implemented in rotor-flux oriented d-q reference frame. The behavior of the current control is improved by the compensation of the stator back electromotive force. The angle θ, determined by a suitable observer, is the phase angle of the rotor flux vector in the stationary reference frame.

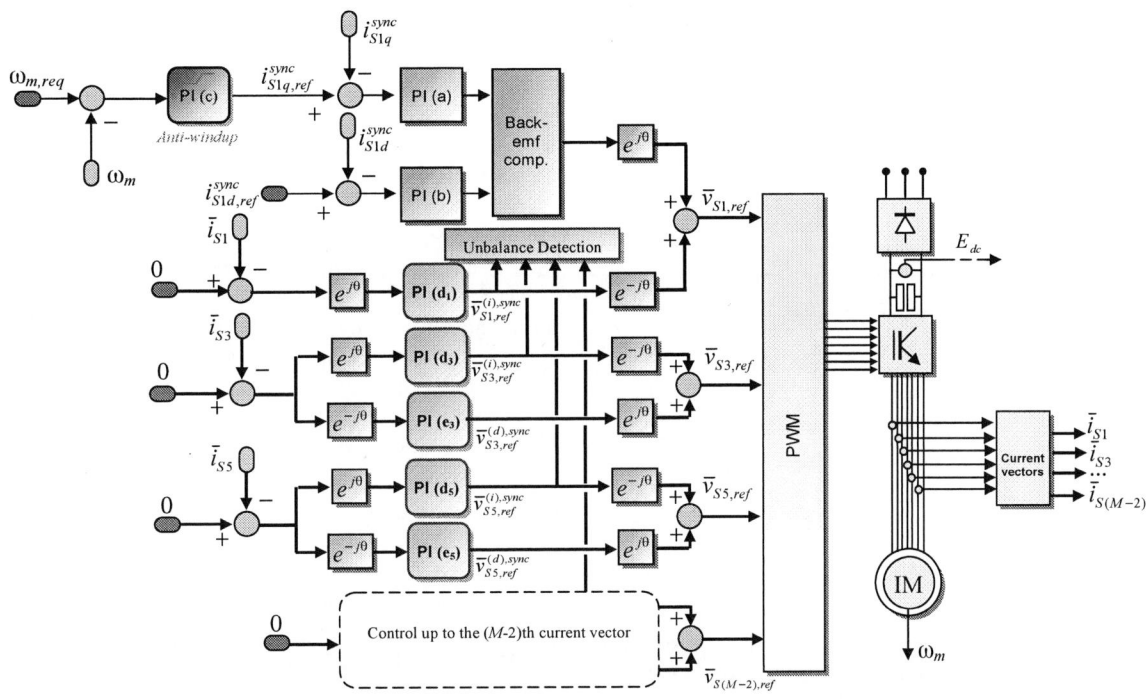

Fig. 1. Block diagram of the proposed control scheme, with the capability to detect and compensate the stator resistance unbalance.

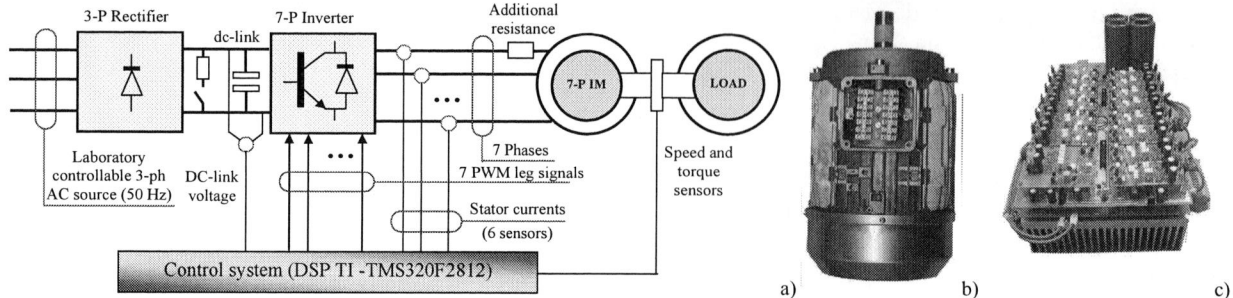

Fig. 2. Experimental set-up. a) Schematic diagram of the test bench and position of the current and voltage sensors. b) Seven phase induction machine. c) Seven phase inverter.

The reference value of i_{S1d}^{sync} in Fig. 1 is equal to the rated magnetizing current of the machine, whereas the reference value of i_{S1q}^{sync}, instead, is determined by the PI regulator (c) on the basis of the speed error.

When the motor is balanced, the current vectors $\bar{i}_{S3}$, $\bar{i}_{S5}$, ..., $\bar{i}_{S(M-2)}$ are theoretically null. However, when an unbalance arises, to ensure a disturbance-free operation of the multiphase machine, the control system should apply suitable voltage to keep these currents equal to zero. However, since the analysis in Section III has shown that each voltage vector contains not only a direct but also an inverse component in steady state operating conditions, for each harmonic component of the air-gap field it is necessary to use a pair of PI regulators implemented in two reference frames rotating in the same and in the opposite direction of the field-oriented reference frame.

For each pair of regulators, the first one drives to zero the direct component of the error signal, and the second one the inverse component. The output voltage vector is obtained by summing the outputs of both regulators of each pair; hence, according to the superposition principle, it can drive to zero the total tracking error.

It is worth noting that the input signal of these regulators is a vector signal, i.e., in the practical implementation it is necessary to decompose the input signal into d- and q-components.

As far as the regulators perform correctly, even in unbalanced operating conditions, it is possible to apply (16), since the current vectors $\bar{i}_{S3}$, $\bar{i}_{S5}$, ..., $\bar{i}_{S(M-2)}$ are approximately zero.

V. Experimental Results

A complete drive system has been built and some experimental tests have been carried out to verify the theoretical analysis. The experimental set-up consists of a 7-phase IGBT inverter and a 4 kW, 7-phase, 4-pole squirrel cage induction machine. A schematic diagram and some pictures of the experimental set-up are shown in Fig. 2. The induction machine has been manufactured by a local company specialized in the construction of electrical motors. The parameters of the induction machine are shown in Table

TABLE I – PARAMETERS OF THE SEVEN-PHASE MACHINE

T_{rated}	=	24	Nm	L_{S1}	=	180	mH
$I_{s,max}$	=	7.5	A$_{(peak)}$	L_{R1}	=	180	mH
$I_{S1d,rated}$	=	3.6	A$_{(peak)}$	M_1	=	175	mH
f_{rated}	=	50	Hz	L_{S3}	=	24	mH
R_S	=	1.3	Ω	L_{R3}	=	24	mH
R_{R1}	=	1.0	Ω	M_3	=	19	mH
R_{R3}	=	0.8	Ω	L_{S5}	=	10	mH
R_{R5}	=	0.6	Ω	L_{R5}	=	10	mH
p	=	2		M_5	=	7	mH

I. The control scheme of Fig. 1 has been implemented on a TMS320F2812 produced by Texas Instruments.

During the experimental tests, one or more external resistors have been added in series with the machine phases to reproduce the effect of high-resistance connections.

Fig. 3 shows the behavior of the multiphase drive when the stator phases are correctly balanced, the speed reference is 500 rpm and the motor delivers approximately the rated torque to the load. As can be seen, the currents are perfectly sinusoidal and have the same amplitude. The voltage vectors $\bar{v}_{S3,ref}$ and $\bar{v}_{S5,ref}$ are negligible. These quantities are shown in Figs. 3(b) and (c), divided by 2/M $I_{S1,ref}$ in such a way that the amplitude of visualized signals directly represents the unbalance resistance.

In the beginning of the time interval shown in Fig. 4, the operating conditions are the same of Fig. 3. After 400 ms, there is an increase of 0.1 Ω in the resistance of phase 1, which corresponds to 7-8% of the nominal value. Despite the unbalance, the waveform of the phase currents does not change, since the control system continues tracking the current references. As a consequence, the motor speed, torque and the Joule losses are not altered. According to the theoretical analysis, the additional resistance of 0.1 Ω leads to unbalances that are near 0.085 Ω in phase 1, and -0.014 Ω for the other phases.

Fig. 5 shows the behavior of the drive in the same operating conditions of Fig. 4, i.e., when a small resistance has been already added to phase 1, during a speed transient from about 400 to about 700 rpm. This test aims to assess the sensitivity of the proposed estimation method to transient operating conditions. The analysis of the first trace of Fig. 5 shows that the estimation of the resistance unbalance ΔR_1

978-1-4799-0024-4/13 $31.00 © 2013 IEEE

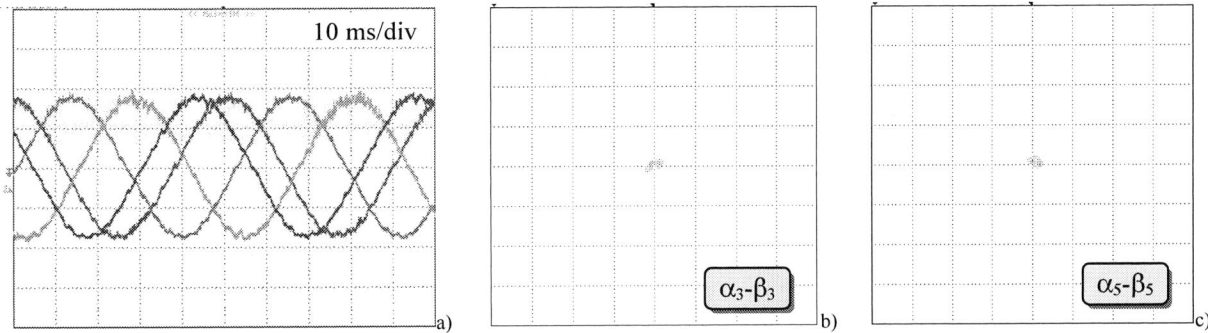

Fig. 3. Behavior of the healthy vector-controlled seven-phase machine rotating at 500 rpm and delivering the rated torque. a) Waveforms of the stator currents of phases 1, 3, 5 and 7 (5 A/div). b) Locus of vector $\bar{v}_{S3,ref}$ on plane α_3-β_3, normalized by dividing by $(2/M)I_{S1,ref}$ (0.2 Ω/div). c) Locus of vector $\bar{v}_{S5,ref}$ on plane α_5-β_5, normalized by dividing by $(2/M)I_{S1,ref}$ (0.2 Ω/div).

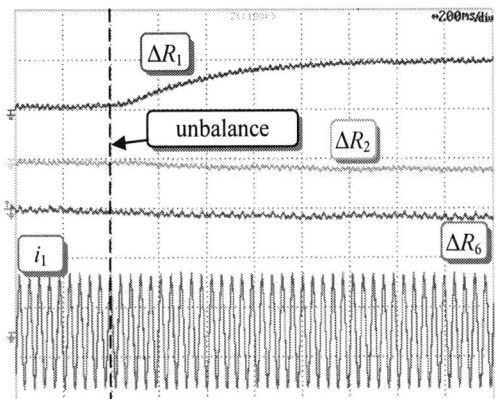

Fig. 4 - Behavior of the estimation algorithm when the resistance of phase 1 is increased by 0.1 Ω (7-8%) and the machine delivers the rated torque to the load. The top traces show the trends of the ΔR_1, ΔR_2, ΔR_6 (0.075 Ω/div), while the bottom trace is the waveform of the current of phase 1 (10 A/div).

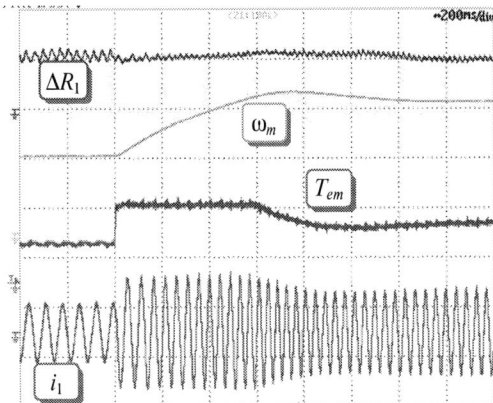

Fig. 5 - Behavior of the estimation algorithm during a speed transient, when the resistance of phase 1 is unbalanced by 0.1 Ω (7-8%). From top to bottom, the traces are the waveform of ΔR_1 (0.075 Ω/div), the rotor speed (250 rpm/div), the electromagnetic torque (15 Nm/div), and the current of phase 1 (10 A/div).

does not appear affected by speed and torque variations, and consequently the estimation algorithm of high-resistance connections seems sufficiently robust.

Finally, Fig. 6 shows the behavior of the estimation algorithm when two phases are unbalanced. Initially, the resistance of phase 2 is increased by 0.1 Ω, then the resistance of phase 1 is increased by the same quantity after 1.6 s.

As can be noted, each time an additional resistance is inserted, the estimated values of the unbalances change, since the average value of the stator resistances varies. The values of the unbalances are in good agreement with the theoretical analysis.

VI. CONCLUSION

This paper investigates the behavior of vector-controlled multiphase induction machines with an odd number of phases when high-resistance connections unbalance the stator resistances.

A control scheme, based on multi-reference frame controllers, with the capability to keep stator currents balanced in presence of this fault, has been proposed. The

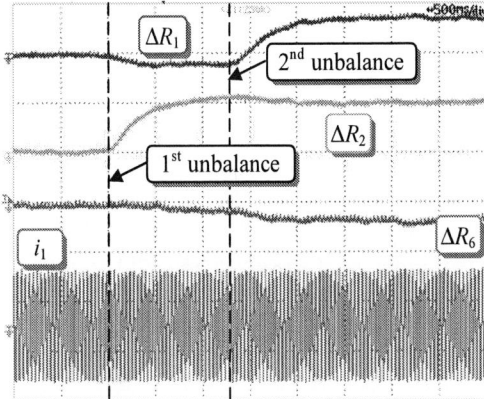

Fig. 6 - Behavior of the estimation algorithm when two resistances are unbalanced, and the machine delivers the rated torque to the load. The resistance of phases 1 and 2 are both increased by 0.1 Ω (7-8%). The top traces show the trends of the ΔR_1, ΔR_2, ΔR_6 (0.075 Ω/div), while the bottom trace is the waveform of the current of phase 1 (10 A/div).

information provided by the PI regulators of the control system can be used to detect any machine dissymmetry that tends to produce negative sequences in the voltage vectors

978-1-4799-0024-4/13 $31.00 © 2013 IEEE

$\bar{v}_{S3}, \bar{v}_{S5}, ..., \bar{v}_{S,M-2}$. When negative sequence components are detected, the method can calculate the resistance unbalance and identify the unbalanced stator phase, under the assumption that the unbalance is only due to a resistance variations. In practical applications the proposed algorithm is suitable to detect a) high-resistance connections of the machine terminals or in the cables feeding the machine, b) high-resistance faults in the stator phases, c) unbalances due to a hot-spot in the feeding cables.

Experimental tests have confirmed the correctness of the theoretical analysis and the feasibility of the control system.

The proposed online technique may improve the efficiency, safety, and reliability of multiphase drives since the maintenance personnel can check the phase connections and the cables by using conventional inspection techniques as soon as they are informed of an incipient fault, whereas the machine behavior is kept unchanged.

APPENDIX: MULTIPLE SPACE VECTORS

To analyze a multiphase system, it is very useful to introduce the definition of multiple space vectors.

For a given set of M (odd number) real variables $x_1, ..., x_k, ..., x_M$ a new set of complex variables, the multiple space vectors $\bar{x}_1, \bar{x}_3, ..., \bar{x}_\rho, ..., \bar{x}_{M-2}$, can be obtained by means of the following symmetrical linear transformations:

$$\bar{x}_\rho = \frac{2}{M} \sum_{k=1}^{M} x_k \, \bar{\alpha}_k^\rho \qquad (\rho = 1, 3, ..., M\text{-}2), \qquad (A1)$$

where

$$\bar{\alpha}_k = e^{j\frac{2\pi}{M}(k-1)}. \qquad (A2)$$

Another important definition is that of zero-sequence component, which is as follows:

$$x_0 = \frac{1}{M} \sum_{k=1}^{M} x_k \,. \qquad (A3)$$

The inverse transformations of (A1) and (A3) are

$$x_k = x_0 + \sum_{\rho=1,3,5,...}^{M-2} \bar{x}_\rho \cdot \bar{\alpha}_k^\rho \qquad (k = 1, 2, ..., M) \qquad (A4)$$

where the symbol " $\cdot$ " represents the dot product, defined as the real part of the product between the first operand and the complex conjugate of the second.

According to (A1) and (A3), a general M-phase system can be represented by $(M-1)/2$ independent space vectors and the zero-sequence component. Alternatively, it is frequently said that the transformation (A1) maps the real variables $x_1, ..., x_M$ into $(M-1)/2$ complex planes α_1-β_1, α_3-β_3, ..., α_{M-2}-β_{M-2}, which are independent of each other.

ACKNOWLEDGMENT

The authors gratefully acknowledge the contributions of Lucchi Elettromeccanica for the prototype of the seven phase induction machine.

VII. REFERENCES

[1] E. Levi, "Multiphase electric machines for variable-speed applications", IEEE Trans. on Industrial Electronics, vol. 55, no. 5, 2008, pp. 1893-1909.

[2] R. Kianinezhad, B.-Nahid Mobarakeh, L. Baghli, F. Betin, G.A. Capolino, "Modeling and control of six-phase symmetrical induction machine under fault condition due to open phases," IEEE Trans. on Industrial Electronics, Vol. 55, No. 5, pp. 1966-1977, May 2008.

[3] C. B. Jacobina, I. S. Freitas, T. M. Oliveira, E. R. C. da Silva, A. M. N. Lima, "Fault tolerant control of five-phase AC motor drive," Proc. of the 35th Annual Power Electronics Specialists Conference, 2004, PESC 04, Aachen, Germany, pp. 3486-3492.

[4] J. Apsley, S. Williamson, "Analysis of multiphase induction machines with winding faults, " IEEE Trans. on Ind. Appl., Vol. 42, No. 2, pp. 465-472, March-Apr. 2006.

[5] A. Tani, M. Mengoni, L. Zarri, G. Serra, D. Casadei, "Control of multiphase induction motors with an odd number of phases under open-circuit phase faults," IEEE Trans. on Power Electronics, Vol. 27, No. 2, Feb. 2012, pp. 565-577.

[6] A. Bellini, F. Filippetti, C. Tassoni, and G. A. Capolino, "Advances in diagnostic techniques for induction machines," IEEE Tran. on Ind. Electron., Vol. 55, N. 12, pp. 4109–4126, Dec. 2008.

[7] R.S. Colby, "Detection of high-resistance motor connections using symmetrical component analysis and neural networks," Proc. of IEEE SDEMPED, pp. 2-6, Atlanta, GA, 2003.

[8] J. Bockstette, E. Stolz, and E.J. Wiedenbrug, "Upstream impedance diagnostics for three phase induction machines," Proc. of IEEE SDEMPED, Cracow, Poland, 2007.

[9] J. Yun, J. Cho, S.B. Lee, J. Yoo, "On-line detection of high-resistance connections in the incoming electrical circuit for induction motors," IEEE Tran. on Ind. Appl., Vol. 45, Issue 2, pp. 694-702, 2009.

[10] J. Yun, K. Lee, K.W Lee, S.B. Lee, J.Y. Yoo, "Detection and classification of stator turn faults and high-resistance electrical connections for induction machines," IEEE Tran. on Ind. Appl., Vol. 45, Issue 2, pp. 666-675, 2009.

[11] L. Zarri, M. Mengoni, Y. Gritli, A. Tani, F. Filippetti, G. Serra, D. Casadei, "Behavior of multiphase induction machines with unbalanced stator windings," Proc. of SDEMPED11, Bologna, Italy, Sept. 5-8, pp. 84-91.

[12] M. Mengoni, L. Zarri, A. Tani, Y. Gritli, G. Serra, F. Filippetti, "Behaviour of Multiphase Induction Machines with Unbalanced Stator Resistances " Proc. of EPE-PEMC 2012/ECCE Europe, 4-6 Sept. 2012, Novi Sad , Serbia, pp.1-8.

[13] L. Zarri, M. Mengoni, Y. Gritli, A. Tani, F. Filippetti, G. Serra, D. Casadei, "Detection and Localization of Stator Resistance Dissymmetry Based on Multireference Frame Controllers in Multiphase Induction Motor Drives", IEEE Trans. on Industrial Electronics, Aug. 2013, Vol. 60, No. 8, pp. 3506 - 3518.

[14] A. M. da Silva, R. J. Povinelli, and N. A. O. Demerdash, "Induction machine broken bar and stator short-circuit fault diagnostics based on three-phase stator current envelopes," IEEE Trans. on Ind. Electron., vol. 55, no. 3, pp. 1310–1318, Mar. 2008.

[15] S. M. A. Cruz and A. J. M. Cardoso, "Multiple reference frames theory: A new method for the diagnosis of stator faults in three-phase induction motors," IEEE Trans. on Energy Conversion., vol. 20, no. 3, pp. 611–619, Sep. 2005.

[16] J. Cusido, L. Romeral, J. A. Ortega, J. A. Rosero, A. Garcia Espinosa, "Fault detection in induction machines using power spectral density in wavelet decomposition," IEEE Trans. on Ind. Electron., vol. 55, no. 2, pp. 633–643, Feb. 2008.

[17] F. Briz, M. W. Degner, J. M. Guerrero, and P. Garcia, "Stator windings fault diagnostics of induction machines operated from inverters and soft-starters using high-frequency negative-sequence currents," IEEE Trans. on Ind. Appl., vol. 45, no. 5, pp. 1637–1646, Sep./Oct. 2009.

Luca Zarri (SM'12) was born in Bologna, Italy, in 1972. He received the M. Sc. in Electrical Engineering, with honors, and the Ph.D. degree from the University of Bologna, Bologna, Italy, in 1998 and 2007, respectively. He worked as a freelance software programmer from 1989 to 1992 and as a plant designer with an engineering company from 1998 to 2002. In 2003 he became a Laboratory Engineer with the Department of Electrical Engineering, University of Bologna. Since 2005 he has been an Assistant Professor with the same department. He is author or co-author of more than 90 scientific papers. His research activity concerns the modulation strategies of innovative converters and the robust control of electric drives. He is a member of the IEEE Industry Applications, IEEE Power Electronics and IEEE Industrial Electronics Societies.

Michele Mengoni (M'13) was born in Forlì, Italy, in 1981. He received the M.Sc and Ph.D. degree in Electrical Engineering, with honors, from the University of Bologna, Bologna, Italy, in 2006 and 2010, respectively. He is currently a fellow researcher at the Department of Electrical Engineering of the University of Bologna. His research interests include sensorless control of induction motors, multiphase drives and ac/ac matrix converters. He is a member of the IEEE Industry Applications Society.

Angelo Tani was born in Faenza, Italy, in 1963. He received the M. Sc. in Electrical Engineering, with honors, from the University of Bologna, Bologna, Italy, in 1988. He joined the Department of Electrical Engineering, University of Bologna, in 1990, where he is currently an Associate Professor. His scientific work is related to electrical machines, motor drives and power electronics. He has authored more than 100 papers published in technical journals and conference proceedings. His current activities include multiphase motor drives, ac/ac matrix converters, and field weakening strategies for induction motor drives.

Yasser Gritli was born in Tunis, Tunisia, in 1975. He received the M.S. degree in Electrical Engineering from the National Engineering School of Tunis (ENIT), Tunisia, in 2006. In 2011, he received the Ph.D.degree in Industrial Informatics from the National Institute of Applied Sciences and Technologies (INSAT), Tunisia. He is currently working toward the Ph.D. in Electrical Engineering at the Department of Electrical Engineering and Information Technology, University of Bologna. His research interests are related to electrical machine and drives, diagnostics of induction motors and renewable energy. His current activities include control and diagnosis of induction machines in wind generator and railway traction systems. Dr. Gritli was the recipient of the best paper Award at the IEEE-SDEMPED 2011.

Fiorenzo Filippetti (M'00) was born in Fano, Italy, in 1945. He received the M.S. degree in electrical engineering from the University of Bologna, Bologna, Italy, in 1970. In 1976, he became Assistant Professor in the Department of Electrical Engineering, University of Bologna, where he is currently a Full Professor of electrical drives. From 1993 to 2002, he was an Adjunct Professor of Electrotechnics and Electrical Drives at the University

of Parma, Parma, Italy. He was a visiting professor at the University Claude Bernard, Centre de Génie Electrique de Lyon (CEGELY), University Claude Bernard, Lyon, France, and the University of Picardie Jules Verne, Amiens, France. In 1998 he held a position at the University Claude Bernard as a member of the Scientific Council of CEGELY. He is a Lecturer for the European Master in Advanced Power Electrical Engineering program recognized by the European Commission in 2004. He has authored or coauthored more than 180 scientific papers published in scientific journals and conference proceedings, and one textbook. He is the holder of one industrial patent. His main research interests include the simulation and modeling of electric circuits and systems, and the study and application of condition-monitoring and fault-detection techniques for ac electrical machines. He was recipient of the Best Paper Awards at the conferences IEEE IAS 2000 and IEEE SDEMPED 2011.

Giovanni Serra (SM'04) received the M. Sc., with honors, in Electrical Engineering from the University of Bologna, Bologna, Italy, in 1975. He joined the Department of Electrical Engineering, University of Bologna, first as a recipient of a Fellowship of the National Research Council, then as a Research Associate, and, since 1987, as an Associate Professor. He is currently Professor of Electrical Machines in the Department of Electrical Engineering. He has authored more than 180 papers published in technical journals and conference proceedings. His fields of interests are electrical machines, electrical drives, and power electronic converters. His current activities include multi-phase drives, direct torque control of ac machines, linear motors, and ac/ac matrix converters. Dr. Serra is a member of the IEEE Industry Applications and IEEE Dielectrics and Electrical Insulation Societies and the Italian Electrotechnical and Electronic Association (AEIT).

Domenico Casadei (SM'04) received the M. Sc., with honors, in Electrical Engineering from the University of Bologna, Italy, in 1974.
He joined the Electrical Engineering Department of the University of Bologna, in 1975, as Research Assistant Professor. He is currently Full Professor of Electrical Drives. His scientific work is related to electrical machines and drives, and power electronics. Prof. Casadei is author or co-author of more than 200 scientific papers, published in technical journals and conference proceedings. His present research areas are vector control of AC drives, multi-phase drives and diagnosis of electrical machines. He has been involved in several research projects with the industry in the same research areas. He is a IEEE Senior Member and a member of the IEEE Industrial Electronics and Power Electronics Societies and a member of the European Power Electronics Society.

Improved Open Switch Fault Detection Based on Normalized Current Analysis in Multiphase Fault Tolerant Converters

Mehdi Salehifar, Manuel Moreno-Eguilaz, Vicent Sala, Ramin Salehi Arashloo, L.Romeral
Department of Electronic Engineering, Universitat Politecnica de Catalunya, BarcelonaTech
Center of Motion Control and Industrial Applications (MCIA)
Barcelona, Spain
mehdi.salehifar@mcia.upc.edu

Abstract – **a new open switch fault detection method based on normalized current analysis is proposed for application in multiphase fault tolerant PMSM drives. Performance characteristics of proposed method are single diagnostic variable, ability to detect open phase fault without using auxiliary variable, ability to detect multiple switch fault, simple diagnostic variable, generality, and robustness in case of high unbalanced current waveforms. Theory of diagnostic method with special multiphase drive application is developed; simulation results using Matlab/Simulink and experimental waveforms are shown to validate effectiveness of the presented fault detection method.**

Keyword – **fault detection, open switch fault, multiphase fault tolerant drive, PMSM**

I. INTRODUCTION

Due to advantages of drive systems, this technology is increasingly used in many applications. There are some applications where a continuous operation of drive is critical to ensure safety criteria. Aerospace, power plants, automotive industry, railway locomotives and military area are among those application areas which need high reliability drive systems. Consequently, developing of fault tolerant systems for drive application has been presented in recent literature. Faulty control methods, fault detection, machine design and fault tolerant power converters are among those interesting topics in research [1]. So, in order to change converter configuration and avoid secondary damages in a fault tolerant converter, a fault detection scheme should be applied which is fast, simple and accurate.

According to a recent survey on reliability of power converters, semiconductor switches are most vulnerable components among others [2]. Regarding faulty types in power semiconductor, open switch or short circuit are two main types; a comprehensive study on aforementioned faulty types, detection methods and performance evaluation has been presented in [3]. Short circuit faults are much more serious than open circuit faults. It should be detected and protected very fast, typically less than 10 µs, to avoid secondary faults. A simple and robust desaturation based short circuit detection has been already implemented in commercial industrial gate drive units.

On the other side, open circuit is less destructive, but it can remain undetected; therefore, secondary faults may happen. Detection for this fault can be implemented at converter level or semiconductor level. A simple and fast fault detection method, based on gate voltage behavior, has been shown in [4]. However, its performance depends on design parameters; moreover it is sensitive to load transient and disturbances. On the other hand, one of main reasons for open switch fault is the control part of the switch, which is implemented in the semiconductor gate drive. Another method based on lower switch voltage measurement was shown in [5]. Although this method is fast and effective, design complexities in case of higher voltage applications are unavoidable.

Open switch fault detection at converter level has also been investigated in literature. This method is less expensive, simpler and more reliable than other methods. However, it is not as fast as switch level detection methods [3]. Current is the main input for such diagnostic techniques. Using this quantity, three different fault detection categories have been presented. First detection scheme is based on load model; according to this method, using load model and input command, converter output current is estimated and compared to real value [6], [7]. Any difference shows a fault signal. Although this method is simple, diagnostic performance mainly depends on estimation process.

Detection methods based on comparison between reference current and real current have also been studied [8], [9]. This approach is simple, general and fast; however it needs reference current for diagnosis purpose which is not available in an open loop system. On the other side, detection performance under high unbalanced current waveforms is poor.

Last detection approach is based on current signal pattern. Different methods have been presented using Park's vector approach, slope method, AC instantaneous frequency, modified normalized DC current method, and wavelet transform. However, there exist some implicit problems such as high complexity, robustness, generality and detection time [3]. Detection based on combination of wavelet transform and a neural network has been shown in [10]. However, computation cost is a main drawback for this method.

978-1-4799-0024-4/13 $31.00 © 2013 IEEE

Recently, a simple approach based on normalized current with respect to Park's vector modulus was presented in [11], where current average value is divided by absolute average value. This method is simple. However, it is limited to three phase systems. On the other side, detection performance under high unbalanced currents is low. Likewise, calculation of Park's vector modulus for multiphase converters is more complex.

To address discussed limitations in literature, a new detection method is presented in this paper which can be applied to any multiphase two level converter. Also, it is more robust and faster than conventional ones. Input current to detection block is nonsinusoidal, and unbalanced current for a multiphase converter which has not been investigated in literature so far. A comparison is done between proposed method and ideas studied in literature in order to validate high performance of presented detection scheme.

Rest of this paper is presented in following order. Detection method and simulated converter topology are explained at part II. Simulation results for different faulty modes are demonstrated at part III, moreover a comparison is done between proposed method and detection scheme in literature. At section IV, presented method is implemented in fault tolerant algorithm of a five phase PMSM; experimental results are shown in part V. A conclusion is made at part VI and finally references are presented.

II. FAULT DIAGNOSTIC METHOD

Multiphase PMSM drive is an effective solution for industrial applications where reliability is of paramount importance, for example, electric and hybrid electric vehicles. This kind of motors is able to remain operational even with two faulty phases [12], [13]. To maintain fault tolerant concept in this motor it is necessary to implement a fault tolerant converter which has a monitoring block to detect and isolate fault (see Fig. 1). A fast, simple and robust fault detection method is an important part for a fault tolerant power converter at PMSM drive. For this reason, a simple diagnostic variable has been presented in literature as [14]:

$$m = \frac{\langle i \rangle}{\langle |i| \rangle} \tag{1}$$

Where $\langle \rangle$ is signal average value at fundamental frequency and $||$ is absolute value. Main advantage of this diagnostic variable is simplicity. However there are two major problems using this method. First one is poor performance in transients and second one is occurrence of false alarms in case of high unbalanced current waveforms. Also, fault detection is not effective in case of open phase fault. To avoid these disadvantages, some auxiliary signals have been proposed, as in [12]. However, using these signals, detection circuit complexity increases.

A. Proposed fault detection method

To overcome drawbacks of conventional methods, a new technique is presented in this paper to extract a single

diagnostic signal which detects fault effectively, being at the same time robust and fast.

One of advantages of the diagnostic signal in (1) is its high immunity to noise, since averaging current automatically eliminates high frequency components. As a result, at first step, a low pass filter is applied to eliminate

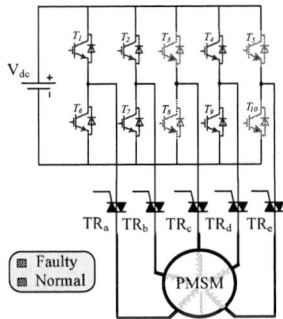

Fig. 1. Power converter configuration

high frequency components. Obviously, a tradeoff should be done between filter delay and noise reduction.

Moreover, in order to avoid false alarms in case of fast transients, it is necessary to normalize the signal. The approach presented in [11] utilizes Park's vector modulus to obtain the stationary components in a three phase drive; according to this method, first Park transformation should be applied. Besides, in case of more than three phases, there is more than one equivalent perpendicular page so that obtaining normalizing factor will be even more difficult. To overcome these drawbacks, a simpler normalizing factor is proposed in this paper as:

$$d_1 = \frac{i(n)}{\sqrt{\dfrac{1}{M} \sum_{j=1}^{M} i_j^2 (n - M)}} \tag{2}$$

Where $i(n\text{-}M)$ shows current samples at sample M before current sample $i(n)$. Period of normalization window is shown in Fig. 2; in contrast to conventional normalization method shown in (1), here normalization is applied with respect to few samples of waveform and not the whole period. According to denominator in (2), current square value is first calculated, and then its average value is obtained. After that, its square root is obtained. Final value is square root mean of current for sampling window shown in Fig. 2. By dividing current to normalizing factor at each phase, current is transferred to a value between -1 and 1, regardless of converter current magnitude.

After normalization, the next step is to develop a fault detection variable, as explained in the following section.

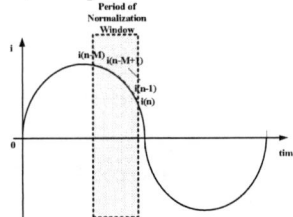

Fig. 2. Normalization window

Here, a new variable is proposed which makes signal less sensitive to transients, and current signals with average dc value while more sensitive to near zero values; it should be noted that current is zero for maximum half a cycle in phase with single faulty switch. Mathematic exponential function shown in (3) has such advantage.

$$d_2 = \exp(-\lambda \times |d_1|), \quad \lambda > 0 \qquad (3)$$

Where λ is a rate parameter. Since normalized ac signal at (2) is between -1 and 1, after applying (3), resulted signal will be 0, while under faulty mode it has a low fundamental oscillatory component with peak value equal to 1. In order to make d_1 signal a suitable value for detection, its average value at four times of fundamental frequency is calculated as in (4). Reason of averaging is to eliminate high frequency components. Also, in order to reduce detection time, filter frequency should be chosen higher. It means that for the same threshold value, filter with fundamental frequency has lower speed than a filter with four times of fundamental frequency. If filter frequency is so high, distortion in detection signal will remain.

$$d_3 = \frac{1}{M} \sum_{j=1}^{M} d_{2j} \qquad (4)$$

Signal d_3 is zero under healthy condition. Moreover, when there is a double switch fault in one phase, this variable is increased up to 1. Here, this advantage is used to detect this kind of fault.

After fault diagnostic, it is important to locate faulty component, since in a fault tolerant system, it can be isolated or replaced with a new component. Here average value of normalized phase current is utilized to locate faulty switch. According to this method, current average value in one fundamental cycle is calculated as follows:

$$\bar{I}_J = \frac{1}{N} \sum_{j=1}^{N} i_j, \quad j\varepsilon\{a,b,c,d,e\} \qquad (5)$$

Where N is the number of samples in one period. If I value is positive, then lower switch is faulty component at corresponding phase, while a negative value is related to upper faulty switch. After fault detection using (4), current polarity is obtained from diagnostic variable given in (5) and multiplied by (4); therefore, unique fault diagnostic signal is calculated as:

$$D = d_3 \times sign(d_1) \qquad (6)$$

A positive D value means that there is a fault at lower switch, whereas a negative means a fault at upper switch.

In other ??? to localize open phase fault, average value of signal d_2 is calculated at fundamental frequency. This value is near 0.5 for single switch fault while it is equal to 1 under open phase fault. So, its average is compared with a threshold, if value is higher than threshold, open phase fault is detected. Total fault detection and localization block is shown in Fig. 3. As it can be seen, first fault diagnostic

variable d_3 is calculated. Then, after fault detection, sub block for fault localization is implemented to determine a faulty switch. Input for this method is phase current. Here, no special transformation is necessary. Moreover, in contrast to available methods in literature, a single variable determines whether there is a faulty component. From the implementation point of view, the exponential operator is a standard practice in digital signal processing. Furthermore, it is much simpler to store its value for a limited data range in a lookup table; this method makes implementation simple for any processor type.

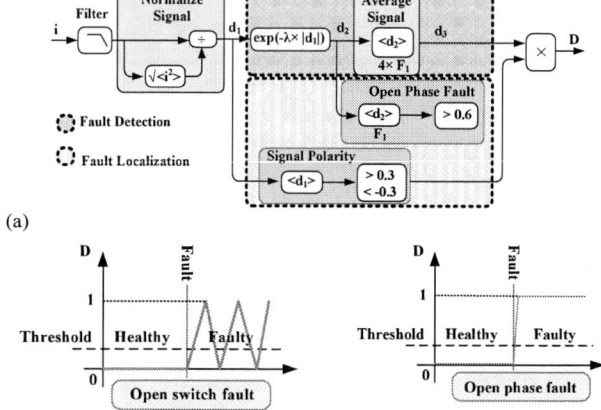

(b)

Fig. 3. Fault detection and localization block (a) fault detection block (b) fault signal

New diagnostic variable D is almost zero under healthy conditions, while it increases up to 1 under phase fault condition. This variable is robust to high load transients and high unbalanced load currents

III. SIMULATION RESULTS

The power converter shown in Fig. 1 has been simulated under different faulty modes using Matlab/Simulink. Load equivalent inductance and resistance are 7 mH and 2 Ω respectively, switching frequency is 2 kHz and dc link voltage is 310 V. Furthermore, in order to make simulation results more realistic, 1 µW white noise is added to current waveforms. Since in multiphase fault tolerant application, converter output current is an unbalanced combination of fundamental and third harmonic. These conditions have also been included in simulation. Current waveforms and detection index of phase currents are shown by a, b, c, d, e respectively. Besides, thresholds for open switch and open phases faults are indicated by Th_1 and Th_2, respectively. Prior to present fault detection, it is important to show transient response of fault detection strategy.

A step load transient from 1 p.u. to 0.05 p.u. is applied at 0.1 s. Simulated current waveforms and diagnostic variable are shown in Fig. 4; as seen, maximum deviation of D is 0.06 Therefore, threshold value should be selected beyond this, here 0.1 is the fault level detection. At multiphase applications, the converter is able to continue operation under simultaneous faults in more than one phase, which is not possible with three phase systems, so this mode is simulated here; where two faults are initiated at T6 and T7 at 0.25 s, as shown in Fig. 5. According to simulation results, faults are detected and lower faulty switch are localized successfully.

Similar to previous case study, two open phase fault can also happen; even under this condition, fault tolerant system is able to maintain continuous operation. This mode is also studied; simulation waveforms are presented in Fig. 6, where two open phase faults are initiated at phases A and E at time 0.24 and 0.28 s, respectively. As shown, fault is detected and its type is also determined. Detection time in this case is smaller. However, it takes half of fundamental cycle to detect the fault type.

In practical applications of five phase converters, it is not able to maintain fault tolerant in case of fault in more than two phases. However, in order to show detection method performance, these operation conditions are also simulated. Open switch fault at two lower switches T6, T7 at time 0.24 s, and two upper switch faults T4, T5 at time 0.28 s are considered. Waveforms are shown in Fig. 7. As seen, faults are detected, moreover faulty switches are successfully localized. Another case is fault at phase A at time 0.24 s, and open phase fault in phase E at time 0.26 s; an intermediate fault in upper switch T2 is initiated at time 0.26 s. Simulated waveforms are shown in Fig. 8. As seen, detection scheme can successfully detect fault and determine fault code.

A. Comparison of proposed method with literature

Here, in order to show the effectiveness of presented method, a comparison is done between detection schemes presented in this paper with methods available in literature. As aforementioned, there are two main problems with presented methods in literature.

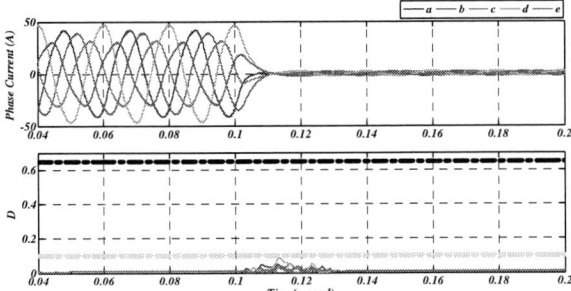

Fig. 4. Current waveforms and diagnostic variables under healthy mode

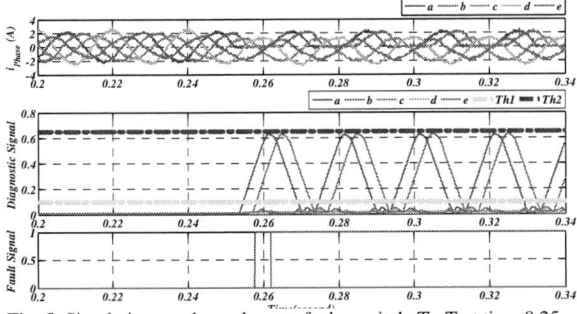

Fig. 5. Simulation results under two faulty switch T_6, T_7 at time 0.25 s

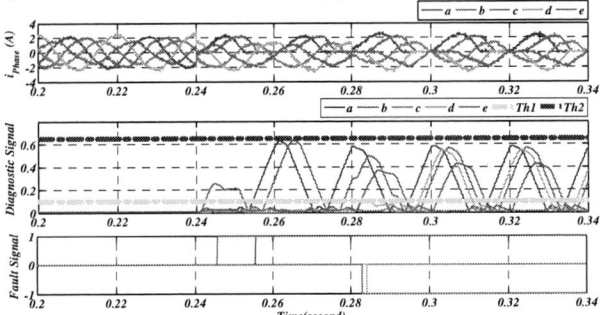

Fig. 6. Simulation results under two faulty phases A at 0.24 s and E at 0.28 s

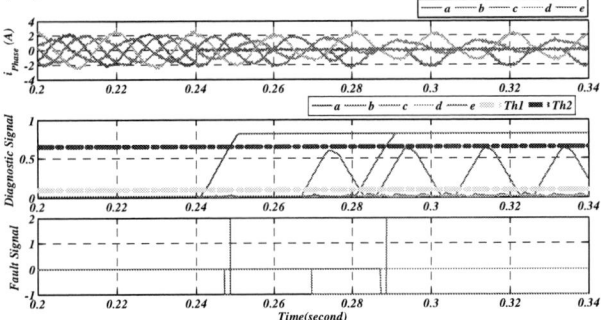

Fig. 7. Simulation results under two faulty switch T_6, T_7 at time 0.24 s and T_4, T_5 at time 0.28 s

Fig. 8. Simulation results under two faulty phases A at 0.24 s, E at 0.28 s and one fault switch T_2 at 0.26 s

First one, these methods have a low performance under load transients in case of multiphase (i.e. more than three phase) nonsinusoidal unbalanced currents. As first case for comparison, the idea presented in [14] is calculated as shown in (1) without using auxiliary variable; average current is normalized with respect to its absolute average value. Second case is utilization of Park's vector modulus as presented in [11] and [15]. Here later idea is developed for a five phase converter as follows. First, current signals are transformed to space vector using Park's transformation shown in (7) and (8).

$$(7)$$

$$T = \frac{2}{5} \begin{bmatrix} 1 & \cos(2\pi/5) & \cos(4\pi/5) & \cos(6\pi/5) & \cos(8\pi/5) \\ 0 & \sin(2\pi/3) & \sin(4\pi/3) & \sin(6\pi/5) & \sin(8\pi/5) \\ 1 & \cos(4\pi/5) & \cos(8\pi/5) & \cos(12\pi/5) & \cos(16\pi/5) \\ 0 & \sin(4\pi/3) & \sin(8\pi/3) & \sin(12\pi/5) & \sin(16\pi/3) \\ \frac{1}{\sqrt{2}} & \frac{1}{\sqrt{2}} & \frac{1}{\sqrt{2}} & \frac{1}{\sqrt{2}} & \frac{1}{\sqrt{2}} \end{bmatrix}$$

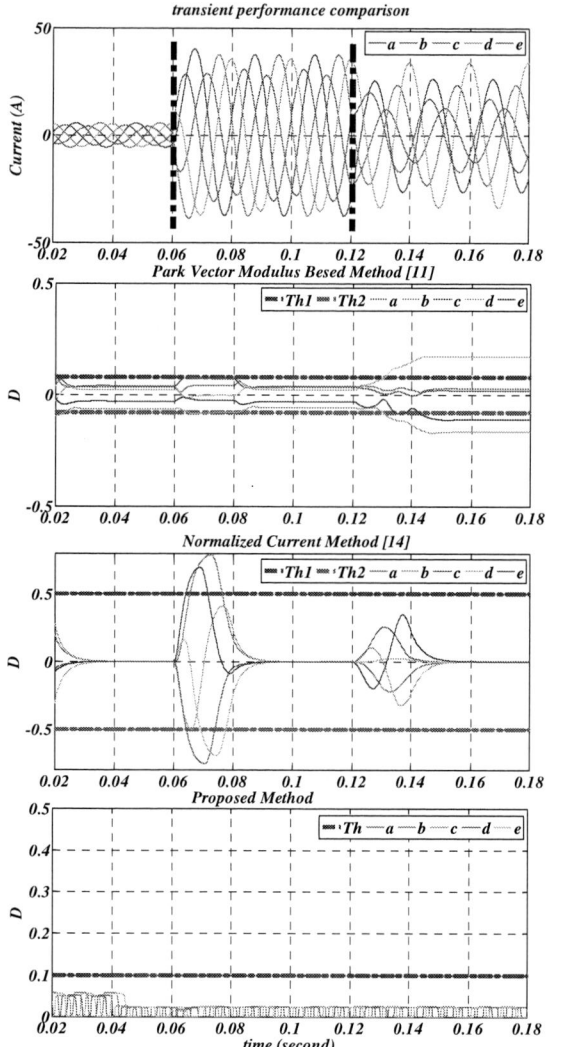

Fig. 9. Performance comparison under load transients and nonsinusoidal unbalanced currents

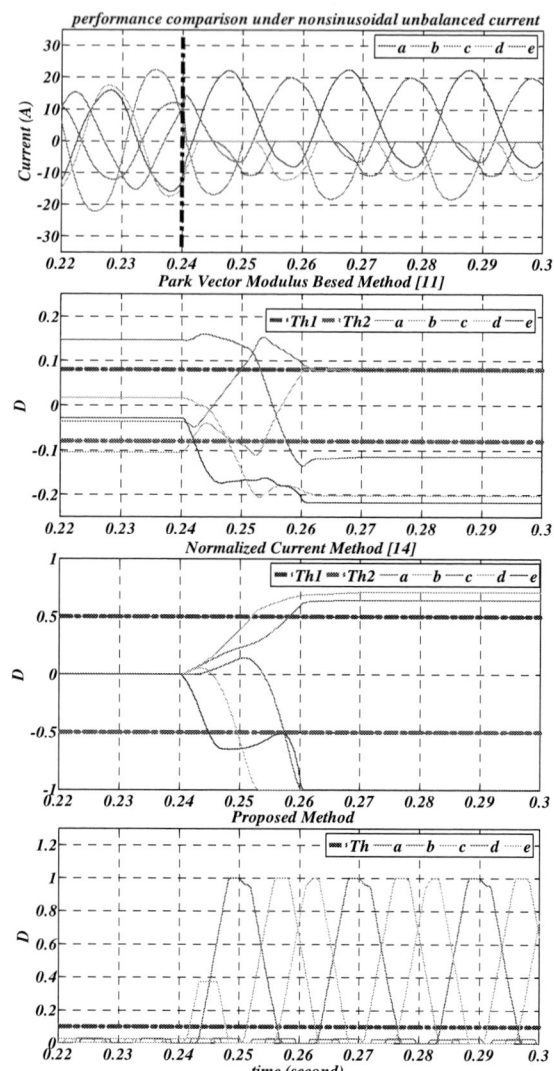

Fig. 10. Performance comparison under high unbalanced currents

$$\begin{bmatrix} i_d \\ i_q \\ i_x \\ i_y \\ i_0 \end{bmatrix} = T \begin{bmatrix} i_a \\ i_b \\ i_c \\ i_d \\ ie \end{bmatrix} \tag{8}$$

Then, input current is normalized with respect to magnitude of Park's vector modules. After that, resultant value is subtracted from absolute value of normalized motor phase currents (i.e. ζ) under balanced sinusoidal waveforms as shown in (8).

$$D = \left\langle \left| \frac{i}{\sqrt{i_d^2 + i_q^2 + i_x^2 + i_y^2}} \right| \right\rangle - \xi \tag{9}$$

It should note that value for a balanced sinusoidal current waveforms is equal to 0.4.

A simulation is done to evaluate transient performance of proposed method and techniques presented in literature. Simulated current waveforms and detection index are shown in Fig. 19. As seen, there are two transients at times 0.06 s and 0.12 s. According waveforms shown Fig. 9 for Park's vector modulus, after 0.12 s., detection variable is increased beyond threshold (i.e. 0.08), consequently a false alarm will be detected in phases b and d. Considering second method, normalized value of phase currents a, c, d, and e exceeds threshold value at 0.07 s. Therefore, a false alarm is initiated in these phases. In contrast to previous methods, the proposed method shows a high performance under both transients.

As second main problem with presented methods in literature, such methods show a low performance in case of an unbalanced current. This reason is explained in the following. In case of fault in each phase, a DC value is added to remaining healthy phases; therefore, average value of healthy phase is equal to this DC component. Under unbalanced load conditions, if there is a fault in phase with highest current amplitude, DC component is big. Therefore, healthy phase average current is close to its absolute average value. So according to (1), detection index is close to one, since numerator and denominator are almost equal. Simulation waveforms for this case are shown in Fig. 10. As seen, the method based on Park's vector has a better performance in this case. However, there are still false alarms. The current average based method in [14], without auxiliary variable, has the lowest performance. Finally, according to simulation waveforms, the proposed method can successfully detect faults.

TABLE I: PERFORMANCE COMPARISON IN TERMS OF DIFFERENT CRITERIA BETWEEN PROPOSED METHOD AND LITERATURE

	[14]	[11]	Proposed method
Transients under Sinusoidal current	*	*	*
Transients Unbalanced nonsinusoidal current			*
Simplicity	*	*	*
Detection speed	*	*	*
robustness			*
flexibility			*
Open phase fault detection	*		*
Multiple fault detection	*	*	*

TABLE II: MOTOR PARAMETERS

Number of Pole Pairs		2
Stator Resistance		0.55 Ω
Stator Inductance	Laa	35 mH
	Lab	10 mH
	Lac	7 mH
Nominal Torque		12 Nm
Nominal Speed		1527 rpm
Permanent Magnet Flux		0.48 Wb
Moment of Inertia		235 µkgm^2

IV. EVALUATION OF DETECTION METHOD PERFORMANCE IN FAULT TOLERANT CONTROL

In order to study performance of proposed method in a practical system, a five phase converter is used to supply a five phase PMSM drive. Such a drive can still maintain continuous operation under one and two faulty phases. A fault tolerant control algorithm (based on a field oriented control) is utilized to operate the machine. Parameters of simulated machine are shown in table II. Reference current values under faulty modes, details of control method and modulation strategy have been presented in [16] and [17].

As first cases study, two open switch fault is considered. Two open switch faults are initiated in lower and upper switches in phases A and B, respectively. Simulated waveforms of phase currents, diagnostic variable D and fault signal are shown in Fig. 11. Threshold value is denoted by *Th*. As seen, current in phase A is positive when fault is initiated at 0.1 second.

So, current waveforms are still non zero. As current starts to cross zero, due to lower open switch fault, it is zero. Considering diagnostic variable waveform in phase A, fault is detected within 2 ms after current value is zero. Same analysis is relevant for upper switch fault in phase B. Moreover, according to fault signal waveforms, fault localization is successfully done.

Open phase fault as common fault in power converters is also studied. Simulated waveforms are shown in Fig. 12.

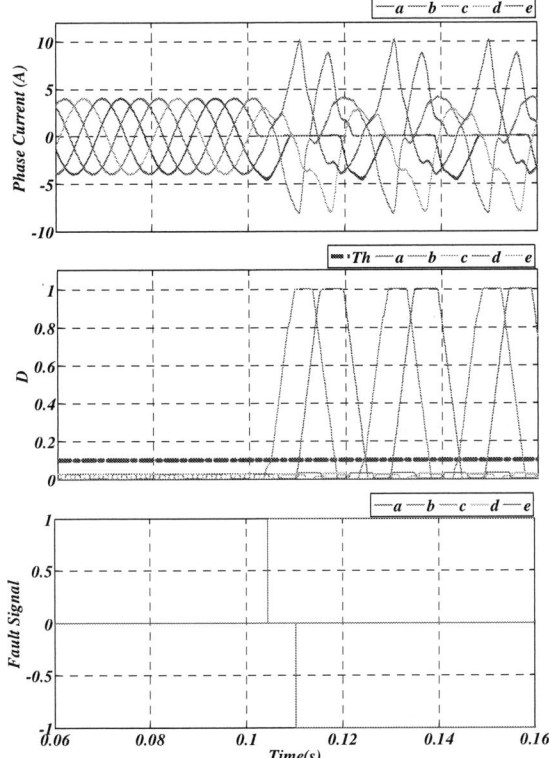

Fig. 11. Fault detection waveforms with two faulty switches

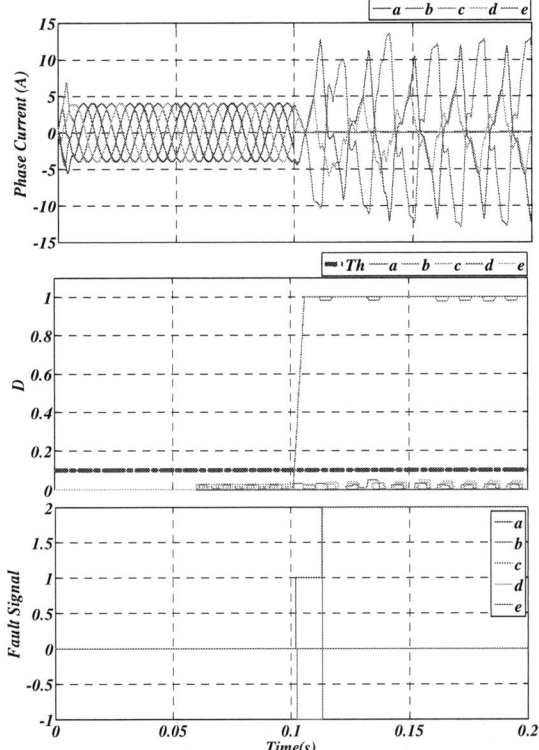

Fig. 12. Fault detection waveforms with two faulty phases

Two open phase faults are initiated in phases A and B at time 0.1 s. As it can be seen, detection method is able to detect and localized faulty type mode with high performance here. As in the previous case, fault is detected in less than two miliseconds. However, fault localization needs more time. This is due to higher threshold value considered for fault detection. It should be noted that to operate fault tolerant converter shown in Fig. 1, it is necessary to only detect fault; since for both single switch and open phase fault, faulty phase is isolated.

To show the high performance of the proposed method, it has been included in fault tolerant algorithm of a five phase PMSM drive. This drive is able to achieve fault tolerant concept with maximum two faulty phases when winding configuration is star with isolated neutral. It has different performance under two adjacent and two nonadjacent faulty phases. Both cases are simulated here.

Considering fault tolerant simulation, a fault is initiated at phase A at time 0.1 s. A second fault is considered at phase B at time 0.23 s. It should be noted that this phase is adjacent to phase A. Simulation waveforms are shown in Fig. 13.

As seen, detection block successfully detects faults. Finally, fault tolerant operation under two nonadjacent faulty phases is simulated. Results are shown in Fig. 14. According to presented waveforms, fault detection and control algorithms are done effectively. So presented simulation waveforms validate high performance of proposed fault detection scheme in fault tolerant drive applications.

V. EXPERIMENTAL RESULTS

Experimental results of a fault tolerant algorithm with the presented detection method are shown to validate theory. This fault tolerant control algorithm of a five phase PMSM with detection method is implemented in a DSPACE model 1005. Experimental setup is shown in Fig. 15. Two case studies are evaluated. First motor is operated in healthy mode. Then, a fault is included randomly in phase A, after that a second fault in phase B which is adjacent to phase A is imposed. Phase currents and torque are shown in Fig. 16. As it can be observed, experimental waveforms validate the high performance of the proposed detection scheme.

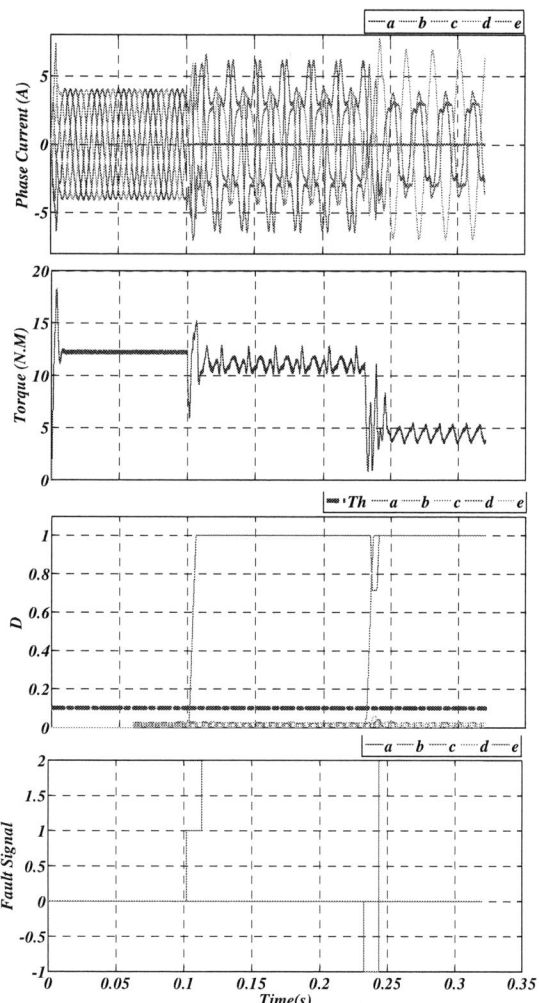

Fig. 13. Fault tolerant waveforms with two adjacent faulty phases

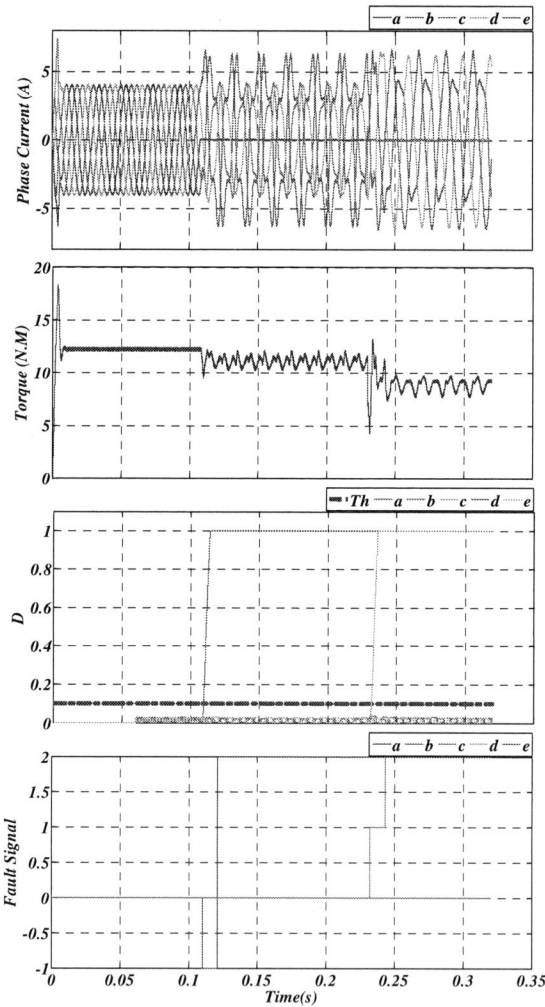

Fig. 14. Fault tolerant waveforms with two nonadjacent faulty phases

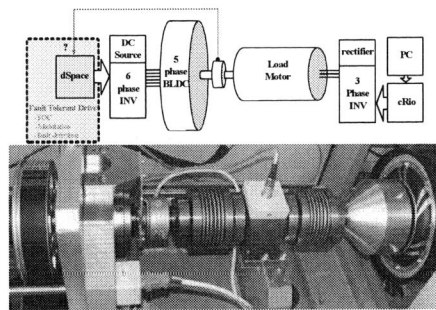

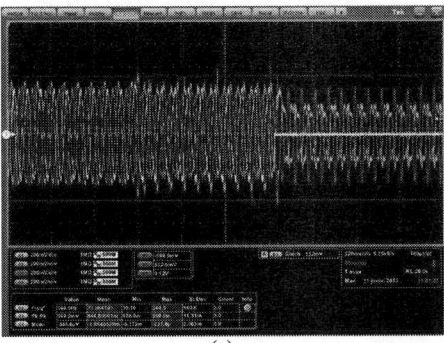

Fig. 15. Experimental setup

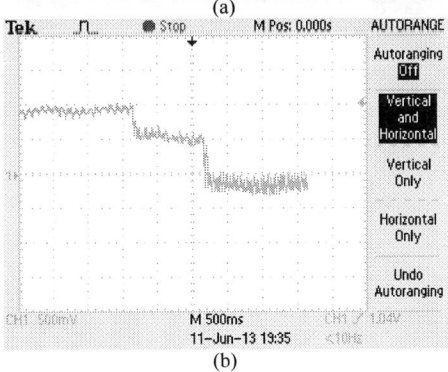

(a)

(b)

Fig. 16. Two adjacent faulty phases (a) phase currents (scale: 1V is equivalent to 16.67 A) (b) torque (scale: 1V is equivalent to 5.56 N.M)

VI. CONCLUSION

A new open circuit fault detection method is proposed which is more robust than conventional method due to normalization and type of operator used for detection. Only a single diagnostic variable is used to detect faults. Presented method was simulated under different faulty modes; a five phase PMSM drive which has nonsinusoidal unbalanced current waveforms under healthy and faulty condition was chosen as application area. According to both simulation and experimental results, successful fault detection and robustness to false alarms validated effectiveness of the proposed method. Average detection times, lower than a quarter of one fundamental cycle were presented; however it depends on fault initiation time.

ACKNOWLEDGEMENT

This work was supported by Spanish Ministry of Economic Affairs and Competiveness under the Research Project TRA 2010-21598-C02-01 and CD2009-00046 Consolidor Project.

REFERENCES

[1] Brian A. Welchko, Thomas A. Lipo, Thomas M. Jahns, and Steven E. Schulz, "Fault Tolerant Three Phase AC Motor Drive Topologies: A Comparison of Features, Cost, and Limitations," *IEEE Trans. Power Electron.*, vol. 19, no. 4, pp. 1108-1116, July 2004.

[2] Shaoyong Yang, Angus Bryant, Philip Mawby, Dawei Xiang, Li Ran, and Peter Tavner, "An Industry-Based Survey of Reliability in Power Electronic Converters," *IEEE Trans. Ind. Appl.*, vol. 47, no. 3, pp. 1441-1451, June 2011.

[3] Bin Lu, and Santosh K. Sharma, "A Literature Review of IGBT Fault Diagnostic and Protection Methods for Power Inverters," *IEEE Trans. Ind. Appl.*, vol. 45, no. 5, pp. 1770-1777, Oct. 2009.

[4] M. A. Rodriguez-Blanco, A. Claudio-Sanchez, D. Theilliol, L. G. Vela Valdes, P. Sibaja-Teran, L. Hernandez-Gonzalez, J. Aguayo "A failure-detection strategy for IGBT based on gate voltage behavior applied to a motor drive system," *IEEE Trans. Ind. Electron.*, vol. 58, no. 5, pp. 1625-1633, May 2011.

[5] Q.-T. An, L.-Z. Sun, K. Zhao and L. Sun, "Switching function model based fast-diagnostic method of open-switch faults in inverters without sensors," *IEEE Trans. Power Electron.*, vol. 26, no. 1, pp. 119-126, Jan. 2011.

[6] Yi Lu Murphey, M. Abul Masrur, Zhi Hang Chen, and Baifang Zhang, "Model-Based Fault Diagnosis in Electric Drives Using Machine Learning," *IEEE/ASME Trans. on Mechatronics*, vol. 11, no. 3, pp. 290 – 303, June 2006.

[7] Shin-Myung Jung, Jin-Sik Park, Hag-Wone Kim, Kwan-Yuhl Cho, and Myung-Joong Youn, "An MRAS-Based Diagnosis of Open-Circuit Fault in PWM Voltage-Source Inverters for PM Synchronous Motor Drive Systems," *IEEE Trans. Power Electron.*, vol. 28, no. 5, pp. 2514-2526, May 2013.

[8] C. Kral and K. Kafka, "Power electronics monitoring for a controlled voltage source inverter drive with induction machines," *IEEE Power Electron. Specialists Conf.*, vol. 1, pp. 213-217, 2000.

[9] J. O. Estima and A. J. M. Cardoso, "A New Algorithm for Real-Time Multiple Open-Circuit Fault Diagnosis in Voltage-Fed PWM Motor Drives by the Reference Current Errors," *IEEE Trans. Ind. Electron.*, vol. 60, no. 8, pp. 3496-3505, Aug. 2013.

[10] F. Charfi, F. Sellami, and K. Al-Haddad, "Fault diagnosis in power system using wavelet transforms and neural networks," *in Proc. IEEE Int. Symp. Ind. Electron.*, pp. 1143–1148, 2006.

[11] Jorge O. Estima and Antonio J. Marques Cardoso, "A New Approach for Real-Time Multiple Open-Circuit Fault Diagnosis in Voltage-Source Inverters," *IEEE Trans. Ind. Appl.*, vol. 47, no. 6, pp. 2487-2494, Dec. 2011.

[12] L. Parsa and H. A. Toliyat, "Fault-tolerant five-phase permanent-magnet motor drives," *IEEE Trans. Ind. Appl.*, vol. 41 , no. 1, pp. 30-37, Feb. 2005.

[13] P. Zheng, Y. Sui, J. Zhao, C. Tong, T.A. Lipo, A. Wang, "Investigation of a Novel Five-Phase Modular Permanent-Magnet In-Wheel Motor," *IEEE Trans. on Magnetics*, vol. 47 , no. 10, pp. 4084–4087, Oct. 2011.

[14] Wojciech Sleszynski, Janusz Nieznanski, and Artur Cichowski, "Open-Transistor Fault Diagnostics in Voltage-Source Inverters by Analyzing the Load Currents," *IEEE Trans. Ind. Electron.*, vol. 56, no. 11, pp. 4681–4688, Nov. 2009.

[15] Nuno M. A. Feire, Jorge O. Estima, and A. J. Marques Cardoso, "Open-Circuit Fault Diagnosis in PMSG Drives for Wind Turbine Applications," *IEEE Trans. Ind. Electron.*, vol. 60, no. 9, pp. 3957-3967, Sep. 2013.

[16] M. Salehifar, R. Salehi Arashloo, M. Moreno-Eguilaz, V. Sala, L. Romeral, "Fault Tolerant Operation of a Five Phase Converter for PMSM Drives," *Applied Power Electronics Conference and Exposition (APEC), 2013 Twenty-Eighth Annual IEEE*, pp: 1177 – 1184, 2013.

[17] R. Salehi Arashloo, M. Salehifar, L. Romeral, "On the Effect of Accessible Neutral Point in Fault Tolerant Five Phase PMSM Drives," *IECON 2012 - 38th Annual Conference on IEEE Industrial Electronics Society*, pp. 1934 – 1939, 2012.

Inter-turn Fault Detection in Five-Phase PMSMs. Effects of the Fault Severity

Harold Saavedra, Jordi-Roger Riba, Luís Romeral

Abstract – **This paper deals with the effects of inter-turn short circuit faults in five-phase permanent magnet synchronous motors (PMSMs). For this purpose a finite-elements model (FEM) of a faulty machine with 1, 2 and 4 inter-turns in short circuit is analyzed. From the results of this model the effects of these fault severities in the stator currents and zero-sequence voltage components (ZSVC) harmonics is analyzed and the possibility of developing a fault diagnosis scheme based on the changes in their spectral content is exposed. Moreover, the effect of the fault severity on the total power losses in the machine is presented. Inter-turn faults generate large circulating currents which may lead to catastrophic failures. Therefore it is very important to know the increase in power losses in the machine due to the occurrence of such faults for applying corrective actions at the precise time once the fault has been diagnosed.**

Index Terms— **fault diagnosis, fault severity, finite-elements method, inter-turn, in-wheel applications, multiphase motors, permanent magnet synchronous machines, power losses, short circuit.**

I. NOMENCLATURE

σ: Electrical conductivity of the ferromagnetic core material in S/m.

d: Lamination thickness in m.

ρ: Lamination mass density in kgm^{-3}.

B_{max}:   Peak value of the magnetic flux density in T.

J: Strength value of the magnetic polarization in T.

W_t: Specific total losses of the ferromagnetic material in Wm^{-3}.

W_h: Specific hysteresis losses in Wm^{-3}.

W_f: Specific eddy currents losses in Wm^{-3}.

W_e: Specific excess or anomalous losses in Wm^{-3}.

P_J: Dissipated power loss by Joule effect in W.

k_h: Experimental coefficient of magnetic losses due to hysteresis in $WsT^{-2}m^{-3}$.

k_e: Experimental coefficient of excess or anomalous losses in $W(Ts^{-1})^{-3/2}m^{-3}$.

K_f: Stacking factor ($0 \bullet K_f \, 1$).

f: Electrical frequency in Hz.

i_a: Stator current in phase a.

i_f: Circulating current in the shorted turns.

N: Total number of turns per phase

n: Number of shorted turns in phase e.

p: Number of poles pairs.

μ: Ratio between the number of shorted turns and the total turns per phase.

V_0:   Zero-sequence voltage component measured between the center point of the stator windings and the dc mid-point of the inverter.

R_s: Stator resistance per phase.

R_f: Resistance that models the insulation failure.

i_f: Fault current through R_f.

j: Current density in Am^{-2}.

E: Electric field strength in Vm^{-1}.

II. INTRODUCTION

THERE is a growing interest in the study of multiphase motors because they are well suited for applications requiring high reliability [1]. This is the case of aerospace, military and automotive industries. Among multiphase-motors, five-phase permanent magnet synchronous motors (PMSMs) are attractive candidates since these motor types offer high-reliability, high power and torque densities as well as high efficiency among other appealing features [2].

According to [3] machine topologies for fault tolerant drives must include physical and magnetic separation among phases by means of the use of concentrated windings, electrical isolation among phases by using separate bridge voltage-fed inverters for each phase, and the machine must be designed to cancel the mutual inductances between phases. Additionally, to produce the rated power during post-fault operation mode, the machine must be over-rated.

Inter-turn short circuits are among the critical faults in electrical machines [4], since they generate large circulating currents (higher than the rated current) which increase power losses, thus producing excessive heating which may lead to catastrophic failures [5]. In [3] it is stated that by shorting the whole phase through the inverter switches, the short circuit may be confined. However, this procedure is ineffective when a single turn is short-circuited due to its low impedance [3]. It is known that the magnitude of the circulating current depends on the number of shorted turns, their position in the slot and the speed of the machine.

In this paper the effects of inter-turn faults in a five-phase PMSM for in-wheel applications are analyzed and they are diagnosed by means of FEM simulations. Different fault

---

This work was supported in part by the Spanish Ministry of Science and Technology under the TRA2010-21598-C02-01 Research Project.

H. Saavedra and L. Romeral are with the Department of Electronic Engineering, Universitat Politècnica de Catalunya, C/Colom 1, 08222 Terrassa, Barcelona, Spain (e-mail: harold.saavedra@mcia.upc.edu; romeral@eel.upc.edu).

J.-R. Riba is with the Department of Electric Engineering, Universitat Politècnica de Catalunya, Escola d'Enginyeria d'Igualada, Plaça Rei 15, 08700 Igualada, Barcelona, Spain (e-mail: riba@ee.upc.edu).

978-1-4799-0024-4/13 $31.00 © 2013 IEEE

severities are studied, i.e. when 1, 2, and 4 turns are short-circuited. To diagnose inter-turn short circuit faults in five-phase PMSMs, this paper proposes the analysis of the fifth harmonic component of the stator currents and the first and third harmonics of the ZSVC. As stated, inter-turn faults may generate excessive heating, thus leading to very severe consequences. The total losses in the machine may modify the operating point of the machine, and therefore may affect the harmonic content of both the stator currents and the ZSVC, especially when the machine operates under rated load or close to rated load. Therefore it is very important to know the increase in power losses in the machine due to the occurrence of such faults, so remedial actions may be applied during the motor design stage and corrective actions may be adopted at the precise time once the fault has been diagnosed. Both, the fault diagnosis stage and the power losses calculations are carried out with the assistance of FEM simulations.

III. THE POWER LOSSES MODEL

Technical literature has paid little attention to analyze changes in power losses due to faults occurrence. However, in [6] the change in machine losses as a result of the unbalanced magnetic field during faulty operation is analyzed. This unbalanced magnetic field may modify the spectral content of the eddy currents. In addition, during faulty operation, the unbalanced stator currents may generate forward and backward rotating harmonic fields which induce eddy currents in the rotor permanent magnets and the iron core [6]. Therefore, the change in machine power losses may severely limit the machine performance as well as the torque and power capabilities.

In this section the power losses calculation system to be used in the FEM model is explained. The dissipated power losses by Joule effect P_J (W) on a conducting volume region may be calculated as

$$P_J = \int_V \vec{j}\vec{E}dV \tag{1}$$

E being the electric field strength and j the current density, which includes the eddy currents induced in conducting regions. Note that eddy currents are calculated in the permanent magnets of the rotor but not in the magnetic circuit, since stator and rotor laminations losses are calculated as in (6). Therefore, Joule losses are calculated from (1) in the stator windings and in the permanent magnets.

The total average power losses per unit volume (or specific losses) W_t in soft magnetic ferromagnetic iron cores may be split into three terms, namely the hysteresis, eddy currents and anomalous or excess losses components, respectively, W_h, W_f and W_e [7,8],

$$W_t = W_h + W_f + W_e \tag{2}$$

The hysteresis losses per unit volume may be expressed as,

$$W_h = k_h B_{max}^2 f \tag{3}$$

According to [7,9], the eddy currents or classical term may be expressed as,

$$W_f = \frac{\pi^2 \sigma d^2}{6} B_{max}^2 f^2 \tag{4}$$

whereas the anomalous power losses term is as follows [9],

$$W_e = 8.67 k_e (B_{max} f)^{3/2} \tag{5}$$

Therefore, by adding (3) to (5), the total specific power losses in the iron core laminations may be expressed as,

$$W_t = \underbrace{k_h B_{max}^2 f}_{\substack{Hysteresis \\ component}} + \underbrace{\frac{\pi^2 \sigma d^2}{6} B_{max}^2 f^2}_{\substack{Eddy\ currents \\ component}} + \underbrace{8.67 k_e (B_{max} f)^{3/2}}_{\substack{Excess\ loss \\ component}} \tag{6}$$

By supposing electrical steel laminations with a stacking factor K_f, (6) leads to the Bertotti fitting,

$$W_t = K_f [k_h B_{max}^2 f + \frac{\pi^2 \sigma d^2}{6} B_{max}^2 f^2 + 8.67 k_e (B_{max} f)^{3/2}] \tag{7}$$

Since the values of σ, d and K_f are known, parameters k_h and k_e may be obtained from the data supplied by the laminations' manufacturer. Therefore, (7) may be applied to compute the power losses in both the stator and rotor steel laminations.

Tables I and II show, respectively, the main physical properties of the rotor permanent magnets and laminations used in both the stator and rotor cores.

TABLE I
DATA OF THE PERMANENT MAGNETS

SmCo$_5$ permanent magnets	
Supplier	ChenYang
Grade	S18
Curie temperature	750 ºC
Max operating temperature	350 ºC
Electrical conductivity, σ	1.11·10^6 Sm^{-1}
Mass density, ρ	8250 kgm^{-3}
Remanence, B_r	0.8 T

TABLE II
DATA OF THE STEEL LAMINATIONS

M235-35A laminations	
Supplier	ArcerlorMittal
Electrical conductivity, σ	1.695·10^6 Sm^{-1}
Density, ρ	7850 kgm^{-3}
Thickness, d	0.35·10^{-3} m
Hysteresis coefficient, k_h	67.38100 WsT^{-2}m^{-3}
Excess losses coefficient, k_e	0.96313 W(Ts^{-1})$^{-3/2}$m^{-3}
Stacking factor, K_f	0.95

In this paper non-oriented electrical steel M235-35A laminations from ArcerlorMittal are analyzed. The specific losses of the M235-35A at f = 164 Hz (rated electrical frequency of the analyzed five-phase PMSM dealt with) as a function of the magnetic polarization and the peak value of the magnetic flux density are shown in Fig.1.

By evaluating two points (for example the points corresponding to $B = 0.52$ T and $B = 1.11$ T) of the curve shown in Fig. 1b, the experimental coefficients k_h and k_e at $f = 164$ Hz (rated speed) are obtained by solving a 2x2 equations system. The experimental coefficients k_h and k_e may also be obtained by applying an optimization algorithm. These values are only valid at this electrical frequency, i.e. at rated speed. Therefore, for variable speed applications, coefficients k_h and k_e must be calculated at each specific speed of the analyzed speed range.

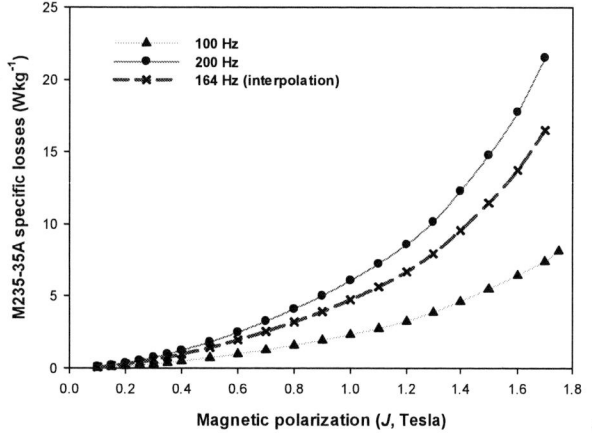

a)

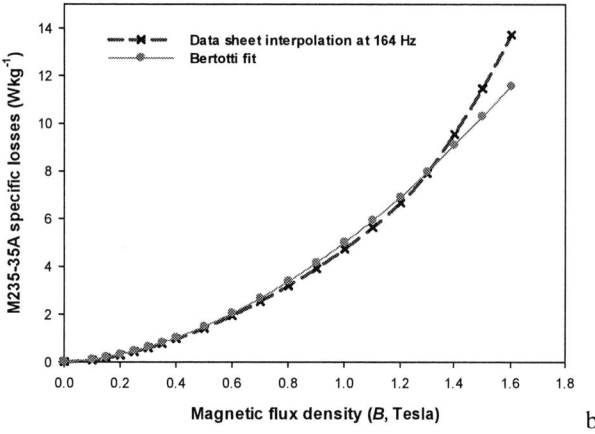

b)

Fig. 1. a) Specific losses of the M235-35A steel laminations at $f = 100$, and 200 Hz (data from ArcerlorMittal catalogue) and interpolation at $f = 164$ Hz as a function of the peak value of the magnetic polarization. b) Specific losses of the M235-35A steel laminations at $f = 164$ Hz as a function of the peak value of the magnetic flux density. Note that $B = \mu_0 H + J$.

The calculation system presented in Fig. 1 is applied in this paper to calculate the values of k_h and k_e which allow obtaining the power losses in both the stator and rotor steel laminations as expressed by (7). Note that the peak value of the magnetic field density B_{max} appearing in (7) is calculated by means of FEM simulations in all points of the analyzed domain.

IV. THE HARMONICS COMPONENTS TO BE ANALYZED

Inter-turn short circuit faults in three-phase PMSMs may be diagnosed by means of the analysis of both the stator currents and the ZSVC spectra [5,10]. In the [5,10], such faults are diagnosed by analyzing the third harmonic component of the stator currents (it arises because of the unbalance effect of inter-turn faults) and the first one of the ZSVC when dealing with three-phase PMSMs. Therefore, when dealing with five-phase PMSMs this paper proposes the analysis of the fifth harmonic component of the stator currents (since the unbalance effect must be reflected in the fifth component) and the first and third harmonics of the ZSVC.

The theoretical justification of the suitability of the analysis of such harmonic components may be found in [10]. As stated in [10], the ZSVC spectrum of a faulty machine must contain a significant first harmonic component and also third and fifth harmonic components because the ZSVC is greatly influenced by the current circulating through the faulty turn. According to [10], the circulating current contains all the odd harmonics since it is mainly due to the induction effect of the rotor magnets. In addition, in the case of a faulty machine there is a link between the circulating current and the phase currents, so the phase currents must contain this chain of odd harmonic components. It is worth noting that at low speed operation the induction effect of the permanent magnets greatly diminishes.

The analysis of the stator currents harmonics is appealing due to its inherent simplicity and low-cost, since no additional hardware is required. However, it is recognized that the stator current harmonics amplitudes may be affected by the PMSM drive current-loops. This is because the motor drive interprets the harmonics of the stator currents as a perturbation, so the current-loops try to compensate such variation [11].

On the other hand, an accessible neutral point of the PMSM is an indispensable requisite to monitor the ZSVC. However, this apparent drawback is highly compatible with fault-tolerant schemes [1]. Moreover, when properly measured, the ZSVC may be decoupled from the motor drive effects [5]. As stated in [10], the analysis of the ZSVC is appealing since it offers improved resolution when compared with the analysis of the stator currents spectra, especially at low speed operation. This feature is of special relevance when dealing with in-wheel drives due to the inherent low-speed range of such applications. The main disadvantage of the ZSVC-based method is that an extra voltage sensor is required.

Fig. 2 shows the electrical connections of the five-phase PMSM and the circuit used to model inter-turn short circuit faults, R_f being the resistance that models the insulation failure and i_f the circulating fault current.

978-1-4799-0024-4/13 $31.00 © 2013 IEEE

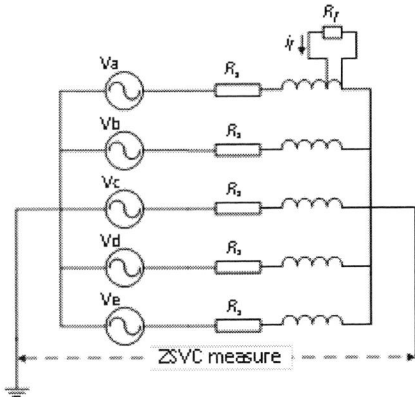

Fig. 2. Connections diagram in the FEM model including the turns in short circuit and the ZSVC measurement.

Note that simulations of the faulty PMSM have been conducted using a fault resistance $R_f = 0.05\ \Omega$, a value in the range of those found in [12-14].

V. THE FEM MODEL

A two-dimensional (2D) FEM model of the analyzed PMSM was carefully prepared in the Flux-Cedrat® environment. The motor was supplied by means of five balanced voltage sources (72° phase shift between two consecutive phase voltages), as detailed in Fig. 2. Simulations were performed with a sampling frequency of 4.9 kHz and 2048 points were simulated with a resolution of 30 points/cycle. Therefore, the FFT of such waveforms outputs 1024 points with a frequency resolution of 2.4 Hz.

Fig. 3 shows the winding distribution of the analyzed PMSM. Note that A+ refers to the conductors in phase A which input the slot, whereas A- refers to conductors which output the slot, so the sequence is A+ A- A+ A- ... in a phase winding.

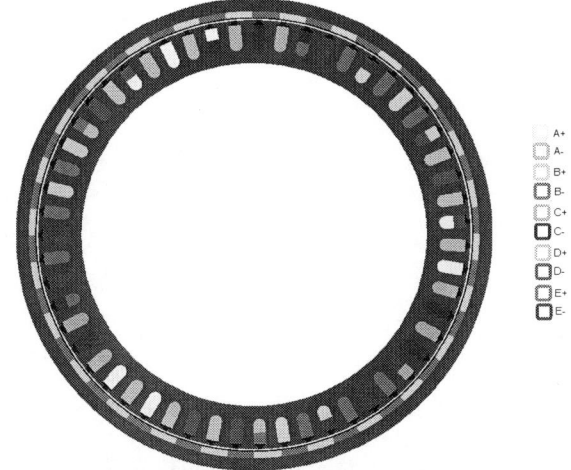

Fig. 3. Winding distribution of the analyzed five-phase PMSM.

Fig. 4 plots the stator laminations, the rotor laminations

with the permanent magnets and a partial view of the 2D-mesh applied in the FEM simulations, respectively.

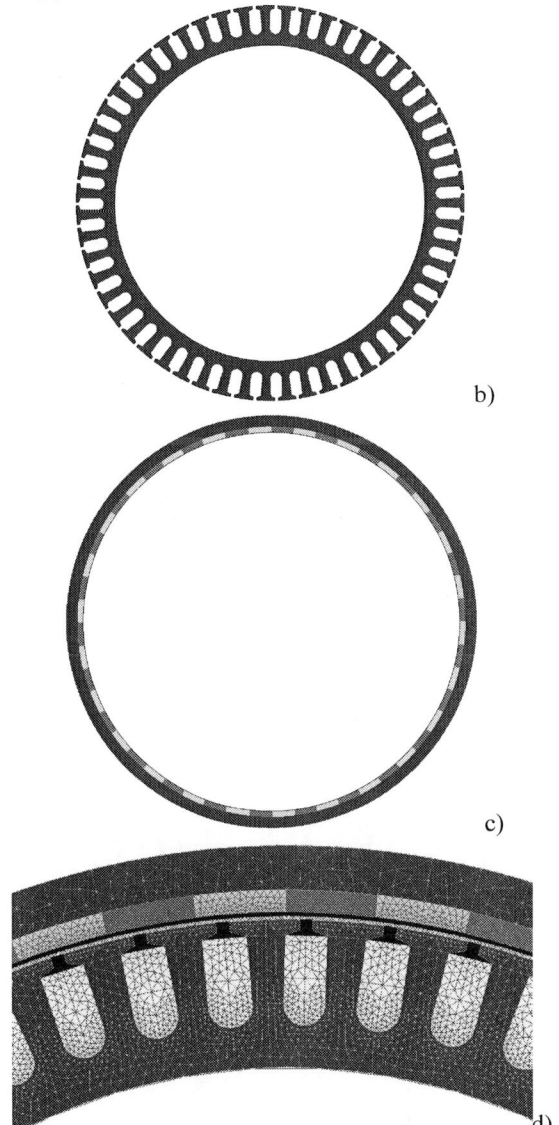

Fig. 4.a) Section of the stator of the analyzed five-phase PMSM. c) Section of the rotor. d) Detail of the 2D-mesh.

The main parameters of the five-phase PMSM analyzed are given in Table III.

TABLE III
MAIN PARAMETERS OF THE ANALYZED PMSM

Outer Rotor Characteristics	
Number of phases	5
Rated torque	23 Nm
Rated speed	379 r/min
Poles pairs	26
Rated phase voltage	18 V
Rated current	13 A
Rated electrical frequency	164 Hz
Resistance per phase	0.1 Ω
Number of turns per phase	110
Number of stator slots	55

VI. RESULTS

In this section the results obtained by means of the 2D-FEM model presented in Section V are discussed. They deal with the healthy five-phase PMSM and three faulty machines, namely ST1, ST2 and ST4, which correspond to 1, 2 and 4 turns in short circuit, respectively.

First, the relative change in the power losses in both the stator and rotor parts due to the occurrence of inter-turn faults is analyzed. Fig. 4 details the power losses in each part of the motor for the healthy machine and the three analyzed fault conditions.

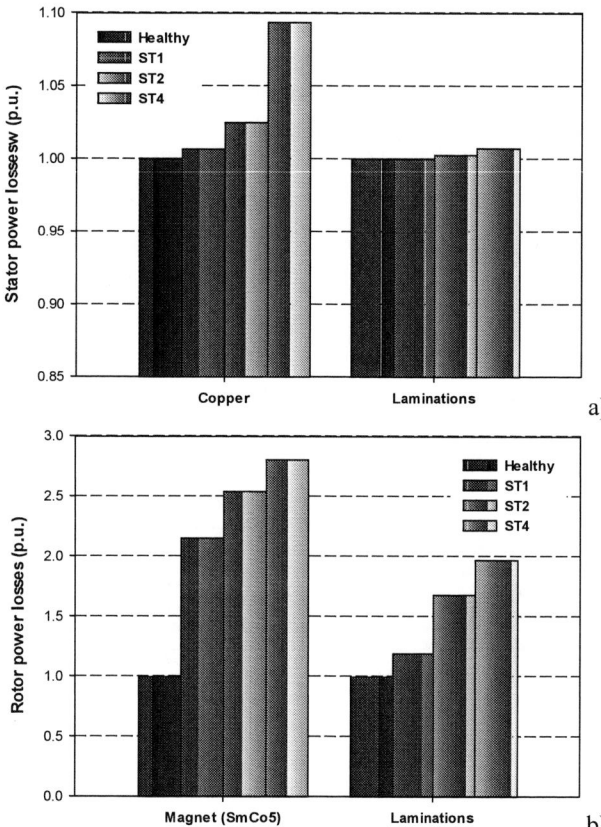

Fig. 4. Power losses in the analyzed five-phase PMSM when operating at rated speed under rated load conditions. a) Stator power losses. b) Rotor power losses.

Results presented in Fig. 4 show the increase of the power losses in both the stator and rotor regions due to inter-turn faults. It clearly shows an increase of power losses as the fault severity increases. Additionally, in the case of copper losses in the stator windings, this increase is mostly concentrated in the region where the short circuit occurs.

Fig. 5 shows the circulating fault current i_f for all analyzed fault severities and their spectra. Results presented in Fig. 5 clearly show that the fault current i_f magnitude increases with a growing number of short-circuited turns, as expected. The increase of i_f with a growing number of turns in short circuit is due to the increase of the voltage induced by the permanent magnets in the shorted turns.

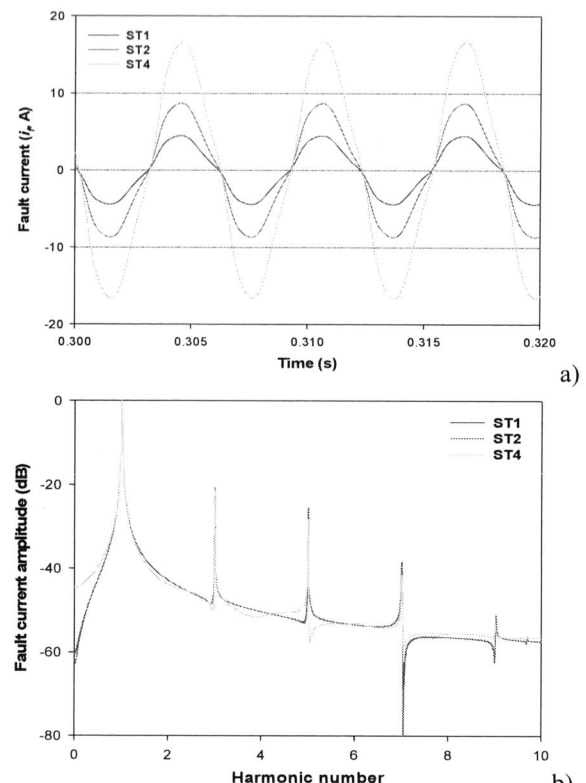

Fig. 5. Circulating fault current i_f simulated at rated speed and rated load under different fault severities. a) Amplitudes. b) Spectra.

Fig. 6 shows the harmonic content of the stator currents of the healthy five-phase PMSM and the three analyzed fault severity degrees.

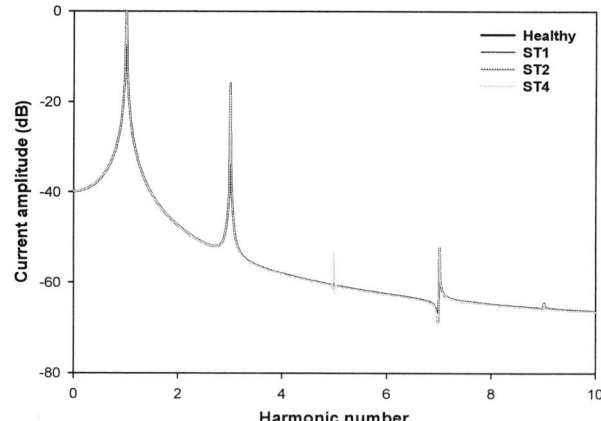

Fig. 6. Stator currents spectra of the healthy and faulty five-phase PMSMs analyzed when running at rated speed under rated load conditions.

Results presented in Fig. 6 show that the fifth harmonic of the stator currents is also present in the spectrum of the circulating fault current i_f as shown in Fig. 5. This suggests that the fault current i_f influences the spectral content of the stator currents.

For better readability, Table IV summarizes the values of

978-1-4799-0024-4/13 $31.00 © 2013 IEEE

the most suitable harmonic components to detect inter-turn faults from the stator currents spectra.

TABLE IV
STATOR CURRENT HARMONICS AMPLITUDES OF THE DIFFERENT
ANALYZED MACHINES

	Current harmonics amplitudes (dB)			
	3rd	5th	7th	9th
Healthy	-15.93	-	-52.26	-44.64
ST1	-15.93	**-57.80**	-52.24	-44.68
ST2	-15.93	**-55.64**	-52.19	-44.77
ST4	-15.93	**-53.41**	-52.00	-44.89

Results presented in Fig. 6 and Table IV show that the fifth stator currents harmonic is the most useful to detect inter-turn faults. It is so because the stator currents spectrum of a healthy machine does not present such harmonic, whereas in the case of all analyzed faulty machines this harmonic is already present in their spectra. Although from the analysis of the 7th and 9th harmonics it is theoretically possible to detect the analyzed fault, in a practical motor this possibility is not feasible due to the small change in their amplitudes when compared with those of a healthy motor.

Fig. 7 shows the harmonic content of ZSVC of the healthy five-phase PMSM and the three analyzed fault severity degrees.

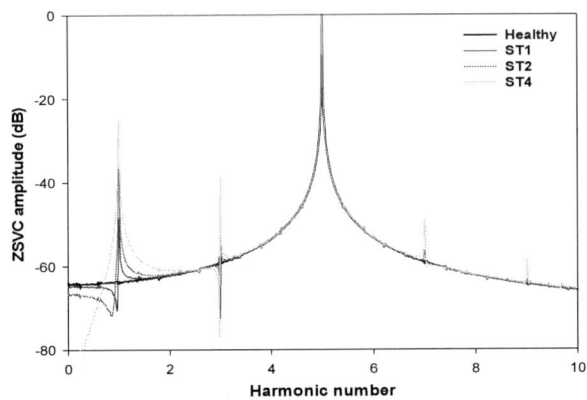

Fig. 7. ZSVC spectra of the healthy and faulty five-phase PMSMs analyzed when running at rated speed under rated load conditions

Table V summarizes the values of the most suitable harmonics components to detect inter-turn faults from the ZSVC spectra.

TABLE V
ZSVC HARMONICS AMPLITUDES OF THE DIFFERENT ANALYZED MACHINES

	ZSVC harmonics amplitudes (dB)			
	1st	3rd	7th	9th
Healthy	-	-	-	-
ST1	**-49.96**	**-59.35**	-59.82	-64.26
ST2	**-37.18**	**-52.63**	-57.53	-63.39
ST4	**-25.21**	**-39.57**	-49.98	-58.46

Results presented in Fig. 7 and Table V show that the healthy machine ZSVC spectrum only contains the fundamental harmonic, i.e. the 5th. Contrarily, the ZSVC spectra of all analyzed faulty cases contain important amplitudes of the 1st and 3rd harmonics, as well a non-

negligible contribution of the 7th and 9th harmonics. These results suggest that the appearance of the 1st and 3rd harmonics in the ZSVC spectrum may be used to detect inter-turn faults. In addition, it seems also feasible to diagnose this type of fault in its early stage, i.e. when only one turn is short-circuited since the 1st ZSVC harmonic of case ST1 has important amplitude, as shown in Table V.

When comparing results presented in Tables IV and V, it is deduced that the analysis of the ZSVC provides better sensitivity. Additionally, in the case of the ZSVC two harmonic components may be used to diagnose inter-turn faults, i.e. the 1st and 3rd, whereas in the case of the analysis of the stator currents the only harmonic susceptible to be analyzed is the 5th.

VII. CONCLUSION

It is recognized that inter-turn faults are among the most potentially harmful faults in electrical machines, thus deteriorating motor efficiency and performance.

This paper has analyzed the effects of such faults when 1, 2 and 4 inter-turns are short-circuited in five-phase PMSMs. It has been proved that inter-turn faults generate an increase in power losses, therefore lowering the PMSM efficiency and changing the machine's operating point.

In this work inter-turn faults have been diagnosed by means of the analysis of the stator currents and the ZSVC spectra. Results presented show better performance and improved sensitivity of the ZSVC-based method when compared with the system based on the analysis of the stator currents spectra. In particular, the analysis of the first and third harmonic frequencies of the ZSVC is recommended to conduct an accurate diagnosis of such faults in five-phase PMSMs, although it is also feasible the analysis of the 7th and 9th harmonics.

VIII. REFERENCES

[1] R. Salehi, M. Salehifar, L. Romeral, "On the Effect of Accessible Neutral Point in Fault Tolerant Five Phase PMSM Drives," in *Proc. of the IEEE 38th Annual Conference on IEEE Industrial Electronics Society, IECON*2012, 2012, pp. 1934-1939.

[2] A.M. El-Refaie, "Fault-tolerant permanent magnet machines: a review," *IET Electr. Power Appl.*, vol. 5, no. 1, pp. 59–74, 2011.

[3] P. Arumugam, T. Hamiti, C. Gerada, "Analytical Modeling of a Vertically Distributed Winding Configuration for Fault Tolerant Permanent Magnet Machines to Suppress Inter-Turn Short Circuit Current Limiting," in *Proc. IEEE International Electric Machines & Drives Conference IEMDC2011*, 2011, pg. 371-376.

[4] W. Tang, G. Liu, and J. Ji, "Winding Turn-to-Turn Faults Detection of Five-Phase Fault-Tolerant Permanent-Magnet Machine Based on Parametric Model," in *Proc. 15th International Conference on Electrical Machines and Systems (ICEMS)*, 2012, pp. 1-6.

[5] J.-C. Urresty, J.-R. Riba, and L. Romeral, "Diagnosis of Interturn Faults in PMSMs Operating Under Nonstationary Conditions by Applying Order Tracking Filtering," *IEEE Trans. Power Electron.*, vol. 28, no. 1, pp. 507-515, Jan. 2013.

[6] T. Raminosoa, C. Gerada, N. Othman, L.D. Lillo, "Rotor losses in fault-tolerant permanent magnet synchronous machines," *IET Electr. Power Appl.*, vol. 5, no. 1, pp. 75–88, 2011.

[7] G. Bertotti, "General Properties of Power Losses in Soft Ferromagnetic Materials," *IEEE Trans. Magn.*, vol. 24, no. 1, pp. 621-630, Jan. 1988.

[8] J. P. Schlegel, N. Sadowski, N. J. Batistela, B. A. T. Iamamura, J. P. A. Bastos, and A. A. Espíndola, "Core Tester Iron Losses Segregation

by Finite Element Modeling," *IEEE Trans. Magn.*, vol. 48, no. 2, pp. 715-718, Feb 2012.

[9] F.A. Fiorillo, A. Novikov, "An Improved Approach to Power Losses in Magnetic Laminations under Nonsinusoidal Induction Waveform," *IEEE Trans. Magn.*, vol. 26, no. 5, pp. 2904-2910, Sept. 1990.

[10] J.–C. Urresty, J.-R. Riba, L. Romeral, "Detection of Stator Winding Inter-Turn Faults in Surface-Mounted Permanent Magnet Synchronous Motors by means of the Zero-Sequence Voltage Component," Electric Power Systems Research, vol. 89, pp. 38-44, Aug. 2012.

[11] W. Le Roux, R. G. Harley, and T. G. Habetler, "Detecting faults in rotors of PM drives," *IEEE Ind. Appl. Mag.*, vol. 14, no. 2, pp. 23–31, March-April 2008.

[12] C. Yeh, R. J. Povinelli, B. Mirafzal, and N. A. O. Demerdash, "Diagnosis of stator winding inter-turn shorts in induction motors fed by PWM-inverter drive systems using a time-series data mining technique," in *Proc. of the International Conference Power System Technology, PowerCon2004, 2004,* pp. 891-896.

[13] A. Sayed-Ahmed, C. Yeh, N.A.O. Demerdash, B. Mirafzal, "Analysis of Stator Winding Inter-Turn Short-Circuit Faults in Induction Machines for Identification of the Faulty Phase," in *Proc. of the IEEE Industry Applications Conference, 2006,* vol. 3, pp. 1519-1524.

[14] L. Romeral, J. –C. Urresty, J. –R. Riba, and A. Garcia, "Modeling of Surface-Mounted Permanent Magnet Synchronous Motors With Stator Winding Interturn Faults," *IEEE Trans. Ind. Electron.*, vol. 58, no. 5, pp. 1576-1585, May 2011.

IX. BIOGRAPHIES

Harold Saavedra received the M.S. degree in electrical engineering in 2009 from the Universidad del Valle, Cali, Colombia. Currently he is a PhD student in the Electronic Engineering Department at the Universitat Politècnica de Catalunya (UPC), in the research group MCIA in Terrassa (Barcelona, Spain). His research interests include signal fault-tolerant systems for multiphase electrical machines, variable-speed drive systems, fault diagnosis systems, signal processing methods and optimization algorithms.

Jordi-Roger Riba (M'09) received the M.S. degree in physics and the Ph.D. degree from the Universitat de Barcelona, Barcelona, Spain, in 1990 and 2000, respectively. In 1992, he joined the Escola d'Enginyeria d'Igualada, Universitat Politècnica de Catalunya, Barcelona, Spain, as a full-time Lecturer, and he joined the Electric Engineering Department in 2001. He belongs to the Motion and Industrial Control Group (MCIA). His research interests include electromagnetic devices modeling, signal processing methods, electrical machines, fault diagnosis in electrical machines and fault detection algorithms.

Luís Romeral (M'98) received the M.S. degree in electrical engineering and the Ph.D. degree from the Universitat Politècnica de Catalunya (UPC), Barcelona, Spain, in 1985 and 1995, respectively.
In 1988, he joined the Electronic Engineering Department, UPC, where he is currently an Associate Professor and the Director of the Motion and Industrial Control Group (MCIA), whose major research activities concern induction and permanent magnet motor drives, enhanced efficiency drives, fault detection and diagnosis of electrical motor drives, and improvement of educational tools. He has developed and taught post-graduate courses on programmable logic controllers, electrical drives and motion control, and sensors and actuators. Dr. Romeral is a Member of the European Power Electronics and Drives Association and the International Federation of Automatic Control.

978-1-4799-0024-4/13 $31.00 © 2013 IEEE

Detection of Coupling Inductor Faults in Three-Phase Adjustable Speed Drives with Direct Power Control-Based Active Front-End Rectifiers

Joaquín G. Norniella, José M. Cano, *Member, IEEE*, Gonzalo A. Orcajo, Carlos H. Rojas,
Joaquín F. Pedrayes, Manés F. Cabanas, *Member, IEEE*, and Manuel G. Melero

Electrical Engineering Department
University of Oviedo
Gijón, Spain
jgnorniella@uniovi.es

Abstract -- Inductors are utilized as the typical filtering option in three-phase adjustable speed drives (ASDs) with an active front-end rectifier (AFE) controlled by the so-called direct power control (DPC) method. Faults in the coupling inductors are one of the possible causes of failure in those systems. In this paper, a new algorithm for the early detection of coupling inductor faults in DPC-based AFEs of ASDs is presented. Most of these faults involve variations of the coil inductance value, as in the case of inter-turn short circuits. The proposed algorithm estimates and tracks the coupling inductance value to achieve a correct diagnosis. Therefore, a major system breakdown can be avoided. The algorithm also provides the identification of the winding where the fault occurs.

Index Terms -- Adjustable speed drive, coupling inductor fault, direct power control, inductance estimation, three-phase active front-end rectifier.

Nomenclature

p, q	Instantaneous active and reactive power.
i_α, i_β	α–β components of ac-current space vector.
e_α, e_β	α–β components of grid voltage space vector.
L_a, L_b, L_c	Rated single-phase coupling inductances.
L	Rated three-phase coupling inductance.
L_A, L_B, L_C	Estimated three-phase coupling inductance considering a fault in a, b or c.
ΔL	Three-phase coupling inductance variation considering a fault in one of the inductors.
$\Delta L_A, \Delta L_B, \Delta L_C$	Estimated three-phase coupling inductance variation considering a fault in a, b or c.
i_a, i_b, i_c	Converter ac-side currents.
i_a', i_b', i_c'	Time derivatives of ac-currents.
v_{dc}	Voltage in the dc-bus.
t, t'	Instants before and after commutation.
S_a, S_b, S_c	Switching states of converter upper switches (1 if closed, 0 if open).
e_a, e_b, e_c	Grid voltages.

I. Introduction

RELIABILITY of ASDs has been a recurrent research topic since decades ago [1]-[3]. Most of the surveys on this field are focused on failures related to the power semiconductor devices included in the ASDs [4]-[6]. As a matter of fact, those devices are ranked as the ones for which reliability in power electronic converters is of most concern [6]. On the other hand, works on coupling inductor faults in ASDs are hardly available in bibliography. This lack of interest may come from the fact that failure rates of industrial power converters caused by fragile components like coupling inductors are relatively low (about 5% according to [6]) when compared to those due to other components. However, an in-service fault of a coupling inductor (e.g., due to an inter-turn short circuit) can involve dramatic consequences for the whole system. In addition to this, the reliability of certain applications against coupling choke failures can be improved to a great extent in a simple fashion, as is the case of the present work. Therefore, the incorporation of an online method in the ASD control system to detect incipient failure of coupling coils can be cost-effective and useful to achieve the overall application fault-tolerance.

Inter-turn short circuits are a common fault in electric components, not only in inductors but also in transformers and rotating machinery [7]. Inter-turn short circuits derive from the degradation of insulation provoked by a wide range of causes, including design/manufacturing defects, power/thermal cycles or environmental causes [6]. It is well-known that inter-turn short circuits make the inductance value of the coil vary. As a matter of fact, methods devoted to detecting this type of failure (e.g., surge tests) take advantage of that variation [8], as is the case of the strategy proposed in this paper.

Coupling filters are indispensable in three-phase AFEs with VSRs controlled by means of the DPC method [9]. In such devices, filters represented by one three-phase or three single-phase inductors suffice for harmonics to comply with regulations. Although the three-phase coil is more economical, three single-phase inductors enable reducing reserve units. The coupling inductance value is required in the DPC method to estimate instantaneous power and grid voltages. The most common option is to use the inductance

---

This work was supported in part by the Spanish Research, Development and Innovation Program under Grant DPI2011-26535.

rated value. However, parameters as the grid interaction or the inductor tolerance, operating point, temperature, etc., lead to mismatches between the considered and actual values of the inductance. Such errors affect the operation of the DPC method negatively, making the use of online algorithms for estimating the coupling inductance in DPC-based strategies advisable. The present paper is partially based on the analytic estimation algorithms proposed by the authors in [10]. Unlike the algorithms in [10], the method presented in this paper enables the estimation of the coupling inductance value even in the case of a faulty coil. Besides, the proposed method calculates the deviation of the faulty inductor value from the rated one.

In this paper, a new method for detecting coupling inductor faults involving inductance variations in DPC-based three-phase AFEs is proposed. The method provides a simple and accurate identification of the phase where the fault occurs. Therefore, a major breakdown can be avoided. Knowing the affected phase enables the rewinding of just the faulty inductor when using three single-phase coils or identifying the flawed winding if one three-phase inductor is used. The method is also a strategy to estimate the coupling inductance value even in the case of a faulty choke. Simulation results obtained by means of Simulink demonstrate the validity of the method.

II. DIRECT POWER CONTROL

The basic diagram of the DPC strategy [9] applied to a three-phase voltage source active rectifier (VSR) included in an ASD is shown in Fig. 1. For the sake of simplicity, a resistor is assumed to be connected to the dc-link.

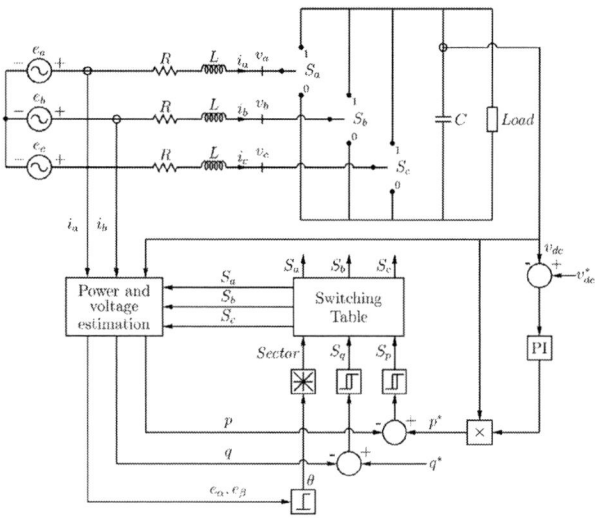

Fig. 1. DPC basic diagram.

The DPC method is based on a hysteresis-band control of p and q. The dc-link voltage tends to be controlled simply by means of a PI controller. The optimal switching states of the converter are stored in a look-up table usually called

switching table. Those states are chosen according to the outputs of the hysteresis controllers utilized for p and q, and to the position of the grid voltage space vector (VSV) in the α-β frame. Because the DPC method has no grid voltage sensors, instantaneous power and grid voltage must be estimated. The grid VSV α-β components and instantaneous active and reactive power are estimated as shown in (1)-(3).

$$\begin{pmatrix} e_\alpha \\ e_\beta \end{pmatrix} = \frac{1}{i_\alpha^2 + i_\beta^2} \begin{pmatrix} i_\alpha & -i_\beta \\ i_\beta & i_\alpha \end{pmatrix} \begin{pmatrix} p \\ q \end{pmatrix} \quad (1)$$

$$p = e_a i_a + e_b i_b + e_c i_c = \\ L_a i_a' i_a + L_b i_b' i_b + L_c i_c' i_c + v_{dc} \left(S_a i_a + S_b i_b + S_c i_c \right) \quad (2)$$

$$q = \left(1/\sqrt{3} \right) \left[(e_b - e_c) i_a + (e_c - e_a) i_b + (e_a - e_b) i_c \right] = \\ \left(1/\sqrt{3} \right) \{ \left(L_b i_b' - L_c i_c' \right) i_a + \left(L_c i_c' - L_a i_a' \right) i_b + \left(L_a i_a' - L_b i_b' \right) i_c - \\ - v_{dc} \left[S_a (i_b - i_c) + S_b (i_c - i_a) + S_c (i_a - i_b) \right] \} \quad (3)$$

In the following sections, the initial inductance values before an inductor fault are considered to be equal to the rated value of the coupling inductors ($L_a = L_b = L_c = L$).

III. DETECTION OF INDUCTOR FAULTS AND COUPLING INDUCTANCE ESTIMATION

Regarding (1)-(3), a correct estimation of the grid VSV α-β components is expected in the DPC method when there are no faults in the coupling inductors. In that case, the trajectory of the estimated grid VSV must be a continuous magnitude due to the continuity of the actual grid voltages. On the other hand, if a fault occurs in one of the chokes, discontinuities in that trajectory can be noticed after commutations. For example, Fig. 2 shows the trajectory of the estimated grid VSV both for the case of healthy inductors and for the case of a fault in the coil of phase a.

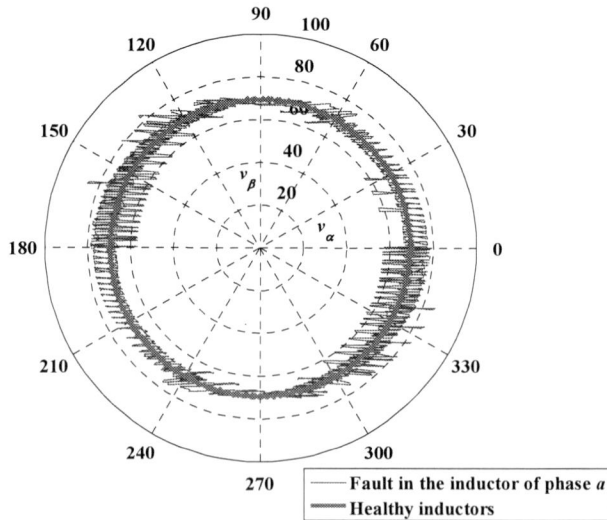

Fig. 2. Effect of a fault in the inductor of phase a on the trajectory of the estimated grid VSV.

The fault in Fig. 2 is modeled as a 15% decrease in the inductance value of the inductor in phase a. Such a value is considered to be significant enough not to be caused by the grid interaction or the inductor tolerance, but by an actual fault in the choke of phase a.

As shown in Fig. 2, the α-β components of the grid VSV must remain constant after a commutation when there are no faults in the coupling inductors (assuming a sufficiently short sampling time). This statement and (1) lead to (4)-(5). These equations are obtained regarding the continuity of ac-current signals; in other words, ac-current signals are assumed to maintain their value after a commutation.

$$e_{at} = e_{at'} \Rightarrow i_\alpha (p_t - p_{t'}) = i_\beta (q_t - q_{t'}) \quad (4)$$

$$e_{\beta t} = e_{\beta t'} \Rightarrow i_\beta (p_t - p_{t'}) = i_\alpha (q_t - q_{t'}) \quad (5)$$

Both p and q must remain constant after a commutation to make (4)-(5) true for any given values of i_α and i_β. This condition leads to (6)-(7).

$$p_t - p_{t'} = 0 \quad (6)$$

$$q_t - q_{t'} = 0 \quad (7)$$

As said before, if a fault occurs in one of the three inductors, a variation of the inductance value (ΔL, normally negative) of the affected inductor is expected [8]. For example, the expressions of p and q in case of a fault in the coupling inductor of phase a can be obtained from (2)-(3) as shown in (8)-(9).

$$p = L(i_a' i_a + i_b' i_b + i_c' i_c) + \Delta L i_a i_a' + v_{dc}(S_a i_a + S_b i_b + S_c i_c) \quad (8)$$

$$q = (1/\sqrt{3})\{(L_b i_b' - L_c i_c')i_a + (L_c i_c' - L_a i_a')i_b + (L_a i_a' - L_b i_b')i_c + \\ + \Delta L(i_c - i_b)i_a' - \\ - v_{dc}[S_a(i_b - i_c) + S_b(i_c - i_a) + S_c(i_a - i_b)]\} \quad (9)$$

The possibility of multiple faults in different phases at the same time is considered to be very unlikely and therefore it is not taken into account in this study. Once a fault takes place in one of the phases, it is detected by the method proposed in this paper. The system is then expected to be suspended until the fault is corrected, preventing other faults from occurring at the same time.

Equations (6)-(7) must continue to be valid even in the aforementioned case of a fault in one of the inductors. Therefore, (8)-(9) can be applied to (6)-(7) leading to (10)-(11). Equations (10)-(11) are obtained considering the continuity of the dc-link voltage and ac-currents, i.e., these magnitudes are assumed to maintain their values after a commutation. Equations (10)-(11) form a system with two unknowns, L_A and ΔL_A, that can be obtained as shown in (12)-(13).

$$p_t - p_{t'} = \\ = [i_a(i_{at}' - i_{at}') + i_b(i_{bt}' - i_{bt}') + i_c(i_{ct}' - i_{ct}')]L_A + \\ + (i_{at}' - i_{at}')i_a \Delta L_A + \\ + v_{dc}[i_a(S_{at} - S_{at'}) + i_b(S_{bt} - S_{bt'}) + i_c(S_{ct} - S_{ct'})] = 0 \quad (10)$$

$$q_t - q_{t'} = \\ = [i_a(i_{bt}' - i_{bt}' - i_{ct}' + i_{ct}') + i_b(i_{ct}' - i_{ct}' - i_{at}' + i_{at}') + i_c(i_{at}' - i_{at}' - i_{bt}' + i_{bt}')]L_A + \\ + (i_{at}' - i_{at}')(i_c - i_b)\Delta L_A - \\ - v_{dc}[(S_{at} - S_{at'})(i_b - i_c) + (S_{bt} - S_{bt'})(i_c - i_a) + (S_{ct} - S_{ct'})(i_a - i_b)] = 0 \quad (11)$$

$$L_A = \frac{v_{dc}[(S_{ct} - S_{ct'}) - (S_{bt} - S_{bt'})]}{(i_{bt}' - i_{bt}') - (i_{ct}' - i_{ct}')} \quad (12)$$

$$\Delta L_A = \frac{L_A[i_a(i_{at}' - i_{at}') + i_b(i_{bt}' - i_{bt}') + i_c(i_{ct}' - i_{ct}')]}{-i_a(i_{at}' - i_{at}')} + \\ + \frac{v_{dc}[i_a(S_{at} - S_{at'}) + i_b(S_{bt} - S_{bt'}) + i_a(S_{ct} - S_{ct'})]}{-i_a(i_{at}' - i_{at}')} \quad (13)$$

Expressions for L_B and ΔL_B, and L_C and ΔL_C, for the cases of a fault in the inductor of phase b or phase c, respectively, can be easily obtained by permuting the subscripts for phases a, b and c in (12)-(13).

Estimations of L_A and ΔL_A should be calculated one sampling time after each commutation because (12)-(13) are not defined at non-commutation instants (time derivatives of ac-currents do not change). In addition to this, L_A estimations are not defined at commutations of the types shown in Table I, as deduced from (12). Similar non-defined estimations can be obtained for L_B and L_C by permuting the subscripts for phases a, b and c in Table I. Similarly, estimations of ΔL_A are not defined at non-commutation instants and at commutations of the types given by Table I, as deduced from (13). Estimations of ΔL_A are also not defined at commutations of the types given by Table II. Similar non-defined estimations can be obtained for ΔL_B and ΔL_C by permuting the subscripts for phases a, b and c in Table I and Table II. Non-defined estimations are inhibited in the proposed method.

TABLE I.   NON-DEFINED ESTIMATIONS FOR L_A AND ΔL_A

Pre-commutation state	Post-commutation state
$(0S_bS_c)$	$(1S_bS_c)$
$(1S_bS_c)$	$(0S_bS_c)$
(S_a00)	(S_a11)
(S_a11)	(S_a00)

TABLE II.   ADDITIONAL NON-DEFINED ESTIMATIONS FOR ΔL_A

Pre-commutation state	Post-commutation state
(S_a01)	(S_a10)
(S_a10)	(S_a01)

Equation (12) and the equivalent expressions for L_B and L_C give an accurate estimation of the coupling inductance when there are no faults in any of the inductors. The average value of the available values of L_A, L_B and L_C (ideally equal to each other) can be considered as the correct estimation of the coupling inductance at each commutation. The time-domain mean value of the successive correct estimations can be supplied online to the control system (according to the selected updating time interval) as an accurate estimation of the inductance value. In the case of healthy inductors, the result of (13) and the equivalent expressions for ΔL_B and ΔL_C must be close to zero. This inductance estimation method is an alternative to the strategy presented by the authors in [10].

On the other hand, when a fault occurs in one of the inductors, only the estimation of the coupling inductance referred to the affected phase (L_A, L_B or L_C) must continue to be correct. Therefore, the fault and the phase where it occurs can be detected by tracking the time-domain mean values of the inductance given by L_A, L_B and L_C. When a simultaneous deviation beyond a specific threshold is detected in two out of the time-domain mean values of L_A, L_B and L_C (e.g., L_B and L_C), it can be considered that the fault is taking place at the other phase (phase a in the example). In this case, the time-domain mean value of the variation of the inductance value corresponding to the faulty inductor (ΔL_A in the example) must reflect the change in its inductance value caused by the fault. If this variation exceeds a specific threshold, the operation of the system can be suspended to avoid a major breakdown. The advantage of knowing the faulty phase enables replacing just the affected inductor when utilizing three single-phase coils or identifying the faulty winding when using one three-phase inductor. If the system is not immediately suspended after the detection of the fault (e.g., due to service continuity causes), the time-domain mean value of the coupling inductance in the faulty phase (L_A in the example) should be supplied online to the control system.

IV. SIMULATION RESULTS

Figure 3 shows the estimations of the coupling inductance value based on (12) and the equivalent expressions for L_B and L_C, respectively, obtained when a fault occurs in the inductor of phase a.

As done before, the fault in Fig. 3 is modeled by means of a 15% decrease in the inductance value of the coil in phase a. According to what is stated in Section III, defined estimations are obtained one sampling time after certain commutations and non-defined estimations are inhibited.

Figure 4 shows the respective time-domain mean values of those estimations. As shown in Fig. 4, the correct value of the coupling inductance after the fault is only maintained in the estimation referred to the faulty choke (L_A), as predicted. According to this, Table III shows, for each phase and after the fault, the steady state error in the time-domain mean value of the inductance estimations with respect to the inductance rated value.

Figure 5 shows the estimations of ΔL_A under the same circumstances as Figs. 3-4. As said before, defined estimations are obtained one sampling time after certain commutations and non-defined estimations are inhibited. Figure 6 demonstrates that the steady state time-domain mean value of ΔL_A after the fault is quite close to the considered -15% variation (-0.66% error).

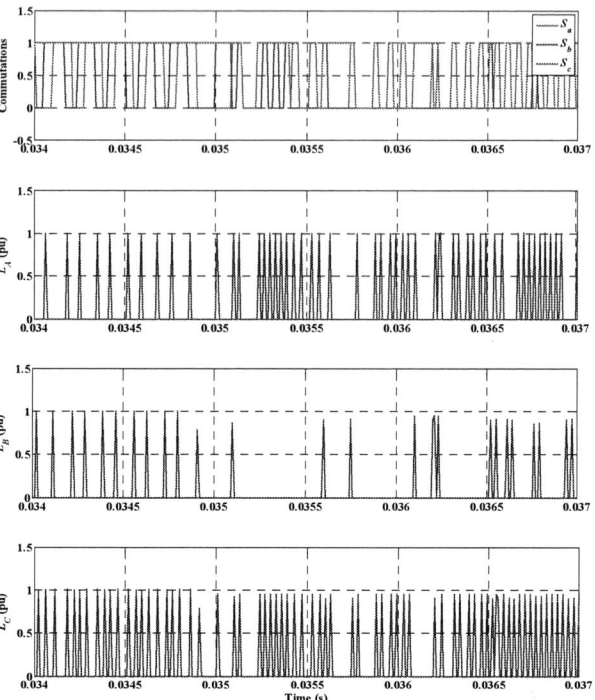

Fig. 3. Commutations and estimation of coupling inductance (L_A, L_B, L_C) – Fault in the inductor of phase a at 0.035 s.

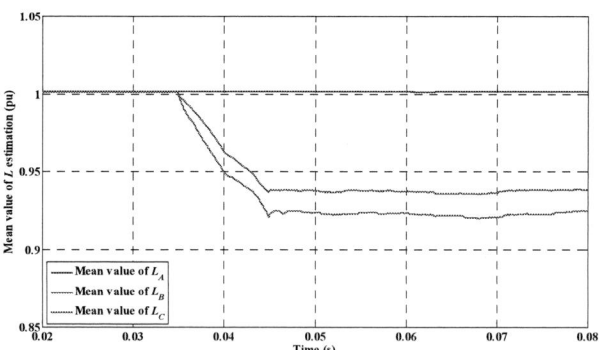

Fig. 4. Mean values of estimations given by L_A, L_B and L_C – Fault in the inductor of phase a at 0.035 s.

TABLE III. ESTIMATION ERRORS AFTER THE FAULT

Phase	Error (%)
a	0.17
b	-6.50
c	-7.50

978-1-4799-0024-4/13 $31.00 © 2013 IEEE

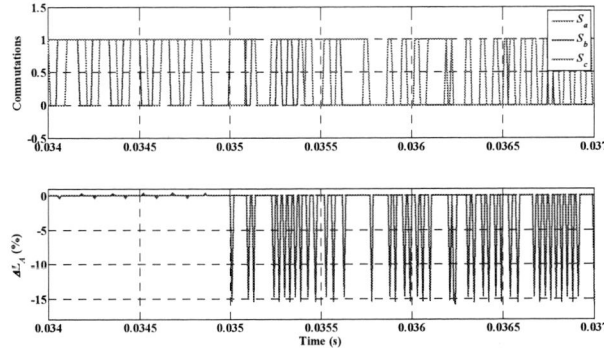

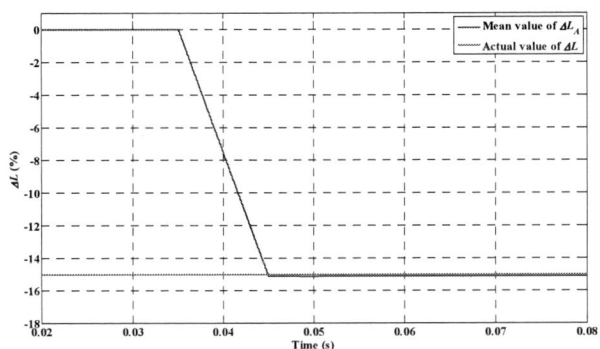

Fig. 5. Commutations and estimations of ΔL_A – Fault in the inductor of phase a at 0.035 s.

Fig. 6. Actual value of ΔL and mean value of ΔL_A – Fault in the inductor of phase a at 0.035 s.

V. Conclusions

In this paper, a new method for detecting faults involving inductance variations of coupling inductors in three-phase AFEs with DPC-based VSRs is proposed. This method enables the estimation of the coupling inductance value in an alternative manner to that of previously existing methods. From the tracking of the coupling inductance estimated values associated with each phase, a fault occurrence in the coupling inductor can be detected. Moreover, the faulty phase is identified and even an accurate estimation of the inductance variation is obtained. The fault detection enables suspending the service to avoid a major breakdown in the system. Besides, knowing the affected phase is a remarkable advantage because it allows replacing only the faulty inductor if three single-phase coils are used or identifying the faulty winding if one three-phase inductor is utilized. Simulation results demonstrate the validity of the proposed strategy.

By means of the setup described in [10], the authors are currently performing successful experimental tests of the proposed method. The results of those tests will be included in further works on the topic. The authors will also discuss in future works the possibility of adapting the proposed method to other control strategies, as the traditional PWM-based ones.

VI. Appendix

The switching table used in this paper coincides with that proposed in [11] and is shown in Table IV.

TABLE IV.   Switching Table

S_p	S_q	Sector											
		1	2	3	4	5	6	7	8	9	10	11	12
1	0	V5	V6	V6	V1	V1	V2	V2	V3	V3	V4	V4	V5
	1	V7	V7	V0	V0	V7	V7	V0	V0	V7	V7	V0	V0
0	0	V6	V1	V1	V2	V2	V3	V3	V4	V4	V5	V5	V6
	1	V1	V2	V2	V3	V3	V4	V4	V5	V5	V6	V6	V1

Parameters used in the simulations performed in this paper are shown below.

Grid voltage RMS	70 V
Coupling inductance	5.12 mH
Grid frequency	50 Hz
DC-link capacitor	1,100 μF
Load resistor	50 Ω
Control and circuit sampling times	10 μs, 0.01 μs
Reference p and q	1,000 W, 0 var
Hysteresis limits on p and q	$\pm$ 20 W, $\pm$ 20 var

VII. Acknowledgment

The authors acknowledge the support provided by the University of Oviedo in the development of this work.

VIII. References

[1] D. Kastha and B. K. Bose, "Investigation of fault modes of voltage-fed inverter system for induction motor drives," *IEEE Trans. Ind. Appl.*, vol. 30, pp. 1028–1038, July/Aug. 1994.

[2] F. W. Fuchs, "Some diagnosis methods for voltage source inverters in variable speed drives with induction machines. A survey," in Proc. IEEE IECON, Roanoke, VA, 2003, vol. 2, pp. 1378–1385.

[3] D. U. Campos-Delgado, E. Palacios, and D. R. Espinoza-Trejo, "Fault tolerant control in variable speed drives: A survey," *IET Elect. Power Appl.*, vol. 2, no. 2, pp. 121–134, Mar. 2008.

[4] B. Lu and S. Sharma, "A literature review of IGBT fault diagnostic and protection methods for power inverters," *IEEE Trans. Ind. Appl.*, vol. 45, no. 5, pp. 1770–1777, Sep./Oct. 2009.

[5] S. Yang, D. Xiang, A. Bryant, P. Mawby, L. Ran, and P. Tavner, "Condition monitoring for device reliability in power electronic converters—A review," *IEEE Trans. Power Electron.*, vol. 25, no. 11, pp. 2734–2752, Nov. 2010.

[6] S. Yang, A. Bryant, P. Mawby, D. Xiang, R. Li, and P. Tavner, "An industry-based survey of reliability in power electronic converters," *IEEE Trans. Ind. Appl.*, vol. 47, no. 3, pp. 1441–1451, May 2011.

[7] Y. Han and Y. H. Song, "Condition monitoring techniques for electrical equipment-a literature survey," IEEE Trans. Power Del., vol. 18, no. 1, pp. 4–13, Jan. 2003.

[8] E. Wiedenbrug, G. Frey, and J. Wilson, "Early intervention: Impulse testing and turn insulation deterioration in electric motors," *IEEE Industry Applications Magazine*, vol. 10, n.o 5, pp. 34–40, Oct. 2004.

[9] P. Noguchi, H. Tomiki, S. Kondo, and I. Takahashi, "Direct power control of PWM converter without power-source voltage sensors," *IEEE Trans. Power Electron.*, vol. 34, no. 3, pp. 473–479, May 1998.

[10] J. G. Norniella, J. M. Cano, G. A. Orcajo, C. H. Rojas, J. F. Pedrayes, M. F. Cabanas, and M. G. Melero, "Analytic and iterative algorithms for online estimation of coupling inductance in direct power control of three-phase active rectifiers," *IEEE Trans. Power Electron.*, vol. 26, no 11, pp. 3298-3307, Nov. 2011.

[11] A. Baktash, A. Vahedi, and M. A. S. Masoum, "Improved switching table for direct power control of three-phase PWM rectifier," in *Australasian Universities Power Engineering Conference*, Perth, Aust., 2007, pp. 1-5.

IX. BIOGRAPHIES

Joaquín G. Norniella was born in Gijón, Spain. He received the M.Sc. and the Ph.D. degrees in electrical engineering from the University of Oviedo in 2005 and 2012, respectively. He is currently working as a Lecturer at the University of Oviedo. His main research interests are in the field of Power Quality and the conditions monitoring of electrical equipment.

José M. Cano (M'98) was born in Oviedo, Spain in 1971. He received the M.Sc. and Ph.D. degrees in electrical engineering from the University of Oviedo, Spain, in 1996 and 2000, respectively. In 1996 he joined the Department of Electrical Engineering, University of Oviedo, where he is currently an Associate Professor. His main research interests are in the field of power quality in industrial power supply networks.

Gonzalo A. Orcajo was born in Gijón, Spain in 1965. He received the M. Sc and Ph. D degrees in electrical engineering from the University of Oviedo, Asturias, Spain, in 1990 and 1998, respectively. In 1992, he joined the Department of Electrical Engineering, University of Oviedo, where he is currently an Associate Professor. His main research interests are in the field of power quality in industrial power systems.

Carlos H. Rojas was born in Caracas Venezuela. He earned his Engineering degree at the Simón Bolivar University - Venezuela in 1994 and his Ph.D. degree at the University of Oviedo, Spain, in 2001, where he is currently an Associate Professor. His main research interests are in the field of FEM modeling of electrical machines, predictive maintenance of electrical equipment and power quality.

Francisco Pedrayes was born in Gijón, Spain, in 1973. He received the M.Sc. degree from the University of Oviedo, Asturias, Spain, in 2001, and he obtained the Ph.D. in electrical engineering in 2007. He is currently working as an Assistant Professor at the University of Oviedo.

Manés F. Cabanas was born in Gijón, Spain, in 1965. He earned his Engineering and Ph.D. degrees at the University of Oviedo in 1991 and 1995, respectively. In 1992 he joined the Department of Electrical Engineering, University of Oviedo, Spain, where he is currently an Associate Professor. His main research interests are in the field of condition monitoring and electrical machinery diagnosis.

Manuel G. Melero was born in Gijón, Spain. He received the M.Sc. and Ph.D. degrees in electrical engineering from the University of Oviedo, Spain, in 1991 and 1998, respectively. In 1991 he joined the Department of Electrical Engineering, University of Oviedo, where he is currently an Associate Professor. His main research interests are in the field of predictive maintenance of rotating machinery.

Study of fault – tolerant inverter

F.khelifi [a,*], B.Nadji [a]

a) Team of Microelectronics and Microsystems, Laboratory electrification of industrial entreprises,
University of Boumerdes, Algeria
b) Team of Infotronic, Laboratory electrification of industrial enterprises,
University of Boumerdes, Algeria

Abstract

This work describes the operation of the fault tolerant inverter. The consequences of the switch defect in a traditional voltage inverter are analyzed. The studied failures are the short-circuit defect and the opening defect. The consequences of transistor defect leading to a final state of short-circuit or open circuit are discussed. It was proposed to implement a topology of fault-tolerant inverter with four-arms. The use of this inverter allows the increase in the availability of the converter. All simulations were done with the PSIM 7.1.2 software.

1. Introduction

The standard topologies representing the quasi totality of the converters on the market, do not tolerate the most common defect: that of the semiconductor short- circuit. To increase the operational availability less penalizing in terms of mass, volume and cost the concept of fault-tolerance consists in introducing one or more active redundancies in the converter.

A great number of industrial processes are based on electric drives at variable speed, in much of these applications; the principal element is a voltage inverter which supplies an asynchronous motor. The inverters are the structures in bridge generally constituted of the electronic switches such as the power transistors.

In Some applications, the high reliability and the availability of equipments are essential. To satisfy these objectives without introducing passive redundancies nor the margins of excessive dimensions, it was proposed to implement a topology of fault-tolerant inverter with four-arms [1]. The PSIM software demo version 7.1 is used for simulation purposes in this paper is downloaded from the link

http://www.powersimtech.com/

2. Dysfunctions mode

2.1 Impact on the system after a switch failure in an inverter

The defect of transistor is caused either by a command defect (propagated by the command itself or emanating from the coupling circuit - command), or by a physical failure of this one. Thereafter, the consequences of transistor defect leading to a final state of short-circuit or open circuit are discussed.

2.1.1. Failure of short-circuit due to the failure of a switch

Whatever the origin of the deffect of short-circuit the switch failure causes the short circuit of the voltage source when the second switch of the cell is commanded at closing. The impedance of the mesh (filter capacitor (C), the faulty cell) is very low, resulting in a strong current of short circuit flowing through the faulty cell [2].

---

* Corresponding author. khelififat@yahoo.fr
Tel: +213 (663)434 588 ; Fax: +213 (24) 818 300

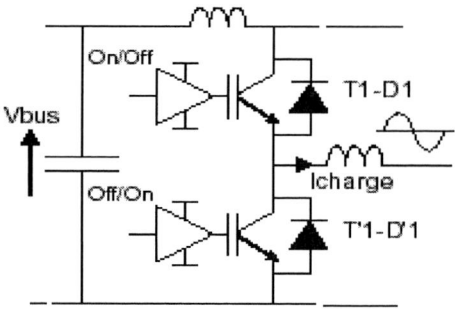

Fig 1. A. generic cell

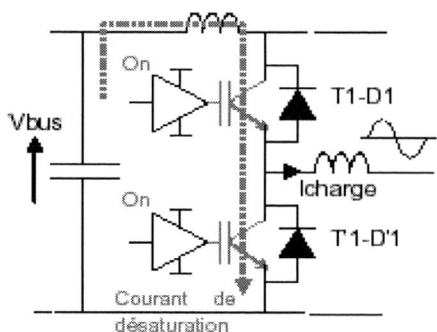

B. Short circuit fault

If nothing is done, one or two power semiconductors (according to the origin of the defect), will pass very quickly at a physical fault. It will not be then possible any more to open them by the command (passage to a physical defect).

If the contribution of energy by the voltage source is not stopped, the temperature will continue to increase until the opening or the explosion of the components.

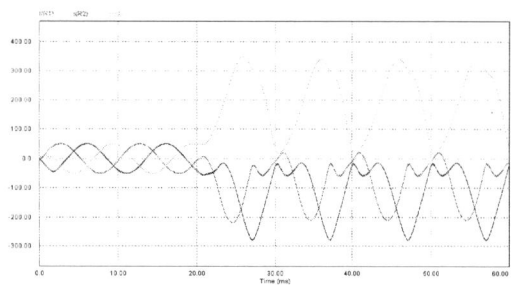

Fig2. Health-wave currents near a fault short-circuit B3

The simulation of the defect of the inverter arm shows that the current of short-circuit is a rectified current type mono-alternation (fig. 2 and 3). Its amplitude reaches three time of the nominal value and it can be even higher according to the operation mode in which it occurs. To protect the healthy components from the faulty cell and to avoid any risk of explosions, it is important to detect very quickly the defect of short-circuit and to open the components of the failing arm before the two switches of the cell are damaged.

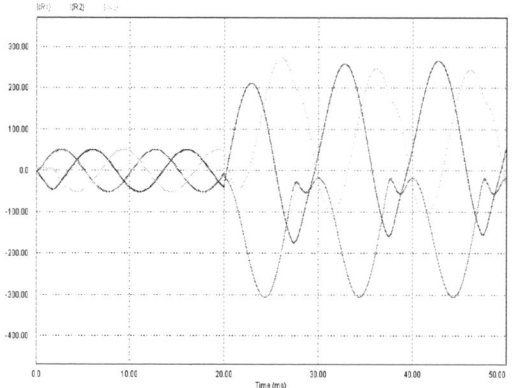

Fig3. Waveforms of the currents after a fault short-circuit B2 and B3

2.2. Failure to initiate a switch

Maintaining the open state of a transistor causes a loss of reversibility in current of cell: in inverter mode, the phase current is unipolar and non-sinusoidal. Indeed, the spontaneous conduction of the anti-parallel diode of failed transistor is conditioned by the sign of the current in phase.

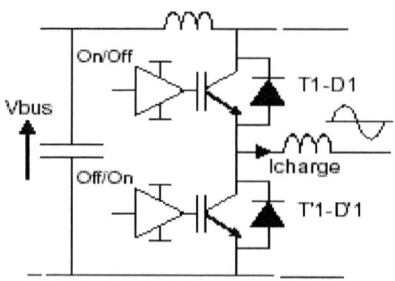

Fig 4 .A. generic cell

978-1-4799-0024-4/13 $31.00 © 2013 IEEE

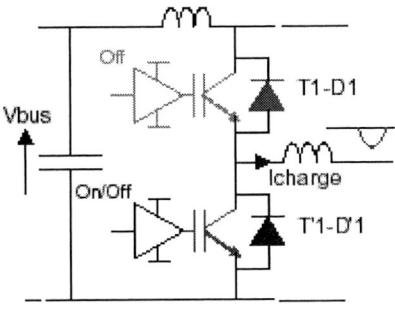

B. Failure to initiate.

For example, if the higher transistor of a cell remains open, the corresponding phase of the machine remains connected to the negative potential of the bus by the diode of bottom. The current in the phase concerned (B3 phase on the fig.5, 6) cancels during half of the modulation period.

As the neutral of the machine is insulated, the currents in the two other phases are also deformed. The distortion of the currents is thus important what implies a significant fluctuating power. Then, the maintenance of the nominal mode requires an increase in the efficacy current.

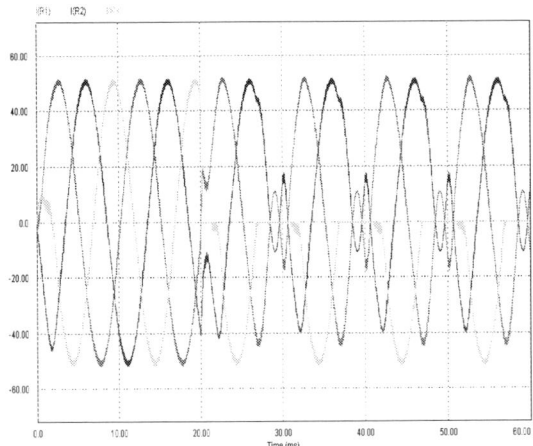

Fig.5. waveforms of the currents after a fault opening in B3

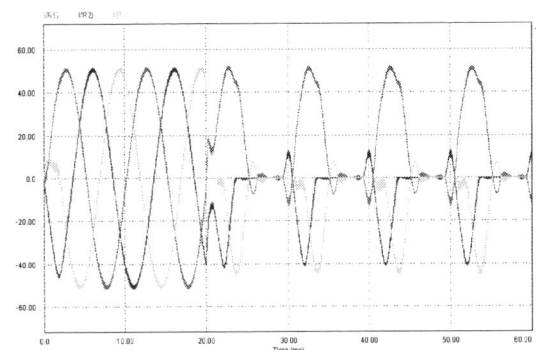

Fig.6. waveforms of the currents after a fault opening in B2 and B3

3. Inverter with four arms

A conventional three-phase inverter topology incorporating isolation switches can tolerate an internal failure. The addition of an internal redundancy to the UPS improves the performance of the operation reconfigured. It is preferable that the additional element is a switching cell rather than a capacitive divider. For reasons of integration as performance: the voltage range that UPS may apply to the machine after reconfiguration is thus more important [3].

To replace a failing arm, the help arm can be connected to each of the 3 phases. Thus after reconfiguration, the operation remains that of a three-phase inverter; no modification of the command algorithm is required and the integrality of nominal power can be provided. However, this topology requires 3 devices of connection and 3 devices of insulation (one by phase), which penalizes the compactness of the converter.

It then seems appropriate to connect the additional arm to the neutral of the machine. The number of connecting devices is reduced to one per phase. However, it is preferable to add also an isolating switch arm of help, to connect it only in the case of defect and thus to avoid any circulation of current in this arm. Conversely, it is possible in normal mode, that the fourth arm plays the role of active filtering of the neutral by vector modulation 4 arms. It is then necessary to isolate the arm if it undergoes failure to allow the reconfiguration of the inverter [4].

978-1-4799-0024-4/13 $31.00 © 2013 IEEE

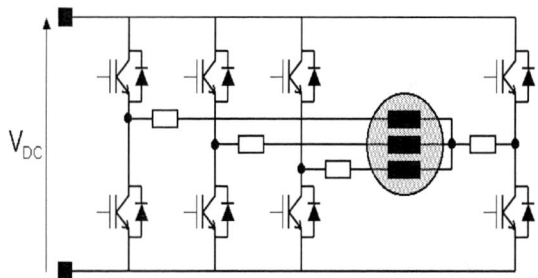

Fig.7. Topology of inverter with help arm connected to neutral

3.1 Isolation procedure of a inverter with 4arms

Following the failure of a semiconductor converter, the corresponding switching cell is disconnected from the phase of the machine. In order to not make the design of the protections elements isolating switch excessive, it is necessary to minimize energy to be dissipated. However, for a failure of type "short circuit" of a transistor, it is not certain that the current of phase vanishes quickly. The proposed strategy is to open all other transistors to cause an intermittent conduction phase motor [5].

However, as the combination of electromotive force (emf) imposes the conduction of diodes and current evolution, it is not certain to obtain the current cancellation before many periods according to the instant at which the failure occurs and the alternation of the phase current. The phase currents can undergo high amplitude oscillations.

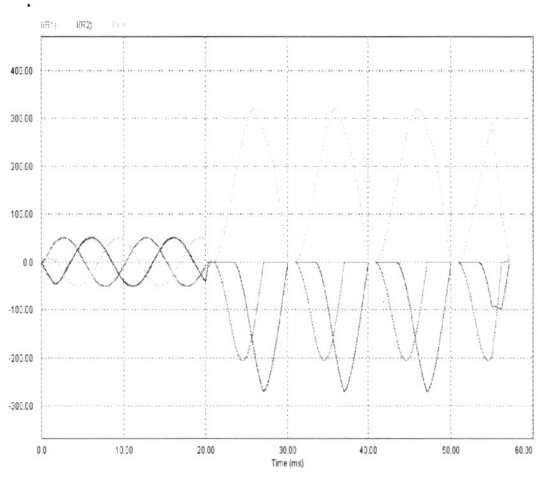

Fig.8. Failure of short-circuit block B3 with B1 and B2

To impose a more fast cancellation of current and avoid the current transient current to the converter and the machine, it is possible to take advantage of the current cancellation in one of the remaining phases to open it. It then remains only one path short-circuits and the current phase can be expressed as the sum of a sinusoid and an exponential decay. It vanishes more quickly(fig.9).

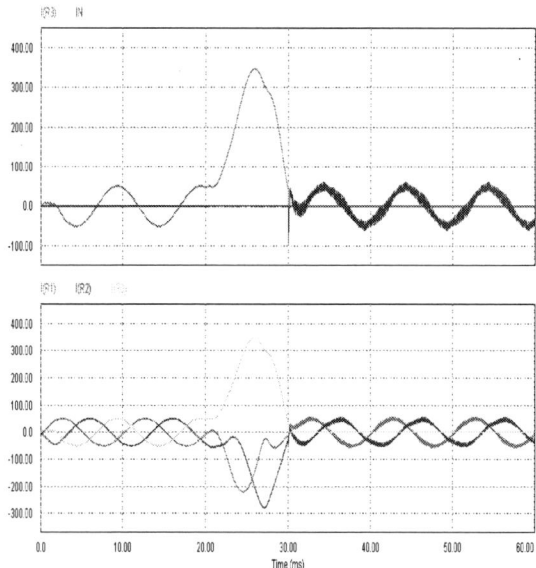

Fig.9. Normal operation, the inverter blocks the appearance of failure, isolation and reconfiguration of the faulty cell (left: B3 opening phase, right: relief arm connection)

Another alternative is to use the help arm to force the cancellation of the phase current. When locking the inverter, the isolation switch connected to the neutral is closed and, in the arm connected to neutral, the transistor is a same level as the faulty of transistor.

The machine winding is thus short-circuited and the current evolution is conditioned by the single emf of the phase connected to the failing arm, it could quickly impose a cancellation (Fig. 12).

If, however, the decrease of the algebraic value of this current does not last until its cancellation, the reconfiguration of isolation switches and help arm as function of the emf allows to ensures a reduced transient regime.

978-1-4799-0024-4/13 $31.00 © 2013 IEEE   536

3.2. Emergency operation of the inverter with four arms

Once the failing arm is isolated, the supervising body connects the help arm to the neutral of the machine. This then ensures the return of two-phase currents through the neutral and plays the role of an active voltage divider to regulate the neutral voltage.

As the converter comprises few components, its reliability is pretty good (MTBF given by the manufacturers that ranging between 4 and 5 years) [6].

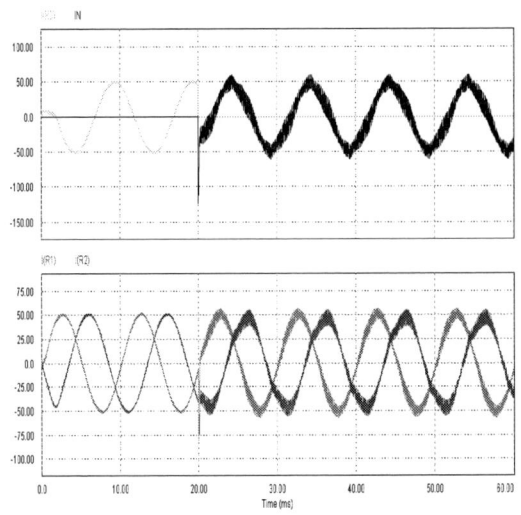

Fig.10. *Normal operation without reconfiguration and modification of the order (two-phase mode 120 °)*

From the viewpoint of the dependability, a defect of opening is much less constraining since the operational safety is assured On the other hand, a failure of a short-circuit of switch does not allow to ensure the safety of the system. Indeed, even if the protections are added to the structure (fuse, smart driver), these cannot avoid the short circuit of the voltage source.

The defect can propagate outside the converter: risk of failure on continuous supply, damage of the mechanical part. Thus, the system is not fault tolerant to the short circuit of a switch and the very fast stopping of the converter is the most adapted strategy.

The addition of the internal redundancy to the inverter allows improving the performance of the reconfigured operation. .

Conclusion

If a defect of opening does not have fatal consequences on the system when a classical inverter is used, it is not the same for a defect of short-circuit. The effects on the voltage source, the motor, the load and the converter can then be dramatic. To remedy this, it must increase the reliability and availability of the converter. The use of the four arms inverter topology with fault tolerance allows the increase in the availability of the converter.

References

[1] MartinAIMÉ «Evaluation and optimization of the band-width of the static inverters Application to the new structures multicellulars ». Thesis de l'INPT, 2000.

[2] Frédéric Richardeau « dependability in electronics of power thesis of synthesis of l'INPT, 2004.

[3] T.Elch-Heb, J.P.Hautier, "Remedial strategy for inverter - induction machine system faults using two-phase operation", EPE, Brighton, 1993.

[4] A.L. Julian, G. Oriti, T.A. Lipo, "Elimination of common-mode voltage in three-phase sinusoidal power converters", IEEE Trans. on Power Electronics, vol. 14, issue 5, Sept. 1999, pp. 982-989.

[5] M. Zhifeng DOU «Sûreté de fonctionnement des convertisseurs ; Nouvelles structures de redondances pour onduleurs sécurisés à tolérance de pannes » thèse de l'université de Toulouse 2011.

[6] Robert A. Hanna, Shiva Prabhu « Medium-voltage ajustable speed drives – Users' an manufacturers' experiences », IEEE IAS, Vol. 33, N°6, Nov/Dec. 1997.

Bearing Fault Detection Using Relative Entropy of Wavelet Components And Artificial Neural Networks

Helder L. Schmitt, Lyvia R. B. Silva, Paulo R. Scalassara*, and Alessandro Goedtel

Abstract—Fault detection in electrical machines have been widely explored by researchers, especially bearing faults that represents about 40% to 60% of the total faults. Since this kind of fault is detectable by particular frequencies at the stator current, it is now a source of investigation. Thus, this work presents a predicability analysis method based on relative entropy measures estimated over reconstructed signals obtained from wavelet-packet decomposition components. The signals were simulated using a real motor current signal with addition of frequency components related to the bearing faults. Using three ANN topologies, these entropy measures are classified in two groups: normal and faulty signals with a high performance rate.

Index Terms—Artificial neural networks, Bearing fault detection, Relative entropy, Wavelet packets

I. INTRODUCTION

Electrical motors are common in factories all around the world due to their versatility and robustness. They can be found throughout the productive process in several applications: ventilators, compressors, pumps, and others. These motors are subject to various noxious factors to their operation, such as environment adverse conditions, misusage, and design problems. These factors can lead to the development of incipient faults, which would result in motor failure [1].

Bearings faults are responsible for a portion of 40% to 60% of all failures [2]–[4]. These faults may be divided in localized faults, which occur in specific bearing elements and generalized faults, which damage the whole bearing.

The presence of a localized bearing fault cause vibrations in known frequencies related to the rotational speed and geometry of the bearing. Techniques for analysis of vibration signals that are often used are Hilbert transform, wavelet-packet transform [5], and Fast Fourier Transform (FFT) [6].

The vibration caused by the presence of bearing faults reflects on the stator current signals of the machine. Techniques of current signals analysis are relatively non-invasive techniques and more economic than vibration signal analysis.

In [7], the authors present a number of analyis techniques based on stator current signals such as artificial neural networks (ANNs), extended park's vector approach, FFT, and wavelet-packet.

The wavelet-packet transform (WPT) is a more powerful tool for the analysis of specific signal bands in comparison to other techniques such as FFT [8]. This is due to the WPT capacity to analyze signals through a size-variable time-frequency windowing, in contrast to the FFT that generates a full spectrum of the entire signal [9], and even the Short Time Fourier Transform (STFT), wherein is used a fixed-size time-frequency windowing [8].

Several researchers used the Wavelet-Packet Decomposition (WPD) for analyses of specific regions of the spectrum, such as [10], where the authors calculated the energy of wavelet components that contained the localized fault frequencies. Also in [6], six fault types were analyzed using energy measures of wavelet components and FFT.

Another approach is the use of predictability analysis of the signals based on measures of information theory such as Shannon entropy and relative entropy. In [11], good results are presented using Shannon entropy measures to discriminate vibration signals of bearing faults of the inner and outer race.

In this work, we analyze the predictability of the signals from measurements of relative entropy between the modeling errors obtained from WPD components. Similar the methodology presented in [12], which is a study conducted to discriminate healthy and pathological larynxes from recorded voice signals.

We also use an ANN for pattern classification, which have improved the signal discrimination, similar to [13], where the classification of statistical measures of continuous wavelet coefficients using Support Vector Machines (SVMs), resulted in signals discrimination close to 100%. Also, in [14], radial basis functions based ANN (RBF) for the fault classification based on energy of WPD components of signals.

For our analyses, we used simulated current signals with bearing localized faults, and ANNs with topologies Multilayer Perceptron (MLP), Kohonen self-organizing maps (KHN), and RBF. We aimed at discriminating the signals in two groups: fault and normal ones.

This paper is structured as follows. In section Theory, we present some theoretical concepts, fault frequencies calculation, a brief review of WPD, and the relative entropy based feature vector. In section Methodology, we detail the proposed method, and, in the section Results and Discussion,

---

* Corresponding author: prscalasssara@utfpr.edu.br.

H. L. Schmitt, L. B. Silva, P. R. Scalassara, and A. Goedtel are with Federal University of Technology - Paraná, Cornélio Procópio, Brazil. (e-mail: helderschmitt@gmail.com, lybiagi@hotmail.com, prscalassara@utfpr.edu.br, agoedtel@utfpr.edu.br).

we present the results of the tests with ANN based classifier using our feature vector. And, in Conclusion, we make our final comments.

II. THEORY

In this section, we discuss some of the theoretical concepts used in this research, starting with the bearing fault frequencies calculation.

A. Fault frequencies calculation

The presence of a fault in a motor bearing causes vibrations in characteristics frequencies related to the bearing element that present the fault. These vibrations affect the air gap eccentricity, generating variations in their length, which, in turn, cause variations in flux density that affect the inductances of the machine producing harmonic in the stator currents [14].

The frequency components related to bearings faults (f_b) are calculated from the stator current in (1) [14], where f_e is the power source frequency, $m = 1, 2, 3, ...$ and f_v is one of the characteristic vibration frequencies.

$$f_b = |f_e \pm m.f_v| \tag{1}$$

The characteristic vibration frequencies (f_v), are given in (2) to (5), respectively for the fault presence in the outer race (2), inner race (3), cage (4), and ball defective elements (5). Where (f_r) is the rotor speed and (ϕ) is ball contact angle [14]. They are calculated based on the geometry of the bearing depicted in Figure 1.

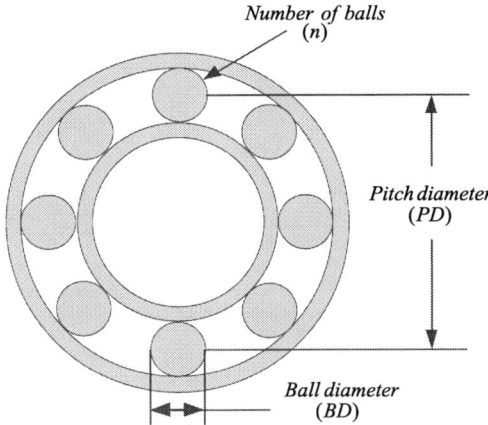

Fig. 1.   Illustration of the bearing geometry, showing the ball diameter (BD), pitch diameter (PD), and the number of the balls (n).

$$f_{or} = \frac{n}{2}.f_r.\left(1 - \frac{BD}{PD}.\cos\phi\right) \tag{2}$$

$$f_{ir} = \frac{n}{2}.f_r.\left(1 + \frac{BD}{PD}.\cos\phi\right) \tag{3}$$

$$f_{cag} = \frac{1}{2}.f_r.\left(1 - \frac{BD}{PD}.\cos\phi\right) \tag{4}$$

$$f_{be} = \frac{PD}{2.BD}.f_r.\left(1 - \left(\frac{BD}{PD}\right)^2.\cos^2\phi\right) \tag{5}$$

B. Wavelet-packet decomposition model

The predictability analysis method proposed in [12], requires the use of prediction models. One such model is based on the wavelet transform which is a representation of a signal with scaled and shifted versions of a single function called wavelet [15].

From a practical point of view, the discrete wavelet decomposition is performed by the convolution of the signal $x[n]$ with a lowpass filter $h_0[n]$ and a highpass filter $g_0[n]$, that splits the signal into approximation coefficients, ($A[n]$), and detail, ($D[n]$), as shown in (6) and (7) respectively [16].

$$A[n] = x[n] * h_0[n] = \sum_{k=0}^{M-1} h_0[k].x[2n-k] \tag{6}$$

$$D[n] = x[n] * g_0[n] = \sum_{k=0}^{M-1} g_0[k].x[2n-k] \tag{7}$$

This is a multilevel process, where they are successively decomposed in (6) and (7). The wavelet-packet transform is a modification where the detail coefficients are also decomposed. Figure 2 depicts this complete decomposition and the frequency domain of the coefficients.

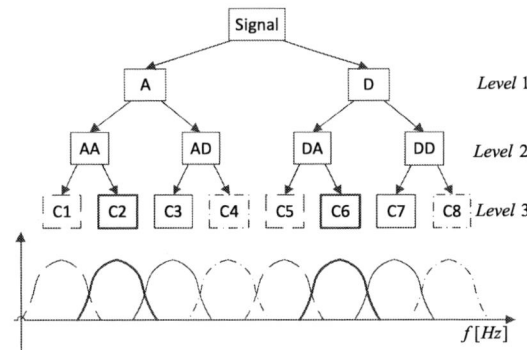

Fig. 2.   Three level wavelet-packet decomposition, showing and the approximation and detail coefficients, and their frequency domain.

C. Feature vector

The relative entropy between two discrete probability density functions (PDFs), $p[n]$ and $q[n]$, is given in (8). The PDFs can be estimated using a nonparametric method based on histogram with a fixed quantization step [17] and (N) bins.

$$D_{p\|q} = \sum_{i=1}^{N} p_i.\log_2 \frac{p_i}{q_i} \tag{8}$$

where p_i and q_i are the original signal and the error of prediction model PDFs, for a N bins histogram.

The feature vector used by the ANN is composed of relative entropy measures between the prediction errors of the phase currents, when used WPD model, according to (9).

$$Features = [D_{e_A\|e_B} \; D_{e_B\|e_C} \; D_{e_C\|e_A}] \qquad (9)$$

D. Artificial Neural Networks

An ANN is a system based on operations performed by virtual objects inspired on biological neurons. They are adaptive systems capable of approximating functions, classify and cluster patterns, identify and optimize control systems, among other applications. These applications are performed on experimental data that represent the operation of the system under study, and do not need to be reprogrammed once training has been completed [18].

The main ANN characteristics are the adaptability based on experience, and learning, as well as the ability of generalization and data organization.

The artificial neurons are responsible for simple operations, collecting and assigning weights to the input signals according to their relevance to the system. The weighting of all system inputs is performed in the synaptic junctions and the neurons response to these inputs is determined by the application of a activation function [18].

The neural network is trained using samples that represent all the aspects of the system's operation. Training algorithms are used in order to adjust the values of the synaptic weights of the neural structures.

Figure 3 illustrates a typical architecture of a feedforward network, for n inputs, n_1 neurons in the hidden layer and m outputs.

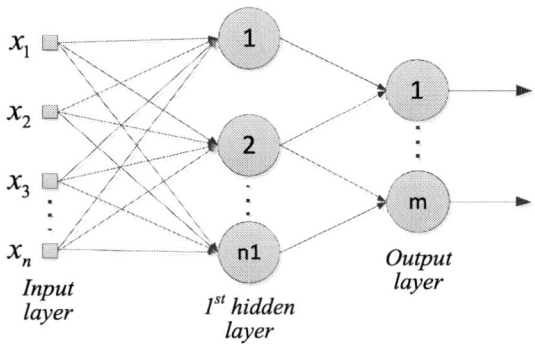

Fig. 3.   Multilayer Perceptron Network.

Supervised training, is the one where the desired network output is known. Examples of topologies are Adaline, MLP, and RBF. In this case, the synaptic weights are adjusted until the errors produced by the network are smaller than a factor of accuracy ε defined by the project specifications.

The same process occurs in unsupervised training, but the desired output is unknown. In these structures, the samples are grouped according to their common characteristics, such as Kohonen self-organizing maps [18].

III. METHODOLOGY

The block diagram of Figure 4 depicts the methodology proposed in this paper.

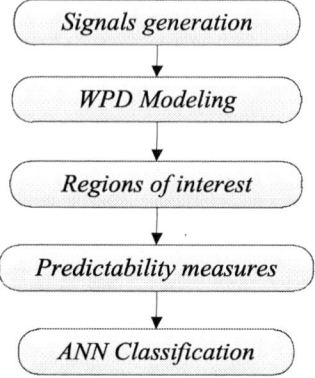

Fig. 4.   Block diagram of the proposed method.

The simulated signals used in this study were created using a base signal without faults, from a real three-phase induction motor operating at nominal voltage and torque conditions. The main parameters of this four pole WEG's motor, 220/380V, IP55, are presented in Table I, as shown in [19].

TABLE I
THREE PHASE INDUCTION MOTOR PARAMETERS

Standard Line - 4 poles - 60 Hz - 220/380 $[V]$	
Power	1 CV
Nominal current	3.02 $[A]$
Stator resistance	7.32 $[\Omega]$
Rotor resistance	2.78 $[\Omega]$
Stator inductance	8.95 $[mH]$
Rotor inductance	5.44 $[mH]$
Mutual inductance	0.141 $[H]$
Moment of Inertia	2.71×10^{-3} $[kg.m^2]$
Synchronous speed	188.49 $[rad/s]$
Slip	3.8%
Nominal torque	4.1 $[N.m]$

The process of signal generation is shown in Figure 5. The faulty signals were obtained from the sum of the basis signal with signals composed by the characteristic frequencies of the analyzed failures and a noisy component, as in [20].

Four groups were created, each with 24 simulated signal, that were based on the parameters of the bearings and the fault frequency. Three groups had signals with faults, one in

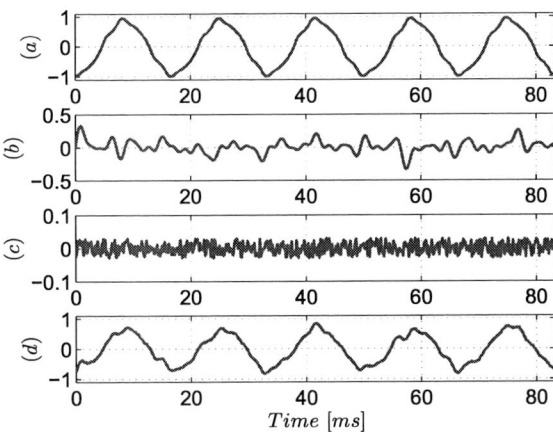

Fig. 5. Signal generation: (a) base signal, (b) fault components signal, (c) random noise, and (d) bearing fault simulated signal.

the inner race, once in the outer race, one in the cage, and the last one composed of normal signals. We considered fault components in the range of 5% to 10% of the maximum amplitude of the current signal, and harmonics components *m* equals to 3 [14], [21].

For the signal analysis, the considered spectral region was determined based on the number of occurrences of the fault frequencies in a single WPD component. Thus, for the modeling of the signals we considered as signal prediction the reconstruction signal using only the fifth component of the seventh WPD level, using a Daubechies (*db4*) wavelet with a frequency band ranging from $93.75Hz$ to $117.18Hz$. This component was chosen because it is related to a region of the spectrum more susceptible to alterations due to the presence of failures.

Based on this modeling, we calculated the relative entropy between prediction error signals of the phases A, B, and C, as in (9), that shows the feature vector. Such feature was created with the intention to highlight the correlations between the motor phase currents.

The ANN classification of the signals into two groups, faulty or normal, were based on MLP, RBF and KHN topologies. Topology details are presented in Table II.

IV. RESULTS AND DISCUSSION

For analysis, the signal we considered the relative entropy measures of only two groups: the normal signals, and the faulty signals (with inner race, outer race, and cage defects). The Figure 6 presents the mean values and standard deviations of the relative entropy of all the signals.

The signals from the faulty groups presented higher relative entropy than the normal signals. High values of these measures indicate that the PDFs of the signals and their prediction errors are more distinct, what results in higher predictability than for the normal signals.

The ANNs topologies presented in the Table II were used for classification of the normal and faulty groups, using the

TABLE II
ANNs TOPOLOGY DETAILS

	Net. 1	Net. 2	Net. 3
Architecture	MLP	RBF	KHN
Training	SP	US+SP	US
Hidden layers	1	1	-
Neurons lay. #1	25	3	-
Neurons output. layer	1	1	5x5
Training algorithm	BP	k-*means* BP	CP
Activation fcn. layer #1	Logistics	Gaussian	-
Output layer	Logistics	Linear	-

(SP) Supervised   (US) Unsupervised

(BP) Backpropagation   (CP) Competitive

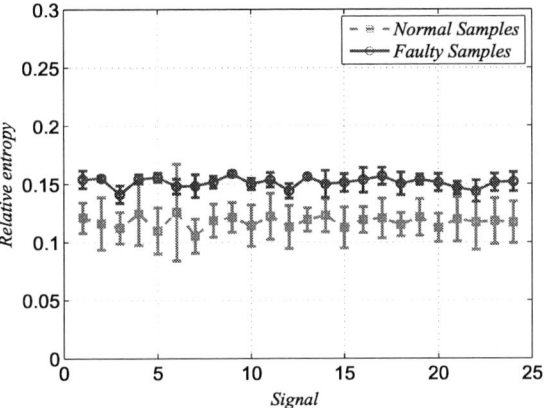

Fig. 6. Mean values and standard deviations of the feature vector measures of all signals.

proposed feature vector. Table III presents the classification performance of the ANNs of Table II.

TABLE III
CLASSIFICATION RESULTS

	Epochs	Required error	Success rate
Net. 1	5164	10^{-7}	100%
Net. 2	358	10^{-7}	95.83%
Net. 3	3798	10^{-10}	70.83%

The success rate is high for the first two ANNs, MLP and RBF, and reasonable for the KHN. The MLP network with one hidden layer obtained 100% of success, therefore may be considered the most appropriate network based on data of this study.

The confusion matrix in Table IV summarizes the classification performance of the classifiers with data validation sets composed with six signals of each group.

TABLE IV
CONFUSION MATRIX

	Net. 1		Net. 2		Net. 3		
	Normal	Faulty	Normal	Faulty	Normal	Faulty	N/C
Normal	6	0	5	1	3	1	2
Inner race faults	0	6	0	6	2	3	1
Outer race Faults	0	6	0	6	1	5	0
Cage faults	0	6	0	6	0	6	0

N/C: Unclassified

Based on Table IV, we see that *Net. 2* only wrongly classified one signal of the normal group, flaging it as a faulty signal. *Net. 3* has not presented a good classification performance of the normal signals and signals with inner race failures, classifying correctly only half of the analyzed signals.

V. CONCLUSION

In this paper, we presented a study of simulated current signals of eletric motors with normal and faulty bearings. The faults were localized on the the outer race, inner race, and in the cage.

When a bearing presents a localized failure, the vibration frequencies related to the failures are reflected in the spectrum of the stator current signals. Analyzing specifically the regions in which these disturbances occur, the frequency components presented in these regions become more unpredictable when compared to the normal signals.

We evaluated the predictability of the signals using measurements of relative entropy between the prediction errors of the signals of motor phases. The predictions were provided by a WPD decomposition model that used the signal reconstruction based only on the fifth component of the seventh level. Only that component was used due to the occurrence of fault characteristic frequencies in its frequency range.

The proposed methodology may be used to analyze electric faults that do not uniformly occur on the three phases. This is due to the evaluation of the predictability using differences between phases, measured by the relative entropy. The method may also be used for energy quality problems such as source voltage unbalance. In such case, we expect occurences of relative entropy differences between phases.

In order to validate the proposed method, further studies should be conducted using experimental signals with localized faults and also generalized faults. In addition, new measures may be considered, such as the Bhattacharyya distance, a symmetric statistical distance measure, the Lempel-Ziv complexity, that assesses the degree of signal periodicity, and the mutual information, that measures the shared information between two signals. This would allow an optimized analysis of more specific components that provide relevant information of the system behavior. Also, new topologies and architectures of the ANNs could be used and MLP performance evaluation using experimental signals.

Finally, a next step is to use reconstructions with several components selected according to the location of faults in the frequency spectra of the signals, what would allow more specific and detailed analyses.

ACKNOWLEDGMENT

This work is supported by the Araucária Foundation for the Support of the Scientific and Technological Development of the State of Paraná, Brazil (Processes Number 06/56093-3 and 338/2012), and the National Council for Technological and Scientific Development - CNPq (Processes number 474290/2008-5, 473576/2011-2, 552269/2011-5) and Capes-DS scolarships.

REFERENCES

[1] J. N. Brito and R. Pederiva, "Hy-nes - a hybrid neural expert system to diagnose problems in induction motors," in *17 th International Congress of Mechanical Engineering*, Sao Paulo, 2003.

[2] A. Bellini, F. Filippetti, C. Tassoni, and G.-A. Capolino, "Advances in diagnostic techniques for induction machines," *IEEE Transactions on Industrial Electronics*, vol. 55, no. 12, pp. 4109 –4126, Dec. 2008.

[3] A. Bonnett and C. Yung, "Increased efficiency versus increased reliability," *IEEE Industry Applications Magazine*, vol. 14, no. 1, pp. 29–36, Jan.-Feb. 2008.

[4] P. Zhang, Y. Du, T. Habetler, and B. Lu, "A survey of condition monitoring and protection methods for medium-voltage induction motors," *IEEE Transactions on Industry Applications*, vol. 47, no. 1, pp. 34–46, Jan.-Feb. 2011.

[5] C. Tang, Q. Miao, and M. Pecht, "Rolling element bearing fault detection: Combining energy operator demodulation and wavelet packet transform," in *Prognostics and System Health Management Conference (PHM-Shenzhen), 2011*, May 2011, pp. 1–6.

[6] O. Seryasat, M. Aliyari Shoorehdeli, F. Honarvar, and A. Rahmani, "Multi-fault diagnosis of ball bearing using fft, wavelet energy entropy mean and root mean square (rms)," in *2010 IEEE International Conference on Systems Man and Cybernetics (SMC)*, Oct. 2010, pp. 4295–4299.

[7] W. Zhou, T. Habetler, and R. Harley, "Stator current-based bearing fault detection techniques: A general review," in *IEEE International Symposium on Diagnostics for Electric Machines, Power Electronics and Drives, 2007. SDEMPED 2007.*, Sep. 2007, pp. 7–10.

[8] N. Mehala and R. Dahiya, "A comparative study of fft, stft and wavelet techniques for induction machine fault diagnostic analysis," in *Proceedings of the 7th WSEAS international conference on Computational intelligence, man-machine systems and cybernetics*, ser. CIMMACS'08, 2008, pp. 203–208.

[9] R. Sharifi and M. Ebrahimi, "Detection of stator winding faults in induction motors using three-phase current monitoring," *ISA Transactions*, vol. 50, no. 1, pp. 14–20, 2011.

978-1-4799-0024-4/13 $31.00 © 2013 IEEE

[10] L. Eren and M. Devaney, "Bearing damage detection via wavelet packet decomposition of the stator current," *IEEE Transactions on Instrumentation and Measurement*, vol. 53, no. 2, pp. 431–436, Apr. 2004.

[11] P. Kankar, S. C. Sharma, and S. Harsha, "Rolling element bearing fault diagnosis using wavelet transform," *Neurocomputing*, vol. 74, no. 10, pp. 1638–1645, 2011.

[12] P. R. Scalassara, C. D. Maciel, and J. C. Pereira, "Predictability analysis of voice signals," *IEEE Engineering in Medicine and Biology Magazine*, vol. 28, pp. 30–34, 2009.

[13] P. Konar and P. Chattopadhyay, "Bearing fault detection of induction motor using wavelet and support vector machines (svms)," *Applied Soft Computing*, vol. 11, no. 6, pp. 4203–4211, 2011.

[14] M. Devaney and L. Eren, "Detecting motor bearing faults," *IEEE Instrumentation Measurement Magazine*, vol. 7, no. 4, pp. 30–50, Dec. 2004.

[15] P. S. R. Diniz, E. A. B. Silva, and S. L. Netto, *Digital Signal Processing: System analysis and design.* Cambridge: Cambridge Univ. Press, 2010.

[16] R. C. Guido, J. F. W. Slaets, R. Koberle, L. O. B. Almeida, and J. C. Pereira, "A new technique to construct a wavelet transform matching a specified signal with applications to digital, real time, spike, and overlap pattern recognition," *Digital Signal Processing*, vol. 16, no. 1, pp. 24–44, 2006.

[17] P. R. Scalassara, C. D. Maciel, J. C. Pereira, S. Oliveira, and D. Stewart, "Problems with nonparametric entropy estimation of voice signals," in *Proceedings of the 20th International Congress on Mechanical Engineering*, 2009, p. 1677.

[18] S. Haykin, *Neural Networks: Comprehensive Foundation.* Prentice Hall, 1999.

[19] T. Santos, A. Goedtel, S. Silva, and M. Suetake, "An ann strategy applied to induction motor speed estimator in closed-loop scalar control," in *XXth International Conference on Electrical Machines (ICEM), 2012*, Sep. 2012, pp. 844–850.

[20] S. McInerny and Y. Dai, "Basic vibration signal processing for bearing fault detection," *IEEE Transactions on Education*, vol. 46, no. 1, pp. 149–156, Fev. 2003.

[21] J. Silva and A. Cardoso, "Bearing failures diagnosis in three-phase induction motors by extended park's vector approach," in *31st Annual Conference of IEEE Industrial Electronics Society, 2005. IECON 2005.*, Nov. 2005, pp. 2591–2596.

Helder L. Schmitt is an undergraduate student of electrical engineering at the Federal University of Technology - Paraná, Cornélio Procópio, Brazil. His research interests are in the field of signal processing, embedded systems, and electrical machines.

Lyvia R. B. Silva is a graduate student at the Federal University of Technology - Paraná, Cornélio Procópio, Brazil. Her research interests are in the field of signal processing, embedded systems, and control systems.

Paulo R. Scalassara is an Assistant Professor at the Federal University of Technology - Paraná, Cornélio Procópio, Brazil. His research interests focuses on the field of signal processing, especially voice analysis. Current projects include predictability analysis and entropy methods applied to discrimination of signals.

Alessandro Goedtel is an Assistant Professor at the Federal University of Technology - Paraná, Cornélio Procópio, Brazil. His research interests are within the fields of electrical machinery, intelligent systems, and power electronics.

Dedicated Hierarchy of Neural Networks applied to Bearings Degradation Assessment

Miguel Delgado, *Member, IEEE*, Giansalvo Cirrincione, *Member, IEEE*, Antonio Garcia Espinosa, *Member, IEEE*, Juan Antonio Ortega, *Member, IEEE*, Humberto Henao, *Senior Member, IEEE*

Abstract -- Condition monitoring schemes, able to deal with different sources of fault are, nowadays, required by the industrial sector to improve their manufacturing control systems. Pattern recognition approaches, allow the identification of multiple system's scenarios by means the relations between numerical features. The numerical features are calculated from acquired physical magnitudes, in order to characterize its behavior. However, only a reduced set of numerical features are used in order to avoid computational performance limitations of the artificial intelligence techniques. In this sense, feature reduction techniques are applied. Classical approaches analyze the features significance from a global data discrimination point of view. This paper, however, proposes a novel and reliable methodology to exploit the information contained in the original features set, by means a dedicated hierarchy of neural networks.

Index Terms-- Ball bearings, Classification algorithms, Curvilinear Component Analysis, Discriminant Analysis, Fault diagnosis, Motor Fault detection, Neural Networks, Time domain analysis, Vibrations.

I. NOMENCLATURE

D	Data space dimension
d	Latent space dimension
D_{ij}	Euclidian distance between i and j points in data space
L_{ij}	Euclidian distance between i and j points in latent space
x_i	The i-th data point from a set
y_i	The i-th projection from a set
α	Learning rate
λ	Distance function threshold

II. INTRODUCTION

THE proposal of reliable health monitoring schemes, applied to electromechanical drives, is a high interest topic in which different scientific communities are focusing its efforts. In this sense, the condition monitoring approaches can be divided in: classical signal processing based structures, led mainly by Industrial Society members [1]-[6], and pattern recognition methodologies, presented by the Computational Intelligence Society members predominately [7]-[9]. Both of them, are based on the characterization of a physical magnitude (such as stator currents, system vibrations and others), under different system's scenarios (operating conditions and considered faults). Afterwards, the identification of the current system status is carried out by the assessment of a new measurement in regard with predefined thresholds or patterns.

Following this procedure, the signal processing approach is based on a direct fault-effect analysis. The behavior of the electromechanical components, in presence of a fault, is analyzed by means of different physical magnitudes. The aim is to emphasize these effects by means of signal processing techniques, and resume the information in specific numerical features. These features will be compared with predefined thresholds during the diagnosis stage. One of the most representative approaches is the detection of characteristic fault-frequencies in the spectral analysis of the stator current or system vibrations [10]-[11]. Also, statistical-time features have been identified in many studies as good parameters for different fault detection schemes [12]. Different works, based on time-frequency techniques, have been also presented, in which non-stationary conditions are considered [13]. Advanced signal processing techniques, such as probabilistic models [14], high-resolution frequency analysis [15] or enhanced wavelets decompositions [16], applied over the measured physical magnitude, have been used also to obtain reliable fault indicators. Most of these methods, however, are based on the analysis of a unique feature in regard with a specific fault. Therefore, although this approach represents the basis of the electromechanical fault diagnosis, additional processing is needed to develop multi-fault detection systems.

In this sense, pattern recognition based methodologies are proposed. This approach allows the identification of different kind of faults under different operating conditions. The so called classifiers generate the corresponding relations between features, in order to distinguish between different system scenarios. A great deal of classification structures has been presented to manage sets of numerical features, such as neural network based approaches [17]-[18]. However, the use of a great deal of numerical features decreases the system performance instead of improve it. The use of a numerous set of features requires a huge amount of system's measurements in order to train properly the artificial intelligence based techniques. Such amount of data is not

---

M. Delgado and J. A. Ortega are with the department of Electronic Engineering, Technical University of Catalonia (UPC), MCIA research center, Rbla. San Nebridi s/n, 08222 Terrassa, Spain (phone: +34-93-739-8518; fax: +34-93-739-8972; e-mails: miguel.delgado@mcia.upc.edu, juan.antonio.ortega@mcia.upc.edu).

A. García is with the department of Electrical Engineering, Technical University of Catalonia (UPC), MCIA research center, Rbla. San Nebridi s/n, 08222 Terrassa, Spain (e-mail: antoni.garcia@mcia.upc.edu).

G. Cirrincione and H, Henao are with the department of Electrical Engineering, University of Picardie (UPJV), LTI 7 Rue du Moulin Neuf, 80000 Amiens, France (e-mails: giansalvo.cirrincione@u-picardie.fr, humberto.henao@u-picardie.fr).

usually available, which will result in a non-well-tuned classifier. However, in an initial numerical feature set, non-significant or redundant information in regard with the considered scenarios is generally present. In this sense, feature reduction techniques are applied first in order to remove and compress the available information. Feature selection and feature extraction variants, such as Principal Component Analysis (PCA) or Sequential Selection respectively [19], have been widely applied. However, classical procedures analyze the features significance from a global data point of view. Hence, only those features with significant discrimination capabilities between all the considered scenarios are chosen. The rest of them, although can be useful to distinguish between a specific pair of scenarios, are removed from the final set due to its low global-significance.

This fact makes mandatory the research towards diagnosis methodologies able to exploit both signal processing and pattern recognition approaches. The contributions of this work include a complete diagnosis methodology, which allows the coexistence of both approaches to maximize the diagnosis capabilities of a proposed set of features. That is, the use of different sets of numerical features depending on the analyzed scenario. A novel dedicated hierarchy of neural networks (dh-NN), is proposed to cover the feature reduction and classification steps, in which different sets of features can be used depending on the discrimination needs. The effectiveness of this condition monitoring scheme has been verified by experimental results in a demanding scenario composed by different bearing faults and severity levels.

The paper is organized as follows: in Section III, an introduction of the feature reduction technique proposed is presented, the Curvilinear Component Analysis, which represents the link between the numerical features and the classification algorithms. Next, in Section IV, the proposed methodology is explained, with special emphasis in the classification structure. Experimental results are shown in Section V. Finally, discussion and conclusions are presented in Section VI.

III. CURVILINEAR COMPONENT ANALYSIS

The feature reduction stage has been implemented usually with linear techniques, such as PCA, Factor Analysis and others. However, this approach has been discussed by many authors emphasizing its limitation dealing with large data sets, because it seeks for a global structure of the data [20]-[21]. In this sense, generally, the characterization of different electromechanical faults under several operating conditions does not result in lineal relations between the calculated numerical features. Concerning with this problem, manifold learning methods has been applied in the last years to preserve the original information in a lower dimensional space [22]-[23]. In this sense, one of the most novel and powerful strategies of non-linear feature reduction is based on distance preservation algorithms. The Curvilinear Component Analysis (CCA) [24]-[27], is a self-organizing neural network. CCA performs the quantization of a data

training set for estimating the corresponding non-linear projection in a lower dimensional space (latent space). Hence, for every pair of different features vectors in the original feature space (data space), a between-points distance D_{ij}, is calculated, $D_{ij} = \|x_i - x_j\|$. The objective is to preserve these distances between the same points in the reduced feature space (latent space), $L_{ij} = \|y_i - y_j\|$, formed by a reduced set of features. In order to face this problem, the CCA technique defines a distance function threshold, λ, in order to determine short and long distances between feature vectors, D_{ij}. By this way, the CCA prioritizes the short distances, which means local distance preservation. The basic procedure of the CCA is shown schematically in Fig.1.

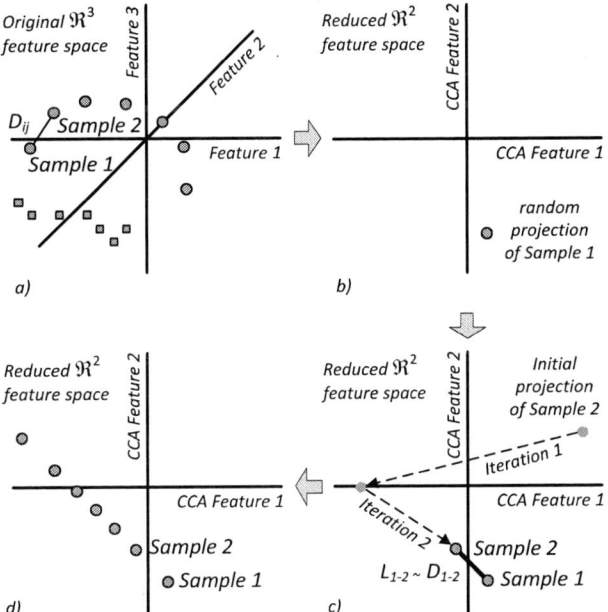

Fig. 1. CCA operation scheme sequence. (a) Seven three-dimensional feature vectors for each of the two classes (circles and squares) represented in a three-dimensional data space. (b) CCA projection of the first feature vector of one operating condition (circles) in the latent space. (c) CCA projection of the second feature vector of the same operating condition (circles) in the latent space. Two iterations are represented until reach $L_{1-2} \sim D_{1-2}$. (d) Resultant CCA projection of the feature vectors corresponding to one operating condition (circles).

The corresponding error function is the *right Bregman divergence* [28]. Indeed, this function penalizes inconsistent long distances and its asymmetry allows a better unfolding of data. The *CCA right Bregman* divergence is given by:

$$E_{Bregman}(p_j) = \lambda^2 p_j \left[e^{-\frac{D_i}{\lambda}} - e^{\frac{L_i}{\lambda}} + (D_i - L_i)\frac{e^{-\frac{L_i}{\lambda}}}{\lambda} \right] \quad (1)$$

being p_j the *j-th* sample to project. The stochastic gradient algorithm for minimizing (1) is then:

$$p_j \leftarrow p_j - \alpha \frac{L_i - D_i}{L_i} e^{-\frac{L_i}{\lambda}} \quad (2)$$

where α and λ are scalars. Due to the CCA properties, the resulting latent space allows interpreting the underlying physical phenomenon of the analyzed faults. In contrast to supervised approaches, the CCA is not influenced by the fault or operating condition to which a pair of measurements belongs. That is, if a pair of measurements, characterized by a specific set of features, exhibits a similar behavior, they are not going to be artificially separated in the latent space. In this sense, the CCA tries to maintain the data topology in a lower-dimensional space, which represents an important property to allow an analysis of the data. In this sense, the CCA is only constrained by the distance preservation.

IV. DIAGNOSIS METHODOLOGY

The proposed methodology is composed of three steps, first, the feature calculation from the vibration signal. Second the feature reduction, to compress and maintain the significant information for diagnosis purposes and, finally, the classification.

In this work four bearing conditions (classes), have been considered, namely: healthy (h), inner race fault (i), outer race fault (o) and ball fault (b). Moreover, three different severity levels (sub-classes) have been considered for each fault: small (i_7, o_7 and b_7), medium (i_{14}, o_{14} and b_{14}) and big single-point defect (i_{21}, o_{21} and b_{21}). Four different steady state operating conditions are taken into account (one speed set point under four different load levels). For each combination of bearing condition, operating condition and severity level, a set of acquisitions is available. The whole set of vibration measurements are divided in training and test set. The training set is used, first, to select the different sets of features, which will be used throughout the hierarchy structure. Second, to train the Hierarchical CCA (h-CCA), and the Hierarchical Multilayer Perceptron (h-MLP), which compose the hd-NN. Then, the test set is evaluated.

A. Feature calculation

From each acquired vibration signal measurement, a set of statistical-time features is computed. This kind of features allows the characterization of the acquired signal. A total of 15 features from time-domain are proposed: mean, maximum value, root mean square (rms), square root mean (srm), standard deviation, variance, root mean square shape factor, square root mean shape factor, crest factor, latitude factor, impulse factor, skewness, kurtosis, normalized 5-th and 6-th moments.

B. Feature reduction by means of selection

All the proposed features exhibit characteristic information of the vibration signal. However, only some of them are really significant in regard with the considered scenarios. These ones, in turn, depend on the considered bearing defects, the appearance of additional sources of vibration and the bearing location among others. In this sense, the most significant features may be different depending on the considered scenario. This fact is considered during the proposed feature reduction stage.

Different techniques can be applied to analyze the feature relevance with regard to the considered diagnosis scenario. Linear Discriminant Analysis (LDA) [29], is one of the classical techniques for feature selection. LDA evaluates quantitatively the discriminant capabilities of the proposed features with regard to classes. This analysis provides information about the contribution of each feature to a well delimited classes' representation. Every two and three combinations of the proposed features, as well as their individual capabilities, have been evaluated. An ordered list, with the most significance features, is obtained. However, although from a global analysis point of view, some features exhibit better discrimination capabilities between classes than others, dedicated feature sets are proposed in this study depending on the discrimination needs. That is, four sets of features are proposed to feed the different layers of the h-MLP: feature set 1 for classification between h, i, b and o, feature set 2 between i_7, i_{14} and i_{21}, feature set 3 between b_7, b_{14} and b_{21} and finally, feature set 4 between o_7, o_{14} and o_{21}. By means of this strategy, the discrimination capabilities are not constrained to a unique set of features, as in classical schemes, but different sets of features can be used depending on the diagnosis stage. Moreover, although the number of features in the feature sets must be predefined initially, the number of features is not increased, which allows to maintain the classification performance.

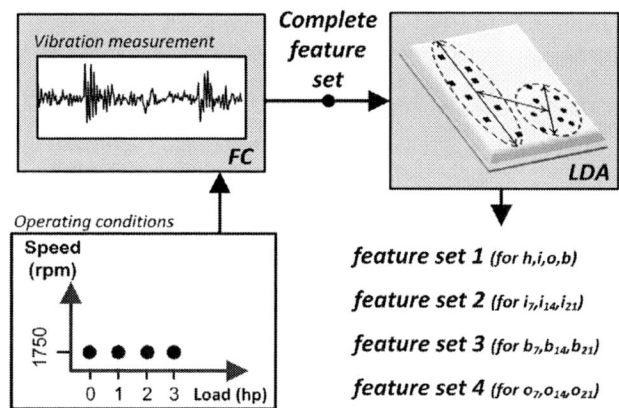

Fig. 2. Generation of the features sets during the training stage by feature selection. The most significant features to distinguish between the predefined classes (h,i,o and b), and sub-classes ($7,14$ and 21 of each fault), are selected by means Linear Discriminant Analysis.

C. Feature reduction by means of extraction

The first level of classification (between h, i, b and o), is more difficult to achieve than the second one (between severities), because of the bigger complexity of the data set distribution. Hence, the all data set has been divided in several subsets in order to improve the CCA data reduction. The hierarchical nature of the classification (classes and subclasses), can be exploited by using a corresponding hierarchy of CCA's, the data reduction problem is so simplified because only one CCA is needed for classify the fault and other three CCA's are needed for estimating the corresponding severity level. During the recall phase, the first feature vector (feature set 1) is fed to CCA$_1$. After this

978-1-4799-0024-4/13 $31.00 © 2013 IEEE

step (bearing fault classification), if a fault bearing is detected, the corresponding feature vector (feature set 2, feature set 3 or feature set 4) is fed to the corresponding CCA (CCA$_2$, CCA$_3$ or CCA$_4$), to assess the fault severity by means of the related neural network classifier. The proposed methodology's structure is shown in Fig. 3.

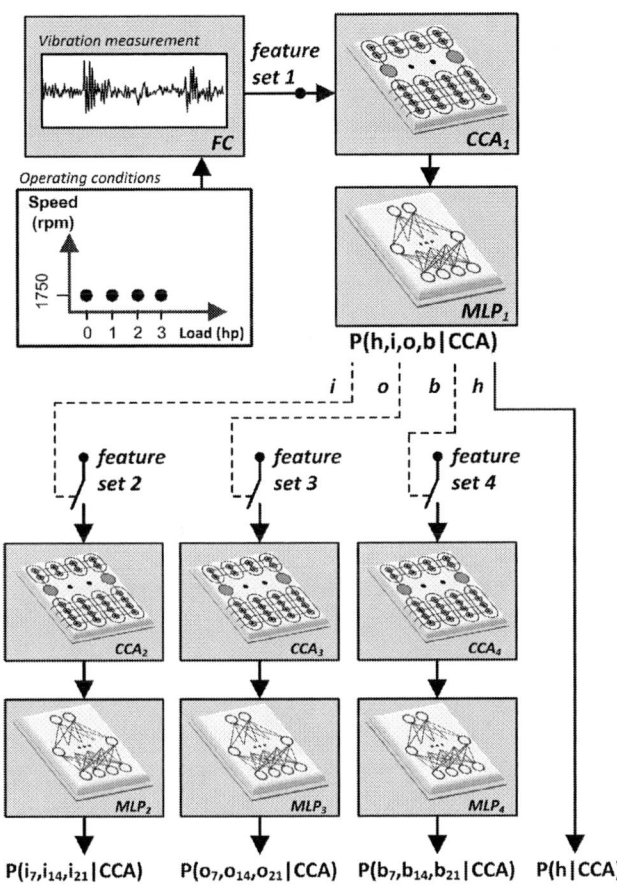

Fig. 3. Proposed diagnosis methodology scheme including feature calculation (FC), feature selection (LDA), feature extraction (*h-CCA*) and classification (*h-NN*). Four set of features are generated depending on the discrimination needs, and probability vectors are obtained in regard with the classification result.

D. Classification

Due to the different number of considered operating conditions (four), and the number of considered bearing fault scenarios (three and healthy) and severity levels (three for each fault scenario), a two-level *h-MLP* is applied. The building block of this architecture is the multilayer perceptron [29]. Each MLP has two layers and the hidden activation function is the hyperbolic tangent. Its training uses the backpropagation rule for the gradient estimation and the scaled conjugate gradient as minimization technique. These blocks are hierarchically organized. This structure allows the classification in two steps: a first neural network classifies a two-dimensional feature vector (resulting from the *CCA$_1$* projection) between four predefined classes (in this application: *h*, *i*, *o* and *b*). If the classification result different from healthy, specific degradation assessment

classifiers (three) are placed in a second level, one for each fault scenario. Then, once the input has been classified in the first neural network, the corresponding second neural network is recalled, and the bearing status and its severity level are obtained. The neural network does not only output the class membership but also its probability. This additional information is important both for assessing the level of confidence of the classification and if risk analysis is required. This information is the output of the neural network because its error function is the cross-entropy error function and the output activation functions are soft-max [29], which are an extension of the sigmoid. The posterior probability for each class is given by the product of the four-class neural network output with the corresponding three-class neural network output. For instance, define as *i* the event *inner race fault* and define the new vector as y_{new}.

$$P(i_7 \mid y_{new}) = P(i_7 \mid i, y_{new})P(i \mid y_{new}) \qquad (3)$$

where $P(i \mid y_{new})$ is the probability of obtaining the event *i* as output of the first four-class classifier, and $P(i_7 \mid i, y_{new})$ is the probability of obtaining event i_7 as output of the corresponding three-*i*-class classifier.

V. EXPERIMENTAL RESULTS

The experimental data comes from the bearing data center [30]. A set of ten ball bearings have been tested, considering inner raceway, rolling element and outer raceway single-point defects as well as healthy bearing. Three severity levels are available for each kind of single-point defect: 0.007 inches, 0.014 inches and 0.021 inches in diameter. Electro-discharge machining was applied during the bearing faults generation. The vibration signal is measured near to the motor bearings at 12kHz. Four operating conditions were considered: one speed set point (1750rpm), and four loads of 0 to 3 hp. The parameters of the bearings under test and the bearing fault conditions of data are shown in Table II and III, respectively. The test bench is shown in Fig. 4.

TABLE II
BEARING PARAMETERS

Type	Outside diameter	Inside diameter	N$_b$	Bd	Pd	$\cos\phi$
SKF6205	2.04 in	0.098 in	9	7.95 in	1.53 in	0.9

TABLE III
CLASSICAL BEARING FAULTS INDICATORS
ANALYZED UNDER RATED CONDITIONS

Bearing condition		Fault specifications	
		Diameter [inches]	Depth [inches]
Healthy (h)		-	-
Inner race fault (i)	i_7	0.007	0.0011
	i_{14}	0.014	0.0011
	i_{21}	0.021	0.0011
Outer race fault (o)	o_7	0.007	0.0011
	o_{14}	0.014	0.0011
	o_{21}	0.021	0.0011
Ball fault (b)	b_7	0.007	0.0011
	b_{14}	0.014	0.0011
	b_{21}	0.021	0.0011

Fig. 4. Test bench composed of a 2 hp motor (left), a torque transducer/encoder (center), a dynamometer (right), and control electronics (not shown). The test bearings support the motor shaft [30].

Following the proposed methodology, the first step is the calculation of the most significant sets of features to discriminate between the four considered scenarios: first h, i, o and b, second i_7, i_{14} and i_{21}, third b_7, b_{14} and b_{21} and fourth, between o_7, o_{14} and o_{21}. In this work, feature sets composed by four features has been selected as a good trade-off between available information in the original set and reduction capabilities needed to visualize it in a final two-dimensional plot. The most significant features to discriminate the considered classes/severities are shown in Table IV, additionally, the most significant features considering the whole data set (classes and sub-classes), are also shown.

TABLE IV
SETS OF THE MOST SIGNIFICANT FEATURES DEPENDING ON THE
CLASSIFICATION SCENARIO.

Classification scenario	Most significant features	
Healthy (h) / Inner race fault (i) / Outer race fault (o) / Ball fault (b)	feature set 1 (fs₁)	
	rms	Shape Factor
	srm	Standard deviation
Inner race fault (i) severities: i_7, i_{14} and i_{21}	feature set 2 (fs₂)	
	rms	Maxim value
	srm	Variance
Outer race fault (o) severities: o_7, o_{14} and o_{21}	feature set 3 (fs₃)	
	rms	Standard deviation
	srm	Variance
Ball fault (b) severities: b_7, b_{14} and b_{21}	feature set 4 (fs₄)	
	rms	Shape factor
	Impulse factor	Crest factor
Whole data set (classes and sub-classes)	feature set 5 (fs₅)	
	rms	Shape Factor
	srm	Variance

Indeed, each of the proposed statistical time features quantifies a specific characteristic of the signal. The combination of different statistical-time features makes possible the identification of the baring status. As it is shown in Fig. 5, the different bearings conditions exhibit different vibration signal characteristics, such as signal variation ratio from the mean, magnitude and others. Therefore, depending on the considered scenarios to distinguish, some statistical time features will be more significant than others.

Thirty measurements are performed for each kind of fault, severity level and operating condition, twenty-five measurements are used for training and five for test purposes. Regarding the proposed methodology, as it has

been mentioned, four CCAs are executed, one for each classification scenario. By means a distributed CCA operation, the projection performance of the second layer CCAs is enhanced due to the lower number of measurements considered. At the same time, the use of different four-dimensional features sets, depending on the classification scenario, allows an increase in the discrimination capabilities and the physical behaviors of the measurements are always preserved.

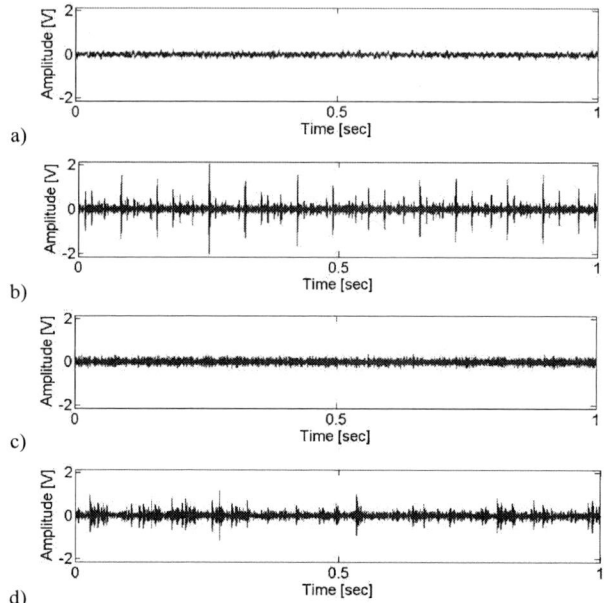

Fig. 5. An example of characteristics vibration signals under different bearing status at the same operating condition (1750rpm speed and 1hp load). (a) Healthy bearing. (b) Inner race fault bearing; 0.007in. (c) Outer race fault bearing; 0.007in. (d) Ball bearing fault; 0.007in.

It can be seen in Fig. 6(a) the resulting CCA₁ projection of the whole training measurements characterized with fs_1. The corresponding dy-dx diagram, Fig. 6(b), relates the distances of the samples in the data space (dx) with the distances in the latent space (dy). It can be seen that most points lie on the bisector. This analysis reveals that the considered bearing faults behave as a set of disconnected manifolds. This fact implies that the use of common reduction techniques as Principal Component Analysis [29], would not be capable to characterize the considered faults.

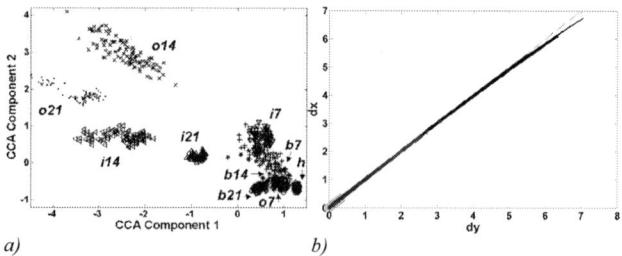

Fig. 6. CCA₁ projection of the whole training measurements characterized with fs_1. (a) CCA₁ projection, α=0.5, λ=1, 50 iterations. ☆ =h, ▽ =i_7, ◁ =i_{14}, △ =i_{21}, + =b_7, ✳ =b_{14}, ◇ =b_{21}, ○ =o_7, × =o_{14} and ∙ =o_{21}. (b) Resulting CCA₁ dy-dx diagram.

978-1-4799-0024-4/13 $31.00 © 2013 IEEE

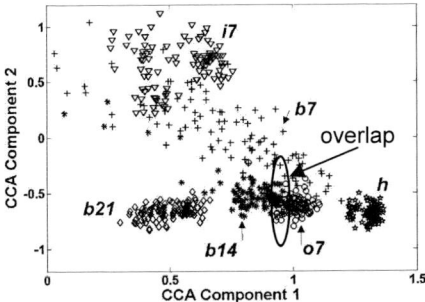

Fig. 7. A detail of CCA₁ projection of the whole training measurements characterized with fs_1. $\alpha=0.5$, $\lambda=1$, 50 iterations, ☆$=h$, $\nabla=i_7$, $+=b_7$, $*=b_{14}$, $\Diamond=b_{21}$ and $O=o_7$.

The resulting MLP₁ projection map, used for the first classification level, is shown in Fig. 8(a). The four classes (h, i, o and b), are well separated, although there is a slight overlapping between o/b. This figure shows the real bearing conditions behavior and how the working conditions and severities degrees influence them. The same procedure has been carried out with the CCA's projections and MLP classifiers in the second layer.

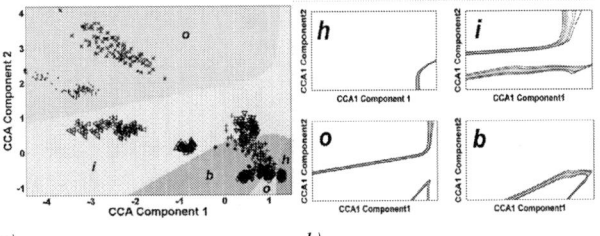

Fig. 8. MLP₁ decision regions resulting from the whole training measurements characterized with fs_1 and projected with CCA₁. (a) h, i, o and b classification regions. ☆$=h$, $\nabla=i_7$, $\triangleleft=i_{14}$, $\triangle=i_{21}$, $+=b_7$, $*=b_{14}$, $\Diamond=b_{21}$, $O=o_7$, $\times=o_{14}$ and $\cdot=o_{21}$. (b) MLP₁ corresponding probability curves, from 100% to 0% of membership's probability.

For checking the generalization properties of the proposed methodology, a test set for the recall phase has been considered. The resulting CCA's projections, and MLP's evaluated with the test measurements are shown from Fig. 8 to Fig. 11.

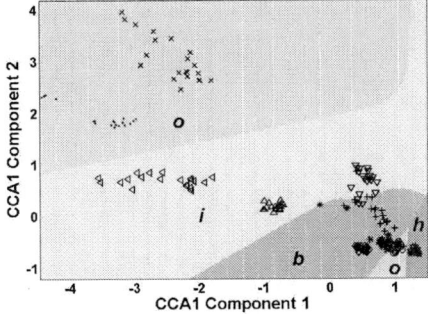

Fig. 9. CCA₁ projection of fs_1 test vectors corresponding to the whole data base and the MLP₁ decision regions between $h/i/o/b$. ☆$=h$, $\nabla=i_7$, $\triangleleft=i_{14}$, $\triangle=i_{21}$, $+=b_7$, $*=b_{14}$, $\Diamond=b_{21}$, $O=o_7$, $\times=o_{14}$ and $\cdot=o_{21}$.

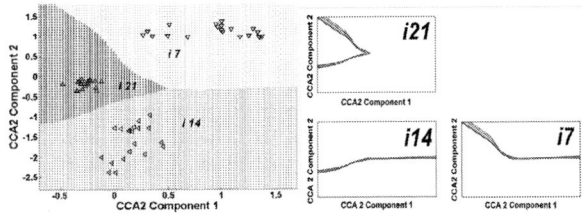

Fig. 10. MLP₂ decision regions resulting from the $i_7/i_{14}/i_{21}$ training measurements characterized with fs_2 and projected with CCA₂. (a) Test measurements characterized with fs_2 and projected with CCA₂, $\nabla=i_7$, $\triangleleft=i_{14}$, $\triangle=i_{21}$. (b) MLP₂ corresponding probability curves, from 100% to 0% of membership's probability.

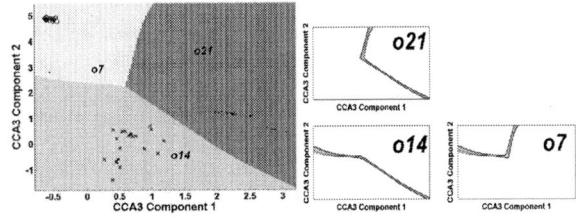

Fig. 11. MLP₃ decision regions resulting from the $o_7/o_{14}/o_{21}$ training measurements characterized with fs_3 and projected with CCA₃. (a) Test measurements characterized with fs_3 and projected with CCA₃, $O=o_7$, $\times=o_{14}$ and $\cdot=o_{21}$. (b) MLP₃ corresponding probability curves, from 100% to 0% of membership's probability.

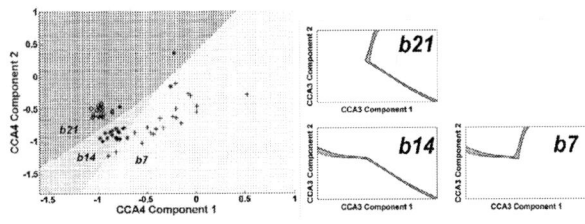

Fig. 12. MLP₄ decision regions resulting from the $o_7/o_{14}/o_{21}$ training measurements characterized with fs_3 and projected with CCA₄. (a) Test measurements characterized with fs_4 and projected with CCA₄, $+=b_7$, $*=b_{14}$, $\Diamond=b_{21}$. (b) MLP₄ corresponding probability curves, from 100% to 0% of membership's probability.

The classification ratio for the test set is 95% approximately. The resulting confusion matrix can be seen in Table V. It is important to notice that all points corresponding to the healthy machine are correctly classified, and only some samples between clusters o/b are misclassified.

TABLE V
CONFUSION MATRIX RESULTING FROM THE
EVALUATION OF THE hd-NN

	h	i_7	i_{14}	i_{21}	o_7	o_{14}	o_{21}	b_7	b_{14}	b_{21}
H	60	0	0	0	0	0	0	0	0	0
i_7	0	19	0	0	1	0	0	0	0	0
i_{14}	0	0	20	0	0	0	0	0	0	0
i_{21}	0	0	0	20	0	0	0	0	0	0
o_7	0	0	0	0	18	0	0	2	0	0
o_{14}	0	0	0	2	0	18	0	0	0	0
o_{21}	0	0	0	0	0	0	20	0	0	0
b_7	0	0	0	0	2	0	0	18	0	0
b_{14}	0	0	0	0	0	0	0	1	17	2
b_{21}	0	0	0	0	0	0	0	0	0	20

In order to compare the obtained performance with classical approaches, the projections and classifications using a unique set of features have been analyzed. Fig. 11(a) shows the MLP decision regions resulting from the training measurements characterized with fs_5 and projected with a CCA, and the projection of the corresponding test vectors. The classification ratio in this first layer is 93%, while the classification ratio in the proposed methodology was 98%. The difference is negligible, which is coherent with the similarity between feature set 1 and 5. However, the performance decreases in the $b_7/b_{14}/b_{21}$ scenario. Fig. 11(b), shows the MLP decision regions resulting from the $b_7/b_{14}/b_{21}$ training measurements characterized with fs_5 and projected with a CCA, and the projection of the corresponding test vectors. The obtained classification ratio decrease until 81% in comparison with the 95% obtained with the fs_4. In this case, fs_4 and fs_5 present important differences, which affect the corresponding sub-classes discrimination.

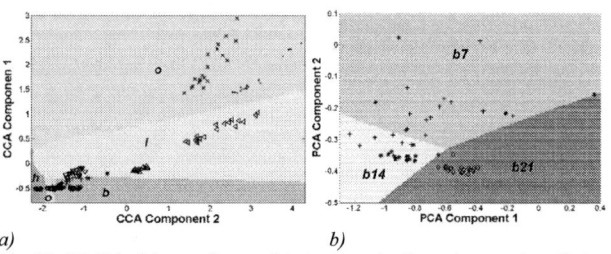

a)   b)

Fig. 11. MLP decision regions and test sets projection using a unique feature set, fs_5. (a) Test vectors corresponding to the whole data base, characterized with fs_5 and projected with a CCA. (b) Test vectors corresponding to the $o_7/o_{14}/o_{21}$ data base, characterized with fs_5 and projected with a CCA, $+=b_7$, $*=b_{14}$, $\diamond=b_{21}$. (b)

Another classical approach is the use of linear feature reduction techniques. Indeed, the feature reduction by lineal techniques, as PCA, exhibits low performances, as it can be seen in Fig. 6(a). The clusters, corresponding to the different classes and sub-classes, are linearly projected. This fact implies, as it can be seen in Fig. 6(b), important overlaps between b_{14}, b_7 and o_7.

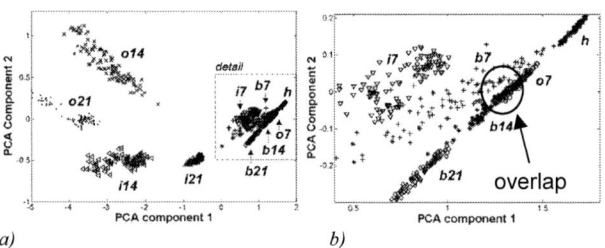

a)   b)

Fig. 6. PCA projection of the whole training measurements characterized with fs_1; $\star=h$, $\triangledown=i_7$, $\triangleleft=i_{14}$, $\triangle=i_{21}$, $+=b_7$, $*=b_{14}$, $\diamond=b_{21}$, $O=o_7$, $\times=o_{14}$ and $\cdot=o_{21}$. (a) Complete training set. (b) Detail of the most critical part of the projection.

VI. Conclusion

The proposed diagnosis methodology, based on a dedicated hierarchy of neural networks (hd-NN) has been introduced. It comprises a hierarchy of curvilinear component projections (h-CCA), and a hierarchy of multilayer perceptrons (h-MLP). The methodology follows

three steps. First, the feature calculation based on the calculation of statistical-time features over the acquired vibration signal. Second, a double feature reduction stage is computed. The whole set of measurements is divided in classes and sub-classes. The most significant sets of features to maximize the discrimination capabilities in each scenario are selected by Linear Discriminant Analysis. Different set of features are obtained. Afterwards, two-dimensional projections are carried out by means of CCAs hierarchically distributed. Third, decision regions are defined by means of NNs also hierarchically distributed. The output is given by the class membership probability, which is suitable for further analysis (e.g. risk). By this way, all the operating conditions as well as the different fault scenarios and its evolution, can be shown in easy and understandable two dimensional space

It should be taken into account that most of the computational cost required by the proposed methodology takes part during the training stage. In this sense, the hd-NN configuration can be carried out in an off-line mode.

Unlike neural and non-neural techniques in the literature, the diagnosis task is shared by two different hierarchies. This is the peculiarity and the power of the proposed method, which allows its application to other complex pattern recognition problems.

VII. Acknowledgment

The authors would like to express their thanks to the Case Western Reserve University (CWS) for their disinterested contribution providing access to the bearing data center analyzed in this paper. This work was supported in part by the Spanish Ministry of Economy and Competitiveness under the TRA2010-21598-C02-01 Research Project.

VIII. References

[1] C. Seungdeog, B. Akin, M. M. Rahimian and H. A. Toliyat; "Implementation of a Fault-Diagnosis Algorithm for Induction Machines Based on Advanced Digital-Signal-Processing Techniques," *IEEE Trans. Ind. Electron.*, vol. 58, no. 3, pp. 937-948, March 2011.

[2] A. Yazidi, H. Henao, G.-A. Capolino, F. Betin and F. Filippetti; "A Web-Based Remote Laboratory for Monitoring and Diagnosis of AC Electrical Machines," *IEEE Trans. Ind. Electron.*, vol. 58, no. 10, pp. 4950,4959, Oct. 2011.

[3] V. Choqueuse, M. E. H. Benbouzid, Y. Amirat and S. Turri; "Diagnosis of Three-Phase Electrical Machines Using Multidimensional Demodulation Techniques," *IEEE Trans. Ind. Electron.*, vol. 59, no. 4, pp. 2014,2023, April 2012.

[4] H. Kum-Kang; R. D. Lorenz and N. J. Nagel; "Gear Fault Diagnostics Integrated in the Motion Servo Drive for Electromechanical Actuators," *IEEE Trans. Ind. Applications*, vol. 48, no. 1, pp. 142,150, Jan.-Feb. 2012.

[5] S. H. Kia, H. Henao and G.-A. Capolino; "Some digital signal processing techniques for induction machines diagnosis in *Proc.* IEEE *Int. SDEMPED*, pp. 322-329, Bologna, Italy, Sept. 2011.

[6] J. Pons-Llinares, V. Climente-Alarcon, F. Vedreno-Santos, J. Antonino-Daviu and M. Riera-Guasp; "Electric machines diagnosis techniques via transient current analysis," in *Proc.* IEEE *IECON*, pp. 3893-3900, Montréal, Canada, Oct. 2012.

[7] M. Seera, C. P. Lim, D. Ishak and H. Singh; "Fault Detection and Diagnosis of Induction Motors Using Motor Current Signature Analysis and a Hybrid FMM–CART Model," *IEEE Trans. Neural Networks and Learning Systems*, vol.23, no. 1, pp. 97-108, Jan. 2012.

[8] W. Wang, "An Intelligent System for Machinery Condition Monitoring," *IEEE Trans. Fuzzy Systems*, vol. 16, no. 1, pp. 110-122, Feb. 2008.

[9] F. Zhao, X. Koutsoukos, H. Haussecker, J. Reich and P. Cheung; "Monitoring and fault diagnosis of hybrid systems," *IEEE Trans. Systems, Man, and Cybernetics, Part B: Cybernetics*, vol. 35, no. 6, pp. 12251240, Dec. 2005.

[10] S. Nandi, H. A. Toliyat, and X Li, "Condition monitoring and fault diagnosis of electrical motors—A review," *IEEE Trans. Energy Conversion*, vol. 20, no. 4, pp. 719-729, 2005.

[11] A. Bellini, F. Filippetti, C. Tassoni, and G. A. Capolino, "Advances in diagnostic techniques for induction machines," *IEEE Trans. Ind. Electron.*, vol. 55, no.12, pp. 4109-4126, 2008.

[12] C. Bianchini, F. Immovilli, M. Cocconcelli, M.; Rubini, R.; Bellini, A., "Fault Detection of Linear Bearings in Brushless AC Linear Motors by Vibration Analysis,", *IEEE Trans. Ind. Electron.*, vol.58, no. 5, pp. 1684-1694, May 2011.

[13] M. Riera-Guasp, M. Pineda-Sanchez, J. Perez-Cruz, R. Puche-Panadero, J. Roger-Folch and J. A. Antonino-Daviu; "Diagnosis of Induction Motor Faults via Gabor Analysis of the Current in Transient Regime," *IEEE Trans. Instrumentation and Measurement*, vol. 61, no. 6, pp. 1583-1596, June 2012.

[14] K. W. Wilson, "Probabilistic inter-disturbance interval estimation for bearing fault diagnosis", in *Proc. IEEE SDEMPED*, pp. 1-6, Cargèse, France, Sept. 2009.

[15] A. Garcia-Perez, R. de Jesus Romero-Troncoso, E. Cabal-Yepez and R. A. Osornio-Rios, "The Application of High-Resolution Spectral Analysis for Identifying Multiple Combined Faults in Induction Motors,"*IEEE Trans. Ind. Electron.*,vol.58,no.5, pp.2002-2010, 2011.

[16] E. C. C. Lau and H. W. Ngan, "Detection of Motor Bearing Outer Raceway Defect by Wavelet Packet Transformed Motor Current Signature Analysis," *IEEE Trans. Instrumentation and Measurement*, vol. 59, no. 10, pp. 2683-2690, 2010.

[17] S. Hamdani, O. Touhami, R. Ibtiouen and M. Fadel; "Neural network technique for induction motor rotor faults classification-dynamic eccentricity and broken bar faults," in *Proc. IEEE SDEMPED*, pp. 626-631, Bologna, Italy, Sept. 2011.

[18] V. N. Ghate and S. V. Dudul; "Cascade Neural-Network-Based Fault Classifier for Three-Phase Induction Motor," *IEEE Trans. Ind. Electron.*, vol. 58, no. 5, pp. 1555-1563, May 2011.

[19] M. Delgado, J. C. Urresty, L. Albiol, J. A. Ortega, A. Garcia, L. Romeral and E. Vidal; "Motor fault classification system including a novel hybrid feature reduction methodology," in *Proc. IEEE IECON*, pp. 2388-2393, Montréal, Canada, Nov. 2011.

[20] J. Yu, "Local and Nonlocal Preserving Projection for Bearing Defect Classification and Performance Assessment," *IEEE Trans Ind. Electron.*, vol. 59, no. 5, pp. 2363-2376, May 2012.

[21] C. Chen, B. Zhang, G. Vachtsevanos and M. Orchard, "Machine Condition Prediction Based on Adaptive Neuro–Fuzzy and High-Order Particle Filtering," *IEEE Trans. Ind. Electron.*, vol. 58, no. 9, pp. 4353-4364, Sept. 2011.

[22] L. A. O Martins, F. L. C. Pádua and P. E. M. Almeida, "Automatic detection of surface defects on rolled steel using Computer Vision and Artificial Neural Networks," *in Proc. IEEE IECON*, pp. 1081-1086, Nov. 2010.

[23] M. I. Chacon-Murguia and S. Gonzalez-Duarte, "An Adaptive Neural-Fuzzy Approach for Object Detection in Dynamic Backgrounds for Surveillance Systems," *IEEE Trans. Ind. Electron.*, vol. 59, no. 8, pp. 3286-3298, Aug. 2012.

[24] G. Cirrincione, M. Delgado, J. A. Ortega and H. Henao, "Bearing fault diagnosis by EXIN CCA," in *Proc. IEEE IJCNN*, Brisbane, Australia, June 2012.

[25] M. Delgado, G. Cirrincione, A. García, J. A. Ortega and H. Henao, "EXIN CCA for Visualization and Classification," in *Proc.* IEEE *ICEM*, Marseille, France, September 2012.

[26] M. Delgado, G. Cirrincione, A. García, J. A. Ortega and H. Henao, "Accurate Bearing Faults Classification based on Statistical-Time Features, Curvilinear Component Analysis and Neural Networks," in *Proc.* IEEE *IECON*, Montréal, Canada, October 2012.

[27] M. Delgado, G. Cirrincione, A. García, J. A. Ortega and H. Henao, "Bearing Faults Detection by a Novel Condition Monitoring Scheme based on Statistical-Time Features and Neural Networks," *IEEE Trans.. Ind. Electron.*, to be published.

[28] J. Sun, M. Crowe and C. Fyfe, "Extending metric multidimensional scaling with Bregman divergences," *Pattern Recognition Journal*, vol. 44, n°5, pp. 1137-1154, 2011.

[29] C. Bishop, *Pattern Recognition and Machine Learning*, Springer; 2007.

[30] K. A. Loparo, "Bearings vibration data set," Case Western Reserve University http://www.eecs.cwru.edu/laboratory/bearing/download.htm

IX. BIOGRAPHIES

Miguel Delgado Prieto (S'08, M'12) received the M.S. degree in Electronics Engineering and the Ph.D. degree in Electronics Engineering from the Universitat Politècnica de Catalunya (UPC), Barcelona, Spain in 2007 and 2012 respectively. From 2004 to 2008 he was a Teaching Assistant in the Electronic Engineering Department of the UPC. In 2008 he joined the Motion and Industrial Control Group (MCIA), where he is currently a Post-Doc Researcher. His research interests include fault detection algorithms, machine learning, signal processing methods and embedded systems.

Giansalvo Cirrincione (M'04) received the "Laurea" degree in electrical engineering from the Politecnico di Torino, Turin, Italy, in 1991, and the Ph.D. degree (with the congratulations of the jury) from the Laboratoire d'Informatique et Signaux de l'Institut National Polytechnique de Grenoble, Grenoble, France, in 1998. In 1999, he was on a post-doc scolarship with the Department SISTA, Leuven University, Leuven, Belgium. Since 2000, he has been an Assistant Professor with the Department of Electrical Engineering, University of Picardie "Jules Verne," Amiens, France. His current research interests include neural networks, data analysis, computer vision, brain models, and system identification.

Antonio Garcia Espinosa (M'05) received the M.S. degree in electrical engineering and the Ph.D. degree from the Universitat Politècnica de Catalunya (UPC), Barcelona, Spain, in 2000 and 2005 respectively. In 2000, he joined the Electric Engineering Department of the UPC, where he is currently a Lecturer. He belongs to the Motion and Industrial Control Group (MCIA). His research interests include electromagnetic devices, electric machines, variable-speed drive systems, and fault-detection algorithms.

Juan Antonio Ortega received the M.S. Telecommunication Engineer and Ph.D. degrees in Electronics from the Technical University of Catalonia (UPC) in 1994 and 1997, respectively. In 1994, he joined the UPC Department of Electronic Engineering as a full time Associate Lecturer. In 1998, he obtained a tenured position as an Associate Professor. Since 1994 he has taught courses of microprocessors and signal processing. From 1994 to 2001 he was with Sensor Systems Group working in the areas of smart sensors, embedded systems, and signal conditioning, acquisition and processing. Since 2001 he belongs to the Motion Control and Industrial Applications research group working in the area of motor current signature analysis. His current research activities include: motor diagnosis, signal acquisition, smart sensors, embedded systems and remote labs. In the last years, he has participated in several Spanish and European funded research projects about these items.

Humberto Henao (M'95–SM'05) received the M.Sc. degree in electrical engineering from the Technological University of Pereira, Pereira, Colombia, in 1983, the M.Sc. degree in power system planning from the Universidad de los Andes, Bogotá, Colombia, in 1986, and the Ph.D. degree in electrical engineering from the Institut National Polytechnique de Grenoble, Grenoble, France, in 1990. From 1987 to 1994, he was a Consultant for companies such as Schneider Industries and GEC Alstom in the Modeling and Control Systems Laboratory, Mediterranean Institute of Technology, Marseille, France. In 1994, he joined the Ecole Supérieure d'Ingénieurs en Electrotechnique et Electronique, Amiens, France, as an Associate Professor. In 1995, he joined the Department of Electrical Engineering, University of Picardie "Jules Verne," Amiens, as an Associate Professor, where he has been a Full Professor since 2010. He is currently the Department Representative for international programs and exchanges (SOCRATES). He also leads the research activities in the field of condition monitoring and diagnosis for power electrical engineering. His main research interests are modeling, simulation, monitoring, and diagnosis of electrical machines and drives.

A Dedicated Application of Artificial Ants for the Condition Monitoring of Induction Motors

A. Soualhi, H. Razik, *senior member, IEEE*, G. Clerc, *senior member, IEEE*

Abstract -- **In the last decade, the field of diagnosis has attracted the attention of many researchers, especially for the detection of faults in induction motors. The condition monitoring of induction motors is generally based on the analysis of signals coming from one or several sensors. This analysis is performed by the motor current signature analysis (MCSA) which is considered as the most popular fault detection technique. This approach considers that a failed component generates a frequency in the motor current spectrum and measuring the amplitude of this frequency can help us to identify and quantify the fault severity. So, the frequency amplitude of the faulty component has to be known. This paper suggests the use of a heuristic technique inspired by the behavior of a colony of ants to track these frequencies. This technique is very easy to implement and converge quickly to a solution. The proposed technique is described and the experimental results illustrate this novel technique.**

Index Terms— **Artificial intelligence, Artificial ants, Broken bars, Fault detection, Induction motors, Monitoring, Motor current spectrum.**

I. INTRODUCTION

Induction motors are widely used in the industry because of their high power-to-weight ratio, low price and an easy maintenance. However, their performances generate constraints such as electrical and mechanical faults leading to the development of various methods for the diagnosis of faults [1]-[5]. These methods can be classified in two categories: model-based methods and the data-based methods. The first category is based on some fundamental understanding of the physics of induction motors which must take into account the disturbances and the model uncertainties. However, these ones are difficult to design and in some cases impossible to extract. In the opposite of model based methods, where a priori knowledge of the system is necessary, data-based methods only require the availability of significant features. These ones can be extracted from signals such as the voltage, current and also the measurement of speed, temperature and vibration in electric motors. However, these features require the establishment of a large number of sensors (flux, torque meters, accelerometer, transducer...) often expensive, sensitive and sometime difficult to place. To correct this drawback, an efficient data-based method should use non-intrusive sensors on the machine and, by analysis, provide the detection of faults. The motor current signature analysis (MCSA) is largely used in the condition monitoring of induction motors [6]-[8]. It is a non-destructive technique which analyzes the current signal in order to identify the frequencies which characterize several kinds of faults like broken rotor bars, eccentricity and bearings failure. Some techniques are based on the quantification of the severity by the analysis of the amplitude of these frequencies. These ones have to be monitored using a postprocessor algorithm based on: NN (neural network) technique [9], PR (pattern recognition) technique [10]. This list is not exhaustive and other techniques may be used, like those inspired from the collective behaviour of decentralized, self-organized systems. This kind of technique is known as Swarm Intelligence (SI) [11]. Most famous such SIs are Ant Colony Optimization (ACO) [12], Stochastic Diffusion Search (SDS) [13] and Particle Swarm Optimization (PSO) [14].

ACO, in its current form, became a powerful source of inspiration for the design of techniques to solve complex problems. Studies have shown that ants are able to solve collectively many complex problems in the nature such as food search, construction of the nest with a structured organization. For this purpose, the Ant Colony Optimization, inspired by the behavior of ants, has been the subject of several studies in different application domains [15]-[17]. The proposed technique is applied to explore the spectral density of the stator current in order to identify broken rotor bars at several levels of load.

This paper is structured as follows: Section II illustrates how it is possible to identify and quantify the fault severity of broken rotor bars by the analysis of the motor current spectrum. Section III gives a description about how the behavior of a colony of ants can be used to localize the characteristic frequencies of broken rotor bars. In section IV, we investigate the described technique on an induction motor of 3 kW operating with less than one broken bar at different levels of load. We conclude in section V with some results to prove the efficiency of this technique.

II. DETECTION OF BROKEN ROTOR BARS

To illustrate our meaning, Fig. 1 presents the spectral density of the stator current of a 3kW induction motor operating with less than one broken bar at 50% of the nominal load. As we can see, lines appear at the frequencies

---

Soualhi. A, Razik. H, Clerc. G are with the Laboratoire Ampère, Université de Lyon, F-69622, Lyon, France ; Université Lyon 1, Villeurbanne, France ; CNRS, UMR 5005. E-mail: [abdenour.soualhi, guy.clerc, hubert.razik]@univ-lyon1.fr

as follows:

$$f_{bb} = (1 \pm 2ks) f_s \qquad (1)$$

where f_{bb} is the faulty lines, k is any integer and f_s is the supply frequency.

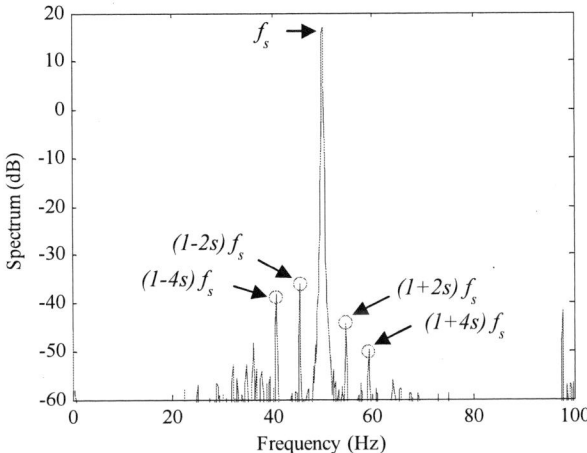

Fig. 1. Spectrum of the stator current of an induction motor operating with less than one broken bar at 50% of full load.

The number of broken bars n is quantified by the following equation which is based on the ratio of the amplitude (I) of the principal faulty lines with the supply frequency.

$$\frac{n}{N_r} \approx \frac{\left(I_{(1-2s)f_s} + I_{(1+2s)f_s} \right)}{I_{f_s}} \qquad (2)$$

where N_r is the number of rotor bars. s represents the slip of the machine and f_s the supply frequency.

In the paper of [18], the authors have noted that the sum of the two sideband components at frequencies depends mainly on the fault degree, therefore, relationship (2) can be used to determine the faulted bars number, if the sum of the amplitude of the two current components is used.

This paper suggests tracking the faulty lines in the stator current of an induction motor induced by the appearance of a problem of broken bars. Some tests have been made using a GA (Genetic Algorithm) and PSO (Particles Swarms Optimization) [19], [20]. The technique described here is based on artificial ants.

III. Ant Colony Optimization for the Detection of Broken Bars

The ability of artificial ants is to solve combinatorial optimization and classification problems using an objective function. One of the first applications of ACO was the problem of minimizing the traveling distance for a salesman in 1997 [21]. As illustrated in Fig. 2, a colony of ants finds the shortest distance/path from a starting point (the nest) towards the source of food thanks to pheromones. It solves this problem according to the highest concentration of pheromone. As the ant lays the same quantity of pheromone on the path chosen, the shortest path has the highest concentration of pheromone knowing that pheromone could evaporate in time.

Let consider Z as the number of ants that will explore the N paths. Starting from the nest, each ant places a quantity of pheromone on a visited path (i) $_{i \in (1,...,N)}$, and then chooses the next path according to the probability $p^t_A(i)$ which depend on the length of the path (i) $d(i)$ and the quantity of pheromones deposited by ants in this path at time t.

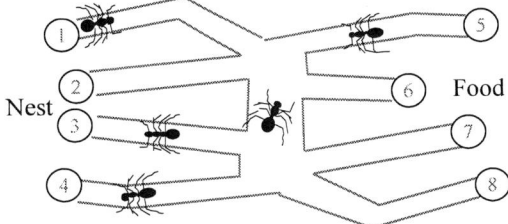

(a) Ants choose paths randomly at the time instant t=0

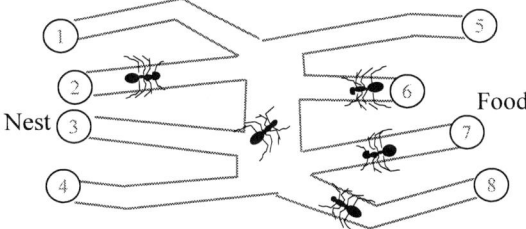

(b) Ants choose paths 2,6,7,8 at the time instant t > 0

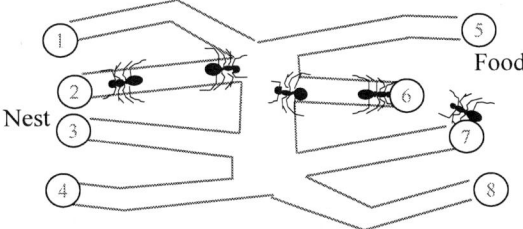

(c) Ants choose paths 2,6 at the time instant t » 0

Fig. 2. Example of the behaviour of real ants to find the shortest way to the source of food.

After returning to the starting point (the nest), the ant lays down a quantity of pheromone. The intensity of this substance depends on the covered distance. When another ant tries to explore other paths, it tends to follow the path which has the highest quantity of pheromone.

$$p^t_A(i) = \frac{\left[\tau^t(i) \right]^\alpha (\eta_i)^\beta}{\sum\limits_{j \in PATH} \left[\tau^t(j) \right]^\alpha (\eta_j)^\beta} \qquad (3)$$
$$_{A \in (1,...,Z)}$$

with
- $A \in (1,..., Z)$ is the Ath ant and Z is the number of ants $Z<N$;
- $d(i)$ is the length of the ith path;
- $\eta_i = 1/d(i)$ is the inverse of the length of the path (i);
- α is a parameter which controls the importance of the pheromone;
- β is a parameter which controls the importance of the path length taken by ants;
- $PATH \in (1,..., N)$ is the set of paths taken by ants;
- $\tau^t(i)$ and $\tau^t(j)$ are the quantities of the pheromone deposited by ants on the path (i) and (j) respectively at time t.

The quantity of the pheromone $\tau(i)$ is updated by:

$$\tau^{t+1}(i) = (1-\rho)\,\tau^t(i) + \Delta\tau(i) \qquad (4)$$

$$\Delta\tau(i) = \sum_{A=1}^{Z} \Delta\tau_A(i) \qquad (5)$$

where
- $\rho \in [0,1]$ is a parameter such that $(1-\rho)$ represents the evaporation of the pheromone;
- $\Delta\tau_A(i)$ is the quantity per unit of length of the pheromone deposited by the Ath ant on the path (i).

$\Delta_A\tau(i)$ is given by:

$$\Delta\tau_A(i) = \begin{cases} \dfrac{Q}{d(i)} & \text{if the } A^{th} \text{ ant has taken} \\ & \quad \text{the path}(i) \\ 0 & \text{otherwise} \end{cases}, \qquad (6)$$

with
- Q : constant related to the quantity of pheromone deposited by ants.

The example described above can be applied to detect broken rotor bars. In this case, the spectral density of the stator current is divided into two parts; the first part, called the nest, includes a number of frequencies less than f_s and the second part, called the source of food, includes a number of frequencies upper than f_s. As shown in Fig. 3, the identification of $(1-2s)\,f_s$ and $(1+2s)\,f_s$ named f_L f_R respectively is related to the quantity of pheromones deposited by ants when they move from the first part of the spectrum of the stator current towards the second part of the spectrum passing through the fundamental frequency (f_s).

The probability that a frequency f_i $i \in (1,..., N)$ corresponds to one of the frequencies f_L and f_R is given by:

$$p_A^t(f_i) = \frac{\left[\tau^t(i)\right]^\alpha (\eta_i)^\beta}{\displaystyle\sum_{j \in PATH} \left[\tau^t(j)\right]^\alpha (\eta_j)^\beta}, \quad \begin{array}{l} A \in (1,...,Z), \\ i \in (1,...,N) \end{array}, \qquad (7)$$

with
- $\eta_{ij} = 1/d(f_i, f_s)$ is the inverse of the distance between f_i and f_s (see Fig. 3);
- $PATH \in (1,..., N)$ is the set of frequencies chosen by ants $Z<N$ and N is the number of frequencies taken into account to identify f_L f_R.

$\Delta_A\tau(i)$ is given by:

$$\Delta\tau_A(i) = \begin{cases} \dfrac{Q}{d(f_i, f_s)} & \text{if the } A^{th} \text{ ant has selected} \\ & \quad \text{the frequency}(f_i) \\ 0 & \text{otherwise} \end{cases} \qquad (8)$$

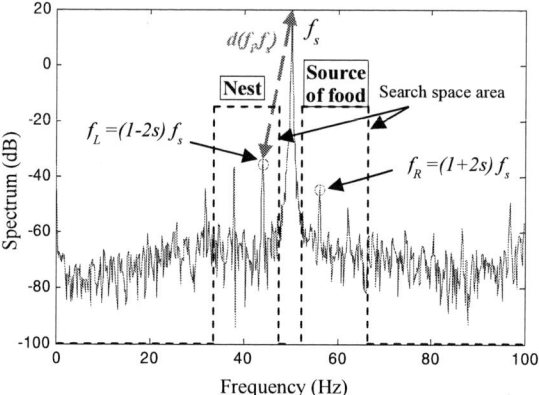

Fig. 3. Spectrum of the stator current of an induction motor operating under one broken bar at 50% of full load.

f_L f_R are obtained by a succession of interactions between the ants. Each ant has behavioural rules which it follows according to the situations that it meets:
- The nest includes frequencies less than f_s, and corresponds to the starting point for ants which leave towards the source of food;
- The source of food includes frequencies upper than f_s;
- The path taken by the ants towards the source of food passing through the fundamental frequency (f_s) corresponds to the choice of f_L f_R;
- Each ant memorizes the frequencies by putting a quantity of pheromone;
- The choice of f_L f_R by an ant depends on the quantity of pheromones deposited by ants previously.

The following steps show how the behaviour of ants can be used to find f_L f_R.

1) Ants start in the nest by choosing randomly Z frequencies. We use here the rand function of MATLAB which returns pseudorandom numbers.

2) The frequencies chosen by ants in the nest are marked by a quantity of pheromones given by:

$$\tau^{t=0}(i) = \begin{cases} \dfrac{Q}{d(f_i, f_s)} & \text{if the } A^{th} \text{ ant has selected} \\ & \text{the frequency } (f_i) \\ 0 & \text{otherwise} \end{cases} \quad (9)$$

These ants move towards the source of food passing through the fundamental frequency (f_s) and chose randomly Z other frequencies.

3) The frequencies chosen by ants in the source of food are marked by a quantity of pheromones given by (9).

4) The frequency f_R is initially identified by (7).

5) Ants choose frequencies situated in the neighbor of f_R.

6) The frequency f_R is chosen once again by (7).

7) Ants identify the frequency f_R by putting a quantity of pheromone given by (4),(5),(8) and move towards the nest.

8) The frequency f_L is chosen according the frequency which has the highest quantity of pheromone $\tau'(i)$.

9) Ants choose frequencies situated in the neighbor of f_L.

10) The frequency f_L is chosen once by (7).

11) Ants identify the frequency f_L by putting a quantity of pheromones given by (4),(5),(8) and move towards the source of food.

12) The frequency f_R is chosen according the frequency which has the highest quantity of pheromone $\tau'(i)$.

13) Repeat step 6 to 13

IV. EXPERIMENTS

This section presents some results obtained on an induction of 3 kW, 5.9A and 2800 rpm with 28 bars, 400V, 2 poles and 50 Hz. The motor is fed by a fixed network

The stator current was measured by a current sensor at 10kHz during 10.3s. Consequently the spectrum has a resolution of 0.15Hz which is enough to make an efficient diagnostic whatever the defect is.

In this application, the current spectrum limited on the range] 0Hz, 100Hz [is considered as represented in Fig. 1 and Fig. 3. Figures 4 to 7 present, respectively, the spectrum of a motor operating with less than one broken bar for 100, 75, 50 and 25 % of the full load. The red curve in these figures represent the characteristic frequencies of broken bars $(f_L \ f_R)$ marked by the highest quantity of the pheromone thanks to artificial ants. The simulation was done with an Intel Xeon 2.8GHz, 8GB RAM.

Table. I . Parameters of the ACO

Number of ants	**Z**	27
Number of iterations		4
Quantity of pheromone	**Q**	30
Importance of the pheromone	α	-0,01
importance of the path length taken by ants	β	1
evaporation of the pheromone	ρ	0,1
Iteration number	-	20

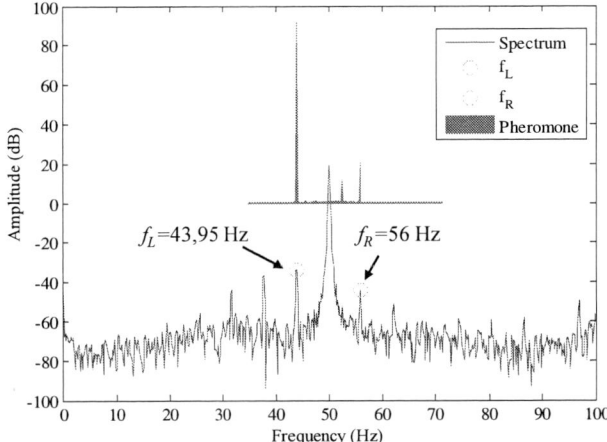

Fig. 4. Spectrum of a motor with less than one broken bar at 100% of the full load.

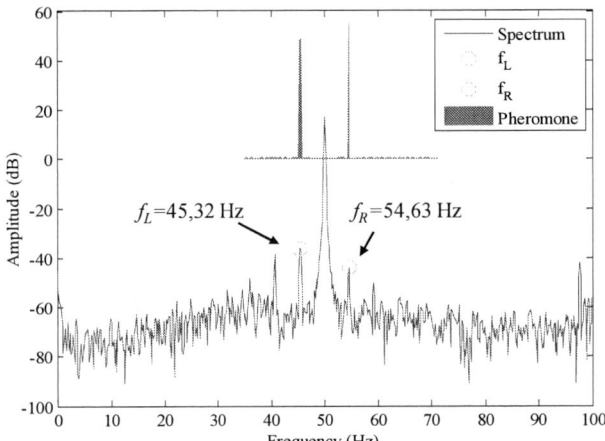

Fig. 5. Spectrum of a motor with less than one broken bar at 75% of the full load.

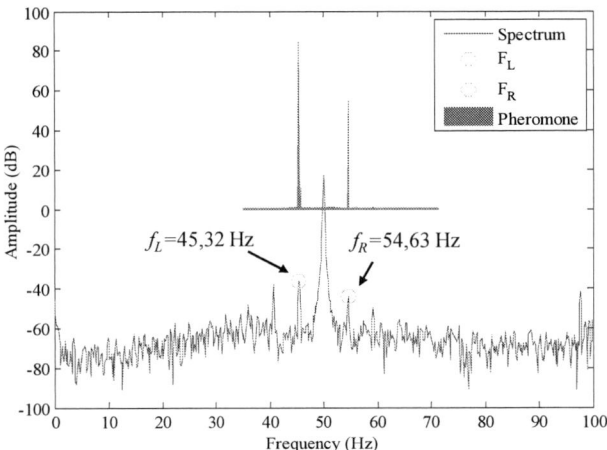

Fig. 6. Spectrum of a motor with less than one broken bar at 50% of the full load.

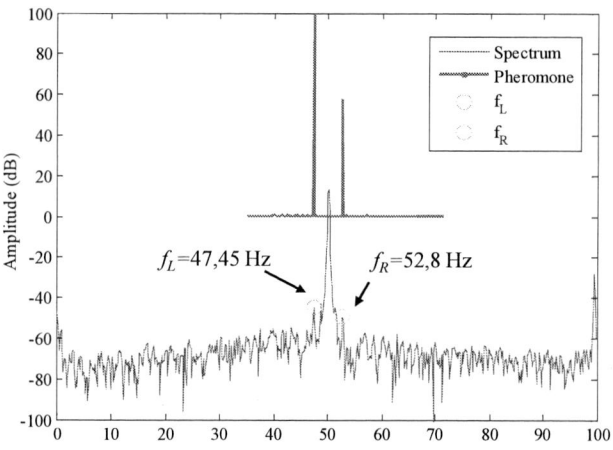

Fig. 7. Spectrum of a motor with less than one broken bar at 25% of the full load.

f_L and f_R are the characteristic frequencies of broken rotor bars. The amplitude (I) of these frequencies allows to equation (2) to quantify the fault severity. It is noticed that the quantity of pheromone deposited by ants in these frequencies cannot quantify the fault severity. This is due to the random choice of the frequencies and the number of iterations used to detect them. The consumed time in an ordinary computer is less than 1s.

V. CONCLUSION

In this paper, we have proposed a technique based on the use of artificial ants to find both defectives lines around the supply frequency. This technique, artificial ants, is simple to implement and moreover, it does not require a huge computational process. So we have presented an application based on the analysis of current spectrum of the induction motor in the case of a broken bar. In the same manner, we can extend this technique to the tracking of faulty lines to other defects in order to make a diagnostic.

VI. REFERENCES

[1] F. Pedrayes, C. H. Rojas, M. F. Cabanas, M. G. Melero, G. A. Orcajo, and J. Cano, "Application of a dynamic model based on a network of magnetically coupled reluctances to rotor fault diagnosis in induction motors," in *Diagnostics for Electric Machines, Power Electronics and Drives (SDEMPED)*, 2007, pp. 241–246.

[2] B. Akin, U. Orguner, H. Toliyat, and M. Rayner, "Phase-sensitive detection of motor fault signatures in the presence of noise," *Industrial Electronics, IEEE Transactions on*, vol. 55, no. 6, pp. 2539–2550, 2008.

[3] S. Choi, B. Akin, M. Rahimian, and H. Toliyat, "Implementation of a fault-diagnosis algorithm for induction machines based on advanced digital-signal-processing techniques," *Industrial Electronics, IEEE Transactions on*, vol. 58, no. 3, pp. 937–948, 2011.

[4] O. Ondel, G. Clerc, E. Boutleux, and E. Blanco, "Fault detection and diagnosis in a set "inverter-induction machine" through multidimensional membership function and pattern recognition,"

Energy Conversion, IEEE Transactions on, vol. 24, no. 2, pp. 431–441, 2009.

[5] A. Bellini, F. Filippetti, C. Tassoni, and G.A. Capolino, "Advances in diagnostic techniques for induction machines," *Industrial Electronics, IEEE Transactions on*, vol. 55, no. 12, pp. 4109–4126, 2008.

[6] J.H. Jung, L. Jong-Jae, and B.H. Kwon, "Online diagnosis of induction motors using mcsa," *Industrial Electronics, IEEE Transactions on*, vol. 53, no. 6, pp. 1842–1852, 2006.

[7] W. Zhou, T. Habetler, and R. Harley, "Stator current-based bearing fault detection techniques: A general review," in *Diagnostics for Electric Machines, Power Electronics and Drives (SDEMPED)*, 2007, pp. 7–10.

[8] L. Frosini and E. Bassi, "Stator current and motor efficiency as indicators for different types of bearing faults in induction motors," *Industrial Electronics, IEEE Transactions on*, vol. 57, no. 1, pp. 244–251, 2010.

[9] S. Guedidi, S. Zouzou, W. Laala, M. Sahraoui, and K. Yahia, "Broken bar fault diagnosis of induction motors using mcsa and neural network," in *Diagnostics for Electric Machines, Power Electronics Drives (SDEMPED)*, 2011, pp. 632–637.

[10] R. Casimir, E. Boutleux, and G. Clerc, "Fault diagnosis in an induction motor by pattern recognition methods," in *Diagnostics for Electric Machines, Power Electronics and Drives (SDEMPED)*, 2003, pp. 294–299.

[11] E. Bonabeau, M. Dorigo, and G. Theraulaz, *Swarm intelligence: from natural to artificial systems*. Oxford University Press, 1999.

[12] M. Dorigo, G. D. Caro, and L. M. Gambardella, "Ant algorithms for discrete optimization," *Artificial Life*, vol. 5, pp. 137–172, 1999.

[13] M. M. al Rifaie, M. J. Bishop, and T. Blackwell, "An investigation into the merger of stochastic diffusion search and particle swarm optimisation," in *Genetic and evolutionary computation (GECCO)*, 2011, pp. 37–44.

[14] J. Kennedy and R. Eberhart, "Particle swarm optimization," in *Neural Networks*, vol. 4, 1995, pp. 1942–1948.

[15] D. Zhao, L. Luo, and K. Zhang, "An improved ant colony optimization for communication network routing problem," in *Bio-Inspired Computing*, 2009, pp. 1–4.

[16] K. Sankar and K. Krishnamoorthy, "Ant colony algorithm for routing problem using rule-mining," in *Computational Intelligence and Computing Research (ICCIC)*, 2010, pp. 1–8.

[17] A. Soualhi, G. Clerc, and H. Razik, "Faults classification of induction machine using an improved ant clustering technique," in *Diagnostics for Electric Machines, Power Electronics Drives (SDEMPED)*, 2011, pp. 316–321.

[18] F. Filippetti, G. Franceschini, C. Tassoni, and P. Vas, "Ai techniques in induction machines diagnosis including the speed ripple effect," in *Industry Applications Conference*, vol. 1, 1996, pp. 655–662.

[19] H. Razik, M. de Rossiter Correa, and E. da Silva, "A novel monitoring of load level and broken bar fault severity applied to squirrel-cage induction motors using a genetic algorithm," *Industrial Electronics, IEEE Transactions on*, vol. 56, no. 11, pp. 4615–4626, 2009.

[20] H. Razik, M. Correa, and E. R. C. Da Silva, "The tracking of induction motor's faulty lines through particle swarm optimization using chaos," in *Industrial Technology (ICIT)*, 2010, pp. 1245–1250.

[21] M. Dorigo and L. Gambardella, "Ant colony system: a cooperative learning approach to the traveling salesman problem," *Evolutionary Computation, IEEE Transactions on*, vol. 1, no. 1, pp. 53–66, 1997.

VII. BIOGRAPHIES

Abdenour Soualhi was born in Sétif, Algeria, on May 16, 1985. He obtained in 2010 the Master Degree in Electrical Engineering from the "Université Claude Bernard Lyon I" at Lyon. He is currently engaged in PhD research in the department of electrical engineering at the laboratory Ampère. His research interests are in the diagnosis and prognosis of faults by means of artificial intelligence techniques.

Hubert Razik (M'98-SM'03) was graduated from the Ecole Normale Supérieure, Cachan, France, in 1987. Since November 2009, he is professor of electrical engineering at the "Université Claude Bernard Lyon I". His fields of research include the modelling, the control the monitoring conditions of multi-phase induction motor.

Guy Clerc (M'90-SM'10) was born in Libourne, France, on November 30, 1960. He received the Engineer's degree and the PhD in electrical engineering from the Ecole Centrale de Lyon, France, in 1984 and 1989, respectively. He is Professor of Universities. He carried out researches on control and diagnosis of induction machines at Ampère - UMR 5005, Villeurbanne, France.

Comparison of Supervised Classification Algorithms Combined with Feature Extraction and Selection : Application to a Turbo-generator Rotor Fault Detection

Alexandre Bacchus, Mélisande Biet, Ludovic Macaire, Yvonnick Le Menach and Abdelmounaïm Tounzi

Abstract—The goal of this paper consists in applying pattern recognition methods to turbo-generators. Previous works have shown that a monitor, thanks to pattern recognition, is practical on asynchronous machines. This procedure has rarely taken advantage of these methods for turbogenerator. The statistical model has been obtained from harmonics extracted from flux probes and from stator current and voltage. For this purpose, the main way is to build a learning matrix to predict the functional state of a new measurement. Finally, three classifiers have been compared : k Nearest Neighbors, Linear Discriminant Analysis and Support Vector Machines. The best classification result is obtained by Linear Discriminant Analysis combined with Factorial Discriminant Analysis achieving a score of 84.6%.

Index Terms—Classification Algorithms, Statistical Analysis, Fault diagnosis, Feature Extraction, Fuzzy logic, Monitoring, Pattern analysis, Support Vector Machines, Turbogenerators

I. NOMENCLATURE

$V_{a,b,c}$ Stator voltage phase a, b and c

$I_{a,b,c}$ Stator current phase a, b and c

σ_i Standard deviation of I_i with $i = \alpha$, β or s (also (α, β))

δ_i Deformation of the normalized characteristic I_i with $i = \alpha$ or β

n Number of prototypes

m Number of test data

p Number of features

This work was supported by EDF R&D, Clamart 92140, France

Alexandre Bacchus is with the L2EP, University of Lille 1, Villeneuve d'Ascq 59650, France. (email : alexandre.bacchus@univ-lille1.fr)

Mélisande Biet is with EDF R&D, Clamart 92141, France. (email : melisande.biet@edf.fr)

Ludovic Macaire is with the LAGIS, UMR CNRS 8219, University of Lille 1, Villeneuve d'Ascq 59650, France. (email : ludovic.macaire@univ-lille1.fr)

Yvonnick Le Menach is with the L2EP, University of Lille 1, Villeneuve d'Ascq 59650, France. (email : yvonnick.le-menach@univ-lille1.fr)

Abdelmounaïm Tounzi is with the L2EP, University of Lille 1, Villeneuve d'Ascq 59650, France. (email : mounaim.tounzi@univ-lille1.fr)

$\mathbf{X}$ $\in \mathbb{R}^{n \times p}$, the learning matrix

$\mathbf{x}$ $\in \mathbb{R}^p$, a prototype (also $\mathbf{x_i}$ $1 \leq i \leq n$)

$\mathbf{Z}$ $\in \mathbb{R}^{m \times p}$, the test matrix

$\mathbf{z}$ $\in \mathbb{R}^p$, a test data

F_i i^{th} feature

$\mathbf{f}$ $\in \mathbb{R}^n$, a column of $\mathbf{X}$

$\omega(\mathbf{x_i})$ The class of $\mathbf{x_i}$

ω_j Class j

$\mathcal{G}$ Set of all classes

K $card(\mathcal{G})$

$\widetilde{\mathbf{x_j}}$ $\in \mathbb{R}^p$, mean of the class j

X Random variable corresponding to the samples

G Random variable corresponding to the classes

$\mu_j(\mathbf{x})$ The membership degree of $\mathbf{x}$ to the class ω_j

$N_k(\mathbf{z})$ k nearest prototypes of $\mathbf{z}$

Σ_T The covariance matrix of the prototypes

Σ_W, Σ_B The within and between classes covariance matrix of the prototypes

Σ_{Wj}, Σ_{Bj} The within and between classes covariance matrix of the class j

II. INTRODUCTION

MOST of the nuclear plants units have been built during the 70-80's. In order to ensure a safe and optimal operating, reflecting the aging equipment, it is necessary to carry out regular preventive maintenance. At the same time, monitoring these turbo-generators is critically important [ ].

In general case, several faults may appear due to the aging of the machine : turn-to-turn failures, cracking issues, mechanical failures onto the shaft, rotor eccentricities, winding faults at the ground etc [ ]. However, the two must common rotor faults are eccentricities and inter turn short circuit in the rotor winding. The first fault induces mechanical unbalance and vibrations and the second one can imply, in the worst case, a rotor ground fault.

Recent works [ ]-[ ], have explored the use of leakage flux, axial shaft voltage, flux probes, rotor current, stator voltage in

978-1-4799-0024-4/13 $31.00 © 2013 IEEE

order to detect rotor faults. Stator voltages and currents remain the most accessible quantities to analyze in order to detect such fault. Furthermore, high power nuclear plant generators are now equipped with flux probes. Thus, we base our diagnosis work on the signal analysis given by these quantities.

To perform the study, it is necessary to have data on the studied defects. However, this cannot be obtained from a real alternator. In our case, we consider a small scale prototype of a nuclear plant generator [9]. It is instrumented so that stator, current and voltage, and the flux from two radial probes located in the air gap can be extracted.

Most of fault detection methods are based on signal threshold. However, it is difficult to determine relevant thresholds. Meanwhile [10]-[11] used several pattern recognition methods for asynchronous machine diagnosis. Biet [1] adapts this strategy to the turbo generator monitoring. Hence, she has developed a diagnosis scheme based on statistical pattern recognition adapted to the global electric and local flux probe measurements.

This scheme is divided into two steps, the off-line learning step followed by the on-line fault detection step. Biet has shown that this approach succeeds in correctly detecting faults for 77% of the cases [1].

The performances reached by this approach strongly depend on the criteria used to select features and on the decision rule followed to assign the test sample. It is the reason why, in this paper, we propose to apply several other feature selection and classification schemes to improve this rate of correct fault detections. In order to select features, we apply filter and wrapper approaches that analyze the class distribution and classification accuracy, respectively. Three classification methods have been used : k Nearest Neighbors (kNN), Linear Discriminant Analysis (LDA) and Support Vector Machines (SVM).

Experimental set-up and the two kinds of faults (eccentricities and short-circuit turns) are presented in section III. Section IV shows the features that are extracted from measurements. Section V describes the rotor fault diagnosis on synchronous machine using supervised classification. Finally, section VI presents feature extraction and selection methods used to compare experimental results obtained by several tested methods.

III. FAULTS DATA FROM AN EXPERIMENTAL SET-UP

Rotor static eccentricities occur when the rotation center is the same as the rotor one but different to the stator one. Fig. 1a gives an explaining scheme on the rotation rotor center with respect to the stator one.

Rotor short-circuited turns (see Fig. 1b) are, generally, due to damage of the insulating material between two turns. Even if this kind of faults is not often serious, a rotor ground fault may occur with its degradation.

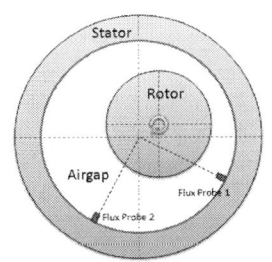

(a) Example of a rotor static eccentricity

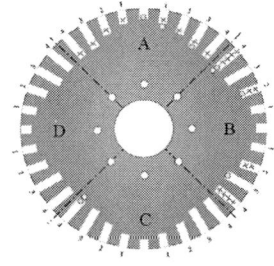

(b) Types of short-turns on 2 poles, each pole is defined by A, B, C or D [2]

Fig. 1.   Machine rotor faults

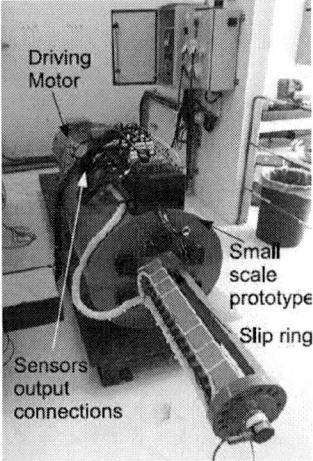

Fig. 2.   Small scale prototype of a turbo-generator

To interpret data related to the studied defects, measures have been carried out on a small scale prototype presented in Fig. 2. It is a 26.7kVA, 50Hz, 4 pole generator presenting 48 stator and 36 rotor slots, which has been adapted in order to, experimentally generate several rotor faults. Thus, this prototype allows us to carry out measures either with static eccentricities and/or with rotor short-circuited turns. Different cases of these defects with varying degrees of severity can be simulated.

The combination of generated eccentricity ("Ecc") and short-circuited turns ("St") leads to 16 functional states (Fs). For each Fs, the machine is studied under five load cases : no loaded (0), two full resistive loads (12.5kW and 25kW) and two inductive ones (10kVA and 20kVA).

Each pair functional state/load corresponds to one of the $K = 80$ classes (16 Fs × 5 loads), that constitutes the subset $\mathcal{G} = \{\omega_1, \omega_2, ..., \omega_{80}\}$. For each of these pairs, measurement

978-1-4799-0024-4/13 $31.00 © 2013 IEEE   559

acquisitions have been achieved at different times.

For example the case healthy with no load matches with the first class ω_1 (see Fig. 3). "Ecc" stands for eccentricity and the number is referring to its gravity. For example, the functional state "Ecc1" is related to a $0.27\,mm$ static eccentricity. "StB2B4" corresponds to short-turns applied to pole B, located in slots 2 and 4. (see Fig. 1b) The Ampere-turns decrease by 22% due to this fault.

The assignation of a test data to one of the 80 classes can be simplified. Because faults are our main concern, only the functional state can be relevant. So we can consider that a correct diagnostic corresponds to assign a data test to one of the 5 classes associated to the true Fs. For example in Fig. 3 by considering only the Healthy state diagnostic, we consider that a data test can be assigned to either ω_1, ω_2, ω_3, ω_4 or ω_5. So, the accuracy obtained with 80 classes (16 Fs $\times$ 5 loads) can also be used to estimate the accuracy with 16 classes (16 Fs).

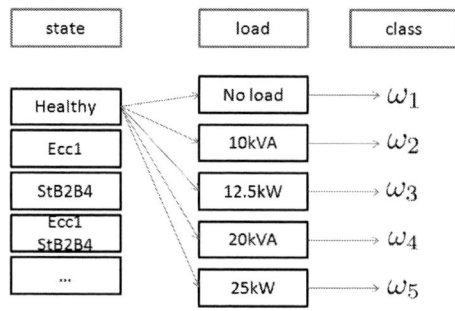

Fig. 3.   Class construction, a class is represented by a pair (Fs/load)

IV. FEATURES EXTRACTED FROM MEASUREMENTS

In a healthy case, magnetic field in the air gap presents a symmetry. This symmetry is no longer ensured when a short-circuited turn occurs and the magnetic flux density differs according to the rotor poles. Then, the measure of the magnetic flux density in the air gap can yield to diagnose this fault. This technique was first described by Albright [12]. In our case, we use its extension given in [3]. Two search coils are located in the air gap shifted by an electrical angle of 180°. The induced voltage in both coils are added. Thus, when the machine is not healthy, the resulting signal is no more nil and when a PSD (Power Spectral Density) is applied, it can be possible to obtain specific harmonics that characterize the different faults (see Fig 4).

Following [1] and [10], stator current and voltage have been processed by the Concordia's transformation. We also

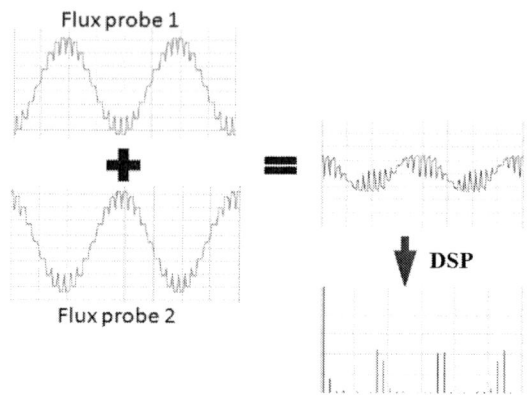

Fig. 4.   Fault detection method with two flux probes

consider the standard deviations σ_α, σ_β and σ_s of I_α, I_β and $\left\|\vec{I}_{\alpha,\beta}\right\|$ respectively, where $\left\|\vec{I}_{\alpha,\beta}\right\| = \sqrt{I_\alpha^2 + I_\beta^2}$ is the current modulus.

Moreover we also use the direct impedance $Z_d = \frac{V_d}{I_d}$, where V_d and I_d are, respectively, the stator voltage and current along the rotor direct axis d. Finally, Table I shows the 33 features that have been extracted from measurements. Note that we analyze 6 global features (current and voltage) and 27 local ones (from flux probe harmonics).

The learning matrix $\mathbf{X} \in \mathbb{R}^{800 \times 33}$ is formed by 80 classes (pair fault/load), each class containing 10 samples called prototypes, and characterized by 33 features. 650 other samples have also been extracted from measures in order to validate diagnosis methods, and constitute the test matrix $\mathbf{Z} \in \mathbb{R}^{650 \times 33}$. Because of technical problems, classes $\omega_{21}... \omega_{25}$ and $\omega_{71...80}$ are not represented by $\mathbf{Z}$.

Table I
SUMMARY OF THE 33 FEATURES USED TO FORM THE LEARNING MATRIX

features	Notations	Definition
F_1	σ_α	Standard deviation
F_2	σ_β	
F_3	σ_s	of I_α, I_β and $\left\|\vec{I}_{\alpha,\beta}\right\|$
F_4	δ_α	Deformation of I_β (I_α)
F_5	δ_β	
F_6	Z_d	Direct impedance
$F_7..F_{21}$	$2^{nd}...16^{th}$	Harmonics from two flux probes
$F_{22}..F_{28}$	$34^{th}...40^{th}$	
$F_{29}..F_{31}$	$70^{th}...72^{th}$	
F_{32}, F_{33}	$75^{th}, 76^{th}$	

978-1-4799-0024-4/13 $31.00 © 2013 IEEE

V. ROTOR FAULT DIAGNOSIS ON TURBOGENERATORS USING SUPERVISED CLASSIFICATION

A. Definition

Pattern recognition consists in the automatic association of a new pattern to a group or a class of objects. Each sample is defined by its characteristics that induce the gathering of samples in classes. These characteristics are also called features, they are the pattern descriptors, and they correspond to the first column in Table I. Moreover, in order to recognize a sample, one has to define a learning matrix $\mathbf{X}$ composed of prototypes. Finally, the goal is to identify the class of a test data, i.e. its load and fault, (see Table 3) using a supervised classification algorithm.

B. Classification

In this part, three well-known supervised classification methods are presented : Linear Discriminant Analysis, k Nearest Neighbor and Support Vector Machines.

1) Linear Discriminant Analysis: Linear Discriminant Analysis (LDA) is a supervised classification method using Bayes' law. The probability that a test data $\mathbf{z}$ belongs to a class ω_j, $j = 1..K$, is based on a Gaussian function :

$$f_j(\mathbf{z}) = \frac{1}{\sqrt{\left|\mathbf{\Sigma_T}^{-1}\right|(2\pi)^{33/2}}} e^{-\frac{1}{2}(\mathbf{z}-\widetilde{\mathbf{x}}_j)^t \mathbf{\Sigma_T}^{-1}(\mathbf{z}-\widetilde{\mathbf{x}}_j)} \quad (1)$$

where $\mathbf{\Sigma_T}$ and $\widetilde{\mathbf{x}}_j$ define, respectively, the total covariance matrix and the mean of class ω_j. In Eq (1) the exponential term is related to Mahalanobis distance (see Eq (6)) and the denominator corresponds to a normalization of $f_j(\mathbf{z})$.

Furthermore, an application of Bayes' law gives

$$Pr(G = \omega_j | X = \mathbf{z}) = \frac{f_j(\mathbf{z})\pi_j}{\sum_{l=1}^{K} f_l(\mathbf{z})\pi_l} \quad (2)$$

where π_j is the prior probability of class ω_j, depending on the number of class prototypes.

Using (1) in the log of (2) , we get the linear discriminant function (see [13]):

$$\delta_j(\mathbf{z}) = -\frac{1}{2}log\left|\mathbf{\Sigma_T}\right| - \frac{1}{2}(\mathbf{z}-\widetilde{\mathbf{x}}_j)^t \mathbf{\Sigma_T}^{-1}(\mathbf{z}-\widetilde{\mathbf{x}}_j) + log\,\pi_j \quad (3)$$

The sample $\mathbf{z}$ is assigned to the class $\omega(\mathbf{z})$ with the highest probability, given by :

$$\omega(\mathbf{z}) = arg \max_{j=1..K} \delta_j(\mathbf{z}) \quad (4)$$

2) k Nearest Neighbors: The k nearest neighbors (kNN) algorithm is a supervised classification method. It assigns a test data to the most represented class among its k nearest prototypes.

The number k has to be experimentally adjusted by the analyst. In this paper, we found that $k = 5$ is often the best choice. To determine the nearest prototypes $N_k(\mathbf{z})$, we use the classic Euclidian and Mahalanobis distances between the data test $\mathbf{z}$ and each prototype $\mathbf{x}$:

$$d_{Euc}(\mathbf{x}, \mathbf{z}) = \sqrt{(\mathbf{x} - \mathbf{z})^t (\mathbf{x} - \mathbf{z})} \quad (5)$$

$$d_{Mah}(\mathbf{x}, \mathbf{z}) = \sqrt{(\mathbf{x} - \mathbf{z})^t \mathbf{\Sigma_T}^{-1} (\mathbf{x} - \mathbf{z})} \quad (6)$$

In (6), high scale variations between two features are weighted thanks to $\mathbf{\Sigma_T}$.

The test data $\mathbf{z}$ is assigned to the class $\omega(\mathbf{z})$ so that :

$$\omega(\mathbf{z}) = arg \max_{j=1..K} \left(\sum_{x \in N_k(\mathbf{z})} h(\omega(\mathbf{x}), \omega_j) \right) \quad (7)$$

where $N_k(\mathbf{z})$ is the k nearest neighbors set of $\mathbf{z}$ and $\forall d, e \in \{1..K\}$ $h(\omega_d, \omega_e)$ is defined by :

$$h(\omega_d, \omega_e) = \begin{cases} 1 & \text{if } \omega_d = \omega_e \\ 0 & \text{otherwise} \end{cases} \quad (8)$$

3) Support Vector Machines: Another issue is to build some margins that separate classes in the feature space. Support vector machines (SVM) are supervised learning models associated with learning algorithms that analyze data and recognize patterns. A SVM model is a representation of the prototypes as points in a feature space, mapped so that the prototypes of the separated classes are divided by a clear gap (or margin) that is as wide as possible. Because the class distribution does not allow us to build these margins, prototypes need to be projected in a higher dimension space than the original one using a kernel function. The standard normal function has been used as kernel, given by :

$$K(\mathbf{x}, \mathbf{x}') = e^{-\frac{1}{p}(\mathbf{x}-\mathbf{x}')^t(\mathbf{x}-\mathbf{x}')} \quad (9)$$

where the penalty coefficient $\frac{1}{p}$ is invariant.

Because this method only performs a binary classifier, we consider a classifier between every available pair of classes, denoted hereafter ω_j and ω_l (one versus one approach).

The solution of the problem induced by the SVM is given by :

$$f^{jl}(\mathbf{z}) = \sum_{\mathbf{x} \in \omega_l \cup \omega_j} \alpha(\mathbf{x})\,\omega(\mathbf{x})\,K(\mathbf{z}, \mathbf{x}) + \beta_0 \quad (10)$$

where $\alpha(\mathbf{x})$ is the Lagrange multiplier associated to the prototype $\mathbf{x}$ and β_0 is a parameter to be estimated.. Let us introduce the score function $\varepsilon^{jl}(\mathbf{z})$:

$$\varepsilon^{jl}(\mathbf{z}) = \begin{cases} \omega_j & \text{if } f^{jl}(\mathbf{z}) \geq 0 \\ \omega_l & \text{otherwise} \end{cases} \quad (11)$$

Finally the classifier is given by :

$$\omega(\mathbf{z}) = arg \max_{j=1..K} \sum_{l \neq j} h(\varepsilon^{jl}(\mathbf{z}), \omega_j) \quad (12)$$

978-1-4799-0024-4/13 $31.00 © 2013 IEEE

where the function h corresponds to (8).

4) Results: We propose to implement these methods and to compare the results. kNN and LDA analyze the data projected in the selected feature space, whereas SVM projects data in an higher dimensional space, assumed to be linear discriminant thanks to a kernel function. kNN takes into account the k prototypes that are the closest ones to each test sample (local decision), while LDA compares the distances between the each test sample and the means of all classes (global decision).

Each method learns the manifold from prototypes and assigns each test data to a class. Therefore the score corresponds to :

$$score = \frac{\text{number of well classified test data} \times 100}{650} \quad (13)$$

According to Table II, for LDA, kNN (with Euclidian distance) and SVM more than 40% (60% for LDA) of the test data have not been correctly classified, which represent almost 260 samples of 650 test data. Based upon these conclusions, one needs to select the feature space in which the clusters can be well discriminated.

Table II
ACCURACY RATES OF $\mathbf{Z}$ INDUCED BY kNN, LDA AND SVM ON $\mathbf{X}$

methods	LDA	kNN	SVM
scores	38.3077%	54.1538%	57.5714%

VI. FEATURE EXTRACTION AND SELECTION METHODS

As pointed out before, features highly affect class distribution and, moreover, the classification accuracy. So it will be interesting to evaluate the discriminating power of these features. Feature extraction and selection methods consist in weighting each feature. In the case of feature selection, these weights are either 0 (feature non selected) or 1.

There exists two kinds of feature selection methods [14]-[13] :

- Wrapper methods use a classifier to evaluate the quality of a feature subset. According to the hold-out partition, the prototypes are divided into two sets : the first is used to learn the classes while the second one is assigned to the classes thanks to a classifier. The accuracy of the second one provides the feature score. As wrapper methods train a new model for each subset, they are very computationally intensive, but usually provide the best performing feature subset for that particular type of model.
- Filter methods use a proxy measure to evaluate a feature subset. This measure is chosen to fast to be computed, whilst evaluating the discriminating power of the feature subset. Many filters provide a feature ranking rather than an explicit best feature subset, and the cut off point in the ranking is determined via cross-validation.

A. Factorial Discriminant Analysis

Factorial Discriminant Analysis (FDA) is a feature extraction method that combines features in order to create a basis where prototypes and test data can be projected. The key point is to solve the following optimization problem :

$$\max \mathbf{a}^t \Sigma_{\mathbf{B}} \, \mathbf{a}, \quad \text{subject to} \quad \mathbf{a}^t \Sigma_{\mathbf{T}} \, \mathbf{a} = 1 \quad (14)$$

where $\mathbf{a} \in \mathbb{R}^p$ contains all the weights of each feature, $\Sigma_{\mathbf{T}} = \Sigma_{\mathbf{B}} + \Sigma_{\mathbf{W}}$ with $\Sigma_{\mathbf{W}} = \sum_{j=1}^{K} \Sigma_{\mathbf{Wi}}$ and $\Sigma_{\mathbf{B}} = \sum_{j=1}^{K} \Sigma_{\mathbf{Bi}}$, where $\Sigma_{\mathbf{Bi}}$ and $\Sigma_{\mathbf{Wi}}$ being, respectively, the between and within covariance matrices of class ω_j. Using the Lagrangian of (14) FDA consists in finding the eigenvalues of $\Sigma_{\mathbf{T}}^{-1} \Sigma_{\mathbf{B}}$.

B. Wrapper Feature Selection Criterion

Intuitively, the accuracy reached by a classifier operating in the examined feature space can be a criterion of quality. Indeed, the goal is to get the best accuracy rate, thus it can be used as a feature evaluation function. In this case one does not introduce test data, but only prototypes thanks to 10 cross validation method. It consists in partitioning prototypes in 10 parts by considering iteratively 90% of data as learning set and 10% for the validation one. Finally one gets 10 accuracy rates, the mean or the maximum inducing the criterion.

C. Filter Feature Selection Criteria

In order to reduce the number of features, one needs to define a feature selection criterion, that evaluates the discriminant power of each feature. Let this criterion be a function $\mathcal{J}$, where Y is a feature subset :

$$\mathcal{J} : \{F_1, ..., F_p\} \rightarrow \mathbb{R} \quad (15)$$
$$Y \mapsto \mathcal{J}(Y) \quad (16)$$

where F_i corresponds to the ith feature, i.e. i^{th} column of prototype matrix $\mathbf{X}$.

The principal goal of the following criterion consists in separating and compacting each class. First, let us introduce the Fisher criterion [1]-[10] that measures the discriminating power of one feature and is defined as :

$$\mathcal{J}_{Fisher}(\{F_i\}) = \sum_{l=1}^{K} \sum_{\substack{j \neq i}}^{K} \frac{\widetilde{x}_{li} - \widetilde{x}_{ji}}{\Sigma_{Wl} - \Sigma_{Wj}} \quad (17)$$

where K is the number of classes and $\widetilde{x}_{li}$ is the mean of class ω_l, according to the feature F_i.

As this criterion only examines one feature, the selection algorithm sorts the features with respect to their Fisher scores. Let us consider Hotelling criterion, coming from Fisher's one. A relevant feature subspace maximizes the numerator of (17), i.e. maximizing the distance between classes, and minimizes the denominator, i.e. minimizing the distance within each class in the feature subset Y.

$$\mathcal{J}_{Hot}(Y) = tr\left(\Sigma_W^{-1}\Sigma_B\right) \quad (18)$$

where $tr(A)$ corresponds to the trace of matrix A.

Semani and al. [15] introduced a new criterion based on fuzzy logic called the "Ambiguities criterion". An ambiguity is due to the projection of all prototypes on axis corresponding to the selected features and the information loss. Each prototype is characterized by a membership degree $\mu_i(x)$ to the class ω_i:

$$\mu : \mathbb{R}^p \to [0,1]^K \quad (19)$$

$$\mathbf{x} \mapsto \begin{bmatrix} \mu_1(\mathbf{x}) & \mu_2(\mathbf{x}) & \dots & \mu_K(\mathbf{x}) \end{bmatrix} \quad (20)$$

$$\mu_j(\mathbf{x}) = \frac{1}{1 + d(\mathbf{x}, \widetilde{\mathbf{x}_j})} \quad (21)$$

where $d(\mathbf{x}, \widetilde{\mathbf{x}_j})$ corresponds to the distance between the prototype $\mathbf{x}$ and the class center $\widetilde{\mathbf{x}_j}$, Euclidian distance has been used. Let us note the t-norm $\forall a, b \in \mathbb{R}$ then $\top(a,b) = min(a,b)$, the t-co norm $\bot(a,b) = max(a,b)$. Then the authors introduced the following operator called $OR-2\ fuzzy$:

$$\overset{2}{\underset{j=1,K}{\bot}} \mu_j = \underset{j=1,K}{\top}\left(\underset{\substack{l=1,K \\ l \neq j}}{\bot} \mu_l\right) \quad (22)$$

It corresponds to the second maximum of the μ_i. Thus the ambiguity measure processed in the feature subset Y is given by

$$A^Y(\mathbf{x}) = \frac{\overset{2}{\underset{j=1,K}{\bot}} \mu_j(\mathbf{x})}{\underset{j=1,K}{\bot} \mu_j(\mathbf{x})} \quad (23)$$

And finally,

$$J_A(Y) = \sum_{i=1}^{n} A^Y(\mathbf{x_i}) \quad (24)$$

A relevant feature subspace minimizes the number of ambiguities and, consequently, $J_A(Y)$.

D. Feature Selection Algorithm

Considering p features, the number possible feature subsets is 2^p (e.g. $p = 33$, $2^{33} = 8.59 \times 10^9$). It is necessary to explore less combinations using some sub-optimal search algorithms. Biet and Casimir et al. used a sequential algorithm called SBS (Sequential Backward Selection) [1]-[10]-[16]. The algorithm of this method is given by Algorithm 1. Whereas SBS iteratively removes each feature starting from a full initial set, SFS (Sequential Forward Selection) iteratively adds each feature by considering an empty initial set [13]. The stopping criterion can be based on the variation between two successive steps of the evaluation function or the classification accuracy.

Algorithm 1 Sequential Backward Selection

$i = 0$
$YB_i = \{F_1, F_2, ..., F_p\}$
Do

$$j_0 = arg \underset{j=1...|YB_i|}{max} \mathcal{J}(YB_i - \{F_j\})$$
$$YB_{i+1} = YB_i - \{F_{j_0}\}$$
$$i = i + 1$$

Until the stop criterion

VII. EXPERIMENTS

A. Results

Table III presents the 3 "best" features, induced by SBS with the Fisher, Hotelling and Ambiguities. In this Table, we indicate the harmonics associated to the selected features. It seems that local measures provided by flux probes are more informative than global ones. In this paper, we focus the study on the pattern recognition view. Consequently, features selected by the previous methods are not interpreted in the physical way.

Table III
FEATURES SELECTED BY SBS BASED ON DIFFERENT CRITERIA

rank of selected features	1st	2nd	3rd
Fisher features	F_7	F_{30}	F_{32}
corresponding harmonics	2nd	71st	75th
Hotelling features	F_{22}	F_7	F_9
corresponding harmonics	34th	2nd	4th
Ambiguities features	F_{17}	F_{16}	F_9
corresponding harmonics	12th	11th	4th

Fig. 5 shows the accuracy rate evolution obtained by kNN, operating in a selected feature space, SBS has been used as feature selection algorithm combined with Hotelling and the ambiguities criteria, for eleven numbers of features. For each method, we compare the results by taking into account either the Euclidian or Mahalanobis distance. Rotor fault accuracy rate corresponds to the case when the fault has been found by the classifier but the load is not necessarily correct, i.e. the 16 class problem as defined in section III. The figures show that the accuracy rate strongly varies when the number of selected features changes. As expected, rotor fault accuracy reaches better percentages than the classic one. Except for the 7 "best" features, Hotelling allows a better accuracy rate than Ambiguities criterion. Although Mahalanobis distance is based on the total covariance matrix, it provides lower results than Euclidian distance. For the classification accuracy rate, corresponding to the 80 class problem, the best value, 77%, is achieved for 6 features using SBS and Hotelling criterion.

978-1-4799-0024-4/13 $31.00 © 2013 IEEE

Even if the load has not been correctly identified, SBS and Ambiguities criterion succeeds in detecting a rotor fault in 87% of cases.

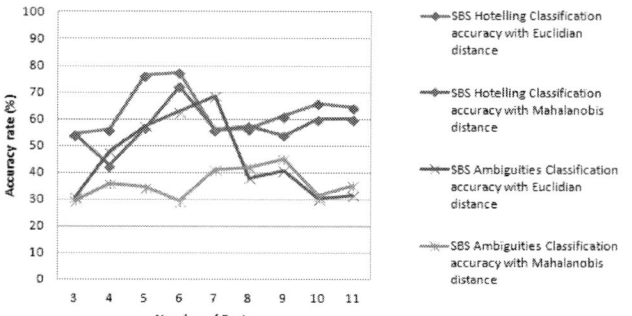

Fig. 5.   Synthesis of the accuracy rates achieved with kNN

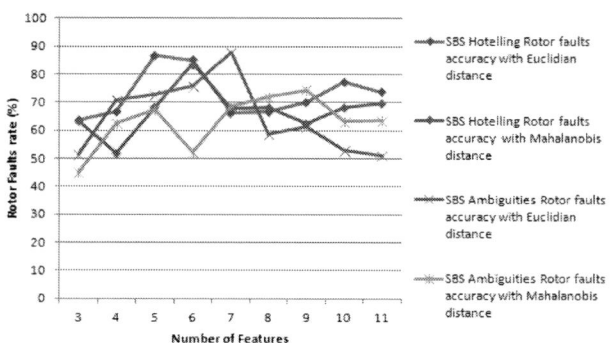

Fig. 6.   Synthesis of the rotor faults rate achieved with kNN

Table IV is a summary of the best accuracy rates induced by all methods presented in this paper. The last column corresponds to the use of the classifier in the first column as the classifier and the selection method score. The number of features, the feature selection method and the percentage associated with cross-validation are not the same for each result. Only the best ones, for each method, are showed in Table IV. The worst rate, 42.5%, corresponds to SVM and FDA. The best accuracy rate of 84.6% has been performed using Linear Discriminant Analysis as the classification method and FDA as feature extraction method. Table V shows the classes that have not been found with the combination of the two last methods. It seems that the no load cases are hard to predict for the unserious faults. We can conclude that when there is no machine load, it is very difficult to identify the flaws or a healthy functional state.

TABLE IV
COMPARISON OF THE ACCURACY RATE OBTAINED BY CLASSIFICATION METHODS COMBINED WITH FEATURE SELECTION AND EXTRACTION

	Extraction	Selection			
		Filter			Wrapper
	FDA	Fisher	Hotelling	Ambi.	Classifier
LDA	84.6%	78.9%	80.6%	66.61%	79.1%
kNN	79%	56.3%	77%	69.1%	76.6%
SVM	42.5%	60.2%	76.8%	68.3%	65.1%

TABLE V
SIMPLIFIED CONFUSION MATRIX REPRESENTING THE CLASSES THAT ARE NOT CORRECTLY FOUND BY LDA AND FDA

classes	fault	load
ω_1	Healthy	no load
ω_6	Ecc1	no load
ω_{11}	Ecc2	no load
ω_{16}	Ecc3	no load
ω_{19}	Ecc3	10kVA
ω_{33}	StB4	25kW
ω_{34}	StB4	10kVA
ω_{38}	StB2B4	25kW
ω_{43}	Ecc4StA1	25kW

B. Discussion

This paper tends to improve the results shown in [ ] since the model is the same. Previously, the k nearest neighbor method has been computed, showing an accuracy rate of 77%. This last percentage results on kNN combined with a feature selection algorithm SBS and Hotelling criterion which induced 6 features. Table IV shows that our work outperforms the preceding one. Indeed, the accuracy obtained by 5 pattern recognition methods are higher than 77%. The best rate achieved by kNN method, SBS and Ambiguities criterion is about 69% in Fig. 6. The main drawback of this last criterion shows that the class of the prototype being labeled is not considered. In such case, the membership should be maximum when it corresponds to the class prototype. Accuracy rates shown in Table IV confirm this result, Ambiguities and Fisher's ones provide lower results in most cases than Hotelling criterion. Fisher's criterion does not consider feature subset as argument, which explains results. The best rate using kNN and SFS in order to build the classifier and select features is almost the same as kNN combined with Hotelling. The selected features depend on the accuracy rate provided by the classifier. Using kNN for both steps of pattern recognition induces the same mistake and does not discriminate classes.

Nevertheless for LDA and kNN, the best rate achieved is based on Factorial Discriminant Analysis. Combining features seems more efficient than selecting features. Finally the most confident classifier, among those presented in this paper, is Linear Discriminant Analysis. Almost all rates achieved by this last are higher or equal to the other classifiers. Best classification accuracy rate is 84.6%, corresponding to 550 test data well classified for 650 ones. Among this test data, only 100 test data are not correctly assigned and 9 classes have not been correctly identified. According to Table V, the no load cases are hard to predict. Because of the high concentration of classes, prediction seems more difficult. By combining LDA with FDA feature extraction scheme, the rotor fault rate corresponding to the 16 class problem is around 91%.

VIII. CONCLUSION

Pattern recognition has proven its efficiency to establish a diagnosis of turbo-generator rotor faults. While a signal extracted from a faulty machine is generally compared to the one given by a healthy one, pattern recognition links a fault to the healthy case and moreover, all other functional states. In addition, the statistical model allows a simple and complete automation of the diagnosis. It is based not only on the difference between two functional states (distance between classes) but on the shape of each class. Furthermore, physic processing followed by pattern recognition is a good way to establish a diagnosis. Nevertheless, algorithms from pattern recognition require a learning step where each class has to be defined as more precisely as possible. This step also selects or combines the available features to build the feature space in which the classifier operates.

Three classification methods have been used for five feature selection and extraction criteria. It should be interesting to compare these results to other pattern recognition methods in order to increase the accuracy rate associated to the data. Moreover global and local measures have been considered as the same by the different methods. Splitting features in two groups may increase the diagnosis performance. Finally, only one validation set has been treated, a cross validation of the two sets would point out the accuracy of our methods.

IX. REFERENCES

[1] M. Biet, "Rotor faults diagnosis using features selection and nearest neighbors rule: Application to a turbogenerator," *IEEE Trans. Industrial Electronics*, vol. 60, no. 9, pp. 4063 – 4073, Sep 2013.

[2] B. A. T. Iamamura, Y. Le Menach, A. Tounzi, N. Sadowski, and E. Guillot, "Study of static and dynamic eccentricities of a synchronous generator using 3d fem," *IEEE Trans. Magnetics*, vol. 46, no. 8, pp. 3516–3519, Aug 2010.

[3] C. Bruzzese, "A virtual instrument for on-line evaluation of alternator's shaft misalignments through icsva (internal current space-vector analysis)," *Proc. 2011 IEEE SDEMPED Conf.*, pp. 55–62, 2011.

[4] D. de Canha, W. A. Cronje, A. S. Meyer, and S. J. Hoffe, "Methods for diagnosing static eccentricity in a synchronous 2 pole generator," *IEEE Lausanne Power Tech*, pp. 2162–2167, Jul 2007.

[5] R. Fiser, D. Makuc, H. Lavric, D. Miljavec, and M. Bugeza, "Modeling, analysis and detection of rotor field winding faults in synchronous generators," *Proc. 2010 Electrical Machines Conf.*, pp. 1–6, 2010.

[6] M. Kiani, W.-J. Lee, R. Kenarangui, and B. Fahimi, "Detection of rotor faults in synchronous generators," *Proc. 2007 IEEE SDEMPED Conf.*, pp. 266–271, 2007.

[7] W. Shuting, L. Heming, L. Yonggang, and W. Yi, "The diagnosis method of generator rotor winding inter-turn short circuit fault based on excitation current harmonics," *Proc. 2003 Power Electronics and Drive Systems Conf.*, vol. 2, pp. 1669–1673, 2003.

[8] G. Stone, M. Sasic, J. Stein, and C. Stinson, "Using magnetic flux monitoring to detect synchronous machine rotor winding shorts," in *Proc. IEEE 2011 Petroleum and Chemical Industry Conf.*, 2011, pp. 1–7.

[9] S. Richard, J.-P. Ducreux, and A. Foggia, "A three dimensional finite element analysis of the magnetic field in the end region of a synchronous generator," in *Proc. IEEE Electric Machines and Drives Conf.*, 1997, pp. WC2/4.1–WC2/4.3.

[10] R. Casimir, E. Boutleux, and G. Clerc, "Fault diagnosis in an induction motor by pattern recognition methods," *Proc. 2003 IEEE SDEMPED Conf.*, pp. 294–299, Aug 2003.

[11] O. Ondel, E. Boutleux, and G. Clerc, "Application of pattern recognition method to the diagnosis in induction machine," *Proc. 2006 on Electrical Machines Conf.*, pp. 2–5, Sep. 2006.

[12] D. Albright, "Interturn short-circuit detector for turbine-generator rotor windings," *IEEE Trans. Power Apparatus and Systems*, vol. 90, no. 2, pp. 478–483, 1971.

[13] R. Tibshirani, J. Friedman, and T. Hastie, *The Elements of Statistical Learning.* Springer, 2009.

[14] A. Jain, R. Duin, and J. Mao, "Statistical pattern recognition: a review," *IEEE Trans. Pattern Analysis and Machine Intelligence*, vol. 22, no. 1, pp. 4–37, 2000.

[15] D. Semani, C. Frelicot, and P. Courtellemont, "Une approche de type filtrage pour la sélection de variables. application à la reconnaissance automatique de poissons," *GRETSI 2005 (french)*, 2005.

[16] M. Kudo and J. Sklansky, "Comparison of algorithms that select features for pattern classifiers," *Pattern Recognition*, vol. 33, no. 1, pp. 25–41, 2000.

Alexandre Bacchus received his Master's degree in numerical analysis from the University of Pierre et Marie Curie (Paris VI) in 2012. He is currently working towards his Ph.D in electrical engineering at the University of Science and Technologies of Lille (USTL), Villeneuve D'Ascq, France, and in the EDF Research and Development division (R&D), Clamart, France. His research interests include the diagnostic of rotor faults in synchronous machines using pattern recognition methods.

Mélisande Biet is graduated from the Hautes Etudes Industrielles, France, in 2004, received an M.Sc. degree in electrical engineering in 2004 from the USTL, and a Ph.D. degree from the USTL, France, in 2007. She is currently with EDF R&D. Her main field of interest now deals with condition monitoring, diagnosis tools and numerical modeling of electrical machines.

Ludovic Macaire is presently full professor in the LAGIS Laboratory at the USTL. His research interests include pattern recognition applied to image segmentation and retrieval.

Yvonnick Le Menach was born in Brittany, France in 1970. He received his PhD in Electrical Engineering in 1999 from the USTL. From 1999 to 2002 he was assistant professor in University of Artois, France. Since 2002, he is working as assistant professor in the L2EP (Laboratory of power electronics and electrical engineering), USTL. His research work concerns the numerical techniques for the Finite Integration Technique and the Finite Element Method applied to the electromagnetic systems.

Abdelmounaïm Tounzi (member IEEE), was born in Casablanca (Morocco) in 1965. He graduated from the University of Nancy, France (M's 1989) and the Institut National Polytechnique de Lorraine, France (Ph.D 1993). From 1993 to 2008, he was an associate professor at the USTL, France and meme of the L2EP. Currently, he is full professor at the same university. His research areas are the design and modeling of electromagnetic systems.

Neural Approach for Bearing Fault Detection in Three Phase Induction Motors

W. S. Gongora, H. V. D. Silva, A. Goedtel, W. F. Godoy, S. A. O. da Silva

Abstract -- **The induction motor has been widely used in various industrial applications. Thus, several studies have presented strategies for the diagnosis and prediction of failures in these motor. One strategy used recently is based on intelligent systems, in particular, artificial neural networks. The purpose of this paper is to present an alternative tool to traditional methods for detection of bearing failures using on a perceptron network with signal analysis in time domain. Experimental results are presented to validate the proposal.**

Index Terms -- **Artificial Neural Networks, Failure prediction, Three phase induction motors.**

I. INTRODUCTION

Current technology context allows us to state that the Three Phase Induction Motor (TIM) is the primary means of transformation of electrical into mechanical energy. Allied to its factors of favoritism already consolidated such as robustness and low cost, it may become great part of industrial applications and draw attention to the production of components and accessories allowing a variety of employment and operation.

As any electrical machine, it requires proper maintenance, once failures compromise entire production causing large losses at industrial process. In [1] it was estimated that the maintenance costs may be listed 15% to 40% of total production. Based on this approach, some studies present maintenance as an important point to be considered and invested in order to minimize process and improve industrial processes performance. Specific techniques present detection models of a particular type of fault by using specialist system [2-5].

This work was supported by Fundação Araucária de Apoio ao Desenvolvimento Científico e Tecnológico do Paraná (Process Nr. 06/56093-3), Conselho Nacional de Desenvolvimento Científico e Tecnológico - CNPq (Process Nr. (Process Nr 474290/2008-5, Nr 473576/2011-2 and Nr 552269/2011-5) and Capes-DS.

W. S. Gongora is with the Department of Electrical Technician, IFPR Institute, Assis Chateaubriand, PR 85935-000 Brazil (wylliam.gongora@ifpr.edu.br).

H. V. D. Silva is with the Department of Electrical Engineering, UNOPAR University, Londrina, PR 86041-120 Brazil (hugolond@gmail.com).

A. Goedtel is with the Department of Electrical Engineering, UTFPR University, Cornélio Procópio, PR 86300-000 Brazil (e-mail: agoedtel@utfpr.edu.br).

W. F. Godoy is with the Department of Electrical Engineering, UTFPR University, Cornélio Procópio, PR 86300-000 Brazil (e-mail: wagnergodoy@utfpr.edu.br).

S. A. O. da Silva is with the Department of Electrical Engineering, UTFPR-CP University, Cornélio Procópio, PR 86300-000 Brazil (e-mail: augus@utfpr.edu.br).

The techniques of Model Predictive Control (MPC) are investigated since the 70s due to the development of preventive maintenance [6]. The traditional maintenance programs are designed to implement the idea of routine services of all machinery and quick responses to unexpected failures. As the opposite, a predictive maintenance program provides specified maintenance duties only when they are actually needed, minimizing to the maximum unexpected failures.

In [7] predictive maintenance is defined as a technique that indicates the actual operating conditions of machines based on data analysis that inform their wear or degradation processes. This technique can be adapted for a non-invasive model indicating the current operating condition of the process based on mechanical, electrical and electromagnetic gathering information.

The treatment of these problems in the three phase induction motor has a set of analysis well explored that is related to problems with bearings that compose the machine. This failure specifically, according to [8], can reach more than 40% of problems that occur in electric motors in general.

Several methods are applied to the fault detection such as mechanical vibration analysis, stator current frequency spectrum, axial flow and others that are showed by [9]. Basically, any kind of change in the bearings of machines cause excessive vibration, and this vibration can cause, among others, harmonic and changes.

The simplest way to detect refers to spectral analysis of the signals obtained from the specific sensor readings coupled to the machine frame at certain points and composed by a combination of acquisitions [10-13]. This procedure is highly reliable considering that each bearing failure has a specific frequency as per treated the aforementioned studies.

However, this method has a considered application cost as it requires specific equipment, installation, time of application and qualified manpower in reading and interpretation of the results. Also, valid procedures which evaluate the consequences that these oscillations produce are able to be incorporated into other types of signals removed from the machine. As per [12], which deals with the analysis of electrical current to the signals in the frequency domain or [14], where the procedure is the analysis of the components positive, negative and zero sequence.

Intelligent systems have been applied to solving various problems of control and drive machine [15-19]. These proposals do not require specific equipment but only

---

978-1-4799-0024-4/13 $31.00 © 2013 IEEE

requires the data directly from the sensors. These systems are capable of classifying determine the presence or absence of faults.

The intelligent systems applied to the diagnosis of machines are based on Artificial Neural Networks (ANN), Fuzzy Logic (FL), Hybrid Systems (HS), among others [20-23]. In the strategies for predicting faults in electrical machines [22] present a comparison of four structures of neural networks able to detect faults in the motor with mathematical analysis and manipulation of stator currents. Among other procedures with their specificities, [24] provides an algorithm able to classify how the two types of motor failures and the severity of the fault itself, based on the readings of the spectrum of the stator currents and machine vibration.

The purpose of this paper is to present a strategy to predict bearing failures in induction motors based on Artificial Neural Networks analysis with signal at time domain, only monitoring the electrical current in the stator power of an induction motor.

This article is organized as follows: Section 2 presents a description of the major faults in electric motors. Section 3 presents aspects of artificial neural networks in an overview. In Section 4 the methodology proposed in this article is presented with experimental results. Finally, in Section 5, the conclusions of the study are presented.

II. ASPECTS RELATED TO FAILURES IN ELECTRIC MOTORS

The induction motors, surely the most widely used in various productive sectors may present malfunctions. This failure can be divided into two major groups: i) electrical faults and ii) mechanical failures. Figure 1 shows a block diagram of the main fault types in which electrical faults are highlighted problems relating to stator winding, rotor winding, which are present in some models of motors; broken bars in the rotor, broken rotor rings; connections among others. Moreover, the mechanical failures may be derived from problems of bearings, eccentricity, wear coupling misalignment among others as reported by [8].

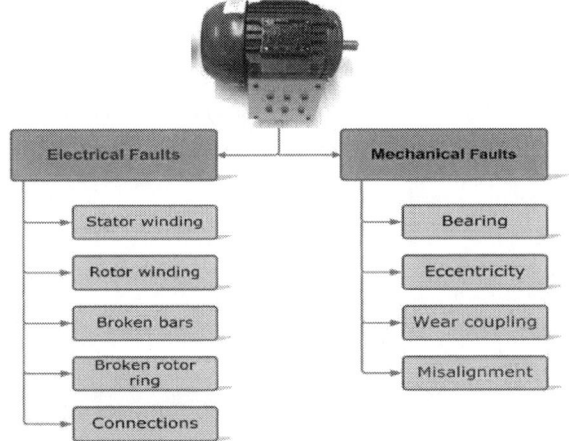

Fig. 1. Failures classification

Of the failures reported in the literature [24-25], it is estimated that the bearings are responsible for approximately 40% of the stops undesired of the electric motors, as can be followed in Figure 2.

Some methods described to predict motor failures are based on non-invasive strategies. This is due to the fact that the mechanical failures can be diagnosed by changes in the signals of the electric current of these motors as reported [26-27], or by analyzing the motor housing vibration [28-11].

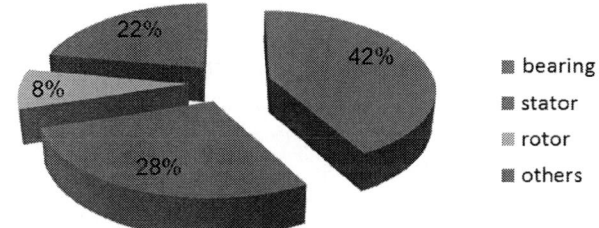

Fig. 2. Possibility of occurrence of failures in induction motors

As an example, Figure 3 (a) shows the normalized currents measured in a motor of 7.5 CV applied to the heating process of a sugar cane grinding with mechanical problems (bearings). This problem has been detected by the conventional method of analysis of mechanical vibration. Even the machine showing such disturbances, still in operation, there was collections of data and can be observed in detail that distortion exists between the motor currents.

After corrective maintainability, new measures of both mechanical vibration and current were taken and no vibrations were noticed. Thus, as showed in Figure 3 (b), it can be inferred the proper operation of the machine by restoring the standard current signal, as it was not observed distortions. Figure 3 (c) presents the motor under analysis in the industrial process.

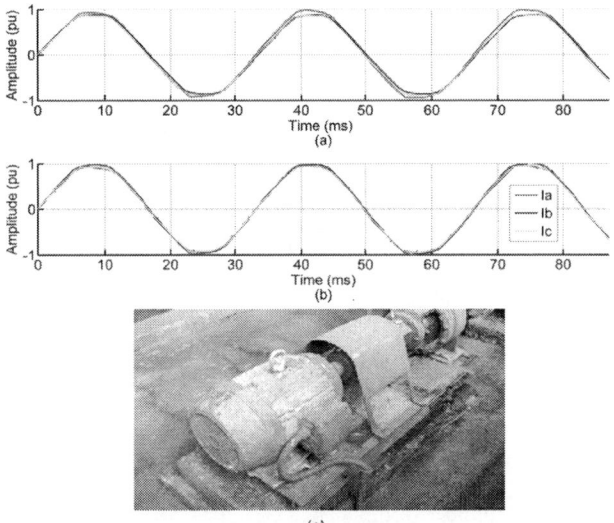

Fig. 3. (a) Current before maintenance (b) Current after maintenance (c) Motor under analysis.

978-1-4799-0024-4/13 $31.00 © 2013 IEEE

Other methods of signal analysis of the motor supply current are based on spectral frequency spectrum, which are characteristic of each mechanical failure [11-29]. However, problems related with electric power quality, which harmonic content is a result of feeding non-linear loads on the same power supply, can influence the analysis of these data and be interpreted as mechanical failures, as per reported in [23].

Thus, in the productive sectors it is observed the use of mechanical vibrations analysis or a combination of two analyzes: mechanical vibrations and the spectrum of the supply signal current of these motors.

III. MODELS OF NEURAL ARTIFICIAL NETWORKS

Identification using artificial neural network has shown promise for the solution of a series of problems involving power systems [30]. More specifically, the use of ANN has provided alternative schemes to handling problems related to electrical machines [22-24]. In this study, ANNs were applied to bearing fault identification in TIM.

For such purpose, a multilayer perceptron network was used, which was trained with a backpropagation algorithm [30]. This training algorithm has two basic steps: the first one, called propagation, applies values to the ANN inputs and verifies the response signal in its output layer.

This value is then compared with the desired signal for that output. The second step occurs in the reverse way, i.e., from the output to the input layer. The error produced by the network is used in the adjustment process of its internal parameters (weights and bias) [30].

The basic element of a neural network is the artificial neuron (Fig. 4), which is also known as the node or processing element.

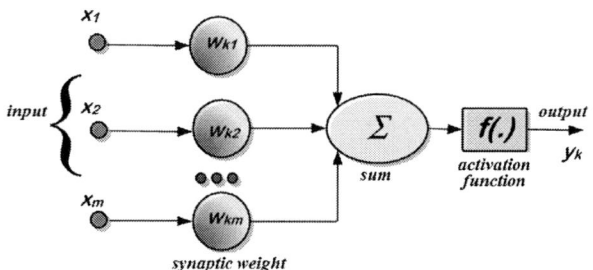

Fig. 4. Representation of the artificial neuron.

The artificial neuron illustrated in Fig. 4 can be modeled mathematically as follows:

$$v_j(k) = \sum_{i=1}^{n} X_i . w_i + b \qquad (1)$$

$$y_j(k) = \varphi_j(v_j(k)) \qquad (2)$$

where:

n is the number of input signals of the neuron;
X_i is the i-th input signal of the neuron;
w_i is the weight associated with the i-th input signal;

b is the threshold associated with the neuron;
$v_j(k)$ is the weighted response (summing junction) of the j-th neuron with respect to the instant k;
$\varphi_j(.)$ is the activation function of the j-th neuron;
$y_j(k)$ is the output signal of the j-th neuron with respect to the instant k.

The adjust of the network weights (w_j), associated with the j-th output neuron, is done by computing of the error signal linked to the k-th iteration or k-th input vector (training example). This error signal is provided by:

$$e_j(k) = d_j(k) - y_j(k) \qquad (3)$$

where $d_j(k)$ is the desired response to the j-th output neuron. Adding all squared errors produced by the output neurons of the network with respect to k-th iteration, we have:

$$E(k) = \frac{1}{2}\sum_{j=1}^{p} e_j^2(k) \qquad (4)$$

where p is the number of output neurons.

For an optimum weight configuration, $E(k)$ is minimized with respect to the synaptic weight w_{ji}. Therefore, the weights associated with the output layer of the network are updated using the following relationship:

$$w_{ji}(k+1) \leftarrow w_{ji}(k) - \eta \frac{\partial E(k)}{\partial w_{ji}(k)} \qquad (5)$$

where w_{ji} is the weight connecting the j-th neuron of the output layer to i-th neuron of the previous layer, and η is a constant that determines the learning rate of the backpropagation algorithm.

The adjustment of weights belonging to the hidden layers of the network is carried out in an analogous way. The necessary steps for adjusting the weights associated with the hidden neurons can be found in [30].

IV. FAILURE IDENTIFICATION USING NEURAL NETWORKS

The proposed work consists in the use of stator current signals of an induction motor presented in time domain to an ANN capable to identify the existence or absence of bearing failures.

Differently from the traditional methods of mechanical vibration analysis, which require the installation of special sensors at various points of the machine in order to acquire the vibration signals, the method proposed in this work uses the data sampled by a digital oscilloscope of four isolated channels TPS 2014 Tecktronix model with current ferrules A622 100 Amp AC/DC.

This unit has a storage capacity which uses a memory card where the signals are recorded as a datasheet of 2500 points.

The sampling rate is variable according to the selector sec/div which is adjusted as a function of the signal displayed on the screen.

978-1-4799-0024-4/13 $31.00 © 2013 IEEE

TABLE II
ARTIFICIAL NEURAL NETWORK PARAMETERS

Type	Network 1	Network 2	Network 3
Architecture	Perceptron	Perceptron	MLP
Training	S	S	S
N° of layers	3	3	2
Neurons in the 1° layer	2	2	4
Neurons in the 2° layer	-	-	2
Training Algorithm	BP	BP+LM	BP
Network function 1° layer	HT	HT	LF
Network function 2° layer	-	-	HT
Output network function	Linear	Linear	Linear

(S) Supervised; (LM) Levenberg Maquardt; (LF) Logistic Function (HT) Hyperbolic Tangent

Based on the collected data and with a proper import routine as per described in Figure 6, these data are handled and evaluated in MATLAB.

The signals are separated by a half cycle and normalized by its peak value to hold then in the time domain disregarding machine scale.

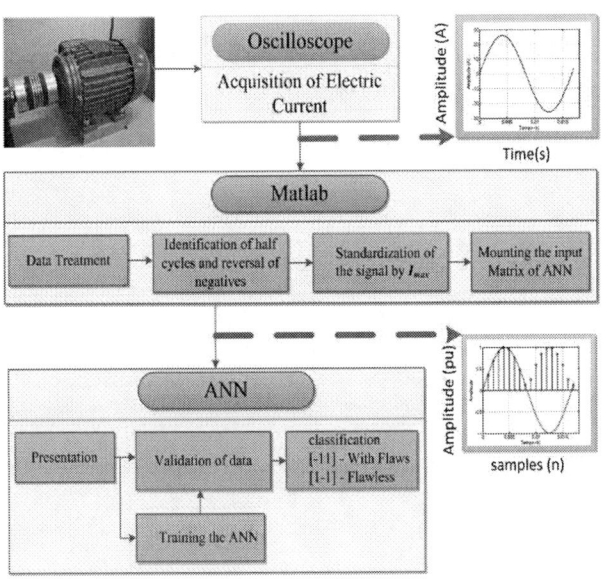

Fig. 6. Data processing routine

In a first step, by using traditional methods of mechanical vibration analysis bearings faults were identified in the motors mentioned in Table 3. Thus, signals of voltage and current were collected at industrial process, during normal operation. It should be emphasized that these machines are connected directly in the line and with no inverter-fed.

TABLE III
MOTORS USED FOR DATA ACQUISITION

Power	Application	Bearing (front)	Bearing (back)
12,5 HP	Mill	6308	6307
7,5 CV	Heating system	6307	6306
1 CV	Laboratory	6204	6203

Motors of 12,5 HP and 7,5CV are used in the first stage of sugar cane milling processes and the 1CV motor, is used in the laboratory. The laboratory motor usage aims to integrate signals increasing the universe of samples with different motors of different powers. Data related with laboratory motor was collected with no bearing failure, being considering a signal with low harmonic distortion. As for the motors used in the milling process, data was collected during normal operation at bearing fault condition.

After proper maintenance, by replacing defective bearings and also cleaning the machine, a further vibration evaluation were done to confirm bearing status, and a new set of data were acquired, assuming that these signals were obtained under normal operating conditions.

A. Input Data Treatment

The current signals of each phase of the various failed motors, were individually collected by the oscilloscope during normal steady state operation. Both motors were assessed before maintenance and again, after replacing the bearings, it was necessary a processing data routine and split training and validation sets of the networks.

The amplitude value of each sample point of the three phase stator current of the motors under analysis are presented as the network input. This method was proposed by [32] where it is considered as the input signal a sinusoidal waveform in continuous time. In this application, each half cycle is divided into a number of samples required to be submitted to the network, thus making the linear signal discretization, as showed in Figure 7.

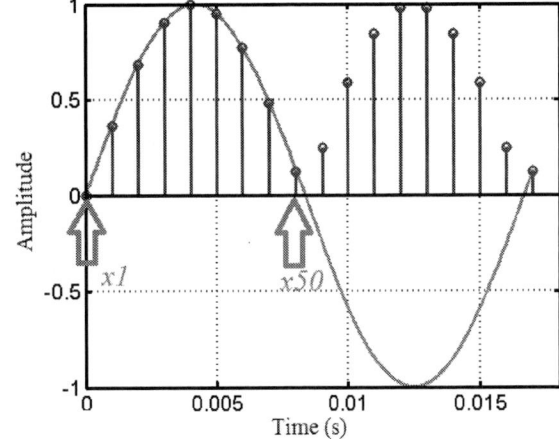

Fig. 7. Input data organization

978-1-4799-0024-4/13 $31.00 © 2013 IEEE

The 2500 current signal points obtained from the oscilloscope with its setting in 10ms/div, totaling 12 half cycles of wave sampled, each with 208 points.

In order to simplify the structure of the proposed network, each half cycle is sampled in 50 points that is capable of recreating the sinusoidal signal.

Due to the current delay and analysis occur for each half cycle, it is considered the absolute values of the sampled signals only getting positive half cycles of the signal as showed in Figure 7.

Another aspect related to the data treatment is the fact that signals representing the currents of a three-phase machine. Thus, it is necessary to mount a column vector with points collected each phase of the collected system so as to disregard the delay between the phases.

Each sample is mounted with the 50 points of the subsequent semi cycle of each phase from one another, thus creating an array of 150 entry points to the network.

When dealing to the use of real signs of electrical currents from machines of different power and different functional states, it is necessary to perform the normalization of these data.

This normalization also performed by processing the data algorithm, consider the peak value of the waveform of each sampled signal. Figure 8 shows the normalized experimental curves.

It is worth mentioning here that the commented standardization satisfies this application and performed work. Differentiated values thereof may be applied to other processes and respond as expected.

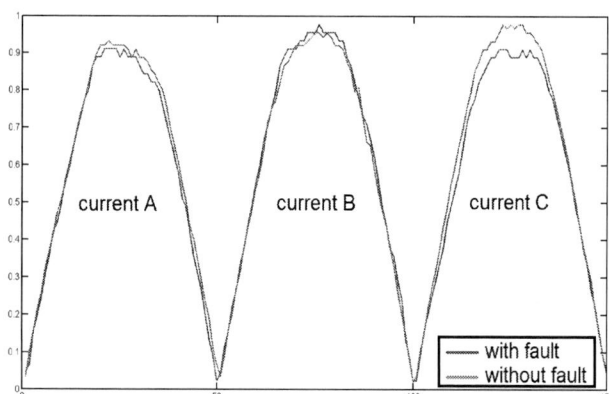

Fig. 8. Standardized inputs

In order to work with the proposed networks only with information collected from real applications without simulation results, sampling vectors were randomly divided into two classes, one for training and one for validation, and they were divided according to Table 4. As can be observed, 26 samples are used for training the network, totaling 66% of the total, and about 33% for validation, totaling 14 samples. The validation samples are compared to the results of the proposed topologies.

TABLE IV
DIVISION OF COLLECTED DATA

Classes	(%)	Samples
Training samples	66	26
Validation samples	33	14

B. Robustness Test - Noise Signal

In order to carry out the network performance evaluation in front of an interference on the presented signs, it was inserted into the sampled signal a noise background. This noise composed of a random value, was considered the worst possible case where its frequency is equals to the frequency discretization of the treated signal. Thus the entire value presented to the network input was altered, which can be noticed in Figure 9 (a).

Thus, it was developed a routine that generates a noise. This routine establishes random values between -1 and 1. These values are attenuated when multiplied by a maximum factor of distortion, thus allowing easy modification and verification of the maximum interference value which the network remain immune.

Considering that the network input is normalized between 0 and 1, when added noise, it suffers changes proportional to the maximum distortion factor by which noise was submitted. This can be observed in Figure 9 (b).

The robustness test procedure was not applied to training step. The insertion of noise was accomplished only to the network validation data. Thus, obtaining the maximum value of interference for which do not alter the network classification result comes close to 3.70% of the normalized value of the input signal.

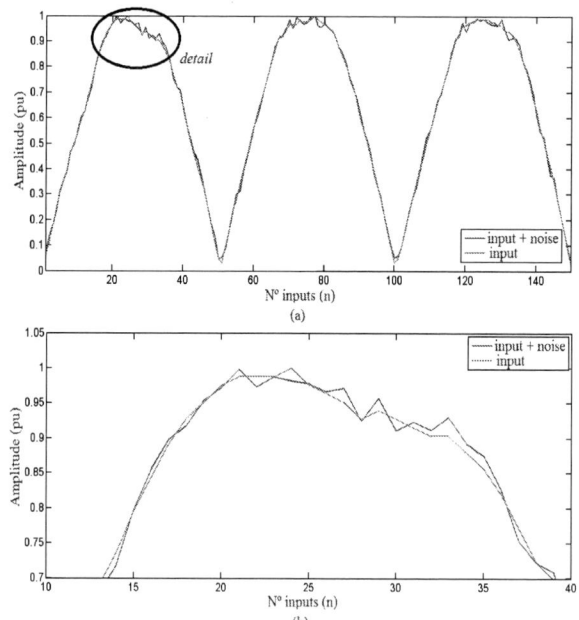

Fig. 9. Noise input

V. Experimental Results

The proposed networks were subjected to training with the same input signals having a learning rate of $\eta = 0.5$. As stopping criterion it was established the mean squared error (MSE), defined for each network as showed in Table 5. The Network 1 converged with 74 epochs, while Network 2 has reached the stopping criterion with 38 epochs. As for the Network 3 it was necessary 7453 epochs, however, after concluding the network training this high number of epochs will not impact on the network performance.

TABLE V
FINAL RESULTS

Type	Net 1	Net 2	Net 3
Training samples	26	26	26
Validation samples	14	14	14
AQE	$1\ 10^{-1}$	$1\ 10^{-8}$	$1\ 10^{-1}$
Learning Coefficient	0,5	0,5	0,5
Epochs	74	38	7453
Positive false	5	0	0
Negative false	0	0	0
Classification error	5/14	0/14	0/14
Accuracy percentage	60,71%	100%	100%

After analyzing the proposed models, it is possible to compare the methods showed in Table 5, whose network architectures are described in Table 2. For Network 1 it was not obtained satisfactory results, with the best performance of 60% for data not presented the network, thus determining a low generalization capacity and infeasibility of its implementation.

In the Network 2, which uses Levenberg-Maquardt algorithm, it was achieved 100% of accuracy for the validation data. When compared with Network 1 is also observed a reduction in training epochs and mean squared error. Network 3 shows a larger structure and also a larger number of neurons applied to the processing. This network showed a percentage of 100% accuracy with results as efficient as the Network 2.

VI. Conclusions

The purpose of this work is achieved with the test of two different topologies able to classify the existence of bearing failure, by analyzing only the stator current of the Three-Phase Induction Motor in the time domain.

Is worth mentioning that the application of such prediction routine can be used in a variety of potencies and also in various operating regimes.

Proposed model estimates the severity of the bearing fault since the network was trained to classify such state, similar to the existing process. This is due to the fact that the signals collected, as well as the traditional methods, identifies the existence of bearing faults in accordance with a range of acceptable vibration not determining the degradation of the piece.

Other possibility is to consider this same ANN to identify other types of faults. This evaluation was not addressed in this paper due to the usage of signals collected in the field.

The proposed methodology presents a direct result of the failure classification and does not use specific equipment such as vibration analyzers and others. Data processing is optimized and applicable to any level of model predictive control. For a possible hardware implementation, proposed networks have compact structures facilitating its application in low-cost processors.

VII. References

[1] L. M. R. Baccarini,"Detection and diagnosis of faults in induction motors". Ph.D. dissertation (in Portuguese), Univ. Federal de Minas Gerais, 2005.

[2] Arkan, M., Perovic, D. K. e Unsworth, P. (2001). Online stator fault diagnosis in induction motors, IEEE Proceedings: Electric Power Applications 148: 537-547.

[3] S.Yang, "A condition-based failure prediction and processing-scheme for preventive maintenance", *Reliability, IEEE Transactions on* 52(3): 373-383, 2003.

[4] S. M. A Cruz, and A. J. M Cardoso, "Diagnosis of stator inter-turn short circuits in dtc induction motor drives", *IEEE Transactions on Industry Applications* 40(5): 1349-1360, 2004.

[5] M. F. S. V. D'Angelo, R. M. Palhares, R. H. C Takahashi, R. H. Loschi, L. M. R Baccarini, and W. M. Caminhas, "Incipient fault detection in induction machine stator winding using a fuzzy-bayesian change point detection approach", *Applied Soft Computing Journal* 11(1): 179 192, 2011.

[6] W. H. Kwong, Introduction to predictive control with MATLAB (in Portuguese), São Carlos: EdUFSCar, 2005.

[7] V. C. Souza, *Organization and Management of Maintenance* (in Portuguese), São Paulo: All Print, 2009.

[8] G. Singh, and S. A. Kazzaz, "Induction machine drive condition monitoring and diagnostic research a survey", *Electric Power Systems Research* 64(2): 145-158, 2003.

[9] C. J. Verucchi, and G. G. Acosta, "Fault detection and diagnosis techniques in induction electrical machines", *Latin America Transactions, IEEE (Revista IEEE America Latina)* 5(1): 41-49, 2007.

[10] T. Omar, N. Lahcene, I. Rachild, and F. Maurice, "Modeling of the induction machine for the diagnosis of rotor defects. Part i. an approach of magnetically coupled multiple circuits", *Industrial Electronics Society, 2005. IECON 2005. 31st Annual Conference of IEEE*, p. 8, 2005.

[11] M. Blodt., P. Granjon, B. Raison, G. Rostaing, "Models for bearing damage detection in induction motors using stator current monitoring" Industrial Electronics, *IEEE Transactions on* 55(4): 1813-1822, 2008.

[12] R. Araujo, R. Rodrigues, H. Paula, and L. Baccarini, " Premature wear of bearings and recurring failures of a three-phase induction motor: Case Study"(in Portuguese), *Industry Applications (INDUSCON), 2010 9th IEEE/IAS International Conference on*, pp. 1-6, 2010.

[13] M. Tsypkin, "Induction motor condition monitoring: Vibration analysis technique - a practical implementation", *Electric Machines Drives Conference (IEMDC), 2011 IEEE International*, pp. 406-411, 2011.

[14] J. Sottile, F. Trutt, and J. L. Kohler, "Experimental investigation of on-line methods for incipient fault detection in induction motors", *Proceedings of the Industry Application Conference* pp. 2682-2687, 2000.

[15] N. Muthuselvan, S. Dash, and P. Somasundaram, "A high performance induction motor drive system using fuzzy logic controller", *TENCON 2006. 2006 IEEE Region 10 Conference*, pp. 1-4, 2006.

[16] A. Goedtel, "Neural speed estimator for three phase induction motors " (in Portuguese), Ph.D. dissertation, São Carlos: Escola de Engenharia de São Carlos, Univ. de São Paulo, 2007.

[17] G.-Y. Li, J.-R Wan, Y.-P. Liu, S. Hong, and C.-H. Yuan, "Permanent magnet synchronous motor direct torque control with zero vector based on intelligent method", *Machine Learning and Cybernetics, 2009 International Conference on*, Vol. 2, pp. 755-760, 2009.

[18] T. H. Santos, A. Goedtel, I. N. Silva, and M. Suetake, "A neural speed estimator in threephase induction motors powered by a driver with scalar control", *Power Electronics Conference (COBEP), 2011 Brazilian*, pp. 44-49, 2011.

[19] Y. Sayouti, A. Abbou, M. Akherraz, and H. Mahmoudi, "Sensor less low speed control with ann mras for direct torque controlled induction motor drive", *Power Engineering, Energy and Electrical Drives (POWERENG), 2011 International Conference on*, pp. 1-5, 2011.

[20] B. Samanta, and K. Al-Balushi, "Artificial neural network based fault diagnostics of rolling element bearings using time-domain features", *Mechanical Systems and Signal Processing* 17(2): 317-328, 2003.

[21] B. Sreejith, A. Verma, and A. Srividya, "Fault diagnosis of rolling element bearing using time-domain features and neural networks", *Industrial and Information Systems, 2008. ICIIS 2008. IEEE Region 10 and the Third international Conference on*, pp. 1-6, 2008.

[22] V. Ghate, and S. Dudul, "Fault diagnosis of three phase induction motor using neural network techniques", *Emerging Trends in Engineering and Technology (ICETET)*, 2009 2nd International Conference on, pp. 922-928, 2009.

[23] M. Seera, C. P. Lim, D. Ishak, and H. Singh, "Fault detection and diagnosis of induction motors using motor current signature analysis and a hybrid fmm - cart model", *Neural Networks and Learning Systems, IEEE Transactions on* 23(1): 97-108, 2012.

[24] C. T. Kowalski, and T. Orlowska-Kowalska, "Neural networks application for induction motor faults diagnosis", *Mathematics and Computers in Simulation* 63(3-5): 435-448, 2003.

[25] T. Han, B.-S. Yang, and Z.-J. Yin, "Feature based fault diagnosis system of induction motors using vibration signal", *Journal of Quality in Maintenance Engineering*, Vol. 13, pp. 163-175, 2007.

[26] R. Schoen, T. Habetler, F. Kamran, and R. Bartheld, "Motor bearing damage detection using stator current monitoring", *Industry Applications Society Annual Meeting, 1994., Conference Record of the 1994 IEEE*, Vol. 1, pp. 110-116, 1994.

[27] W. Saadaoui, and K. Jelassi, "Induction motor bearing damage detection using stator current analysis", *Power Engineering, Energy and Electrical Drives (POWERENG)*, International Conference on, pp. 1-6, 2011.

[28] W. Finley, M. Howdowanec, and W. Holter, "Diagnosing motor vibration problems", *Pulp and Paper Industry Technical Conference*, 2000.

[29] E. Mendel, L. Z. Mariano, I. Drago, S. Loureiro, T. W. Rauber, F. M. Varejão, and R. Batista, " Automatic recognition of patterns bearing failures using vibration signal analysis" (in Portuguese), *XVII Congresso Brasileiro de Automática*, 2008, Juiz de Fora. Anais do XVII Congresso Brasileiro de Automática, 2008 1(1).

[30] S. Haykin, *Redes Neurais: Princípios e Práticas*, 2ª edição, Porto Alegre : Bookman, 2001.

[31] M. Hagan, and M. Menhaj, "Training feedforward networks with the marquardt algorithm", *Neural Networks, IEEE Transactions on* 5(6): 989-993, 1994.

[32] C. F. Nascimento, Jr. A. A. Oliveira, A. Goedtel, and P. J. A. Serni, "Harmonic identification using parallel neural networks in single-phase systems", *Applied Soft Computing* 11(2): 2178-2185.

VIII. .BIOGRAPHIES

Gongora, W. S. was born in Cascavel, Brazil in 1984. He received the B.S degree in control and automation engineering from the Faculdade Assis Gurgacz (2007) and the M.Sc. degree at electrical engineering from Universidade Tecnológica Federal do Paraná (2013). He is a professor at the Federal Institute of Paraná (IFPR), campus Assis Chateaubriand. Working mainly on the themes: artificial neural networks, fault prediction, three-phase induction motor.

Silva, H. V. D. Graduated in Computer Eng. Universidade do Norte do Paraná (2009), Specialization in Automation and Process Control Industrial Technology from the Universidade Tecnológica Federal do Paraná - Campus Cornélio Procópio (2011). He is currently a professor at the Universidade do Norte do Paraná - UNOPAR and masters student in Electrical Engineering from the Universidade Tecnológica Federal do Paraná - Campus Cornélio Procópio

Goedtel, A. was born in Arroio do Meio, Brazil, in 1972. He received the B.S. degree in electrical engineering from the Federal University of Rio Grande do Sul, Porto Alegre, Brazil, in 1997, the M.Sc. degree in industrial engineering from the São Paulo State University (UNESP), São Paulo, Brazil, in 2003, and the Ph.D. degree in electrical engineering from the University of São Paulo (USP), São Paulo, in 2007. Currently, he is an Assistant Professor with . the Federal Technological University of Paraná, Cornélio Procópio, Brazil. His research interests are within the fields of electrical machinery, intelligent systems, and power electronics.

Godoy, W. F. was born in Cornélio Procópio, Brazil, in 1977. He received the B.S. degree in electrical engineering from the University Norte do Paraná, Londrina, Brazil, in 2003, the M.Sc. degree in eletrical engineering from the University of Londrina, Londrina, Brazil, in 2010. Currently, he is an Assistant Professor with the Federal Technological University of Paraná, Cornélio Procópio, Brazil. His research interests are within the fields of electrical machinery, intelligent systems, and electrical maintenance.

Silva, S. A. O. was born in Joaquim Távora, Brazil, in 1964. He received the B. S. and M. S. degrees in electrical engineering from Federal University of Santa Catarina, Florianopolis, Brazil, in 1987 and 1989, respectively. He received the Ph.D. degree in electrical engineering from Federal University of Minas Gerais, Belo Horizonte, Brazil, in 2001. Since 1993, he has been with the Electrical Engineering Department in the Federal Technological University of Parana, where he is currently a Professor of Electrical Engineering. His present research involves UPS systems, Photovoltaic systems, Active power filters, Control systems and Power quality.

978-1-4799-0024-4/13 $31.00 © 2013 IEEE

Early Detection of Unbalance Voltage in Three Phase Induction Motor Based on SVM

D. R. Sawitri, D. A. Asfani, M. H. Purnomo, I. K. E. Purnama, M. Ashari

Abstract -- **Unbalance voltage supply in induction motor is a crucial problem. This paper proposes an original system to detect an unbalance voltage condition in induction motor using Support Vector Machine (SVM). Induction motor current, as a signal due to the unbalance voltage supply in three-phase induction motor, is recorded in lab bench. The features of recorded signals are extracted by wavelet transform and Principal Component Analysis algorithm. Then, the quality of detection is classified by SVM, and the average result of detection is 86%.**

Index Terms--**Unbalance Voltage, Induction Motor, Support Vector Machine (SVM)**

I. NOMENCLATURE

$\underline{i}$ :: Current vector (A)

$\tilde{I}$: Current phasor (A)

p : Instantaneous active power (W)

P_0 : Average active power (W)

qds : Stator oriented reference frame

sn : Negative sequence

sp : Positive sequence

t : Time (s)

T_0 : Average Torque (N/m)

$\tilde{V}$: Voltage phasor (V)

$\underline{v}$: Voltage vector (V)

ω_e : Electrical angular frequency (rad/s).

II. INTRODUCTION

Induction motor is critical device in industrial process. Fault in induction motor causes loss due to disruption of the production process. Beside influenced by internal factor, faults in induction motor are also influenced by external factor. The internal factor, such as inter turn short circuits, broken rotor bar, or the damage of the bearings.

Induction motor faults due to external factor are influenced by the input of power quality. One is the voltage unbalance.

In three phase system, voltage unbalance occurs when the phase or the magnitude of each line and or phase angel is different from a balanced condition. Unbalance voltage will give effect to induction motor performance. [1]. The small unbalance voltage may generate a large negative sequence. This can lead to increase the heat of stator winding and rotor bar. Unfortunately, the heat is not producing the useful power. [2].

Identification of unbalance voltage in induction motor is needed. It can minimize the greater losses. Fault identification in induction motor can be done by analyze the current spectrum. Past research on voltage unbalance is reported by [3]. Voltage unbalance identified based vector Park component and then it is analyzed using Fast Fourier Transform (FFT). The other research for identified voltage unbalance is reported by [4]. This study identified external faults in induction motor include voltage unbalance using Artificial Neural Network (ANN).

Current spectrum has widely been used to detect a fault in induction motor. Motor current signature analysis (MCSA) have been used to detect the broken rotor bar with Discrete Fourier Transform method [5], [6]. Current spectrum analyze is also used to detect short circuits in stator winding [5] and air gap eccentricity [7]. Several previous studies have applied intelligent system to detect fault in induction motor. The methods were applied to include fuzzy logic [8], [9], [10] FFT [8], [10], [11], [12], ANN [8], Support Vector Machine (SVM) [7], [13], [14], and Kalman Filter [15].

Unbalance voltage is difficult to detect by current signal due to the similarity between one type signals with another. However, the small unbalance voltage if not immediately identified may cause large losses. The loss is caused by the energy waste due to the heat generate. If it is not handled immediately then the unbalance voltage may cause damage to the motor, so the production process must be terminate. Unbalance voltage is not easily identified by current spectrum analysis. By applied an intelligent system, it is expected the process of identification become easier. This paper will discuss identification unbalance voltage in induction motor using Support Vector Machine (SVM).

This work was supported in part by Directorate General of Higher Education, Department of Education and Culture, Indonesia

D. R. Sawitri is with Electrical Engineering Departmen, Dian Nuswantoro University, Indonesia (e-mail: dianrs@elect-eng.its.ac.id; drsawitri@gmail.com).

D.A. Asfani, M. H. Purnomo, I. K. E. Purnama, and M. Ashari is with the Department of Electrical Engineering, Institut Teknologi Sepuluh Nopember, Indonesia (e-mail:: anton@ee.its.ac.id; hery@ee.its.ac.id; ketut@ee.its.ac.id; ashari@ee.its.ac.id).

978-1-4799-0024-4/13 $31.00 © 2013 IEEE

III. UNBALANCE VOLTAGE

A. Definition

There are 3 definition of unbalance voltage, which are NEMA definition, IEEE definition, and True definition. These definitions describe as follows [1].

NEMA Definition (National Equipment Manufacturer's Association) about unbalance voltage is also called with Line Voltage Unbalance Rate (*LVUR*).

$$\%LVUR = \frac{\text{max voltage deviation from avg line voltage}}{\text{avg line voltage}}.100$$

IEEE Definition: IEEE definition about unbalance voltage also called with phase voltage unbalance rate (*PVUR*) is defined as:

$$\%PVUR = \frac{\text{max voltage deviation from avg phase voltage}}{\text{avg phase voltage}}.100$$

IEEE definition is similar with NEMA definition. The difference is that IEEE use phase voltage and NEMA use information about voltage line-to-line. Information about phase angel not appears in those equations.

True Definition: Unbalance voltage is defined as ratio of negative sequence voltage with positive sequence voltage. The value is defined as the percentage of voltage unbalance factor (*VUF*).

$$\%VUF = \frac{\text{negative sequence voltage component}}{\text{potitive sequence voltage component}}.100$$

Components voltage of positive and negative sequence is obtained from unbalance of each phase. V_{ab}, V_{bc}, and V_{ca} are described into symmetric component V_{sp} and V_{sn} (from each phase). Two balanced components are definite in (1) and (2).

$$V_{sp} = \frac{V_{ab} + a.V_{bc} + a^2.V_{ca}}{3} \tag{1}$$

$$V_{sn} = \frac{V_{ab} + a^2.V_{bc} + a.V_{ca}}{3} \tag{2}$$

$$a = 1\angle 120^o \ dan \ a^2 = 1\angle 240^o$$

Positive and negative sequences voltage can be used to analyze the behavior induction motor in unbalance state.

In general, effect of unbalance voltage in induction motor according to NEMA, equivalent to that occurring with negative sequence voltage. Rotation occurs in the opposite direction to that the voltage balance [16].

B. Effect of Unbalance Voltage

Unbalance voltage in induction motor with small percentage will result the greater percentage of unbalance current. The temperature will rise. When the motor is an operation, percentage of unbalance voltage will be greater than the balance voltage [16]. Unbalance voltage will be effect in torque and motor power. Unbalance voltage produce the oscillation of power when the supply frequency twice normal condition. This condition causes additional oscillation of torque and causes vibration. [17].

If the motor is supplied with voltage that is not balanced, the voltage and current in steady state describe in (3) and (4).

$$\underline{v}_{qds} = \tilde{V}_{sp} e^{j\omega_e t} + \tilde{V}_{sn}^* e^{-j\omega_e t} \tag{3}$$

$$\underline{i}_{qds} = \tilde{I}_{sp} e^{j\omega_e t} + \tilde{I}_{sn}^* e^{-j\omega_e t} \tag{4}$$

Based on equation of voltage and current, can be defined equation to calculate the active power (5).

$$P = \frac{3}{2}\text{Re}\left[\underline{v}_{qds}\,\underline{i}_{qds}^*\right] \tag{5}$$

With substitution (3) and (4) in (5), obtained the following result:

$$P = \frac{3}{2}\text{Re}\left(\tilde{V}_{sp}\tilde{I}_{sp}^* + \tilde{V}_{sn}^*\tilde{I}_{sn}\right)$$
$$+ \frac{3}{2}\text{Re}\left(\tilde{V}_{sp}\tilde{I}_{sn}e^{j2\omega_e t}\right) + \frac{3}{2}\text{Re}\left(\tilde{V}_{sn}^*\tilde{I}_{sp}^*e^{-j2\omega_e t}\right) \tag{6}$$

In condition such as (6), active power of motor in unbalance voltage condition will cause the components oscillate at a frequency of 2 times the fundamental frequency. The first component will generate the greatest power amplitude. Amplitude depends on positive sequence voltage. Negative sequence voltage will also affect the magnitude of the power amplitude though not as positive sequence.

IV. EXPERIMENTAL SETUP AND DATA GENERATION

A. Data Generation

Three phase induction motor, 0.25 kW, 220/380 V, 1320 rpm, is running in unbalance voltage condition. Current

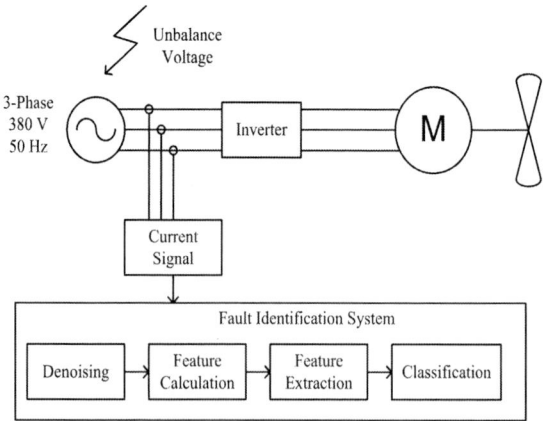

Fig. 1. Experimental Setup for identification of unbalance voltage condition in induction motor

Signal is recorded with oscilloscope 4 channels (Fig. 1). There are 5 conditions of unbalance voltage. The conditions are 1%, 2%, 3%, 4%, and 5%. Each condition consist 3 types of faults (unbalance voltage in phase A, B, and C). Next, the recorded data identified in a system with 4 steps. The steps are denoising signal, feature calculation, feature extraction, and classification.

B. Denoising

Denoising is necessary to remove the noise in the signal. Noise in the signal may raise bias. This paper use combination wavelet and Principal Component Analysis (PCA) [18] as denoising algorithm. The algorithm is shown in Fig. 2.

The process begins with wavelet transform at level 4 for each vector signal. At this stage, signals to be decomposed into detail signals (cDj) and approximation signals (cAj). Frequency band of each wavelet coefficient based on half band wavelet filter are shown in this figure. Using PCA, each coefficient detail and approximation are extracted to obtain the principal components. The principal components perform the simpler signals. Based on result of PCA, it reconstructs a new matrix trough invert wavelet transform. Next, it performs PCA analysis for the new matrix.

Recorded current signals in this experiment have a high noise due to the use of inverter. This noise must be removed to improve the accuracy of identification. Fig. 3 shows 4 types denoising current signal. The example current signal includes 1%, 3%, and 5% unbalance voltage conditions.

C. Feature Calculation

After denoising, feature parameters are calculated based on time domain (T) and frequency domain (F) see Table I.

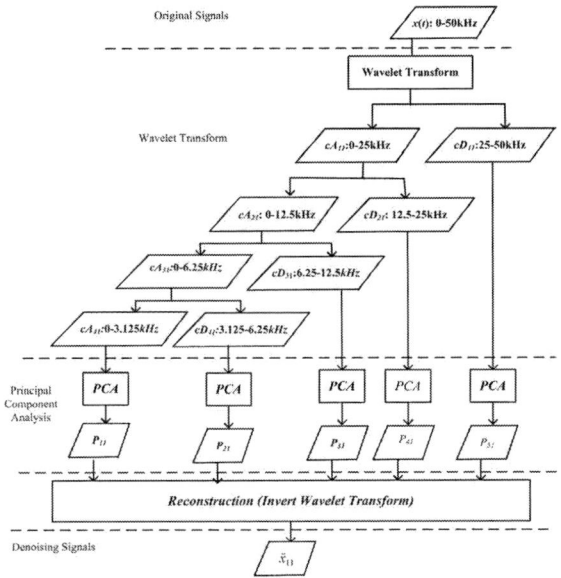

Fig. 2. Denoising algorithm using wavelet and PCA

TABLE I
FEATURE CALCULATION

No	Features	T	F	Equation
1	Mean	√	√	$\mu = \frac{1}{n}\sum_1^n x_i$
2	Variance	√	√	$\sigma^2 = \frac{1}{n-1}\sum_1^n (x_i - \mu)^2$
3	Deviation Standard	√		$\sigma = \sqrt{\frac{1}{n-1}\sum_1^n (x_i - \mu)^2}$
4	Modus	√		Most appear data
5	Median	√		Central value of data
6	Skewness	√		$\tau = \frac{\mu_3}{\sigma^3} = \frac{E(x-\mu)^3}{E\left[(x-\mu)^2\right]^{3/2}}$
7	Kurtosis	√		$K = \frac{\mu_4}{\sigma^4} = \frac{E(x-\mu)^4}{E\left[(x-\mu)^2\right]^2}$
8	Moment	√		$m_r = E(x-\mu)^r$
9	Upper Histogram	√		Max value of histogram
10	Lower Histogram	√		Min value of histogram
11	Linear Rank Correlation	√		$r_s(x_1.x_2) = 1 - \frac{6\sum_{i=1}^n \left(r_i^{x_1} - r_i^{x_2}\right)}{n(n^2-1)}$
12	Entropy	√		$H(X) = -\sum_{i=1}^M p_i \log p_i$
13	Rms	√	√	$I_{rms} = \frac{I_{max}}{\sqrt{2}}$
14	Number of Peak	√	√	Number of peak
15	Average of Peak	√	√	Average of peak
16	Harmonic		√	$I_H = \sqrt{I_2^2 + I_3^2 + \cdots + I_n^2}$ I_F= rms of fundamental current
17	THD		√	$THD = \frac{I_H}{I_F}$
18	Crest Factor		√	$crest\ factor = \frac{peak\ value}{RMS\ value}$
19	Form Factor		√	$form\ factor = \frac{RMS\ value}{Average\ value}$
20	Peak Maximum		√	Maximum of peak

There are 25 feature parameters of each phase (phase A, B, and C). We have 15 conditions and each one has 20 measurements, so the total obtained 300 data calculated. The value of features parameter calculated based statistical value of time domain and frequency domain. Table 1 show the equation of each feature.

D. Feature Extraction

Twenty five features from feature calculation will be selected to find the true feature. This study use Principal Component Analysis (PCA) to extract the features.

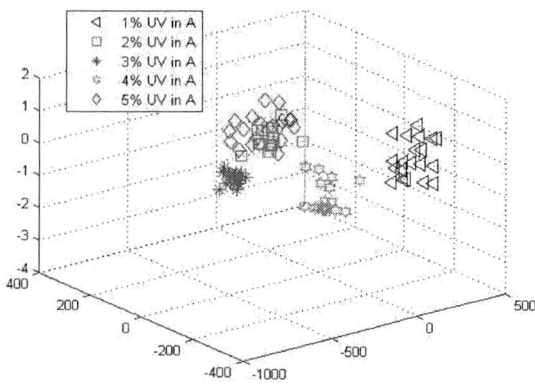

Fig. 4. Scatter Diagram for 3 Largest Principal Components

Feature extraction is done by calculating the eigenvalue and eigenvector from matrix covariance that formed in previous process. By taking the eigenvalue greater that zero, and discard the rest, we obtain the principal component. The final data is composed based on the value of eigenvector.

From the feature calculation process, we have 25 feature parameter based on time domain and frequency domain. Using PCA, only 14 of 25 parameters are selected as principal component. Selection is based on the largest eigenvalue.

Fig. 4 shows the scatter diagram of 3 biggest eigenvalue of principal component for voltage unbalance in phase A. It can be seen the data tends to distribute into groups based on type of fault. The condition of 1% and 2% voltage unbalance looks much closed. This condition is difficult to distinguish. For 3%, 4%, and 5% voltage unbalance seem to cluster so much easier to identify.

E. Classification with Support Vector Machine

In this paper, SVM is used to identify unbalance voltage in induction motor. All the measured data are classified into 15 types of fault. From each type, it is selected 30% as training data and 70% as testing data.

SVM maps the input vectors x into a high-dimensional features space Z through some nonlinear mapping [19]. In this space, an optimal separating hyperplane is constructed.

The optimal hyperplane separated without error and the distance between the closest vector to the hyperplane is maximal (Fig. 5). Suppose the training data

$$(x_i, y_i),...,(x_l, y_l), \quad x \in R^n, \quad y \in \{+1,-1\}$$

Can be separated by a hyperplane:

$$(w, x) + b = 0 \tag{7}$$

where, w and b shall be derived in such a way that unseen data can be classified correctly. To describe the separating hyperplane, the following canonical form can be use:

$$(w.x_i) + b \geq 1 \qquad if \ y = +1,$$

$$(w.x_i) + b \leq -1 \qquad if \ y_i = -1.$$

In the following we use a compact notation for these inequalities:

$$y_i\left[(w.x_i) + b \geq 1, \qquad i = 1,...,n\right] \tag{8}$$

It is easy to check that the optimal hyperplane is the one that satisfies the condition (8) and minimizes.

$$\Phi(w) = \|w\|^2 \tag{9}$$

(The minimization is taken with respect to both vector w and scalar b)

The learning machines construct the decision functions are called Support Vector Machines (SVM), that is nonlinear in the input space,

$$f(x) = sign\left(\sum y_i \alpha_i K(x_i, x) - b\right) \tag{10}$$

The functions are equivalent to linear decision functions in the high-dimensional feature space $\psi_1(x),...,\psi_N(x)$ ($K(x_i, x)$ is a convolution of the inner product for this feature space).

To find the coefficients $\square$i in the separable case (analogously in the non-separable case) it is sufficient to find the maximum of the functional.

$$W(\alpha) = \sum_{i=1}^{n} \alpha_i - \frac{1}{2}\sum_{i,j}^{n} \alpha_i \alpha_j y_i y_j K(x_i, x_j) \tag{11}$$

Subject to the constraints

$$\sum_{i=1}^{n} \alpha_i y_i = 0 \qquad \alpha_i \geq 0, \qquad i = 1,2,...,n \tag{12}$$

K(xi,x) is the Kernel function in input space that equivalent with inner product in feature space. K is a symmetric positive definite function which satisfies Mercer's condition.

$$K(x_i, x) = \sum_{k=1}^{\infty} a_k \psi_k(x_i)\psi_k(x) \qquad a_k > 0 \tag{13}$$

It is necessary and sufficient that the condition
$$\iint K(x_i, x)g(x_i)g(x)\,dx_i dx > 0$$

Be valid for all $g \neq 0$ for which $\int g^2(x_i)dx_i < \infty$

V. RESULT AND DISCUSSION

The unbalance voltage conditions are identified by SVM. Data are classified into 15 conditions. Each condition consist 20 measurements. From each condition, randomly it selected 30% as data training and 70% as data testing. We use 2 types of kernels as machine learning, that are RBF (Radial Basis Function) kernel and MLP (Multi Layer Perceptron) kernel. Table II, show the result of identification.

Simultaneously, 15 types of unbalance voltage can be identified by SVM with strength of identification is 81.57%. MLP kernel gives the better value than RBF kernel, with average 86.00% and 77.13% respectively. Some condition

TABLE I
IDENTIFICATION OF UNBALANCE VOLTAGE BY SVM

No	Types of Fault	Strength of Identification (%)		
		RBF	MLP	Average
1	1% Unbalance Voltage φ A	73.00	84.00	78.50
2	1% Unbalance Voltage φ B	98.00	97.00	97.50
3	1% Unbalance Voltage φ C	40.00	70.00	55.00
4	2% Unbalance Voltage φ A	82.00	99.00	90.50
5	2% Unbalance Voltage φ B	32.00	73.00	52.50
6	2% Unbalance Voltage φ C	90.00	90.00	90.00
7	3% Unbalance Voltage φ A	96.00	99.00	97.50
8	3% Unbalance Voltage φB	100.00	100.00	100.00
9	3% Unbalance Voltage φC	100.00	100.00	100.00
10	4% Unbalance Voltage φ A	100.00	100.00	100.00
11	4% Unbalance Voltage φ B	71.00	84.00	77.50
12	4% Unbalance Voltage φ C	72.00	92.00	82.00
13	5% Unbalance Voltage φ A	61.00	52.00	56.50
14	5% Unbalance Voltage φ B	61.00	54.00	57.50
15	5% Unbalance Voltage φC	81.00	96.00	88.50
	Average	77.13	86.00	81.57

can be identified perfectly with strength of identification 100%. Several other conditions have strength of identification less than 50% for RBF and more than 50% for MLP. Based on this result, it concludes that MLP kernel is more suitable than RBF kernel. From this case, it concludes that SVM with MLP kernel can be identified unbalance voltage with "good" strength of identification. SVM can be used to identify unbalance voltage in induction motor to prevent a greater harm.

By applying this method in industrial applications, losses from damage of the engine due to voltage unbalance can be prevented as early as possible. However, application in industrial process will increase the cost due to additional hardware installation and memory for the learning process.

VI. CONCLUSSION

Unbalance voltage supply in induction motor is a crucial fault. The small percentage of unbalance voltage obtains a large unbalance current. If left uncheck, it will lead to waste in energy and damage the motor. To avoid this, identification unbalance voltage in induction motor with SVM can be done.

Based on experiment in this paper, SVM with MLP kernel can identify unbalance voltage in induction motor with average strength of identification is 86%. This shows SVM with MLP kernel has a good performance to identify the unbalance voltage in induction motor.

VII. ACKNOWLEDGMENT

The authors gratefully acknowledge the contributions of Directorate General of Higher Education, Depart¬ment of Education and Culture, Indonesia.

REFERENCES

[1] P. Pillay and M. Manyage, "Definitions of Voltage Unbalance," *IEEE Power Engineering Review*, vol. 22, no. 11, pp. 49–50, 2002.

[2] G. . Maruthi and K. Panduranga Vittal, "Electrical Fault Detection in Three Phase Squirrel Cage Induction Motor by Vibration Analysis using MEMS Accelerometer," in *Power electronic and Drives System, 2005*, vol. 2, pp. 838–843.

[3] L. El Menzhi and A. Saad, "Induction motor fault diagnosis using voltage Park components of an auxiliary winding - voltage unbalance," 2009, pp. 1–6.

[4] E. M. T. Eldin, H. R. Emara, E. M. Aboul-Zahab, and S. S. Refaat, "Monitoring and Diagnosis of External Faults in Three Phase Induction Motors Using Artificial Neural Network," 2007, pp. 1–7.

[5] I. P. Georgakopoulos, E. D. Mitronikas, and A. N. Safacas, "Detection of Induction Motor Faults in Inverter Drives Using Inverter Input Current Analysis," *IEEE Transactions on Industrial Electronics*, vol. 58, no. 9, pp. 4365–4373, Sep. 2011.

[6] G. Didier, E. Ternisien, O. Caspary, and H. Razik, "A new approach to detect broken rotor bars in induction machines by current spectrum analysis," *Mechanical Systems and Signal Processing*, vol. 21, no. 2, pp. 1127–1142, Feb. 2007.

[7] A. Widodo, B.-S. Yang, D.-S. Gu, and B.-K. Choi, "Intelligent fault diagnosis system of induction motor based on transient current signal," *Mechatronics*, vol. 19, no. 5, pp. 680–689, Aug. 2009.

[8] V. N. Ghate and S. V. Dudul, "Optimal MLP neural network classifier for fault detection of three phase induction motor," *Expert Systems with Applications*, vol. 37, no. 4, pp. 3468–3481, Apr. 2010.

[9] R. Sharifi and M. Ebrahimi, "Detection of stator winding faults in induction motors using three-phase current monitoring," *ISA Transactions*, vol. 50, no. 1, pp. 14–20, Jan. 2011.

[10] A. Çakır, H. Çalış, and E. U. Küçüksille, "Data mining approach for supply unbalance detection in induction motor," *Expert Systems with Applications*, vol. 36, no. 9, pp. 11808–11813, Nov. 2009.

[11] P. J. Tavner, "Review of condition monitoring of rotating electrical machines," *IET Electric Power Applications*, vol. 2, no. 4, p. 215, 2008.

[12] L. Beran, "Thermal effect of short-circuit current in low power induction motors," in *Power Electronics and Motion Control Conference, 2008. EPE-PEMC 2008. 13th*, 2008, pp. 782–786.

[13] K. C. Gryllias and I. A. Antoniadis, "A Support Vector Machine approach based on physical model training for rolling element bearing fault detection in industrial environments," *Engineering Applications of Artificial Intelligence*, vol. 25, no. 2, pp. 326–344, Mar. 2012.

[14] D. Sawitri, I. K. E. Purnama, and M. Ashari, "Detection Of Electrical Faults In Induction Motor Fed by Inverter Using Support Vector Machine and Receiver Operating Characteristic," *Journal of Theoretical and Applied Information Technology*, vol. 40, no. 1, pp. 015–022, Jun. 2012.

[15] F. Karami, J. Poshtan, and M. Poshtan, "Detection of broken rotor bars in induction motors using nonlinear Kalman filters," *ISA Transactions*, vol. 49, no. 2, pp. 189–195, Apr. 2010.

[16] A. H. Bonnett, "The impact that voltage and frequency variations have on AC induction motor performance and life in accordance with NEMA MG-1 standards," in *Pulp and Paper, 1999. Industry Technical Conference Record of 1999 Annual*, pp. 16–26.

[17] G. R. Bossio, C. H. De Angelo, P. D. Donolo, A. M. Castellino, and G. O. Garcia, "Effects of voltage unbalance on IM power, torque and vibrations," 2009, pp. 1–6.

[18] M. Aminghafari, N. Cheze, and J.-M. Poggi, "Multivariate denoising using wavelets and principal component analysis," *Computational Statistics & Data Analysis*, vol. 50, no. 9, pp. 2381–2398, May 2006.

[19] V. N. Vapnik, *The nature of statistical learning theory*. New York: Springer, 1995.

978-1-4799-0024-4/13 $31.00 © 2013 IEEE

BIOGRAPHIES

Dian R. Sawitri is doctoral student in Institut Teknologi Sepuluh Nopember (ITS), Surabaya, Indonesia. She received M.Eng degree in electrical engineering from Gadjah Mada University, Yogyakarta, Indonesia in 2002 and Bachelor Degree from Diponegoro University in 1993. She is currently working in electrical engineering department, Dian Nuswantoro University in Semarang, Indonesia. Her research interest include intelligent system application in fault detection in induction machine, power electronic and power quality.

Dimas Anton Asfani received his B.Eng, M.T degrees in Electrical Engineering from Institut Teknologi Sepuluh Nopember Surabaya, Indonesia, in 2004 and 2006. Ph.D. degree in the Power system Laboratory Kumamoto University in 2012..He was joined Institut Teknologi Sepuluh Nopember, Surabaya, Indonesia in 2005 as lecture and research asistent. His research concentrates mainly on fault detection in Induction machine and Power system protection.

Mauridhi Heri Purnomo received the B.S degree in power system engineering from Institut Teknologi Sepuluh Nopember Surabaya, Indonesia, in 1985. M.Eng and Ph.D degrees in control engineering and Intelligence System from Osaka City Univ. Japan, in 1995 and 1998. He is currently a Professor of Departement of Electrical Enginering. His research interest include fuzzy control, intelligence control and control applications.

I. Ketut Eddy Purnama received Bachelor Degree in the field of Computer System Engineering at The Department of Electrical Engineering, Institut Teknologi Sepuluh Nopember (ITS), Surabaya, Indonesia in 1994. He is currently working at the same department. In 1999, he received M.Sc. degree at The Department of Informatics, Bandung Institute of Technology, Bandung, Indonesia. He continued his studied at The Department of Biomedical, University Medical Center Groningen, University of Groningen, The Netherlands, and graduated in 2007. His interest is in telematics, computer vision, computer graphics, database management system, and telematics for medicine including: telemedicine, mobile-telemedicine, medical image analysis and visualization, medical image sharing and storage.

Mochamad Ashari, received his B.E degree in Electrical Engineering from Institut Teknologi Sepuluh Nopember (ITS), Surabaya, Indonesia, in 1989. The Master and Ph.D degree in Electrical Engineering were obtained from Curtin University, Australia in 1997 and 2001, respectively. He Joined Institut Teknologi Sepuluh Nopember in 1990 and has been a Professor from 2009. His current interest include application of intelligent systems to power electronic, power quality and renewable energy power source.

Exploitation of Induction Machine's High-Frequency Behavior for Online Insulation Monitoring

Peter Nussbaumer, Markus A. Vogelsberger, Thomas M. Wolbank

Abstract -- **The trend to increased energy efficiency and profitable capital expenditure leads to the operation of drive systems at or near their rated values and to increased running time. Thus, down time due to machine breakdown (e.g. in a traction drive) leads to high economic losses for the operator. Furthermore, the breakdown of safety-critical devices may lead to dangerous situations. Due to the above stated reasons the demand for operation reliability is constantly increasing.**

The fast switching of the voltage source inverter (VSI) in adjustable speed drives (ASD) leads to increased stress for the winding insulation. Thus, it gets even more important to monitor the condition of the insulation in inverter-fed drives.

Degradation of the insulation results in an alteration of the machine's high-frequency behavior. The proposed method is capable of detecting such changes by evaluation of the transient current reaction on inverter switching.

Index Terms--**AC motor drives, Fault diagnosis, Induction motor protection, Monitoring, Pulse width modulated inverters, Rotating machine insulation testing, Squirrel cage motors**

I. INTRODUCTION

IN modern traction drives the application of adjustable speed drives consisting of an AC machine (typically induction or permanent magnet synchronous (PMSM) machine) and inverters (nowadays voltage source inverters with IGBTs (Insulated Gate Bipolar Transistors)) is standard. Although electrical machines are generally highly reliable the increased demand of system availability leads to the necessity to implement condition monitoring, fault detection and/or fault tolerant control.

The main causes for machine breakdown have been analyzed in [1] and [2]. The result is that machine breakdown originates in faults that can be classified in three categories – bearing, stator and rotor related faults. According to these investigations the second most common causes are stator related accounting for about 35% of all collected machine breakdowns. Within these stator related faults problems with the insulation leading to short circuit faults account for 70%. Thus, reliable monitoring of the insulation condition allows the shift of maintenance strategy from preventive to predictive. In case of predictive

maintenance, the risk of failure estimated by insulation condition monitoring allows to decide if maintenance (e.g. replacement of the windings before breakdown of the insulation system) is required or not.

Usually breakdown of the insulation is a slowly developing process starting with deterioration of the insulation material and then leading to severe turn-to-turn, phase-to-phase or phase-to-ground short circuits [3]. The exact time of insulation breakdown cannot be determined according to [4]. Therefore only a risk of failure rather than a time to failure can be defined.

The deterioration of the insulation condition is accelerated by different causes. The main cause according to [5] is thermal stress. However, electrical, mechanical and environmental strains lead to deterioration of the insulation material too. Concerning inverter-fed drives the fast rise times of modern switching devices like IGBTs and MOSFETs leads to increased electrical stress of the insulation system as analyzed in [3].

So far, many different insulation fault detection and condition monitoring techniques have been proposed in literature. All methods can be categorized into offline or online approaches. The most industrially accepted methods are applied offline. Thus the machine has to be taken out of service to test its insulation.

The offline partial discharge [8] and offline surge [6] test are able to detect insulation deterioration. Further offline insulation fault detection techniques are the DC conductivity test [7], the insulation resistance (IR) test [4], DC/AC HiPot test [4] and polarization index (PI) test [4].

So far the only industrially accepted online insulation monitoring technique is the online partial discharge (PD) test [9]. However, this test is only applicable for medium to high voltage machines and needs additional measurement hardware and highly sophisticated evaluation software. Many other online insulation fault detection methods have been presented in literature. However, most of them are only able to detect solid stator faults but no insulation deterioration. The approach to use motor current signature analysis (MCSA) for insulation fault detection is presented for example in [10]. The adaption of the surge test for online applicability is proposed in [11]. This test is design for application in mains-fed machines. Evaluation of the transient leakage induction by measuring the current reaction on inverter switching is presented in [12] to detect solid turn-to-turn faults.

The requirements of the presented online insulation monitoring method are applicability for inverter-fed drives, usage of the already available sensors only, the detection of

---

The work to this investigation was supported by the Austrian Science Fund (FWF) under grant number P23496-N24.

P. Nussbaumer and T.M. Wolbank are with the Institute of Energy Systems and Electrical Drives, Vienna University of Technology, 1040 Vienna, Austria (e-mail: peter.nussbaumer@tuwien.ac.at; thomas.wolbank@tuwien.ac.at)

M.A. Vogelsberger is with Bombardier Transportation Austria GmbH, Drives Development Center, PPC, 1220 Vienna, Austria (e-mail: markus.vogelsberger@at.transport.bombardier.com)

insulation deterioration and that no disassembling of the drive is necessary.

This paper briefly presents the developed online insulation monitoring method and compares its application to two different induction machine drive systems with different power ratings in experimental investigations. Furthermore the application of the method for different DC-link voltage values but same machine is compared.

II. HIGH-FREQUENCY BEHAVIOR OF INVERTER-FED DRIVES

The fast voltage rise times of modern switching devices additionally stresses the winding insulation as mentioned in the introduction. The cause for this stress is the occuring transient overvoltage resulting from reflections of the applied voltage pulse at the machine's terminal connections. According to traveling wave theory the mismatch of machine and supply cable impedance leads to these reflections [13]. The machine impedance is by far bigger than the cable impedance. Thus, in theory the voltage pulse is fully reflected (reflection coefficient nearly one) [13]. The reflected voltage pulse leads to an oscillating transient overvoltage with decaying magnitude. The peak value can reach up to twice or even four times (for fast subsequent voltage pulses) the DC-link voltage and the oscillation frequency is in the range of tens kHz to tens MHz [14].

An inverter-fed drive system consists of three main components defining the complex impedance system leading to the voltage pulse reflections:

- Inverter (e.g. voltage source inverter)
- Supply cables
- AC machine (e.g. induction machine)

The elements defining the characteristic of the drive's behavior like stator resistance r_S, stator inductance l_S, cable inductance and resistance per unit length, etc. on the one hand and the parasitic components like phase-to-ground (cable), winding-to-ground (machine), winding-to-winding (machine) and turn-to-turn (machine) capacitance and the inverter's capacitive coupling to ground on the other hand determine the mentioned drive's complex impedance network. Many of the parasitic components are defined and strongly influenced by the insulation system. Thus, these components also influence the oscillating transient voltage. This oscillation is also visible in the current immediately after inverter switching. It is characteristic for the machine's high-frequency behavior.

A change in the machine's insulation system (e.g. due to deterioration) leads to a change in this characteristic oscillation (hf-behavior). The proposed condition monitoring method evaluates the oscillation in the current reaction on voltage switching to detect changes in the machine's hf-behavior. It is preferred to analyze the changes in the current oscillation as current sensors are already available in modern drive systems for machine control.

III. MEASUREMENT PROCEDURE AND SIGNAL PROCESSING

For the evaluation of the hf-oscillation the current has to be acquired with sufficient resolution in time. The excitation of the machine in the hf-range is carried out by application of different changes of the inverter's switching state.

The used transitions of switching states always originate in the lower short circuit (-SC, 000) and end in one of the three positive active switching states (+U, 001; +V, 010; +W, 100). Standard industrial current sensor can be used (bandwidth in the range of 100kHz to 300kHz).

The current's reaction during the switching transition in phase U (-SC to +U) is depicted in Fig. 1.

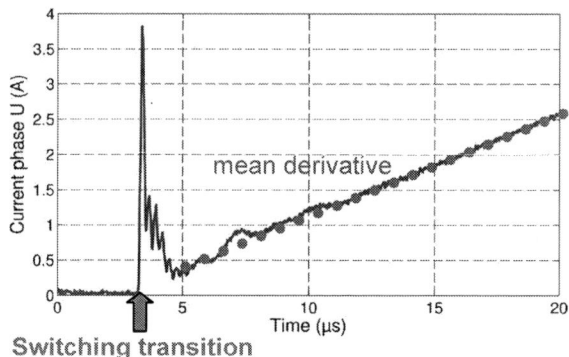

Fig. 1. Current signal with illustrated mean derivative (green, dotted line) and switching instant (red arrow).

The visible current derivative after the switching transition (denoted as mean derivative) is mainly determined by the machine's transient leakage inductance. This transient leakage inductance depends on inherent machine saliencies like rotor slotting and saturation level. Thus, this influence has to be eliminated as only the high-frequency oscillation is of interest. The elimination can be done by simply subtracting the mean derivative after exact detection of the real switching instant. The resulting signal is shown in Fig. 2

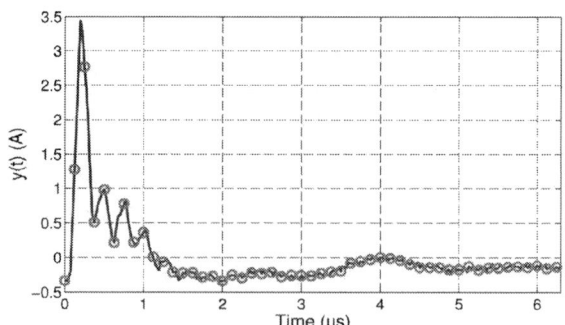

Fig. 2. Current (mean derivative subtracted) after switching transition from lower short circuit (-SC, 000) to +U (001).

The red circles highlight exemplary sampling instants. The transient current oscillation depicted above is further analyzed in the frequency domain. Therefore the depicted signal is transformed by FFT (Fast Fourier Transform). The investigated time window is chosen from the real switching

978-1-4799-0024-4/13 $31.00 © 2013 IEEE

instant until the oscillation has decayed. For the shown signal this time window is chosen to 6.4µs. The used sampling frequency is 40MS/s. Thus the number of sample values equals to $N=256$. The resulting amplitude spectrum after Fourier transform is depicted in Fig. 3 for evaluation in phase U.

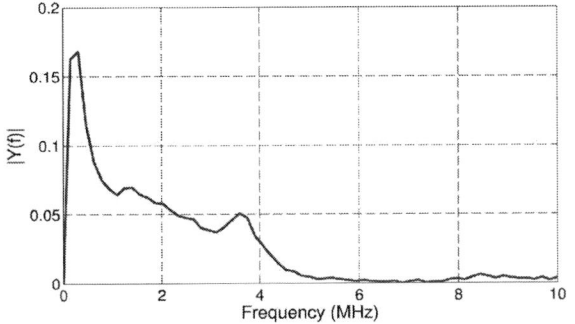

Fig. 3. Amplitude spectrum of measured current in phase U after switching transition from lower short circuit (-SC, 000) to +U (001).

The amplitude spectrum is characteristic for the machine's hf-behavior in phase U. If recorded for healthy machine condition it serves as a reference trace and is compared to later measurements to assess the machine's insulation condition in the proposed monitoring method. To detect changes in the machine's hf-behavior the above described measurement procedure and signal processing is repeated for all three phases. For each of the three phases a reference amplitude spectrum (recorded for healthy machine condition) is stored for later comparison to condition measurements and assessment of the insulation condition.

IV. INSULATION STATE INDICATOR (ISI/SISI)

The measurement and signal processing has been described in the previous section. An insulation state indicator is introduced for the assessment of the insulation condition in one phase. It is based on quantifying the change in the machine's high-frequency behavior by comparison of amplitude spectrum recorded for healthy machine condition (reference) and during condition assessment. The Root Mean Square Deviation (RMSD) is chosen as a comparative value and serves as Insulation State Indicator (ISI) for the respective phase.

$$ISI_{p,k} = RMSD_{p,k}(x_1, x_2) = \sqrt{\frac{\sum_{i=n_{low}}^{n_{high}} \left(Y_{ref,p}(i) - \left|Y_{con,p,k}(i)\right|\right)^2}{n_{high} - n_{low}}} \quad (1)$$

The Fourier transformed signals Y_{ref} and Y_{con} have been obtained by the procedure described in section III. for healthy machine condition and a later condition measurement, respectively. The index p defines the investigated phase (U,V,W). The variables n_{high} and n_{low} define the compared frequency range and depend on sampling rate and investigated window length. The definition of the evaluated frequency range allows separating other influences like cabling or grounding leading to a change in the drive's high-frequency behavior. Due to the

fact that a single measurement's duration is in the range of a few hundred microseconds the procedure can be repeated m-times to increase the accuracy of the method. Thus, the index k (1,2,3,...) denotes the number of the consecutive measurements.

The used reference signal (frequency spectrum for healthy machine condition) is the mean trace calculated from m measurements.

$$Y_{ref}(i) = \frac{\sum_{k=1}^{m}\left|Y_{ref,k}(i)\right|}{m} \quad (2)$$

The variable i identifies the discrete frequency. The frequency is calculated using the sampling rate f_S, the number of samples N and the window length t_{win} according to the following equation

$$f = \frac{i \cdot f_s}{N} ; \ i = 0,1,2,3,...; \ N = t_{win} \cdot f_s \quad (3)$$

The quantity defining the insulation condition in one phase is the insulation state indicator ISI_p calculated from m RMSD values according to equation (1).

$$ISI_p = \frac{\sum_{k=1}^{m} ISI_{p,k}}{m} \quad (4)$$

In a last step of signal processing a Spatial Insulation State Indicator (SISI) is calculated by linear combination of the ISI values of all three phases.

$$SISI = ISI_U + ISI_V \cdot e^{j\frac{2\pi}{3}} + ISI_W \cdot e^{j\frac{4\pi}{3}} \quad (5)$$

Symmetrical changes of the high-frequency behavior (e.g. due to temperature variation, change of cabling,...) are eliminated by this linear combination as these lead to a zero-sequence component. The successful elimination of influences on the insulation state indicator due to the cable has been given in [15] by the authors.

V. EXPERIMENTAL SETUP

The main focus in this paper is to compare the applicability of the above proposed condition monitoring method on induction machines with clear different power rating and insulation systems. Thus two different machines have been investigated. Induction machine IM#1 is a industrial 2-pole machine with 5.5kW and enamel-insulated wire. Induction machine IM#2 on the other hand is a 4-pole 1.4MW machine with fibre-insulation wires. Both machines have a squirrel cage rotor and several tapped windings accessible at the machine's terminal connection block. Thus, it is possible to change the machine's hf-behavior by inserting capacitors between the different taps of the winding. The exemplary scheme of the parasitic capacitances (turn-to-turn C_{t-t}, phase-to-phase C_{ph-ph}, phase-to-ground C_{ph-gnd}) in the machine and this additional capacitor C_{fault} are depicted in Fig. 4. The phase-to-ground capacitances of IM#1 and IM#2 are 1.5nF and 21nF respectively. The additional capacitor in parallel to the turn-to-turn capacitances results in an increase. This is in accordance to the results of increasing capacitance due to insulation deterioration presented in [16].

The measurements, control and signal processing are carried out with a combined system of Real-Time processor, Field Programmable Gate Array (FPGA) and fast sampling ADCs from National Instruments, programmable in LabVIEW.

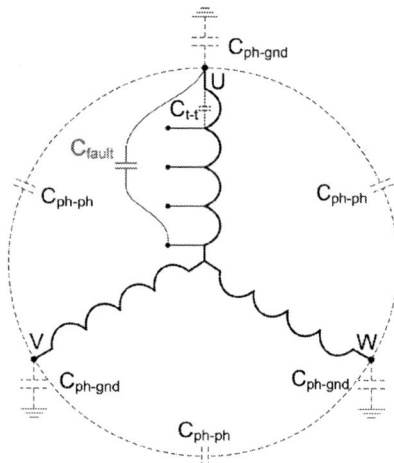

Fig. 4. Additional capacitor C_{fault} inserted in parallel to the full phase winding U, schematically.

VI. EXPERIMENTAL RESULTS

The purpose of the following experimental investigations is to identify differences in the application of the above presented monitoring method between induction machines with different power rating and insulation system.

In a first step the differences in the high-frequency behavior of the two investigated machines is analyzed regarding the amplitude spectrum for health insulation condition. Fig. 5 shows this comparison of the amplitude spectrum normalized on the respective maximum magnitude of each machine.

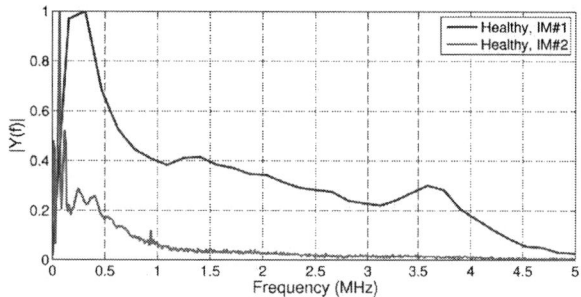

Fig. 5. Normalized Amplitude spectrum of measured current in phase U after switching transition from lower short circuit (000) to +U (001) for induction machine IM#1 (blue) and IM#2 (green).

It is clearly visible that the dominant frequencies for the machine with higher power rating (IM#2) are in a lower

frequency range than the ones of IM#1. The highest magnitude can be identified at 313kHz and 68kHz for induction machine IM#1 and IM#2, respectively.

The sampling rate in both investigations is chosen 40MS/s. The investigated time window length t_{win} is defined to 6.4µs and 102.4µs for IM#1 and IM#2, respectively. The time window depends on the duration of the decaying transient oscillation visible in the current signal. The DC link voltage (voltage pulse magnitude) in case of IM#1 is 440V, in case of IM#2 it is 600V.

In a next investigation an additional capacitor is inserted in parallel to the full phase winding U. The value of the inserted capacitor is chosen with respect to the phase-to-ground capacitance of the two machines IM#1 and IM#2 to 0.5nF and 15nF, respectively. The amplitude spectra for the reference signal in phase U, $Y_{ref,U}$ (according to (2)) is compared to the amplitude during a condition measurement $|Y_{con,U,1}|$. This comparison and the resulting square deviation is depicted in Fig. 6 and Fig. 7 for induction machine IM#1 and IM#2, respectively.

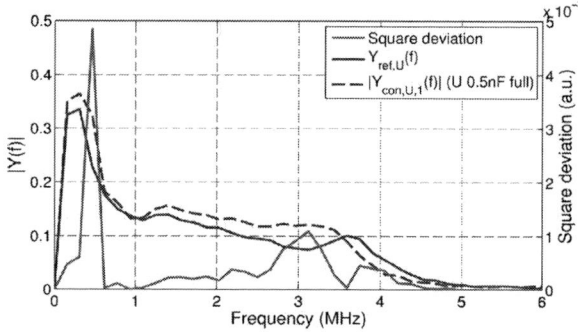

Fig. 6. Reference amplitude spectrum $Y_{ref,U}(f)$ (blue, solid trace), amplitude spectrum of one condition assessment $|Y_{con,U,1}(f)|$ (blue, dashed trace) for 0.5nF capacitor inserted in parallel to the full winding of phase U and calculated square deviation of both traces (green, solid trace); IM#1.

For both investigated machines a dominant change in specific frequency components can be detected.

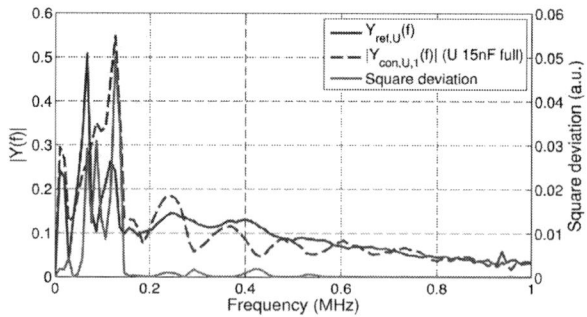

Fig. 7. Reference amplitude spectrum $Y_{ref,U}(f)$ (blue, solid trace), amplitude spectrum of one condition assessment $|Y_{con,U,1}(f)|$ (blue, dashed trace) for 15nF capacitor inserted in parallel to the full winding of phase U and calculated square deviation of both traces (green, solid trace); IM#2, U_{DC}=600V.

For IM#1 this change is at 469kHz. Whereas the most dominant changes for IM#2 can be detected at 68kHz,

88kHz and 127kHz.

Concerning induction machine IM#2 the influence of DC-link voltage on the dominant frequencies and the interesting frequencies in case of changed hf-behavior were investigated next. The results for investigations with DC-link voltage of 600V are already shown in Fig. 7. The results for the same evaluation but a DC-link voltage of 2800V are illustrated in Fig. 8.

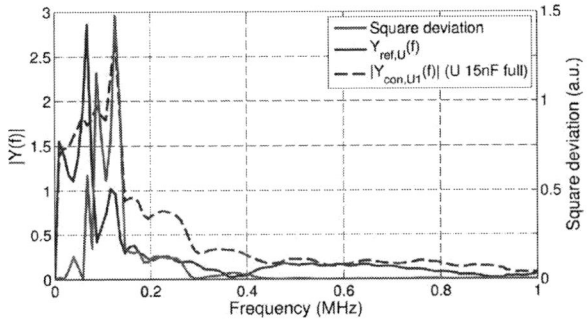

Fig. 8. Reference amplitude spectrum $Y_{ref,U}(f)$ (blue, solid trace), amplitude spectrum of one condition assessment $|Y_{con,U,1}(f)|$ (blue, dashed trace) for 15nF capacitor inserted in parallel to the full winding of phase U and calculated square deviation of both traces (green, solid trace); IM#2, U_{DC}=2800V.

The most dominant frequency component in case of healthy machine condition is identified to 68kHz and the interesting frequency components for changed hf-behavior of the machine are 68kHz, 88kHz and 127kHz. Thus, all regarded frequency component are the same for both scenarios. Therefore the magnitude of the DC-link voltage does not significantly influence the frequency of the transient current oscillation.

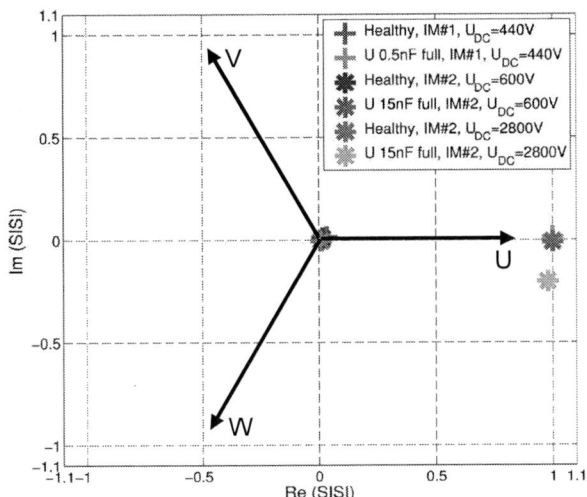

Fig. 9. Normalized spatial insulation state indicator (SISI) for different investigated machines (IM#1 and IM#2) and changed hf-behavior.

At the end of insulation condition assessment using the

proposed monitoring method the spatial insulation state indicator calculated according to equation (5) is regarded. This indicator is depicted in Fig. 9 for the above investigated scenarios. For all scenarios the indicator is normalized on the magnitude of the respective SISI for changed machine's hf-properties. It is shown that in all investigated scenarios the changed machine condition is clearly different to the unchanged ('Healthy') one. The unchanged machine condition always results in a spatial insulations state indictor in or close to the origin of the Gaussian plane. The capacitor has been inserted in phase U for all scenarios. The SISI is pointing in direction of phase axis U. Thus, the phase location of the alteration can be identified. In case of investigated induction machine IM#2 and DC-link voltage of 2800V the angle of the spatial insulation state indicator is approximately 10°. This deviation from the other scenarios (angle equaling ~0°) will be further investigated in future measurements.

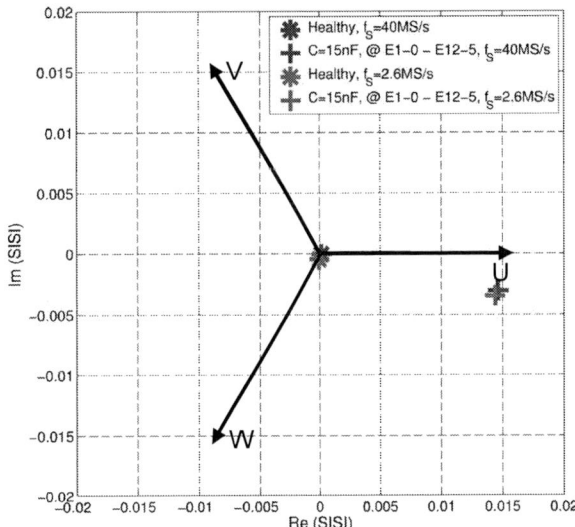

Fig. 10. Spatial insulation state indicator (SISI) for induction machine IM#2 (UDC=2800V); sampling rate f_S equals 40MS/s (blue) and 2.6MS/s (green).

An important parameter concerning the implementation of the proposed condition monitoring technique is the sampling rate needed for accurate calculation of the insulation state indicator. The above presented investigations have been carried out with a sampling rate of 40MS/s. In modern drive systems the sampling rate of the used analog-to-digital converters (ADCs) is lower. Thus, investigations concerning the necessary sampling rate have been carried out on induction machine IM#1. The results have been presented in [17]. The conclusion has been that the necessary sampling rate depends on the frequency range investigated for calculation of the insulation state indicator. The sampling rate has to be 20-times higher than the highest interesting frequency. This results in a necessary sampling rate of 10MS/s in case of induction machine IM#1 as the interesting frequency is about 500kHz. In case of induction machine

IM#2 the highest interesting frequency is lower and can be identified to about 130kHz. According to the above mentioned rule of thumb the necessary sampling rate results to 2.6MS/s. The obtained spatial insulation state indicators for induction machine IM#2 and DC-link voltage of 2800V at different sampling rates (40MS/s and 2.6MS/s) are depicted in Fig. 10. It can be seen that the resulting indicators are almost identical for both investigated scenarios and sampling rates. Thus a reduction of the sampling rate to 10MS/s (IM#1) and 2.6MS/s (IM#2) can be made according to the rule of thumb presented in [17] without deterioration in accuracy.

VII. CONCLUSIONS

Insulation deterioration influences the machine's high-frequency behavior. Thus, a condition monitoring method has been developed and presented that is capable of detecting such changes by evaluation of the transient current reaction immediately after inverters switching. The method has been applied to a small 5.5kW and a bigger 1.4MW induction machine both with different insulation systems. The results have been compared for different machine ratings and different DC-link voltages. It is concluded that the method can be applied to machines with different power ratings as well as insulation systems. Furthermore DC-link voltage does not influence the frequency range investigated during the calculation of the insulation state indicator.

Sampling rate used for current acquisition is a crucial parameter in the design of the measurement hardware needed for the condition monitoring method. Thus, an analysis of the necessary sampling rate has been carried out and compared to previous results. The sampling rate has to be approximately 20-times higher than the maximum frequency component used for the calculation of the insulation state indicator.

VIII. ACKNOWLEDGMENT

The authors want to thank Bombardier Transportion Switzerland, Converter Department (head Markus Jörg, MSc) and especially the team of BT-Powerlab Zürich for the generous support. Special thanks also goes to National Instruments Austria and especially DI Günther Stefan for the generous support and donation to finance the measurement hardware.

IX. REFERENCES

[1] IEEE Committee Report; "Report of large motor reliability survey of industrial and commercial installation, Part I," *IEEE Trans. on Ind. Appl.*, vol.21, no.4, pp.853–864, 1985.

[2] IEEE Committee Report; "Report of large motor reliability survey of industrial and commercial installation, Part II," *IEEE Trans. on Ind. Appl.*, vol.21, no.4, pp.865–872, 1985.

[3] Yang, J; Cho, J.; Lee, S.B.; Yoo, J.-Y.; Kim, H.D.; "An Advanced Stator Winding Insulation Quality Assessment Technique for Inverter-Fed Machines," *IEEE Trans. on Ind. Appl.*, vol.44, no.2, pp.555-564, 2008.

[4] Stone, G.C.; Boulter, E.A.; Culbert, I.; Dhirani, H.; "Electrical Insulation for Rotating Machines," IEEE Press, Wiley & Sons,

2004.

[5] Grubic, S.; Aller, J.M.; Bin Lu; Habetler, T.G.; "A Survey on Testing and Monitoring Methods for Stator Insulation Systems of Low-Voltage Induction Machines Focusing on Turn Insulation Problems," *IEEE Trans. on Industrial Electronics*, vol.55, no.12, pp.4127-4136, 2008.

[6] Wiedenbrug, E.; Frey, G.; Wilson, J.; "Impulse testing and turn insulation deterioration in electric motors," *Annual Pulp and Paper Industry Technical Conference*, pp. 50- 55, 2003.

[7] Schump, D.E.; "Testing to assure reliable operation of electric motors," *Industry Applications Society 37th Annual Petroleum and Chemical Industry Conference*, pp.179-184, 1990.

[8] Stone, G.C.; "Recent important changes in IEEE motor and generator winding insulation diagnostic testing standards," *IEEE Trans. on Ind. Appl.*, vol.41, no.1, pp. 91-100, 2005.

[9] Stone, G.C.; Sedding, H.G.; Costello, M.J.; "Application of partial discharge testing to motor and generator stator winding maintenance," *IEEE Trans. on Ind. Appl.*, vol.32, no.2, pp.459-464, 1996.

[10] Nandi, S.; Toliyat, H.A.; "Novel frequency-domain-based technique to detect stator interturn faults in induction machines using stator-induced voltages after switch-off," *IEEE Trans. on Ind. Appl.*, vol.38, no.1, pp.101-109, 2002.

[11] Grubic, S.; Habetler, T.G.; Restrepo, J.; "A new concept for online surge testing for the detection of winding insulation deterioration," *Energy Conversion Congress and Exposition (ECCE)*, pp.2747-2754, 2010.

[12] Wolbank, T.M.; Wohrnschimmel, R.; "Transient electrical current response evaluation in order to detect stator winding interturn faults of inverter fed ac drives," *Symposium on Diagnostics for Electric Machines, Power Electronics and Drives (SDEMPED)*, pp.1-6, 2001.

[13] Kerkman, R.J.; Leggate, D.; Skibinski, G.L.; "Interaction of drive modulation and cable parameters on AC motor transients," *IEEE Trans. on Ind.Appl.*, vol. 33, pp. 722-731, 1997.

[14] Peroutka, Z. and Kus, V.; "Adverse effects in voltage source inverter-fed drive systems," *Seventeenth Annual IEEE Applied Power Electronics Conference and Exposition, APEC*, pp. 557-563, 2002.

[15] Nussbaumer, P.; Wolbank, T.M.; Vogelsberger, M.A.; "Separation of disturbing influences on induction machine's high-frequency behavior to ensure accurate insulation condition monitoring," *Twenty-Eighth Annual IEEE Applied Power Electronics Conference and Exposition (APEC)*, pp.1158-1163, 2013.

[16] Perisse, F.; Werynski, P.; Roger, D.; "A New Method for AC Machine Turn Insulation Diagnostic Based on High Frequency Resonances," *IEEE Trans. on Diel. and El. Ins.*, vol.14, no.5, pp.1308-1315, 2007.

[17] Nussbaumer, P.; Vogelsberger, M.A.; Wolbank, Th.M.; "Sensitivity Analysis of Insulation State Indicator in Dependence of Sampling Rate and Bit Resolution to Define Hardware Requirements," *International Conference on Industrial Technology (ICIT)*, pp.392-397, 2013.

X. BIOGRAPHIES

Peter Nussbaumer (SM' 2011) received the MSc degree in Power Engineering and the PhD. degree in Electrical Engineering from Vienna University of Technology, Vienna, Austria in 2009 and 2013, respectively. He is currently a Project Assistant at the Institute of Energy Systems and Electrical Drives, Vienna University of Technology and a colleague in the doctoral program Energy Systems 2030. He is engaged in several industrial and scientific projects with focus on electrical drives and electric mobility. His special fields of interest are fault detection, condition monitoring and sensorless control of inverter-fed AC machines.

Markus A. Vogelsberger received the the Dipl.-Ing./M.S. (with honor) and Dr.Techn/Ph.D. (with honor) degrees in Electrical Engineering from Vienna University of Technology, Vienna, Austria, in 2004 and 2009, respectively. He has been a Scientific Research Assistant in the Institute of Electrical Drives and Machines, Vienna University of Technology, where he has been engaged in several industrial and scientific R&D projects in the field of sensorless control, drives systems and power electronics / drive converter.

In 2011, he joined Bombardier Transportation, Vienna, Austria, as a Research & Development Engineer in the PPC-Drives Department.

His field of interest include design of traction motors, power electronics as well as control and simulation of electrical drives.

In particular, currently his focus as R&D-Project Lead for drives is on the research and development of fault detection and condition monitoring strategies for AC traction drives.

Thomas M. Wolbank (M' 1992) studied Industrial Electronics at the Vienna University of Technology in Austria where he also received the Doctor degree in 1996 and the Habilitation in 2004 respectively. At present he is with the Institute of Energy Systems and Electrical Drives at the Vienna University of Technology.

His research interests include saliency-based sensorless control of ac drives, dynamic properties and condition monitoring of inverter fed machines, transient electrical behaviour of ac machines, motor drives and their components, and controlling them by the use of intelligent control algorithms.

He has co-authored some 100 papers in refereed journals and international conferences.

Long-Term Prediction of Bearing Condition by the Neo-Fuzzy Neuron

A. Soualhi, *Student member, IEEE*, G. Clerc, *Senior member, IEEE*, H. Razik, *Senior member, IEEE*, and F. Rivas, *member, IEEE*

Abstract -- **Rolling element bearings are devices used in almost every electrical machine. Therefore, it is important to monitor and track the degradation of bearings. This paper presents a new approach to predict the degradation of bearings by a time series forecasting model called the neo-fuzzy neuron. The proposed approach uses the root mean square extracted from vibration signals as a health indicator. The root mean square is used here as an input of the neo-fuzzy neuron in order to estimate the evolution of bearing's degradation in time. Experimental degradation data provided by the University of Cincinnati is used to validate the proposed approach. A comparative study between the neo-fuzzy neuron and the adaptive neuro-fuzzy inference system is carried out to appraise their prediction capabilities. The experimental results show that the neo-fuzzy model can track the degradation of bearings.**

Index Terms-- **Prognosis, Time domain analysis, Vibration measurement, Feature extraction, Fuzzy neural networks, Artificial intelligence.**

I. INTRODUCTION

Electrical machines are widely used in industrial applications because of their low cost, performance and robustness. However, different kinds of failures can occur during the life of the machine. [1] and [2] reviewed the major causes of failures in electrical machines. They identified broken rotor bars and end ring faults. These defects represent approximately 9% of the electrical machines faults. They identify also inter-turn short circuits in the stator winding and bearing failures which represent respectively 31% and 42% of all faults. Since the bearings is an interface between stator and rotor, the degradation of these devices led to a major breakdown of the motor. Thus, the prognosis of bearings failures becomes very important and advocated.

The prognosis of bearing failures is one of the main approach for ensuring the quality and performance of electrical machines. According to the international standard organization [3], the prognosis of a failure corresponds to the prediction of the remaining useful life and the determination of the future state of the system being tested. For the prediction of the useful reaming time, trends in parameter values, features or changes in probabilities of the bearing

state can be used to forecast the time to failure and to predict the condition of bearings. This is accomplished by a variety of techniques. The most popular techniques are the neural networks, time series analysis (one-step-ahead prediction and multi-step-ahead prediction) and fuzzy logic [4]-[8]. These techniques use the past or the historical information of the system to infer its future state and continually update the prediction. The one-step-ahead prediction using a time series forecasting model like the neural networks and fuzzy systems has been used successfully in the prediction of machine condition degradation [9],[10]. The prediction is provided by a fuzzy system and its parameters are optimized through an artificial neural network. The neural networks learn the behavior of the system from its historical data and therefore do not require a physical model for modeling purposes. These papers use the one step-ahead prediction to forecast at short term the future condition of the system. The one-step-ahead prediction does not provide enough information to know the behavior of the system in the long term case. However, it is possible to use it in order to solve the problem of the long-term prediction. This procedure is known as the multi-step-ahead or the long-term prediction. In this case, the first prediction is given from a time series of past values. The next prediction is given by the previous predicted and past values of the time series. Then, the process is repeated recursively until a certain value.

Neuro-fuzzy systems are widely used to solve significant problems of prediction. However, in real conditions, data processing must be carried out simultaneously with the operation of the process and therefore, time becomes very important. Neuro-fuzzy systems can be in some cases unable to solve the problem mentioned above due to their low rate of convergence during the learning process, and their inability to learn online. In order to take advantage of artificial neural networks, and their ability to handle a great amount of information provided by the fuzzy logic models, a new structure, called the neo-fuzzy neuron was introduced by Takeshi Yamakawa *et al* in [11]-[13]. This structure provides good results for the representation of the behavior of complex systems. The neo-fuzzy neuron is a tool that offers great advantages for modeling complex systems by the simplicity of its structure, which consists of a single neuron, unlike artificial neural networks, where neurons are included, and can be numerous when the system model is very complex.

In this paper, we propose a long-term predictor based on the one-step-ahead prediction and the neo-fuzzy neuron to

---

A.SOUALHI, G. CLERC, H. RAZIK, are with the Laboratoire Ampère, Université de Lyon, F-69622, Lyon, France ; Université Lyon 1, Villeurbanne, France ; CNRS, UMR 5005. E-mail: [abdenour.soualhi, guy.clerc, hubert.razik]@univ-lyon1.fr
F. RIVAS. Universidad de Los Andes, Facultad de Ingeniería Laboratorio de Sistemas Inteligentes, Mérida, Venezuela, e-mail: rivas@ula.ve

provide a long-term estimation of bearing condition. This paper is organized as follows: in the next section, we investigate two methods based on neural networks to solve the long-term prediction problem: the adaptive neuro fuzzy system (ANFIS) predictor and the proposed predictor based on the neo-fuzzy neuron (NFN). In section III, experimental degradation data provided by the University of Cincinnati from Prognostics Center of Excellence are used to validate the proposed approach. Final conclusions are drawn in section IV.

II. LONG TERM PREDICTION PROBLEM

The long-term prediction problem can be formulated as follows: given a time series $\{x_t, x_{t-r}, x_{t-2r}, x_{t-3r}...x_{t-(n-1)r}\}$ of data extracted from a feature, we need to know the value of x_{t+l} such as:

$$x_{t+l} = f(x_t, x_{t-r},..., x_{t-(n-1)r}) \quad (1)$$

where x_{t+l} is the future value of the time series at time $t+l$, l represents the horizon of prediction, r denotes the time series step and n defines the number of previous time steps.

The prediction of x_{t+l} noted $\hat{x}_{t+l}$ can be obtained by the multi-step ahead prediction. This strategy provides an estimation of x_{t+l} from the previous predictions $\{\hat{x}_{t+l-r},...,\hat{x}_{t+l-nr}\}$:

$$\hat{x}_{t+l} = f(\hat{x}_{t+l-r},...,\hat{x}_{t+l-nr}) \quad (2)$$

When l =4, n=4, r=1, this means that the prediction of $\hat{x}_{t+4}$ is obtained from the four previous values of the time series spaced by one-step. In the case of the multi-step ahead prediction, the prediction of $\hat{x}_{t+4}$ can be obtained from the four previous predictions $\{\hat{x}_{t+3}, \hat{x}_{t+2}, \hat{x}_{t+1}, x_t\}$. This is shown in Fig. 1 for r =1, n=4 and l=4.

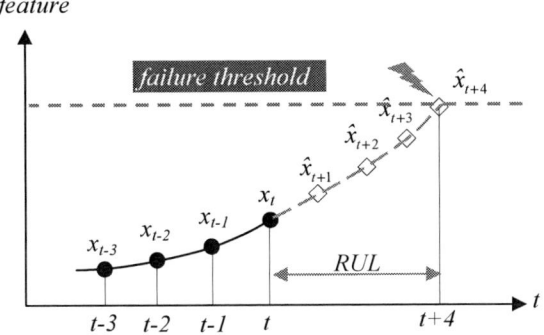

Fig. 1. Prediction of a time series for l =4, n=4, r=1.

The one-step-ahead prediction combined with the neural networks and fuzzy systems can be used for the design of two predictor: the adaptive neuro fuzzy system (ANFIS) predictor and the proposed predictor based on the neo-fuzzy neuron (NFN).

A. ANFIS Predictor

The ANFIS is a fuzzy inference system implemented in the framework of an adaptive neural network. It has been employed successfully for the prediction of machine condition degradation, where the prediction is carried out by a fuzzy system while its parameters are optimized via an artificial neural network [9], [14]. These papers demonstrate that ANFIS is a reliable and robust condition predictor.

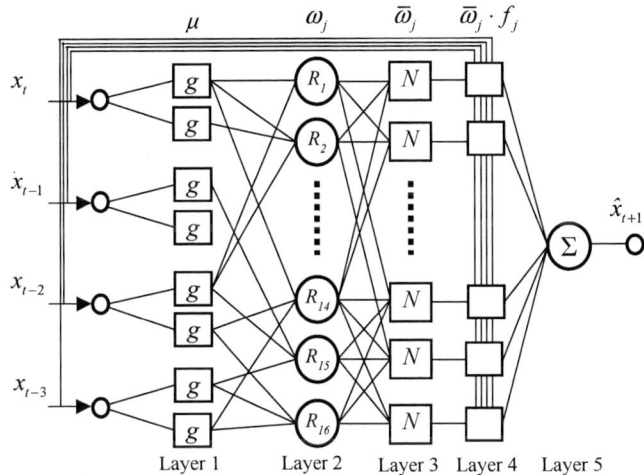

Fig. 2. Architecture of the ANFIS predictor with four inputs and two MFs for each one; g is a Gaussian function.

For simplicity purposes, we assume that the fuzzy inference system has four inputs $\{x_t, x_{t-1}, x_{t-2}, x_{t-3}\}$ with two Membership Functions (MFs) for each input and one output $\hat{x}_{t+1}$ as shown in Fig. 2. ANFIS applies the rules of TSK (Takagi, Sugeno and Kang) form in its architecture. Sixteen TSK fuzzy rules (*if-then*) rules R_j (*j=2⁴*) can be expressed as:
Rule j: *if* x_t is A_0^j and x_{t-1} is A_1^j and x_{t-2} is A_2^j and x_{t-3} is A_3^j , *then*
$\hat{x}_{t+1}^j = c_0^j x_t + c_1^j x_{t-1} + c_2^j x_{t-2} + c_3^j x_{t-3} + c_4^j$, $j = 1,2,...16.$
where (j=*16*) is the number of rules, x_{t-i} (i=0,1,2,3) are the model inputs and $\hat{x}_{t+1}^j$ is the output of the rule R_j. $A_i^j{}_{i=0,1,2,3}$ are a fuzzy sets characterized by membership functions $\mu_{A_i^j}(x_{t-i})$ and $\{c_0^j, c_1^j, c_2^j, c_3^j, c_4^j\}$ are the coefficients of the rule *j*.

The ANFIS structure contains five layers excluding the input layer.
Layer 1: is an adaptive layer which denotes membership functions to each input. In this paper, we choose Gaussian functions as membership functions:

$$\mu_{A_i^j}(x_{t-i}) = exp\left(-\left[(x_{t-i} - m_{ij})/a_{ij}\right]^2\right) \quad (3)$$

where x_{t-i} is the input to node j; $\mu_{A_i^j}$ is the membership function associated with this node; and $\{a_{ij}, m_{ij}\}$ is the parameter set that changes the shapes of the membership function. These parameters are called premise parameters.

Layer 2: is a fixed layer in which each node calculates the firing strength of each rule via multiplication:

$$\omega_j = \mu_{A_0^j}(x_t) \cdot \mu_{A_1^j}(x_{t-1}) \cdot \mu_{A_2^j}(x_{t-2}) \cdot \mu_{A_3^j}(x_{t-3}) \qquad (4)$$

Layer 3: this is the normalization layer which normalizes the strength of all rules:

$$\bar{\omega}_j = \frac{\omega_j}{\sum\limits_{j=1}^{16} \omega_j} \qquad (5)$$

Layer 4: every node j in this layer is an adaptive node with a node function:

$$\bar{\omega}_j f_j = \bar{\omega}_j \cdot \left(c_0^j x_t + c_1^j x_{t-1} + c_2^j x_{t-2} + c_3^j x_{t-3} + c_4^j \right). \qquad (6)$$

After a linear combination of the input signals in the *layer 4,* the output of the ANFIS predictor is given in the *layer 5* by:

$$\hat{x}_{t+1} = \sum_{j=1}^{16} \bar{\omega}_j \cdot \left(c_0^j x_t + c_1^j x_{t-1} + c_2^j x_{t-2} + c_3^j x_{t-3} + c_4^j \right). \qquad (7)$$

The long-term prediction using ANFIS can be formulated as follows:

The prediction of x_{t+1} is given by:

$$\hat{x}_{t+1} = \sum_j \bar{\omega}_j \cdot \left(c_0^j x_t + \ldots + c_{n-1}^j x_{t-n+1} + c_n^j \right) \qquad (8)$$

where t denotes the real value of the input x_t at time t, $\hat{x}_{t+1}$ is the predicted value at time $t+1$.

Based on the first estimate $\hat{x}_{t+1}$, the prediction of x_{t+2} at time $t+2$ is given by:

$$\hat{x}_{t+2} = \sum_j \bar{\omega}_j \cdot \left(c_0^j \hat{x}_{t+1} + \ldots + c_{n-1}^j x_{t-n+2} + c_n^j \right) \qquad (9)$$

where the previous predicted value $\hat{x}_{t+1}$ instead of the real value x_{t+1} is used to carry out the prediction of x_{t+2} at time $t+2$. The long-term prediction can be performed in this way until the predicted value $\hat{x}_{t+l}$ reaches a predefined failure threshold.

B. Neo-Fuzzy Neuron Predictor

The Neo-fuzzy neuron (NFN) is a nonlinear multi-input single output system. Its structure is composed of n-inputs and 1-output as shown in Fig. 3. Among its most important advantages, the high rate of learning and the computational simplicity and also that it is characterized by a set of fuzzy *if-then* rules. The output of the NFN predictor $\hat{x}_{t+1}$ can be expressed as:

$$\hat{x}_{t+1} = \sum_{i=0}^{n-1} f_i(x_{t-i}) \qquad (10)$$

where x_{t-i} is the i-th input $i=(0,1,\ldots,n\text{-}1)$, $\hat{x}_{t+1}$ is the system output. The output of the NFN is the prediction of x_{t+1} at time $t+1$. Structural blocks of neo-fuzzy neuron are

nonlinear synapses which perform transformation of the i-th input in the form:

$$f_i(x_{t-i}) = \sum_{j=1}^{h} \omega_{ji} \cdot \mu_{ji}(x_{t-i}) \qquad (11)$$

where ω_{ji} are the interconnecting weights and $\mu_{ji}(x_{t-i})$ is the membership value fired by the input variables x_{t-i}. The membership functions are triangular functions (see Fig. 4) obtained by normalizing the input variables x_{t-i} ($0 \le x_{t-i} \le 1$). They are expressed in the form:

$$\mu_{ji}(x_{t-i}) = \begin{cases} \dfrac{x_{t-i} - c_{j-1,i}}{c_{ji} - c_{j-1,i}} & , x_{t-i} \in \left[c_{j-1,i}, c_{ji} \right] \\[2ex] \dfrac{c_{j+1,i} - x_{t-i}}{c_{j+1,i} - c_{ji}} & , x_{t-i} \in \left[c_{ji}, c_{j+1,i} \right]. \\[2ex] 0 & \qquad otherwise \end{cases} \qquad (12)$$

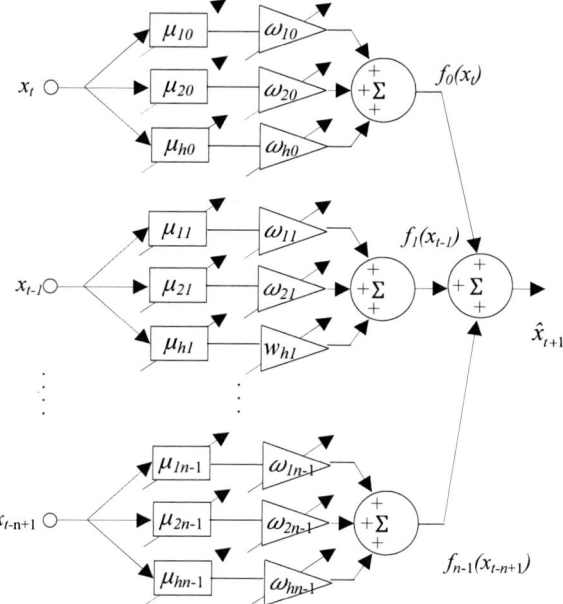

Fig. 3. One-step-ahead prediction by the NFN predictor.

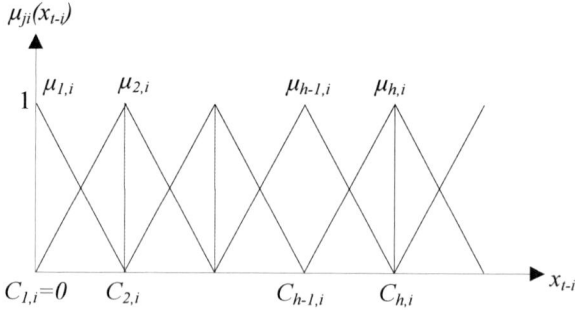

Fig. 4. The triangular membership functions.

So, the output value of the nonlinear synapse can be given by:

$$f_i(x_{t-i}) = \omega_{ji} \cdot \mu_{j,i}(x_{t-i}) + \omega_{j+1,i} \cdot \mu_{j+1,i}(x_{t-i}). \quad (13)$$

The weight ω_{ji} is updated by the incremental updating (Stepwise Training) learning algorithm.

$$\omega_{ji} = -\rho \cdot (\hat{x}_{t+1} - x_{t+1}) \cdot \mu_{ji}(x_{t-i}) \quad (14)$$

where ρ is the learning rate and x_{t+1} is the desired output.

Finally, the long-term prediction using NFN can be formulated as follows:

The prediction of x_{t+1} is given by:

$$\hat{x}_{t+1} = f_0(x_t) + f_1(x_{t-1}) + \ldots + f_{n-1}(x_{t-n+1}) \quad (15)$$

where $\hat{x}_{t+1}$ is the predicted value at time $t+1$, f_i ($i=0,1,\ldots,n-1$) represents the NFN predictor.

Based on the first prediction, the prediction at time $t+2$ is given by:

$$\hat{x}_{t+2} = f_0(\hat{x}_{t+1}) + f_1(x_t) + \ldots + f_{n-1}(x_{t-n+2}) \quad (16)$$

In this case, the prediction of x_{t+l} noted $\hat{x}_{t+l}$ can be obtained by:

$$\hat{x}_{t+l} = f_0(\hat{x}_{t+l-r}) + f_1(x_{t+l-2r}) + \ldots + f_{n-1}(\hat{x}_{t+l-nr}) \quad (17)$$

The use of the NFN and ANFIS for the prediction of bearing condition comprises four steps: data acquisition, feature extraction, training-validation, and prediction.

- Step1. *Data acquisition:* this step is used to collect a set of vibration signals from accelerometers placed on bearings. This set covers a range of historical signals of bearings operated until the occurrence of a failure. These bearings must have the same characteristics and the same operating condition of the bearing being tested.
- Step 2. *Feature extraction:* the root mean square of the vibration signal is used to extract a set of historical data indicating the degradation of bearings.
- Step 3. *Training and validation:* the historical dataset obtained from the previous step is split into two parts: training set and validation set. The training set is used to estimate the parameters of ANFIS and the NFN. The validation set is used to validate the models.
- Step 3. *Prediction:* the NFN and ANFIS are used to estimate the evolution of the root mean square in time.

III. EXPERIMENT

In order to test the proposed approach, bearing run-to-failure tests were performed on a test bench designed by Rexnord Corp. The experimental data were downloaded from the Prognostics Center of excellence [http://ti.arc.nasa.gov/tech/dash/pcoe/prognostic-data-repository/]. The test bench, shown in Fig. 5, is composed of four Rexnord ZA-2115 double row bearings installed on one shaft. The rotation speed was kept constant at 2000 rpm with 6000 lb radial load placed onto the shaft and bearing by a spring mechanism. A magnetic plug installed in the oil feedback pipe collects debris from the oil as evidence of bearing degradation.

The vibration signals of these bearings were recorded from October 2003 to November 2003. The root mean square (RMS) of the vibration signal of bearing 1, 2, 3 and 4 are shown in Fig. 6. The acquisition time is 35 days. The acquisition will stop when the accumulated debris adhered to the magnetic plug exceeds a certain level and causes an electrical switch to stop the test.

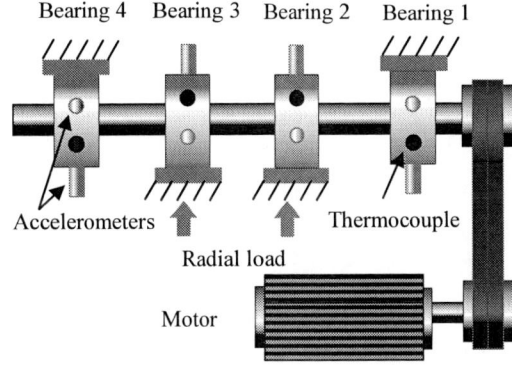

Fig. 5. Bearing test rig [15].

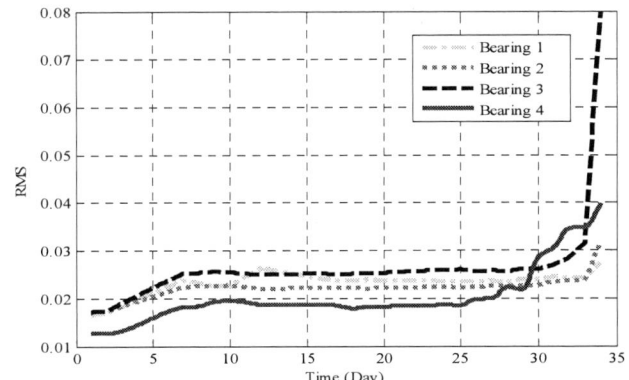

Fig. 6. RMS of bearing 1, 2 , 3 and 4.

The root mean square of the vibration signals of bearing 4 is used to evaluate the performance of the ANFIS and the NFN for the prediction of its evolution in time. For this purpose, we suppose that the vibration data of bearings 1, 2 and 3 are available. These data are used to identify the parameters of ANFIS and the NFN predictor.

The prediction of the future condition of bearing 4 consists of two steps: off-line and on-line. In the off-line step, the RMS of the vibration signals of bearings 1, 2 and 3 are used to train the ANFIS and NFN predictor. In the on-line step, the RMS of the vibration signals of bearing 4 is used as inputs in the ANFIS and the NFN predictor in order to predict its trending in time.

Figures 7(a), (b), (c) show the training and validation of the NFN and the ANFIS model for the RMS of the vibration signal of bearings 1, 2 and 3 respectively. There are four inputs ($n=4$) for each of the ANFIS and the NFN model. Two Gaussian membership functions are chosen for each input of the ANFIS model and fifteenth triangular membership functions are chosen for each input of the NFN model. The parameters of the membership functions (MFs) are adjusted from the training set in order to reduce the error between the output of the model and the actual values of the RMS.

In order to evaluate the prediction performance, the Root Mean Square Error is used:

$$RMSE = \sqrt{\frac{1}{l}\sum_{i=1}^{l}\left(x_{t+i} - \hat{x}_{t+i}\right)}. \qquad (17)$$

where l is the total number of data points, x_{t+i} and $\hat{x}_{t+i}$ are the ith actual and predicted value, respectively.

For the NFN, the difference between the actual values and the predicted values are identical with small RMSE values ranging from $4.28*10^{-4}$ to $3.56*10^{-4}$. These results indicate that the learning capability of the NFN model is excellent. Also, the ANFIS model is trained from the RMS of bearings 1, 2 and 3 with a RMSE of 0.0011, 0.0065 and 0.0011 respectively. These values are higher than those obtained by the NFN model. The reason is the small number of membership functions used in the training process. To obtain a small RMSE of each bearing, the number of MFs in ANFIS must increase. However, this will increase the computational complexity.

With the aim of forecasting the future condition of bearing 4, the RMS of the vibration signals of the first height days of bearing 4, in addition to those of bearings 1, 2 and 3 are used to train the parameters of the ANFIS and the NFN predictor.

Figure 8 shows the prediction given by the NFN and the ANFIS models for bearing 4. The values of the RMSE for these two predictors are summarized in table I. As we can see, the RMSE of the ANFIS model is higher than those of the NFN. The values given by the NFN indicate the condition of bearing 4 with a long-term prediction and a small RMSE. The ANFIS model can be improved by increasing the number of MFs. As we can see in table I, the RMSE of the ANFIS predictor for 2 and 3 membership functions are 0.1219 and 0.0035 respectively.

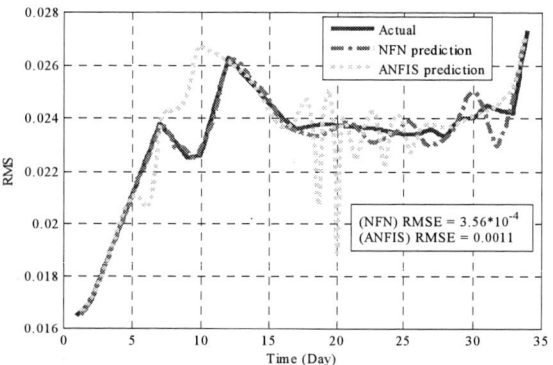

(a)   Training and validation results of the RMS of the vibration signal applied on bearing 1

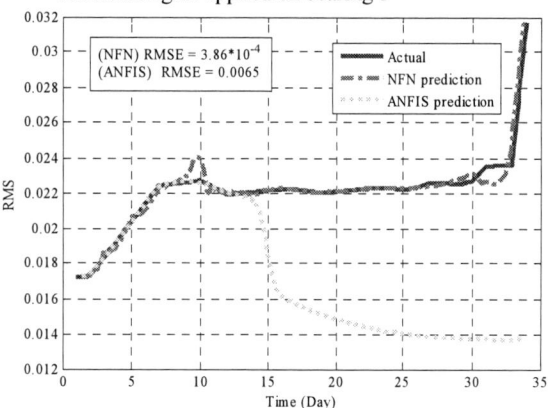

(b)   Training and validation results of the RMS of the vibration signal applied on bearing 2

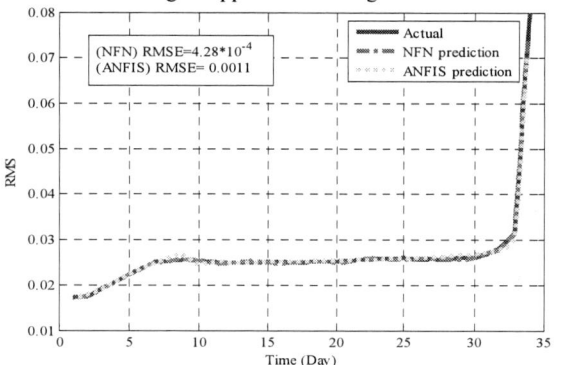

(c)   Training and validation results of the RMS the vibration signal applied on of bearing 3

Fig .7.   Training and validation results of the ANFIS and the NFN.

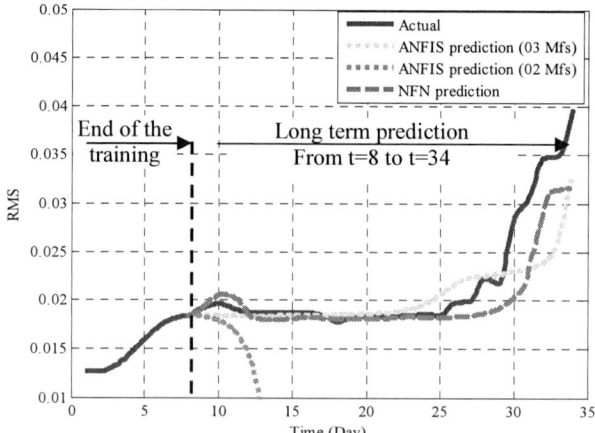

Fig. 8. Predicted results of ANFIS and NFN.

978-1-4799-0024-4/13 $31.00 © 2013 IEEE

Table I
RMSE of ANFIS and NFN

Data type	Training		
	NFN	ANFIS	
RMS of bearing 1	$35.6 \; 10^{-3}$	0.0011	
RMS of bearing 2	$38.6 \; 10^{-3}$	0.0065	
RMS of bearing 3	$42.8 \; 10^{-3}$	0.0011	
	Testing		
	15 MFs	2 MFs	3 MFs
RMS of bearing 4	0.0029	0.1219	0.0035
Computation time (sec)	15	4	75

IV. CONCLUSION

A new approach for long-term prediction of bearing condition based on the neo-fuzzy neuron has been proposed in this paper. The NFN has been combined with the one-step ahead prediction and used as an extrapolation tool in order to predict the evolution of the RMS in time. This approach has been compared with the adaptive neuro-fuzzy system and validated experimentally by analysing bearing data provided by the Center for Intelligent Maintenance Systems, University of Cincinnati. The RMS extracted form bearing vibration signals is considered as a good parameter for tracking the evolution of bearing condition. The obtained results show the efficiency of the proposed approach for the prognosis of bearings.

To apply the proposed approach, a significant amount of past knowledge of the bearing being tested is required because the degradation process must be known in advance and well described. This allows to obtaining a reliable long-term prediction. But on the other hand, limits the applicability of this approach when the availability of historical data is very difficult to obtain.

ACKNOWLEDGEMENT

The authors would like to express their thanks to the "Supervision and maintenance tasks in a shared organizational environment" project under the France-Venezuela Postgraduate Cooperation Program (PCP) and to the Intelligent Maintenance System (IMS) Center and Rexnord Technical services for their database.

V. REFERENCES

[1] P. F. Albrecht, J. C. Appiarius, R. M. McCoy, E. Owen, and D. K. Sharma, "Assessment of the reliability of motors in utility applications - updated," *Energy Conversion, IEEE Transactions on*, vol. 1, pp. 39–46, 1986.

[2] I. C. Report, "Report of large motor reliability survey of industrial and commercial installation, Part I and Part II," *Industry Applications, IEEE Transactions on* , vol. 21, pp. 853-872, 1985.

[3] AFNOR, "Condition monitoring and diagnostics of machines - prognostics - part 1: General guidelines. NF ISO 13381-1," 2005.

[4] S. Zaidi, S. Aviyente, M. Salman, K.-K. Shin, and E. Strangas, "Prognosis of gear failures in dc starter motors using hidden markov models," *Industrial Electronics, IEEE Transactions on*, vol. 58, no. 5, pp. 1695–1706, 2011.

[5] C. Chen, B. Zhang, G. Vachtsevanos, and M. Orchard, "Machine condition prediction based on adaptive neuro-fuzzy and high-order particle filtering," *Industrial Electronics, IEEE Transactions on*, vol. 58, no. 9, pp. 4353–4364, 2011.

[6] E. Strangas, S. Aviyente, J. Neely, and S. Zaidi, "Improving the reliability of electrical drives through failure prognosis," in *Diagnostics for Electric Machines, Power Electronics Drives (SDEMPED)*, 2011, pp. 172–178.

[7] A. K. Mahamad, S. Saon, and T. Hiyama, "Predicting remaining useful life of rotating machinery based artificial neural network." *Computers & Mathematics with Applications*, vol. 60, no. 4, pp. 1078–1087, 2010.

[8] T. Benkedjouh, K. Medjaher, N. Zerhouni, and S. Rechak, "Fault prognostic of bearings by using support vector data description," in *Prognostics and Health Management (PHM)*, 2012, pp. 1–7.

[9] J. Liu, W. Wang, and F. Golnaraghi, "A multi-step predictor with a variable input pattern for system state forecasting," *Mechanical Systems and Signal Processing*, vol. 23, no. 5, pp. 1586 – 1599, 2009.

[10] W. Wang, "An adaptive predictor for dynamic system forecasting," *Mechanical Systems and Signal Processing*, vol. 21, no. 2, pp. 809 – 823, 2007.

[11] T. Yamakawa, E. Uchino, T. Miki, and H. Kusanagi, "A neo fuzzy neuron and its applications to system identification and prediction of the system behavior," *Int Conf on Fuzzy Logic and Neural Networks*, 1992, pp. 477–483.

[12] E. Uchino, and T. Yamakawa, "Soft computing based signal prediction, restoration and filtering," *Fuzzy Logic, Neural Networks and Genetic Algorithms*, 1997, pp. 331–349.

[13] T. Miki, and T. Yamakawa, "Analog implementation of neo-fuzzy neuron and its on-board learning," *Computational Intelligence and Applications*, 1999, pp. 144–149.

[14] F. Zhao, J. Chen, L. Guo, and X. Li, "Neuro-fuzzy based condition prediction of bearing health," *Journal of Vibration and Control*, vol. 15, no. 7, pp. 1079–1091, 2009.

[15] H. Qiu, J. Lee, J. Lin, and G. Yu, "Wavelet filter-based weak signature detection method and its application on rolling element bearing prognostics," *Journal of Sound and Vibration*, vol. 289, no. 45, pp. 1066 – 1090, 2006.

BIOGRAPHIES

Abdenour Soualhi was born in Sétif, Algeria, on May 16, 1985. He obtained in 2010 the Master Degree in Electrical Engineering from the "Université Claude Bernard Lyon I" at Lyon. He is currently engaged in PhD research in the department of electrical engineering at the laboratory Ampère. His research interests are in the diagnosis and prognosis of faults by means of artificial intelligence techniques.

Guy Clerc (M'90-SM'10) was born in Libourne, France, on November 30, 1960. He received the Engineer's degree and the PhD in electrical engineering from the Ecole Centrale de Lyon, France, in 1984 and 1989, respectively. He is Professor of Universities. He carried out researches on control and diagnosis of induction machines at Ampère - UMR 5005, Villeurbanne, France.

Hubert Razik (M'98-SM'03) was graduated from the Ecole Normale Supérieure, Cachan, France, in 1987. Since November 2009, he is professor of electrical engineering at the "Université Claude Bernard Lyon I". His fields of research include the modelling, the control the monitoring conditions of multi-phase induction motor.

Francklin Rivas (M'93) was born in Mérida, Venezuela, on 1969. He received the Engineer's degree, Master and the PhD in Applied Science from the Universidad de Los Andes, Mérida, Venezuela in 1993, 1996 and 2000, respectively. He is Professor and the Director of Artificial Intelligence Laboratory of the Universidad de Los Andes.

Bar Breakage Mechanism and Prognosis in an Induction Motor

Vicente Climente-Alarcon, *Member, IEEE*, Jose Alfonso Antonino-Daviu, *Senior Member, IEEE*, Elias Strangas, *Member, IEEE* and Martin Riera-Guasp, *Senior Member, IEEE*

Abstract— In recent years several methods have been proposed in order to detect broken bars in the cage of induction motors, yielding the necessity of testing their performance in different conditions and their thresholds in assessing incipient cage damage. This paper presents the first results of a fatigue test designed to produce in the most natural way the breakage of a bar in order to determine the effectiveness of different motor current signature analysis. The high accuracy of the transient methods based on the study of the startup current, allows proposing some physical models of the breakage with the aim of establishing the remaining life of the induction motor, once the incipient defect has been detected.

Index Terms—Induction motors, fault diagnosis, rotor broken bar, frequency analysis, time-frequency analysis, fatigue, prognosis

I. INTRODUCTION

Rotor asymmetries, and among them, bar and ring breakages, are responsible for about 10% of the failures of induction motors [1]. Generally, this fault involves big units started under high inertia, with high repair costs and often difficult replacement.

The slow evolution of the defect, that propagates from bar to bar until yielding the machine useless, makes especially suitable the application of condition monitoring and, therefore, the diagnosis of this kind of fault has been a subject of detailed study in the last years, applying, in order to detect it, a wide battery of signal processing methods to magnitudes such as vibrations [2], voltages [3] and currents [1]; prevailing the techniques belonging to this latter group, especially the ones that use the decomposition in frequency of the current, known as Motor Current Signature Analysis (MCSA), which are of widespread use in the industry.

Recently, research has been focused in special aspects of rotor asymmetries, such as the influence of other factors in the traditional MCSA-based diagnosis [4], incipient fault detection [5] and diagnosis in transient regime [6], under which the FFT analysis proves useless. In fact, one of the seminal works on transient current analysis, by Elder et Al.,

The authors are with the Department of Electrical Engineering and Automation, Aalto University, P.O. Box 13000, Aalto, 00076 FINLAND, email: viclial@ieee.org; Instituto de Ingeniería Energética, Universitat Politècnica de València, Camino de Vera s/n, 46022, Valencia, SPAIN, (phone: 0034-963877592, fax: 0034-963877599), e-mails: joanda@die.upv.es, mriera@die.upv.es; and Engineering College, Michigan State University, MI, USA, email: strangas@egr.msu.edu.

pointed out towards combining both of these approaches [7] in the study of direct startup transient since:

[...] Throughout the starting transient period however, (particularly if started Direct on Line) large currents flow in the motor, even under no load conditions. During this short period, the machine is under conditions of severe electrical and mechanical stress [...] under these conditions, machine faults such as high resistance or broken rotor bars could be detected at much earlier stage. [...]

Thus, a fatigue test was designed to prove this hypothesis. An induction motor would suffer heavy startups and stationary periods until this kind of fault developed naturally. The comparison of the resulting rotor asymmetry indicators for the transient regime and the stationary ones would determine which technique detected before the incipient fault.

Furthermore, a particle filtering approach will be used as classification method in order to determine the state of the rotor. Particle filtering (PF), or Sequential Monte Carlo methods [8], has already been successfully applied to fault detection and failure prognosis of mechanical faults due to its versatility [9]. The approximation of the conditional state probability distribution of the machine by a swarm of points having different weights, called particles enables the fusion of the information from different sources and the use of non-linear state models, an improvement over Kalman filters [10].

For these purposes, this work is structured as follows: in section II the experimental procedure used for development of the fault –already introduced in [11]– is completed, a summary of the results are presented in Section III, followed in Section IV by a discussion on them. Section V is dedicated to the fitting of the bar breakage pattern obtained from the motor transient current analysis to known fatigue failure expressions, in order to determine the physical models with application in prognosis. Section VI introduces a state estimation model using the particle filtering approach, and finally, Section VII yields the conclusions.

II. EXPERIMENTAL PROCEDURE

The procedure undertaken to obtain the bar breakage, introduced in [11], is summarized in this section. In addition, the modifications added are presented.

A. Test stand

A one pole pair induction motor whose characteristics are exposed in the Appendix A was coupled to a DC Machine having an inertia of 0.11 kg·m^2 (Fig. 1). Currents in two phases (A and C), vibrations at the middle of the stator, rotation speed and the stator and external temperatures were recorded for each cycle. Fig. 2 shows the first four magnitudes for the test number 80,000. The startup transient lasted seven seconds, being followed by stationary operation until around the 20th second after connection. A plug stopping, lasting six seconds, was added after cycle 60,000 in order to increase the stress in the rotor cage. Further details of the experimental setup and the results after the first 40,000 cycles were presented in [11].

The vibration signals show two resonant rotating frequencies during the acceleration and braking (Fig. 2). A study of the startup by means of higher sensitivity accelerometers and a time-frequency decomposition tool was presented in [12].

Fig. 1. Experimental test stand

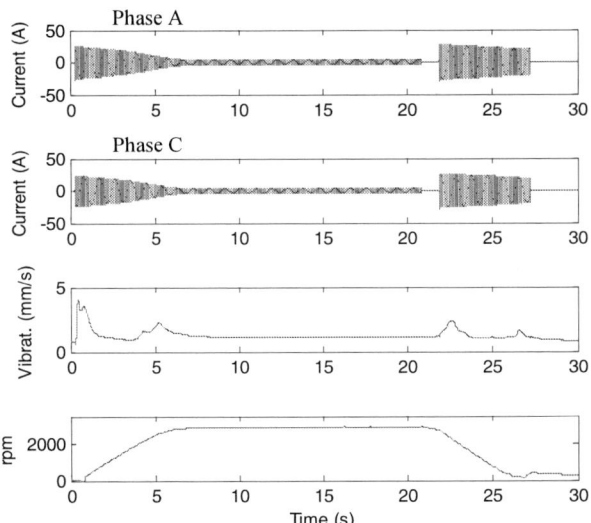

Fig. 2. Signals recorded for test number 80,000. Currents in two phases, vibration and speed (plus stator and external temperature not shown).

B. Stationary indicators computed

As it was presented in [11], several fault indicators, in this case adapted to the detection of a rotor asymmetry, were computed. Three indicators based on the study of the stator current during the last 10 seconds of stationary operation of each test were calculated.

1) The first one compares the amplitude of the low sideband harmonic, LSH, to the amplitude of the fundamental component from the current spectrum (1). In order to increase its accuracy, a Blackman window was used to avoid leakage effects and the Teager-Kaiser algorithm was utilized to enhance the detection of the peaks [13].

$$\gamma_{F1}(dB) = 20 \cdot log\left(\frac{I_1}{I_{LSH}}\right) \tag{1}$$

2) The second indicator modifies (1) also taking into account the amplitude of the upper sideband harmonic (2), as it is recommended in the bibliography [14].

$$\gamma_{F2}(dB) = 20 \cdot log\left(\frac{I_1}{I_{LSH} + I_{USH}}\right) \tag{2}$$

3) Finally, also for the stationary operation, a third indicator was computed. It is based on studying the spectrum of the envelope obtained from the application of the Hilbert transform to the stator current [15]. This enables the suppression of the main current component and thus improved sideband detection. Their energy is compared to the amplitude of the 100 Hz harmonic (3).

$$\gamma_{Hilbert}(dB) = 20 \cdot log\left(\frac{I_{100}}{I_{2fs} + I_{4fs}}\right) \tag{3}$$

C. Transient indicators computed

For transient operation, Wavelet-based indicators (4) and the integration of the energy of the LSH using the Wigner-Ville distribution (5) were computed:

4) The Wavelet indicator [11], [16] takes into account the energy present in a whole band whose limits are determined by the decomposition, by comparing the energy present in one detail or approximation χ to the energy in the original current waveform i, between the limits Ns and Nb, to avoid edge effects (4).

$$\gamma_{\chi}(dB) = 10 \cdot log\left[\frac{\sum_{j=Nb}^{Ns} i_j^2}{\sum_{j=Nb}^{Ns}[\chi(j)]^2}\right] \tag{4}$$

5) The Wigner-Ville indicator, proposed in [17] and improved in [18], however, utilizes this transform to obtain a time-frequency distribution of the energy of the current in a range whose limits can be set freely by use of band pass and notch filters. This time-frequency boxes are selected to isolate prevailing fault harmonics, whose evolution can be tracked by means of the computation of the instantaneous frequency and their energy $e_{if,bw}$ integrated on that track, therefore avoiding

contributions of other harmonics or interferences (5):

$$\gamma_{W,\chi}(dB) = 10 \cdot log\left[\frac{2\sum_{t_c} i_j^2}{e_{if,bw}}\right] \qquad (5)$$

In this case, the selected time-frequency box corresponds to the last part of the evolution of the LSH harmonic, between 0 and 45 Hz during the second part of the startup, in which the effect of cage's resistance is predominant in the rotor currents. Advanced FIR filters, as proposed in [18] have been used for isolating the evolution of this component and the indicator (5) has been normalized, yielding values comparable to the application of (1) and (2) to the analysis of a rotor asymmetry in stationary operation.

III. RESULTS OF THE FATIGUE TEST

The fatigue test comprises 82,265 startups. As no evolution was seen in the indicators, the rotor was progressively weakened. This procedure was carried out removing material from the end rings, in their junction with the rotor bars. In [11] the results of a symmetric lathing process, carried out after startup 33,477, were presented. In spite of that, the prolongation of the test up to 79,007 startups showed no trend in the bar breakage indicators (both for stationary and transient operation), and it was decided to further debilitate one of the bars of the rotor. No procedure was carried out to select a specific one.

A. Modification of the original procedure

In order to proceed with this weakening, two holes were drilled in the junction of a bar to the short circuit ring, inside the groove left by the lathing process. The diameter of the holes drilled, initially 3 mm, was increased to 4 mm (bar section reduced to 14%), and then to 4.5 mm one of them (bar section reduced to 7%), until the bar breakage developed.

This machining made the original casted squirrel cage to become similar to a welded one, since the lathing process increased the cantilevered disposition of the rings, and the holes near the bar produced a configuration alike to a bar protruding from the rotor iron to reach the ring.

B. Indicators behavior

Table I shows the average value of the stationary indicators (1), (2), (3) during the fatigue test. The first epoch corresponds to a healthy rotor. The second spans the operation of the rotor cage after the lathing process, which removed 70% of the material in contact with the bars. Finally, the third epoch covers the final startups of the tests, when one bar was weakened drilling two holes around its junction to the end ring. Each indicator is calculated for both phases A and C. Table II introduces the results of the study of the LSH during the startup using the transient indicators (4), (5).

The difference of value before and after the bar breakage is clear for all the cases. It is remarkable that a clear increase in all indicators appears when the remaining section of the bar was reduced to 14% of the original one (see column designed as "compensation" and (a1) in Fig. 3, 4, 5). This behavior

probably is due to the fact that the LSH produced by the forced incipient fault is combined with the small LSH due to the inherent asymmetry of the cage, performing as a double breakage; but as it is shown in [19] and depending on the relative position of both fault components (incipient and inherent LSH in this case), the resultant LSH can grow or can be compensated reducing its value, as happens in the performed test, yielding higher indicators.

TABLE I
EVOLUTION OF THE STATIONARY INDICATORS DURING THE FATIGUE TEST.

Indicator Phase		First epoch	Second epoch	Third epoch	
				Compensation	After Breakage
Cycles→		1 – 33,477	33,478 – 79,007	80,405 – 80,857	81,216 – 81,653
$\gamma_{F1(1)}$	A	42.8	40.9	49.3	39.8
$\gamma_{F2(2)}$	A	41.3	39.5	46.2	37.5
$\gamma_{F1(1)}$	C	42.5	40.8	46	40.1
$\gamma_{F2(2)}$	C	41	39.3	43.2	37.6
$\gamma_{Hilbert,\,(3)}$	A	30.9	30.2	31.9	28.3
$\gamma_{Hilbert,\,(3)}$	C	31	29.8	30.5	28.4

TABLE II
EVOLUTION OF THE TRANSIENT INDICATORS DURING THE FATIGUE TEST.

Indicator Phase		First epoch	Second epoch	Third epoch	
				Compensation	After Breakage
Cycles→		1- 33,477	33,478 - 79,007	80,405 – 80,857	81,216 – 81,653
$\gamma_{Apr6,\,(4)}$	A	46.2	44	47.2	35.7
$\gamma_{Apr6,\,(4)}$	C	46.2	44	47.4	35.9
$\gamma_{Det7,\,(4)}$	A	46.5	44.5	47.7	36.8
$\gamma_{Det7,\,(4)}$	C	46.5	44.5	47.7	37
$\gamma_{Apr7,\,(4)}$	A	58.1	53.9	57.2	42.2
$\gamma_{Apr7,\,(4)}$	C	58.3	53.9	58.8	42.8
$\gamma_{Det8,\,(4)}$	A	61.1	54.7	59.6	42.7
$\gamma_{Det8,\,(4)}$	C	61.2	54.7	59.9	43.1
$\gamma_{W\text{-}LSH,(5)}$	A	-	43.5	55.7	33.5
$\gamma_{W\text{-}LSH,(5)}$	C	-	43.5	56.2	33.9

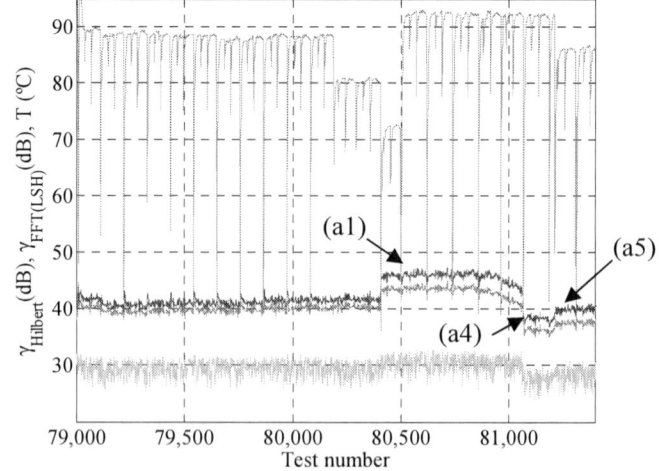

Fig. 3. Stationary rotor asymmetry indicators and stator temperature during the last stage of the fatigue test, phase C. Amber, Hilbert indicator $\gamma_{Hilbert}$ (3); Red, FFT indicator γ_{F2}(2); Blue, FFT indicator γ_{F1}(1). Rose, stator temperature.

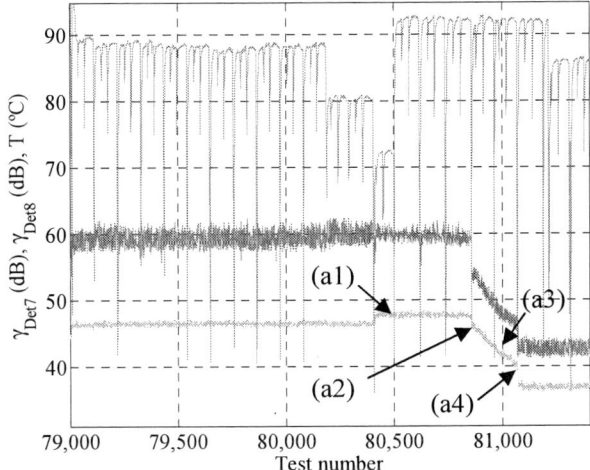

Fig. 4. Transient rotor asymmetry Wavelet indicators and stator temperature during the last stage of the fatigue test, phase A. Orange, Wavelet Detail 7 indicator γ_{Det7}, (4); red, Wavelet Detail 8 indicator γ_{Det8}, (4). Rose, stator temperature.

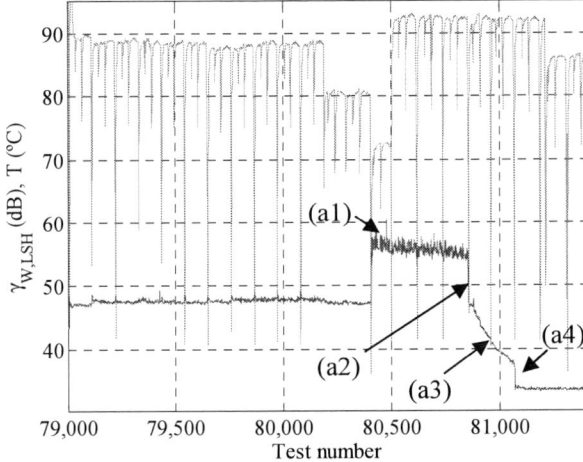

Fig. 5. Transient rotor asymmetry Wigner-Ville-based indicator $\gamma_{W\text{-}LSH}$,(5) (blue) for phase A, and stator temperature during the last stage of the fatigue test (rose).

This result is interesting, since it suggests that an incipient fault, in its initial stage, should be detected through a slight variation of the fault indicators (increase or decrease) and not necessarily through a decrease of these parameters.

Moreover if Fig. 3 showing the stationary indicators is compared to Fig. 4 and 5, which present the transient ones, a much clearer evolution is detected in the last ones once the section of the bar was reduced to 7% of its original value. First of all, the increase of the diameter of one drill from 4 mm to 4.5 mm is clearly presented by a sudden reduction of the indicators (a2). Further, an unambiguous trend in the transient indicators is observed, as the bar weakens (a3). The full breakage occurred after the second startup following a long stop in which the machine reached room temperature (a4).

Thus, the conclusions of the fatigue test support the hypothesis presented in [7]: the study of the direct startup of an induction motor yields a more accurate view of its rotor asymmetry.

C. Bar inspection

Once the rotor was inspected after the indicators stabilized, surprisingly no apparent damage was seen. The bridge between the holes was still in its position. However, a detailed comparison between the pictures taken before and after the breakage (Fig. 6 and Fig. 7) showed some damage, material lost at the junction between the ring and the bar and a crack down the 4 mm drill.

Fig. 6. Bridge between the bar and the end ring before the breakage.

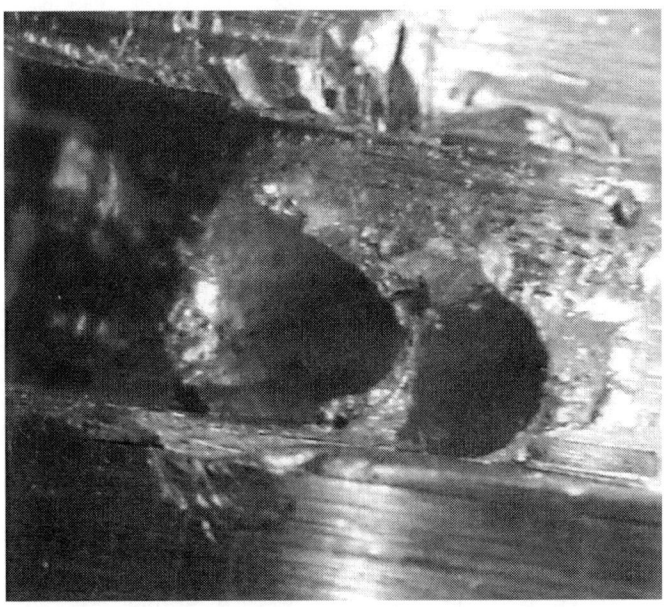

Fig. 7. Bridge between the bar and the ring after the breakage.

Fig. 8 presents a combination of pictures showing this crack (b2) using materials lenses, whose limited depth of field only allow to correctly focusing a few millimeters of the hole at a time. The crack clearly continues deep inside it. The cavities left by the material lost in the upper side of the bar are well seen (b1). Furthermore, an aluminum drop hanging from the hollow where it was formed is also identified (b3).

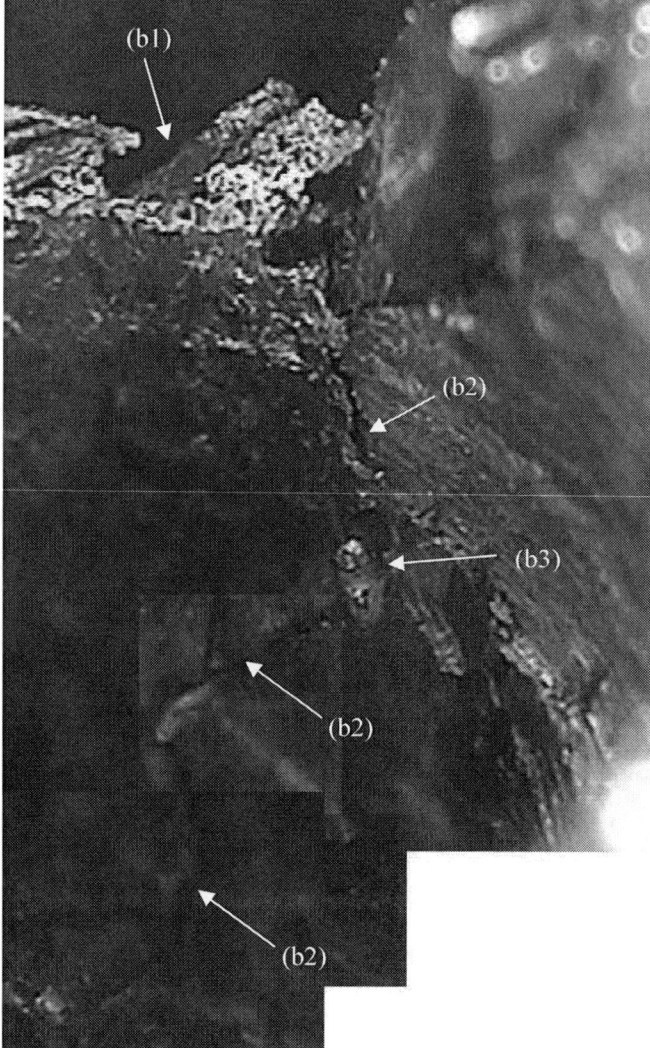

Fig. 8. 4 mm drill after the breakage. A crack and a molten aluminum drop are clearly visible.

IV. DISCUSSION OF THE RESULTS

The bar breakage process achieved during the fatigue test showed a more complicated signature than initially thought.

A. Compensation of a previous rotor asymmetry

It is observed the evolution of the rotor asymmetry indicators contrary to what it is expected, increasing its value as a bar is weakened (Fig. 3, 4, 5 and Tables I and II). This effect is attributed to the compensation of a slight previous asymmetry due to an intrinsic defect in a bar at half the pole pitch of the one that was weakened, as it is indicated in [19].

B. Influence of other factors on the traditional indicators

The disassembling and reassembling of the motor in order to check the breakage, has a clear effect in the stationary indicators that increase their value (Fig. 3, a5), whilst the transient based ones remain unaffected (Fig. 4, 5). The average variation is 1.7 for (1), 1.35 dB for (2) and 0.8 dB for (3). This effect can be a product of a modification of the static

eccentricity as stated in [4]. Therefore, in experiments trying to detect an incipient breakage using stationary indicators, some slight variations of the indicators may be product of unintentional modifications in the position of the rotor within the stator bore when it was extracted in order to artificially cause the damage, and not an effect of the incipient fault.

C. Trends observed during the bar breakage

As the section of the bar is reduced to an estimated 14% of its original area (test number 80,405) and the compensation is achieved, a slight trend in the rotor asymmetry wavelet indicators is appreciated (Fig. 4). In the latter developed method based in the Wigner-Ville distribution, this evolution is wider (Fig. 5 and 9) (c1).

In 452 tests, the indicator is reduced in 1.7 dB (Fig. 9). In addition, the slope increases as this series of tests progresses.

Once the remaining section is reduced to the 7% of the original (test number 80,857), by increasing the diameter of a hole in 0.5 mm (Fig. 3, 4, 5 (a2), Fig. 9 (c2)), the evolution is much faster (a3, c3) until the bar abruptly breaks (a4, c4) after 214 additional tests, that is, between cycles 81,070 and 81,071. There is no further evolution in the indicator (c5).

This behavior well fits into the evolution of a crack: a growing period (c1, c3) and once the section has been reduced enough, a sudden breakage (c4) occurs.

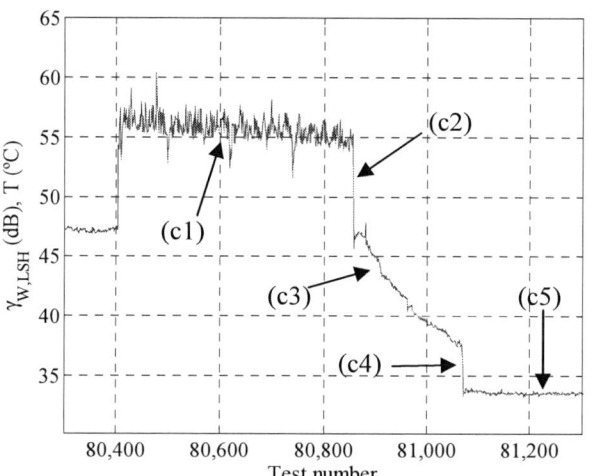

Fig. 9. Wigner-Ville indicator $\gamma_{W\text{-}LSH}$ (5) assessing the energy of the LSH in the startup during the bar weakening and breakage.

It must be noted that although there are signs of creep at the upper part of the remaining bar, with material lost, and in the hole itself (the aluminum drop), the breakage mechanism according to Fig, 9, corresponds to the behavior of a crack propagation.

V. PROPOSED FATIGUE MODELS

The study of the startup transient of the induction machine as the bar breaks, whose results are exposed in Section III and discussed in Section IV, yields accurate enough results on the asymmetry of the rotor cage to adapt the corresponding failure mechanism model, with the aim of carrying out the prognosis of the fault.

A. Currents in bars

The motor was modeled in FLUX™ in order to obtain the currents in the bars and end rings, to be used as an input for the fatigue models. The average currents during each stage are presented in Table III.

In the case one bar is broken; the bars next to this suffer an increase of the average current I_{bb}, also computed by FEM for each period.

TABLE III
AVERAGED CURRENTS IN BARS

Period	I_{RMS} (A)	$I_{bb\ RMS}$ (A)	t(sec)
Startup	735.8	994.2	5.5
Stationary	199.2	235.9	15
Plug stopping	1238.7	1792.6	5.5

B. Low-Cycle (Thermal) Fatigue

In order to model the bar breakage from the data shown in Fig. 9, both evolutions (c1) and (c3) are fit by quadratic equations. To overcome the discontinuity produced by enlarging the hole (c2), a number of tests equal to the necessary to achieve half of the indicator's evolution caused by this modification is estimated for each series. This yields a total of 1,181 tests for the weakening period: 452 in the slow evolution (c1), 213 for the fast evolution until breakage (c3) and 516 needed to extrapolate the evolution between both series (Fig 10).

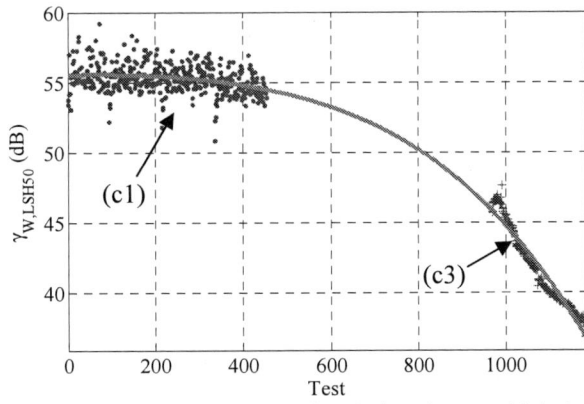

Fig. 10. Interpolation of both series of results in order to model the breakage evolution

Qualitatively, from the relatively rapid development of the failure, below ten thousand cycles, it is inferred that the breakage can be modeled as low cycle fatigue [20]. In addition, the behavior of the transient indicators and the effects that appeared on the surface of the aluminum (Fig. 8, b2), point out to the predominance of stress-strain, caused by thermal expansion and contraction, over creep. Therefore, the equations modeling low cycle (thermal) fatigue and crack propagation could be applied to estimate the remaining life of a bar, once an evolution in the rotor asymmetry indicators based on the study of the transient current has been detected.

A first approximation to the remaining life could be obtained from the Coffin-Manson expression:

$$\Delta\varepsilon_p \cdot N^b = C \qquad (6)$$

where b ranges between 0.5 and 0.7 and C is a constant of the material, $\Delta\varepsilon_p$ is the total plastic range and N is the life expectancy of the sample in cycles. This equation has already been applied to the study of thermal fatigue on an aluminum bar whose ends are constrained [21], a similar case to the breakage of one bar in a welded cage. Following the same steps as in [21] (see Appendix B) and taking into account the characteristics of the Aluminum alloy 1060-O (σ_{str}=70 MPa, σ_{Yield}=28 MPa, ε_f=0.35) [22], it is obtained a value of C equal to 0.265, which yields the following Coffin-Manson equation for estimating the life of the bar:

$$\Delta\varepsilon_p \cdot N^{0.6} = 0.265 \qquad (7)$$

This equation links the plastic deformation achieved in each cycle $\Delta\varepsilon_p$ with the total number of cycles N for an averaged value of b, however, the plastic deformation of the weakened bar is not easy to calculate in this case. Therefore, the fitting will be made to the specific energy dissipated in a section of the bar during a cycle, from the currents in bars computed by FEM analysis and exposed in Table III.

$$K \cdot E_{cycle} \cdot N^{0.6} = 0.265 \qquad (8)$$

For a resistivity of $47.8 \cdot 10^{-9}$ $\Omega \cdot$m obtained from [23] and corrected to a temperature of 100 ºC, and the original section of the bar at the beginning of the evolution, 6 mm², the energies dissipated during the transient periods of the test are:

TABLE IV
ENERGY DISSIPATED IN THE BAR PER MM³

Period	E(joules)
Startup	4
Stationary	0.8
Plug stopping	30.6

As during both transients the energy dissipated is different, in order to adjust the fatigue expressions, it is assumed an average of them, having energy of 17.3 J/mm³. Therefore, for fatigue considerations, there are two cycles in each test, having that average energy for every startup and stop of the motor, so N = 2,362, which yields K = 1.45·10⁻⁴.

This parameterization allows predicting the bar breakage once the incipient fault has been detected. A further extrapolation taking into account the increase of currents in the neighboring bars presented in Table III would establish the remaining life of the machine operating under this fault.

C. Crack propagation

In addition, the rapid evolution indicated by (c3) after the section of the bar had been decreased to a 7% of its original value and its sudden collapse (c4) invites the application of fracture theory to model that behavior.

The Paris equation links the growth rate of a crack of length a to the stress intensity factor ΔK, which is linked to the range of cyclic stress, by means of two constants of the material, C and m:

$$\frac{da}{dN} = C \cdot \Delta K^m \tag{9}$$

For aluminum (m=3, C=10^{-12}) the integration of this equation yields:

$$\left| a_i^{-1/2} - a_0^{-1/2} \right| = -\frac{1}{6} \cdot C \cdot \Delta K^3 \cdot N_f \tag{10}$$

Taking into account the evolution of the Wigner-Ville parameter as evolution of the crack and fitting the expression to the total number of cycles, it gives $\Delta K = 6.4 \cdot 10^2$.

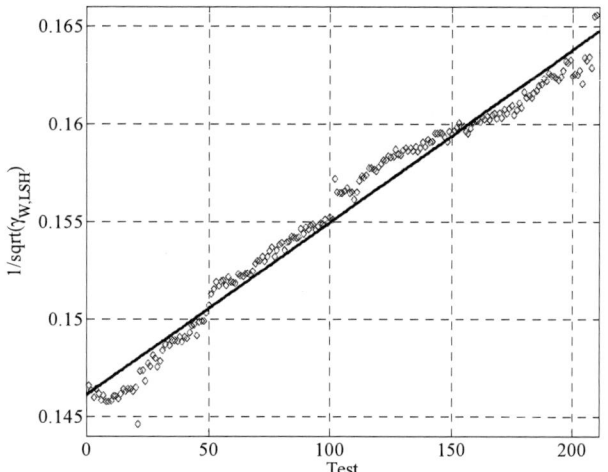

Fig. 11. Fitting of the Paris equation to the evolution of the Wigner-Ville indicator during the final stage of the breakage (c3).

In principle this result links the stress intensity factor to the transient indicators of the bar breakage obtained during a direct startup.

VI. STATE ESTIMATION BY PARTICLE FILTERING

Finally, in this section an implementation of a state estimation model according to the data obtained in the previous sections is applied to the values obtained using the indicator (2) on the current of phase C, which is depicted in Fig. 3.

According precedent results, three states have been defined: healthy, progression of damage and broken bar, numbered 1 to 3 as the severity of the fault increases.

For the definition of the state functions, the rate of the evolution of the failure has been taken into account, considering no evolution for healthy and broken bar and a approximate variation of the state of +1 in one-thousand cycles for the case of damage progression, according to the data obtained from Section V, point B.

The observation equations linking the state to the measurements are defined according to the data obtained from the fatigue test for each state, considered as healthy the values of the indicator above the average value for healthy state minus the standard deviation of those values and broken bar

the average values plus the standard deviation for that state. The progression damage state is defined to a value in-between. The procedure is complemented by a trend detection code that recognizes trends in the data, indicating progressing damage. The probability of each particle according to the measurement, which defines its weight, is derived accordingly, as a normal distribution centered on the detected state.

The results of applying the particle filtering algorithm for the state estimation of the rotor during the breakage are shown in Fig. 12. In this case, 10 particles have been used.

This procedure detects the damage progression at test number 81001, that is, 143 tests after it becomes clearly noticeable in the transient indicators. The final breakage is detected immediately. Some oscillations in the state indicator appear as a consequence of disassembling the motor to inspect the damage, as presented in Section IV point B.

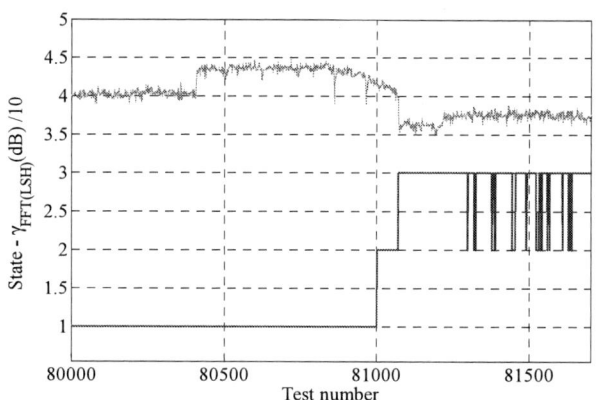

Fig. 12. Results of the state estimation through the particle filtering approach applied to the indicator 2C. Blue line, state (1, healthy; 2 damage progression, 3 broken bar), red line, 2C indicator value divided by 10.

VII. CONCLUSIONS

The results of a fatigue test designed to produce in the most natural way possible a bar breakage in an induction motor confirm the Elder et Al. hypothesis for this fault: the study of a direct transient allows tracking the rotor asymmetry accurately, and thus detecting a breaking bar in an earlier stage.

The result of the fatigue test also suggests that an incipient fault, in its initial stage, should be detected through a slight variation of the fault indicators (increasing or decreasing) and not necessarily through a decrease of these parameters, due to the possible initial compensation of inherent asymmetries, which randomly depends on the position of the defective bar.

The signature of the failure's evolution obtained by means of these transient current analysis permits the fitting of physical models paving the way for carrying out the prognosis of the fault, which will be presented in future works.

APPENDIX A

The characteristics of the motor tested are: star connected, rated voltage (U_n): 400V, rated power (P_n): 1.5 kW, 1 pole pairs, stator rated current (I_{1n}): 3.25 A rated speed (n_n): 2860 rpm.

APPENDIX B

For obtaining the constant C in the Coffin-Manson equation, it is considered that the upper limit of the fatigue test coincides with the failure in a tension test [21], that is, $N=0.25$ and $\Delta\varepsilon_p = \varepsilon_f$, with ε_f equal to 0.6 at 150 °C [22].

VIII. REFERENCES

[1] W.T. Thomson, M. Fenger, "Current signature analysis to detect induction motor faults", *IEEE Ind. Appl. Magazine*, pp. 26-34, July/August 2001.

[2] A. Bellini, C. Concari, G. Franchescini, C. Tassoni, A. Toscani, "Vibrations, currents and stray flux signals to asses induction motors rotor conditions." *IEEE Industrial Electronics, IECON 2006 - 32nd Annual Conference on*, 2006 , pp. 4963 – 4968

[3] A. Khezzar, M. Y. Kaikaa, M. El Kamel Oumaamar, M. Boucherma, H. Razik, "On the Use of Slot Harmonics as a Potential Indicator of Rotor Bar Breakage in the Induction Machine", *IEEE Trans. Ind. Electron.*, Vol.56, No.11, Nov. 2009, pp. 4592-4605.

[4] J.Faiz, B. M. Ebrahimi, H.A. Toliyat, W.S. Abu-Elhaija, Mixed-fault diagnosis in induction motors considering varying load and broken bars location, *Energy Conversion and Management*, Vol. 51, Issue 7, July 2010, pp. 1432–1441

[5] S. H. Kia, H. Henao, G.-A. Capolino, Zoom-MUSIC frequency estimation method for three-phase induction machine fault detection *Industrial Electronics Society, 2005. IECON 2005. 31st Annual Conference of IEEE*, 2005, pp. 2603-2608

[6] S.H. Kia, H. Henao, G. -A. Capolino, Windings monitoring of wound rotor induction machines under fluctuating load conditions, *IECON 2011 - 37th Annual Conference on IEEE Industrial Electronics Society*, 2011, pp. 3459-3465

[7] S. Elder, J. F. Watson, W. T. Thomson, "Fault detection in induction motors as a result of transient analysis", *Electrical Machines and Drives, 1989. Fourth International Conference on*, pp. 182 - 186, 1989

[8] M. E. Orchard, G. J. Vachtsevanos, A particle-filtering approach for on-line fault diagnosis and failure prognosis. *Transactions of the Institute of Measurement and Control*, 2009, 31:221

[9] N. J. Gordon, D. J. Salmond, A. F. M. Smith, Novel approach to nonlinear/non-Gaussian Bayesian state estimation, *IEE Proceedings-F*, Vol. 140, No. 2, April 1993

[10] O. Ondel, E. Boutleux, E. Blanco, G. Clerc, Coupling Pattern Recognition With State Estimation Using Kalman Filter for Fault Diagnosis, *IEEE Trans. Ind. Electron.* Vol. 59, No. 11, 2012

[11] V. Climente-Alarcon, M. Riera-Guasp, J. A. Antonino-Daviu, J. Roger Folch, "Experimental study of the evolution of a bar breakage process in a commercial induction machine", *Electrical Machines, 2008. ICEM 2008. 18th International Conference on*, pp. 1 - 5, 2008

[12] V. Climente-Alarcon, J. Antonino-Daviu, F. Vedreno-Santos, R. Puche-Panadero, Vibration transient detection of broken bars by PSH sidebands, *IEEE Trans. Ind. Appl.* Vol. 49, No. 6, Nov./Dic. 2013.

[13] V. Kandia, Y. Stylianou, "Detection of Creak Clicks of Sperm Whales in low SNR conditions" *IEEE Oceans – Europe* 2005 pp. 1052-1057

[14] A. Bellini, F. Filippetti, C. Tassoni, and G. A. Capolino, "Advances in Diagnostic Techniques for Induction Machines," *IEEE Trans. Ind. Electron.*, Vol. 55, 2008, pp. 4109-4126.

[15] R. Puche-Panadero, M. Pineda-Sanchez, M. Riera-Guasp, J. Roger-Folch, E. Hurtado-Perez, J. Perez-Cruz, Improved Resolution of the MCSA Method Via Hilbert Transform, Enabling the Diagnosis of Rotor Asymmetries at Very Low Slip, *IEEE Trans. Energy Conv.*, Vol. 24, No. 1, 2009, pp: 52 – 59.

[16] M. Riera-Guasp, J. A. Antonino-Daviu, M. Pineda-Sanchez, R. Puche-Panadero, and J. Perez-Cruz, "A General Approach for the Transient Detection of Slip-Dependent Fault Components Based on the Discrete Wavelet Transform," *IEEE Trans. Ind. Electron.*, Vol. 55, 2008, pp. 4167-4180.

[17] V. Climente-Alarcon, J. A. Antonino-Daviu, M. Riera-Guasp, R. Puche-Panadero, L. Escobar, "Application of the Wigner-Ville Distribution for the Detection of Rotor Asymmetries and Eccentricity through High-Order Harmonics", *Electric Power Systems Research*, no. 91, pp.28-36, October 2012

[18] V. Climente-Alarcon, J. Antonino-Daviu, M. Riera-Guasp, M. Vlcek, "Induction Motor Diagnosis by Advanced Notch FIR Filters and the Wigner-Ville Distribution", *IEEE Trans. Ind. Electron.*, (submitted).

[19] M. Riera-Guasp, M. F. Cabanas, J. A. Antonino-Daviu, M. Pineda-Sanchez, C. H. R. Garcia, "Influence of Nonconsecutive Bar Breakages in Motor Current Signature Analysis for the Diagnosis of Rotor Faults in Induction Motors", *IEEE Trans. Energy Conv.*, Vol. 25, 2010, pp. 80-89.

[20] G. R. Halford, Low-Cycle Thermal Fatigue, NASA Technical Memorandum 87225, Lewis Research Center, Cleveland, OH.

[21] E. Velasco, R. Colás, S. Valtierra, J. F. Mojica, A model for thermal fatigue in aluminium casting alloy, *Int. J. Fatigue*, Vol. 17, No. 6, 1995, pp.399-406.

[22] J. G. Kaufman, E. L. Rooy, Aluminum Alloy Castings: Properties, Processes and Applications, ASM International, 2004.

[23] M. Hodowanec, W. R. Finley, Copper Versus Aluminum-Which Construction Is Best? [Induction Motor Rotors], *IEEE Ind. Appl. Magazine*, Vol. 8, 2002, pp. 14-25.

IX. BIOGRAPHIES

Vicente Climente-Alarcon (M'12) received his M.Sc. degrees in Chemical and Industrial Engineering in 2000 and 2011, and his Ph.D. degree in Electrical Engineering in 2012, all from the Universitat Politècnica de València (Spain).
He has worked as Assistant Professor in the School of Industrial Engineering of the mentioned university, on research tasks in the area of condition monitoring of electrical machines, and externally as a consultant in automation and management of power systems. Currently he is carrying out postdoctoral research at the Department of Electrical Engineering and Automation, Aalto University, Espoo, Finland.

Jose Antonino-Daviu (S'04–M'08) received his M.S. and Ph. D. degrees in Electrical Engineering, both from the Universitat Politècnica de València, in 2000 and 2006, respectively. He was working for IBM during 2 years, being involved in several international projects. Currently, he is Associate Professor in the Department of Electrical Engineering of the mentioned University, where he develops his docent and research work. He has been invited professor in Helsinki University of Technology (Finland) in 2005 and 2007 and in Michigan State University (USA) in 2010. He has over 60 publications between international journals, conferences and books. His primary research interests are condition monitoring of electric machines, wavelet theory and its application to fault diagnosis and design and optimization of electrical installations and systems.

Elias G. Strangas (S'74–M'80) received the Dipl. Eng. degree in electrical engineering from the National Technical University, Athens, Greece, in 1976, and the Ph.D. from the University of Pittsburgh, Pittsburgh, PA, in 1980.
He has worked in industry (Schneider Electric) and at the University of Missouri-Rolla from 1983 to 1985. Since 1986, he has been with Michigan State University, where he directs the Electrical Machines and Drives Laboratory. His technical interests are in the design and control of electrical machines and drives and has published extensively in these fields. For the last ten years, he and his students have been working on the fault diagnosis and failure prognosis of electrical machines, using model and data based methods. Their work has been funded by industry and the State and Federal governments.

Martín Riera-Guasp (M'04-SM'12) received the M. Sc. degree in industrial engineering and the Ph.D. degree in electrical engineering from the Universitat Politècnica de València, Valencia (Spain), in 1981 and 1987, respectively.
Currently, he is an Associate Professor with the Department of Electrical Engineering, Polytechnic University of Valencia. His research interests include condition monitoring of electrical machines, applications of the Wavelet Theory to electrical engineering and efficiency in electric power applications.

Time-Frequency Complexity Based Remaining Useful Life (RUL) Estimation for Bearing Faults

Rodney K. Singleton II, Elias G. Strangas and Selin Aviyente

Abstract—Reliable operation of electrical machines depends on the timely detection and diagnosis of faults as well as on prognosis, i.e. estimating the remaining useful life (RUL) of the components. Bearings are the most common components in rotary machines and usually constitute a large portion of the failure cases in these machines. Although there has been a lot of work in the study of bearing life failure mechanisms and modeling, the problem is still far from being solved. In this paper, we introduce a time-frequency feature extraction based method for estimating remaining useful life of bearings from vibration signals. The proposed approach extracts measures that quantify the complexity of time-frequency surfaces corresponding to vibration signals. The extracted features are then tracked through the life time of a bearing using curve fitting and Extended Kalman Filtering algorithms. The proposed methodology is tested on a publicly available bearing data set with known RULs.

Index Terms—Bearing Faults, Prognosis, Remaining Useful Life, Time-Frequency Analysis, Entropy, Extended Kalman Filter

I. INTRODUCTION

The ability to accurately predict the remaining useful life of electromechanical systems is critical for affordable system operation and can also be used to enhance system safety. The theme of condition-based maintenance (CBM) is that maintenance is performed based on an assessment or prediction of the component health instead of its service time, which achieves objectives of cost reduction and safety enhancement. If one can predict the degradation of a component before it actually fails, then it will provide ample time for maintenance engineers to schedule a repair, and to acquire replacement components before the components actually fail. Bearings are of paramount importance to almost all forms of rotating machinery, and are among the most common machine elements. The failures of bearing without warning will result in catastrophic consequences in many situations, such as in helicopters, transportation vehicles, etc. Most of the current maintenance procedure includes periodic visual inspections and replacement of the components at fixed time intervals.

According to early surveys bearing faults represent the most common cause for mechanical failure. Consequently, majority of the proposed fault detection methods are focused on detecting bearing faults. Despite such a variety of approaches, most of them focus on extracting a set of well-established features that indicate bearing surface faults.

R. K. Singleton, E. G. Strangas and S. Aviyente are with the Department of Electrical and Computer Engineering, Michigan State University, East Lansing, MI 48824, USA (e-mail: strangas@egr.msu.edu; aviyente@egr.msu.edu).

This material is based in part upon work supported by the National Science Foundation under Grant No. EECS-1102316 and by the National Science Foundation Graduate Research Fellowship under Grant No. DGE-0802267.

Much work has been done on the diagnosis of bearing faults using data-driven and model-based methods, such as Hidden Markov Modeling and Particle Filtering. It has been shown that by using current signals you can detect the presence of a fault in the bearing as well as diagnose the severity [16], [20]. Further work has been done on the prognosis of bearing faults in order to obtain the Remaining Useful Life (RUL) of bearings, or motors in general. Particle Filters have also been used for the prognosis of fault severity [2], [5], [7], [13]. However, one drawback of using particle filtering is that you must have a reliable physical model for the fault degradation. For most real- life signals and systems, including bearings, a reliable physical model for the degradation process is not available. Kalman Filtering has also been used for fault diagnosis and prognosis, using an n-step ahead Kalman Predictor for data extrapolation [12], [14].

Particularly, with bearing fault analysis, vibration data is used over current signal since vibration data is more robust to operating conditions. Previous studies have shown that bearing diagnostics can also be performed using current signals, but only at certain frequency rates [8]. In this paper, we propose a data-driven method for estimating the remaining useful life of bearings from accelerometer recordings. The proposed method relies on extraction of features from time-frequency distribution of the vibration signals as the signals are nonstationary in nature. The extracted features, entropy and concentration, quantify the spread of the energy across time and frequency and relies on the observation that the vibration signals become more chaotic or impulsive as the severity of the fault progresses. The extracted features are then used by Extended Kalman Filtering to build a degradation model and determine a threshold value for the features from training data. This threshold is then applied to testing data to estimate the RUL as well as a confidence interval associated with the estimated RUL.

II. BACKGROUND

A. Time-Frequency Analysis

In literature, different time-frequency transform methods have been proposed for the analysis of nonstationary signals [4]. The most common approaches include the Short-Time Fourier Transform (STFT), Wigner Distribution, wavelet transform and Cohen's class of time-frequency distributions. Cohen's class of distributions are bilinear time-frequency distributions (TFDs) that are expressed as [1] [4]:

[1] All integrals are from $-\infty$ to ∞ unless otherwise stated.

978-1-4799-0024-4/13 $31.00 © 2013 IEEE

$$C(t,\omega) = \iiint \phi(\theta,\tau)s(u+\tfrac{\tau}{2}) \tag{1}$$
$$s^*(u-\tfrac{\tau}{2})e^{j(\theta u-\theta t-\tau\omega)}\,du\,d\theta\,d\tau,$$

where the function $\phi(\theta,\tau)$ is the kernel function and s is the signal. TFDs represent the energy distribution of a signal over time and frequency, simultaneously. The kernel completely determines the properties of the corresponding TFD.

The major differences between Cohen's class of TFDs compared to other time-frequency representations such as the wavelet transform are the nonlinearity of the distribution, energy preservation and the uniform resolution over time and frequency. The wavelet transform provides a representation over time and scale where the frequency resolution is high at low frequencies and low at high frequencies. Although this property makes wavelet transform attractive in detecting high frequency transients in a given signal, it inherently imposes a non-uniform time-frequency tiling on the analyzed signal and thus results in biased energy representations. Cohen's class of bilinear TFDs on the other hand assumes uniform resolution over the entire time-frequency plane. One popular member of Cohen's class of distributions is Choi-Williams distribution, which offers both high time-frequency resolution and reduced interference for multi-component signals. The Choi-Williams distribution of a signal $s(t)$ is defined as:

$$C(t,\omega) = \iiint \phi(\theta,\tau)s(u+\tfrac{\tau}{2})s^*(u-\tfrac{\tau}{2}) \tag{2}$$
$$e^{j(\theta u-\theta t-\tau\omega)}\,du\,d\theta\,d\tau,$$

where $\phi(\theta,\tau) = \exp(-\frac{(\theta\tau)^2}{\sigma})$ is the kernel function that acts as a filter on the signal's autocorrelation function. This distribution can be thought of as a filtered/smoothed version of the Wigner distribution and the amount of smoothing is controlled by σ. This smoothing removes the cross-terms seen in the Wigner distribution at the expense of reduced resolution.

B. Extended Kalman Filter

Kalman Filters have been used to estimate the state of a system given a finite-length data stream. However, Kalman Filters are only useful for linear degradation models with additive white noise. This proves to be very disadvantageous since most signals in engineering are non-linear. The Extended Kalman Filter (EKF) proposes a solution to this problem by approximating the state using a local linearization of a nonlinear function [6]. In the EKF, the state transition and observation models must be differentiable but not necessarily linear functions.

$$x_k = f(x_{k-1}, u_{k-1}) + w_{k-1} \tag{3}$$
$$z_k = h(x_k) + v_k \tag{4}$$

where x_k is the state, z_k is the observaion, u_k is the input at time sample k, f and h are the nonlinear functions, with w_k and v_k being zero-mean, Gaussian noise with some covariance matrices Q_k and R_k, respectively. Prediction of the state is computed via the following:

$$\hat{x}_{k|k-1} = f(\hat{x}_{k-1|k-1}, u_{k-1}) \tag{5}$$

$$P_{k|k-1} = F_{k-1}P_{k-1|k-1}F_{k-1}^T + Q_{k-1} \tag{6}$$

The estimate is then updated

$$\tilde{y}_k = z_k - h(\hat{x}_{k|k-1}) \tag{7}$$
$$S_k = H_k P_{k|k-1} H_k^T + R_k \tag{8}$$
$$K_k = P_{k|k-1} H_k^T S_k^{-1} \tag{9}$$
$$\hat{x}_{k|k} = \hat{x}_{k|k-1} + K_k \tilde{y}_k \tag{10}$$
$$P_{k|k} = P_{k|k-1} - K_k H_k P_{k|k-1} \tag{11}$$

where F_k and H_k are the local linearizations of the state transition and observation model, respectively, given by:

$$F_{k-1} = \left.\frac{\partial f}{\partial x}\right|_{\hat{x}_{k-1|k-1}, u_{k-1}} \tag{12}$$

$$H_k = \left.\frac{\partial h}{\partial x}\right|_{\hat{x}_{k|k-1}} \tag{13}$$

It is also important to note that the $p(x_k|z_{1:k})$ is approximated by a Gaussian distribution.

C. Information Theoretic Measures on the Time-Frequency Plane

In recent years, there has been an interest in adapting information-theoretic measures to the time-frequency plane in order to quantify signal complexity [1], [3], [10], [19]. The application of information-theoretic measures such as entropy and divergence have made it easier to quantify the complexity of non-stationary signals on the time-frequency plane as well as differentiate between different signals. These measures have been shown to be effective in quantifying the number of signal components. Some of the most desired properties of TFDs are the energy preservation and the marginals. They are given as follows and are satisfied when $\phi(\theta,0) = \phi(0,\tau) = 1 \ \forall \tau,\theta$.

$$\iint C(t,\omega)\,dt\,d\omega = \int |s(t)|^2\,dt = \int |S(\omega)|^2\,d\omega$$
$$\int C(t,\omega)\,d\omega = |s(t)|^2 \ , \int C(t,\omega)\,dt = |S(\omega)|^2. \tag{14}$$

The formulas given above evoke an analogy between a TFD and the probability density function (pdf) of a two-dimensional random variable. This analogy has inspired the adaptation of information-theoretic measures such as entropy to the time-frequency plane. The main difference between TFDs and pdfs is that TFDs are not always positive and thus, not all information measures are well-defined on the time-frequency plane. Another important point is that distributions have to be normalized by their energy before applying any information theoretic measures on them.

The well-known Shannon entropy as applied to TFDs can be defined as:

$$H(C) = -\iint C(t,\omega)\log C(t,\omega)\,dt\,d\omega. \tag{15}$$

This measure is not well-defined when the time-frequency distribution takes on a negative value and can only be applied to positive distributions such as the spectrogram. For this reason, generalized entropy measures such as Rényi entropy have been considered for quantifying the information content of signals on the time-frequency plane [15]. Rényi's generalized entropy of order α for TFDs is:

$$H_\alpha(C) = \frac{1}{1-\alpha} \log_2 \int \int \left(\frac{C(t,\omega)}{\int \int C(u,v)\, du\, dv} \right)^\alpha dt\, d\omega$$

(16)

where $\alpha > 0$ is the order of Rényi entropy, and entropy is well-defined as long as $\int \int C^\alpha(t,\omega) dt\, d\omega > 0$. This condition does not require the TFDs to be positive for all time and frequency points, and is well-defined for a large class of distributions and signals [3]. As $\alpha \to 1$, Rényi entropy approaches the well-known Shannon entropy. Since in actual implementations we are interested in the discrete-time implementation of this measure, we will define Rényi entropy in discrete time-frequency domain as:

$$H_\alpha(C) = \frac{1}{1-\alpha} \log_2 \sum_n \sum_k \left(\frac{C[n,k]}{\sum_{n'} \sum_{k'} C[n',k']} \right)^\alpha + \log_2 \delta_t \delta_\omega$$

(17)

where δ_t and δ_ω are the sampling step size in time and frequency, respectively.

Another measure that is commonly used to quantify the spread of the signals in the time-frequency plane is the concentration measure. Contrary to the entropy, concentration measure is a statistic on how concentrated a signal is in the time-frequency plane and is given below:

$$M[C_{norm}[n,k]] = \left(\sum_n \sum_k |C_{norm}[n,k]|^{\frac{1}{p}} \right)^p$$

(18)

where $p > 1$, and $C_{norm}[n,k]$ must be a probability distribution function characterized by

$$C_{norm}[n,k] = \frac{C[n,k]}{\sum_n \sum_k C[n,k]}$$

(19)

where C[n,k] is the original TFD. Furthermore, $p < 4$ is chosen since higher values of p will emphasize the small energy regions in the TFD [17]. This paper uses p = 2.

III. METHOD

A. Feature Extraction

The vibration signals (see Figs. 1-3) were first transformed into the time-frequency plane using the Choi-Williams distribution. Each sample, which is of variable length, was transformed into the time-frequency (TF) domain with a 256 point FFT. By looking at the transformed signals, we observed that the horizontal accelerometer was more informative of the progression of the fault compared to the vertical one (See Figs. 4 - 6). For the horizontal accelerometer data, a clear trend in the time-frequency distribution across all 6 data sets was observed. At the beginning of each training run, there

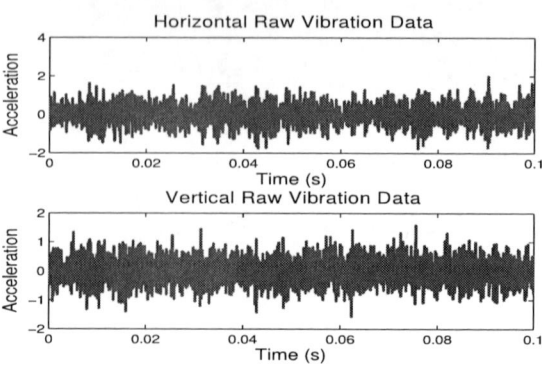

Fig. 1. Raw Data of Initial Vibration Signal

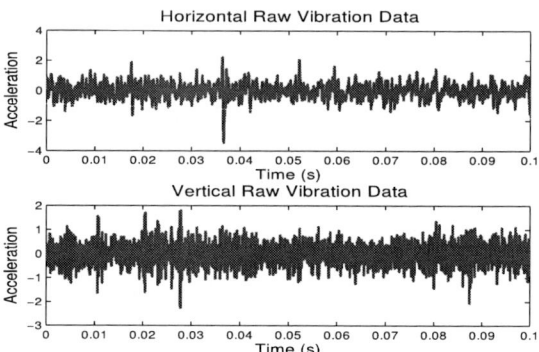

Fig. 2. Raw Data of Intermediate Vibration Signal

was significant energy in the frequency range of 160-200 Hz. As the severity of the fault increased, this band moved up in frequency, around frequency band 236-256 Hz. Finally, as the motor neared failure state, the concentration of energy distributed itself across the entire plot, in a less uniform manner. The progression of the faults, from the first sample, to an intermediate, then finally the last, or failure sample, can be seen in figures 1 - 3, for the raw signals, and figures 4 - 6 in the time-frequency domain, respectively.

Based on these observations, we quantify the spread of energy in the time-frequency plane using the entropy and concentration measures. Since over time, the time frequency energy distribution plot goes from a uniform distribution

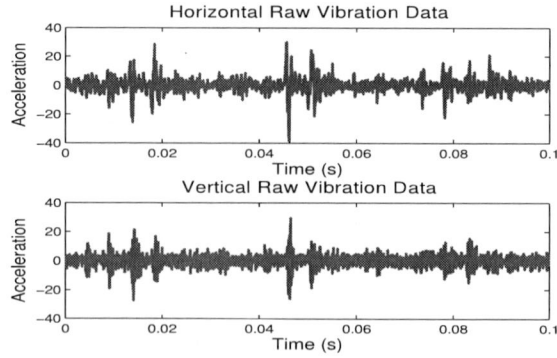

Fig. 3. Raw Data of Final Vibration Signal

978-1-4799-0024-4/13 $31.00 © 2013 IEEE

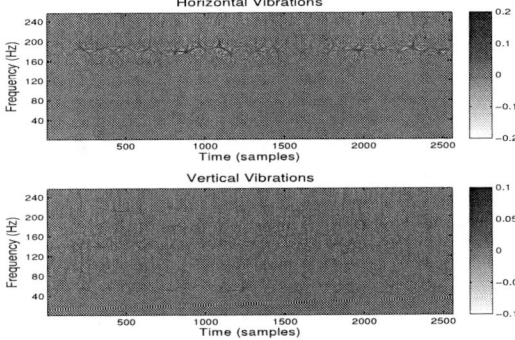

Fig. 4. Choi-WIlliams Transformation of Initial Vibration Signal

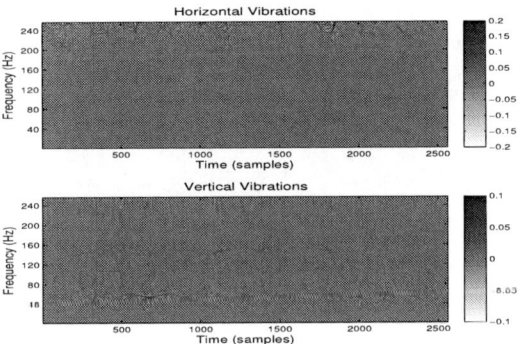

Fig. 5. Choi-WIlliams Transformation of Intermediate Vibration Signal

to a more impulsive distribution near failure state, entropy and concentration measures are suitable. The measures were extracted over selected frequency bands across the samples. Four different frequency bands, 160-200 Hz, 236-256 Hz, 0-40 Hz, and the entire surface 0-256 Hz were considered.

B. Fault Prognosis and RUL Prediction

The extracted features are tracked across time using the Extended Kalman Filter (EKF). First, the trend of the feature across training samples is approximated by an analytical signal, such as a linear, quadratic, or exponential model. Once such an approximation is found, EKF can be used to estimate the parameters of the model by altering equation (3)

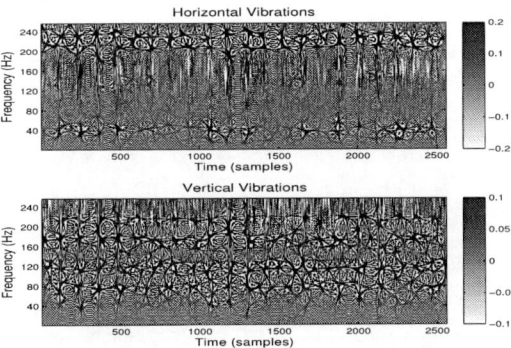

Fig. 6. Choi-WIlliams Transformation of Final Vibration Signal

to track the parameters as well as the overall state, by allowing $\dot{\mathbf{x}} = [\mathbf{x}, \theta]$ [9].

$$[x_k, \theta_k] = f([x_{k-1}, \theta_{k-1}], u_{k-1}) + w_{k-1} \quad (20)$$

where θ is the set of all parameters in the equation. It is also assumed that

$$\theta_k = \theta_{k-1} + n_k \quad (21)$$

where n_k is some zero-mean, Gaussian noise with a covariance M_k. Once the parameters of the model are obtained, the value of the parametric prediction model, obtained through EKF, at the failure sample is extracted as the threshold, γ. The threshold is extracted from each training set individually and then the final threshold for testing is computed as the average across training sets

$$\bar{\gamma} = \frac{1}{K} \sum_{i=1}^{K} \gamma_i \quad (22)$$

where K is the number of training samples, equal to 6 in this case.

In order to predict the RUL of the bearing, the time it will take to reach the failure threshold, $\bar{\gamma}$, is computed. At each sample, k, the degradation equation found by EKF, $q^k(\tau_k)$ is computed by plugging in the estimated parameters, θ_k. Using

$$g^k(\tau_k) = \bar{\gamma} \quad (23)$$

we can solve for τ_k. The RUL, then, simply becomes

$$RUL(k) = \tau_k - k \quad (24)$$

C. Confidence Intervals of RULs

The main focus of this paper is to not only obtain RUL information, but to add confidence to the prediction based on how likely the prediction is. By the end of the algorithm, we have obtained RUL estimates $\{RUL(1), RUL(2), ..., RUL(N)\}$, where N is the number of samples in the test data. From these estimates, we can build a histogram for each test case and feature separately. Confidence intervals are then given in the 95% level rounded to the nearest hundred. This gives the user time estimates in which the motor will fail as well as the likelihood of the bearing failing around a range of time centered at the RUL estimate [18]. This feature is useful to the user because there may be some outlying RUL predictions which may not be realistic. These "outliers" will have a low probability of occurring and the user will be notified about it. For example, in the initial stages of testing the estimated RULs based on just a few sample points may not be very reliable. Similarly, towards the end of the lifetime of the bearing, the accelerometer samples may be noisy making the extracted features unreliable. The RUL predictions with the highest probabilities will be provided to the user.

IV. Data

In this paper, the data provided by FEMTO-ST Institute (Besancon - France, http: //www.femto-st.fr/) was used. Experiments were carried out on a laboratory experimental platform (PRONOSTIA) that enables accelerated degradation of

Fig. 7. Overview of PRONOSTIA set up

bearings under constant and/or variable operating conditions, while gathering online health monitoring data (rotating speed, load force, temperature, vibration). PRONOSTIA is an experimentation platform dedicated to test and validate bearings fault detection, diagnostic and prognostic approaches. The main objective of PRONOSTIA is to provide real experimental data that characterize the degradation of ball bearings along their whole operational life (until their total failure). Data representing 3 different loads were considered (rotating speed and load force): First operating conditions: 1800 rpm and 4000 N; second operating conditions: 1650 rpm and 4200 N; third operating conditions: 1500 rpm and 5000 N. 6 run-to-failure datasets were used to build the prognostics models, and the 11 remaining bearings were used for testing. The characterization of the bearing's degradation is based on two data types of sensors: vibration and temperature. In this paper, we are only using the data from the vibration sensor for prognosis. The vibration sensors consist of two miniature accelerometers positioned at 90^0 to each other; the first is placed on the vertical axis and the second is placed on the horizontal axis. The two accelerometers are placed radially on the external race of the bearing. The acceleration measures are sampled at 25.6 kHz [11]. An overview of the entire PRONOSTIA set up can be found in figure 7.

V. RESULTS

For the selected frequency bands of 160-200 Hz, 236-256 Hz, 0-40 Hz, and 0-256 Hz, the two features discussed above, entropy and concentration measure, were extracted for each recorded sample. Median filtering was performed for smoothing the features using a window size of 3, determined empirically. Out of all the features that were tested, the feature that had the clearest trend across samples was entropy (See Figs. 8 and 9), and in particular, the entropy of the 160-200 Hz band. Across all 6 training sets, this feature looks to have a linear degradation trend which the EKF parameter estimation method can exploit. Median filtering was also performed for smoothing the features using a window size of 3, determined empirically. However, the entropy over the 160-200 Hz band is not exactly linear as seen in figure 10. For feature data extrapolation, two different analytical models were viewed: linear and exponential. The results of the RUL estimations over the 11 testing sets can be found in Table 1.

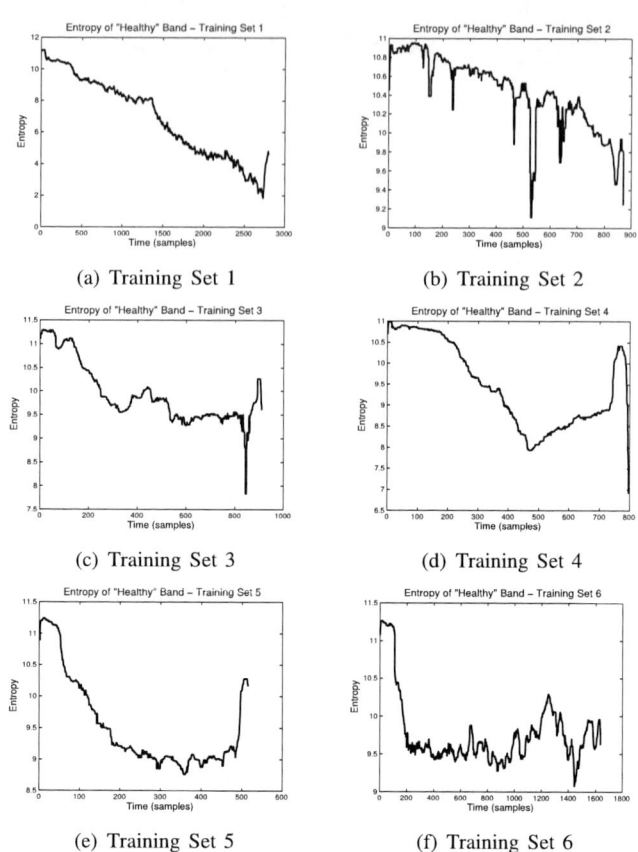

(a) Training Set 1   (b) Training Set 2

(c) Training Set 3   (d) Training Set 4

(e) Training Set 5   (f) Training Set 6

Fig. 8. Median filtered entropy over the "healthy" band across all 6 training sets

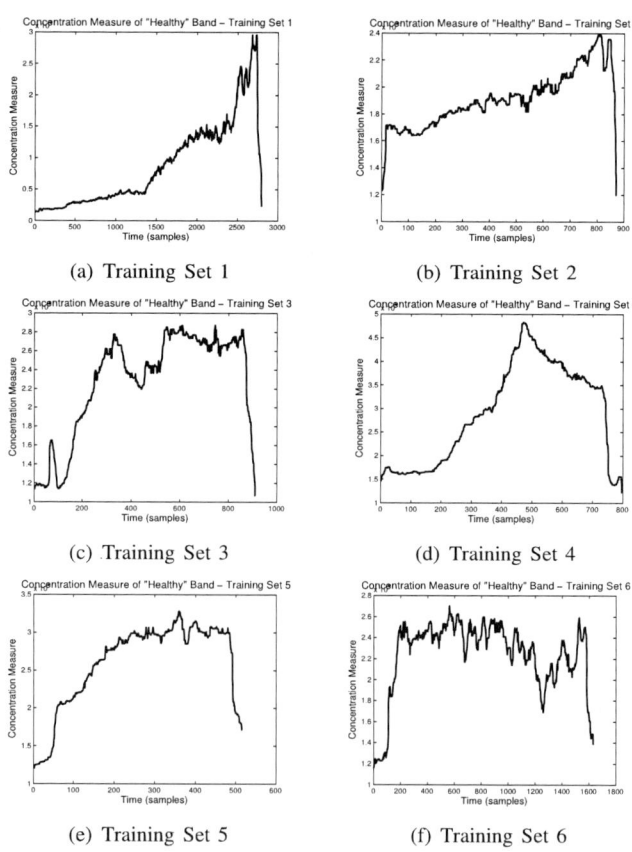

(a) Training Set 1   (b) Training Set 2

(c) Training Set 3   (d) Training Set 4

(e) Training Set 5   (f) Training Set 6

Fig. 9. Median filtered concentration measure over the "healthy" band across all 6 training sets

TABLE I
RUL ESTIMATIONS AND PDF

Test Set	True RUL (s)	Est. RUL (s)	Confidence Int.	Error
1	5730	N/A	N/A	N/A
2	339	1422.5	[200, 1800]	319.62%
3	1610	1117.9	[400, 2300]	30.57%
4	1460	3721.4	[1300, 6800]	154.8%
5	7570	11734.4	[7300, 13000]	55.01%
6	7530	6608.6	[6000, 8400]	12.24%
7	1390	2691.3	[1200, 3100]	93.62%
8	3090	3678.9	[2800, 4200]	19.06%
9	1290	4367.8	[1200, 6900]	238.5%
10	N/A	580	N/A	N/A
11	N/A	820	N/A	N/A

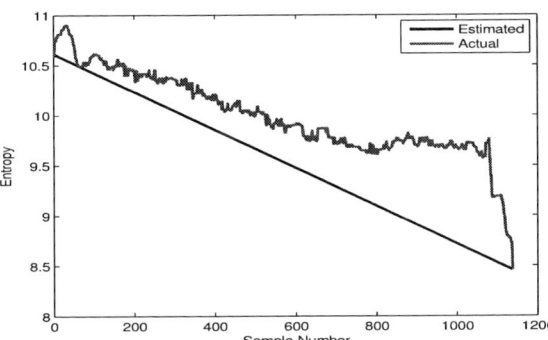

Fig. 10. Linear EKF Parameter Estimation for Bearing 1_5

Out of 11 testing datasets, 3 of them were deemed to already be in failure state because at the time of RUL estimation, the feature was already below the threshold value. These cases are marked as N/A, or not attainable. For the remaining testing sets, the RUL estimation and corresponding confidence intervals were calculated. In particular, we can see the results from bearing 1_5 in the figures 10 and 11. The error can be calculated by

$$error = \frac{estimatedRUL - trueRUL}{trueRUL} * 100 \quad (25)$$

and is also given in Table 1 along with the true estimates. For the 8 testing cases out of 11 that were assigned an RUL, 6 of them had errors less than 100%. One reason for the relatively high errors in RUL estimates is adopting a linear

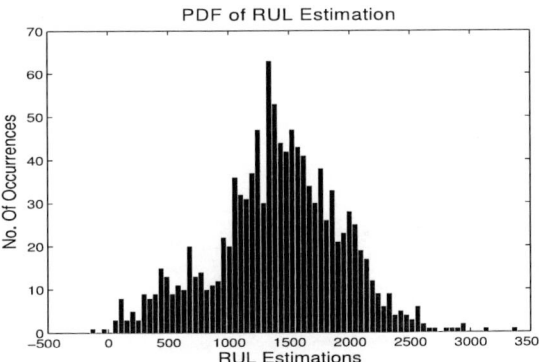

Fig. 11. RUL PDF for Bearing 1_5 in seconds

degradation curve for a nonlinear phenomenon. Furthermore, the RUL estimates at each time point should not be weighted the same since there is less data and more uncertainty at the initial samples. This made it difficult to perform RUL predictions. Even though the RUL estimation error is high for certain test cases, in all of the cases the true RUL lies within the 95% confidence interval computed from the distribution of the estimated RULs. This indicates that not all RUL estimates should be equally weighted and the actual estimate should rather be an interval or a weighted mean based on the reliabilities of each predicted RUL value.

VI. CONCLUSIONS

In this paper, we presented a new method for RUL estimation from bearing data. The proposed method uses complexity based features in the time-frequency plane to predict RUL for bearings as well as to provide how likely a particular outcome is. These RUL notifications, however, are dependent on the extracted features. The methodology requires the extraction of a feature that resembles or can be modeled by an analytical signal across time. This proves to be problematic as the trajectory of the extracted fault is not always a monotonic function and may show fluctuations which makes it harder to fit an analytic signal model. For future work, model-free techniques will be examined to compensate for this drawback. Furthermore, performing prognosis using multiple features instead of one may prove to be more beneficial and improve the reliability of the RUL estimates. Moreover, alternative ways of selecting the threshold, other than the mean, can improve the robustness of the method. The proposed method can also be applied to other types of sensor data such as the current from the bearings.

REFERENCES

[1] A. Abella, J.H. Wright, and A. Gorin. Dialog trajectory analysis. In *Proc ICASSP*, 2004.

[2] M. S. Arulampalam, S. Maskell, N. Gordon, and T. Clapp. A tutorial on particle filters for online nonlinear/non-gaussian bayesian tracking. *Signal Processing, IEEE Transactions on*, 50(2):174–188, 2002.

[3] R. G. Baraniuk, P. Flandrin, A. J. E. M. Janssen, and O. Michel. Measuring time-frequency information content using the Rényi entropies. *IEEE Trans. on Info. Theory*, 47(4):1391–1409, May 2001.

[4] L. Cohen. *Time-Frequency Analysis*. Prentice Hall, New Jersey, 1995.

[5] D. Creal. A survey of sequential monte carlo methods for economics and finance. *Econometric Reviews*, 31(3):245–296, 2012.

[6] AJ Haug. A tutorial on bayesian estimation and tracking techniques applicable to nonlinear and non-gaussian processes. *MITRE Corporation, McLean*, 2005.

[7] W. He, N. Williard, M. Osterman, and M. Pecht. Prognostics of lithium-ion batteries based on dempster–shafer theory and the bayesian monte carlo method. *Journal of Power Sources*, 196(23):10314–10321, 2011.

[8] F. Immovilli, A. Bellini, R. Rubini, and C. Tassoni. Diagnosis of bearing faults in induction machines by vibration or current signals: A critical comparison. *IEEE Transactions on Industry Applications*, 46(4):1350–1359, July-Aug. 2010.

[9] L. Ljung. Asymptotic behavior of the extended kalman filter as a parameter estimator for linear systems. *Automatic Control, IEEE Transactions on*, 24(1):36–50, 1979.

[10] O. Michel, R. G. Baraniuk, and P. Flandrin. Time-frequency based distance and divergence measure. In *Proc. IEEE Int. Symp. Time-Frequency and Time-Scale Analysis*, pages 64–67, 1994.

[11] P. Nectoux, R. Gouriveau, K. Medjaher, E. Ramasso, B. Chebel-Morello, N. Zerhouni, C. Varnier, et al. Pronostia: An experimental platform for bearings accelerated degradation tests. In *Conference on Prognostics and Health Management.*, pages 1–8, 2012.

[12] O. Ondel, E. Boutleux, E. Blanco, and G. Clerc. Coupling pattern recognition with state estimation using kalman filter for fault diagnosis. *Industrial Electronics, IEEE Transactions on*, 59(11):4293–4300, 2012.

[13] M. E. Orchard and G. J. Vachtsevanos. A particle filtering-based framework for real-time fault diagnosis and failure prognosis in a turbine engine. In *Control & Automation, 2007. MED'07. Mediterranean Conference on*, pages 1–6. IEEE, 2007.

[14] K. Qing-chun, X. Ming-qing, and Z. Xin. Fault prognosis modeling of aviation electronic system based on kalman filtering. In *Testing and Diagnosis, 2009. ICTD 2009. IEEE Circuits and Systems International Conference on*, pages 1–4. IEEE, 2009.

[15] A. Rényi. On measures of entropy and information. In *Proceedings 4th Berkeley Symp. Math. Stat. and Prob.*, volume 1, pages 547–561, 1961.

[16] A. Soualhi, G. Clerc, H. Razik, and A. Lebaroud. Fault detection and diagnosis of induction motors based on hidden markov model. In *XXth International Conference on Electrical Machines (ICEM),2012*, pages 1693 –1699, Sept. 2012.

[17] L. Stanković. A measure of some time–frequency distributions concentration. *Signal Processing*, 81(3):621–631, 2001.

[18] E. Sutrisno, H. Oh, A. S. S. Vasan, and M. Pecht. Estimation of remaining useful life of ball bearings using data driven methodologies. In *Prognostics and Health Management (PHM), 2012 IEEE Conference on*, pages 1–7. IEEE, 2012.

[19] W. J. Williams, M. Brown, and A. Hero. Uncertainty, information and time-frequency distributions. In *SPIE-Advanced Signal Processing Algorithms*, volume 1556, pages 144–156, 1991.

[20] S. S. H. Zaidi, S. Aviyente, M. Salman, K. Shin, and E. G. Strangas. Prognosis of gear failures in dc starter motors using hidden markov models. *Industrial Electronics, IEEE Transactions on*, 58(5):1695–1706, 2011.

VII. BIOGRAPHIES

Rodney K. Singleton II received his B.S. in electrical engineering with honors from Michigan State University (MSU) in East Lansing, Michigan in 2010. During his undergraduate career, he completed summer internships at Ford Motor Company and The Johns Hopkins University Applied Physics Laboratory. Currently, he is pursuing his M.S. and Ph.D. in Electrical Engineering, also at MSU, as a GEM and NSF fellow. His graduate research includes the prognosis of bearing faults in electrical motors, as well as remaining useful life estimation of a motor.

Selin Aviyente received her B.S. degree with high honors in electrical and electronics engineering from Bogazici University, Istanbul in 1997. She received her M.S. and Ph.D. degrees, both in Electrical Engineering: Systems, from the University of Michigan, Ann Arbor, in 1999 and 2002, respectively. In August 2002, she joined the Department of Electrical and Computer Engineering at Michigan State University where she is currently an associate professor. Her research focuses on the theory and applications of statistical signal processing, in particular non-stationary signal analysis. She is interested in developing methods for efficient signal representation, detection and classification. She has published over 80 refereed journal articles and conference proceedings on time-frequency analysis, signal detection and classification. She is the recipient of 2005 Withrow Teaching Excellence Award and 2008 NSF CAREER Award.

Elias Strangas received the Dipl. Eng. degree in electrical engineering from the National Technical University of Greece, Athens, Greece, in 1975 and the Ph.D. degree from the University of Pittsburgh, Pittsburgh, PA, in 1980. He was with Schneider Electric (ELVIM), Athens, from 1981 to 1983 and the University of Missouri, Rolla, from 1983 to 1986. Since 1986, he has been with the Department of Electrical and Computer Engineering, Michigan State University, East Lansing, MI, where he heads the Electrical Machines and Drives Laboratory. His research interests include the design and control of electrical machines and drives, finite-element methods for electromagnetics, and fault prognosis and mitigation of electrical drive systems.

Improvements on Lifespan Modeling of the Insulation of Low Voltage Machines with Response Surface and Analysis of Variance

Antoine Picot, David Malec and Pascal Maussion

Abstract- The aim of this paper is to present some improvements of the method for modeling the lifespan of insulation materials in a partial discharge regime. The first step is based on the design of experiments which is well-known for reducing the number experiments and increasing the accuracy of the model. Accelerated aging tests are carried out to determine the lifespan of polyesteimide insulation films under different various stress conditions. A lifespan model is achieved, including an original relationship between the logarithm of the insulation lifespan and that of electrically applied stress and an exponential form of the temperature. The significance of the resulting factor effects are tested through the analysis of variance. Moreover, response surface is helpful to take into account some second order terms in the model and to improve its accuracy. Finally, the model validity is tested with additional points which have not been used for modeling.

Index terms—Diagnosis, dielectrics, insulation, insulation testing, accelerated aging, lifespan estimation, condition monitoring, modeling, aging, response surface, analysis of variance (ANOVA)

I. NOMENCLATURE

E_i Effect of factor number i
E_{ii} Effect of the square of factor number i
E_{ij} Effect of interaction between factor number i and factor number j
n   Number of experiments
F_i Level of factor i, could be +1, -1 or any value between -1 and +1
$\hat{E}$ Effect vector composed of the different E_{ii} terms
X   Experimental matrix
Y   Y is the experimental Weibull's vector
M   Average experimental value
L   Lifespan in minutes
T   Temperature in °C
V   Voltage in V
F   Frequency in Hz
V_A Variance of factor A
N   Number of samples = number of repetitions
k   Number of factors (k=3)
n_0 Number of center points
dof   Degrees of freedom

Authors are with Université de Toulouse ; INPT, UPS ; LAPLACE (LAboratoire PLAsma et Conversion d'Energie); ENSEEIHT, 2 rue Charles Camichel, BP 7122, F-31071 Toulouse cedex 7, France and CNRS; LAPLACE; F-31071 Toulouse, France.

II. INTRODUCTION

Low voltage electric machines are increasingly submitted to heavy electric constraints and their lifespan becomes nowadays a concern. Among many papers, [1] have reported that stator-winding insulation is one of the weakest components in a drive (around 40% of failures). Materials involved in electric ageing affect the insulation system lifespan. Many operating factors such as voltage, frequency, temperature, pressure, etc. could have a dramatic effect as these stresses can synergize as shown in [2]. The existing models of degradation or of lifespan differ slightly from one paper to another but they are all based on physics and include some factors which are specific to the material or the aging mechanism. As examples, [3]-[4] include the physical, thermal and electro-mechanical aspects of the electrical aging process. Additionally, the choice of the lifespan forms remains critical for model accuracy. Various forms can be found in different works such as [5] where they mainly involve a log-based relationship for frequency and voltage and an exponential form for the temperature. Nevertheless, there is no comprehensive model for insulation lifespan prediction under different combined stresses and this phenomenon remains complex and difficult to understand and to model, especially with PWM supply.

Experimental tests are necessary to assess lifespan modeling but full aging tests for all the factors involved in the aging process under nominal conditions could be time consuming. As a consequence, accelerated aging tests which are about to speed up the degradation, are generally performed in order to study and predict the lifespan. IGBT modules in high temperature power cycling are tested in [6] whereas nano-structured enamels on twisted pairs are tested in [7] under severe waveforms but it does not give any lifespan model.

The "Design of Experiments" (DoE) [8] was helpful in many cases either for optimization or modeling purposes. The principle of this methodology is to organize and carry out just the required number of experiments in order to obtain the most accurate information for a specific problem. Its performance has been put in evidence in different applications, especially in chemistry and mechanics, where a high volume of parameters have to be simultaneously optimized. In recent years, DoE have been successfully used in electrical engineering for electrical machine design (induction, reluctance, synchronous...) [9]-[11] or for the control of power electronic devices [12]-[13]. In the field of insulation lifespan modeling, [14] provides a model with some statistical analysis to test which parameters are significant or not. A set of 8 tests are carried out for each combination of the stress factors (Voltage, Temperature and Frequency). To be more cost effective, this paper will try to reduce this number of tests. But, it is important to check whether the results remain statistically significant or not and Analysis of Variance (ANOVA) will be employed.

III. SYSTEM DESCRIPTION AND BASIC DoE

The testbench itself and the tested insulation materials have been fully described in [14]. It consist of steel plates coated with polyesterimide (PEI - thermal class: 180°C) films which are widely used in rotating machine insulation systems (15cm x 9cm with a 90μm coating).

Fig. 1. Tested 90μm coated steel plate (15cm x 9cm)

Eight samples were tested in our experimental setup, shown in Fig.2. Under electrical stress, the steel plate acts as the first electrode and a spherical stainless steel electrode (diameter: 1mm) is the second. Samples were placed in a climatic chamber where the temperature is fully controlled. The lifespan of each sample was measured using a timer (one per sample) which stopped counting as soon as the current increased and crossed a threshold at which the corresponding sample broke down. The faulty specimen was disconnected while the survivors remained under voltage and at the controlled temperature. Accelerated aging tests are carried out in order to relate the applied external stresses (factors) to the insulation lifespan (response). Lifespan data in this paper is presented according to Weibull's statistical processing [15], which is commonly used for breakdown data treatment.

Fig. 2. The experimental setup for accelerated aging tests, climatic chambers and power electronic.

The failure process is driven by several stresses acting simultaneously such as electrical, thermal, mechanical and ambient stresses. The experimental aging conditions of theses accelerated tests were chosen to ensure that the insulation degradation is mainly due to the partial discharges. However, for simplicity and because of their influence, only three major parameters were studied:

1) the square wave applied HVDC (V),
2) the frequency of the applied voltage (F),
3) temperature (T).

TABLE I
LEVELS OF THE THREE STRESS FACTORS

Factors	Level (-1)	Level (+1)	Level (0)
Log (Voltage (kV))	Log(1)	Log(3)	Log(1.73)
Log((Frequency (kHz))	Log(5)	Log(15)	Log(8.7)
Exp(-b.Temperature (°C))	Exp(55b)	Exp(-180b)	Exp(26.7b)

with b= 5.64×10^{-3}

The basic data treatment of the PEI insulation film lifespan relies on the Design of Experiments as written in [16] and [14]. The corresponding plans and results are recalled in Table II and Table III for an easy understanding. Table II gives all the 2^3 (3 factors, 2 levels each) possible combinations between the different factor levels whereas Table III list the sample lifespans (in minutes). They lead to a first model presenting a logarithmic form of electrical stress (i.e. voltage and frequency) and an exponential form of the temperature as expressed in (1). The coefficients of the model are listed in Table IV.

$$
\begin{aligned}
Log(L) \sim\ & M + E_V.log(V) + E_F.log(F) + E_T.exp(-bT) + \\
& E_{FV}.log(V).log(F) + E_{VT}.log(V).exp(-bT) + \\
& E_{FT}.log(F).exp(-bT) + E_{VFT}.log(V).log(F).exp(-bT)
\end{aligned}
\tag{1}
$$

TABLE II
EXPERIMENTAL RESULTS WITH THREE FACTORS AT 2 LEVELS EACH : 8 EXPERIMENTS WITH 8 SAMPLES EACH

Test n°	Log(V) F1	Log(F) F2	e^(-bT) F3	Lifespan (minutes)							
				Sample no. 1	Sample no. 2	Sample no. 3	Sample no. 4	Sample no. 5	Sample no. 6	Sample no. 7	Sample no. 8
1	-1	-1	-1	378	418	568	587	634	642	786	850
2	-1	-1	1	25	29	23	29	26	30	24	31
3	-1	1	-1	169	187	268	268	343	364	162	322
4	-1	1	1	14.5	14.25	13.4	13.8	14.8	15	13	15.5
5	1	-1	-1	26	30	28	33	24	29	28	35
6	1	-1	1	6	6,5	6,2	6,8	5,5	6,5	6	6.3
7	1	1	-1	14	16.21	15.07	15.5	15.28	16	12.5	14.8
8	1	1	1	2.2	2.233	1.133	1.65	2.033	1.516	0.933	1.4

TABLE III
FULL FACTORIAL DESIGN MATRIX FOR THREE FACTORS WITH 2 LEVELS EACH

Test n°	Log(V) M	Log(F) F1	e^(-bT) F2	F3	Log(V).Log(F) I(V.F)	Log(V).e^(-bT) I(V.T)	Log(F).e^(-bT) I(F.T)	Log(V).Log(F).e^(-bT) I(V.F.T)	Log(L) Weibull
1	1	-1	-1	-1	1	1	1	-1	2.824
2	1	-1	-1	1	1	-1	-1	1	1.453
3	1	-1	1	-1	-1	1	-1	1	2.459
4	1	-1	1	1	-1	-1	1	-1	1.167
5	1	1	-1	-1	-1	-1	1	1	1.486
6	1	1	-1	1	-1	1	-1	-1	0.806
7	1	1	1	-1	1	-1	-1	-1	1.187
8	1	1	1	1	1	1	1	1	0.255

Tests at the center of the study domain

Model : Log(L)=f[Log(V), Log(F) and exp(-b.T)]

9	1	0	0	0	0	0	0	0	1.43

The factor effect calculation can be achieved by a simple matrix inversion through the following relationship (2), where (Ê) is the 8x1 effect vector, Y is the 8x1 experimental Weibull's vector of the and X is the 8x8 matrix composed of experiment levels, the dark grey part in Table III.

$$\hat{E} = X^{-1}.Y \qquad (2)$$

TABLE IV
EFFECT VALUES FOR THE LIFESPAN MODEL, VECTOR $\hat{E}$

Model	Effect
M	1.45
Log(V)	-0.53
Log(F)	-0.19
Exp(-bT)	-0.54
$I_{Log(V).Log(F)}$	-0.03
$I_{Log(V).exp(-bT)}$	0.12
$I_{Log(F). exp(-bT)}$	-0.03
$I_{Log(F).Log(F). exp(-bT)}$	-0.05

Once this factors have been calculated as in any kind of regression problem, tests of hypotheses are helpful in measuring their usefulness and whether their contribution to the model are significant or no. The following section describes the analysis of variance which is carried out as test of significance.

IV. SIGNIFICANCE AND REPETITION NUMBER USING ANOVA

A. ANOVA presentation

ANOVA is a widely used statistical model separating the total variability found within a data set into two components: random and systematic factors. It has been demonstrated in [17] that ANalysis Of Variance (ANOVA) is a very interesting technique to assess the statistical significance of the different factor effects in many different applications in the wide field of electrical engineering. A new rotor shape for a high-speed interior permanent-magnet synchronous motor is presented in [18]. Its design is based on a full factorial experimental design and leads to a significant reduction in the amount of permanent magnet employed. Moreover, the design factors of the greatest influence are identified with the help of ANOVA. This method has also been used in [19] in order to evaluate student's performance in tests and exams after supervisory control and data acquisition (SCADA) and robotics experiments in control and automation education. The study put in evidence the usefulness of experimental practice even on low cost experimental setups. Energy production from PV generators is monitored in [20] and ANOVA helps to get concise information about the energy produced by all the inverters for comparison purposes.

B. Significance and ANOVA

Some of the factor effects computed in the ageing model proposed in section I are close to 0. These effects might not be significant and might be only due to the variability of the data, i.e. caused by non-controlled factors. To use ANOVA, two conditions must be verified:
- data must be normally distributed for each factor,
- the different data must be independent.

Variance V_i due to the specific factor i is compared to the variance of the data set, called the residual variance V_r. If the factor is not significant, the variance V_i will be very close to V_r. On the contrary, the variance V_I will be proportionately higher than V_r. The ratio $F_{exp}=V_i/V_r$ is computed and tested using a Fisher-Snedecor test. This method tests the equality of both variances. So, the null hypothesis is to consider the effect of A as non-significant. In this case, the ratio F_{exp} should be less than a threshold F_{lim}. The effect of A is considered as statistically significant if F_{exp} is greater than F_{lim}. The threshold F_{lim} is defined according to the table of upper critical values of the F-distribution depending on the degrees of freedom of the data and the test significance level (generally fixed at 5%). For each test, a residue r_{ij} is computed as the difference between the lifetime obtained for a particular repetition j and the average lifetime obtained for this particular experiment i. The residual variance V_r is computed for all experiments i and repetitions j according to (3):

$$V_r = \frac{\sum r_{ij}^2}{dof_r} \tag{3}$$

where dof_r is the number of degrees of freedom for the residues. If we consider a DoE with N experiments and k repetitions for each test, the degrees of freedom for each test will be equal to k-1 and so the number of degrees of freedom for the residues will be $dof_r=N(k$-$1)$. For each factor i, a variance V_i is computed as follows:

$$V_i = \frac{n \sum E_i^2}{dof_i} \tag{4}$$

where n is the number of tests when factor F_i is to a certain level, E_i is the effect of the factor i and dof_i its number of degrees of freedom. The number of tests n can be computed as the total number of tests $N.k$ divided by the number of levels l_i for the factor i. The number of degrees of freedom is equal to the number of levels of the factor minus 1. The variance V_i can then be expressed in (5):

$$V_i = \frac{N.k \sum E_i^2}{l_i(l_i - 1)} \tag{5}$$

The square of the effect is summed as many times as the number of levels. In the case of an interaction between two factors i and j, the variance V_{ij} is computed as:

$$V_{ij} = \frac{n \sum I_{ij}^2}{dof_{ij}} \tag{6}$$

with I_{ij} the effect of the interaction between factors I and j. The number of repetitions n is calculated as $N/(l_il_j)$ and the number of degrees of freedom dof_{ij} is equal to the product of the number of degrees of freedom for factors i and j. The expression of V_{ij} is then:

$$V_i = \frac{N.k \sum I_{ij}^2}{l_il_j(l_i - 1)(l_j - 1)} \tag{7}$$

The square of the interactions is summed as many times as there are possibilities of combination between the levels of the different factors. For example, if factors i and j are both two-levels factors, there are $4(=2^2)$ possible combinations ($[+1;+1]$, $[+1;-1]$, $[-1;+1]$, $[-1;-1]$).

C. Application to the DoE results

The ANOVA is now applied to the results obtained by the DoE described in section III in order to evaluate the significance of each effect. The normality of the distribution for each experiment is tested using a Shapiro-Wilk test and will not be described here. The Shapiro-Wilk test has been chosen because it is well designed for small populations (between 3 and 5000) as explained in [21]. The independence of the different data is ensured by the fact that each test has been realized independently on different coated steel plates.

DoE has been realized with 3 factors and 2 levels for each. $N=8$ $(=2^3)$ experiments have been set with $k=8$ repetitions for each. The number of degrees of freedom is $dof=1$ for every factor and interaction and $dof_r=8\times7=56$ for the residues. According these degrees of freedom, the theoretical threshold of the Fisher test evaluating the significance of the different factors is $F_{lim}=4.00$. An effect will then be considered as significant if $F_{exp}>F_{lim}$. The results obtained applying the ANOVA on the DoE is summed up in Table V.

TABLE V
SIGNIFICANCE OF THE DOE FACTORS EFFECTS WITH $K=8$ REPETITIONS

Factor	dof	Variance	$F_{exp}=V_i/V_r$	F_{lim}	Significant?
V	1	17.0371	2218.6	4.00	Yes
F	1	2.3610	307.4	4.00	Yes
T	1	17.8524	2324.7	4.00	Yes
V,F	1	0.0209	2.7	4.00	No
V,T	1	0.3286	42.8	4.00	Yes
F,T	1	0.0169	2.2	4.00	No
V,F,T	1	0.1229	16.0	4.00	Yes
Residues	56	0.0077			

Table V confirms that voltage and temperature are the two most significant factors. Consequently, their interaction should be significant too. The effect of the frequency is significant but its interactions with other factors are not. This seems logical because the effect of these interactions were very low, $I(V,F)=-0.0295$ and $I(F,T)=-0.0265$. Nevertheless, the interaction $I(V,F,T)$ is statistically significant according to ANOVA although its effect might seem low, $I(V,F)=-0.0506$. Accordingly, interaction effects between frequency and temperature, frequency and voltage, should not be taken into account in the ageing model.

D. Number of repetitions needed

This method can also be used to predict how many repetitions are needed so a factor effect is significant. The Fisher test can indeed be used to define the minimum number k to have $F_{exp}>F_{lim}$. To do so, the residual variance is supposed known which is often true as it is part of the knowledge of a product due to experience. The minimal effect that must be significant is named E_{mini}. Knowing E_{mini} and V_r, it is then possible to estimate the minimal number of repetitions so E_{mini} is considered as significant by using the relation $V_i/V_r>F_{lim}$. The expression of V_i defined in (4) is injected this relation, for k repetitions, we have:

$$\frac{k}{F_{lim}} > \frac{l_i(l_i-1)V_r}{N\sum E_i^2} \tag{8}$$

F_{lim} depends on dof_i which is fixed by the number of levels for the factor i and $dof_r=N.(k-1)$ with N with the number of experiments. So, F_{lim} depends only on k. This method is applied to the model obtained by the DoE. The effects of interactions between frequency and temperature and between voltage and frequency are not taken into account because it has been demonstrated in the former section that they were not significant. So, the number of experiments of the DoE is now $N=8-2=6$. The minimal effect we want to be significant is $I(V,F,T)=0.0506$. The same residual variance $V_r=0.0077$ than it has be obtained in Table 1 with 8 repetitions is chosen as it is considered as a correct estimation of the residual variance. The results for 2, 3 and 4 repetitions are presented in Table VI.

TABLE VI
ESTIMATION OF THE MINIMUM NUMBER K OF REPETITIONS TO CONSIDER $I(V,F,T)$ AS SIGNIFICANT

k	F_{lim}	k/F_{lim}	$k.V_r/V_i$	Significant ?
4	4.41	0.907	0.499	Yes
3	4.75	0.631	0.499	Yes
2	5.99	0.334	0.499	No

Table II shows that 3 repetitions at least are needed to consider the effect of the interaction between voltage, frequency and temperature as significant. This is an interesting result demonstrating that even if the effect of the double interaction is low, only 3 repetitions could have be done to observe a significant effect, which would reduce the total number of tests to run to 18 instead of 64. Next section will try to see if the model prediction might be improved with second order terms

V. RESPONSE SURFACE METHODOLOGY FOR MODEL IMPROVEMENT

The design of experiments can be seen as a regression method that gives to a model which can include some non-linear relationships between the stress variables. But the resulting model is limited to single factors (voltage V or temperature T for example) or products between them (VT) while some other effects such as the square of the voltage (V^2) could be influent as well and cannot be included in the first order model achieved with DoE. In these cases, Response surface methodology (RSM) is a good candidate to complete the investigations and to provide extended models [22]. RSM has been used in [23] for the multi-objective optimal design of isolated dc–dc converters. For the power loss and the weight of the converter, The impact of different factor such as voltages, currents or model parameters, transformer heat-transfer coefficient for example are taken into account. Besides RSM method in [24] gives an analytical model of some parameters (such as weight, detent force, and thrust force) of a permanent-magnet type transverse flux linear motor (TFLM). The objective functions of a Particle Swarm Optimization (PSO) algorithm for the motor design optimization are based on these model parameters.

In this paper, Response Surface Methodology will be used to improve previous work on lifespan modeling of the insulation of low voltage machines. A specific design is built in order to fit a second order model, which means that the experimental surface is supposed to be fitted on a particular form. The method of least squares enables to estimate the regression coefficients in this multiple linear regression model. The response estimation, , is then given by a second order polynomial, according to (9).

$$\eta = M + \sum_{i=1}^{k} E_i.x_i + \sum_{i=1}^{k} E_{ii}.x_i^2 + \sum_{i=1}^{k-1}\left(\sum_{j=i+1}^{k} E_{ij}.x_i.x_j \right) \quad (9)$$

According to the objective, Central Composite Designs are used. They are specifically designed to solve optimization problems and to give the optimal solution. A central composite design is defined by :

- a complete 2^k factorial design, extracted from Table II and Table III,
- n_0 repetitions on center points, for statistic analysis. The factors will now have 3 levels (-1, 0 and 1),
- 2 axial points on the axis of each parameter generally situated at a distance of from the design center. In our case will be taken as 1 for experimental reasons, i.e. the limitations of the test bench, as level 1 for parameter T corresponds to 180°C which is already the maximum reachable temperature.

It is important to notice that this identification does not require a new complete experimental procedure as the 8 experiments of the DoE in section II are included in this extension of the regression method. It only needs $n_0 + 2k$

additional experiments, 12 experiments in this case. These points contribute to estimation of quadric terms of the fitted model, i.e. give information on the curvature of the model [22]. Figure 3 shows an example of a face-centered cube design for 3 variables with 3 levels each (-1, 0, +1), whereas the corresponding experimental plan can be found in Table VII hereafter.

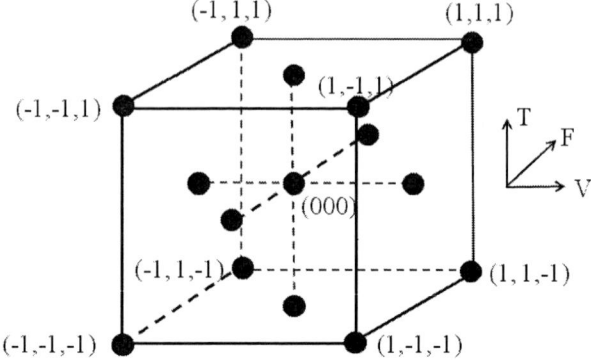

Fig. 3 Face-centered cube design

TABLE VII

RESPONSE SURFACE MATRIX FOR THREE FACTORS WITH 3 LEVELS EACH INCLUDING THE FULL FRACTIONAL PLAN FROM TABLE III

Test n°	M	Log(V) F1	Log(F) F2	$e^{(-bT)}$ F3	$Log(V)^2$ $F1^2$	$Log(F)^2$ $F2^2$	$e^{(-bT)2}$ $F3^2$	Log(V).Log(F) I(V.F)	Log(V).$e^{(-bT)}$ I(V.T)	Log(F). $e^{(-bT)}$ I(F.T)	Log(V). Log(F). e(-bT) I(V.F.T)	Weibull 8 exp Log(L)	Weibull 4 exp Log(L)
1	1	-1	-1	-1	1	1	1	1	1	1	-1	2.824	2.72
2	1	1	-1	-1	1	1	1	-1	-1	1	1	1.486	1.44
3	1	-1	1	-1	1	1	1	-1	1	-1	1	2.46	2.382
4	1	1	1	-1	1	1	1	1	-1	-1	-1	1.187	1.152
5	1	-1	-1	1	1	1	1	1	-1	-1	1	1.453	1.483
6	1	1	-1	1	1	1	1	-1	1	-1	-1	0.806	0.814
7	1	-1	1	1	1	1	1	-1	-1	1	-1	1.167	1.192
8	1	1	1	1	1	1	1	1	1	1	1	0.255	0.27
9	1	-1	0	0	1	0	0	0	0	0	0	1.769	1.751
10	1	1	0	0	1	0	0	0	0	0	0	0.881	0.815
11	1	0	-1	0	0	1	0	0	0	0	0	1.704	1.649
12	1	0	1	0	0	1	0	0	0	0	0	1.314	1.271
13	1	0	0	-1	0	0	1	0	0	0	0	1.985	2.004
14	1	0	0	1	0	0	1	0	0	0	0	0.997	0.939
15	1	0	0	0	0	0	0	0	0	0	0	1.432	1.43
16	1	0	0	0	0	0	0	0	0	0	0	1.444	1.44
17	1	0	0	0	0	0	0	0	0	0	0	1.45	1.39
18	1	0	0	0	0	0	0	0	0	0	0	1.46	1.39
19	1	0	0	0	0	0	0	0	0	0	0	1.42	1.41
20	1	0	0	0	0	0	0	0	0	0	0	1.437	1.413

TABLE VIII
EFFECT VALUES FOR THE LIFESPAN MODEL, VECTOR $\hat{E}$

Model	Effect calculated from 8 exp. for model SRM8	Effect calculated from 4 exp. for model SRM4
M	1.4383	1.4092
Log(V)	-0.5056	-0.5036
Log(F)	-0.1891	-0.1887
exp(-bT)	-0.5265	-0.4997
Log(V)^2	-0.1112	-0.1218
Log(F)^2	0.0725	0.0782
e$^{(-bT)2}$	0.0548	0.0651
I$_{Log(V).Log(F)}$	-0.0249	-0.0256
I$_{Log(V).exp(-bT)}$	0.1315	0.1147
I$_{Log(F).\,exp(-bT)}$	-0.0218	-0.026
I$_{Log(F).Log(F).\,exp(-bT)}$	-0.0412	-0.036

The vector effect $\hat{E}$ is then calculated with the mean square fit, as expressed in (10)

$$\hat{E} = (X^tX)^{-1}.\,X^t.Y \qquad (10)$$

Vector Y could be either the 13° column of Table VII when taking into account the full experimental results or the 14° column of Table VII in case of a reduced set of experiments. Fig. 4 shows for example the non linearity of the response surface associated to model xx, limited to the most influent factors, i.e. V and T.

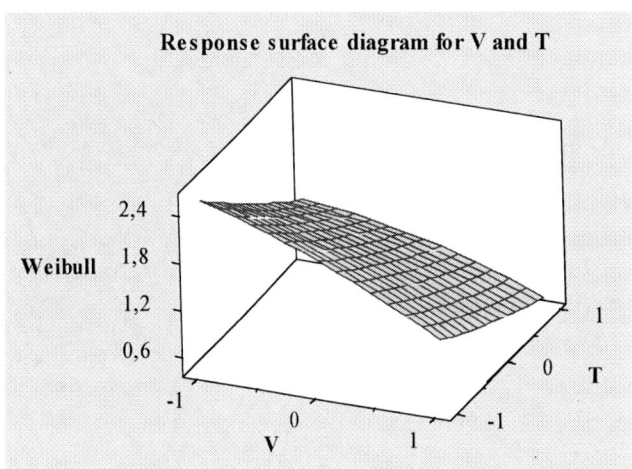

Fig. 4. Response surface diagram for 2 parameters, V and T, out of the 3 parameters involved in the model (V, F and T).

Other experiments inside the experimental domain are carried out such as the 11 points in Table IX, in order to check the models validity. It presents and compares the different results obtained with different models:

- DoE model has been built from 8 experiments focusing only on the main influential factors and interactions,
- SRM8 model relies on 8 experiments and includes only the most influent factors,
- SRM4 model is only based on 4 experiments randomly chosen out of the 8 experiments of SRM8 model and includes only the most influent factors.

It turns out that the use of the SRM increases the model precision and that a limited number of experiments (4 instead of 8) do not dramatically affect the corresponding model precision.

TABLE IX
EXPERIMENTAL POINTS FOR MODEL VERIFICATION I.E. POINTS INSIDE THE EXPERIMENTAL DOMAIN BUT NOT INCLUDED IN TABLE II AND III

	V level	F level	T level	Weibull Exp. Result (8 exp.)	DoE Model (8 exp.)	% Difference	SRM8 Model (8 exp.)	% Difference	SRM4 Model (4 exp.)	% Difference
2kV; 10 kHz; 117.5 °C	0.262	0.262	0.694	1.017	0.920	-9.5%	0.937	-7.9%	0.928	-8.4%
1kV; 5kHz; 20°C	-1	-1	-0.061	2.089	2.206	5.6%	2.137	2.3%	2.098	0.7%
2kV; 10kHz; 100°C	0.262	0.262	0.587	1.069	0.974	-8.9%	0.982	-8.1%	0.970	-9.4%
1.7 kV; 15 kHz; 26.7°C	0	1	0	1.314	1.266	-3.6%	1.321	0.5%	1.298	2.1%
1.7 kV; 8.6 kHz; 180°C	0	0	1	0.997	0.924	-7.4%	0.970	-2.7%	0.978	4.1%
3 kV; 8.7 kHz; 26.7°C	1	0	0	0.881	0.935	6.0%	0.822	-6.7%	0.785	-3.7%
2kV; 10kHz; 62.5°C	0.262	0.262	0.320	1.165	1.109	-4.9%	1.101	-5.5%	1.081	-7.4%
1.5 kV; 7.5 kHz; -17.5°C	-0.262	-0.262	-0.481	1.771	1.915	8.1%	1.901	7.4%	1.858	5.3%
1kV; 8.6 kHz; 26.7°C	-1	-0.013	0.004	1.769	1.976	11.7%	1.833	3.6%	1.791	2.3%
1.73 kV; 5 kHz; 26.7°C	0	-1.000	0	1.704	1.642	-3.7%	1.699	-0.3%	1.675	-1.2%
1.73kV; 8.6kHz; -55°C	0	0	-1	1.986	1.993	0.4%	2.023	1.9%	1.978	-1.1%

978-1-4799-0024-4/13 $31.00 © 2013 IEEE

VI. CONCLUSION

This paper has improved the previous work on lifespan modeling of the insulation of low voltage machines with the help of some statistical tools such as response surface and analysis of variance. It has been shown how the most influential factors can be identified, that the SRM increases modelling accuracy and that an experiment number reduction is possible with low risks.

The final objective of this work is to extend the validity domain of the model, primarily towards low constraint levels, for prognostic purposes. Other constraints such as pressure will be included in future work.

VII. REFERENCES

[1] P.J. Tavner, "Review of condition monitoring of rotating electrical machines", *IET Electr. Power Appl.*, vol. 2, no. 4, pp. 215–247, 2008.

[2] J. Yang, S.B. Lee, J. Yoo, S. Lee,Y. Oh, C. Choi, "A Stator Winding Insulation Condition Monitoring Technique for Inverter-Fed Machines", *IEEE Trans. Power Elec.*, vol. 22, no. 5, pp 2026-2033, 2007.

[3] G. Mazzanti, "The combination of electro-thermal stress, load cycling and thermal transients and its effects on the life of high voltage ac cables", *IEEE Trans. Dielec. and Elec. Insul.* vol. 16, no.4, pp. 1168–1179, 2009.

[4] J.P Crine, "On the interpretation of some electrical aging and relaxation phenomena in solid dielectrics", *IEEE Trans. Dielec. Elec. Insul.*, vol. 12, no. 6, pp. 1089-1107, 2005.

[5] R. Bartnikas, R. Morin, "Multi-stress aging of stator bars with electrical, thermal, and mechanical stresses as simultaneous acceleration factors, *IEEE Trans. Ener. Conv.*, vol. 19, no. 4, pp. 702–714, 2004.

[6] Smet, F. Forest, J. Huselstein, and F. Richardeau, "Diagnosiss for Electric Machines, Power Electronics & Drives", in *Proc. IEEE SDEMPED, 2011*, pp. 278–282.

[7] F. Guastavino, A. Ratto, E. Torello, G. Biondi, G. Loggi, and A. Ceci, "Electrical aging tests on different nanostructured enamels subjected to severe voltage waveforms", *Proc. IEEE SDEMPED, 2011*, pp. 278–282.

[8] R.A. Fisher, *The Design of Experiments*, Edinburgh, Oliver and Boyd, 1935.

[9] I.P. Brown, and R.D. Lorenz, "Induction Machine Design Methodology for Self-Sensing: Balancing Saliencies and Power Conversion Properties", *IEEE Trans. Ind. Appl.*, vol. 47 , no. 1, pp. 79–87, 2011.

[10] M.S. Islam, R. Islam, T. Sebastian, A. Chandy, and S.A. Ozsoylu, "Cogging Torque Minimization in PM Motors Using Robust Design Approach", *IEEE Trans. Ind. App.*, vol. 47, no. 4, pp. 1661–1669, 2011.

[11] S. R. Cove, M. Ordonez, F. Luchino and John E. Quaicoe, "Applying Response Surface Methodology to Small Planar Transformer Winding Design", *IEEE Trans. Ind. Electron.*, vol. 60, no.2, pp 483-493, 2013.

[12] H.M. Hasanien, and S.M Muyeen, "Design Optimization of Controller Parameters Used in Variable Speed Wind Energy Conversion System by Genetic Algorithms", *IEEE Trans. on Sust. Energy*, vol. 3, no. 2, pp. 200 – 208, 2012.

[13] S. Rahmani, N. Mendalek, K. Al-Haddad; "Experimental Design of a Nonlinear Control Technique for Three-Phase Shunt Active Power Filter", *IEEE Trans. Ind. Elec.*, vol. 57, no. 10, pp 3364–3375, 2010.

[14] Lahoud, N.; Faucher J.; Malec, D.; Maussion, P., « Electrical Aging of the Insulation of Low Voltage Machines: Model definition and test with the Design of Experiments", Industrial Electronics, IEEE Transactions on, vol. pp, no. 99, 2013, pp 1

[15] B. Bertsche, *Reliability in Automotive and Mechanical Engineering Determination of Component and System Reliability*, Springer, 2008.

[16] N. Lahoud; J. Faucher; D. Malec, and P. Maussion, "Electrical ageing modeling of the insulation of low voltage rotating machines fed by inverters with the design of experiments (DoE) method", *in Proc. IEEE SDEMPED*, 2011, pp. 272–277.

[17] M. Pillet, *Les plans d'expériences par la méthode Taguchi*, Paris, Les Editions d'Organisation, 2001.

[18] Kim, S-I. I.; Kim, Y-K ; Lee, G-H ; Hong, J-P P, "A Novel Rotor Configuration and Experimental Verification of Interior PM Synchronous Motor for High-Speed Applications", Magnetics, IEEE Transactions on vol. 48 , Issue: 2, 2012 , pp. 843 – 846

[19] Sahin, S.; Isler, Y., "Microcontroller-Based Robotics and SCADA Experiments", IEEE Trans. on Education, vol. PP, no. 99, 2013, p 1

[20] Vergura, S.; Acciani, G.; Amoruso, V.; Patrono, G. E.; Vacca, F., "Descriptive and Inferential Statistics for Supervising and Monitoring the Operation of PV Plants", Industrial Electronics, IEEE Trans. on, vol. 56, 2009 , pp. 4456 – 4464

[21] S. Shapiro and M. Wilk, "An analysis of variance test for normality", Biometrika, vol. 52, no. 3, pp 591-611, 1965.

[22] Myers R.H and Montgomery D.C., *Response Surface Methodology*, J. Wiley and Sons, 2002

[23] Versèle, C.; Deblecker, O.; Lobry, J., "A Response Surface Methodology Approach to Study the Influence of Specifications or Model Parameters on the Multiobjective Optimal Design of Isolated DC–DC Converters", Power Elec.s, IEEE Trans.s on, vol. 27, no. 7, 2012, pp. 3383 – 3395

[24] Hasanien, Hany M., "Particle Swarm Design Optimization of Transverse Flux Linear Motor for Weight Reduction and Improvement of Thrust Force", Industrial Electronics, IEEE Transactions on, vol. 58 , no. 9, 2011 , pp. 4048 – 4056

VII. BIOGRAPHIES

Antoine Picot graduated from National Polytechnique Institut (INP) Grenoble, France in 2006. He received the M degree in signal, image, speech processing in 2006 and his PhD in control and signal processing in 2009 from the INP Grenoble. He is actually associate professor at the INP Toulouse. His research interests are in monitoring and diagnosis of complex systems with signal processing and AI techniques.

David Malec was born in France in 1964. He got the Eng. degree in 1992 from the Conservatoire National des Arts et Métiers and the Ph.D. degree in 1996 from Paul Sabatier University of Toulouse, where he is Professor in Electrical Engineering. His scientific activities deal with the study of solid insulating materials (polymers and ceramics) used in both low and high voltage electrical engineering domains.

Pascal Maussion (member IEEE) got his MSc and PhD in Electrical Engineering in 1985 and 1990 from Institut National Polytechnique (Toulouse, France). He is currently Professor with the University of Toulouse and LAPLACE, Lab. for PLAsma and Conversion of Energy. His research activities deal with control and diagnosis of electrical systems and with the design of experiments for optimisation. He is currently head of Control and Diagnosis group and teaches control and diagnosis.

Diagnosis and Prognosis of In-Service Electric Machine in the Absence of Historic Data Related to Faults and Faults Progression

Syed Sajjad H. Zaidi
National University of Sciences and Technology,
Islamabad, Pakistan
sajjadzaidi@pnec.edu.pk

Abstract—**Extensive work has been presented in the literature related to fault diagnosis and prognosis of machines and related components. Prime focus of the proposed techniques is on either on assembly line checkout of machines or newly installed machines as a large number of methods are based on supervised learning. In this paper, fault diagnosis algorithm of in-service DC starter motor is presented. The proposed approach encompasses on the development of predefined fault progression curves. Features to develop these curves are extracted from machine current in time frequency domain. According to the proposed method, a number of curves are developed each of different order and slope. As the machine fault progresses, the fault features are projected on these curves and the % fault severity is identified. The results are presented and conclusions are made.**

Index Terms—**Electric Machine, Diagnosis, Prognosis, Principle Component Analysis, Pattern Recognition, Time Frequency Analysis, fault estimation**

I. INTRODUCTION

Fault diagnosis and prognosis of electrical machines is an area of great interest and has attracted a huge number of researchers to explore this field. Many techniques related to different types of machines have been presented in literature which propose different variants of diagnostic and prognostic algorithms. All types of machines have been analyzed for diagnosis. Fault diagnosis techniques based on time-frequency domain spectral analysis and pattern recognition were presented in [1]. As fault diagnosis and prognosis have commonalities, intuitively, prognosis should also be possible in the time-frequency domain, using pattern recognition techniques. Abdessalam and Clerc [2] grouped diagnostic approaches in major categories such as signal based and model based. The model based approach uses time-frequency signal representation. One of the popular choices is Hidden Markov Models (HMM), which is a powerful statistical modeling tool, having its main strength in its doubly statistical nature. HMMs are extensively used for speech recognition, hand gesture recognition and text segmentation. It is also used for tool wear detection/prediction [3], [4] and in monitoring bearing faults [5]. In [2] the authors use HMM for the diagnosis of machine faults using features extracted by the time frequency distributions. Most of the current work in field of electrical machines is based on using HMM for diagnosis but not for prognosis. All of these mentioned approaches and another

rich set of techniques have been discussed for machine fault diagnosis and prognosis. However, algorithm procedure of training and testing which is most common in all approaches, has certain limitations. These techniques are more likely to be focused on diagnosis and prognosis on the the assembly line or cases where a large set of the same machine is available. Otherwise training data for a in-service machine is not available.

In this work, the author proposed a time frequency representation based method for the diagnosis and future state estimation of in-service electrical machines. The target platform is selected as a automotive starter with faults in the gears. In Section II, the proposed methodology is presented. In Sections III and IV, time-frequency transformation methods and experimental setup are presented, respectively. Implementation details are given in V. In Section VI, results are presented. Section VII includes conclusions.

II. PROPOSED METHODOLOGY

In this work, diagnosis and prognosis of an in-service electrical machine is proposed. It is assumed the fault progression related data and related curves are not provided. In this case, the manufacture must provide a method suitable for fault diagnosis and predicting remaining useful life. For diagnosis and prognosis of machines, the first question to be answered is about the suitable parameter or signal to be selected for feature extraction. There can be many candidates. Major categories are intrusive such as x rays, vibration measurement, high frequency current injection etc, or non intrusive such as temperature, sound or current. Authors in [6] proved that machine current is an efficient means for feature extraction for diagnosis. As it is a non intrusive method, it is cost effective and an accurate means of feature extraction. Feature extraction is one of the most important aspect in fault diagnosis and prognosis. Diligent efforts are required to collect discriminative and accurate features which are able to exhibit all dynamics within an electromechanical system. Features can be collected in the time domain, frequency domain or time frequency domain. However, time domain features are susceptible to noise and can possess misleading information. Frequency domain features have proved valuable in representing the underlying phenomena in electrical machines and many publication can be referred. However, loss of time in the analyzed data is the inherent issue with frequency domain analysis. Therefore, the time-frequency domain has emerged as the most suitable domain transform to represent transient phenomena related to fault occurrences in electrical machines. In this work, the collected information is transformed to the time frequency domain.

In the spirit of the proposed work, the machine under study is considered to be in-service and historic information related to machine faults is not available. In order to carry out fault diagnosis under the supervised learning method, this data is mandatory. Therefore, an alternate method is being proposed. As data only exist in the present state, new data collected is used datum for diagnosis. A number of samples are collected from machine in the present state, and they are transformed to the time-frequency domain. Representation in the time frequency domain contains large number of redundant features which can tend to inaccurate results. Dimensionality reduction is carried out using principle component analysis of the transformed signal. As the signal belongs to a machine which is in same state, there is only one significant eigenvalue and resultantly only one eigenvector. The mean feature vector of all the samples collected from the machine is projected on the eigenvector. This point is datum base.

Based on this datum base point, a plot can be developed which contains linear regression models having different slopes. Moreover, higher order progression lines can be plotted. Example plots are shown in Fig. 1 & 2.

It is assumed that samples are collected on a regular basis from the machine. On receipt of new each new sample, it is projected on the PCA eigenvector plan. The plan already contains the datum point. With projection of a few samples, the concerned slope can be identified. As long as samples remain within the lower slope, that the machine has not developed any fault. However, if the projection of a new sample follows higher slopes, it is an indication of fault occurrences. As the proposed method falls under the scope of un-supervised learning, the new fault state can only be identified as a percentage increase in the previous state which is the maximum diagnosis algorithm can determine.

However, using the proposed prognosis approach can be more fruitful. Once the concerning slope has been identified, approximate time to failure can be predicted using regression. If the projected samples map themselves along linear slopes, the machine failure stage will not occur soon, whereas if are mapped along the higher order contour, it reflects that failure stage will occur early.

III. THEORETICAL BACKGROUND

In the development of the proposed methodology, theoretical concepts related to the time frequency domain representation, principle component analysis and linear progression are presented in the following subsections.

A. Choi Williams Distribution (CWD)

The Choi-Williams distribution[7] of a signal *s(t)* is defined as:

$$C(t,w) = \int \int \int \phi(\theta,\tau) s(u - \tfrac{\tau}{2}) s^*(u - \tfrac{\tau}{2})$$
$$e^{j(\theta u - \theta t - \tau \omega)} du d\theta d\tau \qquad (1)$$

where $\phi(\theta,\tau) = exp(-\frac{(\theta,\tau)^2}{\sigma})$ is the kernel function that acts as a filter on the signals autocorrelation function. This distribution can be thought of as a filtered/smoothed version of the Wigner distribution [8], and the amount of smoothing is controlled by σ. This smoothing removes the cross-terms seen in the Wigner distribution at the expense of reduced resolution.

B. Principle Component Analysis

Principal Component Analysis(PCA) is a dimensionality reduction method which marks unidentified trends in data. It transforms a signal in a way to prominently represent hidden patterns in it. The transform, also known as Karhunen-Loeve Transform[8], is based on a simplistic approach. The *d*-dimensional mean vector is μ is subtracted from the feature vectors. From the zero mean data, the first principal component is computed as given by Eq. 2

$$w_1 = \underset{\|w\|=1}{\arg \max} Var\{w^T X\} \qquad (2)$$

where w & and X are the principal component vector and feature set respectively. Since the mean is zero, Eq. 2 becomes

$$w_1 = \underset{\|w\|=1}{\arg \max} E\{(w^T X)^2\} \qquad (3)$$

Principal components can be obtained by computing Eigenvalues and Eigenvectors using Singular Value Decomposition. The Eigenvalues represent the significance of the corresponding dimension. The higher the values, the more discriminative the data will be, once projected on that direction.

C. Feature extraction Method

PCA is a non-parametric analysis method which can deal with any type of feature set. Fault currents in time domain generates the most simplistic set of features. Amplitude, variance and mean value can be time domain features for fault analysis. However, time domain features are severely affected by the presence of noise and interference. Frequency domain analysis of complex signals provides more detailed information especially with respect to the distribution of energy in different energy bands. As presented in [6], electrical as well as mechanical faults also appear in the machine current if analyzed in frequency domain. Even more detailed analysis can be preformed in time-frequency domain which is capable of representing transient phenomena occurring in sampled information. In this work, faults in machines are represented in the time-frequency domain which provides a strong set of features; specially in the case of transient fault analysis[9]. The transform used is the Choi William transform as proposed in [10]. High frequency scales obtained from the Choi-Williams distribution of the machine current are utilized as the features for analysis.

978-1-4799-0024-4/13 $31.00 © 2013 IEEE

D. Development of Regression Plots

The features extracted in the time frequency domain are projected on the PCA eigenvectors. As per the proposed procedure, only one eigenvalue is expected be significant. Therefore, only one dimensional data will be processed. It is assumed that the priority of features will increase with the severity of faults. As per the formulation of the problem, information related to the machine in higher fault severity states does not exist. Features computed for the present state of the machine are considered as the initial point. Scaled intensity of the same features is used to give an approximate value of the higher fault severity features. Once the initial point has been computed and the maximum value has been estimated (as five or six times the initial value), linear and higher order curves can be developed having different slopes and different orders respectively. In Fig. 1, linear regression are shown. These curves are developed as follows:

$$\ell_{l_i} = m_i \times t + c_i \qquad (4)$$

$$m_{l_i} = \frac{y_{Maxi} - y_0}{t_{y_{Maxi}} - t_0} \qquad (5)$$

where i=1 ranges from 1 to number of linear slopes; c_i is assumed to be zero for all slopes. In the figure, 10 line having increasing slopes are shown.

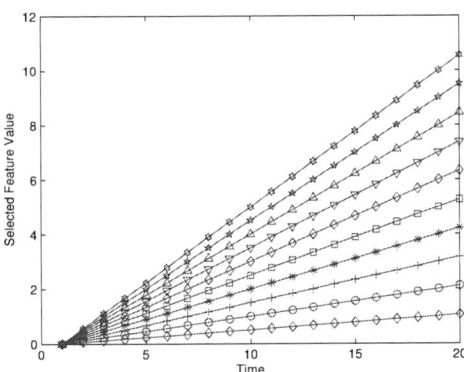

Fig. 1. Linear Fault Progression Curves

In Fig. 2, higher order curves are shown. The curves are developed as follows

$$\ell_{qi} = t^{n_i} + c_i \qquad (6)$$

where i=1 ranges from 1 to number of high order slops and c_i is assumed to be zero for all slopes. In the figure, 10 curves of increasing order are shown.

IV. EXPERIMENTAL SETUP

For this work a laboratory experimental setup was constructed. It comprise complete engine with all accessories and it was housing a starter motor assembly. The starter motor was energized by a 12V battery and was controlled by a PC through electronics comprising of an IGBT switch. The starting signals for the transistor switch were generated by the data acquisition software. A National Instruments card DAQ - 6229 was used

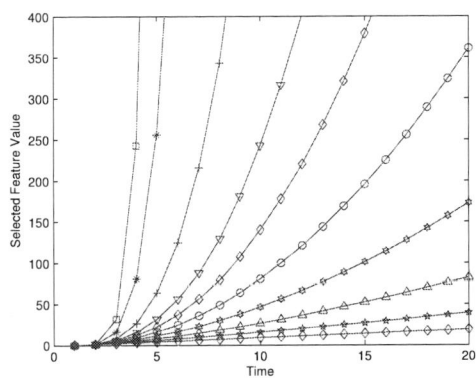

Fig. 2. Non linear Fault Progression Curves

for sampling. As mentioned previously, all phenomena related to electromechanical faults are reflected in the motor current. It is the primary signal for this analysis. A wide bandwidth (50kHz) current sensor LEM HASS-600S was used. A position sensor with an optical pulse counter was used to measure the position of the starter motor gear.

Fig. 3. Hardware Setup

A. Operation of Motor

In this work, the objective was diagnosis and prognosis of a in-service machine with damaged gear teeth. For this purpose automotive DC starter motors were used.

B. Nature of Fault

The DC starter motor turns for approximately $1s$ during each starting attempt. In each attempt, the starter motor gear engages with the flywheel attached to the engine crankshaft. During the compression cycle of each cylinder, the starter motor torque increases, resulting in higher forces on the starter gear teeth. In the case of damaged teeth, the impact and meshing pattern is different, which is reflected in the motor current [6]. Sampled motor currents are shown in Fig. 4. This is a cyclic fault, repeating itself with the frequency of engine compression and expansion. However, in the case

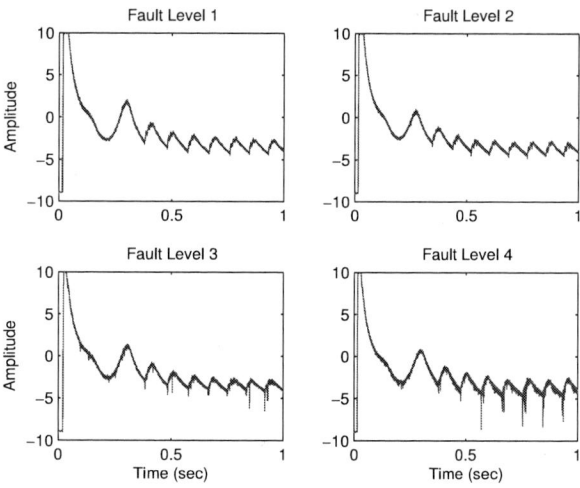

Fig. 4. Sampled Current Signals

of a fault, the signal includes high frequencies, which are generated due to the knock caused by the imperfect meshing of gears. The current is sampled and features are extracted after transforming the samples to the time-frequency domain. Gear faults of increasing intensity were introduced in four machines. Therefore, there are five classes, including one from the healthy machine and four from faulted machines.

V. Implementation

A block diagram of the proposed approach is shown in Fig.5. Data is collected from the in-service machine as described IV.

A. Feature Extraction

In this work, the machine current is the parameter which is used for diagnosis as well as prognosis of faults. Using the Choi Williams transformation method mentioned described in Section-II, a time frequency domain representation of the sampled current is obtained. The Choi Williams transform with 8 bands and 0.2 as the smoothing factor was used. Since the fault representative features mostly lie in the high frequency range, only higher bands were selected, and lower bands were discarded. The bands were placed in the form of vectors to generate feature sets.

B. Development of Regression Plots

Typical diagnosis applications have training and testing subphases. However, in the present scenario, training data is not possible as the machine is already in service and historical data is not available. Therefore, the alternate approach is outlined in Section II. Number of samples of the machine current were collected and their time frequency features were obtained. In this work, 10 samples from the present state were collected. The principle component eigenvector of these samples was computed using the PCA method. The most significant vector was selected, all the samples were projected on it, and the mean position was computed. This mean position is the class

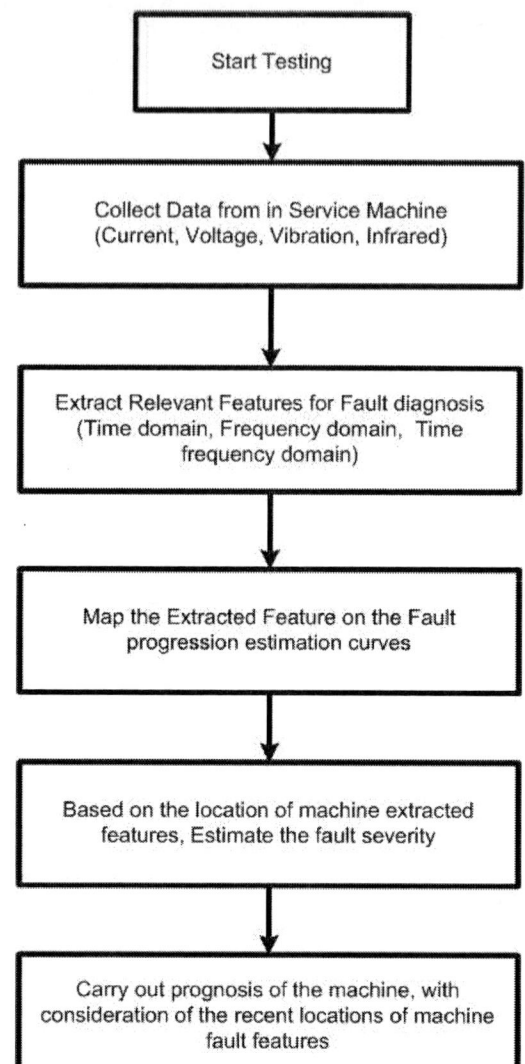

Fig. 5. Development of Fault diagnosis and Prognosis Curves

representative of the present state. These values are considered as the base values. The maximum fault feature value was estimated as five times higher than the base value. Multiple linear and higher order curves were generated keeping the initial point as the base point and final point as the estimated one.

C. Fault Diagnosis

Signals continued to be collected a specific interval. The same process of time frequency transformation, reshaping of frequency bands and feature reduction using PCA is performed. The new feature vectors are projected on the most significant PCA eigenvector and the mean of the projected points is computed and the class is determined.

The new sample class determination, previous class and regression curves are plotted together. If the new class data is projected on the higher slope curve, it can be detected that the fault has occurred. However, since the procedure is based on unsupervised learning, the class cannot be named. The increase

978-1-4799-0024-4/13 $31.00 © 2013 IEEE   618

in fault severity is the only conclusion which can be make using the proposed method.

D. Prognosis

As the data is projected on the plots with pervious having data points and regression curves, estimated time/instances to failure can be predicted by scaling the information obtained by the two projected data points. The severity state which is five times the base class point value is considered as the failure state. Projection of the class representative points on the curve plot will identify the curve of prediction. As long as the new data remains within the same curve, it indicates the faults class is not changed. When new data moves away from the curve, with higher slope or higher order, it means that the failure will occur soon and prognosis estimates can be updated considering the new curves.

VI. RESULTS AND DISCUSSION

In Fig. 6, projected class points and their curves are shown. There are five linear regression curves related to five different fault progression slopes represented as solid lines. The dotted line is the projected data points on the curve plan. It can be

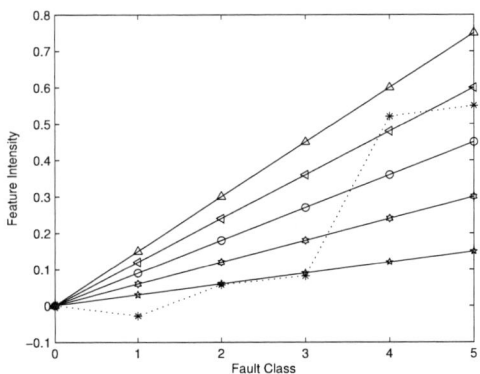

Fig. 6. Fault Features Projected on Linear Curves

observed from these curves and data points the initial trend with low intensity. This is indicative of no or very nominal increase in the fault intensity.

After that data point 3, there is a steep and sudden change in the slope. This indicates a serious increase in the fault severity. Moreover, the slope of the fault progression also changes and prognosis estimates need to be reestablished based on the new regression curve.

As the fault features were generally following linear curves, higher order curves were not used for diagnosis and prognosis of this machine. Notwithstanding these curves can be used when the fault dynamics are quick and severity is changing rapidly.

VII. CONCLUSIONS

In this paper, a proposed method for fault diagnosis and prognosis is presented for an in-service electrical machine with out historical data available. Development of fault regression

curves is proposed. These curves can be linear with increasing slopes or are of higher order depending upon the nature of the fault. These curves are used to detect an increase in fault severity and prognosis for failure.

Fault diagnosis of a machine has been proposed based on information acquired from regression curves. If new data points remain on the same curve, fault severity is not increasing. If the curve is changing however, this indicates an increase in fault severity states.

Prediction of the failure state is estimated by the scaled value of the base data. A low slope curve will decay to the failure state slowly whereas a high slope curve will attain the failure state much earlier. Furthermore, if the failure is catastrophic or its dynamics are very high, the curve will be of higher order which will reach the failure state in a shorter amount of time.

As future work, the regression curve will be used as a dynamic fault model to be used by a predictor such as a Kalman filter, hidden Markov model or particle filter. These estimators require a model of the fault severity dynamics in order to predict the future state. The proposed regression curves, with some effort, can be used to develop such a model. Implementation of this is expected to enhance the prediction confidence.

REFERENCES

[1] E. Strangas, S. Aviyente, and S. Zaidi, "Time frequency analysis for efficient fault diagnosis and failure prognosis for interior permanent magnet AC motors," *IEEE Transactions on Industrial Electronics*, vol. 55, no. 12, pp. 4191–4199, Dec 2008.

[2] L. Abdesselam and G. Clerc, "Diagnosis of induction machine by time frequency representation and hidden Markov modelling," in *IEEE International Symposium on Diagnostics for Electric Machines, Power Electronics and Drives, SDEMPED*, Sep. 2007, pp. 272–276.

[3] L. Heck and J. McClellan, "Mechanical system monitoring using hidden Markov models," in *International Conference on Acoustics, Speech, and Signal Processing, ICASSP*, Apr 1991, pp. 1697–1700 vol.3.

[4] F. Camci and R. Chinnam, "Hierarchical HMMs for autonomous diagnostics and prognostics," in *International Joint Conference on Neural Networks, IJCNN*, 2006, pp. 2445–2452.

[5] H. Ocak, K. A. Loparo, and F. M. Discenzo, "Online tracking of bearing wear using wavelet packet decomposition and probabilistic modeling: A method for bearing prognostics," in *Journal of Sound and Vibration*, vol. 302, no. 4-5, May 2007, pp. 951–961.

[6] W. Thomson and M. Fenger, "Current signature analysis to detect induction motor faults," *IEEE Industry Applications Magazine*, vol. 7, pp. 26–34, 2001.

[7] L. Cohen, *Time-Frequency Analysis*. Prentice Hall, 1995.

[8] R. O. Duda, P. E. Hart, and D. G. Stork, *Pattern Classification*, 2nd ed. Wiley-Interscience, Oct 2000.

[9] W. Zanardelli, E. Strangas, and S. Aviyente, "Failure prognosis for permanent magnet AC drives based on wavelet analysis," in *IEEE International Conference on Electric Machines and Drives*, 2005, pp. 64–70.

[10] L. Cohen, "The scale representation," *IEEE Transactions on Signal Processing*, vol. 41, no. 12, pp. 3275–3292, Dec 1993.

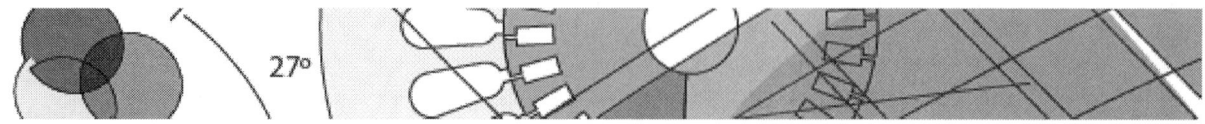

Awards

The 2013 Diagnostics Achievement Award

This award was established to honour innovators who contributed to the technical areas of this committee.

The award is to be presented biennially (once every two years, odd years) to an individual, for outstanding sustained technical contributions and services in the field of monitoring and diagnostics for electrical machines, power electronics, and drives.

The 2013 diagnostics achievement award was given to **Prof. Fiorenzo Filippetti** with the University of Bologna, Italy

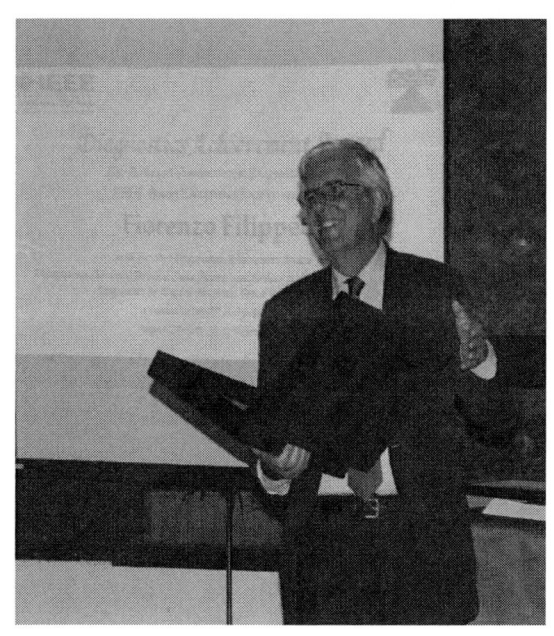

SDEMPED Paper Awards

These awards are presented biennially for up to three technical papers (in no order), for outstanding technical competence displayed in a paper presented at an IEEE SDEMPED conference.

The Technical Program Chair of the SDEMPED Conference will select 10 Prize Paper Award candidates with the highest score from the paper review process.

Voting will be under the leadership of the chair of the award selection committee and the outcome will be based on the scores of the selection committee.

The SDEMPED 2013 Paper Awards were given to the following papers:

-Detection of Stator Slot Magnetic Wedge Failures for Induction Motors without Disassembly

Authors:
Kun Wang Lee,
Jongman Hong,
Doosoo Hyun,
Sang Bin Lee,
Ernesto Wiedenbrug,
Mike Teska and
Chaewoong Lim

- Full Detection of High-Resistance Connections in Multiphase Induction Machines

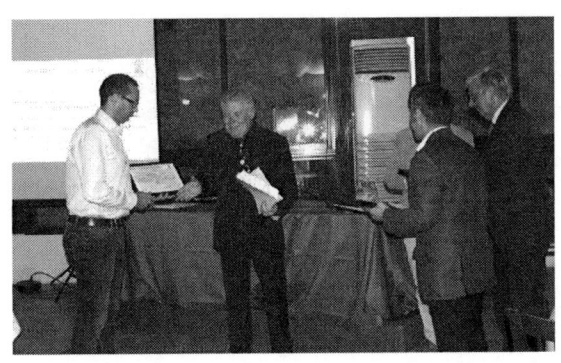

Authors:

Luca Zarri,
Michele Mengoni,
Angelo Tani,
Yasser Gritli,
Giovanni Serra,
Fiorenzo Filippetti,
Domenico Casadei

-Improvements on Lifespan Modeling of the Insulation of Low Voltage Machines with Response Surface and Analysis of Variance

Authors:
Antoine Picot,
David Malec and
Pascal Maussion

Technical Program Summary

Tutorials

- *Testing And Monitoring Of Medium/High Voltage Induction Machines*, Prof. Sang Bin Lee, Korea University, Seoul, Korea
- *Fault Detection And Fault Tolerant Operation Of Adjustable Speed Switched Reluctance And Permanent Magnet Synchronous Motor Drives*, Prof. Babak Fahimi, University of Texas at Dallas
- *Modelling Faults With A Numerical Software*, Dr. Vicente Aucejo, Indielec, Spain

Plenary Sessions

- *Life Estimation Of Copper Bars In Induction Motors: Multi-Physics Approach*, Dr. Pedro Jover Rodríguez, ABB Corporate Research, Sweden
- *Plenary Session 2- Electrical Machines in wind turbines- Maturity of technology*, Clara Mateo Martínez de Albornoz Experiencia Operativa Renovables, Iberdrola ,Dirección Técnica
- *The Reliability of Machine Windings in Adjustable Speed Drives: Stress, Strength, Design, Qualification & Diagnostics*, Dr. Martin Kaufhold ,Siemens AG Industry, Drive Technologies, Large Drives Vice President.

Special Sessions

- *Fault tolerant power converters, electrical machines and drives*, Prof. Fiorenzo Filippetti, Luca Zarri
- *Advanced artificial intelligence approaches applied to fault characterization and classification for electrical machines diagnostic purposes*, Prof. Luis Romeral Martínez, Juan Antonio Ortega Redondo
- **Failure prognosis methods in electrical drives**, Elias Strangas, Hubert Razik,

Sessions

	Sala d'actes	Sala de Graus
Wednesday 28, Afternoon		
14h30	SS1-A1 Fault tolerant Power Converters, Electrical machines and drives (I)	RS1-G1 Power Electronics (I)
17h	RS2-A2 Rotor Faults (I)	RS3-G2 Tools for Diagnostics (I) . Signal Analysis Techniques
Thursday 29, Morning		
8h30	RS4-A3 Rotor Faults (II)	RS5-G3 Tools for Diagnostics (II) . AI Techniques
11h	RS6-A4 Stator Faults (I)	RS7-G4 Adjustable Speed Drives and Power Converters
Thursday29, Afternoon		
14h30	RS8-A5 Stator Faults (II)	RS9-G5 Rotor Faults (III)
17h	RS10-A6 Permanent Magnet Machines	RS11-G6 Tools for Diagnostics (III) . Mechanical faults
Friday 30, Morning		
8h30	RS12-A7 Test for predictive maintenance. Partial Discharge tests	SS4-G7 Failure Prognosis Methods In Electrical Drives
11h	SS3-A8 Advanced Artificial Intelligence Approaches	SS1-G8 Fault tolerant Power Converters, Electrical machines and drives (II)
Friday 30, Afternoon		
14h30	RS13-A9 Rotor Faults (IV)	RS14-G9 Power Electronics II, Power Converters

Committees

General Co-Chairs
Martin Riera-Guasp (SPAIN)
Jose A. Antonino-Daviu (SPAIN)

Technical Program Co-Chairs
Manuel Pineda-Sánchez (SPAIN)
Sang Bin Lee (KOREA)

Honorary Co-Chairs
Gerard-André Capolino (FRANCE)
Giuseppe Buja (ITALY)

Special Sessions Co-Chairs
Joan Pons-Llinares (SPAIN)
Luca Zarri (ITALY)

Publicity Co-Chairs
Vicente Climente-Alarcón (SPAIN)
Antonio J. Marques-Cardoso (PORTUGAL)

Tutorials Co-Chairs
Rubén Puche-Panadero (SPAIN)
Ronald Harley (USA)

Awards Chair
Alberto Bellini (ITALY)

Local Organizing Committee
José Roger-Folch (SPAIN)
Juan Pérez-Cruz (SPAIN)
Francisco Vedreño-Santos (SPAIN)
Miguel Delgado (SPAIN)
Oscar Duque-Pérez (SPAIN)
Manés Fernández-Cabanas (SPAIN)
Antoni Garcia (SPAIN)
Manuel Garcia Melero (SPAIN)
Daniel Moríñigo-Sotelo (SPAIN)
Juan Antonio Ortega (SPAIN)
Jordi-Roger Riba Ruiz (SPAIN)
Jose Luis Romeral Martínez (SPAIN)

International Steering Committee
Gérard-André Capolino (Chair)
Fiorenzo Filippetti (Vice-Chair)
Sang Bin Lee (Secretary)
Manes Fernandez-Cabanas
Alberto Bellini (Awards Chair)
Giuseppe Buja
Thomas Habetler
Ronald Harley
Humberto Henao
Marian Kazmierkowski
Christian Kral
Antonio J. Marques Cardoso
Tadeusz Sobczyk
Elias Strangas
Ernesto Wiedenbrug

SDEMPED 2013 Proceedings

Committees

Special sessions organizers

Fiorenzo	Filippetti
Juan A.	Ortega
Hubert	Razic
Luis	Romeral
Elías	Strangas
Luca	Zarri

Track Chairs

Jose A	Antonino-Daviu
Claudio	Bruzzese
Laurent	Capocchi
Miguel	Delgado Prieto
Oscar	Duque Perez
Manes	F. Cabanas
Fiorenzo	Filippetti
Antoni	García
Manuel	Garcia Melero
Shahin	Hedayati Kia
Sang Bin	Lee
Antonio J.	Marques Cardoso
Daniel	Morinigo Sotelo
Juan A	Ortega
Manuel	Pineda
Joan	Pons Llinares
Ruben	Puche Panadero
Hubert	RAZIK
Jordi R.	Riba Ruiz
Martin	Riera-Guasp
Luis	Romeral
Elias	Strangas
Francisco	Vedreño-Santos
Luca	Zarri

Session Chairs

Jose A.	Antonino-Daviu
Alberto	Bellini
Franck	Betin
Claudio	Bruzzese
Doménico	Casadei
Wenping	Cao
Robert	Cox
Miguel	Delgado Prieto
Oscar	Duque Pérez
Babak	Fahimi
Manés	Fernández-Cabanas
Fiorenzo	Filippetti
Antoni	Garcia Espinosa
Yasser	Gritli
Pedro	Jover Rodríguez
Shahin	Hedayati Kia
Humberto	Henao
Sang Bin	Lee
Antonio J.	Marques Cardoso
Pascal	Maussion
Daniel	Moríñigo Sotelo
Joaquín G.	Norniella
Subhasis	Nandi
Juan A.	Ortega
Manuel	Pineda-Sanchez
Joan	Pons Llinares
Ruben	Puche-Panadero
Remus	Pusca
Hubert	Razik
Jordi R.	Riba Ruiz
Martín	Riera-Guasp
Luis	Romeral
René	Romero-Troncoso
Elías	Strangas
Thomas	Wolbank

SDEMPED 2013 Proceedings

Committees

Tutorial Organizers

Sang Bin Lee
Babak Fahimi
Vicente Aucejo

Keynote Speakers

Pedro Jover Rodríguez
Clara Mateo Martínez de Albornoz
Martin Kaufhold

List of reviewers

Name	institution	Country
A. Goedtel	Federal Technological University Of Paraná	Brazil
A.N. Safacas	University Patras	Greece
Abdelhalim Mayouf	Laboratory Of Research Dimmer	Univ. Of Djelfa
Abdelmalek Khezzar	Laboratoire D'électrotechnique De Constantine	Algeria
Abdenour Soualhi	Ampere Laboratory	France
Aderito Neto Alcaso	Polytechnic Institute Of Guarda	Portugal
Agnieszka Nowak	Abb	Poland
Ahmed Braham	Insat	Tunisia
Alejandro Fernández Gómez	Cracow University Of Technology	Poland
Aleksandrs Andreiciks	Riga Technical University	Latvia
Alessandro Pisano	University Of Cagliari	Italy
Alexandre Bacchus	L2ep University Of Lille 1	Edf R&D
Anders Hultgren	Blekinge Institute Of Technology	Sweden
Anderson Machado	Ctm	Spain
Andre Mendes	University Of Coimbra	Portugal
Andrea Caprara	Techimp Hq Spa	Italy
Angel Sapena	Upv	Spain
Angelo Tani	Bologna University	Italy
Antoine Picot	Enseeiht-Laplace	France
Antoni García	Upc-Mcia	Spain
Antonio J. Marques Cardoso	University Of Beira Interior (Ubi)	Portugal
Antonio J. Marques Cardoso	University Of Beira Interior (Ubi)	Portugal
Arezki Menacer	Electrical Engineering	Algeria
Arkadiusz Dziechciarz	Cracow University Of Technology	Poland
Arshad Jamal	Indian Institute Of Technology	India
Arturo Garcia-Perez	Universidad De Guanajuato	Mexico
Baptiste Trajin	Aeroconseil Blagnac	France
Becharia Nadji	University Of Boumerdes	Algeria
Brice Aubert	Laboratoire Laplace	France
Bruno Baptista	Uc	Portugal
Carlos Platero	Upm	Spain
Chanseung Yang	Korea University	Korea
Chiara Boccaletti	Sapienza University Of Rome	Italy
Ciprian Harlisca	Technical University Of Cluj-Napoca	Romania
Claude Delpha	Laboratoire Des Signaux Et Systèmes (L2s)	France
Claudio Bruzzese	Diaee - University Of Rome Sapienza	Italy
Daniel Morinigo Sotelo	University Of Valladolid	Spain
Daniel Morinigo-Sotelo	Universidad De Valladolid	Spain
Daniel Zurita Millan	Polytechnic University Of Catalonia Mcia	Spain

SDEMPED 2013 Proceedings

List of reviewers

Name	Institution	Country
Dragan Matic	Faculty Of Technical Sciences	Serbia
Elias Strangas	Michigan State University	United States
Elio Usai	University Of Cagliari - Diee	Italy
Emmanuel Boutleux	Ampere - Ecl	France
Fabio Immovilli	University Of Modena And Reggio Emilia	Italy
Francesco Guastavino	University Of Genova	Italy
Francisco Vedreño-Santos	Universidad Politécnica De Valencia	Spain
Francois Philipp	Tu Darmstadt	Germany
Gabriele Rizzoli	University Of Bologna	Italy
George Georgoulas	Tei Of Epirus	Greece
Gerard Champenois	University Of Poitiers	France
Gerasimos Rigatos	Industrial Systems Institute	Greece
Gojko Joksimovic	University Of Montenegro	Montenegro
Gonzalo A. Orcajo Gonzalo	University Of Oviedo	Spain
Goran Stojicic	Vienna University Of Technology	Austria
Guillaume Verez	Greah	France
Guy Clerc	Laboratoire Ampère /Ucbl	France
Hanbo Zheng	Haepc Epri	China
Harold Saavedra	Universidad Politecnica Catalunya	Spain
Hubert Razik	Ampere	France
Ilhem Bouchareb	Electrical Engennering	Algeria
J.M. Anderson	Unc Charlotte	United States
James Ottewill	Abb Corporate Research Center	Poland
Jeremi Regnier	Laplace Laboratory	France
Joan Hernandez	Universitat Politècnica De Catalunya	Spain
Joan Pons Llinares	Universitat Politècnica De València	Spain
Joaquín G. Norniella	University Of Oviedo	Spain
Jochen Immel	Lenze Automation Gmbh	Germany
Jonas Borges Da Silva	Federal University Of Itajubá - Unifei	Brazil
Jordi-Roger Riba Ruiz	Universitat Politècnuica De Catalunya	Spain
Jorge Pleite Guerra	Universidad Carlos Iii De Madrid	Spain
Jose A Antonino-Daviu	Universitat Politecnica De Valencia	Spain
Jose Manuel Cano	University Of Oviedo	Spain
Jose Roger-Folch	Universitat Politecnica De Valencia	Spain
Juan Antonio Ortega	Universitat Politècnica De Catalunya	Spain
Juan Pérez-Cruz	Upv	Spain
Julio-César Urresty	Alstom Power	Spain
Keskes Hassen	Insat	Tunisia
Khaled Yahia	Université De Biskra	Algeria
Klemen Drobnic	University Of Ljubljana	Slovenia
Konstantinos Kampouropoulos	Electric Enginner	Spain
Konstantinos N. Gyftakis	University Of Patras	Greece
Lorand Szabo	Tu Cluj	Romania
Luca Fornasari	Techimp H.Q. Spa	Italy
Luca Zarri	University Of Bologna	Italy

SDEMPED 2013 Proceedings   List of reviewers

Name	Institution	Country
Luis Romeral	Mcia Center - Univ Politècnica De Catalunya	Spain
Maciej Orman	Ph.D. Eng.	Poland
Mahmood Moghadasian	University Of Picardie "Jules Verne"	France
Manes Fernandez Cabanas	Universidad De Oviedo	Spain
Manuel Moreno-Eguilaz	Mcia / Upc	Spain
Manuel Pineda	Universitat Politecnica De Valencia	Spain
Marco Antonio Rodríguez Blanco	Autonomus University Of Carmen City	Mexico
Marco Bonavoglia	Unibo	Italy
Maria Perdomo-Arvizu	Student	Germany
Mario Pacas	University Of Siegen	Germany
Martha Cecilia Amaya	Universidad Del Valle - Colombia	Colombia
Martin Riera-Guasp	Universitat Politecnica De Valencia	Spain
Mehdi Salehifar	Phd Student	Spain
Michele Mengoni	University Of Bologna	Italy
Miguel Delgado	Technical University Of Catalonia (Upc) - Mcia	Spain
Miguel Delgado Prieto	Technical University Of Catalonia (Upc)	Spain
Milan Rapaic	Faculty Of Technical Sciences	Serbia
Mitja Nemec	University Of Ljubljana	Slovenia
Mohamed Benbouzid	University Of Brest	France
Mohamed Drif	University Of Science And Technology Of Oran	Algeria
Mohamed Sahraoui	Electrical Engineering Laboratory Of Biskra	Algeria
Moshen Abadi	University Of Coimbra	Portugal
Mouna Ben Hamed	Tunisia	Tunisia
Muslum Arkan	Inonu University	Turkey
Nabil Ngote	Ecole Nationale De L'industrie Minérale	Morocco
Nayana Mahajan	Yc College Of Engineering	Nagpur
Nirudh Jirasuwankul	Kmitl	Thailand
Nuno Freire	University Of Coimbra/I T	Portugal
O. Touhami	Ecole Nationale Polytechnique D'alger	Algeria
Oliver Magdun	Darmstadt University Of Technology	Germany
Omar Touhami	Ecole Nationale Polytechnique D'alger	Algeria
Osama Al-Naseem	Kuwait University	Kuwait
Oscar Duque	University Of Valladolid	Spain
P.E. Gardel-Sotomayor	Uva And Una	Paraguay
Paolo Sommella	University Of Salerno-D. Of Industrial Engineering	Italy
Pascal Maussion	University Of Toulouse	France
Paulo Costa Branco	Instituto Superior Técnico	Portugal
Pedro Esteban Gardel	Uva And Una	Paraguay
Pedro Rodriguez	Abb	Sweden
Peng Zhang	Marquette University	United States
Peter Thiemann	Fachhochschule Suedwestfalen	Germany
Pu Shi	Glyndwr University	United Kingdom
Ramin Salehi	Universidad Polytecnica De Catalunya	Spain
Rastko Fiser	University Of Ljubljana	Faculty Of Electr. Eng.
Rene Romero-Troncoso	Universidad De Guanajuato	Mexico

List of reviewers

Name	Institution	Country
Robert Cox	University Of North Carolina Charlotte	United States
Robert Hanna	Rpm Engineering Ltd.	Canada
Roque Osornio Rios	Universidad Autonoma De Queretaro	Mexico
Ruben Puche Panadero	Universitat Politecnica De Valencia	Spain
S.N. Dhurvey	Assistant Professor	India
Sachin Kumar	Caterpillar	United States
Saeed Jahdi	University Of Warwick	United Kingdom
Salah Eddine Zouzou	Université De Biskra	Algeria
Salaheddine Ethni, N		United Kingdom
Samir Hamdani	Usthb	Algeria
Sang Bin Lee	Korea University	Korea
Sara Sara	Nust	Pakistan
Sarath Mohan	B-Tech	India
Selin Aviyente	Michigan State University	United States
Sergio Cruz	University Of Coimbra	Portugal
Shahin Hedayati Kia	University Of Picardie "Jules Verne"	France
Sijo Augustine	Indian Institute Of Technology	Madras
Slim Tnani	University Of Poitiers	France
Stefan Grubic	General Electric Company	United States
Subhasis Nandi	University Of Victoria	Canada
Tadeusz Sobczyk	Cracow University Of Technology	Poland
Taner Goktas	Mandatory	Turkey
Thomas Wolbank	Vienna University Of Technology	Austria
Tobias Müller	Fh-Swf	Germany
Valeria Leite	Universidade Federal De Itajubá	Brazil
Vicent Sala	Mcia-Upc	Spain
Vicente Climente Alarcon	Aalto University	Finland
Vincent Choqueuse	University Of Brest	France
Wesley Zanardelli	U.S. Army Tardec	United States
Wilder Herrera	Universidad Del Valle	Colombia
Yasser Gritli	University Of Bologna	Italy
Yassine Maouche	Student Member	Algeria
Z Liu	Newcastle University	United Kingdom
Željko Kanovi?	Faculty Of Technical Sciences	Novi Sad
Zied Lachiri	Insat	Tunisia
Zoran Jelicic	Associate Professor	Univresity Of Novi Sad

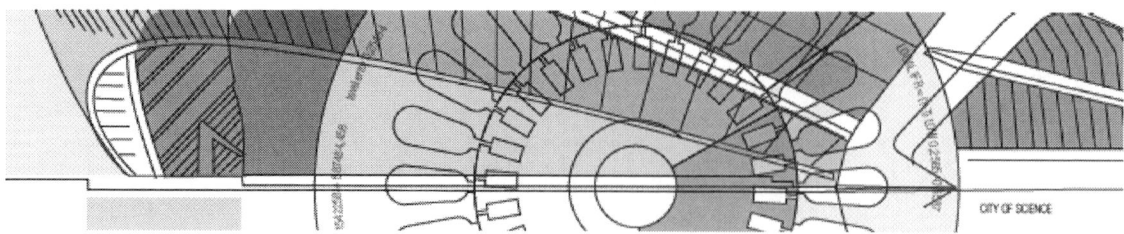

Index of Authors

A

Abadi, Mohsen 497

Ahrend, Ulf 207

Ait-Amar, Sonia 257

Alam, Farhan 428

Alamyal, -Mohamoud 157

Albini, Andrea Albini 371

Alejo, Dominique 329

Al-Naseem, Osama 221

Alvarez Salas, Ricardo 192

Amara, Yacine 342

Ambrozic, Vanja 142

Anderson, Jason M. 1

Andreiciks, Aleksandrs 433

Andrejak, Jean-Marie 77

Andriamalala, Rijaniaina N. 128

Antonino-Daviu, Jose 114, 592

Arkan, Müslüm 122

Arkkio, Antero 215

Asfani, Dimas 288, 573

Ashari, M. 573

B (right column, top)

Athanasopoulos, Dimitrios K. 36, 402

Aubert, Brice 329

Augustine, Sijo 512

Aviyente, Selin 600

B

Bacchus, Alexandre 558

Baptista, Bruno 497

Barakat, Georges 342

Barambones, Oscar 439

Bellini, Alberto 491

Belouchrani, Adel 23

Bennouna, Ouadie 342

Bertilsson, Kent 428

Betin, Franck 469

Biagi, Lyvia 538

Bianchini, Claudio 491

Biet, Mélisande 558

Binder, Andreas 447

Blaabjerg, Frede 433

Blánquez, Francisco R. 177

Index of Authors

Blatt, Sebastien 447

Blázquez, F. 177

Borekci, Selim 28

Bortolozzi, Mauro 477

Boucherma, Mohamed 263, 295

Boukoucha, Abdelaziz 23

Boussaid, Abdelfettah 263, 295

Broniera, Paulo 281

Bruzzese, Claudio 349, 477

Bui, Sonny 485

C

Cabal Yepez, Eduardo 192

Cabanas, Manés F. 527

Cano, José M. 527

Cao, Wenping 9. 269

Capolino, Gérard-André 358, 469

Caprara, Andrea 384

Cardoso, A. J. Marques 249

Casadei, Domenico 57, 505

Caux, Stéphane 329

Chen, Zheng 49, 323

Cirrincione, Giansalvo 544

Clerc, Guy 43, 552, 586

Climente-Alarcon, Vicente 592

Constantin, Alexandru 257

Cox, Robert W. 1

Cruz, Sérgio 497

D

Da Silva, Sérgio A. O. 566

Delgado, Miguel 169, 544

Disselnkoetter, Rolf 207

Drobnic, Klemen 142

Duan, Fang 274

Duque-Perez, Oscar 105, 162

Dziechciarz, Arkadiusz 317

E

El-Sayed, Mohamed A. 221

Eutebach, Thomas 377

F

Fernández Gómez, Alejandro 136, 317

Fernandez-Temprano, Miguel 162

Ferro, Francesco 477

Filippetti, Fiorenzo 57, 505

Fireteanu, Virgiliu 257

Fiser, Rastko 142

Fontes Godoy, Wagner 281, 566

Fornasari, Luca 384

Fornasiero, Emanuele 491

Fournier, Etienne 77

Francois, Bruno 128

Freire, Nuno 249

Frisk, Laura 16

Frosini, Lucia 371

G

G. Melero, Manuel 527

G. Norniella, Joaquín 527

Gadoue, Shady 157

García, Antonio 544

Garcia-Perez, Arturo 105, 192, 233

Garcia-Ramirez, Armando 233

Gardel-Sotomayor, Pedro E. 105, 162

Glessner, Manfred 215

Goedtel, Alessandro 281, 538, 566

Göktas, Taner 122

Gritli, Yasser 57, 505

Gyftakis, Konstantinos N. 36, 302, 402

H

Haddad, Reemon 99

Hamdani, Samir 23, 420

Harlisca, Ciprian 371

Hasni, -Morad 420

Hedayati Kia, Shahin 358

Henao, Humberto 358, 544

Hernández González, Leonardo 241

Hoblos, Ghaleb 342

Hong, Jongman 183

Howey, David 391

Hultgren, Anders 485

Hyun, Doosoo 183

I

Ibtiouen, -Rachid 420

Immel, Jochen 377

Immovilli, Fabio 491

J

Jakovljevic, Boris 64

Jelicic, Zoran 64

Ji, Bing 9, 269

Jirasuwankul, Nirudh 85

K

Kang, Tae-June 114

Kanovic, Željko 64, 407

Kapetina, Mirna 64

Kappatou, Joya C. 36, 302, 402

Kavanagh, Darren 391

Khelif, Samia 309

Khelifi, F. 533

Khezzar, Abdelmalek 263, 295

Krievs, Oskars 433

Kulic, Filip 407

L

Le Menach, Yvonnick 558

Lee, Sang Bin 114, 183

Lee, Kun Wang 183

Lennels, Matz 485

Lim, Chaewoong 183

Linnér, Jörgen 485

Liu, Zheng 269

Lorenzani, Emilio 491

Luleci, Ihsan 28

M

Mabrouk, A.E. 309

Macaire, Ludovic 558

Magdun, Oliver 447

Majid, Abdul 428

Malec, -David 607

Manop, Chalermchat 85

Manuel Moreno, Joan 461, 512

Maouche, Yassine 263, 295

Martinez, Javier 215

Matic, Dragan 64, 407

Maussion, Pascal 77, 607

May Alarcón, Manuel 241

Mazzuca, Teresa 477

Mcculloch, Malcolm 391

Mendes, Andre 497

Mengoni, Michele 57, 505

Mezzarobba, Mario 477

Mishra, Mahesh K. 512

Moghadasian, Mahmood 469

Montanari, Gian Carlo 384

Morinigo-Sotelo, Daniel 105, 162

N

Nadji, B. 533

Nandi, Subhasis 336

Narasamma, N. Lakshmi 512

Nemec, Mitja 142

Nowak, Agnieszka 200

Nussbaumer, Peter 579

O

O'connor, Paul 1

Orcajo, Gonzalo A. 527

Orman, Maciej 200

Oros, Djura 407

Ortega, Juan Antonio 169, 544

Osornio-Rios, Roque A. 105, 233

Ottewill, James 200, 207

Ouaged, S. 23

Oumaamar, M.E.K. 43

Özgüven, Ömer Faruk 122

P

Pacas, Mario 377

Pech Carbonell, Abraham 241

Pedrayes, Joaquín F. 527

Perdomo-Arvizu, Maria 377

Perez-Cruz, Juan 69, 150

Philipp, François 215

Pickert, Volker 9

Picot, Antoine 77, 607

Pineda-Sánchez, Manuel 69, 91, 150

Pinto, Cajetan 200, 207

Pippola, Juha 16

Platero, Carlos 177

Pons-Llinares, Joan 114

Puche-Panadero, Ruben 69, 150

Purnama, I. K. E. 573

Purnomo, Mauridhy 288, 573

Pusca, Remus 257

R

Ranstad, Per 485

Rapaic, Milan 64

Rastko, Zivanovic 274

Razik, Hubert 43, 128, 552, 586

Rebollo, Emilio 177

Régnier, Jérémi 77, 329

Reljic, Dejan 407

Riba, Jordi-Roger 520

Riera-Guasp, Martin 69, 91, 114, 150, 592

Rivas, Francklin 586

Rodriguez, Pedro 207

Rodríguez Blanco, Marco Antonio 241

Roger-Folch, Jose 69, 150

Rojas, Carlos H. 527

Romary, Raphael 257

SDEMPED 2013 Proceedings

Index of Authors

Romeral, Luis 169, 412, 461,512,520

Romero-Troncoso, Rene J. 105, 233

Rossi, Claudio 57

Rzeszucinski, Pawel 207

S

Saavedra, Harold 520

Sadoun, Rabah 23

Sahraoui, Mohammed 309

Sala, Vicent 461, 512

Saleem, Jawad 428

Salehi Arashloo, Ramin 412, 461, 512

Salehifar, Mehdi 412, 461, 512

Salviano Gongora, Wylliam 281, 566

Sapena-Baño, Angel 69, 150

Sargos, Francois-Michel 128

Sawitri, Dian 288, 573

Scala, Giorgio 477

Scalassara, Paulo 538

Schmitt, Helder 538

Sedlacek, Radek 396

Serra, Giovanni 505

Shi, Pu 49, 323

Shi, Dongfeng 365

Singleton, Rodney 600

Sivert, A. 469

Sobczyk, Tadeusz J. 136, 317

Song, Xueguan 269

Soualhi, Abdenour 552, 586

Steiks, Ingars 433

Stojcic, Goran 227

Strangas, Elias 99, 592, 600

Sulowicz, Maciej 207

Szabo, Lorand 371

T

Tan, Zheng 269, 505

Teska, Mike 183

Tessarolo, Alberto 477

Thirumarai Chelvan, -Ilamparithi 336

Tian, Guiyun 269

Tientcheu Yamdeu, Mathias 77

Tomlain, Ján 396

Touhami, Omar 23, 420

Tounzi, Abdelmounaïm 558

V

V. Da Silva, Hugo 566

Vaalasranta, Ilkka 16

Vagapov, Yuriy 49, 323

Vasic, Veran 407

Vázquez Pérez, Amsi 241

Vedral, Josef 396

Vedreño-Santos, Fracisco 91

Verez, Guillaume 342

Villalobos Pina, Francisco Javier 192

Vogelsberger, Markus A. 579

W

Wiedenbrug, Ernesto 183

Wildermuth, Stephan 207

Wolbank, Thomas M. 227, 579

X

Xing, Lei 9

Y

Yang, Chanseung 114

Yazidi, Amine 469

Z

Zahawi, Bashar 157

Zaidi, Syed Sajjad 615

Zarri, Luca 505

Zito, Damiano 477

Zouaoui, Zoubir 49, 323

Zouzou, Salah Eddine 309

Zurita, Daniel 169

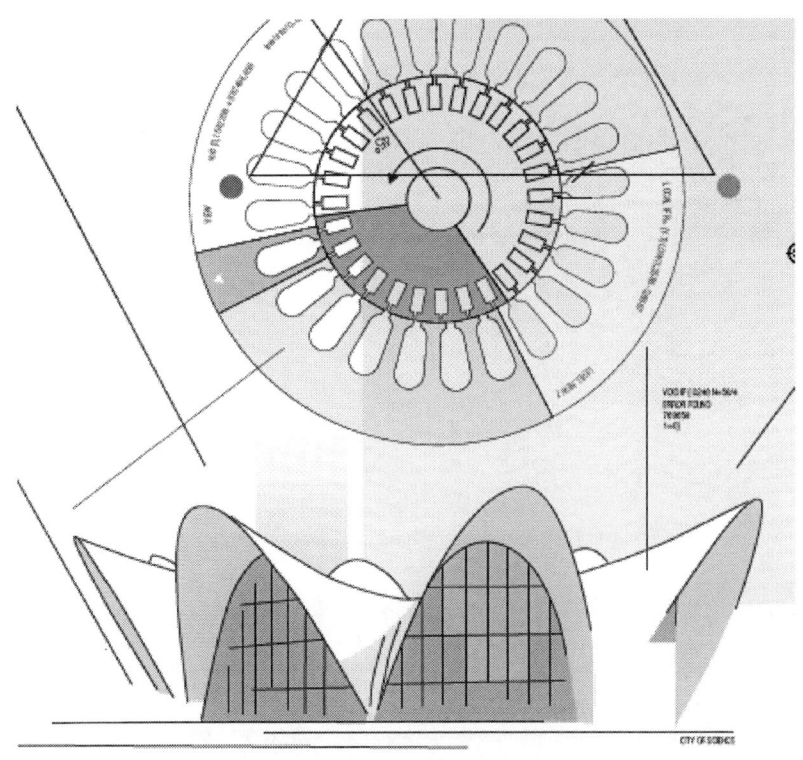

IEEE TRANSACTIONS ON INDUSTRIAL ELECTRONICS
IEEE TRANSACTIONS ON INDUSTRIAL INFORMATICS

Joint Special Section on:

Modern Diagnostics Techniques for Electrical Machines, Power Electronics & Drives

D IAGNOSIS/PROGNOSIS AND CONDITION MONITORING for electrical machines, power electronics, adjustable speed drives and related areas are key elements for modern industrial systems. During the past forty years diagnostics techniques in power electrical engineering have been dedicated to components and many papers have been published on the topic. However, industrial interests have moved towards global systems for which fault detection and predictive maintenance are more difficult to be achieved.

Editors invite original manuscripts presenting recent advances in these fields with special reference to the following topics:

- Electrical machines: Failure detection and location in electrical machines using vibration, audible noise, electrical or mechanical variables, sensors, insulation failures, electrical, mechanical and thermal models

- Power electronics: Diagnostics in power converters using input-output monitoring, thermal and/or electrical measurements on power semiconductors, control supervision, signal processing

- Materials for electrical machines: Insulating and magnetic materials, remaining life models, ageing tests

- Electrical drives: Monitoring and diagnostics for drives using electrical machines (motors and generators), power converters and control systems supervision, computer-based signal processing and data analysis

- Tools for diagnosis/prognosis: Neural networks, fuzzy logic, artificial intelligence, genetic algorithms, expert systems, identification, estimation, observers, signal processing techniques

- Tests for predictive maintenance: Partial discharge analysis and tests, new instruments for diagnostics.

Manuscript Preparation and Submission:

Authors should be aware that Papers will be published in TIE or TII depending on the paper technical content and the space available in both journals. Check carefully the style of the journals described in the guidelines "Information for Authors" in the IEEE- IES web site: http://www.ieee-ies.org/publications.
Please submit your manuscript in electronic form through: https://mc.manuscriptcentral.com/tie-ieee/ or https://mc.manuscriptcentral.com/tii/.

On the submitting page #1 in pop-up menu of manuscript type, select: **"SS on Modern Diagnostics Techniques for EMPED"**, then upload all your manuscript files following the instructions given on the screen.

Corresponding Guest Editor **Prof. Gérard-André Capolino** *Department of Electrical Engineering* *University of Picardie "Jules Verne"* *80039 Amiens, France*	*Guest Editor* **Prof. Jose Antonino-Daviu** *Institute for Energy Engineering* *Universitat Politècnica de València* *46022 Valencia, Spain*	*Guest Editor* **Prof. Martin Riera-Guasp** *Institute for Energy Engineering* *Universitat Politècnica de València* *46022 Valencia, Spain*
EMAIL: gerard.capolino@ieee.org	EMAIL: joanda@die.upv.es	EMAIL: mriera@die.upv.es

Submission management emails: tie-submissions@ieee-ies.org / tii-submissions@ieee-ies.org

Timetable		
Deadline for manuscript submissions:	**Information about manuscript acceptance:**	**Hardcopy publication date:**
31 January, 2014	Summer, 2014	Spring, 2015

Editor in Chief of TIE: Prof. Carlo Cecati, Univ. of L'Aquila, 67100 L'Aquila, Italy, EMAIL: tie@ieee-ies.org. URL: http://tie.ieee-ies.org
Editor in Chief of TII: Prof. Kim F. Man, City University, Honk Kong, HK, EMAIL: tii@ieee-ies.org. URL: http://tii.ieee-ies.org

978-1-4799-0024-4/13 $31.00 © 2013 IEEE